Chapter 4 Systems of Equations and Inequalities

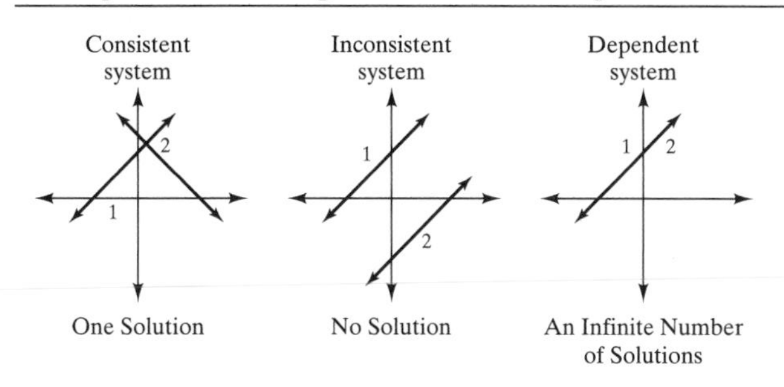

A system of linear equations may be solved: (a) graphically, (b) by the substitution method, (c) by the addition or elimination method, (d) by matrices or (e) by determinants.

$$\begin{vmatrix} a_1 & b_1 \\ a_2 & b_2 \end{vmatrix} = a_1 b_2 - a_2 b_1$$

Cramer's Rule:

Given a system of equations of the form

$$\begin{aligned} a_1 x + b_1 y &= c_1 \\ a_2 x + b_2 y &= c_2 \end{aligned} \text{ then } x = \frac{\begin{vmatrix} c_1 & b_1 \\ c_2 & b_2 \end{vmatrix}}{\begin{vmatrix} a_1 & b_1 \\ a_2 & b_2 \end{vmatrix}} \text{ and } y = \frac{\begin{vmatrix} a_1 & c_1 \\ a_2 & c_2 \end{vmatrix}}{\begin{vmatrix} a_1 & b_1 \\ a_2 & b_2 \end{vmatrix}}$$

Chapter 5 Polynomials and Polynomial Functions

FOIL method to multiply two binomials

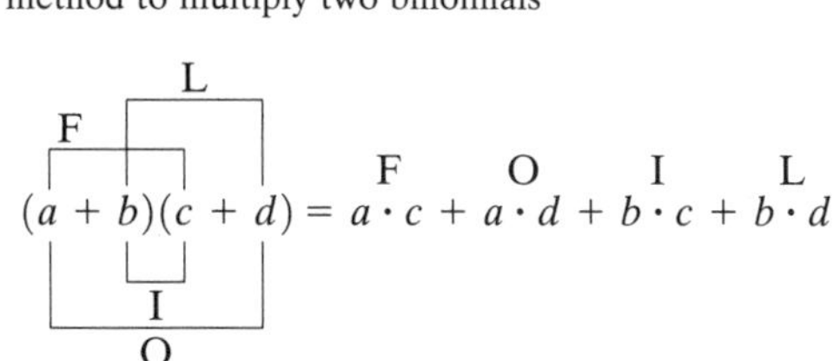

Pythagorean Theorem:

$\text{leg}^2 + \text{leg}^2 = \text{hyp}^2$ or $a^2 + b^2 = c^2$

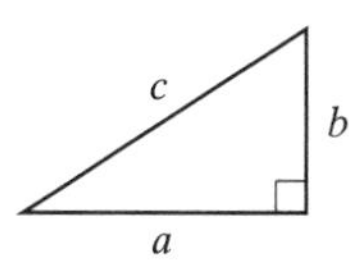

Square of a binomial:

$(a + b)^2 = a^2 + 2ab + b^2.$

$(a - b)^2 = a^2 - 2ab + b^2$

Product of the sum and difference of the same two terms (also called the difference of two squares):

$(a + b)(a - b) = a^2 - b^2$

Perfect square trinomials:

$a^2 + 2ab + b^2 = (a + b)^2, \quad a^2 - 2ab + b^2 = (a - b)^2$

Sum of two cubes:

$a^3 + b^3 = (a + b)(a^2 - ab + b^2)$

Difference of two cubes:

$a^3 - b^3 = (a - b)(a^2 + ab + b^2)$

Standard form of a quadratic equation:

$ax^2 + bx + c = 0, a \neq 0$

Zero-factor property: If $a \cdot b = 0$, then either $a = 0$ or $b = 0$, or both a and $b = 0$.

Chapter 6 Rational Expressions and Equations

To Multiply Rational Expressions:

1. Factor all numerators and denominators.
2. Divide out any common factors.
3. Multiply numerators and multiply denominators.
4. Simplify the answer when possible.

To Divide Fractional Expressions:

Invert the divisor and then multiply the resulting rational expressions.

To Add or Subtract Rational Expressions:

1. Write each fraction with a common denominator.
2. Add or subtract the numerators while keeping the common denominator.
3. When possible factor the numerator and simplify the fraction.

Similar Figures: Corresponding angles and corresponding sides are in proportion.

Proportion: If $\frac{a}{b} = \frac{c}{d}$ then $ad = bc$

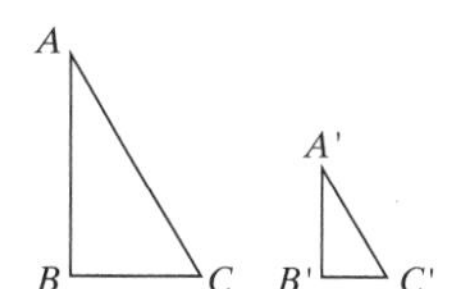

Variation: direct, $y = kx$; inverse, $y = \frac{k}{x}$ joint, $y = kxz$

FIFTH EDITION

INTERMEDIATE ALGEBRA

FOR COLLEGE STUDENTS

ALLEN R. ANGEL
Monroe Community College

PRENTICE HALL
Upper Saddle River, New Jersey 07458

Library of Congress Cataloging-in-Publication Data

Angel, Allen R.
Intermediate algebra for college students / Allen R. Angel. —5th ed.
p. cm.
Includes index.
ISBN 0-13-916321-2
1. Algebra. I. Title.
QA154.2.A53 2000
512.9—dc21 99-14410
CIP

Executive Editor: Karin E. Wagner
Editor-in-Chief: Jerome Grant
Editor-in-Chief, Development: Carol Trueheart
Senior Managing Editor: Linda Mihatov Behrens
Executive Managing Editor: Kathleen Schiaparelli
Assistant Vice President of Production and Manufacturing: David W. Riccardi
Marketing Manager: Eilish Main
Marketing Assistant: Amy Lysik/Vince Jansen
Manufacturing Buyer: Alan Fischer
Manufacturing Manager: Trudy Pisciotti
Editorial Assistant/Supplements Editor: Kate Marks
Associate Editor, Mathematics/Statistics Media: Audra J. Walsh
Art Director: Maureen Eide
Cover Designer: Joseph Sengotta
Interior Designer: Lorraine Castellano
Associate Creative Director: Amy Rosen
Director of Creative Services: Paula Maylahn/Paul Belfanti
Assistant to Art Director: John Christiana
Art Manager: Gus Vibal
Art Editor: Grace Hazeldine
Cover Image: James H. Carmichael, Jr./The Image Bank/Precious "Wentle Trap" Epitonium Scalare (means spiral staircase), W. Pacific
Project Management: Elm Street Publishing Services, Inc.
Photo Researcher: Diana Gongora
Photo Research Administrator: Beth Boyd
Art Studio: Scientific Illustrators
Composition: Prepare, Inc./Emilcomp srl

Printed in the United States of America

10 9 8 7 6 5 4 3 2 1

ISBN 0-13-916321-2

Prentice-Hall International (UK) Limited, *London*
Prentice-Hall of Australia Pty. Limited, *Sydney*
Prentice-Hall Canada, Inc., *Toronto*
Prentice-Hall Hispanoamericana, S.A., *Mexico*
Prentice-Hall of India Private Limited, *New Delhi*
Prentice-Hall of Japan, Inc., *Tokyo*
Prentice-Hall Pte. Ltd., *Singapore*
Editora Prentice-Hall do Brasil, Ltda., *Rio de Janeiro*

To my mother,

Sylvia Angel-Baumgarten

And to the memory of my father, Isaac Angel

Contents

6) Rational Expressions and Equations 357

7) Roots, Radicals, and Complex Numbers 421

8) Quadratic Functions 489

Preface

This book was written for college students who have successfully completed a first course in elementary algebra. My primary goal was to write a book that students can read, understand, and enjoy. To achieve this goal I have used short sentences, clear explanations, and many detailed worked-out examples. I have tried to make the book relevant to college students by using practical applications of algebra throughout the text.

Features of the Text

Four-Color Format Color is used pedagogically in the following ways:

- Important definitions and procedures are color screened.
- Color screening or color type is used to make other important items stand out.
- Artwork is enhanced and clarified with use of multiple colors.
- The four-color format allows for easy identification of important features by students.
- The four-color format makes the text more appealing and interesting to students.

Readability One of the most important features of the text is its readability. The book is very readable, even for those with weak reading skills. Short, clear sentences are used and more easily recognized, and easy-to-understand language is used whenever possible.

Accuracy Accuracy in a mathematics text is essential. To ensure accuracy in this book, mathematicians from around the country have read the pages carefully for typographical errors and have checked all the answers.

Connections Many of our students do not thoroughly grasp new concepts the first time they are presented. In this text we encourage students to make connections. That is, we introduce a concept, then later in the text briefly reintroduce it and build upon it. Often an important concept is used in many sections of the text. Students are reminded where the material was seen before, or where it will be used again. This also serves to emphasize the importance of the concept. Important concepts are also reinforced throughout the text in the Cumulative Review Exercises and Cumulative Review Tests.

Chapter Opening Application Each chapter begins with a real-life application related to the material covered in the chapter. By the time students complete the chapter, they should have the knowledge to work the problem.

Preview and Perspective This feature at the beginning of each chapter explains to the students why they are studying the material and where this material will be used again in other chapters of the book. This material helps students see the connections between various topics in the book, and the connection to real-world situations.

Student's Solution Manual, Videotape, and Software Icons At the beginning of each section, Student's Solution Manual, videotape, and tutorial software icons are displayed. These icons tell the student where material in the section can be found in the Student's Solution Manual, on the videotapes, and in the tutorial software, saving your students time when they want to review this material. Small videotape icons are also placed next to exercises that are worked out on the videotapes.

Keyed Section Objectives Each section opens with a list of skills that the student should learn in that section. The objectives are then keyed to the appropriate portions of the sections with symbols such as 1.

Problem Solving Polya's five-step problem-solving procedure is discussed in Section 2.2. Throughout the book problem solving and Polya's problem-solving procedure are emphasized.

Practical Applications Practical applications of algebra are stressed throughout the text. Students need to learn how to translate application problems into algebraic symbols. The problem-solving approach used throughout this text gives students ample practice in setting up and solving application problems. The use of practical applications motivates students.

Detailed Worked-Out Examples A wealth of examples have been worked out in a step-by-step, detailed manner. Important steps are highlighted in color, and no steps are omitted until after the student has seen a sufficient number of similar examples.

Now Try Exercise In each section, students are asked to work exercises that parallel the examples given in the text. These Now Try Exercises make the students *active*, rather than passive, learners and they reinforce the concepts as students work the exercises.

Study Skills Section Many students taking this course have poor study skills in mathematics. Section 1.1, the first section of this text, discusses the study skills needed to be successful in mathematics. This section should be very beneficial for your students and should help them to achieve success in mathematics.

Helpful Hints The helpful hint boxes offer useful suggestions for problem solving and other varied topics. They are set off in a special manner so that students will be sure to read them.

Avoiding Common Errors Errors that students often make are illustrated. The reasons why certain procedures are wrong are explained, and the correct procedure for working the problem is illustrated. These Avoiding Common Errors boxes will help prevent your students from making those errors we see so often.

Using Your Calculator The Using Your Calculator boxes, placed at appropriate intervals in the text, are written to reinforce the algebraic topics presented in the section and to give the student pertinent information on using the calculator to solve algebraic problems.

Using Your Graphing Calculator Using Your Graphing Calculator boxes are placed at appropriate locations throughout the text. They reinforce the algebraic topics taught and sometimes offer alternate methods of working problems. This book is designed to give the instructor the option of using or not using a graphing calculator in their course. Many Using Your Graphing Calculator boxes contain graphing calculator exercises, whose answers appear in the answer section of the book. The illustrations shown in the Using Your Graphing Calculator boxes are from a Texas Instrument 83 calculator. The Using Your Graphing Calculator boxes are written assuming that the student has no prior graphing calculator experience.

Exercise Sets

The exercise sets are broken into three main categories: Concept/Writing Exercises, Practice the Skills, and Problem Solving. Many exercise sets also contain Challenge Problems and/or Group Activities. Each exercise set is graded in difficulty. The early problems help develop the student's confidence, and then students are eased gradually into the more difficult problems. A sufficient number and variety of examples are given in each section for the student to successfully complete even the more difficult exercises. The number of exercises in each section is more than ample for student assignments and practice. Many exercise sets contain graphing calculator exercises for instructors who wish to assign them.

Concept/Writing Exercises Most exercise sets include exercises that require students to write out the answers in words. These exercises improve students' understanding and comprehension of the material. Many of these exercises involve problem solving, and conceptualization, and help develop better reasoning and critical thinking skills. Writing exercises are indicated by the symbol ✎.

Challenge Problems These exercises, which are part of many exercise sets, provide a variety of problems. Many were written to stimulate student thinking. Others provide additional applications of algebra or present material from future sections of the book so that students can see and learn the material on their own before it is covered in class. Others are more challenging than those in the regular exercise set.

Problem Solving Exercises These exercises have been added to help students become better thinkers and problem solvers. Many of these exercises are applied in nature.

Cumulative Review Exercises All exercise sets (after the first two) contain questions from previous sections in the chapter and from previous chapters. These cumulative review exercises will reinforce topics that were previously covered and help students retain the earlier material, while they are learning the new material. For the students' benefit the Cumulative Review Exercises are keyed to the section where the material is covered.

Group Activities Many exercise sets have group activity exercises that lead to interesting group discussions. Many students learn well in a cooperative learning atmosphere, and these exercises will get students talking mathematics to one another.

Chapter Summary At the end of each chapter is a chapter summary which includes a glossary and important chapter facts.

Review Exercises At the end of each chapter are review exercises that cover all types of exercises presented in the chapter. The review exercises are keyed to the sections where the material was first introduced.

Practice Tests The comprehensive end-of-chapter practice test will enable the students to see how well they are prepared for the actual class test. The Test Item File includes several forms of each chapter test that are similar to the student's practice test. Multiple choice tests are also included in the Test Item File.

Cumulative Review Test These tests, which appear at the end of each chapter, test the students' knowledge of material from the beginning of the book to the end of that chapter. Students can use these tests for review, as well as for preparation for the final exam. These exams, like the cumulative review exercises, will serve to reinforce topics taught earlier.

Answers The *odd answers* are provided for the exercise sets. *All answers* are provided for the Using Your Graphing Calculator Exercises, Cumulative Review Exercises, the Review Exercises, Practice Tests, and the Cumulative Practice Test. *Answers* are not provided for the Group Activity exercises since we want students to reach agreement by themselves on the answers to these exercises.

National Standards

Recommendations of the *Curriculum and Evaluation Standards for School Mathematics*, prepared by the National Council of Teachers of Mathematics, (NCTM) and *Crossroads in Mathematics: Standards for Introductory College Mathematics Before Calculus*, prepared by the American Mathematical Association of Two Year Colleges (AMATYC) are incorporated into this edition.

Prerequisite

The prerequisite for this course is a working knowledge of elementary algebra. Although some elementary algebra topics are briefly reviewed in the text, students should have a basic understanding of elementary algebra before taking this course.

Modes of Instruction

The format and readability of this book lends itself to many different modes of instruction. The constant reinforcement of concepts will result in greater understanding and retention of the material by your students.

The features of the text and the large variety of supplements available make this text suitable for many types of instructional modes including:

- lecture
- self-paced instruction
- modified lecture
- cooperative or group study
- learning laboratory

Changes in the Fifth Edition

When I wrote the fifth edition I considered the many letters and reviews I got from students and faculty alike. I would like to thank all of you who made suggestions for improving the fifth edition. I would also like to thank the many instructors and students who wrote to inform me of how much they enjoyed and appreciated the text.

Some of the changes made in the fifth edition of the text include:

- Real-life chapter-opening applications have been added to each chapter.
- More Using Your Graphing Calculator boxes have been added throughout the text. They are used to provide information on using a graphing calculator. The text is designed so that instructors have the opportunity of using, or not using, a graphing calculator with this book.

- The table of contents has been reorganized to reduce the overlap between the material covered in elementary and intermediate algebra.
- The chapters on Factoring and Polynomials have been combined into one chapter.
- Functions are now integrated throughout the text to help better prepare your students for additional mathematics courses.
- The exercise sets have been rewritten and reorganized. Exercise sets now start with Concept/Writing Exercises, followed by Practice the Skills Exercises, followed by Problem Solving Exercises.
- Problem Solving and George Polya's problem-solving procedure are stressed throughout the book. Problem-solving examples are worked using the following steps: Understand, Translate, Carry out, Check, and Answer.
- Cumulative Review Tests are now at the end of every chapter.
- The Challenge Problem/Group Activity exercises in the previous edition have been broken up into two separate categories, and more exercises have been added to each category.
- The Exercise Sets now have a much greater variety of exercises, and more challenging exercises have been added for those instructors who wish to assign them. The Exercise Sets are graded in level of difficulty.
- Many more problem solving, and thought provoking, exercises have been added to the exercise sets throughout the book.
- Although functions are presented early, certain function topics such as inverse and composite functions are introduced later. This gives students the opportunity to learn and understand functions before being introduced to these more complex topics.
- In the graphing chapters, more emphasis is placed on understanding the meaning of graphs.
- More graphing calculator exercises have been added to the Exercise Sets for those who wish to assign them.
- The Graphing Calculator Corners have been renamed as either Using Your Calculator or Using Your Graphing Calculator. They are colored differently for easy identification.
- The Common Student Error boxes have been renamed Avoiding Common Errors.
- The Exercise Sets have many more real-life applications than in previous editions.
- A more colorful and appealing design results in distinct features being more recognizable. The new exciting design also results in students being more willing to read the text.
- Graphing AIE answers were moved to an appendix in the back of the text. This results in the students' text not having large blocks of empty space in their text.
- Definitions are given in Definition Boxes and Procedures are given in Procedure Boxes.
- In the AIE, Teaching Tips have been added to provide ideas for exploration.
- The Chapter Tests and Cumulative Review Tests have been made more uniform. All Chapter Tests now have 25 problems, and all Cumulative Review Tests now have 20 problems.
- The chapters on Conic Sections and Logarithms have been switched in location.
- Now Try Exercises have been added in each section after many examples. Students are asked to work specific exercises after they read specific examples. Working these exercises reinforces what the student has just learned, and also serves to make students active, rather than passive, learners. The Now Try Exercises are marked in green in the Exercise Sets for easy identification by the student.
- Exercises that are worked on the videotapes are indicated by an icon next to the exercises.
- There are more writing exercises, that is, exercises that require a written answer. Writing exercises are indicated with a pencil icon.

Supplements to the Fifth Edition

For this edition of the book the author has personally coordinated the development of the *Student's Solution Manual* and the *Instructor's Solution Manual*. Experienced mathematics professors who have prior experience in writing supplements, and whose works have been of superior quality, have been carefully selected for authoring the supplements.

For Instructors

Printed Supplements

Annotated Instructor's Edition

- Contains all of the content found in the student edition.
- Answers to all exercises are printed on the same

text page (graphed answers are in a special graphing answer section at the back of the text).

- Teaching Tips throughout the text are placed at key points in the margin.

Instructor's Solutions Manual

- Solutions to even-numbered section exercises.
- Solutions to every (even and odd) exercise found in the Chapter Reviews, Chapter Tests, and Cumulative Reviews.

Instructor's Test Manual

- Two free-response Pretests per chapter.
- Eight Chapter Tests per chapter (3 multiple choice, 5 free response).
- Two Cumulative Review Tests (one multiple choice, one free response) every two chapters.
- Eight Final Exams (3 multiple choice, 5 free response).
- Twenty additional exercises per section for added test exercises if needed.

Media Supplements

TestPro4 Computerized Testing

- Algorithmically driven, text-specific testing program.
- Networkable for administering tests and capturing grades on-line.
- Edit and add your own questions—create nearly unlimited number of tests and drill worksheets.

Companion Web site

- www.prenhall.com/angel
- Links related to the chapter openers at the beginning of each chapter allow students to explore related topics and collect data needed in order to complete application problems.
- Additional links to helpful, generic sites include Fun Math and For Additional Help.
- Syllabus builder management program allows instructor to post course syllabus information and schedule on the Web site.

For Students

Printed Supplements

Student Solution Manual

- Solutions to odd-numbered section exercises.
- Solutions to every (even and odd) exercise found in the Chapter Reviews, Chapter Tests, and Cumulative Reviews.

Student Study Guide

- Includes additional worked-out examples, additional exercises, practice tests and answers.
- Includes information to help students study and succeed in mathematics class.
- Emphasizes important concepts.

New York Times *Themes of the Times*

- Contact your local Prentice Hall sales representative.

How to Study Mathematics

- Contact your local Prentice Hall sales representative.

Internet Guide

- Contact your local Prentice Hall sales representative.

Media Supplements

MathPro4 Computerized Tutorial

- Keyed to each section of the text for text specific tutorial exercises and instruction.
- Includes Warm up exercises and graded Practice Problems.
- Includes video Watch screens.
- Take chapter quizzes.
- Send and receive e-mail from and to your instructor.
- Algorithmically driven and fully networkable.

Videotape Series

- Keyed to each section of the text.
- Step by step solutions to exercises from each section of the text. Exercises from the text that are worked in the videos are marked with a video icon.

Companion Web site

- www.prenhall.com/angel
- Links related to the chapter openers at the beginning of each chapter allow students to explore related topics and collect data needed in order to complete application problems.
- Additional links to helpful, generic sites include Fun Math and For Additional Help.
- Syllabus builder management program allows instructor to post course syllabus information and schedule on the Web site.

Acknowledgments

Writing a textbook is a long and time-consuming project. Many people deserve thanks for encouraging and assisting me with this project. Most importantly I would like to thank my wife Kathy, and sons, Robert and Steven. Without their constant encouragement and understanding, this project would not have become a reality.

I would like to thank Richard Semmler of Northern Virginia Community College, Larry Clar and Donna Petrie of Monroe Community College, and Cindy Trimble and Teri Lovelace of Laurel Technical Services for their conscientiousness and the attention to details they provided in checking pages, artwork, and answers.

I would like to thank students and faculty from around the country for using the fourth edition and offering valuable suggestions for the fifth edition. I was especially pleased in receiving so many letters from students informing me how much they enjoyed using the book. Thank you for your kind words.

I would like to thank my editor at Prentice Hall, Karin Wagner and my production editor Ingrid Mount for their many valuable suggestions and conscientiousness with this project.

I would like to thank the following reviewers and proofreaders for their thoughtful comments and suggestions.

Helen Banes, *Kirkwood Community College*
Jon Becker, *Indiana University Northwest*
Paul Boisvert, *Robert Morris College*
Charlotte Buffington, *New Hampshire Community Technical College—Stratham*
Connie Buller, *Metropolitan Community College—Omaha*
Gerald Busald, *San Antonio College*
Joan Capps, *Raritan Valley Community College*
Mitzi Chaffer, *Central Michigan University*
Larry Clar, *Monroe Community College*
John DeCoursey, *Vincennes University*
Abdollah Hajikandi, *Buffalo State College*
Barney Herron, *Muskegon Community College*
Cheryl Hobneck, *Illinois Valley Community College*
Bruce Hoelter, *Raritan Valley Community College*
Gisele Icore, *Baltimore City Community College*
John Jerome, *Suffolk County Community College—Brentwood*
Patricia Lanz, *Erie Community College—South*
David Lunsford, *Grossmont College*
Chuck Miller, *Albuquerque TVI*
Katherine Nickell, *College of DuPage*
Donna Petrie, *Monroe Community College*
Thomas Pomykalski, *Metropolitan Community College—Omaha*
Shawn Robinson, *Valencia Community College*
Robert Secrist, *Kellogg Community College*
Richard Semmler, *Northern Virginia Community College*
Diane Short, *Southwestern College*
Bettie Truitt, *Black Hawk College*

EMPHASIS ON *Problem Solving*

The fifth edition of the Angel series places a stronger emphasis on problem solving than ever before. Problem solving is now introduced early and incorporated as a theme throughout the texts.

Five-Step Problem-Solving Procedure

The in-text examples demonstrate how to solve each exercise based on Polya's five-step problem-solving procedure: **Understand, Translate, Carry Out, Check,** and **State Answer.**

EXAMPLE 4 Only two-axle vehicles are permitted to cross a bridge that leads to Honeymoon Island State Park. The toll for the bridge is 50 cents for motorcycles and $1.00 for cars and trucks. On Saturday, the toll booth attendant collected a total of $150, and the vehicle counter recorded 170 vehicles crossing the bridge. How many motorcycles and how many cars and trucks crossed the bridge that day?

Solution **Understand and Translate**

Let x = number of motorcycles

y = number of cars and trucks

Since a total of 170 vehicles crossed the bridge, one equation is $x + y = 170$. The second equation comes from the tolls collected.

Tolls from motorcycles	+	tolls from cars and trucks	= 150
$0.50x$	+	$1.00y$	= 150

System of equations $\begin{cases} x + y = 170 \\ 0.50x + 1.00y = 150 \end{cases}$

Carry Out Since the first equation can be easily solved for y, solve this system by substitution. Solving for y in $x + y = 170$ gives $y = 170 - x$. Substitute $170 - x$ for y in the second equation and solve for x.

$0.50x + 1.00y = 1$...

To Solve a System of Equations by the Addition (or Elimination) Method

1. If necessary, rewrite each equation so that the terms containing variables appear on the left side of the equal sign and any constants appear on the right side of the equal sign.
2. If necessary, multiply one or both equations by a constant(s) so that when the equations are added the resulting sum will contain only one variable.
3. Add the equations. This will result in a single equation containing only one variable.
4. Solve for the variable in the equation in step 3.
5. Substitute the value found in step 4 into either of the original equations. Solve that equation to find the value of the remaining variable.
6. Check the values obtained in all original equations.

In step 2 we indicate it may be necessary to multiply one or both equations by a constant. In this text we will use brackets [ ], to indicate that both sides of the equation within the brackets are to be multiplied by some constant.

Procedure Boxes

Important procedures are highlighted and boxed throughout the text, making them easy for students to read and review.

Problem-Solving Exercises

These exercises are designed to help students become better thinkers.

Problem Solving

63. Is the point represented by the ordered pair $\left(\frac{1}{2}, -\frac{2}{23}\right)$ on the graph of the equation $y = \frac{x}{x^2 - 6}$? Explain.

64. Is the point represented by the ordered pair $\left(\frac{1}{2}, \frac{3}{2}\right)$ on the graph of the equation $y = \frac{x^2 - 4}{x - 2}$? Explain.

65. a) Plot the points $A(2, 7)$, $B(2, 3)$, and $C(6, 3)$, and then draw $\overline{AB}$, $\overline{AC}$, and $\overline{BC}$. ($\overline{AB}$ represents the line segment from A to B.)

b) Find the area of the figure.

66. ...Plot the points $A(-4, 5)$, $B(2, 5)$, $C(2, -3)$, and...

b) Estimate the total sales of the three items listed in 1999.

c) In which years were the sales of low/nonfat ice cream greater than $2.5 billion?

d) Does the decrease in the sales of frozen yogurt from 1995 through 2000 appear to be approximately linear? Explain.

68. The following graph shows unemployment rates in Washington, D.C, Maryland, and Virginia from 1990 through 1996.

Unemployment Rates

EMPHASIS ON *Applications*

Each chapter begins with an illustrated, real-world application to motivate students and encourage them to see algebra as an important part of their daily lives. Problems based on real data from a broad range of subjects appear throughout the text, in the end-of-chapter material, and in the exercise sets.

GRAPHS AND FUNCTIONS

CHAPTER

3

3.1 Graphs

3.2 Functions

3.3 Linear Functions: Graphs and Applications

3.4 The Slope–Intercept Form of a Linear Equation

3.5 The Point–Slope Form of a Linear Equation

3.6 The Algebra of Functions

3.7 Graphing Linear Inequalities

Summary
Review Exercises
Practice Test
Cumulative Review Test

Use the Angel Web site at www.prenhall.com/angel to be linked to an internet resource that will help you further explore the following application.

Have you ever dreamed of starting your own business? Before starting it, you should write a business plan which includes, among other things, a projection of the profits of your business. The gross annual profit can be estimated by subtracting your annual expenses from your annual income. On page 172, we project the annual profit of a tire store as a function of the number of tires sold.

Chapter-Opening Applications

New **chapter-opening applications** emphasize the use of mathematics in everyday life, and in the workplace giving students an applied, real-world introduction to the chapter material. The applications are often tied to examples presented in the section, and links in the chapter openers direct students to the Angel Web site.

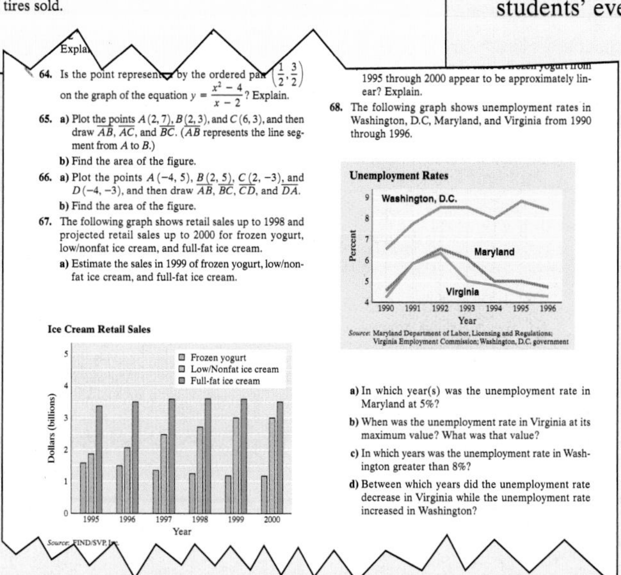

64. Is the point represented by the ordered pair $\left(\frac{1}{2}, \frac{3}{2}\right)$ on the graph of the equation $y = \frac{x^2 - 4}{x - 2}$? Explain.

65. a) Plot the points $A(2, 7)$, $B(2, 3)$, and $C(6, 3)$, and then draw $\overline{AB}$, $\overline{AC}$, and $\overline{BC}$. ($\overline{AB}$ represents the line segment from A to B.)
b) Find the area of the figure.

66. a) Plot the points $A(-4, 5)$, $B(2, 5)$, $C(2, -3)$, and $D(-4, -3)$, and then draw $\overline{AB}$, $\overline{BC}$, $\overline{CD}$, and $\overline{DA}$.
b) Find the area of the figure.

67. The following graph shows retail sales up to 1998 and projected retail sales up to 2000 for frozen yogurt, low/nonfat ice cream, and full-fat ice cream.
a) Estimate the sales in 1999 of frozen yogurt, low/nonfat ice cream, and full-fat ice cream.

1995 through 2000 appear to be approximately linear? Explain.

68. The following graph shows unemployment rates in Washington, D.C, Maryland, and Virginia from 1990 through 1996.

a) In which year(s) was the unemployment rate in Maryland at 5%?
b) When was the unemployment rate in Virginia at its maximum value? What was that value?
c) In which years was the unemployment rate in Washington greater than 8%?
d) Between which years did the unemployment rate decrease in Virginia while the unemployment rate increased in Washington?

Real-World Applications

An abundance of wonderfully updated, **real-world applications** gives students needed practice with practical applications of algebra. Real data is used, and real-world situations emphasize the relevance of the material being covered to students' everyday lives.

EMPHASIS ON *Exercises*

End-of-section exercise sets provide a thorough review of the section material. Each set progresses in difficulty to help students gain confidence and succeed with more difficult exercises.

148 · Chapter 3 · Graphs and Functions

Practice the Skills

List the ordered pairs corresponding to the indicated points.

5.

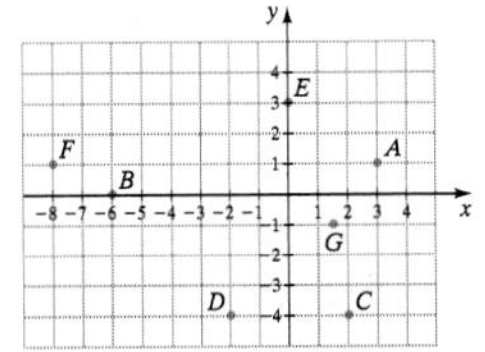

6.

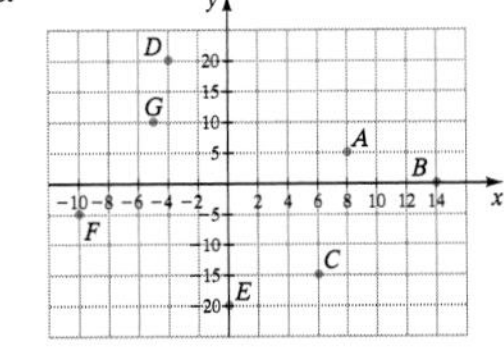

7. Graph the following points on the same axes.
$A(4, 2)$ $B(-6, 2)$ $C(0, -1)$ $D(-2, 0)$

8. Graph the following points on the same axes.
$A(-4, -2)$ $B(3, 2)$ $C(2, -3)$ $D(-3, 3)$

Determine the quadrant in which each point is located.

9. (1, 6) **10.** (−2, 3) **11.** (5, −9) **12.** (24, 116)
13. (−35, 18) **14.** (−24, −8) **15.** (−6, −19) **16.** (8, −120)

Practice the Skills Exercises

Practice the Skills exercises cover all types of exercises presented in the chapter.

Problem-Solving Exercises

These exercises are designed to help students become better thinkers.

Problem Solving

63. Is the point represented by the ordered pair $\left(\frac{1}{2}, -\frac{2}{23}\right)$ on the graph of the equation $y = \frac{x}{x^2 - 6}$? Explain.

64. Is the point represented by the ordered pair $\left(\frac{1}{2}, \frac{3}{2}\right)$ on the graph of the equation $y = \frac{x^2 - 4}{x - 2}$? Explain.

65. a) Plot the points $A(2, 7)$, $B(2, 3)$, and $C(6, 3)$, and then draw $\overline{AB}$, $\overline{AC}$, and $\overline{BC}$. ($\overline{AB}$ represents the line segment from A to B.)
b) Find the area of the figure.

b) Estimate the total sales of the three items listed in 1999.
c) In which years were the sales of low/nonfat ice cream greater than $2.5 billion?
d) Does the decrease in the sales of frozen yogurt from 1995 through 2000 appear to be approximately linear? Explain.

68. The following graph shows unemployment rates in Washington, D.C, Maryland, and Virginia from 1990 through 1996.

Unemployment Rates

Concept/Writing Exercises

New Concept/Writing Exercises encourage students to analyze and write about the concepts they are learning, improving their understanding and comprehension of the material.

Exercise Set 3.1

Concept/Writing Exercises

1. a) What does the graph of any linear equation look like?
b) How many points are needed to graph a linear equation? Explain.

2. What does it mean when a set of points is collinear?

3. When graphing the equation $y = \frac{1}{x}$, what value cannot substituted for x? Explain.

4. What is another name for the Cartesian coordinate system?

EMPHASIS ON *Exercises*

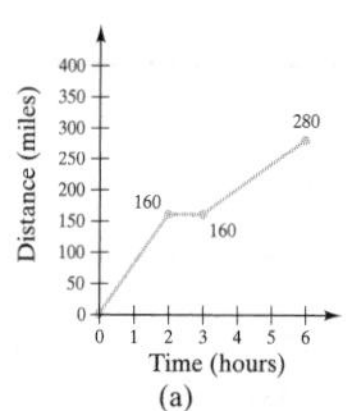

(a)

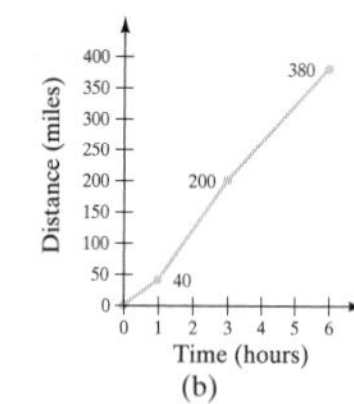

(b)

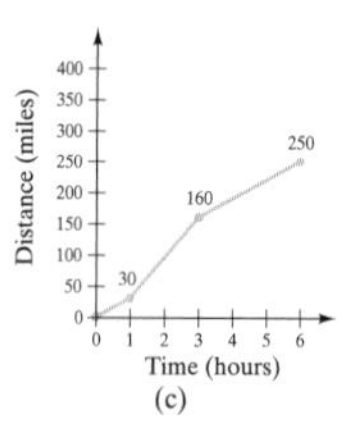

(c)

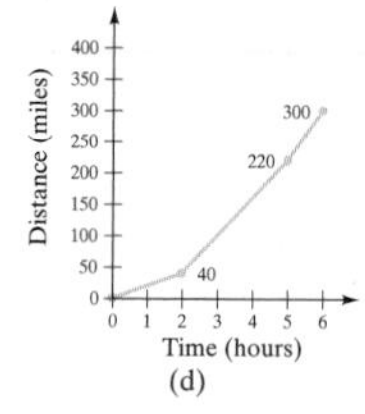

(d)

Use a graphing calculator to graph each function. Make sure you select values for the window that will show the curvature of the graph. Then, if your calculator can display tables, display a table of values in which the x-values extend by units, from 0 to 6.

85. $y = 3x + 5$ **86.** $y = \frac{1}{2}x - 4$ **87.** $y = x^2 + x + 6$ **88.** $y = x^2 - 12$

89. $y = x^3 - 2x + 4$ **90.** $y = 2x^3 - 6x^2 - 1$

Challenge Problems

Graph each equation.

91. $y = |x - 2|$ **92.** $x = y^2 + 2$

Group Activity

Discuss and work Exercises 93–94 as a group.

93. a) Group member 1: Plot the points $(-2, 4)$ and $(6, 8)$. Determine the *midpoint* of the line segment connecting these points.
Group member 2: Follow the above instructions for the points $(-3, -2)$ and $(5, 6)$.
Group member 3: Follow the above instructions for the points $(4, 1)$ and $(-2, 4)$.

b) As a group, determine a formula for the midpoint of the line segment connecting the points (x_1, y_1) and (x_2, y_2). (Note: We will discuss the midpoint formula further in Chapter 10.)

94. Three points on a parallelogram are $A(3, 5)$, $B(8, 5)$, and $C(-1, -3)$.

a) Individually determine a fourth point D that completes the parallelogram.

b) Individually compute the area of your parallelogram.

c) Compare your answers. Did you all get the same answers? If not, why not?

d) Is there more than one point that can be used to complete the parallelogram? If so, give the two points and find the corresponding areas of each parallelogram.

Cumulative Review Exercises

[2.2] **95.** Evaluate $\frac{-b + \sqrt{b^2 - 4ac}}{2a}$ for $a = 2$, $b = 7$, and $c = -15$.

[2.3] **96.** Hertz Automobile Rental Agency charges a daily fee of \$30 plus 14 cents a mile. National Automobile Rental Agency charges a daily fee of \$16 plus 24 cents a mile for the same car. What distance would you have to drive in 1 day to make the cost of renting from Hertz equal to the cost of renting from National?

[2.5] **97.** Solve the inequality $-4 \le \frac{4 - 3x}{2} < 5$. Write the solution in set builder notation.

[2.6] **98.** Find the solution set for the inequality $|3x + 2| > 5$.

Challenge Problems

Challenge Problems stimulate student interest with problems that are conceptually and computationally more demanding.

Group Activities

Group Activities provide students with opportunities for collaborative learning.

Cumulative Review Exercises

Cumulative Review Exercises reinforce previously covered topics. These exercises are keyed to sections where the material is explained.

EMPHASIS ON *Pedagogy*

Preview and Perspective

Every chapter begins with a **Preview and Perspective** to give students an overview of the chapter. The **Preview and Perspective** shows students the connections between the concepts presented in the text and the real world.

> **Preview and Perspective**
>
> Two of the primary goals of this book are to provide you with a good understanding of graphing and of functions. Graphing is heavily used in this course and in other mathematics courses you may take. To reinforce your knowledge of this topic we introduce graphing early and discuss it frequently throughout the book. Many of the exercise sets have graphs taken from newspapers or magazines. The material presented in this chapter may help you understand them better.
>
> In Section 3.2 we introduce the concept of *function.* Functions are a unifying concept used throughout all of mathematics. We also use functions throughout the book to reinforce and expand upon what you learn in this chapter.
>
> Most of you have graphed linear equations before. However, you probably have not graphed the nonlinear equations presented in Section 3.1. Make sure you read this section carefully. We will be using the technique presented in this section when we graph other nonlinear equations later in the book.
>
> In Section 3.7 we graph linear inequalities *in two variables*. We will use the procedures for graphing linear inequalities in two variables again when we graph systems of linear inequalities in Section 4.6.

Numbered Section Objectives

Each section begins with a **list of objectives.** Numbered icons connect the objectives to the appropriate sections of the text.

> **3.1 GRAPHS**
>
> 
>
>
> 1. Plot points in the Cartesian coordinate system.
> 2. Draw graphs by plotting points.
> 3. Graph nonlinear equations.
> 4. Use a graphing calculator.
> 5. Interpret graphs.
>
> 1) **Plot Points in the Cartesian Coordinate System**

In-Text Examples

An abundance of **in-text examples** illustrate the concept being presented and provide a step-by-step annotated solution.

> **EXAMPLE 8** When Jim Herring went to see his mother in Cincinnati, he boarded a Southwest Airlines plane. The plane sat at the gate for 20 minutes, taxied to the runway, and then took off. The plane flew at about 600 miles per hour for about 2 hours. It then reduced its speed to about 300 miles per hour and circled the Cincinnati Airport for about 15 minutes before it came in for a landing. After landing, the plane taxied to the gate and stopped. Which graph in Figure 3.18a–3.18d best illustrates this situation?
>
> a)
>
> 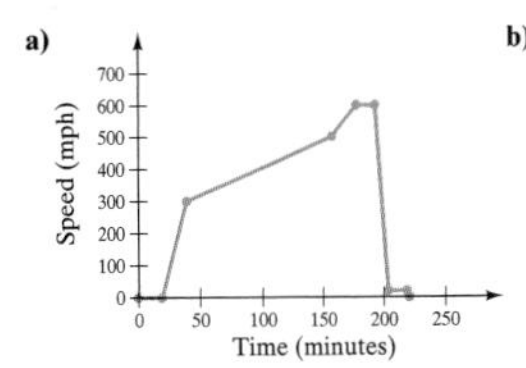
>
>
> b)
>
> 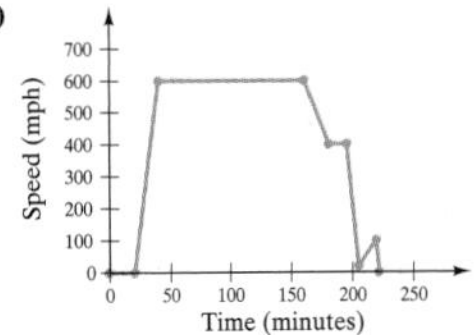
>

EMPHASIS ON *Pedagogy*

Calculator

Using Your Calculator and **Using Your Graphing Calculator** boxes provide more optional exercises for use with technology than in the previous edition as well as keystroke instructions.

Using Your Graphing Calculator Sometimes it may be difficult to estimate the intercepts of a graph accurately. When this occurs, you might want to use a graphing calculator. We demonstrate how in the following example.

EXAMPLE Determine the x- and y-intercepts of the graph $y = 1.3(x - 3.2)$.

Solution Press the Y= key, then assign 1.3 $(x - 3.2)$ to y. Then p... graph the function $y_1 =$ ... igure 3.46

how to find the zeros or roots of a function. On a TI-82 or TI-83 you press the keys 2nd TRACE to get to the CALC menu (which stands for calculate). Then you choose option 2, *zero**. Once the zero feature has been selected, the calculator will display

Left bound?**

At this time, move the cursor along the curve until it is to the *left* of the zero. Then press ENTER.

Avoiding Common Errors

Avoiding Common Errors boxes illustrate common mistakes, explain why certain procedures are wrong, and show the correct methods for working the problems.

Avoiding Common Errors

When graphing nonlinear equations, many students do not plot enough points to get a true picture of the graph. For example, when graphing $y = \frac{1}{x}$ many students consider only integer values of x. Following is a table of values for the equation and two graphs that contain the points indicated in the table.

x	-3	-2	-1	1	2	3
y	$-\frac{1}{3}$	$-\frac{1}{2}$	-1	1	$\frac{1}{2}$	$\frac{1}{3}$

Correct — $y = \frac{1}{x}$

Incorrect

Helpful Hints

Helpful Hints offer useful suggestions for problem solving and various other topics.

HELPFUL HINT

Linear equations that are not solved for y can be written using function notation by solving the equation for y, then replacing y with $f(x)$. For example, the equation $-9x + 3y = 6$ becomes $y = 3x + 2$ when solved for y. We can therefore write $f(x) = 3x + 2$.

Procedures, Important Facts, and Definitions

Procedures and **Important Facts** are presented in boxes throughout the text to make it easy for students to focus on this material and find it when preparing for quizzes and tests. Definitions are set off in **Definition Boxes** for easy reference and review.

The graph of any equation of the form $x = a$ will always be a vertical line for any real number a.

Notice that the graph of $x = -2$ does not represent a function since it does not pass the vertical line test. For $x = -2$ there is more than one value of

Vertical Line Test

If a vertical line can be drawn through any part of the graph and the line intersects another part of the graph, the graph does not represent a function. If a vertical line cannot be drawn to intersect the graph at more than one point, the graph represents a function.

We use the vertical line test to show that Figure 3.23b represents a function and

Now Try Exercises

Now Try Exercises appear after selected examples to reinforce important concepts. **Now Try Exercises** also provide students with immediate practice and make the student an active, rather than passive, learner.

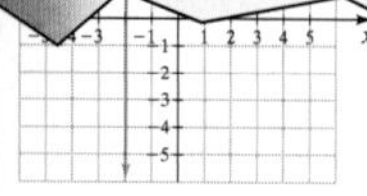

FIGURE 3.40

Notice that the graph of $x = -2$ does not represent a function since it does not pass the vertical line test. For $x = -2$ there is more than one value of y. In fact, when $x = -2$ there are an infinite number of values for y.

NOW TRY EXERCISE 35.

4) **Study Applications of Functions**

Graphs are often used to show the relationship between variables. The axes of a graph do not have to be labeled x and y; they can be any designated variables. Consider the following example.

EXAMPLE 6 The yearly profit, p, of a tire store can be estimated by the function $p(n) = 20n - 30{,}000$, where n is the number of tires sold per year.

a) Draw a graph of profit versus tires sold for up to and including 6000 tires.

b) Estimate the number of tires that must be sold for the company to break even.

c) Estimate the number of tires sold if the company has a $40,000 profit.

Solution **a) Understand** The profit, p, is a function of the number of tires sold, n. The horizontal axis will therefore be labeled number of tires sold...

To the Student

Algebra is a course that cannot be learned by observation. To learn algebra you must become an active participant. You must read the text, pay attention in class, and, most importantly, you must work the exercises. The more exercises you work, the better.

The text was written with you in mind. Short, clear sentences are used, and many examples are given to illustrate specific points. The text stresses useful applications of algebra. Hopefully, as you progress through the course, you will come to realize that algebra is not just another math course that you are required to take, but a course that offers a wealth of useful information and applications.

This text makes full use of color. The different colors are used to highlight important information. Important procedures, definitions, and formulas are placed within colored boxes.

The boxes marked **Helpful Hints** should be studied carefully, for they stress important information. The boxes marked **Avoiding Common Errors** should also be studied carefully. These boxes point out errors that students commonly make, and provide the correct procedures for doing these problems.

Ask your professor early in the course to explain the policy on when the calculator may be used. Pay particular attention to the **Using Your Calculator** boxes. You should also read the **Using Your Graphing Calculator** boxes even if you are not using a graphing calculator in class. You may find the information presented here helps you better understand the algebraic concepts.

Other questions you should ask your professor early in the course include: What supplements are available for use? Where can help be obtained when the professor is not available? Supplements that may be available include: Student's Study Guide, Student's Solutions Manual, tutorial software, and videotapes, including a tape on the study skills needed for success in mathematics.

You may wish to form a study group with other students in your class. Many students find that working in small groups provides an excellent way to learn the material. By discussing and explaining the concepts and exercises to one another you reinforce your own understanding. Once guidelines and procedures are determined by your group, make sure to follow them.

One of the first things you should do is to read Section 1.1, Study Skills Needed for Success in Mathematics. Read this section slowly and carefully, and pay particular attention to the advice and information given. Occasionally, refer back to this section. This could be the most important section of the book. Carefully read the material on doing your homework and on attending class.

At the end of all Exercise Sets (after the first two) are **Cumulative Review Exercises**. You should work these problems on a regular basis, even if they are not assigned. These problems are from earlier sections and chapters of the text, and they will refresh your memory and reinforce those topics. If you have a problem when working these exercises, read the appropriate section of the text or study your notes that correspond to that material. The section of the text where the Cumulative Review Exercise was introduced is indicated in brackets, [ ], to the left of the exercise. After reviewing the material, if you still have a problem, make an appointment to see your professor. Working the Cumulative Review Exercises throughout the semester will also help prepare you to take your final exam.

At the end of each chapter are a **Summary**, a set of **Review Exercises**, and a **Practice Test**. Before each examination you should review this material carefully and take the Practice Test. If you do well on the Practice Test, you should do well on the class test. The questions in the Review Exercises are marked to indicate the section in which that material was first introduced. If you have a problem with a Review Exercise question, reread the section indicated. You may also wish to take the **Cumulative Review Test** that appears at the end of every chapter.

In the back of the text there is an **answer section** which contains the answers to the *odd-numbered* exercises, including the Challenge Problems. Answers to *all* Using Your Graphing Calculator

Exercises, Cumulative Review Exercises, Review Exercises, Practice Tests, and Cumulative Review Tests are provided. Answers to the group exercises are not provided, for we wish students to reach agreement by themselves on answers to these exercises. The answers should be used only to check your work.

I have tried to make this text as clear and error free as possible. No text is perfect, however. If you find an error in the text, or an example or section that you believe can be improved, I would greatly appreciate hearing from you. If you enjoy the text, I would also appreciate hearing from you.

Allen R. Angel

BASIC CONCEPTS

CHAPTER 1

Use the Angel Web site at www.prenhall.com/angel to be linked to an internet resource that will help you further explore the following application.

In 1943, Spearfish, South Dakota experienced one of the most unusual temperature changes ever recorded. In Exercise 133 on page 27, we determine how much the temperature changed during a two minute period. This record change was caused by a Chinook wind.

Preview and Perspective

In this chapter we review many of the concepts that were presented in elementary algebra. We will be using these concepts throughout the course, so you must understand them. In Section 1.1, we discuss study skills that will help you in this and other mathematics courses. Study and apply this material. In Section 1.2, we discuss sets. Sets and their properties are unifying concepts of mathematics and will be used continuously in this and later mathematics courses you may take. We will apply the concepts of the union and intersection of sets when we discuss inequalities in Sections 2.5 and 2.6. The properties of real numbers discussed in Section 1.3 are used throughout the book, often without naming them. Pay particular attention to the definition of absolute value, for it is an important concept. The rules in Section 1.4, "Order of Operations," are used in evaluating expressions and formulas. A knowledge of order of operations is also essential in checking solutions to equations. You must understand the order of operations to be successful in this course. In Section 1.5 we discuss the rules of exponents, which will be used in the chapters on polynomials, rational expressions, and logarithms, as well as in other chapters. In the last section we use scientific notation. This section reinforces the rules of exponents and is important in the sciences and other areas where very large and very small numbers are used.

1.1 STUDY SKILLS FOR SUCCESS IN MATHEMATICS, AND USING A CALCULATOR

1) **Have a positive attitude.**
2) **Prepare for and attend class.**
3) **Prepare for and take examinations.**
4) **Find help.**
5) **Learn to use a calculator.**

You need to acquire certain study skills that will help you to complete this course successfully. These study skills will also help you succeed in any other mathematics courses you may take.

It is important for you to realize that this course is the foundation for more advanced mathematics courses. If you have a thorough understanding of algebra, you will find it easier to be successful in later mathematics courses.

1) Have a Positive Attitude

You may be thinking to yourself, "I hate math" or "I wish I did not have to take this class." You may have heard the term *math anxiety* and feel that you fall into this category. The first thing you need to do to be successful in this course is to change your attitude to a more positive one. You must be willing to give this course and yourself a fair chance.

Based on past experiences in mathematics, you may feel this will be difficult. However, mathematics is something you need to work at. Many of you taking this course are more mature now than when you took previous mathematics courses. Your maturity and your desire to learn are extremely important and can make a tremendous difference in your ability to succeed in mathematics. I believe you can be successful in this course, but you also need to believe it.

2) Prepare for and Attend Class

Preview the Material

Before class, you should spend a few minutes previewing any new material in the textbook. You do not have to understand everything you read yet. Just get a feeling for the definitions and concepts that will be discussed. This quick preview will help you to understand what your instructor is explaining during class. After the material is explained in class, read the corresponding sections of the text slowly and carefully, word by word, as explained below.

Read the Text

A mathematics text is not a novel. Mathematics textbooks should be read slowly and carefully. If you do not understand what you are reading, reread the material. When you come across a new concept or definition, you may wish to underline it so that it stands out. This way, when looking for it later, it will be easier to find. When you come across a worked-out example, read and follow the example very carefully. Do not just skim it. Try working out the example yourself on another sheet of paper. Make notes of anything you do not understand to ask your instructor.

Do the Homework

Two very important commitments that you must make to be successful in this course are to attend class and do your homework regularly. Your assignments must be worked conscientiously and completely. Mathematics cannot be learned by observation. You need to practice what you have heard in class. By doing homework you truly learn the material.

Don't forget to check the answers to your homework assignments. Answers to the odd-numbered exercises are in the back of this book. In addition, the answers to all the cumulative review exercises, end-of-chapter review exercises, practice tests, and cumulative review tests are in the back of the book.

If you have difficulty with some of the exercises, mark them and do not hesitate to ask questions about them in class. You should not feel comfortable until you understand all the concepts needed to work every assigned problem.

When you do your homework, make sure that you write it neatly and carefully. Pay particular attention to copying signs and exponents correctly. Do your homework in a step-by-step manner. This way you can refer back to it later and still understand what was written.

Attend and Participate in Class

You should attend every class. Most instructors will agree that there is an inverse relationship between absences and grades: The more absences you have, the lower your grade will be. Every time you miss a class, you miss important information. If you must miss a class, contact your instructor ahead of time and get the reading assignment and homework.

While in class, pay attention to what your instructor is saying. If you do not understand something, ask your instructor to repeat or explain the material. If you have read the upcoming material before class and have questions that have not been answered, ask your instructor. If you do not ask questions, your instructor will not know that you have a problem understanding the material.

In class, take careful notes. Write numbers and letters clearly so that you can read them later. It is not necessary to write down every word your instructor says. Copy down the major points and the examples that do not appear in the text. You should not be taking notes so frantically that you lose track of what your instructor is saying. It is a mistake to believe that you can copy down material in class without understanding it and then figure it out later.

Study

Study in the proper atmosphere. Study in an area where you are not constantly disturbed so that your attention can be devoted to what you are reading. The area where you study should be well ventilated and well lit. You should have sufficient desk space to spread out all your materials. Your chair should be comfortable. There should be no loud music to distract you from studying.

When studying, you should not only understand how to work a problem, you should also know why you follow the specific steps you do to work the problem. If you do not have an understanding of why you follow the specific process, you will not be able to solve similar problems.

Time Management

It is recommended that students spend at least 2 hours studying and doing homework for every hour of class time. Some students require more time than others. Finding the necessary time to study is not always easy. The following are some suggestions that you may find helpful.

1. Plan ahead. Determine when you will have time to study and do your homework. Do not schedule other activities for these time periods. Try to space these periods evenly over the week.
2. Be organized so that you will not have to waste time looking for your books, pen, calculator, or notes.
3. Use a calculator to perform tedious calculations.
4. When you stop studying, clearly mark where you stopped in the text.
5. Try not to take on added responsibilities. You must set your priorities. If your education is a top priority, as it should be, you may have to cut the time spent on other activities.
6. If time is a problem, do not overburden yourself with too many courses. Consider taking fewer credits. If you do not have sufficient time to study, your understanding and your grades in all of your courses may suffer.

3) Prepare for and Take Examinations

Study for an Exam

If you do some studying each day, you should not need to cram the night before an exam. If you wait until the last minute, you will not have time to seek the help you may need. To review for an exam:

1. Read your class notes.
2. Review your homework assignments.
3. Study the formulas, definitions, and procedures given in the text.

4. Read the Common Student Error boxes and Helpful Hint boxes carefully.
5. Read the summary at the end of each chapter.
6. Work the review exercises at the end of each chapter. If you have difficulties, restudy those sections. If you still have trouble, seek help.
7. Work the practice chapter test.
8. Work the Cumulative Review Test if material from earlier chapters will be included on the test.

Take an Exam

Make sure that you get a good night's sleep the day before the test. If you studied properly, you should not have to stay up late the night before to prepare for the test. Arrive at the exam site early so that you have a few minutes to relax before the exam. If you need to rush to get to the exam, you will start out nervous and anxious. After you receive the exam, do the following:

1. Carefully write down any formulas or ideas that you need to remember.
2. Look over the entire exam quickly to get an idea of its length and to make sure that no pages are missing. You will need to pace yourself to make sure that you complete the entire exam. Be prepared to spend more time on problems worth more points.
3. Read the test directions carefully.
4. Read each problem carefully. Answer each question completely and make sure that you have answered the specific question asked.
5. Starting with number 1, work each question in order. If you come across a question that you are not sure of, do not spend too much time on it. Continue working the questions that you understand. After completing all other questions, go back and finish those questions you were not sure of. Do not spend too much time on any one question.
6. Attempt each problem. You may be able to earn at least partial credit.
7. Work carefully and write clearly so that your instructor can read your work. Also, it is easy to make mistakes when your writing is unclear.
8. Check your work and your answers if you have time.
9. Do not be concerned if others finish the test before you. Do not be disturbed if you are the last to finish. Use all your extra time to check your work.

4) Find Help

Use the Supplements

This text comes with numerous supplements. The student's supplements are listed in the Preface. Find out from your instructor early in the semester which of these supplements are available and which supplements might be beneficial for you to use. Reading supplements should not replace reading the text, but should enhance your understanding of the material. If you miss a class, you may want to review the videotape on the topic you missed before attending the next class.

Seek Help

One thing I stress with my own students is to *get help as soon as you need it!* Do not wait! In mathematics, one day's material is often based on the previous day's material. So, if you don't understand the material today, you may not be able to understand the material tomorrow.

Where should you seek help? There are often a number of places to obtain help on campus. You should try to make a friend in the class with whom you can study. Often you can help one another. You may wish to form a study group with other students in your class. Discussing the concepts and homework with your peers will reinforce your own understanding of the material.

You should not hesitate to visit your instructor when you are having problems with the material. Be sure you read the assigned material and attempt the homework before meeting with your instructor. Come prepared with specific questions to ask.

Often other sources of help are available. Many colleges have a mathematics laboratory or a mathematics learning center where tutors are available to help students. Ask your instructor early in the semester if any tutors are available, and find out where the tutors are located. Then use these tutors as needed.

5) Learn to Use a Calculator

Many instructors require their students to purchase and to use a calculator in class. You should find out as soon as possible which calculator, if any, your instructor expects you to use. If you plan on taking additional mathematics courses, you should determine which calculator will be required in those courses and consider purchasing that calculator for use in this course if its use is permitted by your instructor. Many instructors require a scientific calculator and many others require a graphing calculator.

In this book we provide information about both types of calculators. Always read and save the user's manual for whatever calculator you purchase.

Exercise Set 1.1

Do you know all of the following information? If not, ask your instructor as soon as possible.

1. What is your instructor's name?
2. What are your instructor's office hours?
3. Where is your instructor's office located?
4. How can you best reach your instructor?
5. Where can you obtain help if your instructor is not available?
6. What supplements are available to assist you in learning?
7. Does your instructor recommend or require a specific calculator? If so, which one?
8. When can you use a calculator? Can it be used in class, on homework, on tests?
9. What is your instructor's attendance policy?
10. Why is it important that you attend every class possible?
11. Do you know the name and phone number of a friend in class?
12. For each hour of class time, how many hours outside class are recommended for homework and studying?
13. List what you should do to be properly prepared for each class.
14. Explain how a mathematics textbook should be read.
15. Write a summary of the steps you should follow when taking an exam.
16. Having a positive attitude is very important for success in this course. Are you beginning this course with a positive attitude? It is important that you do!
17. You need to make a commitment to spend the time necessary to learn the material, to do the homework, and to attend class regularly. Explain why you believe this commitment is necessary to be successful in this course.
18. What are your reasons for taking this course?
19. What are your goals for this course?
20. Have you given any thought to studying with a friend or a group of friends? Can you see any advantages in doing so? Can you see any disadvantages in doing so?

1.2 SETS AND OTHER BASIC CONCEPTS

1) **Identify sets.**
2) **Identify and use inequalities.**
3) **Use set builder notation.**
4) **Find the union and intersection of sets.**
5) **Identify important sets of numbers.**

We begin by introducing some important definitions. When a letter is used to represent various numbers it is called a **variable**. For instance, if t = the time, in hours, that a car is traveling, then t is a variable since the time is constantly changing as the car is traveling. We often use the letters x, y, z, and t to represent variables. However, other letters may be used. When presenting properties or rules, the letters a, b, and c are often used as variables.

If a letter represents one particular value it is called a **constant**. For example, if s = the number of seconds in a minute, then s represents a constant because there are always 60 seconds in a minute. The number of seconds in a minute does not vary. In this book, letters representing both variables and constants are italicized.

The term **algebraic expression**, or simply **expression**, will be used often in the text. An expression is any combination of numbers, variables, exponents, mathematical symbols, and mathematical operations.

1) Identify Sets

Sets are used in many areas of mathematics, so an understanding of sets and set notation is important. A **set** is a collection of objects. The objects in a set are called **elements** of the set. Sets are indicated by means of braces, { }, and are often named with capital letters. When the elements of a set are listed within the braces, as illustrated below, the set is said to be in **roster form**.

$$A = \{a, b, c\}$$
$$B = \{\text{yellow, green, blue, red}\}$$
$$C = \{1, 2, 3, 4, 5\}$$

Set A has three elements, set B has four elements, and set C has five elements. The symbol $\in$ is used to indicate that an item is an element of a set. Since 2 is an element of set C, we may write $2 \in C$; this is read "2 is an element of set C."

A set may be finite or infinite. Sets A, B, and C each have a finite number of elements and are therefore *finite sets*. In some sets it is impossible to list all the elements. These are *infinite sets*. The following set, called the set of **natural numbers** or **counting numbers**, is an example of an infinite set.

$$N = \{1, 2, 3, 4, 5, \ldots\}$$

The three dots after the last comma indicate that the set continues in the same manner.

Another important infinite set is the integers. The set of **integers** follows.

$$I = \{\ldots, -4, -3, -2, -1, 0, 1, 2, 3, 4, \ldots\}$$

Notice that the set of integers includes both positive and negative integers and the number 0.

If we write

$$D = \{1, 2, 3, 4, 5, \ldots, 280\}$$

we mean that the set continues in the same manner until the number 280. Set D is the set of the first 280 natural numbers. D is therefore a finite set.

A special set that contains no elements is called the **null set**, or **empty set**, written { } or $\varnothing$. For example, the set of students in your class over the age of 150 is the null or empty set.

2) Identify and Use Inequalities

Before we introduce a second method of writing a set, called *set builder notation*, we will introduce the inequality symbols.

Inequality Symbols

$>$ is read "is greater than."
$\geq$ is read "is greater than or equal to."
$<$ is read "is less than."
$\leq$ is read "is less than or equal to."
$\neq$ is read "is not equal to."

Inequalities can be explained using the real number line (Fig. 1.1).

FIGURE 1.1 −6 −5 −4 −3 −2 −1 0 1 2 3 4 5 6

The number a is greater than the number b, $a > b$, when a is to the right of b on the number line (Fig. 1.2). We can also state that the number b is less than a, $b < a$, when b is to the left of a on the number line. The inequality $a \neq b$ means either $a < b$ or $a > b$.

FIGURE 1.2 Lesser b Greater a

EXAMPLE 1 Insert either $>$ or $<$ in the shaded area between the numbers to make each statement true.

a) 7 ▒ 3 **b)** −6 ▒ −4 **c)** 0 ▒ −2

Solution Draw a number line illustrating the location of all the given values (Fig. 1.3).

FIGURE 1.3 −7 −6 −5 −4 −3 −2 −1 0 1 2 3 4 5 6 7

a) $7 > 3$ Note that 7 is to the right of 3 on the number line.
b) $-6 < -4$ Note that −6 is to the left of −4 on the number line.
c) $0 > -2$ Note that 0 is to the right of −2 on the number line.

NOW TRY EXERCISE 29

Remember that the symbol used in an inequality, if it is true, always points to the smaller of the two numbers.

We use the notation $x > 2$, read "x is greater than 2," to represent *all* real numbers greater than 2. We use the notation $x \leq -3$, read "x is less than or equal to −3," to represent all real numbers that are less than or equal to −3. The notation $-4 \leq x < 3$ means all real numbers that are greater than or equal to −4 and also less than 3. In the inequalities $x > 2$ and $x \leq -3$, the 2 and the −3 are called **endpoints**. In the inequality $-4 \leq x < 3$, the −4 and 3 are the endpoints.

The solutions to inequalities that use either $<$ or $>$ do not include the endpoints, but the solutions to inequalities that use either $\le$ or $\ge$ do include the endpoints. When inequalities are illustrated on the number line, a solid circle is used to show that the endpoint is included in the answer, and an open circle is used to show that the endpoints are not included. Below are some illustrations of how certain inequalities are indicated on the number line.

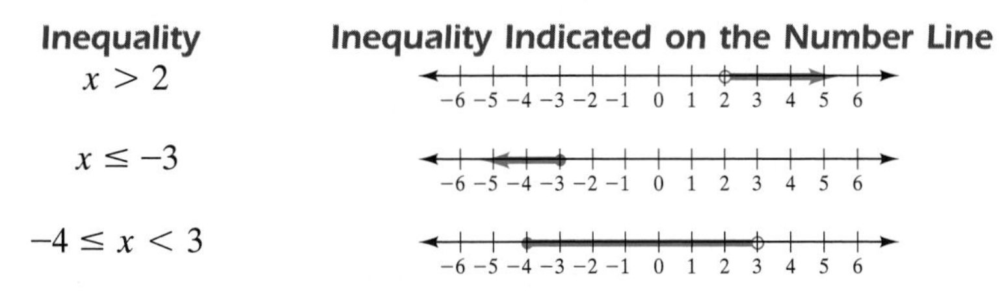

Some students misunderstand the word *between*. The word *between* indicates that the endpoints are not included in the answer. For example, the set of natural numbers between 2 and 6 is $\{3, 4, 5\}$. If we wish to include the endpoints, we can use the word *inclusive*. For example, the set of natural numbers between 2 and 6 inclusive is $\{2, 3, 4, 5, 6\}$.

3) Use Set Builder Notation

Now that we have introduced the inequality symbols we will discuss another method of indicating a set, called **set builder notation**. An example of set builder notation is

$$E = \{x \mid x \text{ is a natural number greater than } 6\}$$

This is read "Set E is the set of all elements x, such that x is a natural number greater than 6." In roster form, this set is written

$$E = \{7, 8, 9, 10, 11, \ldots\}$$

The general form of set builder notation is

$$\{\quad x \quad \mid \quad x \text{ has property } p \quad\}$$

The set of — all elements x — such that — x has the given property

Two condensed ways of writing set $E = \{x \mid x \text{ is a natural number greater than } 6\}$ in set builder notation follow.

$$E = \{x \mid x > 6 \text{ and } x \in N\} \quad \text{or} \quad E = \{x \mid x \ge 7 \text{ and } x \in N\}$$

The set $A = \{x \mid -3 < x \le 4 \text{ and } x \in I\}$ is the set of integers greater than -3 and less than or equal to 4. The set written in roster form is $\{-2, -1, 0, 1, 2, 3, 4\}$. Notice that the endpoint -3 is not included in the set but the endpoint 4 is included.

How do the sets $B = \{x \mid x > 2 \text{ and } x \in N\}$ and $C = \{x \mid x > 2\}$ differ? Can you write each set in roster form? Can you illustrate both sets on the number line? Set B contains only the natural numbers greater than 2, that is, $\{3, 4, 5, 6, \ldots\}$. Set C contains not only the natural numbers greater than 2 but also fractions and decimal numbers greater than 2. If you attempted to write set C in roster form, where would you begin? What is the smallest number greater than 2? Is it 2.1 or 2.01 or 2.001? Since there is no smallest number greater than 2, this set cannot be written in roster form. Below we illustrate these two sets on the number line. We have also illustrated two other sets.

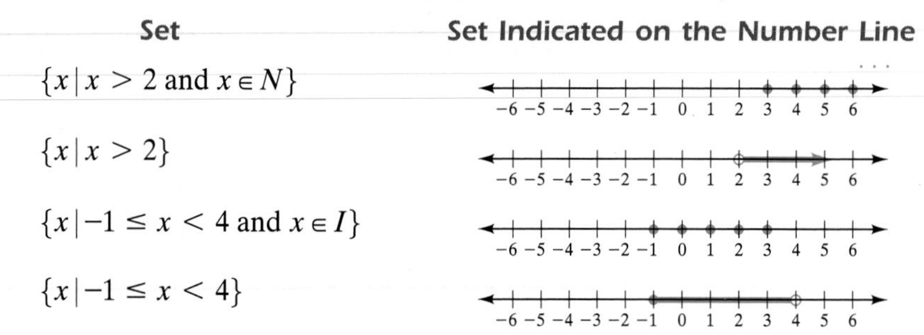

NOW TRY EXERCISE 77

Another method of indicating inequalities, called **interval notation**, will be discussed in Section 2.5.

4) Find the Union and Intersection of Sets

Just as *operations* such as addition and multiplication are performed on numbers, operations can be performed on sets. Two set operations are *union* and *intersection*.

DEFINITION

The **union** of set A and set B, written $A \cup B$, is the set of elements that belong to either set A *or* set B.

The union is formed by combining, or joining together, the elements in set A with those in set B.

Examples of Union of Sets

$$A = \{1, 2, 3, 4, 5\}, \quad B = \{3, 4, 5, 6, 7\}, \quad A \cup B = \{1, 2, 3, 4, 5, 6, 7\}$$

$$A = \{a, b, c, d, e\}, \quad B = \{x, y, z\}, \quad A \cup B = \{a, b, c, d, e, x, y, z\}$$

In set builder notation we can express $A \cup B$ as

$$A \cup B = \{x \mid x \in A \quad \text{or} \quad x \in B\}$$

DEFINITION

The **intersection** of set A and set B, written $A \cap B$, is the set of all elements that are common to both set A *and* set B.

Examples of Intersection of Sets

$$A = \{1, 2, 3, 4, 5\}, \quad B = \{3, 4, 5, 6, 7\}, \quad A \cap B = \{3, 4, 5\}$$

$$A = \{a, b, c, d, e\}, \quad B = \{x, y, z\}, \quad A \cap B = \{\ \}$$

Note that in the last example sets A and B have no elements in common. Therefore, their intersection is the empty set. In set builder notation we can express $A \cap B$ as

NOW TRY EXERCISE 57

$$A \cap B = \{x \mid x \in A \quad \text{and} \quad x \in B\}$$

5 Identify Important Sets of Numbers

At this point we have all the necessary information to discuss important sets of real numbers.

Important Sets of Real Numbers	
Real numbers	$\{x \mid x$ is a point on the number line$\}$
Natural or counting numbers	$\{1, 2, 3, 4, 5, \ldots\}$
Whole numbers	$\{0, 1, 2, 3, 4, 5, \ldots\}$
Integers	$\{\ldots, -3, -2, -1, 0, 1, 2, 3, \ldots\}$
Rational numbers	$\left\{\frac{p}{q} \,\middle|\, p \text{ and } q \text{ are integers, } q \neq 0\right\}$
Irrational numbers	$\{x \mid x$ is a real number that is not rational$\}$

Let us briefly look at the rational, irrational, and real numbers. A **rational number** is any number that can be represented as a quotient of two integers, with the denominator not zero.

Examples of Rational Numbers

$$\frac{3}{5}, \quad \frac{-2}{3}, \quad 0, \quad 1.63, \quad 7, \quad -12, \quad \sqrt{4}$$

Notice that 0, or any other integer, is also a rational number since it can be written as a fraction with a denominator of 1. For example, $0 = \frac{0}{1}$ and $7 = \frac{7}{1}$.

The number 1.63 can be written $\frac{163}{100}$ and is thus a quotient of two integers. Since $\sqrt{4} = 2$ and 2 is an integer, $\sqrt{4}$ is a rational number. *Every rational number when written as a decimal number will be either a repeating or a terminating decimal number.*

Examples of Repeating Decimals	Examples of Terminating Decimals
$\frac{2}{3} = 0.6666\ldots$ The 6 repeats	$\frac{1}{2} = 0.5$
$\frac{1}{7} = 0.142857142857\ldots$ The block 142857 repeats	$\frac{7}{4} = 1.75$

To show that a digit or group of digits repeat, we can place a bar above the digit or group of digits that repeat. For example, we may write

$$\frac{2}{3} = 0.\overline{6} \qquad \text{and} \qquad \frac{1}{7} = 0.\overline{142857}$$

Although $\sqrt{4}$ is a rational number, the square roots of most integers are not. Most square roots will be neither terminating nor repeating decimals when expressed as decimal numbers and are *irrational numbers*. Some irrational numbers are $\sqrt{2}$, $\sqrt{3}$, $\sqrt{5}$, and $\sqrt{6}$. Another irrational number is pi, π. When we give a decimal value for an irrational number, we are giving only an *approximation* of the value of the irrational number. The symbol $\approx$ means "is approximately equal to."

$$\pi \approx 3.14 \qquad \sqrt{2} \approx 1.41$$

The **real numbers** are formed by taking the *union* of the rational numbers with the irrational numbers. Therefore, any real number must be either a rational number or an irrational number. The symbol $\mathbb{R}$ is often used to represent the set of real numbers. Figure 1.4 illustrates various real numbers on the number line.

$-\sqrt{23}$ -3.62 $-\frac{8}{5}$ 0 $\frac{1}{2}$ $\sqrt{2}$ 2 $\frac{20}{7}$ π 4.3

−6 −5 −4 −3 −2 −1 0 1 2 3 4 5 6

FIGURE 1.4

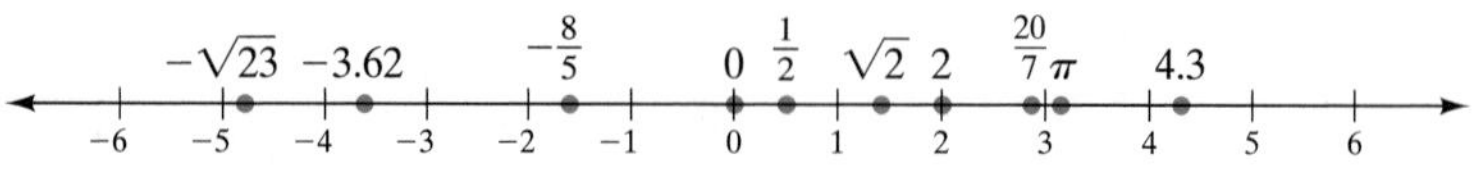

A first set is a **subset** of a second set when every element of the first set is also an element of the second set. For example, the set of natural numbers, $\{1, 2, 3, 4, \ldots\}$ is a subset of the set of whole numbers, $\{0, 1, 2, 3, 4, \ldots\}$, because every element in the set of natural numbers is also an element in the set of whole numbers. Figure 1.5 illustrates the relationships between the various subsets of the real numbers. In Figure 1.5a, you see that the set of natural numbers is a subset of the set of whole numbers, of the set of integers, and of the set of rational numbers. Therefore, every natural number must also be a whole number, an integer, and a rational number. Using the same reasoning, we can see that the set of whole numbers is a subset of the set of integers and of the set of rational numbers, and that the set of integers is a subset of the set of rational numbers.

Using Figure 1.5b we see that the positive integers, 0, and the negative integers form the integers, that the integers and noninteger rational numbers form the rational numbers, and so on.

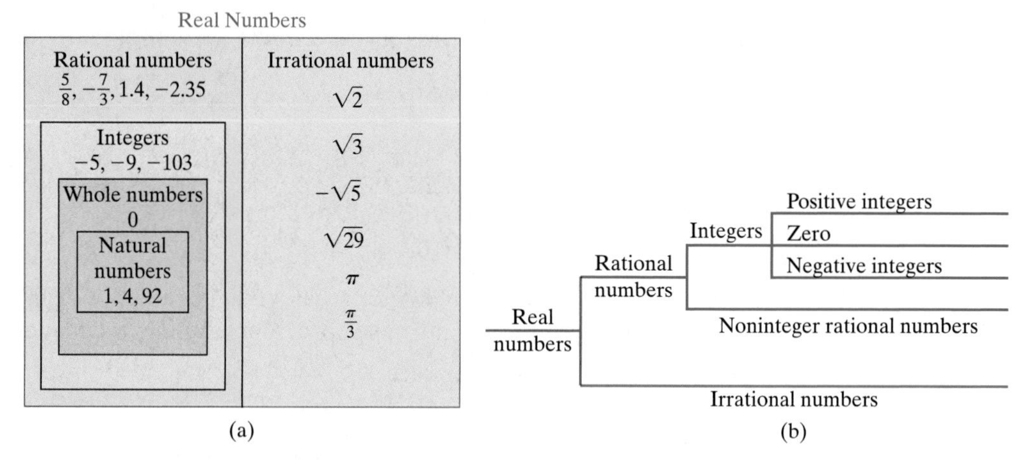

FIGURE 1.5

EXAMPLE 2 Consider the following set:

$$\left\{-4,\ 0,\ \frac{3}{5},\ 2.7,\ \sqrt{3},\ -\sqrt{5},\ \frac{19}{6},\ 18,\ 4.62,\ -23,\ \pi\right\}$$

List the elements of the set that are

a) natural numbers. **b)** whole numbers. **c)** integers.
d) rational numbers. **e)** irrational numbers. **f)** real numbers.

Solution **a)** Natural numbers: 18

b) Whole numbers: 0, 18

c) Integers: −4, 0, 18, −23

d) Rational numbers can be written in the form p/q, $q \neq 0$. Each of the following can be written in this form and is a rational number.

$$-4,\ 0,\ \frac{3}{5},\ 2.7,\ \frac{19}{6},\ 18,\ 4.62,\ -23$$

e) Irrational numbers are real numbers that are not rational. The following numbers are irrational.

$$\sqrt{3},\ -\sqrt{5},\ \pi$$

f) All of the numbers in the set are real numbers. The union of the rational numbers and the irrational numbers forms the real numbers.

$$-4,\ 0,\ \frac{3}{5},\ 2.7,\ \sqrt{3},\ -\sqrt{5},\ \frac{19}{6},\ 18,\ 4.62,\ -23,\ \pi$$

NOW TRY EXERCISE 49

Not all numbers are real numbers. Some numbers that we discuss later in the text that are not real numbers are complex numbers and imaginary numbers.

Exercise Set 1.2

Concept/Writing Exercises

1. What is a variable?
2. What is an algebraic expression?
3. What is a set?
4. What do we call the objects in a set?
5. Is the set of natural or counting numbers a finite or infinite set? Explain.

6. What is the null or empty set?
7. List the 5 inequality symbols and write down how each is read.
8. Give an example of a set that is empty.
9. List the set of integers *between* 4 and 9.
10. List the set of integers *between* 4 and 9 *inclusive*.
11. Explain why every integer is also a rational number.
12. Describe the counting numbers, whole numbers, integers, rational numbers, irrational numbers, and real numbers. Explain the relationships among the sets of numbers.

Indicate whether each statement is true or false.

13. Some rational numbers are integers.
14. Every natural number is a whole number.
15. Every whole number is a natural number.
16. Every integer is a rational number.
17. Every rational number is an integer.
18. The union of the set of rational numbers with the set of irrational numbers forms the set of real numbers.
19. The intersection of the set of rational numbers and the set of irrational numbers is the empty set.
20. The set of natural numbers is a finite set.
21. The set of integers between 1 and 2 is the null set.
22. The set of rational numbers between 1 and 2 is an infinite set.

Practice the Skills

Insert either $<$ or $>$ in the shaded area to make each statement true.

23. $6 \;\square\; 8$
24. $-2 \;\square\; 4$
25. $0 \;\square\; 4$
26. $-1 \;\square\; 1$
27. $-3 \;\square\; -3.5$
28. $4 \;\square\; -3$
29. $-5 \;\square\; -3$
30. $-6 \;\square\; -1$
31. $-4.6 \;\square\; -4.7$
32. $-3.6 \;\square\; -3.2$
33. $1.1 \;\square\; 1.9$
34. $-1.1 \;\square\; -1.9$
35. $-952 \;\square\; -955$
36. $-780 \;\square\; -655$
37. $-\frac{7}{8} \;\square\; -\frac{8}{9}$
38. $-\frac{4}{7} \;\square\; -\frac{5}{9}$

List each set in roster form.

39. $A = \{x \mid 5 < x < 7 \text{ and } x \in N\}$
40. $B = \{x \mid x$ is an even integer between 2 and 8$\}$
41. $C = \{x \mid x$ is an even integer greater than or equal to 6 and less than 10$\}$
42. $D = \{x \mid x > 6 \text{ and } x \in N\}$
43. $E = \{x \mid x < 7 \text{ and } x \in W\}$
44. $F = \{x \mid -\frac{6}{5} \leq x < \frac{15}{4} \text{ and } x \in N\}$
45. $H = \{x \mid x$ is a whole number multiple of 5$\}$
46. $I = \{x \mid x$ is an integer greater than $-5\}$
47. $J = \{x \mid x$ is an integer between -6 and $-4.3\}$
48. $K = \{x \mid x$ is a whole number between 3 and 4$\}$
49. Consider the set $\left\{-3, 4, \frac{1}{2}, \frac{5}{9}, 0, \sqrt{2}, \sqrt{8}, -1.23, \frac{99}{100}\right\}$. List the elements that are:
    a) natural numbers.
    b) whole numbers.
    c) integers.
    d) rational numbers.
    e) irrational numbers.
    f) real numbers.
50. Consider the set $\{2, 4, -5.33, \frac{9}{2}, \sqrt{7}, \sqrt{2}, -100, -7, 4.7\}$. List the elements that are:
    a) whole numbers.
    b) natural numbers.
    c) rational numbers.
    d) integers.
    e) real numbers.
    f) irrational numbers.

Find $A \cup B$ and $A \cap B$ for each set A and B.

51. $A = \{5, 6, 7\}, B = \{6, 7, 8\}$
52. $A = \{2, 4, 6, 8\}, B = \{1, 3, 5, 7\}$
53. $A = \{-2, -4, -5\}, B = \{-1, -2, -4, -6\}$
54. $A = \{-1, 0, 1\}, B = \{0, 2, 4, 6\}$
55. $A = \{\ \}, B = \{0, 1, 2, 3\}$
56. $A = \{2, 4, 6\}, B = \{2, 4, 6, 8, \ldots\}$
57. $A = \{0, 2, 4, 6, 8\}, B = \{1, 3, 5, 7\}$
58. $A = \{1, 3, 5\}, B = \{1, 3, 5, 7, \ldots\}$
59. $A = \{0.1, 0.2, 0.3\}, B = \{0.2, 0.3, 0.4, 0.5, \ldots\}$
60. $A = \left\{1, \frac{1}{2}, \frac{1}{4}, \frac{1}{6}, \ldots\right\}, B = \left\{\frac{1}{4}, \frac{1}{6}, \frac{1}{8}\right\}$

Describe each set.

61. $A = \{1, 2, 3, 4, \ldots\}$
62. $B = \{5, 7, 9, 11, \ldots\}$
63. $C = \{8, 10, 12, \ldots, 30\}$
64. $A = \{a, b, c, d, \ldots, z\}$
65. $B = \{\ldots, -5, -3, -1, 1, 3, 5, \ldots\}$
66. $C = \{$Alabama, Alaska, …, Wyoming$\}$

*In Exercises 67 and 68, **a)** write out how you would read each set; **b)** write the set in roster form.*

67. $A = \{x \mid x < 8 \text{ and } x \in N\}$
68. $B = \{x \mid x$ is one of the last five capital letters in the English alphabet$\}$

Illustrate each set on a number line.

69. $\{x \mid x < 3\}$

70. $\{x \mid x \geq -1\}$

71. $\{x \mid x \geq 6\}$

72. $\{x \mid x \leq -5\}$

73. $\left\{x \,\middle|\, -4 < x \leq \frac{3}{7}\right\}$

74. $\{x \mid -1.67 \leq x < 5.02\}$

75. $\{x \mid x > 2 \text{ and } x \in N\}$

76. $\{x \mid -1.90 \leq x \leq 2.1 \text{ and } x \in I\}$

77. $\left\{x \,\middle|\, x < \frac{40}{9} \text{ and } x \in N\right\}$

78. $\left\{x \,\middle|\, \frac{1}{2} < x \leq \frac{5}{9} \text{ and } x \in N\right\}$

Express in set builder notation each set of numbers indicated on the number line.

79. 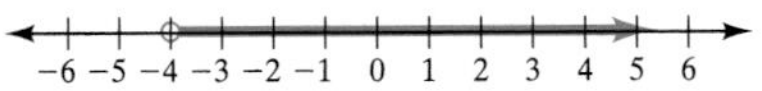

80. 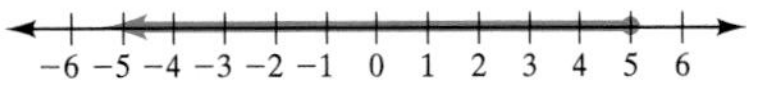

81. 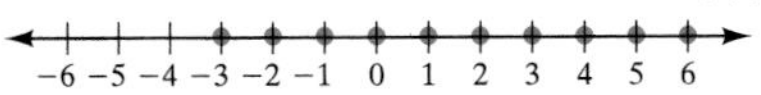

82. 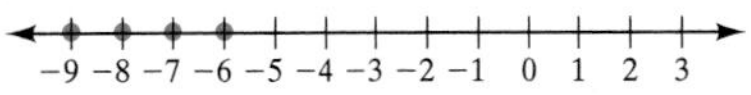

83. 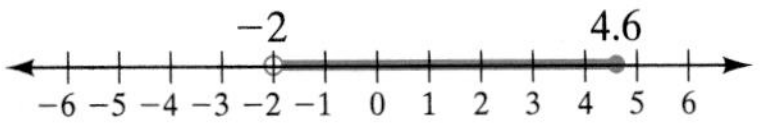

84. 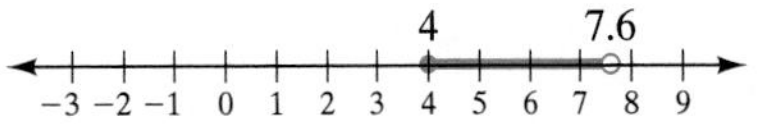

85. 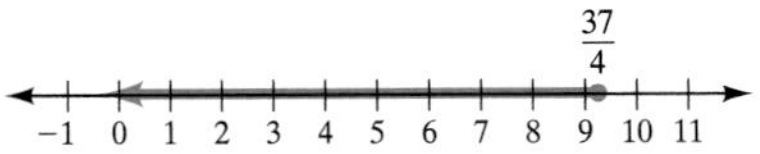

86. 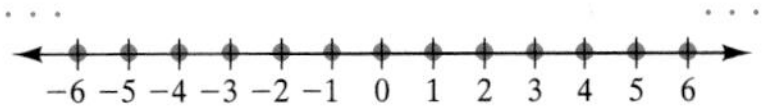

87. 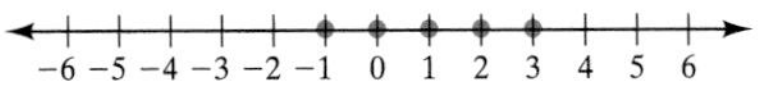

88. 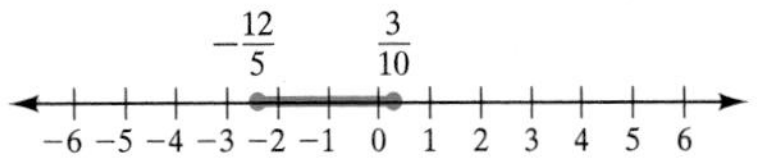

Let N = the set of natural numbers, W = the set of whole numbers, I = the set of integers, Q = the set of rational numbers, H = the set of irrational numbers, and $\mathbb{R}$ = the set of real numbers. Determine whether the first set is a subset of the second set for each pair of sets.

89. N, W

90. W, Q

91. I, Q

92. W, N

93. Q, H

94. $Q, \mathbb{R}$

95. $H, \mathbb{R}$

96. Q, I

Problem Solving

97. Construct a set that contains five rational numbers between 0 and 1.

98. Construct a set that contains five rational numbers between 1 and 2.

99. Determine two sets A and B such that $A \cup B = \{3, 5, 7, 8, 9\}$ and $A \cap B = \{5, 7\}$.

100. Determine two sets A and B such that $A \cup B = \{2, 4, 5, 6, 8, 9\}$ and $A \cap B = \{4, 5, 9\}$.

101. The table shows the top men's and women's brands ranked by repurchase intent for 1996.

a) Find the set of brand names that were listed in *either* the men's or women's category.

b) Find the set of brand names that were in *both* the men's and women's category.

Top Brands in 1996

Men's	Women's
1. Levi's	1. Levi's
2. Lands' End	2. Hanes
3. L.L. Bean	3. Reebok
4. Nike	4. Hanes Her Way
5. Gold Toe	5. L.L. Bean
5. London Fog	6. Arizona
5. Reebok	

SOURCES: Kurt Salmon Associates and NPD Group Inc.

c) Does part **a)** represent the union or intersection of the brands? Explain.

d) Does part b) represent the union or intersection of the brands? Explain.

102. The following charts show the top five network TV advertisers and the top five magazine advertisers in 1996.

Top 10 Network TV Advertisers in 1996 ($ in thousands)	
Company	**Report total**
1 General Motors Corp.	$613,872.7
2 Procter & Gamble Co.	589,467.9
3 Johnson & Johnson	504,776.7
4 PepsiCo Inc.	423,402.5
5 Philip Morris Co.	403,066.9

Top 10 Magazine Advertisers in 1996 ($ in thousands)	
Company	**Report total**
1 General Motors Corp.	$456,433.1
2 Philip Morris Co.	343,147.8
3 Procter & Gamble	280,194.3
4 Ford Motor Co.	280,152.0
5 Chrysler Corp.	269,451.5

a) Find the set of companies that were in the top five in *either* category in 1996.

b) Find the set of companies that were in *both* categories in 1996.

c) Does part a) represent the union or the intersection of the companies? Explain.

d) Does part b) represent the union or the intersection of the companies? Explain.

103. The following graph shows the percentage weight given to different goods and services in the consumer price index.

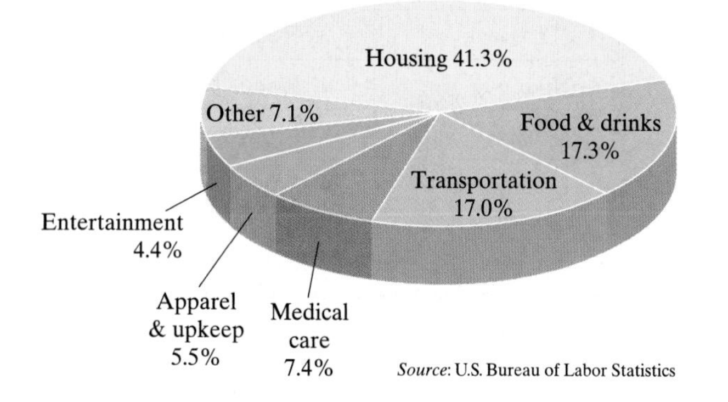

a) List the set of goods and services that have a weight of 17% or greater.

b) List the set of goods and services that have a weight of less than 6%.

104. In the February 1998 issue of Consumer Report, 24 brands of paper towels were tested and ranked. The 10 brands of paper towels that ranked best in overall score are shown in the chart below.

Brand and Model		Price Roll	Price Sheet	Sheets per roll	Overall score (0–10: P F G VG E)
Kleenex	Viva Job Squad	$1.59	3.2 ¢	50	
Kleenex	Viva	0.89	1.6	55	
Bounty	Quilted	0.97	1.5	64	
Bounty	Quilted Big Roll Microwave	1.49	1.6	96	
Bounty	Rinse & Reuse	1.56	1.7	90	
Brawny		0.84	1.4	60	
Green Forest		0.79	1.5	52	
Sparkle		0.75	1.0	72	
Scott Towels		0.72	0.9	80	
So-Dri		0.53	1.0	52	

a) List the set of brands of paper towels that ranked excellent.

b) List the set of brands of paper towels that ranked very good.

c) List the set of brands of paper towels that ranked good.

d) List the set of brands of paper towels that ranked at least good and whose cost per sheet is less than 1.5 cents.

105. The following diagram is called a *Venn diagram*. From the diagram determine the following sets: **a)** A, **b)** B, **c)** $A \cup B$, **d)** $A \cap B$.

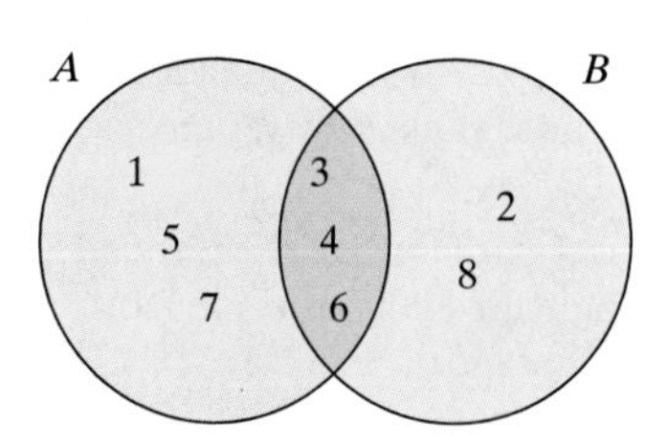

106. Use the following Venn diagram to determine the following sets: **a)** A, **b)** B, **c)** $A \cup B$, **d)** $A \cap B$.

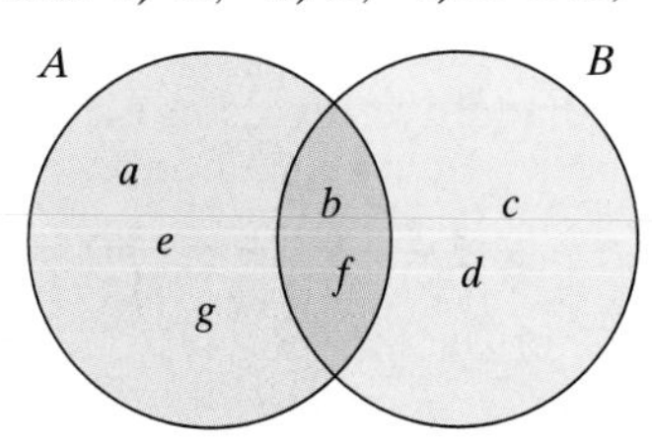

107. a) Explain the difference between the following sets of numbers: $\{x \mid x > 1 \text{ and } x \in N\}$ and $\{x \mid x > 1\}$.

b) Write the first set given in roster form.

c) Can you write the second set in roster form? Explain your answer.

108. Repeat Exercise 107 for the sets $\{x \mid 2 < x < 6 \text{ and } x \in N\}$ and $\{x \mid 2 < x < 6\}$.

Challenge Problems

109. a) Write the decimal numbers equivalent to $\frac{1}{9}$, $\frac{2}{9}$, and $\frac{3}{9}$.

b) Write the fractions equivalent to $0.\overline{4}$, $0.\overline{5}$, and $0.\overline{6}$.

c) What is $0.\overline{9}$ equal to? Explain how you determined your answer.

Group Activity

110. The Venn diagram that follows shows the results of a survey given to 45 people. The diagram shows the number of people in the survey who read the *New York Post*, the *New York Daily News*, and *The Wall Street Journal*.

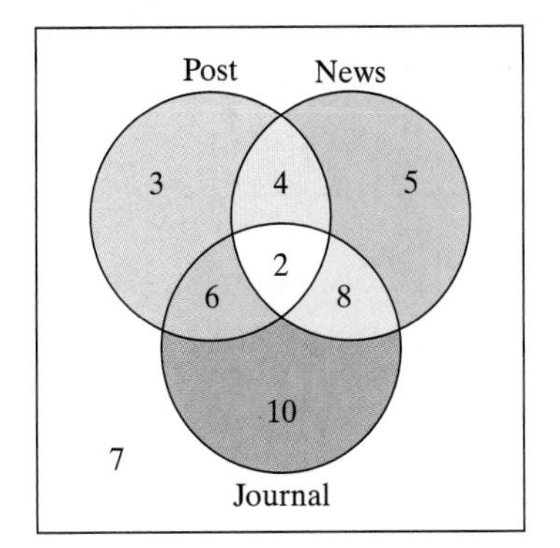

a) Group member 1: Determine the number surveyed who read *both* the *News* and the *Post*, that is, *News* $\cap$ *Post*.

b) Group member 2: Determine the number who read both the *Post* and the *Journal*, that is, *Post* $\cap$ *Journal*.

c) Group member 3: Determine the number who read both the *News* and the *Journal*, that is, *News* $\cap$ *Journal*.

d) Share your answer with the other members of the group and see if the group agrees with your answer.

e) As a group, determine the number of people that read all three papers.

f) As a group, determine the number of people that do not read any of the three papers.

1.3 PROPERTIES OF AND OPERATIONS WITH REAL NUMBERS

1. **Evaluate absolute values.**
2. **Add real numbers.**
3. **Subtract real numbers.**
4. **Multiply real numbers.**
5. **Divide real numbers.**
6. **Use the properties of real numbers.**

To succeed in algebra you must understand how to add, subtract, multiply, and divide real numbers. Before we can explain addition and subtraction of real numbers, we need to discuss absolute value.

Two numbers that are the same distance from 0 on the number line but in opposite directions are called **additive inverses**, or **opposites**, of each other. For example, 3 is the additive inverse of −3, and −3 is the additive inverse of 3. The number 0 is its own additive inverse. The sum of a number and its additive inverse is 0. What are the additive inverses of −56.3 and $\frac{76}{5}$? Their additive inverses are 56.3 and $-\frac{76}{5}$, respectively. Notice that the additive inverse of a positive number is a negative number and the additive inverse of a negative number is a positive number.

Definition

Additive Inverse

For any real number a, its additive inverse is $-a$.

Consider the number −5. Its additive inverse is $-(-5)$. Since we know this number must be positive, this implies that $-(-5) = 5$. This is an example of the double negative property.

Double Negative Property

For any real number a, $-(-a) = a$.

By the double negative property $-(-7.4) = 7.4$ and $-(-\frac{12}{5}) = \frac{12}{5}$.

1) Evaluate Absolute Values

The **absolute value** of a number is its distance from the number 0 on a number line. The symbol | | is used to indicate absolute value.

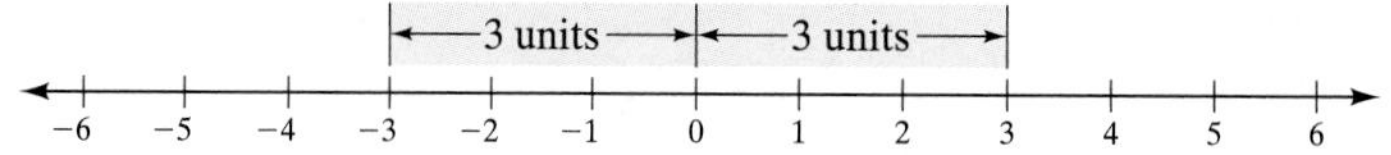

FIGURE 1.6

Consider the numbers 3 and −3 (Fig. 1.6). Both numbers are 3 units from 0 on the number line. Thus

$$|3| = 3 \quad \text{and} \quad |-3| = 3$$

EXAMPLE 1 Evaluate. **a)** $|4|$ **b)** $|-1.6|$ **c)** $|0|$

Solution

a) $|4| = 4$, since 4 is 4 units from 0 on a number line.

b) $|-1.6| = 1.6$, since −1.6 is 1.6 units from 0 on a number line.

c) $|0| = 0$.

The absolute value of any nonzero number will always be a positive number, and the absolute value of 0 is 0.

To find the absolute value of a real number without using a number line, use the following definition.

Definition

Absolute Value

If a represents any real number, then

$$|a| = \begin{cases} a & \text{if } a \geq 0 \\ -a & \text{if } a < 0 \end{cases}$$

The definition of absolute value indicates that the absolute value of any nonnegative number is the number itself, and the absolute value of any negative number is the additive inverse (or opposite) of the number. The absolute value of a number can be found by using the definition, as illustrated below.

$|8.4| = 8.4$ Since 8.4 is greater than or equal to 0, its absolute value is 8.4.

$|0| = 0$ Since 0 is greater than or equal to 0, its absolute value is 0.

$|-12| = -(-12) = 12$ Since -12 is less than 0, its absolute value is $-(-12)$ or 12.

EXAMPLE 2 Evaluate using the definition of absolute value. **a)** $-|5|$ **b)** $-|-6.43|$

Solution **a)** We are finding the opposite of the absolute value of 5. Since the absolute value of 5 is positive, its opposite must be negative.

$$-|5| = -(5) = -5$$

b) We are finding the opposite of the absolute value of -6.43. Since the absolute value of -6.43 is positive, its opposite must be negative.

$$-|-6.43| = -(6.43) = -6.43$$

EXAMPLE 3 Insert $>$, $<$, or $=$ in the shaded area between the two values to make each statement true. **a)** $|-3| \;\square\; |3|$ **b)** $-|-6| \;\square\; |-5|$

Solution **a)** $|-3| = |3|$, since both $|-3|$ and $|3|$ equal 3.

NOW TRY EXERCISE 39

b) $-|-6| < |-5|$, since $-|-6| = -6$ and $|-5| = 5$.

2) Add Real Numbers

We first discuss how to add two numbers with the same sign, either both positive or both negative, and then we will discuss how to add two numbers with different signs, one positive and the other negative.

To Add Two Numbers with the Same Sign (Both Positive or Both Negative)

Add their absolute values and place the common sign before the sum.

The sum of two positive numbers will be a positive number, and the sum of two negative numbers will be a negative number.

EXAMPLE 4 Evaluate $-2 + (-5)$.

Solution Since both numbers being added are negative, the sum will be negative. To find the sum, add the absolute values of these numbers and then place a negative sign before the value.

$$|-2| = 2 \qquad |-5| = 5$$

Now add the absolute values.

$$|-2| + |-5| = 2 + 5 = 7$$

Since both numbers are negative, the sum must be negative. Thus,

$$-2 + (-5) = -7$$

To Add Two Numbers with Different Signs (One Positive and the Other Negative)

Take the difference of the absolute values. The answer is positive if the positive number has the larger absolute value. The answer is negative if the negative number has the larger absolute value.

The sum of a positive number and a negative number may be either positive, negative, or zero. The sign of the answer will be the same as the sign of the number with the larger absolute value.

EXAMPLE 5 Evaluate $5 + (-9)$.

Solution Since the numbers being added are of opposite signs, we find the difference of their absolute values. First we take each absolute value.

$$|5| = 5 \qquad |-9| = 9$$

Now we find the difference, $9 - 5 = 4$. The number -9 has a larger absolute value than the number 5, so their sum is negative.

$$5 + (-9) = -4$$

EXAMPLE 6 Evaluate. **a)** $6.4 + (-8.5)$ **b)** $\frac{5}{8} + \left(-\frac{4}{5}\right)$

Solution **a)** $6.4 + (-8.5) = -2.1$

b) $\frac{5}{8} + \left(-\frac{4}{5}\right) = \frac{25}{40} + \left(-\frac{32}{40}\right) = \frac{25 + (-32)}{40} = \frac{-7}{40} = -\frac{7}{40}$

EXAMPLE 7 The Palau Trench in the Pacific Ocean lies 26,424 feet below sea level. The deepest ocean trench, the Mariana Trench, is 9416 feet deeper than the Palau Trench (Fig. 1.7). Find the depth of the Mariana Trench.

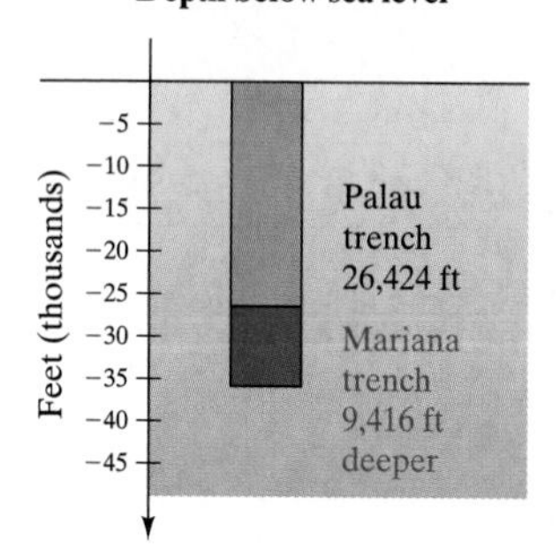

FIGURE 1.7

Solution Consider distance below sea level to be negative. Therefore, the total depth is

$$-26{,}424 + (-9416) = -35{,}840 \text{ feet}$$

or 35,840 feet below sea level.

NOW TRY EXERCISE 135

3) Subtract Real Numbers

Every subtraction problem can be expressed as an addition problem using the following rule.

Subtraction of Real Numbers

$$a - b = a + (-b)$$

To subtract b from a, add the opposite (or additive inverse) of b to a.

For example, $5 - 7$ means $5 - (+7)$. To subtract $5 - 7$, add the opposite of $+7$, which is -7, to 5.

$$5 - 7 = 5 + (-7)$$

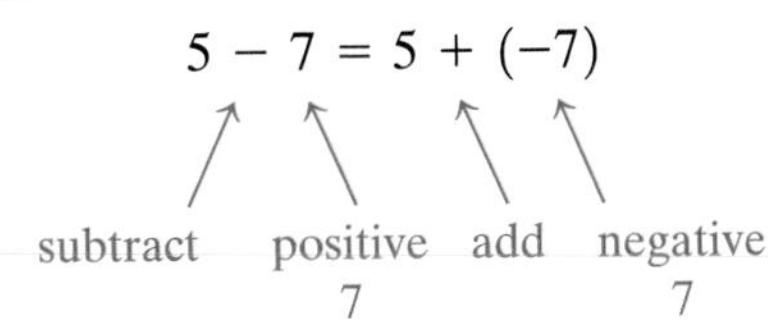

Since $5 + (-7) = -2$, then $5 - 7 = -2$.

EXAMPLE 8 Evaluate. **a)** $6 - 10$ **b)** $-8 - 4$

Solution **a)** $6 - 10 = 6 + (-10) = -4$ **b)** $-8 - 4 = -8 + (-4) = -12$

EXAMPLE 9 Evaluate $8 - (-10)$.

Solution In this problem we are subtracting a negative number. The procedure to subtract remains the same.

$$8 - (-10) = 8 + 10 = 18$$

subtract negative 10 add positive 10

Thus, $8 - (-10) = 18$.

By studying Example 9 and similar problems, we can see that for any real numbers a and b,

$$a - (-b) = a + b$$

We can use this principle to evaluate problems such as $8 - (-10)$ and other problems where we *subtract a negative quantity*.

EXAMPLE 10 Evaluate $-4 - (-12)$.

Solution $-4 - (-12) = -4 + 12 = 8$

EXAMPLE 11 **a)** Subtract 35 from -42. **b)** Subtract $-\frac{3}{5}$ from $-\frac{5}{9}$.

Solution **a)** $-42 - 35 = -77$

NOW TRY EXERCISE 57

b) $-\frac{5}{9} - \left(-\frac{3}{5}\right) = -\frac{5}{9} + \frac{3}{5} = -\frac{25}{45} + \frac{27}{45} = \frac{2}{45}$

EXAMPLE 12 The 1997 Information Please Almanac lists the average high and low temperatures in January for selected cities in the United States. The highest average low temperature, 72.6°, occurs in Honolulu, Hawaii. The lowest average low temperature, −12.6°, occurs in Fairbanks, Alaska, (Fig. 1.8). Find the difference between these two temperatures.

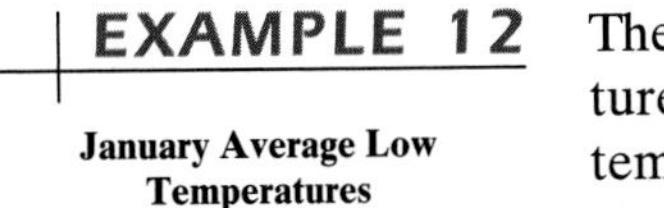

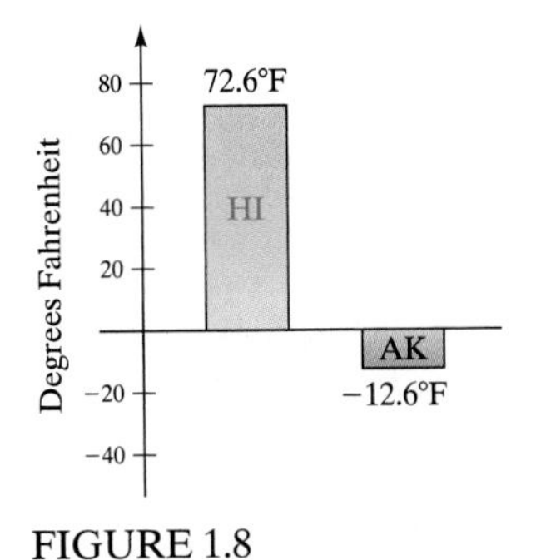

FIGURE 1.8

Solution To find the difference, we subtract.

$$72.6° - (-12.6°) = 72.6° + 12.6° = 85.2°$$

Addition and subtraction are often combined in the same problem, as in the following examples. Unless parentheses are present, if the expression involves only addition and subtraction we add and subtract from left to right. When parentheses are used, we add and subtract within the parentheses first; then we add and subtract from left to right.

EXAMPLE 13 Evaluate $-12 - (4 - 8) - (-3)$.

Solution

$$\begin{aligned} -12 - (4 - 8) - (-3) &= -12 - (-4) + 3 \\ &= -12 + 4 + 3 \\ &= -8 + 3 = -5 \end{aligned}$$

EXAMPLE 14 Evaluate $2 - |-3| + 4 - (6 - |-7|)$.

Solution Begin by replacing the numbers in absolute value signs with their numerical equivalents; then evaluate.

$$\begin{aligned} 2 - |-3| + 4 - (6 - |-7|) &= 2 - 3 + 4 - (6 - 7) \\ &= 2 - 3 + 4 - (-1) \\ &= 2 - 3 + 4 + 1 \\ &= -1 + 4 + 1 \\ &= 3 + 1 = 4 \end{aligned}$$

NOW TRY EXERCISE 69

4 Multiply Real Numbers

The following rules are used in determining the product when two numbers are multiplied.

Multiply Two Real Numbers

1. To multiply two numbers with **like signs**, either both positive or both negative, multiply their absolute values. The answer is **positive**.
2. To multiply two numbers with **unlike signs**, one positive and the other negative, multiply their absolute values. The answer is **negative**.

EXAMPLE 15 Evaluate. **a)** $(4.2)(-1.6)$ **b)** $(-16)\left(-\frac{1}{2}\right)$.

Solution **a)** $(4.2)(-1.6) = -6.72$ *The numbers have unlike signs.*

b) $(-16)\left(-\frac{1}{2}\right) = 8$ *The numbers have like signs, both negative.*

EXAMPLE 16 Evaluate $4(-2)(-3)(1)$.

Solution $4(-2)(-3)(1) = (-8)(-3)(1) = 24(1) = 24$

When multiplying more than two numbers, the product will be *negative* when there is an *odd* number of negative numbers. The product will be *positive* when there is an *even* number of negative numbers.

The multiplicative property of zero indicates that the product of 0 and any number is 0.

Multiplicative Property of Zero

For any number a,

$$a \cdot 0 = 0 \cdot a = 0$$

By the multiplicative property of zero, $5(0) = 0$ and $(-7.3)(0) = 0$.

EXAMPLE 17 Evaluate $9(5)(-2.63)(0)(4)$.

Solution If one or more of the factors is 0, the product is 0. Thus, $9(5)(-2.63)(0)(4) = 0$. Can you explain why the product of any number of factors will be zero if any factor is 0?

5) Divide Real Numbers

The rules for the division of real numbers are very similar to those for multiplication of real numbers.

Divide Two Real Numbers

1. To divide two numbers with **like signs**, either both positive or both negative, divide their absolute values. The answer is **positive**.
2. To divide two numbers with **unlike signs**, one positive and the other negative, divide their absolute values. The answer is **negative**.

EXAMPLE 18 Evaluate. **a)** $-24 \div 6$ **b)** $-6.45 \div (-0.4)$

Solution

a) $\frac{-24}{6} = -4$ *The numbers have unlike signs.*

b) $\frac{-6.45}{-0.4} = 16.125$ *The numbers have like signs.*

EXAMPLE 19 Evaluate $\frac{-3}{8} \div \left|\frac{-2}{5}\right|$.

Solution Since $\left|\frac{-2}{5}\right|$. is equal to $\frac{2}{5}$, we write

$$\frac{-3}{8} \div \left|\frac{-2}{5}\right| = \frac{-3}{8} \div \frac{2}{5}$$

Now invert the divisor and proceed as in multiplication.

$$\frac{-3}{8} \div \frac{2}{5} = \frac{-3}{8} \cdot \frac{5}{2} = \frac{-3 \cdot 5}{8 \cdot 2} = -\frac{15}{16}$$

NOW TRY EXERCISE 85

When the denominator of a fraction is a negative number, we usually rewrite the fraction with a positive denominator. To do this, we use the following fact.

Sign of a Fraction

For any number a and any nonzero number b,

$$\frac{a}{-b} = \frac{-a}{b} = -\frac{a}{b}$$

Thus, when we have a quotient of $\frac{1}{-2}$, we rewrite it as either $\frac{-1}{2}$ or $-\frac{1}{2}$.

6 Use the Properties of Real Numbers

We have already discussed the double negative property and the multiplicative property of zero. Table 1.1 lists other basic properties for the operations of addition and multiplication on the real numbers.

TABLE 1.1

For real numbers *a*, *b*, and *c*	Addition	Multiplication
Commutative property	$a + b = b + a$	$ab = ba$
Associative property	$(a + b) + c = a + (b + c)$	$(ab)c = a(bc)$
Identity property	$a + 0 = 0 + a = a$ (0 is called the **additive identity element**.)	$a \cdot 1 = 1 \cdot a = a$ (1 is called the **multiplicative identity element**.)
Inverse property	$a + (-a) = (-a) + a = 0$ ($-a$ is called the **additive inverse** or **opposite** of a.)	$a \cdot \frac{1}{a} = \frac{1}{a} \cdot a = 1$ ($1/a$ is called the **multiplicative inverse** or **reciprocal** of a, $a \neq 0$.)
Distributive property (of multiplication over addition)	$a(b + c) = ab + ac$	

Note that the commutative property involves a change in *order*, and the associative property involves a change in *grouping*.

The distributive property applies when there are more than two numbers within the parentheses.

$$a(b + c + d + \cdots + n) = ab + ac + ad + \cdots + an$$

This expanded form of the distributive property is called the **extended distributive property**.

EXAMPLE 20 Name each property illustrated.

a) $6 \cdot x = x \cdot 6$ **b)** $(x + 2) + 3y = x + (2 + 3y)$

c) $2x + 3y = 3y + 2x$ **d)** $3x(y + 2) = 3x(y) + 3x(2)$

Solution **a)** Commutative property of multiplication: change of order, $6 \cdot x = x \cdot 6$

b) Associative property of addition: change of grouping, $(x + 2) + 3y = x + (2 + 3y)$

c) Commutative property of addition: change of order, $2x + 3y = 3y + 2x$

d) Distributive property:

$$3x(y + 2) = 3x(y) + 3x(2)$$

In Example 20 **d)** the expression $3x(y) + 3x(2)$ can be simplified to $3xy + 6x$ using the properties of the real numbers. Can you explain why?

NOW TRY EXERCISE 119

EXAMPLE 21 Name each property illustrated.

a) $4 \cdot 1 = 4$ **b)** $x + 0 = x$

c) $4 + (-4) = 0$ **d)** $1(x + y) = x + y$

Solution **a)** Identity property of multiplication **b)** Identity property of addition

c) Inverse property of addition **d)** Identity property of multiplication

EXAMPLE 22 Write the additive inverse (or opposite) and multiplicative inverse (or reciprocal) of each of the following.

a) -3 **b)** $\frac{2}{3}$

Solution **a)** The additive inverse is 3. The multiplicative inverse is $\frac{1}{-3} = -\frac{1}{3}$.

b) The additive inverse is $-\frac{2}{3}$. The multiplicative inverse is $\frac{1}{\frac{2}{3}} = \frac{3}{2}$.

Exercise Set 1.3

Concept/Writing Exercises

1. What are additive inverses or opposites?
2. Give an example of the double negative property?
3. Give the definition of absolute value.
4. Will the absolute value of every real number be a positive number? Explain.

Find the unknown number(s). Explain how you determined your answer.

5. All numbers a such that $|a| = |-a|$
6. All numbers a such that $|a| = a$
7. All numbers a such that $|a| = -a$
8. All numbers a such that $|a| = -3$
9. All numbers a such that $|a| = 5$
10. All numbers x such that $|x - 3| = |3 - x|$
11. Explain in your own words how to add two numbers with different signs.
12. In your own words, explain how to add two numbers with the same sign.
13. In your own words, explain how to subtract real numbers.
14. In your own words, explain how the rules for multiplication and division of real numbers are similar.
15. List two other ways that the fraction $\frac{a}{-b}$ may be written.
16. **a)** Write the commutative property of addition.
    b) In your own words, explain the property.
17. **a)** Write the associative property of multiplication.
    b) In your own words, explain the property.
18. **a)** Write the distributive property of multiplication over addition.
    b) In your own words, explain the property.
19. Using an example, explain why addition is not distributive over multiplication. That is, explain why $a + (b \cdot c) \neq (a + b) \cdot (a + c)$.
20. Give an example of the extended distributive property.

Practice the Skills

Evaluate each absolute value expression.

21. $|3|$ **22.** $|-8|$ **23.** $|-6|$ **24.** $|1.3|$

25. $\left|-\frac{3}{4}\right|$ **26.** $|-7.32|$ **27.** $|0|$ **28.** $-|7|$

29. $-|-7|$ **30.** $-|-9|$ **31.** $-\left|\frac{5}{9}\right|$ **32.** $-\left|-\frac{7}{8}\right|$

Insert <, >, or = in the shaded area to make each statement true.

33. $|6| \;\square\; |-6|$ **34.** $|-9| \;\square\; |8|$ **35.** $-4 \;\square\; |-4|$ **36.** $|-10| \;\square\; -5$

37. $-4 \;\square\; -|4|$ **38.** $|-20| \;\square\; -|24|$ **39.** $|-7| \;\square\; -|3|$ **40.** $-|9| \;\square\; -|11|$

41. $-|-10| \;\square\; |-5|$ **42.** $6 \;\square\; |-12|$ **43.** $|19| \;\square\; |-25|$ **44.** $-|-1| \;\square\; |-6|$

List the values from smallest to largest.

45. $6, 2, -1, |3|, |-5|$

46. $1, -8, |-6|, |13|, -3$

47. $-32, |-7|, 15, -|4|, 4$

48. $-8, -12, -|9|, -|20|, -|-18|$

49. $-2.1, -2, -2.4, |-2.8|, -|2.9|$

50. $-6.1, |-6.3|, -|-6.5|, 6.8, |6.4|$

51. $\frac{1}{3}, \left|-\frac{1}{2}\right|, -2, \left|\frac{3}{5}\right|, \left|-\frac{3}{4}\right|$

52. $\left|-\frac{5}{2}\right|, \frac{3}{5}, |-3|, \left|-\frac{5}{3}\right|, \left|-\frac{2}{3}\right|$

Evaluate each addition and subtraction problem.

53. $4 + (-3)$ **54.** $-4 + 12$ **55.** $-18 + 7$ **56.** $-16 - (-5)$

57. $-14 - (-11)$ **58.** $-6.28 - 3.14$ **59.** $-9.5 - (-3.72)$ **60.** $\frac{5}{6} - \frac{4}{5}$

61. $-\frac{3}{5} - \left(-\frac{2}{9}\right)$ **62.** $-3 - \frac{5}{12}$ **63.** $5 + (-0.43) - 6.97$ **64.** $-|6.4| - 5.3 + 1.4$

65. $8.9 - |8.5| - |17.6|$ **66.** $|9 - 4| - 6$ **67.** $|5 - 12| - |3|$ **68.** $|12 - 5| - |5 - 12|$

69. $-|-3| - |7| + (6 + |-2|)$ **70.** $|-4| - |-4| - |-4 - 4|$ **71.** $\left(\frac{3}{5} + \frac{3}{4}\right) - \frac{1}{2}$ **72.** $\frac{4}{5} - \left(\frac{3}{4} - \frac{2}{3}\right)$

Evaluate each multiplication and division problem.

73. $-4 \cdot 12$ **74.** $(-8)(-7)$ **75.** $-4\left(-\frac{5}{16}\right)$ **76.** $-4\left(-\frac{3}{4}\right)\left(-\frac{1}{2}\right)$

77. $(-1)(-2)(-1)(2)(-3)$ **78.** $(-2.3)(4.9)(-6.2)$ **79.** $(-1.1)(3.4)(8.3)(-7.6)$ **80.** $-16 \div 8$

81. $-80 \div (-8)$ **82.** $36 \div \left(-\frac{1}{4}\right)$ **83.** $-\frac{5}{9} \div \frac{-5}{9}$ **84.** $\left|-\frac{1}{2}\right| \cdot \left|\frac{-3}{4}\right|$

85. $\left|\frac{-4}{7}\right| \div \frac{1}{14}$ **86.** $\left|\frac{3}{8}\right| \div (-2)$ **87.** $\left|\frac{-5}{6}\right| \div \left|\frac{-1}{2}\right|$ **88.** $\frac{-5}{9} \div |-5|$

Evaluate.

89. $4 - 7$ **90.** $-16 - 8$ **91.** $-20 \div (-4)$

92. $-\frac{3}{5} - \frac{5}{9}$ **93.** $4\left(-\frac{8}{5}\right)\left(\frac{5}{2}\right)$ **94.** $8.2 + (-4.9) - (6.8 - 9.4)$

95. $(-2.1)(-22.3)(-8.6)$ **96.** $(4.2)(-1)(-9.6)(3.8)$ **97.** $-16.4 - (-9.6) - 14.8$

98. $9 - (4 - 3) - (-2 - 1)$ **99.** $-|8| \cdot \left|\frac{-1}{2}\right|$ **100.** $-\left|\frac{-12}{5}\right| \cdot \left|\frac{3}{4}\right|$

101. $\left|\frac{-9}{4}\right| \div \left|\frac{-4}{9}\right|$ **102.** $(-|3| + |5|) - (1 - |-9|)$ **103.** $5 - |-7| + 3 - |-2|$

104. $\left(\frac{3}{8} - \frac{4}{7}\right) - \left(-\frac{1}{2}\right)$ **105.** $\left(-\frac{3}{5} - \frac{4}{9}\right) - \left(-\frac{2}{3}\right)$ **106.** $(|-4| - 3) - (3 \cdot |-5|)$

107. $(25 - |32|)(-6 - 5)$ **108.** $\left[(-2)\left|-\frac{1}{2}\right|\right] \div \left|-\frac{1}{4}\right|$

Name each property illustrated.

109. $x + y = y + x$

110. $3(x + 2) = 3x + 6$

111. $x \cdot 0 = 0$

112. $3 \cdot x = x \cdot 3$

113. $(x + 3) + 6 = x + (3 + 6)$

114. $x + 0 = x$

115. $x = 1 \cdot x$

116. $x(y + z) = xy + xz$

117. $5(xy) = (5x)y$

118. $(2x \cdot 3y) \cdot 4y = 2x \cdot (3y \cdot 4y)$

119. $4(x + y + 2) = 4x + 4y + 8$

120. $-(-1) = 1$

121. $5 + 0 = 5$

122. $4 \cdot \frac{1}{4} = 1$

123. $3 + (-3) = 0$

124. $6 \cdot 0 = 0$

125. $x \cdot \frac{1}{x} = 1$

126. $(x + y) = 1(x + y)$

127. $-(-x) = x$

128. $x + (-x) = 0$

List both the additive inverse and the multiplicative inverse for each problem.

129. 4

130. -3

131. $-\frac{2}{3}$

132. $-\frac{3}{7}$

Problem Solving

133. The most unusual temperature change according to the *Guinness Book of World Records* occurred from 7:30 a.m. to 7:32 a.m. on January 22, 1943 in Spearfish, South Dakota. During these two minutes the temperature changed from −4°F to 45°F. Determine the increase in temperature in these two minutes.

134. In New York City, the temperature during a 24-hour period dropped from 46°F to −12°F. Find the change in temperature.

135. A submarine dives 358.9 feet. A short time later the submarine comes up 210.7 feet. Find the submarine's final depth from its starting point. (Consider distance in a downward direction as negative.)

136. Sharon Koch had a balance of −\$32.64 in her checking account when she deposited a check for \$99.38. What is her new balance?

137. Harvey Mutchner signed a contract with a publishing company that called for an advance payment of \$60,000 on the sale of his book *Brooklyn, the Bronx, and Other Places I Love.* When the book is published and sales begin, the publishers will automatically deduct this advance from the author's royalties.

a) Six months after the release of the book, the author's royalties totaled \$47,600 before the advance was deducted. Determine how much money he will receive or owe the publisher.

b) After 1 year, the author's royalties are \$87,500. Determine how much money he will receive or owe to the publishing company.

138. The lowest temperature ever recorded in the United States was −79.8°F on January 23, 1971 in Prospect Creek, Alaska. The lowest temperature in the contiguous states (all states except Alaska and Hawaii) was −69.7°F on January 20, 1954, in Rogers Pass, Montana. Find the difference in these temperatures.

139. In 1999, Joanne Winston made four quarterly estimated income tax payments of \$3000 each. When she completed her year 2000 income tax forms, she found her total tax was \$10,125.

a) Will Joanne be entitled to a refund or will she owe more taxes? Explain.

b) How much of a refund will she receive or how much more in taxes will she owe?

140. On Monday Lisa Coyle purchased 100 shares of Greg Middleton Industries stock for $\$11\frac{1}{4}$ per share. On Friday she sold all 100 shares for $\$10\frac{3}{8}$ per share. What was her gain or loss for this transaction?

141. Write your own realistic word problem that involves subtracting a positive number from a negative number. Indicate the answer to your word problem.

142. Write your own realistic word problem that involves subtracting a negative number from a negative number. Indicate the answer to your problem.

143. The graph to the right shows the balance of payments in 1995 between the federal government and selected states. In some states, there was a large surplus because the federal government spent significantly more in those states than it collected from them in taxes. On the other hand, some states had big deficits, paying much more in taxes than the federal government returned to them in defense spending; grants; payment to the elderly, disabled, and poor; federal employee wages; and so on.

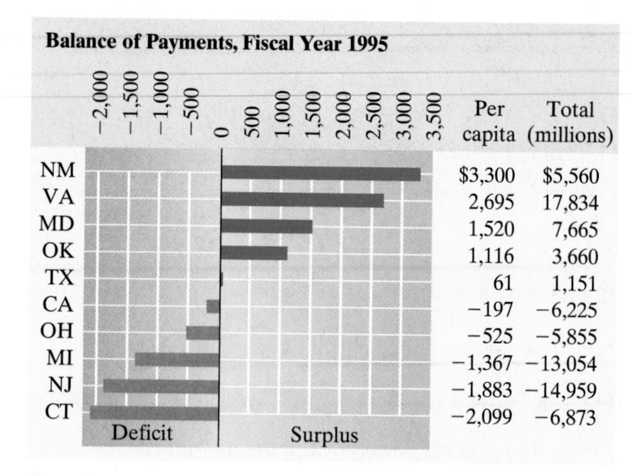

Source: Harvard University, Kennedy School of Government

Determine the difference in the per capita balance of payments between the following states.

a) NM and CT **b)** OK and MI **c)** VA and NJ **d)** TX and CA

144. The Chamber of Commerce in a small town studied the success and failure of small start-up businesses (under 5 employees) in their town. They found the average first-year expenditures and the average first-year incomes, as shown on the chart to the right. Estimate the average first-year profit by subtracting the average first-year expenditures from the average first-year income.

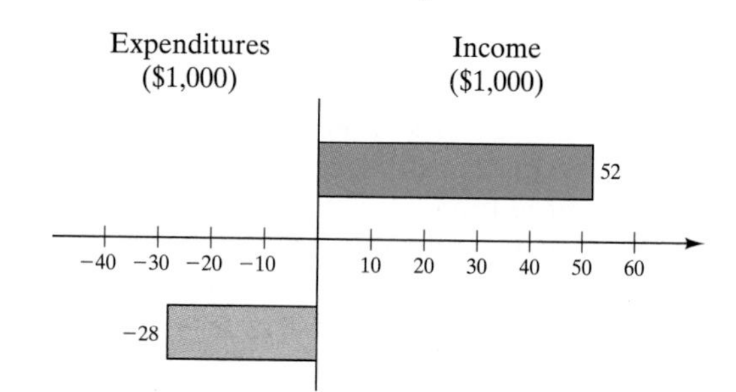

Challenge Problems

145. Evaluate $1 - 2 + 3 - 4 + \cdots + 99 - 100$. (*Hint:* Group in pairs of two numbers.)

146. Evaluate $1 + 2 - 3 + 4 + 5 - 6 + 7 + 8 - 9 + 10 + 11 - 12 + \cdots + 22 + 23 - 24$. (*Hint:* Examine in groups of three numbers.)

147. Evaluate $\dfrac{(1) \cdot |-2| \cdot (-3) \cdot |4| \cdot (-5)}{|-1| \cdot (-2) \cdot |-3| \cdot (4) \cdot |-5|}$.

148. Evaluate $\dfrac{(1)(-2)(3)(-4)(5)\cdots(97)(-98)}{(-1)(2)(-3)(4)(-5)\cdots(-97)(98)}$.

Cumulative Review Exercises

[1.2] **149.** Answer true or false: Every irrational number is a real number.

150. List the set of natural numbers.

151. Consider the set $\{3, 4, -2, \frac{5}{6}, \sqrt{3}, 0\}$. List the elements that are **a)** integers; **b)** rational numbers; **c)** irrational numbers; **d)** real numbers.

152. $A = \{4, 7, 9, 12\}$; $B = \{1, 4, 7, 15\}$. Find **a)** $A \cup B$; **b)** $A \cap B$.

153. Illustrate $\{x \mid -4 < x \leq 6\}$ on a number line.

1.4 ORDER OF OPERATIONS

1. Evaluate exponential expressions.
2. Evaluate square and higher roots.
3. Evaluate expressions using the order of operations.
4. Evaluate expressions containing variables.
5. Evaluate expressions on a graphing calculator.

Before we discuss the order of operations we need to speak briefly about exponents and roots. We will discuss exponents in greater depth in Sections 1.5 and 7.2.

1) Evaluate Exponential Expressions

In a multiplication problem, the numbers or expressions that are multiplied are called **factors**. If $a \cdot b = c$, then a and b are factors of c. For example, since $2 \cdot 3 = 6$, both 2 and 3 are factors of 6. The number 1 is a factor of every number and expression. Can you explain why?

The quantity 3^2 is called an **exponential expression**. In the expression the 3 is called the **base** and the 2 is called the **exponent**. The expression 3^2 is read "three squared" or "three to the second power." Note that

$$3^2 = \underbrace{3 \cdot 3}_{\text{2 factors of 3}}$$

The expression 5^3 is read "five cubed" or "five to the third power." Note that

$$5^3 = \underbrace{5 \cdot 5 \cdot 5}_{\text{3 factors of 5}}$$

In general, the base b to the nth power is written b^n. For any natural number n,

$$b^n = \underbrace{b \cdot b \cdot b \cdot b \cdot \cdots \cdot b}_{n \text{ factors of } b}$$

EXAMPLE 1 Evaluate. **a)** $(0.6)^3$ **b)** $(-2)^5$ **c)** 1^{10} **d)** $\left(\frac{-3}{4}\right)^3$

Solution

a) $(0.6)^3 = (0.6)(0.6)(0.6) = 0.216$

b) $(-2)^5 = (-2)(-2)(-2)(-2)(-2) = -32$

c) $1^{10} = 1$; 1 raised to any power will equal 1. Why?

d) $\left(\frac{-3}{4}\right)^3 = \left(\frac{-3}{4}\right)\left(\frac{-3}{4}\right)\left(\frac{-3}{4}\right) = -\frac{27}{64}$

It is not necessary to write exponents of 1. Whenever we encounter a numerical value or a variable without an exponent, we assume that it has an exponent of 1. Thus, 3 means 3^1, x means x^1, x^3y means x^3y^1, and $-xy$ means $-x^1y^1$.

Students often evaluate expressions containing $-x^2$ incorrectly. The expression $-x^2$ means $-(x^2)$, not $(-x)^2$. Note that -5^2 means $-(5^2) = -(5 \cdot 5) = -25$ while $(-5)^2$ means $(-5)(-5) = 25$. In general, $-x^m$ means $-(x^m)$, not $(-x)^m$. The expression $-x^2$ is read *negative x squared* or *the opposite of x^2*. The expression $(-x)^2$ is read *negative x quantity squared*.

EXAMPLE 2 Evaluate $-x^2$ for each value of x. **a)** 3 **b)** -3

Solution **a)** $-x^2 = -(3)^2 = -9$

b) $-x^2 = -(-3)^2 = -(9) = -9$

EXAMPLE 3 Evaluate $-3^2 + (-2)^4 - 4^3 + (-2^4)$.

Solution First we evaluate each exponential expression. Then we add or subtract, working from left to right.

$$-3^2 + (-2)^4 - 4^3 + (-2^4) = -(3^2) + (-2)^4 - (4^3) + (-2^4)$$
$$= -9 + 16 - 64 + (-16)$$
$$= -9 + 16 - 64 - 16$$
$$= -73$$

NOW TRY EXERCISE 59

Using Your Calculator

Evaluating Exponential Expressions on a Scientific and a Graphing Calculator

On both scientific and graphing calculators the $\boxed{x^2}$ key can be used to square a number. Below we show the sequence of keys to press to evaluate 5^2.

Scientific calculator:

5 $\boxed{x^2}$ 25 ← answer displayed

Graphing calculator:

5 $\boxed{x^2}$ $\boxed{\text{ENTER}}$ 25 ← answer displayed

To evaluate exponential expressions with other exponents, you can use the $\boxed{y^x}$ or $\boxed{\wedge}$ key. Most scientific calculators have a $\boxed{y^x}$* key, whereas graphing calculators use the $\boxed{\wedge}$ key. To evaluate exponential expressions using these keys, first enter the base, then press either the $\boxed{y^x}$ or $\boxed{\wedge}$ key, then enter the exponent. For example, to evaluate 6^4 we do the following:

Scientific calculator:

6 $\boxed{y^x}$ 4 $\boxed{=}$ 1296 ← answer displayed

Graphing calculator:

6 $\boxed{\wedge}$ 4 $\boxed{\text{ENTER}}$ 1296 ← answer displayed

*Some calculators use the $\boxed{x^y}$ or $\boxed{a^b}$ keys instead of the $\boxed{y^x}$ key.

2) Evaluate Square and Higher Roots

The symbol used to indicate a root, $\sqrt{}$, is called a **radical sign.** The number or expression inside the radical sign is called the **radicand**. In $\sqrt{25}$, the radicand is 25. The **principal or positive square root** of a positive number a, written $\sqrt{a}$, is the positive number that when multiplied by itself gives a. Whenever we use the words "square root," we are referring to the "principal square root."

EXAMPLE 4 Evaluate. **a)** $\sqrt{25}$ **b)** $\sqrt{\frac{9}{4}}$ **c)** $\sqrt{0.64}$

Solution **a)** $\sqrt{25} = 5$, since $5 \cdot 5 = 25$

b) $\sqrt{\frac{9}{4}} = \frac{3}{2}$, since $\frac{3}{2} \cdot \frac{3}{2} = \frac{9}{4}$

c) $\sqrt{0.64} = 0.8$, since $(0.8)(0.8) = 0.64$

The square root of 4, $\sqrt{4}$, is a rational number since it is equal to 2. The square roots of other numbers, such as $\sqrt{2}$, $\sqrt{3}$, and $\sqrt{5}$, are irrational numbers. The decimal values of such numbers can never be given exactly since irrational numbers are nonterminating, nonrepeating decimal numbers. The approximate value of $\sqrt{2}$ and other irrational numbers can be found with a calculator.

$$\sqrt{2} \approx 1.414213562 \quad \text{From a calculator}$$

In this section we introduce square roots, cube roots, symbolized by $\sqrt[3]{}$, and higher roots. The number used to indicate the root is called the **index**.

$$\sqrt[n]{a}$$

index → n; radical sign → $\sqrt{}$; radicand → a

The index of a square root is 2. However, we generally do not show the index 2. Therefore, $\sqrt{a} = \sqrt[2]{a}$.

The concept used to explain square roots can be expanded to explain cube roots and higher roots. The cube root of a number a is written $\sqrt[3]{a}$.

$$\sqrt[n]{a} = b \quad \text{if} \quad \underbrace{b \cdot b \cdot b}_{\text{3 factors of } b} = a$$

For example, $\sqrt[3]{8} = 2$, because $2 \cdot 2 \cdot 2 = 8$. The expression $\sqrt[n]{a}$ is read "the nth root of a."

$$\sqrt[n]{a} = b \quad \text{if} \quad \underbrace{b \cdot b \cdot b \cdot \cdots \cdot b}_{n \text{ factors of } b} = a$$

EXAMPLE 5 Evaluate. **a)** $\sqrt[3]{27}$ **b)** $\sqrt[3]{64}$ **c)** $\sqrt[4]{16}$

Solution **a)** $\sqrt[3]{27} = 3$, since $3 \cdot 3 \cdot 3 = 27$

b) $\sqrt[3]{64} = 4$, since $4 \cdot 4 \cdot 4 = 64$

c) $\sqrt[4]{16} = 2$, since $2 \cdot 2 \cdot 2 \cdot 2 = 16$

EXAMPLE 6 Evaluate. **a)** $\sqrt[4]{81}$ **b)** $\sqrt[3]{\frac{1}{27}}$ **c)** $\sqrt[3]{-8}$

Solution **a)** $\sqrt[4]{81} = 3$, since $3 \cdot 3 \cdot 3 \cdot 3 = 81$

b) $\sqrt[3]{\frac{1}{27}} = \frac{1}{3}$, since $\left(\frac{1}{3}\right)\left(\frac{1}{3}\right)\left(\frac{1}{3}\right) = \frac{1}{27}$

c) $\sqrt[3]{-8} = -2$, since $(-2)(-2)(-2) = -8$

Note that in Example 6 **c)** the cube root of a negative number is negative. Why is this so? We will discuss radicals in more detail in Chapter 7.

NOW TRY EXERCISE 25

Using Your Calculator

Evaluating Roots on a Scientific Calculator

The square roots of numbers can be found on calculators with a square-root key, $\boxed{\sqrt{x}}$. To evaluate $\sqrt{25}$ on calculators that have this key, press

25 $\boxed{\sqrt{x}}$ 5 ← answer displayed

Higher roots can be found on calculators that contain either the $\boxed{\sqrt[x]{y}}$ key or the $\boxed{y^x}$ key.* To evaluate $\sqrt[4]{625}$ on a calculator with a $\boxed{\sqrt[x]{y}}$ key, do the following:

625 $\boxed{\sqrt[x]{y}}$ 4 $\boxed{=}$ 5 ← answer displayed

Note that the number within the radical sign (the radicand), 625, is entered, then the $\boxed{\sqrt[x]{y}}$ key is pressed, and then the root (or index) 4 is entered. When the $\boxed{=}$ key is pressed, the answer 5 is displayed.

To evaluate $\sqrt[4]{625}$ on a calculator with a $\boxed{y^x}$ key, use the inverse key as follows:

625 $\boxed{\text{INV}}$ $\boxed{y^x}$ 4 $\boxed{=}$ 5 ← answer displayed

*Calculator keys vary. Some calculators have the $\boxed{x^y}$ or $\boxed{a^b}$ keys instead of the $\boxed{y^x}$ key, and some calculators use a $\boxed{2^{\text{nd}}}$ or $\boxed{\text{shift}}$ key instead of the $\boxed{\text{INV}}$ key.

Using Your Graphing Calculator

Evaluating Roots on a Graphing Calculator

To find the square root on a graphing calculator, use $\sqrt{\ }$. The $\sqrt{\ }$ appears above the $\boxed{x^2}$ key, so you will need to press the $\boxed{2^{\text{nd}}}$ key to evaluate square roots. For example, to evaluate $\sqrt{25}$ press

$\boxed{2^{\text{nd}}}$ $\boxed{x^2}$ 25 $\boxed{\text{ENTER}}$ 5 ⟵ answer displayed

When you press $\boxed{2^{\text{nd}}}$ $\boxed{x^2}$ the Texas Instruments' TI-82 generates $\sqrt{\ }$ and the TI-83 generates $\sqrt{\ }($. Then you insert the radicand and press $\boxed{\text{ENTER}}$. To learn how to find cube and higher roots, refer to your graphing calculator manual. With the TI-83, you can use the $\boxed{\text{MATH}}$ key. When you press this key you get a number of options including 4 and 5, which are shown below.

4: $\sqrt[3]{\ }($

5: $\sqrt[x]{\ }$

Option 4 can be used to find cube roots and option 5 can be used to find higher roots, as shown in the following examples.

EXAMPLE Evaluate $\sqrt[3]{120}$.

Solution

$\boxed{\text{MATH}}$ 4 120 $\boxed{\text{ENTER}}$ 4.932424149 ← answer displayed

(4: select option 4; 120: enter radicand)

To find the root with an index greater than 3, first enter the index, then press the $\boxed{\text{MATH}}$ key, then press option 5.

EXAMPLE Evaluate $\sqrt[4]{625}$.

Solution

4 $\boxed{\text{MATH}}$ 5 625 $\boxed{\text{ENTER}}$ 5 ← answer displayed

(4: index; 5: select option 5; 625: enter radicand)

We will show another way to find roots on a graphing calculator in Section 7.2 when we discuss rational exponents.

NOW TRY EXERCISE 37

3) Evaluate Expressions Using the Order of Operations

You will often have to evaluate expressions containing multiple operations. To do so, follow the *order of operations* indicated below.

Order of Operations

To evaluate mathematical expressions, use the following order:

1. First, evaluate the expressions within grouping symbols, including parentheses, ( ), brackets, [ ], and braces, { }. If the expression contains nested grouping symbol (one pair of grouping symbol within another pair), evaluate the expression in the innermost grouping symbol first.
2. Next, evaluate all terms containing exponents and roots.
3. Next, evaluate all multiplications or divisions in the order in which they occur, working from left to right.
4. Finally, evaluate all additions or subtractions in the order in which they occur, working from left to right.

It should be noted that a *fraction bar* acts as a grouping symbol. Thus, when evaluating expressions containing a fraction bar, we work separately above and below the fraction bar.

Brackets are often used in place of parentheses to help avoid confusion. For example, the expression $7((5 \cdot 3) + 6)$ is easier to follow when written $7[(5 \cdot 3) + 6]$. Remember to evaluate the innermost group first.

EXAMPLE 7 Evaluate $8 + 3 \cdot 5^2 - 7$.

Solution We will use colored shading to indicate the order in which the operations are to be evaluated. Since there are no parentheses, we first evaluate 5^2.

$$8 + 3 \cdot 5^2 - 7 = 8 + 3 \cdot 25 - 7$$

Next, we perform multiplications or divisions from left to right.

$$= 8 + 75 - 7$$

Finally, we perform additions or subtractions from left to right.

$$= 83 - 7$$
$$= 76$$

EXAMPLE 8 Evaluate $10 + \{6 - [4(5 - 2)]\}^2$.

Solution First, evaluate the expression within the innermost parentheses. Then continue according to the order of operations.

$$\begin{aligned} 10 + \{6 - [4(5 - 2)]\}^2 &= 10 + \{6 - [4(3)]\}^2 \\ &= 10 + [6 - (12)]^2 \\ &= 10 + (-6)^2 \\ &= 10 + 36 \\ &= 46 \end{aligned}$$

NOW TRY EXERCISE 77

EXAMPLE 9 Evaluate $\dfrac{6 \div \frac{1}{2} + 5|7 - 3|}{1 + (3 - 5) \div 2}$.

Solution Remember that the fraction bar acts as a grouping symbol. Work separately above the fraction bar and below the fraction bar.

$$\frac{6 \div \frac{1}{2} + 5|7 - 3|}{1 + (3 - 5) \div 2} = \frac{6 \div \frac{1}{2} + 5|4|}{1 + (-2) \div 2}$$

$$= \frac{12 + 20}{1 + (-1)}$$

$$= \frac{32}{0}$$

Since division by 0 is not possible, the original expression is **undefined**.

4 Evaluate Expressions Containing Variables

To evaluate mathematical expressions, we use the order of operations just given. Example 10 is an application problem where we use the order of operations.

EXAMPLE 10 Attendance at NASCAR Winston Cup races has grown consistently over the past 10 years. The approximate attendance at NASCAR races from 1986 through 1996, in millions, can be closely approximated by

$$\text{attendance} = 0.028x^2 + 0.018x + 2.69$$

In the expression on the right side of the equal sign, substitute 1 for x to estimate the attendance in 1987, 2 for x to estimate the attendance in 1988, and so on.

Estimate the NASCAR attendance in **a)** 1987 and **b)** 1996.

Solution **a)** We will substitute 1 for x to evaluate the attendance in 1987.

$$\begin{aligned}\text{attendance} &= 0.028x^2 + 0.018x + 2.69\\ &= 0.028(1)^2 + 0.018(1) + 2.69\\ &= 0.028 + 0.018 + 2.69\\ &= 2.736\end{aligned}$$

The answer is reasonable since it is slightly more than the constant, 2.69. The first two terms add very little to the answer here. In 1987 about 2.736 million, or 2,736,000 people attended NASCAR events.

b) The year 1996 corresponds to the number 10. Therefore, to estimate the attendance in 1996, we substitute 10 for x in the expression.

$$\begin{aligned}\text{attendance} &= 0.028x^2 + 0.018x + 2.69\\ &= 0.028(10)^2 + 0.018(10) + 2.69\\ &= 0.028(100) + 0.18 + 2.69\\ &= 5.67\end{aligned}$$

The answer is reasonable: From the information given we expected to see an increase. The attendance at NASCAR races in 1996 was about 5.67 million, or 5,670,000, people.

NOW TRY EXERCISE 115

EXAMPLE 11 Evaluate $-x^3 - xy - y^2$ when $x = -2$ and $y = 5$.

Solution Substitute −2 for each x and 5 for each y in the expression. Then evaluate.

$$
\begin{aligned}
-x^3 - xy - y^2 &= -(-2)^3 - (-2)(5) - (5)^2 \\
&= -(-8) - (-10) - 25 \\
&= 8 + 10 - 25 \\
&= -7
\end{aligned}
$$

NOW TRY EXERCISE 101

5) Evaluate Expressions on a Graphing Calculator

Throughout this book, the material presented on **graphing calculators** (or graphers) will often reinforce the concepts presented. Therefore, even if you do not have or use a graphing calculator you should read the material related to graphing calculators whenever it appears. You may find that it truly helps your understanding of the concepts. Some graphing calculator information will be given as regular text, and other information about graphing calculators will be given in Using Your Graphing Calculator boxes such as the one on page 32.

The information presented in this book is not meant to replace the manual that comes with your graphing calculator. Because of space limitations in this book, your graphing calculator manual may provide more detailed information on some tasks we discuss. Your manual will also illustrate many other uses for your grapher beyond what we discuss in this course. Keystrokes to use vary from calculator to calculator. When we illustrate keystrokes, they will be for a Texas Instruments 83, (TI-83) calculator. When we display a graphing calculator screen, it will be a TI-83 screen. *We suggest that you carefully read the manual that came with your graphing calculator to determine the sequence of keystrokes to use to accomplish specific tasks.*

Many graphing calculators can store an expression (or equation) and then evaluate the expression for various values of the variable or variables without your having to re-enter the expression each time. This is very valuable in both science and mathematics courses. For example, in Chapter 3 when we graph, we will need to evaluate an expression for various values of the variable.

```
6→X:(2/3)X²+2X-4
                  32
-2.3→X:(2/3)X²+2
X-4
         -5.073333333
```

In the margin we display the screen of a TI-83 graphing calculator showing the expression $\frac{2}{3}x^2 + 2x - 4$ being evaluated for $x = 6$ and $x = -2.3$.

On this calculator screen, $6 \rightarrow X$ shows that we assigned a value of 6 to X. The expression being evaluated, $(2/3)X^2 + 2X - 4$, is shown after the colon. The 32 shown on the right side of the screen (or window) is the value of the expression when $X = 6$. On the next line, on the left side of the screen, we see $-2.3 \rightarrow X$, which shows that a value of -2.3 has been assigned to X. We see that the value of the expression is -5.073333333 when $X = -2.3$. After you have entered the expression to be evaluated it is not necessary to re-enter the expression to evaluate it for a different value of the variable. Read your graphing calculator manual to learn how to evaluate an expression for various values of the variable without having to re-enter the expression each time. On the TI-83, after evaluating an expression for one value of the variable, you can press [2nd] [ENTER] to display the previously assigned value and the expression to be evaluated. Then you can replace the value that was previously assigned to X with the new value to be assigned to X. After doing this and pressing [ENTER] the new answer will be displayed.

The calculator screen displayed above illustrates two important points about graphing calculators.

1. Notice the parentheses around the 2/3. Some graphing calculators interpret $2/3x^2$ as $2/(3x^2)$. To evaluate $\frac{2}{3}x^2$ on such calculators, you must use parentheses around the 2/3. You should learn how your calculator evaluates expressions like $2/3x^2$. *Whenever you are in doubt, use parentheses to prevent possible errors.*

2. In the display, you will notice that the negative sign preceding the 2.3 is slightly smaller and higher than the subtraction sign preceding the 4 in the expression. Graphing calculators generally have both a negative sign key, [(−)], and a subtraction sign key, [−]. You must be sure to use the correct key or you will get an error. The negative sign key is used to enter a negative number. The subtraction key is used to subtract one quantity from another. To enter the expression $-x - 4$ on a graphing calculator, you might press

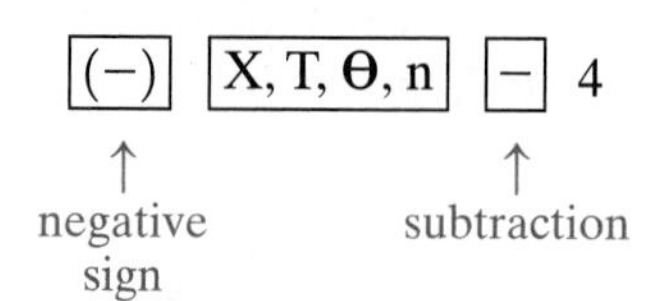

Remember that $-x - 4$ means $-1x - 4$. By beginning with [(−)] you are entering the coefficient −1. Different calculators use different keys to enter the variable x. The key shown after the negative sign is the key used on the TI-83 calculator.

EXAMPLE 12 Video rentals peaked between 1993 and 1994. The number of video rentals, in billions, from 1985 through 1997 can be estimated by

$$\text{video rentals} = -0.04x^2 + 1.09x + 3.55$$

In the expression on the right side of the equal sign, substitute 1 for x to estimate the number of rentals in 1985, 2 for x to estimate the number of rentals for 1986, and so on. Use a graphing calculator, if available, to estimate the number of video rentals in 1985 and 1997.

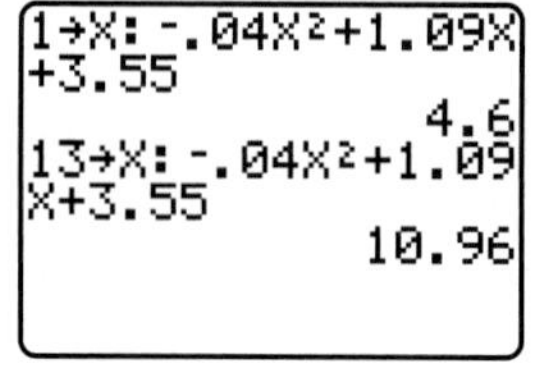

FIGURE 1.9

Solution We first assign x a value of 1, then enter the expression. After we evaluate the expression for $x = 1$, we re-enter the expression and change the value of x to 13 (for 1997). We then evaluate the expression again. Figure 1.9 shows the screen for a TI-83 calculator with the expression evaluated for values of 1 and 13. There were about 4.6 billion video tapes rented in 1985 and about 10.96 billion video tapes rented in 1997.

HELPFUL HINT

Always review your calculator screen to make sure that no keys were pressed incorrectly and no keys were omitted. Note that it is not necessary to enter the 0 before the decimal point in terms such as $-0.04x^2$.

Exercise Set 1.4

Concept/Writing Exercises

1. Consider the expression a^n. **a)** What is the a called? **b)** What is the n called?
2. What is the meaning of a^n?
3. Consider the radical expression $\sqrt[n]{a}$. **a)** What is the n called? **b)** What is the a called?
4. What does it mean if $\sqrt[n]{a} = b$?

5. What is the principal square root of a positive number?
6. Explain why $\sqrt{-4}$ cannot be a real number.
7. Explain why an odd root of a positive number will be positive.
8. Explain why an odd root of a negative number will be negative.
9. In your own words, explain the order of operations to follow when evaluating a mathematical expression.
10. **a)** In your own words, explain step by step how you would evaluate

$$\frac{5 - 18 \div 3^2}{4 - 3 \cdot 2}$$

b) Evaluate the expression.

11. **a)** In your own words, explain step by step how you would evaluate $\{5 - [4 - (3 - 8)]\}^2$.

b) Evaluate the expression.

12. **a)** In your own words, explain step by step how you would evaluate $16 \div 2^2 + 6 \cdot 4 - 24 \div 6$.

b) Evaluate the expression.

Practice the Skills

Evaluate each expression without using a calculator.

13. 4^2 **14.** $(-2)^3$ **15.** $(-5)^2$ **16.** 7^3

17. -4^2 **18.** $\left(\frac{5}{6}\right)^3$ **19.** $-\left(\frac{3}{5}\right)^4$ **20.** $(0.3)^2$

21. $-\left(-\frac{2}{3}\right)^4$ **22.** $\sqrt{36}$ **23.** $-\sqrt{36}$ **24.** $\sqrt{0.64}$

25. $\sqrt[3]{-125}$ **26.** $\sqrt[3]{\frac{-64}{125}}$ **27.** $\sqrt[3]{0.001}$ **28.** $\sqrt[4]{\frac{1}{16}}$

Use a calculator to evaluate each expression. Round answers to the nearest thousandth.

29. $(0.42)^5$ **30.** $-(3.2)^{4.2}$ **31.** $\left(\frac{5}{9}\right)^5$ **32.** $\left(-\frac{3}{7}\right)^6$

33. $-(2.35)^{7.4}$ **34.** $\sqrt{92}$ **35.** $\sqrt[3]{5}$ **36.** $-\sqrt[4]{72.8}$

37. $\sqrt[5]{1246.5}$ **38.** $-\sqrt{\frac{8}{9}}$ **39.** $-\sqrt[3]{\frac{20}{53}}$ **40.** $\sqrt[3]{-\frac{15}{19}}$

Evaluate **a)** x^2 *and* **b)** $-x^2$ *for each given value of x.*

41. 3 **42.** 4 **43.** 1 **44.** −2

45. −1 **46.** −5 **47.** $\frac{1}{3}$ **48.** $-\frac{2}{5}$

Evaluate **a)** x^3 *and* **b)** $-x^3$ *for each given value of x.*

49. 3 **50.** −5 **51.** −3 **52.** −1

53. −2 **54.** −4 **55.** $\frac{2}{3}$ **56.** $-\frac{3}{4}$

Evaluate each expression.

57. $5^2 + 3^2 - 2^2$ **58.** $(-1)^3 - 1^3 + 1^{10} - (-1)^{12}$

59. $-2^2 - 2^3 + 1^{10} + (-2)^3$ **60.** $(-3)^3 - 2^2 - (-2)^2 + (4 - 4)^2$

61. $(0.2)^2 - (1.7)^2 - (3.2)^2$ **62.** $(3.7)^2 - (0.8)^2 + (2.4)^3$

63. $\left(-\frac{1}{2}\right)^3 - \left(\frac{1}{3}\right)^2 - \left(-\frac{2}{3}\right)^2$ **64.** $\left(\frac{3}{4}\right)^2 - \frac{1}{4} - \left(-\frac{3}{8}\right)^2 + \left(\frac{1}{4}\right)^3$

Evaluate each expression.

65. $6 + 4 \cdot 5$

66. $(6^2 - 2) \div (\sqrt{36} - 4)$

67. $20 - 6 \div 3 - 4$

68. $6 \div 2 + 8 - 3^2$

69. $6 \div 3 + 5 \cdot \frac{3}{4}$

70. $24 \cdot 2 \div \frac{1}{3} \div 6$

71. $\frac{3}{4} \div \frac{5}{6} + \frac{1}{2} \cdot \frac{9}{4}$

72. $2[1 - (4 \cdot 5)] + 6^3$

73. $-2 + 5[3 - (2 - 4)]$

74. $-[4(5 - 3)^3] + 2^3$

75. $3[(4 + 6)^2 - \sqrt[3]{8}]$

76. $\{[3(14 \div 7)]^2 - 2\}^2$

77. $\{[(12 - 15) - 3] - 2\}^2$

78. $3\{6 - [(25 \div 5) - 2]\}^3$

79. $4[5(13 - 3) \div (25 \div 5)^2]^2$

80. $\dfrac{15 \div 3 + 2 \cdot 2}{\sqrt{25} \div 5 + 8 \div 2}$

81. $\dfrac{4 - (2 + 3)^2 - 6}{4(3 - 2) - 3^2}$

82. $\dfrac{2(-3) + 4 \cdot 5 - 3^2}{-6 + \sqrt{4}(2^2 - 1)}$

83. $\dfrac{8 + 4 \div 2 \cdot 3 + 4}{5^2 - 3^2 \cdot 2 - 7}$

84. $-2\left|-3 - \frac{2}{3}\right| + 4$

85. $\dfrac{8 - [4 - (3 - 1)^2]}{5 - (-3)^2 + 4 \div 2}$

86. $12 - 15 \div |5| - (|4| - 2)^2$

87. $-2|-3| - \sqrt{36} \div |2| + 3^2$

88. $\dfrac{4 - |-12| \div |3|}{2(4 - |5|) + 9}$

89. $\dfrac{6 - |-4| - 4|6 - 3|}{5 - 6 \cdot 2 \div |-6|}$

90. $-\frac{1}{4}[8 - |-6| \div 3 - 4]^2$

91. $\frac{2}{5}[\sqrt[3]{27} - |-9| + 4 - 3^2]^2$

92. $\dfrac{3(5 - 2)^2}{-3^2} - \dfrac{2(3^2 - 4^2)}{4 - (-2)}$

93. $\dfrac{24 - 5 - 4^2}{|-8| + 4 - 2(3)} + \dfrac{4 - (-3)^2 + |4|}{3^2 - 4 \cdot 3 + |-7|}$

94. $\dfrac{-2 - 8 \div 4^2 \cdot |8|}{|8| - \sqrt{64}} + \dfrac{[(8 - 3)^2 - 4]^2}{2^2 + 16}$

Evaluate each expression for the given value or values.

95. $-3x^2 - 4$ when $x = 1$

96. $5x^2 - 2x + 5$ when $x = 3$

97. $-3x^2 + 6x + 5$ when $x = 5$

98. $3(x - 2)^2$ when $x = \frac{1}{4}$

99. $4(x + 1)^2 - 6x$ when $x = -\frac{5}{6}$

100. $-6x + 3y^2$ when $x = 2, y = 4$

101. $6x^2 + 3y^3 - 5$ when $x = 1, y = -3$

102. $4x^2 - 3y - 5$ when $x = 4, y = -2$

103. $3(a + b)^2 + 4(a + b) - 6$ when $a = 4, b = -1$

104. $-3 - \{2x - [5x - (2x + 1)]\}$ when $x = 3$

105. $-6 - \{x - [2x - (x - 3)]\}$ when $x = 4$

106. $\dfrac{(x - 3)^2}{9} + \dfrac{(y + 5)^2}{16}$ when $x = 4, y = 3$

107. $\dfrac{-b + \sqrt{b^2 - 4ac}}{2a}$ when $a = 6, b = -11, c = 3$

108. $\dfrac{-b - \sqrt{b^2 - 4ac}}{2a}$ when $a = 2, b = 1, c = -10$

Problem Solving

In Exercises 109–113 write an algebraic expression for each problem. Then evaluate the expression for the given value of the variable or variables.

109. Multiply the variable x by 3. To this product add 6. Now square this sum. Find the value of this expression when $x = 3$.

110. Subtract 3 from x. Square this difference. Subtract 5 from this value. Now square this result. Find the value of this expression when $x = -1$.

111. Six is added to the product of 3 and x. This expression is then multiplied by 6. Nine is then subtracted from this product. Find the value of the expression when $x = 3$.

112. The sum of x and y is multiplied by 2. Then 5 is subtracted from this product. This expression is then squared. Find the value of the expression when $x = 2$ and $y = -3$.

113. Three is added to x. This sum is divided by twice y. This quotient is then squared. Finally, 3 is subtracted from this expression. Find the value of the expression when $x = 5$ and $y = 2$.

Use a calculator to answer Exercises 114–121.

114. In October 1997, of the nation's 46.4 million public schoolchildren, about 3.6 million students spoke little or no English. This number is increasing every year. We can find the approximate number of schoolchildren with limited proficiency in English, in millions, by using.

$$\text{number of children} = 0.21x + 1.28$$

where x represents years starting with 1987. Substitute 1 for x to find the number of children with limited English proficiency in 1987, substitute 2 for x to find the number in 1988, and so on.

a) Find the approximate number of schoolchildren with limited English proficiency in 1989.

b) Find the approximate number of schoolchildren with limited English proficiency in 1995.

115. The amount of carbon dioxide (CO_2) gas produced through the burning of fossil fuel is one contributing factor to the greenhouse warming effect of our planet. Since 1905 the amount of CO_2 has been increasing. The total production of CO_2 from all countries except the United States, Canada, and Western Europe (measured in millions of metric tons) can be approximated by

$$CO_2 = 0.073x^2 - 0.39x + 0.55$$

where x represents each 10 year period starting with 1905. Substitute 1 for x to calculate the CO_2 production in 1905, 2 for x to calculate the CO_2 production in 1915, 3 for x in 1925, and so on.

a) Find the approximate amount of CO_2 produced by all countries except the United States, Canada, and Western Europe in 1945.

b) Find the approximate amount of CO_2 produced by all countries except the United States, Canada, and Western Europe in 1995.

116. The number of youths arrested per year for violent crimes increased from 1988 through 1994. Since 1994 the number of youth arrests has been declining. We can use

$$\text{number of youth arrests} = -5.6x^2 + 77x + 241$$

to estimate the arrest rate, per 100,000 youths, from 1988 to 1996. Substitute 1 for x to get the arrest rate in 1988, 2 for x to get the arrest rate in 1989, and so on.

a) The highest arrest rate was in 1994. Find the arrest rate that year.

b) Find the arrest rate in 1996.

117. The number of *latchkey kids*, children who care for themselves while their parents are working, increases with age. The percent of children of different ages, from 5 to 14 years old, who are latchkey kids can be approximated by

$$\text{percent of children} = 0.23x^2 - 1.98x + 4.42$$

The x value represents the age of the children. For example, substitute 5 for x to get the percent of all 5-year-olds who are latchkey kids, substitute 6 for x to get the percent of all 6-year-olds who are latchkey kids, and so on.

a) Find the percent of all 10-year-olds who are latchkey kids.

b) Find the percent of all 14-year-olds who are latchkey kids.

118. The manner in which we receive our news and information is changing rapidly. With television and with the advent of the Internet and ever-increasing computer use, printed material is quickly becoming obsolete. For example, the number of Americans who read a daily newspaper is steadily going down. The percent reading the daily newspaper can be approximated by

$$\text{percent} = -6.2x + 82.2$$

where x represents each 10 year period starting with 1970. Substitute 1 for x to get the percent for 1970, 2 for x to get the percent for 1980, 3 for x to get the percent for 1990, and so forth.

a) Find the percent of U.S. adults who read a daily newspaper in 1970.

b) Assuming that this trend continues, find the percent of U.S. adults who will read a daily newspaper in 2010.

119. As time goes on, computers are getting increasingly more powerful. One measure of this power is the number of instructions per second that the computer can perform. In January 1997, researchers unveiled a supercomputer capable of performing one trillion calculations per second. We can determine the approximate speed of computers, in billions of operations per second, over the past ten years using

$$\text{speed} = 65.3x^3 - 350x^2 + 731x - 444$$

where x represents each two year period starting with 1991. Substitute 1 for x to find the speed in 1991, 2 for x to find the speed in 1993, 3 for x to find the speed in 1995, and so on.

a) Find the speed of the fastest computer in 1991.

b) Find the speed of the fastest computer in 1999.

120. The Congressional Budget Office has projected the U.S. national debt for the next 50 years, and it is not a pretty sight. By 2050, the CBO estimates that the debt will be over 200 times larger than it will be in the year 2000. The projected debt, in trillions of dollars, can be approximated using

$$\text{projected debt} = 0.40x^3 - 4.9x^2 + 20.7x - 19.7$$

where x represents each 5 year period starting with 2000. Substitute 1 for x to get the estimate for the year 2000, 2 for x to get the estimate for 2005, 3 for x for 2010, and so on.

a) Estimate the U.S. national debt in 2030.

b) Estimate the U.S. national debt in 2045.

121. Use of cellular phones is on the rise. We can use

$$\text{number of subscribers} = 0.42x^2 - 3.44x + 5.80$$

to estimate, in millions, the growing numbers of people who used a cell phone from 1983 through 1996. Substitute 1 for x to get the number of subscribers in 1983, 2 for x to get the number of subscribers in 1984, and so on.

a) Find the number of people who used cell phones in 1989.

b) Find the number of people who used cell phones in 1996.

Cumulative Review Exercises

[1.2] **122.** $A = \{a, b, c, d, f\}$, $B = \{b, c, f, g, h\}$. Find **a)** $A \cap B$; **b)** $A \cup B$.

[1.3] *In Exercises 123–125, the letter a represents a real number. For what values of a will each statement be true?*

123. $|a| = |-a|$

124. $|a| = a$

125. $|a| = 4$

126. List from smallest to largest: $-|6|$, -4, $|-5|$, $-|-2|$, 0.

127. Name the following property: $(2 + 3) + 5 = 2 + (3 + 5)$.

1.5 EXPONENTS

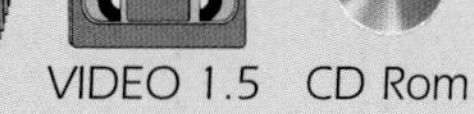

SSM VIDEO 1.5 CD Rom

1. Use the product rule for exponents.
2. Use the quotient rule for exponents.
3. Use the negative exponent rule.
4. Use the zero exponent rule.
5. Use the rule for raising a power to a power.
6. Use the rule for raising a product to a power.
7. Use the rule for raising a quotient to a power.

In the previous section we introduced exponents. In this section we discuss the rules of exponents. We begin with the product rule for exponents.

1 Use the Product Rule for Exponents

Consider the multiplication $x^3 \cdot x^5$. We can simplify this expression as follows:

$$x^3 \cdot x^5 = (x \cdot x \cdot x) \cdot (x \cdot x \cdot x \cdot x \cdot x) = x^8$$

This problem could also be simplified using the **product rule for exponents.***

Product Rule for Exponents

If m and n are natural numbers and a is any real number, then

$$a^m \cdot a^n = a^{m+n}$$

To multiply exponential expressions, maintain the common base and add the exponents.

$$x^3 \cdot x^5 = x^{3+5} = x^8$$

EXAMPLE 1 Simplify. **a)** $3^2 \cdot 3^3$ **b)** $x^3 \cdot x^9$ **c)** $x \cdot x^6$

Solution **a)** $3^2 \cdot 3^3 = 3^{2+3} = 3^5 = 243$ **b)** $x^3 \cdot x^9 = x^{3+9} = x^{12}$

c) $x \cdot x^6 = x^1 \cdot x^6 = x^{1+6} = x^7$

2) Use the Quotient Rule for Exponents

Consider the division $x^7 \div x^4$. We can simplify this expression as follows:

$$\frac{x^7}{x^4} = \frac{\overset{1}{\cancel{x}} \cdot \overset{1}{\cancel{x}} \cdot \overset{1}{\cancel{x}} \cdot \overset{1}{\cancel{x}} \cdot x \cdot x \cdot x}{\underset{1}{\cancel{x}} \cdot \underset{1}{\cancel{x}} \cdot \underset{1}{\cancel{x}} \cdot \underset{1}{\cancel{x}}} = x \cdot x \cdot x = x^3$$

This problem could also be simplified using the **quotient rule for exponents.**

Quotient Rule for Exponents

If a is any nonzero real number and m and n are nonzero integers, then

$$\frac{a^m}{a^n} = a^{m-n}$$

To divide expressions in exponential form, maintain the common base and subtract the exponents.

$$\frac{x^7}{x^4} = x^{7-4} = x^3$$

EXAMPLE 2 Simplify. **a)** $\frac{5^4}{5^2}$ **b)** $\frac{x^5}{x^2}$ **c)** $\frac{y^2}{y^5}$

Solution **a)** $\frac{5^4}{5^2} = 5^{4-2} = 5^2 = 25$ **b)** $\frac{x^5}{x^2} = x^{5-2} = x^3$ **c)** $\frac{y^2}{y^5} = y^{2-5} = y^{-3}$

3) Use the Negative Exponent Rule

Notice in Example 2 **c)** that the answer contains a negative exponent. Let's do part **c)** again by dividing out common factors.

$$\frac{y^2}{y^5} = \frac{\overset{1}{\cancel{y}} \cdot \overset{1}{\cancel{y}}}{\underset{1}{\cancel{y}} \cdot \underset{1}{\cancel{y}} \cdot y \cdot y \cdot y} = \frac{1}{y^3}$$

* The rules given in this section also apply for rational or fractional exponents. Rational exponents will be discussed in Section 7.2. We will review these rules again at that time.

By dividing out common factors and using the result from Example 2 **c)**, we can reason that $y^{-3} = 1/y^3$. This is an example of the negative exponent rule.

Negative Exponent Rule

For any nonzero real number a and any whole number m,

$$a^{-m} = \frac{1}{a^m}$$

EXAMPLE 3 Write each expression without negative exponents.

a) 2^{-3} **b)** $3x^{-2}$ **c)** $\frac{1}{x^{-3}}$

Solution **a)** $2^{-3} = \frac{1}{2^3} = \frac{1}{8}$ **b)** $3x^{-2} = 3 \cdot \frac{1}{x^2} = \frac{3}{x^2}$

c) $\frac{1}{x^{-3}} = 1 \div x^{-3} = 1 \div \frac{1}{x^3} = \frac{1}{1} \cdot \frac{x^3}{1} = x^3$

NOW TRY EXERCISE 25

HELPFUL HINT

In Example 3 **c)** we showed that $\frac{1}{x^{-3}} = x^3$. In general, for any nonzero real number a and any whole number m, $\frac{1}{a^{-m}} = a^m$. When a factor of the numerator or the denominator is raised to any power, the factor can be moved to the other side of the fraction bar provided the sign of the exponent is changed. Thus, for example

$$\frac{2a^{-3}}{b^2} = \frac{2}{a^3b^2} \qquad \frac{a^{-2}b^4}{c^{-3}} = \frac{b^4c^3}{a^2}$$

Generally, we do not leave exponential expressions with negative exponents. *When we indicate that an exponential expression is to be simplified, we mean that your answer should be written without negative exponents.*

EXAMPLE 4 Simplify. **a)** $\frac{3xz^2}{y^{-4}}$ **b)** $4^{-2}x^{-1}y^2$ **c)** $-3^3x^2y^{-3}$

Solution **a)** $\frac{3xz^2}{y^{-4}} = 3xy^4z^2$

b) $4^{-2}x^{-1}y^2 = \frac{1}{4^2} \cdot \frac{1}{x^1} \cdot y^2 = \frac{y^2}{16x}$

c) $-3^3x^2y^{-3} = -(3^3)x^2 \cdot \frac{1}{y^3} = -\frac{27x^2}{y^3}$

NOW TRY EXERCISE 35

4) Use the Zero Exponent Rule

The next rule we will study is the **zero exponent rule**. Any nonzero number divided by itself is 1. Therefore,

$$\frac{x^5}{x^5} = 1$$

By the quotient rule for exponents,

$$\frac{x^5}{x^5} = x^{5-5} = x^0$$

Since $x^0 = \frac{x^5}{x^5}$ and $\frac{x^5}{x^5} = 1$, by the transitive property of equality,

$$x^0 = 1$$

Zero Exponent Rule

If a is any nonzero real number, then

$$a^0 = 1$$

The zero exponent rule illustrates that *any nonzero real number with an exponent of 0 equals 1.* We must specify that $a \neq 0$ because 0^0 is not a real number.

EXAMPLE 5 Simplify (assume that the base is not 0).

a) x^0 **b)** $3x^0$ **c)** $-(a + b)^0$

Solution **a)** $x^0 = 1$ **b)** $3x^0 = 3(1) = 3$ **c)** $-(a + b)^0 = -(1) = -1$

5) Use the Rule for Raising a Power to a Power

Consider the expression $(x^3)^2$. We can simplify this expression as follows:

$$(x^3)^2 = x^3 \cdot x^3 = x^{3+3} = x^6$$

This problem could also be simplified using the rule for **raising a power to a power** (also called the **power rule**).

Raising a Power to a Power (The Power Rule)

If a is a real number and m and n are integers, then

$$(a^m)^n = a^{m \cdot n}$$

To raise an exponential expression to a power, maintain the base and multiply the exponents.

$$(x^3)^2 = x^{3 \cdot 2} = x^6$$

EXAMPLE 6 Simplify. **a)** $(2^3)^2$ **b)** $(y^3)^{-5}$ **c)** $(3^{-2})^3$

Solution **a)** $(2^3)^2 = 2^{3 \cdot 2} = 2^6 = 64$

b) $(y^3)^{-5} = y^{3(-5)} = y^{-15} = \frac{1}{y^{15}}$

c) $(3^{-2})^3 = 3^{-2(3)} = 3^{-6} = \frac{1}{3^6}$ or $\frac{1}{729}$

NOW TRY EXERCISE 75

HELPFUL HINT

Students often confuse the *product rule*

$$a^m \cdot a^n = a^{m+n}$$

with the *power rule*

$$(a^m)^n = a^{m \cdot n}$$

For example, $(x^3)^2 = x^6$, not x^5.

6) Use the Rule for Raising a Product to a Power

Consider the expression $(xy)^2$. We can simplify this expression as follows:

$$(xy)^2 = (xy)(xy) = x \cdot x \cdot y \cdot y = x^2y^2$$

This expression could also be simplified using the rule for **raising a product to a power**.

Raising a Product to a Power

If a and b are real numbers and m is an integer, then

$$(ab)^m = a^m b^m$$

To raise a product to a power, raise all factors within the parentheses to the power outside the parentheses.

EXAMPLE 7 Simplify. **a)** $(-4x^3)^2$ **b)** $(3x^{-2}y^3)^{-3}$

Solution **a)** $(-4x^3)^2 = (-4)^2(x^3)^2 = 16x^6$

b) $(3x^{-2}y^3)^{-3} = 3^{-3}(x^{-2})^{-3}(y^3)^{-3}$ *Raise a power to a power.*

$= \dfrac{1}{3^3} \cdot x^6 \cdot y^{-9}$ *Negative exponent rule, Power rule.*

$= \dfrac{1}{27} \cdot x^6 \cdot \dfrac{1}{y^9}$ *Negative exponent rule.*

$= \dfrac{x^6}{27y^9}$

NOW TRY EXERCISE 89

7) Use the Rule for Raising a Quotient to a Power

Consider the expression $\left(\dfrac{x}{y}\right)^2$. We can simplify this expression as follows:

$$\left(\frac{x}{y}\right)^2 = \frac{x}{y} \cdot \frac{x}{y} = \frac{x \cdot x}{y \cdot y} = \frac{x^2}{y^2}$$

This expression could also be simplified using the rule for **raising a quotient to a power.**

Raising a Quotient to a Power

If a and b are real numbers and m is an integer, then

$$\left(\frac{a}{b}\right)^m = \frac{a^m}{b^m}, \quad b \neq 0$$

To raise a quotient to a power, raise all factors in the parentheses to the exponent outside the parentheses.

EXAMPLE 8 Simplify. **a)** $\left(\frac{2}{x^2}\right)^3$ **b)** $\left(\frac{4x^{-2}}{y^3}\right)^{-2}$

Solution **a)** $\left(\frac{2}{x^2}\right)^3 = \frac{2^3}{(x^2)^3} = \frac{8}{x^6}$

b) $\left(\frac{4x^{-2}}{y^3}\right)^{-2} = \frac{4^{-2}(x^{-2})^{-2}}{(y^3)^{-2}}$ *Raise a quotient to a power.*

$= \frac{4^{-2}x^4}{y^{-6}}$ *Power rule.*

$= \frac{x^4y^6}{4^2}$ *Negative exponent rule.*

$= \frac{x^4y^6}{16}$

NOW TRY EXERCISE 95

Consider $\left(\frac{a}{b}\right)^{-n}$. Using the rule for raising a quotient to a power, we get

$$\left(\frac{a}{b}\right)^{-n} = \frac{a^{-n}}{b^{-n}} = \frac{b^n}{a^n} = \left(\frac{b}{a}\right)^n$$

Using this result, we see that when we have a rational number raised to a negative exponent, we can take the reciprocal of the base and change the sign of the exponent as follows.

$$\left(\frac{5}{9}\right)^{-3} = \left(\frac{9}{5}\right)^3 \qquad \left(\frac{x^2}{y^3}\right)^{-4} = \left(\frac{y^3}{x^2}\right)^4$$

Now we will work some examples that combine a number of properties. Whenever the same variable appears above and below the fraction bar, we generally move the variable with the *lesser exponent*. This will result in the exponent on the variable being positive when the product rule is applied. Examples 9 and 10 illustrate this procedure.

EXAMPLE 9 Simplify. **a)** $\left(\frac{6x^2y^4}{2x^2y}\right)^2$ **b)** $\left(\frac{3x^4y^{-2}}{6xy^3z^{-1}}\right)^{-3}$

Solution Exponential expressions can often be simplified in more than one order. In general, it will be easier to first simplify the expression within the parentheses.

a) $\left(\frac{6x^2y^4}{2x^2y}\right)^2 = (3y^3)^2 = 9y^6$

b) $\left(\dfrac{3x^4 y^{-2}}{6xy^3 z^{-1}}\right)^{-3} = \left(\dfrac{x^4 \cdot x^{-1} z}{2y^3 \cdot y^2}\right)^{-3}$ *Move x, y^{-2}, and z^{-1} to the other side of the fraction bar and change the sign of their exponents.*

$= \left(\dfrac{x^3 z}{2y^5}\right)^{-3}$

$= \left(\dfrac{2y^5}{x^3 z}\right)^{3}$ *Take the reciprocal of the expression inside the parentheses and change the sign of the exponent.*

$= \dfrac{2^3 y^{5\cdot 3}}{x^{3\cdot 3} z^3}$

$= \dfrac{8y^{15}}{x^9 z^3}$

NOW TRY EXERCISE 105

EXAMPLE 10 Simplify $\dfrac{(2p^{-3} q^4)^{-2}}{(p^{-5} q^4)^{-3}}$.

Solution First, use the power rule. Then simplify further.

$\dfrac{(2p^{-3} q^4)^{-2}}{(p^{-5} q^4)^{-3}} = \dfrac{2^{-2} p^6 q^{-8}}{p^{15} q^{-12}}$ *Move 2^{-2}, p^6, and q^{-12} to the other side of the fraction bar and change the signs of their exponents.*

$= \dfrac{q^{-8} \cdot q^{12}}{2^2 p^{15} \cdot p^{-6}}$

$= \dfrac{q^{-8+12}}{4p^{15-6}}$

$= \dfrac{q^4}{4p^9}$

NOW TRY EXERCISE 111

Summary of Rules of Exponents		
For all real numbers a and b and all integers m and n:		
Product rule	$a^m \cdot a^n = a^{m+n}$	
Quotient rule	$\dfrac{a^m}{a^n} = a^{m-n}$,	$a \neq 0$
Negative exponent rule	$a^{-m} = \dfrac{1}{a^m}$,	$a \neq 0$
Zero exponent rule	$a^0 = 1$,	$a \neq 0$
Raising a power to a power	$(a^m)^n = a^{m\cdot n}$	
Raising a product to a power	$(ab)^m = a^m b^m$	
Raising a quotient to a power	$\left(\dfrac{a}{b}\right)^m = \dfrac{a^m}{b^m}$,	$b \neq 0$

Exercise Set 1.5

Concept/Writing Exercises

1. a) Give the product rule for exponents.
   b) Explain the product rule.
2. a) Give the quotient rule for exponents.
   b) Explain the quotient rule.
3. a) Give the negative exponent rule.
   b) Explain the negative exponent rule.
4. a) Give the zero exponent rule.
   b) Explain the zero exponent rule.
5. a) Give the rule for raising a power to a power.
   b) Explain the rule for raising a power to a power.
6. a) Give the rule for raising a product to a power.
   b) Explain the rule for raising a product to a power.
7. a) Give the rule for raising a quotient to a power.
   b) Explain the rule for raising a quotient to a power.
8. What is the exponent on a variable or coefficient if none is shown?
9. If $x^{-1} = 5$, what is the value of x? Explain.
10. If $x^{-1} = y^2$, what is x equal to? Explain.
11. a) Explain the difference between the opposite of x and the reciprocal of x.
    For parts (b) and (c) consider
    $$x^{-1} \qquad -x \qquad \frac{1}{x} \qquad \frac{1}{x^{-1}}$$
    b) Which represent (or are equal to) the *reciprocal* of x?
    c) Which represent the *opposite* (or *additive inverse*) of x?
12. Explain why $-2^{-2} \neq \frac{1}{(-2)^2}$.

Practice the Skills

Evaluate each expression.

13. $3^2 \cdot 3$ 14. $2^2 \cdot 2^3$ 15. $3^4 \cdot 3^0$ 16. $4^2 \cdot 4$

17. $\frac{4^3}{4}$ 18. $\frac{5^5}{5^3}$ 19. $\frac{4^2}{4^0}$ 20. $\frac{6^3}{6^5}$

21. 4^{-2} 22. $(-4)^{-2}$ 23. -4^{-2} 24. $-(-4)^{-2}$

25. 5^{-3} 26. $(-5)^{-3}$ 27. -5^{-3} 28. $-(-5^{-3})$

Simplify each expression and write the answer without negative exponents.

29. $5y^{-3}$ 30. $\frac{1}{x^{-1}}$ 31. $\frac{1}{x^{-4}}$ 32. $\frac{3}{5y^{-2}}$

33. $\frac{2a}{b^{-3}}$ 34. $\frac{6x^4}{y^{-1}}$ 35. $\frac{5m^{-2}n^{-3}}{2}$ 36. $\frac{4x^{-3}}{z^4}$

37. $\frac{5x^{-2}y^{-3}}{z^{-4}}$ 38. $\frac{10ab^5}{2c^{-3}}$ 39. $\frac{6^{-1}x^{-1}}{y}$ 40. $\frac{5^{-1}z}{x^{-1}y^{-1}}$

Evaluate each expression. Assume that all bases represented by variables are nonzero.

41. x^0 42. $5y^0$ 43. $-2x^0$ 44. $-3x^0$

45. $-(a + b)^0$ 46. $3(a + b)^0$ 47. $3x^0 + 4y^0$ 48. $-4(x^0 - 3y^0)$

Simplify each expression and write the answer without negative exponents.

49. $5^3 \cdot 5^{-4}$ 50. $x^2 \cdot x^4$ 51. $x^6 \cdot x^{-2}$ 52. $x^{-4} \cdot x^3$

53. $\frac{3^4}{3^2}$ 54. $\frac{5^2}{5^{-2}}$ 55. $\frac{7^{-5}}{7^{-3}}$ 56. $\frac{x^{-9}}{x^2}$

57. $\frac{x^{-2}}{x}$ 58. $\frac{x^0}{x^{-3}}$ 59. $\frac{5w^{-2}}{w^{-7}}$ 60. $\frac{x^{-3}}{x^{-5}}$

61. $2x^{-4} \cdot 6x^{-3}$ 62. $(-3y^5)(2y^{-4})$ 63. $(-3p^{-2})(-p^3)$ 64. $(2x^{-3}y^{-4})(6x^{-4}y^7)$

65. $(5r^2s^{-2})(-2r^5s^2)$

66. $(-3p^{-4}q^6)(2p^3q)$

67. $(2x^4y^7)(4x^3y^{-5})$

68. $\frac{24x^3y^2}{8xy}$

69. $\frac{27x^5y^{-4}}{9x^3y^2}$

70. $\frac{6x^{-2}y^3z^{-2}}{-2x^4y}$

71. $\frac{9xy^{-4}z^3}{-3x^{-2}yz}$

72. $\frac{(x^{-2})(4x^2)}{x^3}$

Simplify each expression and write the answer without negative exponents.

73. $(2^2)^3$

74. $(3^2)^{-1}$

75. $(2^3)^{-2}$

76. $(x^3)^{-5}$

77. $(x^{-4})^{-2}$

78. $(-x)^2$

79. $(-x)^3$

80. $(-x)^{-3}$

81. $(-2x^{-2})^3$

82. $3(x^4)^{-2}$

83. $\left(\frac{3}{5}\right)^2$

84. $\left(\frac{3}{4}\right)^{-2}$

85. $\left(\frac{2}{7}\right)^{-2}$

86. $\left(\frac{1}{2}\right)^{-3}$

87. $\left(\frac{2x}{3}\right)^{-2}$

88. $(-3x^2y)^4$

89. $(4x^2y^{-2})^2$

90. $(5xy^3)^{-2}$

91. $(2a^3b)^{-3}$

92. $(3x^{-2}y)^{-2}$

93. $(-4x^{-3}y^5)^{-3}$

94. $3(x^2y)^{-4}$

95. $\left(\frac{6x}{y^2}\right)^2$

96. $\left(\frac{3x^2y^4}{z}\right)^3$

97. $\left(\frac{2r^4s^5}{r^2}\right)^3$

98. $\left(\frac{3m^5n^6}{6m^4n^7}\right)^3$

99. $\left(\frac{4xy}{y^3}\right)^{-3}$

100. $\left(\frac{3x^{-2}}{xy}\right)^{-2}$

101. $\left(\frac{4x^{-2}y}{x^{-5}}\right)^3$

102. $\left(\frac{4x^2y}{x^{-5}}\right)^{-3}$

103. $\left(\frac{6x^2y}{3xz}\right)^{-3}$

104. $\left(\frac{5xy}{z^{-2}}\right)^3$

105. $\left(\frac{x^6y^{-2}}{x^{-2}y^3}\right)^2$

106. $\left(\frac{x^2y^{-3}z^4}{x^{-1}y^2z^3}\right)^{-1}$

107. $\left(\frac{4x^{-1}y^{-2}z^3}{2xy^2z^{-3}}\right)^{-2}$

108. $\left(\frac{6x^4y^{-6}z^4}{2xy^{-6}z^{-2}}\right)^{-2}$

109. $\left(\frac{-a^3b^{-1}c^{-3}}{2ab^3c^{-4}}\right)^{-3}$

110. $\frac{(2x^{-1}y^{-2})^{-3}}{(3x^{-1}y^3)^2}$

111. $\frac{(3x^{-4}y^2)^3}{(2x^3y^5)^3}$

112. $\frac{(2xy^2z^{-3})^2}{(3x^{-1}yz^2)^{-1}}$

Problem Solving

Simplify each expression. Assume that all variables represent nonzero integers.

113. $x^{4a} \cdot x^{3a+4}$

114. $y^{4r-2} \cdot y^{-2r+3}$

115. $w^{5b-2} \cdot w^{2b+3}$

116. $d^{5x+3} \cdot d^{-2x-3}$

117. $\frac{x^{2w+3}}{x^{w-4}}$

118. $\frac{y^{5m-1}}{y^{7m-1}}$

119. $(x^{3p+5})(x^{2p-3})$

120. $(s^{2t-3})(s^{-t+5})$

121. $x^{-m}(x^{3m+2})$

122. $y^{3b+2} \cdot y^{2b+4}$

123. $\frac{25m^{a+b}n^{b-a}}{5m^{a-b}n^{a+b}}$

124. $\frac{20x^{c+3}y^{d+4}}{4x^{c-4}y^{d+6}}$

125. **a)** For what values of x is $x^4 > x^3$?
b) For what values of x is $x^4 < x^3$?
c) For what values of x is $x^4 = x^3$?
d) Why can you not say that $x^4 > x^3$?

126. Is 3^{-8} greater than or less than 2^{-8}? Explain.

127. **a)** Explain why $(-1)^n = 1$ for any even number n.
b) Explain why $(-1)^n = -1$ for any odd number n.

128. **a)** Explain why $(-12)^{-8}$ is positive.
b) Explain why $(-12)^{-7}$ is negative.

129. **a)** Is $\left(-\frac{2}{3}\right)^{-2}$ equal to $\left(\frac{2}{3}\right)^{-2}$?
b) Will $(x)^{-2}$ equal $(-x)^{-2}$ for all real numbers x except 0? Explain your answer.

130. **a)** Is $\left(-\frac{2}{3}\right)^{-3}$ equal to $\left(\frac{2}{3}\right)^{-3}$?
b) Will $(x)^{-3}$ equal $(-x)^{-3}$ for any nonzero real numbers x? Explain.
c) What is the relationship between $(-x)^{-3}$ and $(x)^{-3}$ for any nonzero real number x?

Determine what exponents must be placed in the shaded area to make each expression true? Each shaded area may represent a different exponent. Explain how you determined your answer.

131. $\left(\frac{x^2y^{-2}}{x^{-3}y^{\blacksquare}}\right)^2 = x^{10}y^2$

132. $\left(\frac{x^{-2}y^3z}{x^4y^{\blacksquare}z^{-3}}\right)^3 = \frac{z^{12}}{x^{18}y^6}$

133. $\left(\frac{x^{\blacksquare}y^5z^{-2}}{x^4y^{\blacksquare}z}\right)^{-1} = \frac{x^5z^3}{y^2}$

Challenge Problems

We will learn in Section 7.2 that the rules of exponents given in this section also apply when the exponents are rational numbers. Using this information and the rules of exponents, evaluate each expression.

134. $\left(\frac{x^{1/2}}{x^{-1}}\right)^{3/2}$

135. $\left(\frac{x^{5/8}}{x^{1/4}}\right)^{3}$

136. $\left(\frac{x^{4}}{x^{-1/2}}\right)^{-1}$

137. $\frac{x^{1/2}y^{-3/2}}{x^{5}y^{5/3}}$

138. $\left(\frac{x^{1/2}y^{4}}{x^{-3}y^{5/2}}\right)^{2}$

Group Activity

Discuss and answer Exercise 139 as a group.

139. On day 1 you are given a penny. On each following day, you are given double the amount you were given on the previous day.

a) Write down the amounts you would be given on each of the first 6 days.

b) Express each of these numbers as an exponential expression with a base of 2.

c) By looking at the pattern, determine an exponential expression for the number of cents you will receive on day 10.

d) Write a general exponential expression for the number of cents you will receive on day n.

e) Write an exponential expression for the number of cents you will receive on day 30.

f) Calculate the value of the expression in part **e)**. Use a calculator if one is available.

g) Determine the amount found in part **f)** in dollars.

h) Write a general exponential expression for the number of dollars you will receive on day n.

Cumulative Review Exercises

[1.2] **140.** If $A = \{3, 4, 6\}$ and $B = \{1, 2, 5, 8\}$, find **a)** $A \cup B$ and **b)** $A \cap B$.

141. Illustrate the following set on the number line: $\{x \mid -3 \le x < 2\}$.

[1.4] **142.** Evaluate $6 + |12| \div |-3| - 4 \cdot 2^2$.

143. Evaluate $\sqrt[3]{-125}$.

1.6 SCIENTIFIC NOTATION

1) Write numbers in scientific notation.
2) Change numbers in scientific notation to decimal form.
3) Use scientific notation in problem solving.

1) Write Numbers in Scientific Notation

Scientists and engineers often deal with very large and very small numbers. For example, the frequency of an FM radio signal may be 14,200,000,000 hertz (or cycles per second) and the diameter of an atom is about 0.0000000001 meter. Because it is difficult to work with many zeros, scientists often express such numbers with exponents. For example, the number 14,200,000,000 might be written as 1.42×10^{10}, and 0.0000000001 as 1×10^{-10}. Numbers such as 1.42×10^{10} and 1×10^{-10} are in a form called **scientific notation**. In scientific notation, numbers appear as a number greater than or equal to 1 and less than 10 multiplied by some power of 10. When a power of 10 has no numerical coefficient showing, as in 10^5, we assume that the numerical coefficient is 1. Thus, 10^5 means 1×10^5 and 10^{-4} means 1×10^{-4}.

Examples of Numbers in Scientific Notation

3.2×10^6 $\quad 4.176 \times 10^3 \quad$ 2.64×10^{-2}

The following shows the number 32,400 changed to scientific notation.

$$32{,}400 = 3.24 \times 10{,}000$$

$$= 3.24 \times 10^4 \qquad (10{,}000 = 10^4)$$

There are four zeros in 10,000, the same number as the exponent in 10^4. The procedure for writing a number in scientific notation follows.

To Write a Number in Scientific Notation

1. Move the decimal point in the number to the right of the first nonzero digit. This gives a number greater than or equal to 1 and less than 10.
2. Count the number of places you moved the decimal point in step 1. If the original number was 10 or greater, the count is to be considered positive. If the original number was less than 1, the count is to be considered negative.
3. Multiply the number obtained in step 1 by 10 raised to the count (power) found in step 2.

EXAMPLE 1 Write the following numbers in scientific notation.

a) 68,900 **b)** 0.000572 **c)** 0.0074

Solution **a)** The decimal point in 68,900 is to the right of the last zero.

$$68{,}900. = 6.89 \times 10^4$$

The decimal point is moved four places. Since the original number is greater than 10, the exponent is positive.

b) $0.000572 = 5.72 \times 10^{-4}$

The decimal point is moved four places. Since the original number is less than 1, the exponent is negative.

NOW TRY EXERCISE 11

c) $0.0074 = 7.4 \times 10^{-3}$

2) Change Numbers in Scientific Notation to Decimal Form

Occasionally, you may need to convert a number written in scientific notation to its decimal form. The procedure to do so follows.

To Convert a Number in Scientific Notation to Decimal Form

1. Observe the exponent on the base 10.
2. a) If the exponent is positive, move the decimal point in the number to the right the same number of places as the exponent. It may be necessary to add zeros to the number. This will result in a number greater than or equal to 10.

   b) If the exponent is 0, the decimal point in the number does not move from its present position. Drop the factor 10^0. This will result in a number greater than or equal to 1 but less than 10.

   c) If the exponent is negative, move the decimal point in the number to the left the same number of places as the exponent. It may be necessary to add zeros. This will result in a number less than 1.

EXAMPLE 2 Write the following numbers without exponents.

a) 2.1×10^4 **b)** 8.73×10^{-3} **c)** 1.45×10^8

Solution **a)** Move the decimal point four places to the right.

$$2.1 \times 10^4 = 2.1 \times 10{,}000 = 21{,}000$$

b) Move the decimal point three places to the left.

$$8.73 \times 10^{-3} = 0.00873$$

c) Move the decimal point eight places to the right.

$$1.45 \times 10^8 = 145{,}000{,}000$$

NOW TRY EXERCISE 25

Use Scientific Notation in Problem Solving

We can use the rules of exponents when working with numbers written in scientific notation, as illustrated in the following applications.

EXAMPLE 3 On September 30, 1997, the U.S. public debt was approximately \$5,376,000,000,000 (5 trillion 376 billion dollars). The U.S. population on that date was approximately 267,000,000.

a) Find the average U.S. debt for every person in the United States (the per capita debt).

b) On September 30, 1982, the U.S. debt was approximately \$1,142,000,000,000. How much larger was the debt in 1997 than in 1982?

c) How many times greater was the debt in 1997 than in 1982?

Solution **a)** To find the per capita debt, we divide the public debt by the population.

$$\frac{5{,}376{,}000{,}000{,}000}{267{,}000{,}000} = \frac{5.376 \times 10^{12}}{2.67 \times 10^8} \approx 2.01 \times 10^4 \approx 20{,}100$$

Thus, the per capita debt was about \$20,100.

b) We need to find the difference in the debt between 1997 and 1982.

$$\begin{aligned} 5{,}376{,}000{,}000{,}000 - 1{,}142{,}000{,}000{,}000 &= 5.376 \times 10^{12} - 1.142 \times 10^{12} \\ &= (5.376 - 1.142) \times 10^{12} \\ &= 4.234 \times 10^{12} \\ &= 4{,}234{,}000{,}000{,}000 \end{aligned}$$

The U.S. public debt was $4,234,000,000,000 greater in 1997 than it was in 1992.

c) To find out how many times greater the 1997 public debt was, we divide as follows:

$$\frac{5{,}376{,}000{,}000{,}000}{1{,}142{,}000{,}000{,}000} = \frac{5.376 \times 10^{12}}{1.142 \times 10^{12}} \approx 4.71$$

Thus, the public debt in 1997 was about 4.71 times greater than the 1982 public debt.

EXAMPLE 4 The data for the graphs in Figure 1.10 were taken from the 1996 version of the Statistical Abstract of the United States. The graphs show the cumulative state government tax collection in 1970 and 1994. We have given the amounts collected in scientific notation.

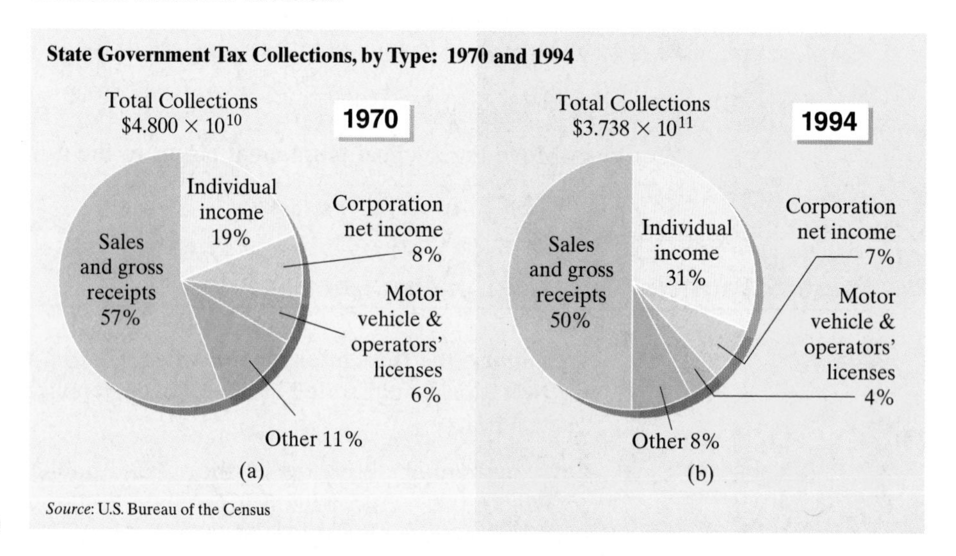

FIGURE 1.10

a) Determine, using scientific notation, how much money was collected from individual income (tax) in 1994.

b) Determine, using scientific notation, how many more dollars were collected, in total taxes, in 1994 than in 1970.

c) Determine, using scientific notation, how many times greater the amount collected in 1994 was than the amount collected in 1970.

d) Describe what you see as major differences in the taxes collected in 1994 versus 1970. Indicate in which sector you see the greatest increase in taxes collected.

Solution **a)** In 1994, 31% of the $\$3.738 \times 10^{11}$ was collected from individual income taxes. In decimal form 31% is 0.31, and it is 3.1×10^{-1} in scientific notation. To determine 31% of $\$3.738 \times 10^{11}$, we multiply using scientific notation as follows:

$$\begin{aligned}\text{income tax collected} &= (3.1 \times 10^{-1})(3.738 \times 10^{11})\\ &= (3.1 \times 3.738)(10^{-1} \times 10^{11})\\ &= 11.5878 \times 10^{-1+11}\\ &= 11.5878 \times 10^{10}\end{aligned}$$

Thus, about 11.5878×10^{10} or $115,878,000,000 was collected from individual income taxes in 1994.

b) In 1970, $\$4.8 \times 10^{10}$ was collected. In 1994, $\$3.738 \times 10^{11}$ was collected. To determine the difference in these amounts we must subtract. We will rewrite 3.738×10^{11} as 37.38×10^{10} so that both numbers will have a power of 10^{10}.

$$\begin{aligned}\text{differences in taxes collected} &= \text{amount collected in 1994} - \text{amount collected in 1970}\\ &= 3.738 \times 10^{11} - 4.80 \times 10^{10}\\ &= 37.38 \times 10^{10} - 4.80 \times 10^{10}\\ &= (37.38 - 4.80) \times 10^{10}\\ &= 32.58 \times 10^{10}\end{aligned}$$

Therefore, $\$32.58 \times 10^{10}$ or $325,800,000,000 more was collected in 1994 than in 1970.

c) To determine the number of times greater the amount collected in 1994 was than in 1970, we need to divide. Can you explain why?

$$\begin{aligned}\text{number of times greater} = \frac{\text{amount collected in 1994}}{\text{amount collected in 1970}} &= \frac{3.738 \times 10^{11}}{4.8 \times 10^{10}}\\ &= \frac{3.738}{4.8} \times 10^{11-10}\\ &= 0.77875 \times 10^{1}\\ &= 7.7875\end{aligned}$$

Thus, the amount collected in 1994 was about 7.8 times greater than the amount collected in 1970.

NOW TRY EXERCISE 75

d) We know from parts **b)** and **c)** that a great deal more taxes were collected in 1994 than in 1970. The graphs show that a smaller percent in 1994 comes from sales tax (50% versus 57%) and a larger percent comes from individual income tax (31% versus 19%). There are smaller changes in the percents of the other sectors.

Using Your Calculator

On a scientific or graphing calculator the product (8,000,000)(400,000) might be displayed as 3.2^{12} or 3.2E12. Both of these represent 3.2×10^{12}, which is 3,200,000,000,000.

To enter numbers in scientific notation on either a scientific or graphing calculator, you generally use the [EE] or [EXP] keys. To enter 4.6×10^{8}, you would press either 4.6 [EE] 8 or 4.6 [EXP] 8. Your calculator screen might then show 4.6^{08} or 4.6E8.

On the TI-82 and TI-83 the EE appears above the [,] key. So to enter (8,000,000)(400,000) in scientific notation you would press

8 [2nd] [,] 6 [×] 4 [2nd] [,] 5 [ENTER] 3.2E12

([2nd] [,] → to get EE; [2nd] [,] → to get EE; 3.2E12 → answer displayed)

Exercise Set 1.6

Concept/Writing Exercises

1. What is the form of a number in scientific notation?
2. Can 1×10^n ever be a negative number for any positive integer n? Explain.
3. Can 1×10^{-n} ever be a negative number for any positive integer n? Explain.
4. Which is greater, 1×10^{-2} or 1×10^{-3}? Explain.

Practice the Skills

Express each number in scientific notation.

5. 7300
6. 900
7. 0.047
8. 0.0000462
9. 19,000
10. 5,260,000,000
11. 0.00000186
12. 0.00000914
13. 5,780,000
14. 0.0000773
15. 0.000101
16. 998,000,000

Express each number without exponents.

17. 6.4×10^3
18. 4×10^7
19. 2.13×10^{-5}
20. 9.64×10^{-7}
21. 3.12×10^{-1}
22. 4.6×10^1
23. 9×10^6
24. 7.3×10^4
25. 2.07×10^5
26. 9.35×10^{-6}
27. 1×10^6
28. 1×10^{-8}

Express each value without exponents.

29. $(5 \times 10^3)(3 \times 10^4)$
30. $(1.6 \times 10^{-2})(4 \times 10^{-3})$
31. $\dfrac{8.4 \times 10^{-6}}{4 \times 10^{-4}}$
32. $\dfrac{25 \times 10^3}{5 \times 10^{-2}}$
33. $\dfrac{5.85 \times 10^4}{4.5 \times 10^{-3}}$
34. $(6.0 \times 10^{-4})(5.0 \times 10^6)$
35. $(8.2 \times 10^5)(1.3 \times 10^{-2})$
36. $(6.3 \times 10^4)(3.7 \times 10^{-8})$
37. $\dfrac{9.2 \times 10^5}{2.3 \times 10^4}$
38. $\dfrac{4.8 \times 10^{-2}}{2.4 \times 10^{-6}}$
39. $(9.1 \times 10^{-4})(6.3 \times 10^{-4})$
40. $\dfrac{6.2 \times 10^{-8}}{3.1 \times 10^{-6}}$

Express each value in scientific notation.

41. $(0.003)(0.00015)$
42. $(230{,}000)(3000)$
43. $\dfrac{1{,}400{,}000}{700}$
44. $\dfrac{20{,}000}{0.005}$
45. $\dfrac{0.0000426}{200}$
46. $\dfrac{0.000012}{0.000006}$
47. $(47{,}000)(35{,}000{,}000)$
48. $\dfrac{0.0000282}{0.00141}$
49. $\dfrac{672}{0.0021}$
50. $\dfrac{0.018}{480}$
51. $\dfrac{0.00153}{0.00051}$
52. $(0.0015)(0.00036)$

Express each value in scientific notation. Round decimal numbers to the nearest thousandth.

53. $(1.23 \times 10^4)(5.67 \times 10^8)$
54. $\dfrac{3.33 \times 10^3}{1.11 \times 10^1}$
55. $(7.23 \times 10^{-3})(1.37 \times 10^5)$
56. $(6.81 \times 10^{-4})(2 \times 10^{-2})$
57. $\dfrac{4.36 \times 10^{-4}}{8.17 \times 10^{-7}}$
58. $\dfrac{8.45 \times 10^{25}}{4.225 \times 10^{15}}$
59. $(3.70 \times 10^{37})(4.15 \times 10^{-30})$
60. $(4.36 \times 10^{-6})(1.07 \times 10^{-6})$
61. $(7.71 \times 10^3)(9.14 \times 10^{-31})$
62. $\dfrac{6.85 \times 10^7}{2.42 \times 10^{15}}$
63. $\dfrac{1.50 \times 10^{35}}{4.5 \times 10^{-26}}$
64. $(3.7 \times 10^5)(1.347 \times 10^{31})$

Problem Solving

65. Explain how you can quickly multiply a number given in scientific notation by **a)** 10; **b)** 100; **c)** 1 million. **d)** Multiply 7.59×10^7 by 1 million. Leave your answer in scientific notation.
66. Explain how you can quickly divide a number given in scientific notation by **a)** 10; **b)** 100; **c)** one million. **d)** Divide 6.58×10^{-4} by 1 million. Leave your answer in scientific notation.

67. During a science experiment you find that the correct answer is 5.25×10^4.

 a) If you mistakenly write the answer as 4.25×10^4, by how much is your answer off?

 b) If you mistakenly write the answer as 5.25×10^5, by how much is your answer off?

 c) Which of the two errors is the more serious? Explain.

68. The distance to the sun is 93,000,000 miles. If a spacecraft travels at a speed of 3100 miles per hour, how long will it take for it to reach the sun?

69. a) Earth completes its 5.85×10^8 mile orbit around the sun in 365 days. Find the distance traveled per day.

 b) Earth's speed is about 8 times faster than a bullet. Estimate the speed of a bullet in miles per hour.

70. We have proof that there are at least 1 sextillion, 10^{21}, stars in the Milky Way.

 a) Write that number without exponents.

 b) How many million stars is this? Explain how you determined your answer to part **b)**.

71. The projected number of people living in the world in 2000 is about 6.09×10^9. The projected number living in the United States is about 2.74×10^8. According to these projections, how many people live outside the United States in 2000?

72. In 1994, computer shipments totaled about $38,400 million. In 1990, computer shipments totaled about $26,000 million.

 a) Write both shipments in scientific notation.

 b) How much larger was the shipment in 1994 than in 1990?

 c) How many times larger was the shipment in 1994 than in 1990?

73. In the United States only about 5% of the 4.2×10^9 pounds of used plastic is recycled annually.

 a) How many pounds are recycled annually?

 b) How many pounds are not recycled annually?

74. If a cubic milliliter of blood contains about 5,900,000 red blood cells, how many red blood cells are contained in 30 cubic milliliters of blood?

75. The top five U.S. airports in 1995, ranked by passengers boarding an airplane (or enplaning) are given in the following circle graph. About 115,700,000,000 or 1.157×10^{11} passengers enplaned in these five airports. The graph also provides the percent of the total number of passengers enplaned for these five airports.

Top Five U.S. Airports in Enplaned Passengers (percent)

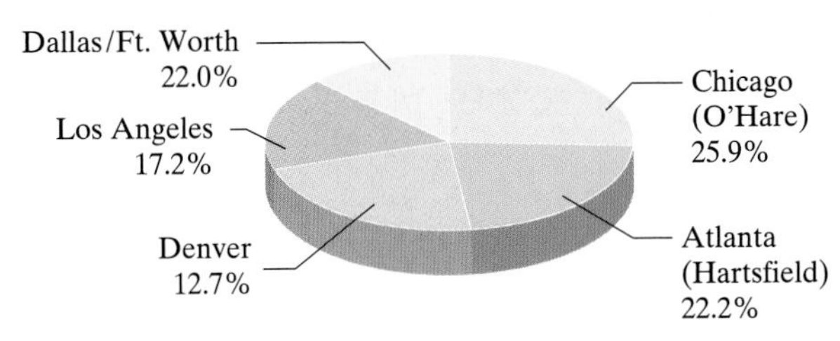

Source: Airports Council International

 a) How many passengers enplaned at Chicago's O'Hare airport?

 b) How many passengers enplaned at Denver International airport?

 c) How many times greater is the number who enplaned at O'Hare than the number who enplaned at Denver International airport?

76. It is estimated that in 2000, the credit card debt by holders of the cards will be $\$6.61 \times 10^{11}$. The table that follows indicates the percent of the total debt by category.

Type of Credit Card	Percent of Debt
Bank	73.5
Retail store	15.0
*Travel & entertainment	5.3
**Others	6.2

* includes American Express and Diners Club.

** includes oil companies, phone companies, air travel cards, Discover, auto rental, and miscellaneous cards.

 a) Write the dollar amount owed in billions of dollars.

 b) Estimate the amount that is owed to banks.

 c) Estimate the amount that is owed to retail stores.

 d) How much greater is the amount that is owed to banks than to retail stores.

77. In 1996, the six most populous countries accounted for 2,945,000,000 people out of the world's total population of 5,772,000,000. The six most populous countries in 1996 are shown in the following graph, along with each country's population.

Six Most Populous Countries (Population in millions)

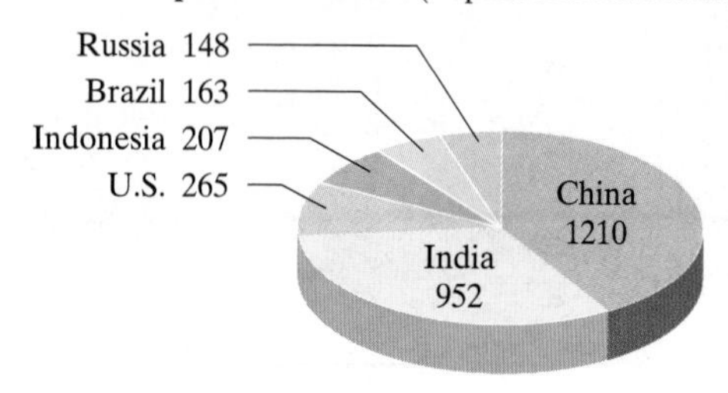

Source: U.S. Census Bureau *Note*: China includes mainland China and Taiwan.

a) How many more people lived in China than in the United States?

b) What percent of the world's population lived in China?

c) If the area of China is 3.70×10^6 square miles, determine China's population density (people per square mile).

d) If the area of the United States 3.62×10^6 square miles, determine the population density in the United States.*

78. The entire span of human history was required for the world's population to reach 5.77×10^9 in 1996. At current rates, the world's population will double in about 40 years.

a) Estimate the world's population in 2036.

b) Assuming 365 days in a year, estimate the average number of additional people added to the world's population each day between 1996 and 2036.

Challenge Problems

79. A *light year* is the distance that light travels in 1 year.

a) Find the number of miles in a light year if light travels at 1.86×10^5 miles *per second.*

b) If Earth is 93,000,000 miles from the sun, how long does it take for light from the sun to reach Earth?

c) Our galaxy, the Milky Way, is about 6.25×10^{16} miles across. If a spacecraft could travel at half the speed of light, how long would it take for the craft to travel from one end of the galaxy to the other?

SUMMARY

Key Words and Phrases

1.2
Between
Constant
Counting numbers
Elements of a set
Endpoints
Expression
Inequalities
Integers
Intersection of sets
Interval notation
Irrational numbers
Natural numbers
Null (or empty) set
Operations
Rational numbers
Real numbers
Roster form
Set
Set builder notation
Subset
Union of sets
Variable
Whole numbers

1.3
Absolute value
Additive inverses
Associative properties
Commutative properties
Distributive property
Opposites
Properties of real numbers

1.5
Base
Exponent
Exponential expression
Factor
Graphing calculator (or grapher)
Index
Order of operations
Principal square root
Radical sign
Radicand
Undefined expression

1.6
Scientific notation

* The area with the greatest population density is Macau, with a population density of 80,426 people per square mile. The country with the greatest population density is Monaco, with a population density of 41,076 people per square mile.

IMPORTANT FACTS

Sets of Numbers

Real numbers	$\{x \mid x \text{ is a point on a number line}\}$
Natural or counting numbers	$\{1, 2, 3, 4, 5, \ldots\}$
Whole numbers	$\{0, 1, 2, 3, 4, \ldots\}$
Integers	$\{\ldots, -3, -2, -1, 0, 1, 2, 3, \ldots\}$
Rational numbers	$\left\{\frac{p}{q} \,\middle|\, p \text{ and } q \text{ are integers}, q \neq 0\right\}$
Irrational numbers	$\{x \mid x \text{ is a real number that is not rational}\}$

Inequalities on the Real Number Line

$\{x \mid x > a\}$

$\{x \mid x \leq a\}$

$\{x \mid a \leq x < b\}$

Properties of the Real Number System

Commutative properties	$a + b = b + a, \quad ab = ba$
Associative properties	$(a + b) + c = a + (b + c), \quad (ab)c = a(bc)$
Identity properties	$a + 0 = 0 + a = a, \quad a \cdot 1 = 1 \cdot a = a$
Inverse properties	$a + (-a) = (-a) + a = 0, \quad a \cdot \frac{1}{a} = \frac{1}{a} \cdot a = 1 \quad (a \neq 0)$
Distributive property	$a(b + c) = ab + ac$
Extended distributive property	$a(b + c + d + \cdots + n) = ab + ac + ad + \cdots + an$
Multiplicative property of 0	$a \cdot 0 = 0 \cdot a = 0$
Double negative property	$-(-a) = a$

Absolute Value: $|a| = \begin{cases} a & a \geq 0 \\ -a & a < 0 \end{cases}$

Exponents and Roots

$$b^n = \underbrace{b \cdot b \cdot b \cdot \cdots \cdot b}_{n \text{ factors of } b} \qquad \sqrt[n]{a} = b \text{ if } \underbrace{b \cdot b \cdot b \cdot \cdots \cdot b}_{n \text{ factors of } b} = a$$

Order of Operations

1. Parentheses or other grouping symbols
2. Exponents and roots
3. Multiplication or division from left to right
4. Addition or subtraction from left to right

Rules for Exponents

1. $a^m \cdot a^n = a^{m+n}$ — *Product rule*
2. $\frac{a^m}{a^n} = a^{m-n}, a \neq 0$ — *Quotient rule*
3. $a^{-m} = \frac{1}{a^m}, a \neq 0$ — *Negative exponent rule*
4. $a^0 = 1, a \neq 0$ — *Zero exponent rule*
5. $(a^m)^n = a^{mn}$ — *Raising a power to a power (Power rule)*
   $(ab)^m = a^m b^m$ — *Raising a product to a power*
   $\left(\frac{a}{b}\right)^m = \frac{a^m}{b^m}, b \neq 0$ — *Raising a quotient to a power*

Review Exercises

[1.2] *List each set in roster form.*

1. $A = \{x \mid x \text{ is a natural number between 2 and 7}\}$
2. $B = \{x \mid x \text{ is a whole number multiple of 3}\}$

Let N = set of natural numbers, W = set of whole numbers, I = set of integers, Q = set of rational numbers, H = set of irrational numbers, ℝ = set of real numbers. Determine whether the first set is a subset of the second set for each pair of sets.

3. N, W
4. $Q, \mathbb{R}$
5. $H, \mathbb{R}$
6. Q, H

Consider the set of numbers $\{-2, 4, 6, \frac{1}{2}, \sqrt{7}, \sqrt{3}, 0, \frac{15}{27}, -\frac{1}{5}, 1.47\}$. *List the elements of the set that are*

7. natural numbers.
8. whole numbers.
9. integers.
10. rational numbers.
11. irrational numbers.
12. real numbers.

Indicate whether each statement is true or false.

13. $\frac{0}{1}$ is not a real number.
14. $0, \frac{3}{5}, -2$, and 4 are all rational numbers.
15. A real number cannot be divided by 0.
16. Every rational number and every irrational number is a real number.

Find A ∪ B and A ∩ B for each set A and B.

17. $A = \{1, 2, 3, 4, 5\}, B = \{2, 3, 4, 5\}$
18. $A = \{3, 5, 7, 9\}, B = \{2, 4, 6, 8\}$
19. $A = \{1, 2, 3, 4, \ldots\}, B = \{2, 4, 6, \ldots\}$
20. $A = \{4, 6, 9, 10, 11\}, B = \{3, 5, 9, 10, 12\}$

Illustrate each set on the number line.

21. $\{x \mid x > 5\}$
22. $\{x \mid x \le -2\}$
23. $\{x \mid -1.3 < x \le 2.4\}$
24. $\{x \mid \frac{2}{3} \le x < 4 \text{ and } x \in N\}$

[1.3] *Insert either <, >, or = in the shaded area between the two numbers to make each statement true.*

25. $-8 \quad 0$
26. $-4 \quad -3.9$
27. $1.06 \quad 1.6$
28. $|-3| \quad 3$
29. $|-4| \quad |-6|$
30. $13 \quad |-5|$
31. $\left|-\frac{2}{3}\right| \quad \frac{3}{5}$
32. $-|-2| \quad -5$

Write the numbers in each list from smallest to largest.

33. $4, -2, -5, |7|$
34. $0, \frac{3}{5}, 2.3, |-3|$
35. $|-7|, |-5|, 3, -2$
36. $-4, -2, -2.1, -|3|$
37. $-4, 6, -|-3|, 5$
38. $|1.6|, |-2.3|, -3, 0$

Name each property illustrated.

39. $3(x + 2) = 3x + 6$
40. $xy = yx$
41. $(x + 3) + 2 = x + (3 + 2)$
42. $a + 0 = a$
43. $(3x)y = 3(xy)$
44. $-(-5) = 5$
45. $3(0) = 0$
46. $x + (-x) = 0$
47. $x \cdot \frac{1}{x} = 1$
48. $(x + y) = 1(x + y)$

[1.3, 1.4] *Evaluate.*

49. $4 - 2^2 + \sqrt{81} \div 3$
50. $-4 \div (-2) + 16 - \sqrt{49}$
51. $(4 - 6) - (-3 + 5) + 12$
52. $3|-2| - (4 - 3) + 2(-3)$
53. $(6 - 9) \div (9 - 6) + 1$
54. $|6 - 3| \div 3 + 4 \cdot 8 - 12$
55. $\sqrt[3]{27} \div 3 + |4 - 2| + 4^2$
56. $3^2 - 6 \cdot 9 + 4 \div 2^2 - 3$
57. $4 - (2 - 9)^0 + 3^2 \div 1 + 3$
58. $4^2 - (2 - 3^2)^2 + 4^3$
59. $-3^2 + 14 \div 2 \cdot 3 - 6$
60. $\{[(9 \div 3)^2 - 1]^2 \div 8\}^3$
61. $\dfrac{8 - 4 \div 2 + 3 \cdot 2}{\sqrt{36} \div 2 - 3}$
62. $\dfrac{-(4 - 6)^2 - 3(-2) + |-6|}{18 - 9 \div 3 \cdot 5}$

Evaluate.

63. Evaluate $2x^2 + 3x + 1$ when $x = 2$

64. Evaluate $4x^2 - 3y^2 + 5$ when $x = 1$ and $y = -\frac{1}{3}$

65. The cost of political campaigning has risen dramatically since 1952. The amount spent, in millions of dollars, on all U.S. elections—including local, state, and national offices; political parties; political action committees; and ballot issues—is approximated by

$$\text{dollars spent} = 50.86x^2 - 316.75x + 541.48$$

where x represents each four year period starting with 1952. Substitute 1 for x to get the amount spent in 1952, 2 for x to get the amount spent in 1956, 3 for x to get the amount spent in 1960, and so on.

a) Find the amount spent for elections in 1976.

b) Find the amount spent for elections in 1996.

66. Railroad traffic has been increasing steadily, and has doubled since 1965. Most of this is due to the increase in trains used to transport goods by container. We can approximate the amount of freight carried in ton-miles (one ton-mile equals one ton of freight hauled 1 mile) by

$$\text{freight hauled} = 14.04x^2 + 1.96x + 712.05$$

where x represents each five year period starting with 1965. Substitute 1 for x to get the amount hauled in 1965, 2 for x to get the amount hauled in 1970, 3 for x for 1975, and so forth.

a) Find the amount of freight hauled by trains in 1980.

b) Find the amount of freight hauled by trains in 1995.

[1.5] *Simplify each expression and write the answer without negative exponents.*

67. $4^2 \cdot 4^1$

68. $x^3 \cdot x^5$

69. $\dfrac{x^6}{x^2}$

70. $\dfrac{y^{12}}{y^3}$

71. $\dfrac{x^4}{x^{-3}}$

72. $x^4 \cdot x^{-7}$

73. $3^{-2} \cdot 3^{-1}$

74. $3x^0$

75. $(3n^2)^2$

76. $\left(\dfrac{2}{3}\right)^{-1}$

77. $\left(\dfrac{3}{4}\right)^{-2}$

78. $\left(\dfrac{x}{y^2}\right)^{-1}$

79. $(7x^2y^5)(-3xy^4)$

80. $(4x^2y^{-3})(2x^{-4}y^2)$

81. $\dfrac{6x^{-3}y^5}{2x^2y^{-2}}$

82. $\dfrac{12x^{-3}y^{-4}}{4x^{-2}y^5}$

83. $\dfrac{a^2b^{-7}c^{-10}}{a^{-3}b^{-3}c^{-4}}$

84. $\dfrac{16p^4q^{-2}r^{-3}}{4p^2q^{-1}r^3}$

85. $\left(\dfrac{5a^2b}{a}\right)^3$

86. $\left(\dfrac{x^5y}{-3y^2}\right)^2$

87. $\left(\dfrac{x^2y}{x^{-1}y^{-3}}\right)^2$

88. $\left(\dfrac{-5x^{-2}y}{z^3}\right)^3$

89. $\left(\dfrac{6xy^3}{z^2}\right)^{-2}$

90. $\left(\dfrac{9m^{-2}n}{3mn}\right)^{-3}$

91. $(-2x^{-3}y^2)^{-4}$

92. $\left(\dfrac{16x^4y^3z^{-2}}{-4x^5y^2z^3}\right)^3$

93. $\left(\dfrac{2x^{-1}y^5z^4}{3x^4y^{-2}z^{-2}}\right)^{-2}$

94. $\left(\dfrac{8x^{-2}y^{-2}z}{-x^4y^{-4}z^3}\right)^{-1}$

[1.6] *Express each number in scientific notation.*

95. 0.0000742

96. 260,000

97. 183,000

98. 0.000001

Simplify each expression and express the answer without exponents.

99. $(25 \times 10^{-3})(1.2 \times 10^6)$

100. $\dfrac{18 \times 10^3}{9 \times 10^5}$

101. $\dfrac{4{,}000{,}000}{0.02}$

102. $(0.004)(500{,}000)$

103. The most successful concert tour of all time, ranked by gross revenue in dollars, was the 1994 Rolling Stones concert tour. The tour's gross revenue was about $\$1.212 \times 10^8$. The third most successful tour was the 1989 Rolling Stones concert tour, with a gross revenue of about $\$9.800 \times 10^7$. (The second most successful tour was the 1994 Pink Floyd concert tour.)

a) How much larger, in dollars, were the gross revenues from the Rolling Stones' 1994 tour than from their 1989 tour?

b) How many times larger was the gross revenue from their 1994 tour than from their 1989 tour?

104. On February 17, 1998, the Voyager 1 spacecraft became the most distant explorer in the solar system, breaking the record of the Pioneer 10. The 20-year-old Voyager 1 has traveled more than 1.04×10^{10} kilometers from Earth (about 70 times the distance from Earth to the sun).

a) Represent 1.04×10^{10} as a decimal number.

b) How many billion kilometers has Voyager 1 traveled?

c) Assuming that Voyager 1 traveled about the same number of kilometers in each of the 20 years, how many kilometers did it average in a year?

d) If 1 kilometer $\approx$ 0.6 miles, how far, in miles, did Voyager 1 travel?

Practice Test

1. List $A = \{x \mid x \text{ is a natural number greater than } 5\}$ in roster form.

Indicate whether each statement is true or false.

2. Every rational number is a real number.

3. The union of the set of rational numbers and the set of irrational numbers is the set of real numbers.

Consider the set of numbers $\{-\frac{3}{5}, 2, -4, 0, \frac{19}{12}, 2.57, \sqrt{8}, \sqrt{2}, -1.92\}$. *List the elements of the set that are*

4. Rational numbers.

5. Real numbers.

Find $A \cup B$ and $A \cap B$ for sets A and B.

6. $A = \{8, 10, 11, 14\}$, $B = \{5, 7, 8, 9, 10\}$

7. $A = \{1, 3, 5, 7, \ldots\}$, $B = \{3, 5, 7, 9, 11\}$

Indicate each set on the number line.

8. $\{x \mid -2.3 \le x < 5.2\}$

9. $\left\{x \,\middle|\, -\frac{5}{2} < x < \frac{6}{5} \text{ and } x \in I\right\}$

10. List from smallest to largest: $|3|, -|4|, -2, 6$.

Name each property illustrated.

11. $(x + y) + 3 = x + (y + 3)$

12. $3x + 4y = 4y + 3x$

Evaluate each expression.

13. $\{4 - [6 - 3(4 - 5)]\}^2 \div (-5)$

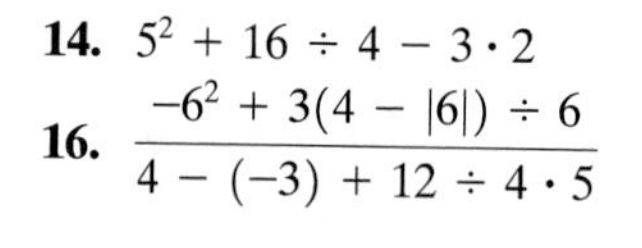

14. $5^2 + 16 \div 4 - 3 \cdot 2$

15. $\dfrac{-3|4 - 8| \div 2 + 4}{-\sqrt{36} + 18 \div 3^2 + 4}$

16. $\dfrac{-6^2 + 3(4 - |6|) \div 6}{4 - (-3) + 12 \div 4 \cdot 5}$

17. Evaluate $-x^2 + 2xy + y^2$ when $x = 2$ and $y = 3$

18. The Federal Fair Labor Services (now dissolved) set a minimum wage of 25 cents per hour for employees in 1940. Ever since then, the minimum wage has increased to keep up with higher costs of living. The minimum wage, in dollars per hour, from 1940 to 1998, can be approximated by

$$\text{minimum wage} = -0.043x^3 + 0.52x^2 - 1.18x + 1.01,$$

where x represents each decade starting with 1940. Substitute 1 for x to get an approximation of the minimum wage in 1940, 2 for x to get an approximation of the minimum wage in 1950, 3 for x for 1960, and so on.

a) Find the minimum wage in 1960.

b) Find the minimum wage in 1990.

Simplify each expression and write the answer without negative exponents.

19. 3^{-2}

20. $\left(\dfrac{3x^{-2}}{y}\right)^2$

21. $\dfrac{3x^2y^3z^2}{9x^5y^{-2}z^{-5}}$

22. $\left(\dfrac{-3x^3y^{-2}}{x^{-1}y^5}\right)^{-3}$

23. Convert 242,000,000 to scientific notation.

24. Simplify $\dfrac{3.12 \times 10^6}{1.2 \times 10^{-2}}$ and write the number without exponents.

25. The total number of people employed in the United States in 1995 was about 1.02×10^8. The graph below shows the men/women breakdown.

a) How many men were employed in the United States?

b) How many women were employed in the United States?

c) How many more men than women were employed?

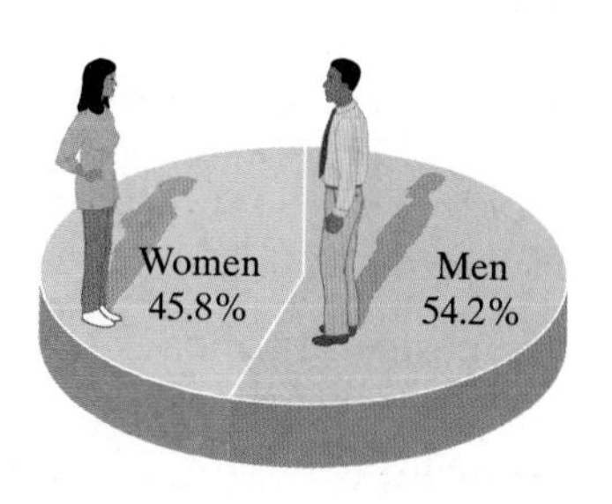

EQUATIONS AND INEQUALITIES

CHAPTER

2

Use the Angel Web site at www.prenhall.com/angel to be linked to an internet resource that will help you further explore the following application

For most people buying a home is the biggest purchase of their life. After negotiating a price for the house, you must also choose a loan. How do you choose a loan? Do you look for the loan which has no application fee, the lowest points, the lowest interest rate, or the one that offers free airline tickets? Is there anything else to consider? On page 90, we compare the costs of two loans by writing equations to describe the cost of each loan and then finding when the costs are the same. While doing this, we learn that the length of time we own the home is a key factor to consider when selecting a loan.

Preview and Perspective

This chapter has two main objectives: solving linear equations and problem solving. In the first section we discuss solving linear equations. We discuss, among other topics, solving equations containing fractions. We will see many more equations containing fractions in Section 6.4.

Algebra is important because it can be used in everyday life. The problem-solving procedure introduced in Section 2.2 will be helpful in writing real-life problems as mathematical expressions and equations. The five-step procedure will be used throughout the book, so you will become very familiar with it. We also work with formulas in Section 2.2. A knowledge of how to evaluate and how to rewrite formulas will prove helpful in many professions.

Changing real-life problems into equations—using the problem-solving procedure—is the emphasis of Sections 2.3 and 2.4. The more you pratice writing equations to represent real-life situations, the better you will become at it. We will emphasize problem solving throughout the book.

Inequalities were introduced in Section 1.2. In Section 2.5 we will learn to solve linear inequalities. In Section 2.6 we will expand on the material presented in Section 2.5 when we solve inequalities containing absolute value.

2.1 SOLVING LINEAR EQUATIONS

1. **Identify the reflexive, symmetric, and transitive properties.**
2. **Combine like terms.**
3. **Solve linear equations.**
4. **Solve equations containing fractions.**
5. **Identify conditional equations, contradictions, and identities.**
6. **Understand the concepts to solve equations.**

1 Identify the Reflexive, Symmetric, and Transitive Properties

In elementary algebra you learned how to solve linear equations. We review these procedures briefly in this section. Before we do so, we need to introduce three useful properties of equality: the *reflexive property*, the *symmetric property*, and the *transitive property*.

Properties of Equality

For all real numbers a, b, and c:

1. $a = a$.	**reflexive property**
2. If $a = b$, then $b = a$.	**symmetric property**
3. If $a = b$ and $b = c$, then $a = c$.	**transitive property**

Examples of the Reflexive Property

$$3 = 3$$

$$x + 5 = x + 5$$

Examples of the Symmetric Property

If $x = 3$, then $3 = x$.

If $y = x + 4$, then $x + 4 = y$.

Examples of the Transitive Property

If $x = a$, and $a = 4y$, then $x = 4y$.

If $a + b = c$ and $c = 4r$, then $a + b = 4r$.

In this book we will often use these properties, without referring to them by name.

2) Combine Like Terms

When an algebraic expression consists of several parts, the parts that are added are called the **terms** of the expression. The expression $3x^2 - 6x - 2$ which can be written $3x^2 + (-6x) + (-2)$, has 3 terms; $3x^2$, $-6x$, and -2. The expression

$$6x^2 - 3(x + y) - 4 + \frac{x + 2}{5}$$

has four terms: $6x^2$, $-3(x + y)$, -4, and $\frac{x + 2}{5}$.

Expression	Terms
$\frac{1}{2}x^2 - 3x - 7$	$\frac{1}{2}x^2$, $-3x$, -7
$-5x^3 + 3x^2y - 2$	$-5x^3$, $3x^2y$, -2
$4(x + 3) + 2x + 5(x - 2) + 1$	$4(x + 3)$, $2x$, $5(x - 2)$, 1

The numerical part of a term that precedes the variable is called its **numerical coefficient** or simply its **coefficient**. In the term $6x^2$, the 6 is the numerical coefficient. When the numerical coefficient is 1 or -1, we generally do not write the numeral 1. For example, x means $1x$, $-x^2y$ means $-1x^2y$, and $(x + y)$ means $1(x + y)$.

Term	Numerical Coefficient
$\frac{2x}{3}$	$\frac{2}{3}$
$-4(x + 2)$	-4
$\frac{x - 2}{3}$	$\frac{1}{3}$
$-(x + y)$	-1

Note that $\frac{x - 2}{3}$ *means* $\frac{1}{3}(x - 2)$ *and* $-(x + y)$ *means* $-1(x + y)$.

When a term consists of only a number, that number is called a **constant**. For example, in the expression $x^2 - 4$, the -4 is a constant.

The **degree of a term** with whole number exponents is the sum of the exponents on the variables in the term. For example, $3x^2$ is a second-degree term, and $-4x$ is a first-degree term $(-4x$ means $-4x^1)$. The number 3 can be written as $3x^0$, so the number 3 (and every other nonzero constant) has degree 0. The term 0 is said to have no degree. The term $4xy^5$ is a sixth-degree term since the sum of the exponents is $1 + 5$ or 6. The term $6x^3y^5$ is an eighth-degree term since $3 + 5 = 8$.

Like terms are terms that have the same variables with the same exponents. For example, $3x$ and $5x$ are like terms, $2x^2$ and $-3x^2$ are like terms, and $3x^2y$ and $-2x^2y$ are like terms. Terms that are not like terms are said to be **unlike terms**. All constants are considered like terms.

To **simplify an expression** means to combine all like terms in the expression. To combine like terms, we can use the distributive property.

Examples of Combining Like Terms

$$5x - 2x = (5 - 2)x = 3x$$

$$3x^2 - 5x^2 = (3 - 5)x^2 = -2x^2$$

$$-7x^2y + 3x^2y = (-7 + 3)x^2y = -4x^2y$$

$$4(x - y) - (x - y) = 4(x - y) - 1(x - y) = (4 - 1)(x - y) = 3(x - y)$$

When simplifying expressions, we can rearrange the terms by using the commutative and associative properties discussed in Chapter 1.

EXAMPLE 1 Simplify. If an expression cannot be simplified, so state.

a) $-2x + 5 + 3x - 7$ **b)** $7x^2 - 2x^2 + 3x + 4$

c) $2x - 3y + 5x - 6y + 3$

Solution **a)** $-2x + 5 + 3x - 7 = \underbrace{-2x + 3x}_{x} + \underbrace{5 - 7}_{-2}$ *Place like terms together.*

This expression simplifies to $x - 2$.

b) $7x^2 - 2x^2 + 3x + 4 = 5x^2 + 3x + 4$

c) $2x - 3y + 5x - 6y + 3 = 2x + 5x - 3y - 6y + 3$ *Place like terms together.*

$= 7x - 9y + 3$

EXAMPLE 2 Simplify $3(x - 2) - [-4(x - 5) + 6]$.

Solution

$3(x - 2) - [-4(x - 5) + 6] = 3(x - 2) - 1[-4(x - 5) + 6]$

$= 3x - 6 - 1[-4x + 20 + 6]$ *Distributive property*

$= 3x - 6 - 1(-4x + 26)$ *Combine like terms.*

$= 3x - 6 + 4x - 26$ *Distributive property*

$= 7x - 32$ *Combine like terms.*

NOW TRY EXERCISE 55

3 Solving Linear Equations

An **equation** is a mathematical statement of equality. *An equation must contain an equal sign* and a mathematical expression on each side of the equal sign.

Examples of Equations

$$x + 4 = -7$$

$$2x^2 - 4 = -3x + 5$$

The numbers that make an equation a true statement are called the **solutions** of the equation. The **solution set** of an equation is the set of real numbers that make the equation true.

Equation	Solution	Solution Set
$2x + 3 = 9$	3	$\{3\}$

Two or more equations with the same solution set are called **equivalent equations**. Equations are generally solved by starting with the given equation and producing a series of simpler equivalent equations.

Example of Equivalent Equations

Equations	Solution Set
$2x + 3 = 9$	$\{3\}$
$2x = 6$	$\{3\}$
$x = 3$	$\{3\}$

In this section we will discuss how to solve **linear equations in one variable**. A linear equation is an equation that can be written in the form $ax + b = c, a \neq 0$.

To solve equations, we use the addition and multiplication properties of equality to isolate the variable on one side of the equal sign.

Addition Property of Equality

If $a = b$, then $a + c = b + c$ for any a, b, and c.

The addition property of equality states that the same number can be added to both sides of an equation without changing the solution to the original equation. Since subtraction is defined in terms of addition, *the addition property of equality also allows us to subtract the same number from both sides of an equation.*

Multiplication Property of Equality

If $a = b$, then $a \cdot c = b \cdot c$ for any a, b, and c.

The multiplication property of equality states that both sides of an equation can be multiplied by the same number without changing the solution. Since division is defined in terms of multiplication, *the multiplication property of equality also allows us to divide both sides of an equation by the same nonzero number.*

To solve an equation, we will often have to use a combination of properties to isolate the variable. Our goal is to get the variable all by itself on one side of the equation (to isolate the variable). A general procedure to solve linear equations follows.

To Solve Linear Equations

1. **Clear fractions**. If the equation contains fractions, eliminate the fractions by multiplying both sides of the equation by the least common denominator.
2. **Simplify each side separately**. Simplify each side of the equation as much as possible. Use the distributive property to clear parentheses and combine like terms as needed.
3. **Isolate the variable term on one side.** Use the addition property to get all terms with the variable on one side of the equation and all constant terms on the other side. It may be necessary to use the addition property a number of times to accomplish this.
4. **Solve for the variable.** Use the multiplication property to get an equation having just the variable (with a coefficient of 1) on one side.
5. **Check.** Check by substituting the value obtained in step 4 back into the original equation.

EXAMPLE 3 Solve the equation $2x + 4 = 9$.

Solution

$$2x + 4 = 9$$

$$2x + 4 - 4 = 9 - 4 \qquad \textit{Subtract 4 from both sides.}$$

$$2x = 5$$

$$\frac{\overset{1}{\cancel{2}}x}{\underset{1}{\cancel{2}}} = \frac{5}{2} \qquad \textit{Divide both sides by 2.}$$

$$x = \frac{5}{2}$$

CHECK:

$$2x + 4 = 9$$

$$\cancel{2}\left(\frac{5}{\cancel{2}}\right) + 4 = 9$$

$$5 + 4 = 9$$

$$9 = 9 \qquad \textit{True}$$

Since the value checks, the solution is $\frac{5}{2}$.

Whenever an equation contains like terms on the same side of the equal sign, combine the like terms before using the addition or multiplication properties.

EXAMPLE 4 Solve the equation $3x - 7 = 5x + 1$.

Solution

$$3x - 7 = 5x + 1$$

$$3x - 3x - 7 = 5x - 3x + 1 \qquad \textit{Subtract 3x from both sides.}$$

$$-7 = 2x + 1$$

$$-7 - 1 = 2x + 1 - 1 \qquad \textit{Subtract 1 from both sides.}$$

$$-8 = 2x$$

$$\frac{-8}{2} = \frac{2x}{2} \qquad \textit{Divide both sides by 2.}$$

$$-4 = x$$

NOW TRY EXERCISE 63

Example 5 contains decimal numbers. We work this problem following the procedure given earlier.

Equation with Decimals

EXAMPLE 5 Solve the equation $4(x - 3.1) = 2.1(x - 4) + 3.5x$.

Solution

$$4(x - 3.1) = 2.1(x - 4) + 3.5x$$

$$4(x) - 4(3.1) = 2.1(x) - 2.1(4) + 3.5x \qquad \textit{Distributive property}$$

$$4x - 12.4 = 2.1x - 8.4 + 3.5x$$

$$4x - 12.4 = 5.6x - 8.4 \qquad \textit{Combine like terms.}$$

$$4x - 12.4 + 8.4 = 5.6x - 8.4 + 8.4 \qquad \textit{Add 8.4 to both sides.}$$

$$4x - 4.0 = 5.6x$$

$$4x - 4x - 4.0 = 5.6x - 4x \qquad \textit{Substract 4x from both sides.}$$

$$-4.0 = 1.6x$$

$$\frac{-4.0}{1.6} = \frac{1.6x}{1.6} \qquad \textit{Divide both sides by 1.6.}$$

$$-2.5 = x$$

The solution is -2.5.

To save space, we will not always show the check of our answers. You should, however, check all your answers. When the equation contains decimal numbers, using a calculator to solve and check the equation may save you some time.

Using Your Calculator

Checking Solutions by Substitution

Solutions to equations may be checked using a calculator. To check, substitute your solution into both sides of the equation to see whether you get the same value (there may sometimes be a slight difference in the last digits). The graphing calculator screen in Figure 2.1 shows that both sides of the equation given in Example 5 equal −22.4 when −2.5 is substituted for x. Thus the solution −2.5 checks.

$$4(x - 3.1) = 2.1(x - 4) + 3.5x$$

$$4(-2.5 - 3.1) = 2.1(-2.5 - 4) + 3.5(-2.5)$$

Exercises

Use your calculator to determine whether the given number is a solution to the equation.

1. $5.2(x - 3.1) = 2.3(x - 5.2)$; 1.4
2. $-2.3(4 - x) = 3.5(x - 6.1)$; 10.125

```
4(-2.5-3.1)
              -22.4      ← Value of the left side of the equation
2.1(-2.5-4)+3.5*
-2.5
              -22.4      ← Value of the right side of the equation
```

FIGURE 2.1

Equation with Nested Parentheses

EXAMPLE 6 Solve the equation $2x + 8 = 3[-2(3x - 5) - 12]$.

Solution First, we use the distributive property and then combine like terms. Since the right side of the equation contains nested parentheses, we work with the inner ones first.

$$2x + 8 = 3[-2(3x - 5) - 12]$$

$$2x + 8 = 3[-6x + 10 - 12] \qquad \text{Distributive property}$$

$$2x + 8 = 3[-6x - 2] \qquad \text{Combine like terms.}$$

$$2x + 8 = -18x - 6 \qquad \text{Distributive property}$$

$$18x + 2x + 8 = -18x + 18x - 6 \qquad \text{Add } 18x \text{ to both sides.}$$

$$20x + 8 = -6$$

$$20x + 8 - 8 = -6 - 8 \qquad \text{Subtract 8 from both sides.}$$

$$20x = -14$$

$$\frac{20x}{20} = \frac{-14}{20} \qquad \text{Divide both sides by 20.}$$

$$x = -\frac{14}{20} = -\frac{7}{10}$$

NOW TRY EXERCISE 87

Notice that the solutions to Examples 5 and 6 are not integers. You should not expect the solutions to equations to always be integer values.

In solving equations, some of the intermediate steps can often be omitted. Now we will illustrate how this may be done.

	Solution		Shortened Solution
a)	$x + 4 = 6$		a) $x + 4 = 6$
	$x + 4 - 4 = 6 - 4$	← *Do this step mentally.*	$x = 2$
	$x = 2$		
b)	$3x = 6$		b) $3x = 6$
	$\frac{3x}{3} = \frac{6}{3}$	← *Do this step mentally.*	$x = 2$
	$x = 2$		

4 Solve Equations Containing Fractions

When an equation contains fractions, we begin by multiplying *both* sides of the equation by the least common denominator. The **least common denominator** (LCD) of a set of denominators (also called the **least common multiple**, LCM) is the smallest number that each of the denominators divides into without remainder. For example, if the denominators of two fractions are 5 and 6, then 30 is the least common denominator since 30 is the smallest number that both 5 and 6 divide into without remainder.

When you multiply both sides of the equation by the LCD, *each term* in the equation will be multiplied by the least common denominator. *After this step is performed, the equation should not contain any fractions.*

EXAMPLE 7 Solve the equation $5 - \frac{2x}{3} = -9$.

Solution The least common denominator is 3. Multiply both sides of the equation by 3 and then use the distributive property on the left side of the equation. *This process will eliminate all fractions from the equation.*

$$5 - \frac{2x}{3} = -9$$

$$3\left(5 - \frac{2x}{3}\right) = 3(-9) \quad \text{Multiply both sides by 3.}$$

$$3(5) - \overset{1}{\cancel{3}}\left(\frac{2x}{\underset{1}{\cancel{3}}}\right) = -27 \quad \text{Distributive Property.}$$

$$15 - 2x = -27$$

$$-2x = -42$$

$$x = 21$$

EXAMPLE 8 Solve the equation $\frac{1}{2}(x + 4) = \frac{1}{3}x$.

Solution Begin by multiplying both sides of the equation by 6, the LCD of 2 and 3.

$$6\left[\frac{1}{2}(x + 4)\right] = 6\left(\frac{1}{3}x\right) \quad \text{Multiply both sides by 6.}$$

$$3(x + 4) = 2x \quad \text{Simplify.}$$

$$3x + 12 = 2x \quad \text{Distributive Property.}$$

$$3x - 2x + 12 = 2x - 2x \quad \text{Subtract } 2x \text{ from both sides.}$$

$$x + 12 = 0$$

$$x + 12 - 12 = 0 - 12 \quad \text{Subtract 12 from both sides.}$$

$$x = -12$$

We will be discussing equations containing fractions further in Section 6.4.

HELPFUL HINT

The equation in Example 8 may also be written as $\frac{x + 4}{2} = \frac{x}{3}$.

Can you explain why?

Equations in one variable may be solved graphically using a graphing calculator. In Section 3.3 we discuss how this is done. You may wish to review that material now.

5 Identify Conditional Equations, Contradictions, and Identities

All equations discussed so far have been true for only one value of the variable. Such equations are **conditional equations**. Some equations are never true and have no solution; these are called **contradictions**. Other equations, called **identities** are always true and have an infinite number of solutions. Table 2.1 summarizes these types of equation and their corresponding number of solutions

TABLE 2.1

Type of linear equation	Solution
Conditional equation	One
Contradiction	None (solution set: ∅)
Identity	Infinite number (solution set: ℝ)

The solution set of a conditional equation contains the solution given in set braces. For example, the solution set to Example 8 is $\{-12\}$. The solution set of a contradiction is the empty set or null set, $\{\ \}$ or $\varnothing$. The solution set of an identity is the set of real numbers, $\mathbb{R}$.

EXAMPLE 9 Determine whether the equation $2x + 1 = 5x + 1 - 3x$ is a conditional equation, a contradiction, or an identity. Give the solution set for the equation.

Solution

$$2x + 1 = 5x + 1 - 3x$$

$$2x + 1 = 2x + 1$$

Since we obtain the same expression on both sides of the equation, it is an identity. This equation is true for all real numbers. Its solution set is $\mathbb{R}$.

NOW TRY EXERCISE 121

EXAMPLE 10 Determine whether $2(3x + 1) = 6x + 3$ is a conditional equation, a contradiction, or an identity. Give the solution set for the equation.

Solution

$$2(3x + 1) = 6x + 3$$

$$6x + 2 = 6x + 3$$

$$6x - 6x + 2 = 6x - 6x + 3 \quad \text{Subtract } 6x \text{ from both sides.}$$

$$2 = 3$$

Since $2 = 3$ is never a true statement, this equation is a contradiction. Its solution set is $\varnothing$.

NOW TRY EXERCISE 115

6) Understand the Concepts to Solve Equations

It is important that you realize that the numbers or variables that appear in equations do not affect the procedures used to solve the equations. In the following example, which does not contain any numbers or letters, we will solve the equation using the concepts and procedures that have been presented.

EXAMPLE 11 Assume in the following equation that $\odot$ represents the variable that we are solving for and all the other symbols represent nonzero real numbers. Solve the equation for $\odot$.

$$\square\odot + \Delta = \#$$

Solution To solve for $\odot$ we need to isolate the $\odot$. We use the addition and multiplication properties to solve for $\odot$.

$$\square\odot + \Delta = \#$$

$$\square\odot + \Delta - \Delta = \# - \Delta \quad \text{Subtract } \Delta \text{ from both sides.}$$

$$\square\odot = \# - \Delta$$

$$\frac{\square\odot}{\square} = \frac{\# - \Delta}{\square} \quad \text{Divide both sides by } \square.$$

$$\odot = \frac{\# - \Delta}{\square}$$

Thus the solution is $\odot = \dfrac{\# - \Delta}{\square}$.

Consider the equation $5x + 7 = 12$. If we let $5 = \square$, $x = \odot$, $7 = \Delta$, and $12 = \#$, the equation has the same form as the equation in Example 11. Therefore, the solution will be of the same form.

Equation	Solution
$\square \odot + \Delta = \#$	$\odot = \dfrac{\# - \Delta}{\square}$
$5x + 7 = 12$	$x = \dfrac{12 - 7}{5} = \dfrac{5}{5} = 1$

If you solve the equation $5x + 7 = 12$, you will see that its solution is 1. Thus the procedure used to solve an equation is not dependent on the numbers or variables given in the equation.

NOW TRY EXERCISE 137

Exercise Set 2.1

Concept/Writing Exercises

1. What are the terms of an expression?
2. Determine the coefficient of each term.

   a) x^2y^3 **b)** $-x^3y^2$ **c)** $-\dfrac{x + 3y}{2}$
3. Determine the coefficient of each term.

   a) $\dfrac{x + y}{4}$ **b)** $-(x + 3)$ **c)** $-\dfrac{3(x + 2)}{5}$
4. How do you find the degree of a term?
5. **a)** What are like terms?

   b) Are $3x$ and $3x^2$ like terms? Explain.
6. What is an equation?
7. Is 4 the solution to the equation $2x + 3 = x + 5$? Explain.
8. Is $\{8\}$ the solution set to the equation $x + 1 = 2x - 7$? Explain.
9. State the addition property of equality.
10. State the multiplication property of equality.
11. **a)** What is an identity?

    b) What is the solution set of an identity?
12. **a)** What is a contradiction?

    b) What is the solution set of a contradiction?
13. **a)** In your own words, explain in a step-by-step manner how you would solve the equation

    $$5x + 2x - 5 = 3(x - 7)$$

    b) Solve the equation.
14. **a)** In your own words, explain step by step how you would solve the equation

    $$\frac{2}{5} = \frac{2}{3}(x + 5)$$

    b) Solve the equation.

Practice the Skills

Name each indicated property.

15. If $x = 7$, then $7 = x$.
16. If $x + 2 = 3$, then $3 = x + 2$.
17. If $x = 3$, and $3 = y$, then $x = y$.
18. If $x + 1 = a$ and $a = 2y$, then $x + 1 = 2y$.
19. $x + 2 = x + 2$
20. If $x = 4$, then $x + 3 = 4 + 3$.
21. If $x = 8$, then $x - 8 = 8 - 8$.
22. If $2x = 4$, then $3(2x) = 3(4)$.
23. If $5x = 4$, then $\frac{1}{5}(5x) = \frac{1}{5}(4)$.
24. If $x + 2 = 4$ then $x + 2 - 2 = 4 - 2$.
25. If $5x = 3$, then $\frac{5x}{5} = \frac{3}{5}$.
26. If $x - 3 = x + y$ and $x + y = z$, then $x - 3 = z$.

Give the degree of each term.

27. $4x$
28. $-6x^2$
29. $3xy$
30. $\frac{1}{2}x^4y$
31. 7
32. -3
33. $-5x$
34. $18x^2y^3$
35. $3x^4y^6z^3$
36. x^4y^6
37. $3x^5y^6z$
38. $-2x^4y^7z^8$

Simplify each expression. If an expression cannot be simplified, so state.

39. $8x + 7 + 7x - 12$
40. $3x^2 + 4x + 5$
41. $5x^2 - 3x + 2x - 5$
42. $6x^2 - 9x + 3 - 4x - 7$
43. $10.6c^2 - 2.3c + 5.9c - 1.9c^2$
44. $7y + 3x - 7 + 4x - 2y$

45. $6b^2 + 6b + 3a$

46. $4x^2 - x^2 - 3x - 3x^2 + 4$

47. $xy + 3xy + y^2 - 2$

48. $3x^2y + 4xy^2 - 2x^2$

49. $8.2(x - 3.4) - 1.2(9.8x + 12.4)$

50. $6(x + 5) + 2\left(x + \frac{2}{3}\right)$

51. $3\left(x + \frac{1}{2}\right) - \frac{1}{3}x + 5$

52. $0.4(x - 3) + 6.5(x - 3) + 4x - 2.3$

53. $4 - [6(3x + 2) - x] + 4$

54. $3(x + y) - 4(x + y) - 3$

55. $4x - [3x - (5x - 4y)] + y$

56. $-2[3x - (2y - 1) - 5x] + y$

57. $5b - \{7[2(3b - 2) - (4b + 9)] - 2\}$

58. $2\{[3a - (2b - 5a)] - 3(2a - b)\}$

59. $-\{[2rs - 3(r + 2s)] - 2(2r^2 - s)\}$

60. $p^2q + 4pq - [-(pq + 4p^2q) + pq]$

Solve each equation.

61. $3x + 5 = 17$

62. $2x + 3 + x = 9$

63. $5x - 9 = 3(x - 2)$

64. $5s - 3 = 2s + 6$

65. $4x - 8 = -4(2x - 3) + 4$

66. $3(x - 5) = 2(x - 6)$

67. $4(x - 3) = 2(x + 9)$

68. $7(x - 1) = 4(x + 2)$

69. $-3(t - 5) = 2(t - 5)$

70. $4(2x - 4) = -2(x + 3)$

71. $3x + 4(x - 2) = 4x - 5$

72. $2 + (3y - 4) = 2(y - 3)$

73. $2 - (x + 5) = 4x - 8$

74. $4x - 2(3x - 7) = 2x - 6$

75. $3y + 2(2y - 1) = 2(y - 6)$

76. $8x + 2(x - 4) = 8x + 10$

77. $5 - 3(2x + 1) = 4x - 8$

78. $3(x - 4) = 6x - (4 - 5x)$

79. $6 - (n + 3) = 3n + 5 - 2n$

80. $8 - 3(2x - 4) = 5 + 3x - 4x$

81. $4(2x - 2) - 3(x + 7) = -4$

82. $6(x + 4) - 4(3x + 3) = 6$

83. $-4(3 - 4x) - 2(x - 1) = 12x$

84. $-4(2z - 6) = -3(z - 4) + z$

85. $-(x - 8) = 6x - 4x - 2(x - 3)$

86. $3(2x - 4) + 3(x + 1) = 9$

87. $5(x - 2) - 14x = -3x - (5 - 4x)$

88. $-[x - (4x - 7)] = 3 - (x - 6)$

***89.** $2[3x - (4x - 6)] = 5(x - 6)$

90. $-z - 6z + 3 = 4 - [6 - z - (3 - 2z)]$

91. $6 - \{4[x - (3x - 4) - x] + 4\} = 2(x + 3)$

92. $3\{[(x - 2) + 4x] - (x - 3)\} = 4 - (x - 12)$

93. $-(3 - x) = 5 - \{6x - [2x - (3x - (5x - 8))]\}$

94. $-3(6 - 4x) = 4 - \{5x - [6x - (4x - (3x + 2))]\}$

Solve each equation. Leave your answer as a fraction if it is not an integer value.

95. $\frac{x}{3} = -12$

96. $\frac{5x + 3}{4} = 2$

97. $\frac{4x - 2}{3} = -6$

98. $\frac{1}{2}(6x - 10) = 7$

99. $\frac{1}{3}x + \frac{1}{2}x = 10$

100. $\frac{1}{4}(x - 2) = \frac{1}{3}(2x + 6)$

101. $4 - \frac{3}{4}x = 7$

102. $x - 2 = \frac{3}{4}(x + 4)$

103. $\frac{1}{2} = \frac{4}{5}x - \frac{1}{4}$

104. $\frac{3}{5} + \frac{5}{2} = 4x + 1$

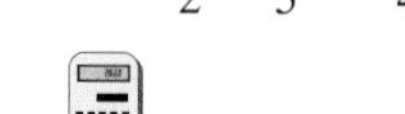

Solve each equation. Round answers to the nearest hundredth.

105. $0.3x = x - 2.7$

106. $0.2(x - 30) = 1.6x$

107. $4.7x - 3.6(x - 1) = 4.9$

108. $3.9x - 5.7(2x - 3) = 9.3$

109. $0.047(3000 - x) = -0.06(x + 900)$

110. $0.05(2000 + 2x) = 0.04(2500 - 6x)$

111. $0.6(500 - 2.4x) = 3.6(2x - 4000)$

112. $3.4t - 12.7t = 4.6(200 + 9t)$

113. $0.04(1000) + 0.2(x + 2000) = 10{,}000$

114. $0.6(14x - 8000) = -0.4(20x + 12{,}000) + 20.6x$

Find the solution set for each exercise. Then indicate whether the equation is conditional, an identity, or a contradiction.

115. $2(x - 3) + x = 3(x + 4)$

116. $4x + 12 - 8x = -6(x - 2) + 2x$

117. $4(2x - 3) + 5 = -6(x - 4) + 12x - 31$

118. $3(2x - 4) = -3(x + 1) + 9$

119. $-4(2 + 4x) + 6x = -(6x + 8) - 4x$

120. $-[4 - (x - 2)] = 2x - 2 - x$

121. $6(x - 1) = -3(2 - x) + 3x$

122. $2(x - 3) + 2x = 4x - 5$

123. $3(x + 4) + 2x = 6 - (x - 3) + 6x$

124. $4(2 - 3x) = -[6x - (8 - 6x)]$

*This exercise also appears on the videotape.

Problem Solving

125. The number of reported runway incidents has increased since 1993. An incident is when an aircraft, vehicle, or person is on a collision course with an aircraft that is landing, taxiing, or taking off. The approximate number of incidents can be found using the equation

$$I = 5x + 23$$

where I = the number of incidents and x is the year starting with 1993. Use $x = 1$ for 1993, $x = 2$ for 1994, etc.

a) Find the number of incidents in 1994.

b) In what year were there 48 incidents?

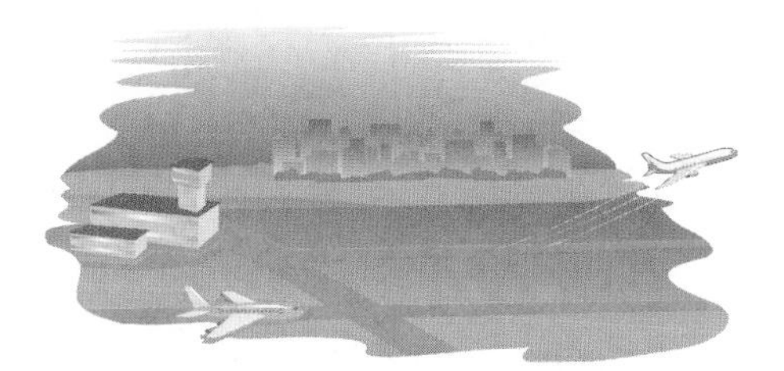

126. Americans now buy 40% of the world's indigestion remedies, and we are buying more every year. The number of remedies bought in the United States each year can be approximated by the equation

$$S = 0.22x + 1.32$$

where S represents total U.S. sales (in billions of dollars) and x is the year starting with 1997. Use $x = 1$ for 1997, $x = 2$ for 1998, and so forth.

a) What were the total U. S. sales of indigestion remedies in 1999?

b) Assuming that this trend continues, in what year will the U.S. sales reach 3.3 billion dollars?

127. Annuities are life insurance contracts that guarantee future payments. One type of annuity, called a variable annuity, is a retirement account that lets someone invest in mutual funds and defer payment of taxes until withdrawals are taken at a later time. The number of variable annuities sold has been growing steadily over the past few years. Variable annuity sales can be approximated by the equation

$$S = 10x + 20$$

where S represents total sales of variable annuities (in billions of dollars) and x is the year starting with 1992. Use $x = 1$ for 1992, $x = 2$ for 1993, etc.

a) Find the total sales of variable annuities in 1995.

b) In what year will annuity sales reach the 100-billion-dollar mark?

128. Florida's public school population is growing much faster than the schools can handle. With enough new students each year to create a new medium-sized school district, overcrowding has become a major problem. The growth in student population can be approximated by the equation

$$S = 62{,}130x + 1{,}758{,}187$$

where S = the state school population and x is the year starting in 1987. In this equation substitute $x = 1$ for 1987, $x = 2$ for 1988, $x = 3$ for 1989, and so on.

a) Find the number of students in 1993.

b) In what year did the student population equal 2,379,487?

129. Consider the equation $x = 4$. Give three equivalent equations. Explain why the equations are equivalent.

130. Consider the equation $2x = 5$. Give three equivalent equations. Explain why the equations are equivalent.

131. Make up an equation that is an identity. Explain how you created the equation.

132. Make up an equation that is a contradiction. Explain how you created the contradiction.

133. Create an equation with three terms to the left of the equal sign and two terms to the right of the equal sign that is equivalent to the equation $3x + 1 = x + 5$.

134. Create an equation with two terms to the left of the equal sign and three terms to the right of the equal sign that is equivalent to the equation $\frac{1}{2}x + 3 = 6$.

135. Consider the equation $-3(x + 2) + 5x + 12 = n$. What real number must n be for the solution of the equation to be 6? Explain how you determined your answer.

136. Consider the equation $2(x + 5) + n = 4x - 8$. What real number must n be for the solution of the equation to be -2? Explain how you determined your answer.

Solve each equation for the given symbol. Assume that the symbol you are solving for represents the variable and that all other symbols represent nonzero real numbers. See Example 11.

137. Solve $*\Delta - \square = \odot$ for Δ.

138. Solve $\odot\square + \Delta = \otimes$ for $\odot$.

139. Solve $\Delta(\odot + \square) = \otimes$ for Δ.

140. Solve $\Delta(\odot + \square) = \otimes$ for $\square$.

Cumulative Review Exercises

[1.3] **141.** **a)** In your own words, explain how to find the absolute value of a number.

b) Write the definition of absolute value.

[1.4] *Evaluate.*

142. **a)** -3^2 **b)** $(-3)^2$

143. $\left(-\frac{3}{4}\right)^3$

144. $\sqrt[3]{-64}$

2.2 PROBLEM SOLVING AND USING FORMULAS

1. **Use the problem-solving procedure.**
2. **Solve for a variable in an equation or formula.**

SSM VIDEO 2.2 CD Rom

1 Use the Problem-Solving Procedure

One of the main reasons for studying mathematics is that we can use it to solve everyday problems. To solve most real-life application problems mathematically, we need to be able to express the problem in mathematical symbols using expressions or equations, so when we do this, we are creating a **mathematical model** of the situation.

In this section we present a problem-solving procedure and discuss formulas. A **formula** is an equation that is a mathematical model of a real-life situation. Throughout the book we will be problem solving. When we do so, we will determine an equation or formula that represents or models the real-life situation.

We will now give the general five-step problem-solving procedure developed by George Polya and presented in his book *How to Solve It*. You can approach any problem by following this general procedure.

Guidelines for Problem Solving

1. **Understand the problem.**
   - Read the problem **carefully** at least twice. In the first reading, get a general overview of the problem. In the second reading, determine (a) exactly what you are being asked to find and (b) what information the problem provides.
   - If possible, make a sketch to illustrate the problem. Label the information given.
   - List the information in a table if it will help in solving the problem.
2. **Translate the problem into mathematical language.**
   - This will generally involve expressing the problem algebraically.
   - Sometimes this involves selecting a particular formula to use, whereas other times it is a matter of generating your own equation. It may be necessary to check other sources for the appropriate formula to use.
3. **Carry out the mathematical calculations necessary to solve the problem.**

continued on next page

4. **Check the answer obtained in step 3.**
   - Ask yourself: "Does the answer make sense?" "Is the answer reasonable?" If the answer is not reasonable, recheck your method for solving the problem and your calculations.
   - Check the solution in the original problem if possible.
5. **Answer the question**. Make sure you have answered the question asked. State the answer clearly.

The following examples show how to apply the guidelines for problem solving. We will sometimes provide the steps in the examples to illustrate the five-step process. However, in some problems it may not be possible or necessary to list every step.

As was stated in step 2 of the problem-solving process—*translate the problem into mathematical language*—we will sometimes need to find and use a *formula*. We will show how to do that in this section. In Section 2.3 we will explain how to develop *equations* to solve real-life application problems.

EXAMPLE 1 Patrice Jones makes a $5000, 6% simple interest loan to her brother Georges for a period of 3 years.

a) At the end of 3 years, what interest will Georges pay Patrice?

b) When Georges settles his loan at the end of 3 years, how much money, in total, must he pay Patrice?

Solution **a) Understand the problem** When a person borrows money using a simple interest loan, the person must repay the simple interest and principal (the original amount borrowed) at the maturity date of the loan. For example, if a simple interest loan is for 3 years, then 3 years after the loan is made, the principal plus the interest must be repaid. We are told in the problem that the simple interest rate is 6% and the loan is for 3 years.

Translate the problem into mathematical language Many business mathematics and investment books include the simple interest formula: interest = principal · rate · time or $i = prt$, which can be used to find the simple interest, i. In the formula, p is the principal, r is the simple interest rate (always changed to decimal form when used in the formula), and t is time. The time and rate must be in the same units. For example, if the rate is 6% per *year*, then the time must be in *years*.

Carry out the mathematical calculations necessary In this problem $p = \$5000$, $r = 0.06$, and $t = 3$. We obtain the simple interest, i, by substituting these values in the simple interest formula.

$$\begin{aligned} i &= prt \\ &= 5000(0.06)(3) \\ &= 900 \end{aligned}$$

Check The answer appears reasonable in that Georges will pay $900 for the use of $5000 of Patrice's money for 3 years.

Answer The simple interest owed is $900.

b) At the end of 3 years Georges must pay the principal he borrowed ($5000) plus the interest determined in part **a)** ($900). (The principal plus interest owed is called the amount owed, A.) Thus when Georges settles his loan, he must pay Patrice $5900.

NOW TRY EXERCISE 67

EXAMPLE 2 Dave Ostrow invests \$1000 in a NationsBank savings account that earns 4% interest compounded quarterly.

a) How much money will he have in his account at the end of 1 year?

b) How much interest has he gained in 1 year?

Solution **a) Understand the problem** Before you can understand the problem you must understand what compound interest is. Compound interest means that you get interest on your investment for one period. Then in the next period you get interest paid on your investment, plus interest paid on the interest that was paid in the first period. This process then continues for each period. In many real-life situations, and in the workplace, you may need to do some research to answer questions asked.

The facts given are that \$1000 is invested and the interest rate is 4% compounded quarterly.

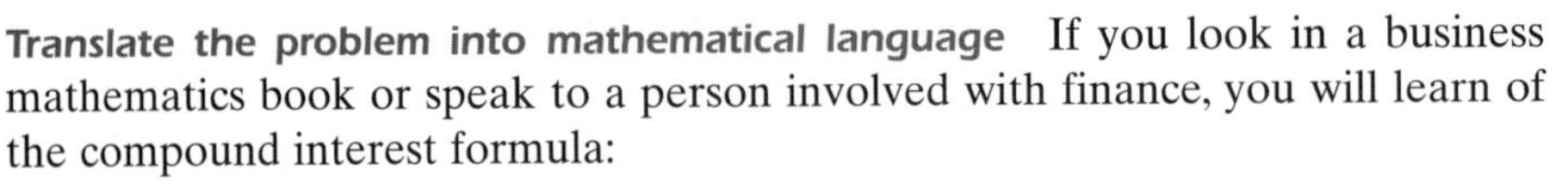

Translate the problem into mathematical language If you look in a business mathematics book or speak to a person involved with finance, you will learn of the compound interest formula:

$$A = p\left(1 + \frac{r}{n}\right)^{nt}$$

The compound interest formula is used by banks to compute the amount (or balance), A, in saving accounts that earns compound interest. In the formula, p represents the principal, r represents the interest rate, n represents the number of compounding periods (the number of times the interest is paid annually), and t represents time in years.

Carry out the mathematical calculations necessary In this problem, $p = \$1000$, $r = 0.04$, $t = 1$, and since the interest is compounded quarterly, $n = 4$. Substitute these values into the formula and evaluate the formula.

$$A = p\left(1 + \frac{r}{n}\right)^{nt}$$

$$= 1000\left(1 + \frac{0.04}{4}\right)^{4(1)}$$

$$= 1000(1 + 0.01)^4$$

$$= 1000(1.01)^4$$

$$= 1000(1.04060401)$$ *From a calculator*

$$= 1040.60$$ *Rounded to the nearest cent*

Check the answer The answer \$1040.60 is reasonable since it is a little larger than the original amount invested.

Answer the question The amount in Dave Ostrow's account at the end of 1 year will be \$1040.60.

b) Understand The interest will be the difference between the original amount invested and the amount in the account at the end of 1 year.

Translate Interest = amount after 1 year − amount originally invested

Carry Out $= 1040.60 - 1000 = 40.60$

Check The answer is reasonable and the arithmetic is easily checked.

NOW TRY EXERCISE 77

Answer The interest gained in the 1-year period was $40.60.

Often a formula contains subscripts. **Subscripts** are numbers (or other variables) placed below and to the right of variables. They are used to help clarify a formula. For example, if a formula contains two velocities, the original velocity and the final velocity, these velocities might be symbolized as V_0 and V_f, respectively. Subscripts are read using the word "sub." For example, V_f is read "V sub f" and x_2 is read "x sub 2." The formula in Example 3 contains subscripts.

EXAMPLE 3 Bonnie Blanchard is in the 28% federal income tax bracket. She is considering making a 4% tax-free investment in municipal bonds. Determine the taxable rate equivalent to a 4% tax-free rate for Bonnie.

Solution **Understand** Some interest we receive, such as from municipal bonds, is tax free. This means that we do not have to pay federal taxes on the interest received. Other interest we receive, such as from savings accounts, is taxable on our federal income tax returns. Paying taxes on the interest has the effect of reducing the amount of money we actually get to keep from the interest. We need to find the taxable interest rate that is equivalent to a 4% tax-free rate for Bonnie (or for anyone in a 28% income tax bracket).

Translate A formula found in many investment books and some government publications that may be used to compare taxable and tax-free rates is

$$T_f = T_a(1 - F)$$

where T_f is the tax free rate, T_a is the taxable rate, and F is the federal income tax bracket. To determine the equivalent taxable interest, T_a, we substitute the appropriate values in the formula and solve for T_a.

Carry Out

$$T_f = T_a(1 - F)$$
$$0.04 = T_a(1 - 0.28)$$
$$0.04 = T_a(0.72)$$
$$\frac{0.04}{0.72} = T_a$$
$$0.05\overline{5} = T_a$$

Check The answer appears reasonable because it is a bit more than 4%, which is what we expected.

NOW TRY EXERCISE 79

Answer The equivalent taxable rate for Bonnie is about 5.56%. After taxes, a taxable investment yielding about 5.56% would give her about the same interest as a 4% tax-free investment.

2) Solve for a Variable in an Equation or Formula

There are many occasions when you might be given an equation or formula that is solved for one variable but you want to solve it for a different variable. Suppose in Example 3 we wanted to find the equivalent taxable rate, T_a, for many different tax-free rates and many different income tax brackets. We could solve each individual problem as was done in Example 3. However, it would be much quicker to solve the formula $T_f = T_a(1 - F)$ for T_a and then substitute the appropriate values in the formula solved for T_a. We will do this in Example 8.

We will begin by solving equations for the variable y. We will need to do this in Chapter 3 when we discuss graphing. Since formulas are equations, the same procedure we use to solve for a variable in an equation will be used for a variable in a formula.

When you are given an equation (or formula) that is solved for one variable and you want to solve it for a different variable, treat each variable in the equation, except the one you are solving for, as if it were a constant. Then *isolate the variable* you are solving for using the procedures similar to those used to solve equations.

EXAMPLE 4 Solve the equation $2x - 3y = 6$ for y.

Solution We will solve for the variable y by isolating the term containing the y on the left side of the equation.

$$2x - 3y = 6$$

$$2x - 2x - 3y = -2x + 6 \qquad \text{Subtract } 2x \text{ from both sides.}$$

$$-3y = -2x + 6$$

$$\frac{-3y}{-3} = \frac{-2x + 6}{-3} \qquad \text{Divide both sides by } -3.$$

$$y = \frac{-1(-2x + 6)}{-1(-3)} \qquad \text{Multiply the numerator and the denominator by } -1.$$

$$y = \frac{2x - 6}{3} \quad \text{or} \quad y = \frac{2}{3}x - 2$$

NOW TRY EXERCISE 29

EXAMPLE 5 Solve the equation $2y - 3 = \frac{1}{2}(x + 3y)$ for y.

Solution Since this equation contains a fraction, we begin by multiplying both sides of the equation by the least common denominator, 2. We then isolate the variable y by collecting all terms containing the variable y on one side of the equation and all other terms on the other side of the equation.

$$2y - 3 = \frac{1}{2}(x + 3y)$$

$$2(2y - 3) = 2\left[\frac{1}{2}(x + 3y)\right] \qquad \text{Multiply both sides by the LCD, 2.}$$

$$4y - 6 = x + 3y \qquad \text{Distributive property}$$

$$4y - 3y - 6 = x + 3y - 3y \qquad \text{Subtract } 3y \text{ from both sides.}$$

$$y - 6 = x$$

$$y - 6 + 6 = x + 6 \qquad \text{Add 6 to both sides.}$$

$$y = x + 6$$

Now let's solve for a variable in a formula. Remember our goal is to isolate the variable for which we are solving. We use the same general procedure we used in Examples 4 and 5.

EXAMPLE 6 The formula for the perimeter of a rectangle is $P = 2l + 2w$, where l is the length and w is the width of the rectangle (see Fig. 2.2). Solve this formula for the width, w.

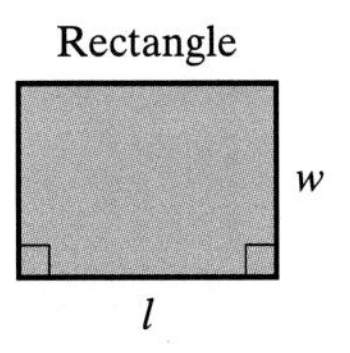

FIGURE 2.2

Solution Since we are solving for w, we must isolate the w on one side of the equation.

$$P = 2l + 2w$$

$$P - 2l = 2l - 2l + 2w \quad \text{Subtract } 2l \text{ from both sides.}$$

$$P - 2l = 2w$$

$$\frac{P - 2l}{2} = \frac{2w}{2} \quad \text{Divide both sides by 2.}$$

$$\frac{P - 2l}{2} = w$$

Thus, $w = \dfrac{P - 2l}{2}$ or $w = \dfrac{P}{2} - \dfrac{2l}{2} = \dfrac{P}{2} - l$.

EXAMPLE 7 A formula used to find the area of a trapezoid is $A = \frac{1}{2}h(b_1 + b_2)$, where h is the height and b_1 and b_2 are the lengths of the bases of the trapezoid (see Fig. 2.3). Solve this formula for b_2.

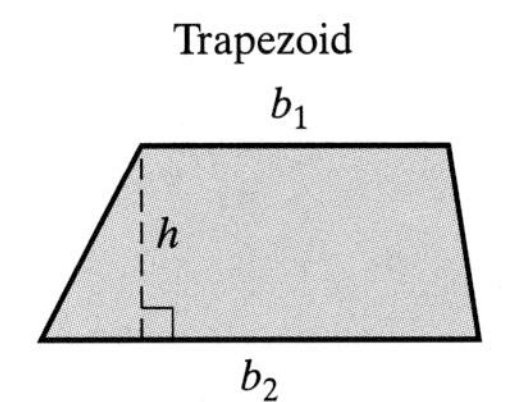

FIGURE 2.3

Solution We begin by multiplying both sides of the equation by 2 to clear fractions.

$$A = \frac{1}{2}h(b_1 + b_2)$$

$$2 \cdot A = 2\left[\frac{1}{2}h(b_1 + b_2)\right] \quad \text{Multiply both sides by 2.}$$

$$2A = h(b_1 + b_2)$$

$$\frac{2A}{h} = \frac{h(b_1 + b_2)}{h} \quad \text{Divide both sides by } h.$$

$$\frac{2A}{h} = b_1 + b_2$$

$$\frac{2A}{h} - b_1 = b_1 - b_1 + b_2 \quad \text{Subtract } b_1 \text{ from both sides.}$$

$$\frac{2A}{h} - b_1 = b_2$$

NOW TRY EXERCISE 57

EXAMPLE 8 In Example 3 we introduced the formula $T_f = T_a(1 - F)$.

a) Solve this formula for T_a.

b) John and Dorothy Czajka are in the 36% income tax bracket. What is the equivalent taxable yield of a 7% tax-free yield?

Solution **a)** We wish to solve this formula for T_a. Therefore we treat all other variables in the equation as if they were constants. Since T_a is multiplied by $(1 - F)$, to isolate T_a we divide both sides of the equation by $1 - F$.

$$T_f = T_a(1 - F)$$

$$\frac{T_f}{1 - F} = \frac{T_a(1 - F)}{1 - F} \qquad \textit{Divide both sides by } 1 - F.$$

$$\frac{T_f}{1 - F} = T_a \quad \text{or} \quad T_a = \frac{T_f}{1 - F}$$

b) Substitute the appropriate values in the formula found in part **a)**.

$$T_a = \frac{T_f}{1 - F}$$

$$T_a = \frac{0.07}{1 - 0.36} = \frac{0.07}{0.64} \approx 0.109$$

NOW TRY EXERCISE 63 Thus, the equivalent taxable yield would be about 10.9%.

Exercise Set 2.2

Concept/Writing Exercises

1. What is a mathematical model?
2. What is a formula?
3. Outline the five-step problem-solving process that we will use to work out problems.
4. When we are solving a formula for a variable we need to isolate the variable. Explain what this means.
5. Consider the equation $10 = 2(4) + 2w$ and the formula $P = 2l + 2w$.

   a) Solve the equation for w.

   b) Solve the formula for w.

   c) Was the procedure used to solve the formula for w any different than the procedure used to solve the equation for w? Explain.

   d) In the formula solved for w in part b) substitute 10 for P and 4 for l, and then find the value of w. How does it compare with your answer in part a)? Explain why this is so.

6. **a)** What are subscripts?

   b) How is x_0 read?

   c) How is v_f read?

Practice the Skills

Evaluate the following formulas for the values given. Use the π *key on your calculator for* π. *Round answers to the nearest hundredth.*

7. $A = lw$ when $l = 7$, $w = 10$ (a formula for finding the area of a rectangle)
8. $C = 2\pi r$ when $r = 10$ (a formula for finding the circumference of a circle)
9. $P = 2l + 2w$ when $l = 15$, $w = 6$ (a formula for finding the perimeter of a rectangle)
10. $V = lwh$ when $l = 10$, $w = 8$, $h = 4$ (a formula for finding the volume of a cube)
11. $A = \pi r^2$ when $r = 8$ (a formula for finding the area of a circle)
12. $\bar{x} = \dfrac{x_1 + x_2 + x_3}{3}$ when $x_1 = 40$, $x_2 = 120$, $x_3 = 80$ (a formula for finding the average of three numbers)
13. $A = \frac{1}{2}h(b_1 + b_2)$ when $h = 10$, $b_1 = 20$, $b_2 = 30$ (a formula for finding the area of a trapezoid)
14. $A = P + Prt$ when $P = 200$, $r = 0.05$, $t = 2$ (a banking formula that yields the total amount in an account after the interest is added)
15. $P_1 = \dfrac{T_1 P_2}{T_2}$ when $T_1 = 250$, $T_2 = 500$, $P_2 = 300$ (a chemistry formula that relates temperature and pressure of gases)
16. $E = a_1 p_1 + a_2 p_2$ when $a_1 = 10$, $p_1 = 0.2$, $a_2 = 100$, $p_2 = 0.3$ (a statistics formula for finding the expected value of an event)
17. $m = \dfrac{y_2 - y_1}{x_2 - x_1}$ when $y_2 = 4$, $y_1 = -3$, $x_2 = -2$, $x_1 = -6$ (a formula for finding the slope of a straight line. We will discuss this formula in Chapter 3.)
18. $F = G\dfrac{m_1 m_2}{r^2}$ when $G = 0.5$, $m_1 = 100$, $m_2 = 200$, $r = 4$ (a physics formula that gives the force of attraction between two masses that are separated by a distance, r)

19. $d = \sqrt{(x_2 - x_1)^2 + (y_2 - y_1)^2}$ when $x_2 = 5, x_1 = -3, y_2 = -6, y_1 = 3$ (a formula for finding the distance between two points on a straight line. We will discuss this formula in Chapter 10.)

20. $R_T = \dfrac{R_1 R_2}{R_1 + R_2}$ when $R_1 = 100, R_2 = 200$ (a formula in electronics for finding the total resistance in a parallel circuit containing two resistors)

21. $x = \dfrac{-b + \sqrt{b^2 - 4ac}}{2a}$ when $a = 2, b = -5, c = -12$ (from the quadratic formula. We will discuss the quadratic formula in Chapter 8.)

22. $x = \dfrac{-b - \sqrt{b^2 - 4ac}}{2a}$ when $a = 2, b = -5, c = -12$ (from the quadratic formula)

23. $A = p\left(1 + \dfrac{r}{n}\right)^{nt}$ when $p = 100,\ r = 0.06,\ n = 1,\ t = 3$ (the compound interest formula; see Example 2)

24. $z = \dfrac{\bar{x} - \mu}{\frac{\sigma}{\sqrt{n}}}$ when $\bar{x} = 80,\ \mu = 70,\ \sigma = 15,\ n = 25$ (a statistics formula for finding the standard, or z score, of a sample mean, $\bar{x}$)

Solve each equation for y (see Examples 4 and 5).

25. $3x + y = 5$
26. $3x + 2y = 8$
27. $2x - y = -5$
28. $2x - 4y = 6$
29. $5x - 3y = -4$
30. $2y = 8x - 3$
31. $\frac{1}{2}x + 2y = 6$
32. $\frac{3}{5}x + \frac{1}{3}y = 1$
33. $3(x - 2) + 3y = 6x$
34. $2(x + 3y) = 4(x - y) + 5$
35. $3x - 5 = 2(3y + 6)$
36. $\frac{1}{5}(x - 2y) = \frac{3}{4}(y + 2)$

Solve each equation for the indicated variable (see Examples 6–8).

37. $d = rt$, for t
38. $C = \pi d$, for d
39. $A = lw$, for l
40. $i = prt$, for t
41. $P = 2l + 2w$, for w
42. $P = 2l + 2w$, for l
43. $V = lwh$, for h
44. $V = \pi r^2 h$, for h
45. $A = \frac{1}{2}bh$, for b
46. $V = \frac{1}{3}lwh$, for l
47. $Ax + By = C$, for y
48. $A = P + Prt$, for r
49. $y = mx + b$, for m
50. $IR + Ir = E$, for R
51. $y - y_1 = m(x - x_1)$, for m
52. $z = \dfrac{x - \mu}{\sigma}$, for σ
53. $z = \dfrac{x - \mu}{\sigma}$, for μ
54. $y = \dfrac{kx}{z}$, for z
55. $P_1 = \dfrac{T_1 P_2}{T_2}$, for T_2
56. $F = \dfrac{mv^2}{r}$, for m
57. $A = \frac{1}{2}h(b_1 + b_2)$, for h
58. $A = \dfrac{x_1 + x_2 + x_3}{n}$, for n
59. $S = \frac{n}{2}(f + l)$ for n
60. $S = \frac{n}{2}(f + l)$, for l
61. $C = \frac{5}{9}(F - 32)$, for F
62. $F = \frac{9}{5}C + 32$, for C
63. $F = \dfrac{km_1m_2}{d^2}$, for m_1
64. $A = \frac{1}{2}h(b_1 + b_2)$, for b_1

Problem Solving

65. Ships at sea measure their speed in terms of knots. For example, when the *Titanic* struck the iceberg, its speed was about 20.5 knots. One knot is one nautical mile per hour. A nautical mile is about 6076 feet. When measuring speed in miles per hour, a mile is 5280 ft.

a) Determine a formula for converting a speed in knots (k) to a speed in miles per hour (m).

b) Explain how you determined this formula.

c) Determine the speed, in miles per hour, at which the *Titanic* struck the iceberg.

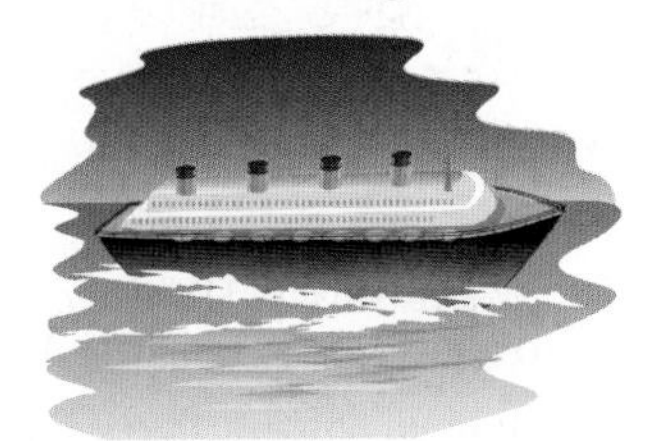

66. **a)** At the time of this writing, $1 of U.S. currency could be exchanged for 8.53 Mexican pesos. Write a formula, using d for U.S. dollars and p for pesos, that can be used to convert from U.S. dollars to pesos.

b) Write a formula that can be used to convert from pesos to U.S. dollars.

c) Explain how you determined your answers to parts **a)** and **b)**.

In Exercises 67–70, we use the simple interest formula $i = prt$. The simple interest formula is used to find the simple interest on an investment or a loan. In the formula, i is the simple interest, p is the principal (or the amount), r is the rate of interest, and t is the time.

67. George Devenney loaned his friend Paul Messelwitz $600 at a simple interest rate of 5% per year for 3 years. Determine the simple interest that Paul must pay George when he repays that loan at the end of 3 years.

68. Dana Frick had to pay $360 interest on a 6% simple interest loan for 2 years. How much had she borrowed?

69. Jim Walsh loaned his friend Donald Tweedt $10,000 dollars to purchase a car. Jim charged Don $4\frac{1}{4}\%$ simple interest. When Don repaid the loan, he repaid Jim the $10,000 plus $1700 interest. What was the length of the loan?

70. Mary Shapiro invested $5000 in a certificate of deposit for 1 year. When she redeemed the certificate she received $5300. What simple interest rate did she receive?

In exercises 71–76, if you are not sure of the formula to use, refer to Appendix A.

71. Lisa Coyle is planning to build a rectangular sandbox for her daughter. She has 38 feet of wood to use for the sides. If the length is to be 11 feet, what is the width to be?

72. George Young, dart-throwing champion in the state of Michigan, practices on a dartboard with concentric circles as shown in the figure.

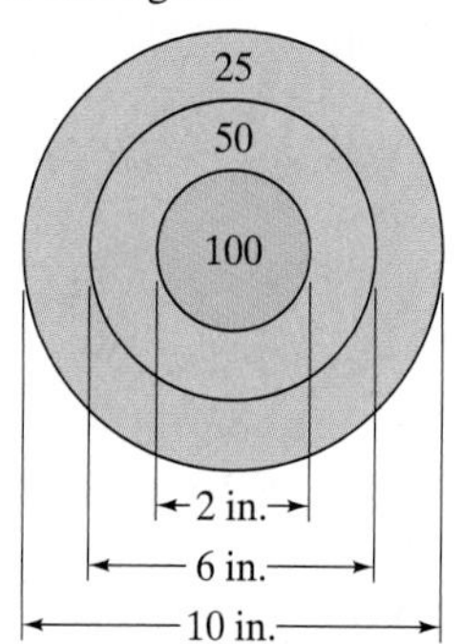

a) Find the area of the circle marked 100.

b) Find the area of the entire dartboard.

73. A helipad in Raleigh, North Carolina, has two concentric circles as shown in the figure.

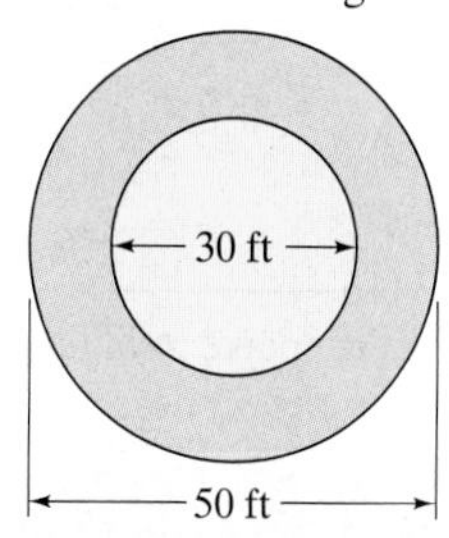

Find the area of the blue region in the figure.

74. Bill Leonard, a construction worker in Georgia, is laying concrete for a driveway. The driveway is to be 15 feet long by 10 feet wide by 6 inches deep.

a) Find the volume of concrete needed in cubic feet.

b) If 1 cubic yard = 27 cubic feet, how many cubic yards of concrete are needed?

c) If the concrete costs $35 per cubic yard, what is the cost of the concrete? Concrete must be purchased in whole cubic yards even if just a portion of a cubic yard is needed.

75. Belen Poltorak has a bucket in which she wishes to mix some detergent. The dimensions of the bucket are shown in the figure.

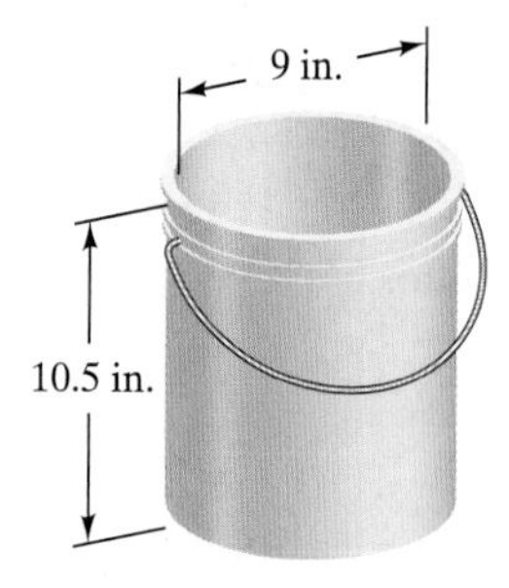

a) Find the capacity of the bucket in cubic inches.

b) If 231 cubic inches = 1 gallon, what is the capacity of the bucket in gallons?

c) If the instructions on the bottle of detergent say to add 1 ounce per gallon of water, how much detergent should Belen add to a full bucket of water?

76. Consider the following two figures. Which has a greater volume, and by how much?

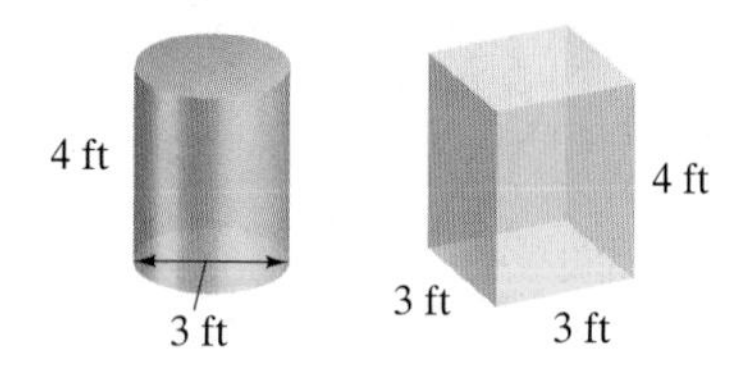

For Exercises 77 and 78, refer to Example 2.

77. Beth Rechsteiner invested $10,000 in a savings account paying 6% interest compounded quarterly. How much money will she have in her account at the end of 2 years?

78. Elizabeth Wood invested $8500 in a savings account paying 6.5% interest compounded monthly. How much money will she have in her account at the end of 4 years?

For Exercises 79 and 80, refer to Example 3.

79. Alfredo Irizany is considering investing $8000 in a $4\frac{1}{4}\%$ tax-free investment or a 5% taxable account. If he is in the 28% tax bracket, which is the better investment?

80. Marissa Felberty is considering investing $9200 in a 6.75% taxable account or in a $5\frac{1}{2}\%$ tax-free account. If she is in the 28% tax bracket, which is the better investment?

81. A person given a stress test is generally told that should the heart rate reach a certain point, the test will be stopped. The maximum allowable heart rate depends on the person's age. The maximum allowable heart rate, m, in beats per minute, can be approximated by the equation $m = -0.875x + 190$, where x represents the patient's age from 1 through 99. Using this mathematical model, find
 a) the maximum heart rate for a 50-year-old.
 b) the age of a person whose maximum heart rate is 160 beats per minute.

82. A nutritionist explains to Jodi Thomas that a person loses weight when they burn more calories than they eat. For example, Jody, a 5'6" woman who weighs 132 pounds, will stay about the same weight with normal exercise if she has a daily diet of 2400 calories. If she burns more than 2400 calories daily, her weight loss can be approximated by the mathematical model $w = 0.02c$, where w is the *weekly* weight loss and c is the number of calories per *day* burned *above* 2400 calories.
 a) Find Jodi's weekly weight loss if she exercises and burns off 2600 calories per day.
 b) How many calories would Jodi needed to burn off in a day to lose two pounds in a week?

83. **a)** A pool and spa maintenance booklet indicates that a spa should be drained and cleaned regularly. The days between cleaning, d, can be determined by dividing the number of gallons in the spa, g, by 3, then dividing this quotient by the average number of daily users, u. Write a formula that can be used to find the number of days between cleaning.
 b) If a hotel spa holds 8600 gallons and has an average of 40 users daily, how often should it be drained and cleaned?

84. Some financial planners are recommending a new rule of thumb for investments. The old rule of thumb was to keep the percentage of stocks in your total portfolio equal to 100 minus your age. The balance was to be put in bonds or bond funds or cash. The new rule of thumb is that the amount to be kept in stocks is found by multiplying a person's age by 80%, then subtracting this number from 100. Construct mathematical models for the percent to be kept in a stock using
 a) the old rule of thumb (use S for percent in stock and a for a person's age).
 b) the new rule of thumb.
 c) Find the percent to be kept in stocks for a 60-year-old under the old rule of thumb.
 d) Find the percent to be kept in stocks for a 60-year-old under the new rule of thumb.

85. The body mass index is a standard way of evaluating a person's body weight in relation to their height. To determine your body mass index (BMI), divide your weight by your height squared—and this is to be done in kilograms and meters. An accurate shortcut for calculating a person's BMI is to multiply a person's weight in pounds by 705, then divide by the square of the person's height, in inches.
 a) Create a formula for finding a person's BMI using kilograms and meters.
 b) Develop a mathematical model for finding a person's BMI when the weight is given in pounds and the height is given in inches.
 c) Determine your BMI.

Challenge Problem

86. Solve the formula $r = \dfrac{s/t}{t/u}$ for **a)** s; **b)** u.

Cumulative Review Exercises

[1.4] 87. Evaluate $-(5-8)^2 + |5-8| - 4^2$.

88. Evaluate $\dfrac{4 - 6 \div 3 + 5^2 - 6 \cdot 4}{5 - |6 \div (-2)|}$

89. Evaluate $6x^2 - 3xy + y^2$ when $x = 2, y = 3$.

[2.1] 90. Solve the equation $\frac{1}{3}x + 4 = \frac{2}{5}(x - 3)$.

2.3 APPLICATIONS OF ALGEBRA

1) **Translate a verbal statement into an algebraic expression or equation.**
2) **Use the problem-solving procedure.**

1) Translate a Verbal Statement into an Algebraic Expression or Equation

The next few sections will present some of the many uses of algebra in real-life situations. Whenever possible, we include other relevant applications throughout the text.

Perhaps the most difficult part of solving a word problem is translating the problem into an equation. This is step 2 in the problem-solving procedure presented in section 2.2. Before we represent problems as equations, we give some examples of phrases represented as algebraic expressions.

Phrase	Algebraic Expression
a number increased by 4	$x + 4$
twice a number	$2x$
5 less than a number	$x - 5$
one-eighth of a number	$\frac{1}{8}x$ or $\frac{x}{8}$
2 more than 3 times a number	$3x + 2$
4 less than 6 times a number	$6x - 4$
3 times the sum of a number and 5	$3(x + 5)$

The variable x was used in these algebraic expressions, but any variable could have been used to represent the unknown quantity.

EXAMPLE 1 Express each phrase as an algebraic expression.

a) the distance, d, increased by 15 miles

b) 3 less than 4 times the area a

c) 4 times a number n is decreased by 5

Solution **a)** $d + 15$ **b)** $4a - 3$ **c)** $4n - 5$

EXAMPLE 2 Write each phrase as an algebraic expression.

a) the cost of purchasing x shirts at \$4 each

b) the distance traveled in t hours at 55 miles per hour

c) the number of cents in n nickels

d) an 8% commission on sales of x dollars

Solution **a)** We can reason like this: one shirt would cost 1(4) dollars; two shirts, 2(4) dollars; three shirts, 3(4) dollars; four shirts, 4(4) dollars, and so on. Continuing this reasoning process, we can see that x shirts would cost $x(4)$ or $4x$ dollars. We can use the same reasoning process to complete each of the other parts.

b) $55t$

c) $5n$

d) $0.08x$ (8% is written as 0.08 in decimal form.)

HELPFUL HINT

A percent is always a percent of some quantity. Therefore, when a percent is listed, it is **always** multiplied by a number or a variable. In the following examples we use the variable c, but any letter could be used to represent the variable.

PHRASE	HOW WRITTEN
6% of a number	$0.06c$
the cost of an item increased by a 7% tax	$c + 0.07c$
the cost of an item reduced by 25%	$c - 0.25c$

Sometimes in a problem two numbers are related to each other. We often represent one of the numbers as a variable and the other as an expression containing that variable. We generally let the less complicated description be represented by the variable and write the second (more complex expression) in terms of the variable. In the following examples, we use x for the variable.

Phrase	One Number	Second Number
Dawn's age now and Dawn's age in 6 years	x	$x + 6$
one number is 4 times the other	x	$4x$
one number is 5 less than the other	x	$x - 5$
a number and the number increased by 7%	x	$x + 0.07x$
a number and the number decreased by 10%	x	$x - 0.10x$
the sum of two numbers is 10	x	$10 - x$
a 6-foot board cut in two lengths	x	$6 - x$
$10,000 shared by two people	x	$10{,}000 - x$

The last three examples may not be obvious. Consider "The sum of two numbers is 10." When we add x and $10 - x$ we get $x + (10 - x) = 10$. When a 6-foot board is cut in two lengths, the two lengths will be x and $6 - x$. For example, if one length is 2 feet, the other must be $6 - 2$ or 4 feet.

EXAMPLE 3 For each relation, select a variable to represent one quantity and express the second quantity in terms of the first.

a) The speed of the second train is 1.2 times the speed of the first.
b) $90 is shared by David and his brother.
c) It takes Tom 3 hours longer than Roberta to complete the task.

d) Hilda has \$4 more than twice the amount of money Hector has.
e) The length of a rectangle is 2 units less than 3 times its width.

Solution **a)** speed of first train, s; speed of second train, $1.2s$
b) amount David has, x; amount brother has, $90 - x$
c) Roberta, t; Tom, $t + 3$
d) Hector, x; Hilda, $2x + 4$
e) Width, w; length, $3w - 2$

The word **is** in a word problem often means **is equal to** and is represented by an equal sign, =.

Verbal Statement	Algebraic Equation
4 less than 3 times a number *is* 5	$3x - 4 = 5$
a number decreased by 4 *is* 3 more than twice the number	$x - 4 = 2x + 3$
the product of two consecutive integers *is* 20	$x(x + 1) = 20$
a number increased by 15% *is* 90	$x + 0.15x = 90$
a number decreased by 12% *is* 38	$x - 0.12x = 38$
the sum of a number and the number increased by 4% *is* 204	$x + (x + 0.04x) = 204$
the cost of renting a VCR for x days at \$15 per day *is* \$120	$15x = 120$

2 Use the Problem-Solving Procedure

There are many types of word problems, and the general problem-solving procedure given in Section 2.2 can be used to solve all types. We now present the five-step problem-solving procedure again so you can easily refer to it. We have included some additional information under step 2, since in this section we are going to emphasize translating word problems into equations.

Problem-Solving Procedure for Solving Application Problems

1. **Understand the problem.** Identify the quantity or quantities you are being asked to find.
2. **Translate the problem into mathematical language** (express the problem as an equation).
   a) Choose a variable to represent one quantity, and **write down exactly what it represents**. Represent any other quantity to be found in terms of this variable.
   b) Using the information from step a), write an equation that represents the word problem.
3. **Carry out the mathematical calculations** (solve the equation).
4. **Check the answer** (using the original wording of the problem).
5. **Answer the question asked.**

Sometimes we will combine or not show some steps in the problem-solving procedure due to space limitations. Even if we do not show a check to a problem, you should always check the problem to make sure that your answer is reasonable and makes sense.

EXAMPLE 4 One AT&T long distance rate plan requires that the customer pay a monthly fee of \$4.95 plus 10 cents per minute, or part thereof, for any long distance call made. A second AT&T plan does not have a monthly fee, but the customer pays 15 cents per minute, or part thereof, for any long distance call made.

a) Rob Sanberg is considering changing to one of these plans. Determine the number of minutes he would need to spend on long distance calls for the two plans to cost the same.

b) If he estimates he will spend 150 minutes on long distance calls per month, which plan would be the least expensive?

Solution **a) Understand** We are given two plans, where one has a monthly fee and the other does not. We are asked to find the *number of minutes* of long distance calls that would result in both plans having the same total cost. To solve the problem, we will set the cost of the two plans equal to one another and solve for the number of minutes.

Translate

$$\text{let } n = \text{number of minutes of long distance calling}$$
$$\text{then } 0.10n = \text{cost for } n \text{ minutes at 10 cents per minute}$$
$$\text{and } 0.15n = \text{cost for } n \text{ minutes at 15 cents per minute}$$

$$\text{cost of first plan} = \text{cost of second plan}$$
$$\text{monthly fee} + \text{calling cost} = \text{calling cost}$$
$$4.95 + 0.10n = 0.15n$$

Carry Out

$$4.95 + 0.10n = 0.15n$$
$$4.95 = 0.05n$$
$$\frac{4.95}{0.05} = n$$
$$99 = n$$

Check If 100 (close to 99) minutes were used per month, both plans would be about \$15 per month. Thus, 99 minutes is a reasonable answer.

Answer If 99 minutes of long distance calls were made per month, the cost of the two plans would be the same.

b) Since Rob plans to spend about 150 minutes per month on long distance calls, the first plan with the lower cost per minute would be better for him. If we substitute 150 for n in both equations, we find that his monthly cost for 150 minutes on the first plan would be \$19.95. On the second plan it would be \$22.50.

NOW TRY EXERCISE 13

EXAMPLE 5 The National Center for Health Statistics reported in an article that the number of divorces and annulments increased dramatically from 1960 to 1985 and that the number has generally been decreasing from 1985 to 1996. The article indicated that there were about 1,150,000 divorces and annulments in the United States in 1996 and that this was a 193% increase over the number in 1960 and a 4% decrease over the number in 1985.

a) Find the number of divorces and annulments in 1960.

b) Find the number of divorces and annulments in 1985.

Solution **a) Understand** We need to find the number of divorces and annulments in 1960. We will use the facts that the number increased by 193% from 1960 to 1996 and that the number of divorces and annulments in 1996 was 1,150,000 to solve this problem.

Translate let x = the number of divorces and annulments in 1960

then $1.93x$ = the increase in divorces and annulments from 1960 to 1996

$$\begin{pmatrix}\text{number of divorces}\\ \text{in 1960}\end{pmatrix} + \begin{pmatrix}\text{increase in divorces}\\ \text{from 1960 to 1996}\end{pmatrix} = \begin{pmatrix}\text{number of divorces}\\ \text{in 1996}\end{pmatrix}$$

$$x \quad + \quad 1.93x \quad = \quad 1{,}150{,}000$$

Carry Out

$$x + 1.93x = 1{,}150{,}000$$
$$2.93x = 1{,}150{,}000$$
$$x \approx 392{,}491$$

Check and Answer The number obtained is less than the number of divorces in 1996, which is what we expected. The number of divorces and annulments in 1960 was about 392,491.

b) Understand We must find the number of divorces and annulments in 1985. We are told that the number decreased by 4% from 1985 to 1996. We will work this part in a manner similar to the way we worked part **a)**.

Translate let x = number of divorces and annulments in 1985

then $0.04x$ = the decrease in the number of divorces and annulments from 1985 to 1996

$$\begin{pmatrix}\text{number of divorces}\\ \text{in 1985}\end{pmatrix} - \begin{pmatrix}\text{decrease in divorces}\\ \text{from 1985 to 1996}\end{pmatrix} = \begin{pmatrix}\text{number of divorces}\\ \text{in 1996}\end{pmatrix}$$

$$x \quad - \quad 0.04x \quad = \quad 1{,}150{,}000$$

Carry Out

$$x - 0.04x = 1{,}150{,}000$$
$$0.96x = 1{,}150{,}000$$
$$x \approx 1{,}197{,}917$$

Check and Answer The answer is greater than 1,150,000, which is what we expected. Therefore the number of divorces and annulments in 1985 was about 1,197,917.

EXAMPLE 6 The format of radio stations often changes. In 1996, there were a total of 10,261 commercial U.S. radio stations. The top five formats accounted for a total of 7268 stations. The top five formats were, in order: country, adult contemporary, talk (talk, news, business, sports), religion (teaching and music), and rock (album, modern, classic). There were 141 more religious stations than rock stations. There were 393 more talk stations than rock stations. The number of adult contemporary stations was 186 less than twice the number of rock stations, and the number of country stations was 112 less than 3 times the number of rock stations. How many of each type of station were there?

Solution **Understand** We are asked to find the number of stations for each of the top five formats. Note that the numbers of stations are all given in relation to the number of rock stations. Therefore, we will let our unknown variable be the number of rock stations. Then we can represent the number of stations with the other formats in terms of the number of rock stations.

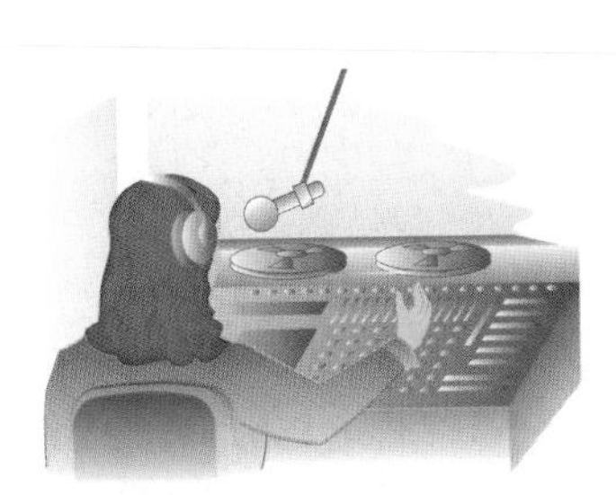

Translate

$$\begin{aligned} \text{let } n &= \text{number of rock stations} \\ \text{then } n + 141 &= \text{number of religous stations} \\ \text{and } n + 393 &= \text{number of talk stations} \\ \text{and } 2n - 186 &= \text{number of adult contemporary stations} \\ \text{and } 3n - 112 &= \text{number of country stations} \end{aligned}$$

$$\begin{pmatrix}\text{number}\\\text{of rock}\end{pmatrix} + \begin{pmatrix}\text{number of}\\\text{religious}\end{pmatrix} + \begin{pmatrix}\text{number}\\\text{of talk}\end{pmatrix} + \begin{pmatrix}\text{number of adult}\\\text{contemporary}\end{pmatrix} + \begin{pmatrix}\text{number of}\\\text{country}\end{pmatrix} = \text{total}$$

$$n + (n + 141) + (n + 393) + (2n - 186) + (3n - 112) = 7268$$

Carry Out

$$\begin{aligned} n + n + 141 + n + 393 + 2n - 186 + 3n - 112 &= 7268 \\ 8n + 236 &= 7268 \\ 8n &= 7032 \\ n &= 879 \end{aligned}$$

Check and Answer There were 879 rock stations and $n + 141$ or $879 + 141$ or 1020 religious stations. The number of talk stations was $n + 393$ or 1272. The number of adult contemporary stations was $2n - 186$ or $2(879) - 186$ or 1572. The number of country stations was $3n - 112$ or 2525. If we add the numbers of each of the five formats, we get $879 + 1020 + 1272 + 1572 + 2525 = 7268$, thus the answer checks.

NOW TRY EXERCISE 27

EXAMPLE 7 Dave Visser took his family to visit the Magic Kingdom at Disneyworld. They stayed for one night at the Holiday Inn in Kissimmee. When they made their hotel reservation, they were quoted a rate of \$70 per night. Their total bill was \$82.10, which included room tax and a \$3 charge for a candy bar (from the in-room bar). Determine the room tax.

Solution **Understand** Their total bill consists of their room rate, the room tax, and the \$3 cost for the candy bar. We need to find the tax rate for the room.

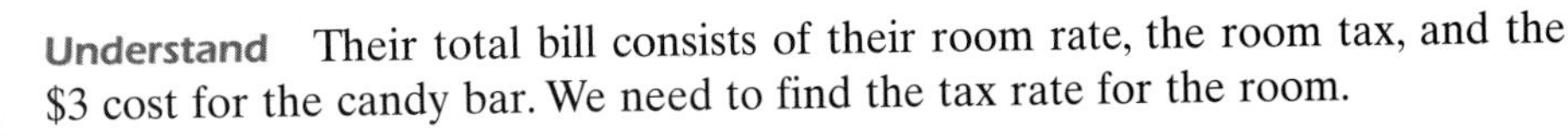

Translate Let t = room tax

$$\text{then } 0.01t = \text{room tax as a decimal}$$

$$\begin{array}{ccccccc} \text{room cost} & + & \text{room tax} & + & \text{candy bar} & = & \text{total} \\ 70 & + & 70(0.01t) & + & 3 & = & 82.10 \end{array}$$

Carry Out

$$\begin{aligned} 70 + 0.7t + 3 &= 82.10 \\ 0.7t + 73 &= 82.10 \\ 0.7t &= 9.10 \\ t &= 13 \end{aligned}$$

Check and Answer If you substitute 13 for t in the equation, you will see that the answer checks. The room tax is 13%.

NOW TRY EXERCISE 23

EXAMPLE 8 The vast majority of the population explosion has taken place in less than one-tenth of 1 percent of human history. When Columbus "discovered" the Americas 508 years ago, global population was small, numbering only 425 million. Much has changed. In just the past 50 years, the world population has increased about 2.4-fold to 6.1 billion.

a) Determine the world population 50 years ago.

b) How many years would it take the world population to double if the world population were to increase steadily at its present rate of 80 million people per year?

Solution **a) Understand** We are told the world population has increased 2.4-fold to 6.1 billion (or 6,100,000,000) in the past 50 years. We need to find the world population 50 years ago.

Translate

$$\text{let } x = \text{world population 50 years ago}$$
$$\text{then } 2.4x = \text{present world population}$$
$$\text{present world population} = 6.1 \text{ billion}$$

Carry Out

$$2.4x = 6.1$$
$$x \approx 2.5$$

Answer Thus, 50 years ago the world population was about 2.5 billion. The increase in the world population in the past 50 years is 6.1 billion − 2.5 billion or 3.6 billion people.

b) Understand We wish to find the number of years for the population to double to 12.2 billion. The growth rate is given in *millions* per year, and the present and future population are given in *billions*. Therefore, when answering part **b)** we will write 80 million as 80,000,000, 6.1 billion as 6,100,000,000, and 12.2 billion as 12,200,000,000.

Translate

$$\text{let } x = \text{number of years for the population to double}$$
$$\text{then } 80{,}000{,}000x = \text{growth in population in } x \text{ years}$$
$$\text{present population} + \text{growth in population} = \text{future population}$$
$$6{,}100{,}000{,}000 + 80{,}000{,}000x = 12{,}200{,}000{,}000$$

Carry Out

$$80{,}000{,}000x = 6{,}100{,}000{,}000$$
$$x = \frac{6{,}100{,}000{,}000}{80{,}000{,}000}$$
$$x = 76.25$$

Answer Thus, the population, if it increased steadily at the rate of 80 million per year, would double in about 76 years.

EXAMPLE 9 Liz Kaster is purchasing her first home and she is considering two banks for a \$60,000 mortgage. Citicorp is charging 8.50% interest with no points for a 30-year loan. (A point is a one-time charge of 1% of the amount of the mortgage.) The monthly mortgage payments for the Citicorp mortgage would be \$461.40. Citicorp is also charging a \$200 application fee. BankAmerica Corporation is charging 8.00% interest with 2 points for a 30-year loan. The monthly mortgage payments for BankAmerica would be \$440.04 and the cost of the points that Liz would need to pay at the time of closing is 0.02 (\$60,000) = \$1200. BankAmerica has waived its application fee.

a) How long would it take for the total payments of the Citicorp mortgage to equal the total payments of the BankAmerica mortgage?

b) If Liz plans to keep her house for 20 years, which mortgage would result in the lower total cost?

Solution **a) Understand** Citicorp is charging a higher interest rate and a small application fee but no points. BankAmerica is charging a lower rate and no application fee but 2 points. We need to determine the number of months when the total payments of the two loans would be equal.

Translate let x = number of months

then $461.40x$ = cost of mortgage payments for x months with the Citicorp mortgage

$440.04x$ = cost of mortgage payments for x months with the BankAmerica Corporation

$$\text{total cost with Citicorp} = \text{total cost with BankAmerica}$$
$$\text{mortage payments} + \text{application fee} = \text{mortage payments} + \text{points}$$
$$461.40x + 200 = 440.04x + 1200$$

Carry Out

$$\begin{aligned} 461.40x + 200 &= 440.04x + 1200 \\ 461.40x &= 440.04x + 1000 \\ 21.36x &= 1000 \\ x &\approx 46.82 \end{aligned}$$

Answer The cost would be the same in about 46.82 months or about 3.9 years.

b) The total cost would be the same at about 3.9 years. Prior to the 3.9 years the cost of the loan with BankAmerica would be more because of the initial \$1200 charge for points. However, after the 3.9 years the cost with BankAmerica would be less because of the lower monthly payment. If we evaluate the total cost with Citicorp over 20 years (240 monthly payments), we obtain \$110,936. If we evaluate the total cost with BankAmerica over 20 years, we obtain \$106,809.60. Therefore, Liz will save \$4126.40 over the 20-year period with BankAmerica.

NOW TRY EXERCISE 29

Exercise Set 2.3

Practice the Skills and Problem Solving

In Exercises 1–45, write an equation that can be used to solve the problem. Find the solution to the problem.

1. Two angles are **supplementary angles** when the sum of their measures is 180°. Angles A and B are supplementary angles. Find the measures of A and B if angle B is 4 times the size of angle A.
2. Angle A and angle B are supplementary angles. Find the measure of each angle if angle A is 30° greater than twice the measure of angle B.
3. The sum of the angles of a triangle is 180°. Find the three angles of a triangle if one angle is 20° greater than the smallest angle and the third angle is twice the smallest angle.
4. Find the three angles of a triangle if one angle is twice the smallest angle and the third angle is 60° greater than the smallest angle.

5. Gary Rodgers is shopping for a new suit. At S & K Menswear, he finds that the sale price of a suit that was reduced by 25% is $187.50. Find the regular price of the suit.

6. Meghan O'Donnell orders magazines through American Educational Services. American Educational Services recently notified Meghan that the price of *Newsweek* was increasing by 12% to $26.88 for 52 issues and that she can beat the increase by renewing her subscription by December 1.

   a) What is the current price of a subscription?

   b) How much will Meghan save by renewing early?

7. It costs Mike Sutton $12.50 a week to wash and dry his clothes at the corner laundry. If a washer and drier cost a total of $940, how many weeks will it take for the laundry cost to equal the cost of a washer and drier? (Disregard energy cost.)

8. Kate Spence buys a monthly bus pass, which entitles the owner to unlimited bus travel, for $40 per month. Without the pass each bus ride costs $1.50. How many rides per month would Kate have to make so that it is less expensive to purchase the pass?

9. The sales tax rate in Denver, Colorado, is 7.3%. What is the maximum price that Shane Stagg can pay for a new car if the total cost of the car, including tax, is not to exceed $22,600?

10. The yearly cost for joining Sam's Club is $35. The Thygesons estimate that they save an average of 3% when compared to other discount stores that do not have a membership fee. How much would the Thygesons have to spend in a year at Sam's Club to recover the yearly membership cost?

11. The cost of renting a truck is $35 a day plus 20 cents per mile. How far can Tanya Richardson drive in 1 day if she has only $80?

12. Scott Montgomery lives in New Jersey and works in New York. He commutes over the George Washington Bridge to go to work. The George Washington Bridge costs $4 for a car going from New Jersey to New York, but there is no cost going from New York to New Jersey. Individuals can purchase a number of different nonrefundable discount ticket books. One, called the All Bridges Book, costs $60 and contains 20 tickets.

   a) How many round trips to New York would Scott need to take to make it worthwhile for him to purchase this ticket book?

   b) If he plans to make 20 trips to New York, how much will he save by purchasing this ticket book?

13. Mr. and Mrs. Ose live on a resort island community attached to the mainland by a toll bridge. The toll is $2 per car going to the island, but there is no toll coming from the island. Island residents can purchase a monthly pass for $20, which permits them to cross the toll bridge from the mainland for only 50 cents each time. How many times a month would the Oses have to go to the island from the mainland for the cost of the monthly pass to equal the regular toll cost?

14. Each week Sandy Ivey receives a flat weekly salary of $240 plus a 12% commission on the total dollar volume of all sales she makes. What must her dollar volume be in a week for her to earn $540?

15. The three largest aircraft carriers in the world, in terms of water displaced, are the USS *George Washington*, the USS *Abraham Lincoln*, and the USS *John C. Stennis*, all of which are Nimitz class carriers. The largest ship of any kind is a Norwegian tanker named *Jahre Viking*, with a water displacement 445% times that of the aircraft carriers named above. If the *Jahre Viking* and one of the aircraft carriers named above displaces 622,608 tons, find the displacement of both the aircraft carrier and the *Jahre Viking*.

16. In 1997, three times as much money was allocated for human space flight as for work on space stations in orbit. Together these two categories accounted for 51.8% of the National Aeronautical and Space Administration total 1997 budget of $13,816 million. Find the amount spent on both human space flight and on space station work in 1997.

17. Rich Rowe makes regular contributions of $5000 annually to the Teachers Insurance and Annuity Association/College Retirement Equity Fund (TIAA/CREF) 403-b retirement plan. He has some of his contribution going to the CREF stock fund and some going to the CREF global equities fund. His contribution to the stock fund is $250 less than twice what he contributes to the global equities fund. How much does he contribute to each fund?

18. According to *Health* magazine, the stress a bone can withstand in pounds per square inch is 6000 pounds

more than 3 times the amount that steel can withstand. If the difference between the amount of stress a bone and steel can withstand is 18,000 pounds per square inch, find the stress that both steel and a bone can withstand.

19. There are 57 major sources of pollen in the United States. These pollen sources are categorized as grasses, weeds, and trees. If the number of weeds is 5 less than twice the number of grasses, and the number of trees is two more than twice the number of grasses, find the number of grasses, weeds, and trees that are major pollen sources.

20. Tabitha McCaun is asked to write a study guide for a textbook. For her work the publishing company is giving her a choice of a one-time payment of $15,000 or a royalty of $3.50 per study guide sold.

a) How many study guides would need to be sold for the total income received by Tabitha to be the same from either choice?

b) If she expects 10,000 copies of the study guide to be sold, which plan should she choose?

21. A security device on a car often reduces the auto insurance premium by 10%. Greg Middleton finds it costs $260 to purchase and install a security device for his car. If the yearly insurance on the car is $460, after how many years would the security device pay for itself?

22. The Midtown Tennis Club offers two payment plans for its members. Plan 1 is a monthly fee of $25 plus $10 per hour of court rental time. Plan 2 has no monthly fee, but court time costs $18.50 per hour. How many hours would Mrs. Levin have to play per month so that plan 1 becomes advantageous?

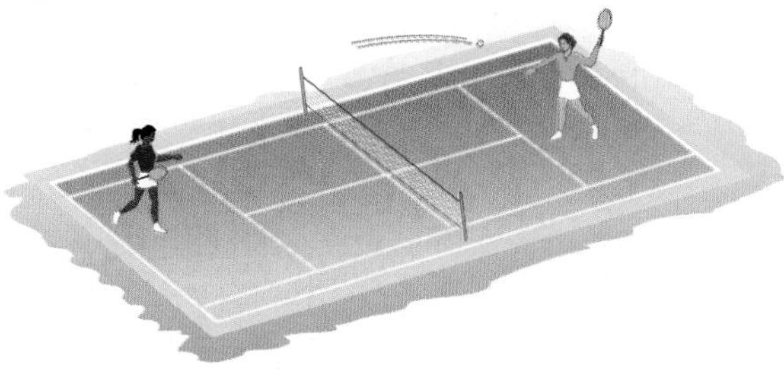

23. After Carol Pharo is seated in a restaurant, she realizes that she only has $15.75. If she must pay 7% sales tax and wishes to leave a 15% tip, what is the maximum price of a lunch she can order?

24. Demetrius Mays bought a bottle of perfume as a gift for his wife. The perfume cost $92 before tax. If the total price including tax was $98.90, find the tax rate.

25. From 1973 to 1997, the median number of hours worked per week increased by 25.1% to 50.8 hours. What was the median number of hours worked per week in 1973?

26. From 1973 to 1997, the median number of hours of leisure time decreased by 25.6% to 19.5 hours. What was the median number of hours of leisure time in 1973?

27. In 1996, the top four makers of personal computers (measured by U.S. shipments) were Compaq, Packard Bell-NEC, IBM, and Dell. Between them, they shipped 10,508,000 computers. IBM shipped 428,000 more computers than Dell. Packard Bell-NEC shipped 812,000 more computers than IBM. Compaq shipped 115,000 less than twice the number of computers shipped by Dell. Determine the number of computers shipped by each of the four manufacturers.

28. The United States and other countries reached an agreement in 1997 to reduce emissions of gases that create the problem of global warming. The focus was on carbon dioxide, which has been accumulating in Earth's atmosphere and trapping heat from the sun. In 1998, the world's five biggest emitters of carbon dioxide (measured in millions of metric tons) were, in order; the United States, China, the Russian Federation, Japan, and Germany. Together these five countries added 11.43 million metric tons of carbon dioxide to the atmosphere. Japan emitted 0.19 million more metric tons than Germany. The Russian Federation emitted 0.36 million metric tons more than twice Germany's emissions. China emitted 1.73 million more metric tons than Germany, and the United States emitted 4.5 million more metric tons than five times Germany's emissions. Determine the emission, in millions of metric tons, of each of the five polluters.

29. The Sanchezes are purchasing a new home and are considering a 30-year \$70,000 mortgage with two different banks. Madison Savings is charging 9.0% with 0 points and First National is charging 8.5% with 2 points. First National is also charging a \$200 application fee, whereas Madison is charging none. The monthly mortgage payments with Madison would be \$563.50 and the monthly mortgage payments with First National would be \$538.30.

    a) After how many months would the total payments for the two banks be the same?

    b) If the Sanchezes plan to keep their house for 30 years, which mortgage would have the lower total cost? (See Example 9)

30. Amy June, a financial planner, is sponsoring dinner seminars. She must pay for the dinners of those attending out of her own pocket. She chooses a restaurant that seats 40 people and charges her \$9.50 per person. If she earns 12% commission of sales made, determine how much in sales she must make from these 40 people

    a) to break even.

    b) to make a profit of \$500.

31. Dung Nguyen is considering refinancing his house at a lower interest rate. He has an 11.875% mortgage, is presently making monthly principal and interest payments of \$510, and has 20 years left on his mortgage. Because interest rates have dropped, Countrywide Mortgage Corporation is offering him a rate of 9.5%, which would result in principal and interest payments of \$420.50 for 20 years. However, to get this mortgage, his closing cost would be \$2500.

    a) How many months after refinancing would he spend the same amount on his new mortgage plus closing cost as he would have on the original mortgage?

    b) If he plans to spend the next 20 years in the house, would he save money by refinancing?

32. Edie Hall is planning to build a rectangular sandbox for her children. She wants its length to be 3 feet more than its width. Find the length and width of the sandbox if only 22 feet of lumber are available to form the frame. Use $P = 2l + 2w$.

33. Craig Campanella, a landscape architect, wishes to fence in two equal areas as illustrated in the figure. If both areas are squares and the total length of fencing used is 91 meters, find the dimensions of each square.

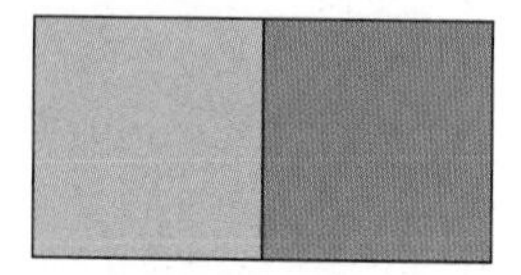

34. Joe Murray wishes to build a bookcase with four shelves (including the top) as shown in the figure. The width of the bookcase is to be 3 feet more than the height. If only 30 feet of wood are available to build the bookcase, what will be the dimensions of the bookcase?

35. Donshay Latrell wishes to fence in three rectangular areas along a river bank as illustrated in the figure. Each rectangle is to have the same dimensions, and the length of each rectangle is to be 1 meter greater than its width (along the river). Find the length and width of each rectangle if the total length of fencing used is 114 meters.

36. Daniella Popoff's farm is divided into three regions. The area of one region is twice as large as the area of the smallest region, and the area of the third region is 4 acres less than three times the area of the smallest region. If the total acreage of the farm is 512 acres, find the area of each of the three regions.

37. During the first week of a going-out-of-business sale, Sam's General Store reduces all prices by 10%. The second week of the sale, Sam's reduces all items by 5 additional dollars. If Sivle Yelserp bought a calculator for \$49 during the second week of the sale, find the original price of the calculator.

38. At a liquidation sale, Quality Photo Company reduces the price of all cameras by $\frac{1}{4}$, and then takes an additional \$10 off. If Holden Brown purchases a Minolta camera for \$290 during this sale, find the original price of the camera.

39. J. P. Richardson sells each of his paintings for \$50. The gallery where he displays his work charges him \$810 a month plus a 10% commission on sales. How many paintings must J. P. sell in a month to break even?

40. The cost of purchasing incandescent bulbs for use over a 9750-hour period is \$9.75. The energy cost for incandescent bulbs over this period is \$73. The cost of one equivalent fluorescent bulb that lasts about 9750 hours is \$20. By using a fluorescent bulb instead of incandes-

cent bulbs for 9750 hours, the total savings of purchase price plus energy cost is $46.75. What is the energy cost for using the fluorescent bulb over this period?

41. Stan Krejecki receives a small clothing allowance from Social Security. He is also provided with a sales tax exclusion number so that when this money is spent on clothing, the 7% sales tax is waived. On all other purchases he must pay the full 7% sales tax. Last week Stan went shopping at Sears, where he used his total clothing allowance to purchase clothing, and he also purchased some other goods on which he paid tax. If he spent a total of $375 before tax and paid a total sales tax of $17.50, find his clothing allowance from Social Security.

42. Approximately 1,500,000 species worldwide have been categorized as either plants, animals, or insects. Insects are often subdivided into beetles and insects that are not beetles. There are about 100,000 more plant than animal species. There are 290,000 more non-beetle insects than animals. The number of beetles is 140,000 less than twice the number of animals. Find the number of animal, plant, non-beetle insect, and beetle species.

43. The five members of the Newton family are going out to dinner with the three members of the Lee family. Before dinner, they decide that the Newtons will pay $\frac{5}{8}$ of the bill (before tip) and the Lees will pay $\frac{3}{8}$ plus the entire 15% tip. If the total bill including the 15% tip comes to $184.60, how much will be paid by each family?

44. To find the average of a set of test grades, we divide the sum of the test grades by the number of test grades. On her first four algebra tests, Paula West's grades were 87, 93, 97, and 96.

a) Write an equation that can be used to determine the grade Paula needs to obtain on her fifth test to have a 90 average.

b) Explain how you determined your equation.

c) Solve the equation and determine the score.

45. Philip Fox's grades on five physics hourly exams are 70, 83, 97, 84, and 74.

a) If the final exam will count twice as much as each hourly exam, what grade does Philip need on the final exam to have an 80 average?

b) If the highest possible grade on the final exam is 100 points, is it possible for Philip to obtain a 90 average? Explain.

46. **a)** Make up your own realistic word problem involving money. Represent this word problem as an equation.

b) Solve the equation and answer the word problem.

47. **a)** Make up your own realistic word problem involving percents. Represent this word problem as an equation.

b) Solve the equation and answer the word problem.

Challenge Problems

48. On Monday Sophia Wagner purchased shares in a money market fund. On Tuesday the value of the shares went up 5%, and on Wednesday the value fell 5%. How much did Sophia pay for the shares on Monday if she sold them on Thursday for $59.85?

49. The Elmers Truck Rental Agency charges $28 per day plus 15 cents a mile. If Denise Duncan rented a small truck for 3 days and the total bill was $121.68, including a 4% sales tax, how many miles did she drive?

Group Activity

Discuss and answer Exercise 50 as a group.

50. **a)** Have each member of the group pick a number. Then multiply the number by 2, add 33, subtract 13, divide by 2, and subtract the number each started with. Record each answer.

b) Now compare answers. If you did not all get the same answer, check each other's work.

c) As a group, explain why this procedure will result in an answer of 10 for any real number n selected.

Cumulative Review Exercises

[1.3] *Evaluate.*

51. $2 + \left|-\frac{3}{5}\right|$

52. $-6.4 - (-3.7)$

53. $\left|-\frac{5}{8}\right| \div |-2|$

54. $5 - |-3| - |7|$

[1.5] **55.** Simplify $(2x^4y^{-6})^{-3}$

2.4 ADDITIONAL APPLICATION PROBLEMS

1) **Solve motion problems.**
2) **Solve mixture problems.**

In this section we discuss two additional types of application problems, motion and mixture problems. We have placed them in the same section because they are solved using similar procedures.

1) Solve Motion Problems

A formula with many useful applications is

amount = rate · time

The "amount" in this formula can be a measure of many different quantities, including distance or length, area, volume, or number of items produced.

When applying this formula, we must make sure that the units are consistent. For example, when speaking about a copier, if the rate is given in copies per *minute*, the time must be given in *minutes*. Problems that can be solved using this formula are called *motion problems* because they involve movement, at a constant rate, for a certain period of time.

A nurse giving a patient an intravenous injection may use this formula to determine the drip rate of the fluid being injected. A company drilling for oil or water may use this formula to determine the amount of time needed to reach its goal.

When the motion formula is used to calculate distance, the word *amount* is replaced with the word *distance* and the formula is called the *distance formula.*

The distance formula is

distance = rate · time

or $d = rt$

When a motion problem has two different rates, it is often helpful to put the information in a table to help analyze the problem.

EXAMPLE 1 The aircraft carrier USS *John F. Kennedy* and the nuclear-powered submarine USS *Memphis* leave from the Puget Sound Naval Yard at the same time heading for the same destination in the Persian Gulf. The aircraft carrier travels at its maximum speed of 34.5 miles per hour (or about 30 knots) and the submarine travels submerged at its maximum speed of 20.2 miles per hour (or about 17.6 knots). The aircraft carrier and submarine are to travel at these speeds until they are 100 miles apart at which time they will receive further communications and instructions from the naval base. How long will it take for the aircraft carrier and submarine to be 100 miles apart? (see Fig. 2.4.)

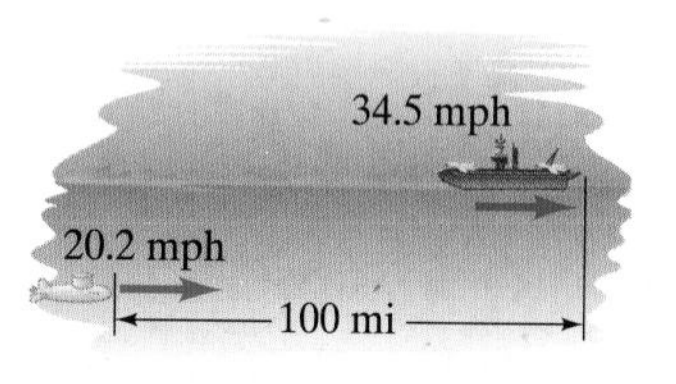

FIGURE 2.4

Solution **Understand** We wish to find out how long it will take for the difference between their distances to be 100 miles. To solve this problem, we will use the distance formula, $d = rt$. When we first introduced the problem-solving procedure, we indicated that to help understand a problem it might be helpful to put the given information in a table, and that is what we will do now.

Let t = time.

	Rate	Time	Distance
Aircraft carrier	34.5	t	$34.5t$
Submarine	20.2	t	$20.2t$

Translate and Carry Out The difference between their distances is 100 miles. Thus,

$$\text{aircraft carrier's distance} - \text{submarine's distance} = 100$$
$$34.5t - 20.2t = 100$$
$$14.3t = 100$$
$$t \approx 6.99$$

Answer The aircraft carrier and submarine will be 100 miles apart in about 7 hours.

EXAMPLE 2 Mrs. Taylor and her daughter Sally, both long distance runners, run together every Saturday starting at 9 A.M. However, because Mrs. Taylor had to attend a homeowners' meeting one Saturday morning, she would not be able to start jogging until 9:30 A.M. She told Sally to start joging slowly, at 4 miles per hour, and she would jog at 6 miles per hour until she caught up with her.

a) Determine the time Mrs. Taylor and Sally will meet.

b) How far from the starting point will they be when they meet?

Solution **a) Understand** Since Mrs. Taylor is going to jog faster, she will cover the same distance in a shorter time period. When they meet, Mrs. Taylor will have jogged $\frac{1}{2}$ hour less than Sally.

$$\text{let } t = \text{time Sally is jogging}$$
$$\text{then } t - \frac{1}{2} = \text{time Mrs. Taylor is jogging}$$

Jogger	Rate	Time	Distance
Sally	4	t	$4t$
Mrs. Taylor	6	$t - \frac{1}{2}$	$6(t - \frac{1}{2})$

Translate When they meet they will both have covered the same distance from the starting point.

$$\text{Sally's distance} = \text{Mrs. Taylor's distance}$$
$$4t = 6\left(t - \frac{1}{2}\right)$$

Carry Out

$$4t = 6t - 3$$
$$-2t = -3$$
$$t = \frac{3}{2}$$

Answer They will meet $1\frac{1}{2}$ hours after Sally begins jogging, or at 10:30 A.M.

b) The distance can be found using either Mrs. Taylor's or Sally's rate. We will use Sally's.

$$d = rt$$
$$= 4\left(\frac{3}{2}\right) = 6$$

NOW TRY EXERCISE 9

They will meet 6 miles from the starting point.

In Example 2, would the answer have changed if we had let t represent the time Mrs. Taylor is jogging rather than the time Sally is jogging? Try it and see.

EXAMPLE 3 A Coca-Cola bottling machine fills and caps bottles. The machine can be run at two different rates. At the faster rate the machine fills and caps 600 more bottles per hour than it does at the slower rate. The machine is turned on for 4.8 hours on the slower rate, then it is changed to the faster rate for another 3.2 hours. During these 8 hours a total of 25,920 bottles were filled and capped. Find the slower rate and the faster rate.

Solution

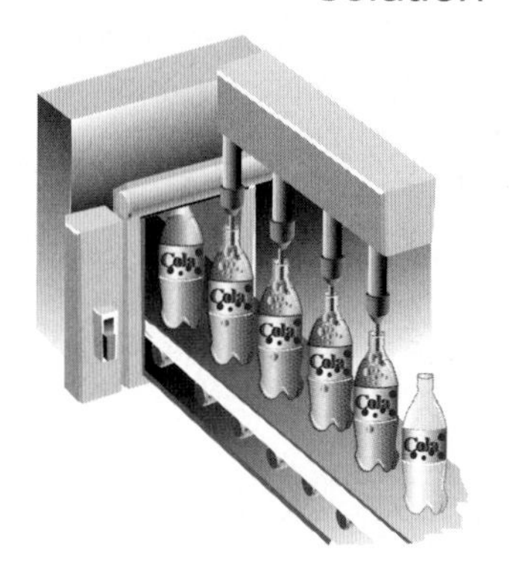

Understand This problem used a number of bottles, an amount, in place of a distance. However, the problem is worked in a similar manner. We will use the formula amount = rate · time. We are given that there are two different rates, and we are asked to find these rates. We will use the fact that the amount filled at the slower rate plus the amount filled at the faster rate equal the total amount filled.

$$\text{let } r = \text{the slower rate}$$
$$\text{then } r + 600 = \text{the faster rate}$$

	Rate	Time	Amount
Slower rate	r	4.8	$4.8r$
Faster rate	$r + 600$	3.2	$3.2(r + 600)$

Translate and Carry Out

$$\text{amount filled at slower rate} + \text{amount filled at faster rate} = 25{,}920$$
$$4.8r + 3.2(r + 600) = 25{,}920$$
$$4.8r + 3.2r + 1920 = 25{,}920$$
$$8r + 1920 = 25{,}920$$
$$8r = 24{,}000$$
$$r = 3000$$

NOW TRY EXERCISE 11

Answer The slower rate is 3000 bottles per hour. The faster rate is $r + 600$ or $3000 + 600 = 3600$ bottles per hour.

2) Solve Mixture Problems

Any problem where two or more quantities are combined to produce a different quantity or where a single quantity is separated into two or more different quantities may be considered a **mixture problem**. As we did when working with motion problems, we will use tables to help organize the information.

EXAMPLE 4 Mr. and Mrs. Glenn van der Hayden won a legal settlement and received \$8000. Some of this money was used for a personal loan made to a friend. The loan was made for a year at 7% simple interest. The balance was put in a certificate of deposit, or CD, that paid $5\frac{1}{4}\%$ simple interest. A year later, while working on their income taxes they found that they received a total of \$458.50 from the two investments, but they did not remember how much money was loaned to their friend. Determine how much was loaned to their friend and how much was invested in the CD.

Solution **Understand and Translate** To work this problem, we will use the **simple interest formula**, interest = principal · rate · time. We are told that part of the investment is made at 7% and part is made at $5\frac{1}{4}\%$ simple interest. We are asked to find the amount invested at each rate.

$$\text{let } p = \text{amount (or principal) invested at } 7\%$$

$$\text{then } 8000 - p = \text{amount invested at } 5\tfrac{1}{4}\%$$

Notice that the sum of the amounts invested at 7% and $5\frac{1}{4}\%$ equals the amount invested, \$8000. We will find the amount invested in each account with the aid of a table.

Investment	Principal	Rate	Time	Interest
Loan	p	0.07	1	$0.07p$
CD	$8000 - p$	0.0525	1	$0.0525(8000 - p)$

Since the total interest from both accounts is \$458.50, we write

$$\text{interest from 7\% loan} + \text{interest from } 5\tfrac{1}{4}\% \text{ CD} = \text{total interest}$$

$$0.07p \quad + \quad 0.0525(8000 - p) \quad = \quad 458.50$$

Carry Out Now we solve the equation.

$$0.07p + 420 - 0.0525p = 458.50$$

$$0.0175p + 420 = 458.50$$

$$0.0175p = 38.50$$

$$p = \frac{38.50}{0.0175} = 2200$$

Answer Therefore, the loan was \$2200, and $8000 - p$ or $8000 - 2200 = \$5800$ was invested in the CD.

NOW TRY EXERCISE 13

EXAMPLE 5 Matt's Hot Dog Stand in Chicago sells hot dogs for \$2.00 each and beef tacos for \$2.25 each. If the sales for the day total \$585.50 and 278 items were sold, how many of each item were sold?

Solution **Understand and Translate** We are asked to find the number of hot dogs and beef tacos sold.

$$\text{let } x = \text{number of hot dogs sold}$$
$$\text{then } 278 - x = \text{number of beef tacos sold}$$

Item	Cost of Item	Number of Items	Total Sales
Hot dogs	2.00	x	$2.00x$
Beef Tacos	2.25	$278 - x$	$2.25(278 - x)$

$$\text{total sales of hot dogs} + \text{total sales of beef tacos} = \text{total sales}$$
$$2.00x + 2.25(278 - x) = 585.50$$

Carry Out

$$2.00x + 625.50 - 2.25x = 585.50$$
$$-0.25x + 625.50 = 585.50$$
$$-0.25x = -40$$
$$x = \frac{-40}{-0.25} = 160$$

Answer Therefore, 160 hot dogs and 278 − 160 or 118 beef tacos were sold. ■

In Example 5 we could have multiplied both sides of the equation by 100 to eliminate the decimal numbers, and then solved the equation.

EXAMPLE 6 Ali Muhammed, a pharmacist, has both 6% and 15% phenobarbital solutions. He receives a prescription for 0.5 liter of an 8% phenobarbital solution. How much of each solution must he mix to fill the prescription?

Solution **Understand and Translate** We are asked to find the amount of each solution to be mixed.

$$\text{let } x = \text{number of liters of 6\% solution}$$
$$\text{then } 0.5 - x = \text{number of liters of 15\% solution}$$

The amount of phenobarbital in a solution is found by multiplying the percent strength of phenobarbital in the solution by the volume of the solution. We will draw a sketch of the problem (see Fig. 2.5) and then construct a table.

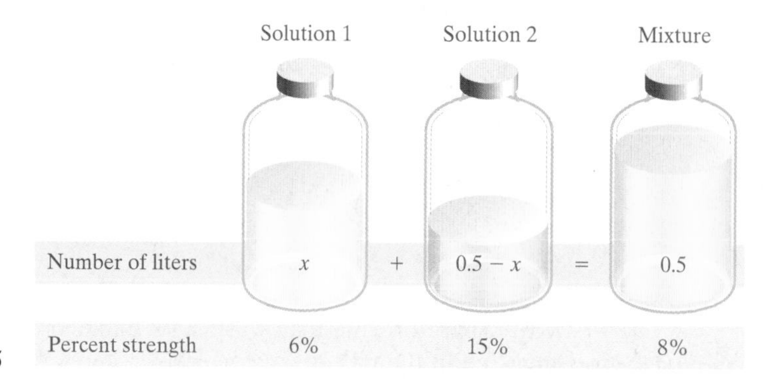

FIGURE 2.5

Solution	Strength of Solution	Number of Liters	Amount of Phenobarbital
1	0.06	x	$0.06x$
2	0.15	$0.5 - x$	$0.15(0.5 - x)$
Mixture	0.08	0.5	$0.08(0.5)$

$$\begin{pmatrix}\text{amount of}\\ \text{phenobarbital}\\ \text{in 6\% solution}\end{pmatrix} + \begin{pmatrix}\text{amount of}\\ \text{phenobarbital}\\ \text{in 15\% solution}\end{pmatrix} = \begin{pmatrix}\text{amount of phenobarbital}\\ \text{in mixture}\end{pmatrix}$$

$$0.06x \quad + \quad 0.15(0.5 - x) \quad = \quad 0.08(0.5)$$

Carry Out

$$0.06x + 0.075 - 0.15x = 0.04$$
$$0.075 - 0.09x = 0.04$$
$$-0.09x = -0.035$$
$$x = \frac{-0.035}{-0.09} = 0.39 \quad \text{(to nearest hundredth)}$$

Ali must mix 0.39 liter of the 6% solution and $0.5 - x$ or $0.5 - 0.39 = 0.11$ liter of the 15% solution to make 0.5 liter of an 8% solution.

NOW TRY EXERCISE 19

Exercise Set 2.4

Practice the Skills and Problem Solving

In Exercises 1–12, write an equation that can be used to solve the motion problem. Solve the equation and answer the question asked.

1. Two friends, Don O'Neal and Judy McElroy, go hiking in the Rocky Mountains. While hiking they come across Bear Lake. They wonder what the distance around the lake is and decide to find out. Don knows he walks at 5 mph and Judy know she walks at 4.5 mph. If they start walking at the same time in opposite directions around the lake, and meet in 1.2 hours, what is the distance around the lake.

2. An earthquake occurs in a desert of California. The shock waves travel outward in a circular path, similar to when a pebble is dropped in a pond. If the P-wave (one kind of shock wave) travels at 2.4 miles per second, how long would it take for the wave to have a diameter of 60 miles? (See the figure.)

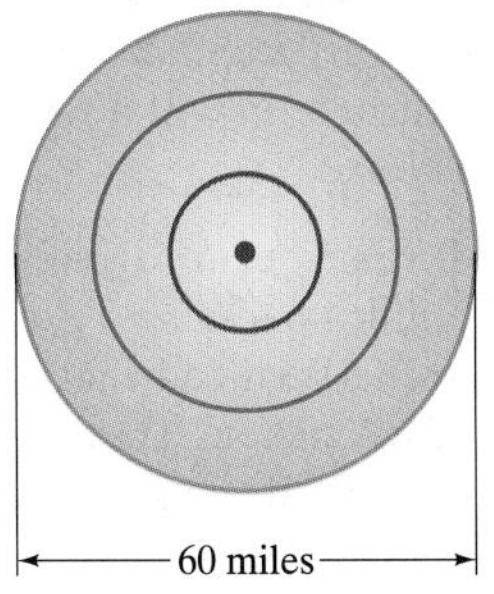

3. Each year Albuquerque, New Mexico, has a hot air balloon festival during which people can obtain rides in hot air balloons. Suppose part of the Diaz family goes in one balloon and other members of the family go in another balloon. Because they fly at different altitudes and carry different weights, one balloon travels at 16 miles per hour while the other travels in the same direction at 14 miles per hour. In how many hours will they be 4 miles apart?

4. A rain forest in Brazil is 150 miles long. The Akala logging company is working on one end, cutting down trees at a rate of 0.15 mile per day. The Okapi logging company is working on the other end, cutting down trees at a rate of 0.10 mile per day. If both logging companies continue at their present pace, how long will it take for the two companies to meet?

5. Often two trains will use the same track. When this happens and the trains are approaching each other, the track's switching device must be engaged so that one train is moved (at least temporarily) to a different set of tracks. A passenger train leaves Norfolk, Virginia, 1.2 hours after a freight train leaves. The passenger train is traveling 18 mph faster than the

freight train on the same tracks in the same direction. If one of the trains did not change tracks, the two trains would be at the same location 3 hours after the passenger train left. Find the planned speed of each train.

6. Two molding machines are turned on at 9 A.M. The older molding machine can produce 40 plastic buckets in 1 hour. The newer machine can produce 50 buckets in 1 hour. How long will it take the two machines to produce a total of 540 buckets?

7. A charity is having a fund-raiser where people's pledges are dependent on the distance that the person on whom they haved pledged travels. This particular fund-raiser is using both joggers and cyclists. Laura Mann, a jogger, and Wayne Seigert, a cyclist, start at the same point at 8 A.M. and travel in the same direction heading for the same point. Wayne's average speed is four times Laura's. In two hours Wayne is 18 miles ahead of Laura.

a) At what rate did Wayne ride?

b) How far did Wayne ride?

c) If Bob Johnson made a $1.50-per-mile pledge on Wayne, and Wayne stops at the distance determined in part **b)**, how much will Bob Johnson have pledged?

8. Lisa Lopez has just had her inground pool refinished. She plans to use two hoses to fill her pool. She knows that the hose with the larger diameter supplies twice as much water as the hose with the smaller diameter. As she is about to fill her pool, she realizes that she loaned the hose with the larger diameter to her sister Ann. She starts filling the pool with the smaller diameter hose. Shortly after, her sister brings over the larger diameter hose. It is hooked up and starts supplying water to the pool 1.5 hours after the smaller diameter hose does. The (volume) markers in the pool show that the volume of water in the pool 3 hours after the larger diameter hose started delivering water is about 1050 gallons. Find the rate of water flow from each hose.

9. Diane Basile hikes down to the bottom of Crayton Canyon, camps overnight, and returns the next day. Her hiking speed down averages 3.4 miles per hour and her return trip averages 1.2 miles per hour. If she spent a total of 12 hours hiking, find

a) how long it took her to reach the bottom of the canyon.

b) the total distance traveled.

10. Carlos de la Lama and Dave Kater are both going to attend a mathematics convention out of town. They wish to arrive at the conference at the same time, but Dave must leave for the conference later than Carlos because he has a meeting to attend. When Carlos has driven 50 miles he gets a call from Dave on his cellular phone, and Dave tells him that he is leaving now. Carlos indicates that he is driving at 55 miles per hour. Dave says that he plans to drive at 70 miles per hour along the same route and should catch up with Carlos in about 3.33 hours.

a) Write the equation used by Dave to determine when he would catch up with Carlos.

b) Solve the equation in part **a)**.

11. Two machines are packing spaghetti into boxes. The smaller machine can package 400 boxes per hour and the larger machine can package 600 boxes per hour. If the larger machine is on for 2 hours before the smaller machine is turned on, how long after the smaller machine is turned on will a total of 15,000 boxes of spaghetti be boxed?

12. Therese Morgan began driving to an airport at a speed of 35 miles per *hour*. Fifteen *minutes* after she left, her parents realized that she had forgotten to take her airline tickets. They then tried to catch up to her in their car. If the parents traveled at 50 miles per hour, how long did it take them to catch Therese?

In Exercises 13–26, write an equation that can be used to solve the mixture problem. Solve each equation and answer the question asked.

13. When Pat Cox was figuring her income taxes, she found that she received a total of \$530 from two simple interest investments. She knows that she invested a total of \$11,000, some in a 4% account and the rest in a 5% account. Determine how much Pat invested in each account.

14. Thea Prettyman invested \$10,000 for 1 year, part at 7% and part at $6\frac{1}{4}\%$. If she earned a total interest of \$656.50, how much was invested at each rate?

15. Bob Davis goes to see an investment counselor because he has just inherited \$10,000. The counselor recommends that Bob should buy stock, and he recommends two in particular: Microsoft and Hilton Hotels. He also thinks that Hilton has the potential to grow faster than Microsoft and that Bob should purchase more shares of Hilton than of Microsoft. Microsoft cost \$108.75 per share and Hilton cost \$27.25 per share. Bob decides to purchase four times as many shares of Hilton than of Microsoft.

a) If only whole shares of stock can be purchased, how many shares of each company can Bob purchase?

b) How much money will Bob have left?

16. J.B. Davis owns a nut shop. He sells almonds for \$6 per pound and walnuts for \$5.20 per pound. He receives a special request from a customer who wants to purchase 30 pounds of a mixture of almonds and walnuts but does not want to pay more than \$165. J. B. used algebra to determine the amount of each to mix so that he could mix the maximum amount of almonds without having the mixture's value exceeding \$165. Determine how many pounds of almonds and walnuts J. B. mixed.

17. Joan Smith is the owner of a Starbucks Coffee Shop. She has many varieties of coffee in the shop including a Kona variety that sells for \$6.20 per pound and an amaretto coffee that sells for \$5.80 per pound. She finds that by mixing these two blends she creates a coffee that sells well. If she uses 18 pounds of amaretto coffee in the blend and wishes to sell the mixture at \$6.10 per pound, how many pounds of the Kona coffee should she mix with the amaretto coffee?

18. Fred Feldon, a chemistry teacher, needs a 10% acetic acid solution for an upcoming chemistry laboratory. When checking the storeroom, he realizes he has only 16 ounces of a 25% acetic acid solution. There is not enough time for him to order more, so he decides to make a 10% acetic acid solution by very carefully adding water to the 25% acetic acid solution. Knowing algebra, he figures out just how much water to add. Determine how much water Fred must add to the 25% solution to reduce it to a 10% solution.

19. Distilled white vinegar that can be purchased in supermarkets generally has a 5% acidity level. Chef Angela Leinenbach makes a special sauerbraten dish that is loved by all. To make her sauerbraten she marinates veal overnight in a special 8% distilled vinegar that she creates herself. To create the 8% solution she mixes a regular 5% vinegar solution with a 12% vinegar solution that she purchases by mail. How many ounces of the 12% vinegar solution should she add to 40 ounces of the 5% vinegar solution to get the 8% vinegar solution?

20. Bill Ding, a mason, has a 50-pound bag of a cement-and-sand mixture. The printing on the bag says that the mixture is 40% sand. For a special project he is doing, he needs the mixture to be 60% sand. He also has a large bag of pure sand available. How much pure sand should Bill mix with the cement-and-sand mixture to raise the amount of sand in the mixture to 60%?

21. Joaquin Gomez is having a holiday party and is making an orange juice–ginger ale–champagne punch. The Brut champagne used is 11.5% alcohol by volume. How much champagne should Joaquin add to 32 ounces of the orange juice–ginger ale combination if the alcohol concentration of the punch is to be 5%?

22. Two acid solutions are available to a chemist. One is a 20% sulfuric acid solution, but the label that indicates the strength of the other sulfuric acid solution is missing. Two hundred milliliters of the 20% acid solution and 100 milliliters of the solution with the unknown strength are mixed together. Upon analysis, the mixture was found to have a 25% sulfuric acid concentration. Find the strength of the solution with the missing label.

23. The Pearlman Nursery sells two types of grass seeds in bulk. The lower-quality seeds have a germination rate of 76%, but the germination rate of the higher-quality seeds is not known. Twelve pounds of the higher-quality seeds are mixed with 16 pounds of the lower-quality seed. If a later analysis of the mixture reveals that the mixture's germination rate was 82%, what is the germination rate of the higher-quality seed?

24. The Agway nursery is selling two types of sunflower seed for bird feed in bulk. The striped sunflower seeds cost \$1.20 per pound, while the black oil sunflower seeds cost \$1.60 per pound. How many pounds of each should the nursery mix to get a 20-pound mixture that sells for \$30?

25. Some states allow a husband and wife to file individual state tax returns (on a single form) even though they file a joint federal return. It is usually to the taxpayer's advantage to do this when both the husband and wife work. The smallest amount of tax owed (or the largest refund) will occur when the husband's and wife's taxable incomes are the same.

    Mr. Juenger's 1999 taxable income was \$28,200 and Mrs. Juenger's taxable income for that year was \$32,450. The Juenger's total tax deductions for the year were \$6400. This deduction can be divided between Mr. and Mrs. Juenger in any way they wish. How should the \$6400 be divided between them to result in each person's having the same taxable income and therefore the greatest refund or least tax?

26. Michael Chang's chain saw uses a fuel mixture of 15 parts gasoline to 1 part oil. How much gasoline must be mixed with a gasoline–oil mixture that is 75% gasoline to make 8 quarts of the mixture to run the chain saw?

In Exercises 27–45, write an equation that can be used to solve the motion or mixture problem. Solve each equation and answer the question asked.

27. Bob and Julie Roe, brother and sister, have met for a family reunion in Branson, Missouri. On Sunday morning they finish breakfast and leave Branson at the same time. Bob is traveling north on Route 65 to Des Moines, Iowa, and Julie is traveling south on Route 65 to Conway, Arkansas. Because of traffic, Julie travels an average of 10 miles per hour faster than Bob. If Bob and Julie are 480 miles apart after 4 hours, find the speed of each car.

28. The Kleins and Pacinos leave their homes at 8 A.M. planning to meet for a picnic at a point between them. If the Kleins travel at 60 miles per hour and the Pacinos travel at 50 miles per hour, and they live 330 miles apart, at what time will they meet?

29. Chuy Carreon invested \$8000 for 1 year, part at 6% and part at 10% simple interest. How much was invested in each account if the same amount of interest was received from each account?

30. The Amityville Fire Department has a number of sump pumps that residents can use to empty flooded basements. Herb Bailey needs to empty his 15,000 gallon inground swimming pool so that it can be refiberglassed, so he calls the fire department. The fire department agrees to lend him two sump pumps to drain water from the pool. One drains 10 gallons of water a minute while the other drains 20 gallons of water per minute. If the pumps are turned on at the same time and remain on until the pool is drained, how long will it take for the pool to be drained?

31. A jetliner flew from Chicago to Los Angeles at an average speed of 500 miles per hour. Then it continued on over the Pacific Ocean to Hawaii at an average speed of 550 miles per hour. If the entire trip covered 5200 miles and the part over the ocean took twice as long as the part over land, how long did the entire trip take?

32. How many quarts of pure antifreeze should Frida Nilforoushan add to 10 quarts of a 20% antifreeze solution to make a 50% antifreeze solution?

33. An Air Force jet is going on a long distance flight and will need to be refueled in midair over the Pacific Ocean. A refueling plane that carries fuel can travel much farther than the jet, but flies at a slower speed. The refueling plane and jet will leave from the same base, but the refueling plane will leave 2 hours before the jet. The jet will fly at 800 mph and the refueling plane will fly at 520 miles per hour.

    a) How long after the jet takes off will the two planes meet?

    b) How far from the base will the refueling take place?

34. Peter Paul Rubin, an artist, sells both large paintings and small paintings at Jackson Square in New Orleans. He sells his small paintings for \$60 and his large paintings for \$180. At the end of the week he determined that the total amount he made by selling 12 paintings was \$1200. Determine the number of small and the number of large paintings that he sold.

35. Kelli Nelson holds two part-time jobs. One job pays \$6.00 per hour and the other pays \$6.50 per hour. Last week she earned a total of \$114 and worked for a total of 18 hours. How many hours did she work at each job?

36. Tim Kent mows part of his lawn in second gear and part in third gear. It took him 2 hours to mow the entire lawn and the odometer on his tractor shows that he covered 13.8 miles while cutting the grass. If he averages 4.2 miles per hour in second gear and 7.8 miles per hour in third gear, how long did he cut in each gear?

37. On a 100-mile trip to their cottage, the Barrs traveled at a steady speed for the first hour. The speed during the second hour of their trip was 16 miles per hour slower than the speed of the first hour. Find their speed during their first hour.

38. Juan Gillespie has 60 ounces of water whose temperature is 92°C. How much water with a temperature of 20°C must he mix in with all the 92°C water to get a mixture whose temperature is 50°C? Neglect heat loss from the water to the surrounding air.

39. Herb Garrett has an 80% methyl alcohol solution. He wishes to make a gallon of windshield washer solution by mixing his methyl alcohol solution with water. If 128 ounces, or a gallon, of windshield washer fluid should contain 6% methyl alcohol, how much of the 80% methyl alcohol solution and how much water must be mixed?

40. Sybil Geraud is making a meat loaf by combining chopped sirloin with veal. The sirloin contains 1.2 grams of fat per ounce, and the veal contains 0.3 gram of fat per ounce. If she wants her 64-ounce mixture to have only 0.8 gram of fat per ounce, find how much sirloin and how much veal she must use.

41. Sundance Dairy has 400 quarts of whole milk containing 5% butterfat. How many quarts of low-fat milk containing 1.5% butterfat should be added to produce milk containing 2% butterfat?

42. Tyrone Whitfield can ride his bike to work in $\frac{3}{4}$ hour. If he takes his car to work, the trip takes $\frac{1}{6}$ hour. If Tyrone drives his car an average of 14 miles per hour faster than he rides his bike, determine the distance he travels to work.

43. A old machine that folds and seals milk cartons can produce 50 milk cartons per minute. A new machine can produce 70 milk cartons per minute. The old machine has made 200 milk cartons when the new machine is turned on. If both machines then continue working, how long after the new machine is turned on will the new machine produce the same total number of milk cartons as the old machine?

44. The salinity (salt content) of the Atlantic Ocean averages 37 parts per thousand. If 64 ounces of water is collected and placed in the sun, how many ounces of pure water would need to evaporate to raise the salinity to 45 parts per thousand? (Only the pure water is evaporated; the salt is left behind.)

45. Two actors, Bruce Willis and Robert DeNiro, are making a western movie. They need to approach each other and be 20 feet apart after 5 seconds of walking. If they both walk at 6 feet per second, how far apart from each other should they be when they start?

46. Two rockets are launched from the Kennedy Space Center. The first rocket, launched at noon, will travel at 8000 miles per hour. The second rocket will be launched some time later and travel at 9500 miles per hour. When should the second rocket be launched if the rockets are to meet at a distance of 38,000 miles from Earth?
    a) Explain how to find the solution to this problem.
    b) Find the solution to the problem.

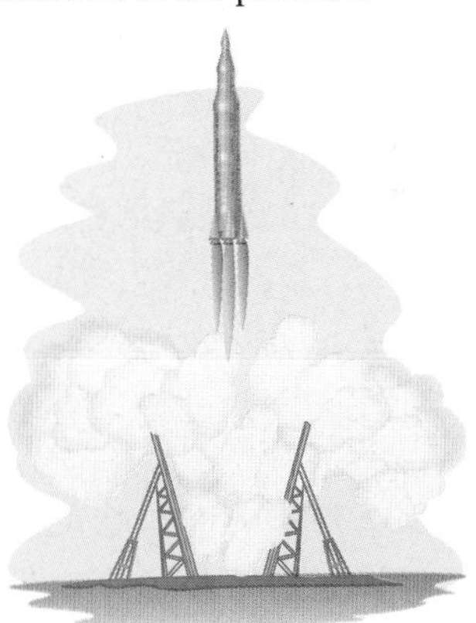

47. **a)** Make up your own realistic motion problem that can be represented as an equation.
    b) Write the equation that represents your problem.
    c) Solve the equation, and then find the answer to your problem.

48. **a)** Make up your own realistic mixture problem that can be represented as an equation.
    b) Write the equation that represents your problem.
    c) Solve the equation, and then find the answer to your problem.

Challenge Problems

49. The Chunnel (the underwater tunnel from Folkstone, England, to Calais, France) is 31 miles long (23.5 miles under water). A person can board France's TGV train (the bullet train) in Paris and travel nonstop through the Chunnel and arrive in London 3 hours later. The TGV averages about 130 miles per hour from Paris to Calais. (It travels up to 185 miles per hour during part of the trip.) It then reduces its speed to an average of 90 miles per hour through the 31-mile Chunnel. When leaving the Chunnel in Folkstone it travels only at an average of about 45 miles per hour for the 68-mile trip from Folkstone to London because of outdated tracks. Using this information, determine the distance from Paris to Calais, France.

50. Two cars labeled A and B are in a 500-lap race. Each lap is 1 mile. The lead car, A, is averaging 125 miles per hour when it reaches the halfway point. Car B is exactly 6.2 laps behind.

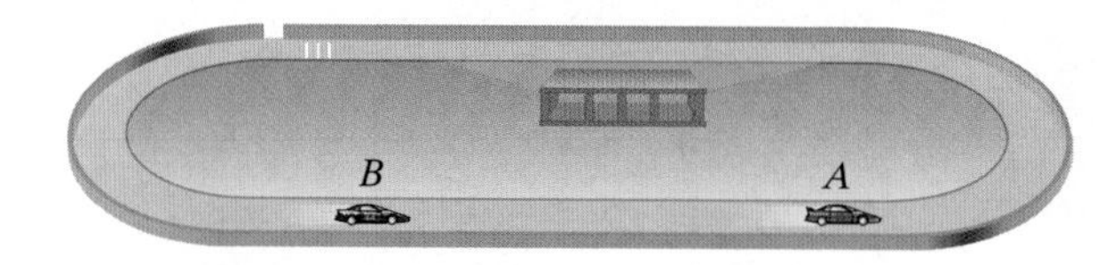

a) Find the average speed of car B.

b) When car A reaches the halfway point, how far behind, in seconds, is car B from car A?

51. The radiator of an automobile has a capacity of 16 quarts. It is presently filled with a 20% antifreeze solution. How many quarts must be drained and replaced with pure antifreeze to make the radiator contain a 50% antifreeze solution?

Cumulative Review Exercises

Solve.

[2.1] **52.** $0.6x + 0.22 = 0.4(x - 2.3)$

53. $\frac{2}{9}x + 3 = x + \frac{1}{5}$

[2.2] **54.** Solve the equation $\frac{3}{5}(x - 2) = \frac{2}{7}(2x + 3y)$ for y.

[2.3] **55.** Hertz Automobile Rental Agency charges $30 per day plus 14 cents a mile. National Automobile Rental Agency charges $16 per day plus 24 cents a mile for the same car. What distance would you have to drive in 1 day to make the cost of renting from Hertz equal to the cost of renting from National?

2.5 SOLVING LINEAR INEQUALITIES

SSM VIDEO 2.5 CD Rom

1) **Solve inequalities.**
2) **Graph solutions on a number line, interval notation, and solution sets.**
3) **Solve compound inequalities involving "and."**
4) **Solve continued inequalities using intersection of sets.**
5) **Solve compound inequalities involving "or."**

1) Solve Inequalities

Inequalities and set builder notation were introduced in Section 1.2. You may wish to review that section now. The inequality symbols follow.*

*$\neq$, not equal to, is also an inequality. $\neq$ means $<$ or $>$. Thus, $2 \neq 3$ means $2 < 3$ or $2 > 3$.

Inequality Symbols	
$>$	is greater than
$\geq$	is greater than or equal to
$<$	is less than
$\leq$	is less than or equal to

A mathematical expression containing one or more of these symbols is called an **inequality**. The direction of the inequality symbol is sometimes called the **order** or **sense** of the inequality.

Examples of Inequalities in One Variable

$2x + 3 \leq 5$ $\quad 4x > 3x - 5$ $\quad 1.5 \leq -2.3x + 4.5$ $\quad \frac{1}{2}x + 3 \geq 0$

To solve an inequality, we must isolate the variable on one side of the inequality symbol. To isolate the variable, we use the same basic techniques used in solving equations.

Properties Used to Solve Inequalities

1. If $a > b$, then $a + c > b + c$.
2. If $a > b$, then $a - c > b - c$.
3. If $a > b$, and $c > 0$, then $ac > bc$.
4. If $a > b$, and $c > 0$, then $\frac{a}{c} > \frac{b}{c}$.
5. If $a > b$, and $c < 0$, then $ac < bc$.
6. If $a > b$, and $c < 0$, then $\frac{a}{c} < \frac{b}{c}$.

The first two properties state that the same number can be added to or subtracted from both sides of an inequality. The third and fourth properties state that both sides of an inequality can be multiplied or divided by any positive real number. The last two properties indicate that **when both sides of an inequality are multiplied or divided by a negative number, the direction of the inequality symbol reverses.**

Example of Multiplication by a Negative Number

$$4 > -2$$

Multiply both sides of the inequality by −1 and reverse the direction of the inequality symbol.

$$-1(4) < -1(-2)$$

$$-4 < 2$$

Example of Division by a Negative Number

$$10 \geq -4$$

Divide both sides of the inequality by −2 and reverse the direction of the inequality symbol.

$$\frac{10}{-2} \leq \frac{-4}{-2}$$

$$-5 \leq 2$$

HELPFUL HINT

Do not forget to reverse the direction of the inequality symbol when multiplying or dividing both sides of the inequality by a negative number.

Inequality	Direction of Inequality Symbol
$-3x < 6$	$\dfrac{-3x}{-3} > \dfrac{6}{-3}$
$-\dfrac{x}{2} > 5$	$(-2)\left(-\dfrac{x}{2}\right) < (-2)(5)$

EXAMPLE 1 Solve the inequality $2x + 6 < 12$.

Solution

$$2x + 6 < 12$$
$$2x + 6 - 6 < 12 - 6$$
$$2x < 6$$
$$\frac{2x}{2} < \frac{6}{2}$$
$$x < 3$$

The solution set is $\{x|x < 3\}$. Any real number less than 3 will satisfy the inequality.

2) Graph Solutions on a Number Line, Interval Notation, and Solution Sets

The solution to an inequality can be indicated on a number line or written as a solution set, as was explained in Section 1.2. The solution can also be written in interval notation, as illustrated. Most instructors have a preferred way to indicate the solution to an inequality.

Recall that *a solid circle on the number line indicates that the endpoint is part of the solution, and an open circle indicates that the endpoint is not part of the solution. In interval notation, brackets,* [ ], *are used to indicate that the endpoints are part of the solution and parentheses,* ( ), *indicate that the endpoints are not part of the solution*. The symbol ∞ is read "infinity"; it indicates that the solution set continues indefinitely. Whenever ∞ is used in interval notation a *parenthesis* must be used on the corresponding side of the interval notation.

Solution of Inequality	Solution Set Indicated on Number Line	Solution Set Represented in Interval Notation
$x > a$		(a, ∞)
$x \geq a$		$[a, \infty)$
$x < a$	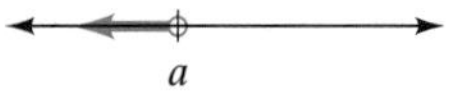	$(-\infty, a)$
$x \leq a$	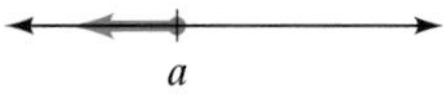	$(-\infty, a]$
$a < x < b$		(a, b)

Inequality	Interval Notation
$a \le x \le b$	$[a, b]$
$a < x \le b$	$(a, b]$
$a \le x < b$	$[a, b)$
$x \ge 5$	$[5, \infty)$
$x < 3$	$(-\infty, 3)$
$2 < x \le 6$	$(2, 6]$
$-6 \le x \le -1$	$[-6, -1]$

EXAMPLE 2 Solve the following inequality and give the solution both on a number line and in interval notation.

$$3(x - 2) \le 5x + 8$$

Solution

$$3(x - 2) \le 5x + 8$$
$$3x - 6 \le 5x + 8$$
$$3x - 5x - 6 \le 5x - 5x + 8$$
$$-2x - 6 \le 8$$
$$-2x - 6 + 6 \le 8 + 6$$
$$-2x \le 14$$
$$\frac{-2x}{-2} \ge \frac{14}{-2}$$
$$x \ge -7$$

Number Line **Interval Notation**

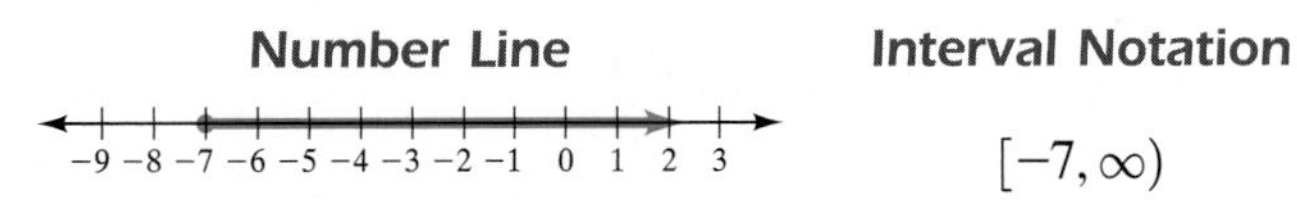

$[-7, \infty)$

NOW TRY EXERCISE 21 The solution set is $\{x | x \ge -7\}$.

In Example 2 we illustrated the solution on a number line, in interval notation, and as a solution set. Your instructor may indicate which form he or she prefers.

EXAMPLE 3 Solve the inequality $\frac{1}{2}(4x + 14) \ge 5x - 6 - 3x$.

Solution

$$\frac{1}{2}(4x + 14) \ge 5x - 6 - 3x$$
$$\frac{1}{2}(4x + 14) \ge 2x - 6$$
$$\left(\frac{1}{2}\right)(4x) + \left(\frac{1}{2}\right)(14) \ge 2x - 6$$
$$2x + 7 \ge 2x - 6$$
$$2x - 2x + 7 \ge 2x - 2x - 6$$
$$7 \ge -6$$

Since 7 is always greater than or equal to −6, the inequality is true for all real numbers. When an inequality is true for all real numbers, the solution set is *the set of all real numbers*, ℝ. The solution set to this example can also be indicated on a number line or given in interval notation.

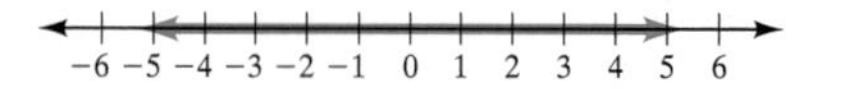

or $(-\infty, \infty)$

If Example 3 had resulted in the expression $7 \le -6$, the inequality would never have been true, since 7 is never less than or equal to −6. When an inequality is never true it has no solution. The solution set of an inequality that has no solution is the *empty or null set*, { } or ∅.

NOW TRY EXERCISE 31

HELPFUL HINT

Generally, when writing a solution to an inequality we write the variable on the left. For example, when solving an inequality, if we obtain $5 \ge y$ we would write the solution as $y \le 5$.

For example,

$-6 < x$ means $x > -6$ (inequality symbol points to −6 in both cases)

$-3 > x$ means $x < -3$ (inequality symbol points to x in both cases)

$a < x$ means $x > a$ (inequality symbol points to a in both cases)

$a > x$ means $x < a$ (inequality symbol points to x in both cases)

EXAMPLE 4 A small single-engine airplane can carry a maximum weight of 1500 pounds. Millie Johnson, the pilot, has to transport boxes weighing 80.4 pounds.

a) Write an inequality that can be used to determine the maximum number of boxes that Millie can safely place on her plane if she weighs 125 pounds.

b) Find the maximum number of boxes that Millie can transport.

Solution **a) Translate** Let n = number of boxes.

$$\text{Millie's weight} + \text{weight of } n \text{ boxes} \le 1500$$
$$125 + 80.4n \le 1500$$

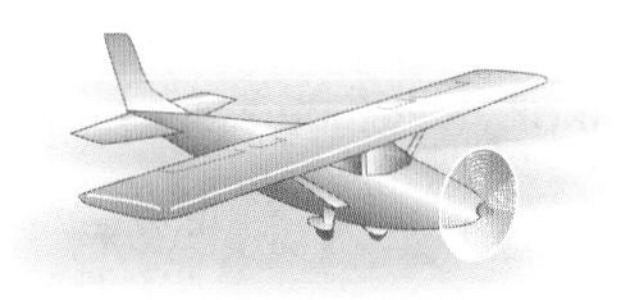

b) Carry Out

$$125 + 80.4n \le 1500$$
$$80.4n \le 1375$$
$$n \le 17.1$$

Answer Therefore, Millie can transport up to 17 boxes per trip.

EXAMPLE 5 A taxi's fare is \$3.50 for the first half-mile and \$2.00 for each additional half-mile. Any additional part of a half-mile is rounded up to the next half-mile.

a) Write an inequality that can be used to determine the maximum distance that Ronesha Jones can travel if she has only \$34.75.

b) Find the maximum distance that Ronesha can travel.

Solution **a) Translate** let x = number of half-miles after the first half-mile

then $2.00x$ = cost of traveling x additional half-miles

$$\text{cost of first half-mile} + \text{cost of additional half-miles} \le \text{total cost}$$
$$3.50 + 2.00x \le 34.75$$

b) Carry Out

$$3.50 + 2.00x \le 34.75$$
$$2.00x \le 31.25$$
$$x \le \frac{31.25}{2.00}$$
$$x \le 15.625$$

Answer Ronesha can travel a distance less than or equal to 15 half-miles after the first half-mile, for a total of 16 half-miles, or 8 miles. If she travels for 16 half-miles after the first, she will owe $3.50 + 2.00(16) = 35.50$ dollars, which is more money than she has.

NOW TRY EXERCISE 65

EXAMPLE 6 For a business to realize a profit, its revenue (or income), R, must be greater than its cost, C. That is, a profit will be obtained when $R > C$ (the company breaks even when $R = C$). A company that produces playing cards has a weekly cost equation of $C = 1525 + 1.7x$ and a weekly revenue equation of $R = 4.2x$, where x is the number of decks of playing cards produced and sold in a week. How many decks of cards must be produced and sold in a week for the company to make a profit?

Solution **Translate and Carry Out** The company will make a profit when $R > C$, or

$$4.2x > 1525 + 1.7x$$
$$2.5x > 1525$$
$$x > \frac{1525}{2.5}$$
$$x > 610$$

Answer The company will make a profit when more than 610 decks are produced and sold in a week.

NOW TRY EXERCISE 67

EXAMPLE 7 The 1997 Internal Revenue Tax Rate Schedule for taxpayers whose filing status is single is duplicated below.

Schedule X–Use if Your Filing Status is Single*			
If the amount on Form 1040, line 38, is: ***Over—***	***But not over—***	**Enter on Form 1040, line 39**	***of the amount over—***
$0	$24,650	.... 15%	$0
24,650	59,750	$3,697.50 + 28%	24,650
59,750	124,650	13,525.50 + 31%	59,750
124,650	271,050	33,644.50 + 36%	124,650
271,050	...	86,348.50 + 39.6%	271,050

* **Caution:** This schedule should only be used when the taxpayer meets certain requirements. In other cases, the **tax table** should be used.

a) Write in interval notation the amounts of taxable income (Form 1040, line 38) that makes up each of the five listed tax brackets, that is, the 15%, 28%, 31%, 36%, and 39.6% tax brackets.

b) Find the tax if a single person's taxable income (line 38) is $25,800.

c) Find the tax if a single person's taxable income is $137,600.

Solution **a)** The words "But not over" mean "less than or equal to." The taxable incomes that make up the five tax brackets are:

$(0, 24{,}650]$ for the 15% tax bracket

$(24{,}650, 59{,}750]$ for the 28% tax bracket

$(59{,}750, 124{,}650]$ for the 31% tax bracket

$(124{,}650, 271{,}050]$ for the 36% tax bracket

$(271{,}050, \infty)$ for the 39.6% tax bracket

b) Translate and Carry Out The tax for a single person with taxable income of $25,800 is $3697.50 plus 28% of the taxable income over $24,650. The taxable income over $24,650 is $25,800 − $24,650 = $1150.

$$\text{tax} = 3697.50 + 0.28(1150) = 3697.50 + 322 = 4019.50$$

Answer Therefore, the tax is $4019.50. This amount will be entered on line 39 of the 1040 tax form.

c) Translate and Carry Out A taxable income of $137,600 places a single taxpayer in the 36% tax bracket. The amount of taxable income over $124,650 is $137,600 − $124,650 = $12,950.

$$\text{tax} = 33{,}644.50 + 0.36(12{,}950) = 33{,}644.50 + 4662 = 38{,}306.50$$

Answer Thus, the tax is $38,306.50.

3 Solve Compound Inequalities Involving "And"

A **compound inequality** is formed by joining two inequalities with the word *and* or *or*.

Examples of Compound Inequalities

$3 < x$ and $x < 5$

$x + 4 > 3$ or $2x - 3 < 6$

$4x - 6 \geq -3$ and $x - 6 < 5$

The solution of a compound inequality using the word *and* is all the numbers that make *both* parts of the inequality true. Consider

$$3 < x \quad \text{and} \quad x < 5$$

What are the numbers that satisfy both inequalities? The numbers that satisfy both inequalities may be easier to see if we graph the solution to each inequality on a number line (see Fig. 2.6). Now we can see that the numbers that satisfy both inequalities are the numbers between 3 and 5. The solution set is $\{x | 3 < x < 5\}$.

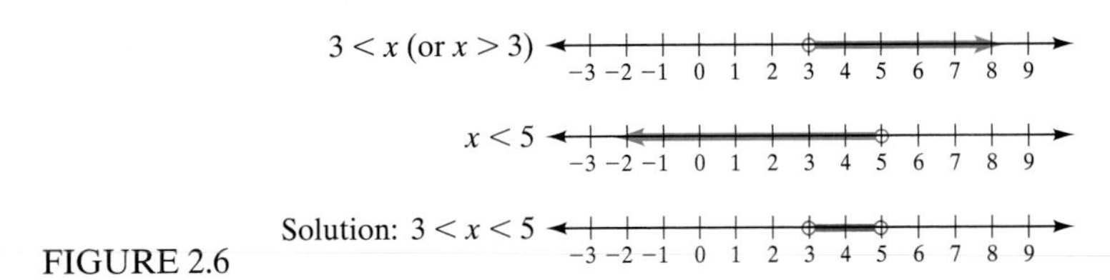

FIGURE 2.6

Recall from Chapter 1 that the intersection of two sets is the set of elements common to both sets. *To find the solution set of an inequality containing the word* **and** *take the* **intersection** *of the solution sets of the two inequalities.*

EXAMPLE 8 Solve $x + 2 \le 5$ and $2x - 4 > -2$.

Solution Begin by solving each inequality separately.

$$x + 2 \le 5 \quad \text{and} \quad 2x - 4 > -2$$

$$x \le 3 \qquad\qquad 2x > 2$$

$$x > 1$$

Now take the intersection of the sets $\{x|x \le 3\}$ and $\{x|x > 1\}$. When we find $\{x|x \le 3\} \cap \{x|x > 1\}$, we are finding the values of x common to both sets. Figure 2.7 illustrates that the solution set is $\{x|1 < x \le 3\}$. In interval notation, the solution is $(1, 3]$.

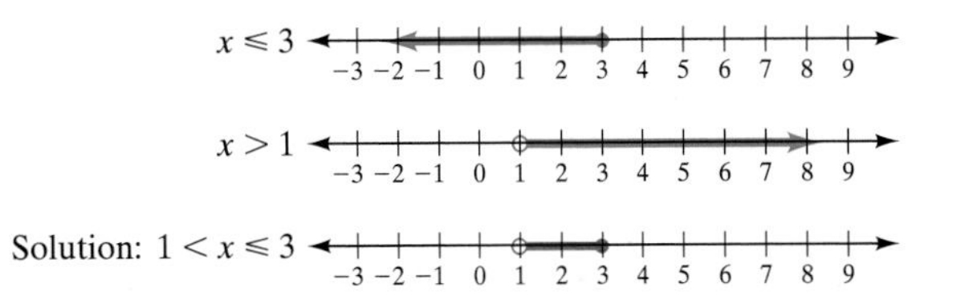

FIGURE 2.7

NOW TRY EXERCISE 57

4) Solve Continued Inequalities

Sometimes a compound inequality using the word *and* can be written in a shorter form. For example, $3 < x$ and $x < 5$ can be written as $3 < x < 5$. The word *and* does not appear when the inequality is written in this form, but it is implied. Inequalities written in the form $a < x < b$ are called **continued inequalities**. The compound inequality $1 < x + 5$ and $x + 5 \le 7$ can be written $1 < x + 5 \le 7$.

EXAMPLE 9 Solve $1 < x + 5 \le 7$.

Solution $1 < x + 5 \le 7$ means $1 < x + 5$ and $x + 5 \le 7$. Solve each inequality separately.

$$1 < x + 5 \quad \text{and} \quad x + 5 \le 7$$

$$-4 < x \qquad\qquad x \le 2$$

Remember that $-4 < x$ means $x > -4$. Figure 2.8 on page 114 illustrates that the solution set is $\{x|-4 < x \le 2\}$. In interval notation, the solution is $(-4, 2]$.

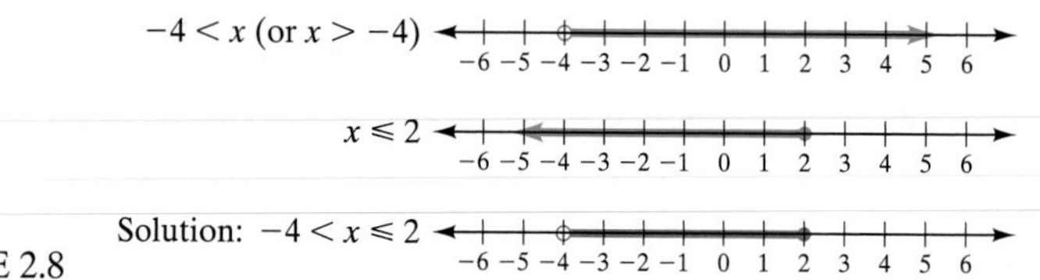

FIGURE 2.8

The inequality in Example 9, $1 < x + 5 \leq 7$, can be solved in another way. We can still use the properties discussed earlier to solve continued inequalities. However, when working with such inequalities, whatever we do to one part we must do to all three parts. In Example 9, we could have subtracted 5 from all three parts to isolate the variable in the middle and solve the inequality.

$$1 < x + 5 \leq 7$$
$$1 - 5 < x + 5 - 5 \leq 7 - 5$$
$$-4 < x \leq 2$$

Note that this is the same solution as obtained in Example 9.

EXAMPLE 10 Solve the inequality $-3 \leq 2x - 7 < 8$.

Solution We wish to isolate the variable x. We begin by adding 7 to all three parts of the inequality.

$$-3 \leq 2x - 7 < 8$$
$$-3 + 7 \leq 2x - 7 + 7 < 8 + 7$$
$$4 \leq 2x < 15$$

Now divide all three parts of the inequality by 2.

$$\frac{4}{2} \leq \frac{2x}{2} < \frac{15}{2}$$
$$2 \leq x < \frac{15}{2}$$

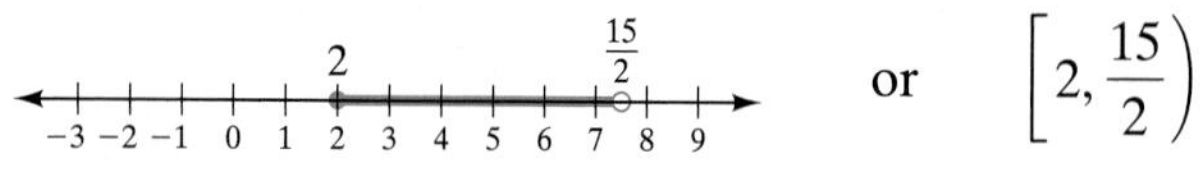

The solution set is $\left\{x \middle| 2 \leq x < \frac{15}{2}\right\}$.

NOW TRY EXERCISE 35

EXAMPLE 11 Solve the inequality $-2 < \frac{4 - 3x}{5} < 8$.

Solution Multiply all three parts by 5 to eliminate the denominator.

$$-2 < \frac{4 - 3x}{5} < 8$$
$$-2(5) < \cancel{5}\left(\frac{4 - 3x}{\cancel{5}}\right) < 8(5)$$
$$-10 < 4 - 3x < 40$$
$$-10 - 4 < 4 - 4 - 3x < 40 - 4$$
$$-14 < -3x < 36$$

Now divide all three parts of the inequality by −3. Remember that when we multiply or divide an inequality by a negative number, the direction of the inequality symbol reverses.

$$\frac{-14}{-3} > \frac{-3x}{-3} > \frac{36}{-3}$$

$$\frac{14}{3} > x > -12$$

Although $\frac{14}{3} > x > -12$ is correct, we generally write continued inequalities with the lesser value on the left. We will, therefore, rewrite the solution as

$$-12 < x < \frac{14}{3}$$

The solution may also be illustrated on a number line, written in interval notation, or written as a solution set.

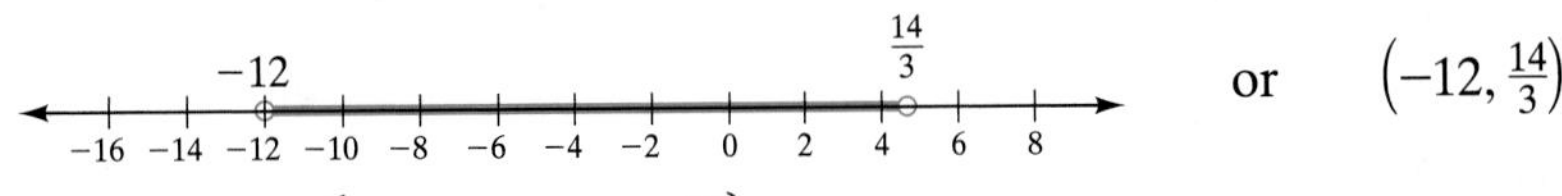

NOW TRY EXERCISE 43

The solution set is $\left\{x \middle| -12 < x < \frac{14}{3}\right\}$.

HELPFUL HINT

You must be very careful when writing the solution to a continued inequality. In Example 11 we can change the solution from

$$\frac{14}{3} > x > -12 \quad \text{to} \quad -12 < x < \frac{14}{3}$$

This is correct since both say that x is greater than −12 and less than $\frac{14}{3}$. Notice that the inequality symbol in both cases is pointing to the smaller number.

In Example 11, had we written the answer $\frac{14}{3} < x < -12$, we would have given the incorrect solution. Remember that the inequality $\frac{14}{3} < x < -12$ means that $\frac{14}{3} < x$ and $x < -12$. There is no number that is both greater than $\frac{14}{3}$ and less than −12. Also, by examining the inequality $\frac{14}{3} < x < -12$, it appears as if we are saying that −12 is a greater number than $\frac{14}{3}$, which is obviously incorrect.

It would also be incorrect to write the answer as

$$\cancel{-12 < x > \frac{14}{3}} \quad \text{or} \quad \cancel{\frac{14}{3} < x > -12}$$

EXAMPLE 12 In an American literature course, an average greater than or equal to 80 and less than 90 will result in a final grade of B. Steve Reinquist received grades of 85, 90, 68, and 70 on his first four exams. For Steve to receive a final grade of B in the course, between which two grades must his fifth (and last) exam fall?

Solution Let x = Steve's last exam grade.

$$80 \le \text{average of five exams} < 90$$

$$80 \le \frac{85 + 90 + 68 + 70 + x}{5} < 90$$

$$80 \le \frac{313 + x}{5} < 90$$

$$400 \le 313 + x < 450$$

$$400 - 313 \le 313 - 313 + x < 450 - 313$$

$$87 \le x < 137$$

Steve would need a minimum grade of 87 on his last exam to obtain a final grade of B. If the highest grade he could receive on the test is 100, is it possible for him to obtain a final grade of A (90 average or higher)? Explain.

5) Solve Compound Inequalities Involving "Or"

The solution to a compound inequality using the word *or* is all the numbers that make *either* of the inequalities a true statement. Consider the compound inequality

$$x > 3 \quad \text{or} \quad x < 5$$

What are the numbers that satisfy the compound inequality? Let's graph the solution to each inequality on the number line (see Fig. 2.9). Note that every real number satisfies at least one of the two inequalities. Therefore, the solution set to the compound inequality is the set of all real numbers, $\mathbb{R}$.

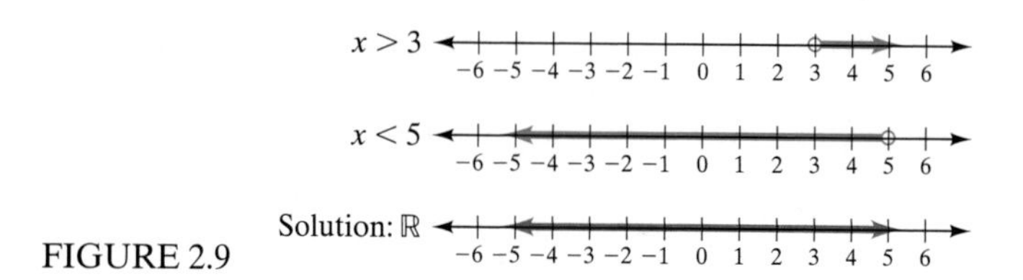

FIGURE 2.9

Recall from Chapter 1 that the *union* of two sets is the set of elements that belongs to *either* of the sets. *To find the solution set of an inequality containing the word* **or**, *take the* **union** *of the solution sets of the two inequalities that comprise the compound inequality.*

EXAMPLE 13 Solve $x + 3 \leq -1$ or $-4x + 3 < -5$.

Solution Solve each inequality separately.

$$\begin{aligned} x + 3 &\leq -1 \quad \text{or} \quad & -4x + 3 &< -5 \\ x &\leq -4 & -4x &< -8 \\ & & x &> 2 \end{aligned}$$

Now graph each solution on number lines and then find the union (Fig. 2.10). The union is $x \leq -4$ or $x > 2$.

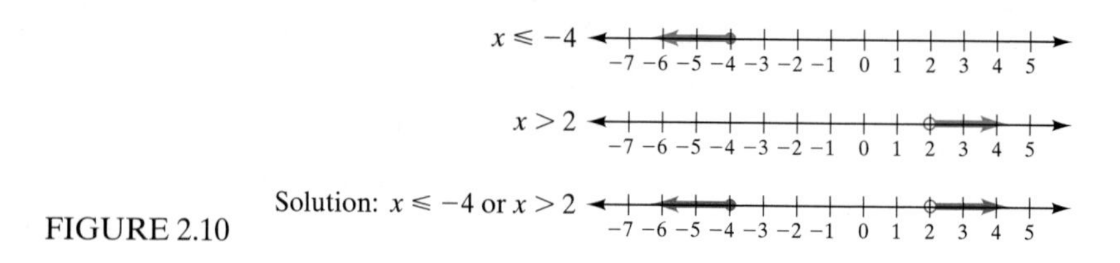

FIGURE 2.10

NOW TRY EXERCISE 59

The solution set is $\{x|x \leq -4\} \cup \{x|x > 2\}$, which can be written as $\{x|x \leq -4 \text{ or } x > 2\}$. In interval notation, the solution is $(-\infty, -4] \cup (2, \infty)$.

We often encounter inequalities in our daily lives. For example, on a highway the minimum speed may be 45 miles per hour and a maximum speed 65 miles per hour. A restaurant may have a sign stating that maximum capacity is 300 people, and the minimum takeoff speed of an airplane may be 125 miles per hour.

HELPFUL HINT

There are various ways to write the solution to an inequality problem. Be sure to indicate the solution to an inequality problem in the form requested by your professor. Examples of various forms follow.

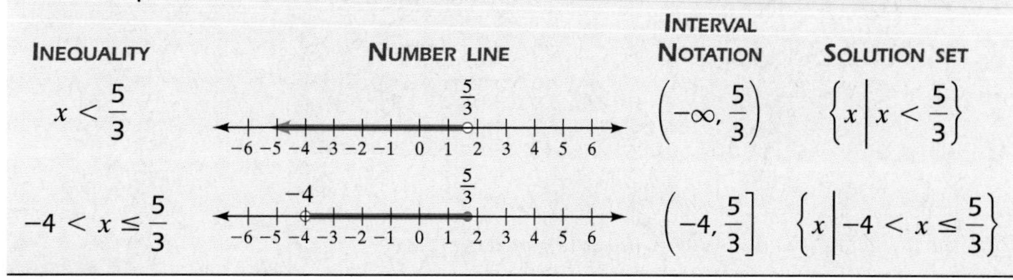

Inequality	Number Line	Interval Notation	Solution Set
$x < \frac{5}{3}$	(number line, −6 to 6, open circle at $\frac{5}{3}$)	$\left(-\infty, \frac{5}{3}\right)$	$\left\{x \middle\| x < \frac{5}{3}\right\}$
$-4 < x \le \frac{5}{3}$	(number line, −6 to 6, open circle at −4, closed circle at $\frac{5}{3}$)	$\left(-4, \frac{5}{3}\right]$	$\left\{x \middle\| -4 < x \le \frac{5}{3}\right\}$

Exercise Set 2.5

Concept/Writing Exercises

1. Explain the difference between $x < 7$ and $x \le 7$.
2. When is it necessary to change the direction of the inequality symbol?
3. **a)** When indicating a solution on a number line, when do you use open circles?
   b) When do you use closed circles?
   c) Give an example of an inequality whose solution on a number line would contain an open circle.
   d) Give an example of an inequality whose solution on a number line would contain a closed circle.
4. What are compound inequalities?
5. Explain why the inequality $4 < x < 2$ is not an acceptable answer.
6. What are continued inequalities?

Practice the Skills

*Express each inequality **a)** using a number line, **b)** in interval notation, and **c)** as a solution set (use set builder notation).*

7. $x < -4$
8. $x > \frac{5}{2}$
9. $x \ge 5.2$
10. $-2 < x < 3$
11. $2 \le x < \frac{12}{5}$
12. $x \ge -\frac{6}{5}$
13. $-7 < x \le -4$
14. $-3 \le x \le 0$

Solve each inequality and graph the solution on the number line.

15. $x + 2 < 8$
16. $2x + 3 > 4$
17. $3 - x < -4$
18. $4x + 3 \le -2x + 9$
19. $4.7x - 5.48 \ge 11.44$
20. $1.4x + 2.2 < 2.6x - 0.2$
21. $4(x - 2) \le 4x - 8$
22. $15.3 > 3(a - 1.4)$
23. $4b - 6 \ge 3(b + 3) + 2b$
24. $-(x - 3) + 4 \le -x + 5$
25. $\frac{y}{3} + \frac{2}{5} \le 4$
26. $2y - 6y + 10 \le 2(-2y + 3)$

Solve each inequality and give the solution in interval notation.

27. $\frac{c + 4}{2} + 9 > c + 2$
28. $4 - 3x < 7 + 2x + 4$
29. $4 + \frac{4x}{3} < 6$

30. $\frac{3y-6}{2} > \frac{2y+5}{6}$ **31.** $\frac{5-6y}{3} \leq 1 - 2y$ **32.** $\frac{3(x-2)}{5} > \frac{5(2-x)}{3}$

33. $-3x + 1 < 3[(x+2) - 2x] - 1$ **34.** $4[x - (3x - 2)] > 3(x+2) - 6$

Solve each inequality and give the solution in interval notation.

35. $1 < x + 3 < 9$ **36.** $-2 \leq x - 5 < 7$ **37.** $-3 < 5x \leq 8$

38. $-2 < -4x < 8$ **39.** $4 \leq 2x - 4 < 7$ **40.** $-12 < 3x - 5 \leq -4$

41. $4.3 < 3.2x - 2.1 \leq 16.46$ **42.** $\frac{1}{2} < 3x + 4 < 6$

Solve each inequality and indicate the solution set.

43. $4 < \frac{4x-2}{2} \leq 12$ **44.** $\frac{3}{5} < \frac{-x-5}{3} < 6$ 45. $6 \leq -3(2x-4) < 12$

46. $-6 < \frac{4-3x}{2} < \frac{2}{3}$ **47.** $0 < \frac{2(x-3)}{5} \leq 12$ **48.** $-15 < \frac{3(x-2)}{5} \leq 0$

Solve each inequality and indicate the solution set.

49. $x < 4$ and $x > 0$ **50.** $x < 4$ or $x > 2$ **51.** $x < 2$ and $x > 4$

52. $x < 3$ or $x > 4$ **53.** $x + 1 < 3$ and $x + 1 > -4$ **54.** $5x - 3 \leq 7$ or $-x + 3 < -5$

Solve each inequality and give the solution in interval notation.

55. $3x - 6 \leq 4$ or $2x - 3 < 5$ **56.** $-x + 6 > -3$ or $4x - 2 < 12$ 57. $4x + 5 \geq 5$ and $3x - 4 \leq 2$

58. $5x - 3 > 10$ and $5 - 3x < -3$ 59. $4 - x < -2$ or $3x - 1 < -1$ **60.** $-x + 3 < 0$ or $2x - 5 \geq 3$

Problem Solving

61. The length, l, plus the width, w, plus the depth, d, of a piece of luggage to be checked on a commercial airplane cannot exceed 61 inches (or else the customer will need to pay an additional fee).

a) Write an inequality that describes this restriction, using l, w, and d as listed above.

b) If Bruce Dodson's luggage is 29 inches long and $21\frac{1}{2}$ inches wide, what is the maximum depth it can have?

62. The length plus the girth of a package to be shipped by United Parcel Service (UPS) can be no larger than 130 inches.

a) Write an inequality that expresses this information, using l for the length and g for the girth.

b) UPS has defined girth as twice the width plus twice the depth. Write an inequality using length, l, width, w, and depth, d, to indicate the maximum allowable dimensions of a package that may be shipped by UPS.

c) If the length of a package is 40 inches and the width of a package is 20.5 inches, find the maximum allowable depth of the package.

In Exercises 63–77, set up an inequality that can be used to solve the problem. Solve the problem and find the desired value.

63. Cal Worth, a janitor, must move a large shipment of books from the first floor to the fifth floor. The sign on the elevator reads "maximum weight 800 pounds." If each box of books weigh 70 pounds, find the maximum number of boxes that Cal should place in the elevator.

64. If the janitor in Exercise 63, weighing 170 pounds, must ride up with the boxes of books, find the maximum number of boxes that can be placed in the elevator.

65. A telephone operator informs a customer in a phone booth that the charge for calling Denver, Colorado, is \$4.25 for the first 3 minutes and 45 cents for each additional minute. Any additional part of a minute will be rounded up to the nearest minute. Find the maximum time the customer can talk if he has only \$9.50.

66. A downtown parking garage in Austin, Texas, charges \$1.25 for the first hour and \$0.75 for each additional hour or part thereof. What is the maximum length of time you can park in the garage if you wish to pay no more than \$3.75?

67. Miriam Ladlow is considering writing and publishing her own book. She estimates her revenue equation as $R = 6.42x$, and her cost equation as $C = 10{,}025 + 1.09x$, where x is the number of books she sells. Find the minimum number of books she must sell to make a profit. See Example 6.

68. Peter Collinge is opening a dry-cleaning store. He estimates his cost equation as $C = 8000 + 0.08x$ and his revenue equation as $R = 1.85x$, where x is the number of garments dry cleaned in a year. Find the minimum number of garments that must be dry cleaned in a year for Peter to make a profit.

69. A for-profit organization can purchase an \$85 per year bulk-mail permit and then send bulk mail at a rate of 25.6 cents per piece. Without the permit, each piece of bulk mail would cost 33 cents. Find the minimum number of pieces of bulk mail that would have to be mailed for it to be financially worthwhile for an organization to purchase the bulk-mail permit.

70. The cost for mailing a package first class is 33 cents for the first ounce and 22 cents for each additional ounce. What is the maximum weight of a package that can be mailed first class for \$5.00?

71. To be eligible to continue her financial assistance for college, Nikita Maxwell can earn no more than \$2000 during her 8-week summer employment. She already earns \$90 per week as a day-care assistant. She is considering adding an evening job at a fast-food restaurant, where she will earn \$6.25 per hour. What is the maximum number of hours she can work at the restaurant without jeopardizing her financial assistance?

72. Pamela Person recently accepted a sales position in Ohio. She can select between two payment plans. Plan 1 is a salary of \$300 per week plus a 10% commission on sales. Plan 2 is a salary of \$400 per week plus an 8% commission on sales. For what amount of weekly sales would Pamela earn more by plan 1?

73. To receive an A in a course, Ray Reynolds must obtain an average of 90 or higher on five exams. If Ray's first four exam grades are 90, 87, 96, and 79, what is the minimum grade that Ray can receive on the fifth exam to get an A in the course?

74. To pass a course, Maria Shepherd needs an average grade of 60 or more. If Maria's grades are 65, 72, 90, 47, and 62, find the minimum grade that Maria can get on her sixth and last exam and pass the course.

75. Calisha Mahoney's grades on her first four exams are 87, 92, 70, and 75. An average greater than or equal to 80 and less than 90 will result in a final grade of B. What range of grades on Calisha's fifth and last exam will result in a final grade of B? Assume a maximum grade of 100.

76. For air to be considered "clean," the average of three pollutants must be less than 3.2 parts per million. If the first two pollutants are 2.7 and 3.42 ppm, what values of the third pollutant will result in clean air?

77. The water acidity in a pool is considered normal when the average pH reading of three daily measurements is greater than 7.2 and less than 7.8. If the first two pH readings are 7.48 and 7.15, find the range of pH values for the third reading that will result in the acidity level being normal.

78. Refer to Example 7. Find the income tax that Ronni Sharone, who is a single taxpayer, will owe if her taxable income is

a) \$26,707.

b) \$83,292.

79. The number of violent crimes in New York City has been dropping since 1990. The number of people receiving public assistance in New York City has also begun dropping. The following graph illustrates this information.

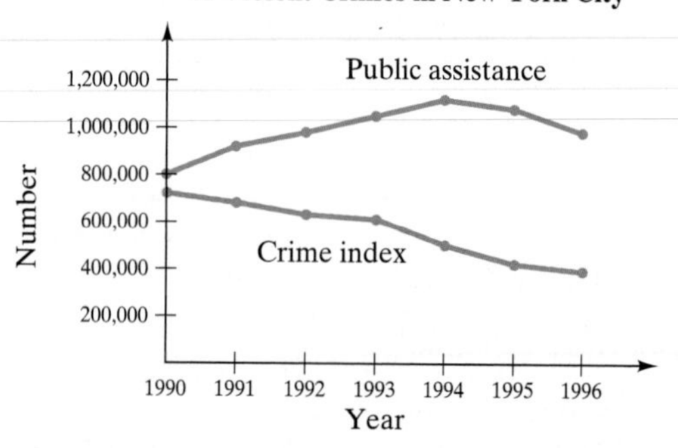

a) In which years was the crime index greater than or equal to 600,000 *and* the number of people on welfare less than or equal to 1,000,000? Explain how you determined your answer.

b) In which years was the crime index less than or equal to 600,000 *or* the number of people on welfare greater than or equal to 900,000? Explain how you determined your answer.

80. The following graph shows the median awards for employee versus employer wrongful termination cases and for discrimination and sexual harassment cases.

Median Awards for Wrongful Termination and Discrimination and Sexual Harassment Cases in the U.S.

Awards: $220,000, $200,000, $180,000, $160,000, $140,000, $120,000, $100,000, $80,000, $60,000, 0

Wrongful termination: 115,168; 152,014; 120,736; 150,000; 150,000; 150,000; 205,794

Discrimination and sexual harassment: 80,000; 83,439; 85,000; 102,500; 78,591; 75,000; 100,050

Year: 1990 1991 1992 1993 1994 1995 1996

a) In which years was the median award for discrimination and sexual harassment cases greater than or equal to $85,000 *and* the median award for wrongful termination greater than or equal to $150,000? Explain how you determined your answer.

b) In which years was the median award for discrimination and sexual harassment cases less than or equal to $85,000 *or* the median award for wrongful termination greater than or equal to $149,385? Explain how you determined your answer.

81. If $a > b$, will a^2 always be greater than b^2? Explain and give an example to support your answer.

82. **a)** Explain the step-by-step procedure used to solve the inequality $a < bx + c < d$ for x (assume that $b > 0$).

b) Solve the inequality for x and write the solution in interval notation.

83. A Blue Cross/Blue Shield insurance policy has a $100 deductible, after which it pays 80% of the total medical cost, c. The customer pays 20% until the customer has paid a total of $500, after which the policy pays 100% of the medical cost. We can describe this policy as follows:

Blue Cross Pays

$$\begin{array}{ll} 0, & \text{if } c \le \$100 \\ 0.80(c - 100), & \text{if } \$100 < c \le \$2100 \\ c - 500, & \text{if } c > \$2100 \end{array}$$

Explain why this set of inequalities describes Blue Cross/Blue Shield's payment plan.

84. Explain why the inequality $a < bx + c < d$ cannot be solved for x unless additional information is given.

Challenge Problems

85. Russell Meek's first five grades in European History were 82, 90, 74, 76, and 68. The final exam for the course is to count one-third in computing the final average. A final average greater than or equal to 80 and less than 90 will result in a final grade of B. What range of final exam grades will result in Russell's receiving a final grade of B in the course? Assume that a maximum grade of 100 is possible.

In Exercises 86–88, **a)** *explain how to solve the inequality, and* **b)** *solve the inequality and give the solution in interval notation.*

86. $x < 3x - 10 < 2x$

87. $x < 2x + 3 < 2x + 5$

88. $x + 3 < x + 1 < 2x$

Cumulative Review Exercises

[1.2] **89.** For $A = \{1, 2, 6, 8, 9\}$ and $B = \{1, 3, 4, 5, 8\}$, find

a) $A \cup B$.

b) $A \cap B$.

90. For $A = \left\{-3, 4, \frac{5}{2}, \sqrt{7}, 0, -\frac{29}{80}\right\}$, list the elements that are

a) counting numbers.

b) whole numbers.

c) rational numbers.

d) real numbers.

[1.3] *Name each illustrated property.*

91. $(3x + 6) + 4y = 3x + (6 + 4y)$

92. $3x + y = y + 3x$

[2.2] **93.** Solve the formula $R = L + (V - D)r$ for V.

2.6 SOLVING EQUATIONS AND INEQUALITIES CONTAINING ABSOLUTE VALUES

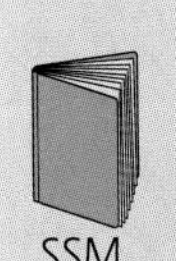

SSM VIDEO 2.6 CD Rom

1. **Understand the geometric interpretation of absolute value.**
2. **Solve equations of the form $|x| = a, a > 0$.**
3. **Solve inequalities of the form $|x| < a, a > 0$.**
4. **Solve inequalities of the form $|x| > a, a > 0$.**
5. **Solve inequalities of the form $|x| > a$ or $|x| < a$ when $a < 0$.**
6. **Solve inequalities of the form $|x| > 0$ or $|x| < 0$.**
7. **Solve equations of the form $|x| = |y|$.**

1 Understand the Geometric Interpretation of Absolute Value

In Section 1.3 we introduced the concept of absolute value. We stated that the absolute value of a number may be considered the distance (without sign) from the number 0 on the number line. The absolute value of 3, written $|3|$, is 3 since it is 3 units from 0 on the number line. Similarly, the absolute value of -3, written $|-3|$, is also 3 since it is 3 units from 0 on the number line.

Consider the equation $|x| = 3$; what values of x make this equation true? We know that $|3| = 3$ and $|-3| = 3$. The solutions to $|x| = 3$ are 3 and -3. When solving the equation $|x| = 3$, we are finding the values whose distance is exactly 3 units from 0 on the number line (see Fig. 2.11a).

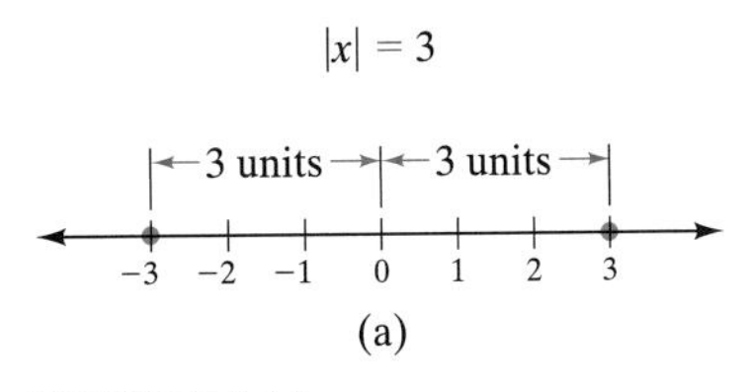

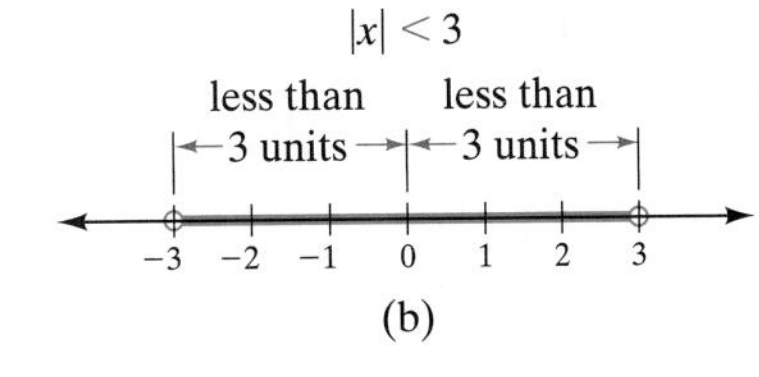

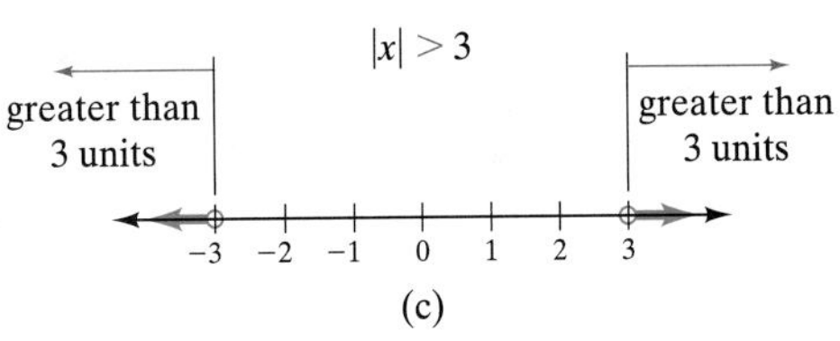

FIGURE 2.11

Now consider the inequality $|x| < 3$. To solve this inequality, we need to find the set of values whose distance is less than 3 units from 0 on the number line. These are the values of x between -3 and 3 (see Fig. 2.11b).

To solve the inequality $|x| > 3$, we need to find the set of values whose distance is greater than 3 units from 0 on the number line. These are the values that are either less than -3 or greater than 3 (see Fig. 2.11c).

In this section we will solve equations and inequalities such as the following:

$$|2x - 1| = 5 \qquad |2x - 1| \le 5 \qquad |2x - 1| > 5$$

The geometric interpretation of $|2x - 1| = 5$ is similar to $|x| = 3$. When solving $|2x - 1| = 5$ we are determining the set of values that result in $2x - 1$ being exactly 5 units away from 0 on the number line.

The geometric interpretation of $|2x - 1| \le 5$ is similar to the geometric interpretation of $|x| \le 3$. When solving $|2x - 1| \le 5$, we are determining the set of values that result in $2x - 1$ being less than or equal to 5 units from 0 on the number line.

The geometric interpretation of $|2x - 1| > 5$ is similar to that of $|x| > 3$. When solving $|2x - 1| > 5$, we are determining the set of values that result in $2x - 1$ being greater than 5 units from 0 on the number line.

We will be solving absolute value equations and inequalities algebraically in the remainder of this section. We will first solve absolute value equations, then we will solve absolute value inequalities. We will end the section by solving absolute value equations where both sides of the equation contain an absolute value, for example, $|x + 3| = |2x - 5|$.

2) Solve Equations of the Form $|x| = a$, $a > 0$

When solving an equation of the form $|x| = a$, $a \geq 0$, we are finding the values that are exactly a units from 0 on the number line. The following procedure may be used to solve such problems.

To Solve Equations of the Form $|x| = a$

If $|x| = a$ and $a > 0$, then $x = a$ or $x = -a$.

EXAMPLE 1 Solve each equation **a)** $|x| = 4$ **b)** $|x| = 0$ **c)** $|x| = -2$

Solution **a)** Using the procedure, we get $x = 4$ or $x = -4$. The solution set is $\{-4, 4\}$.

b) The only real number whose absolute value equals 0 is 0. Thus, the solution set for $|x| = 0$ is $\{0\}$.

c) The absolute value of a number is never negative, so there are no solutions to this equation. The solution set is $\varnothing$.

EXAMPLE 2 Solve the equation $|2w - 1| = 5$.

Solution At first this might not appear to be of the form $|x| = a$. However, if we let $2w - 1$ be x and 5 be a, you will see the equation is of this form. We are looking for the values of w such that $2w - 1$ is exactly 5 units from 0 on the number line. Thus, the quantity $2w - 1$ must be equal to 5 or -5.

$$2w - 1 = 5 \quad \text{or} \quad 2w - 1 = -5$$
$$2w = 6 \qquad\qquad 2w = -4$$
$$w = 3 \qquad\qquad w = -2$$

CHECK:

$w = 3$
$$|2w - 1| = 5$$
$$|2(3) - 1| \stackrel{?}{=} 5$$
$$|6 - 1| \stackrel{?}{=} 5$$
$$|5| \stackrel{?}{=} 5$$
$$5 = 5 \quad \text{True}$$

$w = -2$
$$|2w - 1| = 5$$
$$|2(-2) - 1| \stackrel{?}{=} 5$$
$$|-4 - 1| \stackrel{?}{=} 5$$
$$|-5| \stackrel{?}{=} 5$$
$$5 = 5 \quad \text{True}$$

The solutions 3 and -2 each result in $2w - 1$ being 5 units from 0 on the number line. The solution set is $\{-2, 3\}$.

NOW TRY EXERCISE 21

Consider the equation $|2w - 1| - 3 = 2$. The first step in solving this equation is to isolate the absolute value term. We do this by adding 3 to both sides of the equation. This results in the equation we solved in Example 2.

3) Solve Inequalities of the Form $|x| < a$, $a > 0$

Now let's look at inequalities of the form $|x| < a$. Consider $|x| < 3$. This inequality represents the set of values that are less than 3 units from 0 on a number line (see Figure 2.11b). The solution set is $\{x|-3 < x < 3\}$. The solution set to an inequality of the form $|x| < a$ is the set of values that are *less than a units from 0 on a number line.*

We can use the same reasoning process to solve more complicated problems, as shown in Example 3.

EXAMPLE 3 Solve the inequality $|2x - 3| < 5$.

Solution The solution to this inequality will be the set of values such that the distance between $2x - 3$ and 0 on a number line will be less than 5 units (see Fig. 2.12). Using Figure 2.12, we can see that $-5 < 2x - 3 < 5$.

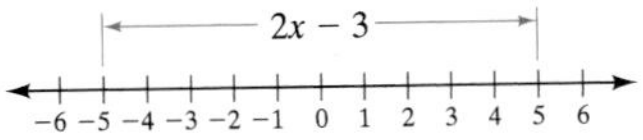

$2x - 3$

−6 −5 −4 −3 −2 −1 0 1 2 3 4 5 6

FIGURE 2.12

Solving, we get

$$-5 < 2x - 3 < 5$$

$$-2 < 2x < 8$$

$$-1 < x < 4$$

The solution set is $\{x|-1 < x < 4\}$. When x is any number between −1 and 4, the expression $2x - 3$ will represent a number that is less than 5 units from 0 on the number line (or a number between −5 and 5).

To solve inequalities of the form $|x| < a$, we can use the following procedure.

To Solve Inequalities of the Form $|x| < a$

If $|x| < a$ and $a > 0$, then $-a < x < a$.

EXAMPLE 4 Solve the inequality $|3x - 4| \leq 5$ and graph the solution on a number line.

Solution Since this inequality is of the form $|x| \leq a$, we write

$$-5 \leq 3x - 4 \leq 5$$

$$-1 \leq 3x \leq 9$$

$$-\frac{1}{3} \leq x \leq 3$$

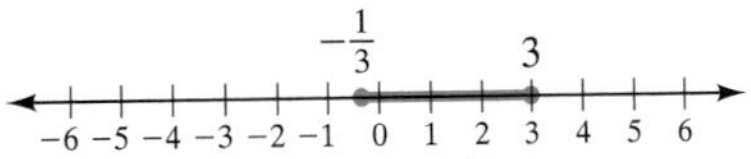

$-\frac{1}{3}$ 3

−6 −5 −4 −3 −2 −1 0 1 2 3 4 5 6

Any value of x greater than or equal to $-\frac{1}{3}$ and less than or equal to 3 would result in $3x - 4$ being less than or equal to 5 units from 0 on a number line.

EXAMPLE 5 Solve the inequality $|4.2 - x| + 1.3 < 3.6$ and graph the solution on a number line.

Solution First isolate the absolute value by subtracting 1.3 from both sides of the inequality. Then solve as in the previous examples.

$$|4.2 - x| + 1.3 < 3.6$$
$$|4.2 - x| < 2.3$$
$$-2.3 < 4.2 - x < 2.3$$
$$-6.5 < -x < -1.9$$
$$-1(-6.5) > -1(-x) > -1(-1.9)$$
$$6.5 > x > 1.9$$
$$\text{or} \quad 1.9 < x < 6.5$$

The solution set is $\{x|1.9 < x < 6.5\}$. The solution in interval notation is (1.9, 6.5).

NOW TRY EXERCISE 37

4 Solve Inequalities of the Form $|x| > a, a > 0$

Now we look at inequalities of the form $|x| > a$. Consider $|x| > 3$. This inequality represents the set of values that are greater than 3 units from 0 on a number line (see Fig. 2.11c). The solution set is $\{x|x < -3 \text{ or } x > 3\}$. The solution set to $|x| > a$ is the set of values that are *greater than a units from* 0 on a number line.

EXAMPLE 6 Solve the inequality $|2x - 3| > 5$ and graph the solution on a number line.

Solution The solution to $|2x - 3| > 5$ is the set of values such that the distance between $2x - 3$ and 0 on a number line will be greater than 5. The quantity $2x - 3$ must either be less than -5 or greater than 5 (see Fig. 2.13).

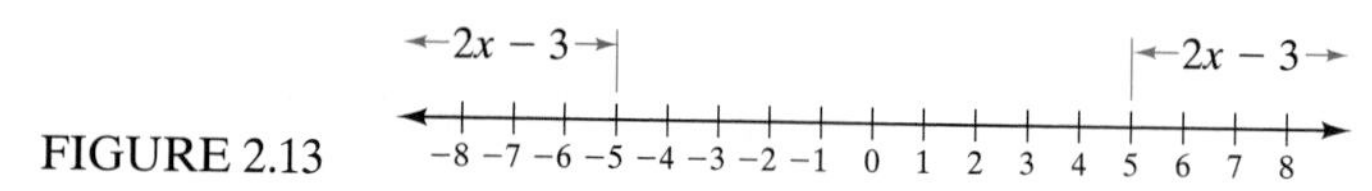

FIGURE 2.13

Since $2x - 3$ must be either less than -5 or greater than 5, we set up and solve the following compound inequality:

$$2x - 3 < -5 \quad \text{or} \quad 2x - 3 > 5$$
$$2x < -2 \qquad\qquad 2x > 8$$
$$x < -1 \qquad\qquad x > 4$$

The solution set to $|2x - 3| > 5$ is $\{x|x < -1 \text{ or } x > 4\}$. When x is any number less than -1 or greater than 4, the expression $2x - 3$ will represent a number that is greater than 5 units from 0 on the number line (or a number less than -5 or greater than 5).

To solve inequalities of the form $|x| > a$, we can use the following procedure.

To Solve Inequalities of the Form $|x| > a$

If $|x| > a$ and $a > 0$, then $x < -a$ or $x > a$.

EXAMPLE 7 Solve the inequality $|2x - 1| \geq 7$ and graph the solution on the number line.

Solution Since this inequality is of the form $|x| \geq a$, we use the procedure given above.

$$2x - 1 \leq -7 \quad \text{or} \quad 2x - 1 \geq 7$$
$$2x \leq -6 \qquad\qquad 2x \geq 8$$
$$x \leq -3 \qquad\qquad x \geq 4$$

−6 −5 −4 −3 −2 −1 0 1 2 3 4 5 6

Any value of x less than or equal to −3, or greater than or equal to 4, would result in $2x - 1$ representing a number that is greater than or equal to 7 units from 0 on the number line. The solution set is $\{x | x \leq -3 \text{ or } x \geq 4\}$. In interval notation, the solution is $(-\infty, -3] \cup [4, \infty)$.

EXAMPLE 8 Solve the inequality $\left|\frac{3x - 4}{2}\right| \geq 9$.

Solution Since this inequality is of the form $|x| \geq a$, we write

$$\frac{3x - 4}{2} \leq -9 \quad \text{or} \quad \frac{3x - 4}{2} \geq 9$$

Now multiply both sides of each inequality by the least common denominator, 2. Then solve each inequality.

$$2\left(\frac{3x - 4}{2}\right) \leq -9 \cdot 2 \quad \text{or} \quad 2\left(\frac{3x - 4}{2}\right) \geq 9 \cdot 2$$
$$3x - 4 \leq -18 \qquad\qquad 3x - 4 \geq 18$$
$$3x \leq -14 \qquad\qquad 3x \geq 22$$
$$x \leq -\frac{14}{3} \qquad\qquad x \geq \frac{22}{3}$$

$-\frac{14}{3}$ $\frac{22}{3}$

−7 −6 −5 −4 −3 −2 −1 0 1 2 3 4 5 6 7 8 9 10

NOW TRY EXERCISE 53

HELPFUL HINT

Some general information about equations and inequalities containing absolute value follows. For real numbers a, b, and c where $a \neq 0$ and $c > 0$:

Form of Equation or Inequality	The Solution Will Be:	Solution on a Number Line:
$\lvert ax + b\rvert = c$	Two distinct numbers, p and q	p q
$\lvert ax + b\rvert < c$	The set of numbers between two numbers, $p < x < q$	p q
$\lvert ax + b\rvert > c$	The set of numbers less than one number or greater than a second number, $x < p$ or $x > q$	p q

5) Solve Inequalities of the Form $|x| > a$ or $|x| < a$ When $a < 0$

We have solved inequalities of the form $|x| < a$ where $a > 0$. Now let us consider what happens in an absolute value inequality when $a < 0$. Consider the inequality $|x| < -3$. Since $|x|$ will always have a value greater than or equal to 0 for any real number x, this inequality can never be true, and the solution is the empty set, $\varnothing$. Whenever we have an absolute value inequality of this type, the solution will be the empty set.

EXAMPLE 9 Solve the inequality $|x - 4| - 3 < -5$.

Solution Begin by adding 3 to both sides of the inequality.

$$|x - 4| - 3 < -5$$
$$|x - 4| < -2$$

Since $|x - 4|$ will always be greater than or equal to 0 for any real number x, this inequality can never be true. Thus, the solution is the empty set, $\varnothing$.

NOW TRY EXERCISE 41

Now consider the inequality $|x| > -3$. Since $|x|$ will always have a value greater than or equal to 0 for any real number x, this inequality will always be true. Since every value of x will make this inequality a true statement, the solution is the set of all real numbers, $\mathbb{R}$. Whenever we have an absolute value inequality of this type, the solution will be the set of all real numbers, $\mathbb{R}$.

EXAMPLE 10 Solve the inequality $|2x + 3| + 4 \geq -7$.

Solution Begin by subtracting 4 from both sides of the inequality.

$$|2x + 3| + 4 \geq -7$$
$$|2x + 3| \geq -11$$

Since $|2x + 3|$ will always be greater than or equal to 0 for any real number x, this inequality is true for all real numbers. Thus, the solution is the set of all real numbers, $\mathbb{R}$.

NOW TRY EXERCISE 59

6) Solve Inequalities of the Form $|x| > 0$ or $|x| < 0$

Now let us discuss inequalities where one side of the inequality is 0. The only value that satisfies the equation $|x - 5| = 0$ is 5, since 5 makes the expression inside the absolute value sign 0. Now consider $|x - 5| \leq 0$. Since the absolute value can never be negative, this inequality is true only when $x = 5$. The inequality $|x - 5| < 0$ has no solution. Can you explain why? What is the solution to $|x - 5| \geq 0$? Since any value of x will result in the absolute value being greater than or equal to 0, the solution is the set of all real numbers, $\mathbb{R}$. What is the solution to $|x - 5| > 0$? The solution is every real number except 5. Can you explain why 5 is excluded from the solution?

EXAMPLE 11 Solve each inequality. **a)** $|x + 3| > 0$ **b)** $|3x - 4| \leq 0$

Solution **a)** The inequality will be true for every value of x except -3. The solution set is $\{x \mid x < -3 \text{ or } x > -3\}$.

b) Determine the number that makes the absolute value equal to 0 by setting the expression within the absolute value equal to 0 and solving for x.

$$3x - 4 = 0$$
$$3x = 4$$
$$x = \frac{4}{3}$$

The inequality will be true only when $x = \frac{4}{3}$. The solution set is $\left\{\frac{4}{3}\right\}$.

7) Solve Equations of the Form $|x| = |y|$

Now we will discuss absolute value equations where an absolute value appears on both sides of the equation. To solve equations of the form $|x| = |y|$, use the procedure that follows.

To Solve Equations of the Form $|x| = |y|$

If $|x| = |y|$, then $x = y$ or $x = -y$.

When solving an absolute value equation with an absolute value expression on each side of the equal sign, the two expressions must have the same absolute value. Therefore, *the expressions must be equal to each other or be opposites of each other.*

EXAMPLE 12 Solve the equation $|z + 3| = |2z - 7|$.

Solution If we let $z + 3$ be x and $2z - 7$ be y, this equation is of the form $|x| = |y|$. Using the procedure given above, we obtain the two equations

$$z + 3 = 2z - 7 \quad \text{or} \quad z + 3 = -(2z - 7)$$

Now solve each equation.

$$z + 3 = 2z - 7 \quad \text{or} \quad z + 3 = -(2z - 7)$$
$$3 = z - 7 \qquad\qquad z + 3 = -2z + 7$$
$$10 = z \qquad\qquad 3z + 3 = 7$$
$$3z = 4$$
$$z = \frac{4}{3}$$

CHECK:

$z = 10$

$$|z + 3| = |2z - 7|$$
$$|10 + 3| \stackrel{?}{=} |2(10) - 7|$$
$$|13| \stackrel{?}{=} |20 - 7|$$
$$|13| \stackrel{?}{=} |13|$$
$$13 = 13 \quad \text{True}$$

$z = \frac{4}{3}$

$$|z + 3| = |2z - 7|$$
$$\left|\frac{4}{3} + 3\right| \stackrel{?}{=} \left|2\left(\frac{4}{3}\right) - 7\right|$$
$$\left|\frac{13}{3}\right| \stackrel{?}{=} \left|\frac{8}{3} - \frac{21}{3}\right|$$
$$\left|\frac{13}{3}\right| \stackrel{?}{=} \left|-\frac{13}{3}\right|$$
$$\frac{13}{3} = \frac{13}{3} \quad \text{True}$$

The solution set is $\left\{10, \frac{4}{3}\right\}$.

EXAMPLE 13 Solve the equation $|4x - 7| = |6 - 4x|$.

Solution

$$\begin{aligned} 4x - 7 &= 6 - 4x & \text{or} \quad 4x - 7 &= -(6 - 4x) \\ 8x - 7 &= 6 & 4x - 7 &= -6 + 4x \\ 8x &= 13 & -7 &= -6 \quad \textit{False} \\ x &= \frac{13}{8} \end{aligned}$$

Since the equation $4x - 7 = -(6 - 4x)$ results in a false statement, the absolute value equation has only one solution. A check will show that the solution set is $\left\{\frac{13}{8}\right\}$.

NOW TRY EXERCISE 63

Summary of Procedures for Solving Equations and Inequalities Containing Absolute Value

For $a > 0$,

If $|x| = a$, then $x = a$ or $x = -a$.

If $|x| < a$, then $-a < x < a$.

If $|x| > a$, then $x < -a$ or $x > a$.

If $|x| = |y|$, then $x = y$ or $x = -y$.

Exercise Set 2.6

Concept/Writing Exercises

1. How do we solve equations of the form $|x| = a$, $a > 0$?
2. What is the solution to $|x| = 0$? Explain your answer.
3. What is the solution to $|x| = -3$? Explain your answer.
4. How do you check to see whether -7 is a solution to $|2x + 3| = 11$? Is -7 a solution?
5. How do we solve inequalities of the form $|x| < a$, $a > 0$?
6. What is the solution to $|x| < 0$? Explain your answer.
7. How do we solve inequalities of the form $|x| > a$, $a > 0$?
8. What is the solution to $|x| > 0$? Explain your answer.
9. Suppose m and n $(m < n)$ are two distinct solutions to the equation $|ax + b| = c$. Indicate the solutions, using both inequality symbols and the number line, to each inequality. (See the Helpul Hint on page 125.)

   a) $|ax + b| < c$

   b) $|ax + b| > c$
10. Explain how to solve an equation of the form $|x| = |y|$.
11. How many solutions will $|ax + b| = k$, $(a \neq 0)$ have if **a)** $k < 0$, **b)** $k = 0$ **c)** $k > 0$?
12. How many solutions are there to the following equations or inequalities if $a \neq 0$ and $k > 0$?

    a) $|ax + b| = k$

    b) $|ax + b| < k$

    c) $|ax + b| > k$
13. Match each absolute value equation or inequality labeled **a)** through **e)** with the graph of its solution set.

    a) $|x| = 4$ A. −6 −5 −4 −3 −2 −1 0 1 2 3 4 5 6

    b) $|x| < 4$ B. −6 −5 −4 −3 −2 −1 0 1 2 3 4 5 6

    c) $|x| > 4$ C. −6 −5 −4 −3 −2 −1 0 1 2 3 4 5 6

    d) $|x| \geq 4$ D. −6 −5 −4 −3 −2 −1 0 1 2 3 4 5 6

    e) $|x| \leq 4$ E. −6 −5 −4 −3 −2 −1 0 1 2 3 4 5 6
14. Match each absolute value equation or inequality, labeled **a)** through **e)**, with its solution set.

    a) $|x| = 7$ A. $\{x \mid x \leq -7 \text{ or } x \geq 7\}$

    b) $|x| < 7$ B. $\{x \mid -7 < x < 7\}$

    c) $|x| > 7$ C. $\{x \mid -7 \leq x \leq 7\}$

    d) $|x| \leq 7$ D. $\{-7, 7\}$

    e) $|x| \geq 7$ E. $\{x \mid x < -7 \text{ or } x > 7\}$

Practice the Skills

Find the solution set for each equation.

15. $|x| = 5$

16. $|y| = 7$

17. $|x| = 12$

18. $|x| = 0$

19. $|x| = -2$

20. $|x + 1| = 5$

21. $|x + 5| = 7$

22. $|3 + y| = \frac{3}{5}$

23. $|2.4 + 0.4x| = 4$

24. $|3.8x - 28.5| = 0$

25. $|5 - 3x| = \frac{1}{2}$

26. $|3(y + 4)| = 12$

27. $\left|\frac{x - 3}{4}\right| = 5$

28. $\left|\frac{3z + 5}{6}\right| - 3 = 6$

29. $\left|\frac{x - 3}{4}\right| + 4 = 4$

30. $\left|\frac{5x - 3}{2}\right| + 2 = 6$

Find the solution set for each inequality.

31. $|y| \le 5$

32. $|x| \le 9$

33. $|x - 7| \le 9$

34. $|7 - x| < 5$

35. $|3z - 5| \le 5$

36. $|x - 3| - 2 < 3$

37. $|2x + 3| - 5 \le 10$

38. $|4 - 3x| - 4 < 11$

39. $|x - 0.4| \le 2.3$

40. $|2x - 3| < -4$

41. $|2x - 6| + 5 \le 2$

42. $\left|\frac{2x - 1}{3}\right| \le \frac{5}{3}$

43. $\left|5 + \frac{3x}{4}\right| < 8$

44. $\left|\frac{x - 3}{2}\right| - 4 \le -2$

45. $|4x - 1| \le 0$

46. $|2x + 3| < 0$

Find the solution set for each inequality.

47. $|x| > 3$

48. $|y| \ge 5$

49. $|x + 4| > 5$

50. $|3x + 1| > 4$

51. $|4 - 3y| \ge 8$

52. $\left|\frac{6 + 2z}{3}\right| > 2$

53. $\left|\frac{5 - 3w}{4}\right| \ge 10$

54. $|2x - 1| - 4 \ge 8$

55. $|0.1x - 0.4| + 0.4 > 0.6$

56. $\left|\frac{2x - 4}{3}\right| - 3 > -5$

57. $\left|\frac{x}{2} + 4\right| \ge 5$

58. $\left|4 - \frac{3x}{5}\right| \ge 9$

59. $|3x + 5| + 2 \ge 2$

60. $|3 - 2x| \ge 0$

61. $|4 - 2x| > 0$

62. $|-8y - 3| > 0$

Find the solution set for each equation.

63. $|2x + 1| = |4x - 9|$

64. $|x - 1| = |2x - 4|$

65. $|6x| = |3x - 9|$

66. $|4x - 2| = |4x - 2|$

67. $\left|\frac{3}{4}x - 2\right| = \left|\frac{1}{2}x + 5\right|$

68. $|3x - 5| = |3x + 5|$

69. $\left|\frac{1}{2}x + \frac{3}{5}\right| = \left|\frac{1}{2}x - 1\right|$

70. $\left|\frac{3}{2}r + 2\right| = \left|\frac{1}{2}r - 3\right|$

Find the solution set for each equation or inequality.

71. $|w| = 7$

72. $|z| \ge 2$

73. $|x - 3| < 5$

74. $|3x - 4| \le -6$

75. $|x + 5| > 9$

76. $|2x - 5| + 3 \le 10$

77. $|4x + 2| = 9$

78. $|2x - 4| + 2 = 10$

79. $|5 + 2x| > 0$

80. $|4 - x| = 5$

81. $|4 + 3x| \le 9$

82. $|2.4x + 4| + 4.9 > 1.9$

83. $|3x - 5| + 4 = 2$

84. $|4 - 2x| - 5 = 5$

85. $\left|\frac{w + 4}{3}\right| - 1 < 3$

86. $\left|\frac{3x + 4}{5}\right| > \frac{7}{5}$

87. $\left|\frac{3x - 2}{4}\right| + 5 \ge 5$

88. $\left|\frac{2x - 4}{5}\right| = 12$

89. $|2x - 8| = \left|\frac{1}{2}x + 3\right|$

90. $\left|\frac{1}{3}y + 3\right| = \left|\frac{2}{3}y - 1\right|$

91. $|2 - 3x| = \left|4 - \frac{5}{3}x\right|$

92. $\left|\frac{3 - 2x}{4}\right| \ge 5$

Problem Solving

93. A spring hanging from a ceiling is bouncing up and down so that its distance, d, above the ground satisfies the inequality $|d - 4| \leq \frac{1}{2}$ ft (see the figure).
a) Solve this inequality for d.
b) Between what distances, measured from the ground, will the spring oscillate?

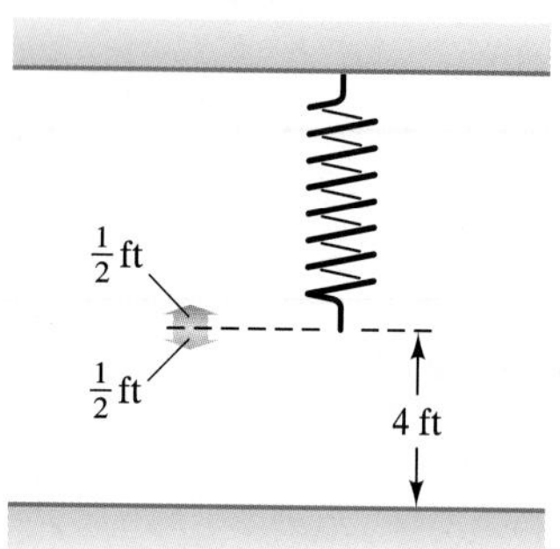

94. A submarine is 160 feet below sea level. It has rock formations above and below it, and should not change its depth by more than 28 feet. Its distance below sea level, d, can be described by the inequality $|d - 160| \leq 28$.
a) Solve this inequality for d.
b) Between what vertical distances, measured from sea level, may the submarine move?

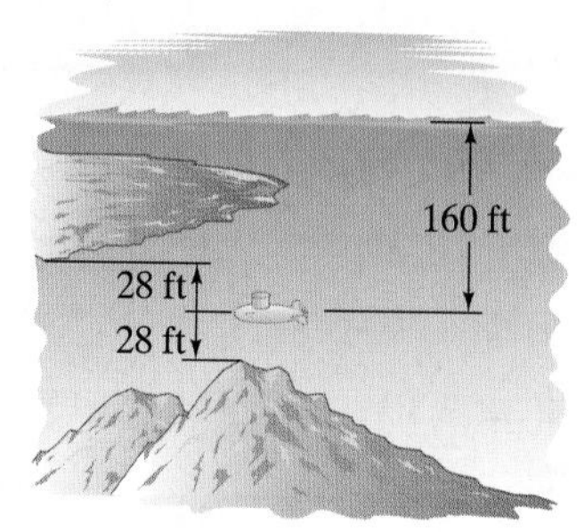

Determine an equation or inequality that has the given solution set.

95. $\{-5, 5\}$

96. $\{x|-5 < x < 5\}$

97. $\{x|x \leq -5 \text{ or } x \geq 5\}$

98. $\{x|-5 \leq x \leq 5\}$

99. For what value of x will the inequality $|ax + b| \leq 0$ be true? Explain.

100. For what value of x will the inequality $|ax + b| > 0$ *not* be true? Explain.

101. a) Explain how to find the solution to the equation $|ax + b| = c$. (Assume that $c > 0$ and $a \neq 0$.)
b) Solve this equation for x.

102. a) Explain how to find the solution to the inequality $|ax + b| < c$. (Assume that $a > 0$ and $c > 0$.)
b) Solve this inequality for x.

103. a) Explain how to find the solution to the inequality $|ax + b| > c$. (Assume that $a > 0$ and $c > 0$.)
b) Solve this inequality for x.

104. a) What is the first step in solving the inequality $-2|3x - 5| \leq -6$?
b) Solve this inequality and give the solution in interval notation.

Determine what values of x will make each equation true. Explain your answer.

105. $|x - 3| = |3 - x|$

106. $|x - 3| = -|x - 3|$

107. $|x| = x$

108. $|x + 2| = x + 2$

Solve. Explain how you determined your answer.

109. $|x + 1| = 2x - 1$

110. $|3x + 1| = x - 3$

111. $|x - 2| = -(x - 2)$

Challenge Problems

Solve by considering the possible signs for x.

112. $|x| + x = 6$

113. $x + |-x| = 6$

114. $|x| - x = 6$

115. $x - |x| = 6$

Group Activity

Discuss and answer Exercise 116 as a group.

116. Consider the equation $|x + y| = |y + x|$.
a) Have each group member select an x value and a y value and determine whether the equation holds. Repeat for two other pairs of x and y values.
b) As a group, determine for what values of x and y the equation is true. Explain your answer.
c) Now consider $|x - y| = -|y - x|$. Under what conditions will this equation be true?

Cumulative Review Exercises

Evaluate.

[1.4] **117.** $\frac{1}{3} + \frac{1}{4} \div \frac{2}{5}\left(\frac{1}{3}\right)^2$

118. $4(x + 3y) - 5xy$ when $x = 1$, $y = 3$

[2.4] **119.** Raul Sanchez swims across a lake averaging 2 miles an hour. Then he turns around and swims back across the lake, averaging 1.6 miles per hour. If his total swimming time was 1.5 hours, what is the width of the lake?

[2.5] **120.** Find the solution set to the inequality $3(x - 2) - 4(x - 3) > 2$.

SUMMARY

Key Words and Phrases

2.1
Coefficient (or numerical coefficient)
Conditional equation
Constant
Contradiction
Degree of a term
Equation
Equivalent equations
Identity
Least common denominator (LCD)
Least common multiple (LCM)
Like terms
Linear equations in one variable
Simplify an expression
Solution set
Solutions of an equation
Terms
Unlike terms

2.2
Formula
Mathematical model
Subscripts
Simple interest formula

2.3
Supplementary angles

2.4
Distance formula
Mixture problem
Motion problem

2.5
And; intersection
Compound inequality
Continued inequality
Inequality
Or; union
Order (or sense) of an inequality

IMPORTANT FACTS

Properties of Equality

Reflexive property: $a = a$

Symmetric property: If $a = b$, then $b = a$.

Transitive property: If $a = b$ and $b = c$, then $a = c$.

Addition property of equality: If $a = b$, then $a + c = b + c$.

Multiplication property of equality: If $a = b$, then $ac = bc$.

To Solve Linear Equations

1. Clear fractions.
2. Simplify each side separately.
3. Isolate the variable term on one side.
4. Solve for the variable.
5. Check.

Problem-Solving Procedure

1. Understand the problem.
2. Translate the problem into mathematical language.
3. Carry out the mathematical calculations necessary to solve the problem.
4. Check the answer obtained in step 3.
5. Answer the question.

continued on next page

Distance Formula	**Simple Interest Formula**
distance = rate · time or $d = rt$	interest = principal · rate · time or $i = prt$
Properties Used to Solve Inequalities	
1. If $a > b$, then $a + c > b + c$.	4. If $a > b$ and $c > 0$, then $\frac{a}{c} > \frac{b}{c}$.
2. If $a > b$, then $a - c > b - c$.	5. If $a > b$ and $c < 0$, then $ac < bc$.
3. If $a > b$ and $c > 0$, then $ac > bc$.	6. If $a > b$ and $c < 0$, then $\frac{a}{c} < \frac{b}{c}$.
Absolute Value, for $a > 0$	
If $\lvert x\rvert = a$, then $x = a$ or $x = -a$.	If $\lvert x\rvert > a$, then $x < -a$ or $x > a$.
If $\lvert x\rvert < a$, then $-a < x < a$.	If $\lvert x\rvert = \lvert y\rvert$, then $x = y$ or $x = -y$.

Review Exercises

[2.1] *State the degree of each term.*

1. $15x^4y^6$ **2.** $6x$ **3.** $-4xyz^5$

Simplify each expression. If an expression cannot be simplified, so state.

4. $x^2 + 3x + 6$ **5.** $x^2 + 2xy + 6x^2 - 4$

6. $3(x + 4) - 3x - 4$ **7.** $2[-(x - y) + 3x] - 5y + 6$

Solve each equation. If an equation has no solution, so state.

8. $\frac{x - 4}{5} = 9 - x$ **9.** $3(x + 2) - 6 = 4(x - 5)$ **10.** $3 + \frac{x}{2} = \frac{5}{6}$

11. $-6 - 2x = \frac{1}{2}(4x + 12) - 12$ **12.** $2\left(\frac{x}{2} - 4\right) = 3\left(x + \frac{1}{3}\right)$ **13.** $3x - 4 = 6x + 4 - 3x$

14. $2(x - 6) = 5 - \{2x - [4(x - 3) - 5]\}$

[2.2] *Evaluate each formula for the given values.*

15. $P = \frac{nRT}{V}$ when $n = 10, R = 100, T = 4, V = 20$

16. $x = \frac{-b + \sqrt{b^2 - 4ac}}{2a}$ when $a = 8, b = 10, c = -3$

17. $h = \frac{1}{2}at^2 + v_0t + h_0$ when $a = -32, v_0 = 60, h_0 = 120, t = 2$

18. $z = \frac{\bar{x} - \mu}{\frac{\sigma}{\sqrt{n}}}$ when $\bar{x} = 60, \mu = 64, \sigma = 5, n = 25$

Solve each equation for the indicated variable.

19. $A = lw$, for l **20.** $A = \pi r^2 h$, for h **21.** $P = 2l + 2w$, for w

22. $d = rt$, for r **23.** $y = mx + b$, for m **24.** $2x - 3y = 5$, for y

25. $P_1V_1 = P_2V_2$, for V_2 **26.** $S = \frac{3a + b}{2}$, for a **27.** $K = 2(d + l)$, for l

[2.3] *Write an equation that can be used to solve each problem. Solve the problem and check your answer.*

28. When the price of a jacket is decreased by 60%, it costs \$20. Find the original price of the jacket.

29. A small town's population is increasing by 350 people per year. If the present population is 4750, how long will it take for the population to reach 5800?

30. Dawn Clark's salary is \$300 per week plus 6% commission of sales. How much in sales must Dawn make to earn \$650 in a week?

31. The one-way bus fare for Sherod Kirby to get to work is \$1.65. A monthly bus pass that provides unlimited free bus travel during the month costs \$27.50. How many *round trips* (to and from work) would Sherod need to make to make the purchase of the bus pass worthwhile?

32. At a going-out-of-business sale, furniture is selling at 40% off the regular price. In addition, green-tagged items are reduced by an additional \$20. If Lalo Broyles purchased a green-tagged item and paid \$120, find the item's regular price.

[2.4] *Solve the following motion and mixture problems.*

33. Tanya Bowlin is a quality control inspector at the Eastman Kodak Company. In a typical 8-hour workday, she inspects 245 rolls of film. What is Tanya's hourly inspection rate?

34. The Sampsons invest \$10,000 in two accounts. One account pays 8% simple interest and the other account pays 5% simple interest. If the total interest for the year is \$680, how much money was invested in each account?

35. Two trains leave Portland, Oregon, at the same time traveling in opposite directions. One train travels at 60 miles per hour and the other at 90 miles per hour. In how many hours will they be 400 miles apart?

36. Space Shuttle 2 takes off 0.5 hour after Shuttle 1 takes off. If Shuttle 2 travels 300 miles per hour faster than Shuttle 1 and overtakes Shuttle 1 exactly 5 hours after Shuttle 2 takes off, find **a)** the speed of Shuttle 1; **b)** the distance from the launch pad when Shuttle 2 overtakes Shuttle 1.

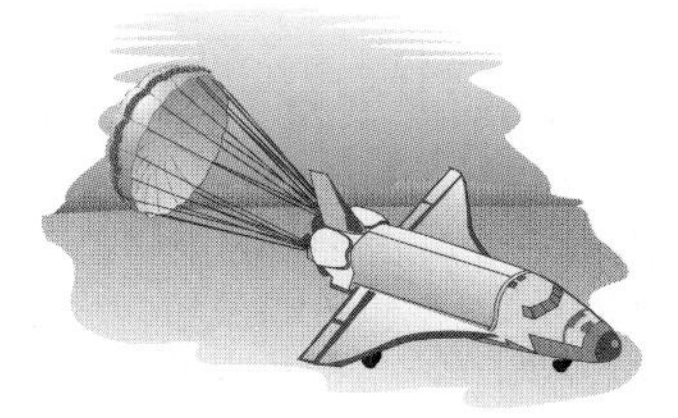

37. Tom Tomlins, the owner of a gourmet coffee shop, has two coffees, one selling for \$6.00 per pound and the other for \$6.80 per pound. How many pounds of each type of coffee should he mix to make 40 pounds of coffee to sell for \$6.50 per pound?

[2.3, 2.4] *Solve.*

38. A blouse has been reduced by 12%. The sale price is \$22. Find the original price.

39. Nicolle Ryba jogged for a distance and then turned around and walked back to her starting point. While jogging she averaged 7.2 miles per hour, and while walking she averaged 2.4 miles per hour. If the total time spent jogging and walking was 4 hours, find **a)** how long she jogged; **b)** the total distance she traveled.

40. Find the three angles of a triangle if one angle measures 25° greater than the smallest angle and the other angle measures 5° less than twice the smallest angle.

41. Two hoses are filling a swimming pool. The hose with the larger diameter supplies 1.5 times as much water as the hose with the smaller diameter. The larger hose is on for 2 hours before the smaller hose is turned on. If 5 hours after the larger hose is turned on there are 3150 gallons of water in the pool, find the rate of flow from each hose.

42. The sum of two consecutive integers is 49. Find the integers.

43. A clothier has two blue dye solutions, both made from the same concentrate. One solution is 6% blue dye and the other is 20% blue dye. How many ounces of the 20% solution must be mixed with 10 ounces of the 6% solution to result in the mixture being a 12% blue dye solution?

44. Ken Reysling invests \$12,000 in two savings accounts. One account is paying 10% simple interest and the other account is paying 6% simple interest. If the same interest is earned on each account how much was invested at each rate?

45. The West Ridge Fitness Center has two membership plans. The first plan is a flat \$40 per month fee plus \$1.00 per visit. The second plan is \$25 per month plus a \$4.00 per visit charge. How many visits would Erick Ruston have to make per month to make it advantageous for him to select the first plan?

46. Two trains leave Philadelphia at the same time along parallel tracks, traveling in opposite directions. The faster train travels 10 miles per hour faster than the slower train. Find the speed of the *faster* train if the trains are 510 miles apart after 3 hours.

[2.5] *Solve the inequality. Graph the solution on a real number line.*

47. $x - 3 \geq 4$

48. $2 - x \leq 5$

49. $2x + 4 > 9$

50. $16 \le 4x - 5$

51. $\frac{4x + 3}{5} > -3$

52. $2(x - 3) > 3x + 4$

53. $-4(x - 2) \le 6x + 4$

54. $\frac{x}{4} \ge 5 - 2x$

Write an inequality that can be used to solve each problem. Solve the inequality and answer the question.

55. A small airplane can carry a maximum load of 1525 pounds if it is to take off safely. If the passengers weigh 468 pounds, how many 80-pound boxes can be safely transported on the plane?

56. Jack Ruimpson, a telephone operator, informs a customer in a phone booth that the charge for calling Omaha, Nebraska, is \$4.50 for the first 3 minutes and 95 cents for each additional minute or any part thereof. How long can the customer talk if he has \$8.65?

57. A fitness center guarantees that customers will lose a minimum of 3 pounds the first week and $1\frac{1}{2}$ pounds each additional week. Find the maximum amount of time needed to lose 27 pounds.

Solve each inequality. Write the solution in interval notation.

58. $1 < x - 4 < 7$

59. $2 \le x + 5 < 8$

60. $3 < 2x - 4 < 8$

61. $-12 < 6 - 3x < -2$

62. $-1 \le \frac{2x - 3}{4} < 5$

63. $-8 < \frac{4 - 2x}{3} < 0$

64. Jekeila Ison's first four exam grades are 94, 73, 72, and 80. If a final average greater than or equal to 80 and less than 90 is needed to receive a final grade of B in the course, what range of grades on the fifth and last exam will result in Jekeila's receiving a B in the course? Assume a maximum grade of 100.

Find the solution set to each compound inequality.

65. $x < 3$ and $2x - 4 > -10$

66. $2x - 1 > 5$ or $3x - 2 \le 7$

67. $4x - 5 < 11$ and $-3x - 4 \ge 8$

68. $\frac{5x - 3}{2} > 7$ or $\frac{2x - 1}{3} \le -3$

[2.5, 2.6] *Find the solution set to each equation or inequality.*

69. $|x| = 4$

70. $|x| < 3$

71. $|x| \ge 4$

72. $|x - 4| = 9$

73. $|x - 2| \ge 5$

74. $|4 - 2x| = 5$

75. $|3 - 2x| < 7$

76. $\left|\frac{2x - 3}{5}\right| = 1$

77. $\left|\frac{x - 4}{3}\right| < 6$

78. $|3x - 4| = |x + 3|$

79. $|2x - 3| + 4 \ge -10$

Solve each inequality. Give the solution in interval notation.

80. $|x + 6| < -1$

81. $3 < 2x - 5 \le 9$

82. $-6 \le \frac{3 - 2x}{4} < 5$

83. $x \le 4$ and $4x - 6 \ge -14$

84. $x - 3 \le 4$ or $2x - 5 > 9$

85. $-10 < 3(x - 4) \le 12$

Practice Test

1. State the degree of the term $-6xy^2z^3$.

Simplify.

2. $2p - 3q + 2pq - 6p(q - 3) - 4p$

3. $4x - \{3 - [2(x - 2) - 5x]\}$

Solve each equation.

4. $3(x - 2) = 4(4 - x) + 5$
5. $\frac{3}{5} - \frac{x}{2} = 4$
6. $-3(x + 3) = 3\{[4 - (2x - 3)] - 4x\}$
7. $7x - 6(2x - 4) = 3 - (5x - 6)$
8. $-\frac{1}{2}(4x - 6) = \frac{1}{3}(3 - 6x) + 2$
9. Find the value of S_n for the given values.

$$S_n = \frac{a_1(1 - r^n)}{1 - r}, a_1 = 3, r = \frac{1}{3}, n = 3$$

10. Solve $c = \frac{a - 3b}{2}$ for b.
11. Solve $A = \frac{1}{2}h(b_1 + b_2)$ for b_2

Write an equation that can be used to solve each problem. Solve the equation and answer the question asked.

12. Find the cost of a set of golf clubs, before tax, if the cost of the clubs plus 7% tax is \$668.75.

13. The cost of renting an automobile is \$35 a day plus 15 cents a mile. How far can Valerie Catching drive in 1 day on \$65?
14. Two joggers start at the same point at the same time and jog in opposite directions. Homer Haines jogs at 4 miles per hour, while Frances Kitchen jogs at $5\frac{1}{4}$ miles per hour. How far apart will they be in $1\frac{1}{4}$ hours?
15. How many liters of 12% salt solution must be added to 10 liters of 25% salt solution to get a 20% salt solution?
16. June Davis has \$12,000 to invest. She places part of her money in a savings account paying 8% simple interest and the balance in a savings account paying 7% simple interest. If the total interest from the two accounts at the end of one year is \$910, find the amount placed in each account.

Solve each inequality and graph the solution on a number line.

17. $4(x - 2) < 3(x - 2) - 5$
18. $\frac{6 - 2x}{5} \geq -12$

Solve each inequality and write the solution in interval notation.

19. $x - 3 \leq 4$ and $2x - 4 > 5$
20. $-4 < \frac{x + 4}{2} < 8$

Find the solution set to the following equations.

21. $|x - 4| = 5$
22. $|2x - 3| = \left|\frac{1}{2}x - 10\right|$

Find the solution set to the following inequalities.

23. $|3x - 2| = 0$
24. $|2x - 3| + 1 > 6$
25. $\left|\frac{2x - 3}{4}\right| \leq \frac{1}{2}$

Cumulative Review Test

1. If $A = \{1, 4, 6, 7, 9, 12\}$ and $B = \{2, 3, 4, 5, 6, 9, 10, 12\}$, find
   a) $A \cup B$.
   b) $A \cap B$.
2. Name each indicated property.
   a) $4x + y = y + 4x$
   b) $(2x)y = 2(xy)$
   c) $2(x + 3) = 2x + 6$
3. Insert $<, >$, or $=$ in the shaded area to make the statement true: $-|-3| \;\square\; |-5|$.

Evaluate.

4. $4 - |-3| - (6 + |-3|)^2$
5. $-4^2 + (-6)^2 \div (2^3 - 2)^2$
6. $x^3 - xy + y^2$ when $x = -3$ and $y = -2$
7. $\frac{8 - \sqrt[3]{27} \cdot 3 \div 9}{|-5| - [5 - (12 \div 4)]^2}$

Simplify.

8. $(2x^4 y^3)^{-2}$

9. $\left(\dfrac{3m^2 n^{-4}}{m^{-3} n^2}\right)^2$

10. $r^{3m-2} \cdot r^{2m-6}$

11. Convert 40,600,000 to scientific notation.

12. Carbonated beverage sales in 1996 were about $\$1.12 \times 10^{10}$. Milk sales in 1996 were about $\$9.25 \times 10^9$. How much more were the sales of carbonated beverages than of milk? Express your answer in scientific notation and as a decimal number.

Solve.

13. $3x - 4 = -2(x - 3) - 9$

14. $1.2(x - 3) = 2.4x - 4.98$

15. $\dfrac{x}{4} - 5 = 3x - \dfrac{1}{3}$

16. $\dfrac{\frac{1}{4}x + 2}{3} = \dfrac{x - 4}{4}$

17. Explain the difference between a conditional linear equation, an identity, and an inconsistent linear equation. Give an example of each.

18. Evaluate the formula $x = \dfrac{-b + \sqrt{b^2 - 4ac}}{2a}$ for $a = 3$, $b = -8$, and $c = -3$.

19. Solve the formula $A = p + prt$ for t.

20. Solve the inequality and give the answer **a)** on a number line, **b)** as a solution set, and **c)** in interval notation.

$$-4 < \frac{5x - 2}{3} < 2$$

Find each solution set.

21. $|4z + 8| = 12$

22. $|2x - 4| - 6 \geq 18$

23. The Computer Tutor has reduced the price of a computer by 20%. Find the original price of the computer if the sale price is $1800.

24. Two cars leave Caldwell, New Jersey, at the same time traveling in opposite directions. The car traveling north is moving 10 miles per hour faster than the car traveling south. If the two cars are 270 miles apart after 3 hours, find the speed of each car.

25. Kimberly Kane has a 20% saltwater solution and a 50% saltwater solution. How much of each solution should she mix to get 2 liters of a 30% saltwater solution?

GRAPHS AND FUNCTIONS

CHAPTER 3

Use the Angel Web site at www.prenhall.com/angel to be linked to an internet resource that will help you further explore the following application.

Have you ever dreamed of starting your own business? Before starting it, you should write a business plan which includes, among other things, a projection of the profits of your business. The gross annual profit can be estimated by subtracting your annual expenses from your annual income. On page 172, we project the annual profit of a tire store as a function of the number of tires sold.

Preview and Perspective

Two of the primary goals of this book are to provide you with a good understanding of graphing and of functions. Graphing is heavily used in this course and in other mathematics courses. To reinforce your knowledge of this topic we introduce graphing early and discuss it frequently throughout the book. Many of the exercise sets have graphs taken from newspapers or magazines. The material presented in this chapter may help you understand them better.

In Section 3.2 we introduce the concept of *function.* Functions are a unifying concept used throughout all of mathematics. We also use functions throughout the book to reinforce and expand upon what you learn in this chapter.

Most of you have graphed linear equations before. However, you probably have not graphed the nonlinear equations presented in Section 3.1. Make sure you read this section carefully. We will be using the technique presented in this section when we graph other nonlinear equations later in the book.

In Section 3.7 we graph linear inequalities *in two variables*. We will use the procedures for graphing linear inequalities in two variables again when we graph systems of linear inequalities in Section 4.6.

3.1 GRAPHS

SSM VIDEO 3.1 CD Rom

1) **Plot points in the Cartesian coordinate system.**
2) **Draw graphs by plotting points.**
3) **Graph nonlinear equations.**
4) **Use a graphing calculator.**
5) **Interpret graphs.**

1) Plot Points in the Cartesian Coordinate System

René Descartes

In this chapter we emphasize graphs and functions. Many algebraic relationships are easier to understand if we can see a visual picture of them. A graph is a picture that shows the relationship between two or more variables in an equation. Before learning how to construct a graph, you must know the Cartesian coordinate system.

The **Cartesian** (or **rectangular**) **coordinate system**, named after the French mathematician and philosopher René Descartes (1596–1650), consists of two axes (or number lines) in a plane drawn perpendicular to each other (Fig. 3.1). Note how the two axes yield four **quadrants**, labeled with capital Roman numerals I, II, III, and IV.

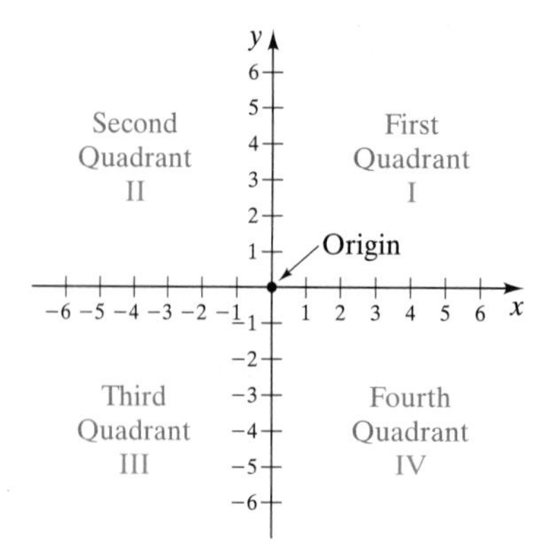

FIGURE 3.1

The horizontal axis is called the ***x*-axis**. The vertical axis is called the ***y*-axis**. The point of intersection of the two axes is called the **origin**. Starting from the origin and moving to the right, the numbers increase; moving to the left, the numbers decrease. Starting from the origin and moving up, the numbers increase; moving down, the numbers decrease.

An **ordered pair** (x, y) is used to give the two **coordinates** of a point. If, for example, the x-coordinate of a point is 2 and the y-coordinate is 3, the ordered pair representing that point is $(2, 3)$. The x-coordinate is always the first coordinate listed in the ordered pair. To plot a point, find the x-coordinate on the x-axis and the y-coordinate on the y-axis. Then suppose there was an imaginary vertical line from the x-coordinate and an imaginary horizontal line from the y-coordinate. The point is placed where the two imaginary lines intersect.

For example, the point corresponding to the ordered pair $(2, 3)$ is plotted in Figure 3.2. The phrase "the point corresponding to the ordered pair $(2, 3)$" is often abbreviated "the point $(2, 3)$." For example, if we write "the point $(-1, 5)$," it means the point corresponding to the ordered pair $(-1, 5)$. The ordered pairs A at $(-2, 3)$, B at $(0, 2)$, C at $(4, -1)$, and D at $(-4, 0)$ are plotted in Figure 3.3.

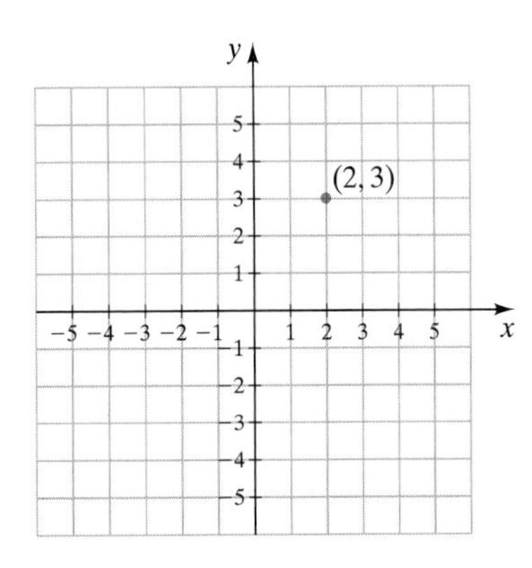

FIGURE 3.2

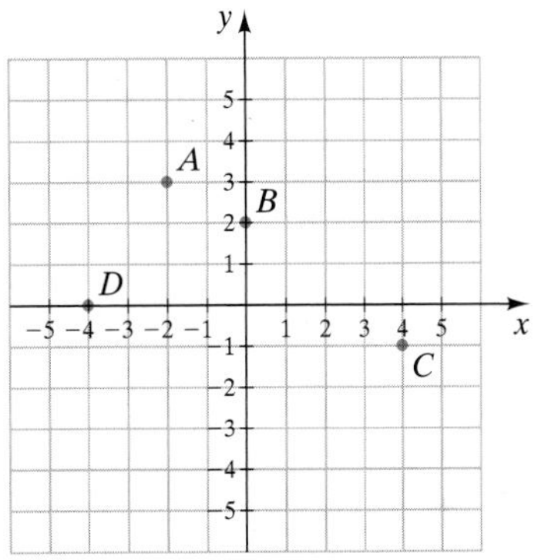

FIGURE 3.3

EXAMPLE 1 Plot the following points on the same axes.

a) $A(4, 2)$ **b)** $B(0, -3)$ **c)** $C(-3, 1)$ **d)** $D(4, 0)$

Solution See Figure 3.4. Notice that when the x-coordinate is 0, as in part **b)**, the point is on the y-axis. When the y-coordinate is 0, as in part **d)**, the point is on the x-axis.

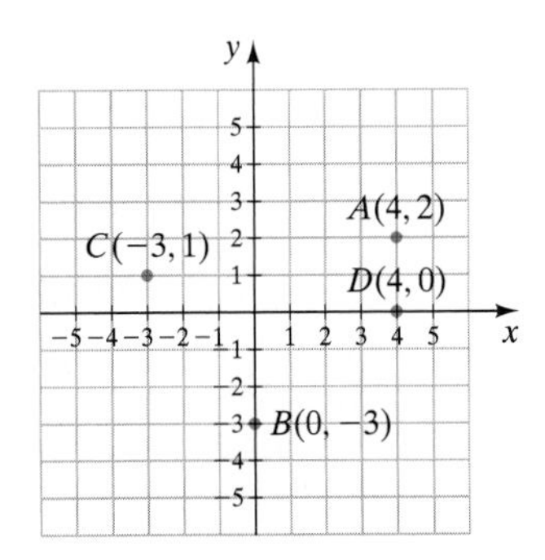

FIGURE 3.4

NOW TRY EXERCISE 7

2 Draw Graphs by Plotting Points

In Chapter 2, we solved equations that contained one variable. Now we will discuss equations that contain two variables. If an equation contains two variables, its solutions are pairs of numbers. *When a solution to an equation is listed as an ordered pair, the first number listed in the pair replaces the variable that occurs first alphabetically.*

EXAMPLE 2 Determine whether the following ordered pairs are solutions of the equation $y = 4x - 2$.

a) $(2, 6)$ **b)** $\left(\frac{1}{2}, 0\right)$ **c)** $(-1, -6)$ **d)** $(3, 12)$

Solution We substitute the first number in the ordered pair for x, and the second number for y. If the substitutions result in a true statement, the ordered pair is a solution. If the substitutions result in a false statement, the ordered pair is not a solution.

a) $y = 4x - 2$

$6 \stackrel{?}{=} 4(2) - 2$

$6 \stackrel{?}{=} 8 - 2$

$6 = 6$

True

b) $y = 4x - 2$

$0 \stackrel{?}{=} 4\left(\frac{1}{2}\right) - 2$

$0 \stackrel{?}{=} 2 - 2$

$0 = 0$

True

c) $y = 4x - 2$

$-6 \stackrel{?}{=} 4(-1) - 2$

$-6 \stackrel{?}{=} -4 - 2$

$-6 = -6$

True

d) $y = 4x - 2$

$12 \stackrel{?}{=} 4(3) - 2$

$12 \stackrel{?}{=} 12 - 2$

$12 = 10$

False

Thus, the ordered pairs $(2, 6)$, $\left(\frac{1}{2}, 0\right)$ and $(-1, -6)$ are solutions to the equation $y = 4x - 2$. The ordered pair $(3, 12)$ is not a solution.

There are many other solutions to the equation in Example 2. In fact, there are an infinite number of solutions. One method that may be used to find solutions to an equation like $y = 4x - 2$ is to substitute values for x and find the corresponding values of y. For example, to find the solution to the equation $y = 4x - 2$ when $x = 0$ we substitute 0 for x and solve for y.

$$y = 4x - 2$$
$$y = 4(0) - 2$$
$$y = 0 - 2$$
$$y = -2$$

Thus, another solution to the equation is $(0, -2)$.

A **graph** is an illustration of the set of points whose coordinates satisfy the equation. Sometimes when drawing a graph, we list some points that satisfy the equation in a table and then plot those points. We then draw a line through the points to obtain the graph. Below is a table of some points that satisfy the equation $y = 4x - 2$. The graph is drawn in Figure 3.5. Note that the equation $y = 4x - 2$ contains an infinite number of solutions and that the line continues indefinitely in both directions (as indicated by the arrows).

x	y	(x, y)
-1	-6	$(-1, -6)$
0	-2	$(0, -2)$
$\frac{1}{2}$	0	$\left(\frac{1}{2}, 0\right)$
2	6	$(2, 6)$

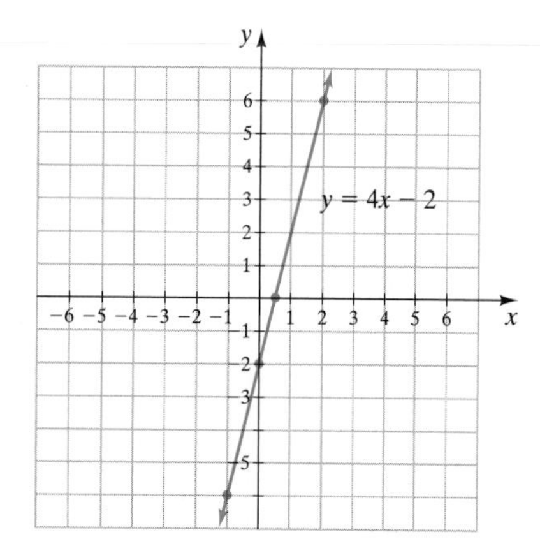

FIGURE 3.5

In Figure 3.5, the four points are in a straight line. Points that are in a straight line are said to be **collinear**. The graph is said to be **linear** because it is a straight line. Any equation whose graph is a straight line is called a **linear equation**. The equation $y = 4x - 2$ is an example of a linear equation. Linear equations are also called **first degree equations** since the greatest exponent that appears on any variable is one. In Examples 3 and 4, we graph linear equations.

EXAMPLE 3 Graph $y = x$.

Solution We first find some ordered pairs that are solutions by selecting values of x and finding the corresponding values of y. We will select 0, some positive values, and some negative values for x. We will also choose numbers close to 0, so that the ordered pairs will fit on the axes, The graph is illustrated in Figure 3.6.

x	y	(x, y)
-2	-2	$(-2, -2)$
-1	-1	$(-1, -1)$
0	0	$(0, 0)$
1	1	$(1, 1)$
2	2	$(2, 2)$

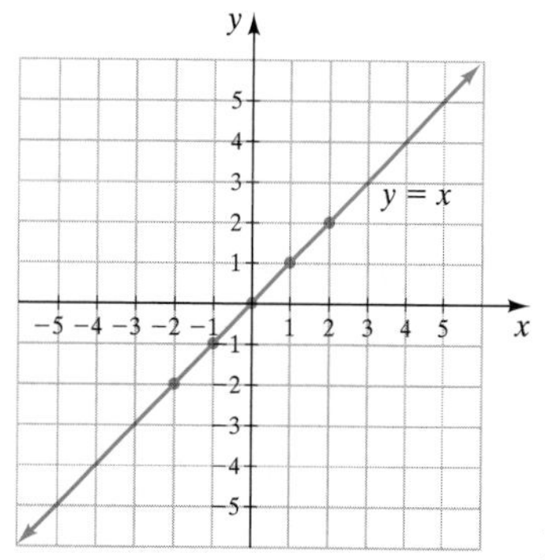

FIGURE 3.6

1. Select values for x.
2. Compute y.
3. Ordered pairs.
4. Plot the points and draw the graph.

EXAMPLE 4 Graph $y = -\frac{1}{2}x + 2$.

Solution We will select some values for x, find the corresponding values of y, and then draw the graph. When we select values for x, we will select some positive values, some negative values, and 0. The graph is illustrated in Figure 3.7. (To conserve space, we will not always list a column in the table for ordered pairs.)

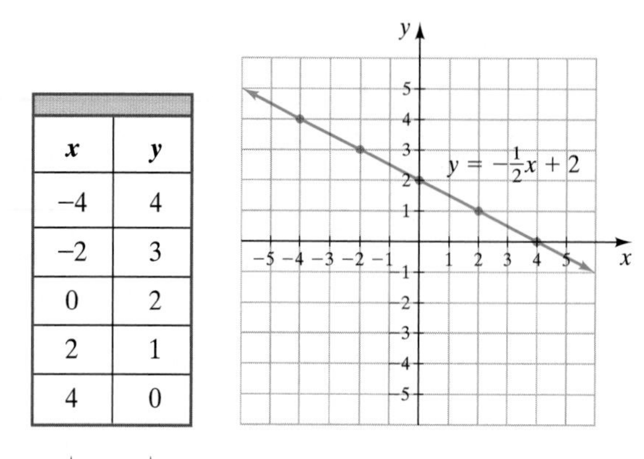

x	y
−4	4
−2	3
0	2
2	1
4	0

FIGURE 3.7

1. Select values for x.
2. Compute y.
3. Plot the points and draw the graph.

NOW TRY EXERCISE 35

Notice in Example 4 that we selected values of x that were multiples of 2 so that we would not have to work with fractions.

If we are asked to graph an equation not solved for y, such as $x + 2y = 4$, our first step will be to solve the equation for y. For example, if we solve $x + 2y = 4$ for y using the procedure discussed in Section 2.2, we obtain

$$x + 2y = 4$$

$$2y = -x + 4 \quad \text{Subtract } x \text{ from both sides.}$$

$$y = \frac{-x + 4}{2} \quad \text{Divide both sides by 2.}$$

or

$$y = -\frac{x}{2} + \frac{4}{2} = -\frac{1}{2}x + 2$$

The resulting equation, $y = -\frac{1}{2}x + 2$ is the same equation we graphed in Example 4. Therefore, the graph of $x + 2y = 4$ is also illustrated in Figure 3.7.

3) Graph Nonlinear Equations

There are many equations whose graphs are not straight lines. Such equations are called **nonlinear equations**. To graph nonlinear equations by plotting points, we follow the same procedure used to graph linear equations. However, since the graphs are not straight lines we may need to plot more points to draw the graph.

EXAMPLE 5 Graph $y = x^2 - 4$.

Solution We select some values for x and find the corresponding values of y. Then we plot the points and connect them with a smooth curve. When we substitute values for x and evaluate the right side of the equation, we follow the order of operations discussed in Section 1.4. For example, if $x = -3$, then $y = (-3)^2 - 4 = 9 - 4 = 5$. The graph is shown in Figure 3.8.

x	y
-3	5
-2	0
-1	-3
0	-4
1	-3
2	0
3	5

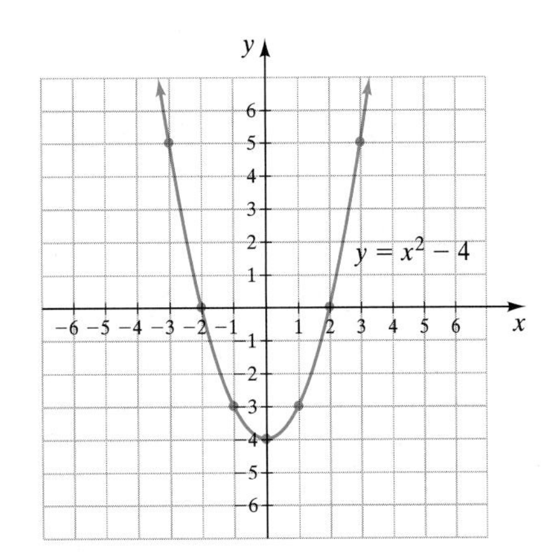

FIGURE 3.8

If we substituted 4 for x, y would equal 12. When $x = 5$, $y = 21$. Notice that this graph rises steeply as x moves away from the origin.

NOW TRY EXERCISE 41

EXAMPLE 6 Graph $y = \dfrac{1}{x}$.

Solution We begin by selecting values for x and finding the corresponding values of y. We then plot the points and draw the graph. Notice that if we substitute 0 for x, we obtain $y = \frac{1}{0}$. Since $\frac{1}{0}$ is undefined, we cannot use 0 as a first coordinate. Instead, we will just select some negative values for x and then some positive values for x. Note for example, that when $x = -\frac{1}{2}$, $y = \frac{1}{-1/2} = -2$. This graph has two branches, one to the left of the y-axis and one to the right of the y-axis, as shown in Figure 3.9.

x	y
-3	$-\frac{1}{3}$
-2	$-\frac{1}{2}$
-1	-1
$-\frac{1}{2}$	-2
$\frac{1}{2}$	2
1	1
2	$\frac{1}{2}$
3	$\frac{1}{3}$

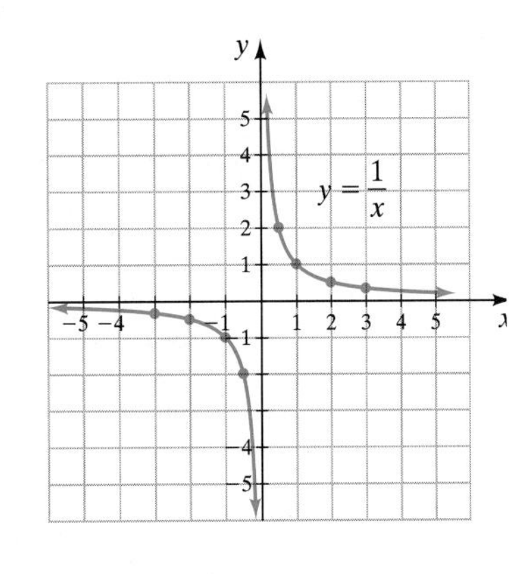

FIGURE 3.9

In the graph for Example 6, notice that for values of x far to the right of 0, or far to the left of 0, the graph approaches the x-axis but does not touch it. For example when $x = 1000$, $y = 0.001$ and when $x = -1000$, $y = -0.001$. Can you explain why y can never have a value of 0?

NOW TRY EXERCISE 51

EXAMPLE 7 Graph $y = |x|$.

Solution Recall that $|x|$ is read "the absolute value of x." Absolute values were discussed in Section 1.3. To graph this absolute value equation, we select some values for x and find the corresponding values of y. For example, if $x = -4$, then $y = |-4| = 4$. Then we plot the points and draw the graph.

Notice that this graph is V-shaped, as shown in Figure 3.10.

x	y
−4	4
−3	3
−2	2
−1	1
0	0
1	1
2	2
3	3
4	4

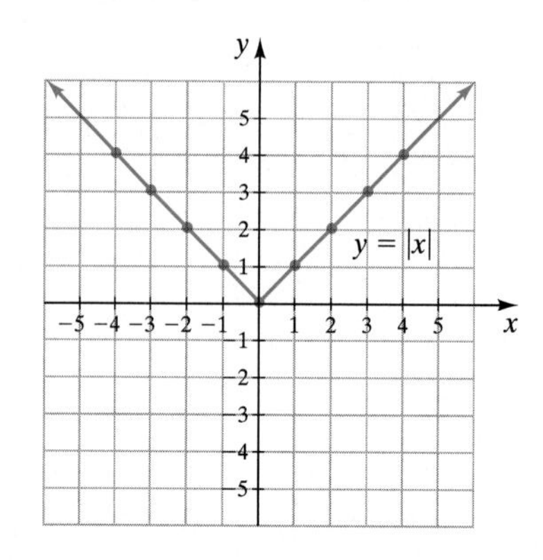

FIGURE 3.10

Avoiding Common Errors

When graphing nonlinear equations, many students do not plot enough points to get a true picture of the graph. For example, when graphing $y = \frac{1}{x}$ many students consider only integer values of x. Following is a table of values for the equation and two graphs that contain the points indicated in the table.

x	−3	−2	−1	1	2	3
y	$-\frac{1}{3}$	$-\frac{1}{2}$	−1	1	$\frac{1}{2}$	$\frac{1}{3}$

Correct

FIGURE 3.11

Incorrect

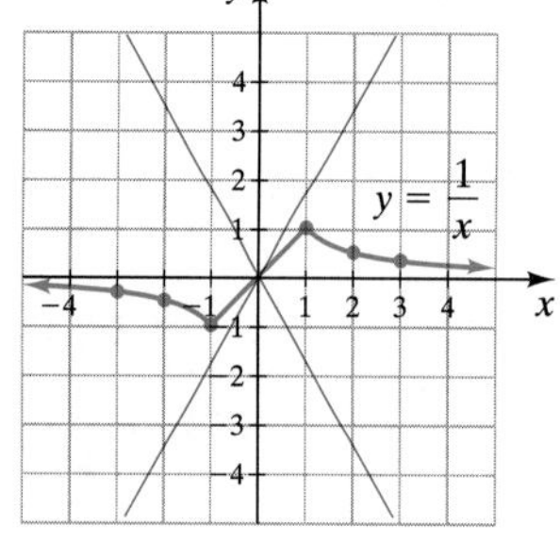

FIGURE 3.12

continues on next page

> If you select and plot fractional values of x near 0, as was done in Example 6, you get the graph in Figure 3.11. The graph in Figure 3.12 cannot be correct because the equation is not defined when x is 0 and therefore the graph cannot cross the y-axis. Whenever you plot a graph that contains a variable in the denominator, select values for the variable that are very close to the value that makes the denominator 0 and observe what happens. For example, when graphing $y = \frac{1}{x - 3}$, you should use values of x close to 3, such as 2.9 and 3.1 or 2.99 and 3.01, and see what values you obtain for y.
>
> Also, when graphing nonlinear equations it is a good idea to consider both positive and negative values. For example, if you used only positive values of x when graphing $y = |x|$, the graph would appear to be a straight line going through the origin, instead of the V-shaped graph shown in Figure 3.10.

4) Use a Graphing Calculator

If an equation is complex, finding ordered pairs can be time consuming. In this section we present a general procedure that can be used to graph equations using a **graphing calculator**.

A primary use of a graphing calculator is to graph equations. A graphing calculator window is the rectangular screen in which a graph is displayed. Figure 3.13 shows a TI-83 calculator window with some information illustrated; Figure 3.14 shows the meaning of the information given in Figure 3.13.

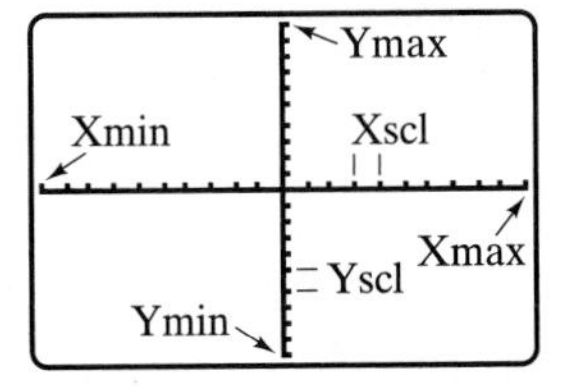

FIGURE 3.13

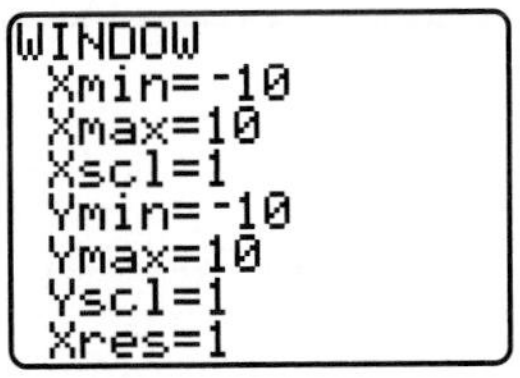

```
WINDOW
 Xmin=-10
 Xmax=10
 Xscl=1
 Ymin=-10
 Ymax=10
 Yscl=1
 Xres=1
```

FIGURE 3.14

The x-axis on the *standard calculator screen* goes from -10 (the minimum value of x, Xmin) to 10 (the maximum value of x, Xmax) with a scale of 1. Therefore each tick mark represents 1 unit $(\text{Xscl} = 1)$. The y axis goes from -10 (the minimum value of y, Ymin) to 10 (the maximum value of y, Ymax) with a scale of 1 $(\text{Yscl} = 1)$.

Since the window is rectangular, the distances between tick marks on the standard window are greater on the horizontal axis than on the vertical axis.

When graphing you will often need to change these window values. Read your graphing calculator manual to learn how to change the window setting. On the TI-82 and TI-83, you press the WINDOW key and then change the settings.

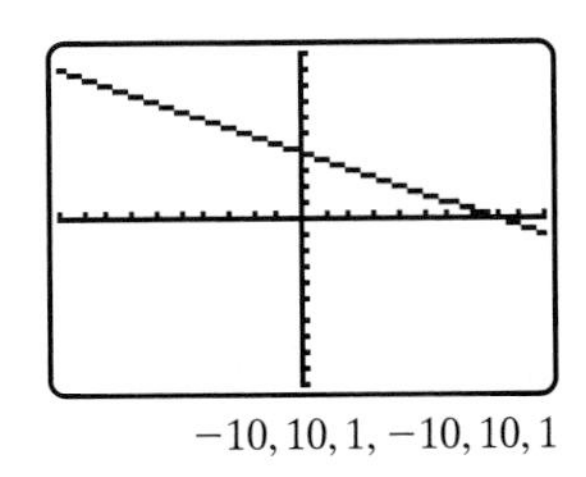

$-10, 10, 1, -10, 10, 1$

FIGURE 3.15

Since the grapher does not display the x- and y-values in the window, we will occasionally list a set of values below the screen. Figure 3.15 shows a TI calculator window with the equation $y = -\frac{1}{2}x + 4$ graphed. Below the window we show six numbers, which represent in order: Xmin, Xmax, Xscl, Ymin, Ymax, and Yscl. The Xscl and Yscl represent the scale on the x- and y-axes, respectively. When we are showing the standard calculator window, we will generally not show these values below the window.

To graph the equation $y = -\frac{1}{2}x + 4$ on a TI-82 or TI-83, you would press

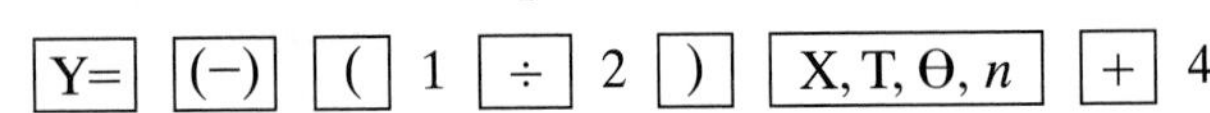

Then when you press GRAPH, the equation is graphed. The X, T, Θ, n key can be used to enter any of the symbols shown on the key. In this book this key will always be used to enter the variable x.

Most graphing calculators offer a **TRACE feature** that allows you to investigate individual points after a graph is displayed. Often the TRACE key is pressed to access this feature. After pressing the TRACE key you can move the flashing cursor along the line by pressing the arrow keys. As the flashing cursor moves along the line, the values of x and y change to correspond with the position of the cursor. Figure 3.16 shows the graph in Figure 3.15 after the TRACE key has been pressed and the right arrow key has been pressed a few times.

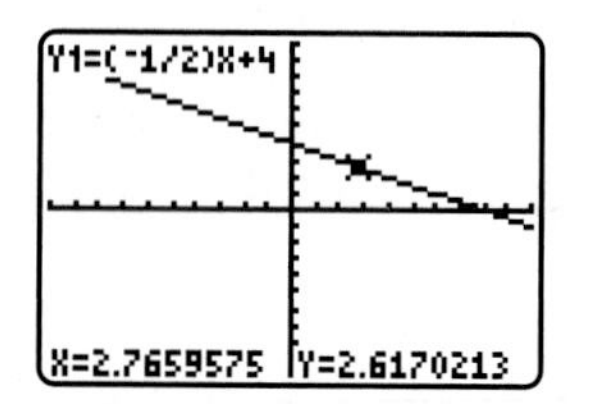

FIGURE 3.16

Many graphing calculators also provide a **TABLE feature**, which will illustrate a table of ordered pairs for any equation entered. On the TI-83, since TABLE appears above the GRAPH key, to obtain a table you press 2nd GRAPH. Figure 3.17 shows a table of values for the equation $y = -\frac{1}{2}x + 4$. You can scroll up and down the table by using the arrow keys.

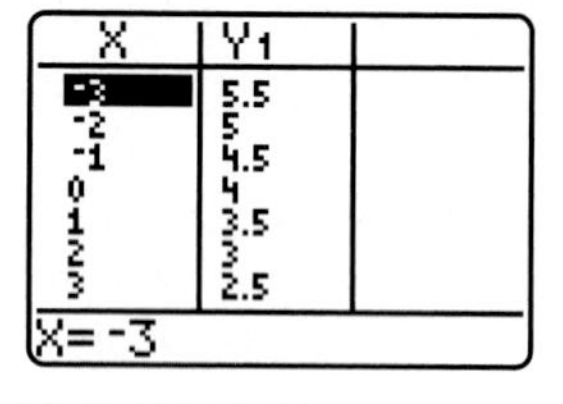

X	Y1
-3	5.5
-2	5
-1	4.5
0	4
1	3.5
2	3
3	2.5

X=-3

FIGURE 3.17

Using TBLSET (for Table setup), you can control the x-values that appear in the table. For example, if you want the table to show values of x in tenths, you could do this using TBLSET.

This section is just a brief introduction to graphing equations, the TRACE feature, and the TABLE feature. You should read your graphing calculator manual to learn how to fully utilize these features.

NOW TRY EXERCISE 85

5 Interpret Graphs

We see many different types of graphs daily in newspapers, in magazines, on television, and so on. Throughout this book, we present a variety of graphs. Since being able to draw and interpret graphs is very important, we will study this further in Section 3.2. In Example 8 you must understand and interpret graphs in order to answer the question.

EXAMPLE 8 When Jim Herring went to see his mother in Cincinnati, he boarded a Southwest Airlines plane. The plane sat at the gate for 20 minutes, taxied to the runway, and then took off. The plane flew at about 600 miles per hour for about 2 hours. It then reduced its speed to about 300 miles per hour and circled the Cincinnati Airport for about 15 minutes before it came in for a landing. After landing, the plane taxied to the gate and stopped. Which graph in Figure 3.18a–3.18d best illustrates this situation?

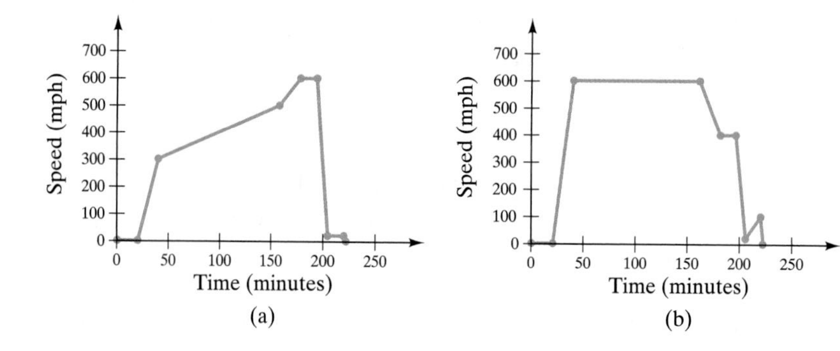

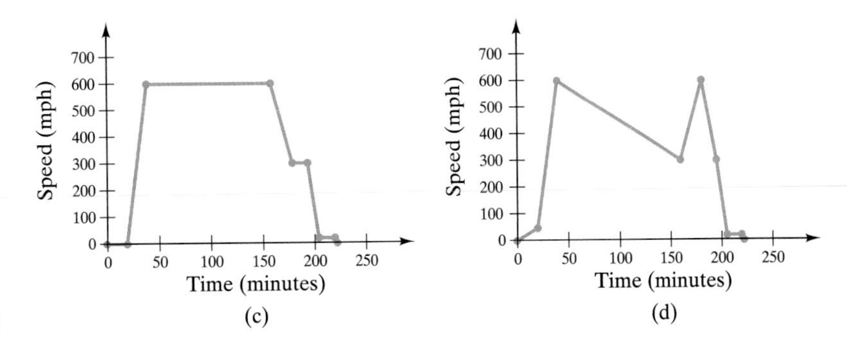

FIGURE 3.18

Solution The graph that depicts the situation described is (c), reproduced with annotations in Figure 3.19. The graph shows speed versus time, with time on the horizontal axis. While the plane sat at the gate for 20 minutes its speed was 0 miles per hour (the horizontal line at 0 from 0 to 20 minutes). After 20 minutes the plane took off, and its speed increased to 600 miles per hour (the near-vertical line going from 0 to 600 mph). The plane then flew at about 600 miles per hour for 2 hours (the horizontal line at about 600 mph). It then slowed down to about 300 miles per hour (the near-vertical line from 600 mph to 300 mph). Next the plane circled at about 300 miles per hour for about 15 minutes (the horizontal line at about 300 mph). The plane then came in for a landing (the near-vertical line from about 300 mph to about 20 mph). It then taxied to the gate (the horizontal line at about 20 mph). Finally it stopped at the gate (the near-vertical line when the speed dropped to 0 mph).

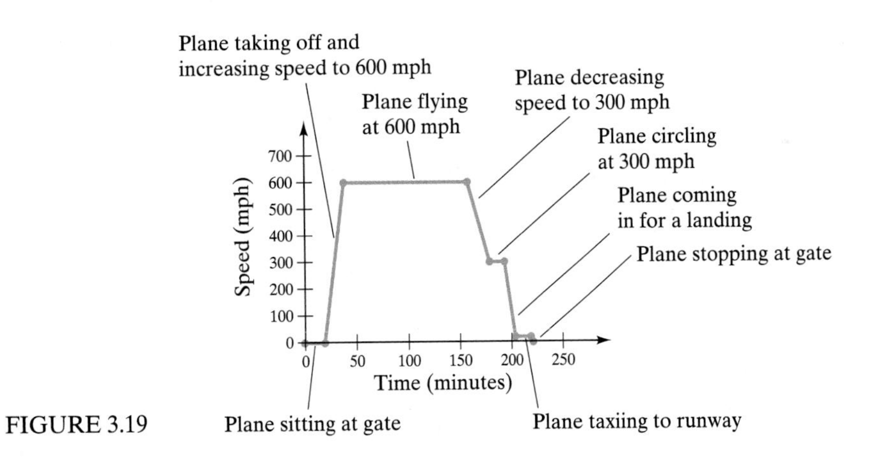

FIGURE 3.19

NOW TRY EXERCISE 77

Exercise Set 3.1

Concept/Writing Exercises

1. a) What does the graph of any linear equation look like?
   b) How many points are needed to graph a linear equation? Explain.
2. What does it mean when a set of points is collinear?
3. When graphing the equation $y = \frac{1}{x}$, what value cannot substituted for x? Explain.
4. What is another name for the Cartesian coordinate system?

Practice the Skills

List the ordered pairs corresponding to the indicated points.

5.

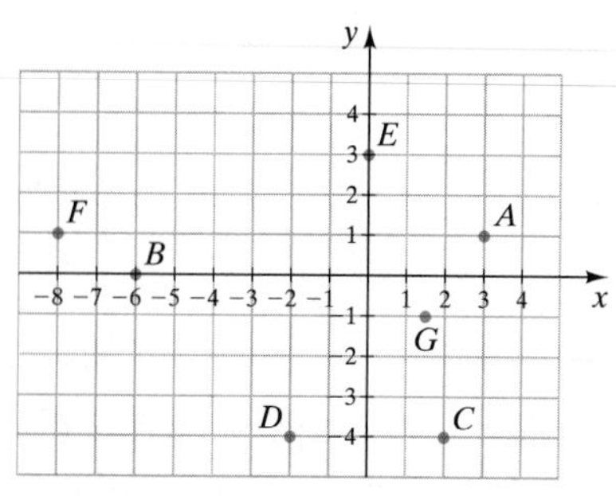

6.

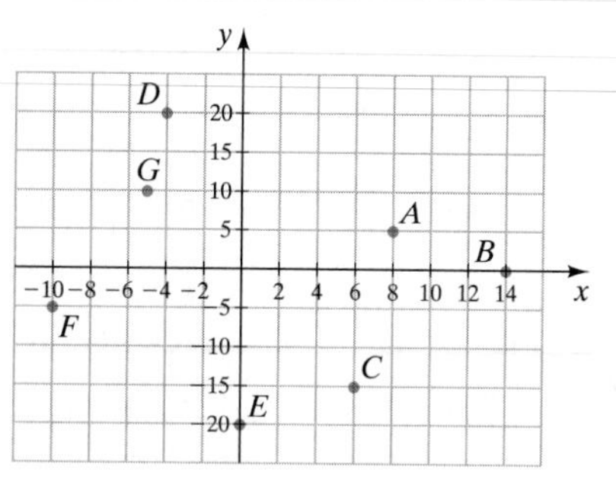

7. Graph the following points on the same axes.
$A(4, 2)$ $B(-6, 2)$ $C(0, -1)$ $D(-2, 0)$

8. Graph the following points on the same axes.
$A(-4, -2)$ $B(3, 2)$ $C(2, -3)$ $D(-3, 3)$

Determine the quadrant in which each point is located.

9. $(1, 6)$
10. $(-2, 3)$
11. $(5, -9)$
12. $(24, 116)$
13. $(-35, 18)$
14. $(-24, -8)$
15. $(-6, -19)$
16. $(8, -120)$

Determine whether the given ordered pair is a solution to the given equation.

17. $(2, 21)$; $y = 6x + 9$
18. $(4, 1)$; $3r + 2s = 9$
19. $(-3, 1)$; $y = |x| - 2$
20. $(-1, 8)$; $y = x^2 - 3x + 4$
21. $(-2, 5)$; $s = 2r^2 - r - 5$
22. $\left(\frac{1}{2}, \frac{5}{2}\right)$; $y = |x - 3|$
23. $(2, 0)$; $2x^2 + y = 8$
24. $(-5, 6)$; $n = 2|m| - 4$
25. $\left(\frac{1}{2}, \frac{3}{2}\right)$; $2x^2 + 4x - y = 0$
26. $\left(-3, \frac{11}{2}\right)$; $2n^2 + 3m = 2$

Graph each equation.

27. $y = x$
28. $y = 2x$
29. $y = \frac{1}{2}x$
30. $y = \frac{1}{3}x$
31. $y = 2x + 4$
32. $y = 4x - 3$
33. $y = -3x - 5$
34. $y = -2x + 2$
35. $y = \frac{1}{2}x + 2$
36. $y = \frac{1}{3}x - 1$
37. $y = -\frac{1}{2}x - 3$
38. $y = -\frac{1}{3}x + 4$
39. $y = x^2$
40. $y = x^2 + 1$
41. $y = -x^2$
42. $y = -x^2 + 1$
43. $y = |x| + 1$
44. $y = |x| - 3$
45. $y = -|x|$
46. $y = |x| - 2$
47. $y = x^3$
48. $y = -x^3$
49. $y = x^3 + 1$
50. $y = \frac{2}{x}$
51. $y = -\frac{1}{x}$
52. $x^2 = 1 - y$
53. $x = |y|$
54. $x = y^2$

In Exercises 55–62, use a calculator to obtain at least eight points that are solutions to the equation. Then graph the equation by plotting the points.

55. $y = x^3 - 2x^2 - 4x$
56. $y = -x^3 + 2x^2 + 4$
57. $y = \frac{1}{x - 1}$
58. $y = \frac{5}{x - 2}$
59. $y = \sqrt{x}$
60. $y = \sqrt{x + 4}$
61. $y = \frac{1}{x^2}$
62. $y = \frac{|x^2|}{2}$

Problem Solving

63. Is the point represented by the ordered pair $\left(\frac{1}{2}, -\frac{2}{23}\right)$ on the graph of the equation $y = \frac{x}{x^2 - 6}$? Explain.

64. Is the point represented by the ordered pair $\left(\frac{1}{2}, \frac{3}{2}\right)$ on the graph of the equation $y = \frac{x^2 - 4}{x - 2}$? Explain.

65. **a)** Plot the points $A(2, 7)$, $B(2, 3)$, and $C(6, 3)$, and then draw $\overline{AB}$, $\overline{AC}$, and $\overline{BC}$. ($\overline{AB}$ represents the line segment from A to B.)
    b) Find the area of the figure.

66. **a)** Plot the points $A(-4, 5)$, $B(2, 5)$, $C(2, -3)$, and $D(-4, -3)$, and then draw $\overline{AB}$, $\overline{BC}$, $\overline{CD}$, and $\overline{DA}$.
    b) Find the area of the figure.

67. The following graph shows retail sales up to 1998 and projected retail sales up to 2000 for frozen yogurt, low/nonfat ice cream, and full-fat ice cream.
    a) Estimate the sales in 1999 of frozen yogurt, low/nonfat ice cream, and full-fat ice cream.

Ice Cream Retail Sales

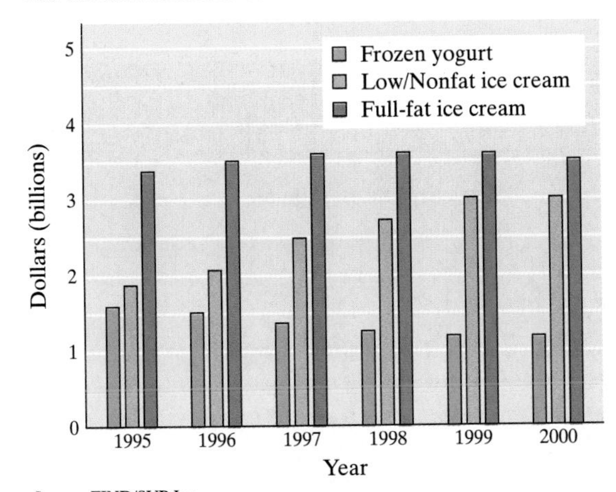

Source: FIND/SVP, Inc.

    b) Estimate the total sales of the three items listed in 1999.
    c) In which years were the sales of low/nonfat ice cream greater than \$2.5 billion?
    d) Does the decrease in the sales of frozen yogurt from 1995 through 2000 appear to be approximately linear? Explain.

68. The following graph shows unemployment rates in Washington, D.C, Maryland, and Virginia from 1990 through 1996.

Unemployment Rates

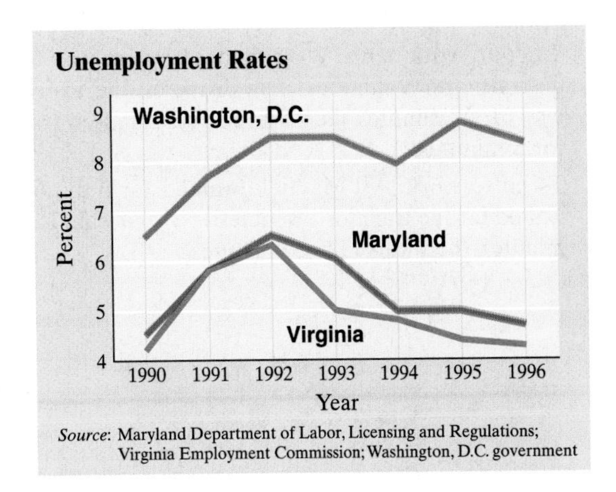

Source: Maryland Department of Labor, Licensing and Regulations; Virginia Employment Commission; Washington, D.C. government

    a) In which year(s) was the unemployment rate in Maryland at 5%?
    b) When was the unemployment rate in Virginia at its maximum value? What was that value?
    c) In which years was the unemployment rate in Washington greater than 8%?
    d) Between which years did the unemployment rate decrease in Virginia while the unemployment rate increased in Washington?

We will discuss many of the concepts introduced in these exercises in Section 3.4.

69. Graph $y = x + 1$, $y = x + 3$, and $y = x - 1$ on the same axes.
    a) What do you notice about the equations and the values where the graphs intersect the y-axis?
    b) Do all equations seem to have the same slant (or slope)?

70. Graph $y = 2x$, $y = 2x - 4$, and $y = 2x + 3$ on the same axes.
    a) What do you notice about the equations and the values where the graphs intersect the y-axis?
    b) Do all equations seem to have the same slant (or slope)?

71. Graph $y = 2x$. Determine the *rate of change* of y with respect to x. That is, by how many units does y change compared to each unit change in x?

72. Graph $y = 3x$. Determine the rate of change of y with respect to x.

73. Graph $y = 3x + 2$. Determine the rate of change of y with respect to x.

74. Graph $y = \frac{1}{2}x$. Determine the rate of change of y with respect to x.

75. The ordered pair $(1, -4)$ represents one point on the graph of a linear equation. If y increases 3 units for each unit increase in x on the graph, find two other solutions to the equation.

76. The ordered pair $(3, -6)$ represents one point on the graph of a linear equation. If y increases 4 units for each unit increase in x on the graph, find two other solutions to the equation.

Match Exercises 77–80 with the corresponding speed versus time graph, labeled a–d below.

77. To go to work, Jamal Washington drove in stop-and-go traffic for 5 minutes, then drove on the expressway for 20 minutes, then drove in stop-and-go traffic for 5 minutes.

78. To go to work, Art Mayfield walked for 3 minutes, waited for the train for 5 minutes, rode the train for 15 minutes, then walked for 7 minutes.

79. To go to work, Katelyn Barth rode her bike uphill for 10 minutes, then rode downhill for 15 minutes, then rode on a level street for 5 minutes.

80. To go to work, Tanya Bates drove on a country road for 10 minutes, then drove on a highway for 12 minutes, then drove in stop-and-go traffic for 8 minutes.

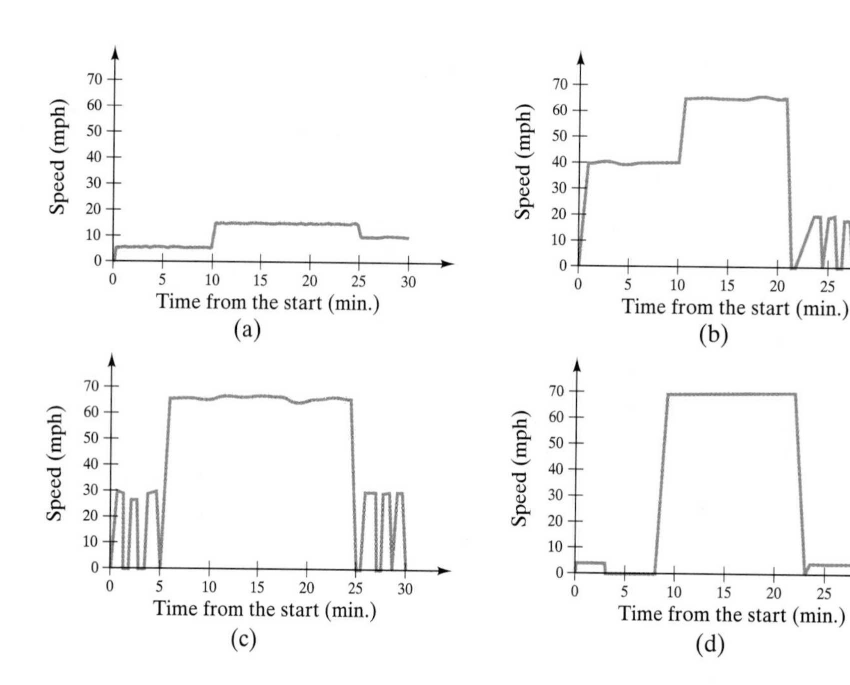

Match Exercises 81–84 with the corresponding distance traveled versus time graphs, labeled a–d.

81. Train A traveled at a speed of 40 mph for 1 hour, then 80 mph for 2 hours, and then 60 mph for 3 hours.

82. Train B traveled at a speed of 20 mph for 2 hours, then 60 mph for 3 hours, and then 80 mph for 1 hour.

83. Train C traveled at a speed of 80 mph for 2 hours, then stayed in a station for 1 hour, and then traveled 40 mph for 3 hours.

84. Train D traveled at 30 mph for 1 hour, then 65 mph for 2 hours, and then 30 mph for 3 hours.

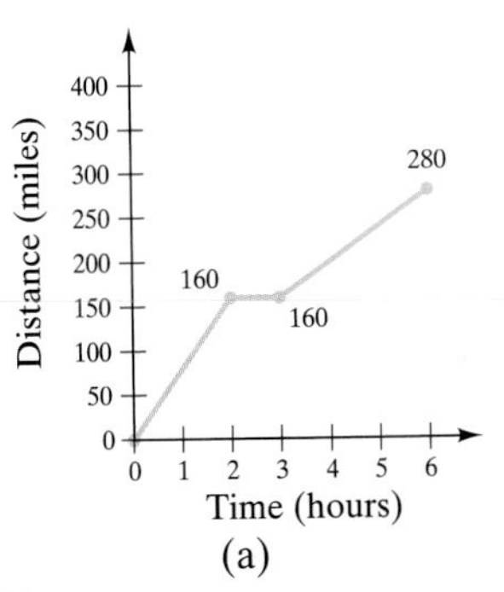

(a)

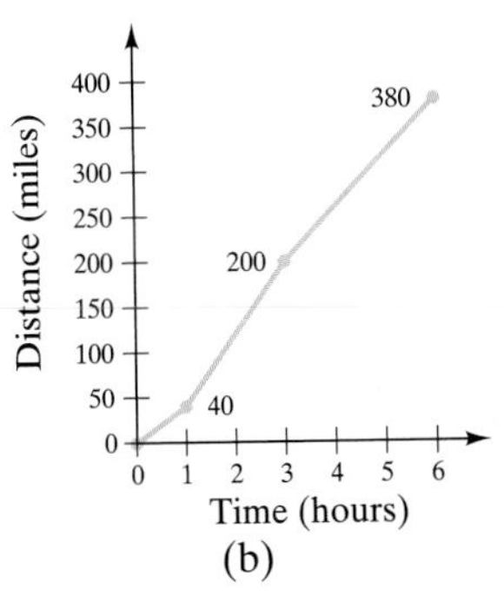

(b)

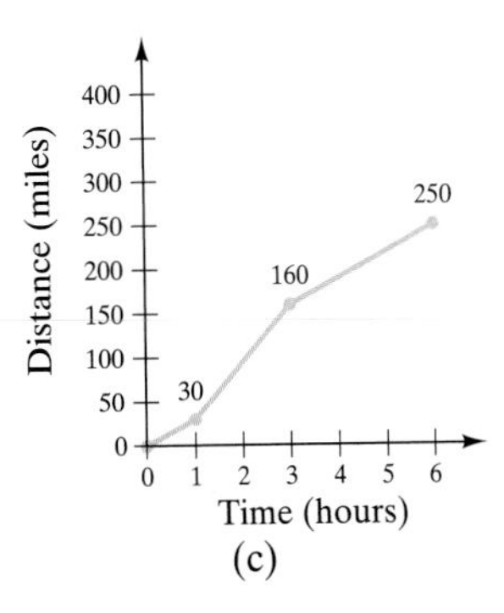

(c)

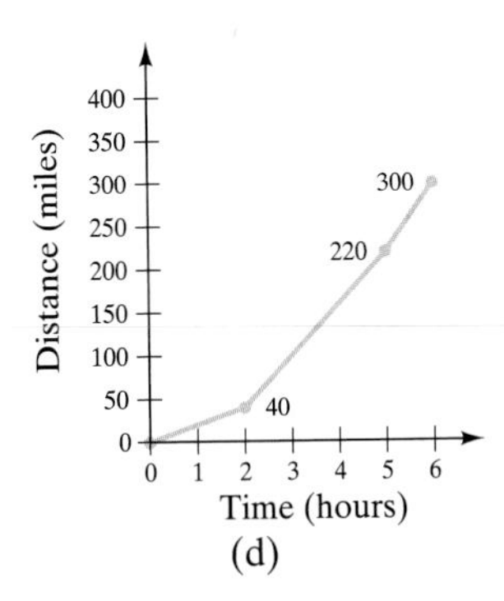

(d)

Use a graphing calculator to graph each function. Make sure you select values for the window that will show the curvature of the graph. Then, if your calculator can display tables, display a table of values in which the x-values extend by units, from 0 to 6.

85. $y = 3x + 5$

86. $y = \frac{1}{2}x - 4$

87. $y = x^2 + x + 6$

88. $y = x^2 - 12$

89. $y = x^3 - 2x + 4$

90. $y = 2x^3 - 6x^2 - 1$

Challenge Problems

Graph each equation.

91. $y = |x - 2|$

92. $x = y^2 + 2$

Group Activity

Discuss and work Exercises 93–94 as a group.

93. a) Group member 1: Plot the points $(-2, 4)$ and $(6, 8)$. Determine the *midpoint* of the line segment connecting these points.
Group member 2: Follow the above instructions for the points $(-3, -2)$ and $(5, 6)$.
Group member 3: Follow the above instructions for the points $(4, 1)$ and $(-2, 4)$.

b) As a group, determine a formula for the midpoint of the line segment connecting the points (x_1, y_1) and (x_2, y_2). (Note: We will discuss the midpoint formula further in Chapter 10.)

94. Three points on a parallelogram are $A(3, 5)$, $B(8, 5)$, and $C(-1, -3)$.

a) Individually determine a fourth point D that completes the parallelogram.

b) Individually compute the area of your parallelogram.

c) Compare your answers. Did you all get the same answers? If not, why not?

d) Is there more than one point that can be used to complete the parallelogram? If so, give the two points and find the corresponding areas of each parallelogram.

Cumulative Review Exercises

[2.2] **95.** Evaluate $\dfrac{-b + \sqrt{b^2 - 4ac}}{2a}$ for $a = 2$, $b = 7$, and $c = -15$.

[2.3] **96.** Hertz Automobile Rental Agency charges a daily fee of $30 plus 14 cents a mile. National Automobile Rental Agency charges a daily fee of $16 plus 24 cents a mile for the same car. What distance would you have to drive in 1 day to make the cost of renting from Hertz equal to the cost of renting from National?

[2.5] **97.** Solve the inequality $-4 \le \dfrac{4 - 3x}{2} < 5$. Write the solution in set builder notation.

[2.6] **98.** Find the solution set for the inequality $|3x + 2| > 5$.

3.2 FUNCTIONS

1. Understand relations.
2. Recognize functions.
3. Use the vertical line test.
4. Understand function notation.
5. Study applications of functions.

1 Understand Relations

In real life we often find that one quantity is related to a second quantity. For example, the amount you spend for oranges is related to the number of oranges you purchase. The speed of a sailboat is related to the speed of the wind. And the income tax you pay is related to the income you earn.

Suppose oranges cost 30 cents apiece. Then one orange costs 30 cents, two oranges cost 60 cents, three oranges cost 90 cents, and so on. We can list this information, or relationship, as a set of ordered pairs by listing the number of oranges first and the cost second. The ordered pairs that represent this situation are $(1, 30)$, $(2, 60)$, $(3, 90)$, and so on. An equation that represents this situation is $c = 30n$, where c is the cost and n is the number of oranges. Since the cost depends on the number of oranges, we say that the cost is the *dependent variable* and the number of oranges is the *independent variable.*

Now consider the equation $y = 2x + 3$. In this equation, the value obtained for y depends on the value selected for x. Therefore x is the *independent variable* and y is the *dependent variable*. Note that in this example, unlike with the oranges, there is no physical connection between x and y. The variable x is the independent variable and y is the dependent variable simply because of their placement in the equation.

For an equation in variables x and y, if the value of y depends on the value of x, then y is the **dependent variable** and x is the **independent variable**. Since related quantities can be represented as ordered pairs, the concept of **relation** can be defined as follows.

Definition

A **relation** is any set of ordered pairs.

2 Recognize Functions

We now develop the idea of a **function**—one of the most important concepts in mathematics. A function is a special type of relation in which each element in one set (called the domain) corresponds to *exactly one* element in a second set (called the range).

Consider the oranges that cost 30 cents apiece that we just discussed. We can illustrate the number of oranges and the cost of the oranges using Figure 3.20.

Notice that each number in the set of numbers of oranges corresponds to (or is mapped to) exactly one number in the set of costs. Therefore, this correspondence is a function. The set consisting of the number of oranges, $\{1, 2, 3, 4, 5, \ldots\}$,

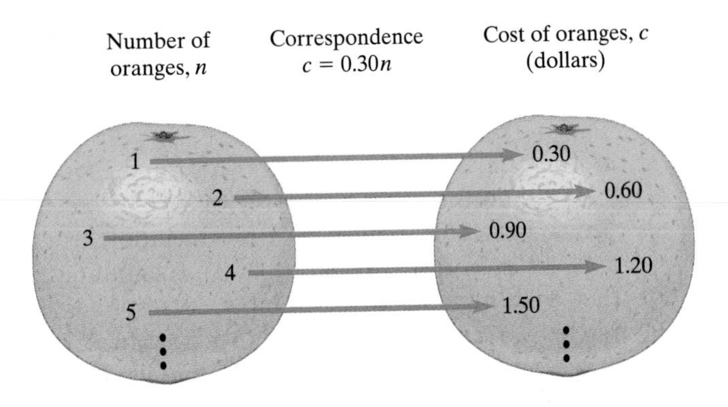

FIGURE 3.20

is called the **domain**. The set consisting of the costs, $\{0.30, 0.60, 0.90, 1.20, 1.50, \ldots\}$, is called the **range**. In general, the set of values for the independent variable is called the **domain**. The set of values for the dependent variable is called the **range**, see Figure 3.21.

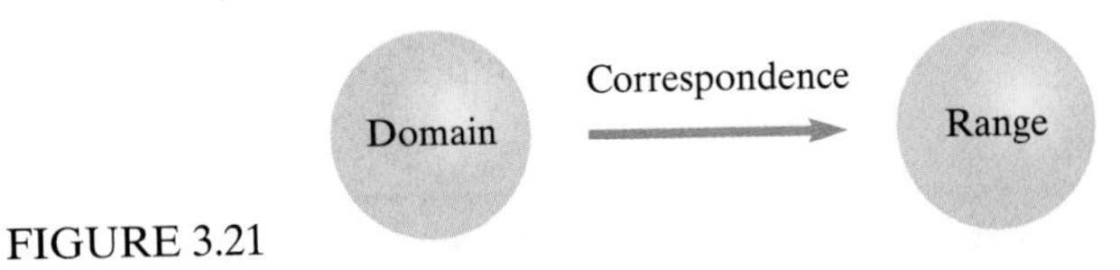

FIGURE 3.21

EXAMPLE 1 Determine whether each correspondence is a function.

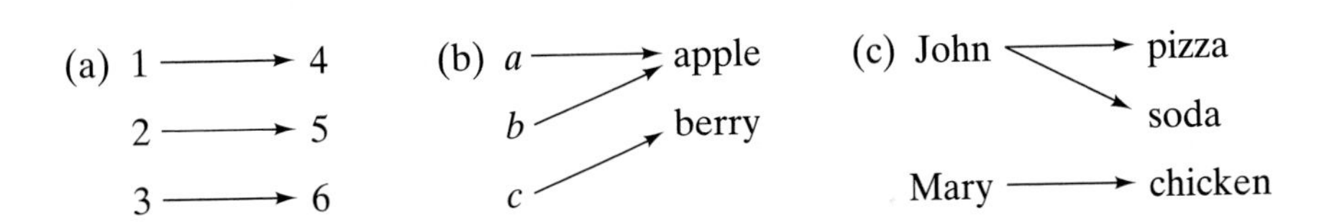

Solution **a)** For a correspondence to be a function, each element in the domain must correspond with exactly one element in the range. Here the domain is $\{1, 2, 3\}$ and the range is $\{4, 5, 6\}$. Since each element in the domain corresponds to exactly one element in the range, this correspondence is a function.

b) Here the domain is $\{a, b, c\}$ and the range is $\{\text{apple}, \text{berry}\}$. Even though the domain has three elements and the range has two elements, each element in the domain corresponds with exactly one element in the range. Thus, this correspondence is a function.

c) Here the domain is {John, Mary} and the range is {pizza, soda, chicken}. Notice that John corresponds to both pizza and soda. Therefore each element in the domain *does not* correspond to exactly one element in the range. Thus, this correspondence is a relation but *not* a function.

NOW TRY EXERCISE 17

Now let us formally define function.

Definition

A **function** is a correspondence between a first set of elements, the domain, and a second set of elements, the range, such that each element of the domain corresponds to *exactly one* element in the range.

EXAMPLE 2 Which of the following relations are functions?

a) $\{(1, 4), (2, 3), (3, 5), (-1, 3) (0, 6)\}$

b) $\{(-1, 3) (4, 2), (3, 1), (2, 6), (3, 5)\}$

Solution **a)** The domain is the set of first coordinates in the set of ordered pairs, $\{1, 2, 3, -1, 0\}$, and the range is the set of second coordinates, $\{4, 3, 5, 6\}$. Notice that when listing the range, we only include the number 3 once, even though it appears in both $(2, 3)$ and $(-1, 3)$. Examining the set of ordered pairs, we see that each number in the domain corresponds with exactly one number in the range. For example, the 1 in the domain corresponds with only the 4 in the range, and so on. No x-value corresponds to more than one y-value. Therefore, this relation *is a function.*

b) The domain is $\{-1, 4, 3, 2\}$ and the range is $\{3, 2, 1, 6, 5\}$. Notice that 3 appears as the first coordinate in two ordered pairs even though it is listed only once in the set of elements that represent the domain. Since the ordered pairs $(3, 1)$ and $(3, 5)$ have *the same first coordinate* and a different second coordinate, each value in the domain does not correspond to exactly one value in the range. Therefore, this relation is *not a function.*

NOW TRY EXERCISE 23

Example 2 leads to an alternate definition of function.

Definition

A **function** is a set of ordered pairs in which no first coordinate is repeated.

If the second coordinate in a set of ordered pairs repeats, the set of ordered pairs may still be a function, as in Example 2 **a)**. However, if two or more ordered pairs contain the same first coordinate, as in Example 2 **b)**, the set of ordered pairs is not a function.

3) Use the Vertical Line Test

The **graph of a function or relation** is the graph of its set of ordered pairs. The two sets of ordered pairs in Example 2 are graphed in Figures 3.22a and 3.22b. Notice that in the function in Figure 3.22a it is not possible to draw a vertical line that intersects two points. We should expect this because, in a function, each x value must correspond to exactly one y value. In Figure 3.22b we *can* draw a vertical line through the points $(3, 1)$ and $(3, 5)$. This shows that each x-value does not correspond to exactly one y-value, and the graph does not represent a function.

This method of determining whether a graph represents a function is called the **vertical line test**.

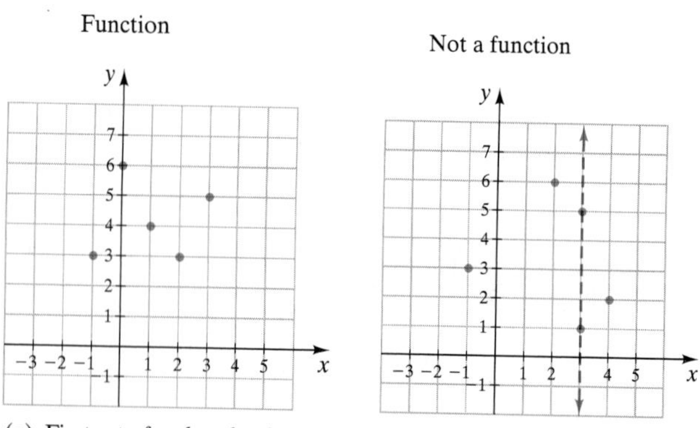

FIGURE 3.22 (a) First set of ordered pairs (b) Second set of ordered pairs

> **Vertical Line Test**
>
> If a vertical line can be drawn through any part of the graph and the line intersects another part of the graph, the graph does not represent a function. If a vertical line cannot be drawn to intersect the graph at more than one point, the graph represents a function.

We use the vertical line test to show that Figure 3.23b represents a function and Figures 3.23a and 3.23c do not represent functions.

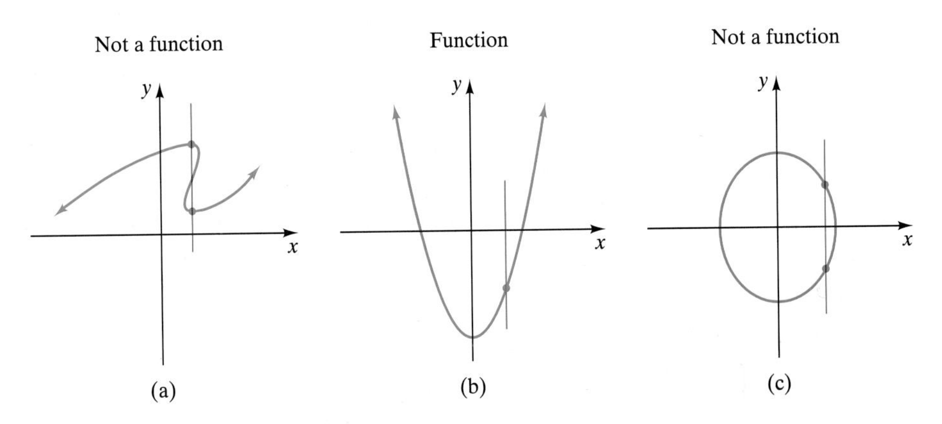

FIGURE 3.23

EXAMPLE 3 Use the vertical line test to determine whether the following graphs represent functions. Determine the domain and range of each function or relation.

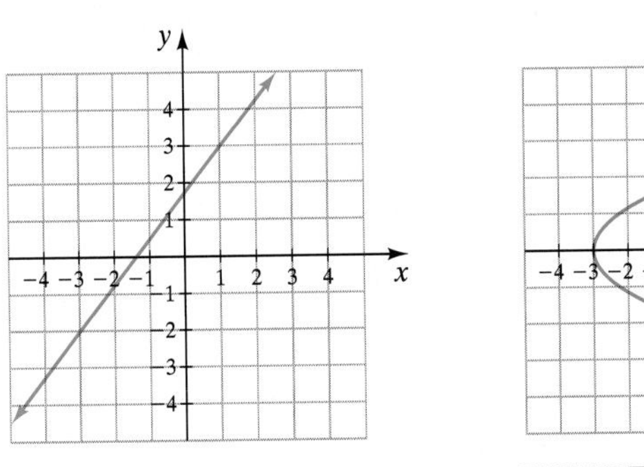

FIGURE 3.24

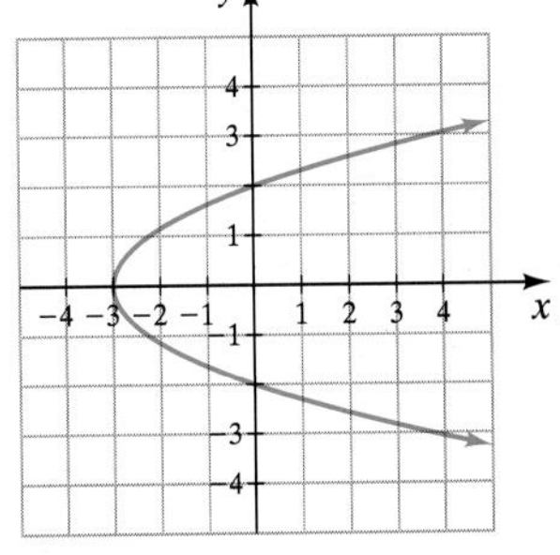

FIGURE 3.25

Solution **a)** A vertical line cannot be drawn to intersect the graph in Figure 3.24 at more than one point. Thus this is the graph of a function. Since the line extends indefinitely in both directions, every value of x will be included in the domain. The domain is the set of real numbers.

$$\text{Domain:}\quad \mathbb{R} \quad \text{or} \quad (-\infty, \infty)$$

The range is also the set of real numbers since all values of y are included on the graph.

$$\text{Range:}\quad \mathbb{R} \quad \text{or} \quad (-\infty, \infty)$$

b) Since a vertical line can be drawn to intersect the graph in Figure 3.25 at more than one point, this is *not* the graph of a function. The domain of this relation is the set of values greater than or equal to −3.

$$\text{Domain:}\quad \{x \mid x \geq -3\} \quad \text{or} \quad [-3, \infty)$$

The range is the set of y-values, which can be any real number.

$$\text{Range: } \mathbb{R} \text{ or } (-\infty, \infty)$$

EXAMPLE 4 Consider the graph shown in Figure 3.26.

a) What member of the range as paired with 4 in the domain?

b) What member (or members) of the domain is (or are) paired with -2 in the range?

c) What is the domain of the function?

d) What is the range of the function?

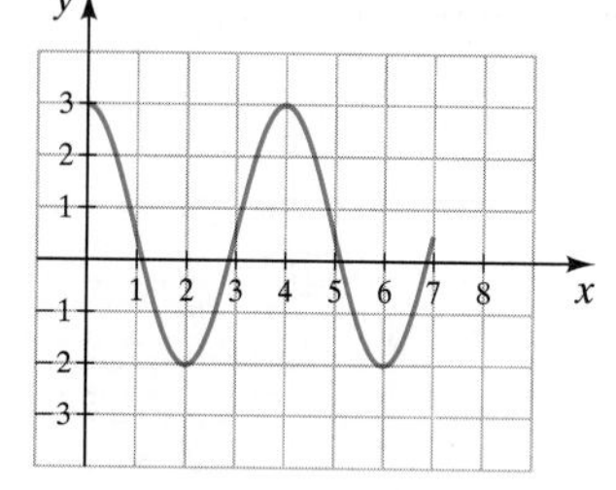

FIGURE 3.26

Solution **a)** The range is the set of y-values. The y-value paired with the x-value of 4 is 3.

b) The domain is the set of x-values. The x-values paired with the y-value of -2 are 2 and 6.

c) The domain is the set of x values, 0 through 7. Thus the domain is

$$\{x \mid 0 \leq x \leq 7\} \text{ or } [0, 7]$$

d) The range is the set of y values, -2 through 3. Thus, the range is

$$\{y \mid -2 \leq y \leq 3\} \text{ or } [-2, 3]$$

NOW TRY EXERCISE 33

EXAMPLE 5 Figure 3.27 illustrates a graph of speed versus time of a man out for a walk and jog. Write a story about the man's outing that corresponds to this function

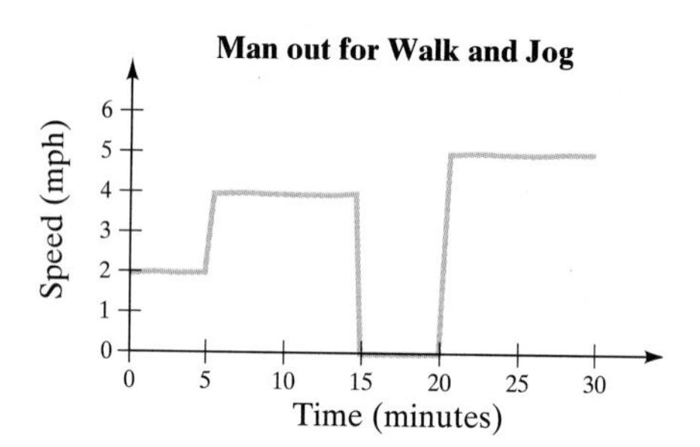

FIGURE 3.27

Solution **Understand** The horizontal axis is time and the vertical axis is speed. When the graph is horizontal it means the person is traveling at the constant speed indicated on the vertical axis. The near-vertical lines that increase with time (or have a positive slope, as will be discussed later) indicate an increase in speed, whereas the near-vertical lines that decrease with time (or have a negative slope) indicate a decrease in speed.

Answer Here is one possible interpretation of the graph. The man walks for about 5 minutes at a speed of about 2 miles per hour. Then the man speeds up to 4 miles per hour and walks fast or jogs at about this speed for about 10 minutes. Then the man slows down and stops, and then rests for about 5 minutes. Finally, the man speeds up to about 5 miles per hour and jogs at this speed for about 10 minutes.

NOW TRY EXERCISE 65

4) Understand Function Notation

In Section 3.1 we graphed a number of equations, as summarized in Table 3.1. If you examine each equation in the table, you will see that they are all functions since their graphs pass the vertical line test.

TABLE 3.1 Example

Section 3.1 Example	Equation Graphed	Graph	Does the Graph Represent a Function?
3	$y = x$		Yes
4	$y = -\frac{1}{2}x + 2$		Yes
5	$y = x^2 - 4$		Yes
6	$y = \frac{1}{x}$		Yes
7	$y = \lvert x \rvert$		Yes

Since the graph of each equation shown represents a function, we may refer to each equation in the table as a function. When we refer to an equation in variables x and y as a function, it means that the graph of the equation satisfies the criteria for a function. That is, each x-value corresponds to exactly one y-value, and the graph of the equation passes the vertical line test.

Not all equations are functions, as you will see in Chapter 10, where we discuss equations of circles and ellipses. However, until we get to Chapter 10, all equations that we discuss will be functions.

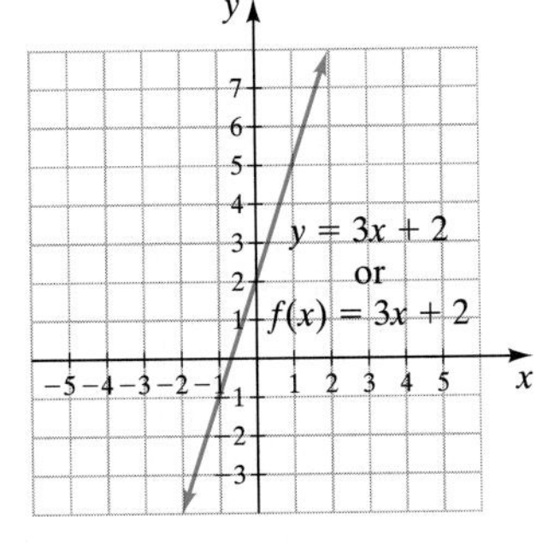

FIGURE 3.28

Consider the equation $y = 3x + 2$. By applying the vertical line test to its graph (Fig. 3.28), we can see that the graph represents a function. When an equation in variables x and y is a function, we often write the equation using *function notation*, $f(x)$, read "f of x". Since the equation $y = 3x + 2$ is a function, and the value of y depends on the value of x, we say that **y is a function of x**. When we are given a linear equation in variables x and y, *that is solved for y,* we can write the equation in function notation by substituting $f(x)$ for y. In this case, we can write the equation in function notation as $f(x) = 3x + 2$. The notation $f(x)$ represents the dependent variable *and does not mean f times x.* Other letters may be used to indicate functions. For example, $g(x)$ and $h(x)$ also represent functions of x, and in Section 5.1 we will use $P(x)$ to represent polynomial functions.

Functions written in function notation are also equations since they contain an equal sign. We may refer to $y = 3x + 2$ as either an equation or a function. Similarly, we may refer to $f(x) = 3x + 2$ as either a function or an equation.

If y is a function of x, the notation $f(5)$, read "f of 5," means the value of y when x is 5. To evaluate a function for a specific value of x, substitute that value for x in the function. For example, if $f(x) = 3x + 2$, then $f(5)$ is found as follows:

$$f(x) = 3x + 2$$

$$f(5) = 3(5) + 2 = 17$$

Therefore, when x is 5, y is 17. The ordered pair (5, 17) would appear on the graph of $y = 3x + 2$.

HELPFUL HINT

Linear equations that are not solved for y can be written using function notation by solving the equation for y, then replacing y with $f(x)$. For example, the equation $-9x + 3y = 6$ becomes $y = 3x + 2$ when solved for y. We can therefore write $f(x) = 3x + 2$.

EXAMPLE 6 If $f(x) = 2x^2 + 3x - 4$, find

a) $f(3)$ **b)** $f(-2)$ **c)** $f(a)$

Solution **a)** $f(x) = 2x^2 + 3x - 4$

$f(3) = 2(3)^2 + 3(3) - 4 = 2(9) + 9 - 4 = 18 + 9 - 4 = 23$

b) $f(-2) = 2(-2)^2 + 3(-2) - 4 = 2(4) - 6 - 4 = 8 - 6 - 4 = -2$

c) To evaluate the function at a, we replace each x in the function with an a.

$$f(x) = 2x^2 + 3x - 4$$

$$f(a) = 2a^2 + 3a - 4$$

EXAMPLE 7 Find each indicated function value.

a) $g(4)$ for $g(r) = r^2 + 2r$

b) $n\left(-\frac{1}{2}\right)$ for $n(t) = 6t^2 - 3t + 4$

c) $h(-3)$ for $h(x) = |x - 4|$

Solution In each part, substitute the indicated value into the function and evaluate the function.

a) $g(4) = 4^2 + 2(4) = 16 + 8 = 24$

b) $n\left(-\frac{1}{2}\right) = 6\left(-\frac{1}{2}\right)^2 - 3\left(-\frac{1}{2}\right) + 4 = 6\left(\frac{1}{4}\right) + \frac{3}{2} + 4 = 7$

NOW TRY EXERCISE 45

c) $h(-3) = |-3 - 4| = |-7| = 7$

5) Study Applications of Functions

Many of the applications that we discussed in Chapter 2 were functions. However, we had not defined a function at that time. Now we examine additional applications of functions.

EXAMPLE 8 The graph in Figure 3.29 shows the number of cases of AIDS in L. A. County from 1984 through 1997.

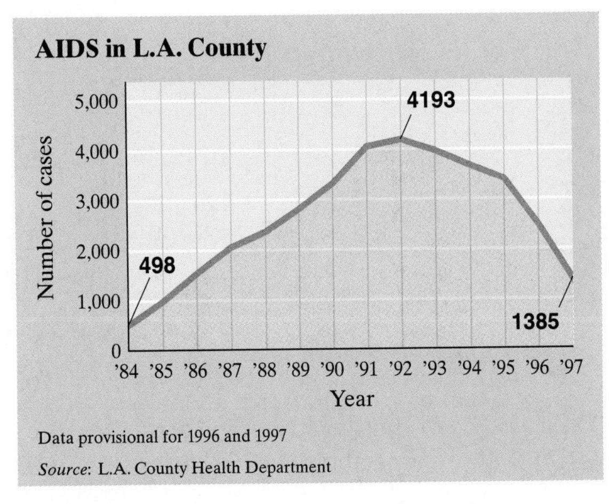

FIGURE 3.29

a) Explain why this graph represents a function.

b) Estimate the number of AIDS cases in Los Angeles County in 1995.

c) Determine the percent increase in the number of AIDS cases from 1984 through 1992.

d) Determine the percent decrease in the number of AIDS cases from 1992 through 1997.

Solution **a)** This graph represents a function because each year corresponds to a single number of AIDS cases. Notice that the graph passes the vertical line test.

b) In 1995 there were about 3400 AIDS cases. We determine this by finding the intersection of the graph with 1995. The answer is found on the vertical axis. If we call the function f, then $f(1995) = 3400$.

c) We will follow our problem-solving procedure to answer this question.

Understand and Translate We are asked to find the percent increase in AIDS cases from 1984 through 1992. To do so, we use the formula

$$\text{percent change (increase or decrease)} = \frac{\begin{pmatrix}\text{amount in}\\ \text{latest period}\end{pmatrix} - \begin{pmatrix}\text{amount in}\\ \text{previous period}\end{pmatrix}}{\text{amount in previous period}}$$

The latest period is 1992 and the previous period is 1984. Substituting the corresponding values, we get

$$\text{percent change} = \frac{4193 - 498}{498}$$

Carry Out

$$= \frac{3695}{498}$$

$$\approx 7.4197$$

$$\approx 742\%$$

Check and Answer Our calculations appear correct. There was about a 742% increase in AIDS cases in Los Angeles County from 1984 through 1992.

d) To find the percent decrease from 1992 through 1997 we follow the same procedure as in part **c)**.

$$\text{percent change} = \frac{\begin{pmatrix}\text{amount in}\\ \text{latest period}\end{pmatrix} - \begin{pmatrix}\text{amount in}\\ \text{previous period}\end{pmatrix}}{\text{amount in previous period}}$$

$$= \frac{1385 - 4193}{4193}$$

$$= \frac{-2808}{4193}$$

$$\approx -0.6697$$

$$\approx -67\%$$

NOW TRY EXERCISE 69

The negative sign preceding the 67% indicates a percent decrease. Thus there was about a 67% decrease in AIDS cases from 1992 through 1997.

EXAMPLE 9 Shopping center space is steadily growing in the United States. In 1980 there were about 12 square feet of shopping center space per capita. There were 13 square feet in 1982, 14 square feet in 1984, 16 square feet in 1986, 17 square feet in 1988, 17 square feet in 1990, 18 square feet in 1992, 19 square feet in 1994, and 20 square feet in 1996 (Source: Deloitte and Touche, LLP).

a) Represent this information on a graph.

b) Using your graph, explain why this set of points represents a function.

c) Using your graph, estimate the number of square feet in 1995.

Solution **a)** The set of points is plotted in Figure 3.30. We placed the year on the horizontal axis and the square feet on the vertical axis. Notice the break in the vertical axis, indicated by ⚡. We do this to show that some numbers have been omitted on the axis to save space.

b) Since each year corresponds to exactly one square footage, this set of points represents a function. Notice that this graph passes the vertical line test.

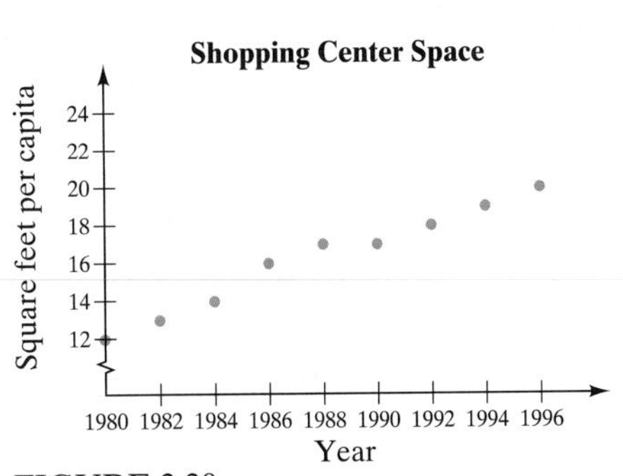

FIGURE 3.30

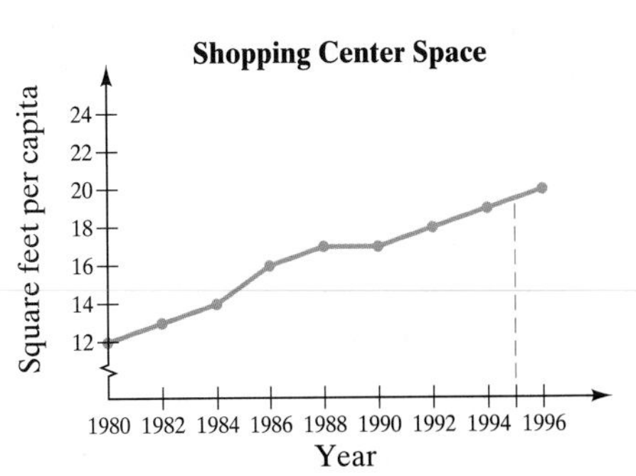

FIGURE 3.31

c) We can connect the points with straight line segments as in Figure 3.31. Then we can estimate from the graph that there were about 19.5 square feet per capita in 1995. If we call the function f, then $f(1995) = 19.5$.

NOW TRY EXERCISE 73

In Section 2.2 we learned to use formulas. Consider the formula for the area of a circle, $A = \pi r^2$. In the formula π is a constant that is approximately 3.14. For each specific value of the radius, r, there corresponds exactly one area, A. Thus the area of a circle is a function of its radius. We may therefore write

$$A(r) = \pi r^2$$

Often formulas are written using function notation, like this.

EXAMPLE 10 The Celsius temperature, C, is a function of the Fahrenheit temperature, F.

$$C(F) = \frac{5}{9}(F - 32)$$

Determine the Celsius temperature that corresponds to 50°F.

Solution We need to find $C(50)$. We do so by substitution.

$$C(F) = \frac{5}{9}(F - 32)$$

$$C(50) = \frac{5}{9}(50 - 32)$$

$$= \frac{5}{9}(18) = 10$$

Therefore, 50°F = 10°C.

In Example 10, F is the independent variable and C is the dependent variable. If we solved the function for F, we would obtain $F(C) = \frac{9}{5}C + 32$. In this formula, C is the independent variable and F is the dependent variable.

NOW TRY EXERCISE 55

Exercise Set 3.2

Concept/Writing Exercises

1. What is a relation?
2. What is a function?
3. Are all relations also functions? Explain.
4. Are all functions also relations? Explain.
5. Explain how to use the vertical line test to determine if a relation is a function.
6. What is the domain of a function?
7. What is the range of a function?
8. What are the domain and range of the function $f(x) = 3x - 2$? Explain your answer.
9. What are the domain and range of a function of the form $f(x) = ax + b, a \neq 0$. Explain your answer.
10. Consider the absolute value function $y = |x|$. What is its domain and range? Explain.
11. What is a dependent variable?
12. What is an independent variable?
13. How is "$f(x)$" read?
14. Are all functions that are given in function notation also equations? Explain.

Practice the Skills

*In Exercises 15–20, **a)** determine if the relation illustrated is a function. **b)** Give the domain and range of each function or relation.*

15. twice number

3 → 6
5 → 10
10 → 20

16. seat number

Bob → 2
Carol → 1
Alice → 3

17. age

Ron → 18
Jayne → 19
Cecilia → 19

18. number squared

4 → 16
5 → 25
6 → 36

19. cost of stamp

1990 → 20
1996 → 32
1999 → 32
1999 → 33

20. absolute value

$|-3|$ → 3
$|3|$ → 3
$|0|$ → 0

*In Exercises 21–28, **a)** determine which of the following relations are also functions. **b)** Give the domain and range of each relation or function.*

21. $\{(1, 4), (2, 2), (3, 5), (4, 3), (5, 1)\}$

22. $\{(1, 1), (4, 4), (3, 3), (2, 2), (4, 1)\}$

23. $\{(3, -1), (5, 0), (1, 2), (4, 4), (2, 2), (7, 5)\}$

24. $\{(-1, 1), (0, -3), (3, 4), (4, 5), (-2, -2)\}$

25. $\{(5, 0), (3, -4), (2, -1), (5, 2), (1, 1)\}$

26. $\{(6, 3), (-3, 4), (0, 3), (5, 2), (3, 5), (2, 5)\}$

27. $\{(0, 3), (1, 3), (2, 2), (1, -1), (2, -7)\}$

28. $\{(3, 5), (2, 5), (1, 5), (0, 5), (-1, 5)\}$

*In Exercises 29–40, **a)** determine whether the graph illustrated represents a function. **b)** Give the domain and range of each function or relation. **c)** Find the value or values of x where $y = 2$.*

29.

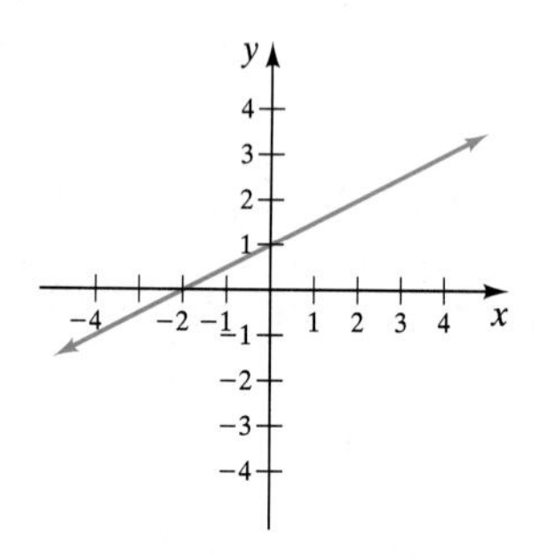

30.

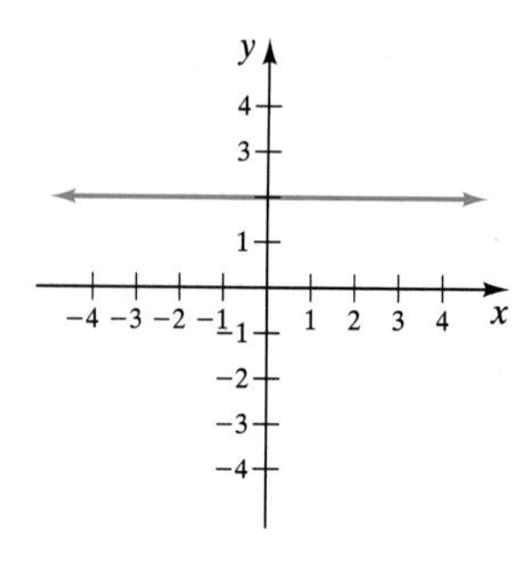

31.

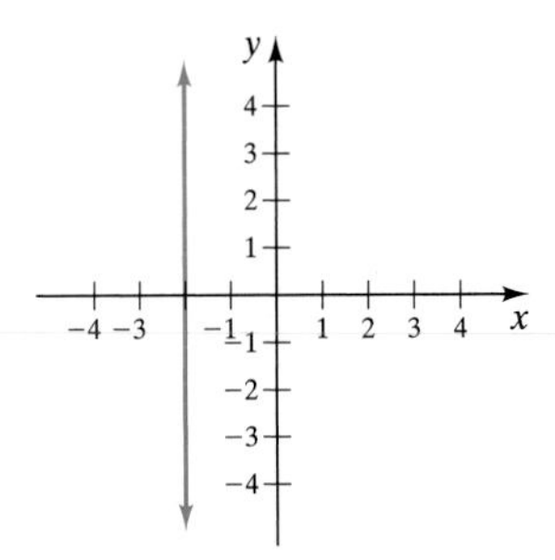

32.

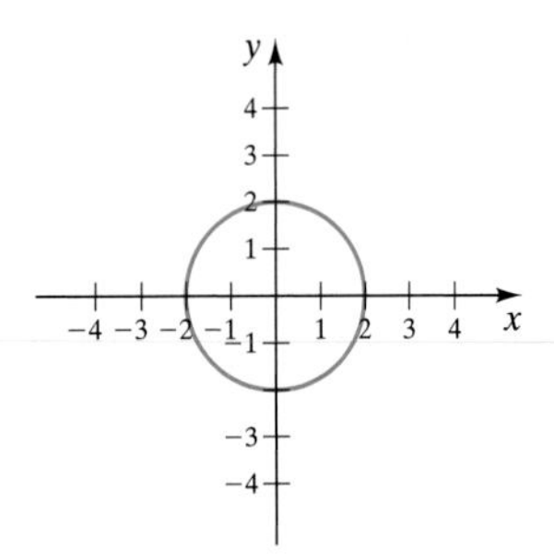

33.

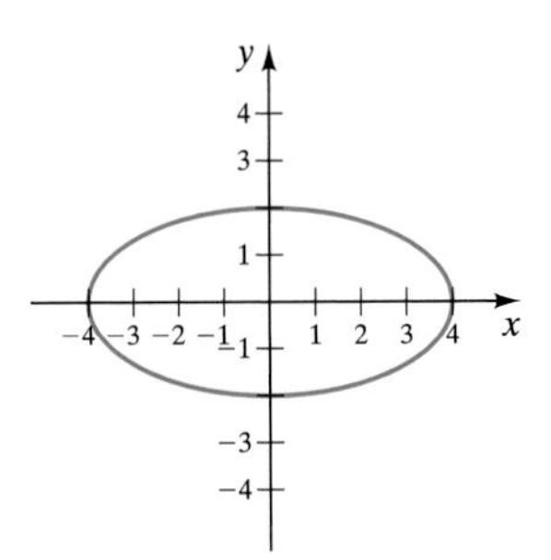

34.

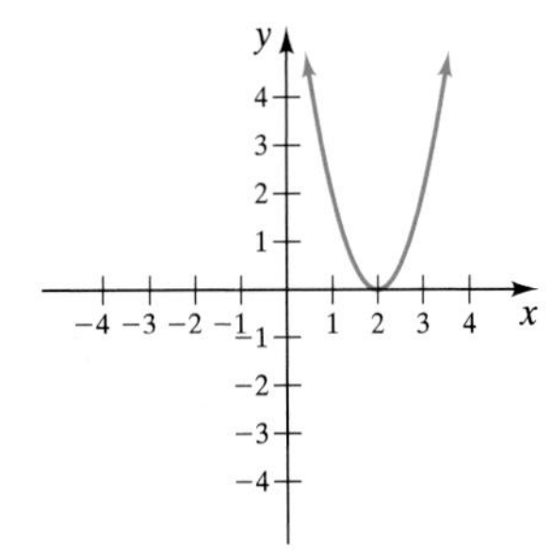

35.

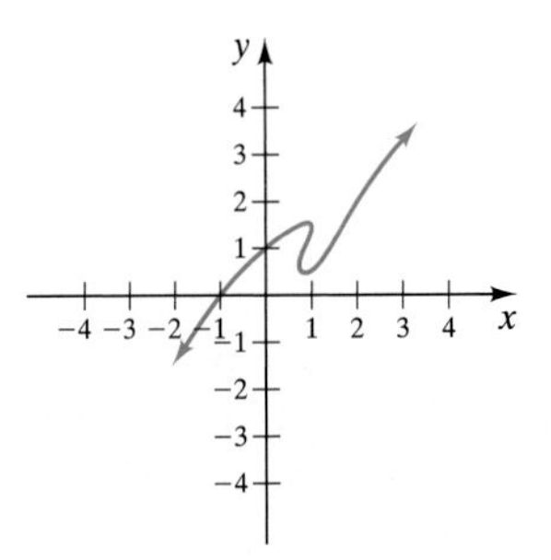

36.

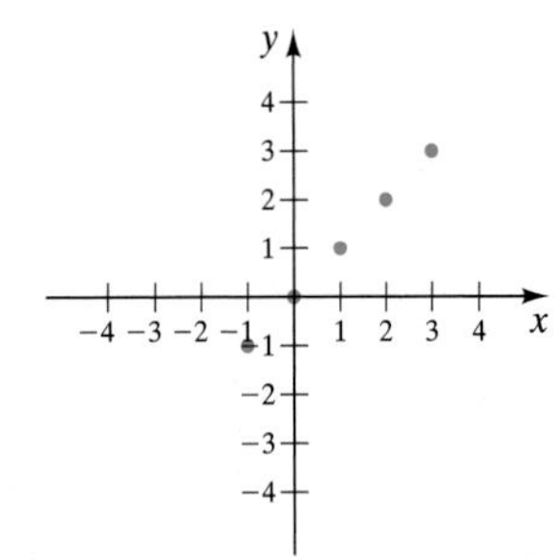

37.

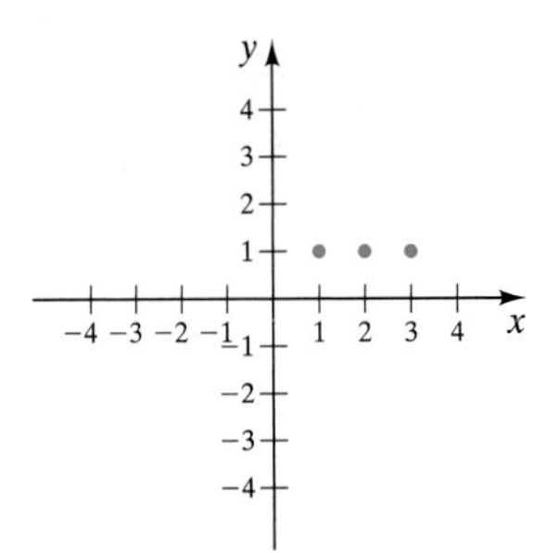

38.

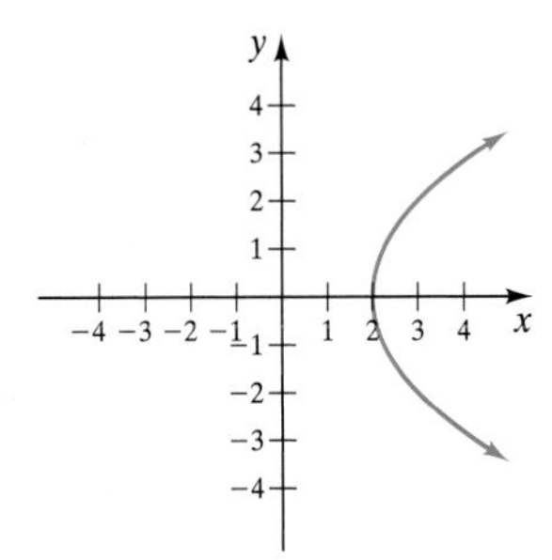

39.

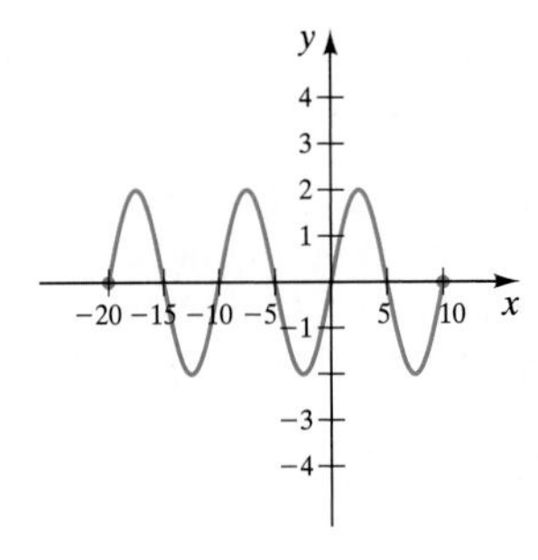

40.

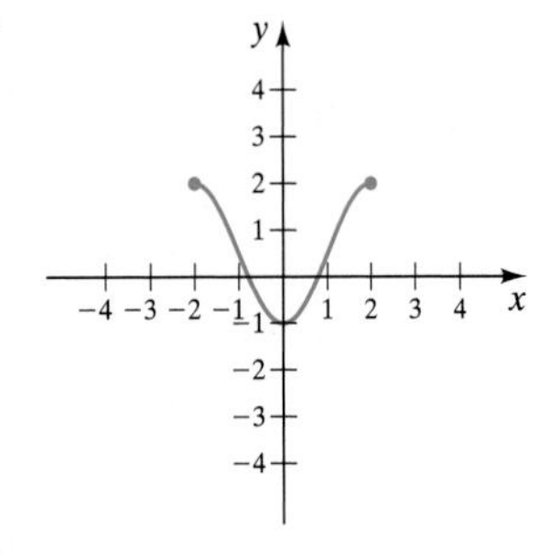

Evaluate each function at the indicated values.

41. $f(x) = 3x + 6$; find
a) $f(5)$;
b) $f(-2)$.

42. $f(x) = \frac{1}{2}x - 4$; find
a) $f(10)$;
b) $f(-4)$.

43. $h(x) = x^2 - x - 6$; find
a) $h(0)$;
b) $h(-1)$.

44. $g(x) = -x^2 - 2x + 3$; find
a) $g(-1)$;
b) $g\left(\frac{1}{2}\right)$.

45. $g(x) = x^3 + 2x^2 - 4$; find
a) $g(2)$;
b) $g(-3)$.

46. $f(x) = 2x^3 - 4x^2 + x - 64$; find
a) $f(0)$;
b) $f(-4)$.

47. $f(x) = |x + 3|$; find
a) $f(-5)$;
b) $f(-12.6)$.

48. $f(x) = |6 - 3x| - 2$; find
a) $f(-1)$;
b) $f(1.3)$.

49. $f(x) = \sqrt{x + 1}$; find
a) $f(3)$;
b) $f(24)$.

50. $f(x) = \sqrt{2x + 7}$; find
a) $f(1)$;
b) $f(9)$.

51. $f(x) = \dfrac{x^2 - 4}{x + 2}$; find
a) $f(2)$;
b) $f(-3)$.

52. $h(x) = \dfrac{x^2 + 4x}{x + 6}$; find
a) $h(-3)$;
b) $h\left(\frac{2}{5}\right)$.

Problem Solving

53. Earlier we introduced the distance formula, $d = rt$. If a car is traveling at a constant 60 miles per hour, then the distance is a function of time, $d(t) = 60t$. Find the distance traveled in **a)** 4 hours and **b)** 12 hours.

54. The formula for the area of a rectangle is $A = lw$. If the width of a rectangle is 8 feet, then the area is a function of its length, $A(l) = 8l$. Find the area when the length is **a)** 4 feet and **b)** 7 feet.

55. The formula for the circumference of a circle is $C = 2\pi r$. The circumference is a function of the radius.
a) Write this formula using function notation.
b) Find the circumference when the radius is 9 feet.

56. The formula for the volume of a sphere is $V = \frac{4}{3}\pi r^3$.
a) Write this formula using function notation.
b) Find the volume of a sphere whose radius is 5 inches

57. The formula for changing Celsius temperature to Fahrenheit temperature is $F = \frac{9}{5}C + 32$. The Fahrenheit temperature is a function of the Celsius temperature.
a) Write this formula using function notation.
b) Find the Fahrenheit temperature that corresponds to 20°C.

58. The formula for the volume of a right circular cylinder is $V = \pi r^2 h$. If the height, h, is 3 feet, then the volume is a function of the radius, r.
a) Write this formula in function notation, where the height is 3 feet.
b) Find the volume if the radius is 2 feet.

59. The temperature, T, in degrees Celsius, in a sauna n minutes after being turned on is given by the function $T(n) = -0.03n^2 + 1.5n + 14$. Find the sauna's temperature after:
a) 3 minutes **b)** 12 minutes

60. The stopping distance, d, in meters for a car traveling v kilometers per hour is given by the function $d(v) = 0.18v + 0.01v^2$. Find the stopping distance for the following speeds:
a) 50 km/hr **b)** 25 km/hr

61. When an air conditioner is turned on maximum in a bedroom at 80°, the temperature, T, in the room after A minutes can be approximated by the function $T(A) = -0.02A^2 - 0.34A + 80$, $0 \le A \le 15$.

a) Estimate the room temperature 4 minutes after the air conditioner is turned on.

b) Estimate the room temperature 12 minutes after the air conditioner is turned on.

62. The number of accidents, n, in 1 month involving drivers x years of age can be approximated by the function $n(x) = 2x^2 - 150x + 4000$. Find the approximate number of accidents in one month that involved:

a) 18-year-olds **b)** 25-year-olds

63. The total number of oranges, T, in a square pyramid whose base is n by n oranges is given by the function

$$T(n) = \frac{1}{3}n^3 + \frac{1}{2}n^2 + \frac{1}{6}n$$

Find the number of oranges if the base is:

a) 6 by 6 oranges **b)** 8 by 8 oranges

64. If the cost of a ticket to a rock concert is increased by x dollars, the estimated increase in revenue, R, in thousands of dollars is given by the function $R(x) = 24 + 5x - x^2$, $x < 8$. Find the increase in revenue if the cost of the ticket is increased by:

a) \$1 **b)** \$4

Review Example 5 before working Exercises 65–68.

65. The following graph shows height above sea level versus time when a man leaves his house and goes for a walk. Write a story that this graph may represent.

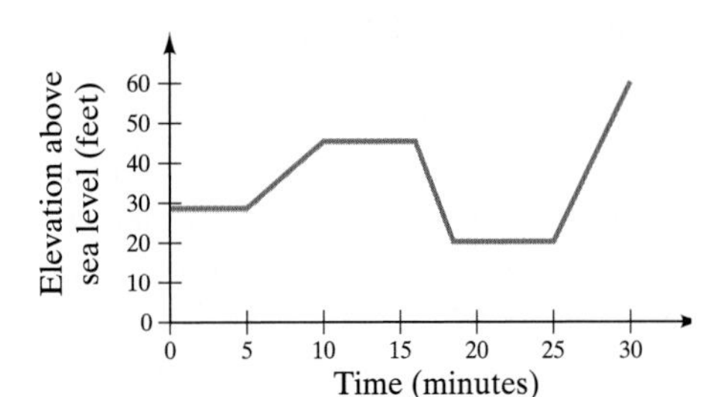

66. The following graph shows the level of water in a bathtub versus time. Write a story that this graph may represent.

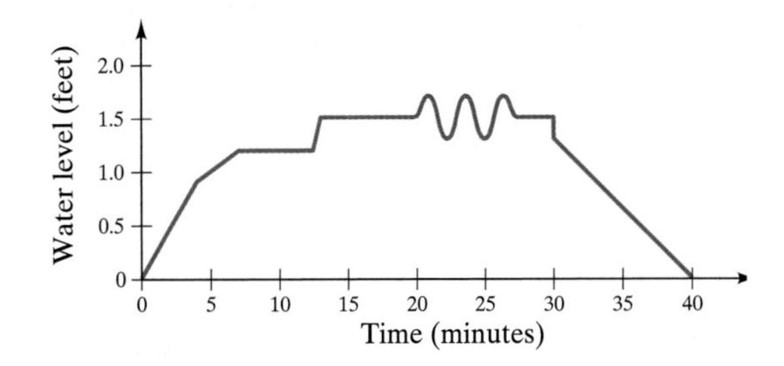

67. The following graph shows the speed of a car versus time. Write a story that this graph may represent.

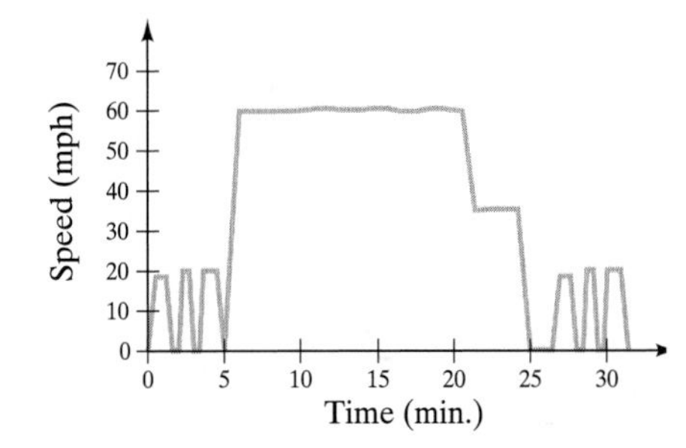

68. The following graph shows the distance traveled by a person in a car versus time. Write a story that this graph may represent.

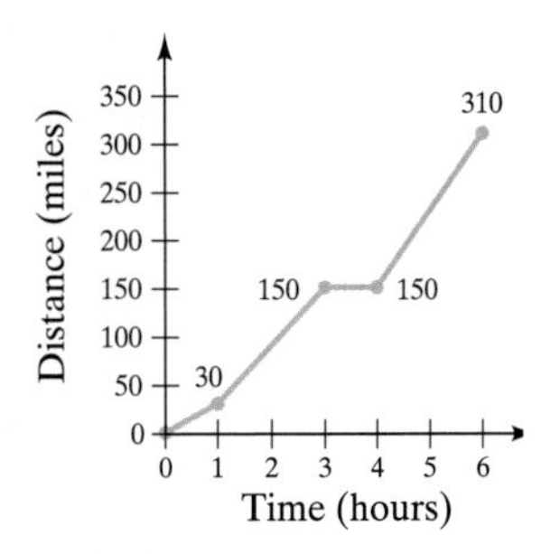

69. The following graph shows the number of active U.S. military personnel, all branches, from 1975 through 1997.

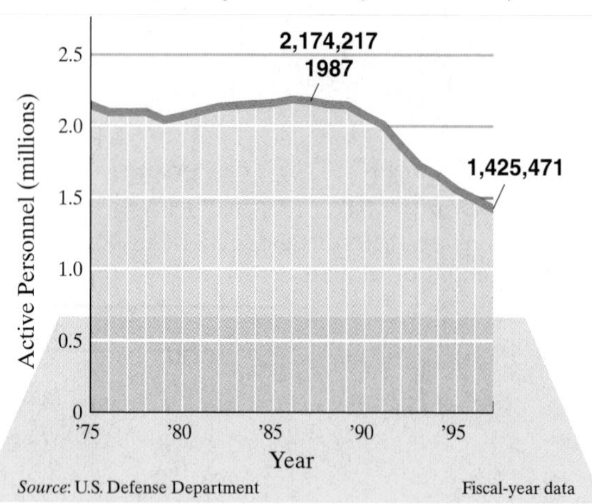

a) Does this graph represent a function? Explain.

b) Describe what the graph shows.

c) Estimate the number of active U.S. military personnel in 1993.

d) Estimate the percent decrease in U.S. military personnel from 1987 through 1997.

70. The following graph shows the change in homicide rates (per 100,000) committed by 14-to-17-year-olds in the United States from 1976 through 1996.

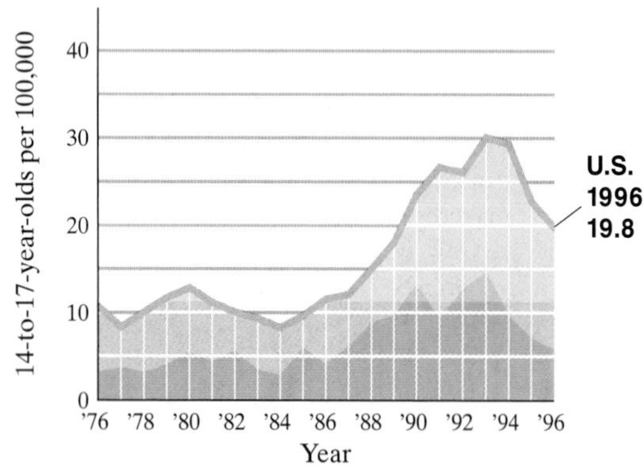

a) Does this graph represent a function? Explain.

b) In this graph, what is the independent variable?

c) If f represents the function, find $f(1993)$.

d) Using the value you determined for $f(1993)$ and the value shown on the graph, determine the percent decrease in the homicide rate from 1993 through 1996.

71. The following graph shows how the number of wireless (cellular) subscribers has been increasing and the avenge monthly bill has been decreasing.

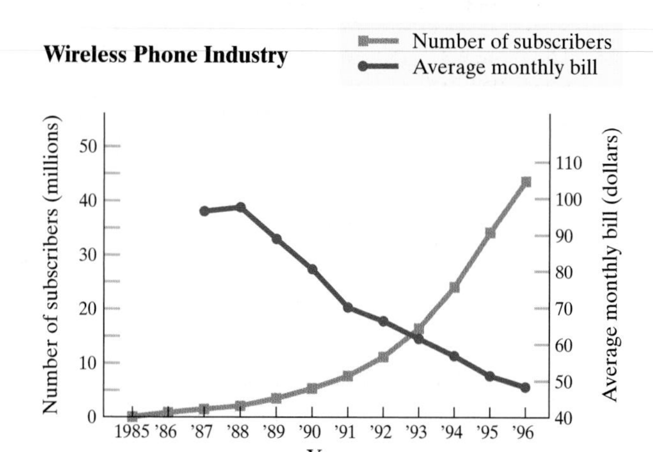

a) Do both graphs represent functions? Explain.

b) Would you say that the graph of the number of subscribers is approximately linear? Explain.

c) Would you say that the graph of the average bill is approximately linear from 1988 through 1996? Explain.

d) If f represents the number of subscribers and t is the year, determine t if $f(t) = 24$ million.

e) If g represents the average monthly bill and t is the year, determine t if $g(t) = \$70$.

72. The average face value of redeemed coupons (in cents) is given in the following bar graph.

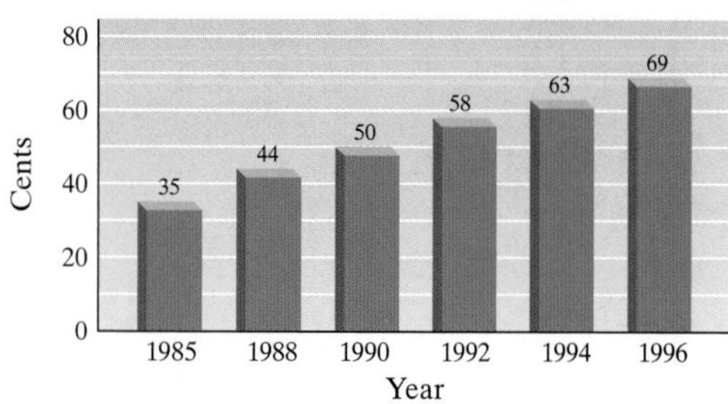

a) Draw a line graph that displays this information.

b) Does the graph you drew in part **a)** appear to be approximately linear? Explain.

c) From the graph you drew, estimate the average value of a coupon redeemed in 1993.

73. The average price of the cost of a 30-second commercial during the Super Bowl has been increasing over the years. The following chart gives the approximate cost of a 30-second commercial for selected years from 1981 through 1997.

Year	Cost ($1000s)
1981	280
1985	500
1989	740
1993	970
1997	1200

a) Draw a line graph that displays this information.

b) Does the graph appear to be approximately linear? Explain.

c) From the graph, estimate the cost of a 30-second commercial in 1995.

74. The average annual household expenditure is a function of the average annual household income. The average expenditure can be estimated by the function

$$f(i) = 0.6i + 5000 \qquad \$3500 \le i \le \$50{,}000$$

where $f(i)$ is the average household expenditure and i is the average household income.

a) Draw a graph showing the relationship between average household income and the average household expenditure.

b) Estimate the average household expenditure for a family whose average household income is $30,000.

75. The price of commodities, like soybeans, is determined by **supply and demand**. If too many soybeans are produced, the supply will be greater than the demand, and the price will drop. If not enough soybeans are produced, the demand will be greater than the supply, and the price of soybeans will rise. Thus the price of soybeans is a function of the number of bushels of soybeans produced. The price of a bushel of soybeans can be estimated by the function

$$f(Q) = -0.00004Q + 4.25, \qquad 10{,}000 \le Q \le 60{,}000$$

where $f(Q)$ is the price of a bushel of soybeans and Q is the annual number of bushels of soybeans produced.

a) Construct a graph showing the relationship between the number of bushels of soybeans produced and the price of a bushel of soybeans.

b) Estimate the cost of a bushel of soybeans if 40,000 bushels of soybeans are produced in a given year.

Group Activity

In many real-life situations, more than one function may be needed to represent a problem. This often occurs where two or more different rates are involved. For example, when discussing federal income taxes, there are different tax rates. When two or more functions are used to represent a problem, the function is called a ***piecewise function****. Following are two examples of piecewise functions and their graphs.*

$$f(x) = \begin{cases} -x + 2, & 0 \le x < 4 \\ 2x - 10, & 4 \le x < 8 \end{cases}$$

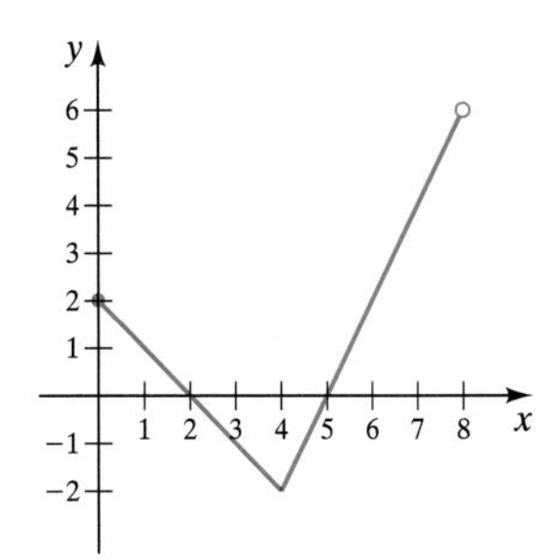

$$f(x) = \begin{cases} 2x - 1, & -2 \le x < 2 \\ x - 2, & 2 \le x < 4 \end{cases}$$

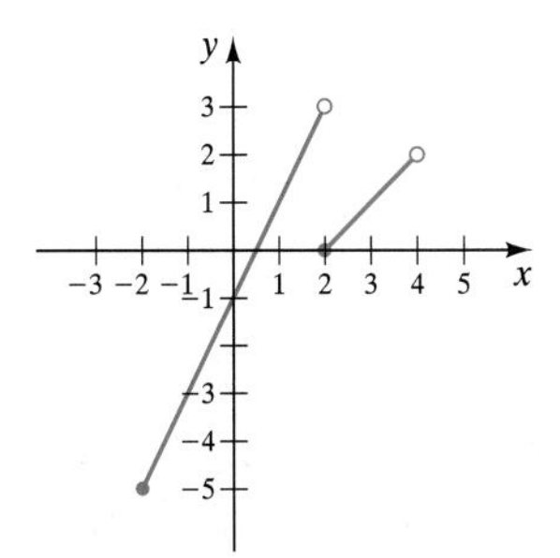

As a group, graph the following piecewise functions.

76. $f(x) = \begin{cases} x + 3, & -1 \le x < 2 \\ 7 - x, & 2 \le x < 4 \end{cases}$

77. $g(x) = \begin{cases} 2x + 3, & -3 < x < 0 \\ -3x + 1, & 0 \le x < 2 \end{cases}$

Cumulative Review Exercises

[2.1] **78.** Solve the equation $3x - 2 = \frac{1}{3}(3x - 3)$.

[2.2] **79.** Solve the following formula for p_2.

$$E = a_1 p_1 + a_2 p_2 + a_3 p_3$$

[2.5] **80.** Solve the inequality $\frac{3}{5}(x - 3) > \frac{1}{4}(3 - x)$ and indicate the solution **a)** on the number line; **b)** in interval notation; and **c)** in set builder notation.

[2.6] **81.** Solve the equation $\left|\frac{x - 4}{3}\right| + 2 = 4$.

3.3 LINEAR FUNCTIONS: GRAPHS AND APPLICATIONS

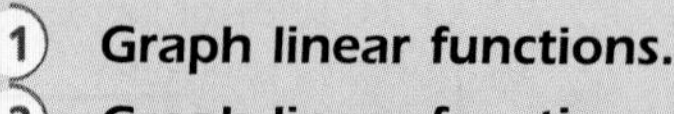

SSM VIDEO 3.3 CD Rom

1. Graph linear functions.
2. Graph linear functions using intercepts.
3. Graph equations of the form $x = a$ and $y = b$.
4. Study applications of functions.
5. Solve linear equations in one variable graphically.

1 Graph Linear Functions

In Section 3.1 we graphed linear equations. To graph the linear equation $y = 2x + 4$, we can make a table of values, plot the points, and draw the graph, as shown in Figure 3.32. Notice that this graph represents a function since it passes the vertical line test.

x	y
−2	0
0	4
1	6

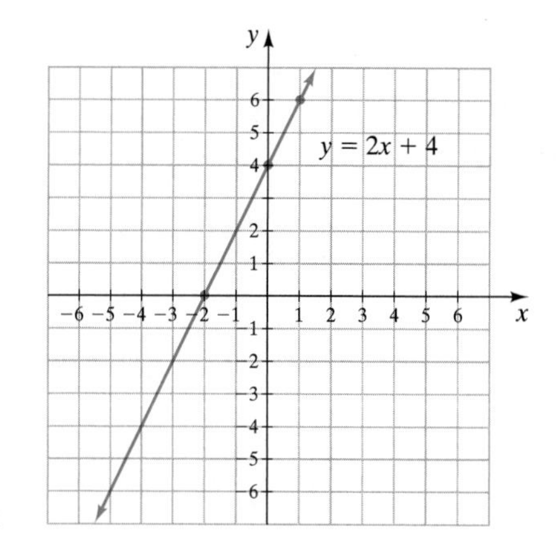

FIGURE 3.32

We may write the equation graphed in Figure 3.32 using function notation as $f(x) = 2x + 4$. This is an example of a linear function. A **linear function** is a function of the form $f(x) = ax + b$. The graph of any linear function is a straight line. The domain of any function is the set of real numbers for which the function is a real number. The domain of any linear function is the set of all real numbers, $\mathbb{R}$: Any real number, x, substituted in a linear function will result in $f(x)$ being a real number. We will discuss domains of functions further in Section 3.6.

To graph a linear function, we treat $f(x)$ as y and follow the same procedure used to graph linear equations.

EXAMPLE 1 Graph $f(x) = -3x + 2$.

Solution We construct a table of values by substituting values for x and finding the corresponding values of $f(x)$ or y. Then we plot the points and draw the graph, as illustrated in Figure 3.33.

x	$f(x)$
−1	5
0	2
2	−4

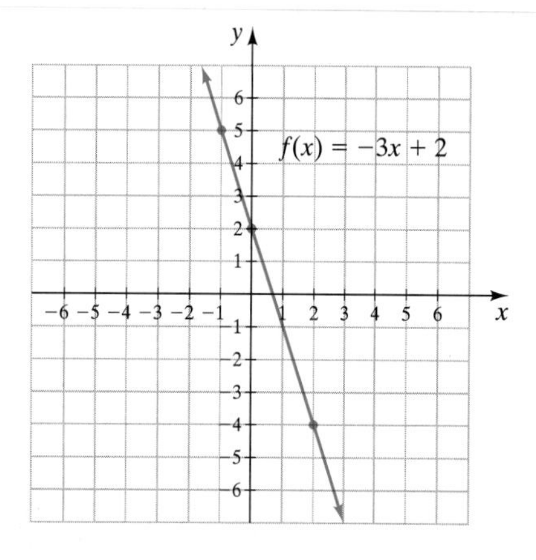

FIGURE 3.33

Note that the vertical axis in Figure 3.33 may also be labeled as $f(x)$ instead of y. In this book we will continue to label it y.

2) Graph Linear Functions Using Intercepts

Linear equations are not always given in the form $y = ax + b$. The equation $2x + 3y = 6$ is an example of a linear equation given in *standard form*.

Definition

The **standard form of a linear equation** is

$$ax + by = c$$

where, a, b, and c are real numbers, and a and b are not both 0.

NOW TRY EXERCISE 13

Examples of Linear Equations in Standard Form

$$2x + 3y = 4 \qquad -x + 5y = -2$$

Sometimes when an equation is given in standard form it may be easier to draw the graph using the x- and y-intercepts. Let's examine two points on the graph shown in Figure 3.32. Note that the graph crosses the x-axis at the point $(-2, 0)$. Therefore, $(-2, 0)$ is called the ***x*-intercept**. Sometimes we say that the x-intercept is *at* −2 (on the x-axis), the x-coordinate of the ordered pair.

The graph crosses the y-axis at the point $(0, 4)$. Therefore, $(0, 4)$ is called the ***y*-intercept**. Sometimes we will say that the y-intercept is *at* 4 (on the y-axis), the y-coordinate of the ordered pair.

Below we explain how the x- and y-intercepts may be determined algebraically.

To Find the x- and y-Intercepts

To find the y-intercept, set $x = 0$ and solve for y.

To find the x-intercept, set $y = 0$ and solve for x.

To graph a linear equation using the x- and y-intercepts, find the intercepts and plot the points. Then draw a straight line through the points. When graphing linear equations using the intercepts you must be very careful. If either of your points is plotted wrong, your graph will be wrong.

EXAMPLE 2 Graph the equation $3y = 6x + 12$ by plotting the x- and y-intercepts.

Solution To find the y-intercept (the point where the graph crosses the y-axis), set $x = 0$ and solve for y.

$$3y = 6x + 12$$
$$3y = 6(0) + 12$$
$$3y = 12$$
$$y = 4$$

The graph crosses the y-axis at $y = 4$. The ordered pair representing the y-intercept is $(0, 4)$.

To find the x-intercept (the point where the graph crosses the x-axis), set $y = 0$ and solve for x.

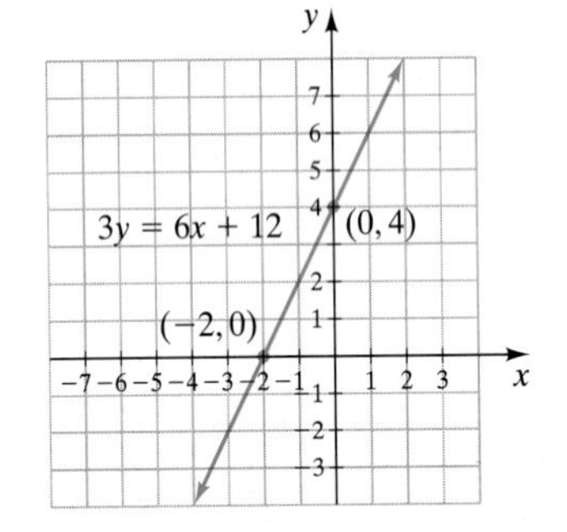

FIGURE 3.34

$$3y = 6x + 12$$
$$3(0) = 6x + 12$$
$$0 = 6x + 12$$
$$-12 = 6x$$
$$-2 = x$$

The graph crosses the x-axis at $x = -2$. The ordered pair representing the x-intercept is $(-2, 0)$. Now plot the intercepts and draw the graph (Fig. 3.34).

NOW TRY EXERCISE 25

EXAMPLE 3 Graph $f(x) = \frac{3}{2}x - 4$ by plotting the x- and y-intercepts.

Solution Treat $f(x)$ the same as y. To find the y-intercept, set $x = 0$ and solve for $f(x)$.

$$f(x) = \frac{3}{2}x - 4$$
$$f(x) = \frac{3}{2}(0) - 4 = -4$$

The y-intercept is $(0, -4)$.

To find the x-intercept, set $f(x) = 0$ and solve for x.

$$f(x) = \frac{3}{2}x - 4$$
$$0 = \frac{3}{2}x - 4$$
$$2(0) = 2\left(\frac{3}{2}x - 4\right) \quad \text{Multiply both sides by 2}$$

$$0 = 2\left(\frac{3}{2}x\right) - 2(4) \qquad \textit{Distributive property}$$

$$0 = 3x - 8$$

$$8 = 3x$$

$$\frac{8}{3} = x$$

The x-intercept is $\left(\frac{8}{3}, 0\right)$. The graph is shown in Figure 3.35.

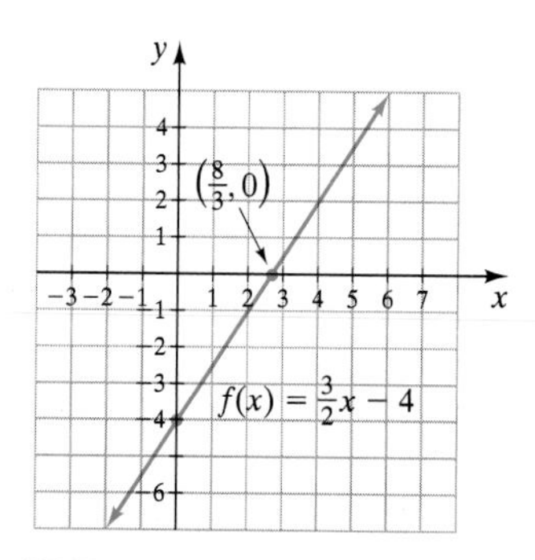

FIGURE 3.35

NOW TRY EXERCISE 19

Using Your Graphing Calculator

Sometimes it may be difficult to estimate the intercepts of a graph accurately. When this occurs, you might want to use a graphing calculator. We demonstrate how in the following example.

EXAMPLE Determine the x- and y-intercepts of the graph $y = 1.3(x - 3.2)$.

Solution Press the [Y=] key, then assign 1.3 $(x - 3.2)$ to y. Then press the [GRAPH] key to graph the function $y_1 = 1.3(x - 3.2)$, as shown in Figure 3.36.

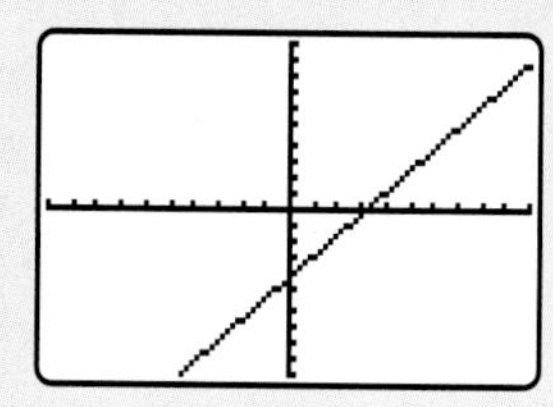

FIGURE 3.36

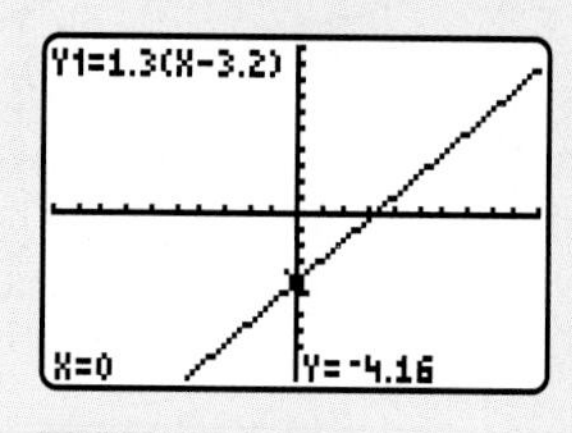

FIGURE 3.37

From the graph it may be difficult to determine the intercepts. One way to find the y-intercept is to use the TRACE feature, which was discussed in Section 3.1. Figure 3.37 shows a TI-83 screen after the [TRACE] key is pressed. Notice it gives the y-intercept is at −4.16.

Some graphing calculators have the ability to find the x-intercepts of a function by pressing just a few keys. A **zero** (or **root**) of a function is a value of x such that $f(x) = 0$. A zero (or root) of a function is the x-coordinate of the x-intercept of the graph of the function. Read your calculator manual to learn how to find the zeros or roots of a function. On a TI-82 or TI-83 you press the keys [2nd] [TRACE] to get to the CALC menu (which stands for calculate). Then you choose option 2, *zero**. Once the zero feature has been selected, the calculator will display

Left bound?**

At this time, move the cursor along the curve until it is to the *left* of the zero. Then press [ENTER]. The calculator now displays

Right bound?

Move the cursor along the curve until it is to the *right* of the zero. Then press [ENTER]. The calculator now displays

Guess?

Now press [ENTER] for the third time and the zero is displayed at the bottom of the screen, as in Figure 3.38. Thus the x-intercept of the function is at 3.2. For practice at finding the intercepts on your calculator, work exercises 61–64.

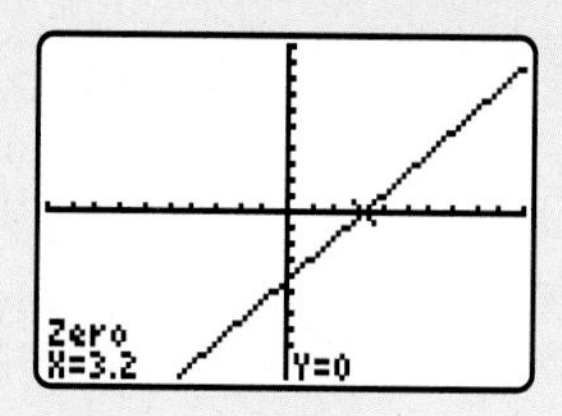

FIGURE 3.38

* The TI-82 uses the word *root* rather than *zero*.

** On the TI-82, the calculator displays Lower Bound? and Upper Bound? rather than Left Bound? and Right Bound?

3) Graph Equations of the Form $x = a$ and $y = b$

Examples 4 and 5 illustrate how equations of the form $x = a$ and $y = b$, where a and b are constants, are graphed.

EXAMPLE 4 Graph the equation $y = 3$.

Solution This equation can be written as $y = 3 + 0x$. Thus, for any value of x selected, y is 3. The graph of $y = 3$ is illustrated in Figure 3.39.

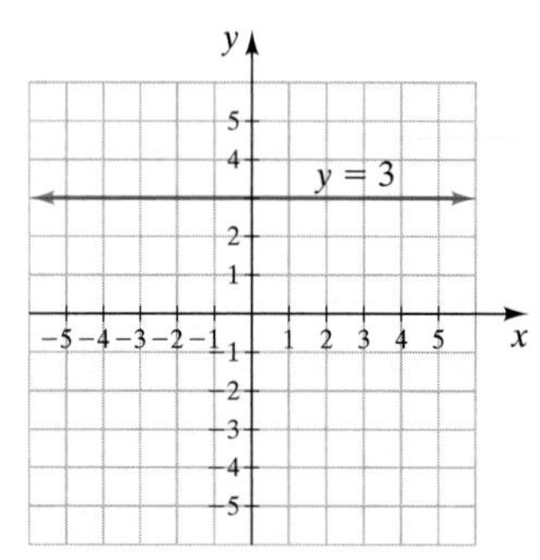

FIGURE 3.39

> The graph of any equation of the form $y = b$ will always be a horizontal line for any real number b.

Notice that the graph of $y = 3$ is a function since it passes the vertical line test. For each value of x selected, the value of y, or the value of the function, is 3. This is an example of a **constant function**. We may write

$$f(x) = 3$$

Any equation of the form $y = b$ or $f(x) = b$, where b represents a constant, is a constant function.

EXAMPLE 5 Graph the equation $x = -2$.

Solution This equation can be written as $x = -2 + 0y$. Thus, for every value of y selected, x will have a value of -2 (Fig. 3.40).

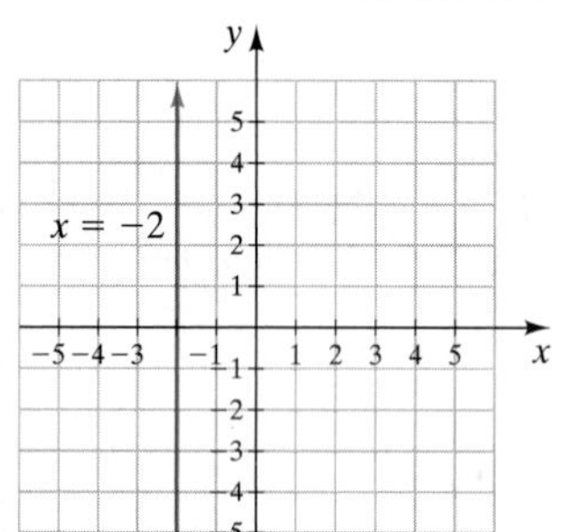

FIGURE 3.40

> The graph of any equation of the form $x = a$ will always be a vertical line for any real number a.

Notice that the graph of $x = -2$ does not represent a function since it does not pass the vertical line test. For $x = -2$ there is more than one value of y. In fact, when $x = -2$ there are an infinite number of values for y.

NOW TRY EXERCISE 35

4) Study Applications of Functions

Graphs are often used to show the relationship between variables. The axes of a graph do not have to be labeled x and y; they can be any designated variables. Consider the following example.

EXAMPLE 6 The yearly profit, p, of a tire store can be estimated by the function $p(n) = 20n - 30{,}000$, where n is the number of tires sold per year.

a) Draw a graph of profit versus tires sold for up to and including 6000 tires.

b) Estimate the number of tires that must be sold for the company to break even.

c) Estimate the number of tires sold if the company has a \$40,000 profit.

Solution **a) Understand** The profit, p, is a function of the number of tires sold, n. The horizontal axis will therefore be labeled number of tires sold (the independent variable) and the vertical axis will be labeled profit (the dependent variable). Since the minimum number of tires that can be sold is 0, negative values do not have to be listed on the horizontal axis. The horizontal axis will therefore go from 0 to 6000 tires.

We will graph this equation by determining and plotting the intercepts.

Translate and Carry Out To find the p-intercept, we set $n = 0$ and solve for $p(n)$.

$$p(n) = 20n - 30{,}000$$

$$p(n) = 20(0) - 30{,}000 = -30{,}000$$

Thus, the p-intercept is $(0, -30{,}000)$

To find the n-intercept, we set $p(n) = 0$ and solve for n.

$$p(n) = 20n - 30{,}000$$

$$0 = 20n - 30{,}000$$

$$30{,}000 = 20n$$

$$1500 = n$$

Thus the n-intercept is $(1500, 0)$.

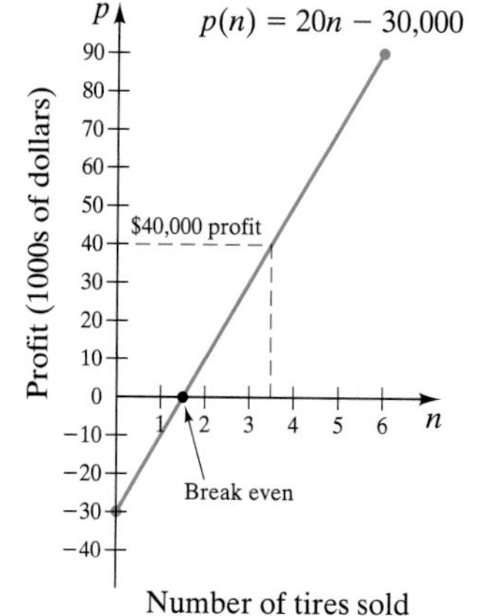

FIGURE 3.41

Answer Now we use the p- and n-intercepts to draw the graph, see Figure 3.41.

b) The break-even point is the number of tires that must be sold for the company to have neither a profit nor a loss. The break-even point is where the graph intersects the n-axis, for this is where the profit, p, is 0. To break even, approximately 1500 tires must be sold.

c) To make $40,000, approximately 3500 tires must be sold (the dashed red line in Figure 3.41).

Sometimes it is difficult to read an exact answer from a graph. To determine the exact number of tires needed to break even in Example 6, substitute 0 for $p(n)$ in the function $p(n) = 20n - 30{,}000$ and solve for n. To determine the exact number of tires needed to obtain a $40,000 profit, substitute 40,000 for $p(n)$ and solve the equation for n.

EXAMPLE 7 Brian Goldenberg is part owner in a newly formed toy company. His monthly salary consists of $200 plus 10% of his sales for that month.

a) Write a function expressing his monthly salary, m, in terms of his sales, s.

b) Draw a graph of his monthly salary for sales up to and including $20,000.

c) If his sales for the month of April are $15,000, what will be Brian Goldenberg's salary for April?

Solution **a)** Brian's monthly salary is a function of sales. His monthly salary, m, consists of $200 plus 10% of the sales, s. Ten percent of s is $0.10s$. Thus the function for finding his salary is

$$m(s) = 200 + 0.10s$$

b) Since monthly salary is a function of sales, sales will be represented on the horizontal axis and monthly salary will be represented on the vertical axis. Since sales can never be negative, the monthly salary can never be negative. Thus both axes will be drawn with only positive numbers. We will draw this graph by plotting points. We select values for s, find the corresponding values

s	m
0	200
10,000	1200
20,000	2200

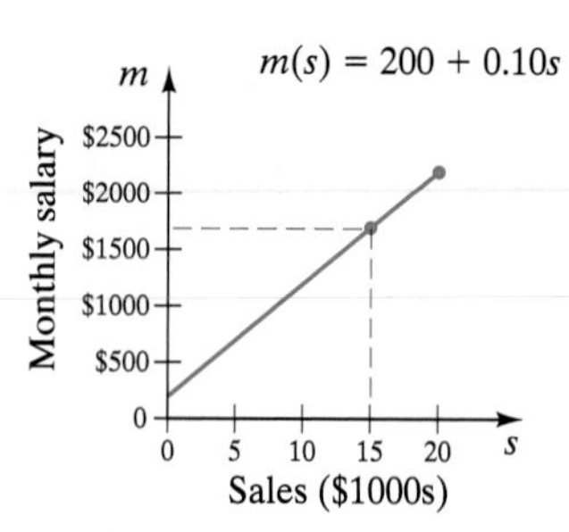

FIGURE 3.42

of m, and then draw the graph. We can select values of s that are between \$0 and \$20,000 (Fig. 3.42).

c) By reading our graph carefully, we can estimate that when his sales are \$15,000, Brian's monthly salary is about \$1700.

NOW TRY EXERCISE 45

5 Solve Linear Equations in One Variable Graphically

Earlier we discussed the graph of $f(x) = 2x + 4$. In Figure 3.43 we illustrate this graph of $f(x)$ along with the graph of $g(x) = 0$. Notice that the two graphs intersect at $(-2, 0)$. We can obtain the x-coordinate of the ordered pair by solving the equation $f(x) = g(x)$. Remember $f(x)$ and $g(x)$ both represent y, and by solving this equation for x we are obtaining the value of x where the y's are equal.

$$\begin{aligned} f(x) &= g(x) \\ \overbrace{2x + 4} &= \overbrace{-0} \\ 2x &= -4 \\ x &= -2 \end{aligned}$$

Note that we obtain -2, the x-coordinate in the ordered pair at the point of intersection.

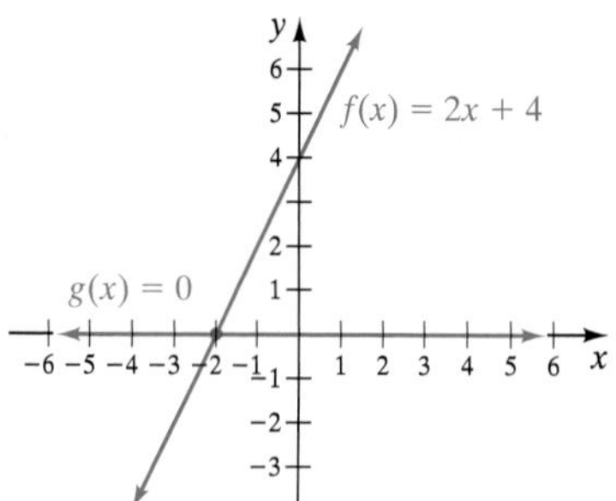

FIGURE 3.43

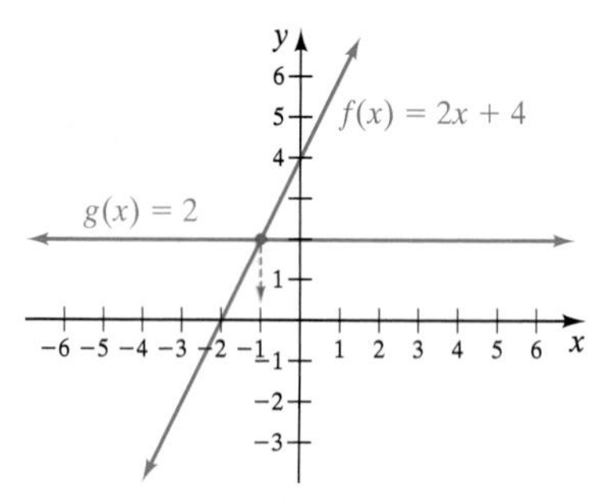

FIGURE 3.44

Now let's find the x-coordinate of the point at which the graphs of $f(x) = 2x + 4$ and $g(x) = 2$ intersect. We solve the equation $f(x) = g(x)$.

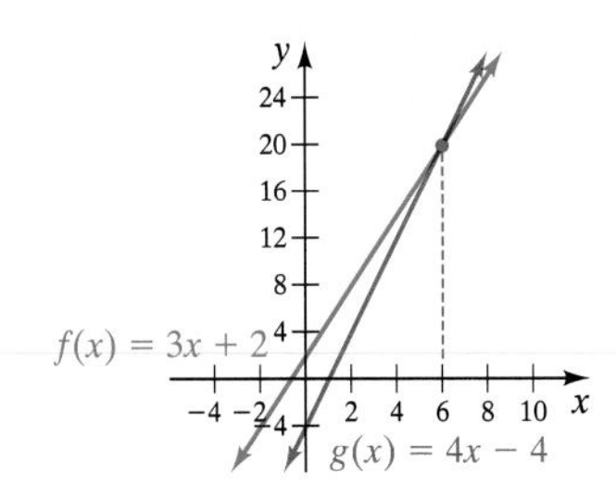

FIGURE 3.45

$$f(x) = g(x)$$

$$2x + 4 = -2$$

$$2x = -2$$

$$x = -1$$

The x-coordinate of the intersection of the two graphs is -1, as shown in Figure 3.44. Notice that $f(-1) = 2(-1) + 4 = 2$.

In general, if we are given an equation in one variable we can regard each side of the equation as a separate function. To obtain the solution to the equation, we can graph the two functions. The x-coordinate of the intersection will be the solution to the equation.

EXAMPLE 8 Find the solution to the equation $3x + 2 = 4x - 4$ graphically.

Solution Let $f(x) = 3x + 2$ and $g(x) = 4x - 4$. The graph of these functions is illustrated in Figure 3.45. The x-coordinate of the point of intersection is 6. Thus, the solution to the equation is 6. Check the solution now.

NOW TRY EXERCISE 57

Using Your Graphing Calculator

In Example 8 we solved an equation in one variable by graphing two functions. In the following example we explain how to find the intersection of two functions on a graphing calculator.

EXAMPLE Use a graphing calculator to find the solution to $2(x + 3) = \frac{1}{2}x + 4$.

Solution Assign $2(x + 3)$ to Y_1 and assign $\frac{1}{2}x + 4$ to Y_2 to get

$$Y_1 = 2(x + 3)$$

$$Y_2 = \frac{1}{2}x + 4$$

Now press the GRAPH key to graph the functions. The graph of the functions is shown in Figure 3.46.

By examining the graph can you determine the x-coordinate of the point of intersection? Is it -1, or -1.5, or some other value? We can determine the point of intersection in a number of different ways. One method involves using the TRACE and ZOOM features. Figure 3.47 shows the window of a TI-83 after the TRACE feature has been used and the cursor has been moved close to the intersection. (Note that pressing the up and down arrows switches the cursor from one function to the other.)

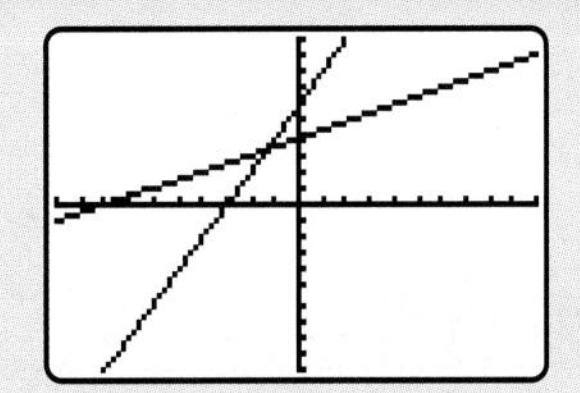

FIGURE 3.46

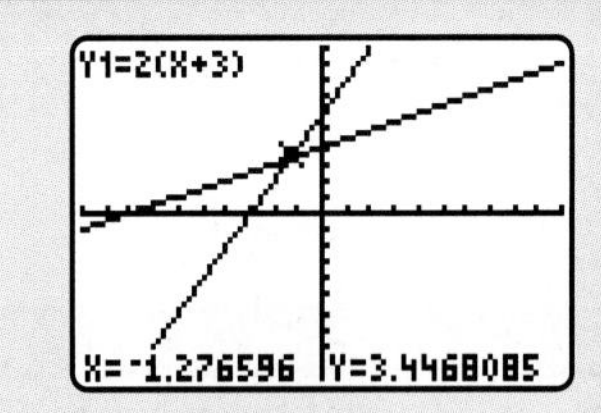

FIGURE 3.47

At the bottom of the screen in Figure 3.47, you see the x- and y-coordinates at the cursor. To get a closer view around the area of the cursor, you can *zoom in* using the ZOOM key. After you zoom in, you can move the cursor closer to the point of intersection and get a better reading (Fig. 3.48). You can do this over and over until you get as accurate an answer as you need. It appears from

continued on next page

Figure 3.48 that the x-coordinate of the intersection is about -1.33.

Graphing calculators can also display the intersection of two graphs from the use of certain keys. The keys to press depend on your calculator. Read your calculator manual to determine how to do this. This procedure is generally quicker and easier to use to find the intersection of two graphs.

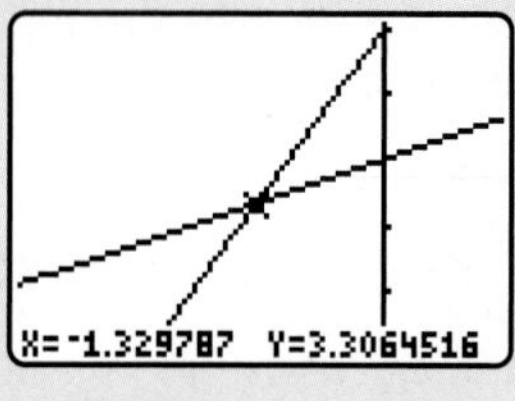

FIGURE 3.48

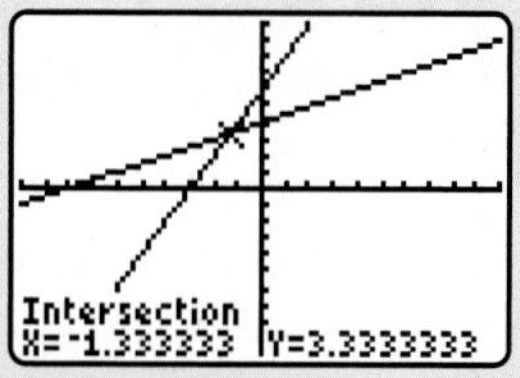

FIGURE 3.49

On the TI-82 or TI-83, select option 5: INTERSECT—from the CALC menu to find the intersection. Once the INTERSECT feature has been selected, the calculator will display

First curve?

At this time, move the cursor along the first curve until it is close to the point of intersection. Then press ENTER. The calculator will next display

Second curve?

with the cursor on the second curve. If the cursor is not close to the point of intersection, move it along this curve until this happens. Then press ENTER. Next the calculator will display

Guess?

Now press ENTER, and the point of intersection will be displayed.

Figure 3.49 shows the window after this procedure has been done. We see that the x-coordinate of the intersection is $-1.333\ldots$ or $-1\frac{1}{3}$ and the y-coordinate of the intersection is $3.333\ldots$ or $3\frac{1}{3}$.

For practice in using a graphing calculator to solve an equation in one variable, work Exercises 57–60.

Exercise Set 3.3

Concept/Writing Exercises

1. What is the standard form of a linear equation?
2. If you are given a linear equation in standard form, and wish to write the equation using function notation, how would you do it?
3. Explain how to find the x- and y-intercepts of the graph of an equation.
4. What terms do graphing calculators use to indicate the x-intercepts?
5. What will the graph of $y = b$ look like for any real number b?
6. What will the graph of $f(x) = b$ look like for any real number b?
7. What will the graph of $x = a$ look like for any real number a?
8. Is the graph of $x = a$ a function? Explain.
9. Explain how to solve an equation in one variable graphically.
10. Explain how to solve the equation $3(x - 2) = 6x + 4$ graphically.

Practice the Skills

Write each equation in standard form.

11. $y = 3x - 2$
12. $2 = 6x - 3y$
13. $2x = 3y - 4$
14. $y = \frac{1}{2}x - 6$
15. $3(x - 2) = 4(y - 5)$
16. $\frac{1}{3}y = 2(x - 3) + 4$

Graph each equation using the x- and y-intercepts.

17. $y = 3x - 6$

18. $y = -2x + 6$

19. $f(x) = 2x + 3$

20. $f(x) = -6x + 5$

21. $y = 4x - 8$

22. $4y + 3x = 12$

23. $4x = 3y - 9$

24. $\frac{1}{2}x + 2y = 4$

25. $30x + 25y = 50$

26. $0.6x - 1.2y = 2.4$

27. $0.25x + 0.50y = 1.00$

28. $-1.6y = 0.4x + 9.6$

29. $120x - 360y = 720$

30. $20x - 240 = -60y$

31. $\frac{1}{3}x + \frac{1}{4}y = 12$

32. $\frac{1}{6}x + \frac{1}{2}y = -1$

Graph each equation.

33. $y = 4$

34. $x = 6$

35. $x = -3$

36. $y = 5$

37. $f(x) = -3$

38. $x = \frac{5}{2}$

39. $g(x) = 0$

40. $x = 0$

Problem Solving

41. Using the distance formula

$$\text{distance} = \text{rate} \cdot \text{time, or } d = rt$$

draw a graph of distance versus time for a constant rate of 50 miles per hour.

42. Using the simple interest formula

$$\text{interest} = \text{principal} \cdot \text{rate} \cdot \text{time, or } i = prt$$

draw a graph of interest versus time for a principal of \$1000 and a rate of 8%.

43. The profit of a bicycle manufacturer can be approximated by the function $p(x) = 60x - 80{,}000$, where x is the number of bicycles produced and sold.

a) Draw a graph of profit versus the number of bicycles sold (for up to 5000 bicycles).

b) Estimate the number of bicycles that must be sold for the company to break even.

c) Estimate the number of bicycles that must be sold for the company to make \$150,000 profit.

44. Jack Fredrick's weekly cost of operating a taxi is \$50 plus 12 cents per mile.

a) Write a function expressing the weekly cost, c, in terms of the number of miles, m.

b) Draw a graph illustrating weekly cost versus the number of miles, up to 200, driven per week.

c) How many miles would Jack have to drive for the weekly cost to be \$70?

d) If the weekly cost is \$60, how many miles did Jack drive?

45. Trudy Belluschi's weekly salary is \$300 plus 10% commission on her weekly sales.

a) Write a function expressing Trudy's weekly salary, s, in terms of her weekly sales, x.

b) Draw a graph of Trudy's weekly salary versus her weekly sales, for up to \$5000 in sales.

c) What is Trudy's weekly salary if her sales were \$4000?

d) If her salary for the week is \$600, what are her weekly sales?

46. Lynn Hicks, a real estate agent, makes \$150 per week plus a 1% sales commission on each property she sells.

a) Write a function expressing her weekly salary, s, in terms of sales, x.

b) Draw a graph of her salary versus her weekly sales, for sales up to \$100,000.

c) If she sells one house per week for \$80,000, what will her weekly salary be?

47. The following graph shows weight, in kilograms, for girls (up to 36 months of age) versus lengths (or heights), in centimeters. The red line is the average weight for all girls of the given length, and the green lines represent the upper and lower limits of the normal range.

a) Explain why the red line represents a function.

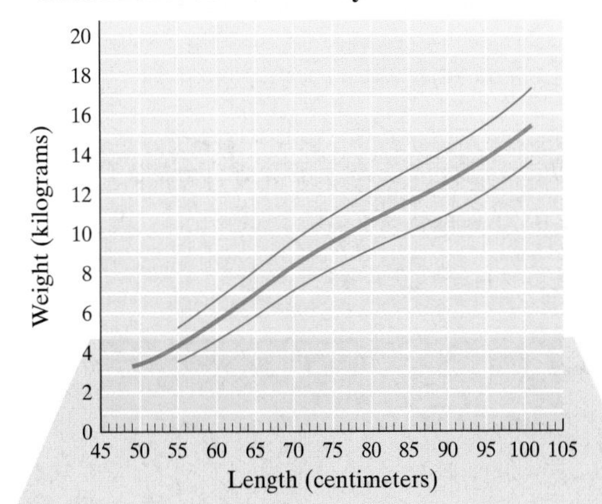

b) What is the independent variable? What is the dependent variable?

c) Is the graph of weight versus length approximately linear?

d) What is the weight in kilograms of the average girl who is 85 centimeter long?

e) What is the average length in centimeters of the average girl with a weight of 7 kilograms?

f) What weights are considered normal for a girl 95 centimeters long?

g) What is happening to the normal range as the lengths increase? Is this what you would expect to happen? Explain.

48. The following graph illustrates the effect of compound interest.

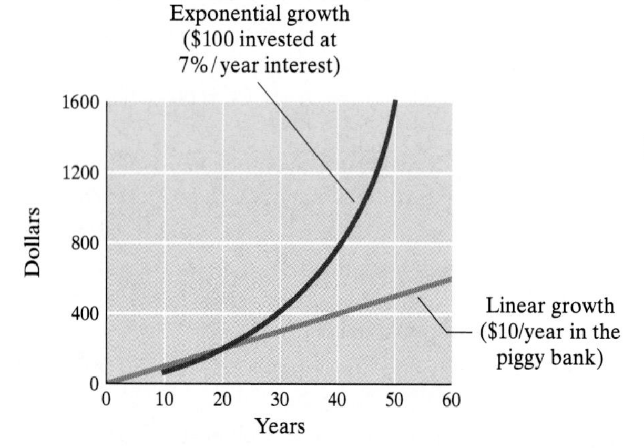

If a child puts $10 each year in a piggy bank, the savings will grow linearly, as shown by the lower curve. If, at age 10, the child invests $100 at 7% interest compounded annually, that $100 will grow exponentially.

a) Explain why both graphs represent functions.

b) What is the independent variable? What is the dependent variable?

c) Using the linear growth curve, determine how long it would take to save $600.

d) Using the exponential growth curve, determine how long after the account is opened would the amount reach $600.

e) Starting at year 20, how long would it take for the money growing at a linear rate to double?

f) Starting at year 20, how long would it take for the money growing exponentially to double? (Exponential growth will be discussed at length in Chapter 9.)

49. When, if ever, will the x- and y-intercepts of a graph be the same? Explain.

50. Write two linear functions whose x- and y-intercepts are both $(0, 0)$.

51. Write a function whose graph will have no x-intercept but will have a y-intercept at $(0, 4)$.

52. Write an equation whose graph will have no y-intercept but will have an x-intercept at $(3, 0)$.

53. If the x- and y-intercepts of a linear function are at 2 and -3, respectively, what will be the new x- and y-intercepts if the graph is moved (or translated) up 3 units?

54. If the x- and y-intercepts of a linear function are -1 and 3, respectively, what will be the new x- and y-intercepts if the graph is moved (or translated) down 4 units?

In Exercises 55 and 56 we give two ordered pairs, which are the x- and y-intercepts of a graph. ***a)*** *Plot the points and draw the line through the points.* ***b)*** *Find the change in y, or the vertical change, between the intercepts.* ***c)*** *Find the change in x, or the horizontal change, between the intercepts.* ***d)*** *Find the ratio of the vertical change to the horizontal change between these two points. Do you know what this ratio represents? (We will discuss this further in Section 3.4.)*

55. $(0, 2)$ and $(-4, 0)$

56. $(3, 5)$ and $(-1, -1)$

Solve each equation for x as done in Example 8. Use a graphing calculator if one is available. If not, draw the graphs yourself.

57. $3(x + 2) = 6x + 12$

58. $-2(x - 2) = 3(x + 6) + 1$

59. $2.4x - 3.6 = 1.6x - 4.8$

60. $2x + \frac{1}{4} = 5x - \frac{1}{2}$

Find the x- and y-intercepts of the graph of each equation on your graphing calculator.

61. $y = 3(x - 1.2)$

62. $3x - 5y = 12$

63. $-4x - 3.2y = 8$

64. $y = \frac{3}{5}x - \frac{1}{2}$

Cumulative Review Exercises

[2.6] *In Exercises 65–67* **a)** *explain the procedure to solve the equation or inequality for x (assume that $b > 0$) and* **b)** *solve the equation or inequality.*

65. $|x - a| = b$

66. $|x - a| < b$

67. $|x - a| > b$

68. Solve the equation $|x - 4| = |2x - 2|$.

3.4 THE SLOPE–INTERCEPT FORM OF A LINEAR EQUATION

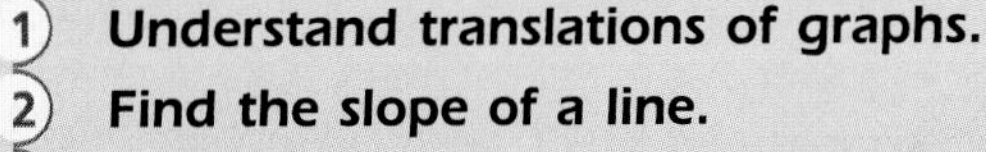

SSM VIDEO 3.4 CD Rom

1. Understand translations of graphs.
2. Find the slope of a line.
3. Recognize slope as a rate of change.
4. Write linear equations in slope–intercept form.
5. Graph linear equations using the slope and the y-intercept.
6. Use the slope–intercept form to construct models from graphs.

1 Understand Translations of Graphs

In this section we discuss the translations of graphs, the concept of slope, and the slope–intercept form of a linear equation.

Consider the three equations

$$y = 2x + 3$$
$$y = 2x$$
$$y = 2x - 3$$

Each equation is graphed in Figure 3.50.

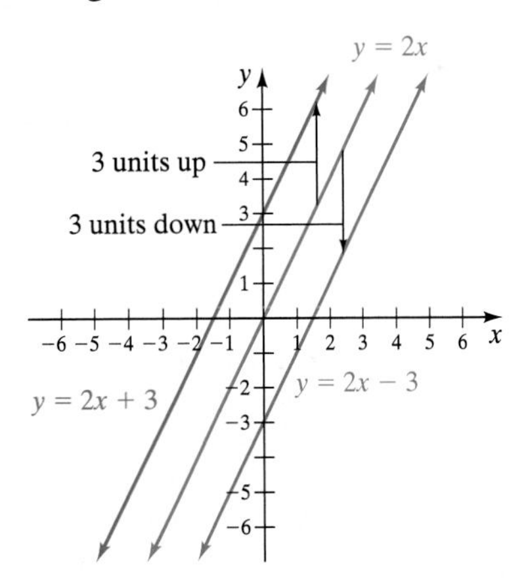

FIGURE 3.50

What are the y-intercepts of $y = 2x + 3$, $y = 2x$ (or $y = 2x + 0$), and $y = 2x - 3$? The y-intercepts are at $(0, 3)$, $(0, 0)$, and $(0, -3)$, respectively. Notice that the graph of $y = 2x + 3$ is the graph of $y = 2x$ shifted, or **translated**, 3 units up and $y = 2x - 3$ is the graph of $y = 2x$ translated 3 units down. All three lines are **parallel**; that is, they do not intersect no matter how far they are extended.

Using this information, can you guess what the y-intercept of $y = 2x + 4$ will be? How about the y-intercept of $y = 2x - \frac{5}{3}$? If you answered $(0, 4)$ and $\left(0, -\frac{5}{3}\right)$, respectively, you answered correctly. In fact, the graph of an equation of the form $y = 2x + b$ will have a y-intercept of $(0, b)$.

Now consider the graphs of the equations $y = -\frac{1}{3}x + 4$, $y = -\frac{1}{3}x$ and $y = -\frac{1}{3}x - 2$, shown in Figure 3.51. The y-intercepts of the 3 lines are $(0, 4)$, $(0, 0)$, and $(0, -2)$, respectively. The graph of $y = -\frac{1}{3}x + b$ will have a y-intercept of $(0, b)$.

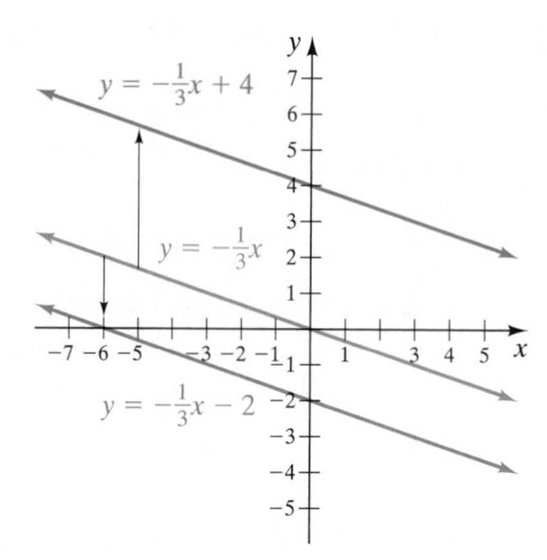

FIGURE 3.51

By looking at the preceding equations, their graphs, and y-intercepts, can you determine the y-intercept of the graph of $y = mx + b$ where m and b are real numbers is $(0, b)$? If you answered $(0, b)$, you answered correctly. In general, the graph of $y = mx + \mathbf{b}$ where m and b are real numbers, has a y-intercept $(0, \mathbf{b})$.

If we look at the graphs in Figure 3.50, we see that the slopes (or slants) of the three lines appear to be the same. If we look at the graphs in Figure 3.51, we see that the slopes of those three lines appear to be the same, but their slope is different from the slope of the three lines in Figure 3.50.

If we consider the equation $y = mx + b$, where the b determines the y-intercept of the line, we can reason that the m is responsible for the slope (or the slant) of the line.

2) Find the Slope of a Line

Now let's speak about slope. The **slope of a line** is the ratio of the vertical change (or rise) to the horizontal change (or run) between any two points on a line. Consider the graph of $y = 2x$ (the blue line in Figure 3.50, repeated in Figure 3.52a). Two points on this line are $(1, 2)$ and $(3, 6)$. Let's find the slope of the line through these points. If we draw a line parallel to the x-axis through the point $(1, 2)$ and a line parallel to the y-axis through the point $(3, 6)$, the two lines intersect at $(3, 2)$, see Figure 3.52b.

From Figure 3.52b we can determine the slope of the line. The vertical change (along the y-axis) is $6 - 2$, or 4 units. The horizontal change (along the x-axis) is $3 - 1$, or 2 units.

$$\text{slope} = \frac{\text{vertical change}}{\text{horizontal change}} = \frac{4}{2} = 2$$

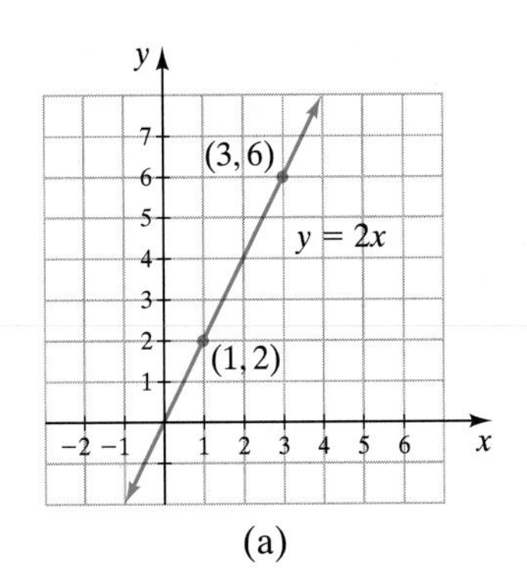

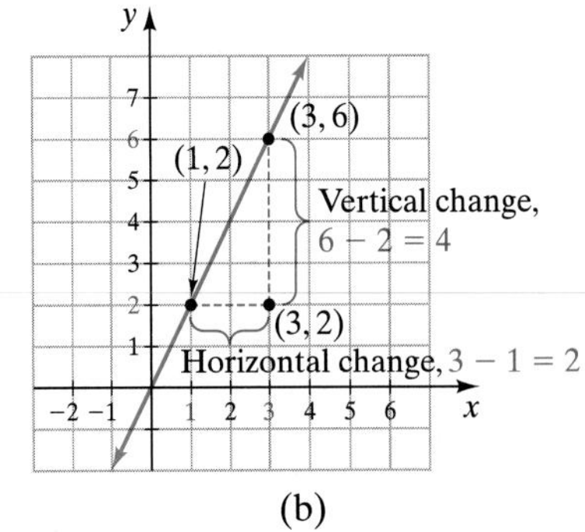

FIGURE 3.52 (a) (b)

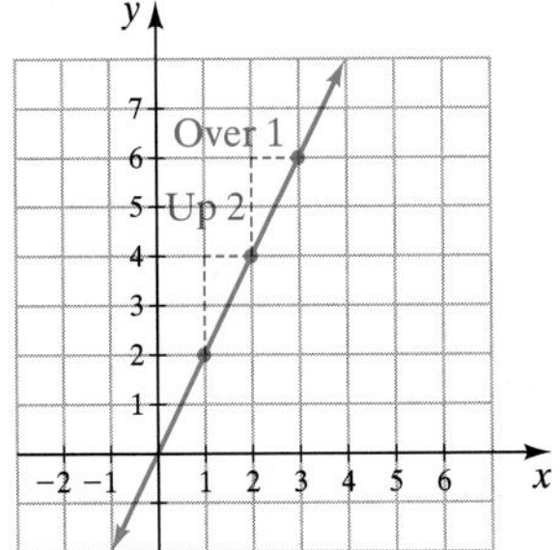

FIGURE 3.53

Thus, the slope of the line through the points (3, 6) and (1, 2) is 2. By examining the line connecting these two points, we can see that for each 2 units the graph moves up the y-axis it moves 1 unit to the right on the x-axis (Fig. 3.53).

We have determined that the slope of the graph of $y = 2x$ is 2. If you were to compute the slope of the other two lines in Figure 3.50, you would find that the graphs of $y = 2x + 3$ and $y = 2x - 3$ also have a slope of 2.

Can you guess what the slope of the graphs of the equations $y = -3x + 2$, $y = -3x$, and $y = -3x - 2$ is? The slope of all three lines is −3. In general, the slope of an equation of the form $y = mx + b$ is m.*

Now let's determine the procedure to find the slope of a line passing through the two points (x_1, y_1) and (x_2, y_2). Consider Figure 3.54. The vertical change can be found by subtracting y_1 from y_2. The horizontal change can be found by subtracting x_1 from x_2.

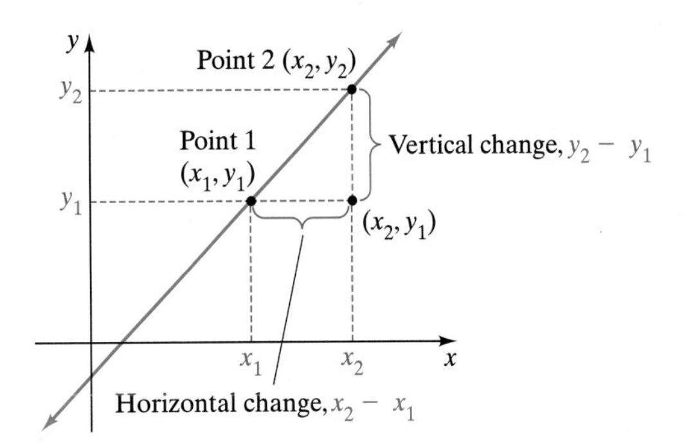

FIGURE 3.54

Definition

The **slope** of the line through the distinct points (x_1, y_1) and (x_2, y_2) is

$$\text{slope} = \frac{\text{change in } y \text{ (vertical change)}}{\text{change in } x \text{ (horizontal change)}} = \frac{y_2 - y_1}{x_2 - x_1}$$

provided that $x_1 \neq x_2$.

*The letter m, is traditionally used for slope. It is believed m comes from the French word *monter*, which means to climb.

It makes no difference which two points on the line are selected when finding the slope of a line. It also makes no difference which point you label (x_1, y_1) or (x_2, y_2). As mentioned before, the letter m is used to represent the slope of a line. The Greek capital letter delta, Δ, is used to represent the words "the change in." Thus, the slope is sometimes indicated as

$$m = \frac{\Delta y}{\Delta x} = \frac{y_2 - y_1}{x_2 - x_1}$$

EXAMPLE 1 Find the slope of the line in Figure 3.55.

Solution Two points on the line are $(-2, 3)$ and $(1, -4)$. Let $(x_2, y_2) = (-2, 3)$ and $(x_1, y_1) = (1, -4)$. Then

$$m = \frac{y_2 - y_1}{x_2 - x_1} = \frac{3 - (-4)}{-2 - 1} = \frac{3 + 4}{-3} = -\frac{7}{3}$$

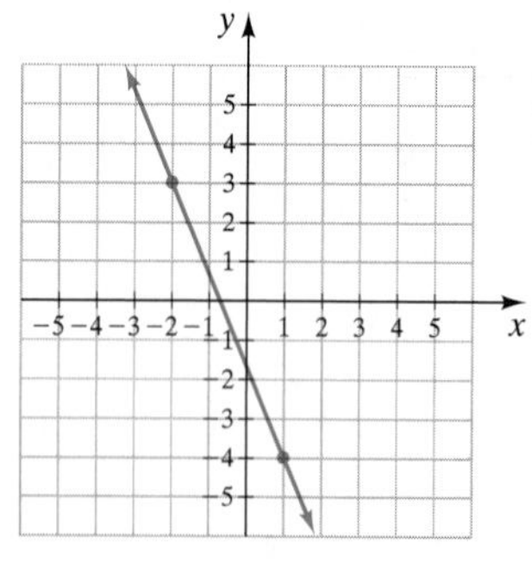
FIGURE 3.55

NOW TRY EXERCISE 37

The slope of the line is $-\frac{7}{3}$. Note that if we had let $(x_1, y_1) = (-2, 3)$ and $(x_2, y_2) = (1, -4)$, the slope would still be $-\frac{7}{3}$. Try it and see.

A line that rises going from left to right (Fig. 3.56a) has a **positive slope**. A line that neither rises nor falls going from left to right (Fig. 3.56b) has **zero slope**. A line that falls going from left to right (Fig. 3.56c) has a **negative slope**.

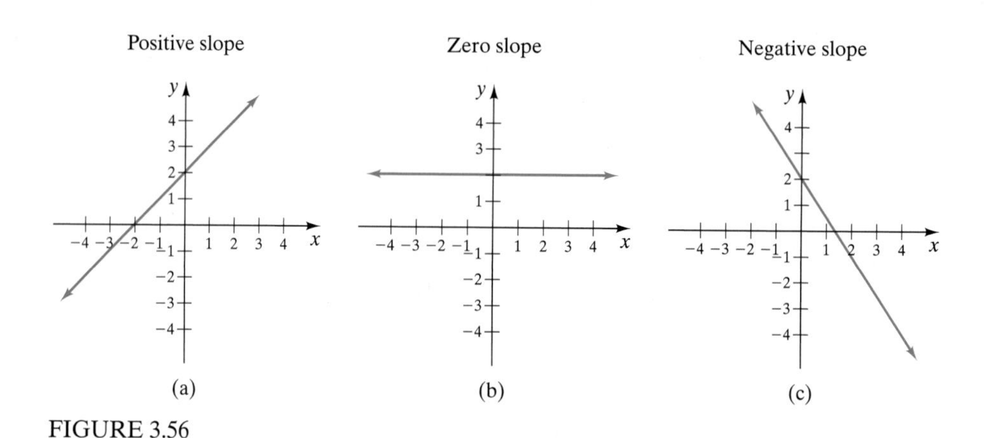

FIGURE 3.56

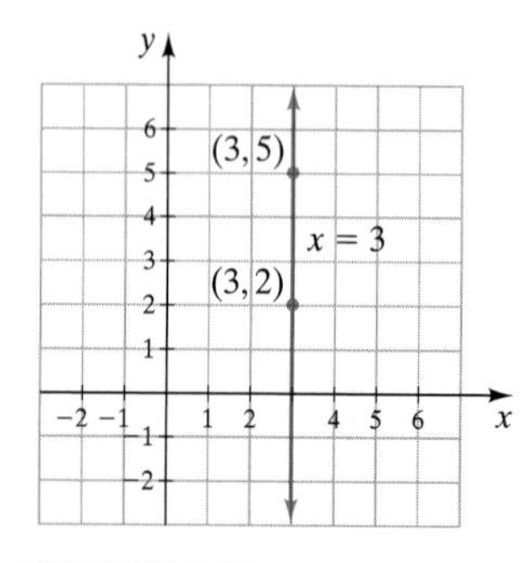

FIGURE 3.57

NOW TRY EXERCISE 19

Consider the graph of $x = 3$ (Fig. 3.57). What is its slope? The graph is a vertical line and goes through the points (3, 2) and (3, 5). Let the point (3, 5) represent (x_2, y_2) and let (3, 2) represent (x_1, y_1). Then the slope of the line is

$$m = \frac{y_2 - y_1}{x_2 - x_1} = \frac{5 - 2}{3 - 3} = \frac{3}{0}$$

Since it is meaningless to divide by 0, we say that the slope of this line is undefined. **The slope of any vertical line is undefined.**

HELPFUL HINT

When students are asked to give the slope of a horizontal or a vertical line, they often answer incorrectly. When asked for the slope of a horizontal line, your response should be "the slope is 0." If you give your answer as "no slope," your instructor may well mark it wrong since these words may have various interpretations. When asked for the slope of a vertical line, your answer should be "the slope is undefined." Again, if you use the words "no slope," this may be interpreted differently by your instructor and marked wrong.

3) Recognize Slope as a Rate of Change

Sometimes it is helpful to describe slope as a *rate of change.* Consider a slope of $\frac{5}{3}$. This means that the y-value increases 5 units for each 3-unit increase in x. Equivalently, we can say that the y-value increases $\frac{5}{3}$ units, or 1.666 ... units, for each 1-unit increase in x. When we give the change in y per unit change in x we are giving the slope as a **rate of change**. When discussing real-life situations or when creating mathematical models, it is often useful to discuss slope as a rate of change.

EXAMPLE 2 The following table of values and the corresponding graph (Fig. 3.58) illustrate the U.S. public debt in billions of dollars from 1910 through 1996.

Year	U.S. Public Debt (billion of dollars)
1910	1.1
1930	16.1
1950	256.1
1970	370.1
1990	3323.3
1996	5207.0

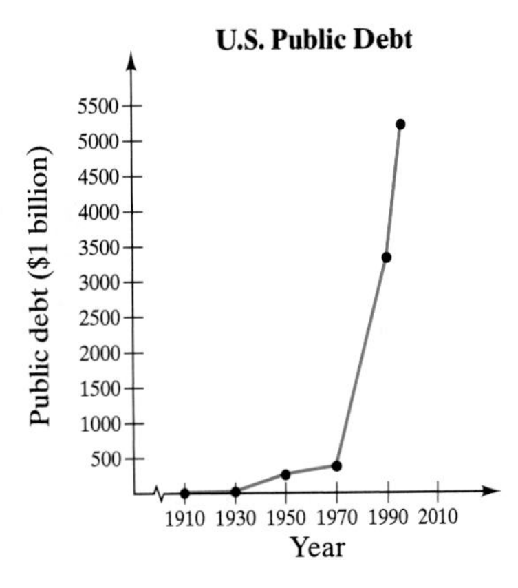

FIGURE 3.58

a) Determine the slope of the line segments between 1910 and 1930 and between 1990 and 1996.

b) Compare the two slopes found in part **a)** and explain what this means in terms of the U.S. public debt.

Solution **a) Understand** To find the slope between any two years, find the ratio of the change in debt to the change in years.

Slope from 1910 to 1930

$$m = \frac{16.1 - 1.1}{1930 - 1910} = \frac{15}{20} = 0.75$$

The U.S. public debt from 1910 to 1930 increased at a rate of $0.75 billion per year.

Slope from 1990 to 1996

$$m = \frac{5207.0 - 3323.3}{1996 - 1990} = \frac{1883.7}{6} = 313.95$$

The U.S. public debt from 1990 to 1996 increased at a rate of \$313.95 billion per year.

b) Slope measures a rate of change. Comparing the slopes for the two periods shows that there was a much greater increase in the average rate of change in the public debt from 1990 to 1996 than from 1910 to 1930. The slope of the line segment from 1990 to 1996 is greater than the slope of any other line segment on the graph. This indicates that the public debt from 1990 to 1996 grew at a faster rate than any other time period illustrated.

NOW TRY EXERCISE 65

4 Write Linear Equations in Slope–Intercept Form

We have already shown that for an equation of the form $y = mx + b$, m represents the slope and b represents the y-intercept. For this reason a linear equation written in the form $y = mx + b$ is said to be in **slope–intercept form.**

Definition

The **slope–intercept form of a linear equation** is

$$y = mx + b$$

where ***m* is the slope** of the line and **(0, *b*) is the *y*-intercept** of the line.

Examples of Equations in Slope–Intercept Form

$$y = 3x - 6 \qquad y = \frac{1}{2}x + \frac{3}{2}$$

Slope ↓ ↓ y-intercept is $(0, b)$

$$y = mx + b$$

Equation	Slope	y-Intercept
$y = 3x - 6$	3	$(0, -6)$
$y = \frac{1}{2}x + \frac{3}{2}$	$\frac{1}{2}$	$\left(0, \frac{3}{2}\right)$

To write an equation in slope–intercept form, solve the equation for y.

EXAMPLE 3 Determine the slope and y-intercept of the equation $-3x + 4y = 8$.

Solution Write the equation in slope–intercept form by solving for y.

$$-3x + 4y = 8$$

$$4y = 3x + 8$$

$$y = \frac{3x + 8}{4}$$

$$y = \frac{3}{4}x + \frac{8}{4}$$

$$y = \frac{3}{4}x + 2$$

NOW TRY EXERCISE 43

The slope is $\frac{3}{4}$; the y-intercept is $(0, 2)$.

5 Graph Linear Equations Using the Slope and the y-Intercept

One reason for studying the slope–intercept form of a line is that it can be useful in drawing the graph of a linear equation, as illustrated in Example 4.

EXAMPLE 4 Graph $2y + 4x = 6$ using the y-intercept and slope.

Solution Begin by solving for y to get the equation in slope–intercept form.

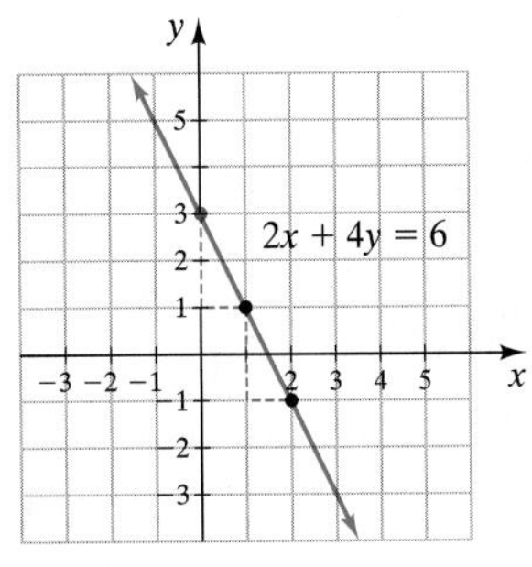

FIGURE 3.59

$$2y + 4x = 6$$
$$2y = -4x + 6$$
$$y = -2x + 3$$

The slope is -2 and the y-intercept is $(0, 3)$. Place a dot at 3 on the y-axis (Fig. 3.59). Then use the slope to obtain a second point. The slope is negative; therefore, the graph must fall as it goes from left to right. Since the slope is -2, the ratio of the vertical change to the horizontal change must be 2 to 1 (remember, 2 means $\frac{2}{1}$). Thus, if you start at $y = 3$ and move down 2 units and to the right 1 unit, we will obtain a second point on the graph.

Continue this process of moving 2 units down and 1 unit to the right to get a third point. Now draw a line through the three points to get the graph.

In Example 4, we chose to move down and to the right to get the second and third points. We could have also chosen to move up and to the left to get the second and third points.

EXAMPLE 5 Graph $f(x) = \frac{4}{3}x - 3$ using the y-intercept and slope.

Solution Since $f(x)$ is the same as y, this function is in slope–intercept form. The y-intercept is $(0, -3)$ and the slope is $\frac{4}{3}$. Place a dot at -3 on the y-axis. Then, since the slope is positive, obtain the second and third points by moving up 4 units and to the right 3 units. The graph is shown in Figure 3.60.

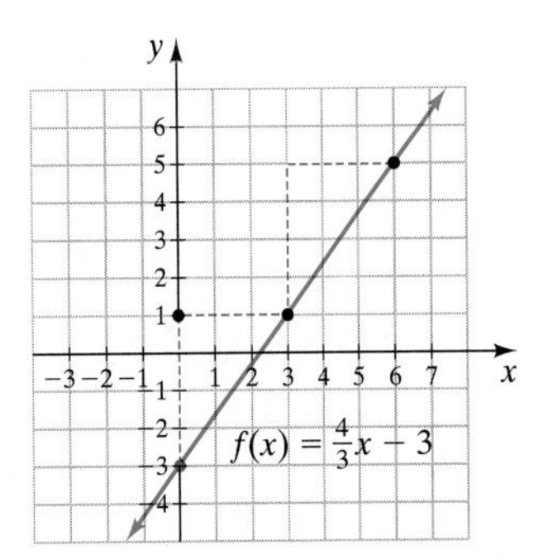

FIGURE 3.60

NOW TRY EXERCISE 49

Use the Slope–Intercept Form to Construct Models from Graphs

Often we can use the slope–intercept form of a linear equation to determine a function that models a real-life situation. Example 6 shows how this may be done.

EXAMPLE 6 Consider the graph in Figure 3.61, which shows the declining number of adults who read the daily newspaper. Notice that the graph is somewhat linear.

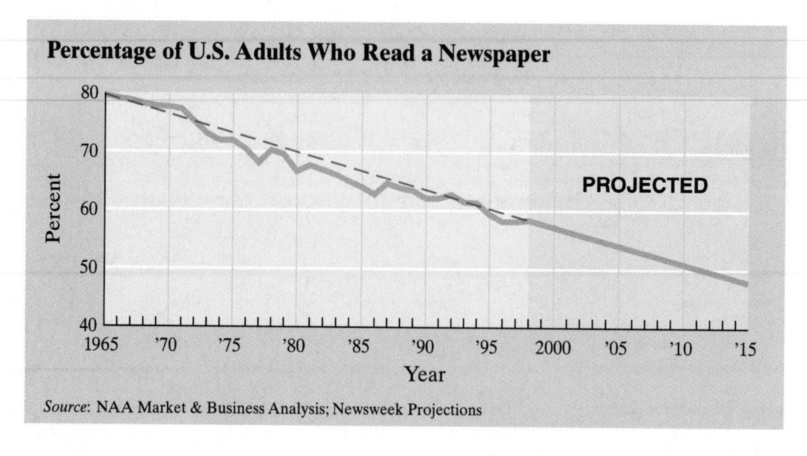

FIGURE 3.61

a) Write a linear function whose graph approximates the graph shown.

b) Estimate the percent of adults who read a daily newspaper in 1983 using the function determined in part **a)**.

Solution **a)** To make the numbers easier to work with, we will select 1965 as a *reference year*. Then we can replace 1965 with 0, 1966 with 1, 1967 with 2, and so on. Then 1998 would be 33, see Figure 3.62.

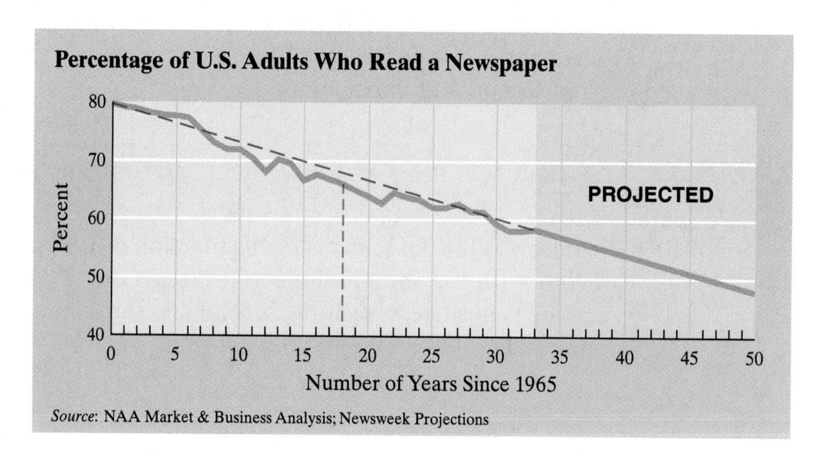

FIGURE 3.62

If we call the vertical axis y and the horizontal axis x, then the y-intercept is 80. The ordered pair that represents the y-intercept is (0, 80). In 1998 it appears that about 57 percent of the adult population read a daily newspaper. Let's select (33, 57) as a second point on the graph of the straight line we drew in Figure 3.62. We designate (33, 57) as (x_2, y_2) and (0, 80) as (x_1, y_1).

$$\text{slope} = \frac{\text{change in percent}}{\text{change in year}} = \frac{y_2 - y_1}{x_2 - x_1} = \frac{57 - 80}{33 - 0} = -\frac{23}{33} \approx -0.70$$

Since the slope is -0.70 and the y-intercept is (0, 80), the equation of the straight line is $y = -0.70x + 80$. This equation in function notation is $f(x) = -0.70x + 80$. To use this function remember that $x = 0$ represents 1965,

$x = 1$ represents 1966, and so on. Note that $f(x)$, the percent, is a function of x, the number of years since 1965.

b) To determine the approximate percent of readers in 1983, we substitute 18 for x in the function.

$$\begin{aligned} f(x) &= -0.70x + 80 \\ f(18) &= -0.70(18) + 80 \\ &= -12.6 + 80 \\ &= 67.4 \end{aligned}$$

NOW TRY EXERCISE 69 Thus, about 67% of adults read a daily newspaper in 1983 (Fig. 3.62).

Exercise Set 3.4

Concept/Writing Exercises

1. Explain how to find the slope of a line from its graph.
2. Explain what it means when the slope of a line is positive.
3. Explain what it means when the slope of a line is negative.
4. What is the slope of a horizontal line? Explain.
5. Why is the slope of a vertical line undefined?
6. When finding the slope of a line, how does the slope change if we interchange (x_1, y_1) and (x_2, y_2)? Explain.
7. Explain how to write an equation given in standard form in slope–intercept form.
8. In the equation $y = mx + b$, what does the m represent? What does the b represent?
9. **a)** What does it mean when a graph is translated up 3 units? moved up 3 units

   b) If the y-intercept of a graph is $(0, -4)$ and the graph is translated up 5 units, what will be its new y-intercept?
10. **a)** What does it mean when a graph is translated down 4 units?

    b) If the y-intercept of a graph is $(0, -3)$ and the graph is translated down 4 units, what will be its new y-intercept?
11. What does it mean when slope is given as a rate of change?
12. Explain how to graph a linear equation using its slope and y-intercept.

Practice the Skills

Find the slope of the line through the given points. If the slope of the line is undefined, so state.

13. $(5, 2)$ and $(2, -4)$
14. $(3, 1)$ and $(5, 4)$
15. $(5, 2)$ and $(1, 4)$
16. $(5, 1)$ and $(2, 4)$
17. $(-1, 4)$ and $(0, 3)$
18. $(2, 3)$ and $(2, -3)$
19. $(3, 2)$ and $(3, -1)$
20. $(8, -4)$ and $(-1, -2)$
21. $(-3, 4)$ and $(-1, 4)$
22. $(2, 6)$ and $(-1, 6)$
23. $(-2, 3)$ and $(7, -3)$
24. $(2, -4)$ and $(-5, -3)$

Solve for the given variable if the line through the two given points is to have the given slope.

25. $(2, a)$ and $(3, 4)$, $m = 1$
26. $(1, 0)$ and $(4, y)$, $m = 3$
27. $(5, b)$ and $(2, -4)$, $m = 2$
28. $(6, 1)$ and $(4, d)$, $m = 3$
29. $(x, 2)$ and $(3, -4)$, $m = 2$
30. $(-2, -3)$ and $(x, 4)$, $m = \frac{1}{2}$
31. $(3, 5)$ and $(x, 3)$, $m = \frac{2}{3}$
32. $(-4, -1)$ and $(x, 2)$, $m = -\frac{3}{5}$

Find the slope of the line in each of the figures. If the slope of the line is undefined, so state. Then write an equation of the given line.

33.

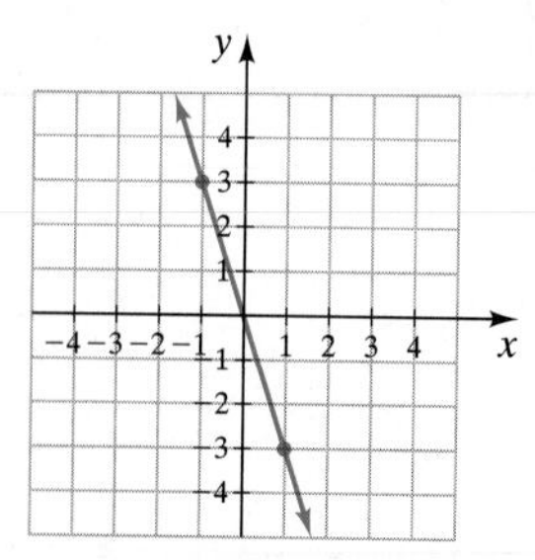

34.

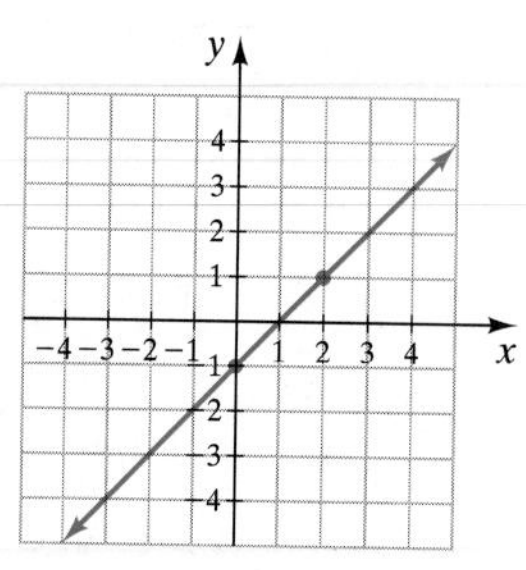

35.

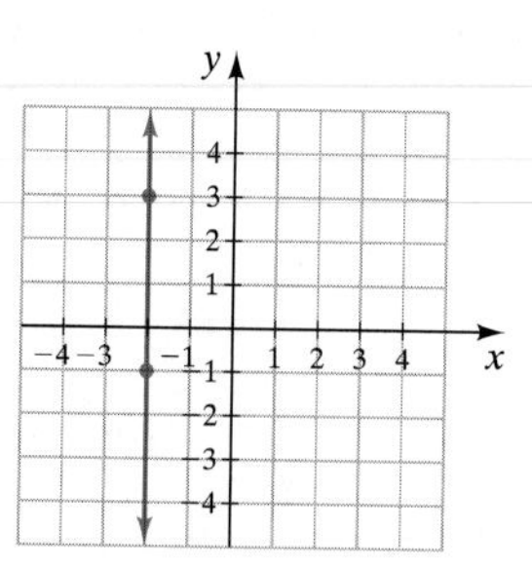

36.

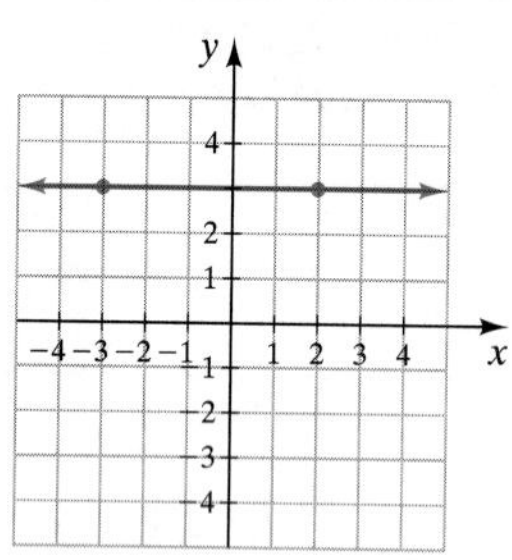

37.

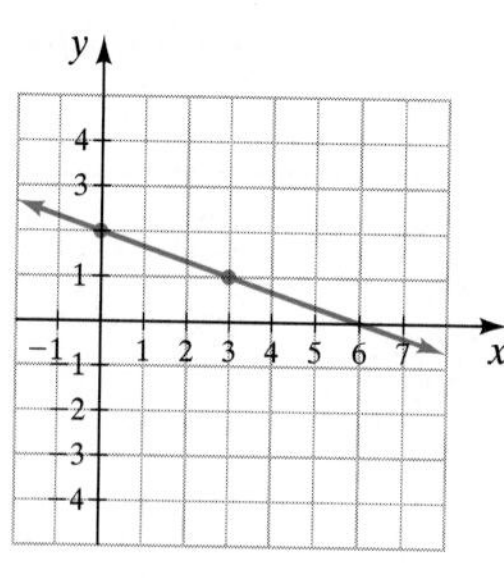

38.

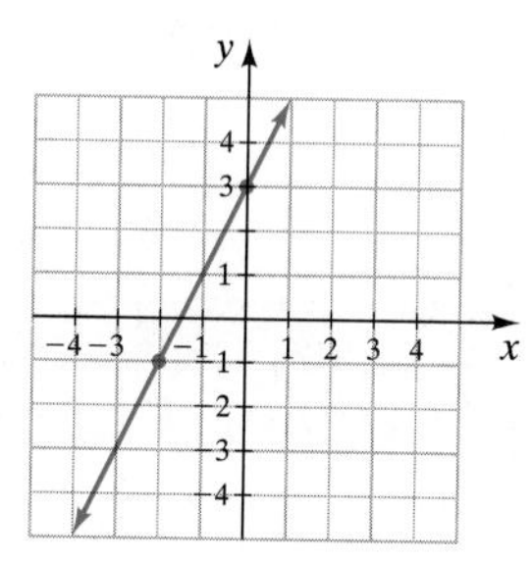

39.

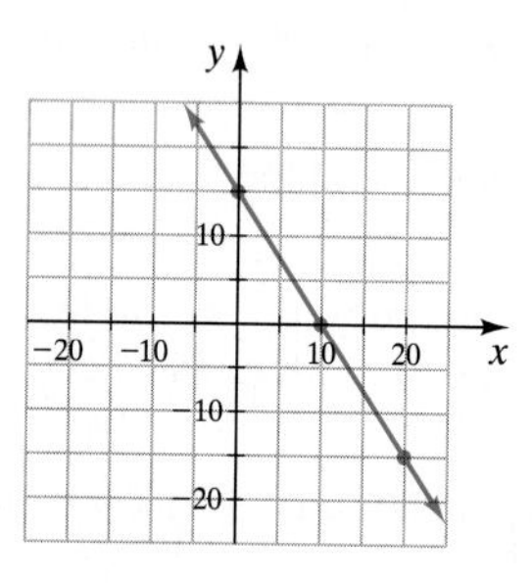

40.

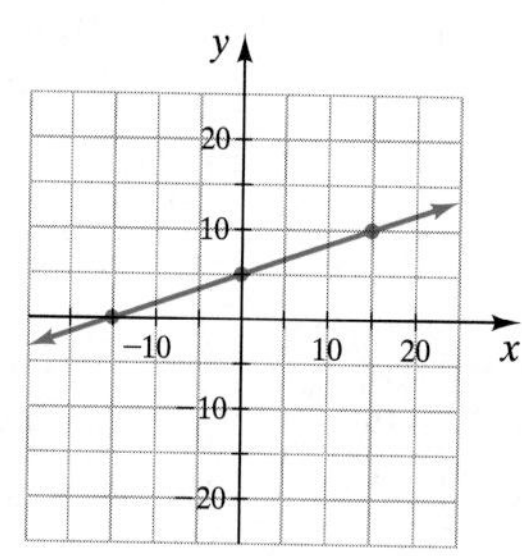

Write each equation in slope–intercept form (if not given in that form). Determine the slope and the y-intercept and use them to draw the graph of the linear equation.

41. $y = -x + 2$

42. $2x + y = 6$

43. $20x - 30y = 60$

44. $5y = 2x - 5$

45. $-50x + 20y = 40$

46. $60x = -30y + 60$

Use the slope and y-intercept to graph each function.

47. $f(x) = 3x - 5$

48. $g(x) = \frac{1}{2}x + 2$

49. $g(x) = -\frac{2}{3}x + 3$

50. $h(x) = -\frac{2}{5}x + 4$

Problem Solving

51. Given the equation $y = mx + b$, for the values of m and b given, match parts **a)–d)** with the appropriate graphs labeled 1–4.

a) $m > 0, b < 0$

b) $m < 0, b < 0$

c) $m < 0, b > 0$

d) $m > 0, b > 0$

1.

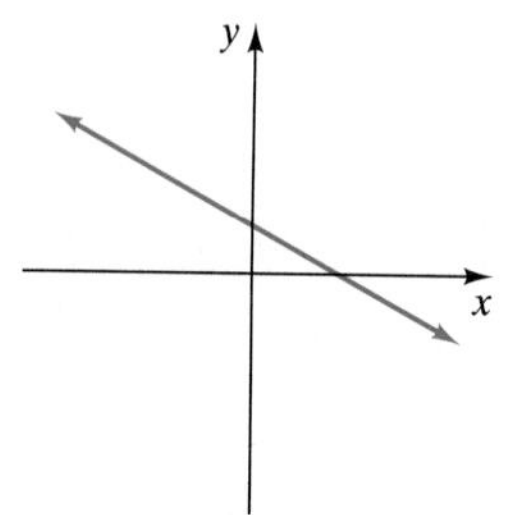

2.

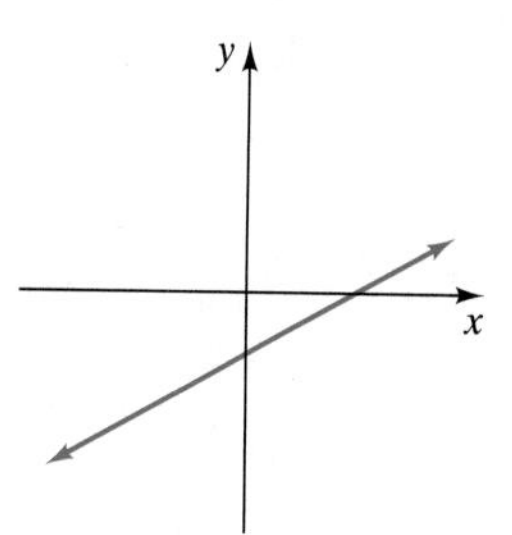

3.

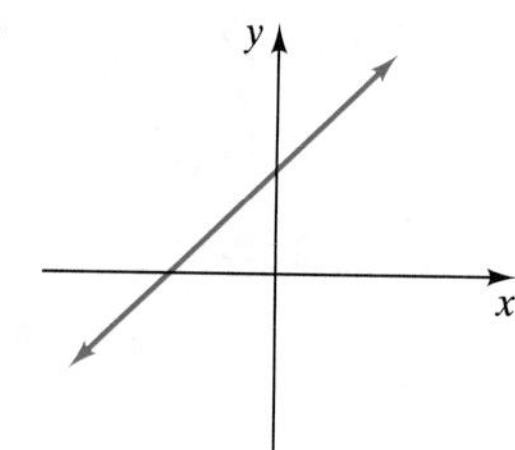

4.

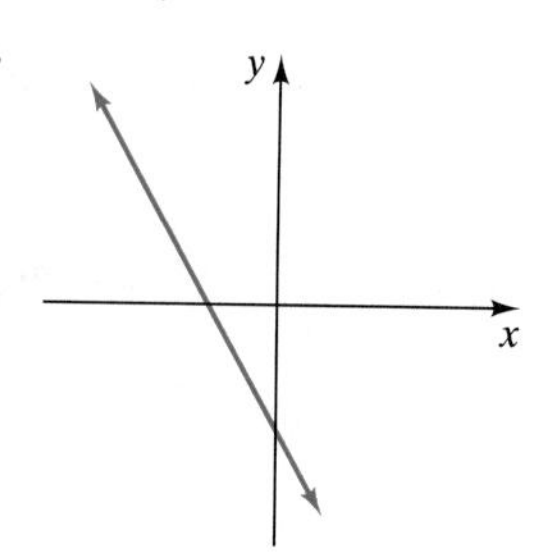

52. Given the equation $y = mx + b$, for the values of m and b given match parts **a)–d)** with the appropriate graphs labeled 1–4.

a) $m = 0, b > 0$ **b)** $m = 0, b < 0$ **c)** m is undefined, x intercept < 0 **d)** m is undefined, x intercept > 0

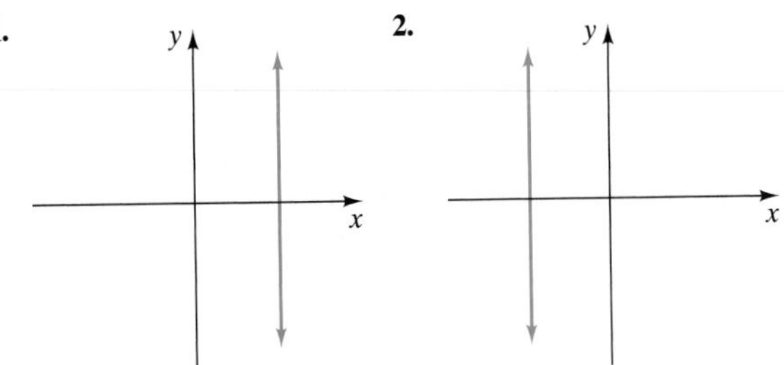

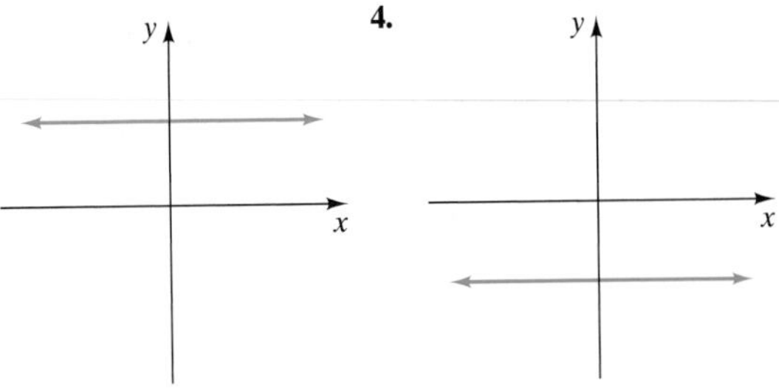

53. We will be discussing parallel lines in the next section. Based on what you have read in this section, explain how you could determine (without graphing) that the graphs of two equations are parallel.

54. How can you determine whether two straight lines are parallel?

55. If one point on a graph is (6, 3) and the slope of the line is $\frac{4}{3}$, find the y-intercept of the graph.

56. If one point on a graph is (−6, 6) and the slope of the line is $-\frac{3}{2}$, find the y-intercept of the graph.

57. In the following graph, the green line is a translation of the blue line.

a) Determine the equation of the blue line.

b) Use the equation of the blue line to determine the equation of the green line.

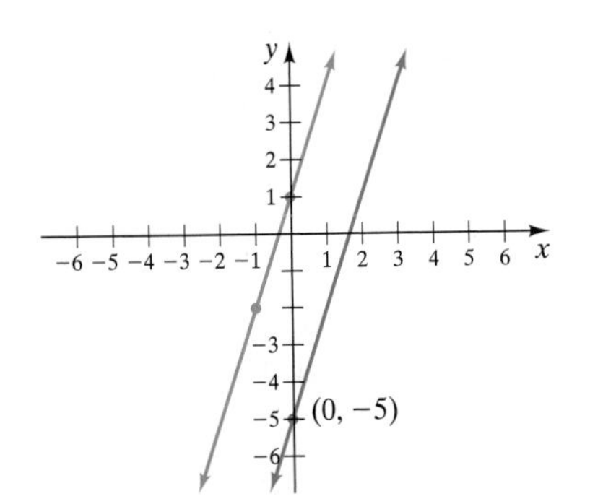

58. In the following graph, the red line is a translation of the blue line.

a) Determine the equation of the blue line.

b) Use the equation of the blue line to determine the equation of the red line.

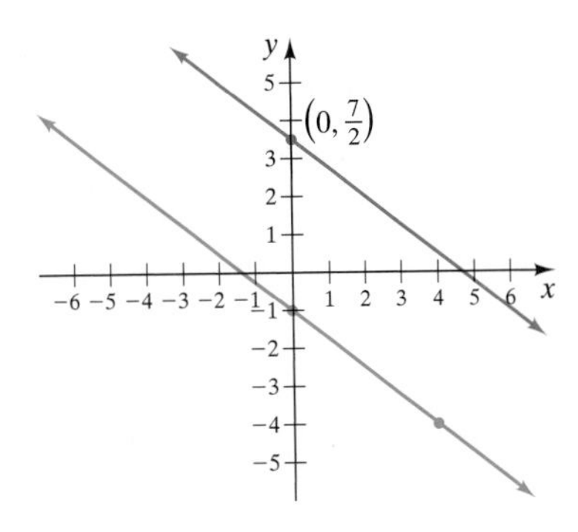

59. The graph of $y = 2x + 1$ is translated up 3 units. Find

a) the slope of the translated graph

b) the y-intercept of the translated graph

c) the equation of the translated graph

60. The graph of $y = \frac{1}{4}x - 3$ is translated down $\frac{5}{2}$ units. Find

a) the slope of the translated graph

b) the y-intercept of the translated graph

c) the equation of the translated graph

61. The graph of $3x - 2y = 6$ is translated down 4 units. Find the equation of the translated graph.

62. The graph of $-3x - 5y = 10$ is translated up 2 units. Find the equation of the translated graph.

63. If a line passes through the points (6, 4) and (−4, 2), find the change of y with respect to a one unit change in x.

64. If a line passes through the points (−3, −4) and (5, 2), find the change of y with respect to a one unit change in x.

65. Starting in 1992 there has been an increase in the number of cigars sold in the United States. The following table gives the sales of cigars, to the nearest thousand, for selected years.

Year	Cigars Sold (1000's)
1977	78
1984	102
1989	81
1996	274

SOURCE: The Cigar Association of America

a) Plot these points on a graph.

b) Connect the points using line segments.

c) Determine the slopes of each of the three line segments.

d) During which period was there the greatest average rate of change? Explain.

66. Each year computers become faster and more powerful. The following table shows the record speed, in billions of operations per second, for supercomputers for selected years.

Year	Operations per Second (billions)
1991	3
1994	143
1996	303
1997	1070

SOURCE: US Dept. of Energy

a) Plot these points on a graph.

b) Connect the points using line segments.

c) Determine the slope of each of the three line segments.

d) During which period was there the greatest average rate of change? Explain.

67. The following bar graph shows the maximum recommended heart rate, in beats per minute, under stress for men of different ages. The bars are connected by a straight line.

a) Use the straight line to determine a function that can be used to estimate the maximum recommended heart rate, h, for $0 \le x \le 50$, where x is the number of years after age 20.

b) Using the function from part **a)**, determine the maximum recommended heart rate for a 34-year-old man.

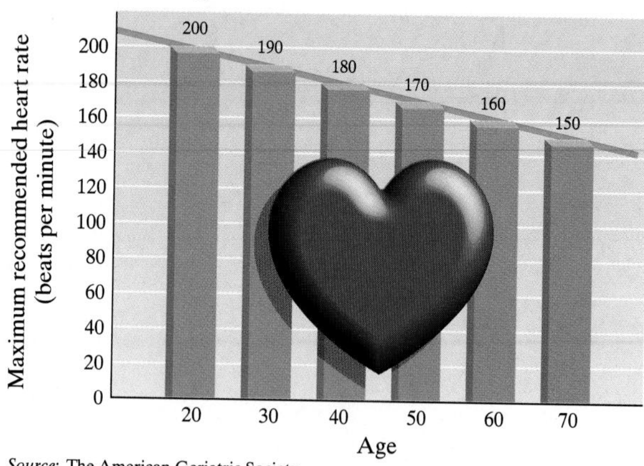

Source: The American Geriatric Society

68. The following graph shows how quality in automobiles has improved from 1990 through 1997.

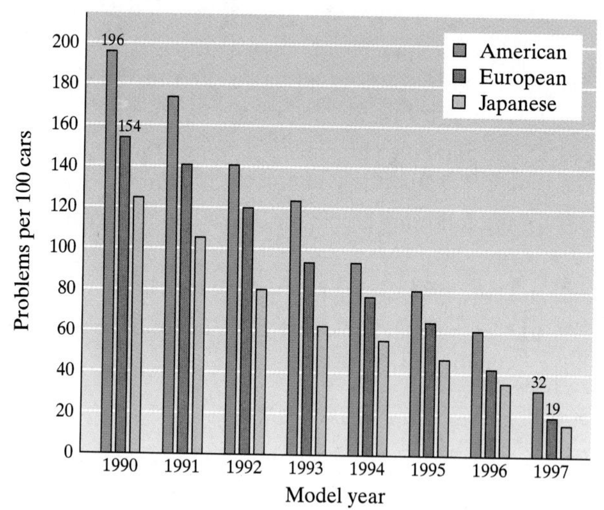

Source: Consumer Reports

a) Determine a function that can be used to estimate the number of problems, P, per 100 cars for American cars from 1990 through 1997. Let t represent the number of years since 1990. (In other words, 1990 would correspond to $t = 0$.)

b) Using the function from part **a)**, determine the number of problems per 100 U.S. cars in 1993. Compare your answer with the graph to see whether the graph supports your answer.

c) Determine a function that can used to estimate the number of problems per 100 cars for European cars from 1990 through 1997.

d) Using the function from part **c)**, determine the number of problems per 100 European cars in 1993. Compare your answer with the graph to see whether the graph supports your answer.

69. Although the quality of U.S. cars has been increasing, the domestic brands' shares, (General Motors, Ford, Chrysler) of U.S. car sales have been dropping, as shown in the following graph.

Domestic Brands' Share of U.S. Car Sales

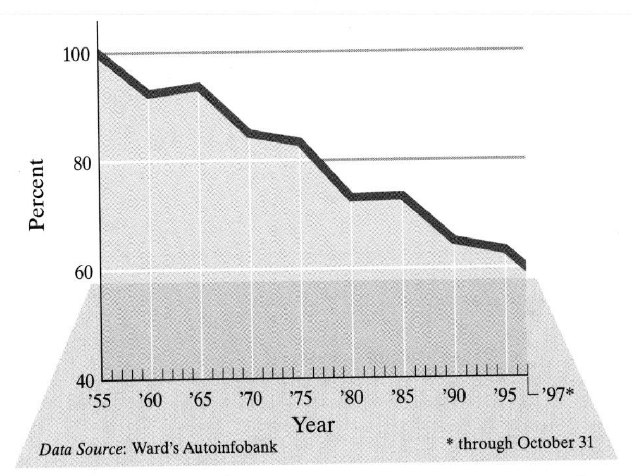

a) Using 1955 as the reference year, determine a function that can be used to estimate the domestic brands' share of U.S. car sales for the years 1955 through 1997. In the function, let t represent the number of years since 1955.

b) Using the function from part **a)**, estimate the percent of sales in 1980. Compare your answer with the graph to see whether the graph supports your answer.

70. The following graph shows that the number of students, n, in millions, with limited proficiency in English has been increasing nationally from 1987 through 1995.

Limited English Abilities

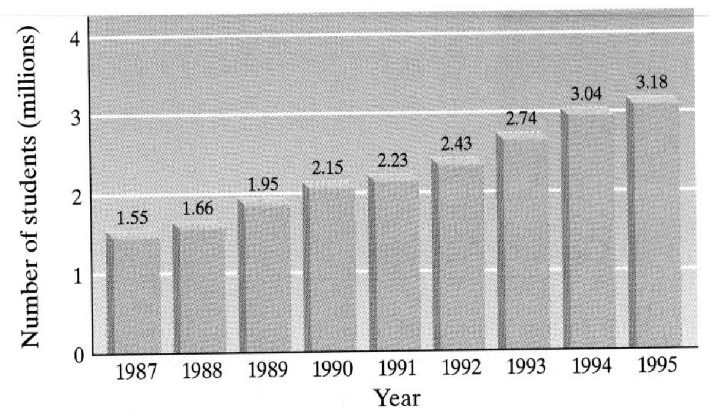

Source: U.S. Department of Education

a) Using 1987 as the reference year, determine a function that can be used to estimate the number of students with limited proficiency in English for the years 1987 through 1995. In the function, let t represent the number of years since 1987.

b) Using the function from part **a)**, estimate the number of students with limited proficiency in English in 1993. Compare your answer with the graph to see whether the graph supports your answer.

Suppose you are attempting to graph the equations shown and you get the screens shown. Explain how you know that you have made a mistake in entering each equation. The standard window setting is used on each graph.

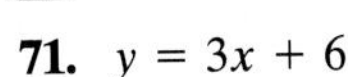

71. $y = 3x + 6$

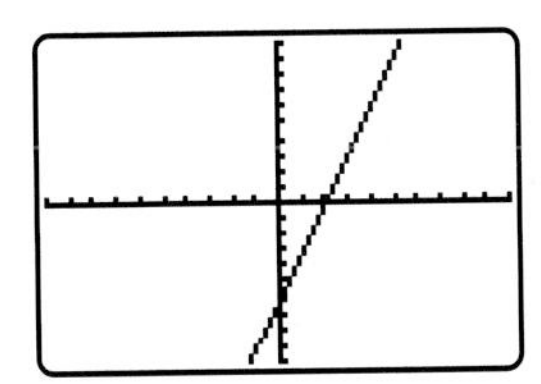

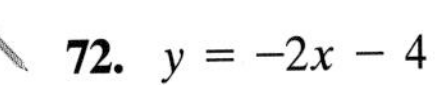

72. $y = -2x - 4$

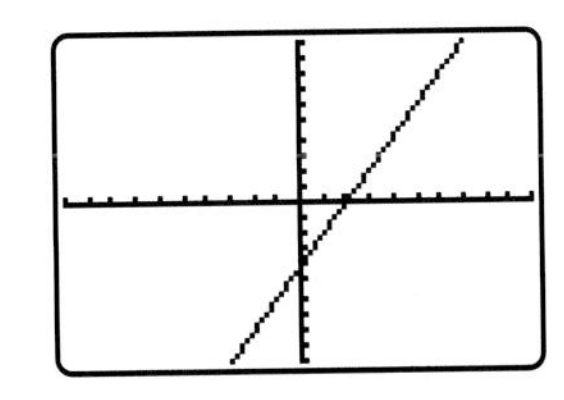

73. $y = \frac{1}{2}x + 4$

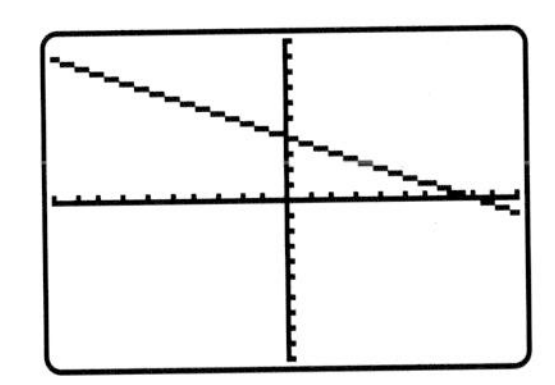

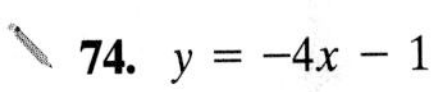

74. $y = -4x - 1$

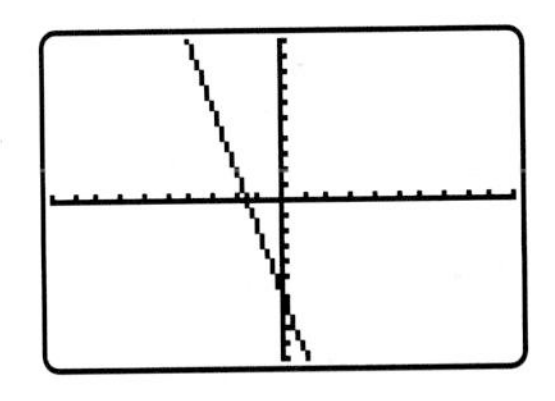

Challenge Problems

75. The photo on the right is the Castle at Chichén Itzá, Mexico. Each side of the castle has a stairway consisting of 91 steps. The steps of the castle are quite narrow and steep, which makes them hard to climb. The total vertical distance of the 91 steps is 1292.2 inches. If a straight line were to be drawn connecting the tips of the steps, the absolute value of the slope of this line would be 2.21875. Find the average height and width of a step.

76. A **tangent line** is a straight line that touches a curve at a single point (the tangent line may cross the curve at a different point if extended). Figure 3.63 shows three tangent lines to the curve at points a, b, and c. Note that the tangent line at point a has a positive slope, the tangent line at point b has a slope of 0, and the tangent line at point c has a negative slope. Now consider the curve in Figure 3.64. Imagine that tangent lines are drawn at all points on the curve except at endpoints a and e. Where on the curve in Figure 3.64 would the tangent lines have a positive slope, a slope of 0, a negative slope?

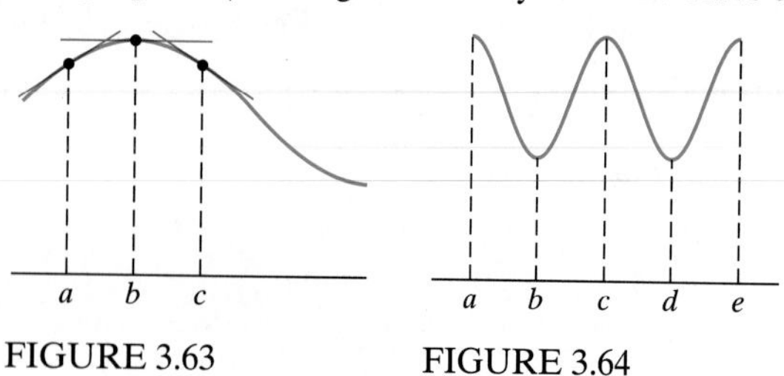

FIGURE 3.63 FIGURE 3.64

Group Activity

77. The following graph from Consumer Reports shows the depreciation on a typical car. The initial purchase price is represented as 100%.

a) Group member 1: Determine the one-year period in which a car depreciates most. Estimate from the graph the percent a car depreciates during this period.

b) Group member 2: Determine between which years the depreciation appears linear or nearly linear.

c) Group member 3: Determine between which two years the depreciation is the lowest.

d) As a group, estimate the slope of the line segment from year 0 to year 1. Explain what this means in terms of rate of change.

Typical Depreciation Curve

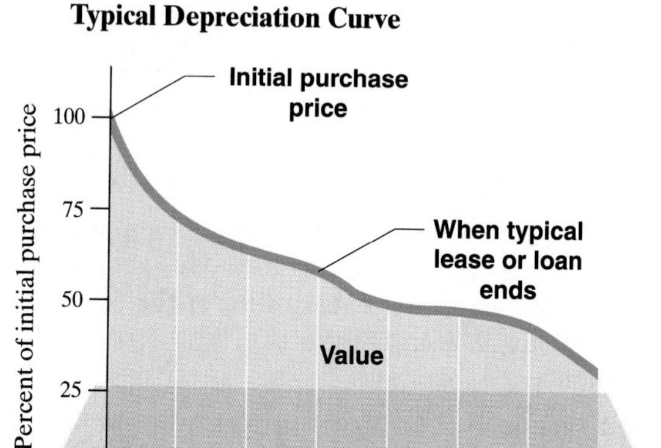

Cumulative Review Exercises

[1.4] **78.** Evaluate $\dfrac{-6^2 - 16 \div 2 \div |-4|}{5 - 3 \cdot 2 - 4 \div 2^2}$.

Solve each equation.

[2.1] **79.** $\frac{3}{4}x + \frac{1}{5} = \frac{2}{3}(x - 2)$

80. $2.6x - (-1.4x + 3.4) = 6.2$

[2.4] **81.** Two trains leave Chicago, Illinois, traveling in the same direction along parallel tracks. The first train leaves 3 hours before the second, and its speed is 15 miles per hour faster than the second. Find the speed of each train if they are 270 miles apart 3 hours after the second train leaves Chicago.

[2.6] **82.** Solve **a)** $|2x + 1| > 3$, **b)** $|2x + 1| < 3$.

3.5 THE POINT–SLOPE FORM OF A LINEAR EQUATION

SSM VIDEO 3.5 CD Rom

1) **Understand the point–slope form of a linear equation.**
2) **Use the point–slope form to construct models from graphs.**
3) **Recognize parallel and perpendicular lines.**

1) Understand the Point–Slope Form of a Linear Equation

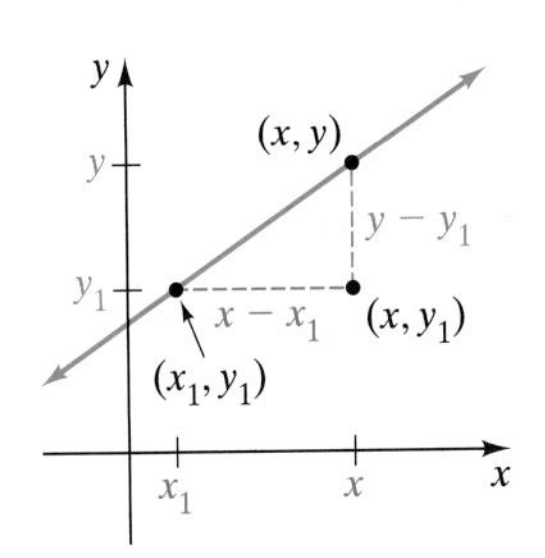

FIGURE 3.65

In the previous section we learned how to use the *slope–intercept form* of a line to determine the equation of a line when the slope and y-intercept of the line are known. In this section we learn how to use the **point–slope form** of a line to determine the equation of a line when the slope and a point on the line are known. The point–slope form can be developed from the expression for the slope between any two points (x, y) and (x_1, y_1) on a line, as shown in Figure 3.65.

$$m = \frac{y - y_1}{x - x_1}$$

Multiplying both sides of the equation by $x - x_1$, we obtain

$$y - y_1 = m(x - x_1)$$

Definition

The **point–slope form of a linear equation** is

$$y - y_1 = m(x - x_1)$$

where m **is the slope** of the line and (x_1, y_1) is a point on the line.

EXAMPLE 1 Write in slope–intercept form the equation of the line that passes through the point (2, 3) and has slope 4.

Solution Since we are given the slope of the line and a point on the line, we can write the equation in point–slope form. We can then solve the equation for y to write the equation in slope–intercept form. The slope, m, is 4. The point on the line is (2, 3); call this (x_1, y_1). Substitute 4 for m, 2 for x_1, and 3 for y_1 in the point–slope form of a line.

$$y - y_1 = m(x - x_1)$$
$$y - 3 = 4(x - 2) \quad \text{Point–slope form}$$
$$y - 3 = 4x - 8$$
$$y = 4x - 5 \quad \text{Slope–intercept form}$$

NOW TRY EXERCISE 5

The graph of $y = 4x - 5$ has a slope of 4 and passes through the point (2, 3).

In Example 1 we used the point–slope form to get the equation of a line when we were given a point on the line and the slope of the line. The point–slope form can also be used to find the equation of a line when we are given two points on the line. We show how to do this in Example 2.

2) Use the Point–Slope Form to Construct Models from Graphs

Now let's look at an application where we use the point–slope form to determine a function that models a given situation.

EXAMPLE 2 An article that appeared in the *Drug Store News* stated that the number of independent drug stores has been decreasing at a linear rate from 1992 through 1997 (while the number of chain drug stores has been increasing at a linear rate during the same time period). In 1992 there were about 23,200 independent drug stores and in 1997 there were about 20,300 independent drug stores (Fig. 3.66).

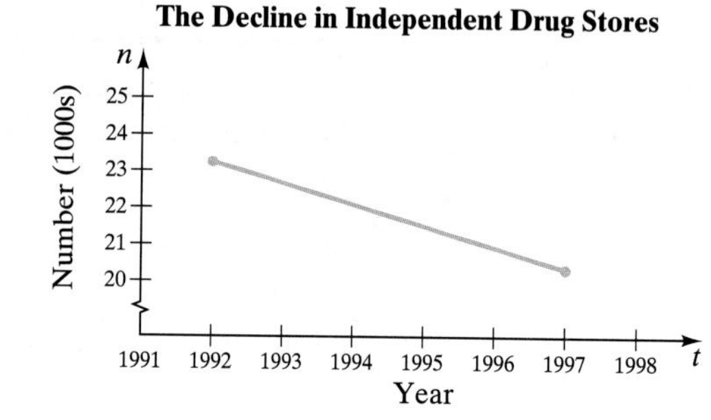

FIGURE 3.66

a) Determine a function that can be used to estimate the number of drug stores from 1992 through 1997.

b) The article indicated that the trend discussed above is expected to continue. Use the function found in part **a)** to estimate the number of independent drug stores in the year 2010.

Solution **a) Understand and Translate** To find the necessary equation, we can use the points (1992, 23,200) and (1997, 20,300). However, using these large values will be awkward. To make the numbers easier to work with, let's select 1991 as a reference year and call it 0. Then 1992 would be represented as year 1, since it is one year after 1991. The year 1993 would be represented as year 2, and so on. Using these values we can develop an equivalent graph (See Fig. 3.67). On the graph we call the horizontal axis t, for time, and the vertical axis n, for the number of drug stores. Therefore our ordered pairs will be of the form (t, n).

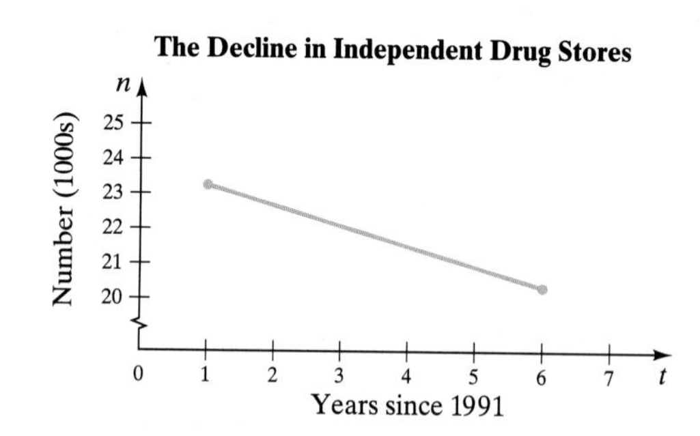

FIGURE 3.67

Instead of using 23,200 and 20,300 for the n-coordinate in the ordered pairs, we can use 23.2 and 20.3, measuring both numbers in thousands. We can therefore use the ordered pairs (1, 23.2) and (6, 20.3) to find the slope of the line and then determine the equation of the line.

Carry Out

$$\text{slope} = \frac{n_2 - n_1}{t_2 - t_1}$$

$$\text{slope} = \frac{20.3 - 23.2}{6 - 1} = \frac{-2.9}{5} = -0.58$$

Now we write the equation using point–slope form. We will use the point (1, 23.2) for (t_1, n_1).

$$n - n_1 = m(t - t_1)$$

$$n - 23.2 = -0.58(t - 1) \quad \text{Point–slope form}$$

$$n - 23.2 = -0.58t + 0.58$$

$$n = -0.58t + 23.78 \quad \text{Slope–intercept form}$$

Answer Since the number of drug stores, n, is a function of time, t, the function we are seeking is

$$n(t) = -0.58t + 23.78$$

b) The year 2010 is 19 years after 1991. Therefore, the number of drug stores in 2010 can be found by substituting 19 for t in the function.

$$n(t) = -0.58t + 23.78$$

$$n(19) = -0.58(19) + 23.78 = 12.76$$

Thus in 2010 if the current trend continues, there will be about 12,760 independent drug stores.

In Example 2 **a)**, would the function have changed if we had selected 1992 instead of 1991 as our reference year? In Example 2 **b)**, would the estimate for the number of independent drug stores in 2010 have changed with a 1992 reference year? See Exercise 57.

NOW TRY EXERCISE 53

3) Recognize Parallel and Perpendicular Lines

Figure 3.68 illustrates two *parallel* lines.

Parallel lines

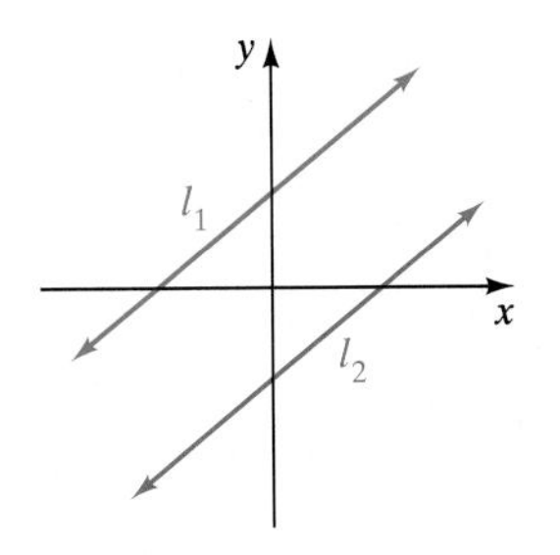

FIGURE 3.68

> Two lines are **parallel** when they have the same slope.

All vertical lines are parallel even though their slope is undefined.

Figure 3.69 illustrates perpendicular lines. Two lines are *perpendicular* when they cross at right (or 90°) angles.

Perpendicular lines

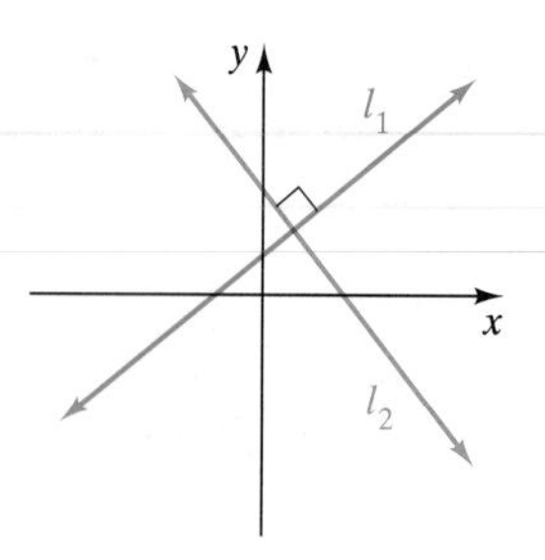

FIGURE 3.69

Two lines are perpendicular when their slopes are negative reciprocals.

For any nonzero number a, its *negative reciprocal* is $\frac{-1}{a}$ or $-\frac{1}{a}$. For example, the negative reciprocal of 2 is $\frac{-1}{2}$ or $-\frac{1}{2}$. The product of any nonzero number and its negative reciprocal is -1.

$$a\left(-\frac{1}{a}\right) = -1.$$

Note that any vertical line is perpendicular to any horizontal line even though the negative reciprocal cannot be applied. (Why not?)

EXAMPLE 3 Two points on l_1 are $(6, 3)$ and $(2, -3)$. Two points on l_2 are $(0, 2)$ and $(6, -2)$. Determine whether l_1 and l_2 are parallel lines, perpendicular lines, or neither.

Solution Determine the slopes of l_1 and l_2.

$$m_1 = \frac{3-(-3)}{6-2} = \frac{6}{4} = \frac{3}{2} \qquad m_2 = \frac{2-(-2)}{0-6} = \frac{4}{-6} = -\frac{2}{3}$$

Since their slopes are different, l_1 and l_2 are not parallel. To see whether the lines are perpendicular, we need to determine whether the slopes are negative reciprocals. If $m_1 m_2 = -1$, the slopes are negative reciprocals and the lines are perpendicular.

$$m_1 m_2 = \frac{3}{2}\left(-\frac{2}{3}\right) = -1$$

NOW TRY EXERCISE 13

Since the product of the slopes equals -1, the lines are perpendicular.

EXAMPLE 4 Consider the equation $2x + 4y = 8$. Determine the equation of the line that has a y-intercept of 5 and is **a)** parallel to the given line and **b)** perpendicular to the given line.

Solution **a)** If we know the slope of a line and its y-intercept, we can use the slope–intercept form, $y = mx + b$, to write the equation. We begin by solving the given equation for y.

$$2x + 4y = 8$$
$$4y = -2x + 8$$
$$y = \frac{-2x + 8}{4}$$
$$y = -\frac{1}{2}x + 2$$

Two lines are parallel when they have the same slope. Therefore, the slope of the line parallel to the given line must be $-\frac{1}{2}$. Since its slope is $-\frac{1}{2}$ and its y-intercept is 5, its equation must be

$$y = -\frac{1}{2}x + 5$$

The graphs of $2x + 4y = 8$ and $y = -\frac{1}{2}x + 5$ are shown in Figure 3.70.

b) Two lines are perpendicular when their slopes are negative reciprocals. We know that the slope of the given line is $-\frac{1}{2}$. Therefore, the slope of the perpen-

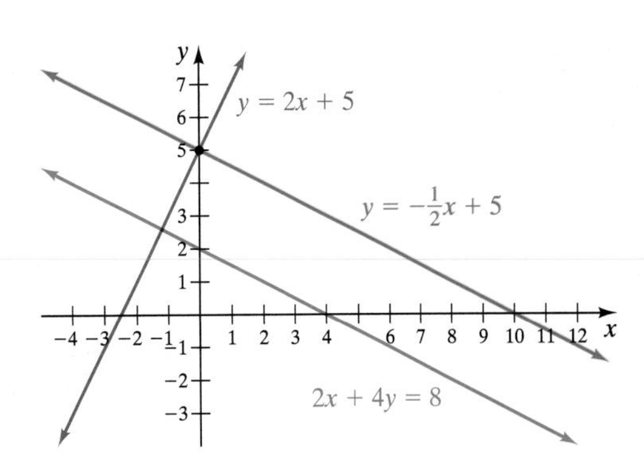

FIGURE 3.70

dicular line must be $-1/\left(-\frac{1}{2}\right)$ or 2. The line perpendicular to the given line has a y-intercept of 5. Thus the equation is

$$y = 2x + 5,$$

Figure 3.70 also shows the graph of $y = 2x + 5$.

EXAMPLE 5 Consider the equation $5y = -10x + 7$.

a) Determine the equation of a line perpendicular to the graph of the equation that passes through $\left(4, \frac{1}{3}\right)$. Write the equation in standard form.

b) Write the equation determined in part **a)** using function notation.

Solution **a)** Determine the slope of the given line by solving the equation for y.

$$5y = -10x + 7$$

$$y = \frac{-10x + 7}{5}$$

$$y = -2x + \frac{7}{5}$$

Since the slope of the given line is -2, the slope of a line perpendicular to it must be $\frac{1}{2}$. The line we are seeking must pass through the point $\left(4, \frac{1}{3}\right)$. Using the point–slope form, we obtain

$$y - y_1 = m(x - x_1)$$

$$y - \frac{1}{3} = \frac{1}{2}(x - 4) \quad \textit{Point–slope form}$$

Now multiply both sides of the equation by the least common denominator, 6, to eliminate fractions.

$$6\left(y - \frac{1}{3}\right) = 6\left[\frac{1}{2}(x - 4)\right]$$

$$6y - 2 = 3(x - 4)$$

$$6y - 2 = 3x - 12$$

Now write the equation in standard form.

$$-3x + 6y - 2 = -12$$

$$-3x + 6y = -10 \quad \textit{Standard form}$$

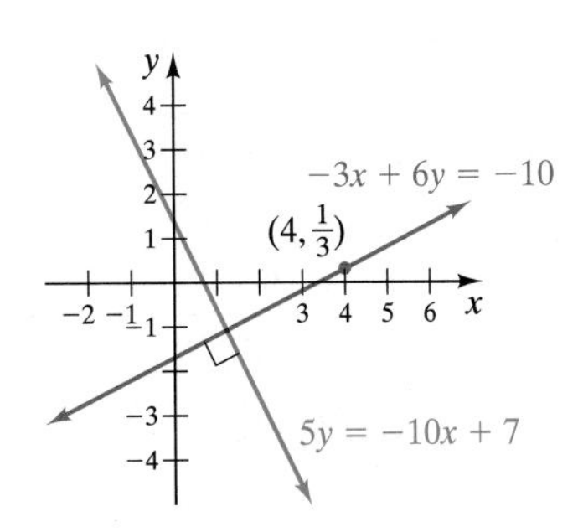

FIGURE 3.71

Note that $3x - 6y = 10$ is also an acceptable answer (Fig. 3.71).

b) To write the equation using function notation, we solve the equation determined in part **a)** for y, then replace y with $f(x)$.

NOW TRY EXERCISE 39

We will leave it to you to show that the function is $f(x) = \frac{1}{2}x - \frac{5}{3}$.

HELPFUL HINT

The following chart summarizes the three forms of a linear equation we have studied and mentions when each may be useful.

Standard form: $ax + by = c$	Useful when finding the intercepts of a graph Will be used in Chapter 4, Systems of Equations
Slope–intercept form: $y = mx + b$	Used to find the slope and y-intercept of a line Used to find the equation of a line given its slope and y-intercept Used to determine if two lines are parallel Used to graph a linear equation
Point–slope form: $y - y_1 = m(x - x_1)$	Used to find the equation of a line when given the slope of a line and a point on the line Used to find the equation of a line when given two points on a line

Exercise Set 3.5

Concept/Writing Exercises

1. Give the point–slope form of a linear equation.
2. How can we determine whether two lines are parallel?
3. How can we determine whether two lines are perpendicular?
4. Why can't the negative reciprocal test be used to determine whether a vertical line is perpendicular to a horizontal line?

Practice the Skills

Use the point–slope form to find the equation of a line with the properties given. Then write the equation in slope–intercept form.

5. Slope = 3, through (2, 4)
6. Slope = −2, through (−4, 5)
7. Through (6, 3) and (5, 2)
8. Through (−4, −2) and (−2, 1)
9. Slope = $\frac{1}{2}$, through (−1, −5)
10. Slope = $-\frac{2}{3}$, through (−1, −2)
11. Through (−4, 6) and (4, −6)
12. Through (4, −2) and (1, 9)

Two points on l_1 and two points on l_2 are given. Determine whether l_1 is parallel to l_2, l_1 is perpendicular to l_2, or neither.

13. l_1: (0, 4) and (2, 8); l_2: (0, −1) and (3, 5)
14. l_1: (3, 4) and (−2, 3); l_2: (0, −3) and (2, −1)
15. l_1: (3, 2) and (−1, −2); l_2: (2, 0) and (3, −1)
16. l_1: (0, 2) and (6, −2); l_2: (4, 0) and (6, 3)
17. l_1: (−1, 3) and (4, 2); l_2: (1, −3) and (4, 2)
18. l_1: (1, 5) and (−2, −1); l_2: (1, −2) and (3, 2)
19. l_1: (2, −2) and (2, 4); l_2: (3, 4) and (5, 4)
20. l_1: (−2, −3) and (4, −3); l_2: (−3, −4) and (−3, 1)

Determine whether the two given lines are parallel, perpendicular, or neither.

21. $y = 3x - 5$
$y = 3x + 1$

22. $2x + 3y = 6$
$y = -\frac{2}{3}x + 5$

23. $4x + 2y = 8$
$8x = 4 - 4y$

24. $3x - 5y = 10$
$3y + 5x = 5$

25. $4x + 2y = 6$
$-x + 4y = 4$

26. $6x + 2y = 8$
$4x - 9 = -y$

27. $y = \frac{1}{2}x - 6$
$-3y = 6x + 9$

28. $2y - 6 = -5x$
$y = -\frac{5}{2}x - 2$

29. $y = 2x - 6$
$x = -2y - 4$

30. $2x + y - 6 = 0$
$6x + 3y = 12$

31. $x - 3y = -9$
$y = 3x + 6$

32. $-4x + 6y = 12$
$2x - 3y = 6$

Find the equation of a line with the properties given. Write the equation in the form indicated.

33. Through (2, 5) parallel to $y = 2x + 4$ (slope–intercept form)

34. Through $\left(\frac{1}{2}, 3\right)$ parallel to $2x + 3y - 9 = 0$ (standard form)

35. Through $\left(\frac{1}{5}, -\frac{2}{3}\right)$ parallel to $-3x = 2y + 6$ (slope– intercept form)

36. Through (2, 3) perpendicular to $y = 2x - 3$ (slope–intercept form)

37. Through $\left(-\frac{2}{3}, -4\right)$ perpendicular to $\frac{1}{2}x = y - 6$ (function notation)

38. With x-intercept (2, 0) and y-intercept (0, 3) (standard form)

39. Through (2, 5) and parallel to the line with x-intercept (1, 0) and y-intercept (0, 3) (function notation)

40. Through (−3, 4) and perpendicular to the line with x-intercept (2, 0) and y-intercept (0, 2) (standard form)

41. Through (6, 2) and perpendicular to the line with x-intercept (2, 0) and y-intercept (0, −3) (slope–intercept form)

42. Through the point (2, 1) parallel to the line through the points (3, 5) and (−2, 3) (function notation)

Problem Solving

43. The number of trademarks registered has grown linearly from 1985 through 1996. In 1985 about 63.1 thousand trademarks were registered. In 1996 the number had grown to 78.7 thousand. Let n be the number of trademarks registered and t the number of years since 1985.

a) Find a linear function $n(t)$ that fits the data.

b) If this trend continues, estimate the number of trademarks that will be registered in 2020.

44. More and more workers are participating in their companies' 401(k) retirement plans. In 1985 about 10.1 million workers participated, while in 1997 about 25.2 million workers participated. The growth in the number of participants in 401(k) plans has been approximately linear.

a) Find a linear function $R(t)$ that fits the data.

b) Use the function in part **a)** to estimate the number of participants in 2000 if this linear trend continues.

45. The number of calories burned in 1 hour on a treadmill is a function of the speed of the treadmill. The average person walking on a treadmill (at 0° incline) at a speed of 2.5 miles per hour will burn about 210 calories. At 6 miles per hour the average person will burn about 370 calories. Let C be the calories burned in 1 hour and s be the speed of the treadmill.

a) Find a linear function $C(s)$ that fits the data.

b) Estimate the calories burned by the average person on a treadmill in 1 hour at a speed of 5 miles per hour.

46. The number of calories burned for 1 hour on a treadmill going at a constant speed is a function of the incline of the treadmill. At 4 miles per hour an average person on a 5% incline will burn 525 calories. At 4 mph on a 15% incline the average person will burn 880 calories. (Most hilly streets have inclines of 10 to 25%.) Let C be the calories burned and d be the degrees of incline of the treadmill (as a decimal number).

a) Find a linear function $C(d)$ that fits the data.

b) Estimate the number of calories burned by the average person in 1 hour on a treadmill going 4 miles per hour and at a 7% incline.

47. The median age for first marriage for both males and females is increasing approximately linearly. In 1966 the median age for males' first marriage was 22.8. In 1995 it was 26.9. (For your information, in 1966 the median age for females' first marriage was 20.5 and in 1995 it was 24.5). Let m be the median age for males' first marriage and t be the the number of years since 1966.

a) Find a function $m(t)$ that fits the data for males.

b) Assuming this trend continues, estimate the median age for males' first marriage in 2005.

48. The number of households with personal computers (PC's) in the United States is expected to increase approximately linearly from 1995 through 2000. The number of households with PC's in the United States in 1995 was 33.2 million. In 2000 it is expected to be 52.8 million. Let h be the number of households with PC's and t be the number of years since 1995.

a) Find a function $h(t)$ that fits the data.

b) Estimate the number of households with PC's in 1998.

49. In the United States the number of farms has been declining approximately linearly from 1975 through 1997 (but their size has been increasing). In 1975 there were about 2.8 million farms; in 1997 there were about 2.1 million farms. Let n be the number of farms and t be the number of years since 1975.

a) Find a function $n(t)$ that fits the data.

b) Assuming this trend continues, estimate the number of farms in 2050.

50. The number of workers per social security beneficiary has been declining approximately linearly since 1970. In 1970 there were 3.7 workers per beneficiary. In 2050 it is projected there will be 2.0 workers per beneficiary. Let W be the workers per social security beneficiary and t be the time since 1970.

a) Find a function $W(t)$ that fits the data.

b) Find the number of workers per beneficiary in 2020.

51. How long will you live? This is a question many people ask as they approach retirement age. The life expectancy of a person, in years, is quite linear from age 30 to age 80. The following chart (a combined chart for men and women) shows life expectancies, that is, the number of additional years a person can expect to live, for various ages.

Age	Life Expectancy
30	55.1
40	45.4
50	36.0
60	27.0
70	18.7
80	11.6

SOURCE: TIAA/CREF

a) Plot these points.

b) In a second color, draw a line segment from (30, 55.1) to (80, 11.6).

c) Determine a function that can be used to estimate the life expectancy, E, as a function of age, a, for ages between 30 and 80.

d) Use the function found in part **c)** to estimate the life expectancy of Gretchen Bertani, who is 65 years of age.

52. The following chart shows the estimated higher education enrollment from 1985 through 1995 and projected enrollment from 1995 through 2005.

Year	Enrollment (in millions)
1985	12.2
1990	13.8
1995	14.3
2000	14.8
2005	15.6

SOURCE: U.S. Department of Education.

a) Plot these points.

b) Draw a line segment from (1985, 12.2) to (2005, 15.6).

c) Determine a function that can be used to estimate higher education enrollment, E, as a function of the number of years since 1985, t.

d) Use the function found in part **c)** to estimate the higher education enrollment in 1999.

53. The following graph shows Superbowl viewers from 1967 through 1995. Notice that the actual graph is not linear, but the general trend of the graph is linear.

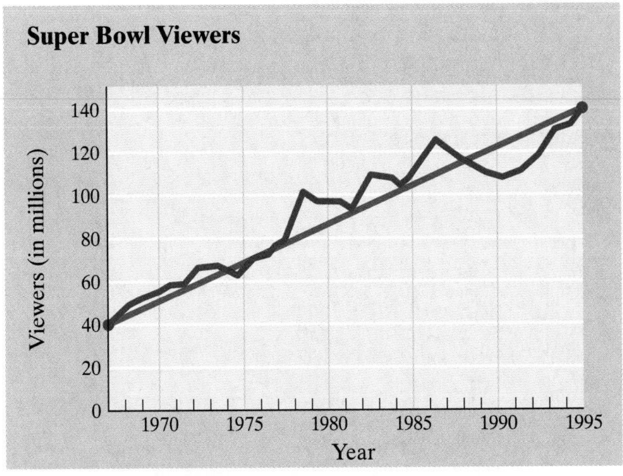

On the graph, we have drawn a red line that can be used to estimate the number of viewers from 1967 through 1995. Let V be the number of viewers in millions and t be the years since 1967.

a) Determine the function $V(t)$ represented by the red line.

b) Assuming this general trend continues, use the function found in part **a)** to estimate the number of viewers in 2000.

54. The following graph shows the global mean temperature of Earth from 1905 through 1995. The graph is not linear, but the general trend of the graph is linear. On the graph, we have drawn a red line that can be used to estimate the global mean temperature. Let T be the temperature and t be the years since 1905.

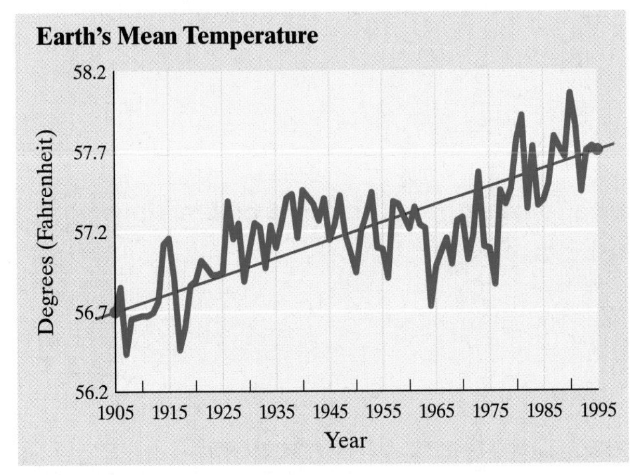

a) Determine the function $T(t)$ represented by the red line.

b) Assuming this trend continues, use the function found in part **a)** to estimate the global mean temperature in 2025.

55. The following graph shows the general trend in the number of take-out and on-premises meals purchased at commercial restaurants per person annually from 1984 through 1996.

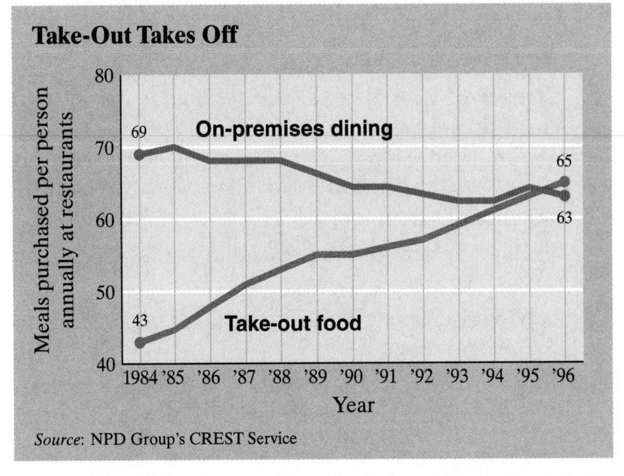

Source: NPD Group's CREST Service

a) Let N be the number of on-premise meals and t be the number of years since 1984. Determine the function $N(t)$ that can be used to approximate the number of on-premise dining meals.

b) Let H be the number of take-out meals and t be the number of years since 1984. Determine the function $H(t)$ that can be used to approximate the number of take-out meals.

56. The following graph shows the changes in Medicare monthly premiums for 38 million Medicare beneficiaries under the previous law and under the new budget passed in 1997.

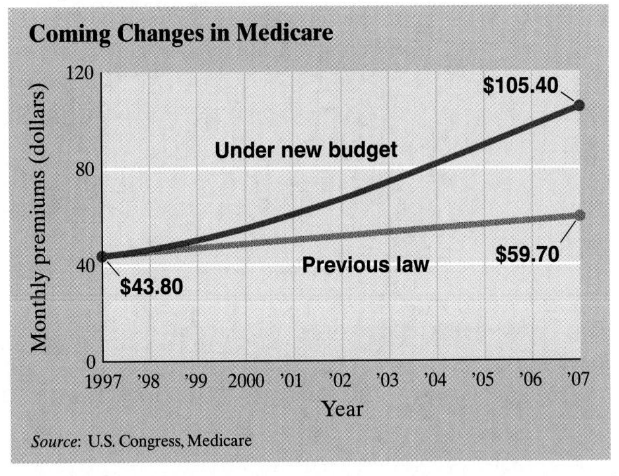

Source: U.S. Congress, Medicare

a) Determine a linear function $P(t)$ that can be used to estimate the monthly premiums under the previous law.

b) Determine a linear function $N(t)$ that can be used to estimate the monthly premium under the new law.

57. **a)** Work Example 2 parts **a)** and **b)** on page 194 using 1992 as the reference year.

b) Did your function in part **a)** change? Explain why or why not.

c) Did the estimate for the number of independent drug stores in 2010 change? Explain why or why not.

Group Activity

58. The following graph shows the growth of the circumference of a girl's head. The red line is the average head circumference of all girls for the given age while the green lines represent the upper and lower limits of the normal range. Discuss and answer the following questions as a group.

    a) Explain why the graph of the average head circumference represents a function.

    b) What is the independent variable? What is the dependent variable?

    c) What is the domain of the graph of the average head circumference? What is the range of the average head circumference graph?

    d) What interval is considered normal for girls of age 18?

    e) For this graph, is head circumference a function of age or is age a function of head circumference? Explain your answer.

    f) Estimate the average girl's head circumference at age 10 and at age 14.

    g) This graph appears to be nearly linear. Determine an equation or function that can be used to estimate the line between (2, 48) and (18, 55).

Head Circumference

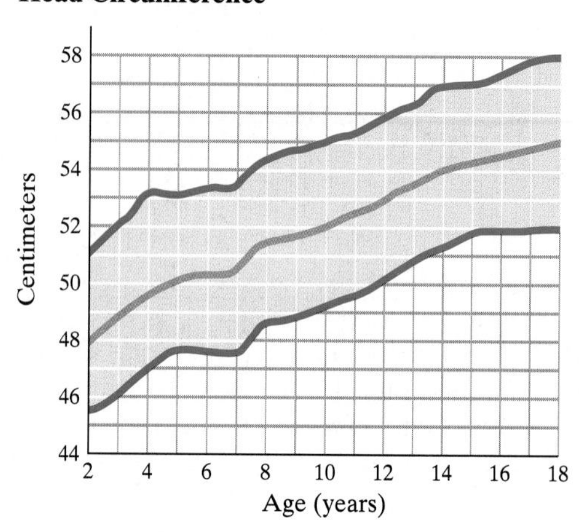

Cumulative Review Exercises

[2.5] **59.** Solve the inequality $4 - \frac{1}{2}x > 2x + 3$ and indicate the solution in interval notation.

60. What must you do when multiplying or dividing both sides of an inequality by a negative number?

[3.2] **61.** **a)** What is a relation? **b)** What is a function. **c)** Draw a graph that is a relation but not a function.

62. Find the domain and range of the function $\{(4, 3), (5, -2), (3, 2), (6, -1)\}$.

3.6 THE ALGEBRA OF FUNCTIONS

SSM VIDEO 3.6 CD Rom

1. Find the sum, difference, product, and quotient of functions.
2. Graph the sum of functions.

1 Find the Sum, Difference, Product, and Quotient of Functions

Let's discuss some ways that functions can be combined. If we let $f(x) = x + 2$ and $g(x) = x^2 + 2x$, we can find $f(5)$ and $g(5)$ as follows.

$$f(x) = x + 2 \qquad g(x) = x^2 + 2x$$

$$f(5) = 5 + 2 = 7 \qquad g(5) = 5^2 + 2(5) = 35$$

If we add $f(x) + g(x)$, we get

$$\begin{aligned} f(x) + g(x) &= (x + 2) + (x^2 + 2x) \\ &= x^2 + 3x + 2 \end{aligned}$$

This new function formed by the sum of $f(x)$ and $g(x)$ is designated as $(f + g)(x)$. Therefore we may write

$$(f + g)(x) = x^2 + 3x + 2$$

We find $(f + g)(5)$ as follows.

$$\begin{aligned} (f + g)(5) &= 5^2 + 3(5) + 2 \\ &= 25 + 15 + 2 = 42 \end{aligned}$$

Notice that

$$\begin{aligned} f(5) + g(5) &= (f + g)(5) \\ 7 + 35 &= 42 \qquad \text{True} \end{aligned}$$

In fact, for any real number substituted for x you will find that

$$f(x) + g(x) = (f + g)(x)$$

Similar notation exists for subtraction, multiplication, and division of functions.

Operations on Functions

If $f(x)$ represents one function, $g(x)$ represents a second function, and x is the domain of both functions, then the following operations on functions may be performed:

Sum of functions: $(f + g)(x) = f(x) + g(x)$

Difference of functions: $(f - g)(x) = f(x) - g(x)$

Product of functions: $(f \cdot g)(x) = f(x) \cdot g(x)$

Quotient of functions: $(f/g)(x) = \dfrac{f(x)}{g(x)}$, provided that $g(x) \neq 0$

EXAMPLE 1 If $f(x) = x^2 + x - 6$ and $g(x) = x - 2$, find

a) $(f + g)(x)$ **b)** $(f - g)(x)$ **c)** $(g - f)(x)$

d) Does $(f - g)(x) = (g - f)(x)$?

Solution To answer parts **a)–c)**, we perform the indicated operation.

a)
$$\begin{aligned} (f + g)(x) &= f(x) + g(x) \\ &= (x^2 + x - 6) + (x - 2) \\ &= x^2 + x - 6 + x - 2 \\ &= x^2 + 2x - 8 \end{aligned}$$

b)
$$\begin{aligned} (f - g)(x) &= f(x) - g(x) \\ &= (x^2 + x - 6) - (x - 2) \\ &= x^2 + x - 6 - x + 2 \\ &= x^2 - 4 \end{aligned}$$

c) $(g - f)(x) = g(x) - f(x)$
$= (x - 2) - (x^2 + x - 6)$
$= x - 2 - x^2 - x + 6$
$= -x^2 + 4$

d) By comparing the answers to parts **b)** and **c)**, we see that

NOW TRY EXERCISE 11

$$(f - g)(x) \neq (g - f)(x)$$

EXAMPLE 2 If $f(x) = x^2 - 4$ and $g(x) = x - 2$, find

a) $(f - g)(6)$ **b)** $(f \cdot g)(4)$ **c)** $(f/g)(8)$

Solution **a)** $(f - g)(x) = f(x) - g(x)$
$= (x^2 - 4) - (x - 2)$
$= x^2 - x - 2$
$(f - g)(6) = 6^2 - 6 - 2$
$= 36 - 6 - 2$
$= 28$

We could have also found the solution as follows:

$$f(x) = x^2 - 4 \qquad g(x) = x - 2$$
$$f(6) = 6^2 - 4 = 32 \qquad g(6) = 6 - 2 = 4$$
$$(f - g)(6) = f(6) - g(6)$$
$$= 32 - 4 = 28$$

b) We will find $(f \cdot g)(4)$ using the fact that

$$(f \cdot g)(4) = f(4) \cdot g(4)$$
$$f(x) = x^2 - 4 \qquad g(x) = x - 2$$
$$f(4) = 4^2 - 4 = 12 \qquad g(4) = 4 - 2 = 2$$

Thus $f(4) \cdot g(4) = 12 \cdot 2 = 24$. Therefore $(f \cdot g)(4) = 24$. We could have also found $(f \cdot g)(4)$ by multiplying $f(x) \cdot g(x)$ and then substituting 4 in the product. We will discuss how to do this in Section 5.2.

c) We will find $(f/g)(8)$ by using the fact that

$$(f/g)(8) = f(8)/g(8)$$
$$f(x) = x^2 - 4 \qquad g(x) = x - 2$$
$$f(8) = 8^2 - 4 = 60 \qquad g(8) = 8 - 2 = 6$$

Then $f(8)/g(8) = 60/6 = 10$. Therefore, $(f/g)(8) = 10$. We could have also found $(f/g)(8)$ by dividing $f(x)/g(x)$ and then substituting 8 in the quotient. We will discuss how to do this in Chapter 5.

NOW TRY EXERCISE 31

Notice that we included the phrase "and x is in the domain of both functions" in the Operations on Functions box on page 203. As we stated earlier, the domain of a function is the set of values that can be used for the independent variable. For example, the domain of the function $f(x) = 2x^2 - 6x + 5$ is all real numbers, because when x is any real number $f(x)$ will also be a real number. The domain of $g(x) = \dfrac{1}{x - 3}$ is all real numbers except 3, because when x is any real number except 3, the function $g(x)$ is a real number. When x is 3, the function is not a real number because $\frac{1}{0}$ is undefined. We will discuss the domain of functions further in Section 6.1.

2) Graph the Sum of Functions

Now we will explain how we can graph the sum, difference, product, or quotient of two functions. Figure 3.72 shows two functions $f(x)$ and $g(x)$.

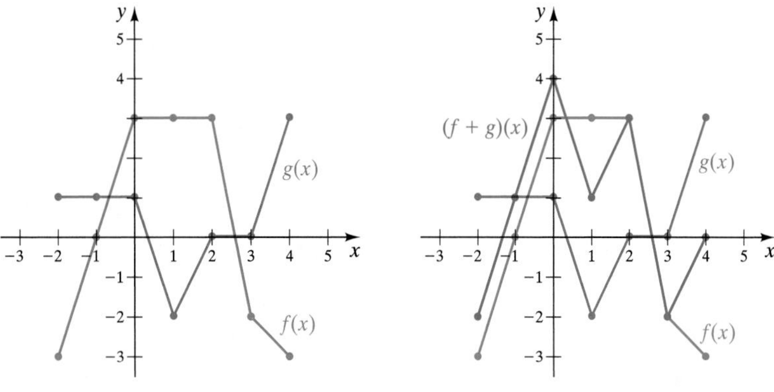

FIGURE 3.72 FIGURE 3.73

To graph the sum of $f(x)$ and $g(x)$, or $(f + g)(x)$, we use $(f + g)(x) = f(x) + g(x)$. The table below gives the integer values of x from -2 to 4, the values of $f(-2)$ through $f(4)$, the values of $g(-2)$ through $g(4)$, and the values of $(f + g)(-2)$ through $(f + g)(4)$. The graph of $(f + g)(x) = f(x) + g(x)$ is illustrated in green in Figure 3.73.

x	$f(x)$	$g(x)$	$(f + g)(x)$
−2	−3	1	$-3 + 1 = -2$
−1	0	1	$0 + 1 = 1$
0	3	1	$3 + 1 = 4$
1	3	−2	$3 + (-2) = 1$
2	3	0	$3 + 0 = 3$
3	−2	0	$-2 + 0 = -2$
4	−3	3	$-3 + 3 = 0$

We could graph the difference, product, or quotient of the two functions using a similar technique. For example, to graph the product function $(f \cdot g)(x)$, we would evaluate $(f \cdot g)(-2)$ as follows:

$$(f \cdot g)(-2) = f(-2) \cdot g(-2)$$
$$= (-3)(1) = -3$$

Thus, the graph of $(f \cdot g)(x)$ would have an ordered pair at $(-2, -3)$. Other ordered pairs would be determined by the same procedure.

NOW TRY EXERCISE 47

In newspapers and magazines we often find graphs that show the sum of two functions. Graphs that show the sum of two functions are generally illustrated in one of two ways. Example 3 shows one way and Example 4 shows the second way.

EXAMPLE 3 Figure 3.74 shows the number of bond funds, stock funds, and total new funds added from 1991 through 1996.

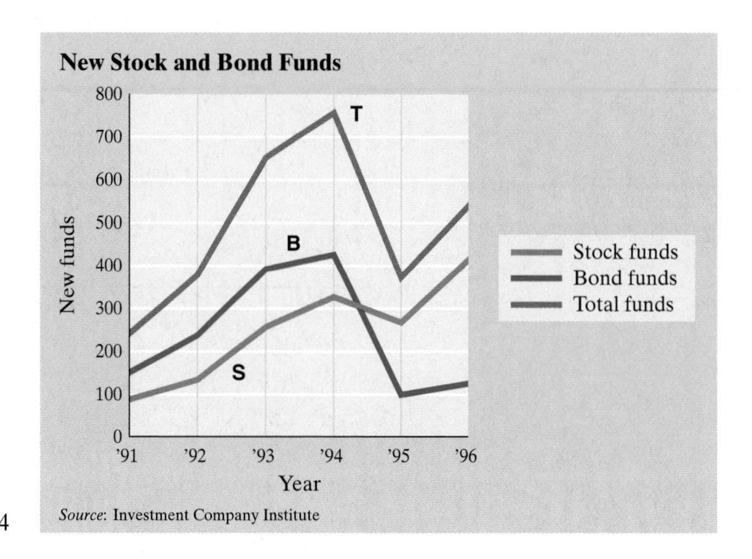

FIGURE 3.74

a) How is the graph of the total new funds, T, determined from the graphs of the stock, S, and bond, B, funds?

b) In which year were the most stock funds added?

c) In which years were there more stock funds than bond funds added?

Solution **a)** In Figure 3.74, the bond and stock graphs are graphed separately on the same axes. The graph for the total new funds is obtained by adding the number of stock funds to the number of bond funds. For example, in 1994 there were about 330 new stock funds and about 425 new bond funds. The sum of these numbers is 755, which is about the total number of funds in 1994. Other points on the total funds graph are found in the same way.

b) The most stock funds were added in 1996. About 415 stock funds were added that year.

c) There were more stock funds than bond funds added in 1995 and 1996.

In Example 4, which also shows a sum of functions, the categories are "piggy backed" on top of one another.

EXAMPLE 4 The graph in Figure 3.75 shows the number of people who received kidney, liver, and other transplants from 1991 through 1996.

a) Find the number people who received kidney transplants in 1996.

b) Find the number of people who received liver transplants in 1996.

c) Find the number of people who received other transplants in 1996.

d) Find the total number of people who received transplants in 1996.

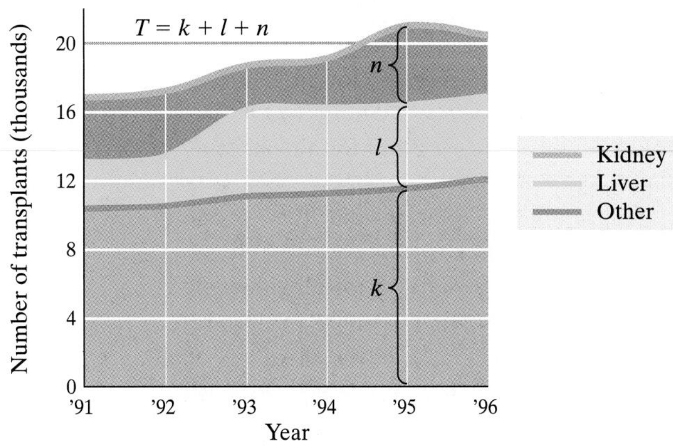

FIGURE 3.75

Solution **a)** In this graph we have labeled the number of kidney transplants k, the number of liver transplants l, and the number of other transplants n. Notice that the total number of transplants, T, is a function of the year, and $T = k + l + n$. By reading the graph, we see that the kidney transplants in 1996 were about 12,000.

b) The gold color on the graph represents the liver transplants. In 1996 the gold shaded area starts at 12,000 and ends at about 17,000. The difference between these two values, $17{,}000 - 12{,}000 = 5000$, represents the number of liver transplants in 1996.

c) The red color on the graph represents the other transplants. In 1996 the red shaded area starts at 17,000 and ends at about 20,500. The difference between these values, $20{,}500 - 17{,}000 = 3500$, represents the number of other transplants.

d) In 1996 the total number of transplants is about 20,500. This can be read directly from the graph. Notice that the total number of transplants = kidney + liver + other.

NOW TRY EXERCISE 55

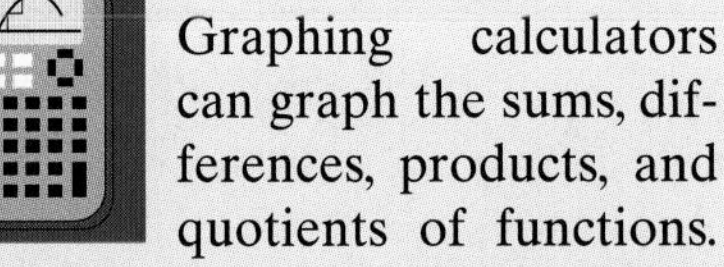

Graphing calculators can graph the sums, differences, products, and quotients of functions. One way to do this is to enter the individual functions. Then, following the instructions that come with your calculator, you can add, subtract, multiply, or divide the functions. For Example, the screen in Figure 3.76 shows a TI-83 ready to graph $Y_1 = x - 3$, $Y_2 = 2x + 4$, and the sum of the functions, $Y_3 = Y_1 + Y_2$. On the TI-83, to get $Y_3 = Y_1 + Y_2$, you press the VARS key. Then you move the cursor to Y-VARS, then you select 1: Function. Next you press 1 to enter Y_1. Next you press +. Then press VARS and go to Y-VARS, and choose 1: Function. Finally press 2 to enter Y_2. Figure 3.77 shows the graphs of the two functions, and the graph of the sum of the functions.

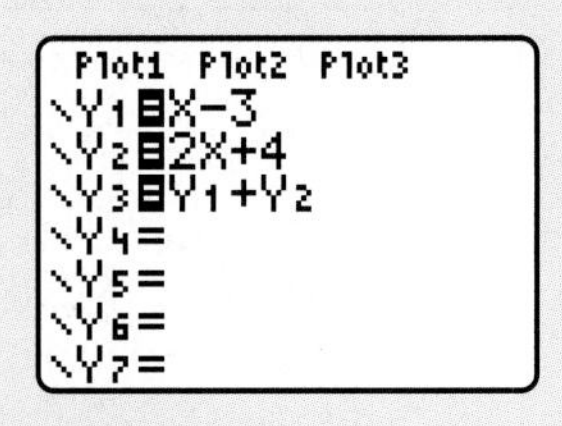

FIGURE 3.76

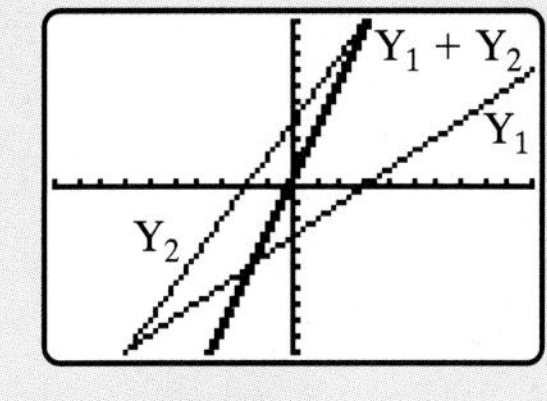

FIGURE 3.77

Exercise Set 3.6

Concept/Writing Exercises

1. Does $f(x) + g(x) = (f + g)(x)$ for all values of x? Explain, and give an example to support your answer.
2. Does $f(x) - g(x) = (f - g)(x)$ for all values of x? Explain, and give an example to support your answer.
3. What restriction is placed on the property $f(x)/g(x) = (f/g)(x)$? Explain.
4. Does $(f + g)(x) = (g + f)(x)$ for all values of x? Explain and give an example to support your answer.
5. Does $(f - g)(x) = (g - f)(x)$ for all values of x? Explain and give an example to support your answer.
6. If $f(4) = 6$ and $g(4) = 2$, find
   a) $(f + g)(4)$ **b)** $(f - g)(4)$
   c) $(f \cdot g)(4)$ **d)** $(f/g)(4)$
7. If $f(-2) = -3$ and $g(-2) = 5$, find
   a) $(f + g)(-2)$ **b)** $(f - g)(-2)$
   c) $(f \cdot g)(-2)$ **d)** $(f/g)(-2)$
8. If $f(5) = 8$ and $g(5) = 0$, find
   a) $(f + g)(5)$ **b)** $(f - g)(5)$
   c) $(f \cdot g)(5)$ **d)** $(f/g)(5)$

Practice the Skills

For each pair of functions, find **a)** $(f + g)(x)$, **b)** $(f + g)(a)$, *and* **c)** $(f + g)(2)$.

9. $f(x) = x + 4,\ g(x) = x^2 - 2x$
10. $f(x) = x^2 - 2x + 3,\ g(x) = x^2 - 4$
11. $f(x) = 2x^2 - 3x + 5,\ g(x) = x^3 - x^2$
12. $f(x) = x^3 - 2x + 3,\ g(x) = 3x^2 - 2x$
13. $f(x) = 4x^3 - 3x^2 - x,\ g(x) = 3x^2 + 4$
14. $f(x) = 3x^2 - x + 4,\ g(x) = 6 - 4x^2$

Let $f(x) = x^2 - 4$ *and* $g(x) = -5x + 3$. *Find the following.*

15. $f(5) + g(5)$
16. $f(2) + g(2)$
17. $f(-2) - g(-2)$
18. $f\left(\frac{1}{2}\right) - g\left(\frac{1}{2}\right)$
19. $f(3) \cdot g(3)$
20. $f(-4) \cdot g(-4)$
21. $f(4)/g(4)$
22. $f(-1)/g(-1)$
23. $g(-3) - f(-3)$
24. $g(6) \cdot f(6)$
25. $g(0)/f(0)$
26. $f(2)/g(2)$

Let $f(x) = 2x^2 - x$ *and* $g(x) = x - 6$. *Find the following.*

27. $(f + g)(x)$
28. $(f + g)(a)$
29. $(f + g)(4)$
30. $(f + g)(-2)$
31. $(f - g)(6)$
32. $(f - g)(0)$
33. $(f \cdot g)(-3)$
34. $(f \cdot g)(-5)$
35. $(f/g)(-1)$
36. $(f/g)(0)$
37. $(g/f)(5)$
38. $(g - f)(3)$
39. $(g - f)(x)$
40. $(g - f)(r)$

Problem Solving

Using the graph to the right, find the value of the following.

41. $(f + g)(2)$
42. $(f - g)(2)$
43. $(f \cdot g)(2)$
44. $(f/g)(2)$
45. $(f - g)(-2)$
46. $(g - f)(0)$
47. $(g/f)(-2)$
48. $(g \cdot f)(0)$

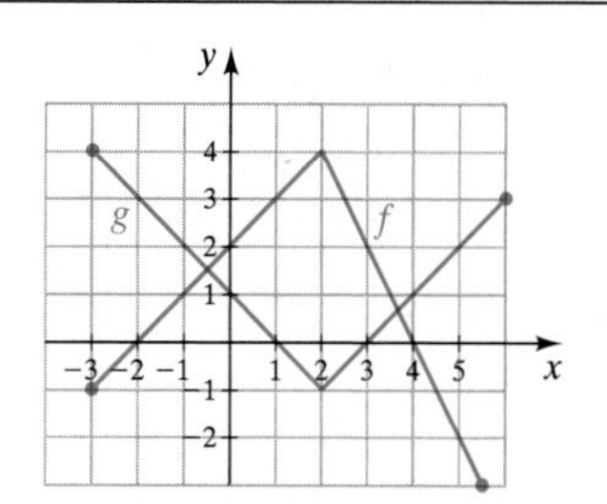

Using the graph below, find the value of the following.

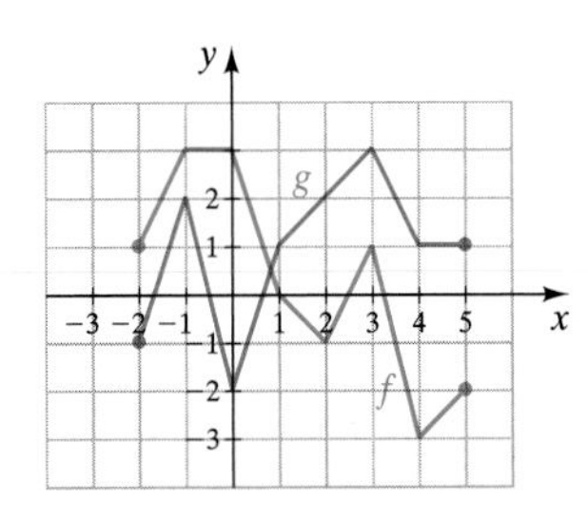

49. $(f + g)(3)$
50. $(f - g)(3)$
51. $(g - f)(2)$
52. $(f \cdot g)(1)$
53. $(f/g)(4)$
54. $(g/f)(5)$
55. $(g \cdot f)(0)$
56. $(g/f)(2)$

57. The following graph shows sales of GM cars and trucks from 1986 through 1996.

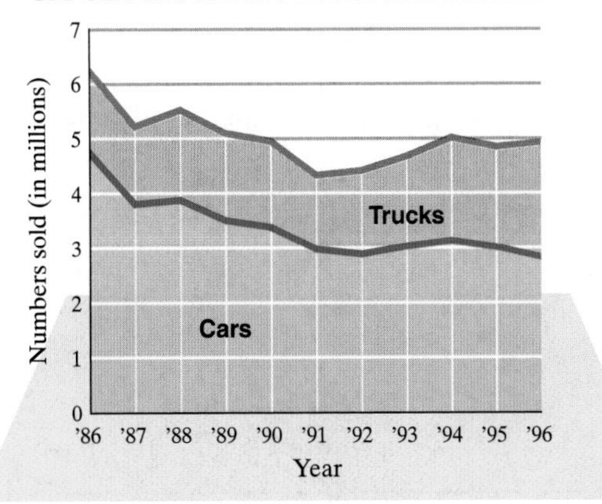

Estimate the number of

a) GM vehicles (total) sold in 1996.

b) GM cars sold in 1986.

c) GM cars sold in 1996.

d) GM trucks sold in 1986.

e) GM trucks sold in 1996.

f) Describe the change in new GM cars and trucks sold from 1986 through 1996.

58. The following graph shows the credit card debt for 1990 and 1995 and projected debt for 2000 and 2005 for Visa, MasterCard, and other credit cards.

a) Estimate the projected debt on other types of credit cards in 2000.

b) Estimate the projected debt on MasterCard in 2000.

c) Estimate the projected debt on Visa in 2000.

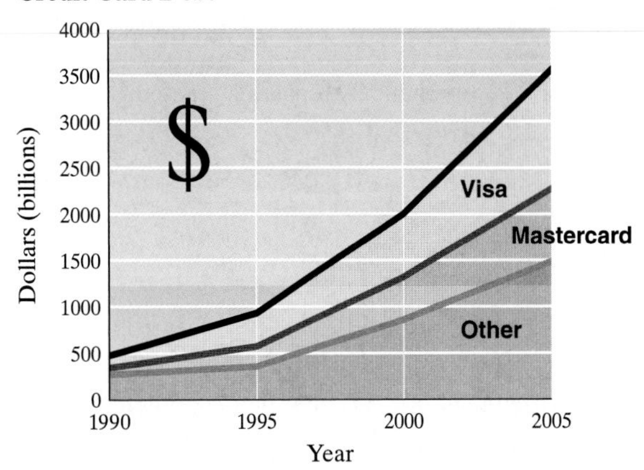

59. The following graph shows U.S., and worldwide shipment of personal computers (PC's) from 1977 through 1998.

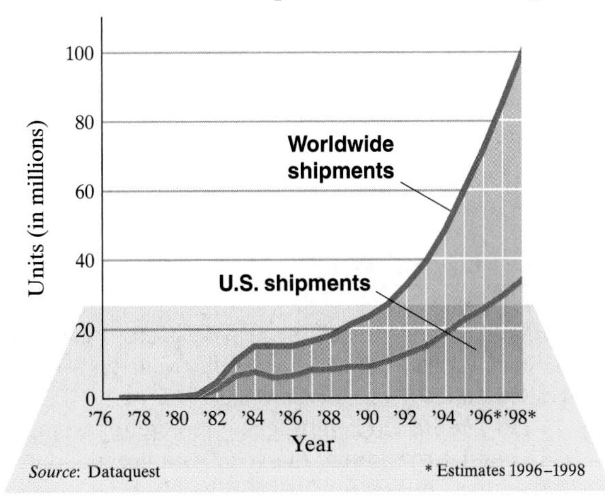

a) Estimate the number of PC's shipped in the United States in 1998.

b) Estimate the number of PC's shipped worldwide in 1998.

c) If U represents U.S. shipments of PC's, W represents worldwide shipments of PC's, and t represents the year, what will the graph of $(W - U)(t)$ represent?

d) Determine the number of PC's shipped outside of the United States in 1998.

60. The following graph shows the number of patents granted for inventions in the United States and worldwide (total) from 1990 through 1996.

a) Estimate the number of patents granted in the United States in 1995.

b) Estimate the number of patents granted worldwide in 1995.

c) Estimate the number of patents granted outside of the United States in 1995.

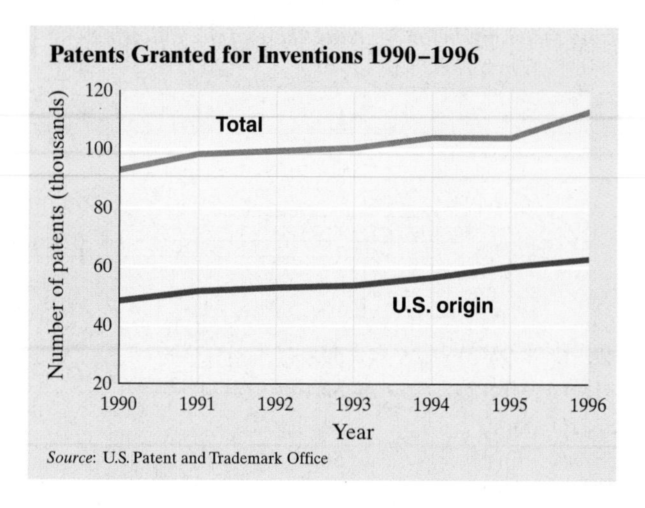

Let f and g represent two functions that are graphed on the same axes.

61. If, at a, $(f + g)(a) = 0$, what must be true about $f(a)$ and $g(a)$?

62. If, at a, $(f \cdot g)(a) = 0$, what must be true about $f(a)$ and $g(a)$?

63. If, at a, $(f - g)(a) = 0$, what must be true about $f(a)$ and $g(a)$?

64. If, at a, $(f - g)(a) < 0$, what must be true about $f(a)$ and $g(a)$?

65. If, at a, $(f/g)(a) < 0$, what must be true about $f(a)$ and $g(a)$?

66. If, at a, $(f \cdot g)(a) < 0$, what must be true about $f(a)$ and $g(a)$?

Graph the following functions on your graphing calculator.

67. $y_1 = 2x + 3$
$y_2 = -x + 4$
$y_3 = y_1 + y_2$

68. $y_1 = x - 3$
$y_2 = 2x$
$y_3 = y_1 - y_2$

69. $y_1 = x$
$y_2 = x + 5$
$y_3 = y_1 \cdot y_2$

70. $y_1 = 2x^2 - 4$
$y_2 = x$
$y_3 = y_1/y_2$

Group Activity

71. The following graph shows sales of passenger cars and light trucks (including sport utility vehicles) in the United States from 1990 through 1995.

Let f represent light trucks, g represent passenger cars, and let t represent the year. As a group, draw a graph that represents $(f + g)(t)$.

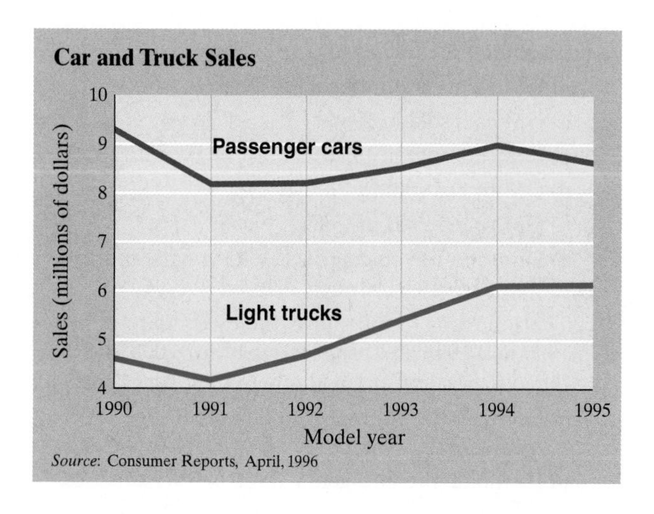

Cumulative Review Exercises

[2.2] **72.** Solve the formula $A = \frac{1}{2}bh$ for h.

[2.3] **73.** The cost of a washing machine, including a 6% sales tax, is \$477. Determine the pre-tax cost of the washing machine.

[3.1] **74.** Graph $y = |x| - 2$.

[3.3] **75.** Graph $3x - 4y = 12$.

3.7 GRAPHING LINEAR INEQUALITIES

1) **Graph linear inequalities in two variables.**

SSM VIDEO 3.7 CD Rom

1) Graph Linear Inequalities in Two Variables

A *linear inequality* results when the equal sign in a linear equation is replaced with an inequality sign.

Examples of Linear Inequalities in Two Variables

$$2x + 3y > 2 \qquad 3y < 4x - 6$$

$$-x - 2y \leq 3 \qquad 5x \geq 2y - 3$$

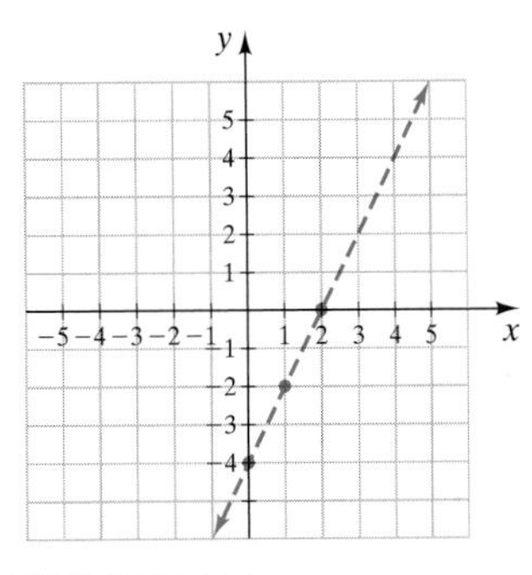

FIGURE 3.78

To Graph a Linear Inequality in Two Variables

1. Replace the inequality symbol with an equal sign.
2. Draw the graph of the equation in step 1. If the original inequality contains a $\geq$ or $\leq$ symbol, draw the graph using a solid line. If the original inequality contains a $>$ or $<$ symbol, draw the graph using a dashed line.
3. Select any point not on the line and determine if this point is a solution to the original inequality. If the point selected is a solution, shade the region on the side of the line containing this point. If the selected point does not satisfy the inequality, shade the region on the side of the line not containing this point.

EXAMPLE 1 Graph the inequality $y < 2x - 4$.

Solution First graph the equation $y = 2x - 4$. Since the original inequality contains a less-than sign, $<$, use a dashed line when drawing the graph (Fig. 3.78). The dashed line indicates that the points on this line are not solutions to the inequality $y < 2x - 4$. Select a point not on the line and determine if this point satisfies the inequality. Often the easiest point to use is the origin, (0, 0).

Check point (0, 0)

$$y < 2x - 4$$

$$0 \overset{?}{<} 2(0) - 4$$

$$0 \overset{?}{<} 0 - 4$$

$$0 < -4 \qquad \textit{False}$$

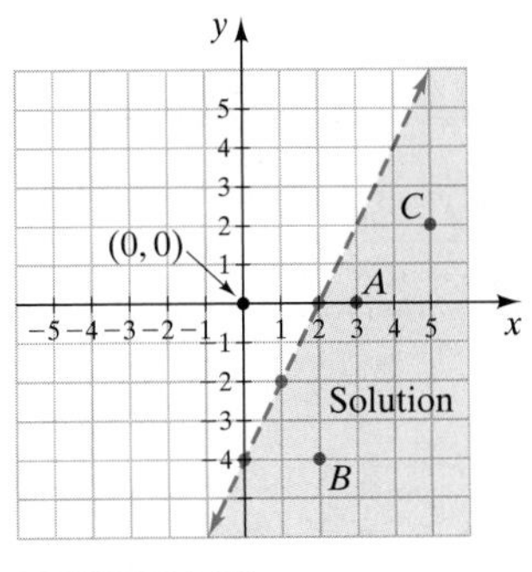

FIGURE 3.79

Since 0 is not less than -4, which is symbolized $0 \not< -4$, the point (0, 0) does not satisfy the inequality. The solution will be all the points on the opposite side of the line from the point (0, 0). Shade in this region (Fig. 3.79). Every point in the shaded area satisfies the given inequality. Let's check a few selected points A, B, and C.

Point A	Point B	Point C
(3, 0)	(2, −4)	(5, 2)
$y < 2x - 4$	$y < 2x - 4$	$y < 2x - 4$
$0 \overset{?}{<} 2(3) - 4$	$-4 \overset{?}{<} 2(2) - 4$	$2 \overset{?}{<} 2(5) - 4$
$0 < 2$ *True*	$-4 < 0$ *True*	$2 < 6$ *True*

EXAMPLE 2 Graph the inequality $y \geq -\frac{1}{2}x$.

Solution First, we graph the equation $y = -\frac{1}{2}x$. Since the inequality is $\geq$, we use a solid line to indicate that the points on the line are solutions to the inequality (Fig. 3.80). Since the point (0, 0) is on the line, we cannot select that point to find the solution. Let's arbitrarily select the point (3, 1).

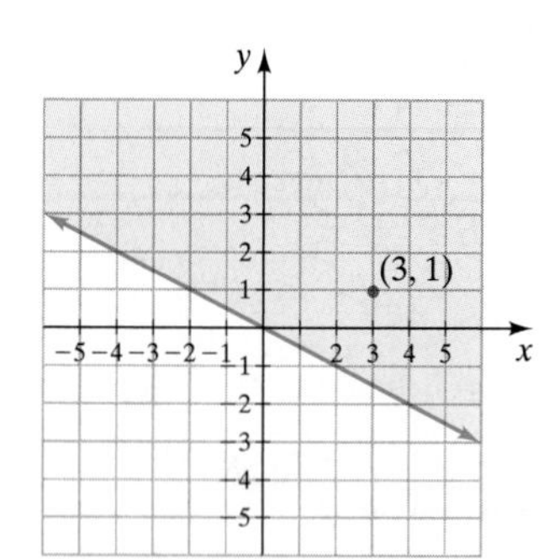

FIGURE 3.80

Check point (3, 1)

$$y \geq -\frac{1}{2}x$$

$$1 \geq -\frac{1}{2}(3)$$

$$1 \geq -\frac{3}{2} \quad \textit{True}$$

Since the point (3, 1) satisfies the inequality, every point on the same side of the line as (3, 1) will also satisfy the inequality $y \geq -\frac{1}{2}x$. Shade this region as indicated. Every point in the shaded region as well as every point on the line satisfies the inequality.

EXAMPLE 3 Graph the inequality $3x - 2y < -6$.

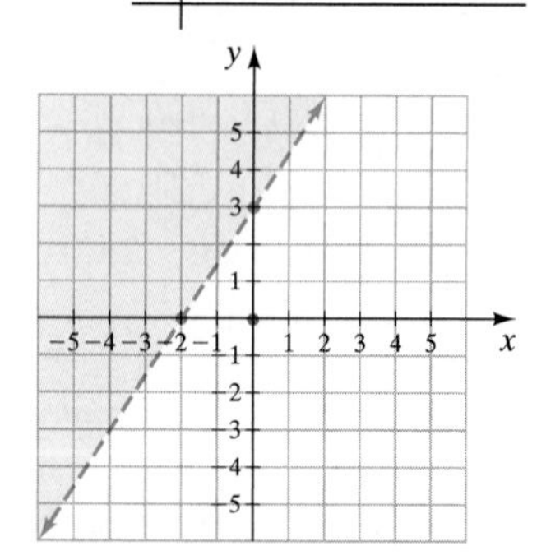

FIGURE 3.81

Solution First, we graph the equation $3x - 2y = -6$. Since the inequality is $<$, we use a dashed line when drawing the graph (Fig. 3.81). Substituting the checkpoint (0, 0) into the inequality results in a false statement.

Check point (0, 0)

$$3x - 2y < -6$$

$$3(0) - 2(0) \overset{?}{<} -6$$

$$0 < -6 \quad \textit{False}$$

NOW TRY EXERCISE 23

The solution is, therefore, that part of the plane that does not contain the origin.

Using Your Graphing Calculator

Graphers can also display graphs of inequalities. The procedure to display the graphs varies from calculator to calculator. In Figure 3.82, we show the graph of $y > 2x + 3$. Read your graphing calculator manual and learn how to display graphs of inequalities.

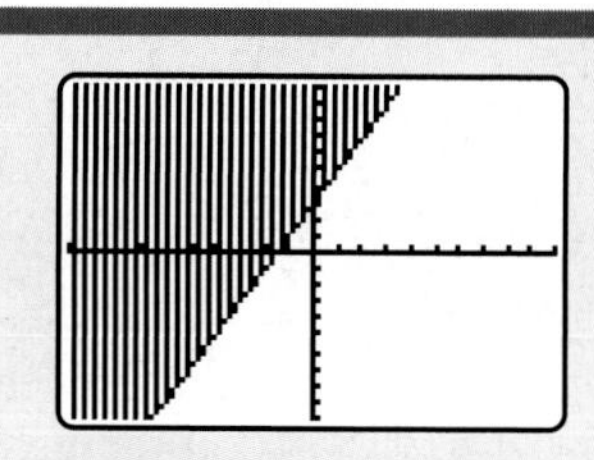
FIGURE 3.82

Exercise Set 3.7

Concept/Writing Exercises

1. When graphing an inequality containing $\geq$ or $\leq$, why are points on the line solutions to the inequality?

2. When graphing an inequality containing $>$ or $<$, why are points on the line not solutions to the inequality?

3. When can $(0, 0)$ not be used as a check point?

4. When graphing a linear inequality of the form $y > ax + b$ where a and b are real numbers, will the solution always be above the line? Explain.

Practice the Skills

Graph each inequality.

5. $x > 1$
6. $x \geq \frac{5}{2}$
7. $y < -2$
8. $y < x$
9. $y \geq 2x$
10. $y > -2x$
11. $y < 2x + 1$
12. $y \geq 3x - 1$
13. $y < -3x + 4$
14. $y \geq 2x + 4$
15. $y \geq \frac{1}{2}x - 3$
16. $y < 3x + 5$
17. $y \leq \frac{1}{3}x + 6$
18. $y > 6x + 1$
19. $y \leq -3x + 5$
20. $y \leq \frac{2}{3}x + 3$
21. $2x + y < 4$
22. $3x - 4y \leq 12$
23. $2x \leq 5y + 10$
24. $-x - 2y > 4$

Problem Solving

25. The average daily volume on the New York Stock Exchange has been increasing approximately linearly from 1990 through 1996. In 1990 the average daily volume was about 150 million shares, and in 1996 it was about 400 million shares.
    a) Draw a graph that fits this data.
    b) On the graph, darken the part of the graph where the average daily volume is less than or equal to 300 million shares.
    c) Estimate the year when the number of shares first exceeded 300 million shares.

26. The percent of working mothers with children has been increasing approximately linearly from 1975 through 1996. The percent of working mothers with children under 6 years old increased from 39.0% in 1975 to 62.3% in 1996.
    a) Draw a graph that fits this data.
    b) On the graph, darken the part of the graph where the percent of working mothers with children under 6 years old is greater than or equal to 50%.
    c) Estimate the first year in which the number of working mothers with children under 6 years old was greater than or equal to 50%?

27. The number of active lawyers has been growing approximately linearly in the United States from 1970 through 1996. In 1970 there were about 327 thousand lawyers and in 1996 there were about 946 thousand lawyers.
    a) Draw a graph that fits this data.
    b) On the graph, darken the part of the graph where the number of lawyers is greater than or equal to 500 thousand.
    c) Estimate the first year that the number of lawyers exceeded 500 thousand?

28. The number of miles of roadway in the United States increased at an approximately linear rate from 1990 through 1996. In 1990 there about 3.80 million miles of roadway. In 1996 there were 3.92 million miles of roadway.
    a) Draw a graph that fits this data.
    b) On the graph, darken the part of the graph where the number of miles of roadway is less than or equal to 3.85 million.
    c) Estimate the first year in which there were more than 3.85 million miles of roadway.

29. **a)** Graph $f(x) = 2x - 4$.
    b) On the graph, shade the region bounded by $f(x)$, $x = 2$, $x = 4$, and the x-axis.

30. **a)** Graph $g(x) = -x + 4$.
    b) On the graph, shade the region bounded by $g(x)$, $x = 1$, and the x- and y-axes.

Challenge Problems

Graph each inequality.

31. $y < |x|$

32. $y \geq x^2$

33. $y < x^2 - 4$

Cumulative Review Exercises

[2.1] **34.** Solve the equation $4 - \frac{5x}{3} = -6$.

[2.2] **35.** If $C = \bar{x} + Z\frac{\sigma}{\sqrt{n}}$, find C when $\bar{x} = 80$, $Z = 1.96$, $\sigma = 3$ and $n = 25$.

[2.3] **36.** El Gigundo Department Store is going out of business. The first week all items are being reduced by 10%. The second week all items are being reduced by an additional \$2. If during the second week Sean Russell purchases a CD for \$12.15, find the original cost of the CD.

[3.3] **37.** Write an equation of the line that passes through the point $(6, -2)$ and is perpendicular to the line $2x - y = 4$.

SUMMARY

Key Words and Phrases

3.1

Cartesian coordinate system
Collinear points
Coordinates
First-degree equation
Graph
Graphing calculator
Linear equation
Midpoint
Nonlinear equation
Ordered pair
Origin
Quadrant
Rectangular coordinate system
TABLE feature
TRACE feature
Window of a graphing calculator
x-axis
y-axis

3.2

Dependent variable
Domain
Function
Function notation
Graph of a function or relation
Independent variable
Piecewise function
Range
Relation
Supply and demand
Vertical line test
y is a function of x

3.3

Constant function
Linear function
Root
Standard form of a linear equation
x-intercept
y-intercept
Zero
ZOOM feature

3.4

Negative slope
Parallel lines
Positive slope
Rate of change
Slope of a line
Slope–intercept form of a linear equation
Tangent line
Translated graph
Zero slope

3.5

Negative reciprocal
Perpendicular line
Point–slope form of a linear equation

3.6

Difference of functions
Product of functions
Quotient of functions
Sum of functions

3.7

Linear inequality

IMPORTANT FACTS

Slope of a Line $m = \frac{\Delta y}{\Delta x} = \frac{y_2 - y_1}{x_2 - x_1}$

Forms of a Linear equation

Standard form: $ax + by = c$
Slope–intercept form: $y = mx + b$
Point–slope form: $y - y_1 = m(x - x_1)$
To find the x-intercept, set $y = 0$ and solve the equation for x.
To find the y-intercept, set $x = 0$ and solve the equation for y.
To write an equation in slope–intercept form, solve the equation for y.

continued on next page

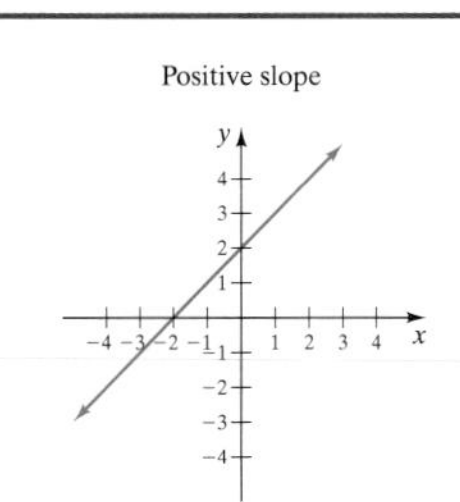

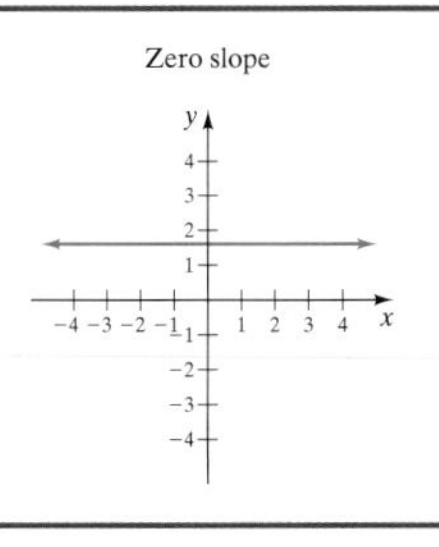

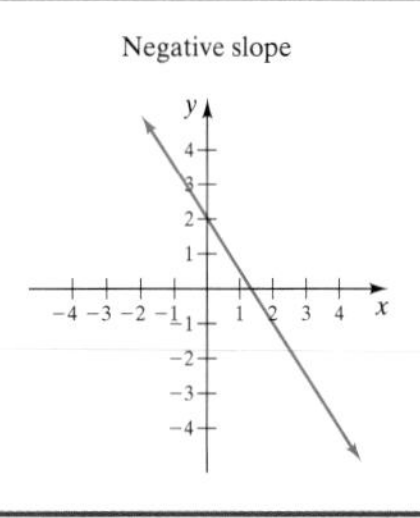

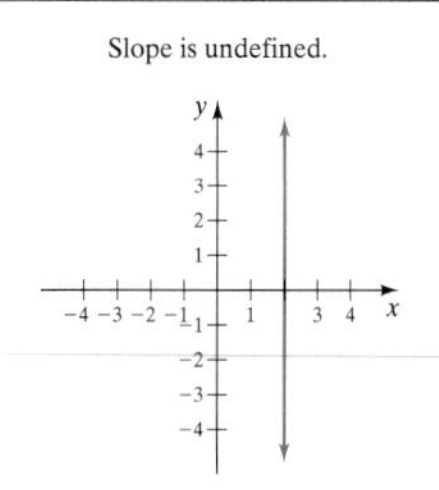

Operations on Functions

Sum of functions: $(f + g)(x) = f(x) + g(x)$
Difference of functions: $(f - g)(x) = f(x) - g(x)$
Product of functions: $(f \cdot g)(x) = f(x) \cdot g(x)$

Quotient of functions: $(f/g)(x) = \dfrac{f(x)}{g(x)}, g(x) \neq 0$

Review Exercises

[3.1] **1.** Plot the ordered pairs on the same axes.

a) $A\ (5, 3)$ **b)** $B\ (0, 4)$ **c)** $C\left(5, \frac{1}{2}\right)$ **d)** $D\ (-4, 3)$ **e)** $E\ (-6, -1)$ **f)** $F\ (-2, 0)$

Graph each equation.

2. $y = 4x$
3. $y = -3x + 4$
4. $y = \frac{3}{2}x - 3$
5. $y = -\frac{1}{2}x + 2$
6. $y = x^2$
7. $y = x^2 - 1$
8. $y = |x|$
9. $y = |x| - 1$
10. $y = x^3$
11. $y = x^3 + 4$

[3.2] **12.** Define function.

13. Is every relation a function? Is every function a relation? Explain.

Determine whether the following relations are functions. Explain your answers.

14. $a \to 2$, $b \to 1$, $c \to 3$

15. $\{(4, 2), (6, 3), (5, -1), (4, 0), (-2, 5)\}$

For Exercises 16–19, ***a)*** *determine whether the following relations are functions;* ***b)*** *determine the domain and range of each.*

16.

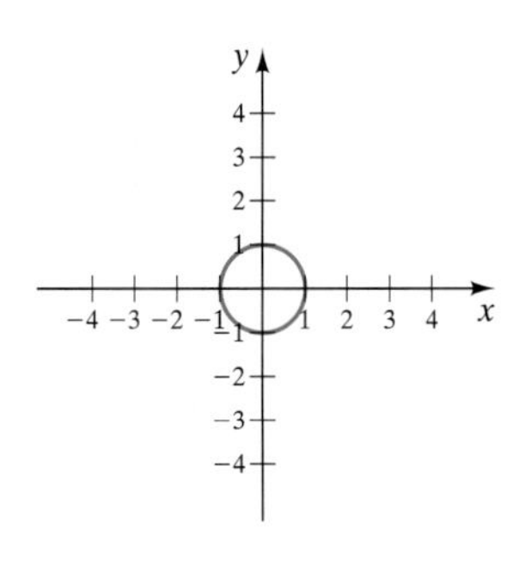

17.

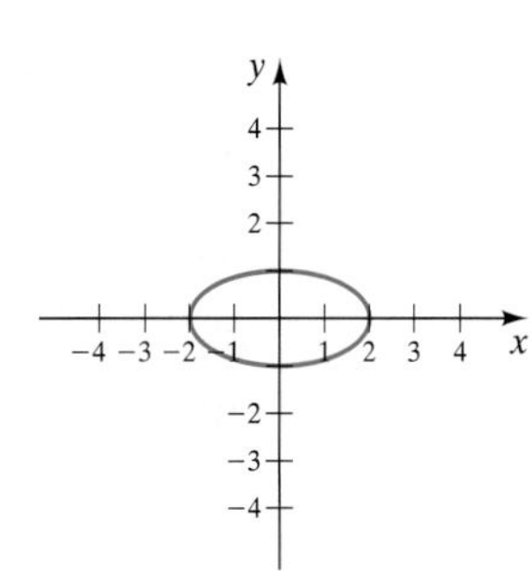

18.

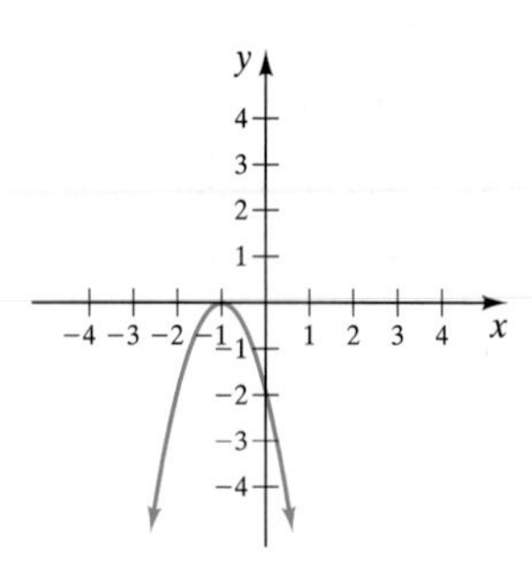

19.

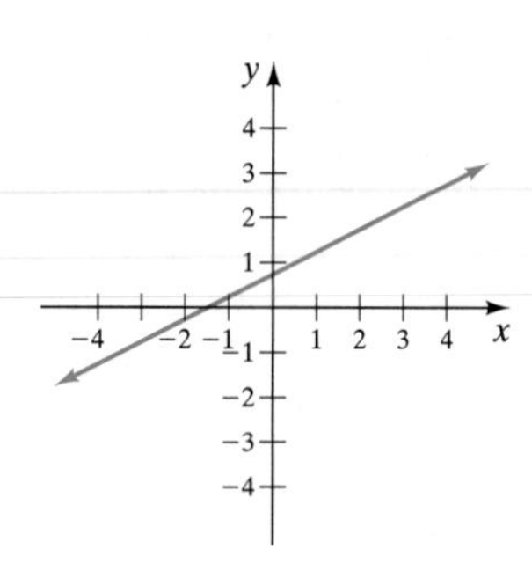

20. If $f(x) = x^2 + 2x - 5$, find **a)** $f(3)$ and **b)** $f(a)$.

21. If $g(t) = t^3 - 5t + 2$, find **a)** $g(-3)$ and **b)** $g(4)$.

22. Jane Covillion goes for a ride in a car. The following graph shows the car's speed as a function of time. Make up a story that corresponds to this graph.

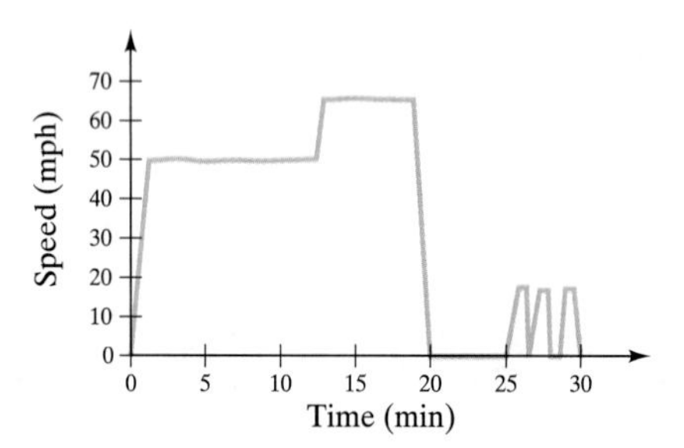

23. The number of baskets of apples, N, that are produced by x trees in a small orchard ($x \leq 100$) is given by the function $N(x) = 40x - 0.2x^2$. How many baskets of apples are produced by

a) 20 trees? **b)** 50 trees?

24. If a ball is dropped from the top of a 100-foot building, its height above the ground, h, at any time, t, can be found by the function $h(t) = -16t^2 + 100$, $0 \leq t \leq 2.5$. Find the height of the ball at

a) 1 second. **b)** 2 seconds.

[3.3] *Graph each equation using intercepts.*

25. $f(x) = \frac{1}{2}x - 4$

26. $\frac{2}{3}x = \frac{1}{4}y + 20$

Graph each equation or function.

27. $f(x) = 4$

28. $x = -2$

29. The yearly profit, p, of a bagel company can be estimated by the function $p(x) = 0.1x - 5000$, where x is the number of bagels sold per year.

a) Draw a graph of profits versus bagels sold for up to 250,000 bagels.

b) Estimate the number of bagels that must be sold for the company to break even.

c) Estimate the number of bagels sold if the company has a $20,000 profit.

30. Draw a graph illustrating the interest on a $12,000 loan for a 1-year period for various interest rates up to 20%. Use interest = principal · rate · time.

[3.4] *Determine the slope and y-intercept of each equation.*

31. $y = -x + 5$

32. $f(x) = -4x + \frac{1}{2}$

33. $3x + 5y = 12$

34. $9x + 7y = 15$

35. $x = -2$

36. $f(x) = 6$

Determine the slope of the line through the two given points.

37. $(4, 6), (5, -1)$

38. $(-2, 3)\ (4, 1)$

Find the slope of each line. If the slope is undefined, so state. Then write the equation of the line.

39.

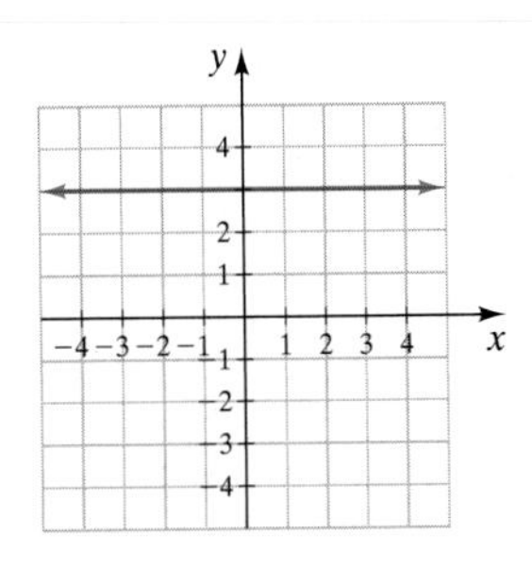

40.

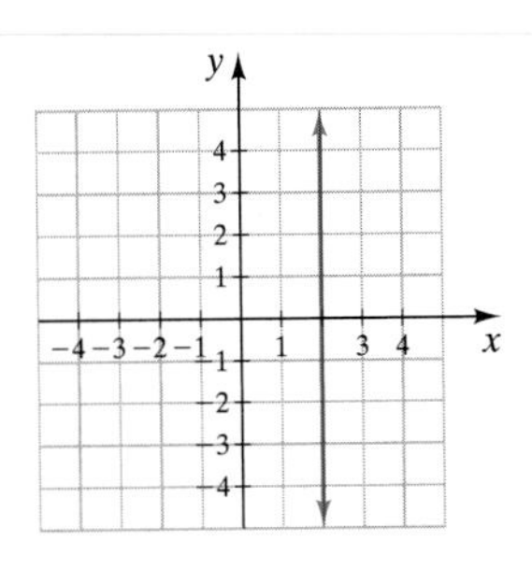

41.

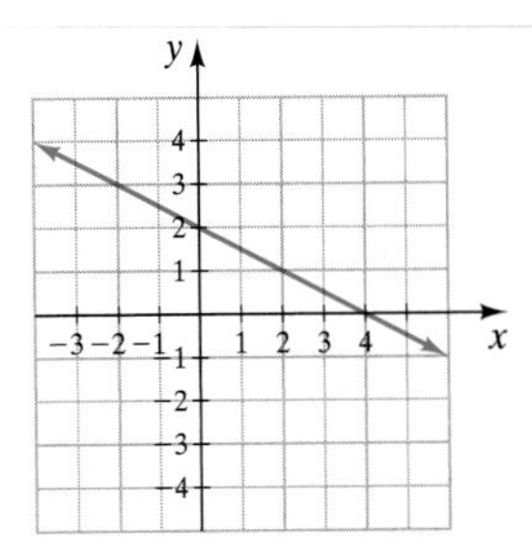

42. If a graph passes through $(3, 5)$ and $(-2, 8)$, find the change in y with respect to a one unit of change in x.

43. If one point on a graph is $(-6, -8)$ and the slope is $\frac{4}{3}$, find the y-intercept of the graph.

44. The following chart shows how the number of murders has been declining in New Orleans.

a) Plot each point and draw line segments from point to point.

b) Compute the slope of the line segments.

Year	Number of Murders
1994	437
1995	360
1996	349
1997	192

45. The following graph shows the number of Social Security beneficiaries from 1980 projected through 2070. Use the slope–intercept form to find the function $n(t)$ (represented by the straight line) that can be used to represent this data.

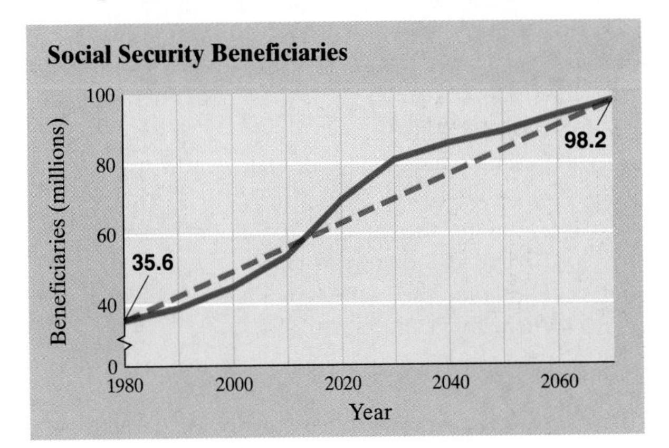

[3.5] *Determine whether the two given lines are parallel, perpendicular, or neither.*

46. $y = 3x - 6$
$6y = 18x + 6$

47. $2x - 3y = 9$
$-3x - 2y = 6$

48. $4x - 2y = 10$
$-2x + 4y = -8$

Find the equation of the line with the properties given. Write each answer in slope–intercept form.

49. Slope $= -\frac{2}{3}$, through $(3, 2)$

50. Through $(4, 3)$ and $(2, 1)$

51. Through $(-6, 2)$ parallel to $y = 3x - 4$

52. Through $(4, -2)$ parallel to $2x - 5y = 6$

53. Through $(-3, 1)$ perpendicular to $y = \frac{3}{5}x + 5$

54. Through $(4, 2)$ perpendicular to $4x - 2y = 8$

Two points on l_1 and two points on l_2 are given. Determine whether l_1 is parallel to l_2, l_1 is perpendicular to l_2, or neither.

55. l_1: $(4, 3)$ and $(0, -3)$; l_2: $(1, -1)$ and $(2, -2)$

56. l_1: $(3, 2)$ and $(2, 3)$; l_2: $(4, 1)$ and $(1, 4)$

57. l_1: $(4, 0)$ and $(1, 3)$; l_2: $(5, 2)$ and $(6, 3)$

58. l_1: $(-3, 5)$ and $(2, 3)$; l_2: $(-4, -2)$ and $(-1, 2)$

59. The number of bank failures declined approximately linearly from a high of 206 in 1989 to a low of 11 in 1994. Let b be the number of bank failures and t be the years since 1989.

a) Find a function $b(t)$ that fits this data.

b) Estimate the number of bank failures in 1990.

60. The number of prisoners has been growing approximately linearly from 1986 through 1996, as shown by the red line in the following figure. Let p be the number of prisoners in thousands, and t be the years since 1986.

a) Find a function that fits this data.

b) Using the function in part **a)**, estimate the number of prisoners in 1993.

c) Does the graph support your answer in part **b)**?

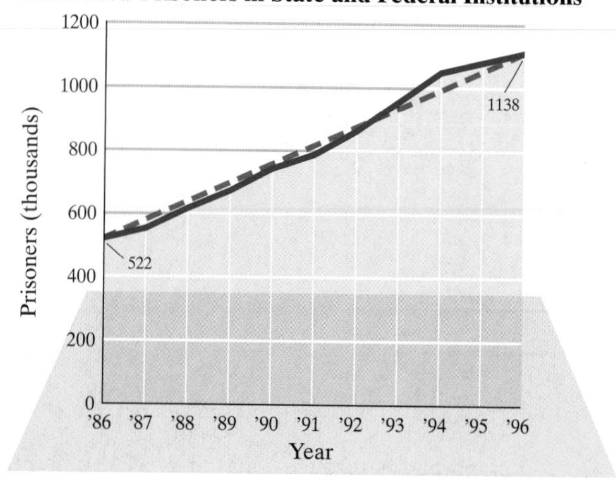

[3.6] *Given $f(x) = x^2 - 3x + 4$ and $g(x) = 2x - 5$, find the following.*

61. $(f + g)(x)$ **62.** $(f + g)(3)$ **63.** $(g - f)(x)$ **64.** $(g - f)(-1)$

65. $(f \cdot g)(-1)$ **66.** $(f \cdot g)(5)$ **67.** $(f/g)(1)$ **68.** $(f/g)(2)$

69. The following graph shows the number of deaths from Alzeimer's disease.

a) Estimate the number of male deaths from Alzheimer's disease in 1994.

b) Estimate the number of female deaths from Alzheimer's disease in 1994.

c) Estimate the total number of deaths from Alzheimer's disease in 1994.

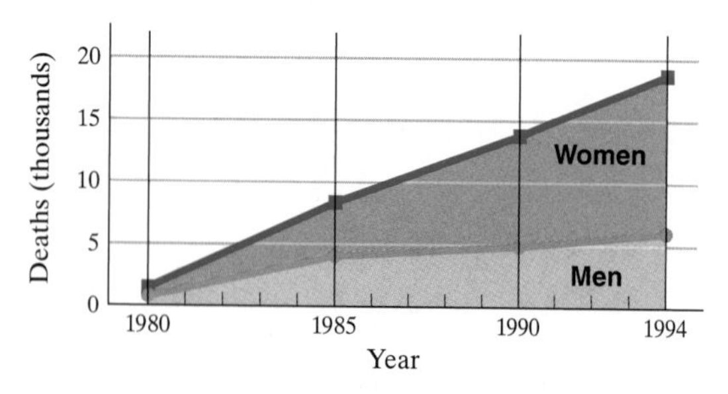

70. The following graph shows the number of doctors and lawyers in the United States from 1970 through 1995. Let d = number of doctors and l = number of lawyers. Estimate

a) $d(1995)$ **b)** $l(1995)$ **c)** $(d + l)(1995)$

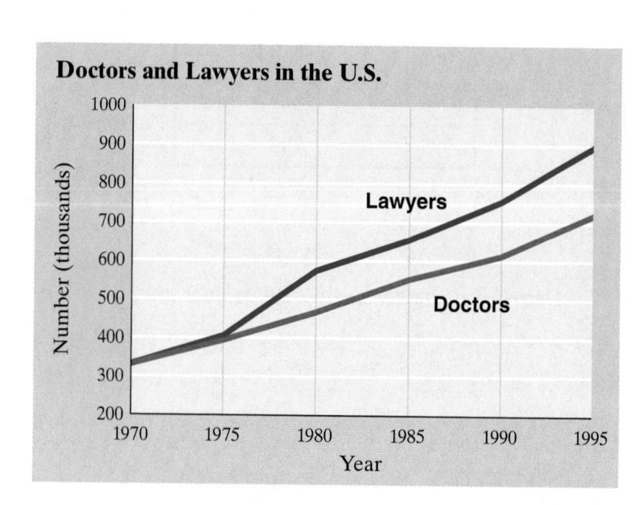

[3.7] *Graph each inequality.*

71. $y \geq -3$ **72.** $x < 4$ **73.** $y \leq 4x - 3$ **74.** $y < \frac{1}{3}x - 2$

Practice Test

1. Graph $y = 4x - 2$.
2. Graph $y = x^2$.
3. Graph $y = x^3 - 1$.
4. Graph $y = |x|$.
5. Define function.
6. Is the following set of ordered pairs a function? Explain your answer.

$$\{(3, 1), (-2, 6), (4, 6), (5, 2), (6, 3)\}$$

Determine whether the following relations are functions. Give the domain and range of the relation or function.

7.

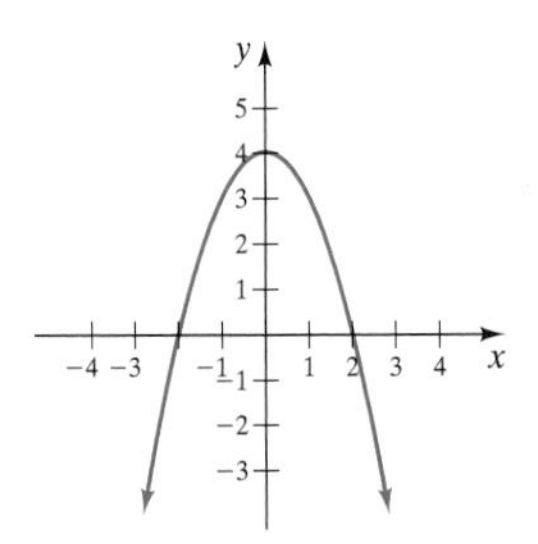

8.

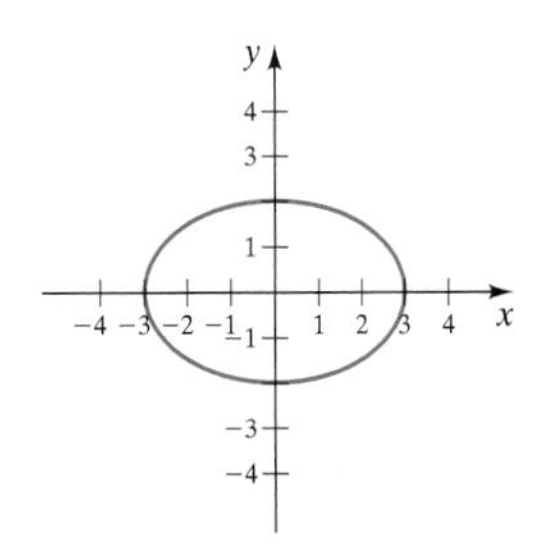

9. If $f(x) = 3x^2 - 6x + 2$, find $f(-2)$.

In Exercises 10 and 11, graph using x- and y-intercepts.

10. $100y + 200x = 400$
11. $\frac{1}{2}y = -\frac{1}{3}x + 4$
12. Graph $f(x) = -3$.
13. Graph $x = 4$.
14. The yearly profit, p, for Zico Publishing Company on the sales of a particular book can be estimated by the function $p(x) = 10.2x - 50{,}000$, where x is the number of books produced and sold.
    a) Draw a graph of profit versus books sold for up to 30,000 books.
    b) Use function $p(x)$ to estimate the number of books that must be sold for the company to break even.
    c) Use function $p(x)$ to estimate the number of books that the company must sell to make a \$100,000 profit.
15. Determine the slope and y-intercept of the graph of the equation $4x - 3y = 9$.
16. Determine the slope of the line through the points $(-6, 2)$ and $(4, -1)$.
17. Determine the function represented by the red line on the graph to the right that can be used to estimate the projected U.S. population, p, from 2000 through 2050. Let 2000 be the reference year so that 2000 is represented by $t = 0$.
18. Determine whether the graphs of the two equations are parallel, perpendicular, or neither. Explain your answer.

$$3x - 6y = -4$$
$$-6x - 3y = 10$$

19. Find the equation of the line through $(3, -4)$ that is perpendicular to the graph of $3x - 2y = 6$.
20. The number of deaths from AIDS grew approximately linearly from 1985 through 1994. In 1985 there were about 6851 deaths from AIDS; while in 1994 there were 47,761 deaths. Fortunately, since 1994 the number of AIDS deaths has been dropping, with a dramatic drop between 1995 and 1996.
    a) Let n be the number of AIDS deaths and t be the years since 1985. Determine a function that fits the data from 1985 through 1994.
    b) Use the function in part **a)** to estimate the number of AIDS deaths in 1990.

U.S. Population Projections 2000–2050

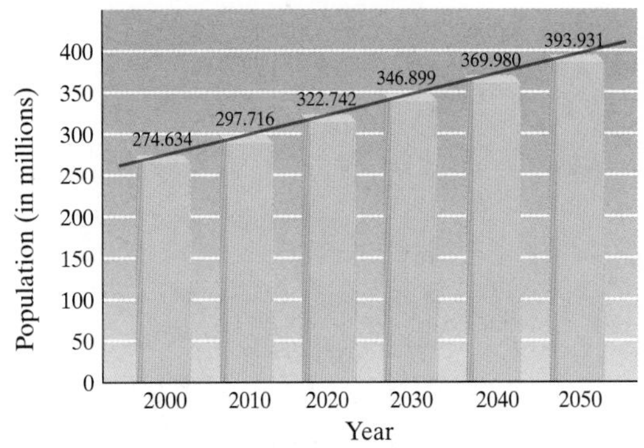

Source: U.S. Bureau of the Census

If $f(x) = 2x^2 - x$ *and* $g(x) = x - 5$, *find*

21. $(f + g)(3)$

22. $(f/g)(-1)$

23. $f(a)$

24. The following graph shows paper usage in 1995 and projected paper usage from 1995 through 2015.

a) Estimate the total number of tons of paper to be used in 2010.

b) Estimate the number of tons of paper to be used by businesses in 2010.

c) Estimate the number of tons of paper to be used for reference, print media, and household use in 2010.

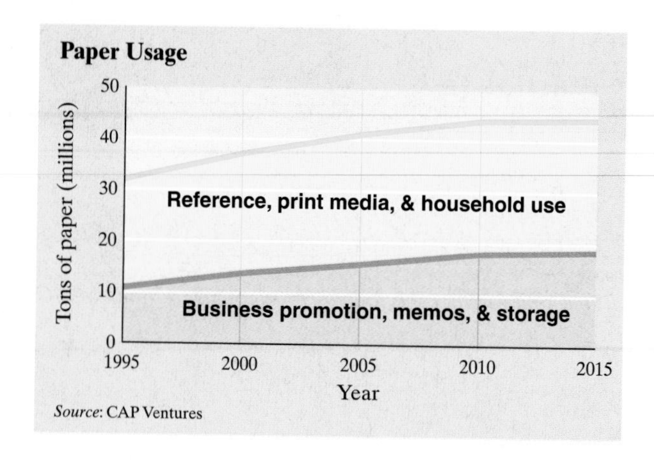

25. Graph $y < 3x - 2$.

Cumulative Review Test

1. For $A = \{2, 4, 6, 8\}$ and $B = \{1, 2, 3, 4, 5, 6\}$, find **a)** $A \cap B$ and **b)** $A \cup B$.

2. Consider the set $\{-6, -4, \frac{1}{3}, 0, \sqrt{3}, 4.67, \frac{37}{2}, -\sqrt{2}\}$. List the elements of the set that are

a) natural numbers

b) real numbers

3. Evaluate $2 - \{3[6 - 4(6^2 \div 4)]\}$.

Simplify.

4. $\left(\frac{4x^2}{y^{-3}}\right)^2$

5. $\left(\frac{2x^4y^{-2}}{4xy^3}\right)^3$

6. The total number of passengers traveling between U.S. cities in 1995 was 1.23×10^9 passengers. The following graph shows a breakdown of how passengers were transported.

a) How many passengers traveled by air?

b) How many traveled by bus?

c) How many more passengers traveled by air than by bus?

Passengers in 1995

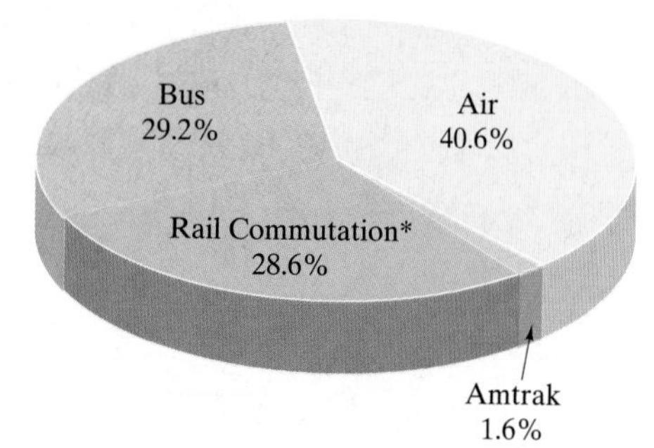

*Travel by rail on a regular basis at reduced rates.

In Exercises 7 and 8, solve the following equations.

7. $4(x - 3) - 2 = 4[x - (-3 + x)]$

8. $\frac{4}{5} - \frac{x}{3} = 10$

9. Simplify $5x - \{4 - [2(x - 4)] - 5\}$.

10. Solve $A = \frac{1}{2}h(b_1 + b_2)$ for b_1.

11. How many liters of 10% salt water solution must be mixed with 8 liters of 6% solution to get a 9% solution?

12. Solve the inequality $3(x - 4) < 6(2x + 3)$.

13. Solve the inequality $-4 < 3x - 7 < 8$.

14. Find the solution set of $|2x - 3| > 4$.

15. Find the solution set of $|2x - 4| = \left|\frac{1}{2}x - 2\right|$.

16. Graph $2x + 4y = 10$.

17. Determine whether the graphs of the two given equations are parallel, perpendicular, or neither.

$$2x - 5y = 6$$
$$5x - 2y = 9$$

18. a) Determine whether the following graph is a function. **b)** Find the domain and range of the graph.

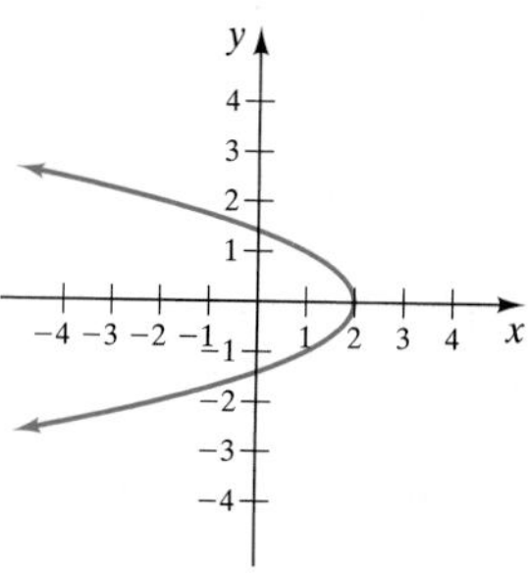

If $f(x) = x^2 + 3x - 2$ *and* $g(x) = 4x - 6$, *find*

19. $(f + g)(x)$

20. $(f \cdot g)(4)$

SYSTEMS OF EQUATIONS AND INEQUALITIES

Use the Angel Web site at www.prenhall.com/angel to be linked to an internet resource that will help you further explore the following application.

Business owners strive to run their companies at full capacity. Mathematics can be used to find the most effective way to distribute resources. On page 247, we solve a system of equations to find how resources at a boat manufacturing plant should be allocated so it operates at full capacity. The field of mathematics which analyzes business operations is called operations research.

Preview and Perspective

In this chapter we discuss solving systems of linear equations by the following methods: by graphing, by substitution, by the addition method, by using matrices, and by using determinants and Cramer's rule. Do not be surprised if you cover only a few techniques for solving systems. Often, time does not permit covering all.

We also discuss solving systems of equations containing three equations in three variables by a variety of techniques. There are many real-life applications of systems of equations, as illustrated in Section 4.3.

In Section 4.6 we solve systems of *linear inequalities* graphically. We use the techniques that were presented in Section 3.7 to graph inequalities.

Businesses often need to consider the relationships that exist between variables. For example, a business considers items such as cost of materials, cost of labor, cost of transportation, sale price of the item manufactured, and a host of other items. Businesses relate these variables to each other as systems of equations. These systems are often solved by computer. However, individuals must first construct the system of equations to be solved, as we will learn to do in this chapter.

4.1 SOLVING SYSTEMS OF LINEAR EQUATIONS IN TWO VARIABLES

1. Solve systems of linear equations graphically.
2. Solve systems of linear equations by substitution.
3. Solve systems of linear equations using the addition method.

It is often necessary to find a common solution to two or more linear equations. We refer to these equations as a **system of linear equations**. For example,

$$\left.\begin{aligned}(1)\ y &= x + 5\\ (2)\ y &= 2x + 4\end{aligned}\right\}\quad \text{System of linear equations}$$

A **solution to a system of equations** is an ordered pair or pairs that satisfy all equations in the system. The only solution to the system above is (1, 6).

Check in Equation (1)	Check in Equation (2)
(1, 6)	(1, 6)
$y = x + 5$	$y = 2x + 4$
$6 \stackrel{?}{=} 1 + 5$	$6 \stackrel{?}{=} 2(1) + 4$
$6 = 6$ *True*	$6 = 6$ *True*

The ordered pair (1, 6) satisfies *both* equations and is the solution to the system of equations.

A system of equations may consist of more than two equations. If a system consists of three equations in three variables, such as x, y, and z, the solution will be an **ordered triple** of the form (x, y, z). If the ordered triple (x, y, z) is a solution to the system, it must satisfy all three equations in the system. Systems with three variables are discussed in Section 4.2. Systems of equations may have more than three variables, but we will not discuss them in this book.

1) Solve Systems of Linear Equations Graphically

To solve a system of linear equations in two variables graphically, graph both equations in the system on the same axes. The solution to the system will be the ordered pair (or pairs) common to both lines, or the point of intersection of both lines in the system.

When two lines are graphed, three situations are possible, as illustrated in Figure 4.1. In Figure 4.1a, lines 1 and 2 intersect at exactly one point. This system of equations has *exactly one solution*. This is an example of a *consistent* system of equations. A **consistent system of equations** is a system of equations that has a solution.

Lines 1 and 2 of Figure 4.1b are different but parallel lines. The lines do not intersect, and this system of equations has *no solution*. This is an example of an *inconsistent* system of equations. An **inconsistent system of equations** is a system of equations that has no solution.

In Figure 4.1c, lines 1 and 2 are actually the same line. In this case, every point on the line satisfies both equations and is a solution to the system of equations. This system has *an infinite number of solutions*. This is an example of a *dependent* system of equations. In a dependent system of linear equations both equations represent the same line. A **dependent system of equations** is a system of equations that has an infinite number of solutions. *Note that a dependent system is also a consistent system since it has solutions.*

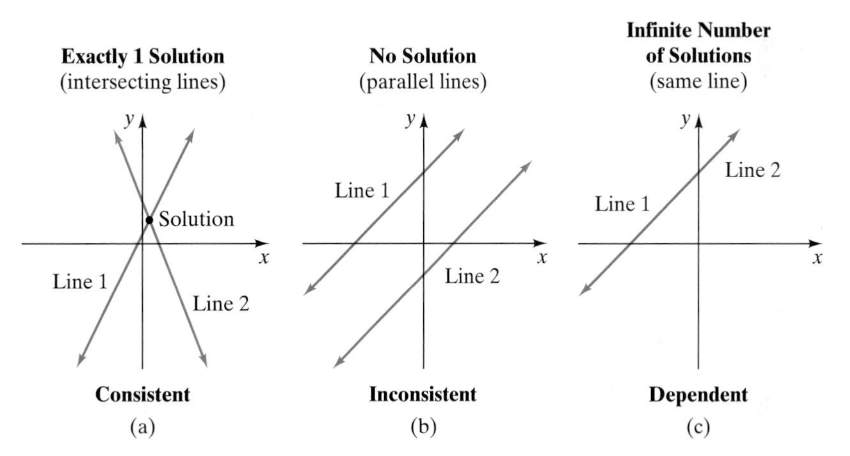

FIGURE 4.1

We can determine if a system of linear equations is consistent, inconsistent, or dependent by writing each equation in slope-intercept form and comparing the slopes and y-intercepts. If the slopes of the lines are different (Fig. 4.1a), the system is consistent. If the slopes are the same but the y-intercepts different (Fig. 4.1b), the system is inconsistent, and if both the slopes and the y-intercepts are the same (Fig. 4.1c), the system is dependent.

EXAMPLE 1 Without graphing the equations, determine whether the following system of equations is consistent, inconsistent, or dependent.

$$2x + y = 3$$
$$4x + 2y = 12$$

Solution Write each equation in slope–intercept form.

$$2x + y = 3 \qquad\qquad 4x + 2y = 12$$
$$y = -2x + 3 \qquad\qquad 2y = -4x + 12$$
$$y = -2x + 6$$

Since both equations have the same slope, -2, and different y-intercepts, the lines are parallel lines. Therefore, the system is inconsistent and has no solution.

NOW TRY EXERCISE 17

EXAMPLE 2 Solve the following system of equations graphically.

$$y = x + 2$$
$$y = -x + 4$$

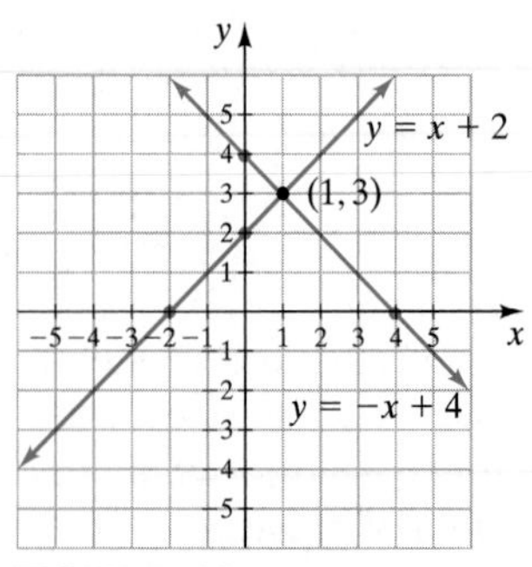

FIGURE 4.2

NOW TRY EXERCISE 25

Solution Graph both equations on the same axes (Fig. 4.2).

The solution is the point of intersection of the two lines, (1, 3).

The system of equations in Example 2 could be represented in function notation as

$$f(x) = x + 2$$
$$g(x) = -x + 4$$

Using Your Graphing Calculator

In the Using Your Graphing Calculator box on page 175, Section 3.3, we discussed using a graphing calculator to find the intersection of two graphs. Let's use the information provided in that Graphing Calculator Corner to solve a system of equations.

EXAMPLE Use your graphing calculator to solve the system of equations. Round the solution to the nearest hundredth.

$$-2.6x - 5.2y = -15.3$$
$$-8.6x + 3.7y = -12.5$$

Solution First solve each equation for y.

$$\begin{aligned} -2.6x - 5.2y &= -15.3 \\ -2.6x &= 5.2y - 15.3 \\ -2.6x + 15.3 &= 5.2y \\ \frac{-2.6x + 15.3}{5.2} &= y \end{aligned}$$

$$\begin{aligned} -8.6x + 3.7y &= -12.5 \\ 3.7y &= 8.6x - 12.5 \\ y &= \frac{8.6x - 12.5}{3.7} \end{aligned}$$

Now let $y_1 = \dfrac{-2.6x + 15.3}{5.2}$ and $y_2 = \dfrac{8.6x - 12.5}{3.7}$. The graphs y_1 and y_2 are illustrated in Figure 4.3.

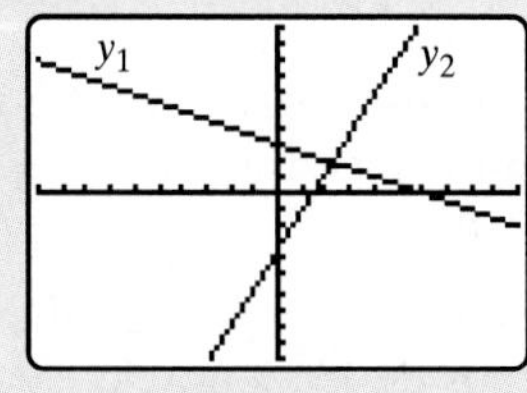

FIGURE 4.3

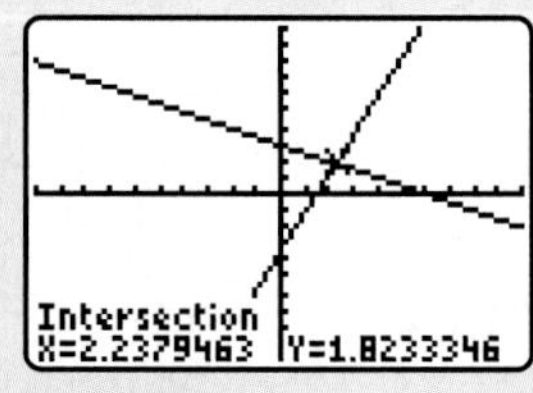

FIGURE 4.4

Figure 4.4 shows that the intersection of the two graphs occurs at (2.24, 1.82), rounded to the nearest hundredth.

Use your graphing calculator to determine the solution to each system. Round your answers to the nearest hundredth.

1. $2x + 3y = 8$
$-3x + 4y = -5$

2. $5x - 6y = 9$
$-3x + 5y = 8$

3. $3.4x - 5.6y = 10.2$
$5.8x + 1.4y = -33.6$

4. $-2.3x + 7.9y = 88.3$
$-5.3x - 2.7y = -16.5$

2) Solve Systems of Linear Equations by Substitution

Often, an exact solution to a system of equations may be difficult to find on a graph. A graphing calculator may not give an exact answer either. When an exact answer is necessary, the system should be solved algebraically, either by substitution or by addition (elimination) of equations. We discuss **substitution** first.

To Solve a Linear System of Equations by Substitution

1. Solve for a variable in either equation. (If possible, solve for a variable with a numerical coefficient of 1 to avoid working with fractions.)
2. Substitute the expression found for the variable in step 1 into the other equation. This will result in an equation containing only one variable.
3. Solve the equation obtained in step 2 to find the value of this variable.
4. Substitute the value found in step 3 into the equation from step 1. Solve the equation to find the remaining variable.
5. Check your solution in *all* equations in the system.

EXAMPLE 3 Solve the following system of equations by substitution.

$$y = 2x + 5$$
$$y = -4x + 2$$

Solution Since both equations are already solved for y, we can substitute $2x + 5$ for y in the second equation and then solve for the remaining variable, x.

$$2x + 5 = -4x + 2$$
$$6x + 5 = 2$$
$$6x = -3$$
$$x = -\frac{1}{2}$$

Now find y by substituting $-\frac{1}{2}$ for x in either of the original equations. We will use the first equation.

$$y = 2x + 5$$
$$= 2\left(-\frac{1}{2}\right) + 5 \qquad \text{Substitute } -\tfrac{1}{2} \text{ for } x.$$
$$= -1 + 5 = 4$$

A check will show that the solution is $\left(-\frac{1}{2}, 4\right)$.

EXAMPLE 4 Solve the following system of equations by substitution.

$$2x + y = 11$$
$$x + 3y = 18$$

Solution Begin by solving for one of the variables in either of the equations. You may solve for either of the variables; however, if you solve for a variable with a numerical coefficient of 1, you may avoid working with fractions. In this system the y-term in $2x + y = 11$ and the x-term in $x + 3y = 18$ both have numerical coefficient 1.

Let's solve for y in $2x + y = 11$.

$$2x + y = 11$$
$$y = -2x + 11$$

Next, substitute $-2x + 11$ for y in the *other equation*, $x + 3y = 18$, and solve for the remaining variable, x.

$$x + 3y = 18$$
$$x + 3(-2x + 11) = 18 \quad \text{Substitute } -2x + 11 \text{ for } y.$$
$$x - 6x + 33 = 18$$
$$-5x + 33 = 18$$
$$-5x = -15$$
$$x = 3$$

Finally, substitute $x = 3$ in the equation $y = -2x + 11$ and solve for y.

$$y = -2x + 11$$
$$y = -2(3) + 11 = 5$$

The solution is the ordered pair (3, 5). Check this solution.

If, when solving a system of equations by either substitution or the addition method, you arrive at an equation that is false, such as $5 = 6$ or $0 = 3$, the system is inconsistent and has no solution. If you obtain an equation that is always true, such as $6 = 6$ or $0 = 0$, the system is dependent and has an infinite number of solutions.

NOW TRY EXERCISE 41

HELPFUL HINT

Students sometimes successfully solve for one of the variables and forget to solve for the other. Remember that a solution must contain a numerical value for each variable in the system.

3 Solve Systems of Linear Equations Using the Addition Method

A third and often the easiest method of solving a system of equations is the **addition** (or elimination) **method**. The object of this process is to obtain two equations whose sum will be an equation containing only one variable. Keep in mind that your immediate goal is to obtain one equation containing only one unknown.

EXAMPLE 5 Solve the following system of equations using the addition method.

$$x + y = 6$$
$$2x - y = 3$$

Solution Note that one equation contains $+y$ and the other contains $-y$. By adding the equations, we can eliminate the variable y and obtain one equation containing only one unknown, x.

$$\begin{aligned} x + y &= 6 \\ 2x - y &= 3 \\ \hline 3x \quad &= 9 \end{aligned}$$

Now solve for the remaining variable, x.

$$\frac{3x}{3} = \frac{9}{3}$$

$$x = 3$$

Finally, solve for y by substituting 3 for x in either of the original equations.

$$x + y = 6$$

$$3 + y = 6 \qquad \text{Substitute 3 for } x.$$

$$y = 3$$

A check will show that the solution is (3, 3).

To Solve a Linear System of Equations Using the Addition (or Elimination) Method

1. If necessary, rewrite each equation in standard form, that is, with the terms containing variables on the left side of the equal sign and the constant on the right side of the equal sign.
2. If necessary, multiply one or both equations by a constant(s) so that when the equations are added, the sum will contain only one variable.
3. Add the respective sides of the equations. This will result in a single equation containing only one variable.
4. Solve for the variable in the equation obtained in step 3.
5. Substitute the value found in step 4 in either of the original equations. Solve that equation to find the value of the remaining variable.
6. Check your solution in all equations in the system.

In step 2 of the procedure, we indicate that it may be necessary to multiply both sides of an equation by a constant. To help avoid confusion, we will number our equations using parentheses, such as (*eq.* 1) or (*eq.* 2).

In Example 6 we will solve the same system we solved in Example 4, but this time we will use the addition method.

EXAMPLE 6 Solve the following system of equations using the addition method.

$$2x + y = 11 \qquad (eq.\ 1)$$

$$x + 3y = 18 \qquad (eq.\ 2)$$

Solution The object of the addition process is to obtain two equations whose sum will be an equation containing only one variable. To eliminate the variable x, we multiply (*eq.* 2) by -2 and then add the two equations.

$$2x + y = 11 \qquad (eq.\ 1)$$

$$-2x - 6y = -36 \qquad (eq.\ 2) \qquad \text{Multiplied by } -2$$

Now add.

$$\begin{aligned} 2x + y &= 11 \\ -2x - 6y &= -36 \\ \hline -5y &= -25 \\ y &= 5 \end{aligned}$$

Now solve for x by substituting 5 for y in either of the original equations.

$$\begin{aligned} 2x + y &= 11 \\ 2x + 5 &= 11 \qquad \text{Substitute 5 for } y \\ 2x &= 6 \\ x &= 3 \end{aligned}$$

The solution is (3, 5). Note that we could have first eliminated the variable y by multiplying (*eq.* 1) by −3 and then adding.

Sometimes both equations must be multiplied by different numbers in order for one of the variables to be eliminated. This procedure is illustrated in Example 7.

EXAMPLE 7 Solve the following system of equations using the addition method.

$$\begin{aligned} 4x + 3y &= 7 \qquad (eq.\ 1) \\ 3x - 7y &= -3 \qquad (eq.\ 2) \end{aligned}$$

Solution We can eliminate the variable x by multiplying (*eq.* 1) by −3 and (*eq.* 2) by 4.

$$\begin{aligned} -12x - 9y &= -21 \qquad (eq.\ 1) \text{ Multiplied by } -3 \\ 12x - 28y &= -12 \qquad (eq.\ 2) \text{ Multiplied by } 4 \\ \hline -37y &= -33 \qquad \text{Sum of equations} \\ y &= \frac{33}{37} \end{aligned}$$

We can now find x by substituting $\frac{33}{37}$ for y in one of the original equations and solving for x. If you try this, you will see that, although it can be done, it gets messy. An easier method to solve for x is to go back to the original equations and eliminate the variable y.

$$\begin{aligned} 28x + 21y &= 49 \qquad (eq.\ 1) \text{ Multiplied by } 7 \\ 9x - 21y &= -9 \qquad (eq.\ 2) \text{ Multiplied by } 3 \\ \hline 37x \qquad &= 40 \qquad \text{Sum of equations} \\ x &= \frac{40}{37} \end{aligned}$$

The solution is $\left(\frac{40}{37}, \frac{33}{37}\right)$.

NOW TRY EXERCISE 67

In Example 7, the same solution could be obtained by multiplying (*eq.* 1) by 3 and (*eq.* 2) by −4 and then adding. Try it now and see.

EXAMPLE 8 Solve the following system of equations using the addition method.

$$\begin{aligned} 0.2x + 0.1y &= 1.1 \qquad (eq.\ 1) \\ \frac{x}{18} + \frac{y}{6} &= 1 \qquad (eq.\ 2) \end{aligned}$$

Solution When a system of equations contains fractions or decimal numbers, it is generally best to *clear*, or remove, the fractions or decimals. In (*eq.* 1), if we multiply both sides of the equation by 10 we obtain

$$10(0.2x) + 10(0.1y) = 10(1.1)$$
$$2x + y = 11 \qquad (eq. 3)$$

In (*eq*. 2), if we multiply both sides of the equation by the least common denominator, 18, we obtain

$$18\left(\frac{x}{18}\right) + 18\left(\frac{y}{6}\right) = 18(1)$$
$$x + 3y = 18 \qquad (eq. 4)$$

The system of equations is now simplified to

$$2x + y = 11 \qquad (eq. 3)$$
$$x + 3y = 18 \qquad (eq. 4)$$

This is the same system of equations we solved in Example 6. Thus the solution to this system is (3, 5), the same as obtained in Example 6.

EXAMPLE 9 Solve the following system of equations using the addition method.

$$2x + y = 3 \qquad (eq. 1)$$
$$4x + 2y = 12 \qquad (eq. 2)$$

Solution

$$-4x - 2y = -6 \qquad (eq. 1) \text{ Multiplied by } -2$$
$$4x + 2y = 12 \qquad (eq. 2)$$
$$0 = 6 \qquad \text{False}$$

NOW TRY EXERCISE 57

Since 0 = 6 is a false statement, this system has no solution. The system is inconsistent and the graphs of these equations are parallel lines.

EXAMPLE 10 Solve the following system of equations using the addition method.

$$x - \frac{1}{2}y = 2$$
$$y = 2x - 4$$

Solution First, align the x- and y-terms on the left side of the equation.

$$x - \frac{1}{2}y = 2 \qquad (eq. 1)$$
$$2x - y = 4 \qquad (eq. 2)$$

Now proceed as in previous examples.

$$-2x + y = -4 \qquad (eq. 1) \text{ Multiplied by } -2$$
$$2x - y = 4 \qquad (eq. 2)$$
$$0 = 0 \qquad \text{True}$$

Since 0 = 0 is a true statement, the system is dependent and has an infinite number of solutions. Both equations represent the same line. Notice that if you multiply both sides of (*eq*. 1) in the solution by 2 you obtain (*eq*. 2) in the solution.

We have illustrated three methods that can be used to solve a system of linear equations: graphing, substitution, and the addition method. When you are given a system of equations, which method should you use to solve the system? When you need an exact solution, graphing may not be the best to use. Of the two algebraic methods, the addition method may be the easiest to use if there are no numerical coefficients of 1 in the system. If one or more of the variables has a coefficient of 1, you may wish to use either method. We will present a fourth method, using matrices, in Section 4.4 and a fifth method, using determinants, in Section 4.5.

Exercise Set 4.1

Concept/Writing Exercises

1. What is a solution to a system of linear equations?
2. What is the solution to a system of linear equations in three variables called?
3. What is a consistent system of equations?
4. What is a dependent system of equations?
5. What is an inconsistent system of equations?
6. Explain how to determine the solution of a linear system graphically.
7. Explain how you can determine, without graphing or solving, whether a system of two linear equations is consistent, inconsistent or dependent.
8. When solving a system of linear equations using the addition (or elimination) method, what is the object of the process?
9. When solving a linear system by addition, how can you tell if the system is inconsistent?
10. When solving a linear system by addition, how can you tell if the system is dependent?

Practice the Skills

Determine which, if any, of the given ordered pairs or ordered triples satisfy the system of linear equations.

11. $y = 2x + 4$
$y = 2x - 1$
a) $(0, 4)$ **b)** $(3, 10)$

12. $2x - 3y = 6$
$y = \frac{2}{3}x - 2$
a) $(3, 0)$ **b)** $(3, -2)$

13. $0.5s = -0.5r + 2$
$2s = -2r + 8$
a) $(2, 5)$ **b)** $(1, 3)$

14. $3m - 4n = 8$
$2n = \frac{3}{2}m - 4$
a) $\left(-\frac{1}{3}, -\frac{9}{4}\right)$ **b)** $(0, -2)$

15. $x + 2y - z = -5$
$2x - y + 2z = 8$
$3x + 3y + 4z = 5$
a) $(1, 3, -2)$ **b)** $(1, -2, 2)$

16. $4x + y - 3z = 1$
$2x - 2y + 6z = 11$
$-6x + 3y + 12z = -4$
a) $(2, -1, -2)$ **b)** $\left(\frac{1}{2}, 2, 1\right)$

Write each equation in slope–intercept form. Without graphing the equations, state whether the system of equations is consistent, inconsistent, or dependent. Also indicate whether the system has exactly one solution, no solution, or an infinite number of solutions.

17. $3y = -x + 6$
$x - 2y = 1$

18. $3y = 2x + 3$
$y = \frac{2}{3}x - 2$

19. $y = \frac{1}{3}x + 4$
$3y = x + 12$

20. $2x - 3y = 4$
$3x - 2y = -2$

21. $3x - 3y = 9$
$2x - 2y = -4$

22. $2x = 3y + 4$
$6x - 9y = 12$

23. $y = \frac{3}{2}x + \frac{1}{2}$
$3x - 2y = -\frac{1}{2}$

24. $x - y = 3$
$\frac{1}{2}x - 2y = -6$

Determine the solution to each system of equations graphically. If the system is inconsistent or dependent, so state.

25. $y = x + 5$
$y = -x + 3$

26. $y = 2x + 4$
$y = -3x - 6$

27. $y = 4x - 1$
$2y = 8x + 6$

28. $y = -2x - 1$
$x + 2y = 4$

29. $2x + 3y = 6$
$4x = -6y + 12$

30. $x + y = 1$
$3x - y = -5$

31. $x + 3y = 4$
$x = 1$

32. $2x - 5y = 10$
$y = \frac{2}{5}x - 2$

33. $y = -5x + 5$
$y = 2x - 2$

34. $4x - y = 9$
$x - 3y = 16$

35. $2x - y = -4$
$2y = 4x - 6$

36. $y = -\frac{1}{3}x - 1$
$3y = 4x - 18$

Find the solution to each system of equations by substitution.

37. $x + 2y = 5$
$x = 2y + 1$

38. $y = x + 2$
$2y = -x - 2$

39. $x + y = 10$
$x = y$

40. $2x + y = 4$
$2y = 6 - 4x$

41. $2r + s = 4$
$2r + s + 6 = 0$

42. $y = 2x + 4$
$y = -\frac{3}{4}$

43. $x = \frac{1}{2}$
$x + \frac{1}{3}y + 6 = 0$

44. $y = \frac{1}{3}x - 2$
$x - 3y = 6$

45. $a - \frac{1}{2}b = 2$
$b = 2a - 4$

46. $y = 2x - 13$
$-4x - 7 = 9y$

47. $x = y + 4$
$3x + 7y = -18$

48. $5x - 2y = -7$
$5 = y - 3x$

49. $5x - 4y = -7$
$x - \frac{3}{5}y = -2$

50. $m + 2n = 4$
$m + \frac{1}{2}n = 4$

51. $\frac{1}{2}x - \frac{1}{3}y = 2$
$\frac{1}{4}x + \frac{2}{3}y = 6$

52. $\frac{1}{2}x + \frac{1}{3}y = 13$
$\frac{1}{5}x + \frac{1}{8}y = 5$

Solve each system of equations using the addition method.

53. $x + y = 0$
$x - y = 4$

54. $x - y = 12$
$x + y = 2$

55. $3x + 2y = 15$
$x - 2y = -7$

56. $3x + 3y = 18$
$x + y = 4$

57. $3p + q = 6$
$-6p - 2q = 10$

58. $2r + s = 14$
$-3r + s = -2$

59. $2x + y = 6$
$3x - 2y = 16$

60. $4x - 3y = 8$
$-2x + 5y = 14$

61. $2a - 5b = 13$
$5a + 3b = 17$

62. $4x = 2y + 6$
$y = 2x - 3$

63. $3y = 2x + 4$
$3y = 2x + 4$

64. $5x + 4y = 10$
$-3x - 5y = 7$

65. $2x - y = 8$
$3x + y = 6$

66. $3x + 4y = 2$
$2x = -5y - 1$

67. $3x - 4y = 5$
$2x = 5y - 3$

68. $4x + 5y = 3$
$2x - 3y = 4$

69. $0.2x + 0.5y = 1.6$
$-0.3x + 0.4y = -0.1$

70. $0.15x - 0.40y = 0.65$
$0.60x + 0.25y = -1.1$

71. $2.1m - 0.6n = 8.4$
$-1.5m - 0.3n = -6.0$

72. $-0.25x + 0.10y = 1.05$
$-0.40x - 0.625y = -0.675$

73. $\frac{1}{2}x - \frac{1}{3}y = 1$
$\frac{1}{4}x - \frac{1}{9}y = \frac{2}{3}$

74. $\frac{1}{3}x = 4 - \frac{1}{4}y$
$3x = 4y$

75. $\frac{1}{5}x + \frac{1}{5}y = 4$
$\frac{2}{3}x - y = \frac{8}{3}$

76. $\frac{2}{3}x - 4 = \frac{1}{2}y$
$x - 3y = \frac{1}{3}$

Problem Solving

77. **a)** Write a system of equations that would be most easily solved by substitution.
b) Explain why substitution would be the easiest method to use.
c) Solve the system by substitution.

78. **a)** Write a system of equations that would be most easily solved using the addition method.
b) Explain why the addition method would be the easiest method to use.
c) Solve the system using the addition method.

79. As the following graph shows, from 1995 through 1997 Netscape's Web browser has been losing market shares to Microsoft Explorer.

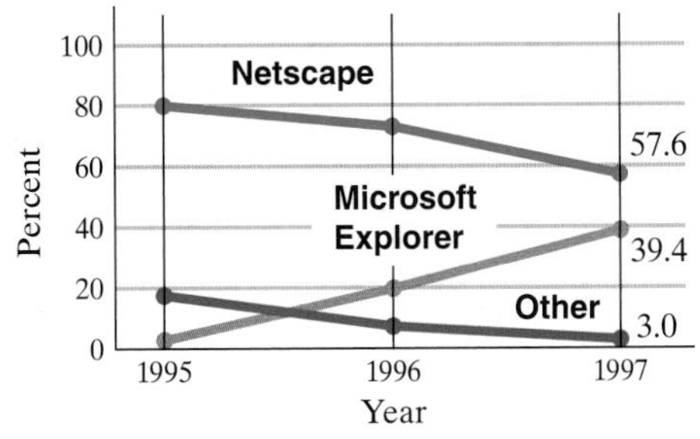

Source: Bloomberg, Dataquest

The percent market share for Netscape can be approximated by the function $N(t) = -11.2t + 80$, where t is years since 1995. The percent market share for Microsoft Explorer can be approximated by the function $M(t) = 18.7t + 2$, where t is years since 1995. If the trend continues, determine the year that the market share of Microsoft Explorer will equal the market share of Netscape.

80. The U.S. market shares of General Motors (GM), Ford, and Chrysler from 1992 through 1997 are shown below.

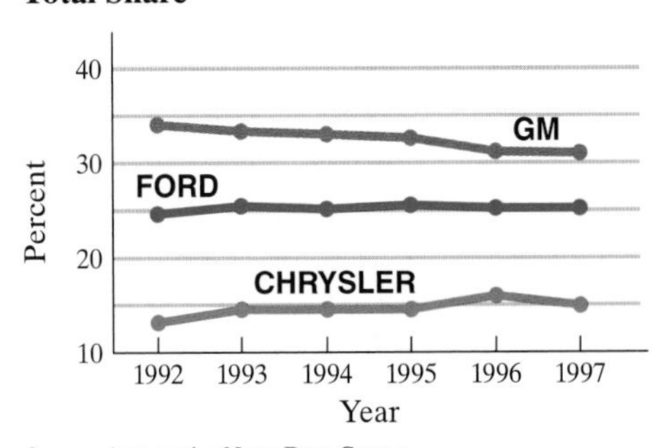

Source: Automotive News Data Center

The market share for GM can be approximated by the function $G(t) = -0.5t + 34$ and the market share for Ford can be approximated by $F(t) = 25.2$, where t is the number of years since 1992. Assuming the present trend continues, determine when the market share for GM and Ford will be equal.

81. Explain how you can tell by observation that the following system is inconsistent.

$$3x - 2y = 6$$
$$-6x + 4y = -8$$

82. Explain how you can tell by observation that the following system is dependent.

$$5x - 3y = 4$$
$$-10x + 6y = -8$$

83. The solutions of a system of linear equations are $(2, 4)$ and $(-2, 6)$.

a) How many other solutions does the system have? Explain.

b) Find the slope of the line containing $(2, 4)$ and $(-2, 6)$. Determine the equation of the line containing these points. Then determine the y-intercept.

c) Is this graph a function? Explain.

84. The solutions of a system of linear equations are $(3, 4)$ and $(3, -2)$.

a) How many other solutions does the system have? Explain.

b) Find the slope of the line containing $(3, 4)$ and $(3, -2)$. Determine the equation of the line containing these points.

c) Is this graph a function? Explain.

85. Construct a system of equations that is dependent. Explain how you created your system.

86. Construct a system of equations that is inconsistent. Explain how you created your system.

In Exercises 87 and 88, **a)** *create a system of linear equations that has the solution indicated and* **b)** *explain how you determined your solution.*

87. $(2, 5)$

88. $(-3, 4)$

89. The solution to the following system of equations is $(2, -3)$. Find A and B.

$$Ax + 4y = -8$$
$$3x - By = 21$$

90. The solution to the following system of equations is $(-5, 3)$ Find A and B.

$$3x + Ay = -3$$
$$Bx - 2y = -16$$

91. If $(2, 6)$ and $(-1, -6)$ are two solutions of $f(x) = mx + b$, find m and b.

92. If $(3, -5)$ and $(-2, 10)$ are two solutions of $f(x) = mx + b$, find m and b.

93. Suppose you graph a system of two linear equations on your graphing calculator, but only one line shows in the window. What are two possible explanations for this?

94. Suppose you graph a system of linear equations on your graphing caluclator and get the following.

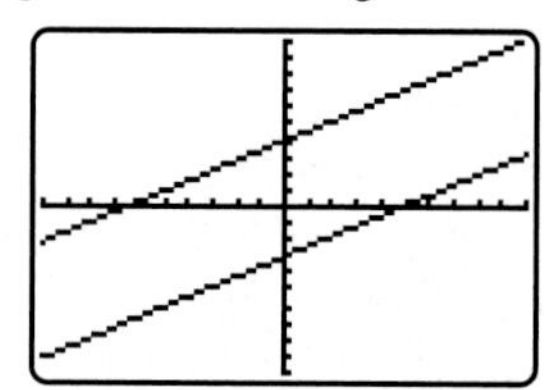

a) By observing the window, can you be sure that this system is inconsistent? Explain.

b) What can you do on your graphing calculator to determine whether the system is inconsistent?

Challenge Problems

Solve each system of equations

95. $\dfrac{x+2}{2} - \dfrac{y+4}{3} = 4$

$\dfrac{x+y}{2} = \dfrac{1}{2} + \dfrac{x-y}{3}$

96. $\dfrac{5x}{2} + 3y = \dfrac{9}{2} + y$

$\dfrac{1}{4}x - \dfrac{1}{2}y = 6x + 12$

Solve each system of equations. Hint: $\dfrac{3}{a} = 3 \cdot \dfrac{1}{a} = 3x$ *if* $x = \dfrac{1}{a}$.

97. $\dfrac{3}{a} + \dfrac{4}{b} = -1$

$\dfrac{1}{a} + \dfrac{6}{b} = 2$

98. $\dfrac{6}{x} + \dfrac{1}{y} = -1$

$\dfrac{3}{x} - \dfrac{2}{y} = -3$

By solving for x and y, determine the solution to each system of equations. In all equations, a ≠ 0 and b ≠ 0. The solution will contain either a, b, or both letters.

99. $4ax + 3y = 19$

$-ax + y = 4$

100. $ax = 2 - by$

$-ax + 2by - 1 = 0$

Group Activity

Discuss and answer Exercise 101 as a group.

101. The graph below appeared in both the *Journal of the American Medical Association* and *Scientific American*. The red line indicates the long-term trend of firearms deaths through 1992, and the purple line indicates the long-term trend in motor vehicle deaths through 1988. The black lines indicate the short-term trends in deaths from firearms and motor vehicles.

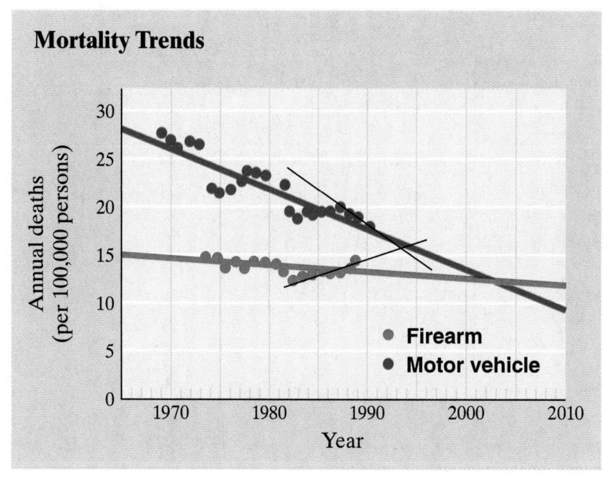

a) Discuss the long-term trend in motor vehicle deaths.

b) Discuss the long-term trend in firearms deaths.

c) Discuss the short-term trend in motor vehicle deaths compared with the long-term trend in motor vehicle deaths.

d) Discuss the short-term trend in firearms deaths compared with the long-term trend in firearms deaths.

e) Using the long-term trends, estimate when the number of deaths from firearms will equal the number of deaths from motor vehicles.

f) Repeat part **e)** using the short-term trends

g) Determine a function, $M(t)$, that can be used to estimate the number of deaths (long term) from motor vehicles from 1965 through 2010.

h) Determine a function, $F(t)$, that can be used to estimate the number of deaths (long term) from firearms from 1965 through 2010

i) Solve the system of equations formed in parts **g)** and **h)**. Does the solution agree with the solution in part **e)**? If not, explain why.

Cumulative Review Exercises

[1.2] **102.** Explain the difference between a rational number and an irrational number.

[1.2] **103. a)** Are all rational numbers real numbers?

b) Are all irrational numbers real numbers?

[2.2] **104.** Find all numbers such that $|x - 4| = |4 - x|$.

[2.2] **105.** Evaluate $A = p\left(1 + \frac{r}{n}\right)^t$, when $p = 500, r = 0.08, n = 2$, and $t = 1$.

[3.5] **106.** Is the relation below a function? Explain your answer. $\{(-3, 4), (7, 2), (-4, 5), (5, 0), (-3, 2)\}$

4.2 SOLVING SYSTEMS OF LINEAR EQUATIONS IN THREE VARIABLES

1) **Solve systems of linear equations in three variables.**
2) **Learn the geometric interpretation of a system of equations in three variables.**
3) **Recognize inconsistent and dependent systems.**

1) Solve Systems of Linear Equations in Three Variables

The equation $2x - 3y + 4z = 8$ is an example of a linear equation in three variables. The solution to a linear equation in three variables is an *ordered triple* of the form (x, y, z). One solution to the equation given is $(1, 2, 3)$. Check now to verify that $(1, 2, 3)$ is a solution to the equation.

To solve systems of linear equations with three variables, we can use either substitution or the addition method, both of which were discussed in Section 4.1.

EXAMPLE 1 Solve the following system of equations by substitution.

$$x = 4$$
$$2x + y = 20$$
$$-x + 4y + 2z = 24$$

Solution Since we know that $x = 4$, we substitute 4 for x in the equation $2x + y = 20$ and solve for y.

$$2x + y = 20$$
$$2(4) + y = 20$$
$$8 + y = 20$$
$$y = 12$$

Now we substitute $x = 4$ and $y = 12$ in the last equation and solve for z.

$$-x + 4y + 2z = 24$$
$$-(4) + 4(12) + 2z = 24$$
$$-4 + 48 + 2z = 24$$
$$44 + 2z = 24$$
$$2z = -20$$
$$z = -10$$

Check: $x = 4$, $y = 12$, $z = -10$. The solution must be checked in *all three* original equations.

$x = 4$	$2x + y = 20$	$-x + 4y + 2z = 24$
$4 = 4$ *True*	$2(4) + 12 \stackrel{?}{=} 20$	$-(4) + 4(12) + 2(-10) \stackrel{?}{=} 24$
	$20 = 20$ *True*	$24 = 24$ *True*

The solution is the ordered triple (4, 12, −10). Remember that the ordered triple lists the x-value first, the y-value second, and the z-value third.

Not every system of linear equations in three variables can be solved by substitution. When such a system cannot be solved using substitution, we can find the solution by the addition method, as illustrated in Example 2.

EXAMPLE 2 Solve the following system of equations using the addition method.

$$3x + 2y + z = 4 \quad (eq.\ 1)$$
$$2x - 3y + 2z = -7 \quad (eq.\ 2)$$
$$x + 4y - z = 10 \quad (eq.\ 3)$$

Solution To solve this system of equations, we must first obtain two equations containing the same two variables. We do so by selecting two equations and using the addition method to eliminate one of the variables. For example, by adding (*eq.* 1) and (*eq.* 3) the variable z will be eliminated. Then we use a different pair of equations [either (*eq.* 1) and (*eq.* 2) or (*eq.* 2) and (*eq.* 3)] and use the addition method to eliminate the *same* variable that was eliminated previously. If we multiply (*eq.* 1) by -2 and add it to (*eq.* 2), the variable z will again be eliminated. We will then have two equations containing only two unknowns. Let us begin by adding (*eq.* 1) and (*eq.* 3).

$$\begin{array}{rl} 3x + 2y + z = 4 & (eq.\ 1) \\ x + 4y - z = 10 & (eq.\ 3) \\ \hline 4x + 6y \quad = 14 & \text{Sum of equations, } (eq.\ 4) \end{array}$$

Now let's use a different pair of equations and again eliminate the variable z.

$$\begin{array}{rl} -6x - 4y - 2z = -8 & (eq.\ 1) \text{ Multiplied by } -2 \\ 2x - 3y + 2z = -7 & (eq.\ 2) \\ \hline -4x - 7y \quad = -15 & \text{Sum of equations, } (eq.\ 5) \end{array}$$

We now have a system consisting of two equations with two unknowns, (*eq.* 4) and (*eq.* 5). If we add these two equations, the variable x will be eliminated.

$$\begin{array}{rl} 4x + 6y = 14 & (eq.\ 4) \\ -4x - 7y = -15 & (eq.\ 5) \\ \hline -y = -1 & \text{Sum of equations} \\ y = 1 & \end{array}$$

Next we substitute $y = 1$ in either one of the two equations containing only two variables [(*eq.* 4) or (*eq.* 5)] and solve for x.

$$\begin{array}{rl} 4x + 6y = 14 & (eq.\ 4) \\ 4x + 6(1) = 14 & \text{Substitute 1 for } y \text{ in } (eq.\ 4) \\ 4x + 6 = 14 & \\ 4x = 8 & \\ x = 2 & \end{array}$$

Finally, we substitute $x = 2$ and $y = 1$ in any of the original equations and solve for z.

$$\begin{array}{rl} 3x + 2y + z = 4 & (eq.\ 1) \\ 3(2) + 2(1) + z = 4 & \text{Substitute 2 for } x \text{ and 1 for } y \text{ in } (eq.\ 1) \\ 6 + 2 + z = 4 & \\ 8 + z = 4 & \\ z = -4 & \end{array}$$

The solution is the ordered triple $(2, 1, -4)$. Check this solution in *all three* original equations.

In Example 2 we chose first to eliminate the variable z by using (*eq.* 1) and (*eq.* 3) and then (*eq.* 1) and (*eq.* 2). We could have elected to eliminate either variable x or variable y first. For example, we could have eliminated variable x by multiplying (*eq.* 3) by -2 and then adding it to (*eq.* 2). We could also eliminate variable x by multiplying (*eq.* 3) by -3 and then adding it to (*eq.* 1). Try solving the system in Example 2 by first eliminating the variable x.

EXAMPLE 3 Solve the following system of equations.

$$\begin{aligned} 2x - 3y + 2z &= -1 \quad (eq.\ 1)\\ x + 2y \qquad\quad &= 14 \quad (eq.\ 2)\\ x \qquad\ - 3z &= -5 \quad (eq.\ 3) \end{aligned}$$

Solution The third equation does not contain y. We will therefore work to obtain another equation that does not contain y. We will use $(eq.\ 1)$ and $(eq.\ 2)$ to do this.

$$\begin{aligned} 4x - 6y + 4z &= -2 \quad (eq.\ 1) \text{ Multiplied by 2}\\ 3x + 6y + 4z &= 42 \quad (eq.\ 2) \text{ Multiplied by 3}\\ \hline 7x + 6y + 4z &= 40 \quad \text{Sum of equations, } (eq.\ 4) \end{aligned}$$

We now have two equations containing only the variables x and z.

$$\begin{aligned} 7x + 4z &= 40 \quad (eq.\ 4)\\ x - 3z &= -5 \quad (eq.\ 3) \end{aligned}$$

Let's now eliminate the variable x.

$$\begin{aligned} 7x + 4z &= 40 \quad (eq.\ 4)\\ -7x + 21z &= 35 \quad (eq.\ 3) \text{ Multiplied by } -7\\ \hline 25z &= 75 \quad \text{Sum of equations}\\ z &= 3 \end{aligned}$$

Now we solve for x by using one of the equations containing only the variables x and z We substitute 3 for z in $(eq.\ 3)$.

$$\begin{aligned} x - 3z &= -5 \quad (eq.\ 3)\\ x - 3(3) &= -5 \quad \text{Substitute 3 for } z \text{ in } (eq.\ 3)\\ x - 9 &= -5\\ x &= 4 \end{aligned}$$

Finally, we solve for y using any of the original equations that contains y.

$$\begin{aligned} x + 2y &= 14 \quad (eq.\ 2)\\ 4 + 2y &= 14 \quad \text{Substitute 4 for } x \text{ in } (eq.\ 2)\\ 2y &= 10\\ y &= 5 \end{aligned}$$

The solution is the ordered triple (4, 5, 3).

Check: $(eq.\ 1)$

$$\begin{aligned} 2x - 3y + 2z &= -1\\ 2(4) - 3(5) + 2(3) &\overset{?}{=} -1\\ 8 - 15 + 6 &\overset{?}{=} -1\\ -1 &= -1 \end{aligned}$$

True

$(eq.\ 2)$

$$\begin{aligned} x + 2y &= 14\\ 4 + 2(5) &\overset{?}{=} 14\\ 4 + 10 &\overset{?}{=} 14\\ 14 &= 14 \end{aligned}$$

True

$(eq.\ 3)$

$$\begin{aligned} x - 3z &= -5\\ 4 - 3(3) &\overset{?}{=} -5\\ 4 - 9 &\overset{?}{=} -5\\ -5 &= -5 \end{aligned}$$

True

NOW TRY EXERCISE 15

HELPFUL HINT

If an equation in a system contains fractions, eliminate the fractions by multiplying each term in the equation by the least common denominator. Then continue to solve the system. If, for example, one equation in the system is $\frac{3}{4}x - \frac{5}{8}y + z = \frac{1}{2}$, multiply both sides of the equation by 8 to obtain the equivalent equation $6x - 5y + 8z = 4$.

2) Learn the Geometric Interpretation of a System of Equations in Three Variables

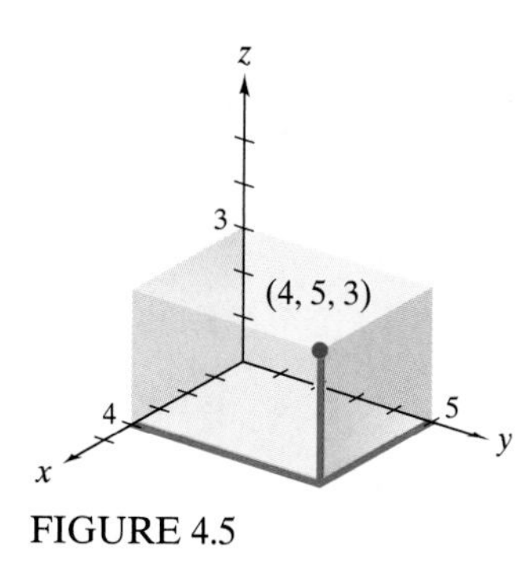

FIGURE 4.5

When we have a system of linear equations in two variables, we can find its solution graphically using the Cartesian coordinate system. A linear equation in three variables, x, y, and z, can be graphed on a coordinate system with three axes drawn perpendicular to each other (see Fig. 4.5).

A point plotted in this three-dimensional system would appear to be a point in space. If we were to graph an equation such as $x + 2y + 3z = 4$, we would find that its graph would be a plane, not a line. In Example 3 we indicated the solution to be the ordered triple (4, 5, 3). This means that the three planes, one from each of the three given equations, all intersect at the point (4, 5, 3). Figure 4.5 shows the location of this point of intersection of the three planes. The drawing in Exercise 38 illustrates three planes intersecting at a point.

3) Recognize Inconsistent and Dependent Systems

We discussed inconsistent and dependent systems of equations in Section 4.1. Systems of linear equations in three variables may also be inconsistent or dependent. When solving a system of linear equations in three variables, if you obtain a false statement like $3 = 0$, the system is inconsistent and has no solution. This means that at least two of the planes are parallel, so the three planes cannot intersect. (See Exercises 35 and 36.)

When solving a system of linear equations in three variables, if you obtain the true statement, $0 = 0$, it indicates that the system is dependent and has an infinite number of solutions. This may happen when all three equations represent the same plane or when the intersection of the planes is a line, as in the drawing in Exercise 37. Examples 4 and 5 illustrate an inconsistent system and a dependent system, respectively.

EXAMPLE 4 Solve the following system of equations.

$$x + 2y + 3z = 0 \quad (eq.\ 1)$$
$$2x + 4y + 6z = 2 \quad (eq.\ 2)$$
$$3x + 6y - 4z = 3 \quad (eq.\ 3)$$

Solution We will begin by eliminating x from (eq. 1) and (eq. 2).

$$-2x - 4y - 6z = 0 \quad (eq.\ 1)\ \text{Multiplied by } -2$$
$$2x + 4y + 6z = 2 \quad (eq.\ 2)$$
$$0 = 2 \quad \text{False}$$

Since we obtained the false statement $0 = 2$, this system is inconsistent and has no solution.

NOW TRY EXERCISE 29

EXAMPLE 5 Solve the following system of equations.

$$x - y + z = 1 \quad (eq.\ 1)$$
$$x + 2y - z = 1 \quad (eq.\ 2)$$
$$x - 4y + 3z = 1 \quad (eq.\ 3)$$

Solution We will begin by eliminating the variable x from $(eq.\ 1)$ and $(eq.\ 2)$ and then from $(eq.\ 1)$ and $(eq.\ 3)$.

$$-x + y - z = -1 \quad (eq.\ 1) \text{ Multiplied by } -1$$
$$x + 2y - z = 1 \quad (eq.\ 2)$$
$$3y - 2z = 0 \quad \text{Sum of equations, } (eq.\ 4)$$

$$x - y + z = 1 \quad (eq.\ 1)$$
$$-x + 4y - 3z = -1 \quad (eq.\ 3) \text{ Multiplied by } -1$$
$$3y - 2z = 0 \quad \text{Sum of equations, } (eq.\ 5)$$

Now we eliminate the variable y using $(eq.\ 4)$ and $(eq.\ 5)$.

$$-3y + 2z = 0 \quad (eq.\ 4) \text{ Multiplied by } -1$$
$$3y - 2z = 0 \quad (eq.\ 5)$$
$$0 = 0 \quad \text{True}$$

Since we obtained the true statement $0 = 0$, this system is dependent and has an infinite number of solutions.

Recall from Section 4.1 that systems of equations that are dependent are also consistent since they have a solution.

NOW TRY EXERCISE 31

Exercise Set 4.2

Concept/Writing Exercises

1. What will be the graph of an equation such as $2x + 3y + 4z = 5$?
2. Assume that the solution to a system of equations in three variables is (2, 3, 4). What does this mean geometrically?

Practice the Skills

Solve by substitution.

3. $x = 3$
   $x + 2y = 7$
   $-3x - y + 4z = 9$
4. $2x + 3y = 9$
   $4x - 6z = 12$
   $y = 5$
5. $5x - 6z = -17$
   $3x - 4y + 5z = -1$
   $2z = -6$
6. $2x - 5y = 12$
   $-3y = -9$
   $2x - 3y + 4z = 8$
7. $x + 2y = 6$
   $3y = 9$
   $x + 2z = 12$
8. $x - y + 5z = -4$
   $3x - 2z = 6$
   $4z = 2$

Solve using the addition method.

9. $x + y + z = 3$
   $x - z = -5$
   $2x - y + 2z = 0$
10. $x - 2y = 2$
    $2x + 3y = 11$
    $-y + 4z = 7$
11. $x - 2z = -5$
    $-y + 3z = 3$
    $-2x + z = 4$
12. $x - 3y = 13$
    $2y + z = 1$
    $y - 2z = 11$
13. $3p + 2q = 11$
    $4q - r = 6$
    $2p + 2r = 2$
14. $-4s + 3t = 16$
    $2t - 2u = 2$
    $-s + 6u = -2$
15. $p + q + r = 4$
    $p - 2q - r = 1$
    $2p - q - 2r = -1$
16. $x - 2y + 3z = -7$
    $2x - y - z = 7$
    $-x + 3y + 2z = -8$
17. $2x - 2y + 3z = 5$
    $2x + y - 2z = -1$
    $4x - y - 3z = 0$

18. $\begin{aligned}2x - y - z &= 4\\ 4x - 3y - 2z &= -2\\ 8x - 2y - 3z &= 3\end{aligned}$

19. $\begin{aligned}2r + 2s - 3t &= 12\\ -2r + s + t &= -11\\ 3r + 4s + 2t &= 4\end{aligned}$

20. $\begin{aligned}a + 2b + 2c &= 1\\ 2a - b + c &= 3\\ 4a + b + 2c &= 0\end{aligned}$

21. $\begin{aligned}2a + 2b - c &= 2\\ 3a + 4b + c &= -4\\ 5a - 2b - 3c &= 5\end{aligned}$

22. $\begin{aligned}x + y + z &= 0\\ -x - y + z &= 0\\ -x + y + z &= 0\end{aligned}$

23. $\begin{aligned}-x + 3y + z &= 0\\ -2x + 4y - z &= 0\\ 3x - y + 2z &= 0\end{aligned}$

24. $\begin{aligned}-\frac{1}{4}x + \frac{1}{2}y - \frac{1}{2}z &= -2\\ \frac{1}{2}x + \frac{1}{3}y - \frac{1}{4}z &= 2\\ \frac{1}{2}x - \frac{1}{2}y + \frac{1}{4}z &= 1\end{aligned}$

25. $\begin{aligned}\frac{2}{3}x + y - \frac{1}{3}z &= \frac{1}{3}\\ \frac{1}{2}x + y + z &= \frac{5}{2}\\ \frac{1}{4}x - \frac{1}{4}y + \frac{1}{4}z &= \frac{3}{2}\end{aligned}$

26. $\begin{aligned}x - \frac{2}{3}y - \frac{2}{3}z &= -2\\ \frac{2}{3}x + y - \frac{2}{3}z &= \frac{1}{3}\\ -\frac{1}{4}x + y - \frac{1}{4}z &= \frac{3}{4}\end{aligned}$

27. $\begin{aligned}0.2x + 0.3y + 0.3z &= 1.1\\ 0.4x - 0.2y + 0.1z &= 0.4\\ -0.1x - 0.1y + 0.3z &= 0.4\end{aligned}$

28. $\begin{aligned}0.3x - 0.4y + 0.2z &= 1.6\\ -0.1x - 0.2y + 0.3z &= 0.9\\ -0.2x - 0.1y - 0.3z &= -1.2\end{aligned}$

Determine whether the following systems are inconsistent, dependent, or neither.

29. $\begin{aligned}2x + y + 2z &= 1\\ x - 2y - z &= 0\\ 3x - y + z &= 2\end{aligned}$

30. $\begin{aligned}-x + y + z &= 0\\ x - y + z &= 0\\ 3x - 3y - z &= 0\end{aligned}$

31. $\begin{aligned}3x - 4y + z &= 4\\ x + 2y + z &= 4\\ -6x + 8y - 2z &= -8\end{aligned}$

32. $\begin{aligned}2p - 4q + 6r &= 8\\ -p + 2q - 3r &= 6\\ 3p + 4q + 5r &= 8\end{aligned}$

33. $\begin{aligned}x + 3y + 2z &= 6\\ x - 2y - z &= 8\\ -3x - 9y - 6z &= -4\end{aligned}$

34. $\begin{aligned}2x - 2y + 4z &= 2\\ -3x + y &= -9\\ 2x - y + z &= 5\end{aligned}$

Problem Solving

An equation in three variables, x, y, and z, represents a plane. Consider a system of equations consisting of three equations in three variables. Answer the following questions.

35. If the three planes are parallel to one another as illustrated in the figure, how many points will be common to all three planes? Is the system consistent or inconsistent? Explain your answer.

36. If two of the planes are parallel to each other and the third plane intersects each of the other two planes, how many points will be common to all three planes? Is the system consistent or inconsistent? Explain your answer.

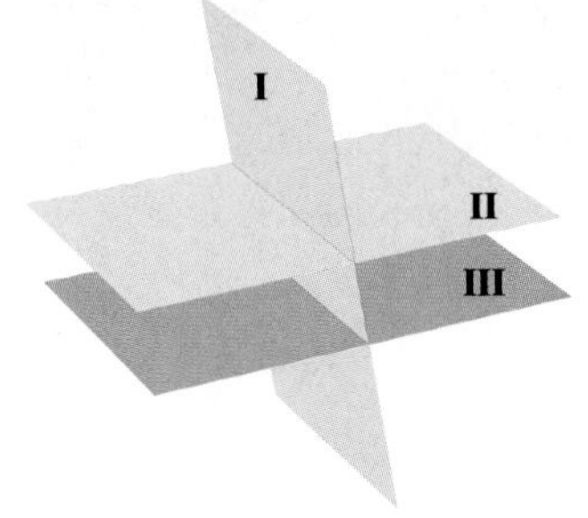

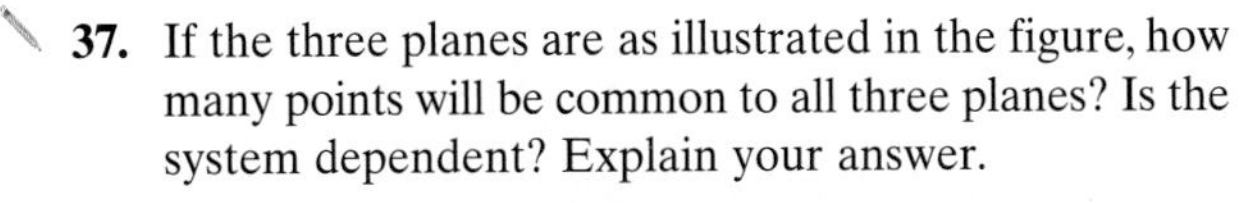

37. If the three planes are as illustrated in the figure, how many points will be common to all three planes? Is the system dependent? Explain your answer.

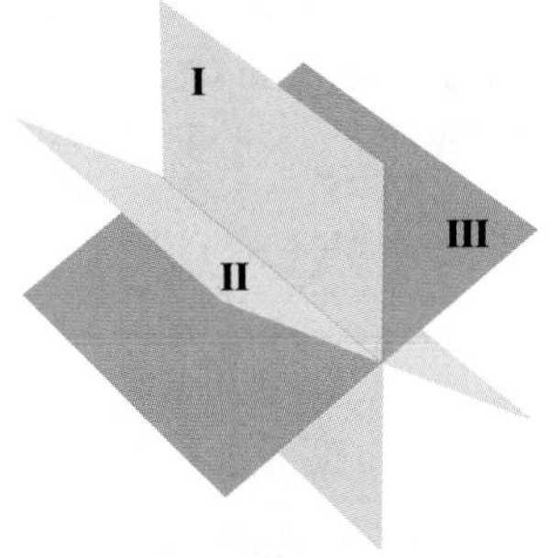

38. If the three planes are as illustrated in the figure, how many points will be common to all three planes? Is the system consistent or inconsistent? Explain your answer.

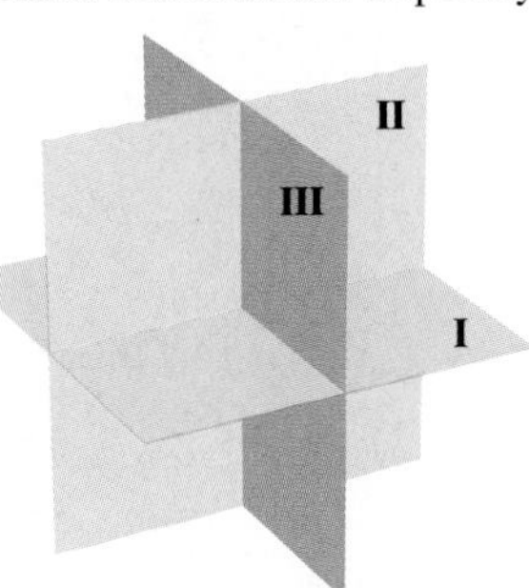

39. Is it possible for a system of linear equations in three variables to have exactly **a)** no solution, **b)** one solution, **c)** two solutions? Explain your answer.

40. In a system of linear equations in three variables, if the graphs of two equations are parallel planes, is it possible for the system to be **a)** consistent, **b)** dependent, **c)** inconsistent? Explain your answer.

41 Three solutions to the equation $Ax + By + Cz = -2$ are $(1, 2, -1)$, $(1, 1, -3)$ and $(2, 3, -2)$. Find the values of A, B, and C and write the equation using the numerical values found.

42. Three solutions to the equation $Ax + By + Cz = 14$ are $(3, -1, 2)$, $(2, -2, 1)$ and $(-5, 3, -24)$. Find the values of A, B, and C and write the equation using the numerical values found.

Write a system of linear equations in three variables that has the given solution. Explain how you determined your answer.

43. $(3, 1, 6)$

44. $(-2, 5, 3)$

45. **a)** Find the values of a, b, and c such that the points $(1, -1)$, $(-1, -5)$, and $(3, 11)$ lie on the graph of $y = ax^2 + bx + c$.

b) Find the quadratic equation whose graph passes through the three points indicated. Explain how you determined your answer.

46. **a)** Find the values of a, b, and c such that the points $(1, 7)$, $(-2, -5)$, and $(3, 5)$ lie on the graph of $y = ax^2 + bx + c$.

b) Find the quadratic equation whose graph passes through the three points indicated. Explain how you determined your answer.

Challenge Problems

Find the solution to the following system of equations,

47.
$$\begin{aligned} 3a + 2b - c &= 0 \\ 2a + 2c + d &= 5 \\ a + 2b - d &= -2 \\ 2a - b + c + d &= 2 \end{aligned}$$

48.
$$\begin{aligned} 3p + 4q &= 11 \\ 2p + r + s &= 9 \\ q - s &= -2 \\ p + 2q - r &= 2 \end{aligned}$$

Cumulative Review Exercises

[2.4] **49.** Phillipa Willis and her son Cameron go cross-country skiing. Phillipa averages 5 miles per hour, and Cameron averages 3 miles per hour. If Cameron begins $\frac{1}{6}$ hour before his mother, **a)** how long after Cameron starts skiing will his mother catch up with him?

b) How far from the starting point will they be when they meet?

[2.6] *Determine each solution set.*

50. $\left|4 - \frac{2x}{3}\right| > 5$

51. $\left|\frac{3x - 4}{2}\right| - 1 < 5$

52. $\left|2x - \frac{1}{2}\right| = -5$

See Exercise 49.

4.3 SYSTEMS OF LINEAR EQUATIONS: APPLICATIONS AND PROBLEM SOLVING

1) Use systems of equations to solve applications.
2) Use linear systems in three variables to solve applications.

SSM VIDEO 4.3 CD Rom

1) Use Systems of Equations to Solve Applications

Many of the applications solved in earlier chapters using only one variable can now be solved using two variables. Following are some examples showing how applications can be represented by systems of equations.

EXAMPLE 1 The graph in Figure 4.6 indicates that the percent of males in the workforce is constantly decreasing while the percent of females is constantly increasing. The function $m(t) = -0.25t + 85.4$, where t = years since 1955, can be used to estimate the percent of males in the workforce, and the function $w(t) = 0.52t + 35.7$ can be used to estimate the percent of women in the workforce. If this trend continues, determine when the percent of women in the workforce will equal the percent of men.

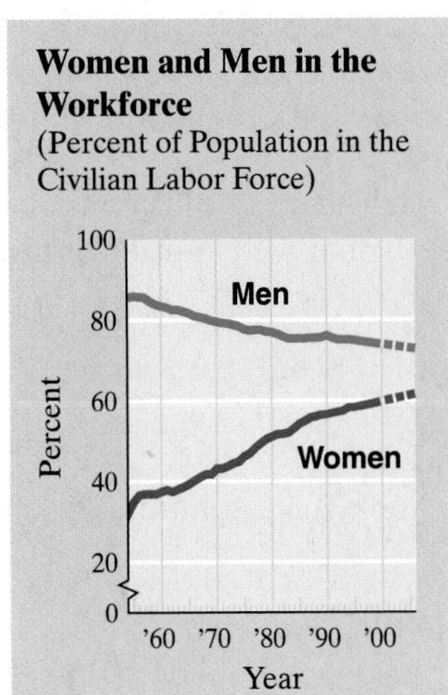

FIGURE 4.6

Solution **Understand and Translate** Consider the two functions given above as the system of equations. To determine when the percent of women will equal the percent of men, we can set the two functions equal to one another and solve for time, t.

Carry Out

$$\text{percent of women} = \text{percent of men}$$
$$0.52t + 35.7 = -0.25t + 85.4$$
$$0.77t = 49.7$$
$$t \approx 64.5$$

Answer Thus the percent of women in the workforce will equal the percent of men about 64.5 years after 1955. Since $1955 + 64.5 = 2019.5$, the percents will be equal in the year 2019.

NOW TRY EXERCISE 37

EXAMPLE 2 Will Ethridge is on The Austin Rowing (crew) Team. The team practices in many different locations to better prepare for the competition. One Sunday, while practicing on the Colorado River, the team rowed an average of 25.6 miles per hour with the current and 18.8 miles per hour against the current. Find the team's rowing speed in still water.

Solution **Understand** When the team is traveling with the current the boat's speed will be the boat's speed in still water plus the current. When traveling against the current, the boat's speed will be the boat's speed in still water minus the current.

Translate Let s = speed of boat in still water

c = current

The system of equations is:

speed of boat going with the current: $s + c = 25.6$

speed of boat going against the current: $s - c = 18.8$

Carry Out We will add the two equations to eliminate the variable c.

$$\begin{array}{r} s + c = 25.6 \\ s - c = 18.8 \\ \hline 2s \quad = 44.4 \\ s = 22.2 \end{array}$$

The speed of the boat is 22.2 miles per hour in still water. Now let's find the current.

$$\begin{aligned} s + c &= 25.6 \\ 22.2 + c &= 25.6 \\ c &= 3.4 \end{aligned}$$

Answer The current is 3.4 miles per hour and the speed of the boat in still water is 22.2 miles per hour.

EXAMPLE 3 Yamil Bermudez, a salesman at Hancock Appliances, receives a weekly salary plus a commission, which is a percentage of his sales. One week, on sales of \$3000, his total take-home pay was \$850. The next week, on sales of \$4000, his total take-home pay was \$1000. Find his weekly salary and his commission rate.

Solution **Understand** Yamil's take-home pay consists of his weekly salary plus commission. We are given information about two specific weeks that we can use to find his weekly salary and his commission rate.

Translate Let s = his weekly salary

r = his commission rate

In week 1 his commission on \$3000 is $3000r$, and in week 2 his commission on \$4000 is $4000r$. Thus the system of equations is

salary + commission = take-home salary

$$\left.\begin{array}{ll} \text{1st week} & s + 3000r = 850 \\ \text{2nd week} & s + 4000r = 1000 \end{array}\right\} \text{System of equations}$$

Carry Out

$$\begin{array}{rl} -s - 3000r = -850 & \text{1st week multiplied by } -1 \\ s + 4000r = 1000 & \text{2nd week} \\ \hline 1000r = 150 & \text{Sum of equations} \end{array}$$

$$r = \frac{150}{1000}$$

$$r = 0.15$$

Yamil's commission rate is 15%. Now we find his weekly salary by substituting 0.15 for r in either equation.

$$\begin{aligned} s + 3000r &= 850 \\ s + 3000(0.15) &= 850 \qquad \text{Substitute 0.15 for } r \text{ in the 1st week equation.} \\ s + 450 &= 850 \\ s &= 400 \end{aligned}$$

NOW TRY EXERCISE 9

Answer Yamil's weekly salary is \$400 and his commission rate is 15%.

EXAMPLE 4 A Coast Guard cutter in a San Diego marina is sent information that a boat that has smuggled drugs has left the marina 2 hours earlier traveling in open waters at 20 miles per hour (or about 17.4 knots). The Coast Guard cutter whose top speed is 32 mph (or about 27.8 knots), leaves immediately in chase of the smugglers.

a) How long after the smugglers leave will the Coast Guard cutter catch them?

b) When it catches them, how far away from the marina will they be?

Solution

a) Understand When the Coast Guard cutter catches the drug smugglers, they will both have traveled the same distance. The Coast Guard cutter will have traveled that distance in 2 hours less time since it left 2 hours after the drug smugglers. We will use the formula distance = rate · time to solve this problem.

Translate Let x = time traveled by the drug smugglers

and y = time traveled by the Coast Guard

We will set up a table to organize the information.

	Rate	Time	Distance
Smugglers	20	x	$20x$
Coast Guard	32	y	$32y$

Since both the smugglers and Coast Guard cover the same distance, we write

$$\text{distance of drug smugglers} = \text{distance of Coast Guard}$$

$$20x = 32y$$

Our second equation comes from the fact that the Coast Guard cutter is traveling for two hours less than the smugglers. Therefore $y = x - 2$.

$$\text{system of equations} \quad \begin{cases} 20x = 32y \\ y = x - 2 \end{cases}$$

Carry Out Since $y = x - 2$, substitute $x - 2$ for y in the first equation then solve for x.

$$\begin{aligned} 20x &= 32y \\ 20x &= 32(x - 2) \\ 20x &= 32x - 64 \\ -12x &= -64 \\ x &= \frac{-64}{-12} = 5\frac{1}{3} \end{aligned}$$

Answer The Coast Guard cutter will catch the smugglers $5\frac{1}{3}$ hours after the smugglers leave.

b) We can use either distance equation from the table to find the distance traveled from the marina. Let's use the distance equation of the smugglers

$$d = 20x = 20(5.3\overline{3}) = 106.\overline{6}$$

Thus when the Coast Guard cutter catches the smugglers, they will have traveled 106.67 miles from the marina.

EXAMPLE 5 Martina Sanchez, a neuroscientist, is working on an AIDS vaccine. She needs to mix a 15% sodium–iodine solution with a 40% sodium–iodine solution to get 6 liters of a 25% sodium–iodine solution. How many liters of the 15% solution and of the 40% solution will she need to mix?

Solution **Understand** To solve this problem we use the fact that the amount of sodium–iodine in a solution is found by multiplying the percent strength of the solution by the number of liters (the volume) of the solution. Martina needs to mix a 15% solution and a 40% solution to obtain 6 liters of a solution whose strength, 25%, is between the strengths of the two solutions being mixed.

Translate Let x = number of liters of the 15% solution

y = number of liters of the 40% solution

We will draw a sketch (Fig. 4.7) and then set up a table to help analyze the problem.

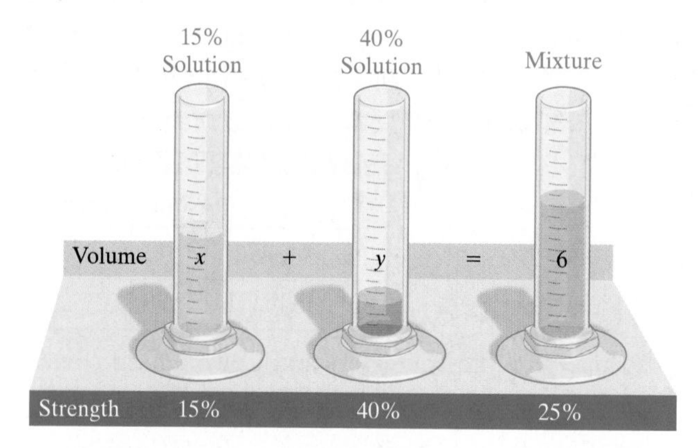

FIGURE 4.7

Solution	Strength of Solution	Number of Liters	Amount of Sodium–Iodine
15% Solution	0.15	x	$0.15x$
40% Solution	0.40	y	$0.40y$
Mixture	0.25	6	$0.25(6)$

Since the sum of the volumes of the 15% solution and the 40% solution is 6 liters, our first equation is

$$x + y = 6$$

The second equation comes from the fact that the solutions are mixed.

$$\begin{pmatrix}\text{amount of}\\ \text{sodium–iodine}\\ \text{in 15\% solution}\end{pmatrix} + \begin{pmatrix}\text{amount of}\\ \text{sodium–iodine}\\ \text{in 40\% solution}\end{pmatrix} = \begin{pmatrix}\text{amount of}\\ \text{sodium–iodine}\\ \text{in mixture}\end{pmatrix}$$

$$0.15x \quad + \quad 0.40y \quad = \quad 0.25(6)$$

The system of equations is, therefore,

$$x + y = 6$$
$$0.15x + 0.40y = 0.25(6)$$

Carry Out Solving $x + y = 6$ for y we get $y = -x + 6$. Substituting $-x + 6$ for y in the second equation gives us

$$0.15x + 0.40y = 0.25(6)$$
$$0.15x + 0.40(-x + 6) = 0.25(6)$$
$$0.15x - 0.40x + 2.4 = 1.5$$
$$-0.25x + 2.4 = 1.5$$
$$-0.25x = -0.9$$
$$x = \frac{-0.9}{-0.25} = 3.6$$

Therefore, Martina must use 3.6 liters of the 15% solution. Since the two solutions must total 6 liters, she must use $6 - 3.6$ or 2.4 liters of the 40% solution.

In Example 5, the equation $0.15x + 0.40y = 0.25(6)$ could have been simplified by multiplying both sides of the equation by 100. This would give the equation $15x + 40y = 25(6)$ or $15x + 40y = 150$. Then the system of equations would be $x + y = 6$ and $15x + 40y = 150$. If you solve this system, you should obtain the same solution. Try it and see.

NOW TRY EXERCISE 13

2) Use Linear Systems in Three Variables to Solve Applications

Now let us look at some applications that involve three equations and three variables.

EXAMPLE 6 Tiny Tots Toys must borrow \$25,000 to pay for an expansion. They are not able to obtain a loan for the total amount from a single bank, so they take out three loans from three different banks. They borrowed some of the money at a bank that charged them 8% interest. At the second bank, they borrowed \$2000 more than one-half the amount borrowed from the first bank. The interest rate at the second bank is 10%. The balance of the \$25,000 is borrowed from a third bank where they paid 9% interest. The total annual interest Tiny Tots Toys pays for the three loans is \$2220. How much did they borrow at each rate?

Solution **Understand** We are asked to determine how much was borrowed at each of the three different rates. Therefore, this problem will contain three variables, one for each amount borrowed. Since the problem will contain three variables, we will need to determine three equations to use in our system of equations.

Translate Let x = amount borrowed at first bank

y = amount borrowed at second bank

z = amount borrowed at third bank

Since the total amount borrowed is \$25,000 we know that

$$x + y + z = 25{,}000$$ *Total amount borrowed is \$25,000.*

At the second bank, Tiny Tots Toys borrowed \$2000 more than one-half the money borrowed from the first bank. Therefore, our second equation is

$$y = \frac{1}{2}x + 2000$$ *Second, y, is \$2000 more than $\frac{1}{2}$ of first, x.*

Our last equation comes from the fact that the total annual interest charged by the three banks is \$2220. The interest at each bank is found by multiplying the interest rate by the amount borrowed.

$$0.08x + 0.10y + 0.09z = 2220$$ *Total interest is \$2220.*

Thus, our system of equation is

$$x + y + z = 25{,}000 \qquad (1)$$

$$y = \frac{1}{2}x + 2000 \qquad (2)$$

$$0.08x + 0.10y + 0.09z = 2220 \qquad (3)$$

Both sides of equation (2) can be multiplied by 2 to remove fractions.

$$2(y) = 2\left(\tfrac{1}{2}x + 2000\right)$$

$$2y = x + 4000 \qquad \text{Distributive Property}$$

$$-x + 2y = 4000 \qquad \text{Subtract } x \text{ from both sides}$$

The decimals in equation (3) can be removed by multiplying both sides of the equation by 100. This gives

$$8x + 10y + 9z = 222{,}000$$

Our simplified system of equations is therefore

$$\begin{aligned} x + y + z &= 25{,}000 && (eq.\ 1)\\ -x + 2y &= 4000 && (eq.\ 2)\\ 8x + 10y + 9z &= 222{,}000 && (eq.\ 3) \end{aligned}$$

Carry Out There are various ways of solving this system. Let's use ($eq.$ 1) and ($eq.$ 3) to eliminate the variable z.

$$\begin{aligned} -9x - 9y - 9z &= -225{,}000 && (eq.\ 1) \text{ Multiplied by } -9\\ 8x + 10y + 9z &= 222{,}000 && (eq.\ 3)\\ \hline -x + y &= -3000 && \text{Sum of equations, } (eq.\ 4) \end{aligned}$$

Now we use ($eq.$ 2) and ($eq.$ 4) to eliminate the variable x and solve for y.

$$\begin{aligned} -x + y &= -3000 && (eq.\ 4)\\ x - 2y &= -4000 && (eq.\ 2) \text{ Multiplied by } -1\\ \hline -y &= -7000 && \text{Sum of equations}\\ y &= 7000 \end{aligned}$$

Now that we know the value of y we can solve for x.

$$\begin{aligned} -x + 2y &= 4000 && (eq.\ 2)\\ -x + 2(7000) &= 4000 && \text{Substitute 7000 for } y \text{ in } (eq.\ 2)\\ -x + 14{,}000 &= 4000\\ -x &= -10{,}000\\ x &= 10{,}000 \end{aligned}$$

Finally, we solve for z.

$$\begin{aligned} x + y + z &= 25{,}000 && (eq.\ 1)\\ 10{,}000 + 7000 + z &= 25{,}000\\ 17{,}000 + z &= 25{,}000\\ z &= 8000 \end{aligned}$$

Answer Tiny Tots Toys borrowed \$10,000 at 8%, \$7000 at 10%, and \$8000 at 9% interest.

EXAMPLE 7 Hobson, Inc., has a small manufacturing plant that makes three types of inflatable boats: one-person, two-person, and four-person models. Each boat requires the service of three departments: cutting, assembly, and packaging. The cutting, assembly, and packaging departments are allowed to use a total of 380, 330, and 120 person-hours per week, respectively. The time requirements for each boat and department are specified in the following table. Determine how many of each type of boat Hobson's must produce each week for its plant to operate at full capacity.

	Time (person-hr)		
Department	**One-Person Boat**	**Two-Person Boat**	**Four-Person Boat**
Cutting	0.6	1.0	1.5
Assembly	0.6	0.9	1.2
Packaging	0.2	0.3	0.5

Solution **Understand** We are told that three different types of boats will be produced and we are asked to determine the number of each type to be produced. Since this problem involves three amounts to be found, the system will contain three equations in three variables.

Translate We will use the information given in the table.

Let x = number of one-person boats

y = number of two-person boats

z = number of four-person boats

The total number of cutting hours for the three types of boats must equal 380 person-hours.

$$0.6x + 1.0y + 1.5z = 380$$

The total number of assembly hours must equal 330 person-hours.

$$0.6x + 0.9y + 1.2z = 330$$

The total number of packaging hours must equal 120 person-hours.

$$0.2x + 0.3y + 0.5z = 120$$

Therefore, the system of equations is

$$\begin{aligned} 0.6x + 1.0y + 1.5z &= 380 \\ 0.6x + 0.9y + 1.2z &= 330 \\ 0.2x + 0.3y + 0.5z &= 120 \end{aligned}$$

Multiplying each equation in the system by 10 will eliminate the decimal numbers and give a simplified system of equations.

$$\begin{aligned} 6x + 10y + 15z &= 3800 \quad (eq.\ 1) \\ 6x + 9y + 12z &= 3300 \quad (eq.\ 2) \\ 2x + 3y + 5z &= 1200 \quad (eq.\ 3) \end{aligned}$$

Carry Out Let's first eliminate the variable x using (*eq.* 1) and (*eq.* 2), and then (*eq.* 1) and (*eq.* 3).

$$\begin{aligned} 6x + 10y + 15z &= 3800 \quad (eq.\ 1) \\ -6x - 9y - 12z &= -3300 \quad (eq.\ 2) \text{ Multiplied by } -1 \\ \hline y + 3z &= 500 \quad \text{Sum of equations, } (eq.\ 4) \end{aligned}$$

$$\begin{aligned} 6x + 10y + 15z &= 3800 \quad (eq.\ 1) \\ -6x - 9y - 15z &= -3600 \quad (eq.\ 3) \text{ Multiplied by } -3 \\ \hline y &= 200 \quad \text{Sum of equations, } (eq.\ 5) \end{aligned}$$

Note that when we added the last two equations, both variables x and z were eliminated at the same time. Now we know the value of y and can solve for z.

$$\begin{aligned} y + 3z &= 500 \quad (eq.\ 4) \\ 200 + 3z &= 500 \quad \text{Substitute 200 for } y. \\ 3z &= 300 \\ z &= 100 \end{aligned}$$

Finally, we find x.

$$\begin{aligned} 6x + 10y + 15z &= 3800 \quad (eq.\ 1) \\ 6x + 10(200) + 15(100) &= 3800 \\ 6x + 2000 + 1500 &= 3800 \\ 6x + 3500 &= 3800 \\ 6x &= 300 \\ x &= 50 \end{aligned}$$

Answer Hobson's should produce 50 one-person boats, 200 two-person boats, and 100 four-person boats per week.

NOW TRY EXERCISE 49

Exercise Set 4.3

Practice the Skills/Problem Solving

1. An article in the January 1998 *Newsweek* entitled "Hollywood's New Math" discusses the movie *Men in Black*. The article indicates that Tommy Lee Jones, an actor, received a base salary of \$7 million and 5% of the gross while director Steven Spielberg received a nominal salary (use \$0) plus 20% of the gross. (Will Smith, an actor, received \$5 million.)

   a) How much would the movie have to gross for Spielberg's total income from the movie to equal that of Jones?
   b) At the time the article was written, the movie had grossed about \$600 million worldwide. How much did Jones receive in total?
   c) How much did Spielberg receive in total?
2. In 1997, France was the most visited country in the world and the United States was the second most visited. A total of about 116 million people visited one of the two countries. If 18 million more people visited France than the United States, how many people visited each country?
3. A nutritionist finds that a large order of fries at McDonalds has more fat than their quarter-pound hamburger. The fries have 4 grams more than 3 times the amount of fat that the hamburger has. The difference in the fat content between the fries and the hamburger is 46 grams. Find the fat content of the hamburger and of the fries. (Note: A maximum of 50 grams of fat is recommended daily.)

4. A study in the *Journal of the American Medical Association* indicated that the more educated people (male or female) are, the more likely they are to get tension headaches. Men with a graduate school degree are 11% more than twice as likely to get tension headaches than men who only graduated from grade school. The difference between the percents is 30%. Find the percent of male grade school graduates and male graduate school graduates that get tension headaches.

5. The cost of going to a baseball game increases yearly. The fan cost index, FCI, includes the cost of four average tickets, two small beers, four small sodas, four hot dogs, parking, two programs, and two caps. There is a large spread between the FCI for a family of four going to a Montreal Expos game (the team with the smallest FCI) and an Atlanta Braves game (the largest FCI). (The San Francisco Giants have the closest to the average FCI of \$104.51.) The FCI for the Atlanta Braves is \$26.68 less than twice the FCI for the Montreal Expos. The difference between the FCI of Atlanta and that of Montreal is \$53.74. Find the FCI for the Braves and for the Expos.

6. Two angles are **complementary angles** if the sum of their measures is 90°. If the larger of two complementary angles is 15° more than 2 times the smaller angle, find the two angles.

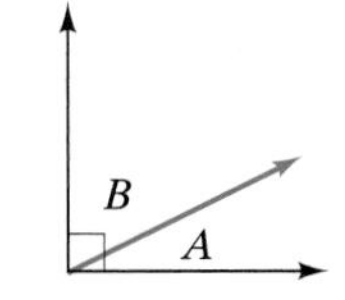

Angles A and B are complementary angles.

7. Two angles are **supplementary angles** if the sum of their measures is 180°. Find the two supplementary angles if one angle is 28° less than 3 times the other.

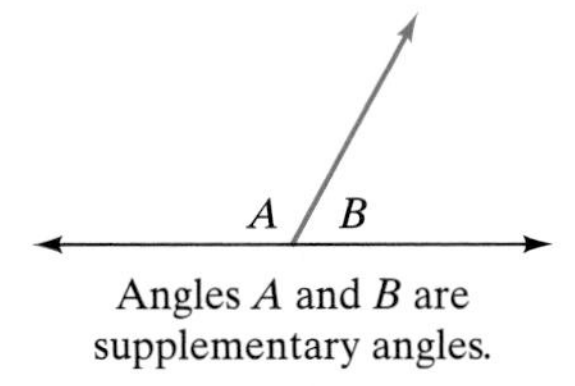

Angles A and B are supplementary angles.

8. Paul Martin, a pilot for the Federal Express Company, noticed that on his flight from the Federal Express headquarters in Memphis, Tennessee, to Omaha, Nebraska, his plane flew at 610 miles per hour with a tailwind (flying with the wind). As soon as he landed and dropped off his mail he flew back to Memphis with a headwind (flying against the wind) at 560 miles per hour. Assuming that the wind speed and direction had not changed and that the plane's speed going to and coming from Memphis would have been the same if it were not for the wind, find the plane's speed in still air and the speed of the wind.

9. Donna Lopez, a book salesperson, earns a weekly salary plus a commission on sales. One week her salary on sales of \$4000 was \$660. The next week her salary on sales of \$6000 was \$740. Find her weekly salary and her commission rate.

10. Trina Smith's truck rental agency charges a daily fee plus a mileage fee. Mel Dobson was charged \$85 for 2 days and 100 miles and Sarah Schwartz was charged \$165 for 3 days and 400 miles. What is the agency's daily fee, and what is their mileage fee?

11. John Tweeddale, director of marketing for a publishing company, wishes to have some advertisement fliers printed. When investigating prices he finds that a printer charges a fixed charge with an additional charge for each flier printed. John finds that the total cost for 1000 fliers is \$550 and the total cost for 2000 fliers is \$800. Find the fixed charge and the additional per-flier charge.

12. Bill Bilderback, a druggist, needs 1000 milliliters of a 10% phenobarbital solution. He has only 5% and 25% phenobarbital solution available. How many milliliters of each solution should he mix to obtain the desired solution?

13. Hy-Pro Spray Cleaner is a concentrated cleaner that is meant to be mixed with water at various concentrations to do a wide variety of chores. Dave Visser had previously made large volumes of 20% Hy-Pro solution (for the kitchen and bathroom cleaning) and 4% Hy-Pro solution (for carpet cleaning). Now Dave needs a 10% solution for general cleaning, but he is out of the concentrate. How much of the 20% solution and how much of the 4% solution should he mix to get 10 gallons of the 10% solution?

14. Birdseed costs \$0.59 a pound and sunflower seeds costs \$0.89 a pound. Angela Leinenbachs' pet store wishes to make a 40-pound mixture of birdseed and sunflower seeds that sells for \$0.76 per pound. How many pounds of each type of seed should she use?

15. Round-Up Concentrate Grass and Weed Killer consists of an 18% active ingredient glyphosate (and 82% inactive ingredients). The concentrate is to be mixed with water and the mixture applied to weeds. If the final mixture is to contain 0.9% active ingredient, how much concentrate and how much water should be mixed to make 200 gallons of the final mixture?

16. Scott's Winterizer Lawn Fertilizer is 22% nitrogen. Schultz's Lime with Lawn Fertilizer is 4% nitrogen. William Weaver, owner of Weaver's Nursery, wishes to mix these two fertilizers to make 400 pounds of a special 10% nitrogen mixture for mid-season lawn feeding. How much of each fertilizer should he mix?

17. Jimmy Stephen's Wing House sells both regular orders and jumbo orders of buffalo wings. The regular order of wings costs \$5.99 and the jumbo order of wings costs \$8.99. One Saturday a total of 134 orders of wings were taken and the receipts from the wings was \$1024.66. How many regular orders and how many jumbo orders were filled?

18. John Bronson runs a grocery store. He wishes to mix 30 pounds of coffee to sell for a total cost of \$170. To obtain the mixture, he will mix coffee that sells for \$5.20 per pound with coffee that sells for \$6.30 per pound. How many pounds of each coffee should he use?

19. Laura Curless owns a dairy. She has milk that is 5% butterfat and skim milk without butterfat. How much 5% milk and how much skim milk should she mix to make 100 gallons of milk that is 3.5% butterfat?

20. Mr. and Mrs. McAdams invest a total of \$10,000 in two savings accounts. One account pays 5% interest and the other 6%. Find the amount placed in each account if they receive a total of \$540 in interest after 1 year. Use interest = principal · rate · time.

21. Steve Gable's recipe for quiche lorraine calls for 2 cups (16 ounces) of light cream that is 20% butter fat. It is often difficult to find light cream with 20% butter fat at the supermarket. What is commonly found is heavy cream, which is 36% butter fat, and half-and-half which is 10.5% butter fat. How much of the heavy cream and how much of the half-and-half should Steve mix to obtain the mixture necessary for the recipe?

22. Professor David Gesell invested \$30,000, part at 9% and part at 5%. If he had invested the entire amount at 6.5%, his total annual interest would be the same as the sum of the annual interest received from the two other accounts. How much was invested at each interest rate?

23. The Friendly Face Fruit Juice Company sells apple juice for 8.3 cents an ounce and raspberry juice for 9.3 cents an ounce. The company wishes to market and sell 8-ounce cans of apple–raspberry juice for 8.7 cents an ounce. How many ounces of each should be mixed?

24. Two cars start at the same point in Alexandria, Virginia, and travel in opposite directions. One car travels 5 miles per hour faster than the other car. After 4 hours, the two cars are 420 miles apart. Find the speed of each car.

25. In an article published in the *Journal of Comparative Physiology and Psychology*, J. S. Brown discusses how we often approach a situation with mixed emotions. For example, when a person is asked to give a speech, he may be a little apprehensive about his ability to do a good job. At the same time, he would like the recognition that goes along with making the speech. J. S. Brown performed an experiment on trained rats. He placed their food in a metal box. He used that same box to administer small electrical shocks to the mice. Therefore, the rats "wished" to go into the box to receive food, yet did not "wish" to go into the box for fear of receiving a small shock. Using the appropriate apparatus, Brown arrived at the following relationships:

pull (in grams) toward food

$$= -\frac{1}{5}d + 70 \qquad 30 < d < 172.5$$

pull (in grams) away from shock

$$= -\frac{4}{3}d + 230 \qquad 30 < d < 172.5$$

where d is the distance in centimeters from the box (and food).

a) Using the substitution method, find the distance at which the pull toward the food equals the pull away from the shock.

b) If the rat is placed 100 cm from the box (or food), what will the rat do?

26. The Centerville School District is having a retirement party and as one of the gifts is giving each retiree a bag containing a mixture of almonds and walnuts. They need 30 pounds of the mixture. Jason Black, a member of the retirement party planning committee, goes to the bulk food department of Safeway, where almonds are selling for \$6.50 per pound and walnuts are selling for \$5.90 per pound. How many pounds of each should the clerk mix if Jason indicates that he can only spend \$6.30 per pound, but he wants as many almonds in the mixture as possible?

27. Rita Pendegrass has an ice cream cone stand in Deming Park in Terre Haute, Indiana. Her small cones sell for \$1.00 and her large cones sell for \$1.50, including tax. At the end of the day she realizes that she made a total of 260 sales. How many of each size cone was sold if her receipts for the day totaled \$299?

28. June Dawn is traveling from just north of Williamsport, Pennsylvania, to North Miami Beach, Florida, where she will spend the winter. Route 15, the road from Williamsport to Route 83, is not all highway but she does average 50 miles per hour on this route. Once she reaches Route 83 (and then Route 95) she averages 70 miles per hour. After 7 hours of driving she decides to stop in Florence, South Carolina, a distance of 430 miles from her starting point. She was hoping to make it to her friends' house in Waterborough, South Carolina, a distance of 500 miles from her starting point, but she stopped driving when it turned dark. Later in her motel room she figured out that if she had averaged 10 miles per hour more for each leg of the trip she would have made it to her friends' house. How many hours did she travel at each speed?

29. Two photocopy machines are used to make large quantities of copies at Kinko's. The slower machine produces 75 copies per minute and the faster machine produces 120 copies per minute. The faster machine was in operation for 3 minutes before the slower machine was started. If they both continue copying together until they produced a total of 1335 copies, find the length of time both machines were in operation.

30. Some states allow a husband and wife to file individual state tax returns (on a single form) even though they file a joint federal return. It is usually to the taxpayer's advantage to do this when both the husband and wife work. The smallest amount of tax owed (or the largest refund) will occur when the husband's and wife's taxable incomes are the same.

Mr. Clar's 2000 taxable income was \$26,200 and Mrs. Clar's income for that year was \$22,450. The Clars' total tax deduction for the year was \$12,400. This deduction can be divided between Mr. and Mrs. Clar in any way they wish. How should the \$12,400 be divided between them to result in each person having the same taxable income and therefore the lowest tax owed?

31. An automobile radiator has a capacity of 16 liters. How much pure antifreeze must be added to a mixture of water and antifreeze that is 18% antifreeze to make a mixture of 20% antifreeze that can be used to fill the radiator?

32. Animals in an experiment are on a strict diet. Each animal is to receive, among other nutrients, 20 grams of protein and 6 grams of carbohydrates. The scientist has only two food mixes available of the following compositions.

Mix	Protein (%)	Carbohydrate (%)
Mix A	10	6
Mix B	20	2

How many grams of each mix should be used to obtain the right diet for a single animal?

33. A company makes two models of chairs. Information about the construction of the chairs is given in the following table.

Model	Time to Assemble	Time to Paint
Model A	1 hr	0.5 hr
Model B	3.2 hr	0.4 hr

On a particular day the company allocated 46.4 person-hours for assembling and 8.8 person-hours for painting. How many of each chair can be made?

34. By weight, one alloy of brass is 70% copper and 30% zinc. Another alloy of brass is 40% copper and 60% zinc. How many grams of each of these alloys need to be melted and combined to obtain 300 grams of a brass alloy that is 60% copper and 40% zinc?

35. Tom Johnson and Melissa Acino start traveling from Oklahoma City, Oklahoma, heading south on Route 35. When Melissa reaches the Dallas/Ft. Worth, Texas, area, a distance of 150 miles, Tom had only reached Denton, Texas, a distance of 120 miles. If Melissa averaged 15 miles per hour faster than Tom, find the average speed of each car.

36. Kamil Abduhl tries to exercise every day. He walks at 3 mph and then jogs at 5 mph. If it takes him 0.9 hr to travel a total of 3.5 mi, how long does he jog?

37. Each year the number of independent drugstores decreases linearly while the number of chain drug stores increases linearly, as shown in the graph. The number of independent drug stores, I, in thousands, can be estimated by the function $I(t) = -1.2t + 27.2$, where t = number of years since 1992. The number of chain drug stores, C, in thousands, can be estimated by the function $C(t) = 0.24t + 17.1$. Assuming this trend continues, in what year will the number of chain drug stores equal the number of independent drug stores?

Chain vs Independent Drugstores

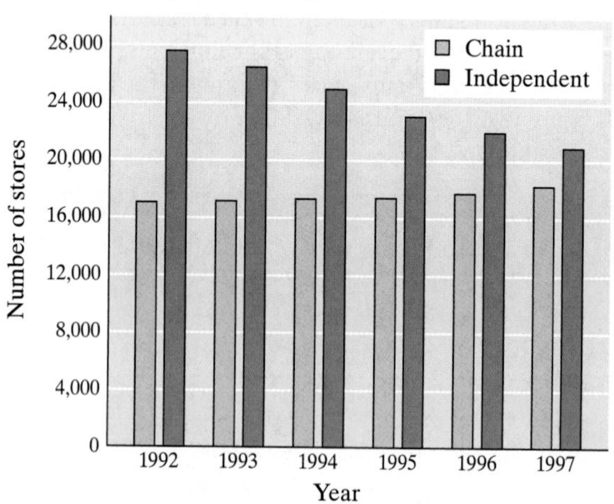

Source: Drug Store News/Chain Store Guides

38. At present in the United States, more men than women receive doctorate degrees annually. But, according to the U.S. Dept. of Education, change is under way. It is expected that the number of men receiving doctorate degrees will fall linearly while the number of women receiving doctorate degrees will continue to increase linearly. The number of men receiving doctorate degrees, D, in thousands, can be approximated by the function $D(t) = -0.86t + 27.1$, where t is the number of years since 1999. The number of women receiving doctorate degrees, W, in thousands, can be approximated by the function $W(t) = 0.43t + 17.5$. Estimate the year when the number of women receiving doctorate degrees will equal the number of men receiving doctorate degrees.

39. On March 30, 1997, AT&T had the following two plans for long distance calling.

Plan 1 15¢ per minute anytime

Plan 2 $4.95 monthly service and 10¢ per minute anytime

a) Represent this information as a system of equations.

b) Graph the system for up to 300 minutes.

c) From the graph, estimate the number of minutes a person would need to spend on the phone for the cost of the two plans to be the same.

d) Solve this system algebraically. If your answer does not agree with your answer in part **c)**, explain why.

40. At a local copy center two plans are available.

Plan 1: 10 cents per copy

Plan 2: an annual charge of $120 plus 4 cents per copy

a) Represent this information as a system of equations.

b) Graph the system of equations for up to 4000 copies made.

c) From the graph, estimate the number of copies a person would have to make in a year for the two plans to have the same total cost.

d) Solve the system algebraically. If your answer does not agree with your answer in part **c)**, explain why.

In Exercises 41–50, **a)** *express the problem as a linear system in three variables, and* **b)** *solve the problem.*

41. A 141-man crew is standard on a Los Angeles class submarine. The number of chief petty officers (enlisted) is four more than the number of commissioned officers. The number of other enlisted men is three less than eight times the number of commissioned officers. Determine the number of commissioned officers, chief petty officers, and other enlisted men on the submarine.

42. The average American household receives 24 pieces of mail each week. The number of bills and statements is two less than twice the number of pieces of personal mail. The number of advertisements is two more than five times the number of pieces of personal mail. How many pieces of personal mail, bills and statements, and advertisements does the

average family get each week? (*Source:* Arthur D. Little, Inc.).

43. One of Princess Diana's goals was to reduce land mines throughout the world. An article in the Chicago Tribune indicated that there are about 100 million land mines buried worldwide and that about 26,000 people are killed or wounded annually from land mines. The countries that have the largest number of land mines are, in order, Iran, Angola, and Iraq. The total number in these three countries is estimated to be 41 million. The number of land mines in Iran is about 14 million less than 3 times the number in Iraq. The number in Angola is about 5 million less than twice the number in Iraq. Find the estimated number of land mines in Iraq, Angola and Iran.

44. Three kinds of tickets are available for a George Strait concert. The up-front main floor tickets are the most expensive; the seats farther back on the main floor are the second most expensive; and the balcony seats are the least expensive. The up-front main floor seats are twice as expensive as the balcony seats. The balcony seats are \$10 less than the main floor seats in the back and \$30 less than the up-front main floor seats. Find the price of each seat.

45. The sum of the measures of the angles of a triangle is 180°. The smallest angle of a triangle is $\frac{2}{3}$ of the middle-sized angle. The largest angle is 30° less than 3 times the middle-sized angle. Find the measure of each angle.

46. Marion Monrow received a check for \$10,000. She decided to divide the money (not equally) into three different investments. She placed part of her money in a savings account paying 3% interest. The second amount, which was twice the first amount, she placed in a certificate of deposit paying 5% interest. She placed the balance in a money market fund that yielded 6% interest. If Marion's total interest over the period of 1 year was \$525.00, how much was placed in each account?

47. A 10% solution, a 12% solution, and a 20% solution of hydrogen peroxide are to be mixed to get 8 liters of a 13% solution. How many liters of each must be mixed if the volume of the 20% solution must be 2 liters less than the volume of the 10% solution?

48. An 8% solution, a 10% solution, and a 20% solution of sulfuric acid are to be mixed to get 100 milliliters of a 12% solution. If the *volume of acid* from the 8% solution is to equal half the *volume of acid* from the other two solutions, how much of each solution is needed?

49. Donaldson Furniture Company produces three types of rocking chairs: the children's model, the standard model, and the executive model. Each chair is made in three stages: cutting, construction, and finishing. The time needed for each stage of each chair is given in the following chart. During a specific week the company has available a maximum of 154 hours for cutting, 94 hours for construction, and 76 hours for finishing. Determine how many of each chair the company should make to be operating at full capacity.

Stage	Children's	Standard	Executive
Cutting	5 hr	4 hr	7 hr
Construction	3 hr	2 hr	5 hr
Finishing	2 hr	2 hr	4 hr

50. By volume, one alloy is 60% copper, 30% zinc, and 10% nickel. A second alloy has percentages 50, 30, and 20, respectively, of the three metals. A third alloy is 30% copper and 70% nickel. How much of each alloy must be mixed so that 100 pounds of the resulting alloy is 40% copper, 15% zinc, and 45% nickel?

51. In electronics it is necessary to analyze current flow through paths of a circuit. In three paths (A, B, and C) of a circuit, the relationships are the following:

$$I_A + I_B + I_C = 0$$

$$-8I_B + 10I_C = 0$$

$$4I_A - 8I_B = 6$$

where I_A, I_B, and I_C represent the current in paths A, B, and C, respectively. Determine the current in each path of the circuit.

52. In physics we often study the forces acting on an object. For three forces, F_1, F_2, and F_3, acting on a beam, the following equations were obtained.

$$3F_1 + F_2 - F_3 = 2$$

$$F_1 - 2F_2 + F_3 = 0$$

$$4F_1 - F_2 + F_3 = 3$$

Find the three forces.

Group Activity

Discuss and answer Exercise 53 as a group.

53. A *nonlinear system of equations* is a system of equations containing at least one equation which is not linear. (Nonlinear systems of equations will be discussed in Chapter 10.) The graph shows a nonlinear system of equations. The curves represent speed versus time for two cars.

a) Are the two curves functions? Explain.

b) Discuss the meaning of this graph.

c) At time $t = 0.5$ hr, which car is traveling at a greater speed? Explain your answer.

d) Assume the two cars start at the same position and are traveling in the same direction. Which car, A or B, traveled farther in 1 hour? Explain your answer.

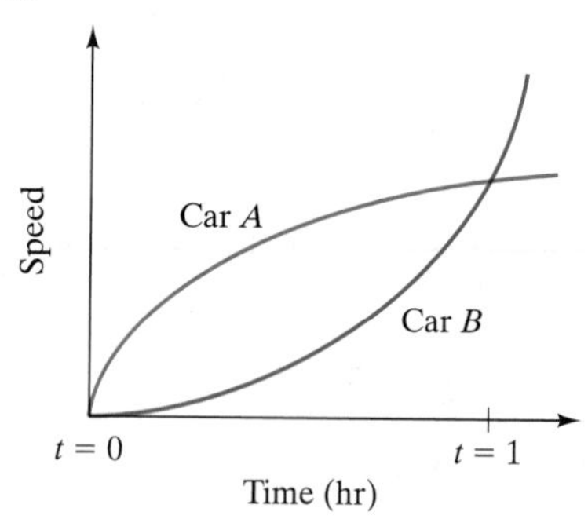

Cumulative Review Exercises

[1.4] **54.** Evaluate $\frac{1}{2}x + \frac{2}{5}xy + \frac{1}{8}y$ when $x = -2$, $y = 5$.

55. Solve $4 - 2[(x - 5) + 2x] = -(x + 6)$

56. Explain how to determine whether a graph is a function.

57. Write an equation of the line that passes through points $(6, -4)$ and $(2, -8)$.

4.4 SOLVING SYSTEMS OF EQUATIONS USING MATRICES

SSM VIDEO 4.4 CD Rom

1. Write an augmented matrix.
2. Solve systems of linear equations.
3. Solve systems of linear equations in three variables.
4. Recognize inconsistent and dependent systems.

1 Write an Augmented Matrix

A **matrix** is a rectangular array of numbers within brackets. The plural of *matrix* is **matrices**. Examples of matrices are

$$\begin{bmatrix} 4 & 6 \\ 9 & -2 \end{bmatrix} \qquad \begin{bmatrix} 5 & 7 & 2 \\ -1 & 3 & 4 \end{bmatrix}$$

The numbers inside the brackets are referred to as **elements** of the matrix.

The matrix on the left contains 2 rows and 2 columns and is called a 2 by 2 (2×2) matrix. The matrix on the right contains 2 rows and 3 columns and is a 2 by 3 (2×3) matrix. The number of rows is the first dimension given, and the number of columns is the second dimension given when describing the dimensions of a matrix. A **square matrix** has the same number of rows as columns. Thus, the matrix on the left is a square matrix.

In this section we will use matrices to solve systems of linear equations. The first step in solving a system of two linear equations using matrices is to write each equation in the form $ax + by = c$. The next step is to write the **augmented matrix**, which is made up of two smaller matrices separated by a vertical line. The numbers on the left of the vertical line are the coefficients of the variables in the system of equations, and the numbers on the right are the constants. For the system of equations

$$a_1 x + b_1 y = c_1$$
$$a_2 x + b_2 y = c_2$$

the augmented matrix is written

$$\left[\begin{array}{cc|c} a_1 & b_1 & c_1 \\ a_2 & b_2 & c_2 \end{array}\right]$$

Following is a system of equations and its augmented matrix.

System of Equations **Augmented Matrix**

$$-x + \frac{1}{2}y = 4 \qquad\qquad \left[\begin{array}{cc|c} -1 & \frac{1}{2} & 4 \\ -3 & -5 & -\frac{1}{2} \end{array}\right]$$
$$-3x - 5y = -\frac{1}{2}$$

Notice that the bar in the augmented matrix separates the numerical coefficients from the constants. Since the matrix is just a shortened way of writing the system of equations, we can solve a linear system using matrices in a manner very similar to solving a system of equations using the addition method.

2) Solve Systems of Linear Equations

To solve a system of two linear equations using matrices, we rewrite the augmented matrix in **triangular form**,

$$\left[\begin{array}{cc|c} 1 & a & p \\ 0 & 1 & q \end{array}\right]$$

where the a, p and q are constants. From this type of augmented matrix we can write an equivalent system of equations. This matrix represents the linear system

$$\begin{aligned} 1x + ay &= p \\ 0x + 1y &= q \end{aligned} \qquad \text{or} \qquad \begin{aligned} x + ay &= p \\ y &= q \end{aligned}$$

For example,

$$\left[\begin{array}{cc|c} 1 & 3 & 4 \\ 0 & 1 & 2 \end{array}\right] \quad \text{represents} \quad \begin{aligned} x + 3y &= 4 \\ y &= 2 \end{aligned}$$

Note that the system above on the right can be easily solved by substitution. Its solution is $(-2, 2)$.

We use **row transformations** to rewrite the augmented matrix in triangular form. We will use two row transformation procedures.

Procedures for Row Transformations

1. All the numbers in a row may be multiplied (or divided) by any nonzero real number. (This is the same as multiplying both sides of an equation by a nonzero real number.)
2. All the numbers in a row may be multiplied by any nonzero real number. These products may then be added to the corresponding numbers in any other row. (This is equivalent to eliminating a variable from a system of equations using the addition method.)

Generally, when changing an element in the augmented matrix to 1 we use row transformation procedure 1, and when changing an element to 0 we use row transformation procedure 2. *Work by columns starting from the left.* Start with the first column, first row.

EXAMPLE 1 Solve the following system of equations using matrices.

$$2x - 3y = 10$$
$$2x + 2y = 5$$

Solution First we write the augmented matrix.

$$\left[\begin{array}{cc|c} 2 & -3 & 10 \\ 2 & 2 & 5 \end{array}\right]$$

Our goal is to obtain a matrix of the form $\left[\begin{array}{cc|c} 1 & a & p \\ 0 & 1 & q \end{array}\right]$. We begin by using row transformation procedure 1 to change the 2 in the first column, first row, to 1. To do so, we multiply the first row of numbers by $\frac{1}{2}$. (We abbreviate this multiplication as $\frac{1}{2}R_1$ and place it to the right of the matrix in the same row where the operation *was* performed. This may help you follow the process more clearly.)

$$\left[\begin{array}{cc|c} 2(\frac{1}{2}) & -3(\frac{1}{2}) & 10(\frac{1}{2}) \\ 2 & 2 & 5 \end{array}\right] \quad \frac{1}{2}R_1$$

This gives

$$\left[\begin{array}{cc|c} 1 & -\frac{3}{2} & 5 \\ 2 & 2 & 5 \end{array}\right]$$

The next step is to obtain 0 in the first column, second row. At present 2 is in this position. We do this by multiplying the numbers in row one by −2, and adding the products to the numbers in row two. (This is abbreviated $-2R_1 + R_2$.)

The numbers in the first row multiplied by −2 are

$$1(-2) \quad -\frac{3}{2}(-2) \quad 5(-2)$$

Now we add these products to their respective numbers in the second row. This gives

$$\left[\begin{array}{cc|c} 1 & -\frac{3}{2} & 5 \\ 2 + 1(-2) & 2 + \left(-\frac{3}{2}\right)(-2) & 5 + 5(-2) \end{array}\right] \quad -2R_1 + R_2$$

Now we have

$$\left[\begin{array}{cc|c} 1 & -\frac{3}{2} & 5 \\ 0 & 5 & -5 \end{array}\right]$$

To obtain 1 in the second column, second row, we multiply the second row of numbers by $\frac{1}{5}$.

$$\left[\begin{array}{cc|c} 1 & -\frac{3}{2} & 5 \\ 0\left(\frac{1}{5}\right) & 5\left(\frac{1}{5}\right) & -5\left(\frac{1}{5}\right) \end{array}\right] \quad \frac{1}{5}R_2$$

$$\left[\begin{array}{cc|c} 1 & -\frac{3}{2} & 5 \\ 0 & 1 & -1 \end{array}\right]$$

The matrix is now in the form we are seeking. The equivalent triangular system of equations is

$$x - \frac{3}{2}y = 5$$

$$y = -1$$

Now we can solve for x using substitution.

$$x - \frac{3}{2}y = 5$$

$$x - \frac{3}{2}(-1) = 5$$

$$x + \frac{3}{2} = 5$$

$$x = \frac{7}{2}$$

NOW TRY EXERCISE 17 A check will show that the solution to the system is $\left(\frac{7}{2}, -1\right)$.

3 Solve Systems of Linear Equations in Three Variables

Now we will use matrices to solve a system of three linear equations in three variables. We use the same row transformation procedures used when solving a system of two linear equations. Our goal is to obtain an augmented matrix in the triangular form

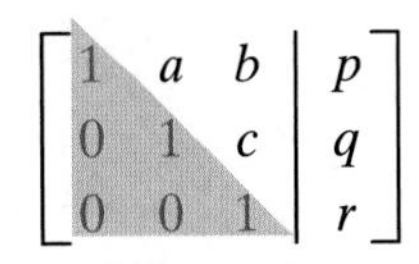

$$\left[\begin{array}{ccc|c} 1 & a & b & p \\ 0 & 1 & c & q \\ 0 & 0 & 1 & r \end{array}\right]$$

where p, q, and r are constants. This matrix represents the following system of equations.

$$\begin{aligned} 1x + ay + bz &= p \\ 0x + 1y + cz &= q \\ 0x + 0y + 1z &= r \end{aligned} \qquad \text{or} \qquad \begin{aligned} x + ay + bz &= p \\ y + cz &= q \\ z &= r \end{aligned}$$

When constructing the augmented matrix, *work by columns, from the left-hand column to the right-hand column. Always complete one column before moving to the next column. In each column, first obtain the 1 in the indicated position, then obtain the zeros*. Example 2 illustrates this procedure.

EXAMPLE 2 Use matrices to solve the following system of equations.

$$\begin{aligned} x + 2y + z &= 0 \\ 2x - y + 2z &= 10 \\ x + 3y - 3z &= -14 \end{aligned}$$

Solution First write the augmented matrix.

$$\left[\begin{array}{ccc|c} 1 & 2 & 1 & 0 \\ 2 & -1 & 2 & 10 \\ 1 & 3 & -3 & -14 \end{array}\right]$$

Our next step is to use row transformations to change the first column to $\begin{smallmatrix} 1 \\ 0 \\ 0 \end{smallmatrix}$. Since the number in the first column, first row is already a 1, we will work with the 2 in the first column, second row. Multiplying the numbers in the first row by -2 and adding those products to the respective numbers in the second row will result in the 2 changing to 0. The matrix is now

$$\left[\begin{array}{ccc|c} 1 & 2 & 1 & 0 \\ 0 & -5 & 0 & 10 \\ 1 & 3 & -3 & -14 \end{array}\right] \quad -2R_1 + R_2$$

Continuing down the first column, we now change the 1 in the third row to 0. By multiplying the numbers in the first row by -1 and adding the products to the third row, we get

$$\left[\begin{array}{ccc|c} 1 & 2 & 1 & 0 \\ 0 & -5 & 0 & 10 \\ 0 & 1 & -4 & -14 \end{array}\right] \quad -1R_1 + R_3$$

Now we work with the second column. We wish to change the numbers in the second column to the form $\begin{smallmatrix} a \\ 1 \\ 0 \end{smallmatrix}$ where a represents a number. We start by changing the -5 in the second row to 1 by multiplying the numbers in the second row by $-\frac{1}{5}$. This gives

$$\left[\begin{array}{ccc|c} 1 & 2 & 1 & 0 \\ 0 & 1 & 0 & -2 \\ 0 & 1 & -4 & -14 \end{array}\right] \quad -\tfrac{1}{5}R_2$$

Continuing down the second column, we now change the 1 in the third row to 0 by multiplying the numbers in the second row by -1 and adding those products to the third row. This gives

$$\left[\begin{array}{ccc|c} 1 & 2 & 1 & 0 \\ 0 & 1 & 0 & -2 \\ 0 & 0 & -4 & -12 \end{array}\right] \quad -1R_2 + R_3$$

Now we work with the third column. We wish to change the numbers in the third column to the form $\begin{matrix} b \\ c \\ 1 \end{matrix}$ where b and c represent numbers. We must change the -4 in the third row to 1. We can do this by multiplying the numbers in the third row by $-\frac{1}{4}$. This results in

$$\left[\begin{array}{ccc|c} 1 & 2 & 1 & 0 \\ 0 & 1 & 0 & -2 \\ 0 & 0 & 1 & 3 \end{array}\right] \quad -\tfrac{1}{4}R_3$$

This matrix is now in the desired form. From this matrix we obtain the system of equations

$$\begin{aligned} x + 2y + z &= 0 \\ y + 0z &= -2 \\ z &= 3 \end{aligned}$$

From the second equation we see that $y = -2$. Now we solve for x by substituting -2 for y and 3 for z.

$$\begin{aligned} x + 2y + z &= 0 \\ x + 2(-2) + 3 &= 0 \\ x - 1 &= 0 \\ x &= 1 \end{aligned}$$

NOW TRY EXERCISE 31 The solution is $(1, -2, 3)$. Check this solution.

4 Recognize Inconsistent and Dependent Systems

When solving a system of two equations, if you obtain an augmented matrix in which one row of numbers on the left side of the vertical line are all zeros but a zero does not appear in the same row on the right side of the vertical line, the system is inconsistent and has no solution. For example, a system of equations that yields the following augmented matrix is an inconsistent system.

$$\left[\begin{array}{cc|c} 1 & 2 & 5 \\ 0 & 0 & 4 \end{array}\right] \longleftarrow \text{Inconsistent}$$

The second row of the matrix represents the equation

$$0x + 0y = 4$$

which is never true.

If you obtain a matrix in which a 0 appears across an entire row, the system of equations is dependent. For example, a system of equations that yields the following augmented matrix is a dependent system.

$$\left[\begin{array}{cc|c} 1 & -3 & -2 \\ 0 & 0 & 0 \end{array}\right] \longleftarrow \text{Dependent}$$

The second row of the matrix represents the equation

$$0x + 0y = 0$$

which is always true.

Similar rules hold for systems with three equations.

$$\left[\begin{array}{ccc|c} 1 & 2 & 4 & 5 \\ 0 & 0 & 0 & -1 \\ 0 & 1 & -2 & 3 \end{array}\right] \longleftarrow \text{Inconsistent system}$$

$$\left[\begin{array}{ccc|c} 1 & 3 & -1 & 2 \\ 0 & 0 & 0 & 0 \\ 0 & 4 & 1 & -3 \end{array}\right] \longleftarrow \text{Dependent system}$$

NOW TRY EXERCISE 25

Using Your Graphing Calculator

Many graphing calculators have the ability to work with matrices. Such calculators have the ability to perform row operations on matrices. These graphing calculators can therefore be used to solve systems of equations using matrices.

Read the instruction manual that came with your graphing calculator to see if it can handle matrices. If so, learn how to use your graphing calculator to solve systems of equations using matrices.

Exercise Set 4.4

Concept/Writing Exercises

1. What is a square matrix?
2. Explain how to construct an augmented matrix.
3. If you obtain the following augmented matrix when solving a system of equations, what would be your next step in completing the process? Explain.

$$\left[\begin{array}{cc|c} 1 & 3 & 6 \\ 0 & -1 & 4 \end{array}\right]$$

4. If you obtained the following augmented matrix when solving a system of equations, what would be your next step in completing the process? Explain your answer.

$$\left[\begin{array}{ccc|c} 1 & 3 & 7 & -1 \\ 0 & -1 & 5 & 3 \\ 2 & 4 & 6 & 8 \end{array}\right]$$

5. When solving a system of equations using matrices, how will you know if the system is **a)** dependent, **b)** inconsistent?
6. When solving a system of linear equations by matrices, if two rows are identical, will the system be consistent, dependent, or inconsistent?

Practice the Skills

Perform each row transformation indicated and write the new matrix.

7. $\left[\begin{array}{cc|c} -8 & 4 & 10 \\ 3 & 5 & -1 \end{array}\right]$ multiply numbers in the first row by $-\frac{1}{8}$

8. $\left[\begin{array}{cc|c} 10 & -3 & 12 \\ 5 & -8 & -\frac{1}{2} \end{array}\right]$ multiply numbers in the second row by $-\frac{1}{8}$

9. $\left[\begin{array}{ccc|c} 4 & 0 & 3 & 8 \\ 5 & -7 & 2 & 14 \\ -1 & 3 & 5 & 12 \end{array}\right]$ multiply numbers in the second row by $-\frac{1}{7}$

10. $\left[\begin{array}{ccc|c} 1 & 4 & -6 & 10 \\ 0 & 2 & 4 & -4 \\ 3 & -5 & 2 & -6 \end{array}\right]$ multiply numbers in the third row by $\frac{1}{2}$

11. $\left[\begin{array}{cc|c} 1 & 3 & 12 \\ -3 & 8 & -6 \end{array}\right]$ multiply numbers in the first row by 3 and add the product to the second row

12. $\left[\begin{array}{cc|c} 1 & 5 & 6 \\ \frac{1}{2} & 10 & -4 \end{array}\right]$ multiply numbers in the first row by $-\frac{1}{2}$ and add the product to the second row

13. $\left[\begin{array}{ccc|c} 1 & 0 & 8 & \frac{1}{4} \\ 5 & 2 & 2 & -2 \\ 6 & -3 & 1 & 0 \end{array}\right]$ multiply numbers in the first row by -5 and add the product to the second row

14. $\left[\begin{array}{ccc|c} 1 & 2 & -1 & 6 \\ 0 & 1 & 5 & 0 \\ 0 & 0 & 2 & 4 \end{array}\right]$ multiply numbers in the second row by -2 and add the product to the first row

Solve each system using matrices.

15. $x + 3y = 3$
$-x + y = -3$

16. $2x - y = -6$
$3x + y = 1$

17. $x - 4y = -1$
$3x - 2y = 7$

18. $4x - y = 25$
$2x - 2y = 20$

19. $-3a + 6b = 3$
$4a - 2b = -1$

20. $-3x + 5y = -22$
$4x - 2y = 20$

21. $2x - 5y = -6$
$-4x + 10y = 12$

22. $-x - 5y = -10$
$2x - 3y = 7$

23. $12x + 10y = -14$
$4x - 3y = -11$

24. $4r + 2s = -10$
$-2r + s = -7$

25. $-3x + 6y = 5$
$2x - 4y = 8$

26. $6x - 3y = 9$
$-2x + y = -3$

27. $9x - 8y = 4$
$-3x + 4y = -1$

28. $2x - 3y = 3$
$-3x + 9y = -3$

29. $10m = 8n + 15$
$16n = -15m - 2$

30. $8x = 9y + 4$
$24x + 6y = 1$

Solve each system using matrices.

31. $x + y - 3z = -1$
$2x + y - z = 3$
$-x + 2y - z = -3$

32. $x - 2y + 3z = -2$
$2x - y - z = 0$
$3x + 2y - 3z = -2$

33. $x + 2y = 5$
$y - z = -1$
$2x - 3z = 0$

34. $4a + 3b = 10$
$2a - b = 10$
$-2a + c = -9$

35. $2x + 12y + 4z = 0$
$-4x + 9y - z = 11$
$-x - 12y + 3z = -2$

36. $3x - 5y + 2z = 8$
$-x - y - z = -3$
$3x - 2y + 4z = 10$

37. $6x - 2y + 8z = 26$
$-3x + y - 4z = -13$
$x + 2y + 3z = 14$

38. $6x - 10y + 2z = -4$
$2x - 15y - z = -4$
$x + 5y + 3z = 6$

39. $4p - q + r = 4$
$-6p + 3q - 2r = -5$
$2p + 5q - r = 7$

40. $-4r + 3s - 6t = 14$
$4r + 2s - 2t = -3$
$2r - 5s - 8t = -23$

41. $2x - 4y + 3z = -12$
$3x - y + 2z = -3$
$-4x + 8y - 6z = 10$

42. $3x - 2y + z = -1$
$12x - 10y - 3z = 2$
$-9x + 8y - 4z = 5$

43. $5x - 3y + 4z = 22$
$-x - 15y + 10z = -15$
$-3x + 9y - 12z = -6$

44. $9x - 4y + 5z = -2$
$-9x + 5y - 10z = -1$
$9x + 3y + 10z = 1$

Problem Solving

45. When solving a system of linear equations using matrices, if two rows of matrices are switched, will the solution to the system change? Explain.

46. You can tell whether a system of two equations in two variables is consistent, dependent, or inconsistent by comparing the slopes and y-intercepts of the graphs of the equations. Can you tell, without solving, if a system of three equations in three variables is consistent, dependent, or inconsistent? Explain.

Solve using matrices.

47. In a triangular cross section of a roof, the largest angle is 55° greater than the smallest angle. The largest angle is 20° greater than the remaining angle. Find the measure of each angle.

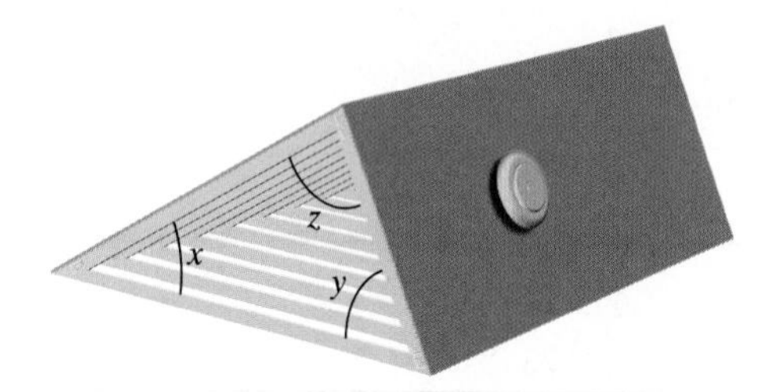

48. A right angle is divided into three smaller angles. The largest of the three angles is twice the smallest. The remaining angle is 10° greater than the smallest angle. Find the measure of each angle.

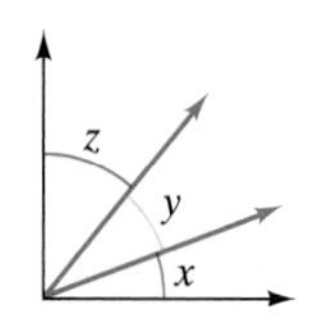

49. Sixty-five percent of the world's bananas are controlled by Chiquita, Dole, or Del Monte (all American companies). Chiquita, the largest, controls 12% more bananas than Del Monte. Dole, the second largest, controls 3% less than twice the percent that Del Monte controls. Determine the percents to be placed in each sector of the circle graph shown.

World's Bananas

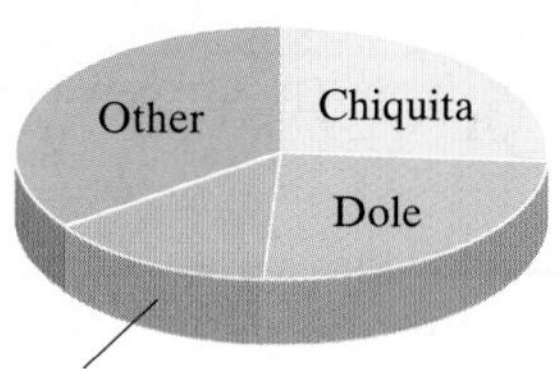

50. The following graph shows that in 1996 Netscape and Microsoft controlled major shares of the Web browser market. Netscape controlled 25% more than Microsoft and Microsoft controlled twice the amount of all others. Determine the percent to be placed in each sector of the circle graph shown.

Web Browser Market

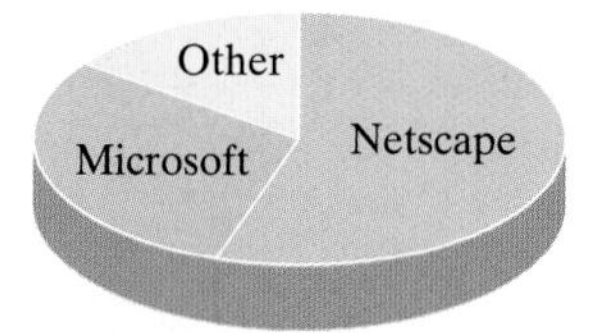

Cumulative Review Exercises

[1.2] **51.** $A = \{1, 2, 4, 6, 9\}$; $B = \{3, 4, 5, 6, 10\}$. Find **a)** $A \cup B$; **b)** $A \cap B$.

[2.5] **52.** Indicate the inequality $-2 < x \leq 4$ **a)** on a number line, **b)** as a solution set, and **c)** in interval notation.

[3.2] **53.** What does a graph represent?

[3.4] **54.** If $f(x) = -2x^2 + 4x - 6$, find $f(-5)$.

4.5 SOLVING SYSTEMS OF EQUATIONS USING DETERMINANTS AND CRAMER'S RULE

SSM VIDEO 4.5 CD Rom

1) Evaluate a determinant of a 2 × 2 matrix.
2) Use Cramer's rule.
3) Evaluate a determinant of a 3 × 3 matrix.
4) Use Cramer's rule with systems in three variables.

1) Evaluate a Determinant of a 2 x 2 Matrix

We have discussed various ways of solving a system of linear equations, including: graphing, substitution, the addition (or elimination) method, and matrices. A system of linear equations may also be solved using determinants.

Associated with every square matrix is a number called its **determinant**. For a 2×2 matrix, its determinant is defined as follows.

Definition

The **determinant** of a 2×2 matrix $\begin{bmatrix} a_1 & b_1 \\ a_2 & b_2 \end{bmatrix}$ is denoted $\begin{vmatrix} a_1 & b_1 \\ a_2 & b_2 \end{vmatrix}$ and is evaluated as

$$\begin{vmatrix} a_1 & b_1 \\ a_2 & b_2 \end{vmatrix} = a_1 b_2 - a_2 b_1$$

EXAMPLE 1 Evaluate each determinant.

a) $\begin{vmatrix} 4 & 6 \\ -3 & 2 \end{vmatrix}$ **b)** $\begin{vmatrix} -3 & 4 \\ 1 & 5 \end{vmatrix}$

Solution **a)** $a_1 = 4,\ a_2 = -3,\ b_1 = 6,\ b_2 = 2$

$$\begin{vmatrix} 4 & 6 \\ -3 & 2 \end{vmatrix} = 4(2) - (-3)(6) = 8 + 18 = 26$$

b) $$\begin{vmatrix} -3 & 4 \\ 1 & 5 \end{vmatrix} = (-3)(5) - (1)(4) = -15 - 4 = -19$$

The value of the determinant in **a)** is 26 and the value of the determinant in **b)** is -19.

NOW TRY EXERCISE 5

2 Use Cramer's Rule

If we begin with the equations

$$a_1 x + b_1 y = c_1$$

$$a_2 x + b_2 y = c_2$$

we can use the addition method to show that

$$x = \frac{c_1 b_2 - c_2 b_1}{a_1 b_2 - a_2 b_1} \quad \text{and} \quad y = \frac{a_1 c_2 - a_2 c_1}{a_1 b_2 - a_2 b_1}$$

(see Challenge Problem 55). Notice that the *denominators* of x and y are both $a_1 b_2 - a_2 b_1$. Following is the determinant that yields this denominator. We have labeled this denominator D.

$$D = \begin{vmatrix} a_1 & b_1 \\ a_2 & b_2 \end{vmatrix} = a_1 b_2 - a_2 b_1$$

The *numerators* of x and y are different. Following are two determinants, labeled D_x and D_y, that yield the numerators of x and y.

$$D_x = \begin{vmatrix} c_1 & b_1 \\ c_2 & b_2 \end{vmatrix} = c_1 b_2 - c_2 b_1 \qquad D_y = \begin{vmatrix} a_1 & c_1 \\ a_2 & c_2 \end{vmatrix} = a_1 c_2 - a_2 c_1$$

We use determinants D, D_x, and D_y in Cramer's rule. **Cramer's rule** can be used to solve systems of linear equations.

Cramer's Rule for Systems of Linear Equations

For a system of linear equations of the form

$$a_1 x + b_1 y = c_1$$
$$a_2 x + b_2 y = c_2$$

$$x = \frac{\begin{vmatrix} c_1 & b_1 \\ c_2 & b_2 \end{vmatrix}}{\begin{vmatrix} a_1 & b_1 \\ a_2 & b_2 \end{vmatrix}} = \frac{D_x}{D} \quad \text{and} \quad y = \frac{\begin{vmatrix} a_1 & c_1 \\ a_2 & c_2 \end{vmatrix}}{\begin{vmatrix} a_1 & b_1 \\ a_2 & b_2 \end{vmatrix}} = \frac{D_y}{D}, \quad D \neq 0$$

HELPFUL HINT

The elements in determinant D are the numerical coefficients of the x and y terms in the two given equations, listed in the same order they are listed in the equations. To obtain the determinant D_x from determinant D, replace the coefficients of the x terms (the values in the first column) with the constants of the two given equations. To obtain the determinant D_y from determinant D, replace the coefficients of the y terms (the values in the second column) with the constants of the two given equations.

EXAMPLE 2 Use Cramer's rule to solve the following system.

$$2x - 4y = 8$$
$$3x + 5y = -10$$

Solution Both equations are given in the desired form, $ax + by = c$. When labeling a, b, and c, we will refer to $2x - 4y = 8$ as equation 1 and $3x + 5y = -10$ as equation 2 (in the subscripts).

$$\begin{array}{c} a_1 \quad b_1 \quad c_1 \\ \downarrow \quad \downarrow \quad \downarrow \\ 2x - 4y = 8 \\ 3x + 5y = -10 \\ \uparrow \quad \uparrow \quad \uparrow \\ a_2 \quad b_2 \quad c_2 \end{array}$$

We now find D, D_x, and D_y.

$$D = \begin{vmatrix} a_1 & b_1 \\ a_2 & b_2 \end{vmatrix} = \begin{vmatrix} 2 & -4 \\ 3 & 5 \end{vmatrix} = 2(5) - 3(-4) = 22$$

$$D_x = \begin{vmatrix} c_1 & b_1 \\ c_2 & b_2 \end{vmatrix} = \begin{vmatrix} 8 & -4 \\ -10 & 5 \end{vmatrix} = 8(5) - (-10)(-4) = 0$$

$$D_y = \begin{vmatrix} a_1 & c_1 \\ a_2 & c_2 \end{vmatrix} = \begin{vmatrix} 2 & 8 \\ 3 & -10 \end{vmatrix} = 2(-10) - (3)(8) = -44$$

Now we find the value of x and y.

$$x = \frac{D_x}{D} = \frac{0}{22} = 0$$

$$y = \frac{D_y}{D} = \frac{-44}{22} = -2$$

Thus, the solution is $x = 0$, $y = -2$ or the ordered pair $(0, -2)$. A check will show that this ordered pair satisfies both equations.

NOW TRY EXERCISE 19

When determinant $D = 0$, Cramer's rule does not apply since division by 0 is not possible. You may then use a different method to solve the system. Or you may evaluate D_x and D_y.

> If $D = 0$, $D_x = 0$, $D_y = 0$, then the system is dependent.
> If $D = 0$ and either $D_x \neq 0$ or $D_y \neq 0$, then the system is inconsistent.

3 Evaluate a Determinant of a 3 x 3 Matrix

For the determinant

$$\begin{vmatrix} a_1 & b_1 & c_1 \\ a_2 & b_2 & c_2 \\ a_3 & b_3 & c_3 \end{vmatrix}$$

the **minor determinant** of a_1 is found by crossing out the elements in the same row and column in which the element a_1 appears. The remaining elements form the minor determinant of a_1. The minor determinants of other elements are found similarly.

$$\begin{vmatrix} \not{a_1} & \not{b_1} & \not{c_1} \\ \not{a_2} & b_2 & c_2 \\ \not{a_3} & b_3 & c_3 \end{vmatrix} \qquad \begin{vmatrix} b_2 & c_2 \\ b_3 & c_3 \end{vmatrix} \qquad \text{Minor determinant of } a_1$$

$$\begin{vmatrix} \not{a_1} & b_1 & c_1 \\ \not{a_2} & \not{b_2} & \not{c_2} \\ \not{a_3} & b_3 & c_3 \end{vmatrix} \qquad \begin{vmatrix} b_1 & c_1 \\ b_3 & c_3 \end{vmatrix} \qquad \text{Minor determinant of } a_2$$

$$\begin{vmatrix} \not{a_1} & b_1 & c_1 \\ \not{a_2} & b_2 & c_2 \\ \not{a_3} & \not{b_3} & \not{c_3} \end{vmatrix} \qquad \begin{vmatrix} b_1 & c_1 \\ b_2 & c_2 \end{vmatrix} \qquad \text{Minor determinant of } a_3$$

To evaluate determinants of a 3×3 matrix, we use minor determinants. The following box shows how such a determinant may be evaluated by **expansion of the minors of the first column.**

Expansion of the Determinant by the Minors of the First Column

Minor determinant of a_1 ↓   Minor determinant of a_2 ↓   Minor determinant of a_3 ↓

$$\begin{vmatrix} a_1 & b_1 & c_1 \\ a_2 & b_2 & c_2 \\ a_3 & b_3 & c_3 \end{vmatrix} = a_1\begin{vmatrix} b_2 & c_2 \\ b_3 & c_3 \end{vmatrix} - a_2\begin{vmatrix} b_1 & c_1 \\ b_3 & c_3 \end{vmatrix} + a_3\begin{vmatrix} b_1 & c_1 \\ b_2 & c_2 \end{vmatrix}$$

EXAMPLE 3 Evaluate $\begin{vmatrix} 4 & -2 & 6 \\ 3 & 5 & 0 \\ 1 & -3 & -1 \end{vmatrix}$ using expansion by the minors of the first column.

Solution We will follow the procedure given in the box.

$$\begin{vmatrix} 4 & -2 & 6 \\ 3 & 5 & 0 \\ 1 & -3 & -1 \end{vmatrix} = 4\begin{vmatrix} 5 & 0 \\ -3 & -1 \end{vmatrix} - 3\begin{vmatrix} -2 & 6 \\ -3 & -1 \end{vmatrix} + 1\begin{vmatrix} -2 & 6 \\ 5 & 0 \end{vmatrix}$$

$$= 4[5(-1) - (-3)0] - 3[(-2)(-1) - (-3)6] + 1[(-2)0 - 5(6)]$$

$$= 4(-5 + 0) - 3(2 + 18) + 1(0 - 30)$$

$$= 4(-5) - 3(20) + 1(-30)$$

$$= -20 - 60 - 30$$

$$= -110$$

NOW TRY EXERCISE 11 The determinant has a value of −110.

4) Use Cramer's Rule with Systems in Three Variables

Cramer's rule can be extended to systems of equations in three variables as follows.

Cramer's Rule for a System of Equations in Three Variables

To evaluate the system

$$a_1x + b_1y + c_1z = d_1$$
$$a_2x + b_2y + c_2z = d_2$$
$$a_3x + b_3y + c_3z = d_3$$

with

$$D = \begin{vmatrix} a_1 & b_1 & c_1 \\ a_2 & b_2 & c_2 \\ a_3 & b_3 & c_3 \end{vmatrix} \qquad D_x = \begin{vmatrix} d_1 & b_1 & c_1 \\ d_2 & b_2 & c_2 \\ d_3 & b_3 & c_3 \end{vmatrix}$$

$$D_y = \begin{vmatrix} a_1 & d_1 & c_1 \\ a_2 & d_2 & c_2 \\ a_3 & d_3 & c_3 \end{vmatrix} \qquad D_z = \begin{vmatrix} a_1 & b_1 & d_1 \\ a_2 & b_2 & d_2 \\ a_3 & b_3 & d_3 \end{vmatrix}$$

then

$$x = \frac{D_x}{D} \qquad y = \frac{D_y}{D} \qquad z = \frac{D_z}{D}, \qquad D \neq 0$$

Note that the denominators of the expressions for x, y, and z are all the same determinant, D. Note that the d's replace the a's, the numerical coefficients of the x-terms, in D_x. The d's replace the b's, the numerical coefficients of the y-terms, in D_y. And the d's replace the c's, the numerical coefficients of the z-terms, in D_z.

EXAMPLE 4 Solve the following system of equations using determinants.

$$3x - 2y - z = -6$$
$$2x + 3y - 2z = 1$$
$$x - 4y + z = -3$$

Solution

$$\begin{array}{llll} a_1 = 3 & b_1 = -2 & c_1 = -1 & d_1 = -6 \\ a_2 = 2 & b_2 = 3 & c_2 = -2 & d_2 = 1 \\ a_3 = 1 & b_3 = -4 & c_3 = 1 & d_3 = -3 \end{array}$$

We will use expansion of the minor determinants of the first column to evaluate D, D_x, D_y, and D_z.

$$D = \begin{vmatrix} 3 & -2 & -1 \\ 2 & 3 & -2 \\ 1 & -4 & 1 \end{vmatrix} = 3\begin{vmatrix} 3 & -2 \\ -4 & 1 \end{vmatrix} - 2\begin{vmatrix} -2 & -1 \\ -4 & 1 \end{vmatrix} + 1\begin{vmatrix} -2 & -1 \\ 3 & -2 \end{vmatrix}$$

$$= 3(-5) - 2(-6) + 1(7)$$

$$= -15 + 12 + 7 = 4$$

$$D_x = \begin{vmatrix} -6 & -2 & -1 \\ 1 & 3 & -2 \\ -3 & -4 & 1 \end{vmatrix} = -6\begin{vmatrix} 3 & -2 \\ -4 & 1 \end{vmatrix} - 1\begin{vmatrix} -2 & -1 \\ -4 & 1 \end{vmatrix} + (-3)\begin{vmatrix} -2 & -1 \\ 3 & -2 \end{vmatrix}$$

$$= -6(-5) - 1(-6) - 3(7)$$

$$= 30 + 6 - 21 = 15$$

$$D_y = \begin{vmatrix} 3 & -6 & -1 \\ 2 & 1 & -2 \\ 1 & -3 & 1 \end{vmatrix} = 3\begin{vmatrix} 1 & -2 \\ -3 & 1 \end{vmatrix} - 2\begin{vmatrix} -6 & -1 \\ -3 & 1 \end{vmatrix} + 1\begin{vmatrix} -6 & -1 \\ 1 & -2 \end{vmatrix}$$

$$= 3(-5) - 2(-9) + 1(13)$$

$$= -15 + 18 + 13 = 16$$

$$D_z = \begin{vmatrix} 3 & -2 & -6 \\ 2 & 3 & 1 \\ 1 & -4 & -3 \end{vmatrix} = 3\begin{vmatrix} 3 & 1 \\ -4 & -3 \end{vmatrix} - 2\begin{vmatrix} -2 & -6 \\ -4 & -3 \end{vmatrix} + 1\begin{vmatrix} -2 & -6 \\ 3 & 1 \end{vmatrix}$$

$$= 3(-5) - 2(-18) + 1(16)$$

$$= -15 + 36 + 16 = 37$$

We found that $D = 4, D_x = 15, D_y = 16$, and $D_z = 37$. Therefore,

$$x = \frac{D_x}{D} = \frac{15}{4} \qquad y = \frac{D_y}{D} = \frac{16}{4} = 4 \qquad z = \frac{D_z}{D} = \frac{37}{4}$$

The solution to the system is $\left(\frac{15}{4}, 4, \frac{37}{4}\right)$. Note the ordered triple lists x, y, and z in this order.

When we have a system of equations in three variables in which one or more equations are missing a variable, we insert the variable with a coefficient of 0. Thus,

$$\begin{array}{rl} 2x - 3y + 2z &= -1 \\ x + 2y &= 14 \\ x - 3z &= -5 \end{array} \quad \text{is written} \quad \begin{array}{rl} 2x - 3y + 2z &= -1 \\ x + 2y + 0z &= 14 \\ x + 0y - 3z &= -5 \end{array}$$

NOW TRY EXERCISE 25

HELPFUL HINT

When evaluating determinants, if any two rows (or columns) are identical, or identical except for opposite signs, the determinant has a value of 0. For example,

$$\begin{vmatrix} 5 & -2 \\ 5 & -2 \end{vmatrix} = 0 \quad \text{and} \quad \begin{vmatrix} 5 & -2 \\ -5 & 2 \end{vmatrix} = 0$$

$$\begin{vmatrix} 5 & -3 & 4 \\ 2 & 6 & 5 \\ 5 & -3 & 4 \end{vmatrix} = 0 \quad \text{and} \quad \begin{vmatrix} 5 & -3 & 4 \\ -5 & 3 & -4 \\ 6 & 8 & 2 \end{vmatrix} = 0$$

As with determinants of a 2×2 matrix, when determinant $D = 0$, Cramer's rule does not apply since division by 0 is not defined. You may then use a different method to solve the system. Or you may evaluate D_x, D_y, and D_z.

If $D = 0$, $D_x = 0$, $D_y = 0$, and $D_z = 0$, then the system is dependent.
If $D = 0$ and $D_x \neq 0$, $D_y \neq 0$, or $D_z \neq 0$, then the system is inconsistent.

In the previous section we mentioned that some graphing calculators have the ability to handle matrices. Graphing calculators with matrix capabilities can also find determinants of square matrices. Read your graphing calculator manual to learn if your calculator can find determinants. If so, learn how to do so on your calculator.

Exercise Set 4.5

Concept/Writing Exercises

1. Explain how to evaluate a 2×2 determinant.

2. Explain how to evaluate a 3×3 determinant by expansion of the minors of the first column.

3. Explain how you can determine whether a system of three linear equations is dependent using determinants.

4. Explain how you can determine whether a system of three linear equations is inconsistent using determinants.

Practice the Skills

Evaluate each determinant.

5. $\begin{vmatrix} 5 & 3 \\ -1 & 4 \end{vmatrix}$

6. $\begin{vmatrix} -1 & 3 \\ 5 & 6 \end{vmatrix}$

7. $\begin{vmatrix} \frac{1}{2} & 3 \\ 2 & -4 \end{vmatrix}$

8. $\begin{vmatrix} 5 & -\frac{2}{3} \\ -1 & 0 \end{vmatrix}$

9. $\begin{vmatrix} 3 & 2 & 0 \\ 0 & 5 & 3 \\ -1 & 4 & 2 \end{vmatrix}$

10. $\begin{vmatrix} 5 & -1 & 3 \\ 0 & 4 & 6 \\ 0 & 5 & -2 \end{vmatrix}$

11. $\begin{vmatrix} 2 & 3 & 1 \\ 1 & -3 & -6 \\ -4 & 5 & 9 \end{vmatrix}$

12. $\begin{vmatrix} 5 & -8 & 6 \\ 3 & 0 & 4 \\ -5 & -2 & 1 \end{vmatrix}$

Solve each system of equations using determinants.

13. $x + 2y = 5$
    $x - 2y = 1$

14. $3x - 2y = 4$
    $3x + y = -2$

15. $x - 2y = -1$
    $x + 3y = 9$

16. $2m - n = 1$
    $3m - 3n = 9$

17. $3a + 4b = 8$
    $2a - 3b = 9$

18. $6x + 3y = -4$
    $9x + 5y = -6$

19. $2x = y + 5$
    $6x + 2y = -5$

20. $x + 5y = 3$
    $2x + 10y = 6$

21. $3r = -4s - 6$
    $3s = -5r + 1$

22. $5x - 5y = 3$
$x - y = -2$

23. $6.3x - 4.5y = -9.9$
$-9.1x + 3.2y = -2.2$

24. $-1.1x + 8.3y = 36.5$
$3.5x + 1.6y = -4.1$

Solve each system of equations using determinants.

25. $x + y - z = -2$
$x + z = 5$
$2x - y + 2z = 11$

26. $2x - y + 3z = 0$
$x + 2y - z = 5$
$2y + z = 1$

27. $-x + y = 1$
$y - z = 2$
$x + z = -2$

28. $-x + 2y + 3z = -1$
$-3x - 3y + z = 0$
$2x + 3y + z = 2$

29. $2x + 2y + 2z = 0$
$-x - 3y + 7z = 15$
$3x + y + 4z = 21$

30. $x - 2y + 3z = 4$
$2x - y + z = -5$
$x + y - z = -2$

31. $a - b + 2c = 3$
$a - b + c = 1$
$2a + b + 2c = 2$

32. $2x + y - 2 = 0$
$3x + 2y + z = 3$
$x - 3y - 5z = 5$

33. $a + 2b + c = 1$
$a - b + c = 1$
$2a + b + 2c = 2$

34. $2r + s - 2t = -4$
$r + s + t = 1$
$r + s + 2t = 3$

35. $1.1x + 2.3y - 4.0z = -9.2$
$-2.3x + 4.6z = 6.9$
$-8.2y - 7.5z = -6.8$

36. $4.6y - 2.1z = 24.3$
$-5.6x + 1.8y = -5.8$
$2.8x - 4.7y - 3.1z = 7.0$

37. $x + y + z = 1$
$2x + 2y + 2z = 2$
$3x + 3y + 3z = 3$

38. $x - 2y + z = 2$
$4x - 6y + 2z = 1$
$2x - 3y + z = 0$

39. $4x - 3y + 8z = 12$
$2x - \frac{3}{2}y + 4z = 11$
$x - 5z = -10$

40. $\frac{1}{2}x + y + z = 8$
$2x - y + \frac{1}{2}z = 3$
$x + y + z = 9$

41. $0.2x - 0.1y - 0.3z = -0.1$
$0.2x - 0.1y + 0.1z = -0.9$
$0.1x + 0.2y - 0.4z = 1.7$

42. $0.6u - 0.4v + 0.5w = 3.1$
$0.5u + 0.2v + 0.2w = 1.3$
$0.1u + 0.1v + 0.1w = 0.2$

Problem Solving

43. Given a determinant of the form $\begin{vmatrix} a_1 & b_1 \\ a_2 & b_2 \end{vmatrix}$, how will the value of the determinant change if the a's are switched with the b's, $\begin{vmatrix} b_1 & a_1 \\ b_2 & a_2 \end{vmatrix}$? Explain your answer.

44. Given a determinant of the form $\begin{vmatrix} a_1 & b_1 \\ a_2 & b_2 \end{vmatrix}$, how will the value of the determinant change if the a's are switched with each other and the b's are switched with each other, $\begin{vmatrix} a_2 & b_2 \\ a_1 & b_1 \end{vmatrix}$? Explain your answer.

45. If all the elements in one row or one column of a 2×2 determinant are zero, what is the value of the determinant?

46. In a 2×2 determinant, if the rows are the same, what is the value of the determinant?

47. Given a 3×3 determinant, if all the elements in one row are multiplied by -1, will the value of the determinant change? Explain.

48. If all the elements in one row or one column of a 3×3 determinant are zero, what is the value of the determinant?

49. Given a 3×3 determinant, if the first and second rows are switched, will the value of the determinant change? Explain.

50. In a 3×3 determinant, if any two rows are the same, can you make a generalization about the value of the determinant?

Solve for the given letter.

51. $\begin{vmatrix} 4 & 6 \\ -2 & y \end{vmatrix} = 32$

52. $\begin{vmatrix} b-2 & -4 \\ b+3 & -6 \end{vmatrix} = 14$

53. $\begin{vmatrix} 3 & x & -2 \\ 0 & 5 & -6 \\ -1 & 4 & -7 \end{vmatrix} = -31$

54. $\begin{vmatrix} 4 & 7 & y \\ 3 & -1 & 2 \\ 4 & 1 & 5 \end{vmatrix} = -35$

Challenge Problems

55. Use the addition method to solve the following system for **a)** x and **b)** y.

$$a_1x + b_1y = c_1$$
$$a_2x + b_2y = c_2$$

Cumulative Review Exercises

[2.5] **56.** Solve the inequality $3(x - 2) < \frac{4}{5}(x - 4)$ and indicate the solution in interval notation.

Graph $3x + 4y = 8$ *using the indicated method.*

[3.2] **57.** By plotting points

58. Using the x- and y-intercepts

[3.3] **59.** Using the slope and y-intercept

4.6 SOLVING SYSTEMS OF LINEAR INEQUALITIES

SSM VIDEO 4.6 CD Rom

1. Solve systems of linear inequalities.
2. Solve linear programming problems.
3. Solve systems of linear inequalities containing absolute value.

1 Solve Systems of Linear Inequalities

In Section 3.7 we showed how to graph linear inequalities in two variables. In Section 4.1 we learned how to solve systems of equations graphically. In this section we show how to solve **systems of linear inequalities** graphically.

To Solve a System of Linear Inequalities

Graph each inequality on the same axes. The solution is the set of points whose coordinates satisfy all the inequalities in the system.

EXAMPLE 1 Determine the solution to the following system of inequalities.

$$x + y \leq 6$$
$$y > 2x - 3$$

Solution First graph the inequality $x + y \leq 6$ (Fig. 4.8). Now on the same axes graph the inequality $y > 2x - 3$ (Fig. 4.9). The solution is the set of points common to the graphs of both inequalities. It is the part of the graph that contains both shadings. The dashed line is not part of the solution, but the part of the solid line that satisfies both inequalities is.

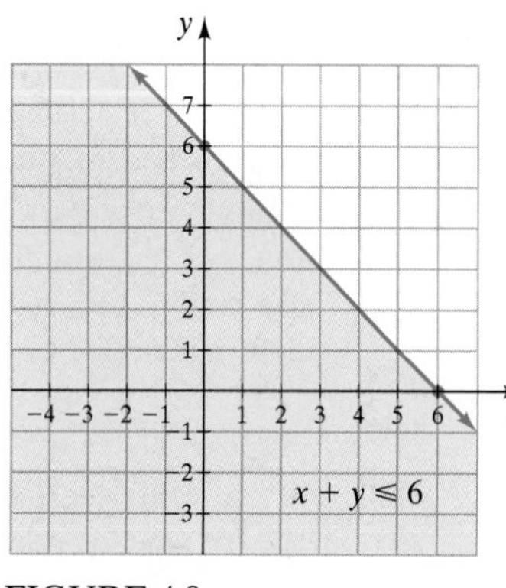

FIGURE 4.8

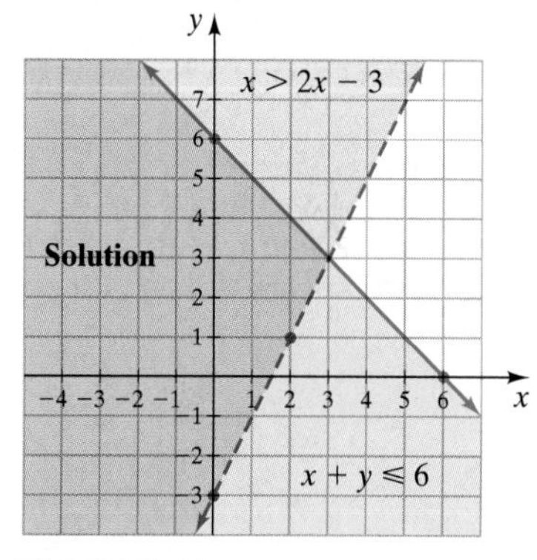

FIGURE 4.9

EXAMPLE 2 Determine the solution to the following system of inequalities.

$$2x + 3y \geq 4$$
$$2x - y > -6$$

Solution Graph $2x + 3y \geq 4$ (see Fig. 4.10). Graph $2x - y > -6$ on the same set of axes (Fig. 4.11). The solution is the part of the graph with both shadings and the part of the solid line that satisfies both inequalities.

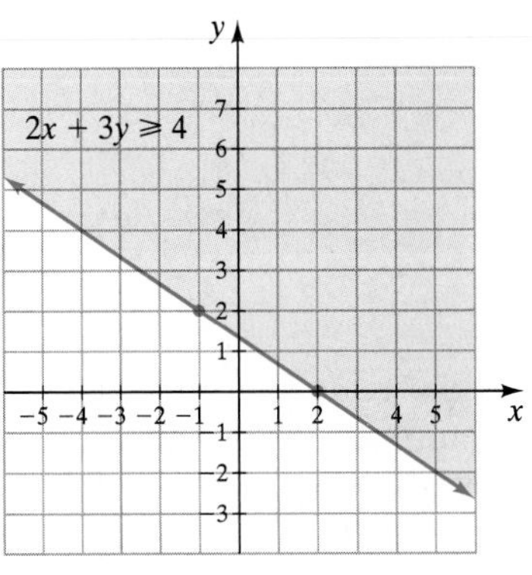

FIGURE 4.10

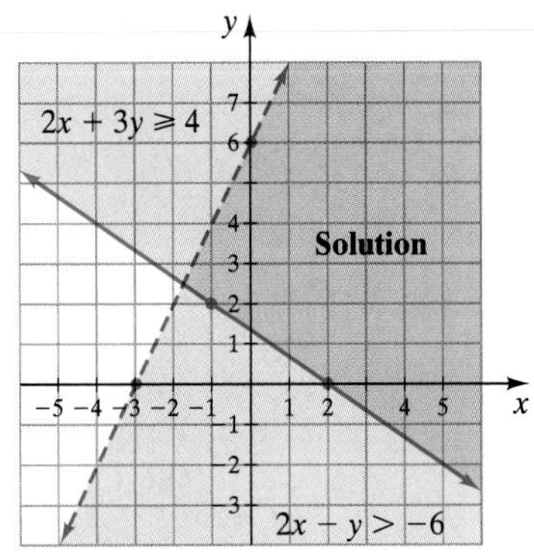

FIGURE 4.11

NOW TRY EXERCISE 9

EXAMPLE 3 Determine the solution to the following system of inequalities.

$$y < 4$$
$$x > -2$$

Solution The solution is illustrated in Figure 4.12.

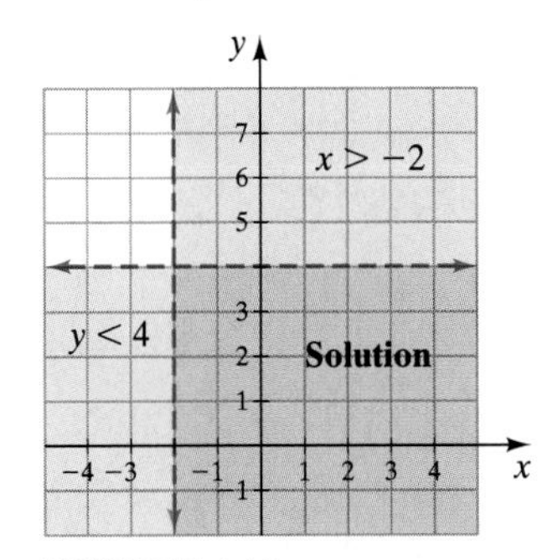

FIGURE 4.12

NOW TRY EXERCISE 13

2 Solve Linear Programming Problems

There is a mathematical process called **linear programming** for which you often have to graph more than two linear inequalities on the same axes. These inequalities are called **constraints**. The following two examples illustrate how to determine the solution to a system of more than two inequalities.

EXAMPLE 4 Determine the solution to the following system of inequalities.

$$x \geq 0$$
$$y \geq 0$$
$$2x + 3y \leq 12$$
$$2x + y \leq 8$$

Solution The first two inequalities, $x \geq 0$ and $y \geq 0$, indicate that the solution must be in the first quadrant because that is the only quadrant where both x and y are positive. Figure 4.13 illustrates the graphs of the four inequalities.

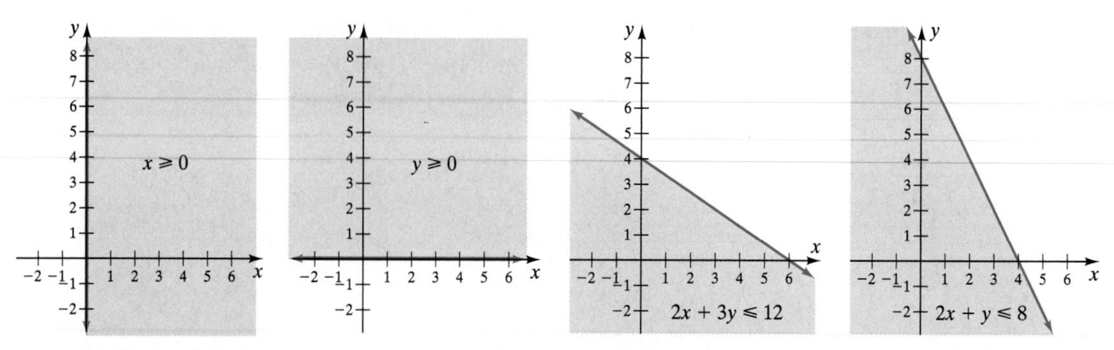

FIGURE 4.13

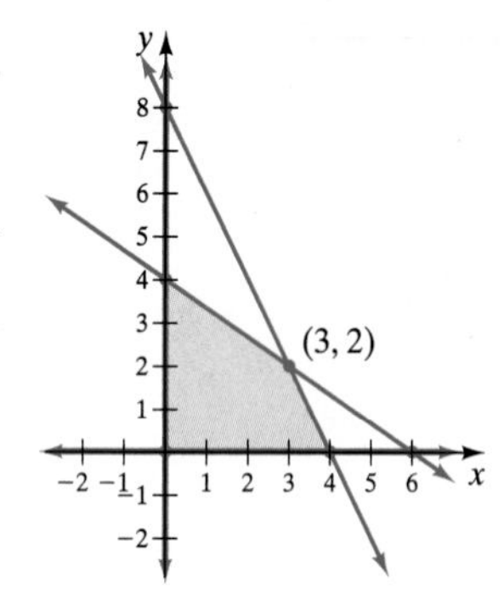

Figure 4.14 illustrates the graphs on the same axes and the solution to the system of inequalities. Note that every point in the shaded area and every point on the lines that form the polygonal region is part of the answer.

FIGURE 4.14

EXAMPLE 5 Determine the solution to the following system of inequalities.

$$\begin{aligned} x &\geq 0 \\ y &\geq 0 \\ x &\leq 15 \\ 8x + 8y &\leq 160 \\ 4x + 12y &\leq 180 \end{aligned}$$

Solution The first two inequalities indicate that the solution must be in the first quadrant. The third inequality indicates that x must be a value less than or equal to 15. Figure 4.15a indicates the graphs of the last three equations. Figure 4.15b indicates the solution to the system of inequalities.

NOW TRY EXERCISE 21

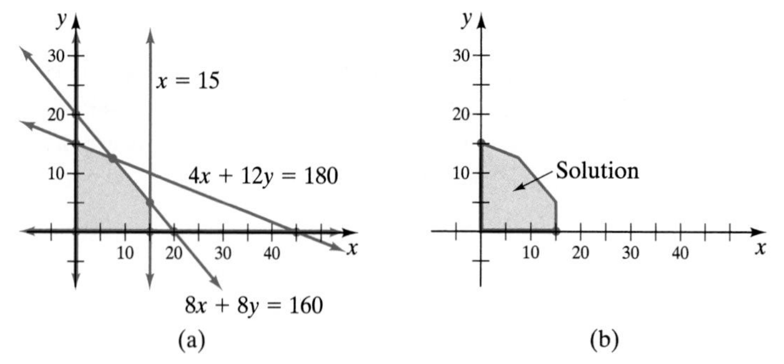

FIGURE 4.15

3) Solve Systems of Linear Inequalities Containing Absolute Value

Now we will graph *systems of linear inequalities containing absolute value* in the Cartesian coordinate system. Before we do some examples, let us recall the rules for absolute value inequalities that we learned in Section 2.6. Recall that

If $|x| < a$ and $a > 0$, then $-a < x < a$.
If $|x| > a$ and $a > 0$, then $x < -a$ or $x > a$.

EXAMPLE 6 Graph $|x| < 3$ in the Cartesian coordinate system.

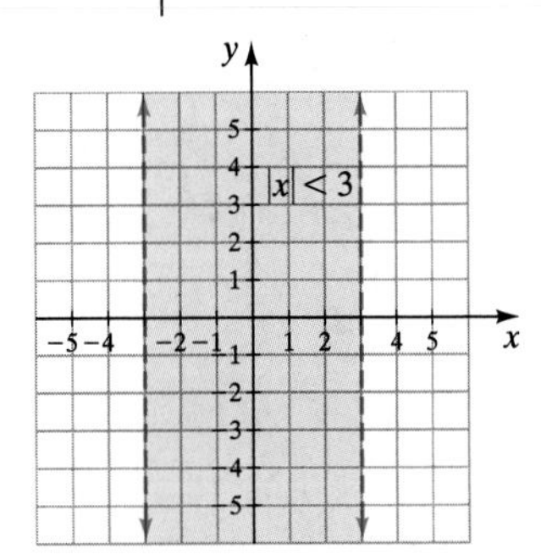

Solution From the rules given, we know that $|x| < 3$ means $-3 < x < 3$. We draw dashed vertical lines through -3 and 3 and shade the area between the two (Fig. 4.16).

FIGURE 4.16

EXAMPLE 7 Graph $|y + 1| > 3$ in the Cartesian coordinate system.

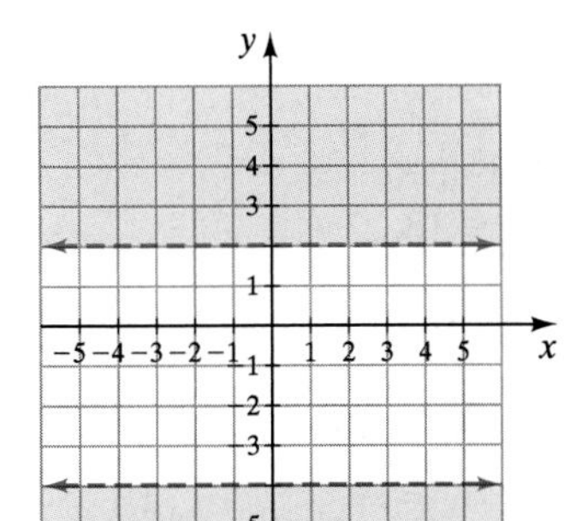

FIGURE 4.17

Solution From the rules given above, we know that $|y + 1| > 3$ means $y + 1 < -3$ or $y + 1 > 3$. First we solve each inequality.

$$y + 1 < -3 \quad \text{or} \quad y + 1 > 3$$

$$y < -4 \qquad\qquad y > 2$$

Now we graph both inequalities and take the *union* of the two graphs. The solution is the shaded area in Figure 4.17.

EXAMPLE 8 Graph the following system of inequalities.

$$|x| < 3$$
$$|y + 1| > 3$$

Solution We draw both inequalities on the same axes. Therefore, we combine the graph drawn in Example 6 with the graph drawn in Example 7 (see Fig. 4.18). The points common to both inequalities form the solution to the system.

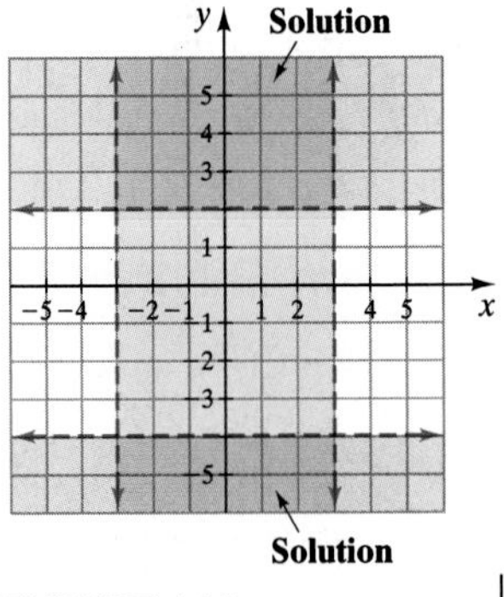

FIGURE 4.18

NOW TRY EXERCISE 37

Exercise Set 4.6

Concept/Writing Exercises

1. Explain how to find the solution to a system of linear inequalities graphically.

2. If one inequality contains < and the other inequality contains ≥, is the point of intersection of the related system of linear equations in the solution set? Explain.

Practice the Skills

Determine the solution to each system of inequalities.

3. $x - y > 2$
$y < -2x + 3$

4. $y \geq 3x - 2$
$y > -4x$

5. $y \leq x - 4$
$y < -2x + 4$

6. $2x + 3y < 6$
$4x - 2y \geq 8$

7. $y < x$
$y \geq 3x + 2$

8. $-x + 3y \geq 6$
$-2x - y > 4$

9. $4x - 2y < 6$
$y \leq -x + 4$

10. $y \leq 3x + 4$
$y > 2$

11. $-4x + 5y < 20$
$x \geq -3$

12. $3x - 4y \leq 6$
$y > -x + 4$

13. $x \leq 4$
$y \geq -2$

14. $x \geq 0$
$x - 3y < 6$

15. $5x + 2y > 10$
$3x - y > 3$

16. $3x + 2y > 8$
$x - 5y < 5$

17. $-2x > y + 4$
$-x < \frac{1}{2}y - 1$

18. $y \leq 3x - 2$
$\frac{1}{3}y < x + 1$

19. $y < 3x - 4$
$6x \geq 2y + 8$

20. $\frac{1}{2}x + \frac{1}{2}y \geq 2$
$2x - 3y \leq -6$

Determine the solution to each system of inequalities. Use the method discussed in Examples 4 and 5.

21. $x \geq 0$
$y \geq 0$
$5x + 4y \leq 20$
$x + 2y \leq 6$

22. $x \geq 0$
$y \geq 0$
$3x + 2y \leq 10$
$2x + 5y \leq 15$

23. $x \geq 0$
$y \geq 0$
$x + y \leq 6$
$7x + 4y \leq 28$

24. $x \geq 0$
$y \geq 0$
$8x + 3y \leq 24$
$2x + 3y \leq 12$

25. $x \geq 0$
$y \geq 0$
$7x + 4y \leq 24$
$2x + 5y \leq 20$

26. $x \geq 0$
$y \geq 0$
$5x + 4y \leq 16$
$x + 6y \leq 18$

27. $x \geq 0$
$y \geq 0$
$x \leq 4$
$x + y \leq 6$
$x + 2y \leq 8$

28. $x \geq 0$
$y \geq 0$
$x \leq 15$
$40x + 25y \leq 1000$
$5x + 30y \leq 900$

29. $x \geq 0$
$y \geq 0$
$x \leq 15$
$30x + 25y \leq 750$
$10x + 40y \leq 800$

Determine the solution to each system of inequalities.

30. $|x| < 3$
$y > x$

31. $|y| > 2$
$y \leq x + 3$

32. $|x| > 1$
$y \leq 3x + 2$

33. $|y| < 4$
$y \geq -2x + 2$

34. $|x| \leq 3$
$|y| > 2$

35. $|x| \geq 1$
$|y| \geq 2$

36. $|x| < 2$
$|y| \geq 3$

37. $|x + 2| < 3$
$|y| > 4$

38. $|x - 3| \geq 2$
$x + y < 5$

39. $|x - 2| > 1$
$y > -2$

40. $|x - 3| \leq 4$
$|y + 2| \leq 1$

Problem Solving

41. Is it possible for a system of linear inequalities to have no solution? Explain. Make up an example to support your answer.

42. Is it possible for a system of two linear inequalities to have exactly one solution? Explain. If you answer yes, make up an example to support your answer.

Without graphing, determine the number of solutions in each indicated system of inequalities. Explain your answers.

43. $2x + y < 6$
$2x + y > 6$

44. $3x - y \leq 4$
$3x - y > 4$

45. $5x - 2y \leq 3$
$5x - 2y \geq 3$

46. $2x - y < 7$
$3x - y < 4$

47. $5x - 3y > 5$
$5x - 3y > 6$

48. $x + y \leq 0$
$x - y \geq 0$

Challenge Problems

Determine the solution to each system.

49. $y < |x|$
$y < 4$

50. $y \geq |x - 2|$
$y \leq -|x - 2|$

Cumulative Review Exercises

[2.2] **51.** A formula for levers in physics is $f_1 d_1 + f_2 d_2 = f_3 d_3$. Solve this formula for f_2.

[3.2] *State the domain and range of each function.*

52. $\{(4, 3), (5, -2), (-1, 2), (0, -5)\}$ **53.** $f(x) = \frac{2}{3}x - 4$ **54.**

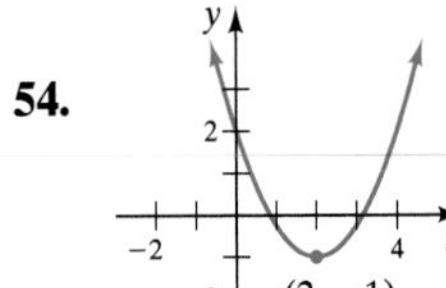

SUMMARY

Key Words and Phrases

4.1
Addition (or elimination) method
Consistent system of equations
Dependent system of equations
Inconsistent system of equations
Ordered triple
Solution to a system of equations
Substitution
System of linear equations

4.2
Geometric interpretation of a system of linear equations in three variables

4.3
Complementary angles
Supplementary angles

4.4
Augmented matrix
Elements
Matrix
Row transformations
Square matrix
Triangular form

4.5
Cramer's rule
Determinant
Expansion of the determinant by minors
Minor determinant

4.6
Constraints
Linear programming
Systems of linear inequalities
Systems of linear inequalities containing absolute value

IMPORTANT FACTS

Augmented Matrices

The matrix $\left[\begin{array}{cc|c} 1 & a & p \\ 0 & 1 & q \end{array}\right]$ represents the system $\begin{aligned} x + a &= p \\ y &= q \end{aligned}$

The matrix $\left[\begin{array}{ccc|c} 1 & a & b & p \\ 0 & 1 & c & q \\ 0 & 0 & 1 & r \end{array}\right]$ represents the system $\begin{aligned} x + ay + bz &= p \\ y + cz &= q \\ z &= r \end{aligned}$

Value of a Second-Order Determinant

$$\begin{vmatrix} a_1 & b_1 \\ a_2 & b_2 \end{vmatrix} = a_1 b_2 - a_2 b_1$$

Cramer's Rule:

For a system of equations of the form $\begin{aligned} a_1 x + b_1 y &= c_1 \\ a_2 x + b_2 y &= c_2 \end{aligned}$

$$x = \frac{\begin{vmatrix} c_1 & b_1 \\ c_2 & b_2 \end{vmatrix}}{\begin{vmatrix} a_1 & b_1 \\ a_2 & b_2 \end{vmatrix}} = \frac{D_x}{D} \quad \text{and} \quad y = \frac{\begin{vmatrix} a_1 & c_1 \\ a_2 & c_2 \end{vmatrix}}{\begin{vmatrix} a_1 & b_1 \\ a_2 & b_2 \end{vmatrix}} = \frac{D_y}{D}, \quad D \neq 0$$

continued on next page

Value of a Third-Order Determinant

Minor determinant of a_1 ↓ Minor determinant of a_2 ↓ Minor determinant of a_3 ↓

$$\begin{vmatrix} a_1 & b_1 & c_1 \\ a_2 & b_2 & c_2 \\ a_3 & b_3 & c_3 \end{vmatrix} = a_1\begin{vmatrix} b_2 & c_2 \\ b_3 & c_3 \end{vmatrix} - a_2\begin{vmatrix} b_1 & c_1 \\ b_3 & c_3 \end{vmatrix} + a_3\begin{vmatrix} b_1 & c_1 \\ b_2 & c_2 \end{vmatrix}$$

Cramer's Rule:

For a system of equations of the form

$$a_1x + b_1y + c_1z = d_1$$
$$a_2x + b_2y + c_2z = d_2$$
$$a_3x + b_3y + c_3z = d_3$$

$$x = \frac{\begin{vmatrix} d_1 & b_1 & c_1 \\ d_2 & b_2 & c_2 \\ d_3 & b_3 & c_3 \end{vmatrix}}{\begin{vmatrix} a_1 & b_1 & c_1 \\ a_2 & b_2 & c_2 \\ a_3 & b_3 & c_3 \end{vmatrix}} = \frac{D_x}{D}, \quad y = \frac{\begin{vmatrix} a_1 & d_1 & c_1 \\ a_2 & d_2 & c_2 \\ a_3 & d_3 & c_3 \end{vmatrix}}{\begin{vmatrix} a_1 & b_1 & c_1 \\ a_2 & b_2 & c_2 \\ a_3 & b_3 & c_3 \end{vmatrix}} = \frac{D_y}{D}, \quad z = \frac{\begin{vmatrix} a_1 & b_1 & d_1 \\ a_2 & b_2 & d_2 \\ a_3 & b_3 & d_3 \end{vmatrix}}{\begin{vmatrix} a_1 & b_1 & c_1 \\ a_2 & b_2 & c_2 \\ a_3 & b_3 & c_3 \end{vmatrix}} = \frac{D_z}{D}, \quad D \neq 0$$

Review Exercises

[4.1] *Write each equation in slope–intercept form. Without graphing or solving the system of equations, state whether the system of linear equations is consistent, inconsistent, or dependent. Also indicate whether the system has exactly one solution, no solution, or an infinite number of solutions.*

1. $x + 2y = 8$
$3x + 6y = 12$

2. $y = -3x - 6$
$2x + 3y = 8$

3. $y = \frac{1}{2}x + 4$
$x + 2y = 8$

4. $6x = 4y - 8$
$4x = 6y + 8$

Determine the solution to each system of equations graphically.

5. $y = x + 3$
$y = 2x + 5$

6. $x = -2$
$y = 3$

7. $2x + 2y = 8$
$2x - y = -4$

8. $2y = 2x - 6$
$\frac{1}{2}x - \frac{1}{2}y = \frac{3}{2}$

Find the solution to each system of equations by subsitution.

9. $y = 2x + 1$
$y = 3x - 2$

10. $y = -x + 5$
$y = 2x - 1$

11. $a = 2b - 8$
$2b - 5a = 0$

12. $3x + y = 17$
$\frac{1}{2}x - \frac{3}{4}y = 1$

Find the solution to each system of equations using the addition method.

13. $x + y = 6$
$x - y = 10$

14. $x + 2y = -3$
$2x - 2y = 6$

15. $2x + 3y = 4$
$x + 2y = -6$

16. $0.6x + 0.5y = 2$
$0.25x - 0.2y = 1.65$

17. $4r - 3s = 8$
$2r + 5s = 8$

18. $-2m + 3n = 15$
$3m + 3n = 10$

19. $x + \frac{2}{5}y = \frac{9}{5}$
$x - \frac{3}{2}y = -2$

20. $2x + 2y = 8$
$y = 4x - 3$

21. $y = -\frac{3}{4}x + \frac{5}{2}$
$x + \frac{5}{4}y = \frac{7}{2}$

22. $2x - 5y = 12$
$x - \frac{4}{3}y = -2$

23. $2x + y = 4$
$x + \frac{1}{2}y = 2$

24. $2x = 4y + 5$
$2y = x - 6$

[4.2] *Determine the solution to each system using substitution or the addition method.*

25. $x + 2y = 12$
$4x = 8$
$3x - 4y + 5z = 20$

26. $3x + 4y - 5z = 10$
$4x + 2z = 16$
$2z = -4$

27. $x + 5y + 5z = 6$
$3x + 3y - z = 10$
$x + 3y + 2z = 5$

28. $-x - y - z = -6$
$2x + 3y - z = 7$
$-3x + y + z = -6$

29. $3y - 2z = -4$
$3x - 5z = -7$
$2x + y = 6$

30. $3a + 2b - 5c = 19$
$2a - 3b + 3c = -15$
$5a - 4b - 2c = -2$

31. $x - y + 3z = 1$
$-x + 2y - 2z = 1$
$x - 3y + z = 2$

32. $-2x + 2y - 3z = 6$
$4x - y + 2z = -2$
$2x + y - z = 4$

[4.3] *Express each problem as a system of linear equations and use the method of your choice to find the solution to the problem.*

33. By coincidence, Bob Edward's niece and brother have the same birthday. If Bob's brother is 4 times as old as his niece and the difference in their ages is 18 years, how old are his niece and brother?

34. A plane can travel 600 miles per hour with the wind and 530 miles per hour against the wind. Find the speed of the wind and the speed of the plane in still air.

35. James Curtis has two acid solutions, as illustrated. How much of each must he mix to get 6 liters of a 40% acid solution?

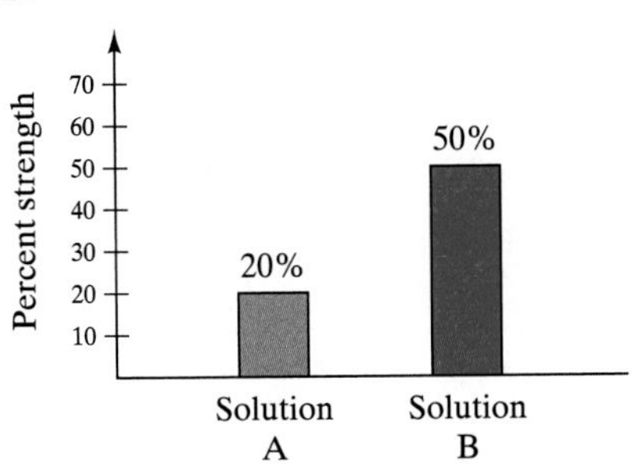

36. The admission at an ice hockey game is $15 for adults and $11 for children. A total of 650 tickets were sold. Determine how many children's tickets and how many adult tickets were sold if a total of $8790 was collected.

37. John Glenn was the first American astronaut to go into orbit around the earth. Many years later he returned to space. The second time he returned to space he was 5 years younger than twice his age when he went into space for the first time. The sum of his ages for both times he was in space is 118. Find his age each time he was in space.

38. Mary Weiner has a total of $40,000 invested in three different savings accounts. She has some money invested in one account that gives 7% interest. The second account has $5000 less than the first account and gives 5% interest. The third account gives 3% interest. If the total annual interest that Mary receives in a year is $2300, find the amount in each account.

[4.4] *Solve each system of equations using matrices.*

39. $-4x + 9y = 7$
$5x + 6y = -3$

40. $2x - 3y = 4$
$2x = y - 2$

41. $y = 2x - 4$
$4x = 2y + 8$

42. $2x - y - z = 5$
$x + 2y + 3z = -2$
$3x - 2y + z = 2$

43. $3a - b + c = 2$
$2a - 3b + 4c = 4$
$a + 2b - 3c = -6$

44. $x + y + z = 3$
$3x + 2y = 1$
$y - 3z = -10$

[4.5] *Solve each system of equations using determinants.*

45. $5x + 6y = 14$
$x - 3y = 7$

46. $3x + 5y = -2$
$5x + 3y = 2$

47. $4m + 3n = 2$
$7m - 2n = -11$

48. $r + s + t = 8$
$r - s - t = 0$
$r + 2s + t = 9$

49. $x + 2y - 4z = 17$
$2x - y + z = -9$
$2x - y - 3z = -1$

50. $y + 3z = 4$
$-x - y + 2z = 0$
$x + 2y + z = 1$

[4.6] *Graph the solution to each system of inequalities.*

51. $-x + 3y > 6$
$2x - y \leq 2$

52. $5x - 2y \leq 10$
$3x + 2y > 6$

53. $y > 2x + 3$
$y < -x + 4$

54. $x > -2y + 4$
$y < -\frac{1}{2}x - \frac{3}{2}$

Determine the solution to the system of inequalities.

55. $x \geq 0$
$y \geq 0$
$x + y \leq 6$
$4x + y \leq 8$

56. $x \geq 0$
$y \geq 0$
$2x + y \leq 6$
$4x + 5y \leq 20$

57. $|x| \leq 3$
$|y| > 2$

58. $|x| > 4$
$|y - 2| \leq 3$

Practice Test

1. Define **a)** a consistent system of equations, **b)** a dependent system of equations, **c)** an inconsistent system of equations.

Determine, without solving the system, whether the system of equations is consistent, inconsistent, or dependent. State whether the system has exactly one solution, no solution, or an infinite number of solutions.

2. $4x + 3y = -6$
$6y = 8x + 4$

3. $5x + 3y = 9$
$2y = -\frac{10}{3}x + 6$

4. $5x - 4y = 6$
$-10x + 8y = -10$

Solve each system of equations by the method indicated.

5. $y = 3x - 2$
$y = -2x + 8$
graphically

6. $y = -x + 6$
$y = 2x + 3$
graphically

7. $y = 6x - 12$
$y = 4x - 8$
substitution

8. $3x + y = 8$
$x - 2y = 5$
substitution

9. $4x - 3y = 10$
$2x + y = 5$
addition

10. $0.3x = 0.2y + 0.4$
$-1.2x + 0.8y = -1.6$
addition

11. $\frac{3}{2}a + b = 6$
$a - \frac{5}{2}b = -4$
addition

12. $x + y + z = 2$
$-2x - y + z = 1$
$x - 2y - z = 1$
addition

13. Write the augmented matrix for the following system of equations.

$$4x - 5y + 3z = 2$$
$$2x - y - 2z = 4$$
$$3x + 2y - z = -3$$

14. Consider the following augmented matrix.

$$\left[\begin{array}{ccc|c} 6 & -2 & 4 & 4 \\ 4 & 3 & 5 & 6 \\ 2 & -1 & 4 & -3 \end{array}\right]$$

Show the results obtained by multiplying the elements in the third row by -2 and adding the products to their corresponding elements in the second row

Solve each system of equations using matrices

15. $x - 5y = -2$
$3x - y = 8$

16. $x - 2y + z = 7$
$-2x - y - z = -7$
$3x - 2y + 2z = 15$

Evaluate each determinant.

17. $\begin{vmatrix} 4 & 6 \\ -2 & 5 \end{vmatrix}$

18. $\begin{vmatrix} 8 & 2 & -1 \\ 3 & 0 & 5 \\ 6 & -3 & 4 \end{vmatrix}$

Solve each system of equations using determinants and Cramer's rule.

19. $5x - 2y = -13$
$2x + y = 11$

20. $2x - y - z = -3$
$3x - 2y - 2z = -5$
$-x + y + 2z = 4$

Express each problem as a system of linear equations and use the method of your choice to find the solution to the problem.

21. Max Wells has cashews that sell for \$7 a pound and peanuts that sell for \$5.50 a pound. How much of each must he mix to get 20 pounds of a mixture that sells for \$6.00 per pound?

22. Tyesha Blackwell, a chemist, has 6% and 15% solutions of sulfuric acid. How much of each solution should she mix to get 10 liters of a 9% solution?

23. The sum of three numbers is 25. The greatest number is 3 times the smallest number. The remaining number is 1 more than twice the smallest number. Find the three numbers.

Determine the solution to each system of inequalities.

24. $3x + 2y < 9$
$-2x + 5y \leq 10$

25. $|x| > 3$
$|y| \leq 1$

Cumulative Review Test

1. Evaluate $24 \div 4[2 - (5 - 2)]^2 - 6$.
2. Consider the set of numbers

$$\left\{\frac{1}{2}, -4, 9, 0, \sqrt{3}, -4.63, 1\right\}.$$

List the elements of the set that are **a)** natural numbers; **b)** rational numbers; **c)** real numbers.

3. Write the following numbers in order from smallest to largest.

$$-1, |-4|, \frac{3}{4}, \frac{5}{8}, -|-8|, |-10|$$

Solve.

4. $-[3 - 2(x - 4)] = 3(x - 6)$
5. $\frac{1}{3}x = \frac{3}{5}x + 4$
6. $|4x - 3| + 2 = 10$
7. Solve the formula $R = 3(a + b)$ for b.
8. Find the solution set of the inequality.

$$0 < \frac{3x - 2}{4} \leq 8$$

9. Simplify $\left(\frac{3x^2y^{-2}}{y^3}\right)^{-2}$
10. Graph $2y = 3x - 8$.
11. Write in slope–intercept form the equation of the line that is parallel to the graph of $2x - 3y = 8$ and passes through the point $(2, 3)$.
12. Graph the inequality $6x - 3y < 12$.
13. Determine which of the following graphs are functions. Explain.

a)

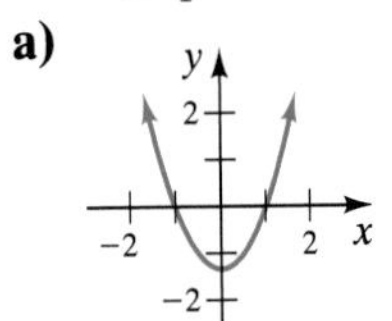

b)

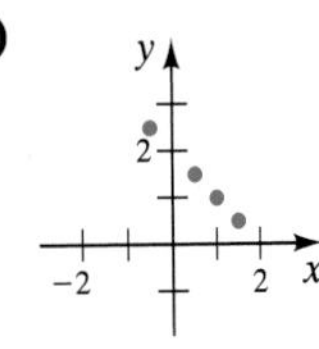

c)

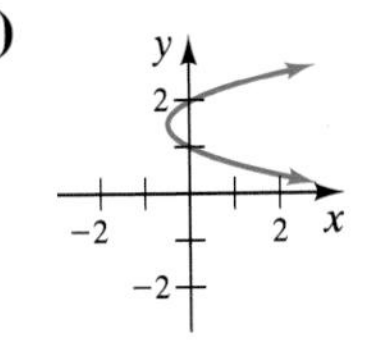

14. If $f(x) = x^2 - 3x + 4$, find **a)** $f(-2)$ and **b)** $f(c)$.

Solve each system of equations.

15. $3x + y = 6$
$y = 2x + 1$

16. $5x + 4y = 10$
$3x + 5y = -7$

17. $x - 2y = 0$
$2x + z = 7$
$y - 2z = -5$

18. If the largest angle of a triangle is 9 times the measure of the smallest angle, and the middle-sized angle is 70° greater than the measure of the smallest angle, find the measure of the three angles.

19. Dawn Davis speed-walks at 4 miles per hour and Judy Bolin jogs at 6 miles per hour. Dawn begins walking $\frac{1}{2}$ hour before Judy starts jogging. If Judy jogs on the same path that Dawn speed-walks, how long after Judy begins jogging will she catch up to Dawn?

20. There are two different prices of seats at a rock concert. The higher-priced seats sell for \$20 and the less expensive seats sell for \$16. If a total of 1000 tickets are sold and the total ticket sales are \$18,400, how many of each type of ticket are sold?

POLYNOMIALS AND POLYNOMIAL FUNCTIONS

CHAPTER 5

Use the Angel Web site at www.prenhall.com/angel to be linked to an internet resource that will help you further explore the following application.

The Internet is changing the way we make airline reservations and purchases. To meet the needs of their customers in a rapidly expanding environment, travel companies use mathematical models to predict the number of Internet clients they can expect. On page 284, we evaluate a polynomial function which predicts the annual online travel revenue. As the online travel revenue increases, the kinds of employees needed will also change. What are some qualities that you think an online travel company will look for when hiring new employees?

Preview and Perspective

In this chapter we discuss polynomials, polynomial functions, and factoring. In the first few sections we add, subtract, multiply, and divide polynomials and polynomial functions. Since graphing is such an important part of this course, make sure you understand the graphs of polynomial functions (objective 3, Section 5.1).

After discussing polynomials, we turn our attention to factoring. *You must have a thorough understanding of factoring to work the problems in many of the remaining chapters.* In Section 5.8 we explain how to solve quadratic equations by factoring and show how applications are solved using factoring. Pay particular attention to how to use factoring to find the x-intercepts of a quadratic function (objective 4, Section 5.8). We will refer back to this topic later in the course.

5.1 ADDITION AND SUBTRACTION OF POLYNOMIALS

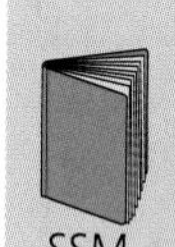

SSM VIDEO 5.1 CD Rom

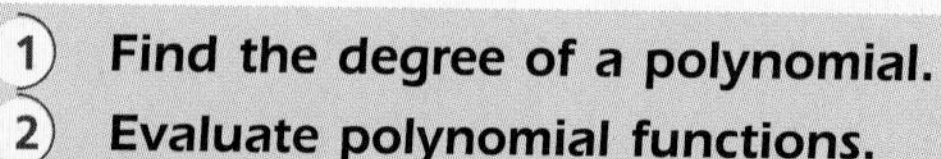

1) Find the degree of a polynomial.
2) Evaluate polynomial functions.
3) Understand graphs of polynomial functions.
4) Add and subtract polynomials.

1) Find the Degree of a Polynomial

Recall from Chapter 2 that the parts that are added or subtracted in a mathematical expression are called **terms.** The **degree of a term** with whole number exponents is the sum of the exponents of the variables, if there are variables. Nonzero constants have degree 0, and the term 0 has no degree.

A **polynomial** is a finite sum of terms in which all variables have whole number exponents and no variable appears in a denominator. The expression $3x^2 + 2x + 6$ is an example of a *polynomial in one variable*, x. The expression $x^2y - 2x + 3$ is an example of a *polynomial in two variables*, x and y. The expressions $x^{1/2}$ and $\frac{1}{x}$ (or x^{-1}) are *not* polynomials because the variables do not have whole number exponents. The expression $\frac{1}{x-1}$ is *not* a polynomial because a variable appears in the denominator.

The **leading term** of a polynomial is the term of highest degree. The **leading coefficient** is the coefficient of the leading term.

EXAMPLE 1 For each polynomial give the number of terms, the degree of the polynomial, the leading term, and the leading coefficient.

a) $2x^4 - 3x^2 + 6x - 4$ **b)** $2x^2y^4 - 6xy^3 + 3xy^2z^4$

Solution We will organize the answers in tabular form.

Polynomial	Number of Terms	Degree of Polynomial	Leading Term	Leading Coefficient
a) $2x^4 - 3x^2 + 6x - 4$	4	4 (from $2x^4$)	$2x^4$	2
b) $2x^2y^4 - 6xy^3 + 3xy^2z^4$	3	7 (from $3xy^2z^4$)	$3xy^2z^4$	3

NOW TRY EXERCISE 29

Polynomials are classified according to the number of terms they have, as indicated in the following chart.

Polynomial Type	Description	Examples
Monomial	A polynomial of one term	$4x^2, 6x^2y, 3, -2xyz^5, 7$
Binomial	A polynomial of two terms	$x^2 + 1, 2x^2 - y, 6x^3 - 5y^2$
Trinomial	A polynomial of three terms	$x^3 + 6x - 4, x^2y - 6x + y^2$

Polynomials containing more than three terms are not given specific names. *Poly* is a prefix meaning *many*. A polynomial is called **linear** if it is of degree 0 or 1. A polynomial in one variable is called **quadratic** if it is of degree 2, and **cubic** if it is of degree 3.

Type of Polynomial	Examples
Linear	$2x - 4, \;\; 5$
Quadratic	$3x^2 + x - 6, \;\; 4x^2 - 6$
Cubic	$-4x^3 + 3x^2 + 5, \;\; 2x^3 + 6x$

The polynomials $2x^3 + 4x^2 - 6x + 3$ and $4x^2 - 3xy + 5y^2$ are examples of polynomials in **descending order** of the variable x because the exponents on the variable x descend (or get lower) as the terms go from left to right. Polynomials are often written in descending order of a given variable.

EXAMPLE 2 Write each polynomial in descending order of the variable x.

a) $3x + 4x^2 - 6$ **b)** $xy - 6x^2 + 3y^2$

Solution **a)** $3x + 4x^2 - 6 = 4x^2 + 3x - 6$

b) $xy - 6x^2 + 3y^2 = -6x^2 + xy + 3y^2$

2 Evaluate Polynomial Functions

The expression $2x^3 + 6x^2 + 3$ is a polynomial. If we write $P(x) = 2x^3 + 6x^2 + 3$, then we have a polynomial function. In a **polynomial function** the expression used to describe the function is a polynomial. To evaluate a polynomial function, we use substitution, just as we did to evaluate other functions in Chapter 3.

EXAMPLE 3 For the polynomial function $P(x) = 4x^3 - 6x^2 - 2x + 8$, find

a) $P(0)$ **b)** $P(3)$ **c)** $P(-2)$

Solution

a)
$$P(x) = 4x^3 - 6x^2 - 2x + 8$$
$$P(0) = 4(0)^3 - 6(0)^2 - 2(0) + 8$$
$$= 0 - 0 - 0 + 8 = 8$$

b)
$$P(3) = 4(3)^3 - 6(3)^2 - 2(3) + 8$$
$$= 4(27) - 6(9) - 6 + 8 = 56$$

c)
$$P(-2) = 4(-2)^3 - 6(-2)^2 - 2(-2) + 8$$
$$= 4(-8) - 6(4) + 4 + 8 = -44$$

NOW TRY EXERCISE 35

Businesses, governments, and other organizations often need to track and make projections about things such as sales, profits, changes in the population, effectiveness of new drugs, and so on. To do so, they often use graphs and functions. Example 4 gives one such example.

EXAMPLE 4 The bar graph in Figure 5.1 shows the projected online travel revenue (travel plans made on the Internet) in billions of dollars from 1996 through 2002. A polynomial function that can be used to approximate the revenues is $R(t) = 0.18t^2 + 0.37t + 0.28$, where t is years since 1996, $0 \le t \le 6$, and R is revenue in billions of dollars.

a) Using the function, estimate the online travel revenue in 1996.

b) Using the function, estimate the revenue in 2000.

Solution **a) Understand** We first need to determine what value of t to substitute into the function. Since t is years since 1996, the year 1996 corresponds to $t = 0$. Thus to estimate the revenue in 1996, we evaluate $R(0)$.

Online Travel Revenues

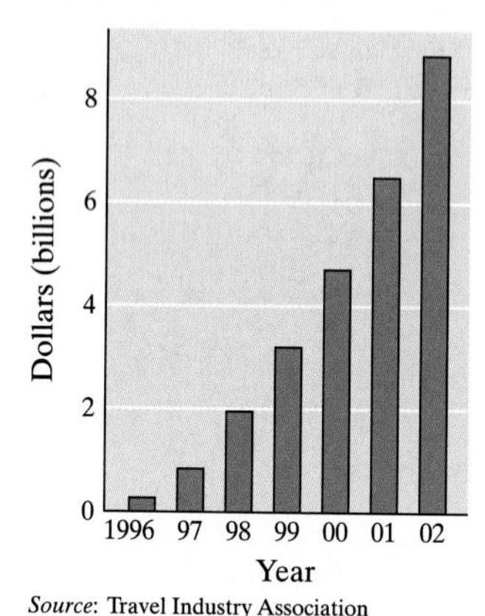

Source: Travel Industry Association of America

FIGURE 5.1

Translate and Carry Out

$$\begin{aligned} R(t) &= 0.18t^2 + 0.37t + 0.28 \\ R(0) &= 0.18(0)^2 + 0.37(0) + 0.28 \\ &= 0 + 0 + 0.28 \\ &= 0.28 \end{aligned}$$

Check and Answer Thus the revenue in 1996 is about \$0.28 billion. The graph supports this answer.

b) Understand The year 2000 is 4 years since 1996 $(2000 - 1996 = 4)$. Therefore to estimate the revenue in 2000, we evaluate $R(4)$.

Translate and Carry Out

$$\begin{aligned} R(t) &= 0.18t^2 + 0.37t + 0.28 \\ R(4) &= 0.18(4)^2 + 0.37(4) + 0.28 \\ &= 2.88 + 1.48 + 0.28 \\ &= 4.64 \end{aligned}$$

Check and Answer The estimated revenue in 2000 is about \$4.64 billion. This is consistent with the amount shown on the graph for the year 2000.

NOW TRY EXERCISE 91

3 Understand Graphs of Polynomial Functions

The graphs of all polynomial functions are smooth, continuous curves. Figure 5.2 shows a graph of a quadratic polynomial function. The graphs of all quadratic polynomial functions with a *positive leading coefficient* will have the shape of the graph in Figure 5.2.

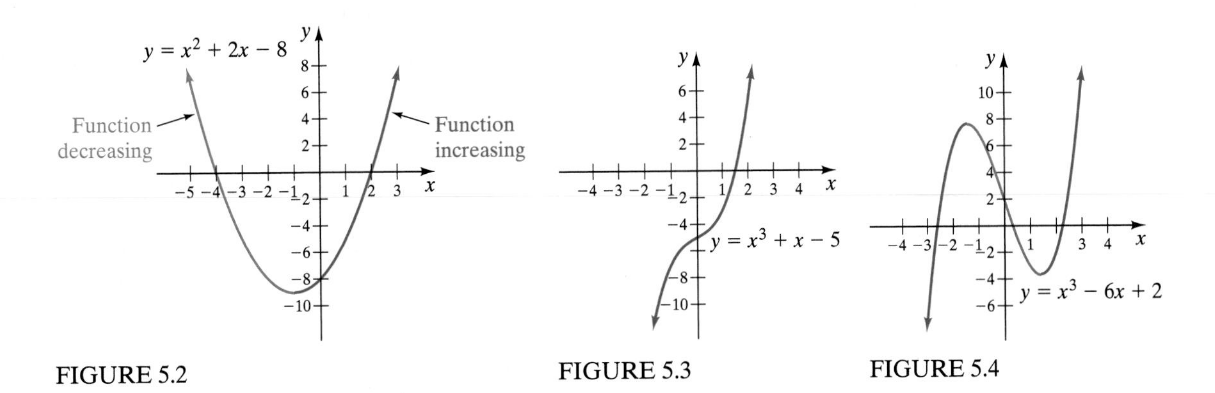

FIGURE 5.2

FIGURE 5.3

FIGURE 5.4

The graph of a cubic polynomial function with a *positive leading coefficient* may have the shape of the graph in either Figure 5.3 or Figure 5.4. Notice that *whenever the leading coefficient in a polynomial function is positive, the polynomial function will increase (or move upward as x increases—the green part of the curve) to the right of some value of x*. For example, in Figure 5.2, the graph continues increasing to the right of $x = -1$. In Figure 5.3 the graph is continuously increasing, and in Figure 5.4 the graph is increasing to the right of about $x = 1.4$.

Polynomial functions with a negative leading coefficient will decrease (or move downward as x increases—the red part of the curve) to the right of some value of x. A quadratic polynomial function with a negative leading coefficient is shown in Figure 5.5, and cubic polynomial functions with negative leading coefficients are shown in Figure 5.6 and Figure 5.7. In Figure 5.5 the quadratic function is decreasing to the right of $x = 2$. In Figure 5.6 the cubic function is continuously decreasing, and in Figure 5.7 the cubic function is decreasing to the right of about $x = 1.2$.

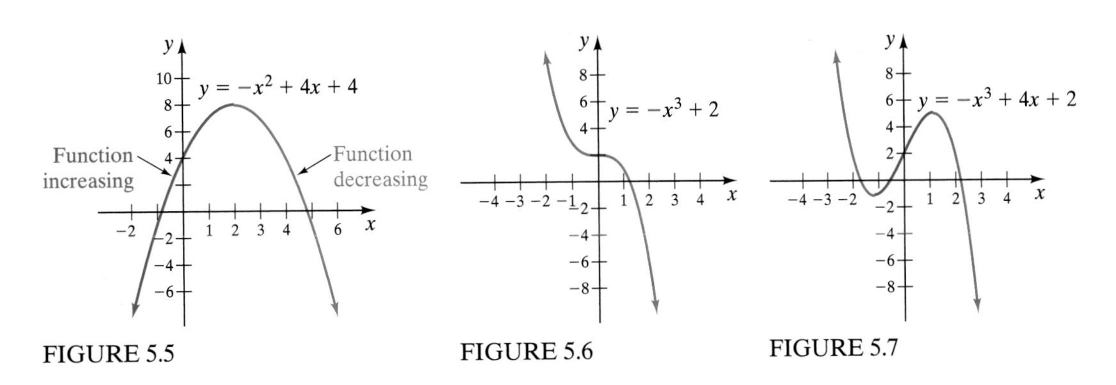

FIGURE 5.5

FIGURE 5.6

FIGURE 5.7

Why does the leading coefficient determine whether a function will increase or decrease to the right of some value of x? The leading coefficient is the coefficient of the term with the greatest exponent on the variable. As x increases, this term will eventually dominate all the other terms in the function. So if the coefficient of this term is positive, the function will *eventually* increase as x increases. If the leading coefficient is negative, the function will *eventually* decrease as x increases. This information, along with checking the y-intercept of the graph, can be useful in determining whether a graph is correct or complete. Read the Using Your Graphing Calculator box that follows, even if you are not using a graphing calculator. Also work Exercises 87 through 90.

Using Your Graphing Calculator

Whenever you graph a polynomial function on your grapher, make sure your screen shows every change in direction of your graph. For example, suppose you graph $y = 0.1x^3 - 2x^2 + 5x - 8$ on your grapher. Using the standard window, you get the graph shown in Figure 5.8.

$y = 0.1x^3 - 2x^2 + 5x - 8$

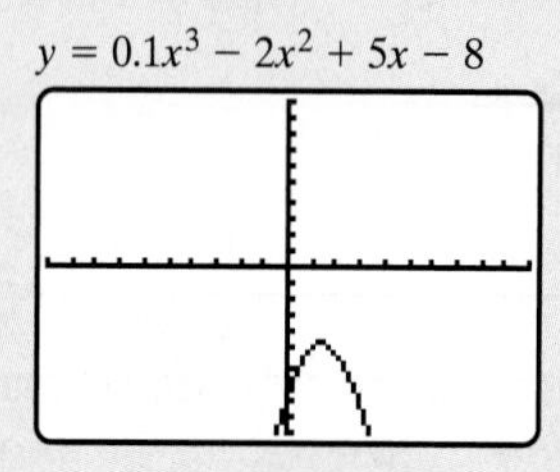

FIGURE 5.8

$y = 0.1x^3 - 2x^2 + 5x - 8$

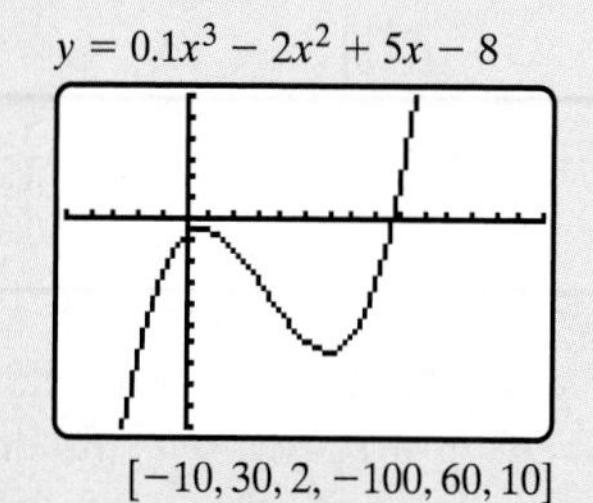

[−10, 30, 2, −100, 60, 10]

FIGURE 5.9

However, from our preceding discussion you should realize that since the leading coefficient, 0.1, is positive, the graph must increase to the right of some value of x. The graph in Figure 5.8 does not show this. If you change your window as shown in Figure 5.9, you will get the graph shown. Now you can see how the graph increases to the right of about $x = 12$. When graphing, the y-intercept is often helpful in determining the values to use for the range. Recall that to find the y-intercept, we set $x = 0$ and solve for y. For example, if graphing

$$y = 4x^3 + 6x^2 + x - 180,$$

the y-intercept will be at -180.

Exercises

Use your grapher to graph each polynomial. Make sure your window shows all changes in direction of the graph.

1. $y = 0.2x^3 + 5.1x^2 - 6.2x + 9.3$
2. $y = 4.1x^3 - 19.6x^2 + 5.4x - 60.2$

4 Add and Subtract Polynomials

When we found sums and differences of functions in Section 3.6 we added and subtracted polynomials, although they were not called polynomials at that time. To *add or subtract polynomials,* remove parentheses if any are present, then combine like terms.

EXAMPLE 5 Simplify $(4x^2 - 6x + 3) + (2x^2 + 5x - 1)$.

Solution

$$(4x^2 - 6x + 3) + (2x^2 + 5x - 1)$$
$$= 4x^2 - 6x + 3 + 2x^2 + 5x - 1 \quad \text{Remove the parentheses.}$$
$$= \underbrace{4x^2 + 2x^2}\ \underbrace{- 6x + 5x}\ \underbrace{+ 3 - 1} \quad \text{Rearrange terms.}$$
$$= 6x^2 \quad -x \quad +2 \quad \text{Combine like terms.}$$

EXAMPLE 6 Simplify $(3x^2y - 4xy + y) + (x^2y + 2xy + 3y - 2)$.

Solution

$$(3x^2y - 4xy + y) + (x^2y + 2xy + 3y - 2)$$
$$= 3x^2y - 4xy + y + x^2y + 2xy + 3y - 2 \quad \text{Remove the parentheses.}$$
$$= \underbrace{3x^2y + x^2y}\ \underbrace{- 4xy + 2xy}\ \underbrace{+ y + 3y} - 2 \quad \text{Rearrange terms.}$$
$$= 4x^2y \quad -2xy \quad +4y \quad -2 \quad \text{Combine like terms.}$$

HELPFUL HINT

Recall that $-x$ means $-1 \cdot x$. Thus $-(2x^2 - 4x + 6)$ means $-1(2x^2 - 4x + 6)$, and the distributive property applies. When you subtract one polynomial from another, the *signs of every term* of the polynomial being subtracted must change. For example,

$$\begin{aligned} x^2 - 6x + 3 - (2x^2 - 4x + 6) &= x^2 - 6x + 3 - 1(2x^2 - 4x + 6) \\ &= x^2 - 6x + 3 - 2x^2 + 4x - 6 \\ &= -x^2 - 2x - 3 \end{aligned}$$

EXAMPLE 7 Subtract $(-x^2 - 2x + 3)$ from $(x^3 + 4x + 6)$.

Solution $(x^3 + 4x + 6) - (-x^2 - 2x + 3)$

$= (x^3 + 4x + 6) - 1(-x^2 - 2x + 3)$ *Insert 1.*

$= x^3 + 4x + 6 + x^2 + 2x - 3$ *Distributive Property.*

$= x^3 + x^2 + 4x + 2x + 6 - 3$ *Rearrange terms.*

$= x^3 + x^2 + 6x + 3$ *Combine like terms.*

EXAMPLE 8 Simplify $x^2y - 4xy^2 + 5 - (2x^2y - 3y^2 + 4)$.

Solution $x^2y - 4xy^2 + 5 - 1(2x^2y - 3y^2 + 4)$ *Insert 1.*

$= x^2y - 4xy^2 + 5 - 2x^2y + 3y^2 - 4$ *Distributive Property.*

$= x^2y - 2x^2y - 4xy^2 + 3y^2 + 5 - 4$ *Rearrange terms.*

$= -x^2y - 4xy^2 + 3y^2 + 1$ *Combine like terms.*

Note that $-x^2y$ and $-4xy^2$ are not like terms since the variables have different exponents. Also, $-4xy^2$ and $3y^2$ are not like terms since $3y^2$ does not contain the variable x.

NOW TRY EXERCISE 45

Exercise Set 5.1

Concept/Writing Exercises

1. What are terms of a mathematical expression?
2. What is the degree of a nonzero constant?
3. What is a polynomial?
4. What is the leading term of a polynomial?
5. What is the leading coefficient of a polynomial?
6. **a)** How do you determine the degree of a term?
   b) What is the degree of $6x^4y^3z$?
7. **a)** How do you determine the degree of a polynomial?
   b) What is the degree of $-4x^4 + 6x^3y^4 + z^5$?
8. What does it mean when a polynomial is in descending order of the variable x?
9. **a)** When is a polynomial linear?
   b) Give an example of a linear polynomial.
10. **a)** When is a polynomial quadratic?
    b) Give an example of a quadratic polynomial.
11. **a)** When is a polynomial cubic?
    b) Give an example of a cubic polynomial.
12. When one polynomial is being subtracted from another, what happens to the signs of all the terms of the polynomial being subtracted?

13. Write a fifth-degree trinomial in x in descending order that lacks fourth-, third-, and second-degree terms.

14. Write a seventh-degree polynomial in y in descending order that lacks fifth-, third-, and second degree terms.

Practice the Skills

Determine whether each expression is a polynomial. If the polynomial has a specific name, for example, "monomial" or "binomial," give the name. If the expression is not a polynomial, explain why it is not.

15. $5y$ **16.** $5x^2 - 6x + 9$ **17.** -10 **18.** $5x^{-3}$

19. $8x^2 - 2x + 8y^2$ **20.** $3x^{1/2} + 2xy$ **21.** $-2x^2 + 5x^{-1}$ **22.** $2xy + 5y^2$

Write each polynomial in descending order of the variable x. If the polynomial is already in descending order, so state. Give the degree of each polynomial.

23. $-8 - 4x - x^2$

24. $2x + 4 - x^2$

25. $6y^2 + 3xy + 10x^2$

26. $-4 + x - 3x^2 + 4x^3$

27. $-2x^4 + 5x^2 - 4$

28. $5xy^2 + 3x^2y - 6 - 2x^3$

*Give **a)** the degree of each polynomial and **b)** its leading coefficient.*

29. $x^4 + 3x^6 - 2x - 10$

30. $-2x^4 + 6x^5 - x^7 + 5x^3$

31. $4x^2y^3 + 6xy^4 + 9xy^5$

32. $-a^4b^3c^2 + 2a^8b^9c^4 - 8a^7c^{20}$

33. $-\frac{1}{2}m^4n^5p^8 + \frac{3}{5}m^3p^6 - \frac{5}{9}n^4p^6q$

34. $-0.6x^2y^3z^2 - 2.9xyz^9 - 1.1x^8y^4$

Evaluate each polynomial function at the given value.

35. Find $P(2)$ if $P(x) = x^2 - 6x + 4$.

36. Find $P(-1)$ if $P(x) = 4x^2 - 6x + 12$.

37. Find $P\left(\frac{1}{2}\right)$ if $P(x) = 2x^2 - 3x - 6$.

38. Find $P\left(\frac{1}{3}\right)$ if $P(x) = \frac{1}{2}x^3 - x^2 + 6$.

39. Find $P(0.4)$ if $P(x) = 0.2x^3 + 1.6x^2 - 2.3$.

40. Find $P(-1.2)$ if $P(x) = -1.6x^3 - 4.6x^2 - 0.1x$.

Simplify.

41. $(x^2 + 5x - 2) + (6x - 7)$

42. $(5b^2 - 4b + 9) - (2b^2 - 2b - 5)$

43. $(x^2 - 6x + 3) - (2x + 5)$

44. $(x - 4) - (3x^2 - 4x + 6)$

45. $(4y^2 + 6y - 3) - (2y^2 + 6)$

46. $(5n^2 - 7) + (2n^2 + 3n + 12)$

47. $\left(-\frac{5}{9}a + 8\right) + \left(-\frac{2}{3}a^2 - \frac{1}{4}a - 5\right)$

48. $(6y^2 - 6y + 4) - (-2y^2 - y + 7)$

49. $(1.4x^2 + 0.6x - 8.3) - (4.9x^2 + 3.7x + 19.2)$

50. $(-12.4x^2y - 6.2xy + 9.3y^2) - (-5.3x^2y + 1.6xy - 10.4y^2)$

51. $\left(-\frac{1}{3}x^3 + \frac{1}{4}x^2y + 3xy^2\right) + \left(-x^3 - \frac{1}{2}x^2y + xy^2\right)$

52. $\left(-\frac{3}{5}xy^2 + \frac{5}{8}\right) - \left(-\frac{1}{2}xy^2 + \frac{3}{5}\right)$

53. $(3a - 6b + 5c) - (-2a + 4b - 8c)$

54. $(6r + 5s - t) + (-3r - 2s - 5t)$

55. $(3a^2b - 6ab + 5b^2) - (4ab - 6b^2 - 5a^2b)$

56. $(3x^2 - 5y^2 - 2xy) - (4x^2 + 3y^2 - 5xy)$

57. $(3r^2 - 5t^2 + 2rt) + (-6rt + 2t^2 - r^2)$

58. $(a^2 - b^2 + 5ab) + (-3b^2 - 2ab + a^2)$

59. $6x^2 - 2x - [3x - (4x^2 - 6)]$

60. $3xy^2 - 2x - [-(4xy^2 + 3x) - 5xy]$

61. $5w - 6w^2 - [(3w - 2w^2) - (4w + w^2)]$

62. $-[-(5r^2 - 3r) - (2r - 3r^2) - 2r^2]$

63. Subtract $(4x - 6)$ from $(3x + 5)$.

64. Subtract $(-x^2 + 3x + 5)$ from $(4x^2 - 6x + 2)$.

65. Add $-2x^2 + 4x - 12$ and $-x^2 - 2x$.

66. Subtract $(5x^2 - 6)$ from $(2x^2 - 4x + 8)$.

67. Subtract $0.2a^2 - 3.9a + 26.4$ from $-4.2a^2 - 9.6a$.

68. Add $6x^2 + 3xy$ and $-2x^2 + 4xy + 3y$.

69. Subtract $\left(5x^2y + \frac{5}{9}\right)$ from $\left(-\frac{1}{2}x^2y + 6xy^2 + \frac{3}{5}\right)$.

70. Subtract $(6x^2y + 3xy)$ from $(2x^2y + 12xy)$.

Simplify. Assume that all exponents represent natural numbers.

71. $(3x^{2r} - 2x^r + 6) + (2x^{2r} - 6x^r - 3)$

72. $(6x^{2r} - 5x^r + 4) + (2x^{2r} - 2x^r + 3)$

73. $(x^{2s} - 6x^s + 4) - (2x^{2s} - 4x^s - 3)$

74. $(5a^{2m} - 6a^m + 4) - (2a^{2m} + 7)$

75. $(7b^{4n} - 3b^{2n} - 1) - (5b^{3n} - b^{2n})$

76. $(-3r^{3a} + r^a - 6) - (-2r^{3a} - 5r^{2a} + 6)$

Problem Solving

77. Is the sum of the two trinomials always a trinomial? Explain, and give an example to support your answer.

78. Is the sum of two binomials always a binomial? Explain, and give an example to support your answer.

79. Is the sum of two quadratic polynomials always a quadratic polynomial? Explain, and give an example to support your answer.

80. Is the difference of two cubic polynomials always a cubic polynomial? Explain, and give an example to support your answer.

81. The area of a circle is a function of its radius, where $A(r) = \pi r^2$. Find the area of a circle if its radius is 6 inches. Use the $\boxed{\pi}$ key on your calculator.

82. The value of a sphere is a function of its radius where $V(r) = \frac{4}{3}\pi r^3$. A circular balloon is being blown up. Find its volume when its radius is 4 inches.

83. When an object is dropped from the Empire State Building (height 1250 feet), the object's height, h, in feet, from the ground at time, t, in seconds, after being dropped can be determined by

$$h = P(t) = -16t^2 + 1250$$

Find the distance an object is from the ground 6 seconds after being dropped.

84. The number of ways that the first-, second-, and third-place winners in a spelling bee can be selected from n participants is given by $P(n) = n^3 - 3n^2 + 2n$. If there are 6 participants, how many ways can the winner and first and second runners-up be selected?

The profit of a company is found by subtracting its cost from its revenue. In Exercises 85 and 86, $R(x)$ represents the company's revenue when selling x items and $C(x)$ represents the company's cost when producing x items. ***a)*** *Determine a profit function $P(x)$.* ***b)*** *Evaluate $P(x)$ when $x = 100$.*

85. $R(x) = 2x^2 - 60x + 4000, C(x) = 8050 - 420x$

86. $R(x) = 5.5x^2 - 80.3x + 5790.4,$
$C(x) = 1.2x^2 + 16.3x + 12{,}040.6$

*In Exercises 87–90, determine which of the graphs—**a), b),** or **c)**—is the graph of the given equation. Explain how you determined your answer.*

87. $y = x^2 + 3x - 4$

a)

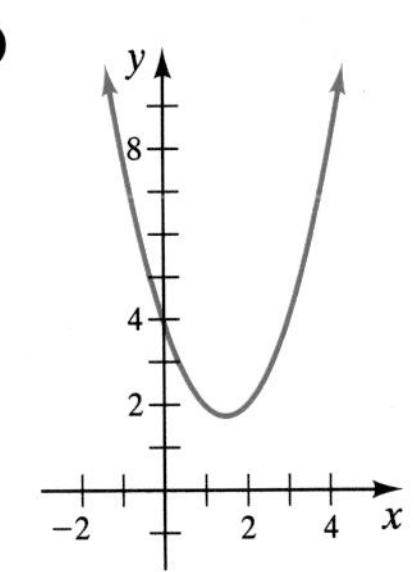

b)

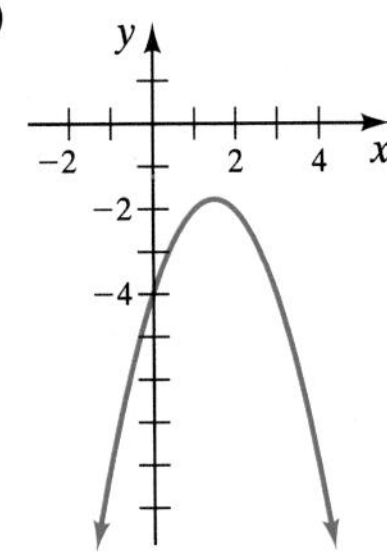

c)

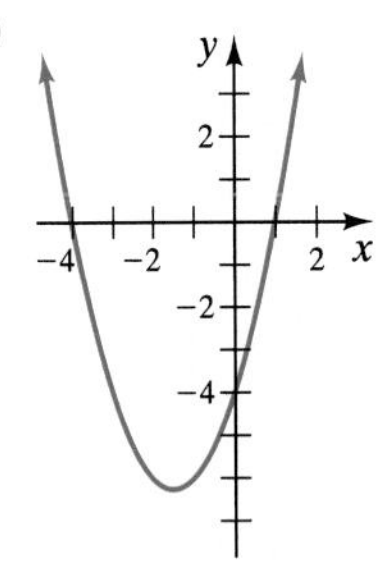

88. $y = x^3 + 2x^2 - 4$

a)

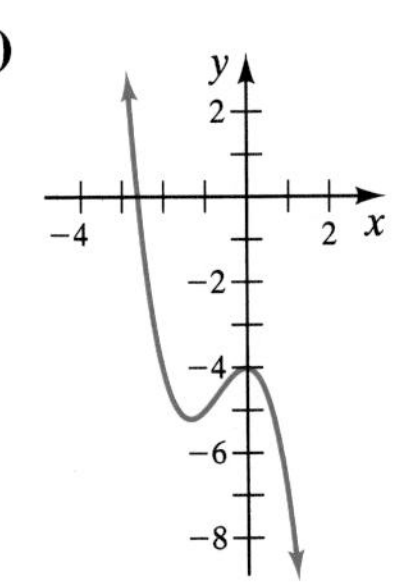

b)

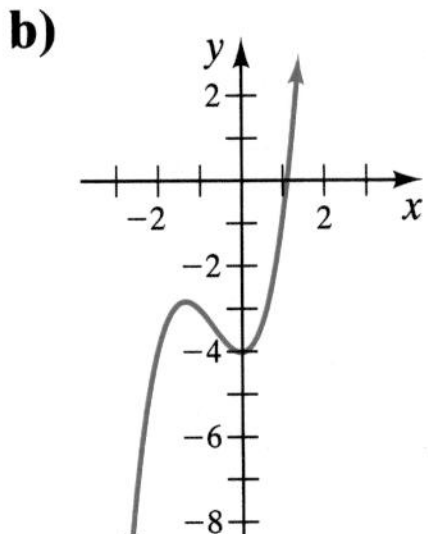

c)

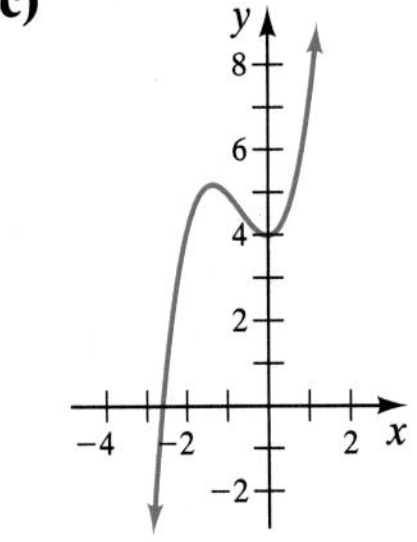

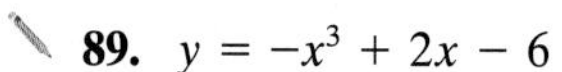

89. $y = -x^3 + 2x - 6$

a)
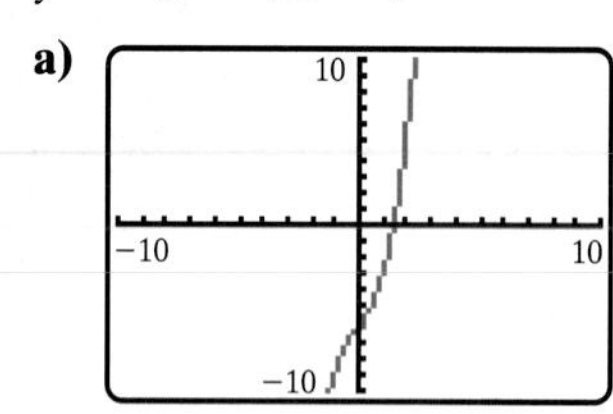

b)
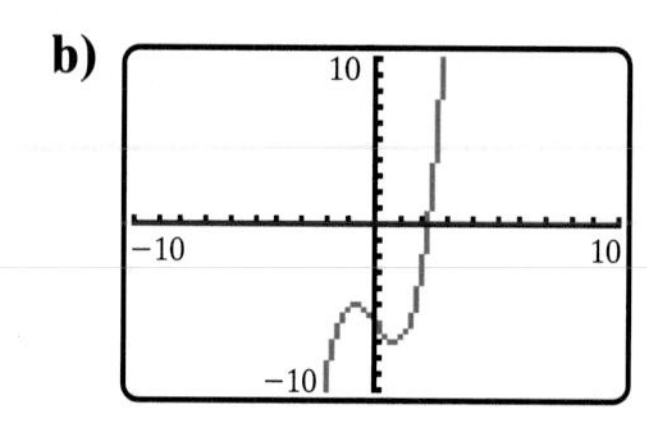

c)

90. $y = x^3 + 4x^2 - 5$

a)
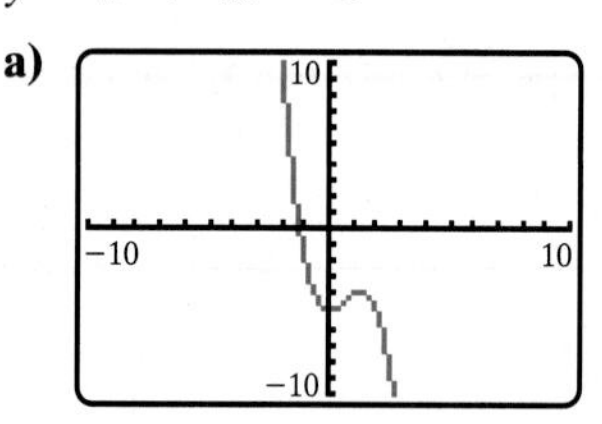

b)
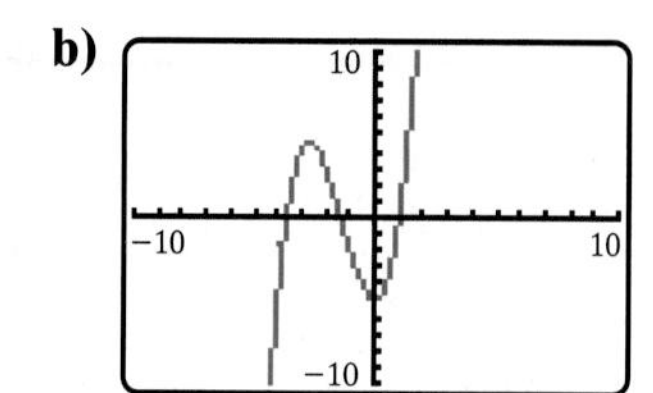

c)
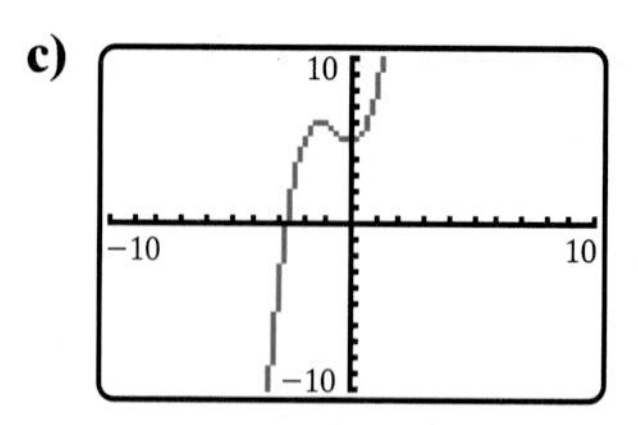

91. The chart shows the projected growth in the sale of digital cameras from 1996 through 2003. The function $s(t) = 45.4t^2 - 26.31t + 261.9$, where t is the number of years since 1996 and $0 \le t \le 7$, can be used to estimate the sales, in thousands, of digital cameras.

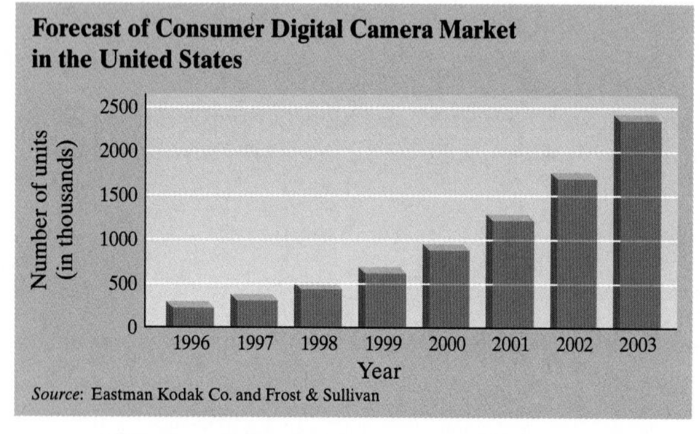

a) Using the function, estimate the number of digital cameras sold in 2000.

b) Compare your answer from part **a)** to the bar graph. Does the graph support your answer?

92. Inflation erodes our purchasing power. Because of inflation, we will pay more for the same goods in the future than we pay for them today. The function $C(t) = 0.31t^2 + 0.59t + 9.61$, where t is years since 1997, approximates the cost, in thousands of dollars, for purchasing in the future what \$10,000 would purchase in 1997. This function is based on a 6% annual inflation rate and $0 \le t \le 25$. Estimate the cost in 2012 for goods that cost \$10,000 in 1997.

93. Each year the number of centenarians, people 100 years old and older, increases. The U.S. Bureau of the Census predicts that by 2050 there will be more than 800,000 centenarians. The number of centenarians, N, in thousands, from 1995 to 2050 can be approximated by the function $N(t) = 0.28t^2 - 2.84t + 78.97$, where t is the number of years since 1995 and $0 \le t \le 55$.

a) Estimate the number of centenarians in 2020.

b) Estimate the number of centenarians in 2045.

94. Data provided by the American Medical Association shows that the number of medical school applications, A, in thousands, in the United States from 1986 through 1997 can be closely approximated by the function $A(t) = -0.11t^3 + 1.83t^2 - 6.19t + 31.82$, where t = years since 1986 and $0 \le t \le 11$.

a) Estimate the number of medical school applications in 1989.

b) Estimate the number of medical school applications in 1995.

Answer Exercises 95 and 96 using a graphing calculator if you have one. If you do not have a graphing calculator, draw the graphs in part ***a)*** *by plotting points. Then answer parts* ***b)*** *through* ***e)***.

95. a) Graph

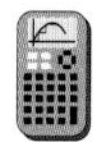

$$y_1 = x^3$$

$$y_2 = x^3 - 3x^2 - 3$$

b) In both graphs, for values of $x > 3$, do the functions increase or decrease as x increases?

c) When the leading term of a polynomial function is x^3, the polynomial must increase for $x > a$, where a is some real number greater than 0. Explain why this must be so.

d) In both graphs, for values of $x < -3$, do the functions increase or decrease as x decreases?

e) When the leading term of a polynomial function is x^3, the polynomial must decrease for $x < a$, where a is some real number less than 0. Explain why this must be so.

96. **a)** Graph

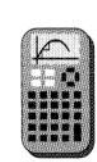

$$y_1 = x^4$$
$$y_2 = x^4 - 6x^2$$

b) In each graph, for values of $x > 3$, are the functions increasing or decreasing as x increases?

c) When the leading term of a polynomial function is x^4, the polynomial must increase for $x > a$, where a is some real number greater than 0. Explain why this must be so.

d) In each graph, for values of $x < -3$, are the functions increasing or decreasing as x decreases?

e) When the leading term of a polynomial function is x^4, the polynomial must increase for $x < a$, where a is some real number less than 0. Explain why this must be so.

Challenge Problems

*Determine which of the graphs—**a)**, **b)**, or **c)**—is the graph of the given equation. Explain how you determined your answer.*

97. $y = -x^4 + 3x^3 - 5$

a)

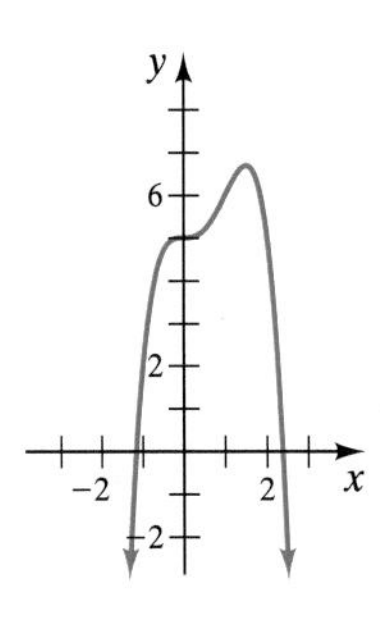

b)

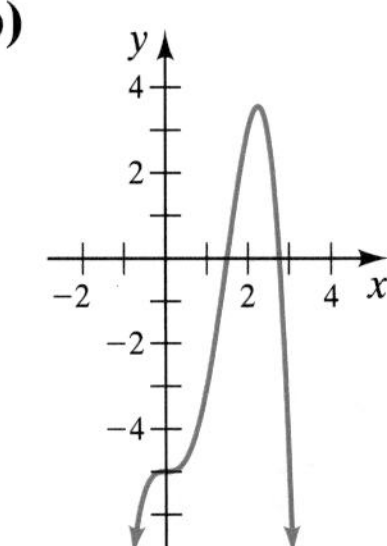

c)

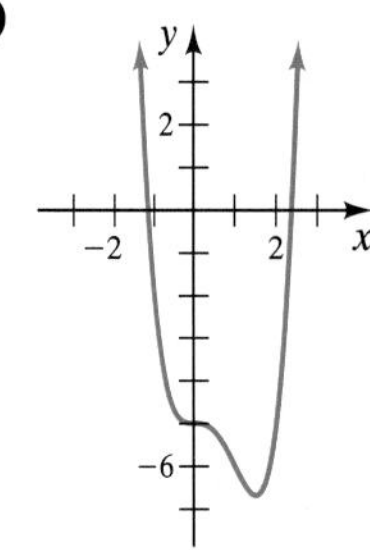

98. $y = 2x^4 + 9x^2 - 5$

a)

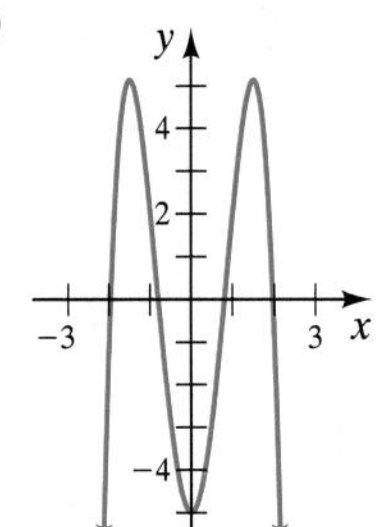

b)

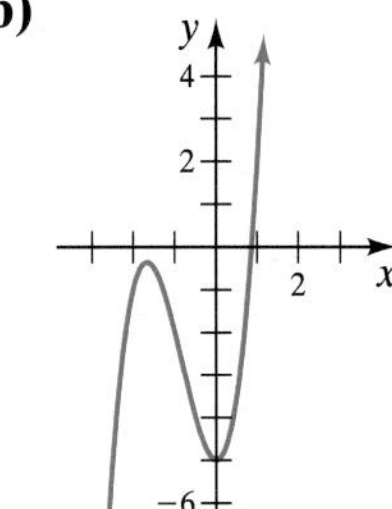

c)

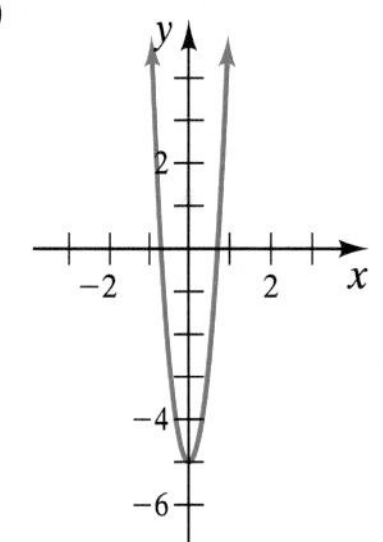

Group Activity

Discuss and answer Exercises 99 and 100 as a group.

99. If the leading term of a polynomial function is $3x^3$, which of the following could possibly be the graph of the polynomial? Explain. Consider what happens for large positive values of x and for large negative values of x.

a)

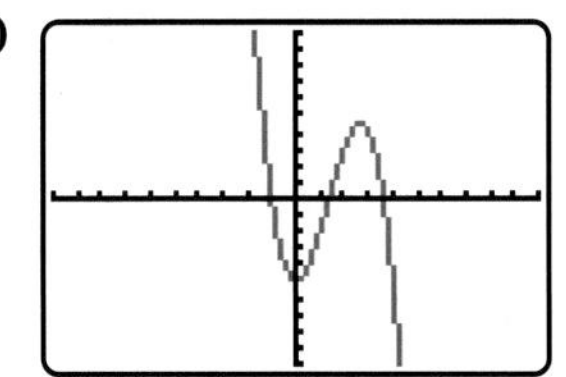

b)

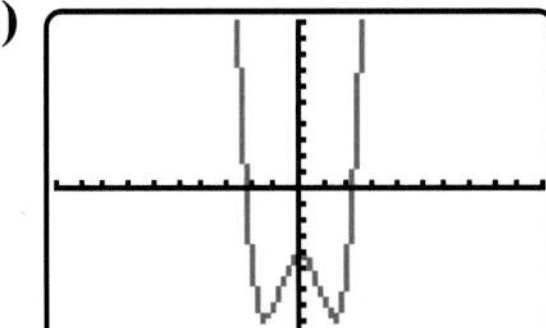

c)

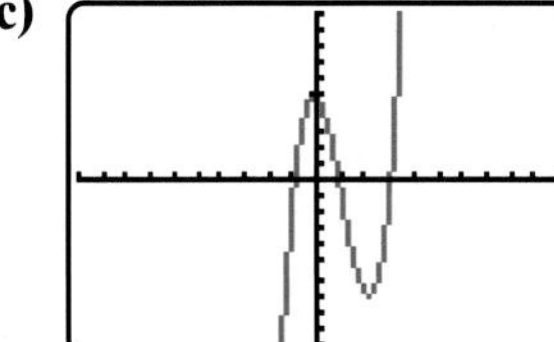

100. If the leading term of a polynomial is $-2x^4$, which of the following could possibly be the graph of the polynomial? Explain.

a) **b)** **c)**

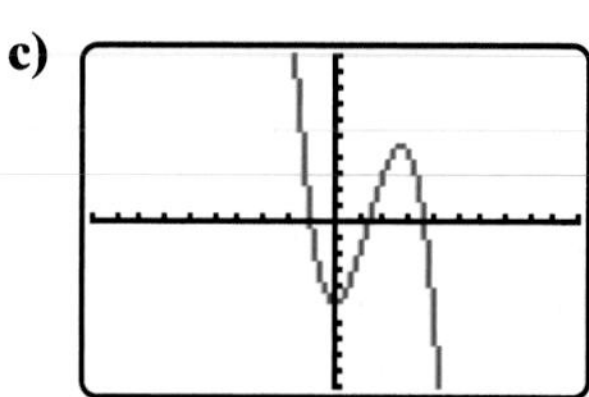

Cumulative Review Exercises

Consider the system of equations

$$x - 4y = -16$$
$$2x + 3y = -10$$

[4.1] **101.** Solve using the addition method

[4.4] **102.** Solve using matrices.

[4.5] **103.** Solve using determinants

[4.3] **104.** The sum of three numbers is 12. If the sum of the two smaller numbers equals the largest number, and the largest number is four less than twice the middle-sized number, find the three numbers.

5.2 MULTIPLICATION OF POLYNOMIALS

SSM VIDEO 5.2 CD Rom

1. **Multiply a monomial by a polynomial.**
2. **Multiply a binomial by a binomial.**
3. **Multiply a polynomial by a polynomial.**
4. **Find the square of a binomial.**
5. **Find the product of the sum and difference of the same two terms.**
6. **Find the product of polynomial functions.**

1 Multiply a Monomial by a Polynomial

In Section 3.6 we added and subtracted functions, but we did not multiply polynomial functions. After this section you will be able to find the product of functions, that is, $(f \cdot g)(x)$.

To multiply polynomials, you must remember that *each term of one polynomial must multiply each term of the other polynomial.* This results in monomials multiplying monomials. To multiply monomials, we use the product rule of exponents, $a^m \cdot a^n = a^{m+n}$, which was presented in Section 1.5.

EXAMPLE 1 Multiply. **a)** $(3x^2y)(4x^5y^3)$ **b)** $(-2a^4b^7)(-3a^8b^3c)$

Solution **a)**
$$\begin{aligned}(3x^2y)(4x^5y^3) &= 3 \cdot 4 \cdot x^2 \cdot x^5 \cdot y \cdot y^3\\ &= 12x^{2+5}y^{1+3}\\ &= 12x^7y^4\end{aligned}$$

b)
$$\begin{aligned}(-2a^4b^7)(-3a^8b^3c) &= (-2)(-3)a^4 \cdot a^8 \cdot b^7 \cdot b^3 \cdot c\\ &= 6a^{4+8}b^{7+3}c\\ &= 6a^{12}b^{10}c\end{aligned}$$

Recall that expressions that are multiplied are called factors. In Example 1a) both $3x^2y$ and $4x^5y^3$ are *factors* of the product $12x^7y^4$.

When multiplying a monomial by a binomial, we can use the distributive property. When multiplying a monomial by a polynomial that contains more than two terms we can use the **expanded form of the distributive property**.

Distributive Property, Expanded Form

$$a(b + c + d + \cdots + n) = ab + ac + ad + \cdots + an$$

EXAMPLE 2 Multiply. **a)** $3x^2\left(\frac{1}{6}x^3 - 5x^2\right)$ **b)** $2xy(3x^2y + 6xy^2 + 4)$

Solution **a)** $3x^2\left(\frac{1}{6}x^3 - 5x^2\right) = 3x^2\left(\frac{1}{6}x^3\right) - 3x^2(5x^2) = \frac{1}{2}x^5 - 15x^4$

b) $2xy(3x^2y + 6xy^2 + 4) = (2xy)(3x^2y) + (2xy)(6xy^2) + (2xy)(4)$

$= 6x^3y^2 + 12x^2y^3 + 8xy$

NOW TRY EXERCISE 13

2) Multiply a Binomial by a Binomial

Consider multiplying $(a + b)(c + d)$. Treating $(a + b)$ as a single term and using the distributive property, we get

$$\begin{aligned}(a + b)(c + d) &= (a + b)\,c + (a + b)\,d \\ &= ac + bc + ad + bd\end{aligned}$$

When multiplying a binomial by a binomial, each term of the first binomial must be multiplied by each term of the second binomial, then all like terms are combined.

Binomials can be multiplied vertically as well as horizontally.

EXAMPLE 3 Multiply $(3x + 2)(x - 5)$.

Solution List the binomials in descending order of the variable one beneath the other. It makes no difference which one is placed on top. Then multiply each term of the top binomial by each term of the bottom binomial, as shown. Remember to align like terms so that they can be added.

$$\begin{array}{lrl} & 3x + 2 & \\ & x - 5 & \\ \hline -5(3x + 2) \longrightarrow & -15x - 10 & \text{Multiply the top binomial by } -5. \\ x(3x + 2) \longrightarrow & 3x^2 + 2x & \text{Multiply the top binomial by } x. \\ \hline & 3x^2 - 13x - 10 & \text{Add like terms in columns.} \end{array}$$

In Example 3 the binomials $3x + 2$ and $x - 5$ are *factors* of the trinomial $3x^2 - 13x - 10$.

The FOIL Method

A convenient way to multiply two binomials is called the FOIL method. To multiply two binomials using the FOIL method, list the binomials side by side. The word FOIL indicates that you multiply the First terms, Outer terms, Inner terms, and Last terms of the two binomials. This procedure is illustrated in Example 4, where we multiply the same two binomials we multiplied in Example 3.

EXAMPLE 4 Multiply $(3x + 2)(x - 5)$ using the FOIL method.

Solution

F L

$(3x + 2)(x - 5)$

I

O

$$\begin{array}{ccccccccc} & \text{F} & & \text{O} & & \text{I} & & \text{L} & \\ & (3x)(x) & + & (3x)(-5) & + & (2)(x) & + & (2)(-5) & \\ = & 3x^2 & - & 15x & + & 2x & - & 10 & = 3x^2 - 13x - 10 \end{array}$$

We performed the multiplications following the FOIL order. However, any order could be followed as long as each term of one binomial is multiplied by each term of the other binomial. We use FOIL rather than OILF or any other combination of letters because it is easier to remember.

NOW TRY EXERCISE 19

3) Multiply a Polynomial by a Polynomial

When multiplying a trinomial by a binomial or a trinomial by a trinomial, every term of the first polynomial must be multiplied by every term of the second polynomial. It is helpful for aligning terms to place each polynomial in descending order, if not given that way.

EXAMPLE 5 Multiply $x^2 + 2 - 3x$ by $2x^2 - 3$.

Solution Since the trinomial is not in descending order, rewrite it as $x^2 - 3x + 2$.

Place the longer polynomial on top, then multiply. Make sure you align like terms as you multiply so that the terms can be added more easily.

$$\begin{array}{rrl} & x^2 - 3x + 2 & \textit{Trinomial written in descending order} \\ & 2x^2 - 3 & \\ \hline -3(x^2 - 3x + 2) \longrightarrow & -3x^2 + 9x - 6 & \textit{Multiply top expression by } -3. \\ 2x^2(x^2 - 3x + 2) \rightarrow & 2x^4 - 6x^3 + 4x^2 \qquad\quad & \textit{Multiply top expression by } 2x^2. \\ \hline & 2x^4 - 6x^3 + x^2 + 9x - 6 & \textit{Add like terms in columns.} \end{array}$$

EXAMPLE 6 Multiply $3x^2 + 6xy - 5y^2$ by $x + 3y$.

Solution

$$\begin{array}{rrl} & 3x^2 + 6xy - 5y^2 & \\ & x + 3y & \\ \hline 3y(3x^2 + 6xy - 5y^2) \longrightarrow & 9x^2y + 18xy^2 - 15y^3 & \textit{Multiply top expression by } 3y. \\ x(3x^2 + 6xy - 5y^2) \rightarrow & 3x^3 + 6x^2y - 5xy^2 \qquad\quad & \textit{Multiply top expression by } x. \\ \hline & 3x^3 + 15x^2y + 13xy^2 - 15y^3 & \textit{Add like terms in columns.} \end{array}$$

NOW TRY EXERCISE 31

4) Find the Square of a Binomial

Now we will study some special formulas. We must often *square a binomial*, so we have special formulas for doing so.

Square of a Binomial

$$(a + b)^2 = a^2 + 2ab + b^2$$
$$(a - b)^2 = a^2 - 2ab + b^2$$

If you forget the formulas, you can easily derive them by multiplying $(a + b)(a + b)$ and $(a - b)(a - b)$.

Examples 7 and 8 illustrate the use of the square of a binomial formula.

EXAMPLE 7 Expand. **a)** $(3x + 5)^2$ **b)** $(4x^2 - 3y)^2$

Solution

a) $(3x + 5)^2 = (3x)^2 + 2(3x)(5) + (5)^2$
$= 9x^2 + 30x + 25$

b) $(4x^2 - 3y)^2 = (4x^2)^2 - 2(4x^2)(3y) + (3y)^2$
$= 16x^4 - 24x^2y + 9y^2$

Squaring binomials, as in Example 7, can also be done using the FOIL method.

Avoiding Common Errors

Remember the middle term when squaring a binomial.

CORRECT	INCORRECT
$(x + 2)^2 = (x + 2)(x + 2)$	~~$(x + 2)^2 = x^2 + 4$~~
$= x^2 + 4x + 4$	
$(x - 3)^2 = (x - 3)(x - 3)$	~~$(x - 3)^2 = x^2 + 9$~~
$= x^2 - 6x + 9$	

EXAMPLE 8 Expand $[x + (y - 1)]^2$.

Solution This problem looks more complicated than the previous example, but it is worked the same way as any other square of a binomial. Treat x as the first term and $(y - 1)$ as the second term. Use the formula twice.

$$[x + (y - 1)]^2 = (x)^2 + 2(x)(y - 1) + (y - 1)^2$$
$$= x^2 + (2x)(y - 1) + y^2 - 2y + 1$$
$$= x^2 + 2xy - 2x + y^2 - 2y + 1$$

None of the six terms are like terms, so no terms can be combined. Note that $(y - 1)^2$ is also the square of a binomial and was expanded as such.

NOW TRY EXERCISE 45

5 Find the Product of the Sum and Difference of the Same Two Terms

Below we multiply $(x + 6)(x - 6)$ using the FOIL method.

$$(x + 6)(x - 6) = x^2 - 6x + 6x - (6)(6) = x^2 - 6^2$$

Note that the outer and inner products add to 0. By examining this example, we see that the product of the sum and difference of the same two terms is the difference of the squares of the two terms.

Product of the Sum and Difference of the Same Two Terms

$$(a + b)(a - b) = a^2 - b^2$$

In other words, to multiply two binomials that differ only in the sign between their two terms, subtract the square of the second term from the square of the first term. Note that $a^2 - b^2$ represents a **difference of two squares**.

EXAMPLE 9 Multiply $\left(3x + \frac{2}{5}\right)\left(3x - \frac{2}{5}\right)$.

Solution This is a product of the sum and difference of the same two terms. Therefore,

$$\left(3x + \frac{2}{5}\right)\left(3x - \frac{2}{5}\right) = (3x)^2 - \left(\frac{2}{5}\right)^2 = 9x^2 - \frac{4}{25}$$

EXAMPLE 10 Multiply $(5x + y^3)(5x - y^3)$.

Solution $(5x + y^3)(5x - y^3) = (5x)^2 - (y^3)^2 = 25x^2 - y^6$

EXAMPLE 11 Multiply $[4x + (3y + 2)][4x - (3y + 2)]$.

Solution We treat $4x$ as the first term and $3y + 2$ as the second term. Then we have the sum and difference of the same two terms.

$$\begin{aligned}[4x + (3y + 2)][4x - (3y + 2)] &= (4x)^2 - (3y + 2)^2 \\ &= 16x^2 - (9y^2 + 12y + 4) \\ &= 16x^2 - 9y^2 - 12y - 4\end{aligned}$$

NOW TRY EXERCISE 49

6 Find the Product of Polynomial Functions

Earlier we learned that for functions $f(x)$ and $g(x)$, $(f \cdot g)(x) = f(x) \cdot g(x)$. Let's work one example involving multiplication of polynomial functions now.

EXAMPLE 12 Let $f(x) = x + 4$ and $g(x) = x - 2$. Find

a) $f(3) \cdot g(3)$ **b)** $(f \cdot g)(x)$ **c)** $(f \cdot g)(3)$

Solution **a)** Both $f(x)$ and $g(x)$ are polynomial functions since the expressions to the right of the equal signs are polynomials.

$$f(x) = x + 4 \qquad g(x) = x - 2$$

$$f(3) = 3 + 4 = 7 \qquad g(3) = 3 - 2 = 1$$

$$f(3) \cdot g(3) = 7 \cdot 1 = 7$$

b) From Section 3.6 we know that

$$\begin{aligned}(f \cdot g)(x) &= f(x) \cdot g(x) \\ &= (x + 4)(x - 2) \\ &= x^2 - 2x + 4x - 8 \\ &= x^2 + 2x - 8\end{aligned}$$

c) To evaluate $(f \cdot g)(3)$, substitute 3 for each x in $(f \cdot g)(x)$.

$$(f \cdot g)(x) = x^2 + 2x - 8$$
$$(f \cdot g)(3) = 3^2 + 2(3) - 8$$
$$= 9 + 6 - 8 = 7$$

Note that in part **c)** we found $(f \cdot g)(3) = 7$ and in part **a)** we found $f(3) \cdot g(3) = 7$. Thus $(f \cdot g)(3) = f(3) \cdot g(3)$, which is what we expected from what we learned in Section 3.6.

NOW TRY EXERCISE 69

In Example 12 we found that if $f(x) = x + 4$ and $g(x) = x - 2$, then $(f \cdot g)(x) = x^2 + 2x - 8$. The graphs of $y = f(x) = x + 4$, $y = g(x) = x - 2$, and $y = (f \cdot g)(x) = x^2 + 2x - 8$, are shown in Figure 5.10.

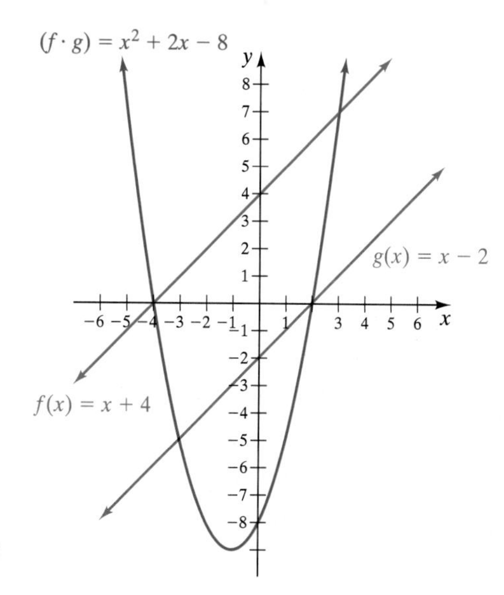

FIGURE 5.10

We see from the graphs that $f(3) = 7$, $g(3) = 1$, and $(f \cdot g)(3) = 7$, which is what we expected from Example 12. Every point on $y = x^2 + 2x - 8$ can be determined the same way. For example, $f(-4) = 0$ and $g(-4) = -6$. Since $0(-6) = 0$, $(f \cdot g)(-4) = 0$. Also $f(2) = 6$ and $g(2) = 0$, thus $(f \cdot g)(2) = 6 \cdot 0 = 0$. Notice in Figure 5.10 that the product of two linear functions gives a quadratic function.

Exercise Set 5.2

Concept/Writing Exercises

1. **a)** Explain how to multiply two binomials using the FOIL method.
   b) Make up two binomials and multiply them using the FOIL method.
   c) Multiply the same two binomials using LIOF (last, inner, outer, first).
   d) Compare the results of parts **b)** and **c)**. If they are not the same, explain why.

2. **a)** Explain how to multiply a monomial by a polynomial.
   b) Multiply $3x(4x^2 - 6x - 5)$ using your procedure from part **a)**.

3. **a)** Explain how to multiply a polynomial by a polynomial.
   b) Multiply $4 + x$ by $x^2 - 6x + 3$ using your procedure from part **a)**.

4. **a)** Explain how to expand $(2x - 3)^2$ using the formula for the square of a binomial.
   b) Expand $(2x - 3)^2$ using your procedure from part **a)**.
5. **a)** What is meant by the product of the sum and difference of the same two terms?
   b) Give an example of a problem that is the product of the sum and difference of the same two terms.
   c) How do you multiply the product of the sum and difference of the same two terms?
   d) Multiply the example you gave in part **b)** using your procedure from part **c)**.
6. Will the product of two binomials always be **a)** binomial? **b)** trinomial? Explain.
7. Will the product of two first-degree polynomials always be a second-degree polynomial? Explain.
8. **a)** Given $f(x)$ and $g(x)$, explain how you would find $(f \cdot g)(x)$.
   b) If $f(x) = x - 2$ and $g(x) = x + 2$, find $(f \cdot g)(x)$.

Practice the Skills

Multiply.

9. $(4xy)(6xy^4)$
10. $(-2xy^4)(3x^4y^6)$
11. $\left(\frac{5}{9}x^2y^5\right)\left(\frac{1}{5}x^5y^3z^2\right)$
12. $2y^3(3y^2 + 2y - 6)$
13. $-3x^2y(-2x^4y^2 + 3xy^3 + 4)$
14. $3x^4(2xy^2 + 5x^7 - 6y)$
15. $\frac{2}{3}yz(3x + 4y - 9y^2)$
16. $\frac{1}{2}x^2y(4x^5y^2 + 3x - 6y^2)$
17. $0.3a^5b^4(9.5a^6b - 4.6a^4b^3 + 1.2ab^5)$
18. $4.6m^2n(1.3m^4n^2 - 2.6m^3n^3 + 5.9n^4)$

Multiply the following binomials.

19. $(4x - 6)(3x - 5)$
20. $(x - y)(x + y)$
21. $(4 - x)(3 + 2x^2)$
22. $\left(\frac{1}{2}x + 2y\right)\left(2x - \frac{1}{3}y\right)$
23. $\left(\frac{2}{5}x - \frac{1}{5}z\right)\left(\frac{1}{3}x + z\right)$
24. $(3xy^2 + y)(4x - 3xy)$
25. $(2.3a - 1.4b)(5.6a + 4.2b)$
26. $(4.6r - 5.8s)(0.2r - 2.3s)$

Multiply the following polynomials.

27. $(x^2 - 3x + 2)(x - 4)$
28. $(7x - 3)(-2x^2 - 4x + 1)$
29. $(x - 2)(4x^2 + 9x - 2)$
30. $(5x^3 + 4x^2 - 6x + 2)(x + 5)$
31. $(a - 3b)(2a^2 - ab + 2b^2)$
32. $(x^3 - 2x^2 + 5x - 6)(2x^2 - 3x + 4)$
33. $(3x - 1)^3$
34. $(x - 2)^3$
35. $(2a^2 - 6a + 3)(3a^2 - 5a - 2)$
36. $(3m^2 - 2m + 4)(m^2 - 3m - 5)$
37. $(5r^2 - rs + 2s^2)(2r^2 - s^2)$
38. $(4a^2 - 5ab + b^2)(a^2 - 2b^2)$

Multiply using either the formula for the square of a binomial or for the product of the sum and difference of the same two terms.

39. $(x + 2)(x + 2)$
40. $(2x - 1)(2x + 1)$
41. $(2x - 3y)^2$
42. $(2x + 5y)^2$
43. $(5m^2 + 2n)(5m^2 - 2n)$
44. $(3x^2 - 4y)(3x^2 + 4y)$
45. $[y + (4 - 2x)]^2$
46. $[5x + (2y + 3)]^2$
47. $[4 - (x - 3y)]^2$
48. $[(x + y) + 4]^2$
49. $[a + (b + 2)][a - (b + 2)]$
50. $[(x - 2y) - 3]^2$

Multiply.

51. $(3y + 4)(2y - 3)$
52. $-\frac{3}{5}x^2y\left(-\frac{2}{3}xy^4 + \frac{1}{9}xy + 3\right)$
53. $\left(2x - \frac{3}{4}\right)\left(2x + \frac{3}{4}\right)$
54. $(4x - 5y)^2$
55. $\frac{2}{3}x^2y^4\left(\frac{3}{5}xy^3 - \frac{1}{4}x^4y + 2xy^3z^5\right)$
56. $-\frac{3}{5}xy^3z^2\left(-xy^2z^5 - 5xy + \frac{1}{6}xz^7\right)$
57. $(x + 3)(2x^2 + 4x - 3)$
58. $(5x + 4)(x^2 - x + 4)$
59. $(2x - 3y)(3x^2 + 4xy - 2y^2)$
60. $(3w^2 + 4)(3w^2 - 4)$
61. $(x + 3)^3$
62. $(2m + n)(3m^2 - mn + 2n^2)$
63. $[w + (3x + 4)][w - (3x + 4)]$
64. $[(3m + 2) + n][(3m + 2) - n]$

65. $(a + b)(a - b)(a^2 - b^2)$

66. $(2a + 3)(2a - 3)(4a^2 + 9)$

67. $(x - 4)(6 + x)(2x - 8)$

68. $(3x - 5)(5 - 2x)(3x + 8)$

For the functions given, find **a)** $(f \cdot g)(x)$ *and* **b)** $(f \cdot g)(4)$.

69. $f(x) = x - 5, g(x) = x + 4$

70. $f(x) = 3x - 5, g(x) = 2x^2 - 4$

71. $f(x) = 3x^2 + 2, g(x) = 4 - x$

72. $f(x) = -x^2 + 3x, g(x) = x^2 + 2$

73. $f(x) = 2x^2 + 6x - 4, g(x) = 5x + 3$

74. $f(x) = -x^2 + 2x + 7, g(x) = x^2 - 1$

Problem Solving

In Exercises 75 and 76, **a)** *find the area of the rectangle by finding the area of the four sections and adding them, and* **b)** *multiply the two sides and compare the product with your answer to part* **a)**.

75.

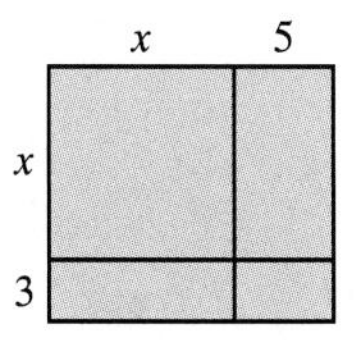

76.

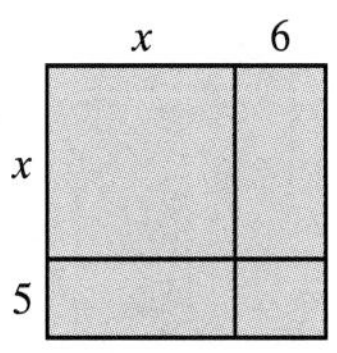

Write a polynomial expression for the area of each figure. All angles in the figures are right angles.

77.

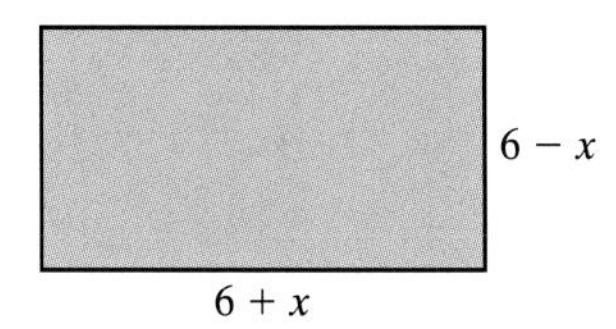

78.

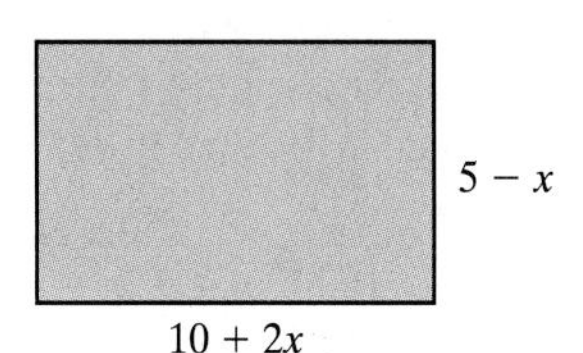

79.

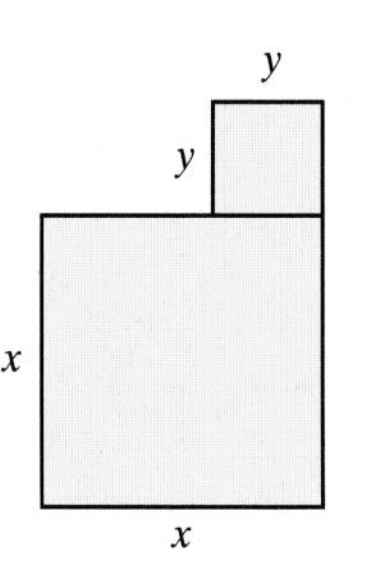

80.

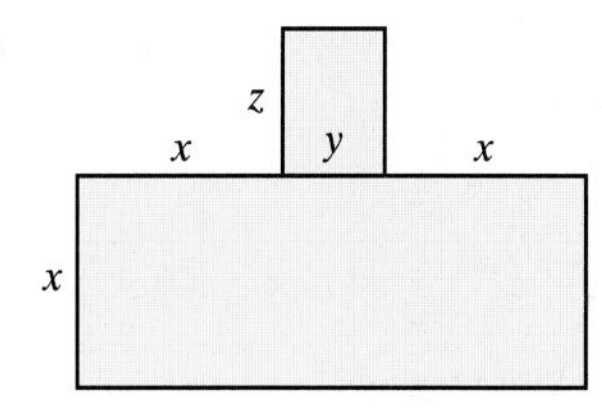

In Exercises 81 and 82, **a)** *write a polynomial expression for the area of the shaded portion of the figure.* **b)** *The area of the shaded portion is indicated above each figure. Find the area of the larger and smaller rectangles.*

81.

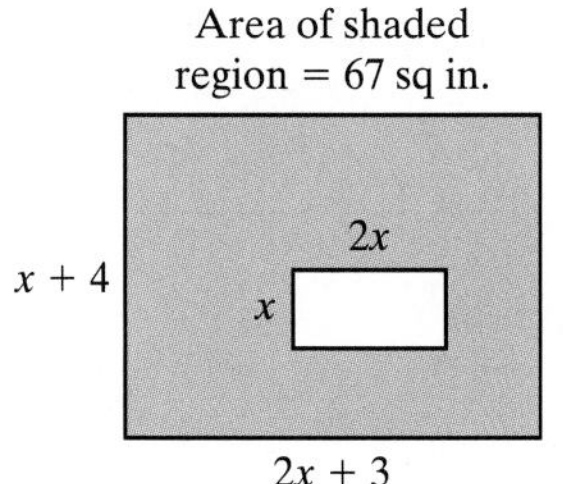

82.

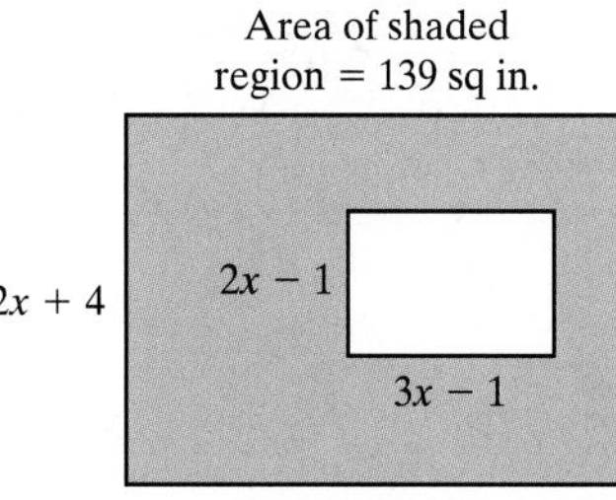

83. Write two binomials whose product is $x^2 - 36$. Explain how you determined your answer.

84. Write two binomials whose product is $4x^2 - 9$. Explain how you determined your answer.

85. Write two binomials whose product is $x^2 + 12x + 36$. Explain how you determined your answer.

86. Write two binomials whose product is $4y^2 - 12y + 9$. Explain how you determined your answer.

87. Consider the expression $a(x - n)^3$. Write this expression as a product of factors.

88. Consider the expression $P(1 - r)^4$. Write this expression as a product of factors.

89. The expression $(a + b)^2$ can be represented by the following figure.

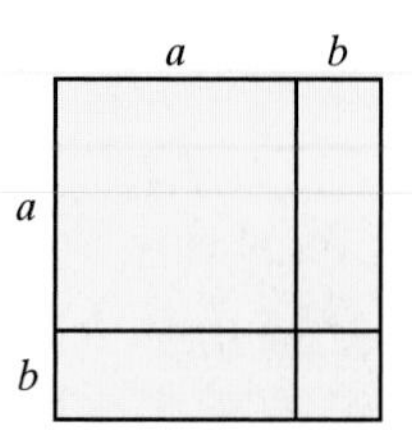

a) Explain why this figure represents $(a + b)^2$.

b) Find $(a + b)^2$ using the figure by finding the area of each of the four parts of the figure, then adding the areas together.

c) Simplify $(a + b)^2$ by multiplying $(a + b)(a + b)$.

d) How do the answers in parts **b)** and **c)** compare? If they are not the same explain why.

90. The expression $(a + b)^3$ can be represented by the following figure.

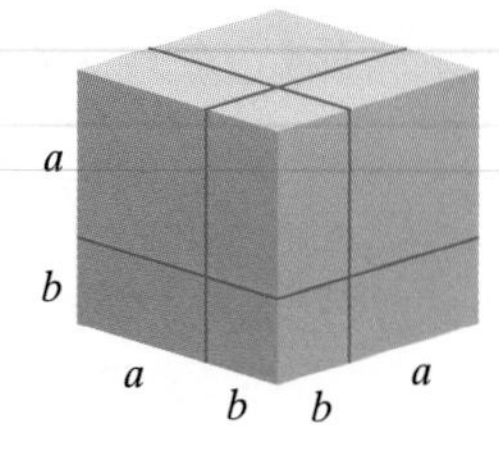

a) Explain why this figure represents $(a + b)^3$.

b) Find $(a + b)^3$ by adding the volume of the eight parts of the figure.

c) Simplify $(a + b)^3$ by multiplying.

d) How do the answers in parts **b)** and **c)** compare? If they are not the same, explain why.

91. The compound interest formula is

$$A = P\left(1 + \frac{r}{n}\right)^{nt}$$

where A is the amount, P is the principal invested, r is the annual rate of interest, n is the number of times the interest is compounded annually, and t is the time in years.

a) Simplify this formula for $n = 1$.

b) Find the value of A if $P = \$1000$, $n = 1$, $r = 6\%$, and $t = 2$ years.

92. Use the formula given in Exercise 91 to find A if $P = \$4000$, $n = 2$, $r = 8\%$, and $t = 2$ years.

93. If $f(x) = x^2 + 3x + 4$, find $f(a + b)$ by substituting $(a + b)$ for each x in the function.

94. If $f(x) = 2x^2 - x + 3$, find $f(a + b)$.

Simplify. Assume that all variables represent natural numbers.

95. $3x^t(5x^{2t-1} + 4x^{3t})$

96. $5k^{r+2}(4k^{r+2} - 3k^r + k)$

97. $(6x^m - 5)(2x^{2m} - 3)$

98. $(x^{3n} - y^{2n})(x^{2n} + 2y^{4n})$

99. $(y^{a-b})^{a+b}$

100. $(a^{m+n})^{m+n}$

Perform each polynomial multiplication.

101. $(x - 3y)^4$

102. $(2a - 4b)^4$

103. a) Explain how a multiplication in one variable, such as $(x^2 + 2x + 3)(x + 2) = x^3 + 4x^2 + 7x + 6$, may be checked using a graphing calculator.

b) Check the multiplication given in part **a)** using your grapher.

104. a) Show that the multiplication $(x^2 - 4x - 5)(x - 1) \neq x^3 + 6x^2 - 5x + 5$ using your grapher.

b) Multiply $(x^2 - 4x - 5)(x - 1)$.

c) Check your answer in part **b)** on your grapher.

Challenge Problems

Multiply.

105. $[(y + 1) - (x + 2)]^2$

106. $[(a - 2) - (a + 1)]^2$

Cumulative Review Exercises

[3.2] **107.** Is every function a relation? Is every relation a function? Explain the difference between a function and a relation.

[3.3] [3.5] **108.** Write **a)** the standard form of a linear equation; **b)** the slope–intercept form of a linear equation; **c)** the point–slope form of a linear equation.

[4.2] **109.** Solve the following system of equations.

$$\begin{aligned} -2x + 3y + 4z &= 17 \\ -5x - 3y + z &= -1 \\ -x - 2y + 3z &= 18 \end{aligned}$$

[4.3] **110.** A postage meter can stamp both first-class postage, 33 cents, and bulk postage, 22.8 cents, on envelopes. If after 1 day the meter indicates that 550 envelopes were stamped and the total cost of the postage was $140.70, how many first-class and how many bulk-mail letters were stamped?

5.3 DIVISION OF POLYNOMIALS AND SYNTHETIC DIVISION

SSM VIDEO 5.3 CD Rom

1) **Divide a polynomial by a monomial.**
2) **Divide a polynomial by a binomial.**
3) **Divide polynomials using synthetic division.**
4) **Use the Remainder Theorem.**

1) Divide a Polynomial by a Monomial

In division of polynomials, division by zero is not permitted. When we are given a division problem containing a variable in the denominator, *we always assume that the denominator is nonzero.*

To divide a polynomial by a monomial, we use the fact that

$$\frac{A+B}{C} = \frac{A}{C} + \frac{B}{C}$$

If the polynomial has more than two terms, we expand this procedure.

To Divide a Polynomial by a Monomial

Divide each term of the polynomial by the monomial.

EXAMPLE 1 Divide $\dfrac{4x^2 - 8x - 3}{2x}$.

Solution

$$\frac{4x^2 - 8x - 3}{2x} = \frac{4x^2}{2x} - \frac{8x}{2x} - \frac{3}{2x}$$
$$= 2x - 4 - \frac{3}{2x}$$

EXAMPLE 2 Divide $\dfrac{4y - 6x^4y^3 - 3x^5y^2 + 5x}{2xy^2}$.

Solution

$$\frac{4y - 6x^4y^3 - 3x^5y^2 + 5x}{2xy^2} = \frac{4y}{2xy^2} - \frac{6x^4y^3}{2xy^2} - \frac{3x^5y^2}{2xy^2} + \frac{5x}{2xy^2}$$
$$= \frac{2}{xy} - 3x^3y - \frac{3x^4}{2} + \frac{5}{2y^2}$$

NOW TRY EXERCISE 15

2) Divide a Polynomial by a Binomial

We divide a polynomial by a binomial in much the same way as we perform long division.

EXAMPLE 3 Divide $\dfrac{x^2 + 7x + 10}{x + 2}$.

Solution Rewrite the division problem as

$$x + 2\overline{)x^2 + 7x + 10}$$

Divide x^2 (the first term in $x^2 + 7x + 10$) by x (the first term in $x + 2$).

$$\frac{x^2}{x} = x$$

Place the quotient, x, above the term containing x in the dividend.

$$\begin{array}{r} x \\ x + 2\overline{)x^2 + 7x + 10} \end{array}$$

Next, multiply the x by $x + 2$ as you would do in long division and place the product under the dividend, aligning like terms.

Times

$$\begin{array}{r} x \\ x + 2\overline{)x^2 + 7x + 10} \\ \text{Equals} \quad x^2 + 2x \longleftarrow x(x+2) \end{array}$$

Now subtract $x^2 + 2x$ from $x^2 + 7x$.

$$\begin{array}{r} x \\ x + 2\overline{)x^2 + 7x + 10} \\ \underline{-(x^2 + 2x)} \\ 5x \end{array}$$

Now bring down the next term, +10.

$$\begin{array}{r} x \\ x + 2\overline{)x^2 + 7x + 10} \\ \underline{x^2 + 2x} \\ 5x + 10 \end{array}$$

Divide $5x$ by x.

$$\frac{5x}{x} = +5$$

Place +5 above the constant in the dividend and multiply 5 by $x + 2$. Finish the problem by subtracting.

Times

$$\begin{array}{r} x + 5 \\ x + 2\overline{)x^2 + 7x + 10} \\ \underline{x^2 + 2x} \\ 5x + 10 \\ \text{Equals} \quad \underline{5x + 10} \longleftarrow 5(x+2) \\ 0 \longleftarrow \text{remainder} \end{array}$$

Thus, $\dfrac{x^2 + 7x + 10}{x + 2} = x + 5$. There is no remainder.

In Example 3, there was no remainder. Therefore, $x^2 + 7x + 10 = (x + 2)(x + 5)$. Note that $x + 2$ and $x + 5$ are both *factors* of $x^2 + 7x + 10$. In a division problem if there is no remainder, then both the divisor and quotient are factors of the dividend.

When writing an answer in a division problem when there is a remainder, write the remainder over the divisor and add this expression to the quotient. For example, suppose that the remainder in Example 3 was 4. Then the answer would be written $x + 5 + \frac{4}{x + 2}$. If the remainder in Example 3 was -7, the answer would be written $x + 5 + \frac{-7}{x + 2}$, which we would simplify to $x + 5 - \frac{7}{x + 2}$.

EXAMPLE 4 Divide $\frac{6x^2 - 5x + 5}{2x + 3}$.

Solution In this example we will not show the change of sign in the subtractions.

$$
\begin{array}{r l}
3x - 7 & \\
2x + 3\overline{)6x^2 - 5x + 5} & \\
6x^2 + 9x & \leftarrow 3x(2x + 3) \\
\hline
-14x + 5 & \\
-14x - 21 & \leftarrow -7(2x + 3) \\
\hline
26 & \leftarrow \text{remainder}
\end{array}
$$

Thus, $\frac{6x^2 - 5x + 5}{2x + 3} = 3x - 7 + \frac{26}{2x + 3}$.

When dividing a polynomial by a binomial, the answer may be *checked* by multiplying the divisor by the quotient and then adding the remainder. You should obtain the polynomial you began with. To check Example 4, we do the following:

$$
\begin{aligned}
(2x + 3)(3x - 7) + 26 &= 6x^2 - 5x - 21 + 26 \\
&= 6x^2 - 5x + 5
\end{aligned}
$$

NOW TRY EXERCISE 25

Since we got the polynomial we began with, our division is correct.

When you are dividing a polynomial by a binomial, you should list both the polynomial and binomial in descending order. If a term of any degree is missing, it is often helpful to include that term with a numerical coefficient of 0. For example, when dividing $(6x^2 + x^3 - 4) \div (x - 2)$, we rewrite the problem as $(x^3 + 6x^2 + 0x - 4) \div (x - 2)$ before beginning the division.

EXAMPLE 5 Divide $(4x^2 - 12x + 3x^5 - 17)$ by $(-2 + x^2)$.

Solution Write both the dividend and divisor in descending powers of the variable x. This gives $(3x^5 + 4x^2 - 12x - 17) \div (x^2 - 2)$. Where a power of x is missing, add that power of x with a coefficient of 0, then divide.

$$
\begin{array}{r l}
3x^3 \qquad\qquad + 6x + 4 & \\
x^2 + 0x - 2\overline{)3x^5 + 0x^4 + 0x^3 + 4x^2 - 12x - 17} & \\
3x^5 + 0x^4 - 6x^3 \qquad\qquad\qquad\qquad & \leftarrow 3x^3(x^2 + 0x - 2) \\
\hline
6x^3 + 4x^2 - 12x \qquad & \\
6x^3 + 0x^2 - 12x \qquad & \leftarrow 6x(x^2 + 0x - 2) \\
\hline
4x^2 + 0x - 17 & \\
4x^2 + 0x - 8 & \leftarrow 4(x^2 + 0x - 2) \\
\hline
-9 & \leftarrow \text{remainder}
\end{array}
$$

In obtaining the answer we performed the divisions

$$\frac{3x^5}{x^2} = 3x^3 \qquad \frac{6x^3}{x^2} = 6x \qquad \frac{4x^2}{x^2} = 4$$

The quotients $3x^3$, $6x$, and 4 were placed above their like terms in the dividend. The answer is $3x^3 + 6x + 4 - \frac{9}{x^2 - 2}$. You should check this answer for yourself by multiplying the divisor by the quotient and adding the remainder.

NOW TRY EXERCISE 41

3) Divide Polynomials Using Synthetic Division

When a polynomial is divided by a binomial of the form $x - a$, the division process can be greatly shortened by a process called **synthetic division**. Consider the following examples. In the example on the right, we use only the numerical coefficients.

$$\begin{array}{r} 2x^2 + 5x - 4 \\ x - 3\overline{)2x^3 - x^2 - 19x + 15} \\ \underline{2x^3 - 6x^2} \qquad\qquad \\ 5x^2 - 19x \qquad \\ \underline{5x^2 - 15x} \qquad \\ -4x + 15 \\ \underline{-4x + 12} \\ 3 \end{array} \qquad\qquad \begin{array}{r} 2 + 5 - 4 \\ 1 - 3\overline{)2 - 1 - 19 \quad 15} \\ \underline{2 - 6} \qquad\qquad \\ 5 - 19 \qquad \\ \underline{5 - 15} \qquad \\ -4 \quad 15 \\ \underline{-4 \quad 12} \\ 3 \end{array}$$

Note that the variables do not play a role in determining the numerical coefficients of the quotient. This division problem can be done more quickly and easily using synthetic division.

Following is an explanation of how we use synthetic division. Consider again the division

$$\frac{2x^3 - x^2 - 19x + 15}{x - 3}$$

1. Write the dividend in descending powers of x. Then list the numerical coefficients of each term in the dividend. If a term of any degree is missing, place 0 in the appropriate position to serve as a placeholder. In the problem above the numerical coefficients of the dividend are

$$2 \quad -1 \quad -19 \quad 15$$

2. When dividing by a binomial of the form $x - a$, place a to the left of the line of numbers from step 1. In this problem we are dividing by $x - 3$; thus, $a = 3$. We write

$$\underline{3|} \quad 2 \quad -1 \quad -19 \quad 15$$

3. Leave some space under the row of coefficients, then draw a horizontal line. Bring down the first coefficient on the left as follows:

$$\begin{array}{r|rrrr} \underline{3} & 2 & -1 & -19 & 15 \\ & & & & \\ \hline & 2 & & & \end{array}$$

4. Multiply the 3 by the number brought down, the 2, to get 6. Place the 6 under the next coefficient, the −1. Then add −1 + 6 to get 5.

3	2	−1	−19	15
		6		
	2	5		

5. Multiply the 3 by the sum 5 to get 15. Place 15 under −19. Then add to get −4. Repeat this procedure as illustrated.

3	2	−1	−19	15
		6	15	−12
	2	5	−4	3

In the last row, the first three numbers are the numerical coefficients of the quotient, as shown in the long division. The last number, 3, is the remainder obtained by long division. The quotient must be one less than the degree of the dividend since we are dividing by $x - 3$. The original dividend was a third-degree polynomial. Therefore, the quotient must be a second-degree polynomial. Use the first three numbers from the last row as the coefficients of a second-degree polynomial in x. This gives $2x^2 + 5x - 4$, which is the quotient. The last number, 3, is the remainder. Therefore,

$$\frac{2x^3 - x^2 - 19x + 15}{x - 3} = 2x^2 + 5x - 4 + \frac{3}{x - 3}$$

EXAMPLE 6 Use synthetic division to divide.

$$(6 - x^2 + x^3) \div (x + 2)$$

Solution First, list the terms of the dividend in descending order of x.

$$(x^3 - x^2 + 6) \div (x + 2)$$

Since there is no first-degree term, insert 0 as a placeholder when listing the numerical coefficients. Since $x + 2 = x - (-2)$, $a = -2$.

−2	1	−1	0	6	
		−2	6	−12	
	1	−3	6	−6	← Remainder

Since the dividend is a third-degree polynomial, the quotient must be a second degree polynomial. The answer is $x^2 - 3x + 6 - \dfrac{6}{x + 2}$.

EXAMPLE 7 Use synthetic division to divide.

$$(3x^4 + 11x^3 - 20x^2 + 7x + 35) \div (x + 5)$$

Solution

−5	3	11	−20	7	35	
		−15	20	0	−35	
	3	−4	0	7	0	← Remainder

Since the dividend is of the fourth degree, the quotient must be of the third degree. The quotient is $3x^3 - 4x^2 + 0x + 7$ with no remainder. This can be simplified to $3x^3 - 4x^2 + 7$.

In Example 7, since there was no remainder, both $x + 5$ and $3x^3 - 4x + 7$ are *factors* of $3x^4 + 11x^3 - 20x^2 + 7x + 35$. Furthermore, since both are factors,

NOW TRY EXERCISE 53

$$(x + 5)(3x^3 - 4x^2 + 7) = 3x^4 + 11x^3 - 20x^2 + 7x + 35$$

4 Use the Remainder Theorem

In Example 6, when we divided $x^3 - x^2 + 6$ by $x + 2$, we found that the remainder was -6. If we write $x + 2$ as $x - (-2)$ and evaluate the polynomial function $P(x) = x^3 - x^2 + 6$ at -2, we obtain -6.

$$P(x) = x^3 - x^2 + 6$$
$$P(-2) = (-2)^3 - (-2)^2 + 6 = -8 - 4 + 6 = -6$$

Is it just a coincidence that $P(-2)$, the value of the function at -2, is the same as the remainder when the function $P(x)$ is divided by $x - (-2)$? The answer is no. It can be shown that for any polynomial function $P(x)$, the value of the function at a, $P(a)$, has the same value as the remainder when $P(x)$ is divided by $x - a$.

To obtain the remainder when a polynomial $P(x)$ is divided by a binomial of the form $x - a$, we can use the **Remainder Theorem**.

Remainder Theorem

If the polynomial $P(x)$ is divided by $x - a$, the remainder is equal to $P(a)$.

EXAMPLE 8 Use the Remainder Theorem to find the remainder when $3x^4 + 6x^3 - 2x + 4$ is divided by $x + 4$.

Solution First we write the divisor $x + 4$ in the form $x - a$. Since $x + 4 = x - (-4)$, we evaluate $P(-4)$.

$$\begin{aligned} P(x) &= 3x^4 + 6x^3 - 2x + 4 \\ P(-4) &= 3(-4)^4 + 6(-4)^3 - 2(-4) + 4 \\ &= 3(256) + 6(-64) + 8 + 4 \\ &= 768 - 384 + 8 + 4 = 396 \end{aligned}$$

Thus, when $3x^4 + 6x^3 - 2x + 4$ is divided by $x + 4$, the remainder is 396.

Using synthetic division, we will show that the remainder in Example 8 is indeed 396.

-4	3	6	0	-2	4	
		-12	24	-96	392	
	3	-6	24	-98	396	← Remainder

If we were to graph the polynomial $P(x) = 3x^4 + 6x^3 - 2x + 4$, the value of $P(x)$, or y, at $x = -4$ would be 396.

EXAMPLE 9 Use the Remainder Theorem to determine whether $x - 5$ is a factor of $6x^2 - 25x - 25$.

Solution Let $P(x) = 6x^2 - 25x - 25$. If $P(5) = 0$, then the remainder of $(6x^2 - 25x - 25)/(x - 5)$ is 0 and $x - 5$ is a factor of the polynomial. If $P(5) \neq 0$, then there is a remainder and $x - 5$ is not a factor.

$$\begin{aligned} P(x) &= 6x^2 - 25x - 25 \\ P(5) &= 6(5)^2 - 25(5) - 25 \\ &= 6(25) - 25(5) - 25 \\ &= 150 - 125 - 25 = 0 \end{aligned}$$

Since $P(5) = 0$, $x - 5$ is a factor of $6x^2 - 25x - 25$. Note that $6x^2 - 25x - 25 = (x - 5)(6x + 5)$.

NOW TRY EXERCISE 67

Exercise Set 5.3

Concept/Writing Exercises

1. a) Explain how to divide a polynomial by a monomial.

   b) Divide $\dfrac{5x^4 - 6x^3 - 4x^2 - 12x + 7}{3x}$ using the procedure you gave in part **a)**.

2. a) Explain how to divide a trinomial in x by a binomial in x.

   b) Divide $2x^2 - 12 + 5x$ by $x + 4$ using the procedure you gave in part **a)**.

3. A trinomial divided by a binomial has a remainder of 0. Is the quotient a factor of the trinomial? Explain.

4. a) Explain how the answer may be checked when dividing a polynomial by a binomial.

   b) Use your explanation in part **a)** to check whether the following division is correct.

   $$\frac{8x^2 + 2x - 15}{4x - 5} = 2x + 3$$

   c) Check to see whether the following division is correct.

   $$\frac{6x^2 - 23x + 14}{3x - 4} = 2x - 5 - \frac{6}{3x - 4}$$

5. When dividing a polynomial by a polynomial, before you begin the division, what should you do to the polynomials?

6. Explain why $\dfrac{x - 1}{x}$ is not a polynomial.

7. a) Describe in your own words how to divide a polynomial by $(x - a)$ using synthetic division.

   b) Divide $x^2 + 3x - 4$ by $x - 5$ using the procedure in part **a)**.

8. a) State the Remainder Theorem in your own words.

   b) Find the remainder when $x^2 - 6x - 4$ is divided by $x - 1$, using the procedure you stated in part **a)**.

Practice the Skills

Divide.

9. $\dfrac{6x + 8}{2}$
10. $\dfrac{3x + 6}{2}$
11. $\dfrac{4x^2 + 2x}{2x}$
12. $\dfrac{5y^3 + 6y^2 + 3y}{3y}$
13. $\dfrac{12x^2 - 4x - 8}{4}$
14. $\dfrac{15y^6 + 5y^2}{5y^4}$
15. $\dfrac{4x^5 - 6x^4 + 12x^3 - 8x^2}{4x^2}$
16. $\dfrac{6x^2y - 9xy^2}{3xy}$
17. $\dfrac{4x^2y^2 - 8xy^3 + 3y^4}{2y^2}$
18. $\dfrac{15x^{12} - 5x^9 + 30x^6}{5x^6}$
19. $\dfrac{6x^2y - 12x^3y^2 + 9y^3}{2xy^2}$
20. $\dfrac{a^2b^2c - 6abc^2 + 5a^3b^5}{2abc^2}$
21. $\dfrac{3xyz + 6xyz^2 - 9x^3y^5z^7}{6xy}$
22. $\dfrac{6abc^3 - 5a^2b^3c^4 + 8ab^5c}{3ab^2c^3}$

Divide.

23. $\dfrac{x^2 + 4x + 3}{x + 1}$
24. $\dfrac{x^2 + 7x + 10}{x + 5}$
25. $\dfrac{2x^2 + 13x + 15}{x + 5}$
26. $\dfrac{6x^2 + 16x + 8}{3x + 2}$
27. $\dfrac{2a^2 + a - 9}{a - 2}$
28. $\dfrac{2x^2 + x - 1}{2x + 5}$
29. $\dfrac{6x^2 + x - 2}{2x - 1}$
30. $\dfrac{8c^2 + 6c - 25}{4c + 9}$
31. $\dfrac{4r^2 - 9}{2r - 3}$
32. $\dfrac{16s^2 - 9}{4s + 3}$
33. $\dfrac{9x^3 - 3x^2 - 3x + 4}{3x + 2}$
34. $\dfrac{-2x^2 - 11x - 7}{x + 5}$
35. $\dfrac{-x^3 - 6x^2 + 2x - 3}{x - 1}$
36. $\dfrac{x^3 + 3x^2 + 5x + 4}{x + 1}$
37. $\dfrac{4y^3 + 12y^2 + 7y - 3}{2y + 3}$
38. $(2x^3 + 6x - 4) \div (x + 4)$
39. $(4x^3 - 5x) \div (2x - 1)$
40. $\dfrac{3t^4 - 9t^3 + 13t^2 - 11t + 4}{t^2 - 2t + 1}$
41. $\dfrac{3x^5 + 4x^2 - 12x - 8}{x^2 - 2}$
42. $\dfrac{4x^5 - 18x^3 + 8x^2 + 18x - 12}{2x^2 - 3}$
43. $\dfrac{3x^4 + 4x^3 - 32x^2 - 5x - 20}{3x^3 - 8x^2 - 5}$
44. $\dfrac{2x^4 - 8x^3 + 19x^2 - 33x + 15}{x^2 - x + 5}$
45. $\dfrac{3x^4 + 4x^3 - 32x^2 - 5x - 20}{x + 4}$
46. $\dfrac{2x^5 + 2x^4 - 3x^3 - 15x^2 + 18}{2x^2 - 3}$

Use synthetic division to divide.

47. $(x^2 + x - 6) \div (x - 2)$

48. $(x^2 + 5x - 6) \div (x + 6)$

49. $(x^2 + 5x - 12) \div (x - 3)$

50. $(2x^2 - 9x + 15) \div (x - 6)$

51. $(3x^2 - 7x - 10) \div (x - 4)$

52. $(x^3 + 6x^2 + 4x - 7) \div (x + 5)$

53. $(4x^3 - 3x^2 + 2x) \div (x - 1)$

54. $(x^3 - 7x^2 - 13x + 5) \div (x - 2)$

55. $(3x^3 + 7x^2 - 4x + 12) \div (x + 3)$

56. $(3x^4 - 25x^2 - 20) \div (x - 3)$

57. $(y^4 - 1) \div (y - 1)$

58. $(x^4 + 16) \div (x + 4)$

59. $(y^5 + y^4 - 10) \div (y + 1)$

60. $(z^5 + 4z^4 - 10) \div (z + 1)$

61. $(3x^3 + 2x^2 - 4x + 1) \div \left(x - \frac{1}{3}\right)$

62. $(8x^3 - 6x^2 - 5x + 3) \div \left(x + \frac{3}{4}\right)$

63. $(2x^4 - x^3 + 2x^2 - 3x + 1) \div \left(x - \frac{1}{2}\right)$

64. $(9y^3 + 9y^2 - y + 2) \div \left(y + \frac{2}{3}\right)$

Determine the remainder for the following divisions using the Remainder Theorem. If the divisor is a factor of the dividend, so state.

65. $(4x^2 - 5x + 4) \div (x - 2)$

66. $(-2x^2 + 3x - 2) \div (x + 3)$

67. $(x^3 - 2x^2 + 4x - 8) \div (x - 2)$

68. $(-3x^3 + 4x - 12) \div (x + 4)$

69. $(-2x^3 - 6x^2 + 2x - 4) \div \left(x - \frac{1}{2}\right)$

70. $(-5x^3 - 6) \div \left(x - \frac{1}{5}\right)$

Problem Solving

71. The area of a rectangle is $6x^2 - 8x - 8$. If the length is $2x - 4$, find the width.

72. The area of a rectangle is $15x^2 - 29x - 14$. If the width is $5x + 2$, find the length.

In Exercises 73–74, how many times greater is the area or volume of the figure on the right than the figure on the left? Explain how you determined your answer.

73.

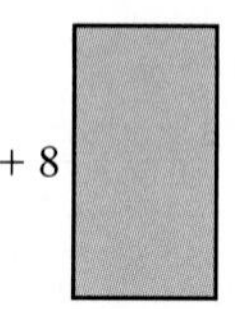

74.

75. Is it possible to divide a binomial by a monomial and obtain a monomial as a quotient? Explain.

76. **a)** Is the sum, difference, and product of two polynomials always a polynomial?

b) Is the quotient of two polynomials always a polynomial? Explain.

77. Explain how you can determine using synthetic division if an expression of the form $x - a$ is a factor of a polynomial in x.

78. Given $P(x) = ax^2 + bx + c$ and a value d such that $P(d) = 0$, explain why d is a solution to the equation $ax^2 + bx + c = 0$.

79. If $\dfrac{P(x)}{x - 4} = x + 2$, find $P(x)$.

80. If $\dfrac{P(x)}{2x + 4} = x - 3$, find $P(x)$.

81. If $\dfrac{P(x)}{x - 4} = x + 3 + \dfrac{4}{x - 4}$, find $P(x)$.

82. If $\dfrac{P(x)}{2x - 3} = 2x - 1 - \dfrac{3}{2x - 3}$, find $P(x)$.

Divide.

83. $\dfrac{2x^3 - x^2y - 7xy^2 + 2y^3}{x - 2y}$

84. $\dfrac{x^3 + y^3}{x + y}$

Divide. The answers contain fractions.

85. $\dfrac{2x^2 + 2x - 2}{2x - 3}$

86. $\dfrac{3x^3 - 5}{3x - 2}$

87. The volume of the box that follows is $2r^3 + 4r^2 + 2r$. Find w in terms of r.

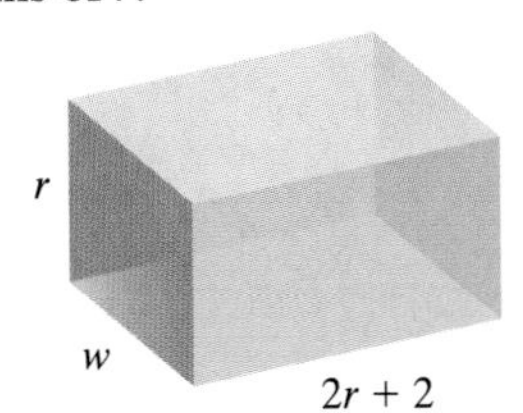

88. The volume of the box that follows is $6a^3 + a^2 - 2a$. Find b in terms of a.

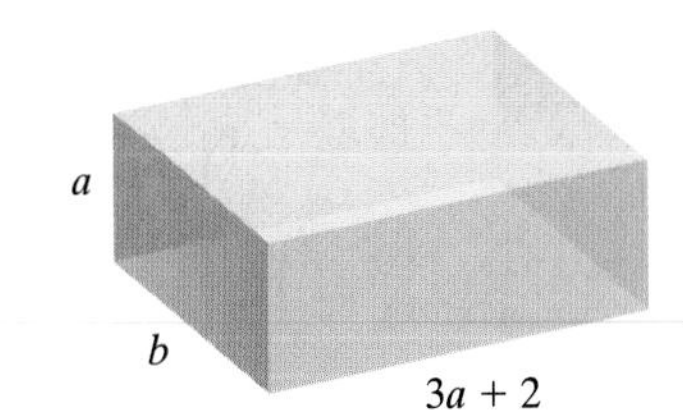

89. When a polynomial is divided by $x - 3$, the quotient is $x^2 - 3x + 4 + \dfrac{2}{x - 3}$. What is the polynomial? Explain how you determined your answer.

90. When a polynomial is divided by $2x - 3$, the quotient is $2x^2 + 6x - 5 + \dfrac{5}{2x - 3}$. What is the polynomial? Explain how you determined your answer.

Divide. Assume that all variables in the exponents are natural numbers.

91. $\dfrac{4x^{n+1} + 2x^n - 3x^{n-1} - x^{n-2}}{2x^n}$

92. $\dfrac{3x^n + 6x^{n-1} - 2x^{n-2}}{2x^{n-1}}$

93. Is $x - 1$ a factor of $x^{100} + x^{99} + \cdots + x^1 + 1$? Explain.

94. Is $x + 1$ a factor of $x^{100} + x^{99} + \cdots + x^1 + 1$? Explain.

95. Is $x + 1$ a factor of $x^{99} + x^{98} + \cdots + x^1 + 1$? Explain.

96. Divide $(0.2x^3 - 4x^2 + 0.32x - 0.64)$ by $(x - 0.4)$.

97. Synthetic division can be used to divide polynomials by binomials of the form $ax - b$, $a \neq 1$. To perform this division, divide $ax - b$ by a to obtain $x - \dfrac{b}{a}$. Then place $\dfrac{b}{a}$ to the left of the numerical coefficients of the polynomial. Work the problem as explained previously. After summing the numerical values below the line, divide all of them, except the remainder, by a. Then write the quotient of the problem using these numbers.

a) Use this procedure to divide $(9x^3 + 9x^2 + 5x + 12)$ by $(3x + 5)$.

b) Explain why we do not divide the remainder by a.

Cumulative Review Exercises

[1.5] **98.** Simplify $(4x^{-2}y^3)^{-2}$.

[2.5] **99.** Solve the inequality and graph the solution on the number line.

$$-1 < \frac{4(3x - 2)}{3} \leq 5$$

[3.2] **100.** If $f(x) = \dfrac{1}{2}x + \dfrac{3}{7}$, find $f\left(-\dfrac{2}{3}\right)$.

[3.3] **101.** Graph the equation $20x - 60y = 120$.

5.4 FACTORING A MONOMIAL FROM A POLYNOMIAL AND FACTORING BY GROUPING

SSM VIDEO 5.4 CD Rom

1. **Find the greatest common factor.**
2. **Factor a monomial from a polynomial.**
3. **Factor a common binomial factor.**
4. **Factor by grouping.**

Factoring is the opposite of multiplying. To factor an expression means to write it as a product of other expressions. For example, in Section 5.2 we learned to perform the following multiplications:

$$3x^2(6x + 3y + 5x^3) = 18x^3 + 9x^2y + 15x^5$$

and

$$(6x + 3y)(2x - 5y) = 12x^2 - 24xy - 15y^2$$

In this section we learn how to determine the *factors* of a given expression. For example, we will learn how to perform each factoring illustrated below.

$$18x^3 + 9x^2y + 15x^5 = 3x^2(6x + 3y + 5x^3)$$

and

$$12x^2 - 24xy - 15y^2 = (6x + 3y)(2x - 5y)$$

1) Find the Greatest Common Factor

To factor a monomial from a polynomial, we factor the **greatest common factor** (GCF) from each term in the polynomial. The GCF is the product of the factors common to all terms in the polynomial. For example, the GCF for $6x + 15$ is 3 since 3 is the largest number that is a factor of both $6x$ and 15. To factor, we use the distributive property.

$$6x + 15 = 3(2x + 5)$$

The 3 and the $2x + 5$ are *factors* of the polynomial $6x + 15$.

Consider the terms x^3, x^4, x^5, and x^6. The GCF of these terms is x^3, since x^3 is the highest power of x that divides all four terms. Note that the GCF of a collection of terms contains the *lowest power of the common variable.*

EXAMPLE 1 Find the GCF of the following terms.

a) y^{12}, y^4, y^9, y^7 **b)** x^3y^2, xy^4, x^4y^5 **c)** $6x^2y^3, 9x^3y^4, 24x^4$

Solution **a)** Note that y^4 is the lowest power of y that appears in any of the four terms. The GCF is, therefore, y^4.

b) The lowest power of x that appears in any of three terms is x (or x^1). The lowest power of y that appears in any of the three terms is y^2. Thus, the GCF of the three terms is xy^2.

c) The GCF is $3x^2$. Since y does not appear in $24x^4$, it is not part of the GCF.

EXAMPLE 2 Find the GCF of the following terms.

$$6(x - 3)^2, 5(x - 3), 18(x - 3)^4$$

Solution The three numbers 6, 5, and 18 have no common factor other than 1. The lowest power of $(x - 3)$ in any of the three terms is $(x - 3)$. Thus, the GCF of the three terms is $(x - 3)$.

2) Factor a Monomial from a Polynomial

When we factor a monomial from a polynomial, we are factoring out the greatest common factor. *The first step in any factoring problem is to determine and then factor out the GCF.*

To Factor a Monomial from a Polynomial

1. Determine the greatest common factor of all terms in the polynomial.
2. Write each term as the product of the GCF and another factor.
3. Use the distributive property to *factor out* the GCF

EXAMPLE 3 Factor $15x^4 - 5x^3 + 20x^2$.

Solution The GCF is $5x^2$. Write each term as the product of the GCF and another product. Then factor out the GCF.

$$\begin{aligned} 15x^4 - 5x^3 + 20x^2 &= 5x^2 \cdot 3x^2 - 5x^2 \cdot x + 5x^2 \cdot 4 \\ &= 5x^2(3x^2 - x + 4) \end{aligned}$$

To check the factoring process, multiply the factors using the distributive property. The product should be the expression with which you began. For instance, in Example 3,

CHECK:
$$\begin{aligned} 5x^2(3x^2 - x + 4) &= 5x^2(3x^2) + 5x^2(-x) + 5x^2(4) \\ &= 15x^4 - 5x^3 + 20x^2 \end{aligned}$$

EXAMPLE 4 Factor $20x^3y^3 + 6x^2y^4 - 12xy^5$.

Solution The GCF is $2xy^3$. Write each term as the product of the GCF and another product. Then factor out the GCF.

$$\begin{aligned} 20x^3y^3 + 6x^2y^4 - 12xy^5 &= 2xy^3 \cdot 10x^2 + 2xy^3 \cdot 3xy - 2xy^3 \cdot 6y^2 \\ &= 2xy^3(10x^2 + 3xy - 6y^2) \end{aligned}$$

NOW TRY EXERCISE 19

CHECK: $2xy^3(10x^2 + 3xy - 6y^2) = 20x^3y^3 + 6x^2y^4 - 12xy^5$

When the leading coefficient of a polynomial is negative, we generally factor out a common factor with a negative coefficient. This results in the leading coefficient of the remaining polynomial being positive.

EXAMPLE 5 Factor. **a)** $-12x - 18$ **b)** $-2x^3 + 6x^2 - 18x$

Solution Since the leading coefficients in parts **a)** and **b)** are negative, we factor out common factors with a negative coefficient.

a) $-12x - 18 = -6(2x + 3)$ *Factor out -6.*

b) $-2x^3 + 6x^2 - 18x = -2x(x^2 - 3x + 9)$ *Factor out $-2x$.*

EXAMPLE 6 When an object is thrown upward with a velocity of 32 feet per second from the top of a 160-foot-tall building, its distance, d, from the ground at any time, t, can be determined by the function $d(t) = -16t^2 + 32t + 160$.

a) Evaluate the distance from the ground after 3 seconds—that is, find $d(3)$.

b) Factor out the GCF from the right side of the function.

c) Evaluate $d(3)$ in factored form.

d) Compare your answers to parts **a)** and **c)**.

Solution

a)
$$\begin{aligned} d(t) &= -16t^2 + 32t + 160 \\ d(3) &= -16(3)^2 + 32(3) + 160 \quad \text{Substitute 3 for } t. \\ &= -16(9) + 96 + 160 \\ &= 112 \end{aligned}$$

b) Factor -16 from the three terms on the right side of the equal sign.

$$d(t) = -16(t^2 - 2t - 10)$$

c) $d(t) = -16(t^2 - 2t - 10)$

$$\begin{aligned} d(3) &= -16[3^2 - 2(3) - 10] && \text{Substitute 3 for } t. \\ &= -16(9 - 6 - 10) \\ &= -16(-7) \\ &= 112 \end{aligned}$$

d) The answers are the same. You may find the calculations in part **c)** easier than the calculations in part **a)**.

NOW TRY EXERCISE 59

3) Factor a Common Binomial Factor

Sometimes factoring involves factoring a binomial as the greatest common factor, as illustrated in Examples 7 through 9.

EXAMPLE 7 Factor $3x(5x - 2) + 4(5x - 2)$.

Solution The GCF is $(5x - 2)$. Factoring out the GCF gives

$$3x(5x - 2) + 4(5x - 2) = (3x + 4)(5x - 2)$$

In Example 7, we could have also used the commutative property of multiplication to write the expression as

$$(5x - 2)3x + (5x - 2)4 = (5x - 2)(3x + 4)$$

The factored forms $(3x + 4)(5x - 2)$ and $(5x - 2)(3x + 4)$ are equivalent and both are correct.

EXAMPLE 8 Factor $9(2x - 5) + 6(2x - 5)^2$.

Solution The GCF is $3(2x - 5)$. Rewrite each term placing the factors in the GCF together.

$$\begin{aligned} 9(2x - 5) + 6(2x - 5)^2 &= 3(2x - 5) \cdot 3 + 3(2x - 5) \cdot 2(2x - 5) \\ &= 3(2x - 5)[3 + 2(2x - 5)] \\ &= 3(2x - 5)[3 + 4x - 10] && \text{Distributive property.} \\ &= 3(2x - 5)(4x - 7) && \text{Simplify.} \end{aligned}$$

EXAMPLE 9 Factor $(2x - 5)(a + b) - (x - 1)(a + b)$.

Solution The binomial $a + b$ is the GCF of the two terms. We therefore factor it out.

$$\begin{aligned} (2x - 5)(a + b) - (x - 1)(a + b) &= (a + b)[(2x - 5) - (x - 1)] && \text{Factor out } (a + b). \\ &= (a + b)(2x - 5 - x + 1) && \text{Simplify.} \\ &= (a + b)(x - 4) && \text{Factors.} \end{aligned}$$

NOW TRY EXERCISE 43

4) Factor by Grouping

When a polynomial contains *four terms*, it may be possible to factor the polynomial by grouping. To **factor by grouping**, remove common factors from groups of terms. This procedure is illustrated in the following example.

EXAMPLE 10 Factor $ax + ay + bx + by$.

Solution There is no factor (other than 1) common to all four terms. However, a is common to the first two terms and b is common to the last two terms. Factor a from the first two terms and b from the last two terms.

$$ax + ay + bx + by = a(x + y) + b(x + y)$$

Now $(x + y)$ is common to both terms. Factor out $(x + y)$.

$$a(x + y) + b(x + y) = (a + b)(x + y)$$

Thus, $ax + ay + bx + by = (a + b)(x + y)$.

To Factor Four Terms by Grouping

1. Arrange the four terms into two groups of two terms each. Each group of two terms must have a GCF.
2. Factor the GCF from each group of two terms.
3. If the two terms formed in step 2 have a GCF, factor it out.

EXAMPLE 11 Factor $x^3 - 5x^2 + 2x - 10$ by grouping.

Solution There are no factors common to all four terms. However, x^2 is common to the first two terms and 2 is common to the last two terms.

$$\begin{aligned} x^3 - 5x^2 + 2x - 10 &= x^2(x - 5) + 2(x - 5) \\ &= (x^2 + 2)(x - 5) \end{aligned}$$

In Example 11 $(x - 5)(x^2 + 2)$ is also an acceptable answer. Would the answer to Example 11 change if we switch the order of $2x$ and $-5x^2$? Let's try it in Example 12.

EXAMPLE 12 Factor $x^3 + 2x - 5x^2 - 10$.

Solution Factor x from the first two terms and -5 from the last two terms.

$$\begin{aligned} x^3 + 2x - 5x^2 - 10 &= x(x^2 + 2) - 5(x^2 + 2) \\ &= (x - 5)(x^2 + 2) \end{aligned}$$

NOW TRY EXERCISE 51

Notice that we got the same results in Examples 11 and 12.

HELPFUL HINT

When factoring four terms by grouping, if the *first* and *third* terms are positive, you must factor a positive expression from both the first two terms and the last two terms to obtain a factor common to the remaining two terms (see Example 11). If the *first* term is positive and the *third* term is negative, you must factor a positive expression from the first two terms and a negative expression from the last two terms to obtain a factor common to the remaining two terms (see Example 12).

The first step in any factoring problem is to determine whether all the terms have a common factor. If so, begin by factoring out the common factor. For instance, to factor $x^4 - 5x^3 + 2x^2 - 10x$, we first factor out x from each term. Then we factor the remaining four terms by grouping, as was done in Example 11.

$$x^4 - 5x^3 + 2x^2 - 10x = x(x^3 - 5x^2 + 2x - 10)$$ Factor out the GCF, x, from all four terms.

$$= x(x - 5)(x^2 + 2)$$ Factors from Example 11

Exercise Set 5.4

Concept/Writing Exercises

1. What is the first step in *any* factoring problem?
2. What is the greatest common factor of the terms of an expression?
3. **a)** Explain how to find the greatest common factor of the terms of a polynomial.
   b) Using your procedure from part **a)**, find the greatest common factor of the polynomial
   $$6x^2y^5 - 2x^3y + 12x^9y^3$$
   c) Factor the polynomial in part **b)**.
4. When a term of a polynomial is itself the GCF, what is written in place of that term when the GCF is factored out? Explain.
5. **a)** Explain how to factor a polynomial of four terms by grouping.
   b) Factor $6x^3 - 2xy^3 + 3x^2y^2 - y^5$ by your procedure from part **a)**.
6. What is the first step when factoring $-x^2 + 8x - 15$? Explain your answer.
7. Determine the GCF of the following terms:
   $$x^4y^6, x^3y^5, xy^6, x^2y^4$$
   Explain how you determined your answer.
8. Determine the GCF of the following terms:
   $$12(x - 4)^3, 6(x - 4)^4, 3(x - 4)^5$$
   Explain how you determined your answer.

Practice the Skills

Factor out the greatest common factor.

9. $8n + 8$
10. $12x + 9$
11. $6x^2 + 3x - 9$
12. $16x^2 - 12x - 6$
13. $8p^2 - 6p + 4$
14. $12w^3 - 8w^2 - 6w$
15. $7x^5 - 9x^4 + 3x^3$
16. $45y^{12} + 30y^{10}$
17. $-24y^{15} + 9y^3 - 3y$
18. $38x^4 - 16x^5 - 9x^3$
19. $3x^2y + 6x^2y^2 + 3xy$
20. $24a^2b^2 + 16ab^4 + 64ab^3$
21. $40a^5b^4c - 8a^4b^2c^2 + 4a^2c$
22. $36xy^2z^3 + 36x^3y^2z + 9x^2yz$
23. $9p^4q^5r - 3p^2q^2r^2 + 6pq^5r^3$
24. $24x^6 + 8x^4 - 4x^3y$
25. $-52x^2y^2 - 16xy^3 + 26z$
26. $14y^3z^5 - 28y^3z^6 - 9xy^2z^2$

Factor out a factor with a negative coefficient.

27. $-6x + 2$
28. $-15m - 5$
29. $-w^2 + 7w - 5$
30. $-s^2 - 5s + 8$
31. $-3r^2 - 6r + 9$
32. $-12t^2 - 48t - 36$
33. $-6r^4s^3 + 4r^2s^4 + 2rs^5$
34. $-5m^6r^3 - 10m^4r^4 + 25mr^7$
35. $-a^4b^2c + 5a^3bc^2 + a^2b$
36. $-20x^5y^3z - 4x^4yz^2 - 8x^2y^5$

Factor.

37. $x(a - 2) + 1(a - 2)$
38. $x(b - 3) - 4(b - 3)$
39. $4a(x - 1) - 3(x - 1)$
40. $3y(a + 3) - 2(a + 3)$
41. $(x - 2)(3x + 5) - (x - 2)(5x - 4)$
42. $(x + 4)(x + 3) + (x - 1)(x + 3)$
43. $(2a + 4)(a - 3) - (2a + 4)(2a - 1)$
44. $(6b - 1)(b + 4) + (6b - 1)(2b + 5)$
45. $x^2 + 3x - 5x - 15$
46. $x^2 + 3x - 2x - 6$
47. $8n^2 - 4n - 20n + 10$
48. $12x^2 + 30x + 8x + 20$
49. $ax + ay + bx + by$
50. $ac + ad + bc + bd$
51. $x^3 - 3x^2 + 4x - 12$
52. $2x^3 + 4x^2 - 5x - 10$
53. $10m^2 - 12mn - 25mn + 30n^2$
54. $12x^2 - 9xy + 4xy - 3y^2$
55. $6x^3 + 18x^2 - 12x - 36$
56. $2a^4 - 2a^3 - 5a^2 + 5a$
57. $a^5 - a^4 + a^3 - a^2$
58. $b^4 - b^3 - b + b^2$

Problem Solving

59. When a flare is shot upward with a velocity of 128 feet per second, its height, h, in feet, above the ground at t seconds can be found by the function $h(t) = -16t^2 + 128t$.

 a) Find the height of the flare 3 seconds after it was shot.

 b) Express the function with the right side in factored form.

 c) Evaluate $h(3)$ using the factored form from part **b)**.

60. When a basketball player shoots a jump shot, the height, h, in feet, of the ball above the ground at any time t, under certain circumstances, may be found by the function $h(t) = -16t^2 + 20t + 8$.

 a) Find the height of the ball at 1 second.

 b) Express the function with the right side in factored form.

 c) Evaluate $h(1)$ using the factored form in part **b)**.

61. The area of the skating rink with semicircular ends shown is $A = \pi r^2 + 2rl$.

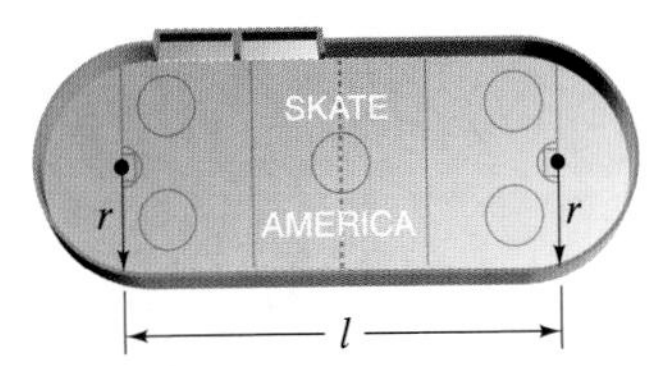

 a) Find A when $r = 20$ feet and $l = 40$ feet.

 b) Write the area A in factored form.

 c) Find A when $r = 20$ feet and $l = 40$ feet using the factored form in part **b)**.

62. The formula for finding the area of a parallelogram may be given as $A = \frac{1}{2}hb_1 + \frac{1}{2}hb_2$. Express this formula in factored form.

63. When the 2000 cars came out, the list price of one model increased by 6% over the list price of the 1999 model. Then in a special sale, the prices of all 2000 cars were reduced by 6%. The sale price can be represented by $(x + 0.06x) - 0.06(x + 0.06x)$, where x is the list price of the 1999 model.

 a) Factor out $(x + 0.06x)$ from each term.

 b) Is the sale price more or less than the price of the 1999 model?

Read Exercise 63 before working Exercises 64–66.

64. A dress is reduced by 10%, then the sale price is reduced by another 10%.

 a) Write an expression for the final price of the dress.

 b) How does the final price compare with the regular price of the dress? Use factoring in obtaining your answer.

65. The price of a Toro lawn mower is increased by 15%. Then at a 4th of July sale the price is reduced by 20%.

 a) Write an expression for the final price of the lawn mower.

 b) How does the sale price compare with the regular price? Use factoring in obtaining your answer.

66. In which of the following, **a)** or **b)**, will the final price be lower, and by how much?

 a) Decreasing the price of an item by 6%, then increasing that price by 8%.

 b) Increasing the price of an item by 6%, then decreasing that price by 8%.

Factor.

67. $5a(3x - 2)^5 + 4(3x - 2)^4$

68. $4p(2r - 3)^7 - 3(2r - 3)^6$

69. $4x^2(x - 3)^3 - 6x(x - 3)^2 + 4(x - 3)$

70. $12(p + 2q)^4 - 40(p + 2q)^3 + 12(p + 2q)^2$

71. $ax^2 + 2ax - 3a + bx^2 + 2bx - 3b$

72. $6a^2 - a^2c + 18a - 3ac + 6ab - abc$

Factor. Assume that all variables in the exponents represent natural numbers.

73. $x^{6m} - 2x^{4m}$

74. $x^{2mn} + x^{4mn}$

75. $3x^{4m} - 2x^{3m} + x^{2m}$

76. $r^{y+4} + r^{y+3} + r^{y+2}$

77. $a^r b^r + c^r b^r - a^r d^r - c^r d^r$

78. $6a^k b^k - 2a^k c^k - 9b^k + 3c^k$

79. **a)** Does $6x^3 - 3x^2 + 9x = 3x(2x^2 - x + 3)$?

b) If the above factoring is correct, what should be the value of $6x^3 - 3x^2 + 9x - [3x(2x^2 - x + 3)]$ for any value of x? Explain.

c) Select a value for x and evaluate the expression in part **b)**. Did you get what you expected? If not, explain why.

80. **a)** Determine whether the following factoring is correct.

$$3(x-2)^2 - 6(x-2) = 3(x-2)[(x-2)-2]$$
$$= 3(x-2)(x-4)$$

b) If the above factoring is correct, what should be the value of $3(x-2)^2 - 6(x-2) - [3(x-2)(x-4)]$ for any value of x? Explain.

c) Select a value for x and evaluate the expression in part **b)**. Did you get what you expected? If not, explain why.

81. Consider the factoring $8x^3 - 16x^2 - 4x = 4x(2x^2 - 4x - 1)$.

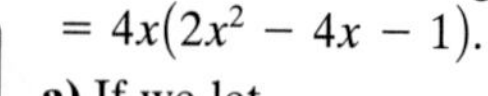

a) If we let

$$y_1 = 8x^3 - 16x^2 - 4x$$
$$y_2 = 4x(2x^2 - 4x - 1)$$

and graph each function, what should happen? Explain.

b) On your graphing calculator, graph y_1 and y_2 as given in part **a)**.

c) Did you get the results you expected?

d) When checking a factoring process by this technique, what does it mean if the graphs do not overlap? Explain.

82. Consider the factoring $2x^4 - 6x^3 - 8x^2 = 2x^2(x^2 - 3x - 4)$.

a) Enter $y_1 = 2x^4 - 6x^3 - 8x^2$

$y_2 = 2x^2(x^2 - 3x - 4)$

in your calculator.

b) If you use the TABLE feature of your calculator, how would you expect the table of values for y_1 to compare with the table of values for y_2? Explain.

c) Use the TABLE feature to show the values for y_1 and y_2 for values of x from 0 to 6.

d) Did you get the results you expected?

e) When checking a factoring process using the table feature, what does it mean if the values of y_1 and y_2 are different?

Cumulative Review Exercises

[1.4] **83.** Evaluate $\dfrac{\left(\left|\frac{1}{2}\right| - \left|-\frac{1}{3}\right|\right)^2}{-\left|\frac{1}{3}\right| \cdot \left|-\frac{2}{5}\right|}$.

[2.5] **84.** Solve the inequality $-4 < \dfrac{6-3x}{2} \le 5$ and give the solution in set builder notation.

[2.3] **85.** Kara Gutterman invested a total of \$10,000 in two savings accounts. One account earned 5% simple and the other account earned 6% simple interest annually. If the total interest earned from both accounts in 1 year is \$560, how much was invested in each account? Solve using only one variable.

[4.3] **86.** Solve Exercise 85 using two variables.

[3.3] **87.** Give the equation of the line that is parallel to the line $3x - 4y = 12$ and has a y-intercept of $(0, -2)$.

5.5 FACTORING TRINOMIALS

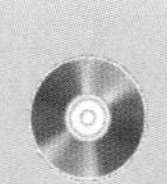

SSM VIDEO 5.5 CD Rom

1. Factor trinomials of the form $x^2 + bx + c$.
2. Factor out a common factor.
3. Factor trinomials of the form $ax^2 + bx + c$, $a \neq 1$, using trial and error.
4. Factor trinomials of the form $ax^2 + bx + c$, $a \neq 1$, using grouping.
5. Factor trinomials using substitution.

1 Factor Trinomials of the Form $x^2 + bx + c$

In this section we learn how to factor trinomials of the form $ax^2 + bx + c$, $a \neq 0$.

Trinomials	Coefficients
$3x^2 + 2x - 5$	$a = 3,\quad b = 2,\quad c = -5$
$-\frac{1}{2}x^2 - 4x + 3$	$a = -\frac{1}{2},\quad b = -4,\quad c = 3$

To Factor Trinomials of the Form $x^2 + bx + c$ (note: $a = 1$)

1. Find two numbers (or factors) whose product is c and whose sum is b.
2. The factors of the trinomial will be of the form

$$(x + \square)(x + \square)$$

One factor determined in step 1 — *Other factor determined in step 1*

If the numbers determined in step 1 are, for example, 3 and −5, the factors would be written $(x + 3)(x - 5)$. This procedure is illustrated in the following examples.

EXAMPLE 1 Factor $x^2 - x - 12$.

Solution $a = 1, b = -1, c = -12$. We must find two numbers whose product is c, which is −12, and whose sum is b, which is −1. We begin by listing the factors of −12, trying to find a pair whose sum is −1.

Factors of −12	Sum of Factors
$(1)(-12)$	$1 + (-12) = -11$
$(2)(-6)$	$2 + (-6) = -4$
$(3)(-4)$	$3 + (-4) = -1$
$(4)(-3)$	$4 + (-3) = 1$
$(6)(-2)$	$6 + (-2) = 4$
$(12)(-1)$	$12 + (-1) = 11$

The numbers we are seeking are 3 and −4 because their product is −12 and their sum is −1. Now we factor the trinomial using the 3 and −4.

$$x^2 - x - 12 = (x + 3)(x - 4)$$

One factor of −12 — Other factor of −12

Notice in Example 1 that we listed all the factors of −12. However, after the two factors whose product is c and whose sum is b are found, there is no need to go further in listing the factors. The factors were listed here to show, for example, that $(2)(-6)$ is a different set of factors than $(-2)(6)$. Note that as the positive factor increases the sum of the factors increases.

HELPFUL HINT

Consider the factors $(2)(-6)$ and $(-2)(6)$ and the sums of these factors.

Factors	Sum of Factors
$2(-6)$	$2 + (-6) = -4$
$-2(6)$	$-2 + 6 = 4$

Notice that if the signs of each number in the product are changed, the sign of the sum of factors is changed. We can use this fact to more quickly find the factors we are seeking. If, when seeking a specific sum, you get the opposite of that sum, change the sign of each factor to get the sum you are seeking.

EXAMPLE 2 Factor $x^2 - 7x + 6$.

Solution We must find two numbers whose product is 6 and whose sum is −7. Since the product is positive and the sum is negative, both numbers must be negative. Why? The numbers are −1 and −6.

Factors of 6	Sum of Factors
$(-1)(-6)$	$-1 + (-6) = -7$

Therefore,

$$x^2 - 7x + 6 = (x - 1)(x - 6)$$

Since the factors may be placed in any order, $(x - 6)(x - 1)$ is also an acceptable answer.

NOW TRY EXERCISE 15

HELPFUL HINT

Checking Factoring

Factoring problems can be checked by multiplying the factors obtained. If the factoring is correct, you will obtain the polynomial you started with. To check Example 2, we will multiply the factors using the FOIL method.

$$(x + 1)(x - 6) = x^2 - 6x + x - 6 = x^2 - 5x - 6$$

Since the product of the factors is the trinomial we began with, our factoring is correct. You should always check your factoring.

The procedure used to factor trinomials of the form $x^2 + bx + c$ can be used on other trinomials, as in the following example.

EXAMPLE 3 Factor $x^2 + 2xy - 15y^2$.

Solution We must find two numbers whose product is −15 and whose sum is 2. The two numbers are 5 and −3.

Factors of −15	Sum of Factors
$15(-3)$	$5 + (-3) = 2$

Since the last term of the trinomial contains y^2, the second term of each factor must contain y.

$$x^2 + 2xy - 15y^2 = (x + 5y)(x - 3y)$$

CHECK:
$$(x + 5y)(x - 3y) = x^2 - 3xy + 5xy - 15y^2$$
$$= x^2 + 2xy - 15y^2$$

2) Factor Out a Common Factor

The first step when factoring any trinomial is to determine whether all three terms have a common factor. If so, factor out that common factor. Then factor the remaining polynomial.

EXAMPLE 4 Factor $3x^4 - 6x^3 - 72x^2$

Solution The factor $3x^2$ is common to all three terms of the trinomial. Factor it out first.

$$3x^4 - 6x^3 - 72x^2 = 3x^2(x^2 - 2x - 24) \quad \text{Factor out } 3x^2.$$

The $3x^2$ that was factored out is a part of the answer but plays no further part in the factoring process. Now continue to factor $x^2 - 2x - 24$. Find two numbers whose product is -24 and whose sum is -2. The numbers are -6 and 4.

$$3x^2(x^2 - 2x - 24) = 3x^2(x - 6)(x + 4)$$

NOW TRY EXERCISE 41 Therefore, $3x^4 - 6x^3 - 72x^2 = 3x^2(x - 6)(x + 4)$.

3 Factor Trinomials of the Form $ax^2 + bx + c$, $a \neq 1$, Using Trial and Error

Now we will look at some examples of factoring trinomials of the form

$$ax^2 + bx + c, \quad a \neq 1$$

Two methods of factoring this type of trinomial will be illustrated. The first method, trial and error, involves trying various combinations until the correct combination is found. The second method makes use of factoring by grouping, a procedure that was presented in Section 5.4

Let's first discuss the trial-and-error method of factoring trinomials. This procedure is sometimes called the FOIL (or Reverse FOIL) method. As an aid in our explanation, we will multiply $(2x + 3)(x + 1)$ using the FOIL method.

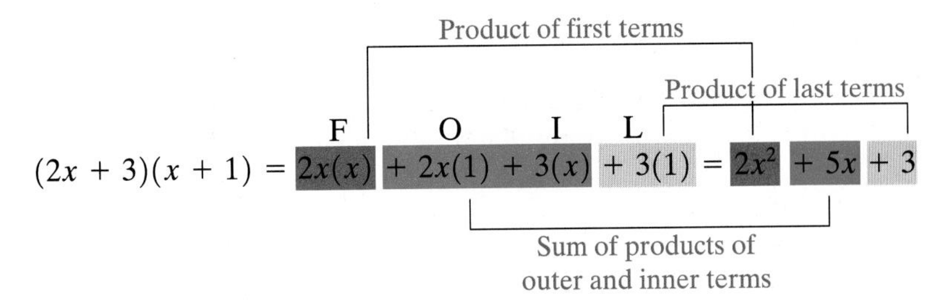

Therefore, if you are factoring the trinomial $2x^2 + 5x + 3$, you should realize that the product of the first terms of the factors must be $2x^2$, the product of the last terms must be 3, and the sum of the products of the outer and inner terms must be $5x$.

To factor $2x^2 + 5x + 3$, we begin as shown here.

$$2x^2 + 5x + 3 = (2x \qquad)(x \qquad) \quad \text{The product of first terms is } 2x^2.$$

Now we fill in the second terms using positive integers whose product is 3. Only positive integers will be considered since the product of the last terms is positive, and the sum of the products of the outer and inner terms is also positive. The two possibilities are

$$\left.\begin{array}{l}(2x + 1)(x + 3)\\(2x + 3)(x + 1)\end{array}\right\} \text{The product of the last terms is 3.}$$

To determine which factoring is correct, we find the sum of the products of the outer and inner terms. If either has a sum of $5x$, the middle term of the trinomial, that factoring is correct.

$(2x + 1)(x + 3) = 2x^2 + 6x + x + 3 = 2x^2 + 7x + 3$ *Wrong middle term*

$(2x + 3)(x + 1) = 2x^2 + 2x + 3x + 3 = 2x^2 + 5x + 3$ *Correct middle term*

Therefore, the factors of $2x^2 + 5x + 3$ are $2x + 3$ and $x + 1$. Thus,

$$2x^2 + 5x + 3 = (2x + 3)(x + 1)$$

Note that if we had begun factoring by writing

$$2x^2 + 5x + 3 = (x \qquad)(2x \qquad)$$

we could have also obtained the correct factors.

Following are guidelines for the **trial and error** method of factoring a trinomial where $a \neq 1$ and the three terms have no common factors.

To Factor Trinomials of the Form $ax^2 + bx + c$, $a \neq 1$, Using Trial and Error

1. Write all pairs of factors of the coefficient of the squared term, a.
2. Write all pairs of factors of the constant, c.
3. Try various combinations of these factors until the correct middle term, bx, is found.

EXAMPLE 5 Factor $3x^2 - 13x + 10$.

Solution First we determine that the three terms have no common factor. Next we determine that a is 3 and the only factors of 3 are 1 and 3. Therefore, we write

$$3x^2 - 13x + 10 = (3x \qquad)(x \qquad)$$

The number 10 has both positive and negative factors. However, since the product of the last terms must be positive (+10), and the sum of the products of the outer and inner terms must be negative (−13), the two factors of 10 must both be negative. (Why?) The negative factors of 10 are $(-1)(-10)$ and $(-2)(-5)$. Below is a list of the possible factors. We look for the factors that give us the correct middle term, $-13x$.

Possible Factors	Sum of Products of Outer and Inner Terms	
$(3x - 1)(x - 10)$	$-31x$	
$(3x - 10)(x - 1)$	$-13x$	⟵ *Correct middle term*
$(3x - 2)(x - 5)$	$-17x$	
$(3x - 5)(x - 2)$	$-11x$	

Thus, $3x^2 - 13x + 10 = (3x - 10)(x - 1)$.

The following Helpful Hint is very important. Study it carefully.

HELPFUL HINT

Factoring by Trial and Error

When factoring a trinomial of the form $ax^2 + bx + c$, the sign of the constant term, c, is very helpful in finding the solution. If $a > 0$, then:

1. When the constant term, c, is positive, and the numerical coefficient of the x term, b, is positive, both numerical factors will be positive.

   Example $x^2 + 7x + 12 = (x + 3)(x + 4)$
   ↑ Positive ↑ Positive ↑ Positive ↑ Positive

2. When c is positive and b is negative, both numerical factors will be negative.

   Example $x^2 - 5x + 6 = (x - 2)(x - 3)$
   ↑ Negative ↑ Positive ↑ Negative ↑ Negative

Whenever the constant, c, is positive (as in the two examples above) the sign in both factors will be the same as the sign in the x term of the trinomial.

3. When c is negative, one of the numerical factors will be positive and the other will be negative.

   Example $x^2 + x - 6 = (x + 3)(x - 2)$
   ↑ Negative ↑ Positive ↑ Negative

EXAMPLE 6 Factor $8x^2 + 8x - 30$.

Solution First we check to see whether the three terms have a common factor. We notice that 2 can be factored out.

$$8x^2 + 8x - 30 = 2(4x^2 + 4x - 15)$$

The factors of 4, the leading coefficient, are $4 \cdot 1$ and $2 \cdot 2$. Therefore the factoring will be of the form $(4x \quad)(x \quad)$ or $(2x \quad)(2x \quad)$. It makes no difference whether you start with the first set of factors or the last set of factors. We will generally start with the medium-sized factors first, so we will start with $(2x \quad)(2x \quad)$. If using these factors does not give our answer, we will work with the other set of factors. The factors of -15 are $(1)(-15)$, $(3)(-5)$, $(5)(-3)$, and $(15)(-1)$. We want our middle term to be $4x$.

Possible Factors	Sum of Products of Outer and Inner Terms
$(2x + 1)(2x - 15)$	$-28x$
$(2x + 3)(2x - 5)$	$-4x$
$(2x + 5)(2x - 3)$	$4x$

Since we found the set of factors that gives the correct x-term, we can stop. Thus

$$8x^2 + 8x - 30 = 2(2x + 5)(2x - 3)$$

In Example 6, if we compare the second and third set of factors we see they are the same except for the signs of the second terms. Notice that when the signs of the second term in each factor are switched the sum of the products of the outer and inner terms also changes sign.

Using Your Graphing Calculator

The graphing calculator can be used to check factoring problems. To check the factoring in Example 6,

$$8x^2 + 8x - 30 = 2(2x + 5)(2x - 3)$$

we let $y_1 = 8x^2 + 8x - 30$ and $y_2 = 2(2x + 5)(2x - 3)$. Then we use the TABLE feature to compare results, as in Figure 5.11.

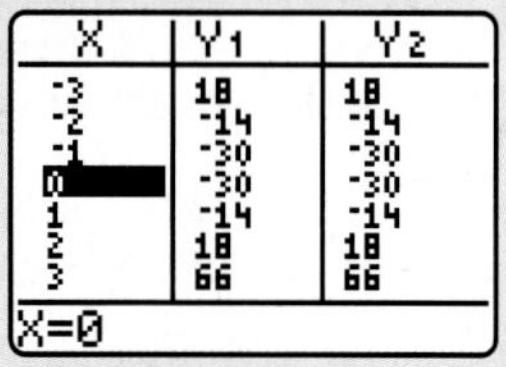

X	Y1	Y2
-3	18	18
-2	-14	-14
-1	-30	-30
0	-30	-30
1	-14	-14
2	18	18
3	66	66

X=0

FIGURE 5.11

Since y_1 and y_2 have the same values for each value of x, a mistake has not been made. This procedure can only tell you if a mistake has been made; it cannot tell you if you have factored completely. For example, $8x^2 + 8x - 30$ and $(4x + 10)(2x - 3)$ will give the same set of values.

Exercises

Use your grapher to determine whether each trinomial is factored correctly.

1. $30x^2 + 37x - 84 \stackrel{?}{=} (6x - 7)(5x + 12)$
2. $72x^2 + 20x - 35 \stackrel{?}{=} (9x - 5)(8x + 7)$

EXAMPLE 7 Factor $6x^2 - 11xy - 10y^2$.

Solution The factors of 6 are either $6 \cdot 1$ or $2 \cdot 3$. Therefore, the factors of the trinomial may be of the form $(6x \qquad)(x \qquad)$ or $(2x \qquad)(3x \qquad)$. We will begin with the middle-sized factors; thus we write

$$6x^2 - 11xy - 10y^2 = (2x \qquad)(3x \qquad)$$

The factors of -10 are $(-1)(10)$, $(1)(-10)$, $(-2)(5)$, and $(2)(-5)$. Since there are eight factors of -10, there will be eight pairs of possible factors to try. Can you list them? The correct factorization is

$$6x^2 - 11xy - 10y^2 = (2x - 5y)(3x + 2y)$$

In Example 7 we were fortunate to find the correct factors by using the form $(2x \qquad)(3x \qquad)$. If we had not found the correct factors using these, we would have tried $(6x \qquad)(x \qquad)$.

When factoring a trinomial whose leading coefficient is negative, we start by factoring out a negative number. For example,

$$\begin{aligned} -24x^3 - 60x^2 + 36x &= -12x(2x^2 + 5x - 3) \quad \text{Factor out } -12x. \\ &= -12x(2x - 1)(x + 3) \end{aligned}$$

and

$$-3x^2 + 8x + 16 = -1(3x^2 - 8x - 16) \quad \text{Factor out } -1.$$
$$= -(3x + 4)(x - 4)$$

NOW TRY EXERCISE 35

4) Factor Trinomials of the Form $ax^2 + bx + c$, $a \neq 1$, Using Grouping

Now we will discuss the **grouping** method of factoring trinomials of the form $ax^2 + bx + c, a \neq 1$.

To Factor Trinomials of the Form $ax^2 + bx + c$, $a \neq 1$, Using Grouping

1. Find two numbers whose product is $a \cdot c$ and whose sum is b.
2. Rewrite the middle term, bx, using the numbers found in step 1.
3. Factor by grouping.

EXAMPLE 8 Factor $2x^2 - 5x - 12$.

Solution We see that $a = 2$, $b = -5$, and $c = -12$. We must find two numbers whose product is $a \cdot c$ or $2(-12) = -24$, and whose sum is b, -5. The two numbers are -8 and 3 because $(-8)(3) = -24$ and $-8 + 3 = -5$. Now rewrite the middle term, $-5x$, using $-8x$ and $3x$.

$$2x^2 - 5x - 12 = 2x^2 \overbrace{- 8x + 3x}^{-5x} - 12$$

Now factor by grouping as explained in Section 5.4. Factor out $2x$ from the first two terms and 3 from the last two terms.

$$2x^2 - 5x - 12 = 2x(x - 4) + 3(x - 4)$$
$$= (2x + 3)(x - 4) \quad \text{Factor out } (x - 4).$$

Note in Example 8 that we wrote $-5x$ as $-8x + 3x$. As we show below, the same factors would be obtained if we wrote $-5x$ as $3x - 8x$. Therefore, it makes no difference which factor is listed first when factoring by grouping. Below we factor x from the first two terms and -4 from the last two terms.

$$2x^2 - 5x - 12 = 2x^2 \overbrace{+ 3x - 8x}^{-5x} - 12$$
$$= x(2x + 3) - 4(2x + 3)$$
$$= (x - 4)(2x + 3) \quad \text{Factor out } (2x + 3).$$

EXAMPLE 9 Factor $12x^2 - 19xy + 5y^2$.

Solution We must find two numbers whose product is $(12)(5) = 60$ and whose sum is -19. Since the product of the numbers is positive and their sum is negative, the two numbers must both be negative. (Why?)

The two numbers are -15 and -4 because $(-15)(-4) = 60$ and $-15 + (-4) = -19$. Now rewrite the middle term, $-19xy$, using $-15xy$ and $-4xy$. Then factor by grouping.

$$12x^2 - 19xy + 5y^2 = 12x^2 \overbrace{- 15xy - 4xy}^{-19xy} + 5y^2$$
$$= 3x(4x - 5y) - y(4x - 5y)$$
$$= (3x - y)(4x - 5y)$$

NOW TRY EXERCISE 27

Try Example 9 again, this time writing $-19xy$ as $-4xy - 15xy$. If you do it correctly, you should get the same factors.

It is important for you to realize that not every trinomial can be factored by the methods presented in this section. In Sections 8.1 and 8.2 we will give some procedures that can be used to factor polynomials that cannot be factored using only the integers (or over the set of integers). A polynomial that cannot be factored (over a specific set of numbers) is called a **prime polynomial.**

EXAMPLE 10 Factor $2x^2 + 6x + 5$.

Solution When you try to factor this polynomial, you will see that it cannot be factored using either trial and error or grouping. This polynomial is prime over the set of integers.

5) Factor Trinomials Using Substitution

Sometimes a more complicated trinomial can be factored by substituting one variable for another. The next two examples illustrate **factoring using substitution.**

EXAMPLE 11 Factor $y^4 - y^2 - 6$.

Solution If we can rewrite this expression in the form $ax^2 + bx + c$, it will be easier to factor. Since $(y^2)^2 = y^4$, if we substitute x for y^2, the trinomial becomes

$$y^4 - y^2 - 6 = (y^2)^2 - y^2 - 6$$
$$= x^2 - x - 6 \quad \text{Substitute } x \text{ for } y^2.$$

Now factor $x^2 - x - 6$.

$$= (x + 2)(x - 3)$$

Finally, substitute y^2 in place of x to obtain

$$= (y^2 + 2)(y^2 - 3) \quad \text{Substitute } y^2 \text{ for } x.$$

Thus, $y^4 - y^2 - 6 = (y^2 + 2)(y^2 - 3)$. Note that x was substituted for y^2, and then y^2 was substituted back for x.

EXAMPLE 12 Factor $3z^4 - 17z^2 - 28$.

Solution Let $x = z^2$. Then the trinomial can be written

$$3z^4 - 17z^2 - 28 = 3(z^2)^2 - 17z^2 - 28$$
$$3x^2 - 17x - 28 \quad \text{Substitute } x \text{ for } z^2.$$
$$= (3x + 4)(x - 7) \quad \text{Factor.}$$

Now substitute z^2 for x.

$$= (3z^2 + 4)(z^2 - 7) \quad \text{Substitute } z^2 \text{ for } x.$$

NOW TRY EXERCISE 55

Thus, $3z^4 - 17z^2 - 28 = (3z^2 + 4)(z^2 - 7)$.

EXAMPLE 13 Factor $2(x + 5)^2 - 5(x + 5) - 12$.

Solution We will again use a substitution, as in Examples 11 and 12. By substituting $a = x + 5$ in the equation, we obtain

$$2(x + 5)^2 - 5(x + 5) - 12$$

$$= 2a^2 - 5a - 12 \qquad \text{Substitute } a \text{ for } (x + 5).$$

Now factor $2a^2 - 5a - 12$.

$$= (2a + 3)(a - 4)$$

Finally, replace a with $x + 5$ to obtain

$$= [2(x + 5) + 3][(x + 5) - 4] \qquad \text{Substitute } (x + 5) \text{ for } a.$$

$$= [2x + 10 + 3][x + 1]$$

$$= (2x + 13)(x + 1)$$

Thus, $2(x + 5)^2 - 5(x + 5) - 12 = (2x + 13)(x + 1)$. Note that a was substituted for $x + 5$, and then $x + 5$ was substituted back for a.

In Examples 11 and 12 we used x in our substitution, whereas in Example 13 we used a. The letter selected does not affect the final answer.

NOW TRY EXERCISE 59

Exercise Set 5.5

Concept/Writing Exercises

1. When factoring any trinomial, what should the first step always be?
2. On a test, Kim Clark wrote the following factoring and did not receive full credit. Explain why Kim's factoring is not complete.
   $$15x^2 - 21x - 18 = (5x + 3)(3x - 6)$$
3. **a)** Explain in your own words the step-by-step procedure to factor $6x^2 - x - 12$.
   b) Factor $6x^2 - x - 12$ using the procedure you explained in part **a)**.
4. **a)** Explain in your own words the step-by-step procedure to factor $8x^2 - 26x + 6$.
   b) Factor $8x^2 - 26x + 6$ using the procedure you explained in part **a)**.

When factoring a trinomial of the form $ax^2 + bx + c$, what will be the signs between the terms in the binomials factors if:

5. $a > 0, b > 0$, and $c > 0$
6. $a > 0, b > 0$, and $c < 0$
7. $a > 0, b < 0$, and $c < 0$
8. $a > 0, b < 0$, and $c > 0$

Practice the Skills

Factor each trinomial completely.

9. $x^2 + 7x + 6$
10. $p^2 - 3p - 10$
11. $y^2 - 12y + 11$
12. $w^2 - 10w + 9$
13. $x^2 - 16x + 64$
14. $x^2 - 34x + 64$
15. $x^2 - 13x - 30$
16. $-a^2 + 18a - 45$
17. $x^2 - 4xy + 3y^2$

18. $x^2 - 6xy + 8y^2$
19. $z^2 - 7yz + 10y^2$
20. $-5x^2 - 20x - 15$
21. $4x^2 + 12x - 16$
22. $x^2 - 12xy - 45y^2$
***23.** $x^3 - 3x^2 - 18x$
24. $x^3 + 11x^2 - 42x$
25. $5p^2 - 8p + 3$
26. $4w^2 + 13w + 3$
27. $3x^2 - 3x - 6$
28. $-3x^2 - 14x + 5$
29. $6x^2 - 13x - 63$
30. $4x^2 + 4xy - 3y^2$
31. $30x^2 - 71x + 35$
32. $6x^3 + 5x^2 - 4x$
33. $32x^2 - 22x + 3$
34. $8x^2 - 8xy - 6y^2$
35. $18w^2 + 18wz - 8z^2$
36. $35x^2 + 13x - 12$
37. $8x^2 + 30xy - 27y^2$
38. $9x^2 - 104x - 48$
39. $100b^2 - 90b + 20$
40. $x^5y - 3x^4y - 18x^3y$
41. $a^3b^5 - a^2b^5 - 12ab^5$
42. $a^3b + 2a^2b - 35ab$
43. $3b^4c - 18b^3c^2 + 27b^2c^3$
44. $6p^3q^2 - 24p^2q^3 - 30pq^4$
45. $8m^8n^3 + 4m^7n^4 - 24m^6n^5$
46. $18x^2 + 9x - 20$
47. $30x^2 - x - 20$
48. $36x^2 - 23x - 8$
49. $8x^4y^4 + 24x^3y^4 - 32x^2y^4$
50. $5a^3b^2 - 8a^2b^3 + 3ab^4$

Factor each trinomial completely.

51. $x^4 + x^2 - 6$
52. $x^4 - 3x^2 - 10$
53. $x^4 + 5x^2 + 6$
54. $x^4 - 2x^2 - 15$
55. $6a^4 + 5a^2 - 25$
56. $(2x + 1)^2 + 2(2x + 1) - 15$
57. $4(x + 1)^2 + 8(x + 1) + 3$
58. $(2x + 3)^2 - (2x + 3) - 6$
59. $6(a + 2)^2 - 7(a + 2) - 5$
60. $6(p - 5)^2 + 11(p - 5) + 3$
61. $a^2b^2 - 8ab + 15$
62. $x^2y^2 + 10xy + 24$
63. $3x^2y^2 - 2xy - 5$
64. $3p^2q^2 + 11pq + 6$
65. $2a^2(5 - a) - 7a(5 - a) + 5(5 - a)$
66. $2y^2(y + 2) + 13y(y + 2) + 15(y + 2)$
67. $2x^2(x - 3) + 7x(x - 3) + 6(x - 3)$
68. $3x^2(x - 2) + 5x(x - 2) - 2(x - 2)$
69. $y^4 - 7y^2 - 30$
70. $3z^4 - 14z^2 - 5$
71. $x^2(x + 3) + 3x(x + 3) + 2(x + 3)$
72. $x^2(x - 1) - x(x - 1) - 30(x - 1)$
73. $5a^5b^2 - 8a^4b^3 + 3a^3b^4$
74. $2x^2y^6 + 3xy^5 - 9y^4$

Problem Solving

75. If the factors of a polynomial are $(2x + 3y)$ and $(x - 4y)$, find the polynomial. Explain how you determined your answer.

76. If the factors of a polynomial are 3, $(4x - 5)$, and $(2x - 3)$, find the polynomial. Explain how you determined your answer.

77. If we know that one factor of the polynomial $x^2 + 3x - 18$ is $x - 3$, how can we find the other factor? Find the other factor.

78. If we know that one factor of the polynomial $x^2 - xy - 6y^2$ is $x - 3y$, how can we find the other factor? Find the other factor.

79. **a)** Which of the following do you think would be more difficult to factor by trial and error? Explain your answer.

$30x^2 + 23x - 40$ or $49x^2 - 98x + 13$

b) Factor both trinomials.

80. **a)** Which of the following do you think would be more difficult to factor by trial and error? Explain your answer.

$48x^2 + 26x - 35$ or $35x^2 - 808x + 69$

b) Factor both trinomials.

81. Find all integer values of b for which $2x + bx - 5$ is factorable.

82. Find all integer values of b for which $3x^2 + bx - 7$ is factorable.

83. If $x^2 + bx + 5$ is factorable, what are the only two possible values of b? Explain.

84. If $x^2 + bx + c$ is factorable and c is a prime number, what are the only two possible factors of b? Explain.

*This exercise also appears on the videotape.

Consider the trinomial $ax^2 + bx + c$. We will learn later in the course that if the expression $b^2 - 4ac$, called the ***discriminant****, is not a perfect square, the trinomial cannot be factored over the set of integers.* ***Perfect squares*** *are 1, 4, 9, 16, 25, 49, and so on. The square root of a perfect square is a whole number. For Exercises 85–88,* ***a)*** *find the value of $b^2 - 4ac$.* ***b)*** *If $b^2 - 4ac$ is a perfect square, factor the polynomial; if $b^2 - 4ac$ is not a perfect square, indicate that the polynomial cannot be factored.*

85. $x^2 - 8x + 15$

86. $6y^2 - 5y - 6$

87. $x^2 - 4x + 6$

88. $3t^2 - 6t + 2$

89. Construct a trinomial of the form $x^2 + (c + 1)x + c$, where c is a real number, that is factorable.

90. Construct a trinomial of the form $x^2 - (c + 1)x + c$, where c is a real number, that is factorable.

Factor completely. Assume that the variables in the exponents represent positive integers.

91. $4a^{2n} - 4a^n - 15$

92. $a^2(a + b) - 2ab(a + b) - 3b^2(a + b)$

93. $x^2(x + y)^2 - 7xy(x + y)^2 + 12y^2(x + y)^2$

94. $3m^2(m - 2n) - 4mn(m - 2n) - 4n^2(m - 2n)$

95. $x^{2n} + 3x^n - 10$

96. $9r^{4y} + 3r^{2y} - 2$

97. Consider $x^2 + 2x - 8 = (x + 4)(x - 2)$.

a) Explain how you can check this factoring using graphs on your graphing calculator.

b) Check the factoring as you explained in part **a)** to see whether it is correct.

98. Consider $6x^3 - 11x^2 - 10x = x(2x - 5)(3x + 2)$.

a) Explain how you can check this factoring using the TABLE feature of a graphing calculator.

b) Check the factoring as you explained in part **a)** to see whether it is correct.

Cumulative Review Exercises

[1.6] **99.** Write the following quotient in scientific notation.

$$\frac{36{,}000{,}000}{0.0004}$$

[3.4] **100.** What is the slope of a horizontal line? Explain.

101. What is the slope of a vertical line? Explain.

[3.5] **102.** Find the equation of the line through $(5, -2)$ that is perpendicular to the graph of $2x - 3y = 6$.

[5.1] **103.** Simplify $2x^2y - 6xy^2 - (3x^2y + 2xy^2 - 6)$.

5.6 SPECIAL FACTORING FORMULAS

SSM VIDEO 5.6 CD Rom

1. Factor the difference of two squares.
2. Factor perfect square trinomials.
3. Factor the sum and difference of two cubes.

1 Factor the Difference of Two Squares

In this section we present some special formulas for factoring the difference of two squares, perfect square trinomials, and the sum and difference of two cubes. It will be to your advantage to memorize these formulas.

The expression $x^2 - 9$ is an example of the difference of two squares.

$$x^2 - 9 = (x)^2 - (3)^2$$

To factor the difference of two squares, it is convenient to use the **difference of two squares formula**. This formula was first presented in Section 5.2.

Difference of Two Squares

$$a^2 - b^2 = (a + b)(a - b)$$

EXAMPLE 1 Factor the following using the difference of two squares formula.

a) $x^2 - 16$ **b)** $16x^2 - 9y^2$

Solution Rewrite each expression as a difference of two squares. Then use the formula.

a) $x^2 - 16 = (x)^2 - (4)^2$
$= (x + 4)(x - 4)$

b) $16x^2 - 9y^2 = (4x)^2 - (3y)^2$
$= (4x + 3y)(4x - 3y)$

EXAMPLE 2 Factor the following differences of squares.

a) $x^6 - y^4$ **b)** $2z^4 - 162x^6$

Solution Rewrite each expression as a difference of two squares. Then use the formula.

a) $x^6 - y^4 = (x^3)^2 - (y^2)^2$
$= (x^3 + y^2)(x^3 - y^2)$

b) $2z^4 - 162x^6 = 2(z^4 - 81x^6)$
$= 2[(z^2)^2 - (9x^3)^2]$
$= 2(z^2 + 9x^3)(z^2 - 9x^3)$

NOW TRY EXERCISE 13

EXAMPLE 3 Factor $x^4 - 16y^4$.

Solution

$$x^4 - 16y^4 = (x^2)^2 - (4y^2)^2$$
$$= (x^2 + 4y^2)(x^2 - 4y^2)$$

Note that $(x^2 - 4y^2)$ is also a difference of two squares. We use the difference of two squares formula a second time to obtain

$$= (x^2 + 4y^2)[(x)^2 - (2y)^2]$$
$$= (x^2 + 4y^2)(x + 2y)(x - 2y)$$

EXAMPLE 4 Factor $(x - 5)^2 - 9$ using the formula for the difference of two squares.

Solution First we express $(x - 5)^2 - 9$ as a difference of two squares.

$$(x - 5)^2 - 9 = (x - 5)^2 - 3^2$$
$$= [(x - 5) + 3][(x - 5) - 3]$$
$$= (x - 2)(x - 8)$$

NOW TRY EXERCISE 21

Note: **It is not possible to factor the sum of two squares of the form $a^2 + b^2$ over the set of real numbers.**

For example, it is not possible to factor $x^2 + 4$ since $x^2 + 4 = x^2 + 2^2$, which is a sum of two squares.

2) Factor Perfect Square Trinomials

In Section 5.2 we saw that

$$(a + b)^2 = a^2 + 2ab + b^2$$
$$(a - b)^2 = a^2 - 2ab + b^2$$

If we reverse the left and right sides of these two formulas, we obtain two *special factoring formulas.*

Perfect Square Trinomials

$$a^2 + 2ab + b^2 = (a + b)^2$$
$$a^2 - 2ab + b^2 = (a - b)^2$$

These two trinomials are called **perfect square trinomials** since each is the square of a binomial. *To be a perfect square trinomial, the first and last terms must be the squares of some expression and the middle term must be twice the product of the first and last terms.* When you are given a trinomial to factor, determine whether it is a perfect square trinomial before you attempt to factor it by the procedures explained in Section 5.5. If it is a perfect square trinomial, you can factor it using the formulas given on the previous page.

Examples of Perfect Square Trinomials

$$y^2 + 6y + 9 \quad \text{or} \quad y^2 + 2(y)(3) + 3^2$$

$$9a^2b^2 - 24ab + 16 \quad \text{or} \quad (3ab)^2 - 2(3ab)(4) + 4^2$$

$$(r + s)^2 + 6(r + s) + 9 \quad \text{or} \quad (r + s)^2 + 2(r + s)(3) + 3^2$$

Now let's factor some perfect square trinomials.

EXAMPLE 5 Factor $x^2 - 8x + 16$.

Solution Since the first and last terms, x^2 and 4^2, are squares, this trinomial might be a perfect square trinomial. To determine whether it is, take twice the product of x and 4 to see if you obtain $8x$.

$$2(x)(4) = 8x$$

Since $8x$ is the middle term and since the sign of the middle term is negative, factor as follows:

NOW TRY EXERCISE 25

$$x^2 - 8x + 16 = (x - 4)^2$$

EXAMPLE 6 Factor $9x^4 - 12x^2 + 4$.

Solution The first term is a square, $(3x^2)^2$, as is the last term, 2^2. Since $2(3x^2)(2) = 12x^2$, we factor as follows:

$$9x^4 - 12x^2 + 4 = (3x^2 - 2)^2$$

EXAMPLE 7 Factor $(a + b)^2 + 6(a + b) + 9$.

Solution The first term, $(a + b)^2$, is a square. The last term, 9 or 3^2, is a square. The middle term is $2(a + b)(3) = 6(a + b)$. Therefore, this is a perfect square trinomial. Thus,

NOW TRY EXERCISE 35

$$(a + b)^2 + 6(a + b) + 9 = [(a + b) + 3]^2 = (a + b + 3)^2$$

EXAMPLE 8 Factor $x^2 - 6x + 9 - y^2$.

Solution Since $x^2 - 6x + 9$ is a perfect square trinomial, we write

$$(x - 3)^2 - y^2$$

Now $(x - 3)^2 - y^2$ is a difference of two squares; therefore

$$\begin{aligned}(x - 3)^2 - y^2 &= [(x - 3) + y][(x - 3) - y] \\ &= (x - 3 + y)(x - 3 - y)\end{aligned}$$

NOW TRY EXERCISE 41

Thus, $x^2 - 6x + 9 - y^2 = (x - 3 + y)(x - 3 - y)$.

The polynomial in Example 8 has four terms. In Section 5.4 we learned to factor polynomials with four terms by grouping. If you study Example 8, you will see that no matter how you arrange the four terms they cannot be arranged so that the first two terms have a common factor and the last two terms have a common factor. Whenever a polynomial of four terms cannot be factored by grouping, try to rewrite three of the terms as the square of a binomial and then factor using the difference of two squares formula.

EXAMPLE 9 Factor $4a^2 + 12ab + 9b^2 - 25$.

Solution We first notice that this polynomial of four terms cannot be factored by grouping. We next look to see if three terms of the polynomial can be expressed as the square of a binomial. Since this can be done, we write the three terms as the square of a binomial. We complete our factoring using the difference of two squares formula.

$$\begin{aligned} 4a^2 + 12ab + 9b^2 - 25 &= (2a + 3b)^2 - 5^2 \\ &= [(2a + 3b) + 5][(2a + 3b) - 5] \\ &= (2a + 3b + 5)(2a + 3b - 5) \end{aligned}$$

3) Factor the Sum and Difference of Two Cubes

Earlier in this section we factored the difference of two squares. Now we will factor the sum and difference of two cubes. Consider the product of $(a + b)(a^2 - ab + b^2)$.

$$\begin{array}{rrrr} & a^2 & - ab & + b^2 \\ & & a & + b \\ \hline & a^2b & - ab^2 & + b^3 \\ a^3 & - a^2b & + ab^2 & \\ \hline a^3 & & & + b^3 \end{array}$$

Thus, $a^3 + b^3 = (a + b)(a^2 - ab + b^2)$. Using multiplication, we can also show that $a^3 - b^3 = (a - b)(a^2 + ab + b^2)$. Formulas for factoring **the sum and the difference of two cubes** appear in the following box.

Sum of Two Cubes

$$a^3 + b^3 = (a + b)(a^2 - ab + b^2)$$

Difference of Two Cubes

$$a^3 - b^3 = (a - b)(a^2 + ab + b^2)$$

EXAMPLE 10 Factor $x^3 + 27$.

Solution Rewrite $x^3 + 27$ as a sum of two cubes, $x^3 + 3^3$. Let x correspond to a and 3 to b. Then factor using the sum of two cubes formula.

$$\begin{aligned} a^3 + b^3 &= (a + b)(a^2 - a\ b + b^2) \\ &\quad \downarrow \ \downarrow \ \downarrow \quad \downarrow\downarrow \quad \downarrow \\ x^3 + 3^3 &= (x + 3)[x^2 - x(3) + 3^2] \\ &= (x + 3)(x^2 - 3x + 9) \end{aligned}$$

Thus, $x^3 + 27 = (x + 3)(x^2 - 3x + 9)$.

EXAMPLE 11 Factor $27x^3 - 8y^6$.

Solution We first observe that $27x^3$ and $8y^6$ have no common factors other than 1. Since we can express both $27x^3$ and $8y^6$ as cubes, we can factor using the difference of two cubes formula.

$$\begin{aligned} 27x^3 - 8y^6 &= (3x)^3 - (2y^2)^3 \\ &= (3x - 2y^2)[(3x)^2 + (3x)(2y^2) + (2y^2)^2] \\ &= (3x - 2y^2)(9x^2 + 6xy^2 + 4y^4) \end{aligned}$$

NOW TRY EXERCISE 53

Thus, $27x^3 - 8y^6 = (3x - 2y^2)(9x^2 + 6xy^2 + 4y^4)$.

EXAMPLE 12 Factor $8y^3 - 64x^6$.

Solution First factor out 8, which is common to both terms.

$$8y^3 - 64x^6 = 8(y^3 - 8x^6)$$

Next factor $y^3 - 8x^6$ by writing it as a difference of two cubes.

$$\begin{aligned} 8(y^3 - 8x^6) &= 8[(y)^3 - (2x^2)^3] \\ &= 8(y - 2x^2)[y^2 + y(2x^2) + (2x^2)^2] \\ &= 8(y - 2x^2)(y^2 + 2x^2y + 4x^4) \end{aligned}$$

Thus, $8y^3 - 64x^6 = 8(y - 2x^2)(y^2 + 2x^2y + 4x^4)$.

EXAMPLE 13 Factor $(x - 2)^3 + 64$.

Solution Write $(x - 2)^3 + 64$ as a sum of two cubes, then use the sum of two cubes formula to factor.

$$\begin{aligned} (x - 2)^3 + (4)^3 &= [(x - 2) + 4][(x - 2)^2 - (x - 2)(4) + (4)^2] \\ &= (x - 2 + 4)(x^2 - 4x + 4 - 4x + 8 + 16) \\ &= (x + 2)(x^2 - 8x + 28) \end{aligned}$$

NOW TRY EXERCISE 61

HELPFUL HINT

The square of a binomial has a 2 as part of the middle term of the trinomial.

$$(a + b)^2 = a^2 + 2ab + b^2$$
$$(a - b)^2 = a^2 - 2ab + b^2$$

The sum or the difference of two cubes has a factor similar to the trinomial in the square of the binomial. However, the middle term does not contain a 2.

$$a^3 + b^3 = (a + b)(a^2 - ab + b^2)$$
$$a^3 - b^3 = (a - b)(a^2 + \underbrace{ab}_{\text{not } 2ab} + b^2)$$

Exercise Set 5.6

Concept/Writing Exercises

1. a) Explain in your own words how to factor the difference of two squares.
   b) Factor $x^2 - 16$ using the procedure you explained in part **a)**.
2. Explain why a sum of two squares, $a^2 + b^2$, cannot be factored over the set of real numbers.
3. Explain in your own words how to determine whether a trinomial is a perfect square trinomial.
4. a) Explain in your own words how to factor a perfect square trinomial.
   b) Factor $x^2 + 12x + 36$ using the procedure you explained in part **a)**.
5. Give the formula for factoring the sum of two cubes.
6. Give the formula for factoring the difference of two cubes.

Practice the Skills

Use the difference of two squares formula or the perfect square trinomial formula to factor each polynomial.

7. $x^2 - 81$ **8.** $x^2 - 9$ **9.** $x^2 - 16$
10. $1 - 4x^2$ **11.** $1 - 9a^2$ **12.** $x^2 - 36y^2$
13. $25 - 16y^4$ **14.** $x^6 - 144y^4$ **15.** $\frac{1}{16} - x^2$
16. $\frac{1}{49} - b^2$ **17.** $a^2b^2 - 49c^2$ **18.** $4a^2c^2 - 16x^2y^2$
19. $0.4x^2 - 0.9$ **20.** $100 - (x + y)^2$ **21.** $36 - (x - 6)^2$
22. $(2x + 3)^2 - 9$ **23.** $a^2 - (3b + 2)^2$ **24.** $a^2 + 2ab + b^2$
25. $x^2 + 10x + 25$ **26.** $25 - 10t + t^2$ **27.** $4 + 4x + x^2$
28. $y^2 - 8y + 16$ **29.** $4x^2 - 20xy + 25y^2$ **30.** $9y^2 + 6yz + z^2$
31. $0.81x^2 - 0.36x + 0.04$ **32.** $25a^2b^2 - 20ab + 4$ **33.** $w^4 + 16w^2 + 64$
34. $0.25x^2 - 0.40x + 0.16$ **35.** $(x + y)^2 + 2(x + y) + 1$ **36.** $(x + 1)^2 + 6(x + 1) + 9$
37. $a^4 - 2a^2b^2 + b^4$ **38.** $(w - 3)^2 + 8(w - 3) + 16$ **39.** $x^2 + 6x + 9 - y^2$
40. $9 - (x^2 - 8x + 16)$ **41.** $25 - (x^2 + 4x + 4)$ **42.** $a^2 + 2ab + b^2 - 16c^2$
43. $9a^2 - 12ab + 4b^2 - 9$ **44.** $x^4 - 6x^2 + 9$ **45.** $(x + y)^2 - (x - y)^2$
46. $(4a - 3b)^2 - (2a + 5b)^2$

Factor using the sum or difference of two cubes formula.

47. $x^3 - 27$ **48.** $y^3 + 125$ **49.** $x^3 + y^3$
50. $x^3 - 8a^3$ **51.** $64 - a^3$ **52.** $w^3 - 216$
53. $27y^3 - 8x^3$ **54.** $5x^3 + 40y^3$ **55.** $24x^3 - 81y^3$
56. $y^6 + x^9$ **57.** $5x^3 - 625y^3$ **58.** $16y^6 - 250x^3$
59. $(x + 1)^3 + 1$ **60.** $(x - 3)^3 + 8$ **61.** $(x - y)^3 - 27$
62. $(2x + y)^3 - 64$ **63.** $b^3 - (b + 3)^3$ **64.** $(m - n)^3 - (m + n)^3$

Factor using a special factoring formula.

65. $121y^4 - 49x^2$ **66.** $a^4 - 4b^4$ **67.** $16y^2 - 81x^2$
68. $49 - 64x^2y^2$ **69.** $25x^4 - 81y^6$ **70.** $(x + y)^2 - 16$
71. $x^3 - 64$ **72.** $2a^2 - 24a + 72$ **73.** $9x^2y^2 + 24xy + 16$
74. $a^4 + 12a^2 + 36$ **75.** $a^4 + 2a^2b^2 + b^4$ **76.** $8y^3 - 125x^6$
77. $x^2 - 2x + 1 - y^2$ **78.** $4r^2 + 4rs + s^2 - 9$ **79.** $(x + y)^3 + 1$
80. $9x^2 - 6xy + y^2 - 4$ **81.** $(m + n)^2 - (2m - n)^2$ **82.** $(r + p)^3 + (r - p)^3$

Problem Solving

*In Exercises 83– 87, **a)** find the area or volume of the shaded figure by subtracting the smaller area or volume from the larger. The formula to find the area or volume is given under the figure. **b)** Write the expression obtained in part **a)** in factored form. Part of the GCF in Exercises 84, 86, and 87 is π.*

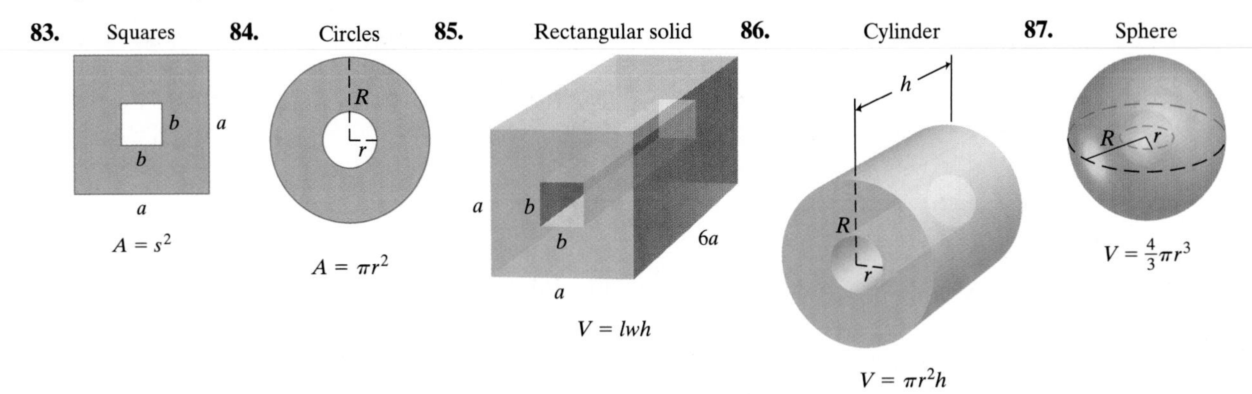

88. A circular hole is cut from a cube of wood, as shown in the figure.

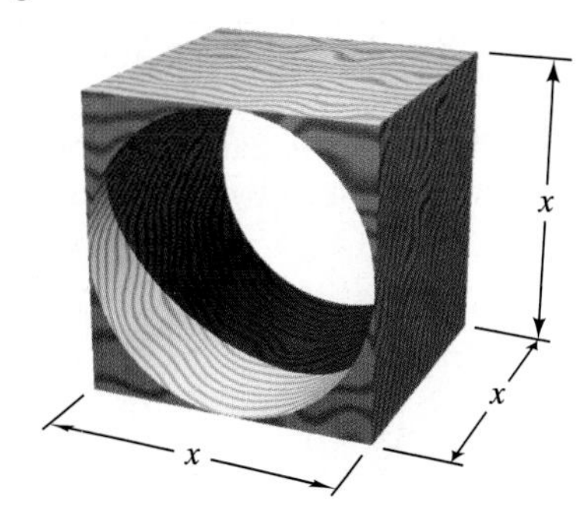

a) Write an expression in factored form in terms of x for the cross-sectional area of the remaining wood.

b) Write an expression in factored form in terms of x for the volume of the remaining wood.

89. Find two values of b that will make $4x^2 + bx + 9$ a perfect square trinomial. Explain how you determined your answer.

90. Find two values of c that will make $16x^2 + cx + 4$ a perfect square trinomial. Explain how you determined your answer.

91. Find the value of c that will make $25x^2 + 20x + c$ a perfect square trinomial. Explain how you determined your answer.

92. Find the value of d that will make $49x^2 - 42x + d$ a perfect square trinomial. Explain how you determined your answer.

93. A formula for the area of a square is $A = s^2$, where s is a side. If a square has an area $A(x) = 25x^2 - 30x + 9$,

$A(x) = 25x^2 - 30x + 9$

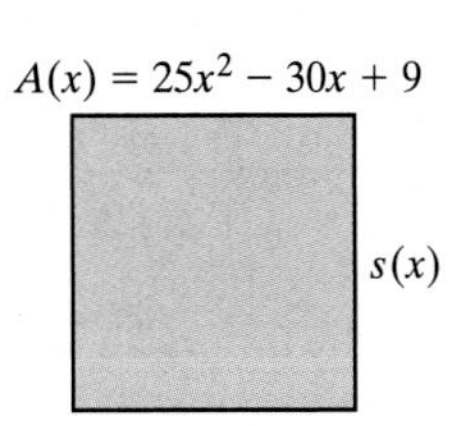

a) explain how to find the length of side x, $s(x)$,

b) find $s(x)$, and

c) find $s(2)$.

94. The formula for the area of a circle is $A = \pi r^2$, where r is the radius. If a circle has an area of $A(x) = 9\pi x^2 + 12\pi x + 4\pi$,

$A(x) = 9\pi x^2 + 12\pi x + 4\pi$

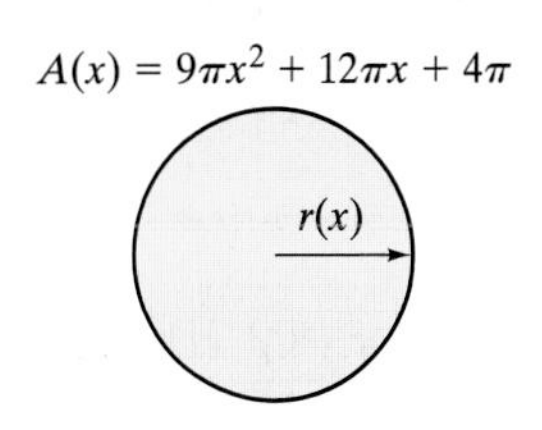

a) explain how to find the radius, $r(x)$, **b)** find $r(x)$, and **c)** find $r(4)$.

95. Factor $x^4 + 64$ by writing the expression as $(x^4 + 16x^2 + 64) - 16x^2$, which is a difference of two squares.

96. Factor $x^4 + 4$ by adding and subtracting $4x^2$. (See Exercise 95)

97. If $P(x) = x^2$, use the difference of two squares to simplify $P(a + h) - P(a)$.

98. If $P(x) = x^2$, use the difference of two squares to simplify $P(a + 1) - P(a)$.

99. The figure shows how we *complete the square*. The sum of the areas of the 3 parts of the square that are shaded in blue is

$$x^2 + 4x + 4x \quad \text{or} \quad x^2 + 8x$$

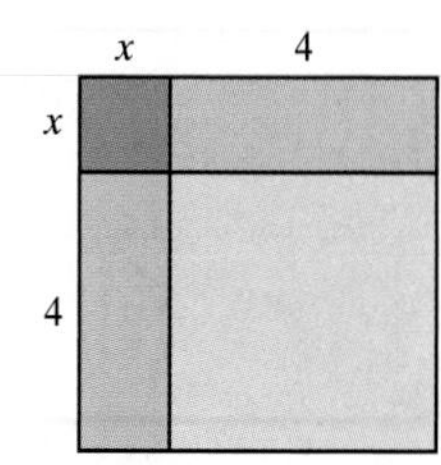

a) Find the area of the fourth part (in red) to complete the square.

b) Find the sum of the areas of the four parts of the square.

c) This process has resulted in a perfect square trinomial in part **b)**. Write this perfect square trinomial as the square of a binomial.

100. Factor $(m - n)^3 - (9 - n)^3$.

Factor completely.

101. $64x^{4a} - 9y^{6a}$

102. $16p^{8w} - 49p^{6w}$

103. $a^{2n} - 16a^n + 64$

104. $144r^{8k} + 48r^{4k} + 4$

105. $x^{3n} - 8$

106. $27x^{3m} + 64x^{6m}$

In Exercises 107 and 108, use your graphing calculator to check the factoring. Explain your answers.

107. $2x^2 - 18 \stackrel{?}{=} 2(x + 3)(x - 3)$

108. $8x^3 + 27 \stackrel{?}{=} 2x(4x^2 + 5x + 9)$

Challenge Problem

109. The expression $x^6 - 1$ can be factored using either the difference of two squares or the difference of two cubes. At first the factors do not appear the same. But with a little algebraic manipulation they can be shown to be equal. Factor $x^6 - 1$ using **a)** the difference of two squares and **b)** the difference of two cubes. **c)** Show that these two answers are equal by factoring the answer obtained in part **a)** completely. Then multiply the two binomials by the two trinomials.

Group Activity

Discuss and answer Exercise 110 as a group.

110. Later in the book we will need to construct perfect square trinomials. Examine some perfect square trinomials with a leading coefficient of 1.

a) Explain how b and c are related if the trinomial

$x^2 + bx + c$ is a perfect square trinomial.

b) Construct a perfect square trinomial if the first two terms are $x^2 + 6x$.

c) Construct a perfect square trinomial if the first two terms are $x^2 - 10x$.

d) Construct a perfect square trinomial if the first two terms are $x^2 - 14x$.

Cumulative Review Exercises

[1.2] **111.** Consider the set of elements $\{-2, \frac{5}{9}, -1.67, 0, \sqrt{3}, -\sqrt{6}, 3, 6\}$. List the elements that are

a) counting numbers.

b) rational numbers.

c) irrational numbers.

d) real numbers.

Write each set in roster form.

112. $\{x | 4 \le x < 8 \text{ and } x \in N\}$

113. $\{x | x > 3 \text{ and } x \in N\}$

[2.2] **114.** Given the formula $z = \dfrac{x - \bar{x}}{\dfrac{s}{\sqrt{n}}}$

find the value of z when $x = 15, \bar{x} = 13, s = 3$, and $n = 9$.

[2.3] **115.** The length of a rectangular hallway is 2 feet greater than twice its width. Find the length and width of the hallway if its perimeter is 22 feet.

5.7 A GENERAL REVIEW OF FACTORING

1) **Factor polynomials using a combination of techniques.**

1) Factor Polynomials Using a Combination of Techniques

We have presented a number of factoring methods. Now we will combine problems and techniques from the previous sections.

A general procedure to factor any polynomial follows.

To Factor a Polynomial

1. Determine whether all the terms in the polynomial have a greatest common factor other than 1. If so, factor out the GCF.
2. If the polynomial has two terms, determine whether it is a difference of two squares or a sum or difference of two cubes. If so, factor using the appropriate formula.
3. If the polynomial has three terms, determine whether it is a perfect square trinomial. If so, factor accordingly. If it is not, factor the trinomial using trial and error, grouping, or substitution as explained in Section 5.5.
4. If the polynomial has more than three terms, try factoring by grouping. If that does not work, see if three of the terms are the square of a binomial.
5. As a final step, examine your factored polynomial to see if any factors listed have a common factor and can be factored further. If you find a common factor, factor it out at this point.

The following examples illustrate how to use the procedure.

EXAMPLE 1 Factor $3x^4 - 27x^2$.

Solution First, check for a greatest common factor other than 1. Since $3x^2$ is common to both terms, factor it out.

$$3x^4 - 27x^2 = 3x^2(x^2 - 9) = 3x^2(x + 3)(x - 3)$$

Note that $x^2 - 9$ is factored as a difference of two squares.

EXAMPLE 2 Factor $3x^2y^2 - 24xy^2 + 48y^2$.

Solution Begin by factoring the GCF, $3y^2$, from each term.

$$3x^2y^2 - 24xy^2 + 48y^2 = 3y^2(x^2 - 8x + 16) = 3y^2(x - 4)^2$$

Note that $x^2 - 8x + 16$ is a perfect square trinomial. If you did not recognize this, you would still obtain the correct answer by factoring the trinomial into $(x - 4)(x - 4)$.

NOW TRY EXERCISE 27

EXAMPLE 3 Factor $24x^2 - 6xy + 16xy - 4y^2$.

Solution As always, begin by determining if all the terms in the polynomial have a common factor. In this example 2 is common to all terms. Factor out the 2; then factor the remaining four-term polynomial by grouping.

$$\begin{aligned} 24x^2 - 6xy + 16xy - 4y^2 &= 2(12x^2 - 3xy + 8xy - 2y^2) \\ &= 2[3x(4x - y) + 2y(4x - y)] \\ &= 2(3x + 2y)(4x - y) \end{aligned}$$

EXAMPLE 4 Factor $10a^2b - 15ab + 20b$.

Solution

$$10a^2b - 15ab + 20b = 5b(2a^2 - 3a + 4)$$

Since $2a^2 - 3a + 4$ cannot be factored, we stop here.

EXAMPLE 5 Factor $2x^4y + 54xy$.

Solution

$$\begin{aligned} 2x^4y + 54xy &= 2xy(x^3 + 27) \\ &= 2xy(x + 3)(x^2 - 3x + 9) \end{aligned}$$

NOW TRY EXERCISE 19

Note that $x^3 + 27$ was factored as a sum of two cubes.

EXAMPLE 6 Factor $6x^2 - 3x + 6y^2 - 9$.

Solution First, we factor 3 from all four terms.

$$6x^2 - 3x + 6y^2 - 9 = 3(2x^2 - x + 2y^2 - 3)$$

Now we determine whether the four terms within parentheses can be factored by grouping. Since these four terms cannot be factored by grouping, we check to see whether any three of the terms can be written as the square of a binomial. No matter how we rearrange the terms, this cannot be done. We conclude that this expression cannot be factored further. Thus,

$$6x^2 - 3x + 6y^2 - 9 = 3(2x^2 - x + 2y^2 - 3)$$

EXAMPLE 7 Factor $3x^2 - 18x + 27 - 3y^2$.

Solution Factor out 3 from all four terms.

$$3x^2 - 18x + 27 - 3y^2 = 3(x^2 - 6x + 9 - y^2)$$

Now try factoring by grouping. Since the four terms within parentheses cannot be factored by grouping, check to see whether any three of the terms can be written as the square of a binomial. Since this can be done, express $x^2 - 6x + 9$ as $(x - 3)^2$ and then use the difference of two squares formula. Thus,

$$\begin{aligned} 3x^2 - 18x + 27 - 3y^2 &= 3[(x - 3)^2 - y^2] \\ &= 3[(x - 3 + y)(x - 3 - y)] \\ &= 3(x - 3 + y)(x - 3 - y) \end{aligned}$$

NOW TRY EXERCISE 43

Exercise Set 5.7

Concept/Writing Exercises

1. Explain the possible procedures that may be used to factor a polynomial of **a)** two terms; **b)** three terms; **c)** four terms.
2. What is the first step in the factoring process?

Practice the Skills

Factor each polynomial completely.

3. $4x^2 + 4x - 48$
4. $3x^2 - 24x + 48$
5. $10s^2 + 19s - 15$
6. $6x^3y^2 + 10x^2y^3 + 8x^2y^2$
7. $-8r^2 + 26r - 15$
8. $3x^3 - 12x^2 - 36x$
9. $0.4x^2 - 0.036$
10. $0.5x^2 - 0.08$
11. $5x^5 - 45x$
12. $6x^2y^2z^2 - 24x^2y^2$
13. $3x^6 - 3x^5 - 12x^5 + 12x^4$
14. $2x^2y + 6xy^2 - 10xy^2 - 30y^2$
15. $5x^4y^2 + 20x^3y^2 - 15x^3y^2 - 60x^2y^2$
16. $6x^2 - 15x - 9$
17. $x^4 - x^2y^2$
18. $4x^3 + 108$
19. $x^7y^2 - x^4y^2$
20. $x^4 - 16$
21. $x^5 - 16x$
22. $12x^2y^2 + 33xy^2 - 9y^2$
23. $4x^6 + 32y^3$
24. $12x^4 - 6x^3 - 6x^3 + 3x^2$
25. $2(a + b)^2 - 18$
26. $12x^3y^2 + 4x^2y^2 - 40xy^2$
27. $6x^2 + 36xy + 54y^2$
28. $3x^2 - 30x + 75$
29. $(x + 2)^2 - 4$
30. $4y^4 - 36x^6$
31. $(2a + b)(2a - 3b) - (2a + b)(a - b)$
32. $pq + 6q + pr + 6r$
33. $(y + 3)^2 + 4(y + 3) + 4$
34. $b^4 + 2b^2 + 1$
35. $45a^4 - 30a^3 + 5a^2$
36. $(x + 1)^2 - (x + 1) - 6$
37. $x^3 + \dfrac{1}{27}$
38. $8y^3 - \dfrac{1}{8}$
39. $3x^3 + 2x^2 - 27x - 18$
40. $6y^3 + 14y^2 + 4y$
41. $a^3b - 16ab^3$
42. $x^6 + y^6$
43. $81 - (x^2 + 2xy + y^2)$
44. $x^2 - 2xy + y^2 - 25$
45. $24x^2 - 34x + 12$
46. $40x^2 + 52x - 12$
47. $16x^2 - 34x - 15$
48. $7(a - b)^2 + 4(a - b) - 3$
49. $x^4 - 81$
50. $(x + 2)^2 - 12(x + 2) + 36$
51. $5bc - 10cx - 6by + 12xy$
52. $16y^4 - 9y^2$
53. $3x^4 - x^2 - 4$
54. $x^2 + 16x + 64 - 100y^2$
55. $y^2 - (x^2 - 8x + 16)$
56. $4a^3 + 32$
57. $2(y + 4)^2 + 5(y + 4) - 12$
58. $x^6 - 11x^3 + 30$
59. $a^2 + 12ab + 36b^2 - 16c^2$
60. $y - y^3$
61. $6x^4y + 15x^3y - 9x^2y$
62. $4x^2y^2 + 12xy + 9$
63. $x^4 - 2x^2y^2 + y^4$
64. $6r^2s^2 + rs - 1$

Problem Solving

*Match Exercises 65–72 with the items labeled **a)** through **h)**.*

65. $a^2 + b^2$
66. $a^2 + 2ab + b^2$
67. $a^3 + b^3$
68. $a^2 - b^2$
69. $a^3 - b^3$
70. $a^2 - 2ab + b^2$
71. a factor of $a^3 + b^3$
72. a factor of $a^3 - b^3$

a) $(a + b)(a^2 - ab + b^2)$
b) $(a - b)^2$
c) $a^2 - ab + b^2$
d) $(a + b)^2$
e) not factorable
f) $(a - b)(a^2 + ab + b^2)$
g) $a^2 + ab + b^2$
h) $(a + b)(a - b)$

In Exercises 73–76, ***a)*** *write an expression for the shaded area of the figure, and* ***b)*** *write the expression in factored form.*

73.

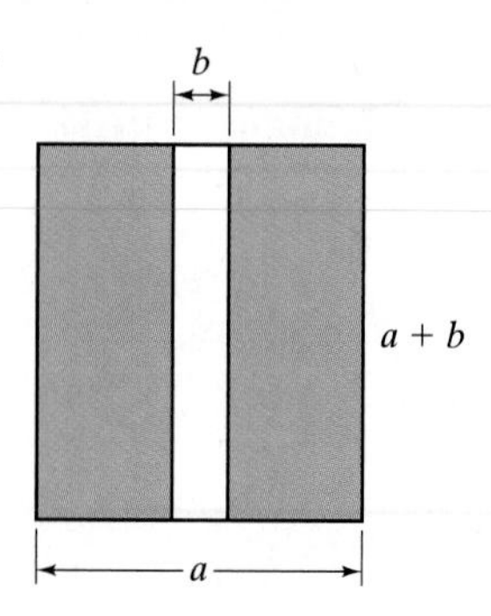

74.

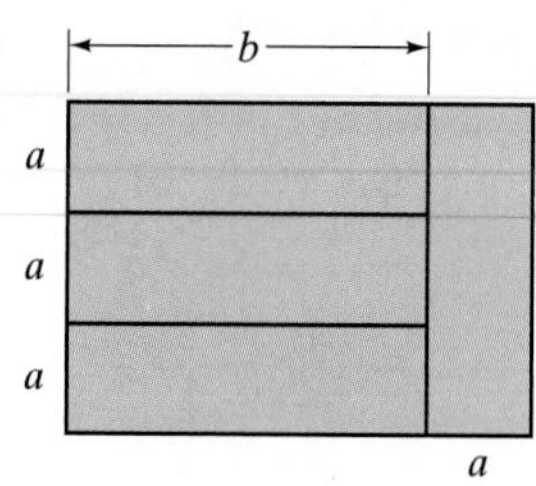

75.

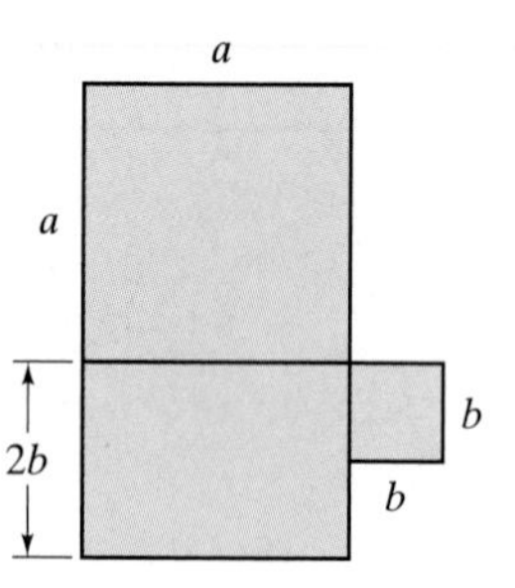

76.

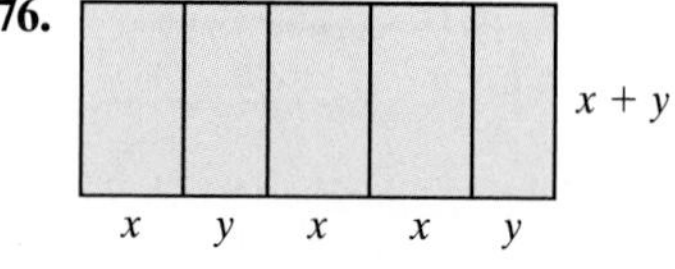

77. **a)** Write an expression for the surface area of the four sides of the box shown (omit top and bottom).

b) Write the expression in factored form.

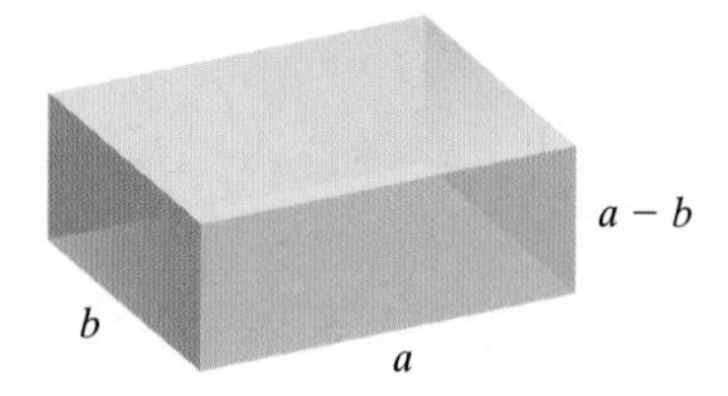

78. Explain how the formula for factoring the *difference* of two cubes can be used to factor $x^3 + 27$.

79. **a)** Explain how to construct a perfect square trinomial.

b) Construct a perfect square trinomial and then show its factors.

Challenge Problems

We have worked only with integer exponents in this chapter. However, fractional exponents may also be factored out of an expression. The expressions below are not polynomials. ***a)*** *Factor out the variable with the lowest (or most negative) exponent from each expression. (Fractional exponents are discussed in Section 7.2.)* ***b)*** *Factor completely.*

80. $x^{-2} - 5x^{-3} + 6x^{-4}$, factor out x^{-4}

81. $x^{-3} - 2x^{-4} - 3x^{-5}$, factor out x^{-5}

82. $x^{5/2} + 3x^{3/2} - 4x^{1/2}$, factor out $x^{1/2}$

83. $5x^{1/2} + 2x^{-1/2} - 3x^{-3/2}$, factor out $x^{-3/2}$

Cumulative Review Exercises

[2.1] *Solve.*

84. $4(x - 2) = 3(x - 4) - 4$

85. $-5(x - 2) + 3 = -5x - 6$

Find the solution set for each inequality.

[2.5] **86.** $4(x - 3) < 6(x - 4)$.

[2.6] **87.** $|2x - 3| > -4$.

5.8 POLYNOMIAL EQUATIONS

1) Use the zero-factor property to solve equations.
2) Use factoring to solve equations.
3) Use factoring to solve applications.
4) Use factoring to find the x-intercepts of a quadratic function.

Whenever two polynomials are set equal to each other, we have a **polynomial equation**.

Examples of Polynomial Equations

$$x^2 + 2x = x - 5$$
$$y^3 + 3y - 2 = 0$$
$$4x^4 + 2x^2 = -3x + 2$$

The **degree of a polynomial equation** is the same as the degree of its highest term. For example, the three equations above have degree 2, 3, and 4, respectively. A second-degree equation in one variable is often called a **quadratic equation**.

Examples of Quadratic Equations

$$3x^2 + 6x - 4 = 0$$
$$5x = 2x^2 - 4$$
$$(x + 4)(x - 3) = 0$$

Any quadratic equation can be written in **standard form**.

Standard Form of a Quadratic Equation

$$ax^2 + bx + c = 0, \quad a \neq 0$$

where a, b, and c are real numbers.

Before going any further, make sure that you can convert each of the three quadratic equations given above to standard form, with $a > 0$.

1) Use the Zero-Factor Property to Solve Equations

To solve equations using factoring, we use the **zero-factor property**.

Zero-Factor Property

For all real numbers a and b, if $a \cdot b = 0$, then either $a = 0$ or $b = 0$, or both a and $b = 0$.

The zero-factor property states that *if the product of two factors equals zero, one (or both) of the factors must be zero.*

EXAMPLE 1 Solve the equation $(x + 5)(x - 3) = 0$.

Solution Since the product of the factors equals 0, according to the zero-factor property, one or both factors must equal zero. Set each factor equal to 0 and solve each equation separately.

$$x + 5 = 0 \qquad \text{or} \qquad x - 3 = 0$$

$$x = -5 \qquad\qquad\qquad x = 3$$

Thus, if x is either -5 or 3, the product of the factors is 0.

Check: $\mathbf{x = -5}$

$$(x + 5)(x - 3) = 0$$
$$(-5 + 5)(-5 - 3) \stackrel{?}{=} 0$$
$$0(-8) \stackrel{?}{=} 0$$
$$0 = 0 \quad \textit{True}$$

$\mathbf{x = 3}$

$$(x + 5)(x - 3) = 0$$
$$(3 + 5)(3 - 3) \stackrel{?}{=} 0$$
$$8(0) \stackrel{?}{=} 0$$
$$0 = 0 \quad \textit{True}$$

2) Use Factoring to Solve Equations

Following is a procedure that can be used to obtain the solution to an equation by factoring.

To Solve an Equation by Factoring

1. Use the addition property to remove all terms from one side of the equation. This will result in one side of the equation being equal to 0.
2. Combine like terms in the equation and then factor.
3. Set each factor *containing a variable* equal to zero, solve the equations, and find the solutions.
4. Check the solutions in the *original* equation.

EXAMPLE 2 Solve the equation $2x^2 = 12x$.

Solution First, make the right side of the equation equal to 0 by subtracting $12x$ from both sides of the equation. Then factor the left side of the equation.

$$2x^2 - 12x = 0$$

$$2x(x - 6) = 0$$

Now set each factor equal to zero and solve for x.

$$2x = 0 \qquad \text{or} \qquad x - 6 = 0$$

$$x = 0 \qquad\qquad\qquad x = 6$$

A check will show that the numbers 0 and 6 both satisfy the equation $2x^2 = 12x$.

Avoiding Common Errors

The zero-factor property can be used only when one side of the equation is equal to 0.

CORRECT	INCORRECT
$(x - 4)(x + 3) = 0$	~~$(x - 4)(x + 3) = 2$~~
$x - 4 = 0$ or $x + 3 = 0$	~~$x - 4 = 2$ or $x + 3 = 2$~~
$x = 4 \qquad x = -3$	~~$x = 6 \qquad x = -1$~~

In the incorrect process illustrated on the right, the zero-factor property cannot be used since the right side of the equation is not equal to 0. Example 3 shows how to solve such problems correctly.

EXAMPLE 3 Solve the equation $(x - 1)(3x + 2) = 4x$.

Solution Since the right side of the equation is not 0, we cannot use the zero-factor property yet. Instead, we begin by multiplying the factors on the left side of the equation. Then we subtract $4x$ from both sides of the equation to obtain 0 on the right side. Then we can factor and solve the equation.

$$(x - 1)(3x + 2) = 4x$$

$$3x^2 - x - 2 = 4x \qquad \text{Multiply the factors.}$$

$$3x^2 - 5x - 2 = 0 \qquad \text{Make one side 0.}$$

$$(3x + 1)(x - 2) = 0 \qquad \text{Factor the trinomial.}$$

$$3x + 1 = 0 \quad \text{or} \quad x - 2 = 0 \qquad \text{Zero-factor property.}$$

$$3x = -1 \qquad x = 2 \qquad \text{Solve the equations.}$$

$$x = -\frac{1}{3}$$

The solutions are $-\frac{1}{3}$ and 2. Check these values in the original equation. 

EXAMPLE 4 Solve the equation $3x^2 + 2x - 12 = -7x$.

Solution

$$3x^2 + 2x - 12 = -7x$$

$$3x^2 + 9x - 12 = 0 \qquad \text{Make one side 0.}$$

$$3(x^2 + 3x - 4) = 0 \qquad \text{Factor out 3.}$$

$$3(x + 4)(x - 1) = 0 \qquad \text{Factor the trinomial.}$$

$$x + 4 = 0 \quad \text{or} \quad x - 1 = 0 \qquad \text{Zero-factor property.}$$

$$x = -4 \qquad x = 1 \qquad \text{Solve for } x.$$

Since the factor 3 does not contain a variable, we do not have to set it equal to zero. Only the numbers -4 and 1 satisfy the equation $3x^2 + 2x - 12 = -7x$.

NOW TRY EXERCISE 31

HELPFUL HINT

When solving an equation whose leading term has a negative coefficient, we generally make the coefficient positive by multiplying both sides of the equation by -1. This makes the factoring process easier, as in the following example.

$$-x^2 + 5x + 6 = 0$$
$$-1(-x^2 + 5x + 6) = -1 \cdot 0$$
$$x^2 - 5x - 6 = 0$$

Now we can solve the equation $x^2 - 5x - 6 = 0$ by factoring.

$$(x - 6)(x + 1) = 0$$
$$x - 6 = 0 \quad \text{or} \quad x + 1 = 0$$
$$x = 6 \qquad\qquad x = -1$$

The numbers 6 and -1 both satisfy the original equation, $-x^2 + 5x + 6 = 0$.

The equations in Examples 1 through 4 were all quadratic equations that were rewritten in the form $ax^2 + bx + c = 0$ and solved by factoring. Other methods that can be used to solve quadratic equations include completing the square and the quadratic formula; we discuss these methods in Chapter 8.

The zero-factor property can be extended to three or more factors, as illustrated in Example 5.

EXAMPLE 5 Solve the equation $2x^3 + 5x^2 - 3x = 0$.

Solution First factor, then set each factor containing an x equal to 0.

$$2x^3 + 5x^2 - 3x = 0$$
$$x(2x^2 + 5x - 3) = 0 \qquad \textit{Factor out } x.$$
$$x(2x - 1)(x + 3) = 0 \qquad \textit{Factor the trinomial.}$$
$$x = 0 \quad \text{or} \quad 2x - 1 = 0 \quad \text{or} \quad x + 3 = 0 \qquad \textit{Zero-factor property}$$
$$2x = 1 \qquad x = -3 \qquad \textit{Solve for } x.$$
$$x = \frac{1}{2}$$

The numbers $0, \frac{1}{2}$, and -3 are all solutions to the equation.

Note that the equation in Example 5 is not a quadratic equation because the exponent in the leading term is 3, not 2. This is a **cubic**, or **third-degree, equation**.

EXAMPLE 6 For the function $f(x) = 2x^2 - 13x - 16$, find all values of a for which $f(a) = 8$.

Solution First we rewrite the function as $f(a) = 2a^2 - 13a - 16$. Since $f(a) = 8$, we write

$$2a^2 - 13a - 16 = 8 \qquad \textit{Set } f(a) \textit{ equal to 8.}$$
$$2a^2 - 13a - 24 = 0 \qquad \textit{Make one side 0.}$$
$$(2a + 3)(a - 8) = 0 \qquad \textit{Factor the trinomial.}$$
$$2a + 3 = 0 \quad \text{or} \quad a - 8 = 0 \qquad \textit{Zero-factor property}$$
$$2a = -3 \qquad a = 8 \qquad \textit{Solve for } a.$$
$$a = -\frac{3}{2}$$

NOW TRY EXERCISE 59

If you check these answers, you will find that $f(-\frac{3}{2}) = 8$ and $f(8) = 8$.

3) Use Factoring to Solve Applications

Now let us look at some applications that use factoring in their solution.

EXAMPLE 7 At an exhibition a large canvas tent is to have an entrance in the shape of a triangle (see Fig. 5.12).

Find the base and height of the entrance if the height is to be 3 feet less than twice the base and the total area of the entrance is 27 square feet.

FIGURE 5.12

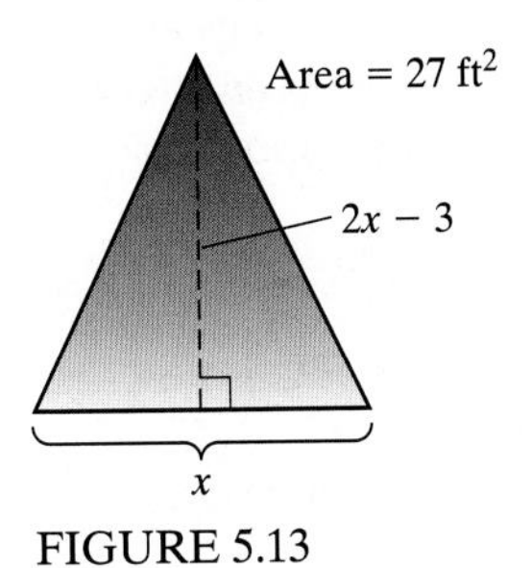

FIGURE 5.13

Solution **Understand** Let's draw a picture of the entrance and label it with the given information (Fig. 5.13).

Translate We use the formula for a triangle to solve the problem.

$$A = \frac{1}{2}(\text{base})(\text{height})$$

$$27 = \frac{1}{2}(x)(2x - 3) \qquad \textit{Substitute expressions for base and height.}$$

Carry out $$2(27) = 2\left[\frac{1}{2}(x)(2x - 3)\right] \qquad \textit{Multiply both sides by 2 to remove fractions.}$$

$$54 = x(2x - 3)$$

$$54 = 2x^2 - 3x$$

$$\text{or} \quad 2x^2 - 3x - 54 = 0 \qquad \textit{Make one side 0.}$$

$$(2x + 9)(x - 6) = 0 \qquad \textit{Factor the trinomial.}$$

$$2x + 9 = 0 \quad \text{or} \quad x - 6 = 0 \qquad \textit{Zero-factor property}$$

$$2x = -9 \qquad x = 6 \qquad \textit{Solve for } x.$$

$$x = -\frac{9}{2}$$

Answer Since the dimensions of a geometric figure cannot be negative, we can eliminate $x = -\frac{9}{2}$ as an answer to our problem. Therefore,

$$\text{base} = x = 6 \text{ feet}$$

$$\text{height} = 2x - 3 = 2(6) - 3 = 9 \text{ feet}$$

EXAMPLE 8 Fireworks are set up on top of a 384-ft mountain overlooking a lake. The fireworks will be shot upward at a speed of 32 feet per second. The height, h, in feet, of the fireworks shell casing above the lake at any time, t, in seconds, can be found by the function $h(t) = -16t^2 + 32t + 384$. Find the time it takes for the shell casing to hit the water after being launched.

Solution **Understand** We will draw a picture to help analyze the problem (see Fig. 5.14). When the shell casing strikes the water, its height above the water is 0 feet.

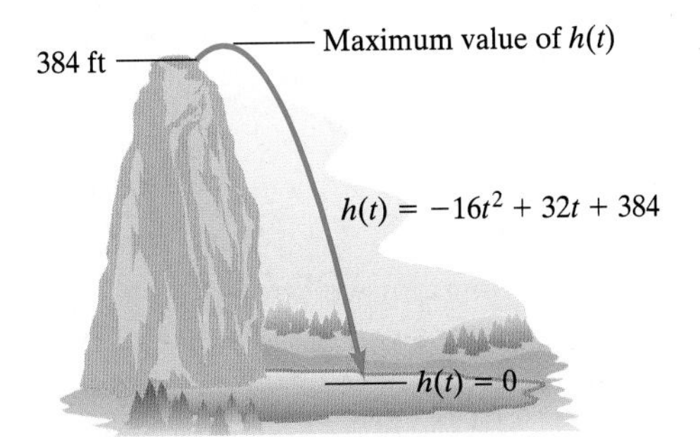

FIGURE 5.14

Translate To solve the problem we need to find the time, t, when $h(t) = 0$. To do so we set the given function equal to 0 and solve for t.

$$-16t^2 + 32t + 384 = 0 \qquad \text{Set } h(t) = 0.$$

Carry Out

$$-16(t^2 - 2t - 24) = 0$$
$$-16(t + 4)(t - 6) = 0$$
$$t + 4 = 0 \quad \text{or} \quad t - 6 = 0$$
$$t = -4 \qquad\qquad t = 6$$

Answer Since t is the number of seconds, -4 is not a possible answer. The shell casing will strike the water in 6 seconds.

NOW TRY EXERCISE 87

Pythagorean Theorem

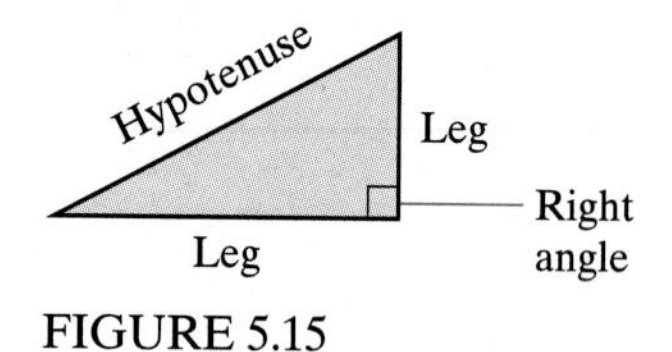

FIGURE 5.15

Our next application uses the Pythagorean Theorem. Consider a right triangle (see Fig. 5.15). The two shorter sides of a right triangle are called the **legs** and the side opposite the right angle is called the **hypotenuse**. The **Pythagorean theorem** expresses the relationship between the legs of the triangle and its hypotenuse.

Pythagorean Theorem

The square of the length of the hypotenuse of a right triangle is equal to the sum of the squares of the lengths of the two legs; that is, $\text{leg}^2 + \text{leg}^2 = \text{hyp}^2$. If a and b represent the lengths of the legs and c represents the length of the hypotenuse, then

$$a^2 + b^2 = c^2$$

EXAMPLE 9

Thuy Thi Thack places a guy wire on a tree to help it grow straight. The location of the stake and where the wire attaches to the tree are given in Figure 5.16. Find the length of the wire. Notice that the length of the wire is the hypotenuse.

Solution

Understand To solve this problem we use the Pythagorean Theorem. From the figure, we see that the legs are x and $x + 1$ and the hypotenuse is $x + 2$.

FIGURE 5.16

Translate

$$\text{leg}^2 + \text{leg}^2 = \text{hyp}^2 \qquad \textit{Pythagorean Theorem}$$
$$x^2 + (x + 1)^2 = (x + 2)^2 \qquad \textit{Substitute expressions for legs and hypotenuse.}$$

Carry Out

$$x^2 + x^2 + 2x + 1 = x^2 + 4x + 4 \qquad \textit{Square terms.}$$
$$2x^2 + 2x + 1 = x^2 + 4x + 4 \qquad \textit{Simplify.}$$
$$x^2 - 2x - 3 = 0 \qquad \textit{Make one side 0.}$$
$$(x - 3)(x + 1) = 0 \qquad \textit{Factor.}$$
$$x - 3 = 0 \quad \text{or} \quad x + 1 = 0 \qquad \textit{Solve.}$$
$$x = 3 \qquad\qquad x = -1$$

Answer From the diagram we know that x cannot be a negative value. Therefore the only possible answer is 3. The stake is placed 3 feet from the tree. The wire attaches to the tree $x + 1$ or 4 feet from the ground. The length of the wire is $x + 2$ or 5 feet.

NOW TRY EXERCISE 81

Using Your Graphing Calculator

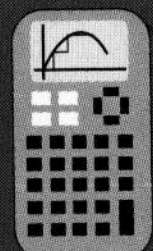

The applications given in this section and the exercise set have been written so that the quadratic equations are factorable. In real life, quadratic equations generally are not factorable (over the set of integers) and need to be solved in other ways. We will discuss methods to solve quadratic equations that are not factorable in Sections 8.1 and 8.2.

You can find approximate solutions to quadratic equations that are not factorable using your graphing calculator. Consider the following real-life example.

EXAMPLE The number of relay antennas for cellular phones in the United States has been increasing. From 1996 through 2002 the number of cellular relay antennas, N, in thousands, in the United States can be closely approximated by the function

$$N(t) = -1.45t^2 + 21.88t + 25.44$$

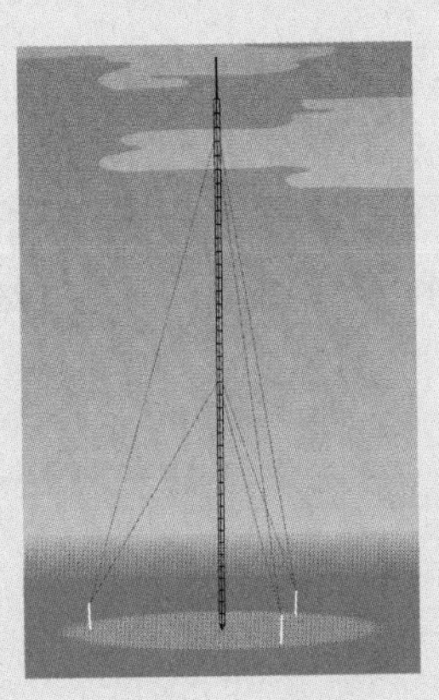

where t is the number of years since 1996. Determine the year in which the number of cellular antennas reached 80 thousand.

Solution **Understand and Translate** To answer this question we need to set the function $N(t)$ equal to 80 and solve for t.

$$-1.45t^2 + 21.88t + 25.44 = 80 \qquad \text{Set } N(t) = 80.$$

We cannot solve this equation by factoring, but we can solve it using a graphing calculator. To do so, let's call one side of the equation y_1 and the other side y_2.

$$y_1 = -1.45x^2 + 21.88x + 25.44$$
$$y_2 = 80$$

Carry out Now graph the two functions on your graphing calculator and use the TRACE and ZOOM keys, or other keys (for example the CALC key with option 5, *intersect*, on the TI-82 and TI-83) to obtain your answer. Figure 5.17. Illustrates the screen of the TI-83, showing that the x-coordinate of the intersection of the equations is about $x = 3.1520$.

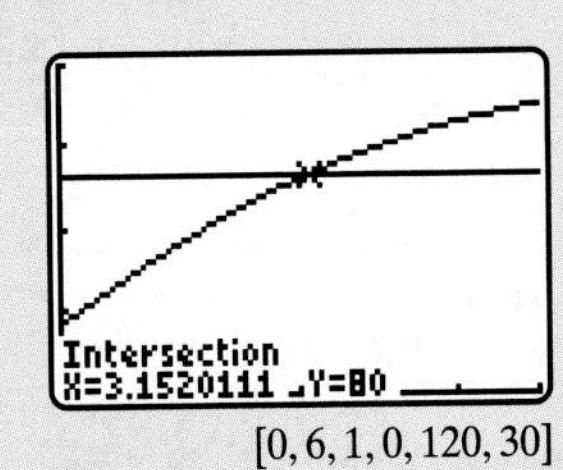

FIGURE 5.17 [0, 6, 1, 0, 120, 30]

Answer Therefore, there will be about 80,000 relay antennas about 3 years after 1996, or in 1999.

4 Use Factoring to Find the x-Intercepts of a Quadratic Function

Consider the graph in Figure 5.18.

At the x-intercepts the value of the function, or y, is 0. Thus if we wish to find the **x-intercepts of a graph**, we can set the function equal to 0 and solve for x.

EXAMPLE 10 Find the x-intercepts of the graph of $y = x^2 - 2x - 8$.

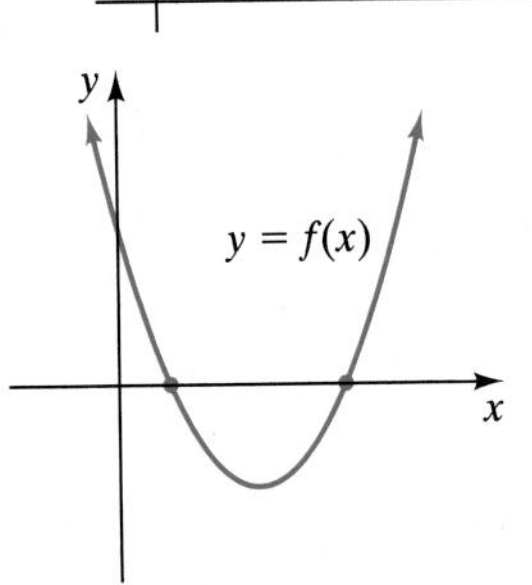

FIGURE 5.18

Solution At the x-intercepts y has a value of 0. Thus to find the x-intercepts we write

$$x^2 - 2x - 8 = 0$$
$$(x - 4)(x + 2) = 0$$
$$x - 4 = 0 \quad \text{or} \quad x + 2 = 0$$
$$x = 4 \qquad\qquad x = -2$$

The solutions of $x^2 - 2x - 8 = 0$ are 4 and -2. The x-intercepts of the graph of $y = x^2 - 2x - 8$ are $(-2, 0)$ and $(4, 0)$, as illustrated in Figure 5.19.

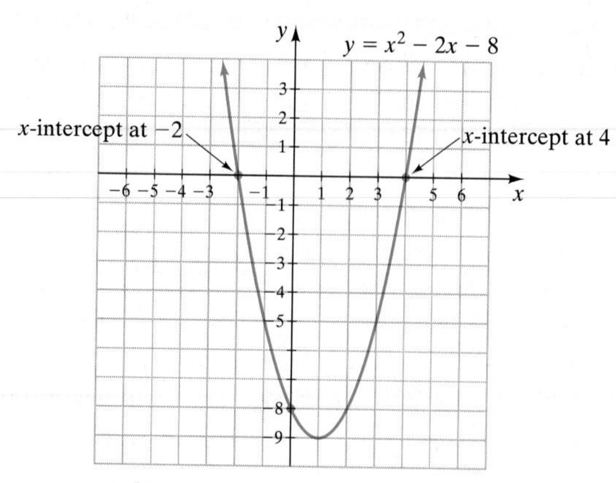

FIGURE 5.19

If we know the x-intercepts of a graph, we can work backward to find the equation of the graph. Read the Using Your Graphing Calculator box that follows to learn how this is done.

NOW TRY EXERCISE 65

Using Your Graphing Calculator

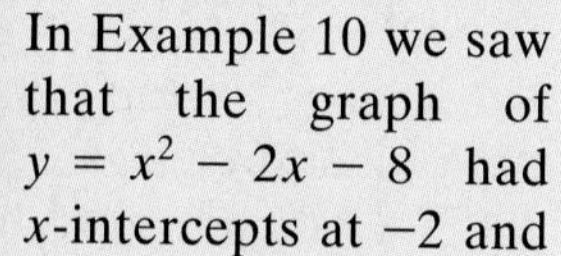

In Example 10 we saw that the graph of $y = x^2 - 2x - 8$ had x-intercepts at -2 and 4. If we know the x-intercepts of a graph, we can find equations whose graphs have those intercepts. For example,

x-Intercepts at	**Factors**	**Equation**
-2 and 4	$(x + 2)(x - 4)$	$y = (x + 2)(x - 4)$ or $y = x^2 - 2x - 8$

Note that other equations may have graphs with the same x-intercepts. For example, the graph of $y = 2(x^2 - 2x - 8)$ or $y = 2x^2 - 4x - 16$ also has x-intercepts at 4 and -2. In fact, the graph of $y = a(x^2 - 2x - 8)$, for any nonzero real number a, will have x-intercepts at 4 and -2.

Consider the graph in Figure 5.20.

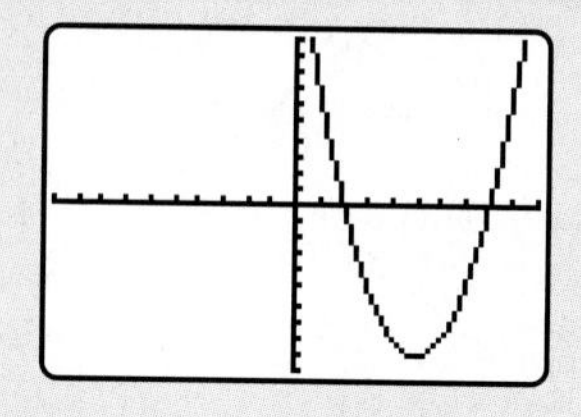

FIGURE 5.20

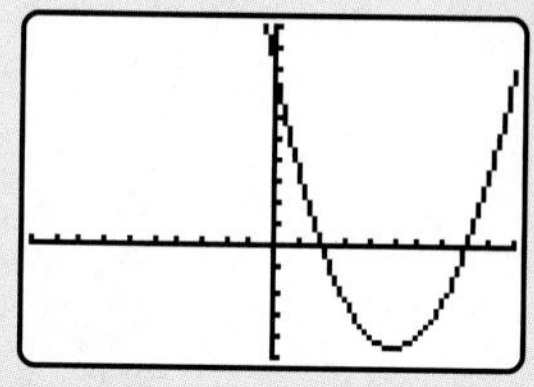

$[-10, 10, 1, -10, 20, 2]$

FIGURE 5.21

If we assume the intercepts are integer values, then the x-intercepts are at 2 and 8. Therefore,

x-Intercepts at	**Factors**	**Possible Equation of Graph**
2 and 8	$(x - 2)(x - 8)$	$y = (x - 2)(x - 8)$ or $y = x^2 - 10x + 16$

If we change the window, we see that the y-intercept of the graph is at 16 (see Fig. 5.21). Therefore, $y = x^2 - 10x + 16$ is the equation of the graph. If, for example, the y-intercept of the graph was at 32, then the equation of the graph would be $y = 2(x^2 - 10x + 16)$ or $y = 2x^2 - 20x + 32$.

Exercises

Write the equation of each graph illustrated. Assume that all x-intercepts are integer values and that the standard window is shown.

1.

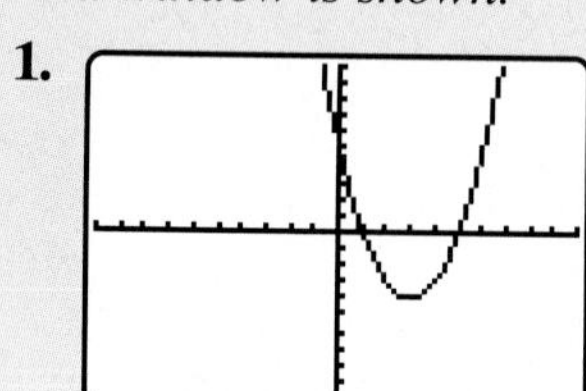

2.

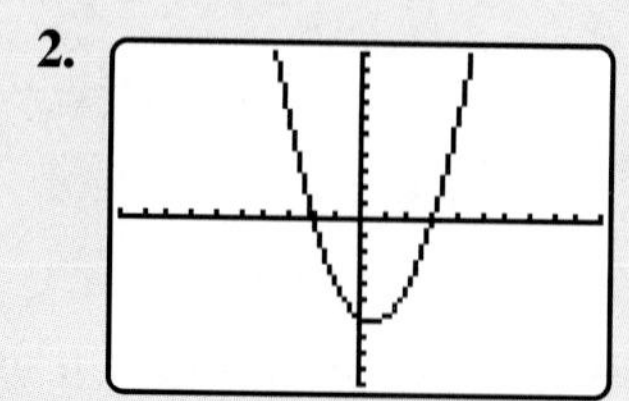

3.

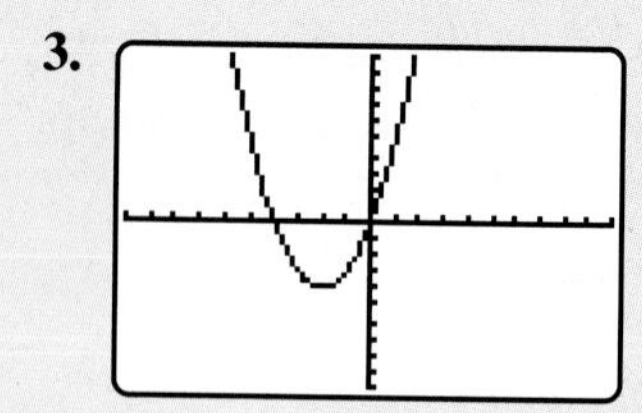

Exercise Set 5.8

Concept/Writing Exercises

1. How do you determine the degree of a polynomial function?
2. What is a quadratic equation?
3. What is the standard form of a quadratic equation?
4. **a)** Explain the zero-factor property in your own words.
   b) Solve the equation $(3x - 7)(2x + 3) = 0$ using the zero-factor property.
5. **a)** Explain why the equation $(x + 3)(x + 4) = 2$ *cannot* be solved by writing $x + 3 = 2$ or $x + 4 = 2$.
   b) Solve the equation $(x + 3)(x + 4) = 2$.
6. When a constant is factored out of an equation, why is it not necessary to set that constant equal to 0 when solving the equation?
7. **a)** Explain in your own words how to solve a polynomial equation using factoring.
   b) Solve the equation $-x - 20 = -12x^2$ using the procedure in part **a)**.
8. **a)** What is the first step in solving the equation $-x^2 + 2x + 35 = 0$?
   b) Solve the equation in part **a)**.
9. **a)** What are the two shorter sides of a right triangle called?
   b) What is the longest side of a right triangle called?
10. Give the Pythagorean Theorem and explain its meaning.
11. If the graph of $y = x^2 + 10x + 16$ has x-intercepts at -8 and -2, what is the solution to the equation $x^2 + 10x + 16 = 0$? Explain.
12. If the solutions to the equation $2x^2 - 15x + 18 = 0$ are $\frac{3}{2}$ and 6, what are the x-intercepts of the graph of $y = 2x^2 - 15x + 18$? Explain.

Practice the Skills

Solve.

13. $x(x + 5) = 0$
14. $3x(x - 5) = 0$
15. $5x(x + 9) = 0$
16. $2(x + 3)(x - 5) = 0$
17. $(2x + 5)(x - 3)(3x + 6) = 0$
18. $x(2x + 3)(x - 5) = 0$
19. $4x - 12 = 0$
20. $9x - 27 = 0$
21. $-x^2 + 12x = 0$
22. $x^2 + 4x = 0$
23. $9x^2 = -18x$
24. $x^2 + 6x + 5 = 0$
25. $x^2 + x - 12 = 0$
26. $x(x + 6) = -9$
27. $x(x - 12) = -20$
28. $3y^2 - 2 = -y$
29. $-z^2 - 3z = -18$
30. $3x^2 = -21x - 18$
31. $3x^2 - 6x - 72 = 0$
32. $x^3 = 3x^2 + 18x$
33. $x^3 + 19x^2 = 42x$
34. $3x^2 - 9x - 30 = 0$
35. $2y^2 + 22y + 60 = 0$
36. $-16x - 3 = -12x^2$
37. $-7x - 10 = -6x^2$
38. $-28x^2 + 15x - 2 = 0$
39. $-2y^2 + 24y - 22 = 0$
40. $3x^3 - 8x^2 - 3x = 0$
41. $z^3 + 16z^2 = -64z$
42. $3p^2 = 22p - 7$
43. $4x^3 + 4x^2 - 48x = 0$
44. $x^2 - 25 = 0$
45. $6x^2 = 16x$
46. $4x^2 = 9$
47. $25x^3 - 16x = 0$
48. $2x^4 - 32x^2 = 0$
49. $(x + 4)^2 - 16 = 0$
50. $(2x + 5)^2 - 9 = 0$
51. $(x - 7)(x + 5) = -20$
52. $(x + 1)^2 = 3x + 7$
53. $6a^2 - 12 - 4a = 19a - 32$
54. $(x - 4)^2 - 4 = 0$
55. $(b - 1)(3b + 2) = 4b$
56. $2(a^2 - 3) - 3a = 2(a + 3)$
57. $2x^3 + 16x^2 + 30x = 0$
58. $18x^3 - 15x^2 = 12x$
59. For $f(x) = 3x^2 + 7x + 9$, find all values of a for which $f(a) = 7$.
60. For $f(x) = 4x^2 - 11x$, find all values of a for which $f(a) = -6$.
61. For $g(x) = 10x^2 - 31x + 19$, find all values of a for which $g(a) = 4$.
62. For $g(x) = 6x^2 + x - 6$, find all values of a for which $g(a) = -5$.
63. For $r(x) = 4x^2 - 19x$, find all values of a for which $r(a) = 30$.
64. For $r(x) = 4x^2 - 4x - 19$, find all values of a for which $r(a) = -4$.

Use factoring to find the x-intercepts of the graphs of each equation (see Example 10).

65. $y = x^2 + 2x - 24$
66. $y = x^2 - 14x + 48$
67. $y = x^2 + 14x + 49$
68. $y = 15x^2 - 14x - 8$
69. $y = 6x^3 - 23x^2 + 20x$
70. $y = 12x^3 - 39x^2 + 30x$

Problem Solving

Determine the x-intercepts of each graph, then match the equation with the appropriate graph labeled **a)–d)**.

71. $y = x^2 - 5x + 6$

72. $y = x^2 - x - 6$

73. $y = x^2 + 5x + 6$

74. $y = x^2 + x - 6$

a)

b)

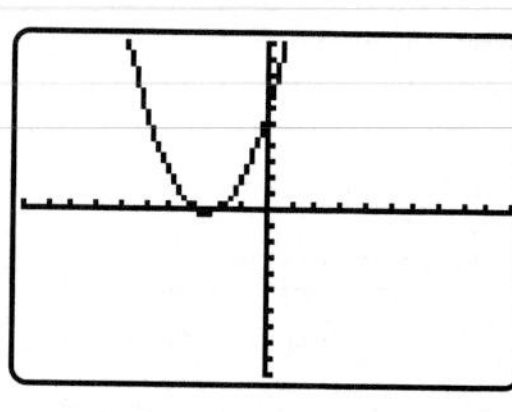

c)

d)

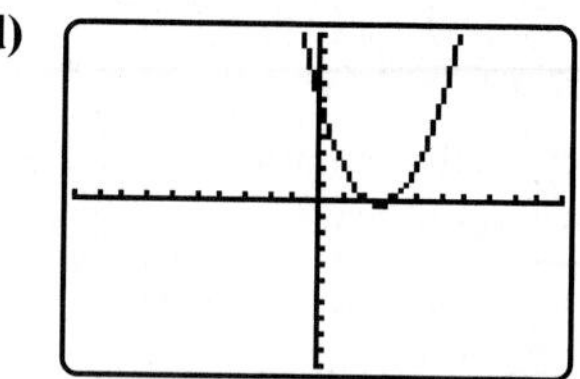

Write an equation whose graph will have x-intercepts at the given values.

75. 4 and −3

76. $\frac{3}{2}$ and 6

77. $-\frac{5}{2}$ and 3

78. −0.4 and 2.6

79. A coffee table is rectangular. If the length of its surface area is 1 foot greater than twice its width and the surface area of the table top is 10 square feet, find its length and width.

80. The floor of a shed has an area of 54 square feet. Find the length and width if the length is 3 feet less than twice its width.

81. A sailboat sail is triangular with a height 6 feet greater than its base. If the sail's area is 80 square feet, find its base and height.

82. A triangular tent has a height that is 4 feet less than its base. If the area of a side is 70 square feet, find the base and height of the tent.

83. The outside dimensions of a picture frame are 28 cm and 23 cm. The area of the picture is 414 cm. Find the width of the frame.

84. Jessyca Nino Aquino's garden is surrounded by a uniform-width walkway. The garden and the walkway together cover an area of 320 square feet. If the dimensions of the garden are 12 feet by 16 feet, find the width of the walkway.

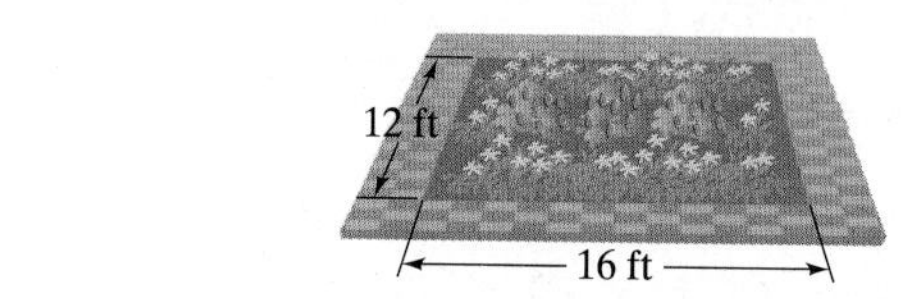

85. Nola Akala's rectangular vegetable garden is 20 feet by 30 feet. In addition to mulching her garden, she wants to put mulch around the outside of her garden in a uniform width. If she has enough mulch to cover an area of 936 square feet, how wide should the mulch border be?

86. Dan Currier has a square garden. He adds a 2-foot-wide walkway around his garden. If the total area of the walkway and garden is 196 square feet, find the dimensions of the garden.

87. In a building at Navy Pier in Chicago, a water fountain jet shoots short spurts of water over a walkway. The water spurts reach a maximum height, then come down into a pond of water on the other side of the walkway. The height above the fountain jet, h, of a spurt of water t seconds after leaving the jet can be found by the function $h(t) = -16t^2 + 32t$. Find the time it takes for the spurt of water to return to the jet's height; that is, when $h(t) = 0$.

88. A model rocket will be launched from a hill 80 feet above sea level. The launch site is next to the ocean (sea level), and the rocket will fall into the ocean. The rocket's distance, s, above sea level at any time, t, is found by the equation $s = -16t^2 + 64t + 80$. Find the time it takes for the rocket to strike the ocean.

89. A tent has wires attached to it to help stabilize it. The wire is attached to the ground 12 feet from the tent. The length of wire used is 8 feet greater than the height from the ground where the wire is attached. How long is the wire?

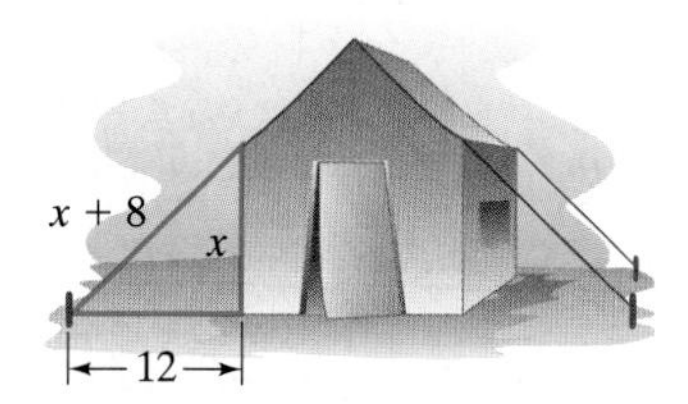

90. Speakers are being wired where the walls meet the ceiling in the corners of a rectangular room. The wires are to go through the attic and be connected at point A, as shown in the figure. If the width of the room is 12 feet and the distance from the corner to point A is 3 feet less than twice the distance from point A to the wall, find the length of wire in the attic needed to wire one speaker.

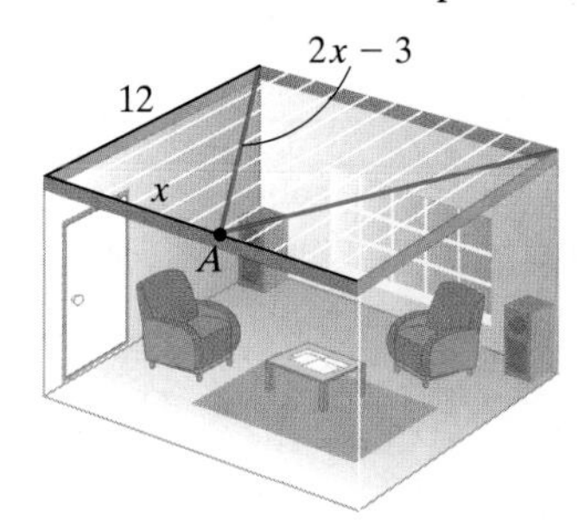

91. The Energy Conservatory Bicycle Shop has a monthly revenue equation $R(x) = 3x^2 - 200x + 450$ and a monthly cost equation $C(x) = x^2 - 75x + 150$, where x is the number of bicycles sold. Find the number of bicycles that must be sold for the company to break even, that is, where revenue equals costs.

92. Edith Hall makes silk plants and sells them to various outlets. Her company has a revenue equation $R(x) = 2x^2 - 29x + 100$ and a cost equation $C(x) = x^2 - 10x + 250$, where x is the number of plants sold. Find the number of plants that must be sold for the company to break even.

93. Phoenix Ta is making a box by cutting out 2 in. by 2 in. squares from a square piece of cardboard and folding up the edges to make a 2-inch-high box.

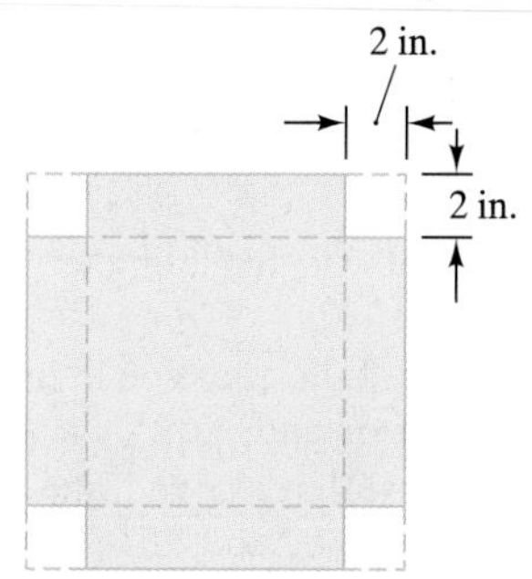

What size piece of cardboard does Phoenix need to make a 2-inch-high box with a volume of 162 cubic inches?

94. A rectangular box is to be formed by cutting squares from each corner of a rectangular piece of tin and folding up the sides. The box is to be 3 inches high, the length is to be twice the width, and the volume of the box is to be 96 cubic inches. Find the length and width of the box.

95. A solid cube with dimensions a^3 has a rectangular solid with dimensions ab^2 removed.

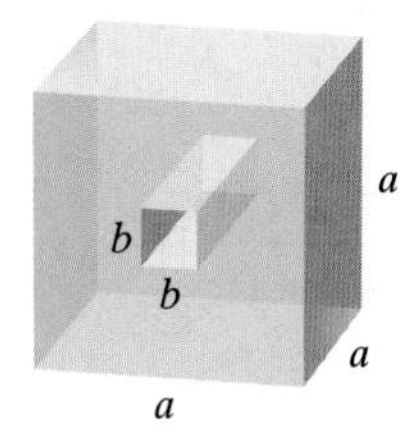

a) Write a formula for the remaining volume, V.
b) Factor the right side of the formula in part **a)**.
c) If the volume is 1620 cubic inches and a is 12 in., find b.

96. A circular steel blade has a hole cut out of its center.

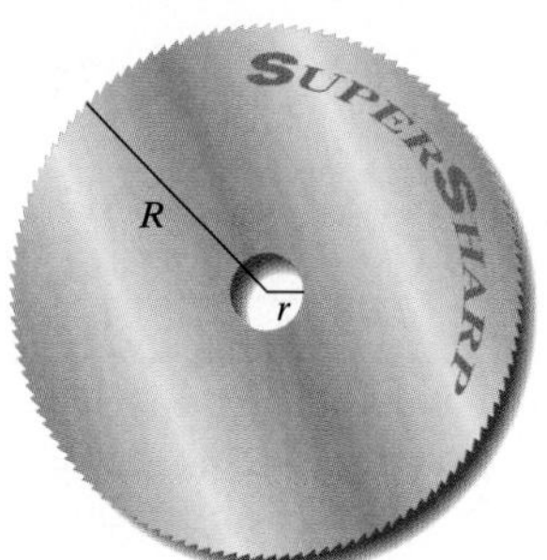

a) Write a formula for the remaining area of the blade.
b) Factor the right side of the formula in part **a)**.
c) Find A if $R = 10$ cm and $r = 3$ cm.

97. Consider the following graph of a quadratic function.

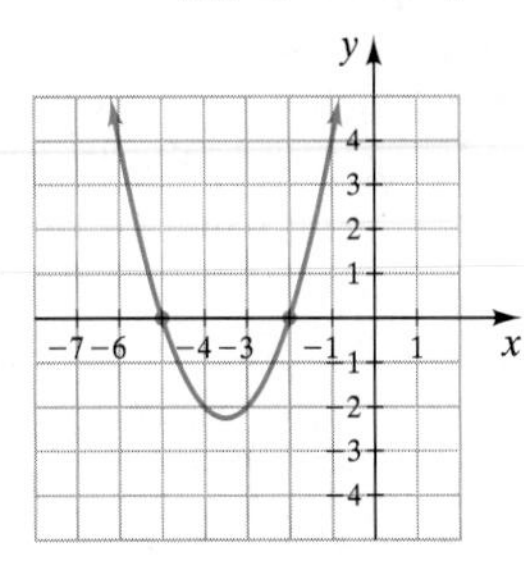

a) Write a quadratic function that has the x-intercepts indicated.

b) Write a quadratic equation in one variable that has solutions of -2 and -5.

c) How many different quadratic functions can have x-intercepts of -2 and -5? Explain.

d) How many different quadratic equations in one variable can have solutions of -2 and -5? Explain.

98. The graph of the equation $y = x^2 + 4$ is illustrated below.

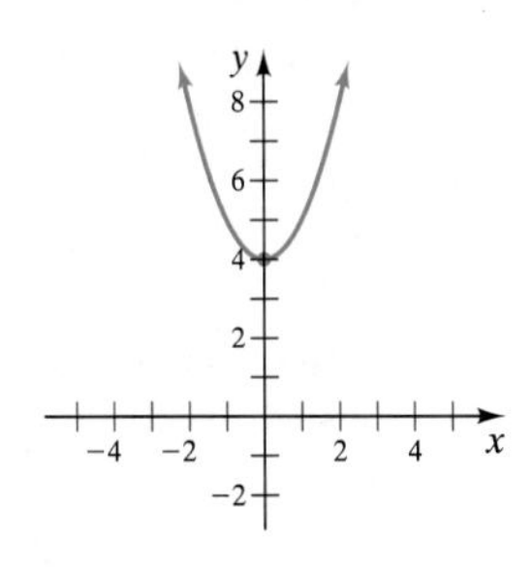

a) How many x-intercepts does the graph have?

b) How many real solutions does the equation $x^2 + 4 = 0$ have? Explain your answer.

99. Consider the quadratic function $P(x) = ax^2 + bx + c, a > 0$.

a) The graph of this type of function may have no x-intercepts, one x-intercept, or two x-intercepts. Sketch each of these possibilities.

b) How many possible real solutions may the equation $ax^2 + bx + c = 0, a > 0$ have? Explain your answer to part **b)** by using the sketches in part **a)**.

100.  A typical car's stopping distance on dry pavement, d, in feet, can be approximated by the function $d(s) = 0.034s^2 + 0.56s - 17.11$, where s is the speed of the car before braking and $60 \leq s \leq 80$ miles per hour. How fast is the car going if it requires 190 ft to stop after the brakes are applied?

101. A typical car's stopping distance on wet pavement, d, in feet, can be approximated by the function $d(s) = -0.31s^2 + 59.82s - 2180.22$, where s is the speed of the car before braking and $60 \leq s \leq 80$ miles per hour. How fast is the car going if it requires 545 feet for the car to stop after the brakes are applied?

Challenge Problems

Solve.

102. $x^4 - 5x^2 + 4 = 0$

103. $x^4 - 13x^2 = -36$

104. $x^6 - 9x^3 + 8 = 0$

Group Activity

In more advanced mathematics courses you may need to solve an equation for y' (read "y prime"). When doing so, treat the y' as a different variable from y. Individually solve each equation for y'. Compare your answers and as a group obtain the correct answers.

105. $xy' + yy' = 1$

106. $xy - xy' = 3y' + 2$

107. $2xyy' - xy = x - 3y'$

Cumulative Review Exercises

[2.3] **108.** Two distance runners, Carmen Kahill and Bob Bruin, run the same course with Bob running 1.2 miles per hour faster than Carmen. Bob finishes in 4 hours and Carmen finishes in 5 hours.

a) What is the rate of each?

b) How long is the course?

[3.1] **109.** Graph $f(x) = x^3 + 3$.

[4.1] **110.** Solve the following system of equations.

$$3x + 5y = 9$$
$$2x - y = 6$$

[4.6] **111.** Find the solution to the following system of inequalities.

$$2y > 6x + 12$$
$$\tfrac{1}{2}y < \tfrac{3}{2}x + 2$$

[5.3] **112.** Divide $(6x^2 - x - 12) \div (2x - 3)$.

SUMMARY

Key Words and Phrases

5.1
Addition of polynomials
Cubic polynomial
Degree of a term
Descending order
Leading coefficient
Leading term
Linear polynomial
Polynomial
Polynomial function
Quadratic polynomial
Subtraction of polynomials
Terms

5.2
Difference of two squares
Expanded form of the distributive property
Factors of a trinomial
FOIL method
Multiplication of polynomials
Square of a binomial

5.3
Division of polynomials
Remainder Theorem
Synthetic division

5.4
Factor a monomial from a polynomial
Factor by grouping
Greatest common factor

5.5
Factor by grouping
Factor by trial and error
Factor trinomials
Factor using substitution
Prime polynomial

5.6
Difference of two cubes
Difference of two squares
Perfect square trinomials
Special factoring formulas
Sum of two cubes

5.8
Cubic equation
Degree of a polynomial equation
Equations quadratic in form
Hypotenuse of a right triangle
Leg of a right triangle
Polynomial equation
Pythagorean Theorem
Quadratic equation
Solution to a polynomial equation
Standard form of a quadratic equation
x-intercepts of a graph
Zero-factor property

IMPORTANT FACTS

FOIL Method to Multiply Binomials

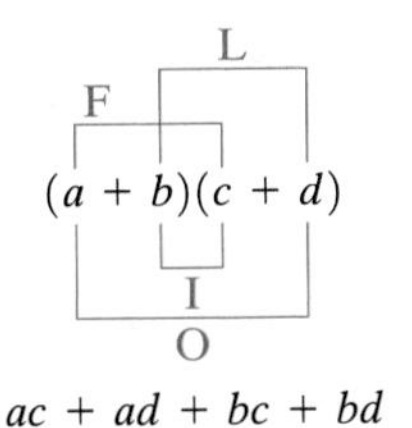

$ac + ad + bc + bd$

F—Multiply First terms.
O—Multiply Outer terms.
I—Multiply Inner terms.
L—Multiply Last terms.

Special Product Formulas

$(a + b)^2 = a^2 + 2ab + b^2$
$(a - b)^2 = a^2 - 2ab + b^2$ } *Square of a binomial*

$(a + b)(a - b) = a^2 - b^2$ *Product of sum and difference of same two terms (or difference of two squares)*

Remainder Theorem

If the polynomial $P(x)$ is divided by $x - a$, the remainder is equal to $P(a)$.

Special Factoring Formulas

$a^2 - b^2 = (a + b)(a - b)$ *Difference of two squares*

$a^2 + 2ab + b^2 = (a + b)^2$
$a^2 - 2ab + b^2 = (a - b)^2$ } *Perfect square trinomials*

$a^3 + b^3 = (a + b)(a^2 - ab + b^2)$ *Sum of two cubes*

$a^3 - b^3 = (a - b)(a^2 + ab + b^2)$ *Difference of two cubes*

Note: The sum of two squares, $a^2 + b^2$, cannot be factored over the set of real numbers.

continued on next page

Standard Form of a Quadratic Equation: $ax^2 + bx + c = 0 \quad a \neq 0.$

Zero-Factor Property: If $a \cdot b = 0$, then either $a = 0$ or $b = 0$, or both a and $b = 0$.

Pythagorean Theorem:

$\text{leg}^2 + \text{leg}^2 = \text{hyp}^2$ or $a^2 + b^2 = c^2$

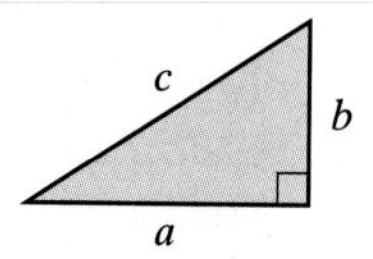

Review Exercises

[5.1] *Determine whether each expression is a polynomial. If the expression is a polynomial,* **a)** *give the special name of the polynomial if it has one,* **b)** *write the polynomial in descending order of the variable x, and* **c)** *give the degree of the polynomial.*

1. $4x^2 - 3 + 5x$

2. $x^2 - y^2 + xy$

3. $-3 - 9x^2y + 6xy^3 + 2x^4$

4. $3x^2 + 6x^{-1} + 4$

[5.1–5.3] *Perform each indicated operation.*

5. $(7x^2 + 3x - 6) - (5x^2 - 7x - 9)$

6. $4x(x^2 + 2x + 3)$

7. $(x + 5)^2$

8. $\dfrac{21y^3 + 6y}{3y}$

9. $(5x + 3)(5x - 3)$

10. $(3xy + 1)(2x + 3y)$

11. $(6x^2 - 11x + 3) \div (3x - 1)$

12. $(2x^3 - 4x^2 - 3x) - (4x^2 - 3x + 9)$

13. $-2xy^2(x^3 + x^2y^5 - 6y)$

14. $(3x - 2y)^2$

15. $(3x^2y + 6xy - 5y^2) - (4y^2 + 3xy)$

16. $(5xy - 6)(5xy + 6)$

17. $\dfrac{9xy - 6y^2 + 3y}{3y}$

18. $(2x - 5y^2)(2x + 5y^2)$

19. $(x^2 + x - 17) \div (x - 3)$

20. $\dfrac{4x^3y^2 + 8x^2y^3 + 12xy^4}{8xy^3}$

21. $[(x + 3y) + 2]^2$

22. $[(x + 3y) + 2][(x + 3y) - 2]$

23. $(-6xy + 6y^2 - 3x) - (y^2 + 3xy + 6x)$

24. $(3x^2 + 4x - 6)(2x - 3)$

25. $(4x^4 - 7x^2 - 5x + 4) \div (2x - 1)$

26. $(4x^3 + 6x - 5)(x + 3)$

27. $(4x^3 + 12x^2 + x - 10) \div (2x + 3)$

28. $(x^2y + 6xy + y^2)(x + y)$

29. Find $P(-3)$ if $P(x) = 3x^2 - 7x + 9$.

30. Find $P(4)$ if $P(x) = -x^3 - 7x^2 + 6x + 3$

The following graph shows Social Security receipts and outlays from 1997 through 2025.

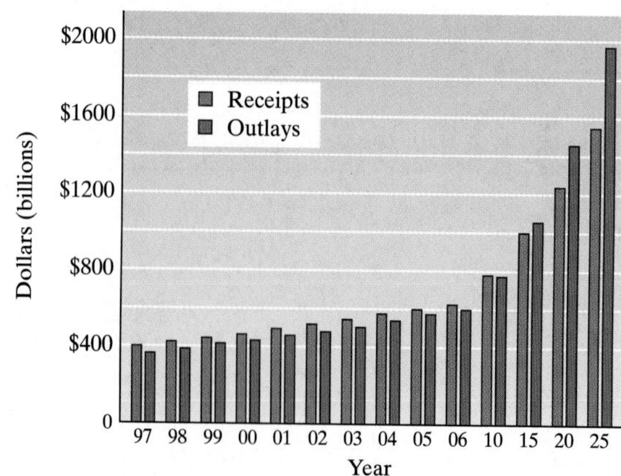

31. The function $R(t) = 0.78t^2 + 20.28t + 385.0$, where t is years since 1997 and $0 \le t \le 28$, gives an approximation of Social Security receipts, $R(t)$, in billions of dollars.

a) Using the function provided, estimate the receipts in 2010.

b) Compare your answer in part **a)** with the graph. Does the graph support your answer?

32. The function $G(t) = 1.74t^2 + 7.32t + 383.91$, where t is years since 1997 and $0 \le t \le 28$, gives an approximation of Social Security outlays, $G(t)$, in billions of dollars.

a) Using the function provided, estimate the outlays in 2010.

b) Compare your answer in part **a)** with the graph. Does the graph support your answer?

For each pair of functions, find **a)** $(f \cdot g)(x)$ *and* **b)** $(f \cdot g)(3)$

33. $f(x) = x + 2, g(x) = x - 3$

34. $f(x) = 2x - 4, g(x) = x^2 - 3$

35. $f(x) = x^2 + x - 3, g(x) = x - 2$

36. $f(x) = x^2 - 2, g(x) = x^2 + 2$

[5.3] *Use synthetic division to obtain each quotient.*

37. $(3x^3 - 2x^2 + 10) \div (x - 3)$

38. $(2y^5 - 10y^3 + y - 1) \div (y + 1)$

39. $(x^5 - 20) \div (x - 2)$

40. $(2x^3 + x^2 + 5x - 3) \div \left(x - \frac{1}{2}\right)$

Determine the remainder of each division problem using the Remainder Theorem. If the divisor is a factor of the dividend, so state.

41. $(x^2 - 4x + 6) \div (x - 3)$

42. $(2x^3 - 6x^2 + 3x) \div (x + 4)$

43. $(3x^3 - 6) \div \left(x - \frac{1}{3}\right)$

44. $(2x^4 - 6x^2 - 8) \div (x + 2)$

[5.4] *Factor out the greatest common factor in each expression.*

45. $12x^2 + 4x + 8$

46. $60x^4 + 6x^9 - 18x^5y^2$

47. $4x(2x - 1) + 3(2x - 1)^2$

48. $12xy^4z^3 + 6x^2y^3z^2 - 15x^3y^2z^3$

Factor by grouping.

49. $5x^2 - xy + 20xy - 4y^2$

50. $12x^2 - 8xy + 15xy - 10y^2$

51. $(3x - y)(x + 2y) - (3x - y)(5x - 7y)$

52. $3a^4 - 12a^2b + 9a^2b - 36b^2$

[5.5] *Factor each trinomial.*

53. $x^2 + 8x + 15$

54. $-x^2 + 12x + 45$

55. $x^2 - 15xy - 54y^2$

56. $8x^3 + 10x^2 - 25x$

57. $4x^3 - 9x^2 + 5x$

58. $12x^3 + 61x^2 + 5x$

59. $x^4 - x^2 - 20$

60. $(x + 5)^2 + 10(x + 5) + 24$

[5.6] *Use a special factoring formula to factor.*

61. $x^2 - 36$

62. $x^4 - 81$

63. $(x - 3)^2 - 4$

64. $4x^2 - 12x + 9$

65. $9y^2 + 24y + 16$

66. $w^4 - 16w^2 + 64$

67. $a^2 + 6ab + 9b^2 - 4c^2$

68. $x^3 - 8$

69. $8x^3 + 27$

70. $(x + 1)^3 - 8$

In Exercises 71 and 72, **a)** *find the area or volume by subtracting the smaller area(s) or volume from the larger and* **b)** *write the expression obtained in part* **a)** *in factored form.*

71.

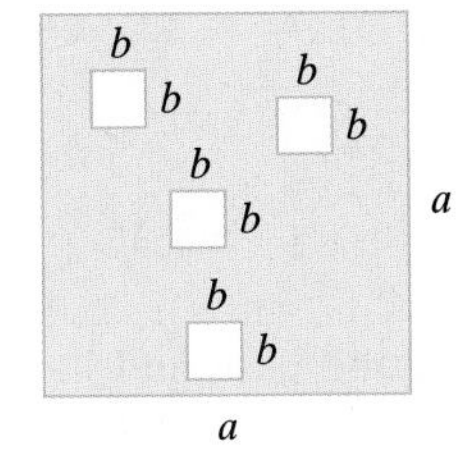

72.

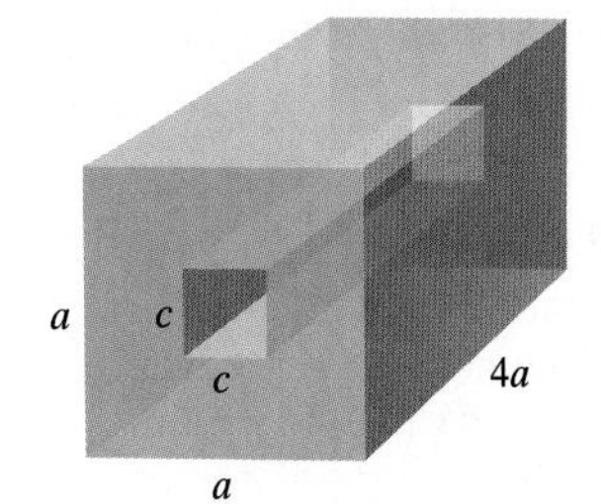

[5.4–5.7] *Factor completely.*

73. $x^2y^2 - 2xy^2 - 15y^2$

74. $3x^3 - 18x^2 + 24x$

75. $3x^3y^4 + 18x^2y^4 - 6x^2y^4 - 36xy^4$

76. $3y^5 - 27y$

77. $2x^3y + 16y$

78. $5x^4y + 20x^3y + 20x^2y$

79. $6x^3 - 21x^2 - 12x$

80. $x^2 + 10x + 25 - y^2$

81. $3x^3 + 24y^3$

82. $x^2(x + 4) + 3x(x + 4) - 4(x + 4)$

83. $4(2x + 3)^2 - 12(2x + 3) + 5$

84. $4x^4 + 4x^2 - 3$

85. $(x - 1)x^2 - (x - 1)x - 2(x - 1)$

86. $9ax - 3bx + 12ay - 4by$

87. $6p^2q^2 - 5pq - 6$

88. $9x^4 - 12x^2 + 4$

89. $4y^2 - (x^2 + 4x + 4)$

90. $6(2a + 3)^2 - 7(2a + 3) - 3$

91. $6x^4y^4 + 9x^3y^4 - 27x^2y^4$

92. $x^3 - \frac{8}{27}y^6$

[5.7] *In Exercise 93–96,* ***a)*** *write an expression for the shaded area of the figure;* ***b)*** *write the expression in part* ***a)*** *in factored form.*

93.

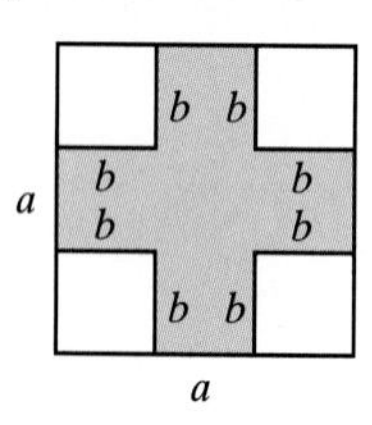

94.

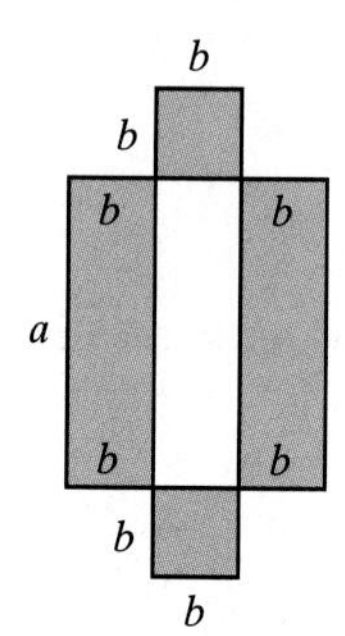

95.

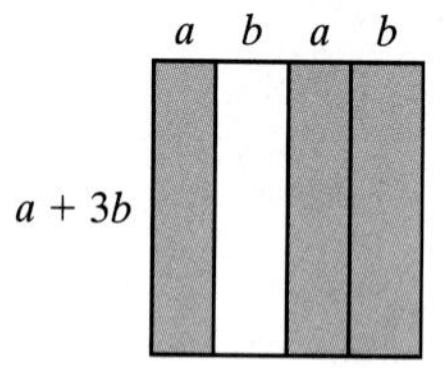

96.

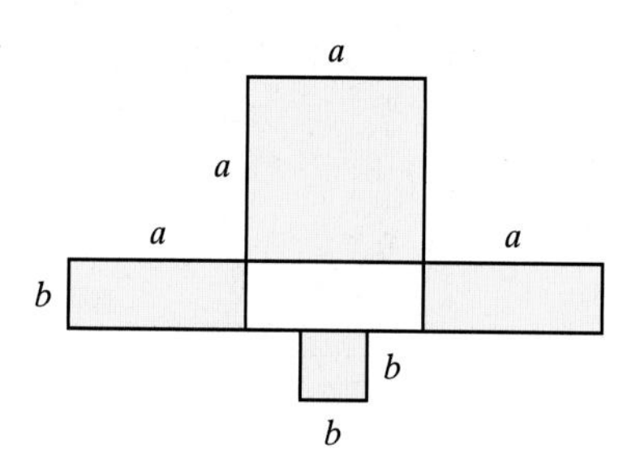

[5.8] *Solve.*

97. $(x - 5)(3x + 2) = 0$

98. $2x^2 = 3x$

99. $15x^2 + 20x = 0$

100. $x^2 - 2x - 24 = 0$

101. $x^2 + 8x + 15 = 0$

102. $x^2 = -2x + 8$

103. $3x^2 + 21x + 30 = 0$

104. $x^3 - 6x^2 + 8x = 0$

105. $12x^3 - 13x^2 - 4x = 0$

106. $8x^2 - 3 = -10x$

107. $4x^2 = 16$

108. $x(x + 3) = 2(x + 4) - 2$

Use factoring to find the x-intercepts of the graph of each equation.

109. $y = 2x^2 - 2x - 60$

110. $y = 20x^2 - 49x + 30$

111. $y = 14x^2 - 41x - 28$

Write an equation that can be used to solve the problem. Solve the equation and answer the question.

112. The area of Lois Heater's rectangular carpet is 63 square feet. Find the length and width of the carpet if the length is 2 feet greater than the width.

113. The base of a large triangular sign is 3 feet more than twice the height. Find the base and height if the area of the triangle is 22 square feet.

114. One square has a side 4 inches longer than the side of a second square. If the area of the larger square is 81 square inches, find the length of a side of each square.

115. A rocket is projected upward from the top of a 144-foot-tall building with a velocity of 128 feet per second. The rocket's distance from the ground, s, at any time, t, in seconds, is given by the formula $s = -16t^2 + 128t + 144$. Find the time it takes for the rocket to strike the ground.

116. Two guy wires are attached to a telephone pole to help stabilize it.

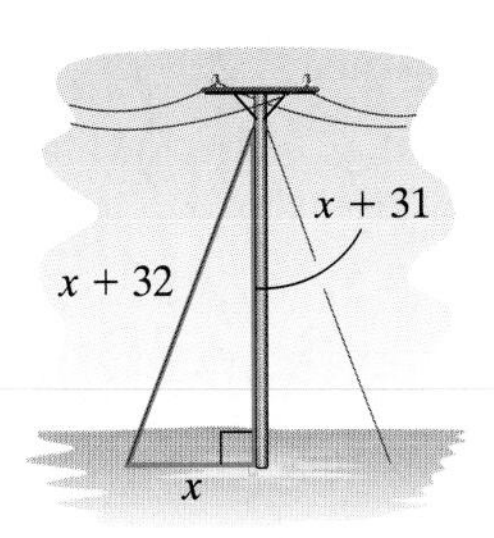

The wire is attached to the ground x feet from the base of the pole. The height of the pole is $x + 31$ and the length of the wire is $x + 32$. Find x.

Practice Test

1. a) Give the specific name of the following polynomial.

$$-4x^2 + 2x - 6x^4$$

b) Write the polynomial in descending powers of the variable x.

c) State the degree of the polynomial.

d) What is the leading coefficient of the polynomial?

Perform each operation.

2. $(12x^6 - 6xy^2 + 15) \div 3x$

3. $(3x + y)(y - 2x)$

4. $(2x^2 + 3xy - 6y^2)(2x + y)$

5. $(2x^2 - 7x + 10) \div (2x + 3)$

6. $(6x^2y + 3y^2 + 5x) - (4x^2y + 2x - 4y^2)$

7. $(2x^3 - x^2 + 5x - 7) \div (x - 3)$

8. $3x^2y^4(-2x^5y^2 + 6x^2y^3 - 3x)$

9. Use synthetic division to obtain the quotient.

$$(3x^4 - 12x^3 - 60x + 4) \div (x - 5)$$

10. Use the Remainder Theorem to find the remainder when $2x^3 - 6x^2 - 5x + 4$ is divided by $x + 3$.

Factor completely.

11. $9x^3y^2 + 12x^2y^5 - 27xy^4$

12. $3x^3 - 6x^2 - 9x$

13. $2x^2 + 4xy + 3xy + 6y^2$

14. $5(x - 2)^2 + 15(x - 2)$

15. $27x^3y^6 - 8y^6$

16. $(x + 3)^2 + 2(x + 3) - 3$

17. $2x^4 + 5x^2 - 18$

18. If $f(x) = 3x - 4$ and $g(x) = x - 5$, find **a)** $(f \cdot g)(x)$ and **b)** $(f \cdot g)(2)$

Solve.

19. $4x^2 - 18 = 21x$

20. $x^3 + 4x^2 - 5x = 0$

21. Solve $b_n = b_1 + ac - c$ for c.

22. Use factoring to find the x-intercepts of the graph of the equation $y = 8x^2 + 10x - 3$.

23. Find an equation whose graph has x-intercepts of 2 and 6.

24. The area of a triangle is 28 square meters. If the base of the triangle is 2 meters greater than 3 times the height, find the base and height of the triangle.

25. A baseball is projected upward from the top of a 448-foot-tall building with an initial velocity of 48 feet per second. The distance, s, of the baseball from the ground at any time, t, in seconds, is given by the equation $s = -16t^2 + 48t + 448$. Find the time it takes for the baseball to strike the ground.

Cumulative Review Test

1. Evaluate $\dfrac{\sqrt[3]{27} - \sqrt[3]{-8} + |-4|}{3^0 - 12 \div 3 \div 4 - 8}$.

2. Simplify $\left(\dfrac{8x^{-2}y^3}{4xy^{-1}}\right)^2$.

3. Simplify $(2p^4q^3)(3pq^4)^3$

4. Solve the equation $\frac{1}{3}(x - 6) = \frac{3}{4}(2x - 1)$.

5. Solve the formula $3P = \dfrac{2L - W}{4}$ for L.

6. Graph $f(x) = 3x - 4$.

7. Graph the inequality $2x - y \leq 6$.

8. Indicate whether the following sets of ordered pairs are functions. Explain your answer.

   a) $\{(0, 1), (3, -2), (-2, 6), (5, 6)\}$

   b) $\{(1, 2), (3, 4), (5, 6), (1, 0)\}$

9. Solve the following system of equations.

$$3x - 2y = 8$$

$$2x - 5y = 10$$

10. Graph the equation $y = |x| - 2$.

11. Simplify $3x^2 - 4x - 6 - (5x - 4x^2 - 6)$.

12. Multiply $(x^2 - 3x - 6)(2x - 5)$.

13. Divide $\dfrac{9x^3y^5 - 8x^2y^4 - 12xy}{3x^2y}$.

14. If $f(x) = 3x^3 - 6x^2 - 4x + 3$, find $f(2)$.

Factor.

15. $x^4 - 3x^3 + 2x^2 - 6x$

16. $12x^2y - 27xy + 6y$

17. $y^4 + 2y^2 - 24$

18. $8x^3 - 27y^6$

19. Copy World charges 15 cents a page for making a master copy from material that must be hand-fed into the copier. After the master copy is made, they can make additional copies from the master copy for 5 cents a page. Cecil Rock has a manuscript to be copied, but since the pages must be hand-fed into the copier, he has a master copy made and six additional copies of the manuscript made from the master copy. If his total bill before tax is \$279, how many pages are in the manuscript?

20. Joseph Santo's first four test grades are 68, 72, 90, and 86. What range of grades on his fifth test will result in an average greater than or equal to 70 and less than 80?

RATIONAL EXPRESSIONS AND EQUATIONS

CHAPTER 6

Use the Angel Web site at www.prenhall.com/angel to be linked to an internet resource that will help you further explore the following application.

Sometimes the context of a problem makes it seem more difficult than it really is. If this happens to you, try to write a similar problem using a context with which you are more familiar. For instance, on page 405, you are asked to determine how far a space station is from NASA headquarters by writing an equation based on the relative times it takes two shuttles traveling at different speeds to reach the station. Although the specific numbers will be different, the problem is the same as determining how far your college is from your home based on the relative time it takes you and your parents to get to the school if you are driving at different speeds.

6.1 THE DOMAINS OF RATIONAL FUNCTIONS AND MULTIPLICATION AND DIVISION OF RATIONAL EXPRESSIONS

1. Find the domains of rational functions.
2. Simplify rational expressions.
3. Multiply rational expressions.
4. Divide rational expressions.

1 Find the Domains of Rational Functions

To understand rational expressions, you must have a thorough understanding of the factoring techniques discussed in Chapter 5. **A rational expression** is an expression of the form p/q, where p and q are polynomials and $q \neq 0$.

Examples of Rational Expressions

$$\frac{2}{3}, \quad \frac{x+3}{x}, \quad \frac{x^2+4x}{x-3}, \quad \frac{x}{x^2-4}$$

Note that the denominator of a rational expression cannot equal 0 because division by 0 is undefined. In the expression $(x+3)/x$, x cannot equal 0, since the denominator would then equal 0. In $(x^2+4x)/(x-3)$, x cannot equal 3 because that would result in the denominator having a value of 0. What values of x cannot be used in the expression $x/(x^2-4)$? If you answered 2 and -2, you answered correctly.

Whenever we write a rational expression containing a variable in the denominator, we always assume that the value or values of the variable that make the denominator 0 are excluded.

In Section 5.1 we discussed polynomial functions. Now we introduce rational functions. A **rational function** is a function of the form $f(x) = p/q$ or $y = p/q$, where p and q are polynomials and $q \neq 0$.

Examples of Rational Functions

$$f(x) = \frac{1}{x} \qquad y = \frac{x^2+9}{x+3} \qquad T(a) = \frac{a^2}{a^2-4}$$

The **domain** of a rational function will be the set of values that can be used to replace the variable. For example, in the rational function $f(x) = (x+2)/(x-3)$, the domain will be all real numbers except 3, written $\{x \mid x \text{ is a real number and } x \neq 3\}$. If x were 3, the denominator would be 0, and division by 0 is undefined.

EXAMPLE 1 For the given functions $f(x)$ and $g(x)$, find the domain of $(f/g)(x)$.

a) $f(x) = x$, $g(x) = x^2 - 4$

b) $f(x) = x - 2$, $g(x) = x^2 + 2x - 8$

Solution **a)** Since $f(x)$ and $g(x)$ are polynomial functions, each of their domains is the set of all real numbers. The domain of the quotient of functions $(f/g)(x)$ will therefore be all real numbers for which the denominator of the quotient is not equal to 0. From Section 3.6 we know that

$$(f/g)(x) = \frac{f(x)}{g(x)}$$

Therefore, $(f/g)(x) = \frac{x}{x^2 - 4}$ *Substitute expressions for $f(x)$ and $g(x)$.*

$= \frac{x}{(x+2)(x-2)}$ *Factor the denominator.*

From this factored form we see that x cannot be 2 or -2. Therefore the domain is all real numbers except 2 or -2. The domain may be expressed as $\{x \mid x \text{ is a real number and } x \neq 2 \text{ and } x \neq -2\}$.

b) $(f/g)(x) = \frac{f(x)}{g(x)}$

$= \frac{x-2}{x^2 + 2x - 8}$ *Substitute expressions for $f(x)$ and $g(x)$.*

$= \frac{x-2}{(x-2)(x+4)}$ *Factor the denominator.*

Notice that the $x - 2$ in the numerator will divide out with the $x - 2$ in the denominator. However, when determining the domain of the quotient of functions, we do so *before* we simplify the expression. Since the denominator cannot be 0, x cannot have values of 2 or -4. The domain is $\{x \mid x \text{ is a real number and } x \neq 2 \text{ and } x \neq -4\}$.

NOW TRY EXERCISE 19

Using Your Graphing Calculator

Although graphing rational functions is beyond the scope of this course, if you have a graphing calculator, you may wish to experiment by graphing some rational functions. This will give you some idea of the wide variety of graphs of rational functions.

If you graph $y = \frac{x^2}{x^2 - 4}$ on a graphing calculator, the display might look like that in Figure 6.1.

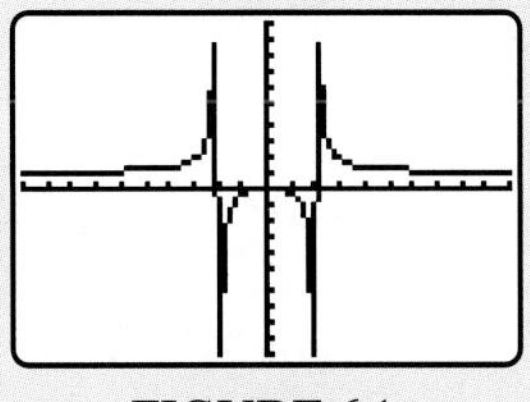

FIGURE 6.1

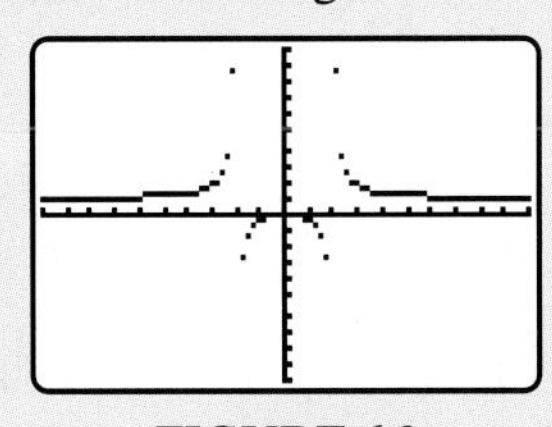

FIGURE 6.2

Notice what appear to be vertical lines at $x = -2$ and $x = 2$, the values of x where the function is undefined. This calculator is in a mode called *connected mode*. When a calculator is in connected mode, it connects all points it plots, going from the point with the smallest x-coordinate to the next larger one. Just to the left of -2, the value of y is a very large positive number, and just to the right of -2 the value of y is a very large negative number. The vertical line is the calculator's attempt to connect the point with this very large positive y-value to the point with this very large negative y-value. A similar situation occurs at $x = 2$.

You may sometimes wish to have your calculator in *dot mode*. When the calculator is in dot mode it displays unconnected points that have been calculated. Read the manual that comes with your calculator to learn how to change from connected to dot mode, and vice versa. The graph in Figure 6.2 shows the same graph as in Figure 6.1 except that this time the calculator is in dot mode.

2) Simplify Rational Expressions

When we work problems containing rational expressions, we must make sure that we write the answer in lowest terms. A rational expression is **simplified** when the numerator and denominator have no common factors other than 1.

The fraction $\frac{6}{9}$ is not simplified because the 6 and 9 both contain the common factor of 3. When the 3 is factored out, the simplified fraction is $\frac{2}{3}$.

$$\frac{6}{9} = \frac{\overset{1}{\cancel{3}} \cdot 2}{\underset{1}{\cancel{3}} \cdot 3} = \frac{2}{3}$$

The rational expression $\frac{ab - b^2}{2b}$ is not simplified because both the numerator and denominator have a common factor, b. To simplify this expression, factor b from each term in the numerator; then divide it out.

$$\frac{ab - b^2}{2b} = \frac{\cancel{b}(a - b)}{2\cancel{b}} = \frac{a - b}{2}$$

Thus $\frac{ab - b^2}{2b}$ becomes $\frac{a - b}{2}$ when simplified.

To Simplify Rational Expressions

1. Factor both numerator and denominator as completely as possible.
2. Divide both the numerator and the denominator by any common factors.

EXAMPLE 2 Simplify $\frac{x^2 + 2x - 3}{x + 3}$.

Solution Factor the numerator; then divide out the common factor.

$$\frac{x^2 + 2x - 3}{x + 3} = \frac{\cancel{(x + 3)}(x - 1)}{\cancel{x + 3}} = x - 1$$

The rational expression simplifies to $x - 1$.

When the terms in a numerator differ only in sign from the terms in a denominator, we can factor out -1 from either the numerator or denominator. *When -1 is factored from a polynomial, the sign of each term in the polynomial changes.* For example,

$$-2x + 3 = -1(2x - 3) = -(2x - 3)$$
$$6 - 5x = -1(-6 + 5x) = -(5x - 6)$$
$$-3x^2 + 5x - 6 = -1(3x^2 - 5x + 6) = -(3x^2 - 5x + 6)$$

EXAMPLE 3 Simplify $\frac{27x^3 - 8}{2 - 3x}$.

Solution

$$\frac{27x^3 - 8}{2 - 3x} = \frac{(3x)^3 - (2)^3}{2 - 3x}$$

Write the numerator as a difference of two cubes.

$$= \frac{(3x - 2)(9x^2 + 6x + 4)}{2 - 3x}$$

Factor; recall that $a^3 - b^3 = (a - b)(a^2 + ab + b^2)$.

$$= \frac{\cancel{(3x - 2)}(9x^2 + 6x + 4)}{-1\cancel{(3x - 2)}}$$

Factor -1 from the denominator and divide out common factors.

$$= -(9x^2 + 6x + 4)$$
$$= -9x^2 - 6x - 4$$

NOW TRY EXERCISE 33

Avoiding Common Errors

INCORRECT INCORRECT

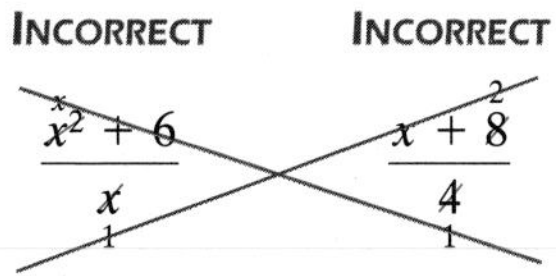

Remember that you can divide out only common **factors**. Only when expressions are *multiplied* can they be factors. Neither of the expressions above can be simplified from their original form.

CORRECT

$$\frac{x^2 - 4}{x - 2} = \frac{(x + 2)\cancel{(x - 2)}}{\cancel{x - 2}}$$

$$= x + 2$$

INCORRECT

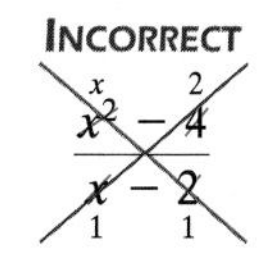

3) Multiply Rational Expressions

Now that we know how to simplify a rational expression, we can discuss multiplication of rational expressions.

To Multiply Rational Expressions

To multiply rational expressions, use the following rule

$$\frac{a}{b} \cdot \frac{c}{d} = \frac{a \cdot c}{b \cdot d}, \quad b \neq 0, d \neq 0$$

To multiply rational expressions, follow these steps.

1. Factor all numerators and denominators as far as possible.
2. Divide out any common factors.
3. Multiply using the above rule.
4. Simplify the answer when possible. (This step will not be necessary if step 2 is done completely!)

If all common factors were factored out in step 2, you will be unable to simplify the answer in step 4. However, if you missed a common factor in step 2, you can factor it out in step 4.

EXAMPLE 4 Multiply **a)** $\frac{x - 5}{4x} \cdot \frac{x^2 - 2x}{x^2 - 7x + 10}$ **b)** $\frac{2x - 5}{x - 4} \cdot \frac{x^2 - 8x + 16}{5 - 2x}$

Solution a) $\frac{x - 5}{4x} \cdot \frac{x^2 - 2x}{x^2 - 7x + 10} = \frac{\cancel{x - 5}}{4\cancel{x}} \cdot \frac{\cancel{x}\cancel{(x - 2)}}{\cancel{(x - 2)}\cancel{(x - 5)}}$ *Factor; divide out common factors.*

$$= \frac{1}{4}$$

b) $\frac{2x-5}{x-4} \cdot \frac{x^2-8x+16}{5-2x} = \frac{2x-5}{x-4} \cdot \frac{(x-4)(x-4)}{5-2x}$ Factor.

$= \frac{\cancel{2x-5}}{\cancel{x-4}} \cdot \frac{\cancel{(x-4)}(x-4)}{-1\cancel{(2x-5)}}$ Factor −1 from denominator; divide out common factors.

$= \frac{x-4}{-1}$

$= -(x-4)$ or $-x+4$ or $4-x$

EXAMPLE 5 Multiply $\frac{x^2-y^2}{x+y} \cdot \frac{x+2y}{2x^2-xy-y^2}$.

Solution $\frac{x^2-y^2}{x+y} \cdot \frac{x+2y}{2x^2-xy-y^2} = \frac{\cancel{(x+y)}\cancel{(x-y)}}{\cancel{x+y}} \cdot \frac{x+2y}{(2x+y)\cancel{(x-y)}}$ Factor; divide out common factors.

$= \frac{x+2y}{2x+y}$

NOW TRY EXERCISE 47

EXAMPLE 6 Multiply $\frac{ab-ac+bd-cd}{ab+ac+bd+cd} \cdot \frac{b^2+bc+bd+cd}{b^2+bd-bc-cd}$.

Solution Factor both numerators and denominators by grouping; then divide out common factors.

$$\frac{ab-ac+bd-cd}{ab+ac+bd+cd} \cdot \frac{b^2+bc+bd+cd}{b^2+bd-bc-cd}$$

$= \frac{a(b-c)+d(b-c)}{a(b+c)+d(b+c)} \cdot \frac{b(b+c)+d(b+c)}{b(b+d)-c(b+d)}$ Factor by grouping.

$= \frac{\cancel{(a+d)}\cancel{(b-c)}}{\cancel{(a+d)}\cancel{(b+c)}} \cdot \frac{\cancel{(b+d)}\cancel{(b+c)}}{\cancel{(b-c)}\cancel{(b+d)}} = 1$ Factor completely; divide out common factors.

4 Divide Rational Expressions

Now let's discuss division of rational expressions.

To Divide Rational Expressions

To divide rational expressions, use the following rule:

$$\frac{a}{b} \div \frac{c}{d} = \frac{a}{b} \cdot \frac{d}{c} = \frac{a \cdot d}{b \cdot c}, \quad b \neq 0, c \neq 0, d \neq 0$$

To divide rational expressions, invert the divisor (the second or bottom fraction) and proceed as when multiplying rational expressions.

EXAMPLE 7 Divide $\frac{12x^4}{5y^3} \div \frac{3x^5}{10y}$.

Solution $\frac{12x^4}{5y^3} \div \frac{3x^5}{10y} = \frac{\overset{4}{\cancel{12}}\,\overset{1}{\cancel{x^4}}}{\underset{1}{\cancel{5}}\,\underset{y^2}{\cancel{y^3}}} \cdot \frac{\overset{2}{\cancel{10}}\,\overset{1}{\cancel{y}}}{\underset{1}{\cancel{3}}\,\underset{x}{\cancel{x^5}}}$ Invert divisor; divide out common factors.

$= \frac{4 \cdot 2}{y^2x} = \frac{8}{xy^2}$

EXAMPLE 8 Divide **a)** $\frac{x^2 - 9}{x + 4} \div \frac{x - 3}{x + 4}$ **b)** $\frac{12a^2 - 22a + 8}{3a} \div \frac{3a^2 + 2a - 8}{2a^2 + 4a}$

Solution **a)**

$$\frac{x^2 - 9}{x + 4} \div \frac{x - 3}{x + 4} = \frac{x^2 - 9}{x + 4} \cdot \frac{x + 4}{x - 3} \qquad \text{Invert divisor.}$$

$$= \frac{(x + 3)\cancel{(x - 3)}}{\cancel{x + 4}} \cdot \frac{\cancel{x + 4}}{\cancel{x - 3}} \qquad \text{Factor; divide out common factors.}$$

$$= x + 3$$

b)

$$\frac{12a^2 - 22a + 8}{3a} \div \frac{3a^2 + 2a - 8}{2a^2 + 4a}$$

$$= \frac{12a^2 - 22a + 8}{3a} \cdot \frac{2a^2 + 4a}{3a^2 + 2a - 8} \qquad \text{Invert divisor.}$$

$$= \frac{2(6a^2 - 11a + 4)}{3a} \cdot \frac{2a(a + 2)}{(3a - 4)(a + 2)} \qquad \text{Factor.}$$

$$= \frac{2\cancel{(3a - 4)}(2a - 1)}{3\cancel{a}} \cdot \frac{2\cancel{a}\cancel{(a + 2)}}{\cancel{(3a - 4)}\cancel{(a + 2)}} \qquad \text{Factor further; divide out common factors.}$$

$$= \frac{4(2a - 1)}{3}$$

EXAMPLE 9 Divide $\frac{x^4 - y^4}{x - y} \div \frac{x^2 + xy}{x^2 - 2xy + y^2}$.

Solution

$$\frac{x^4 - y^4}{x - y} \div \frac{x^2 + xy}{x^2 - 2xy + y^2}$$

$$= \frac{x^4 - y^4}{x - y} \cdot \frac{x^2 - 2xy + y^2}{x^2 + xy} \qquad \text{Invert divisor.}$$

$$= \frac{(x^2 + y^2)(x^2 - y^2)}{x - y} \cdot \frac{(x - y)(x - y)}{x(x + y)} \qquad \text{Factor.}$$

$$= \frac{(x^2 + y^2)\cancel{(x + y)}\cancel{(x - y)}}{\cancel{x - y}} \cdot \frac{(x - y)(x - y)}{x\cancel{(x + y)}} \qquad \text{Factor further; divide out common factors.}$$

$$= \frac{(x^2 + y^2)(x - y)^2}{x}$$

NOW TRY EXERCISE 61

Exercise Set 6.1

Concept/Writing Exercises

1. **a)** What is a rational expression?
   b) Give your own example of a rational expression.
2. Explain why $\frac{\sqrt{x}}{x + 1}$ is not a rational expression.
3. **a)** What is a rational function?
   b) Give your own example of a rational function.
4. Explain why $f(x) = \frac{2}{\sqrt{x + 3}}$ is not a rational function.
5. **a)** What is the domain of a rational function?
   b) What is the domain of $f(x) = \frac{3}{x^2 - 9}$?
6. **a)** Explain how to simplify a rational expression.
   b) Using the procedure stated in part **a)**, simplify $\frac{6x^2 + 7x - 20}{4x^2 - 25}$

7. **a)** Explain how to simplify a rational expression where the numerator and denominator differ only in sign.
b) Using the procedure explained in part **a)**, simplify

$$\frac{3x^2 - 2x - 8}{-3x^2 + 2x + 8}$$

8. **a)** Explain how to multiply rational expressions.
b) Using the procedure given in part **a)**, multiply

$$\frac{6a^2 + a - 1}{3a^2 + 2a - 1} \cdot \frac{3a^2 + 4a + 1}{6a^2 + 5a + 1}$$

9. **a)** Explain how to divide rational expressions.
b) Using the procedure given in part **a)**, divide

$$\frac{r + 2}{r^2 + 7r + 12} \div \frac{(r + 2)^2}{r^2 + 5r + 6}$$

10. Consider $f(x) = \frac{x}{x}$. Will $f(x) = 1$ for all values of x? Explain.

Practice the Skills

Determine the values that are excluded in the following expressions.

11. $\frac{3x}{2x - 8}$
12. $\frac{3}{x^2 - 64}$
13. $\frac{4}{2x^2 - 15x + 25}$
14. $\frac{2}{(x - 3)^2}$
15. $\frac{x - 3}{x^2 + 4}$
16. $\frac{-2}{16 - r^2}$

Determine the domain of each function.

17. $f(p) = \frac{p + 1}{p - 4}$
18. $f(z) = \frac{-2}{-8z + 15}$
19. $y = \frac{5}{x^2 + x - 6}$
20. $y = \frac{x - 3}{x^2 + 4x - 21}$
21. $f(a) = \frac{3a^2 - 6a + 4}{2a^2 + 3a - 2}$
22. $f(x) = \frac{4 - 2x}{x^3 + 9x}$

Simplify each rational expression.

23. $\frac{x - xy}{x}$
24. $\frac{x^2 - 2x}{x}$
25. $\frac{5x^2 - 10xy}{25x}$
26. $\frac{4x^2y + 12xy + 18x^3y^3}{8xy^2}$
27. $\frac{5r - 2}{2 - 5r}$
28. $\frac{4x^2 - 9}{2x^2 - x - 3}$
29. $\frac{p^2 - 2p - 24}{6 - p}$
30. $\frac{4x^2 - 16x^4 + 6x^5y}{8x^3y}$
31. $\frac{a^2 - 3a - 10}{a^2 + 5a + 6}$
32. $\frac{y^2 - 10yz + 24z^2}{y^2 - 5yz + 4z^2}$
33. $\frac{8x^3 - 125y^3}{2x - 5y}$
34. $\frac{x(x - 3) + x(x - 4)}{2x - 7}$
35. $\frac{(x + 1)(x - 3) + (x + 1)(x - 2)}{2(x + 1)}$
36. $\frac{(2x - 5)(x + 4) - (2x - 5)(x + 1)}{3(2x - 5)}$
37. $\frac{xy - yw + xz - zw}{xy + yw + xz + zw}$
38. $\frac{a^2 + 3a - ab - 3b}{a^2 - ab + 5a - 5b}$
39. $\frac{a^3 - b^3}{a^2 - b^2}$
40. $\frac{x^2 + 2x - 3}{x^3 + 27}$

Multiply or divide as indicated. Simplify all answers.

41. $\frac{2x}{3y} \cdot \frac{y^3}{6}$
42. $\frac{16x^2}{y^4} \cdot \frac{5x^2}{4y^2}$
43. $\frac{9x^3}{4} \div \frac{3}{16y^2}$
44. $\frac{80m^4}{49x^5y^7} \cdot \frac{14x^{12}y^5}{25m^5}$
45. $\frac{3 - r}{r - 3} \cdot \frac{r - 5}{5 - r}$
46. $\frac{2a + 2b}{3} \div \frac{a^2 - b^2}{a - b}$
47. $\frac{p^2 + 3p - 10}{p + 5} \cdot \frac{1}{p - 2}$
48. $\frac{x^2 + 3x - 10}{5x} \cdot \frac{x^2 - 3x}{x^2 - 5x + 6}$
49. $\frac{x^2 + 10x + 21}{x + 7} \div (x^2 - 5x - 24)$
50. $(x - 3) \div \frac{x^2 + 3x - 18}{x}$
51. $\frac{x^2 - 9x + 14}{x^2 - 5x + 6} \div \frac{x^2 - 5x - 14}{x + 2}$
52. $\frac{1}{x^2 - 17x + 30} \div \frac{1}{x^2 + 7x - 18}$

53. $\dfrac{a-b}{9a+9b} \div \dfrac{a^2-b^2}{a^2+2a+1}$

54. $\dfrac{2x+4y}{x^2+4xy+4y^2} \cdot \dfrac{2x^2+7xy+6y^2}{4x^2+14xy+12y^2}$

55. $\dfrac{3x^2-x-4}{4x^2+5x+1} \cdot \dfrac{2x^2-5x-12}{6x^2+x-12}$

56. $\dfrac{6x^3-x^2-x}{2x^2+x-1} \cdot \dfrac{x^2-1}{x^3-2x^2+x}$

57. $\dfrac{x+2}{x^3-8} \cdot \dfrac{(x-2)^2}{x^2+4}$

58. $\dfrac{x^2-y^2}{x^2-2xy+y^2} \div \dfrac{x+y}{(x-y)^2}$

59. $\dfrac{x^4-y^8}{x^2+y^4} \div \dfrac{x^2-y^4}{3x^2}$

60. $\dfrac{(x^2-y^2)^2}{(x^2-y^2)^3} \div \dfrac{x^2+y^2}{x^4-y^4}$

61. $\dfrac{2x^4+4x^2}{6x^2+14x+4} \div \dfrac{x^2+2}{3x^2+x}$

62. $\dfrac{8a^3-1}{4a^2+2a+1} \div \dfrac{a-1}{(a-1)^2}$

63. $\dfrac{r^2-9}{r^3-27} \div \dfrac{r^2+6r+9}{r^2+3r+9}$

64. $\dfrac{(a-b)^3}{a^3-b^3} \cdot \dfrac{a^2-b^2}{(a-b)^2}$

65. $\dfrac{2x^3-7x^2+3x}{x^2+2x-3} \cdot \dfrac{x^2+3x}{(x-3)^2}$

66. $\dfrac{4x+y}{5x+2y} \cdot \dfrac{25x^2-5xy-6y^2}{20x^2-7xy-3y^2}$

67. $\dfrac{3r^2+17rs+10s^2}{6r^2+13rs-5s^2} \div \dfrac{6r^2+rs-2s^2}{6r^2-5rs+s^2}$

68. $\dfrac{ac-ad+bc-bd}{ac+ad+bc+bd} \cdot \dfrac{pc+pd-qc-qd}{pc-pd+qc-qd}$

69. $\dfrac{2p^2+2pq-pq^2-q^3}{p^3+p^2+pq^2+q^2} \div \dfrac{p^3+p+p^2q+q}{p^3+p+p^2+1}$

70. $\dfrac{x^3-4x^2+x-4}{x^4-x^3+x^2-x} \cdot \dfrac{2x^3+2x^2+x+1}{2x^3-8x^2+x-4}$

Problem Solving

71. Make up a rational expression that is undefined at $x = 2$ and $x = -3$. Explain how you determined your answer.

72. Make up a rational expression that is undefined at $x = 4$ and $x = -2$. Explain how you determined your answer.

73. Consider the rational function $f(x) = \dfrac{1}{x}$. Explain why this function can never equal 0.

74. Consider the rational function $g(x) = \dfrac{2}{x-3}$. Explain why this function can never equal 0.

75. Consider the rational function $f(x) = \dfrac{x-4}{x^2-4}$. For what value of x, if any, will this function **a)** equal 0; **b)** be undefined? Explain.

76. Consider the function $f(x) = \dfrac{x-4}{x^2-81}$. For what value of x, if any, will this function **a)** equal 0; **b)** be undefined? Explain.

77. Give a function that is undefined at $x = 3$ and $x = -1$ and has a value of 0 at $x = 2$. Explain how you determined your answer.

78. Give a function that is undefined at $x = -4$ and $x = -2$ and has a value of 0 at $x = 3$. Explain how you determined your answer.

Determine the polynomial to be placed in the shaded area to give a true statement. Explain how you determined your answer.

79. $\dfrac{\square}{x^2+2x-15} = \dfrac{1}{x-3}$

80. $\dfrac{\square}{3x+4} = x-3$

81. $\dfrac{y^2-y-20}{\square} = \dfrac{y+4}{y+1}$

82. $\dfrac{\square}{6p^2+p-15} = \dfrac{2p-1}{2p-3}$

Determine the polynomial to be placed in the shaded area to give a true statement. Explain how you determined your answer.

83. $\dfrac{x^2-x-12}{x^2+2x-3} \cdot \dfrac{\square}{x^2-2x-8} = 1$

84. $\dfrac{x^2-4}{(x+2)^2} \cdot \dfrac{2x^2+x-6}{\square} = \dfrac{x-2}{2x+5}$

85. $\dfrac{x^2-9}{2x^2+3x-2} \div \dfrac{2x^2-9x+9}{\square} = \dfrac{x+3}{2x-1}$

86. $\dfrac{4r^2-r-18}{\square} \div \dfrac{4r^3-9r^2}{6r^2-9r+3} = \dfrac{3(r-1)}{r^2}$

87. Consider the triangle below. If its area is $a^2 + 2ab - 3b^2$ and its base is $a + 3b$, find its height h. Use area $= \frac{1}{2}$(base)(height).

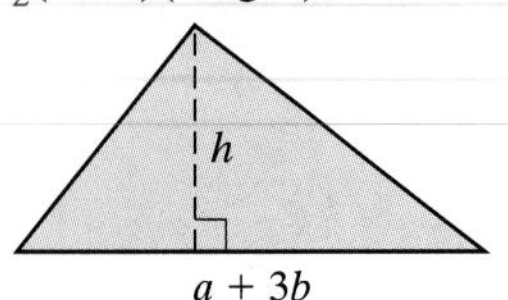

88. Consider the trapezoid below. If its area is $a^2 - b^2$, find its height, h. Use area $= \frac{1}{2}$(height)(base 1 + base 2).

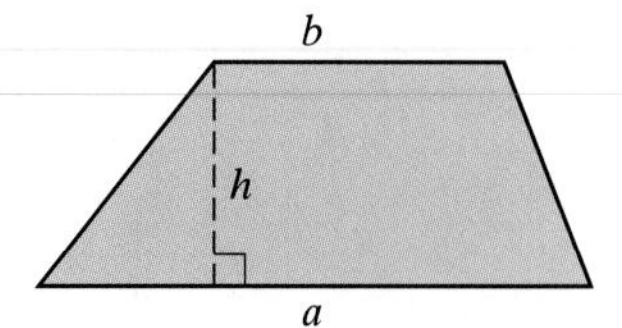

Perform each indicated operation.

89. $\left(\dfrac{2x^2 - 3x - 14}{2x^2 - 9x + 7} \div \dfrac{6x^2 + x - 15}{3x^2 + 2x - 5}\right) \cdot \dfrac{6x^2 - 7x - 3}{2x^2 - x - 3}$

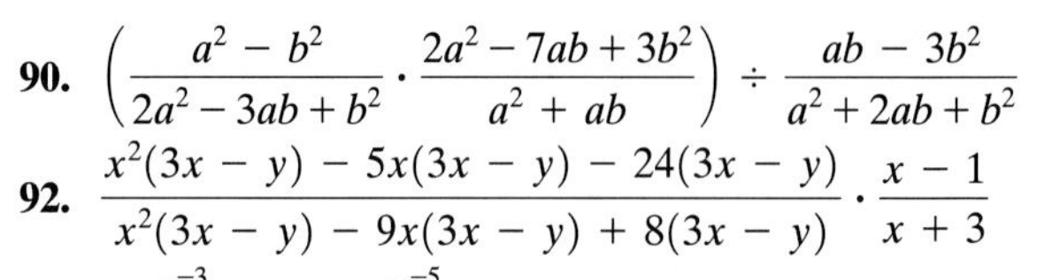

90. $\left(\dfrac{a^2 - b^2}{2a^2 - 3ab + b^2} \cdot \dfrac{2a^2 - 7ab + 3b^2}{a^2 + ab}\right) \div \dfrac{ab - 3b^2}{a^2 + 2ab + b^2}$

91. $\dfrac{5x^2(x - 1) - 3x(x - 1) - 2(x - 1)}{10x^2(x - 1) + 9x(x - 1) + 2(x - 1)} \cdot \dfrac{2x + 1}{x + 3}$

92. $\dfrac{x^2(3x - y) - 5x(3x - y) - 24(3x - y)}{x^2(3x - y) - 9x(3x - y) + 8(3x - y)} \cdot \dfrac{x - 1}{x + 3}$

93. $\dfrac{(x - p)^n}{x^{-2}} \div \dfrac{(x - p)^{2n}}{x^{-4}}$

94. $\dfrac{x^{-3}}{(a - b)^r} \div \dfrac{x^{-5}}{(a - b)^{r+2}}$

Simplify.

95. $\dfrac{x^{5y} + 3x^{4y}}{3x^{3y} + x^{4y}}$

96. $\dfrac{m^{2x} - m^x - 2}{m^{2x} - 4}$

For Exercises 97–100.

a) *Determine the domain of the function.*

b) *Graph the function in connected mode.*

c) *Is the function increasing, decreasing, or remaining the same as x gets closer and closer to 2, approaching 2 from the* left side?

d) *Is the function increasing, decreasing, or remaining the same as x gets closer and closer to 2, approaching 2 from the* right side?

97. $f(x) = \dfrac{1}{x - 2}$

98. $f(x) = \dfrac{x}{x - 2}$

99. $f(x) = \dfrac{x^2}{x - 2}$

100. $f(x) = \dfrac{x - 2}{x - 2}$

Group Activity

101. Consider the rational function $f(x) = \dfrac{1}{x}$.

a) As a group, determine the domain of the function.

b) Have each group member individually complete the following table for the function.

x	−10	−1	−0.5	−0.1	−0.01	0.01	0.1	0.5	1	10
y										

c) Compare your answers to part **b)** and agree on the correct values.

d) As a group, draw the graph of $f(x) = \dfrac{1}{x}$. Consider what happens to the function as x gets closer and closer to 0, approaching 0 from both the left and right sides.

e) Can this graph ever have a value of 0? Explain your answer.

102. Consider the rational function $f(x) = \dfrac{x^2 - 4}{x - 2}$.

a) As a group, determine the domain of this function.

b) Have each member of the group individually complete the following table for the function.

x	−2	−1	0	1	1.9	1.99	2.01	2.1	3	4	5	6
y												

c) Compare your answers to part **b)** and agree on the correct table values.

d) As a group draw the graph of $f(x) = \dfrac{x^2 - 4}{x - 2}$. Is the function defined when $x = 2$?

e) Can this graph ever have a value of 0? If so, for what value(s) of a is $f(a) = 0$?

Cumulative Review Exercises

[1.5] **103.** Simplify $\dfrac{4x^{-3}y^4}{12x^{-2}y^3}$

[2.2] **104.** Solve the formula $V = \frac{4}{3}\pi r^2 h$ for h.

[2.5] **105.** Solve the inequality $-4 < 3x - 4 < 8$. Write the solution in interval notation.

[3.4] **106.** Find the slope and y-intercept of the graph of the equation $3(y - 4) = -(x - 2)$.

[4.2] **107.** Solve the following system of equations using the addition method.

$$x + 2y = 4$$
$$2y = 6x + 6$$

[5.1] **108.** Simplify
$3x^2y - 4xy + 2y^2 - (3xy + 6y^2 + 2x)$.

6.2 ADDITION AND SUBTRACTION OF RATIONAL EXPRESSIONS

SSM VIDEO 6.2 CD Rom

1. **Add and subtract expressions with a common denominator.**
2. **Find the least common denominator (LCD).**
3. **Add and subtract expressions with unlike denominators.**

1 Add and Subtract Expressions with a Common Denominator

When adding (or subtracting) two rational expressions with a common denominator, we add (or subtract) the numerators while keeping the common denominator.

To Add or Subtract Rational Expressions

To add or subtract rational expressions, use the following rules.

Addition	Subtraction
$\dfrac{a}{c} + \dfrac{b}{c} = \dfrac{a + b}{c}, \quad c \neq 0$	$\dfrac{a}{c} - \dfrac{b}{c} = \dfrac{a - b}{c}, \quad c \neq 0$

To add or subtract rational expressions with a common denominator,

1. Add or subtract the expressions using the rules given above.
2. Simplify the expression if possible.

EXAMPLE 1 Add. **a)** $\dfrac{3}{x + 2} + \dfrac{x - 4}{x + 2}$ **b)** $\dfrac{x^2 + 3x - 2}{(x + 5)(x - 2)} + \dfrac{4x + 12}{(x + 5)(x - 2)}$

Solution **a)** Since the denominators are the same, we add the numerators and keep the common denominator.

$$\frac{3}{x + 2} + \frac{x - 4}{x + 2} = \frac{3 + (x - 4)}{x + 2} \qquad \text{Add numerators.}$$
$$= \frac{x - 1}{x + 2}$$

b)
$$\frac{x^2 + 3x - 2}{(x + 5)(x - 2)} + \frac{4x + 12}{(x + 5)(x - 2)} = \frac{x^2 + 3x - 2 + (4x + 12)}{(x + 5)(x - 2)} \qquad \text{Add numerators.}$$
$$= \frac{x^2 + 7x + 10}{(x + 5)(x - 2)} \qquad \text{Combine terms.}$$
$$= \frac{\cancel{(x + 5)}(x + 2)}{\cancel{(x + 5)}(x - 2)} \qquad \text{Factor; divide out common factors.}$$
$$= \frac{x + 2}{x - 2}$$

When subtracting rational expressions, be sure to subtract the entire numerator of the fraction being subtracted. Study the Avoiding Common Errors box that follows very carefully.

Avoiding Common Errors

The error presented here is sometimes made by students. Study the information presented so that you will not make this error.

How do you simplify this problem?

$$\frac{4x}{x-2} - \frac{2x+1}{x-2}$$

CORRECT

$$\frac{4x}{x-2} - \frac{2x+1}{x-2} = \frac{4x-(2x+1)}{x-2} = \frac{4x-2x-1}{x-2} = \frac{2x-1}{x-2}$$

INCORRECT

$$\frac{4x}{x-2} - \frac{2x+1}{x-2} = \frac{4x-2x+1}{x-2} = \frac{2x+1}{x-2}$$

The procedure on the right side is incorrect because the *entire numerator*, $2x + 1$, must be subtracted from $4x$. Instead, only $2x$ was subtracted. Note that **the sign of each term** (not just the first term) **in the numerator of the fraction being subtracted must change**. Note that $-(2x + 1) = -2x - 1$, by the distributive property.

EXAMPLE 2 Subtract $\frac{3a}{a-6} - \frac{a^2-4a+6}{a-6}$.

Solution

$$\frac{3a}{a-6} - \frac{a^2-4a+6}{a-6} = \frac{3a-(a^2-4a+6)}{a-6}$$ *Subtract numerators.*

$$= \frac{3a-a^2+4a-6}{a-6}$$

$$= \frac{-a^2+7a-6}{a-6}$$ *Combine terms.*

$$= \frac{-(a^2-7a+6)}{a-6}$$ *Factor out −1.*

$$= \frac{-\cancel{(a-6)}(a-1)}{\cancel{a-6}}$$ *Factor; divide out common factors.*

$$= -(a-1)$$

NOW TRY EXERCISE 7

2) Find the Least Common Denominator

To add or subtract two numerical fractions with *unlike denominators*, we must first obtain a common denominator. Obtaining a common denominator may involve writing numerical values as products of prime numbers. A **prime number** is a number greater than 1 that has only two divisors, itself and 1. Some prime numbers are 2, 3, 5, 7, 11, 13, and 17. Following we show how the numbers 36 and 48 are written as a product of prime numbers,

$$36 = 2 \cdot 2 \cdot 3 \cdot 3 = 2^2 \cdot 3^2$$

$$48 = 2 \cdot 2 \cdot 2 \cdot 2 \cdot 3 = 2^4 \cdot 3$$

We may need to write numerical coefficients as products of prime numbers to find the **least common denominator** of a rational expression.

To Find the Least Common Denominator (LCD) of Rational Expressions

1. Write each nonprime coefficient (other than 1) that appears in a denominator as a product of prime numbers.
2. Factor each denominator completely. Any factors that occur more than once should be expressed as powers. For example, $(x + 5)(x + 5)$ should be expressed as $(x + 5)^2$.
3. List all different factors (other than 1) that appear in any of the denominators. When the same factor appears in more than one denominator, write the factor with the highest power that appears.
4. The least common denominator is the product of all the factors found in step 3.

EXAMPLE 3 Find the LCD of each expression.

a) $\dfrac{3}{5x} - \dfrac{2}{x^2}$ **b)** $\dfrac{1}{18x^3y} + \dfrac{5}{27x^2y^3}$ **c)** $\dfrac{3}{x} - \dfrac{2y}{x+5}$

Solution **a)** The factors that appear in the denominators are 5 and x. List each factor with its highest power. The LCD is the product of these factors.

Highest power of x

$$\text{LCD} = 5 \cdot x^2 = 5x^2$$

b) The numerical coefficients written as products of prime numbers are $18 = 2 \cdot 3^2$ and $27 = 3^3$. The variable factors are x and y. Using the highest powers of the factors, we obtain the LCD.

$$\text{LCD} = 2 \cdot 3^3 \cdot x^3y^3 = 54x^3y^3$$

c) The factors are x and $(x + 5)$. Note that the x in the second denominator, $x + 5$, is not a factor of that denominator since the operation is addition rather than multiplication.

$$\text{LCD} = x(x + 5)$$

EXAMPLE 4 Find the LCD of each expression.

a) $\dfrac{3}{2x^2 - 4x} + \dfrac{x^2}{x^2 - 4x + 4}$ **b)** $\dfrac{5x}{x^2 - x - 12} - \dfrac{6x^2}{x^2 - 7x + 12}$

Solution **a)** Factor both denominators.

$$\frac{3}{2x^2 - 4x} + \frac{x^2}{x^2 - 4x + 4} = \frac{3}{2x(x-2)} + \frac{x^2}{(x-2)^2}$$

The factors are 2, x, and $x - 2$. Multiply the factors raised to the highest power that appears for each factor.

$$\text{LCD} = 2 \cdot x \cdot (x - 2)^2 = 2x(x - 2)^2$$

b) Factor both denominators.

$$\frac{5x}{x^2 - x - 12} - \frac{6x^2}{x^2 - 7x + 12} = \frac{5x}{(x + 3)(x - 4)} - \frac{6x^2}{(x - 3)(x - 4)}$$

Multiply the factors.

$$\text{LCD} = (x + 3)(x - 4)(x - 3)$$

Note that although $(x - 4)$ is a common factor of each denominator, the highest power of the factor that appears in either denominator is 1.

3) Add and Subtract Expressions with Unlike Denominators

The procedure used to add or subtract rational expressions with unlike denominators is given below.

To Add or Subtract Rational Expressions with Unlike Denominators

1. Determine the LCD.
2. Rewrite each fraction as an equivalent fraction with the LCD. This is done by multiplying both the numerator and denominator of each fraction by any factors needed to obtain the LCD.
3. Leave the denominator in factored form, but multiply out the numerator.
4. Add or subtract the numerators while maintaining the LCD.
5. When possible, factor the numerator and reduce fractions.

EXAMPLE 5 Add. **a)** $\frac{3}{x} + \frac{5}{y}$ **b)** $\frac{5}{4a^2} + \frac{3}{14ab^3}$

Solution **a)** First we determine the LCD.

$$\text{LCD} = xy$$

Now we write each fraction with the LCD. We do this by multiplying *both* numerator and denominator of each fraction by any factors needed to obtain the LCD.

In this problem, the fraction on the left must be multiplied by $\frac{y}{y}$ and the fraction on the right must be multiplied by $\frac{x}{x}$.

$$\frac{3}{x} + \frac{5}{y} = \frac{y}{y} \cdot \frac{3}{x} + \frac{5}{y} \cdot \frac{x}{x} = \frac{3y}{xy} + \frac{5x}{xy}$$

By multiplying both the numerator and denominator by the same factor, we are in effect multiplying by 1, which does not change the value of the fraction, only its appearance. Thus, the new fraction is equivalent to the original fraction.

Now we add the numerators while leaving the LCD alone.

$$\frac{3y}{xy} + \frac{5x}{xy} = \frac{3y + 5x}{xy} \quad \text{or} \quad \frac{5x + 3y}{xy}$$

Therefore, $\frac{3}{x} + \frac{5}{y} = \frac{5x + 3y}{xy}$.

b) The LCD is $28a^2b^3$. We must write each fraction with the denominator $28a^2b^3$. To do this, we multiply the fraction on the left by $\frac{7b^3}{7b^3}$ and the fraction on the right by $\frac{2a}{2a}$.

$$\frac{5}{4a^2} + \frac{3}{14ab^3} = \frac{7b^3}{7b^3} \cdot \frac{5}{4a^2} + \frac{3}{14ab^3} \cdot \frac{2a}{2a}$$ *Multiply to obtain LCD.*

$$= \frac{35b^3}{28a^2b^3} + \frac{6a}{28a^2b^3}$$

$$= \frac{35b^3 + 6a}{28a^2b^3}$$ *Add numerators.*

NOW TRY EXERCISE 27

EXAMPLE 6 Subtract $\frac{x + 2}{x - 4} - \frac{x + 3}{x + 4}$.

Solution The LCD is $(x - 4)(x + 4)$. Write each fraction with the denominator $(x - 4)(x + 4)$.

$$\frac{x + 2}{x - 4} - \frac{x + 3}{x + 4} = \frac{x + 4}{x + 4} \cdot \frac{x + 2}{x - 4} - \frac{x + 3}{x + 4} \cdot \frac{x - 4}{x - 4}$$ *Multiply to obtain LCD.*

$$= \frac{(x + 4)(x + 2)}{(x + 4)(x - 4)} - \frac{(x + 3)(x - 4)}{(x + 4)(x - 4)}$$

$$= \frac{x^2 + 6x + 8}{(x + 4)(x - 4)} - \frac{x^2 - x - 12}{(x + 4)(x - 4)}$$ *Multiply binomials in numerators.*

$$= \frac{x^2 + 6x + 8 - (x^2 - x - 12)}{(x + 4)(x - 4)}$$ *Subtract numerators.*

$$= \frac{x^2 + 6x + 8 - x^2 + x + 12}{(x + 4)(x - 4)}$$

$$= \frac{7x + 20}{(x + 4)(x - 4)}$$ *Combine terms.*

EXAMPLE 7 Add $\frac{4}{x - 3} + \frac{x + 5}{3 - x}$.

Solution Note that each denominator is the opposite, or additive inverse, of the other. (The terms of one denominator differ only in sign from the terms of the other denominator.) When this special situation arises, we can multiply the numerator and denominator of either one of the fractions by −1 to obtain the LCD.

$$\frac{4}{x-3} + \frac{x+5}{3-x} = \frac{4}{x-3} + \frac{-1}{-1} \cdot \frac{(x+5)}{(3-x)}$$ *Multiply to obtain LCD.*

$$= \frac{4}{x-3} + \frac{-x-5}{x-3}$$

$$= \frac{4-x-5}{x-3}$$ *Add numerators.*

$$= \frac{-x-1}{x-3}$$ *Combine terms.*

NOW TRY EXERCISE 31

EXAMPLE 8 Subtract $\dfrac{3x+4}{2x^2-5x-12} - \dfrac{2x-3}{5x^2-18x-8}$.

Solution Factor the denominator of each expression.

$$\frac{3x+4}{2x^2-5x-12} - \frac{2x-3}{5x^2-18x-8} = \frac{3x+4}{(2x+3)(x-4)} - \frac{2x-3}{(5x+2)(x-4)}$$

The LCD is $(2x+3)(x-4)(5x+2)$.

$$\frac{3x+4}{(2x+3)(x-4)} - \frac{2x-3}{(5x+2)(x-4)}$$

$$= \frac{5x+2}{5x+2} \cdot \frac{3x+4}{(2x+3)(x-4)} - \frac{2x-3}{(5x+2)(x-4)} \cdot \frac{2x+3}{2x+3}$$ *Multiply to obtain LCD.*

$$= \frac{15x^2+26x+8}{(5x+2)(2x+3)(x-4)} - \frac{4x^2-9}{(5x+2)(2x+3)(x-4)}$$

$$= \frac{15x^2+26x+8-(4x^2-9)}{(5x+2)(2x+3)(x-4)}$$ *Subtract numerators.*

$$= \frac{15x^2+26x+8-4x^2+9}{(5x+2)(2x+3)(x-4)}$$

$$= \frac{11x^2+26x+17}{(5x+2)(2x+3)(x-4)}$$ *Combine terms.*

EXAMPLE 9 Perform the indicated operations.

$$\frac{x-1}{x-2} - \frac{x+1}{x+2} + \frac{x-6}{x^2-4}$$

Solution First, factor x^2-4. The LCD of the three fractions is $(x+2)(x-2)$.

$$\frac{x-1}{x-2} - \frac{x+1}{x+2} + \frac{x-6}{x^2-4}$$

$$= \frac{x-1}{x-2} - \frac{x+1}{x+2} + \frac{x-6}{(x+2)(x-2)}$$

$$= \frac{x+2}{x+2} \cdot \frac{x-1}{x-2} - \frac{x+1}{x+2} \cdot \frac{x-2}{x-2} + \frac{x-6}{(x+2)(x-2)}$$ *Multiply to obtain LCD.*

$$= \frac{x^2 + x - 2}{(x + 2)(x - 2)} - \frac{x^2 - x - 2}{(x + 2)(x - 2)} + \frac{x - 6}{(x + 2)(x - 2)}$$

$$= \frac{x^2 + x - 2 - (x^2 - x - 2) + (x - 6)}{(x + 2)(x - 2)}$$ *Subtract and add numerators.*

$$= \frac{x^2 + x - 2 - x^2 + x + 2 + x - 6}{(x + 2)(x - 2)}$$

$$= \frac{3x - 6}{(x + 2)(x - 2)}$$ *Combine terms.*

$$= \frac{3\cancel{(x - 2)}}{(x + 2)\cancel{(x - 2)}}$$ *Factor; divide out common factors.*

$$= \frac{3}{x + 2}$$

NOW TRY EXERCISE 55

Using Your Graphing Calculator

In Example 9 we found that

$$\frac{x - 1}{x - 2} - \frac{x + 1}{x + 2} + \frac{x - 6}{x^2 - 4} = \frac{3}{x + 2}$$

Suppose we let

$$y_1 = \frac{x - 1}{x - 2} - \frac{x + 1}{x + 2} + \frac{x - 6}{x^2 - 4}$$

$$y_2 = \frac{3}{x + 2}$$

If we use the TABLE feature on a graphing calculator, how will the values of y_1 and y_2 compare? The function y_1 is not defined at -2 and 2. The function y_2 is not defined at -2. For all values of x other than -2 and 2, the values of y_1 and y_2 should be the same if we have not made a mistake. Following is a table of values for y_1 and y_2, for values of x from -3 to 3.

X	Y1	Y2
-3	-3	-3
-2	ERROR	ERROR
-1	3	3
0	1.5	1.5
1	1	1
2	ERROR	.75
3	.6	.6

X= -3

The graphs of y_1 and y_2 are shown in Figures 6.3 and 6.4, respectively. We illustrated the graphs in this format (rather than on a graphing calculator screen) to show more detail. The open circle on the graph in Figure 6.3 is not shown on a grapher. Notice that the graph of y_1 has an open circle at 2 because y_1 is not defined at $x = 2$. Since y_2 is defined at $x = 2$, Figure 6.4 does not include this open circle. Neither graph is defined at $x = -2$.

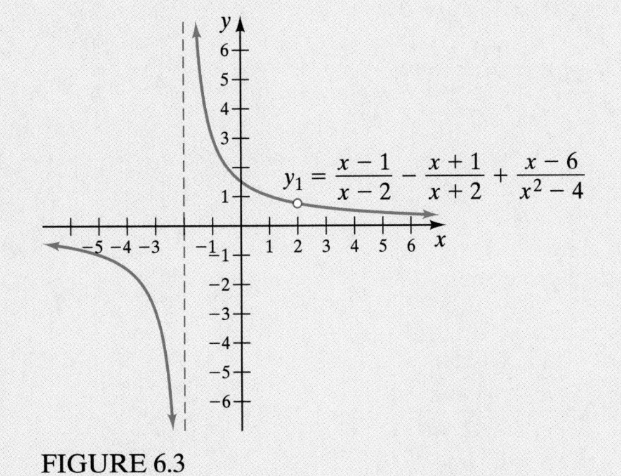

FIGURE 6.3

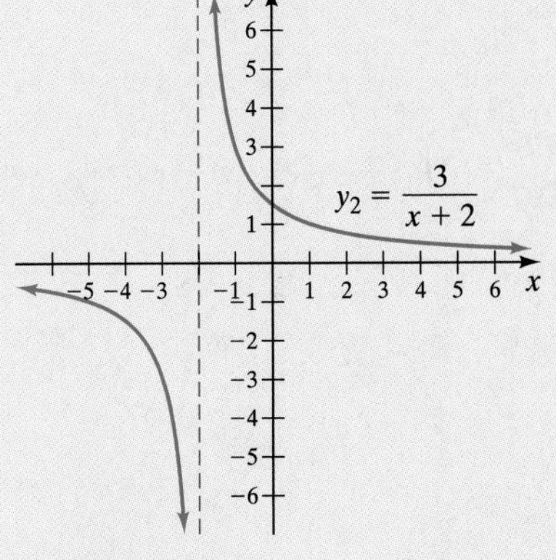

FIGURE 6.4

Exercise Set 6.2

Concept/Writing Exercises

1. a) What is the least common denominator of two or more rational expressions?
   b) Explain how to find the LCD.
   c) Using the procedure you gave in part b), find the LCD of

$$\frac{5}{64x^2 - 121} \text{ and } \frac{1}{8x^2 - 27x + 22}$$

2. a) Explain how to add or subtract two rational expressions.
   b) Add $\frac{4}{x+2} + \frac{x}{3x^2 - 4x - 20}$ following the procedure you gave in part a).

In Exercises 3 and 4, ***a)*** *explain why the subtraction is not correct, and* ***b)*** *show the correct subtraction.*

3. $\frac{x^2 - 4x}{(x+3)(x-2)} - \frac{x^2 + x - 2}{(x+3)(x-2)} \neq \frac{x^2 - 4x - x^2 + x - 2}{(x+3)(x-2)}$

4. $\frac{x - 5}{(x+4)(x-3)} - \frac{x^2 - 6x + 5}{(x+4)(x-3)} \neq \frac{x - 5 - x^2 - 6x + 5}{(x+4)(x-3)}$

Practice the Skills

Add or subtract.

5. $\frac{5x - 6}{x - 2} + \frac{2x - 5}{x - 2}$

6. $\frac{-2x + 6}{x^2 + x - 6} + \frac{3x - 3}{x^2 + x - 6}$

7. $\frac{x^2 - 2}{x^2 + 6x - 7} - \frac{-4x + 19}{x^2 + 6x - 7}$

8. $\frac{-x^2}{x^2 + 5xy - 14y^2} + \frac{x^2 + xy + 7y^2}{x^2 + 5xy - 14y^2}$

9. $\frac{3r^2 + 15r}{r^3 + 2r^2 - 8r} + \frac{2r^2 + 5r}{r^3 + 2r^2 - 8r}$

10. $\frac{x^3 - 12x^2 + 45x}{x(x - 8)} - \frac{x^2 + 5x}{x(x - 8)}$

11. $\frac{3x^2 - x}{2x^2 - x - 21} + \frac{3x - 8}{2x^2 - x - 21} - \frac{x^2 - x + 27}{2x^2 - x - 21}$

12. $\frac{2x^2 + 8x - 15}{2x^2 - 13x + 20} - \frac{2x + 10}{2x^2 - 13x + 20} - \frac{3x - 5}{2x^2 - 13x + 20}$

Find the least common denominator

13. $\frac{4x}{x + 3} + \frac{6}{x + 2}$

14. $\frac{-4}{8x^2y^2} + \frac{7}{5x^4y^5}$

15. $\frac{x + 3}{16x^2y} - \frac{x^2}{3x^3}$

16. $\frac{4}{(r - 7)(r + 3)} - \frac{x + 8}{r - 7}$

17. $5z^2 + \frac{9z}{z - 4}$

18. $\frac{b^2 + 3}{18b} - \frac{b - 7}{12(b + 5)}$

19. $\frac{a - 2}{a^2 - 5a - 24} + \frac{3}{a^2 + 11a + 24}$

20. $\frac{3x - 5}{6x^2 + 13xy + 6y^2} + \frac{3}{3x^2 + 5xy + 2y^2}$

21. $\frac{3}{x^2 + 3x - 4} - \frac{4}{4x^2 + 5x - 9} + \frac{x + 2}{4x^2 + 25x + 36}$

22. $\frac{x}{2x^2 - 7x + 3} + \frac{x - 3}{4x^2 + 4x - 3} - \frac{x^2 + 1}{2x^2 - 3x - 9}$

Add or subtract.

23. $\frac{2}{3x} + \frac{2}{x}$

24. $\frac{6}{x^2} + \frac{3}{2x}$

25. $\frac{5}{6y} + \frac{3}{4y^2}$

26. $\frac{3x}{4y} + \frac{5}{6xy}$

27. $\frac{5}{8x^4y} - \frac{1}{5x^2y^3}$

28. $\frac{3}{4xy^3} + \frac{1}{6x^2y}$

29. $\frac{4x}{3xy} + 2$

30. $\frac{8}{b - 2} + \frac{3x}{2 - b}$

31. $\frac{2a}{a - b} - \frac{a}{b - a}$

32. $\frac{b}{a - b} + \frac{a + b}{b}$

33. $\dfrac{x}{x^2 - 9} - \dfrac{4(x-3)}{x+3}$

34. $\dfrac{4x}{x-4} + \dfrac{x+4}{x+1}$

35. $\dfrac{2m+1}{m-5} - \dfrac{4}{m^2 - 3m - 10}$

36. $\dfrac{x}{x+1} + \dfrac{1}{x^2 + 2x + 1}$

37. $\dfrac{-x^2 + 5x}{(x-5)^2} + \dfrac{x+1}{x-5}$

38. $\dfrac{x}{x^2 + 2x - 8} + \dfrac{x+2}{x^2 - 3x + 2}$

39. $\dfrac{4}{(2p-3)(p+4)} - \dfrac{3}{(p+4)(p-4)}$

40. $\dfrac{5x}{x^2 - 9x + 8} - \dfrac{3(x+2)}{x^2 - 6x - 16}$

41. $5 - \dfrac{x-1}{x^2 + 3x - 10}$

42. $\dfrac{3x}{2x-3} + \dfrac{3x+6}{2x^2 + x - 6}$

43. $\dfrac{3a-4}{4a+1} + \dfrac{3a+6}{4a^2 + 9a + 2}$

44. $\dfrac{7}{3q^2 + q - 4} + \dfrac{9q+2}{3q^2 - 2q - 8}$

45. $\dfrac{x-y}{x^2 - 4xy + 4y^2} + \dfrac{x-3y}{x^2 - 4y^2}$

46. $\dfrac{x+2y}{x^2 - xy - 2y^2} - \dfrac{y}{x^2 - 3xy + 2y^2}$

47. $\dfrac{2r}{r-4} - \dfrac{2r}{r+4} + \dfrac{64}{r^2 - 16}$

48. $\dfrac{4}{p+1} + \dfrac{3}{p-1} + \dfrac{p+2}{p^2 - 1}$

49. $\dfrac{x^2+2}{x^2 - x - 2} + \dfrac{1}{x+1} - \dfrac{x}{x-2}$

50. $\dfrac{2}{x^2 - 16} + \dfrac{x+1}{x^2 + 8x + 16} + \dfrac{3}{x-4}$

51. $\dfrac{x}{3x+4} + \dfrac{3x+2}{x-5} - \dfrac{7x^2 + 24 + 28}{3x^2 - 11x - 20}$

52. $\dfrac{4}{3x-2} - \dfrac{1}{x-4} + 5$

53. $\dfrac{x}{x^2 - 10x + 24} - \dfrac{3}{x-6} + 1$

54. $3 - \dfrac{4}{8r^2 + 2r - 15} + \dfrac{r+2}{4r-5}$

55. $\dfrac{3}{5x+6} + \dfrac{x^2 - x}{5x^2 - 4x - 12} - \dfrac{4}{x-2}$

56. $\dfrac{3}{x^2 - 13x + 36} + \dfrac{4}{2x^2 - 7x - 4} + \dfrac{1}{2x^2 - 17x - 9}$

57. $\dfrac{m}{6m^2 + 13mn + 6n^2} + \dfrac{2m}{4m^2 + 8mn + 3n^2}$

58. $\dfrac{(x-y)^2}{x^3 - y^3} + \dfrac{1}{x^2 + xy + y^2}$

59. $\dfrac{5r - 2s}{25r^2 - 4s^2} - \dfrac{2r - s}{10r^2 - rs - 2s^2}$

60. $\dfrac{6}{(2r-1)^2} + \dfrac{2}{2r-1} - 4$

61. $\dfrac{2}{2x+3y} - \dfrac{4x^2 - 6xy + 9y^2}{8x^3 + 27y^3}$

62. $\dfrac{4}{4x-5y} - \dfrac{3x^2 + 2y^2}{64x^3 - 125y^3}$

Problem Solving

63. When two rational expressions are being added or subtracted, should the numerators of the expressions being added or subtracted be factored? Explain.

64. Are the fractions $\dfrac{x-3}{5-x}$ and $-\dfrac{x-3}{x-5}$ equivalent? Explain.

65. Are the fractions $\dfrac{8-x}{3-x}$ and $\dfrac{x-8}{x-3}$ equivalent? Explain.

66. If $f(x)$ and $g(x)$ are both rational functions, will $(f+g)(x)$ always be a rational function?

67. If $f(x) = \dfrac{x+2}{x-3}$ and $g(x) = \dfrac{x}{x+4}$, find

a) the domain of $f(x)$

b) the domain of $g(x)$

c) $(f+g)(x)$

d) the domain of $(f+g)(x)$

68. If $f(x) = \dfrac{x+1}{x^2 - 4}$ and $g(x) = \dfrac{x}{x-2}$, find

a) the domain of $f(x)$

b) the domain of $g(x)$

c) $(f+g)(x)$

d) the domain of $(f+g)(x)$

The dashed red lines in the figures below are called **asymptotes**. *The asymptotes are not a part of the graph but are used to show values that the graph approaches, but does not touch. In Exercises 69 and 70, determine the domain and range of the rational function shown.*

69.

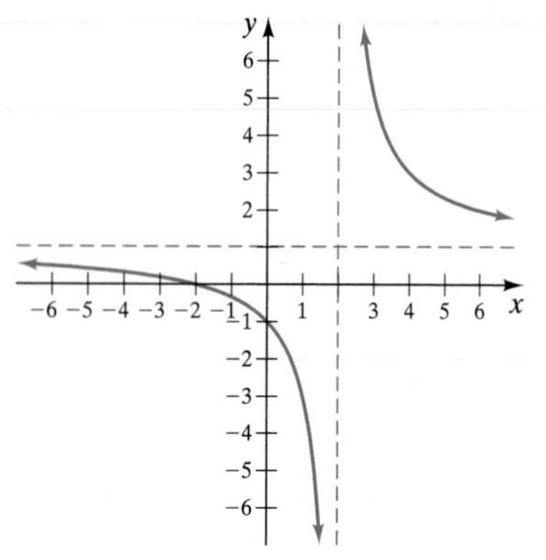

70.

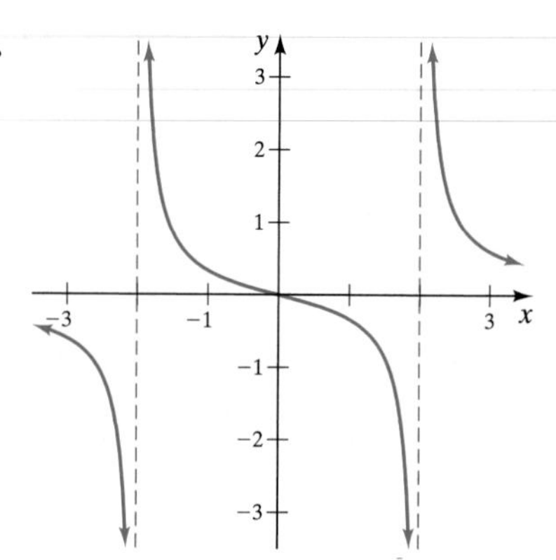

For $f(x) = \dfrac{x}{x^2 - 4}$ *and* $g(x) = \dfrac{3}{x^2 + x - 6}$, *find the following.*

71. $(f + g)(x)$

72. $(f - g)(x)$

73. $(f \cdot g)(x)$

74. $(f/g)(x)$

75. Show that $\dfrac{a}{b} + \dfrac{c}{d} = \dfrac{ad + bc}{bd}$.

76. Show that $x^{-1} + y^{-1} = \dfrac{x + y}{xy}$.

Consider the rectangles given below. Find **a)** *the perimeter, and* **b)** *the area.*

77.

$\dfrac{a + b}{a}$ (top side), $\dfrac{a - b}{a}$ (left side)

78.

$\dfrac{a + 2b}{b}$ (top side), $\dfrac{-a + 2b}{b}$ (left side)

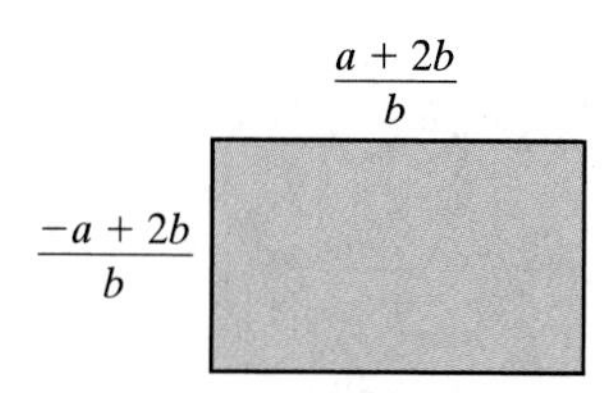

Determine the polynomial to be placed in the shaded area to give a true statement. Explain how you determined your answer.

79. $\dfrac{5x^2 - 6}{x^2 - x - 1} - \dfrac{\square}{x^2 - x - 1} = \dfrac{-2x^2 + 6x - 12}{x^2 - x - 1}$

80. $\dfrac{r^2 - 6}{r^2 - 5r + 6} - \dfrac{\square}{r^2 - 5r + 6} = \dfrac{1}{r - 2}$

Perform the indicated operations.

81. $\left(3 + \dfrac{1}{x + 3}\right)\left(\dfrac{x + 3}{x - 2}\right)$

82. $\left(\dfrac{3}{r + 1} - \dfrac{4}{r - 2}\right)\left(\dfrac{r - 2}{r + 10}\right)$

83. $\left(\dfrac{5}{a - 5} - \dfrac{2}{a + 3}\right) \div (3a + 25)$

84. $\left(\dfrac{x^2 + 4x - 5}{2x^2 + x - 3} \cdot \dfrac{2x + 3}{x + 1}\right) - \dfrac{2}{x + 2}$

85. $\left(\dfrac{x + 5}{x - 3} - x\right) \div \dfrac{1}{x - 3}$

86. $\left(\dfrac{x + 5}{x^2 - 25} + \dfrac{1}{x + 5}\right)\left(\dfrac{2x^2 - 13x + 15}{4x^2 - 6x}\right)$

87. The weighted average of two values a and b is given by $a\left(\dfrac{x}{n}\right) + b\left(\dfrac{n - x}{n}\right)$, where $\dfrac{x}{n}$ is the weight given to a and $\dfrac{n - x}{n}$ is the weight given to b.

a) Express this sum as a single fraction.

b) On exam a you received a grade of 60 and on exam b you received a grade of 92. If exam a counts $\frac{2}{5}$ of your final grade and exam b counts $\frac{3}{5}$, determine your weighted average.

88. Show that $\left(\dfrac{x}{y}\right)^{-1} + \left(\dfrac{y}{x}\right)^{-1} + (xy)^{-1} = \dfrac{x^2 + y^2 + 1}{xy}$.

Perform the indicated operation.

89. $(a - b)^{-1} + (a - b)^{-2}$

90. $\left(\dfrac{a - b}{a}\right)^{-1} - \left(\dfrac{a + b}{a}\right)^{-1}$

Use your graphing calculator to determine whether the following additions are correct.

91. $\dfrac{x-3}{x+4}+\dfrac{x}{x^2-2x-24}\stackrel{?}{=}\dfrac{x^2-10x+18}{(x+4)(x-6)}$

92. $\dfrac{x-2}{x^2-25}+\dfrac{x-2}{2x^2+17x+35}\stackrel{?}{=}\dfrac{3x^2-4x-4}{(x+5)(x-5)(2x+7)}$

Challenge Problem

93. Express each sum as a single fraction.

a) $1+\dfrac{1}{x}$

b) $1+\dfrac{1}{x}+\dfrac{1}{x^2}$

c) $1+\dfrac{1}{x}+\dfrac{1}{x^2}+\cdots+\dfrac{1}{x^5}$

d) $1+\dfrac{1}{x}+\dfrac{1}{x^2}+\cdots+\dfrac{1}{x^n}$

Cumulative Review Exercises

[2.3] **94.** The price of a suit is increased by 20%. The price is then decreased by \$20. If the suit sells for \$196, find the original price of the suit.

[2.4] **95.** A bottling machine fills and caps bottles at a rate of 80 per minute. Then the machine is slowed down and fills and caps bottles at a rate of 60 per minute. If the sum of the two time periods was 14 minutes and the number of bottles filled and capped at the higher rate was the same as the number filled and capped at the lower rate, determine **a)** how long the machine was used at the faster rate, and **b)** the total number of bottles filled and capped over the 14-minute period.

[5.3] **96.** Divide $\dfrac{9x^4y^6-3x^3y^2-5xy^5}{3xy^4}$.

97. Divide $\dfrac{6x^2+5x-4}{3x+4}$.

6.3 COMPLEX FRACTIONS

SSM VIDEO 6.3 CD Rom

1) Recognize complex fractions.
2) Simplify complex fractions by multiplying by a common denominator.
3) Simplify complex fractions by simplifying the numerator and denominator.

1) Recognize Complex Fractions

A **complex fraction** is one that has a fractional expression in its numerator or its denominator or both its numerator and denominator.

Examples of Complex Fractions

$$\frac{\frac{2}{3}}{5},\quad \frac{\frac{x+1}{x}}{3x},\quad \frac{\frac{x}{y}}{x+1},\quad \frac{\frac{a+b}{a}}{\frac{a-b}{b}},\quad \frac{3+\frac{1}{x}}{\frac{1}{x^2}+\frac{3}{x}}$$

The expression above the **main fraction line** is the numerator, and the expression below the main fraction line is the denominator of the complex fraction.

$$\frac{\dfrac{a+b}{a}}{\dfrac{a-b}{b}}$$

$\dfrac{a+b}{a}$ ⟵ numerator of complex fraction

⟵ main fraction line

$\dfrac{a-b}{b}$ ⟵ denominator of complex fraction

We will explain two methods that can be used to simplify complex fractions. To simplify a complex fraction means to write the expression without a fraction in its numerator and its denominator.

2) Simplify Complex Fractions by Multiplying by a Common Denominator

The first method involves multiplying both the numerator and denominator of the complex fraction by a common denominator.

To Simplify a Complex Fraction by Multiplying by a Common Denominator

1. Find the least common denominator of all fractions appearing within the complex fraction. This is the LCD of the complex fraction.
2. Multiply both the numerator and denominator of the complex fraction by the LCD of the complex fraction found in step 1.
3. Simplify when possible.

In step 2, you are actually multiplying the complex fraction by $\frac{\text{LCD}}{\text{LCD}}$, which is equivalent to multiplying the fraction by 1.

EXAMPLE 1 Simplify $\dfrac{\dfrac{2}{x^2}-\dfrac{3}{x}}{\dfrac{x}{5}}$.

Solution The denominators in the complex fraction are x^2, x, and 5. Therefore, the LCD of the complex fraction is $5x^2$. Multiply the numerator and denominator by $5x^2$.

$$\frac{\dfrac{2}{x^2}-\dfrac{3}{x}}{\dfrac{x}{5}}=\frac{5x^2\left(\dfrac{2}{x^2}-\dfrac{3}{x}\right)}{5x^2\left(\dfrac{x}{5}\right)}$$

Multiply the numerator and denominator by $5x^2$.

$$=\frac{5x^2\left(\dfrac{2}{x^2}\right)-5x^2\left(\dfrac{3}{x}\right)}{5x^2\left(\dfrac{x}{5}\right)}$$

Distributive property.

$$= \frac{10 - 15x}{x^3} \qquad \textit{Simplify.}$$

$$= \frac{5(2 - 3x)}{x^3}$$

EXAMPLE 2 Simplify $\dfrac{a + \dfrac{1}{b}}{b + \dfrac{1}{a}}$.

Solution Multiply the numerator and denominator of the complex fraction by its LCD, ab.

$$\frac{a + \dfrac{1}{b}}{b + \dfrac{1}{a}} = \frac{ab\left(a + \dfrac{1}{b}\right)}{ab\left(b + \dfrac{1}{a}\right)} \qquad \textit{Multiply the numerator and denominator by ab.}$$

$$= \frac{a^2b + a}{ab^2 + b} \qquad \textit{Distributive property.}$$

$$= \frac{a(ab + 1)}{b(ab + 1)} = \frac{a}{b}$$

NOW TRY EXERCISE 13

EXAMPLE 3 Simplify $\dfrac{a^{-1} + ab^{-2}}{ab^{-2} - a^{-2}b^{-1}}$.

Solution First rewrite each expression without negative exponents.

$$\frac{a^{-1} + ab^{-2}}{ab^{-2} - a^{-2}b^{-1}} = \frac{\dfrac{1}{a} + \dfrac{a}{b^2}}{\dfrac{a}{b^2} - \dfrac{1}{a^2b}}$$

$$= \frac{a^2b^2\left(\dfrac{1}{a} + \dfrac{a}{b^2}\right)}{a^2b^2\left(\dfrac{a}{b^2} - \dfrac{1}{a^2b}\right)} \qquad \textit{Multiply the numerator and denominator by } a^2b^2\textit{, the LCD of the complex fraction.}$$

$$= \frac{a^2b^2\left(\dfrac{1}{a}\right) + a^2b^2\left(\dfrac{a}{b^2}\right)}{a^2b^2\left(\dfrac{a}{b^2}\right) - a^2b^2\left(\dfrac{1}{a^2b}\right)} \qquad \textit{Distributive property.}$$

$$= \frac{ab^2 + a^3}{a^3 - b}$$

NOW TRY EXERCISE 37

3) Simplify Complex Fractions by Simplifying the Numerator and Denominator

Complex fractions can also be simplified as follows:

To Simplify a Complex Fraction by Simplifying the Numerator and the Denominator

1. Add or subtract as necessary to get one rational expression in the numerator.
2. Add or subtract as necessary to get one rational expression in the denominator.
3. Invert the denominator of the complex fraction and multiply by the numerator of the complex fraction.
4. Simplify when possible.

Example 4 will show how Example 1 can be simplified by this second method.

EXAMPLE 4 Simplify $\dfrac{\dfrac{2}{x^2} - \dfrac{3}{x}}{\dfrac{x}{5}}$.

Solution Subtract the fractions in the numerator to get one rational expression in the numerator. The common denominator of the fractions in the numerator is x^2.

$$\frac{\dfrac{2}{x^2} - \dfrac{3}{x}}{\dfrac{x}{5}} = \frac{\dfrac{2}{x^2} - \dfrac{3}{x} \cdot \dfrac{x}{x}}{\dfrac{x}{5}} \qquad \textit{Obtain common denominator in numerator.}$$

$$= \frac{\dfrac{2}{x^2} - \dfrac{3x}{x^2}}{\dfrac{x}{5}}$$

$$= \frac{\dfrac{2 - 3x}{x^2}}{\dfrac{x}{5}}$$

$$= \frac{2 - 3x}{x^2} \cdot \frac{5}{x} \qquad \textit{Invert denominator and multiply.}$$

$$= \frac{5(2 - 3x)}{x^3}$$

This is the same answer obtained in Example 1.

HELPFUL HINT

Some students prefer to use the second method when the complex fraction consists of a single fraction over a single fraction, such as

$$\frac{\dfrac{x+3}{18}}{\dfrac{x-8}{6}}$$

For more complex fractions, many students prefer the first method because you do not have to add fractions.

NOW TRY EXERCISE 7

Exercise Set 6.3

Concept/Writing Exercises

1. What is a complex fraction?

2. We have indicated two procedures for evaluating complex fractions. Which procedure do you prefer? Why?

Practice the Skills

Simplify.

3. $\dfrac{1-\dfrac{x}{y}}{x}$

4. $\dfrac{\dfrac{x^2y}{4}}{\dfrac{2}{x}}$

5. $\dfrac{\dfrac{15a}{b^2}}{\dfrac{b^3}{5}}$

6. $\dfrac{\dfrac{10x^2y}{3z^3}}{\dfrac{5xy}{9z^5}}$

7. $\dfrac{\dfrac{36x^4}{5y^4z^5}}{\dfrac{9xy^2}{15z^5}}$

8. $\dfrac{x+\dfrac{1}{y}}{\dfrac{x}{y}}$

9. $\dfrac{x-\dfrac{x}{y}}{\dfrac{1+x}{y}}$

10. $\dfrac{\dfrac{9}{x}+\dfrac{3}{x^2}}{3+\dfrac{1}{x}}$

11. $\dfrac{\dfrac{2}{a}+\dfrac{1}{2a}}{a+\dfrac{a}{2}}$

12. $\dfrac{3-\dfrac{1}{y}}{2-\dfrac{1}{y}}$

13. $\dfrac{\dfrac{x}{y}-\dfrac{y}{x}}{\dfrac{x+y}{x}}$

14. $\dfrac{\dfrac{a^2}{b}-b}{\dfrac{b^2}{a}-a}$

15. $\dfrac{\dfrac{1}{m}+\dfrac{2}{m^2}}{2+\dfrac{1}{m^2}}$

16. $\dfrac{\dfrac{a}{b}-2}{\dfrac{-a}{b}+2}-1$

17. $\dfrac{\dfrac{x^2-y^2}{x}}{\dfrac{x+y}{x^3}}$

18. $\dfrac{\dfrac{4x+8}{3x^2}}{\dfrac{4x}{6}}$

19. $\dfrac{\dfrac{a}{a+1}-1}{\dfrac{2a+1}{a-1}}$

20. $\dfrac{\dfrac{x}{4}-\dfrac{1}{x}}{1+\dfrac{x+4}{x}}$

21. $\dfrac{1+\dfrac{x}{x+1}}{\dfrac{2x+1}{x-1}}$

22. $\dfrac{\dfrac{1}{x-1}+1}{\dfrac{1}{x+1}-1}$

23. $\dfrac{\dfrac{a+1}{a-1}+\dfrac{a-1}{a+1}}{\dfrac{a+1}{a-1}-\dfrac{a-1}{a+1}}$

24. $\dfrac{\dfrac{a-2}{a+2}-\dfrac{a+2}{a-2}}{\dfrac{a-2}{a+2}+\dfrac{a+2}{a-2}}$

25. $\dfrac{\dfrac{5}{5-x}+\dfrac{6}{x-5}}{\dfrac{3}{x}+\dfrac{2}{x-5}}$

26. $\dfrac{\dfrac{2}{m}+\dfrac{1}{m^2}+\dfrac{3}{m-1}}{\dfrac{2}{m-1}}$

27. $\dfrac{\dfrac{3}{x^2} - \dfrac{1}{x} + \dfrac{2}{x - 2}}{\dfrac{1}{x}}$

28. $\dfrac{\dfrac{2}{a^2 - 3a + 2} + \dfrac{2}{a^2 - a - 2}}{\dfrac{2}{a^2 - 1} + \dfrac{2}{a^2 + 4a + 3}}$

29. $\dfrac{\dfrac{1}{x^2 + 5x + 4} + \dfrac{2}{x^2 + 2x - 8}}{\dfrac{2}{x^2 - x - 2} + \dfrac{1}{x^2 - 5x + 6}}$

30. $\dfrac{\dfrac{2}{x^2 + x - 20} + \dfrac{3}{x^2 - 6x + 8}}{\dfrac{2}{x^2 + 3x - 10} + \dfrac{3}{x^2 + 2x - 24}}$

Simplify.

31. $3a^{-2} + b$

32. $4a^{-2} + b^{-1}$

33. $(a^{-1} + b^{-1})^{-1}$

34. $\dfrac{a^{-1} + b^{-1}}{\dfrac{1}{ab}}$

35. $\dfrac{a^{-1} + 1}{b^{-1} - 1}$

36. $\dfrac{\dfrac{a}{b} + a^{-1}}{\dfrac{b}{a} + a^{-1}}$

37. $\dfrac{x^{-1} - y^{-1}}{x^{-1} + y^{-1}}$

38. $\dfrac{x^{-2} + \dfrac{1}{x}}{x^{-1} + x^{-2}}$

39. $\dfrac{a^{-1} + b^{-1}}{(a + b)^{-1}}$

40. $\dfrac{3a^{-1} - b^{-1}}{(a - b)^{-1}}$

41. $2x^{-1} - (3y)^{-1}$

42. $\dfrac{\dfrac{5}{x} + \dfrac{1}{y}}{(x - y)^{-1}}$

43. $\dfrac{\dfrac{2}{xy} - \dfrac{3}{y} + \dfrac{5}{x}}{3x^{-1} - 2y^{-2}}$

44. $\dfrac{4m^{-1} + 3n^{-1} + (2mn)^{-1}}{\dfrac{5}{m} + \dfrac{3}{n}}$

Problem Solving

45. The efficiency of a jack, E, is given by the formula

$$E = \frac{\frac{1}{2}h}{h + \frac{1}{2}}$$

where h is determined by the pitch of the jack's thread.

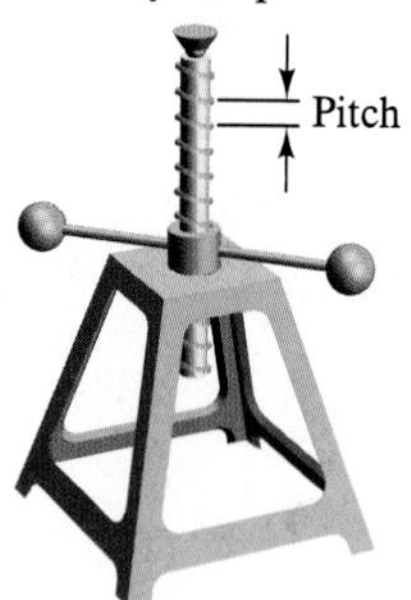

Determine the efficiency of a jack whose values of h are:

a) $\dfrac{2}{3}$ **b)** $\dfrac{4}{5}$

46. If two resistors with resistances R_1 and R_2 are connected in parallel, their combined resistance, R_T, can be found from the formula

$$R_T = \frac{1}{\frac{1}{R_1} + \frac{1}{R_2}}$$

Simplify the right side of the formula.

47. If three resistors with resistances R_1, R_2, and R_3 are connected in parallel, their combined resistance, R_T, can be found from the formula

$$R_T = \frac{1}{\frac{1}{R_1} + \frac{1}{R_2} + \frac{1}{R_3}}$$

Simplify the right side of this formula.

48. A formula used in the study of optics is

$$f = (p^{-1} + q^{-1})^{-1}$$

where p is the object's distance from a lens, q is the image distance from the lens, and f is the focal length of the lens. Express the right side of the formula without any negative exponents.

49. If $f(x) = \dfrac{1}{x}$, find $f(f(a))$.

50. If $f(x) = \dfrac{1}{x + 1}$, find $f(f(a))$.

For each function, find ***a)*** $f(x+h)$, ***b)*** $f(x+h)-f(x)$, *and* ***c)*** $\dfrac{f(x+h)-f(x)}{h}$.

51. $f(x)=\dfrac{1}{x}$

52. $f(x)=\dfrac{3}{x-1}$

53. $f(x)=\dfrac{2}{x^2}$

54. $f(x)=\dfrac{5}{x^2}$

Challenge Problems

Simplify.

55. $\dfrac{1}{2a+\dfrac{1}{2a+\dfrac{1}{2a}}}$

56. $\dfrac{1}{x+\dfrac{1}{x+\dfrac{1}{x+1}}}$

57. $\dfrac{1}{2+\dfrac{1}{2+\dfrac{1}{2}}}$

Cumulative Review Exercises

[1.4] **58.** Evaluate $\dfrac{\left|-\frac{3}{9}\right|-\left(-\frac{5}{9}\right)\cdot\left|-\frac{3}{8}\right|}{|-5-(-3)|}$.

[2.6] **59.** Find the solution set to the inequality

$$\left|\frac{4-2x}{3}\right|\geq 3.$$

[3.7] **60.** Graph the inequality $6y-3x<12$.

[5.3] **61.** Use synthetic division to divide.

$$x^3-7x^2-13x+9\div(x-2)$$

6.4 SOLVING RATIONAL EQUATIONS

SSM VIDEO 6.4 CD Rom

1. Solve rational equations.
2. Check solutions.
3. Solve proportions.
4. Solve problems involving rational functions.
5. Solve applications using rational expressions.
6. Solve for a variable in a formula containing rational expressions.

1 Solve Rational Equations

In Sections 6.1 through 6.3 we presented techniques to add, subtract, multiply, and divide rational expressions. In this section we present a method for solving rational equations. A **rational equation** is an equation that contains at least one rational expression.

To Solve Rational Equations

1. Determine the LCD of all rational expressions in the equation.
2. Multiply *both* sides of the equation by the LCD. This will result in every term in the equation being multiplied by the LCD.
3. Remove any parentheses and combine like terms on each side of the equation.
4. Solve the equation using the properties discussed in earlier sections.
5. Check the solution in the *original* equation.

In step 2, we multiply both sides of the equation by the LCD to eliminate fractions. In some examples we will not show the check to save space.

EXAMPLE 1 Solve $\frac{x}{4} + \frac{1}{2} = \frac{x-1}{2}$.

Solution Multiply both sides of the equation by the LCD, 4. Then use the distributive property, which results in each term in the equation being multiplied by the LCD.

$$4\left(\frac{x}{4} + \frac{1}{2}\right) = \frac{x-1}{2} \cdot 4 \qquad \textit{Multiply both sides by 4.}$$

$$4\left(\frac{x}{4}\right) + 4\left(\frac{1}{2}\right) = 2(x-1) \qquad \textit{Distributive Property.}$$

$$x + 2 = 2x - 2$$

$$2 = x - 2$$

$$4 = x$$

2) Check Solutions

Whenever a variable appears in any denominator, you must check your apparent solution in the original equation. When checking, if an apparent solution makes any denominator equal to zero, that value is not a solution to the equation. Such values are called **extraneous roots** or **extraneous solutions.** An extraneous root is a number obtained when solving an equation that is not a solution to the original equation.

EXAMPLE 2 Solve $2 - \frac{4}{x} = \frac{1}{3}$.

Solution Multiply both sides of the equation by the LCD, $3x$.

$$3x\left(2 - \frac{4}{x}\right) = \left(\frac{1}{3}\right) \cdot 3x \qquad \textit{Multiply both sides by } 3x.$$

$$3x(2) - 3\cancel{x}\left(\frac{4}{\cancel{x}}\right) = \left(\frac{1}{\cancel{3}}\right)\cancel{3}x \qquad \textit{Distributive Property.}$$

$$6x - 12 = x$$

$$5x - 12 = 0$$

$$5x = 12$$

$$x = \frac{12}{5}$$

CHECK:

$$2 - \frac{4}{x} = \frac{1}{3}$$

$$2 - \frac{4}{(12/5)} \stackrel{?}{=} \frac{1}{3} \qquad \textit{Substitute } \frac{12}{5} \textit{ for } x.$$

$$2 - \frac{20}{12} \stackrel{?}{=} \frac{1}{3}$$

$$\frac{1}{3} = \frac{1}{3} \qquad \textit{True}$$

EXAMPLE 3 Solve $x + \dfrac{12}{x} = -7$.

Solution

$$x \cdot \left(x + \frac{12}{x}\right) = -7 \cdot x \qquad \textit{Multiply both sides by the LCD, } x.$$

$$x(x) + x\left(\frac{12}{x}\right) = -7x \qquad \textit{Distributive Property.}$$

$$x^2 + 12 = -7x$$

$$x^2 + 7x + 12 = 0$$

$$(x + 3)(x + 4) = 0$$

$$x + 3 = 0 \quad \text{or} \quad x + 4 = 0$$

$$x = -3 \qquad\qquad x = -4$$

NOW TRY EXERCISE 31 Checks of -3 and -4 will show that they are both solutions to the equation.

EXAMPLE 4 Solve $\dfrac{2x}{x^2 - 4} + \dfrac{1}{x - 2} = \dfrac{2}{x + 2}$.

Solution First factor the denominator $x^2 - 4$, then find the LCD.

$$\frac{2x}{(x + 2)(x - 2)} + \frac{1}{x - 2} = \frac{2}{x + 2}$$

The LCD is $(x + 2)(x - 2)$. Multiply both sides of the equation by the LCD, and then use the distributive property. This process will eliminate the fractions from the equation.

$$(x + 2)(x - 2) \cdot \left[\frac{2x}{(x + 2)(x - 2)} + \frac{1}{x - 2}\right] = \frac{2}{x + 2} \cdot (x + 2)(x - 2)$$

$$\cancel{(x + 2)}\cancel{(x - 2)} \cdot \frac{2x}{\cancel{(x + 2)}\cancel{(x - 2)}} + (x + 2)\cancel{(x - 2)} \cdot \frac{1}{\cancel{x - 2}} = \frac{2}{\cancel{x + 2}} \cdot \cancel{(x + 2)}(x - 2)$$

$$2x + (x + 2) = 2(x - 2)$$

$$3x + 2 = 2x - 4$$

$$x + 2 = -4$$

$$x = -6$$

NOW TRY EXERCISE 35 A check will show that -6 is the solution.

EXAMPLE 5 Solve the equation $\dfrac{22}{2p^2 - 9p - 5} - \dfrac{3}{2p + 1} = \dfrac{2}{p - 5}$.

Solution Factor the denominator, then determine the LCD.

$$\frac{22}{(2p + 1)(p - 5)} - \frac{3}{2p + 1} = \frac{2}{p - 5}$$

Multiply both sides of the equation by the LCD, $(2p + 1)(p - 5)$.

$$\cancel{(2p + 1)}\cancel{(p - 5)} \cdot \frac{22}{\cancel{(2p + 1)}\cancel{(p - 5)}} - \cancel{(2p + 1)}(p - 5) \cdot \frac{3}{\cancel{2p + 1}} = (2p + 1)\cancel{(p - 5)} \cdot \frac{2}{\cancel{p - 5}}$$

$$22 - 3(p - 5) = 2(2p + 1)$$

$$22 - 3p + 15 = 4p + 2$$

$$37 - 3p \stackrel{?}{=} 4p + 2$$

$$35 = 7p$$

$$5 = p$$

The solution appears to be 5. However, since a variable appears in a denominator, this solution must be checked.

CHECK:

$$\frac{22}{2p^2 - 9p - 5} - \frac{3}{2p + 1} = \frac{2}{p - 5}$$

$$\frac{22}{2(5)^2 - 9(5) - 5} - \frac{3}{2(5) + 1} \stackrel{?}{=} \frac{2}{5 - 5} \qquad \textit{Substitute 5 for p.}$$

$$\text{Undefined} \longrightarrow \frac{22}{0} - \frac{3}{11} = \frac{2}{0} \longleftarrow \text{Undefined}$$

Since 5 makes a denominator 0 and division by 0 is undefined, 5 is an extraneous solution. Therefore, you should write **"no solution"** as your answer.

In Example 5 the only possible solution is 5. However, when $p = 5$ the denominator of $2/(p - 5)$ is 0. Therefore, 5 cannot be a solution. We did not actually have to show the complete check, but we did so in this example for clarity.

NOW TRY EXERCISE 39

HELPFUL HINT

Remember, whenever you solve an equation where a variable appears in any denominator, you must check the apparent solution to make sure it is not an extraneous solution. If the apparent solution makes any denominator 0, then it is an extraneous solution and not a true solution to the equation.

3) Solve Proportions

Proportions are equations of the form $\dfrac{a}{b} = \dfrac{c}{d}$. Proportions are one type of rational equation. Proportions may be solved by *cross multiplication* as follows.

If $\dfrac{a}{b} = \dfrac{c}{d}$, then $ad = bc$, $b \neq 0$, $d \neq 0$. Proportions may also be solved by

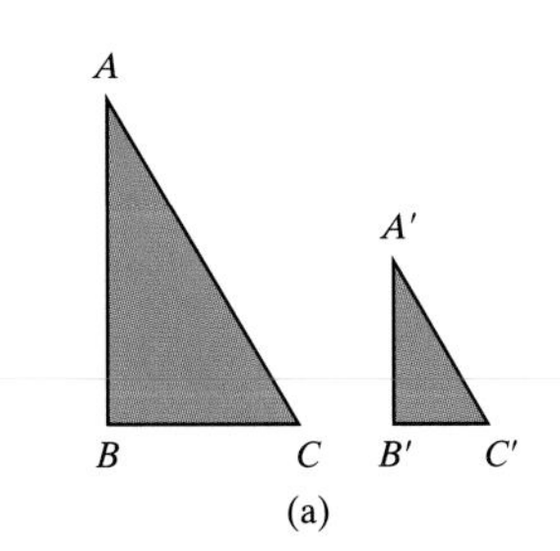

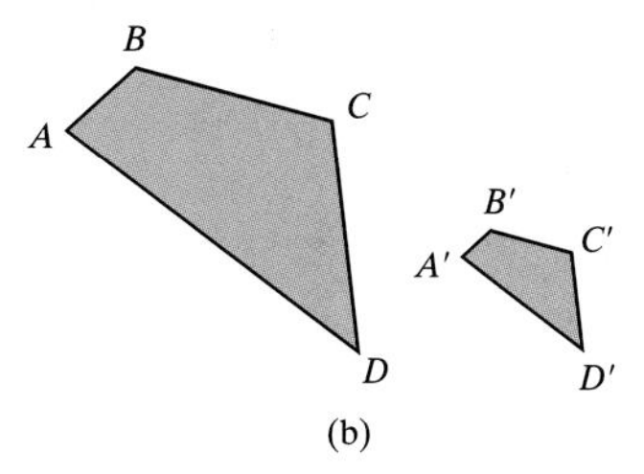

FIGURE 6.5

multiplying both sides of the proportion by the least common denominator. In Examples 6 and 7 we solve proportions by multiplying both sides by the LCD. We then ask you to determine the solutions, if possible, using cross multiplication. *When solving a proportion where the denominator of one or more of the ratios contains a variable, you must check to make sure that your solution is not extraneous.*

Proportions are often used when working with similar figures. **Similar figures** are figures whose corresponding angles are equal and whose corresponding sides are in proportion. Figure 6.5 illustrates two sets of similar figures.

In Figure 6.5a), the ratio of the length of side AB to the length of side BC is the same as the ratio of the length of side $A'B'$ to the length of side $B'C'$. That is,

$$\frac{AB}{BC} = \frac{A'B'}{B'C'}$$

In a pair of similar figures, if the length of a side is unknown, it can often be found by using proportions, as illustrated in Example 6.

EXAMPLE 6 Triangles ABC and $A'B'C'$ in Figure 6.6 are similar figures. Find the length of sides AB and $B'C'$.

Solution We can set up a proportion and then solve for x. Then we can find the lengths.

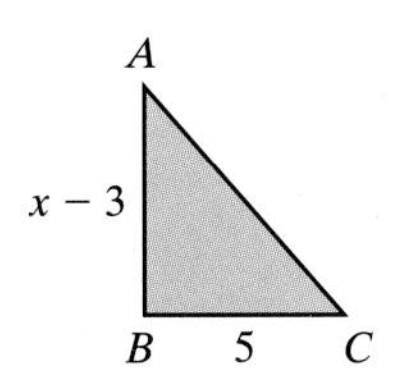

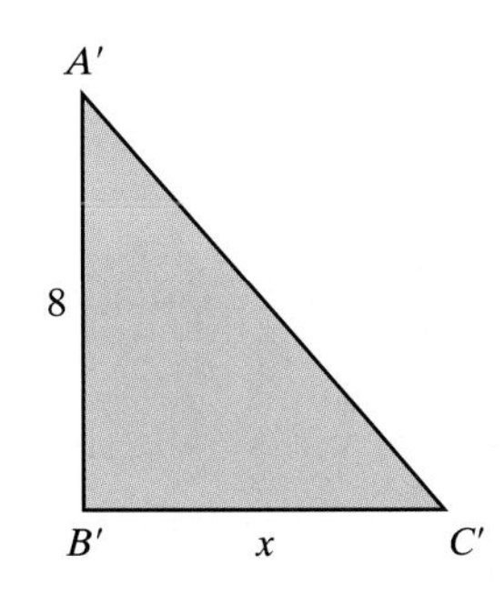

FIGURE 6.6

$$\frac{AB}{BC} = \frac{A'B'}{B'C'}$$

$$\frac{x-3}{5} = \frac{8}{x}$$

$$5x \cdot \frac{x-3}{5} = \frac{8}{x} \cdot 5x \qquad \text{Multiply both sides by the LCD, } 5x.$$

$$x(x-3) = 8 \cdot 5$$

$$x^2 - 3x = 40$$

$$x^2 - 3x - 40 = 0$$

$$(x-8)(x+5) = 0 \qquad \text{Factor the trinomial.}$$

$$x - 8 = 0 \quad \text{or} \quad x + 5 = 0$$

$$x = 8 \qquad\qquad x = -5$$

Since the length of the side of a triangle cannot be a negative number, −5 is not a possible answer. Substituting 8 for x, we see that the length of side $B'C'$ is 8 and the length of side AB is $8 - 3$ or 5.

CHECK:

$$\frac{AB}{BC} = \frac{A'B'}{B'C'}$$

$$\frac{5}{5} \stackrel{?}{=} \frac{8}{8}$$

$$1 = 1 \qquad \text{True}$$

The answer to Example 6 could also be obtained using cross multiplication. Solve Example 6 using cross multiplication now.

EXAMPLE 7 Solve $\frac{x^2}{x-4} = \frac{16}{x-4}$.

Solution This equation is a proportion. We will solve this equation by multiplying both sides of the equation by the LCD, $x - 4$.

$$(x-4) \cdot \frac{x^2}{x-4} = \frac{16}{x-4} \cdot (x-4)$$

$$x^2 = 16$$

$$x^2 - 16 = 0$$

$$(x+4)(x-4) = 0 \qquad \text{Factor the difference of squares.}$$

$$x + 4 = 0 \quad \text{or} \quad x - 4 = 0$$

$$x = -4 \qquad\qquad x = 4$$

CHECK:

$x = -4$

$$\frac{x^2}{x-4} = \frac{16}{x-4}$$

$$\frac{(-4)^2}{-4-4} \stackrel{?}{=} \frac{16}{-4-4}$$

$$\frac{16}{-8} \stackrel{?}{=} \frac{16}{-8}$$

$$-2 = -2 \qquad \text{True}$$

$x = 4$

$$\frac{x^2}{x-4} = \frac{16}{x-4}$$

$$\frac{4^2}{4-4} \stackrel{?}{=} \frac{16}{4-4}$$

$$\frac{16}{0} \stackrel{?}{=} \frac{16}{0} \qquad \longleftarrow \text{Undefined.}$$

Since $x = 4$ results in a denominator of 0, 4 is *not* a solution to the equation. It is an extraneous root. The only solution to the equation is -4.

In Example 7, what would you obtain if you began by cross multiplying? Try it and see.

4) Solve Problems Involving Rational Functions

Now we will work a problem that involves a rational function.

EXAMPLE 8 Consider the function $f(x) = x - \frac{2}{x}$. Find all a for which $f(a) = 1$.

Solution Since $f(a) = a - \frac{2}{a}$, we need to find all values for which $a - \frac{2}{a} = 1, a \neq 0$. We begin by multiplying both sides of the equation by a, the LCD.

$$a \cdot \left(a - \frac{2}{a}\right) = a \cdot 1$$
$$a^2 - 2 = a$$
$$a^2 - a - 2 = 0$$
$$(a - 2)(a + 1) = 0$$
$$a - 2 = 0 \quad \text{or} \quad a + 1 = 0$$
$$a = 2 \qquad\qquad a = -1$$

CHECK:

$$f(x) = x - \frac{2}{x}$$
$$f(2) = 2 - \frac{2}{2} = 2 - 1 = 1$$
$$f(-1) = -1 - \frac{2}{(-1)} = -1 + 2 = 1$$

For $a = 2$ or $a = -1$, $f(a) = 1$.

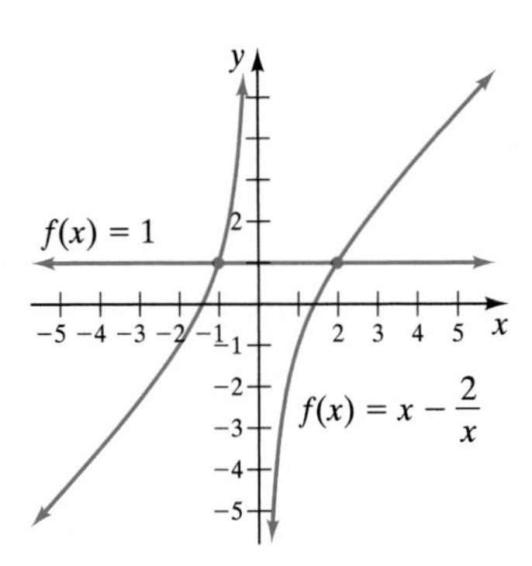

FIGURE 6.7

We used $f(x) = x - \frac{2}{x}$ in Example 8. Figure 6.7 shows the graph of $f(x) = x - \frac{2}{x}$. In this course you will not have to graph functions like this. We illustrate this graph to reinforce the answer obtained in Example 8.

Notice the function is undefined at $x = 0$. Also notice that when $x = -1$ or $x = 2$, it appears that $f(x) = 1$. This was what we should have expected from the results obtained in Example 8.

Example 8 also could have been solved using your graphing calculator by setting $y_1 = x - \frac{2}{x}$ and $y_2 = 1$ and finding the x-coordinate of the intersection of the two graphs.

5) Solve Applications Using Rational Expressions

Now let us look at an application of rational equations.

EXAMPLE 9 In electronics the total resistance, R_T, of resistors connected in a parallel circuit is determined by the formula

$$\frac{1}{R_T} = \frac{1}{R_1} + \frac{1}{R_2} + \frac{1}{R_3} + \cdots + \frac{1}{R_n}$$

where $R_1, R_2, R_3, \ldots, R_n$ are the resistances of the individual resistors (measured in ohms) in the circuit. Find the total resistance if two resistors, one of 200 ohms and the other of 300 ohms, are connected in a parallel circuit.

Solution Since there are only two resistances, use the formula

$$\frac{1}{R_T} = \frac{1}{R_1} + \frac{1}{R_2}$$

Let $R_1 = 200$ ohms and $R_2 = 300$ ohms; then

$$\frac{1}{R_T} = \frac{1}{200} + \frac{1}{300}$$

Multiply both sides of the equation by the LCD, $600R_T$.

$$600R_T \cdot \frac{1}{R_T} = 600R_T\left(\frac{1}{200} + \frac{1}{300}\right)$$

$$\overset{}{600R_T} \cdot \frac{1}{R_T} = \overset{3}{600}R_T\left(\frac{1}{200}\right) + \overset{2}{600}R_T\left(\frac{1}{300}\right)$$

$$600 = 3R_T + 2R_T$$

$$600 = 5R_T$$

$$R_T = \frac{600}{5} = 120$$

Thus, the total resistance of the parallel circuit is 120 ohms. Notice that the resistors actually have less resistance when connected in a parallel circuit than separately.

6) Solve for a Variable in a Formula Containing Rational Expressions

Sometimes you may need to solve for a variable in a formula where the variable you are solving for occurs in more than one term. When this happens it may be possible to solve for the variable by using factoring. To do so, collect all the terms containing the variable you are solving for on one side of the equation and all other terms on the other side of the equation. Then factor out the variable you are solving for. This process is illustrated in Examples 10 through 12.

EXAMPLE 10 A formula used in the study of optics is $\frac{1}{p} + \frac{1}{q} = \frac{1}{f}$. In the formula, p is the distance of an object from a lens or a mirror, q is the distance of the image from the lens or mirror, and f is the focal length of the lens or mirror. For people wearing glasses, the image distance is the distance from the lens to their retina. Solve this formula for f.

Solution Our goal is to isolate the variable f. We begin by multiplying both sides of the equation by the least common denominator, pqf, to eliminate fractions.

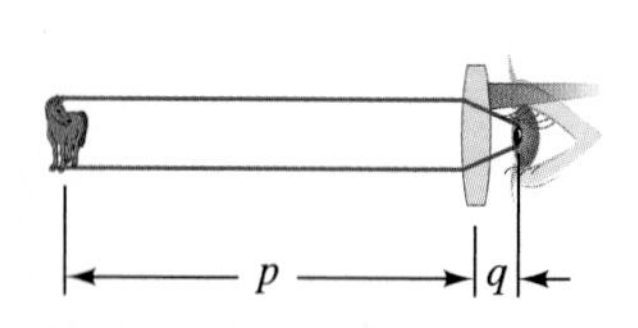

$$\frac{1}{p} + \frac{1}{q} = \frac{1}{f}$$

$$pqf\left(\frac{1}{p} + \frac{1}{q}\right) = pqf\left(\frac{1}{f}\right) \quad \text{Multiply both sides by LCD, } pqf.$$

$$pqf\left(\frac{1}{p}\right) + pqf\left(\frac{1}{q}\right) = pqf\left(\frac{1}{f}\right) \quad \text{Distributive Property.}$$

$$qf + pf = pq \quad \text{Simplify.}$$

$$f(q + p) = pq \quad \text{Factor out } f.$$

$$\frac{f(q + p)}{q + p} = \frac{pq}{q + p} \quad \text{Divide both sides by } q + p.$$

$$f = \frac{pq}{q + p} \quad \text{or} \quad f = \frac{pq}{p + q}$$

NOW TRY EXERCISE 65

EXAMPLE 11 A formula used in banking is $A = P + Prt$, where A represents the amount that must be repaid to the bank when P dollars are borrowed at simple interest rate, r, for time, t, in years. Solve this equation for P.

Solution Since both terms containing P are by themselves on the right side of the equation, we factor P from both terms.

$$A = P + Prt \quad \text{P is in both terms.}$$

$$A = P(1 + rt) \quad \text{Factor out P.}$$

$$\frac{A}{1 + rt} = \frac{P(1 + rt)}{1 + rt} \quad \text{Divide both sides by } 1 + rt \text{ to isolate } P.$$

$$\frac{A}{1 + rt} = P$$

Thus $P = \dfrac{A}{1 + rt}$.

EXAMPLE 12 A formula used for levers in physics is $d = \dfrac{fl}{f + w}$. Solve this formula for f.

Solution We begin by multiplying both sides of the formula by $f + w$ to clear fractions. Then we will rewrite the expression with all terms containing f on one side of the equal sign and all terms not containing f on the other side of the equal sign.

$$d = \frac{fl}{f + w}$$

$$d(f + w) = \frac{fl}{\cancel{(f + w)}} \cancel{(f + w)} \quad \text{Multiply by } f + w \text{ to clear fractions.}$$

$$d(f + w) = fl$$

$$df + dw = fl \quad \text{Distributive property}$$

$$df - df + dw = fl - df \quad \text{Isolate terms containing f on the right side of the equations.}$$

$$dw = fl - df$$

$$dw = f(l - d) \quad \text{Factor out f.}$$

$$\frac{dw}{l - d} = \frac{f\cancel{(l - d)}}{\cancel{l - d}} \quad \text{Isolate f by dividing both sides by } l - d.$$

Thus, $f = \dfrac{dw}{l - d}$.

NOW TRY EXERCISE 67

Exercise Set 6.4

Concept/Writing Exercises

1. What is an extraneous root?
2. Under what circumstances is it necessary to check your answers for extraneous roots?
3. Consider the equation $\frac{x}{4} - \frac{x}{3} = 2$ and the expression $\frac{x}{4} - \frac{x}{3} + 2$.
   a) What is the first step in solving the equation? Explain what effect the first step will have on the equation.
   b) Solve the equation.
   c) What is the first step in simplifying the expression? Explain what effect the first step has when simplifying the expression.
   d) Simplify the expression.
4. Consider the equation $\frac{x}{2} - \frac{x}{3} = 3$ and the expression $\frac{x}{2} - \frac{x}{3} + 3$.
   a) What is the first step in solving the equation?

b) Solve the equation.

c) What is the first step in simplifying the expression? Explain what effect the first step has when simplifying the expression.

d) Simplify the expression.

5. What are similar figures?

6. a) Explain in your words how to solve a rational equation.

b) Solve $\frac{3}{x-4}+\frac{1}{x+4}=\frac{4}{x^2-16}$ following your procedure in part **a)**.

Practice the Skills

Solve each equation and check your solution.

7. $\frac{18}{3b}=\frac{-6}{2}$

8. $\frac{4x+5}{6}=\frac{7}{2}$

9. $\frac{1}{4}=\frac{z+1}{8}$

10. $\frac{a}{5}=\frac{a-3}{2}$

11. $\frac{6x+7}{10}=\frac{2x+9}{6}$

12. $\frac{n}{10}=9-\frac{n}{5}$

13. $\frac{x}{3}-\frac{3x}{4}=-\frac{5x}{12}$

14. $\frac{2}{8}+\frac{3}{4}=\frac{w}{5}$

15. $\frac{3}{4}-x=2x$

16. $\frac{2}{y}+\frac{1}{2}=\frac{5}{2y}$

17. $\frac{3}{r}+\frac{5}{3r}=1$

18. $\frac{x}{4}-\frac{x}{6}=\frac{1}{4}$

19. $\frac{x-1}{x-5}=\frac{4}{x-5}$

20. $\frac{2x+3}{x+1}=\frac{3}{2}$

21. $\frac{5y-2}{7}=\frac{15y-2}{28}$

22. $\frac{2}{x+1}=\frac{1}{x-2}$

23. $\frac{5.6}{-p-6.2}=\frac{2}{p}$

24. $\frac{4.5}{y-3}=\frac{6.9}{y+3}$

25. $\frac{m+1}{m+10}=\frac{m-2}{m+4}$

26. $\frac{x-3}{x+1}=\frac{x-6}{x+5}$

27. $x-\frac{4}{3x}=-\frac{1}{3}$

28. $\frac{b}{2}-\frac{4}{b}=-\frac{7}{2}$

29. $\frac{2x-1}{3}-\frac{x}{4}=\frac{7.4}{6}$

30. $x+\frac{3}{x}=\frac{12}{x}$

31. $x+\frac{6}{x}=-5$

32. $\frac{15}{x}+\frac{9x-7}{x+2}=9$

33. $2-\frac{5}{2b}=\frac{2b}{b+1}$

34. $\frac{3z-2}{z+1}=4-\frac{z+2}{z-1}$

35. $\frac{1}{w-3}+\frac{1}{w+3}=\frac{-5}{w^2-9}$

36. $c-\frac{c}{3}+\frac{c}{5}=26$

37. $\frac{8}{x^2-9}=\frac{2}{x-3}-\frac{4}{x+3}$

38. $\frac{2}{w-5}=\frac{22}{2w^2-9w-5}-\frac{3}{2w+1}$

39. $\frac{y}{2y+2}+\frac{2y-16}{4y+4}=\frac{2y-3}{y+1}$

40. $\frac{3}{x+3}+\frac{5}{x+4}=\frac{12x+19}{x^2+7x+12}$

41. $\frac{1}{x+2}+\frac{1}{x-2}=\frac{4}{x^2-4}$

42. $\frac{4r-1}{r^2+5r-14}=\frac{1}{r-2}-\frac{2}{r+7}$

43. $\frac{5}{x^2+4x+3}+\frac{2}{x^2+x-6}=\frac{3}{x^2-x-2}$

44. $\frac{2}{x^2+2x-8}-\frac{1}{x^2+9x+20}=\frac{4}{x^2+3x-10}$

For each pair of similar figures, find the length of the two unknown sides (that is, those two sides whose lengths involve the variable x).

45.

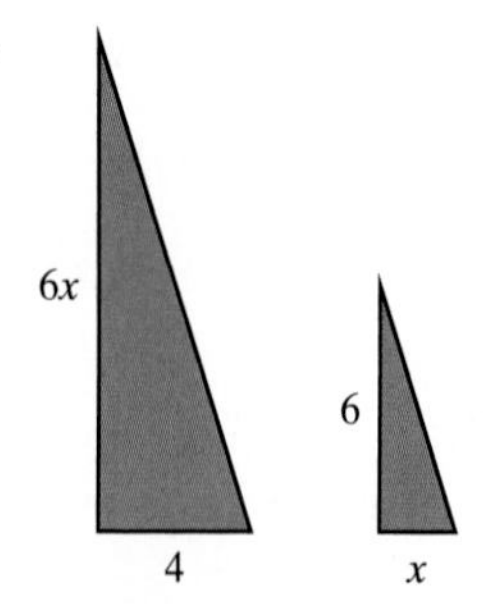

46.

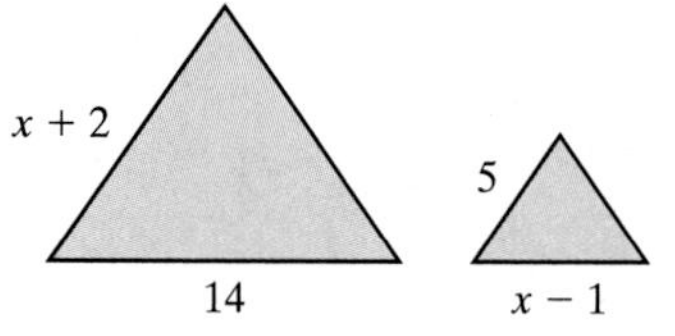

47.

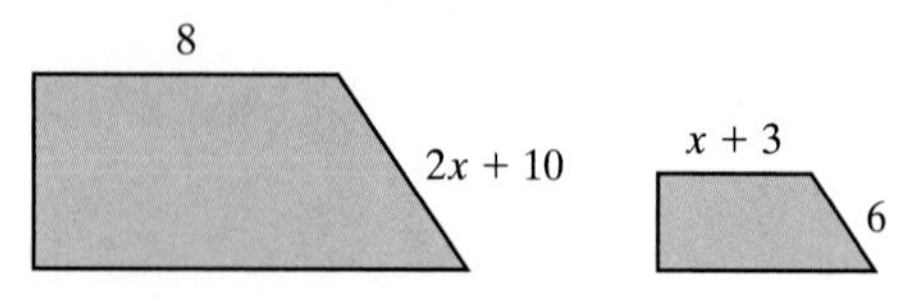

48.

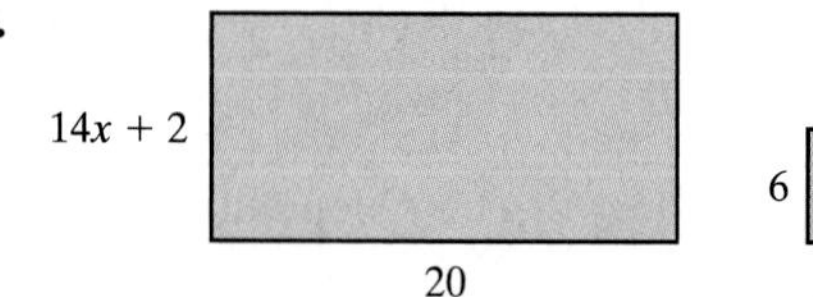

For each rational function given, find all values a for which f(a) has the indicated value.

49. $f(x) = 2x - \frac{4}{x}, f(a) = -2$

50. $f(x) = \frac{x - 2}{x + 5}, f(a) = \frac{3}{5}$

51. $f(x) = 3x - \frac{5}{x}, f(a) = -14$

52. $f(x) = \frac{6}{x} + \frac{6}{2x}, f(a) = 6$

53. $f(x) = \frac{x + 1}{x + 3}, f(a) = \frac{2}{3}$

54. $f(x) = \frac{4}{x} - \frac{3}{2x}, f(a) = 4$

Solve each formula for the indicated variable.

55. $\frac{V_1}{V_2} = \frac{P_2}{P_1}$, for P_2 (chemistry)

56. $T_a = \frac{T_f}{1 - f}$, for f (investment formula)

57. $\frac{V_1}{V_2} = \frac{P_2}{P_1}$, for V_2 (chemistry)

58. $m = \frac{y - y_1}{x - x_1}$, for y (slope)

59. $S = \frac{a}{1 - r}$, for r (mathematics)

60. $m = \frac{y - y_1}{x - x_1}$, for x_1 (slope)

61. $z = \frac{x - \bar{x}}{s}$, for s (statistics)

62. $z = \frac{x - \bar{x}}{s}$, for x (statistics)

63. $d = \frac{fl}{f + w}$, for w (physics)

64. $\frac{1}{p} + \frac{1}{q} = \frac{1}{f}$, for p (optics)

65. $\frac{1}{p} + \frac{1}{q} = \frac{1}{f}$, for q (optics)

66. $\frac{1}{R_T} = \frac{1}{R_1} + \frac{1}{R_2}$, for R_T (electronics)

67. $at_2 - at_1 + v_1 = v_2$, for a (physics)

68. $2P_1 - 2P_2 - P_1P_c = P_2P_c$, for P_c (economics)

69. $a_n = a_1 + nd - d$, for d (mathematics)

70. $S_n - S_nr = a_1 - a_1r^n$, for S_n (mathematics)

*Simplify each expression in **a)** and solve each equation in **b)**.*

71. a) $\frac{2}{x - 2} + \frac{3}{x^2 - 4}$ **b)** $\frac{2}{x - 2} + \frac{3}{x^2 - 4} = 0$

72. a) $\frac{4}{x - 3} + \frac{5}{2x - 6} + \frac{1}{2}$ **b)** $\frac{4}{x - 3} + \frac{5}{2x - 6} = \frac{1}{2}$

73. a) $\frac{b + 3}{b} - \frac{b + 4}{b + 5} - \frac{15}{b^2 + 5b}$ **b)** $\frac{b + 3}{b} - \frac{b + 4}{b + 5} = \frac{15}{b^2 + 5b}$

74. a) $\frac{4x + 3}{x^2 + 11x + 30} - \frac{3}{x + 6} + \frac{2}{x + 5}$ **b)** $\frac{4x + 3}{x^2 + 11x + 30} - \frac{3}{x + 6} = \frac{2}{x + 5}$

Problem Solving

75. What restriction must be added to the statement "If $ac = bc$, then $a = b$"? Explain.

76. Consider $\frac{x - 2}{x - 5} = \frac{3}{x - 5}$.

a) Solve the equation.

b) If you subtract $\frac{3}{x - 5}$ from both sides of the equation, you get $\frac{x - 2}{x - 5} - \frac{3}{x - 5} = 0$. Simplify the difference on the left side of the equation and solve the equation.

c) Use the information obtained in parts **a)** and **b)** to construct another equation that has no solution.

77. Below are two graphs. One is the graph of $f(x) = \frac{x^2 - 9}{x - 3}$ and the other is the graph of $g(x) = x + 3$. Determine which graph is $f(x)$ and which graph is $g(x)$. Explain how you determined your answer.

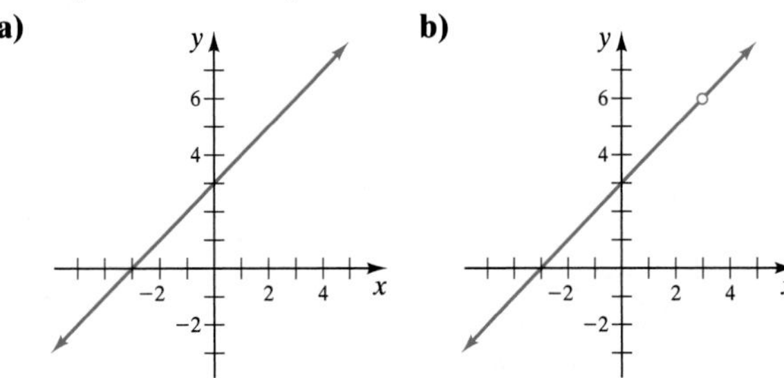

78. The formula $T_a = \dfrac{T_f}{1 - f}$ can be used to find the equivalent taxable yield, T_a, of a tax-free investment, T_f. In this formula f is the individual's federal income tax bracket. Dave Clar is in the 28% income tax bracket.

a) Determine the equivalent taxable yield of a 6% tax-free investment for Dave.

b) Solve this equation for T_f.

c) Determine the equivalent tax-free yield for a 9% taxable investment for Dave.

79. When a homeowner purchases a homeowner's insurance policy where the dwelling is insured for less than 80% of the replacement value, the insurance company will not reimburse the homeowner in total for their loss. The following formula is used to determine the insurance company pay-out, I, when the dwelling is insured for less than 80% of the replacement value.

$$I = \frac{AC}{0.80R}$$

In the formula A is the amount of insurance carried, C is the cost of repairing the damaged area, and R is the replacement value of the home. (There are some exceptions to when this formula is used.)

a) Suppose Jan Burdett had a fire in her kitchen that caused $10,000 worth of damage. If she carried $50,000 of insurance on a home with $100,000 replacement value, what would the insurance company pay for the repairs?

b) Solve this formula for R, the replacement value.

80. Average velocity is defined as a change in distance divided by a change in time, or

$$v = \frac{d_2 - d_1}{t_2 - t_1}$$

This formula can be used when an object at distance d_1 at time t_1 travels to a distance d_2 at time t_2.

a) Assume $t_1 = 2$ hours, $d_1 = 124$ miles, $t_2 = 5$ hours, and $d_2 = 220$ miles. Find the average velocity.

b) Solve the formula for t_2.

81. Average acceleration is defined as a change in velocity divided by a change in time, or

$$a = \frac{v_2 - v_1}{t_2 - t_1}$$

This formula can be used when an object at velocity v_1 at time t_1 accelerates (or decelerates) to velocity v_2 at time t_2.

a) Assume $v_1 = 20$ feet per minute, $t_1 = 20$ minutes, $v_2 = 60$ feet per minute, and $t_2 = 22$ minutes. Find the average acceleration. The units will be ft/min^2.

b) Solve the formula for t_1.

82. The *rate of discount*, P, expressed as a fraction or decimal, can be found by the formula

$$P = 1 - \frac{R - D}{R}$$

where R is the regular price of an item and D is the discount (the amount saved off the regular price).

a) Determine the rate of discount on an umbrella with a regular price of $11.99 that is on sale for $6.99.

b) Solve the formula above for D.

c) Solve the formula above for R.

83. A formula for break-even analysis in economics is

$$Q = \frac{F + D}{R - V}$$

This formula is used to determine the number of units, Q, in an apartment building that must be rented for an investor to break even. In the formula, F is the monthly fixed expenses for the entire building, D is the monthly debt payment on the building, R is the rent per unit, and V is the variable expense per unit.

Assume an investor is considering investing in ABC Properties, a 50-unit building. Each two-bedroom apartment can be rented for $500 per month. Variable expenses are estimated to be $200 per month per unit, fixed expenses are estimated to be $2500 per month, and the monthly debt payment is $8000. How many apartments must be rented for ABC to break even?

Refer to Example 9 for Exercises 84–86.

84. What is the total resistance in the circuit if resistors of 500 ohms and 750 ohms are connected in parallel?

85. What is the total resistance in the circuit if resistors of 300 ohms, 500 ohms, and 3000 ohms are connected in parallel?

86. Three resistors of identical resistance are to be connected in parallel. What should be the resistance of each resistor if the circuit is to have a total resistance of 700 ohms?

Refer to Example 10 for Exercises 87 and 88.

87. In a slide or movie projector, the film acts as the object whose image is projected on a screen. If a 100-mm-focal length (or 0.10 meter) lens is to project an image on a screen 7.5 meters away, how far from the lens should the film be?

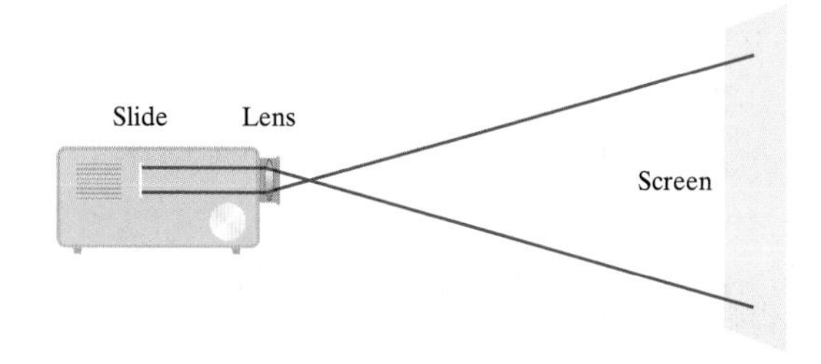

88. A diamond ring is placed 20.0 cm from a concave (curved in) mirror whose focal length is 15.0 cm. Find the position of the image (or the *image distance*).

89. Some investments, such as certain municipal bonds and municipal bond funds, are not only federally tax free but are also state and county or city tax free. When you wish to compare a taxable investment, T_a, with an investment that is federal, state, and county tax free, T_f, you can use the formula

$$T_a = \frac{T_f}{1 - [f + (s + c)(1 - f)]}$$

In the formula, s is your state tax bracket, c is your county or local tax bracket, and f is your federal income tax bracket. Howard Levy, who lives in Detroit, Michigan, is in a 4.6% state tax bracket, a 3% city tax bracket, and a 33% federal tax bracket. He is choosing between the Fidelity Michigan Triple *Tax-Free* Money Market Portfolio yielding 6.01% and the Fidelity *Taxable* Cash Reserve Money Market Fund yielding 7.68%.

a) Using his tax brackets, determine the taxable equivalent of the 6.01% tax free yield.

b) Which investment should Howard make? Explain your answer.

90. The synodic period of Mercury is the time required for swiftly moving Mercury to gain one lap on Earth in their orbits around the sun. If the orbital periods (in Earth days) of the two planets are designated P_m and P_e, Mercury will be seen on the average to move $1/P_m$ of a revolution per day, while Earth moves $1/P_e$ of a revolution per day in pursuit. Mercury's daily gain on Earth is $(1/P_m) - (1/P_e)$ of a revolution, so that the time for Mercury to gain one complete revolution on Earth, the synodic period s, may be found by the formula

$$\frac{1}{s} = \frac{1}{P_m} - \frac{1}{P_e}$$

If P_e is 365 days and P_m is 88 days, find the synodic period in Earth days.

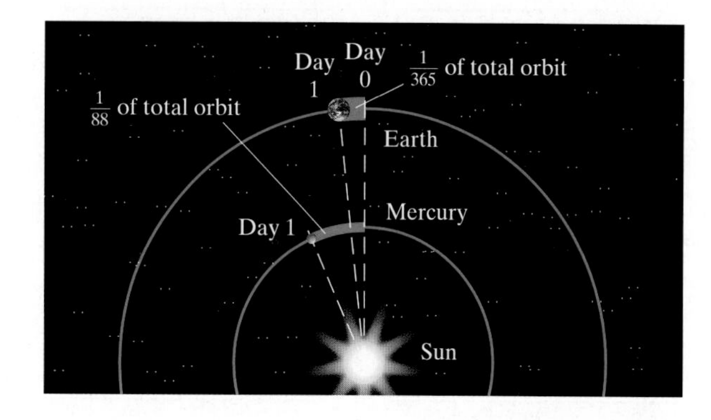

Challenge Problems

91. Make up an equation that cannot have 4 or −2 as a solution. Explain how you determined your answer.

92. Make up an equation containing the sum of two rational expressions in the variable x whose solution is the set of *real numbers*. Explain how you determined your answer.

93. Make up an equation in the variable x containing the sum of two rational expressions whose solution is the set of all real numbers except 0. Explain how you determined your answer.

Group Activity

94. An 80-mm focal length lens is used to focus an image on the film of a camera. The maximum distance allowed between the lens and the film plane is 120 mm.

a) Group member 1: Determine how far ahead of the film the lens should be if the object to be photographed is 10.0 meters away.

b) Group member 2: Repeat part **a)** for a distance of 3 meters away.

c) Group member 3: Repeat part **a)** for a distance of 1 meter away.

d) Individually, determine the closest distance that an object could be photographed sharply.

e) Compare your answers to see whether they appear reasonable and consistent.

Cumulative Review Exercises

[2.5] **95.** Solve the inequality $-2 \leq 4 - 2x < 6$.

[3.2] **96.** If $f(x) = \frac{1}{2}x^2 - 3x + 4$, find $f(5)$.

[5.2] **97.** Multiply $(3x^2 - 6x - 4)(2 - 3x)$.

[5.6] **98.** Factor $8x^3 - 64y^6$.

6.5 RATIONAL EQUATIONS: APPLICATIONS AND PROBLEM SOLVING

SSM VIDEO 6.5 CD Rom

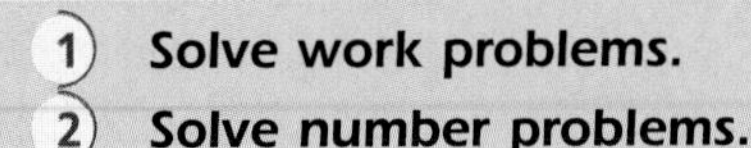

1) Solve work problems.
2) Solve number problems.
3) Solve motion problems.

Some applications of equations with rational expressions were illustrated in Section 6.4 and Exercise Set 6.4. In this section we examine some additional applications. We study work problems first.

1) Solve Work Problems

Problems where two or more machines or people work together to complete a certain task are sometimes referred to as **work problems.** To solve work problems we use the fact that the part of the work done by person 1 (or machine 1) plus the part of the work done by person 2 (or machine 2) is equal to the total amount of work done by both people (or both machines), or 1 (for 1 whole task completed).

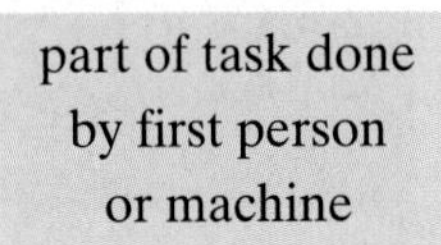

part of task done by first person or machine	+	part of task done by second person or machine	=	1 (one whole task) (completed)

To determine the part of the task done by each person or machine we use the formula

part of task completed = rate · time

This formula is very similar to the formula *amount* = *rate* · *time* that was discussed in Section 2.4.

Let's now discuss how to determine the rate. If, for example, John can do a particular task in 5 hours, he could complete $\frac{1}{5}$ of the task in 1 hour. Thus, his rate is $\frac{1}{5}$ of the task per hour. If Kishi can do a job in 6 hours, her rate is $\frac{1}{6}$ of the job per hour. Similarly, if Maria can do a job in x minutes, her rate is $\frac{1}{x}$ of the job per minute. *In general, if a person or machine can complete a task in x units of time, the rate is* $\frac{1}{x}$.

EXAMPLE 1 Saif and Alejandro Bardini both work for a botanical garden. The botanical garden is adding a number of floral designs around its grounds. Saif, who has more experience, can plant the flowers and make the design by himself in 3 hours. It takes Alejandro 4 hours working by himself to make the same design. If Saif and Alejandro are assigned to work together, how long will it take them to make the design?

Solution **Understand** We need to find the time for both Saif and Alejandro working together to make the floral design. Let x = time, in hours, for Saif and Alejandro to make the floral design together. We will construct a table to assist us in finding the part of the task completed by each person.

Worker	Rate of Work	Time Worked	Part of Task Completed
Saif	$\frac{1}{3}$	x	$\frac{x}{3}$
Alejandro	$\frac{1}{4}$	x	$\frac{x}{4}$

Translate

$$\left(\begin{array}{c}\text{part of floral design made}\\ \text{by Saif in } x \text{ hours}\end{array}\right) + \left(\begin{array}{c}\text{part of floral design made}\\ \text{by Alejandro in } x \text{ hours}\end{array}\right) = 1 \text{ (entire floral design)}$$

$$\frac{x}{3} + \frac{x}{4} = 1$$

Carry Out We multiply both sides of the equation by the LCD, 12. Then we solve for x, the number of hours.

$$12\left(\frac{x}{3} + \frac{x}{4}\right) = 12 \cdot 1 \quad \text{Multiply by the LCD, 12.}$$

$$12\left(\frac{x}{3}\right) + 12\left(\frac{x}{4}\right) = 12 \quad \text{Distributive Property.}$$

$$4x + 3x = 12$$

$$7x = 12$$

$$x = \frac{12}{7} \quad \text{1.71 hours (to the nearest hundredth)}$$

Answer Saif and Alejandro together can make the floral design in about 1.71 hours. This answer is reasonable because this time is less than it takes either person to make the design by himself.

NOW TRY EXERCISE 5

EXAMPLE 2 Jim and Joy Love just moved into a house that has a Jacuzzi bathtub. When they turn on the tap water to fill the tub, the water is cloudy from lack of use by the previous owner. They wish to run as much water through the tub as possible until the water clears. To accomplish this they turn on the cold water tap (to save energy, they don't use the hot water tap) and they open the drain of the tub. The cold water tap can fill the tub in 7.6 minutes and the drain can empty the tub in 10.3 minutes. If the drain is open and the cold water faucet is turned on, how long will it take before the water fills the tub?

Solution **Understand** As water from the faucet is filling the tub, water going down the drain is emptying the tub. Thus, the faucet and drain are working against each other. Let x = amount of time needed to fill the tub.

	Rate of Work	Time Worked	Part of Tub Filled or Emptied
Faucet filling tub	$\frac{1}{7.6}$	x	$\frac{x}{7.6}$
Drain emptying tub	$\frac{1}{10.3}$	x	$\frac{x}{10.3}$

Translate Since the faucet and drain are working against each other, we will *subtract* the part of the water being emptied from the part of water being added to the tub.

$$\begin{pmatrix}\text{part of tub filled}\\ \text{in } x \text{ minutes}\end{pmatrix} - \begin{pmatrix}\text{part of tub emptied}\\ \text{in } x \text{ minutes}\end{pmatrix} = 1 \text{ (whole tub filled)}$$

$$\frac{x}{7.6} - \frac{x}{10.3} = 1$$

Carry Out Using a calculator, we can determine that the LCD is $(7.6)(10.3) = 78.28$. Now we multiply both sides of the equation by 78.28 to remove fractions.

$$78.28\left(\frac{x}{7.6} - \frac{x}{10.3}\right) = 78.28(1)$$

$$\overset{10.3}{\cancel{78.28}}\left(\frac{x}{\cancel{7.6}}\right) - \overset{7.6}{\cancel{78.28}}\left(\frac{x}{\cancel{10.3}}\right) = 78.28(1)$$

$$10.3x - 7.6x = 78.28$$

$$2.7x = 78.28$$

$$x \approx 28.99$$

NOW TRY EXERCISE 13

Answer The tub will fill in about 29 minutes.

EXAMPLE 3 Patty Jones and Mike Prenzton work in the sporting goods department at Sears, where they assemble bicycles. When Patty and Mike work together, they can assemble a bicycle in 20 minutes. When Patty assembles a bike by herself, it takes her 36 minutes. How long would it take Mike to assemble the bike by himself?

Solution **Understand** Let x = amount of time for Mike to assemble the bike by himself. We know that when working together they can assemble the bike in 20 minutes. We organize this information in the table that follows.

Worker	Rate of Work	Time Worked	Part of Bicycle Completed
Patty	$\frac{1}{36}$	20	$\frac{20}{36}$
Mike	$\frac{1}{x}$	20	$\frac{20}{x}$

Translate

$$\begin{pmatrix}\text{part of bicycle}\\ \text{assembled by Patty}\end{pmatrix} + \begin{pmatrix}\text{part of bicycle}\\ \text{assembled by Mike}\end{pmatrix} = 1$$

$$\frac{20}{36} + \frac{20}{x} = 1$$

Carry Out

$$36x\left(\frac{20}{36} + \frac{20}{x}\right) = 36x \cdot 1$$

$$36x\left(\frac{20}{36}\right) + 36x\left(\frac{20}{x}\right) = 36x$$

$$20x + 720 = 36x$$

$$720 = 16x$$

$$45 = x$$

NOW TRY EXERCISE 11 **Answer** Mike can assemble a bike by himself in 45 minutes.

2) Solve Number Problems

Now let us look at a **number problem**, where we must find a number described in relation to one or more other numbers.

EXAMPLE 4 When the reciprocal of 3 times a number is subtracted from 1, the result is the reciprocal of twice the number. Find the number.

Solution **Understand** Let x = unknown number. Then $3x$ is 3 times the number, and $\frac{1}{3x}$ is the reciprocal of 3 times the number. Twice the number is $2x$, and $\frac{1}{2x}$ is the reciprocal of twice the number.

Translate

$$1 - \frac{1}{3x} = \frac{1}{2x}$$

Carry Out

$$6x\left(1 - \frac{1}{3x}\right) = 6x \cdot \frac{1}{2x} \quad \text{Multiply by the LCD, } 6x.$$

$$6x(1) - 6x\left(\frac{1}{3x}\right) = 6x\left(\frac{1}{2x}\right)$$

$$6x - 2 = 3$$

$$6x = 5$$

$$x = \frac{5}{6}$$

NOW TRY EXERCISE 21 **Answer** A check will verify that the number is $\frac{5}{6}$.

3) Solve Motion Problems

The last type of problem we will look at is **motion problems.** Recall that we discussed motion problems in Section 2.4. In that section we learned that distance = rate · time. Sometimes it is convenient to solve for the time when solving motion problems.

$$\text{time} = \frac{\text{distance}}{\text{rate}}$$

EXAMPLE 5 Sally McPherson owns a single-engine Cessna airplane. When making her preflight plan, she finds that there is a 20 mile per hour wind moving from east to west at the altitude she plans to fly. If she travels west (with the wind), she will be able to travel 400 miles in the same amount of time that she would be able to travel 300 miles flying east (against the wind). Assuming that if it were not for the wind, the plane would fly at the same speed going east or west, find the speed of the plane in still air.

Solution **Understand** Let x = speed of the plane in still air. We will set up a table to help answer the question.

Wind (20mph)

West East

Flying with the wind, 400 miles

Flying against the wind, 300 miles

Plane	Distance	Rate	Time
Against wind	300	$x - 20$	$\frac{300}{x-20}$
With wind	400	$x + 20$	$\frac{400}{x+20}$

Translate Since the times are the same, we set up and solve the following equation:

$$\frac{300}{x-20} = \frac{400}{x+20}$$

Carry Out

$$300(x + 20) = 400(x - 20) \quad \text{Cross multiply.}$$
$$300x + 6000 = 400x - 8000$$
$$6000 = 100x - 8000$$
$$14{,}000 = 100x$$
$$140 = x$$

NOW TRY EXERCISE 29

Answer The speed of the plane in still air is 140 miles per hour.

EXAMPLE 6 Milt McGowen rides his bike to and from work from his home in Fayetteville, Arkansas. Going to work, he rides mostly downhill and averages 15 miles per hour. Coming home, mostly uphill, he averages only 6 miles per hour. If it takes him $\frac{1}{2}$ hour longer to ride home than to ride to work, how far is his work from his home?

Solution **Understand** In this problem the times going and returning are not equal. Milt's time returning is $\frac{1}{2}$ hour longer than his time going. Therefore, to make the times equal, we must add $\frac{1}{2}$ hour to his time going (or subtract $\frac{1}{2}$ hour from his time returning).

Let x = the distance from his home to his workplace.

Ride	Distance	Rate	Time
Going	x	15	$\frac{x}{15}$
Returning	x	6	$\frac{x}{6}$

Translate time going $+ \frac{1}{2}$ hour $=$ time returning

$$\frac{x}{15} + \frac{1}{2} = \frac{x}{6}$$

Carry Out

$$30\left(\frac{x}{15}\right) + 30\left(\frac{1}{2}\right) = 30\left(\frac{x}{6}\right) \quad \textit{Multiply by the LCD, 30.}$$

$$2x + 15 = 5x$$

$$15 = 3x$$

$$5 = x$$

NOW TRY EXERCISE 41

Answer Therefore, Milt lives 5 miles from his workplace.

EXAMPLE 7 Dawn Puppel lives in Buffalo, New York, and travels to college in South Bend, Indiana. The speed limit on some of the roads she travels is 55 miles per hour, while on others it is 65 miles per hour. The total distance traveled by Dawn is 490 miles. If Dawn follows the speed limits and the total trip takes 8 hours, how long did she drive at 55 miles per hour and how long did she drive at 65 miles per hour?

Solution **Understand and Translate** let $x =$ number of miles driven at 55 mph

then $490 - x =$ number of miles driven at 65 mph

Speed Limit	Distance	Rate	Time
55 mph	x	55	$\frac{x}{55}$
65 mph	$490 - x$	65	$\frac{490 - x}{65}$

Since the total time is 8 hours, we write

$$\frac{x}{55} + \frac{490 - x}{65} = 8$$

Carry Out The LCD of 55 and 65 is 715.

$$715\left(\frac{x}{55} + \frac{490 - x}{65}\right) = 715 \cdot 8$$

$$715\left(\frac{x}{55}\right) + 715\left(\frac{490 - x}{65}\right) = 5720$$

$$13x + 11(490 - x) = 5720$$

$$13x + 5390 - 11x = 5720$$

$$2x + 5390 = 5720$$

$$2x = 330$$

$$x = 165$$

Answer The number of miles driven at 55 mph is 165 miles. Then the time driven at 55 mph is $\frac{165}{55} = 3$ hours, and the time driven at 65 mph is $\frac{490 - 165}{65} = \frac{325}{65} = 5$ hours.

Notice that in Example 7 the answer to the question was not the value obtained for x. The value obtained was a distance, and the question asked us to find the time. *When working word problems, you must read and work the problems very carefully and make sure you answer the questions that were asked.*

Exercise Set 6.5

Practice the Skills and Problem Solving

1. Two brothers take exactly the same time to paint a wall. If they paint the wall together, will the total time needed to paint the wall be less than $\frac{1}{2}$ the time, equal to $\frac{1}{2}$ the time, or greater than $\frac{1}{2}$ the time it takes each brother separately to paint the wall? Explain.

2. Two tractors, a larger one and a smaller one, work together to level a field. In the same amount of time, the larger tractor levels more land than the smaller one. Will the smaller tractor take more or less than twice the time working alone than the two take working together? Explain.

Write an equation that can be used to solve each work problem. Then answer the question asked.

3. Jason La Rue can shampoo the carpet on the main floor of the Sheraton Hotel in 3 hours. Tom Lockheart can shampoo the same carpet in 6 hours. If they work together, how long will it take them to shampoo the carpet?.

4. At the Merck Corporation it takes one computer 3 hours to print checks for its employees and a second computer 5 hours to complete the same job. How long will it take the two computers working together to complete the job?

5. Paula Wentz and Mike Abernathy are friends and neighbors, and each owns a farm in Cabrillo, California. Paula can plow a specific section used for corn in 4 hours using her John Deere tractor. Mike can plow the same acreage in 6 hours using his Caterpillar tractor. One Saturday Paula has to catch a plane but wishes to first plow the section of her farm used for corn. Since she has a time concern, she asks Mike to help her. How long will it take them working together to plow the section?

6. A $\frac{1}{2}$-inch-diameter hose can fill a swimming pool in 8 hours. A $\frac{4}{5}$-inch-diameter hose can fill the same pool in 5 hours. How long will it take to fill the pool when both hoses are used?

7. A conveyor belt operating at full speed can fill a tank with topsoil in 3 hours. When a valve at the bottom of the tank is opened, the tank will empty in 4 hours. If the conveyor belt is operating at full speed and the valve at the bottom of the tank is open, how long will it take to fill the tank?

8. An oil refinery has large tanks to hold oil. Each tank has an inlet valve and an outlet valve. The tank can be filled with oil in 20 hours when the inlet valve is wide open and the outlet valve is closed. The tank can be emptied in 25 hours when the outlet valve is wide open and the inlet valve is closed. If a new tank is placed in operation and both the inlet valve and outlet valve are wide open, how long will it take to fill the tank?

9. Waunetta Verduin and O'Shea Donald work as individual contractors for Alpha Tile Company. Both are hired by Alpha Tile to lay tile floors for its customers. Waunetta can lay a specific kitchen floor in 12 hours. If Waunetta and O'Shea work together on the kitchen floor, they can complete the job in 8 hours. How long will it take O'Shea to tile the kitchen floor working by himself?

10. Dr. Indiana Jones and his father, Dr. Henry Jones, are archeologists working on a dig near the Forum in Rome. Indiana and his father working together can unearth a specific plot of land in 2.6 months. Indiana can unearth the entire area by himself in 3.9 months. How long would it take Henry to unearth the entire area by himself?

11. Alan MacDonald of New Zealand holds the record for sheep shearing.* He can shear a small herd of sheep in 60 minutes. Mr. MacDonald and Mr. Guesner together can shear the same herd of sheep in 50 minutes. How long would it take Mr. Guesner to shear the herd by himself?

12. William Wade and Sally Sitongia work for General Telephone. Together it takes them 1.8 hours to dig a trench where a wire is to be laid. If William can dig the trench by himself in 3.2 hours, how long would it take Sally by herself to dig the trench?

13. When only the cold water valve is opened, a washtub will fill in 8 minutes. When only the hot water valve is opened, the washtub will fill in 12 minutes. When the drain of the washtub is open, it will drain completely in 7 minutes. If both the hot and cold water valves are open and the drain is open, how long will it take for the washtub to fill?

14. A large tank is being used on Hank Martel's farm to irrigate the crops. The tank has two inlet pipes and one outlet pipe. The two inlet pipes can fill the tank in 10 and 12 hours, respectively. The outlet pipe can empty the tank in 15 hours. If the tank is empty, how long would it take to fill the tank when all three valves are open?

15. The Rushville fire department uses 3 pumps to remove water from flooded basements. One pump can remove all the water from a flooded basement in 6 hours. The second pump can remove the same amount of water in 5 hours, and the third pump requires only 4 hours to remove the water. If all 3 pumps work together to remove the water from the flooded basement, how long will it take to empty the basement?

16. It takes Frank Lee twice as long as Beth Barns to knit an afgan. If together they knit an afghan in 12 hours, how long would it take Beth to knit the afghan by herself?

17. A roofer, Vic Triola, requires 15 hours to put a new roof on a house. Anna Divis, the roofer's apprentice, can reroof the house by herself in 20 hours. After working alone on a roof for 6 hours, Vic leaves for another job. Anna takes over and completes the job. How long will it take Anna to complete the job?

18. Two pipes are used to fill an oil tanker. When the larger pipe is used alone, it takes 60 hours to fill the tanker. When the smaller pipe is used alone, it takes 80 hours to fill the tanker. The large pipe begins filling the tanker. After 20 hours the large pipe is closed down and the smaller pipe is opened. How much longer will it take to finish filling the tanker using only the smaller pipe?

Write an equation that can be used to solve each number problem. Then answer the question asked.

19. What number multiplied by the numerator and added to the denominator of the fraction $\frac{4}{3}$ makes the resulting fraction $\frac{5}{2}$?

20. What number added to the numerator and multiplied by the denominator of the fraction $\frac{3}{2}$ makes the resulting fraction $\frac{1}{8}$?

21. One number is twice another. The sum of their reciprocals is $\frac{3}{4}$. Find the numbers.

22. The sum of the reciprocals of two consecutive integers is $\frac{11}{30}$. Find the two integers.

23. The sum of the reciprocals of two consecutive even integers is $\frac{5}{12}$. Find the two integers.

24. When a number is added to both the numerator and denominator of the fraction $\frac{5}{7}$, the resulting fraction is $\frac{4}{5}$. Find the number added.

25. When 3 is added to twice the reciprocal of a number, the sum is $\frac{31}{10}$. Find the number.

26. The reciprocal of 3 less than a certain number is twice the reciprocal of 6 less than twice the number. Find the number(s).

27. If 3 times a number is added to twice the reciprocal of the number, the answer is 5. Find the number(s).

28. If 3 times the reciprocal of a number is subtracted from twice the reciprocal of the square of the number, the difference is -1. Find the number(s).

* Alan MacDonald sheared 805 sheep in 9 hours (an average of 40.2 seconds per sheep).

Write an equation that can be used to solve each motion problem. Then answer the question asked

29. The Amtrac Auto Train travels nonstop from Lorton, Virginia (just south of Washington, D.C.) to Sanford, Florida (just north of Orlando). Wanda Garner wishes to bring two cars down to Florida for the winter, so she decides to send one by auto train and drive the other down. The train travels 600 kilometers in the time it takes for her to drive 400 kilometers. If the average speed of the train is 40 kilometers greater than the speed of Wanda's car, find the speeds of the train and car.

30. Richard Zucker's boat, the Phantom, is going 20 miles per hour in the still water of a bay. He decides to keep the throttle the same, but to ride to a nearby river. Once in the river, he finds that it takes his boat the same amount of time to travel 3.1 miles downstream (with the current) as it does to travel 2.2 miles upstream (against the current). Find the current.

31. The moving sidewalk at Chicago's O'Hare International Airport moves at a speed of 2.0 feet per second. Walking on the moving sidewalk, Nancy Killian walks 120 feet in the same time that it takes her to walk 52 feet without the moving sidewalk. How fast does Nancy walk?

32. The moving sidewalk at the Denver International Airport moves at a speed of 1.8 feet per second. Nathan Nearpass walks 100 feet on the moving sidewalk, then turns around on the moving sidewalk and walks at the same speed 40 feet in the opposite direction. If the time walking in each direction was the same, find Nathan's walking speed.

33. Ruth and Jerry Nardella go for an outing in Memorial Park in Houston, Texas. Ruth jogs while Jerry rollerblades. Jerry rollerblades 2.9 miles per hour faster than Ruth jogs. When Jerry has rollerbladed 5.7 miles, Ruth has jogged 3.4 miles. Find Ruth's jogging speed.

34. Jan Burdett and Hector Diaz go on a cross-country skiing trail in Park City, Utah. On the trail, Hector averages 8 miles per hour while Jan averages 6 miles per hour. If it takes Jan $\frac{1}{2}$ hour longer to ski the trail and return to their starting point, how long is the trail?

35. Ersnia Jacques rented a boat and traveled to a docking area on the Great Barrier Reef, a distance of 40 kilometers. She then snorkeled for twice the length of time that it took her to get to the docking station. The total time spent traveling to the reef and snorkeling was 4 hours. Find the average speed of the boat, in kilometers per hour, to the docking station.

36. Ray Packerd starts out on a boating trip at 8 A.M. Ray's boat can go 20 miles per hour in still water.

a) How far downstream can Ray go if the current is 5 miles per hour and he wishes to go down and back in 4 hours?

b) At what time must Ray turn back?

37. At a football game in Green Bay, Wisconsin, the Packers have the ball on their own 20 yard line, which is 80 yards from the opposing team's end zone and the distance they must travel for a touchdown. Brett Favre passed the ball to Tyrone Davis, who catches the ball and runs for a touchdown. Assume that the ball, when passed, traveled at 14.7 yards per second and that after Tyrone caught the ball, he ran untouched at 5.8 yards per second into the end zone. If the play, from the time Brett released the ball to the time Tyrone reached the end zone, took 10.6 seconds, how far from where Brett threw the ball did Tyrone catch it? Assume that the ball was thrown straight down the center of the field and the run continued down the center of the field (see the diagram on page 405).

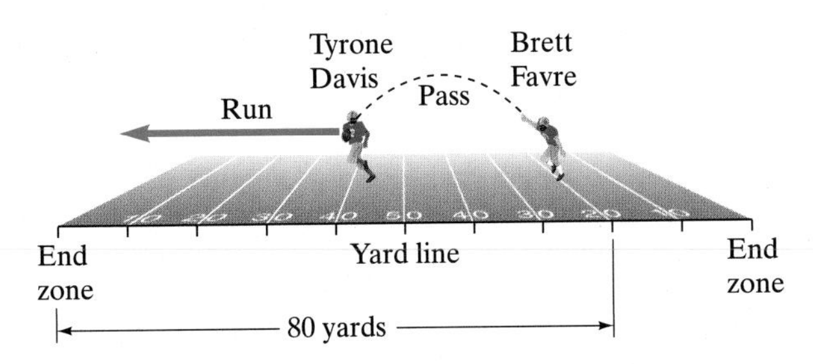

38. In one day, Pauline Shannon drove from Front Royal, Virginia, to Asheville, North Carolina, along the scenic Skyline Drive and Blue Ridge Parkway, a distance of 492 miles. For part of the trip she drove at a steady rate of 50 miles per hour, but in some of the more scenic areas she drove at a steady rate of 35 miles per hour. If the total time of the trip was 11.13 hours, how far did she travel at each speed?

39. The number 4 train in the New York City subway system goes from Woodlawn/Jerome Avenue in the Bronx to Flatbush Avenue in Brooklyn. The one-way distance between these two stops is 24.2 miles. On this route, two tracks run parallel to each other, one for the local train and the other for the express train. The local and express trains leave Woodlawn/Jerome Avenue at the same time. When the express reaches the end of the line at Flatbush Avenue, the local is at Wall Street, 7.8 miles from Flatbush. If the express averages 5.2 miles per hour faster than the local, find the speeds of the two trains.

40. Each morning, Ron Lucky takes his horse, Beauty, for a ride on Pfeiffer Beach in Big Sur, California. He typically rides Beauty for 5.4 miles, then walks Beauty 2.3 miles. His speed when riding is 4.2 times his speed when walking. If his total outing takes 1.5 hours, find the rate at which he walks Beauty.

41. A train and car leave from the Pasadena, California, railroad station at the same time headed for the state fair in Sacramento. The car averages 50 miles per hour and the train averages 70 miles per hour. If the train arrives at the fair 2 hours ahead of the car, find the distance from the railroad station to the state fair.

42. A train and a plane leave from Boston at the same time for Columbia, South Carolina, 900 miles away. If the speed of the plane is 5 times the speed of the train, and the plane arrives 12 hours before the train, find the speeds of the train and the plane.

43. Brothers Jim and Pete Crandell are long-distance swimmers. Jim averages 3.6 miles per hour and Pete averages 2.4 miles per hour. The brothers start swimming at the same time across Lake Mead to a point on the other side of the lake. If Jim arrives 0.2 hour ahead of Pete, find the distance they swam. (Of course each swimmer was accompanied by a boat for safety reasons.)

44. In a 30-mile sailboat race the winning boat, the Buccaneer, finishes 10 minutes ahead of the second place boat, the Raven. If the Buccaneer's average speed was 2 miles per hour faster than the Raven's, find the average speed of the Buccaneer.

45. Two friends drive from Dallas going to El Paso, a distance of 600 miles. Mary Ann Zilke travels by highway and arrives at the same destination 2 hours ahead of Carla Canola, who took a different route. If the average speed of Mary Ann's car was 10 miles per hour faster than Carla's car, find the average speed of Mary Ann's car.

46. Kathy Angel took a helicopter ride to the top of the Mt. Cook Glacier in New Zealand. The flight to the top of the glacier covered a distance of 60 kilometers. Kathy stayed on top of the glacier for $\frac{1}{2}$ hour. She then flew to the town of Te Anu, 140 kilometers away. The helicopter averaged 20 kilometers per hour faster going to Te Anu than on the flight up to the top of the glacier. The total time involved in the outing was 2 hours. Find the average speed of the helicopter going to the glacier.

47. Two rockets are to be launched at the same time from NASA headquarters in Houston, Texas, and are to meet at a space station many miles from Earth. The first rocket is to travel at 20,000 miles per hour and the second rocket will travel at 18,000 miles per hour. If the first rocket is scheduled to reach the space station 0.6 hour before the second rocket, how far is the space station from NASA headquarters?

48. A trip up Mt. Pilatus, near Lucerne, Switzerland, involves riding an inclined railroad up to the top of the mountain, spending time at the top, then coming down the opposite side of the mountain in an aerial tram. A van or bus then returns you to your starting point 18 kilometers away, where your vehicle is parked.

The distance traveled up the mountain is 7.5 kilometers and the distance traveled down the mountain is 8.7 kilometers. The speed coming down the mountain is 1.2 times the speed going up. It takes 36 minutes for the van to return you to your vehicle. If the Lieblichs stayed at the top of the mountain for 3 hours and the total time of their outing (from when they started going up the mountain until they were returned to their car) took 9 hours, find the speed, in kilometers per hour, of the inclined railroad.

49. Make up your own work problem and find the solution.

50. Make up your own motion problem and find the solution.

51. Make up your own number problem and find the solution.

Challenge Problem

52. An officer flying a California Highway Patrol aircraft determines that a car 10 miles ahead of her is speeding at 90 miles per hour.

a) If the aircraft is traveling 450 miles per hour, how far will the car have traveled in the time it takes the aircraft to reach it?

b) How long, in minutes, will it take for the aircraft to reach the car?

c) If the pilot wishes to reach the car in exactly 1 minute, how fast must the airplane fly?

Cumulative Review Exercises

[5.1] **53.** Subtract $\frac{1}{2}x^2 - 3x^2 + 2xy - \left(\frac{3}{5}xy + 6y^2\right)$.

[5.2] **54.** Multiply $(4x^2 - 6x - 1)(3x - 4)$.

[5.3] **55.** Divide $(12x^2 + 7x + 12) \div (3x - 2)$.

[5.5] **56.** Factor $8x^2 + 26x + 15$.

6.6 VARIATION

1) **Solve direct variation problems.**
2) **Solve inverse variation problems.**
3) **Solve joint variation problems.**
4) **Solve combined variation problems.**

In Sections 6.4 and 6.5 we saw many applications of equations containing rational expressions. In this section we see still more.

1) Solve Direct Variation Problems

Many scientific formulas are expressed as variations. A **variation** is an equation that relates one variable to one or more other variables using the operations of multiplication or division (or both operations). There are essentially three types of variation problems: direct, inverse, and joint variation.

In **direct variation** the two related variables will both increase together or both decrease together; that is, as one increases so does the other, and as one decreases so does the other.

Consider a car traveling at 30 miles an hour. The car travels 30 miles in 1 hour, 60 miles in 2 hours, and 90 miles in 3 hours. Notice that as the time increases, the distance traveled increases.

The formula used to calculate distance traveled is

$$\text{distance} = \text{rate} \cdot \text{time}$$

Since the rate is a constant, 30 miles per hour, the formula can be written

$$d = 30t$$

We say that distance varies directly as time or that distance is directly proportional to time. This is an example of a direct variation.

Direct Variation

If a variable y varies directly as a variable x, then

$$y = kx$$

where k is the **constant of proportionality** (or the variation constant).

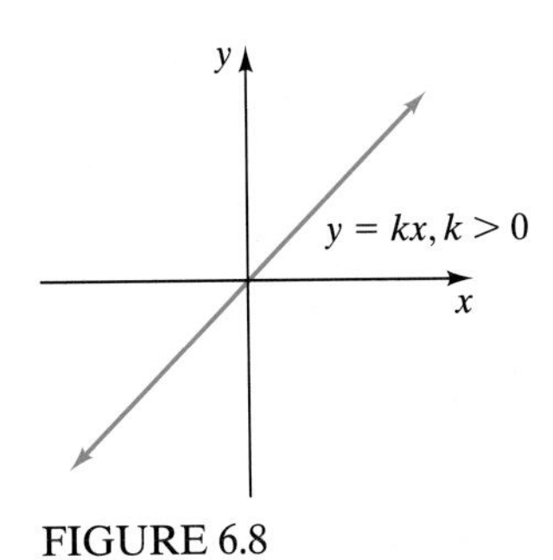

FIGURE 6.8

The graph of $y = kx$, $k > 0$, is always a straight line that goes through the origin (see Fig. 6.8). The slope of the line depends on the value of k. The greater the value of k, the greater the slope.

EXAMPLE 1 The circumference of a circle, C, is directly proportional to (or varies directly as) its radius, r. Write the equation for the circumference of a circle if the constant of proportionality, k, is 2π.

Solution

$$C = kr \text{ (} C \text{ varies directly as } r\text{)}$$

$$C = 2\pi r \text{ (constant of proportionality is } 2\pi\text{)}$$

EXAMPLE 2 The amount of the drug theophylline given to patients is directly proportional to the patient's mass, in kilograms.

a) Write this variation as an equation.

b) If 200 mg is given to a boy whose mass is 40 kg, find the constant of proportionality.

c) How much of the drug should be given to a patient whose mass is 62 kg?

Solution **a)** We are told this is a direct variation. That is, the greater a person's mass the more of the drug that will need to be given. We therefore set up a direct variation.

$$a = km$$

b) Understand and Translate To determine the value of the constant of proportionality, we substitute the given values for the amount, a, and mass, m. We then solve for k

$$a = km$$

$$200 = k(40) \quad \text{Substitute the given values.}$$

Carry Out

$$5 = k$$

Answer Thus $k = 5$ mg. Five milligrams of the drug should be given for each kilogram of a person's mass.

c) Understand and Translate Now that we know the constant of proportionality we can use it to determine the amount of the drug to use for a person's mass. We set up the variation and substitute the values of k and m.

$$a = km$$

$$a = 5(62) \quad \text{Substitute the given values.}$$

Carry Out $$a = 310$$

Answer Thus 310 mg of theophylline should be given to a person whose mass is 62 kg.

EXAMPLE 3 y varies directly as the square of z. If y is 80 when z is 20, find y when z is 90.

Solution Since y varies directly as the *square of* z, we begin with the formula $y = kz^2$. Since the constant of proportionality is not given, we must first find k using the given information.

$$y = kz^2$$

$$80 = k(20)^2 \quad \text{Substitute the given values.}$$

$$80 = 400k \quad \text{Solve for } k$$

$$\frac{80}{400} = \frac{400k}{400}$$

$$0.2 = k$$

We now use $k = 0.2$ to find y when z is 90.

$$y = kz^2$$

$$y = 0.2(90)^2 \quad \text{Substitute the given values.}$$

$$y = 1620$$

NOW TRY EXERCISE 29 Thus, when z equals 90, y equals 1620.

2) Solve Inverse Variation Problems

A second type of variation is **inverse variation.** When two quantities vary inversely, it means that as one quantity increases, the other quantity decreases, and vice versa.

To explain inverse variation, we again use the formula, distance = rate · time. If we solve for time, we get time = distance/rate. Assume that the distance is fixed at 120 miles; then

$$\text{time} = \frac{120}{\text{rate}}$$

At 120 miles per hour it would take 1 hour to cover this distance. At 60 miles an hour, it would take 2 hours. At 30 miles an hour, it would take 4 hours. Note that as the rate (or speed) decreases the time increases, and vice versa.

The equation above can be written

$$t = \frac{120}{r}$$

This equation is an example of an inverse variation. The time and rate are inversely proportional. The constant of proportionality is 120.

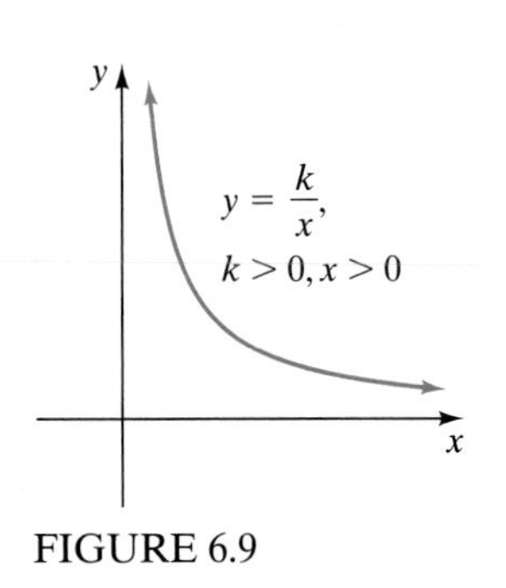

FIGURE 6.9

Inverse Variation

If a variable y varies inversely as a variable x, then

$$y = \frac{k}{x} \qquad (\text{or} \quad xy = k)$$

where k is the constant of proportionality.

Two quantities vary inversely, or are inversely proportional, when as one quantity increases the other quantity decreases. The graph of $y = k/x$, for $k > 0$ and $x > 0$, will have the shape illustrated in Figure 6.9. The graph of an inverse variation is not defined at $x = 0$ because 0 is not in the domain of the function $y = k/x$.

EXAMPLE 4 The amount of time, t, it takes a block of ice to melt in water is inversely proportional to the water's temperature, T.

a) Write this variation as an equation.

b) If a block of ice takes 10 minutes to melt in 75°F water, determine the constant of proportionality.

c) Determine how long it will take a block of ice of the same size to melt in 80°F water.

Solution **a)** The hotter the water temperature, the shorter the time for the block of ice to melt. The inverse variation is

$$t = \frac{k}{T}$$

b) Understand and Translate To determine the constant of proportionality, we substitute the values for the temperature and time and solve for k.

$$t = \frac{k}{T}$$

$$10 = \frac{k}{75} \qquad \textit{Substitute the given values.}$$

Carry Out

$$750 = k$$

Answer The constant of proportionality is 750.

c) Understand and Translate Now that we know the constant of proportionality we can use it to determine how long it will take for the same size block of ice to melt in 80°F water. We set up the proportion, substitute the values for k and T, and solve for t.

$$t = \frac{k}{T}$$

$$t = \frac{750}{80} \qquad \textit{Substitute the given values.}$$

Carry Out

$$t = 9.375$$

Answer It will take 9.375 minutes for the block of ice to melt in the 80°F water.

EXAMPLE 5 The illuminance, I, of a light source varies inversely as the square of the distance, d, from the source. Assuming that the illuminance is 75 units at a distance of 6 meters, find a formula that expresses the relationship between the illuminance and the distance.

Solution **Understand and Translate** Since the illuminance varies inversely as the *square* of the distance, the general form of the equation is

$$I = \frac{k}{d^2} \qquad (\text{or} \quad Id^2 = k)$$

To find k, we substitute the given values for I and d.

$$75 = \frac{k}{6^2} \qquad \text{Substitute the given values}$$

Carry Out

$$75 = \frac{k}{36} \qquad \text{Solve for } k.$$

$$(75)(36) = k$$

$$2700 = k$$

NOW TRY EXERCISE 47

Answer A formula is $I = \dfrac{2700}{d^2}$.

Solve Joint Variation Problems

One quantity may vary as a product of two or more other quantities. This type of variation is called **joint variation.**

Joint Variation

If y varies jointly as x and z, then

$$y = kxz$$

where k is the constant of proportionality.

EXAMPLE 6 The area, A, of a triangle varies jointly as its base, b, and height, h. If the area of a triangle is 48 square inches when its base is 12 inches and its height is 8 inches, find the area of a triangle whose base is 15 inches and height is 20 inches.

Solution **Understand and Translate** First write the joint variation; then substitute the known values and solve for k.

$$A = kbh$$

$$48 = k(12)(8) \qquad \text{Substitute the given values.}$$

Carry Out

$$48 = k(96) \qquad \text{Solve for } k.$$

$$\frac{48}{96} = k$$

$$k = \frac{1}{2}$$

Now solve for the area of the given triangle.

$$A = kbh$$

$$= \frac{1}{2}(15)(20) \qquad \text{Substitute the given values.}$$

$$= 150$$

NOW TRY EXERCISE 51

Answer The area of the triangle is 150 square inches.

Summary of Variations

DIRECT	INVERSE	JOINT
$y = kx$	$y = \dfrac{k}{x}$	$y = kxz$

4) Solve Combined Variation Problems

Often in real-life situations one variable varies as a combination of variables. The following examples illustrate the use of **combined variations.**

EXAMPLE 7 The owners of the Freeport Pretzel Shop find that their weekly sales of pretzels, S, varies directly as their advertising budget, A, and inversely as their pretzel price, P. When their advertising budget is \$400 and the price is \$1, they sell 6200 pretzels.

a) Write an equation of variation expressing S in terms of A and P. Include the value of the constant.

b) Find the expected sales if the advertising budget is \$600 and the price is \$1.20.

Solution **a) Understand and Translate** We begin with the equation

$$S = \frac{kA}{P}$$

$$6200 = \frac{k(400)}{1} \qquad \textit{Substitute the given values.}$$

Carry Out

$$6200 = 400k \qquad \textit{Solve for } k.$$

$$15.5 = k$$

Answer Therefore, the equation for the sales of pretzels is $S = \dfrac{15.5A}{P}$.

b) Understand and Translate Now that we know the combined variation equation, we can use it to determine the expected sales for the given values.

$$S = \frac{15.5A}{P}$$

$$= \frac{15.5(600)}{1.20} \qquad \textit{Substitute the given values.}$$

Carry Out

$$= 7750$$

Answer They can expect to sell 7750 pretzels.

EXAMPLE 8 The electrostatic force, F, of repulsion between two positive electrical charges is jointly proportional to the two charges, q_1 and q_2, and inversely proportional to the square of the distance, d, between the two charges. Express F in terms of q_1, q_2, and d.

Solution

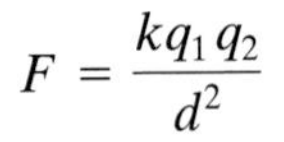

$$F = \frac{kq_1q_2}{d^2}$$

NOW TRY EXERCISE 53

Exercise Set 6.6

Concept/Writing Exercises

1. **a)** Explain what it means when two items vary directly.
   b) Give your own example of two quantities that vary directly.
   c) Write the direct variation for your example in part **b)**.
2. **a)** Explain what it means when two items vary inversely.
   b) Give your own example of two quantities that vary inversely.
   c) Write the inverse variation for your example in part **b)**.
3. What is meant by joint variation?
4. What is meant by combined variation?

Use your intuition to determine whether the variation between the indicated quantities is direct or inverse.

5. The speed and distance traveled by a car in a specified time period.

6. The number of pages a person can read in a fixed period of time and her reading speed.

7. The speed of a runner and the time it takes her to cross the finish line in a race.

8. A person's adjusted gross income and his federal income tax rate.

9. The time it takes an ice cube to melt in water and the temperature of the water.

10. The diameter of a hose and volume of water coming from the hose.

11. The distance between two cities on a map and the actual distance between the two cities.

12. The shutter opening of a camera and the amount of sunlight that reaches the film.

13. A person's weight (due to Earth's gravity) and his distance from Earth.

14. The cubic-inch displacement in liters and the horsepower of the engine.

15. The volume of a balloon and its radius.

16. The length of a board and the force needed to break the board at the center.

17. The light illuminating an object and the distance the light is from the object.

18. The number of calories eaten and the amount of exercise required to burn off those calories.

Practice the Skills

*For Exercises 19–26, **a)** write the variation and **b)** find the quantity indicated.*

19. x varies directly as y. Find x when $y = 12$ and $k = 6$.

20. C varies directly as the square of Z. Find C when $Z = 9$ and $k = \frac{3}{4}$.

21. y varies directly as R. Find y when $R = 180$ and $k = 1.7$.

22. x varies inversely as y. Find x when $y = 25$ and $k = 5$.

23. R varies inversely as W. Find R when $W = 160$ and $k = 8$.

24. L varies inversely as the square of P. Find L when $P = 4$ and $k = 100$.

25. A varies directly as B and inversely as C. Find A when $B = 12$, $C = 4$, and $k = 3$.

26. A varies jointly as R_1 and R_2 and inversely as the square of L. Find A when $R_1 = 120$, $R_2 = 8$, $L = 5$, and $k = \frac{3}{2}$.

*For Exercises 27–34, **a)** write the variation and **b)** find the quantity indicated.*

27. x varies directly as y. If x is 9 when y is 18, find x when y is 36.

28. Z varies directly as W. If Z is 7 when W is 28, find Z when W is 140.

29. y varies directly as the square of R. If y is 5 when $R = 5$, find y when R is 10.

30. S varies inversely as G. If S is 12 when G is 0.4, find S when G is 5.

31. C varies inversely as J. If C is 7 when J is 0.7, find C when J is 12.

32. x varies inversely as the square of P. If $x = 10$ when P is 6, find x when $P = 20$.

33. F varies jointly as M_1 and M_2 and inversely as d. If F is 20 when $M_1 = 5$, $M_2 = 10$, and $d = 0.2$, find F when $M_1 = 10$, $M_2 = 20$, and $d = 0.4$.

34. F varies jointly as q_1 and q_2 and inversely as the square of d. If F is 8 when $q_1 = 2$, $q_2 = 8$, and $d = 4$, find F when $q_1 = 28$, $q_2 = 12$, and $d = 2$.

Problem Solving

35. Assume a varies directly as b. If b is doubled, how will it affect a? Explain.

36. Assume a varies directly as b^2. If b is doubled, how will it affect a? Explain.

37. Assume y varies inversely as x. If x is doubled, how will it affect y? Explain.

38. Assume y varies inversely as a^2. If a is doubled, how will it affect y? Explain.

39. The recommended dosage, d, of the antibiotic drug vancomycin is directly proportional to a person's weight, w. If Carmen Brown, who is 132 pounds is given 2376 mg, find the recommended dosage for Bill Glenn, who is 172 pounds.

40. When converting American dollars into Mexican pesos there is a direct variation between dollars, d, and pesos, p. If on October 1, 1998, Paul Ricardo exchanged \$425 for 3761.25 pesos, how many pesos would he get if he converted \$925 to pesos?

41. The time, t, it takes a car to finish a Winston Cup race is directly proportional to the number of laps, l. If Jeff Gordon covers 62 laps in the Watkins Glen International in 96 minutes, how long will it take him to cover the total 90 laps (assuming he continues his 62-lap pace for the remainder of the race)?

42. The time, t, required to build a brick wall varies inversely as the number of people, n, working on it. If it takes 8 hours for five bricklayers to build a wall, how long will it take eight bricklayers to complete the job?

43. When a car travels at a constant speed, the distance traveled, d, is directly proportional to the time, t. If a car travels 150 miles in 2.5 hours, how far will the same car travel in 4 hours?

44. The time, t, it takes a runner to cover a specified distance is inversely proportional to the runner's speed. If Nhat Chung runs at an average of 6 miles per hour, he will finish a race in 1.4 hours. How long will it take Leif Lundgren, who runs at 5 miles per hour, to finish the same race?

45. The volume of a gas, V, varies inversely as its pressure, P. If the volume, V, is 800 cc when the pressure is 200 millimeters (mm) of mercury, find the volume when the pressure is 25 mm of mercury.

46. Hooke's law states that the length a spring will stretch, S, varies directly with the force (or weight), F, attached to the spring. If a spring stretches 1.4 inches when 20 pounds is attached, how far will it stretch when 10 pounds is attached?

47. When a tennis player serves the ball, the time it takes for the ball to hit the ground in the service box is inversely proportional to the speed the ball is traveling. If Pete Sampras serves at 122 miles per hour, it takes 0.21 seconds for the ball to hit the ground after striking his racquet. How long will it take the ball to hit the ground if he serves at 80 miles per hour?

48. When a ball is pitched in a professional baseball game, the time, t, it takes for the ball to reach home plate varies inversely with the speed, s, of the pitch.* A ball pitched at 90 miles per hour takes 0.459 seconds to reach the plate. How long will it take a ball pitched at 75 miles per hour to reach the plate?

49. On Earth the weight of an object, w, varies directly with its mass, m. If an elephant that weighs 7040 pounds has a mass of 3200 kilograms, find the mass of a koala weighing 18 pounds.

50. The intensity, I, of light received at a source varies inversely as the square of the distance, d, from the source. If the light intensity is 20 foot-candles at 15 feet, find the light intensity at 12 feet.

51. The monthly mortgage payment, P, you pay on a mortgage varies jointly as the interest rate, r, and the amount of the mortgage, m. If the monthly mortgage payment on a \$50,000 mortgage at a 7% interest rate is \$332.50, find the monthly payment on a \$66,000 mortgage at 7%.

52. The weight, w, of an object in Earth's atmosphere varies inversely with the square of the distance, d, between the object and the center of Earth. A 140-pound person standing on Earth is approximately 4000 miles from Earth's center. Find the weight (or gravitational force of attraction) of this person at a distance 100 miles from Earth's surface.

53. The weekly videotape rentals, R, at Busterblock Video vary directly with their advertising budget, A, and inversely with the daily rental price, P. When their advertising budget is \$400 and the rental price is \$2 per day, they rent 4600 tapes per week. How many tapes would they rent per week if they increased their advertising budget to \$500 and raised their rental price to \$2.50?

54. The wattage rating of an appliance, W, varies jointly as the square of the current, I, and the resistance, R. If the wattage is 1 watt when the current is 0.1 ampere and the resistance is 100 ohms, find the wattage when the current is 0.4 ampere and the resistance is 250 ohms.

55. The electrical resistance of a wire, R, varies directly as its length, L, and inversely as its cross-sectional area A. If the resistance of a wire is 0.2 ohm when the length is 200 feet and its cross-sectional area is 0.05 square inch, find the resistance of a wire whose length is 5000 feet with a cross-sectional area of 0.01 square inch.

---

*A ball slows down on its way to the plate due to wind resistance. For a 95 mph pitch, the ball is about 8 mph faster when it leaves the pitcher's hand than when it crosses the plate.

56. The number of phone calls between two cities during a given time period, N, varies directly as the populations p_1 and p_2 of the two cities and inversely to the distance, d, between them. If 100,000 calls are made between two cities 300 miles apart and the populations of the cities are 60,000 and 200,000, how many calls are made between two cities with populations of 125,000 and 175,000 that are 450 miles apart?

57. In a specific region of the country, the amount of a customer's water bill, W, is directly proportional to the average daily temperature for the month, T, the lawn area, A, and the square root of F, where F is the family size, and inversely proportional to the number of inches of rain, R.

In one month, the average daily temperature is 78°F, and the number of inches of rain is 5.6. If the average family of four who have 1000 square feet of lawn pay $68 for water, estimate the water bill in the same month for the average family of six who have 1500 square feet of lawn.

58. An article in the magazine *Outdoor and Travel Photography* states, "If a surface is illuminated by a point-source of light, the intensity of illumination produced is inversely proportional to the square of the distance separating them. In practical terms, this means that foreground objects will be grossly overexposed if your background subject is properly exposed with a flash. Thus direct flash will not offer pleasing results if there are any intervening objects between the foreground and the subject."

If the subject you are photographing is 4 feet from the flash, and the illumination on this subject is $\frac{1}{16}$ of the light of the flash, what is the intensity of illumination on an intervening object that is 3 feet from the flash?

59. One of Newton's laws states that the force of attraction, F, between two masses is directly proportional to the masses of the two objects, m_1 and m_2, and inversely proportional to the square of the distance, d, between the two masses.

a) Write the formula that represents Newton's law.

b) What happens to the force of attraction if one mass is doubled, the other mass is tripled, and the distance between the objects is halved?

60. The pressure, P, in pounds per square inch (psi) on an object x feet below the sea is 14.70 psi plus the product of a constant of proportionality, k, and the number of feet, x, the object is below sea level (see the figure). The 14.70 represents the weight, in pounds, of the column of air (from sea level to the top of the atmosphere) standing over a 1-inch by 1-inch square of ocean. The kx represents the weight, in pounds, of a column of water 1 inch by 1 inch by x feet.

a) Write a formula for the pressure on an object x feet below sea level.

b) If the pressure gauge in a submarine 60 feet deep registers 40.5 psi, find the constant k.

c) A submarine is built to withstand a pressure of 160 psi. How deep can the submarine go?

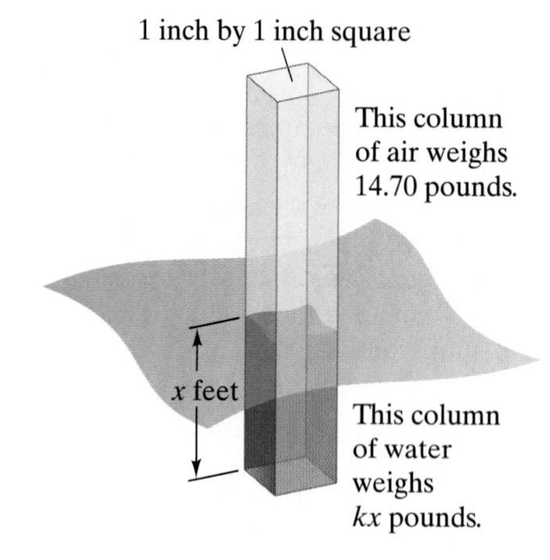
1 inch by 1 inch square
This column of air weighs 14.70 pounds.
x feet
This column of water weighs kx pounds.

Cumulative Review Exercises

[3.4] **61.** Find the variable d if the line through the two given points is to have the given slope: $(5, 1)$ and $(-4, d)$, $m = \frac{2}{3}$.

[3.7] **62.** Graph $2y < 3x - 6$.

[4.3] **63.** Daniel Wilcox, a salesman, earns a base weekly salary plus a commission on his sales. The first week in February, on sales of $5000, his total income was $550. The second week in February, on sales of $8000, his total income was $640. Find his weekly salary and his commission rate.

[4.6] **64.** Graph the inequality $|x - 2| < 4$ in the Cartesian coordinate system.

SUMMARY

Key Words and Phrases

6.1
Domain
Rational expression
Rational function
Simplified (or reduced to lowest terms)

6.2
Least common denominator (LCD)
Prime number

6.3
Complex fraction
Main fraction line

6.4
Extraneous roots or solutions
Rational equation
Similar figures

6.5
Motion problem
Number problem
Work problem

6.6
Combined variations
Constant of proportionality
Direct variation
Inverse variation
Joint variation
Variation

IMPORTANT FACTS

Types of Variation

Direct	Inverse	Joint
$y = kx$	$y = \dfrac{k}{x}$	$y = kxz$

Review Exercises

[6.1] *Determine the values of the variable that must be excluded in each rational expression.*

1. $\dfrac{3}{x - 4}$ **2.** $\dfrac{x}{x + 1}$ **3.** $\dfrac{-2x}{x^2 + 5}$

Determine the domain of each rational function.

4. $y = \dfrac{5}{(x + 3)^2}$ **5.** $f(x) = \dfrac{x + 6}{x^2}$ **6.** $f(x) = \dfrac{x^2 - 2}{x^2 - 3x - 10}$

Simplify each expression.

7. $\dfrac{x^2 + xy}{x + y}$ **8.** $\dfrac{x^2 - 9}{x + 3}$ **9.** $\dfrac{4 - 5x}{5x - 4}$

10. $\dfrac{x^2 + 2x - 3}{x^2 + x - 6}$ **11.** $\dfrac{2x^2 - 6x + 5x - 15}{2x^2 + 7x + 5}$ **12.** $\dfrac{a^3 - 8b^3}{a^2 - 4b^2}$

[6.2] *Find the least common denominator of each expression.*

13. $\dfrac{6x}{x + 1} - \dfrac{3}{x}$ **14.** $\dfrac{9x - 3}{x + y} - \dfrac{4x + 7}{x^2 - y^2}$

15. $\dfrac{19x - 5}{x^2 + 2x - 35} + \dfrac{3x - 2}{x^2 + 9x + 14}$ **16.** $\dfrac{3}{(x + 2)^2} - \dfrac{6(x + 3)}{x^2 - 4} - \dfrac{4x}{x + 1}$

[6.1, 6.2] *Perform each indicated operation.*

17. $\frac{15x^2y^3}{3z} \cdot \frac{6z^3}{5xy^3}$

18. $\frac{1}{x-2} \cdot \frac{2-x}{2}$

19. $\frac{16x^2y^4}{xz^5} \div \frac{2x^2y^4}{x^4z^{10}}$

20. $\frac{4}{2x} + \frac{x}{x^2}$

21. $\frac{4x+4y}{x^2y} \cdot \frac{y^3}{8x}$

22. $\frac{4x^2-11x+4}{x-3} - \frac{x^2-4x+10}{x-3}$

23. $\frac{a-2}{a+3} \cdot \frac{a^2+4a+3}{a^2-a-2}$

24. $\frac{x^2-5x-2}{6x^2-11x-35} - \frac{x^2-7x+5}{6x^2-11x-35}$

25. $\frac{5x}{3xy} - \frac{4}{x^2}$

26. $6 + \frac{x}{x+2}$

27. $5 - \frac{3}{x+3}$

28. $\frac{x^2-y^2}{x-y} \cdot \frac{x+y}{xy+x^2}$

29. $\frac{1}{a^2+8a+15} \div \frac{3}{a+5}$

30. $\frac{a+c}{c} - \frac{a-c}{a}$

31. $\frac{4x^2+8x-5}{2x+5} \cdot \frac{x+1}{4x^2-4x+1}$

32. $(x+3) \div \frac{x^2-4x-21}{x-7}$

33. $\frac{x^2-3xy-10y^2}{6x} \div \frac{x+2y}{12x^2}$

34. $\frac{2}{3x} - \frac{3x}{3x-6}$

35. $\frac{x-4}{x-5} - \frac{3}{x+5}$

36. $\frac{x+3}{x^2-9} + \frac{2}{x+3}$

37. $\frac{1}{a-3} \cdot \frac{a^2-2a-3}{a^2+3a+2}$

38. $\frac{4x^2-16y^2}{9} \div \frac{(x+2y)^2}{12}$

39. $\frac{4}{(x+2)(x-3)} - \frac{4}{(x-2)(x+2)}$

40. $\frac{2x^2+10x+12}{(x+2)^2} \cdot \frac{x+2}{x^3+5x^2+6x}$

41. $\frac{x+2}{x^2-x-6} + \frac{x-3}{x^2-8x+15}$

42. $\frac{x+5}{x^2-15x+50} - \frac{x-2}{x^2-25}$

43. $\frac{y^4-x^6}{x^3-y^2} \div (y^2-x^3)$

44. $\frac{1}{x+3} - \frac{2}{x-3} + \frac{6}{x^2-9}$

45. $\frac{x^3+27}{4x^2-4} \div \frac{x^2-3x+9}{(x-1)^2}$

46. $\frac{x-4}{x-5} - \frac{3}{x+5} - \frac{10}{x^2-25}$

47. $\left(\frac{x^2-8x+16}{2x^2-x-6} \cdot \frac{2x^2-7x-15}{x^2-2x-24}\right) \div \frac{x^2-9x+20}{x^2+2x-8}$

48. $\left(\frac{x^2-x-56}{x^2+14x+49} \cdot \frac{x^2+4x-21}{x^2-9x+8}\right) + \frac{3}{x^2+8x-9}$

49. If $f(x) = \frac{x-4}{x+3}$ and $g(x) = \frac{x}{x+5}$, find

a) the domain of $f(x)$

b) the domain of $g(x)$

c) $(f+g)(x)$

d) the domain of $(f+g)(x)$

50. If $f(x) = \frac{x}{x^2-9}$ and $g(x) = \frac{x+2}{x-3}$, find

a) the domain of $f(x)$

b) the domain of $g(x)$

c) $(f+g)(x)$

d) the domain of $(f+g)(x)$

[6.3] *Simplify each complex fraction.*

51. $\frac{\frac{15xy}{6z}}{\frac{3x}{z^2}}$

52. $\frac{x+\frac{1}{y}}{y^2}$

53. $\frac{\frac{4}{x}+\frac{2}{x^2}}{6-\frac{1}{x}}$

54. $\frac{a^{-1}+2}{a^{-1}+\frac{1}{a}}$

55. $\frac{x^{-2}+\frac{1}{x}}{\frac{1}{x^2}-\frac{1}{x}}$

56. $\frac{\frac{1}{x^2-3x-18}+\frac{2}{x^2-2x-15}}{\frac{3}{x^2-11x+30}+\frac{1}{x^2-9x+20}}$

[6.4] *Solve each equation.*

57. $\frac{3}{x} = \frac{8}{24}$

58. $\frac{5.6}{a} = \frac{14.6}{7.3}$

59. $\frac{x+3}{5} = \frac{9}{5}$

60. $\frac{x}{3.4} = \frac{x-4}{5.2}$

61. $\frac{3x+4}{5} = \frac{2x-8}{3}$

62. $\frac{x}{4.8} + \frac{x}{2} = 1.7$

63. $\frac{4}{x} - \frac{1}{6} = \frac{1}{x}$

64. $\frac{1}{x-2} + \frac{1}{x+2} = \frac{1}{x^2-4}$

65. $\frac{x-3}{x-2} + \frac{x+1}{x+3} = \frac{2x^2+x+1}{x^2+x-6}$

66. $\frac{x}{x^2-9} + \frac{2}{x+3} = \frac{4}{x-3}$

67. Solve $I = \frac{nE}{R+nr}$ for E.

68. Solve $S = \frac{a - ar^n}{1-r}$ for a.

In Exercises 69 and 70, use the formula $\frac{1}{R_T} = \frac{1}{R_1} + \frac{1}{R_2} + \cdots + \frac{1}{R_n}$, *where* $R_1, R_2, \ldots, R_n$ *are the resistances of the individ-*

69. Three resistors of 200, 400, and 1200 ohms, respectively, are connected in parallel. Find the total resistance of the circuit.

70. Two resistors are to be connected in parallel. One is to contain twice the resistance of the other. What should be the resistance of each resistor if the circuit's total resistance is to be 600 ohms?

In Exercises 71 and 72, use the formula $\frac{1}{p} + \frac{1}{q} = \frac{1}{f}$, *where p represents the distance of the object from the mirror, q represents the distance of the image from the mirror, and f represents the focal length of the mirror.*

71. What is the focal length of a curved mirror if the object distance is 12 centimeters and the image distance is 4 centimeters?

72. The focal length of a curved mirror is 10 centimeters. Find the object's distance from the lens if the image distance is twice the object's distance.

[6.5] *Write an equation that can be used to solve each problem. Then answer the question asked.*

73. It takes Dan Moore 3 hours to mow Mr. Lee's lawn. It takes Kim Stevens 4 hours to mow the same lawn. How long will it take them working together to mow Mr. Lee's lawn?

74. Annette Cello and Pete Baird are both copy editors for a publishing company. Together they can edit a 500-page manuscript in 40 hours. If Annette by herself can edit the manuscript in 75 hours, how long will it take Pete to edit the manuscript by himself?

75. What number multiplieÏd by the numerator and added to the denominator of the fraction $\frac{5}{8}$ makes the result equal to 1?

76. When the reciprocal of twice a number is subtracted from 1, the result is the reciprocal of 3 times the number. Find the number.

77. Danesha Neville's motorboat can travel 15 miles per hour in still water. Traveling with the current of a river, the boat can travel 20 miles in the same time it takes to go 10 miles against the current. Find the current.

78. A small plane and a car start from the same location, at the same time, heading toward the same town 450 miles away. The speed of the plane is 3 times the speed of the car. The plane arrives at the town 6 hours ahead of the car. Find the speeds of the car and the plane.

[6.6] *Find each indicated quantity.*

79. A is directly proportional to the square of C. If A is 5 when C is 5, find A when $C = 10$.

80. W is directly proportional to L and inversely proportional to A. If $W = 80$ when $L = 100$ and $A = 20$, find W when $L = 50$ and $A = 40$.

81. z is jointly proportional to x and y and inversely proportional to the square of r. If $z = 12$ when $x = 20$, $y = 8$, and $r = 8$, find z when $x = 10$, $y = 80$, and $r = 3$.

82. The scale of a map is 1 inch to 60 miles. How large a distance on the map represents 300 miles?

83. An electric company charges $0.162 per kilowatthour. What is the electric bill if 740 kilowatthours are used in a month?

84. The distance, d, an object drops in free fall is directly proportional to the square of the time, t. If an object falls 16 feet in 1 second, how far will an object fall in 5 seconds?

85. The area, A, of a circle varies directly with the square of its radius, r. If the area is 78.5 when the radius is 5, find the area when the radius is 8.

86. The time, t, for an ice cube to melt is inversely proportional to the temperature of the water it is in. If it takes an ice cube 1.7 minutes to melt in 70°F water, how long will it take an ice cube of the same size to melt in 50°F water?

Practice Test

1. Determine the values that are excluded in the expression $\dfrac{x+4}{x^2-3x-28}$.

2. Determine the domain of the function $f(x)=\dfrac{x^2-4x}{8x^2+10x-25}$.

Simplify each expression.

3. $\dfrac{8x^7y^2+16x^2y+18x^3y^3}{2x^2y}$

4. $\dfrac{4x^2+4xy-15y^2}{4x^2-4xy-3y^2}$

Perform each indicated operation.

5. $\dfrac{3xy^4}{6x^2y^3}\cdot\dfrac{2x^2y^4}{x^5y^7}$

6. $\dfrac{a^2-9a+14}{a-2}\cdot\dfrac{a^2-4a-21}{(a-7)^2}$

7. $\dfrac{x^2-9y^2}{3x+6y}\div\dfrac{x+3y}{x+2y}$

8. $\dfrac{x^3+y^3}{x+y}\div\dfrac{x^2-xy+y^2}{x^2+y^2}$

9. $\dfrac{5}{x}+\dfrac{3}{2x^2}$

10. $\dfrac{x-5}{x^2-16}-\dfrac{x-2}{x^2+2x-8}$

11. $\dfrac{x+1}{4x^2-4x+1}+\dfrac{3}{2x^2+5x-3}$

12. $\dfrac{m}{12m^2+4mn-5n^2}+\dfrac{2m}{12m^2+28mn+15n^2}$

13. $\dfrac{r^2-16}{r^3-64}\div\dfrac{r^2+2r-8}{r^2+4r+16}$

If $f(x)=\dfrac{x-3}{x+5}$ *and* $g(x)=\dfrac{x}{2x+3}$, *find*

14. $(f+g)(x)$

15. $(f-g)(x)$

16. the domain of $(f+g)(x)$

Simplify.

17. $\dfrac{\frac{1}{x}+\frac{1}{y}}{\frac{1}{x}-\frac{1}{y}}$

18. $\dfrac{x+\frac{x}{y}}{x^{-1}+y^{-1}}$

Solve each equation.

19. $\dfrac{x}{3}-\dfrac{x}{4}=5$

20. $\dfrac{x}{x-8}+\dfrac{6}{x-2}=\dfrac{x^2}{x^2-10x+16}$

21. Solve the formula $I=\dfrac{2R}{w+2s}$ for s.

22. The wattage rating of an appliance, W, varies jointly as the square of the current, I, and the resistance, R. If the wattage is 10 when the current is 1 ampere and the resistance is 1000 ohms, find the wattage when the current is 0.5 ampere and the resistance is 300 ohms.

23. W varies jointly as P and Q and inversely as the square of T. If $W=6$ when $P=20$, $Q=8$, and $T=4$, find W if $P=30$, $Q=4$, and $T=8$.

24. Kris Murphy can level a 1-acre field in 8 hours on his tractor. Heather Meridy can level a 1-acre field in 5 hours on her tractor. How long will it take them to level a 1-acre field if they work together?

25. Cameron Barnette and Ashley Elliot start jogging at the same time at the beginning of a trail. Cameron averages 8 miles per hour, while Ashley averages 5 miles per hour. If it takes Ashley $\frac{1}{2}$ hour longer than Cameron to reach the end of the trail, how long is the trail?

Cumulative Review Test

1. Illustrate the set $\left\{x \middle| -\frac{5}{3} < x \le \frac{19}{4}\right\}$.

2. Evaluate $-3x^3 - 2x^2y + \frac{1}{2}xy^2$ when $x = 2$ and $y = \frac{1}{2}$.

3. Simplify $\left(\frac{4a^3b^{-2}}{2a^{-2}b^{-3}}\right)^{-2}$.

4. In 1998, the federal government's total receipts were about $\$1.631 \times 10^{12}$. The percent received from various sources is shown in the graph.

Federal Government Total Receipts

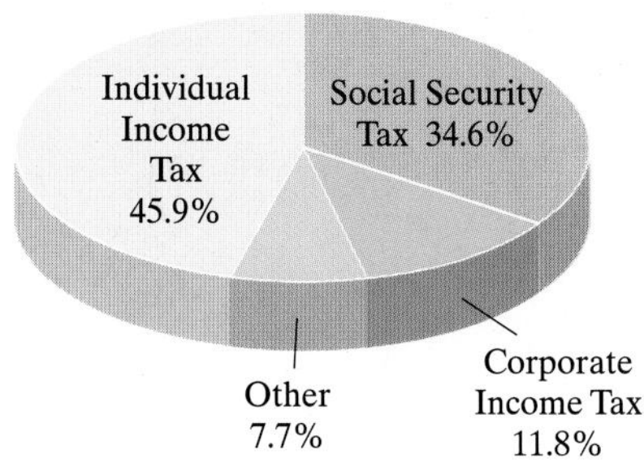

Source: U.S. Office of Management and Budget

(Write your answers in scientific notation.)

a) How much income tax was collected from individuals in 1998?

b) How much income tax was collected from corporations in 1998?

c) How much more was collected from individual income tax than from corporate income tax?

5. Solve the equation
$4[3x - 2(2x - 4)] = -[6x - 4 - (3x + 8)]$

6. Evaluate $\frac{-b + \sqrt{b^2 - 4ac}}{2a}$ when $a = 1, b = -14$, and $c = 48$.

7. Conchita Gonzales, a financial planner, offers her clients two payment plans. Plan 1 is 6% of the client's assets under management and plan 2 is an annual fee of \$500 plus 4% of the client's assets under management.

a) How much in assets would a client have to have under management for the two plans to have the same fee?

b) If a client has \$60,000 assets under management, which plan would give the lower fee?

8. A bottling machine can be run at two different speeds. At the faster speed the machine fills 400 bottles per hour more than it does at the slower speed. The machine runs for 3.6 hours at the faster speed, then it is changed to the slower speed and runs for 4.4 hours. During the 8-hour period a total of 25,440 bottles were filled. Find the faster speed, in bottles filled per hour.

9. Graph $y = x^2 - 2$.

10. Graph $y = |x| + 2$.

11. Indicate which of the following relations are functions.

a) $\{(4, 2), (-5, 3), (6, 3), (5, 3)\}$

b) **c)**

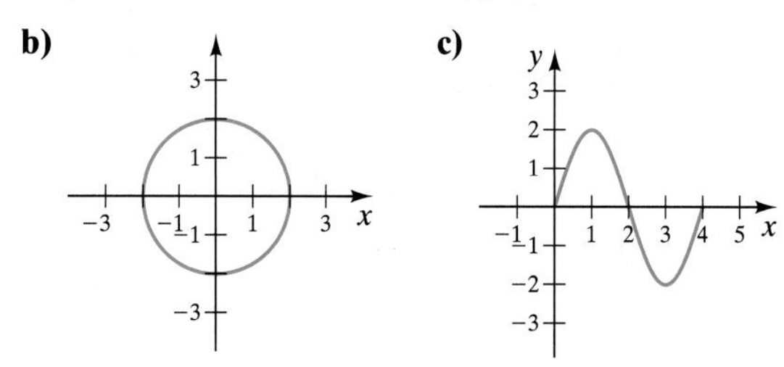

12. Find the equation of the line whose graph has a y-intercept of $(0, -2)$ and is parallel to the graph of $3x - 5y = 7$.

13. Find the equation of the line whose graph passes through $(3, -2)$ and is perpendicular to the graph of $4x - 3y = 12$.

14. The following graph shows the number of boys and girls participating in Florida high school sports for school years from 1987 through 1996. Let $G(t)$ represent the number of girls as a function of time and $B(t)$ represent the number of boys as a function of time.

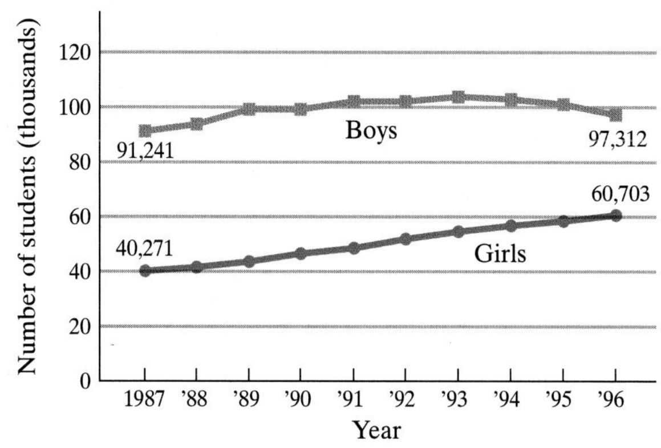

Source: Florida High School Activities Association

a) Determine $(G + B)(96)$.

b) Determine $(B - G)(96)$.

c) Explain how you would graph $(G + B)(t)$.

15. Solve the the following system of equations.

$$2a - b - 2c = -1$$
$$a - 2b - c = 1$$
$$a + b + c = 4$$

16. John Shumack has a 10% sulfuric acid solution and a 20% sulfuric acid solution. How much of each solution should he mix to get 4 liters of a 12% sulfuric acid solution?

17. Given $f(x) = x^2 - 11x + 30$ and $g(x) = x - 5$, find

a) $(f \cdot g)(x)$

b) $(f/g)(x)$

c) the domain of $(f/g)(x)$

18. Factor $24p^3q + 16p^2q - 30pq$.

19. Solve $\frac{5}{c} = 2 - \frac{2c}{c + 1}$.

20. Solve $\frac{4}{r + 5} + \frac{1}{r + 3} = \frac{2}{r^2 + 8r + 15}$.

ROOTS, RADICALS, AND COMPLEX NUMBERS

CHAPTER

7

Use the Angel Web site at www.prenhall.com/angel to be linked to an internet resource that will help you further explore the following application.

The length of a skid mark depends on the speed of the car when braking began and the road conditions. When determining the cause of an accident, investigators use a formula which relates the length of the skid mark and a numerical description of the road conditions to find the speed of the car when braking began. On page 445 you will use this formula to determine speed of a car which leaves an 80-foot skid mark.

Preview and Perspective

In this chapter we explain how to add, subtract, multiply, and divide radical expressions. We also introduce imaginary numbers and complex numbers.

In Section 7.1 we graph square root and cube root functions. In Section 7.2 we change expressions from radical form to exponential form, and vice versa. The rules of exponents discussed in Section 1.5 still apply to rational exponents and we use those rules again here.

In Section 7.4 we discuss rationalizing the denominator, which removes radicals from a denominator. Make sure that you understand the three requirements for a radical expression to be simplified, as discussed in Section 7.4.

In Section 7.6 we discuss how to solve equations that contain radical expressions. We will use these procedures again in Chapters 8 and 10. Section 7.6 also illustrates applications of radical equations.

Imaginary numbers and complex numbers are introduced in Section 7.7. These numbers play a very important role in higher mathematics courses. We will be using imaginary and complex numbers throughout Chapter 8.

7.1 ROOTS AND RADICALS

SSM

VIDEO 7.1

CD Rom

1) **Find square roots.**
2) **Find cube roots.**
3) **Understand odd and even roots.**
4) **Evaluate radicals using absolute value.**

In this chapter we expand on the concept of radicals introduced in Chapter 1. In the expression $\sqrt{x}$, the $\sqrt{}$ is called the **radical sign**. The expression within the radical sign is called the **radicand**.

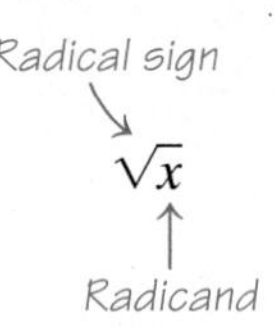

The entire expression, including the radical sign and radicand, is called the **radical expression**. Another part of the radical expression is its index. The **index** (plural indices) gives the "root" of the expression. Square roots have an index of 2. The index of a square root is generally not written. Thus,

$$\sqrt{x} \quad \text{means} \quad \sqrt[2]{x}$$

1) Find Square Roots

Every positive number has two square roots, a principal or positive square root and a negative square root. For any positive number x, the positive square root is written $\sqrt{x}$, and the negative square root is written $-\sqrt{x}$.

Number	Principal or Positive Square Root	Negative Square Root
25	$\sqrt{25}$	$-\sqrt{25}$
10	$\sqrt{10}$	$-\sqrt{10}$

Definition

The **principal square root** of a positive number a, written $\sqrt{a}$, is the *positive* number b such that $b^2 = a$.

Examples

$\sqrt{25} = 5$ since $5^2 = 5 \cdot 5 = 25$

$\sqrt{0.36} = 0.6$ since $(0.6)^2 = (0.6)(0.6) = 0.36$

$\sqrt{\frac{4}{9}} = \frac{2}{3}$ since $\left(\frac{2}{3}\right)^2 = \left(\frac{2}{3}\right)\left(\frac{2}{3}\right) = \frac{4}{9}$

In this book whenever we use the words *square root* we will be referring to the principal or positive square root. Thus if you are asked to find the value of $\sqrt{25}$, your answer will be 5.

In Chapter 1 we indicated that a rational number is one that can be represented as either a terminating or repeating decimal number. If you use the square root key on your calculator, $\boxed{\sqrt{}}$, to evaluate the three examples above you will find they are all terminating or repeating decimal numbers. Thus they are all *rational numbers*. Many radicals, such as $\sqrt{2}$ and $\sqrt{10}$ are not rational numbers. When evaluating $\sqrt{2}$ and $\sqrt{10}$ on a calculator, the results are nonterminating, nonrepeating decimal numbers. Thus $\sqrt{2}$ and $\sqrt{10}$ are *irrational numbers*.

Radical	Calculator Results	
$\sqrt{2}$	1.414213562	*Nonterminating, nonrepeating decimal*
$\sqrt{10}$	3.16227766	*Nonterminating, nonrepeating decimal*

Now let's consider $\sqrt{-25}$. Since the square of any real number will always be greater than or equal to 0, there is no real number that when squared equals −25. For this reason, $\sqrt{-25}$ is *not a real number*. Since the square of any real number cannot be negative, *the square root of a negative number is not a real number*. If you evaluate $\sqrt{-25}$ on a calculator you will get an error message. We will discuss numbers like $\sqrt{-25}$ later in this chapter.

Radical	Calculator Results	
$\sqrt{-25}$	Error	*$\sqrt{-25}$ is not a real number.*
$\sqrt{-2}$	Error	*$\sqrt{-2}$ is not a real number.*

The Square Root Function

When graphing square root functions, functions of the form $f(x) = \sqrt{x}$, we must always remember that the radicand, x, cannot be negative. Thus the domain of $f(x) = \sqrt{x}$ is $\{x \mid x \geq 0\}$, or $[0, \infty)$ in interval notation. To graph $f(x) = \sqrt{x}$, we can select some convenient values of x and find the corresponding values of $f(x)$ or y and then plot the ordered pairs, as shown in Figure 7.1.

x	y
0	0
1	1
4	2
9	3

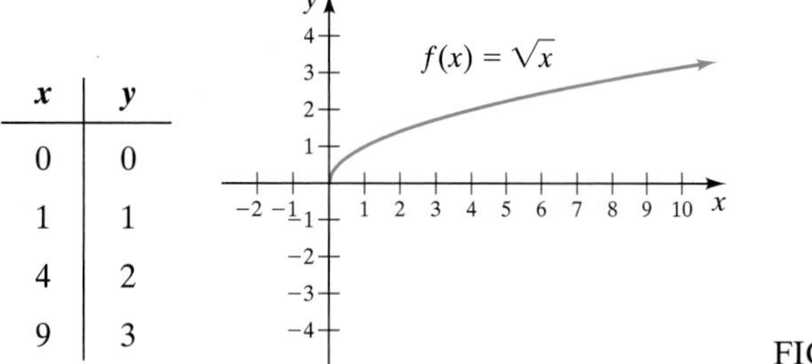

FIGURE 7.1

Since the value of $f(x)$ can never be negative, the range of $f(x) = \sqrt{x}$ is $\{y \mid y \geq 0\}$ or $[0, \infty)$.

By studying Figure 7.1, do you think you can you graph $g(x) = -\sqrt{x}$? (See Exercise 94.) What about the graph of $h(x) = \sqrt{x - 4}$? For the graph of $h(x) = \sqrt{x - 4}$, you would only select values of $x \geq 4$ since the radicand cannot be negative. The domain of $f(x) = \sqrt{x - 4}$ is $\{x | x \geq 4\}$ or $[4, \infty)$.

To evaluate radical functions, it may be necessary to use a calculator.

EXAMPLE 1 For each function, find the indicated value.

a) $f(x) = \sqrt{12x - 20}$, $f(3)$ **b)** $g(r) = -\sqrt{3r + 19}$, $g(-4)$

Solution **a)** $f(3) = \sqrt{12(3) - 20}$ *Substitute 3 for x.*

$= \sqrt{16}$

$= 4$

b) $g(-4) = -\sqrt{3(-4) + 19}$ *Substitute −4 for r.*

$= -\sqrt{7}$

≈ -2.645751311 *From a calculator*

2) Find Cube Roots

The index of a cube root is 3. We introduced cube roots in Section 1.4, where we explained how to find cube roots on a calculator. You may wish to review that material now.

Definition

The **cube root** of a number a, written $\sqrt[3]{a}$, is the number b such that $b^3 = a$.

Examples

$\sqrt[3]{8} = 2$ since $2^3 = 8$

$\sqrt[3]{-27} = -3$ since $(-3)^3 = -27$

For each real number, there is only one cube root. The cube root of a positive number is positive and the cube root of a negative number is negative. The cube root function $f(x) = \sqrt[3]{x}$ has all real numbers as its domain.

EXAMPLE 2 For each function, find the indicated value.

a) $f(x) = \sqrt[3]{10x + 14}$, $f(5)$ **b)** $g(r) = \sqrt[3]{12x - 20}$, $g(-4)$

Solution **a)** $f(5) = \sqrt[3]{10(5) + 14}$ *Substitute 5 for x.*

$= \sqrt[3]{64}$

$= 4$

b) $g(-4) = \sqrt[3]{12(-4) - 20}$ *Substitute −4 for r.*

$= \sqrt[3]{-68}$

≈ -4.081655102 *From a calculator*

NOW TRY EXERCISE 67

The Cube Root Function

Figure 7.2 shows the graph of $y = \sqrt[3]{x}$. To obtain this graph, we substituted values for x and found the corresponding values of $f(x)$ or y.

x	y
−8	−2
−1	−1
0	0
1	1
8	2

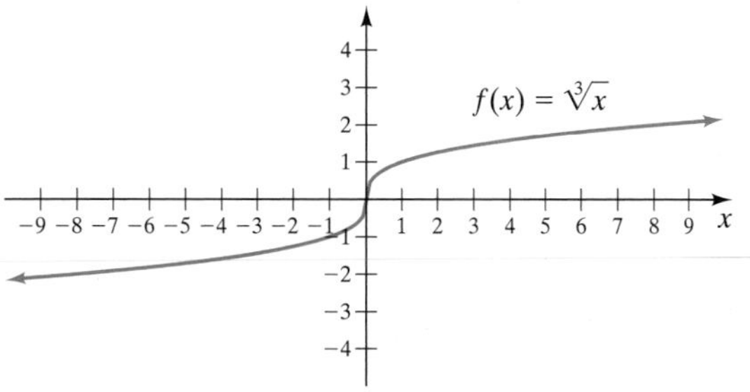

$f(x) = \sqrt[3]{x}$

FIGURE 7.2

Notice that both the domain and range are all real numbers, ℝ. We will ask you to graph cube root functions on your graphing calculator in the exercise set.

3) Understand Odd and Even Roots

Up to this point we have discussed square and cube roots. Other radical expressions have different indices. For example, in the expression $\sqrt[5]{xy}$, (read "the fifth root of xy") the index is 5 and the radicand is xy.

Radical expressions that have indices of 2, 4, 6, ... or any even integer are **even roots**. Square roots are even roots since their index is 2. Radical expressions that have indices of 3, 5, 7, ... or any odd integer are **odd roots**.

Even Indices

The nth root of a, $\sqrt[n]{a}$, where n is an *even index* and a is a nonnegative real number, is the nonnegative real number b such that $b^n = a$.

Examples of Even Roots

$\sqrt{9} = 3$ since $3^2 = 3 \cdot 3 = 9$

$\sqrt[4]{16} = 2$ since $2^4 = 2 \cdot 2 \cdot 2 \cdot 2 = 16$

$\sqrt[6]{729} = 3$ since $3^6 = 3 \cdot 3 \cdot 3 \cdot 3 \cdot 3 \cdot 3 = 729$

Any real number when raised to an even power results in a positive real number. Thus, *when the index of a radical is even, the radicand must be nonnegative for the radical to be a real number.*

HELPFUL HINT

There is an important difference between $-\sqrt[4]{16}$ and $\sqrt[4]{-16}$. The number $-\sqrt[4]{16}$ is the opposite of $\sqrt[4]{16}$. Since $\sqrt[4]{16} = 2$, $-\sqrt[4]{16} = -2$. However, $\sqrt[4]{-16}$ is not a real number since no real number when raised to the fourth power equals −16.

$$-\sqrt[4]{16} = -(\sqrt[4]{16}) = -2$$

$\sqrt[4]{-16}$ is not a real number

Odd Indices

The nth root of a, $\sqrt[n]{a}$, where n is an *odd index* and a is *any real number*, is the real number b such that $b^n = a$.

Examples of Odd Roots

$\sqrt[3]{8} = 2$ since $2^3 = 2 \cdot 2 \cdot 2 = 8$

$\sqrt[3]{-8} = -2$ since $(-2)^3 = (-2)(-2)(-2) = -8$

$\sqrt[5]{243} = 3$ since $3^5 = 3 \cdot 3 \cdot 3 \cdot 3 \cdot 3 = 243$

$\sqrt[5]{-243} = -3$ since $(-3)^5 = (-3)(-3)(-3)(-3)(-3) = -243$

An odd root of a positive number is a positive number, and an odd root of a negative number is a negative number.

It is important to realize that a radical with an even index must have a nonnegative radicand if it is to be a real number. A radical with an odd index will be a real number with any real number as its radicand. Note that $\sqrt[n]{0} = 0$, regardless of whether n is an odd or even index.

EXAMPLE 3 Indicate whether each radical expression is a real number. If the expression is a real number, find its value.

a) $\sqrt[4]{-81}$ **b)** $-\sqrt[4]{81}$ **c)** $\sqrt[5]{-32}$ **d)** $-\sqrt[5]{-32}$

Solution **a)** Not a real number. Even roots of negative numbers are not real numbers.

b) Real number, $-\sqrt[4]{81} = -(\sqrt[4]{81}) = -(3) = -3$

c) Real number, $\sqrt[5]{-32} = -2$ since $(-2)^5 = -32$

NOW TRY EXERCISE 21

d) Real number, $-\sqrt[5]{-32} = -(-2) = 2$

Table 7.1 summarizes the information about even and odd roots.

TABLE 7.1

	***n* is even**	***n* is odd**
$a > 0$	$\sqrt[n]{a}$ is a positive real number.	$\sqrt[n]{a}$ is a positive real number.
$a < 0$	$\sqrt[n]{a}$ is not a real number.	$\sqrt[n]{a}$ is a negative real number.
$a = 0$	$\sqrt[n]{0} = 0$	$\sqrt[n]{0} = 0$

4 Evaluate Radicals Using Absolute Value

You may think that $\sqrt{a^2} = a$, but this is not necessarily true. Below we evaluate $\sqrt{a^2}$ for $a = 2$ and $a = -2$. You will see that when $a = -2$, $\sqrt{a^2} \neq a$.

$a = 2$: $\sqrt{a^2} = \sqrt{2^2} = \sqrt{4} = 2$ Note that $\sqrt{2^2} = 2$

$a = -2$: $\sqrt{a^2} = \sqrt{(-2)^2} = \sqrt{4} = 2$ Note that $\sqrt{(-2)^2} \neq -2$

By examining these examples and other examples we could make up, we can reason that $\sqrt{a^2}$ *will always be a positive real number* for any nonzero real number a. Recall from Section 1.3 that the *absolute value* of any real number a, or $|a|$, is also a positive number for any nonzero number. We use these facts to reason that

For any real number a,

$$\sqrt{a^2} = |a|$$

This indicates that the principal square root of a^2 is the absolute value of a.

EXAMPLE 4 Use absolute value to evaluate. **a)** $\sqrt{5^2}$ **b)** $\sqrt{0^2}$ **c)** $\sqrt{(-71)^2}$

Solution **a)** $\sqrt{5^2} = |5| = 5$ **b)** $\sqrt{0^2} = |0| = 0$ **c)** $\sqrt{(-71)^2} = |-71| = 71$

When simplifying a square root, if the radicand contains a variable and we are not sure that the radicand is positive, we need to use absolute value signs when simplifying.

EXAMPLE 5 Simplify. **a)** $\sqrt{(x+2)^2}$ **b)** $\sqrt{9x^2}$ **c)** $\sqrt{25x^8}$ **d)** $\sqrt{x^2 - 6x + 9}$

Solution Each square root has a radicand that contains a variable. Since we do not know the value of the variable, we do not know whether the radicand is positive or negative. Therefore we must use absolute value signs when simplifying.

a) $\sqrt{(x+2)^2} = |x+2|$

b) Write $9x^2$ as $(3x)^2$, then simplify.

$$\sqrt{9x^2} = \sqrt{(3x)^2} = |3x|$$

c) Write $25x^8$ as $(5x^4)^2$, then simplify.

$$\sqrt{25x^8} = \sqrt{(5x^4)^2} = |5x^4|$$

NOW TRY EXERCISE 59

d) Notice that $x^2 - 6x + 9$ is a perfect square trinomial. Write the trinomial as the square of a binomial, then simplify.

$$\sqrt{x^2 - 6x + 9} = \sqrt{(x-3)^2} = |x-3|$$

If you have a square root whose radicand contains a variable and are given instructions like "Assume all variables represent positive values and the radicand is nonnegative," then it is not necessary to use the absolute value sign when simplifying.

EXAMPLE 6 Simplify. Assume all variables represent positive values and the radicand is nonnegative.

a) $\sqrt{36x^2}$ **b)** $\sqrt{81r^4}$ **c)** $\sqrt{4x^2 - 12xy + 9y^2}$

Solution

a) $\sqrt{36x^2} = \sqrt{(6x)^2} = 6x$ — *Write $36x^2$ as $(6x)^2$.*

b) $\sqrt{81r^4} = \sqrt{(9r^2)^2} = 9r^2$ — *Write $81r^4$ as $(9r^2)^2$.*

c) $\sqrt{4x^2 - 12xy + 9y^2} = \sqrt{(2x-3y)^2}$ — *Write $4x^2 - 12xy + 9y^2$ as $(2x-3y)^2$.*

$= 2x - 3y$

We only need to be concerned about adding absolute value signs when discussing square (and other even) roots. We do not need to be concerned with absolute value signs when the index is odd.

Exercise Set 7.1

Concept/Writing Exercises

1. **a)** How many square roots does every positive real number have? Name them.
   b) Find all square roots of the number 36.
   c) In this text, when we refer to "the square root," which square root are we referring to?
   d) Find the square root of 36.
2. **a)** What are even roots? Give an example of an even root.
   b) What are odd roots? Give an example of an odd root.
3. Explain why $\sqrt{-49}$ is not a real number.
4. Will a radical expression with an odd index and a real number as the radicand always be a real number? Explain your answer.
5. Will a radical expression with an even index and a real number as the radicand always be a real number? Explain your answer.
6. **a)** To what is $\sqrt{a^2}$ equal?
   b) To what is $\sqrt{a^2}$ equal if we know $a \geq 0$?

Practice the Skills

Evaluate each radical expression if it is a real number. Use a calculator to approximate irrational numbers to the nearest hundredth. If the expression is not a real number, so state.

7. $\sqrt{49}$
8. $\sqrt[3]{27}$
9. $\sqrt[3]{-64}$
10. $\sqrt[5]{32}$
11. $\sqrt[3]{125}$
12. $\sqrt[4]{81}$
13. $\sqrt{-9}$
14. $\sqrt[6]{64}$
15. $\sqrt[3]{216}$
16. $\sqrt[5]{-1}$
17. $\sqrt[3]{-343}$
18. $\sqrt[3]{343}$
19. $\sqrt[4]{16}$
20. $\sqrt[4]{-16}$
21. $\sqrt{-36}$
22. $-\sqrt{-25}$
23. $-\sqrt[5]{102.4}$
24. $\sqrt{1600}$
25. $\sqrt{\frac{25}{9}}$
26. $\sqrt[3]{\frac{1}{8}}$
27. $\sqrt[4]{-81}$
28. $\sqrt[4]{-50}$
29. $\sqrt[5]{16.2}$
30. $-\sqrt{92.6}$

Use absolute value to evaluate.

31. $\sqrt{4^2}$
32. $\sqrt{(-6)^2}$
33. $\sqrt{(-1)^2}$
34. $\sqrt{(-17)^2}$
35. $\sqrt{(43)^2}$
36. $\sqrt{(-96)^2}$
37. $\sqrt{(235.23)^2}$
38. $\sqrt{(-147.23)^2}$
39. $\sqrt{(-0.03)^2}$
40. $\sqrt{(-57)^2}$
41. $\sqrt{\left(-\frac{162}{5}\right)^2}$
42. $\sqrt{\left(\frac{40}{9}\right)^2}$

Write as an absolute value.

43. $\sqrt{(a-9)^2}$
44. $\sqrt{(x-7)^2}$
45. $\sqrt{(x-3)^2}$
46. $\sqrt{(3x^2-y)^2}$
47. $\sqrt{(3x+5)^2}$
48. $\sqrt{(x^2-3x+4)^2}$
49. $\sqrt{(6-3x)^2}$
50. $\sqrt{(4-5x^2)^2}$
51. $\sqrt{(y^2-4y+3)^2}$
52. $\sqrt{(x^2-3x)^2}$
53. $\sqrt{(8a-b)^2}$
54. $\sqrt{(3w^4-4w)^2}$

Use absolute value to simplify. You may need to factor first.

55. $\sqrt{a^8}$
56. $\sqrt{r^{12}}$
57. $\sqrt{a^{24}}$
58. $\sqrt{x^{104}}$
59. $\sqrt{a^2-8a+16}$
60. $\sqrt{a^2+2ab+b^2}$
61. $\sqrt{x^2-8x+16}$
62. $\sqrt{9a^2-12ab+4b^2}$

Find the indicated value of each function. Use your calculator to approximate irrational numbers. Round irrational numbers to the nearest thousandth.

63. $f(x)=\sqrt{5x-6}, f(2)$
64. $f(x)=\sqrt{14x-36}, f(4)$
65. $f(x)=\sqrt{64-8x}, f(-3)$
66. $f(x)=\sqrt[3]{8x-8}, f(2)$
67. $f(c)=\sqrt[3]{9c^2-4}, f(4)$
68. $g(x)=\sqrt[4]{16x-5}, g(4)$
69. $g(x)=\sqrt[3]{-3x^2+6x-1}, g(-3)$
70. $h(r)=\sqrt[3]{6r^2-20r-12}, h(-2)$

Problem Solving

71. Select a value for x for which $\sqrt{(2x-1)^2} \neq 2x-1$.

72. Select a value for x for which $\sqrt{(5x-3)^2} \neq 5x-3$.

73. For what values of x will $\sqrt{(x-1)^2} = x-1$? Explain how you determined your answer.

74. For what values of x will $\sqrt{(x+4)^2} = x+4$? Explain how you determined your answer.

75. For what values of x will $\sqrt{(2x-6)^2} = 2x-6$? Explain how you determined your answer.

76. For what values of x will $\sqrt{(4x-4)^2} = 4x-4$? Explain how you determined your answer.

77. **a)** For what values of a is $\sqrt{a^2} = |a|$?
b) For what values of a is $\sqrt{a^2} = a$?

78. Under what circumstances is the expression $\sqrt[n]{x}$ not a real number?

79. Explain why the expression $\sqrt[n]{x^n}$ is a real number for any real number x.

80. Under what circumstances is the expression $\sqrt[n]{x^m}$ not a real number?

81. Find the domain of $\frac{\sqrt{x+3}}{\sqrt[3]{x+3}}$. Explain how you determined your answer.

82. Find the domain of $\frac{\sqrt[3]{x-2}}{\sqrt[6]{x+4}}$. Explain how you determined your answer.

By considering the domains of the functions in Exercises 83 through 86, match each function with its graph, labeled ***a)*** *through* ***d)***.

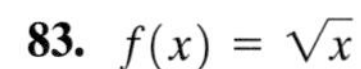

83. $f(x) = \sqrt{x}$

84. $f(x) = \sqrt{x^2}$

85. $f(x) = \sqrt{x - 5}$

86. $f(x) = \sqrt{x + 5}$

a)

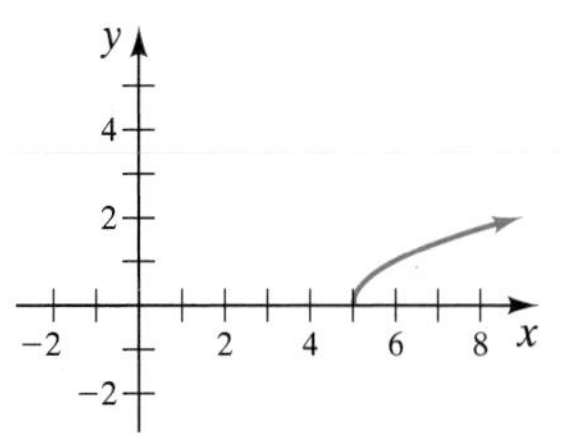

b)

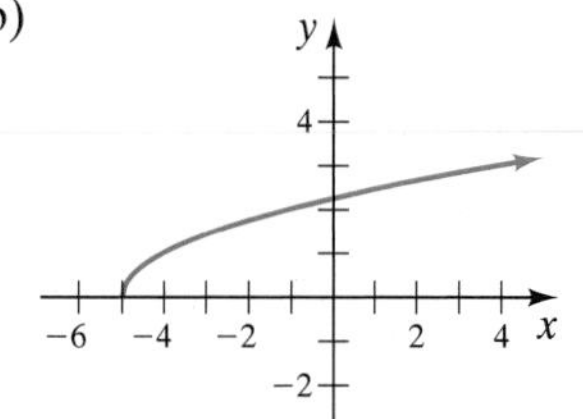

c)

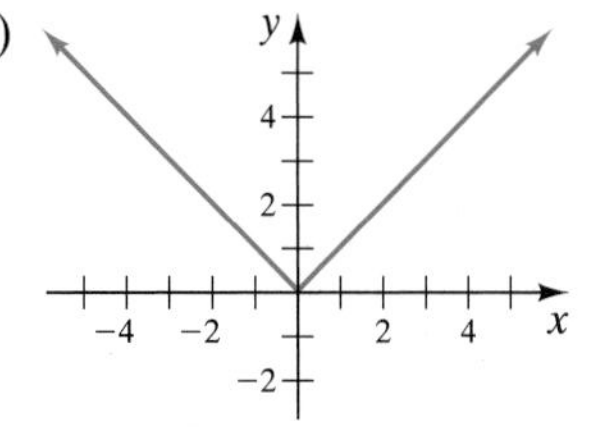

d)

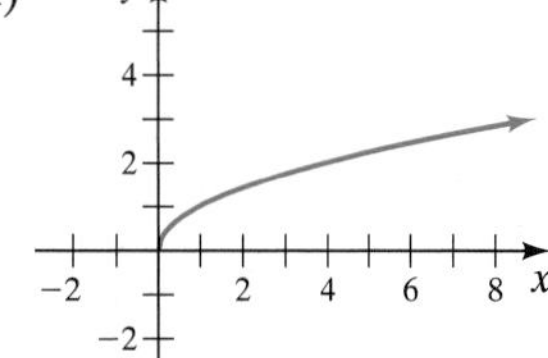

87. Give a radical function whose domain is $\{x | x \geq 4\}$.

88. Give a radical function whose domain is $\{x | x \leq 2\}$.

89. If $f(x) = -\sqrt{x}$, can $f(x)$ ever be **a)** greater than 0, **b)** equal to 0, **c)** less than 0? Explain.

90. If $f(x) = \sqrt{x + 5}$, can $f(x)$ ever be **a)** less than 0, **b)** equal to 0, **c)** greater than 0? Explain.

91. The velocity, V, of an object, in feet per second, after it has fallen a distance, h, in feet, can be found by the formula $V = \sqrt{64.4h}$. A pile driver is a large mass that is used as a hammer to drive pilings into soft earth to support a building or other structure.

With what velocity will the hammer hit the piling if it falls from **a)** 20 feet above the top of the piling, **b)** 40 feet above the top of the piling?

92. Scripps Institute of Oceanography in La Jolla, California, developed the formula for relating wind speed, u, in knots, with the height, H, in feet, of the waves the wind produces in certain areas of the ocean. This formula is

$$u = \sqrt{\frac{H}{0.026}}$$

If waves produced by a storm have a height of 15 feet, what is the wind speed producing the waves?

93. Graph $f(x) = \sqrt{x + 1}$.

94. Graph $g(x) = -\sqrt{x}$.

95. Graph $g(x) = \sqrt{x} + 1$.

96. Graph $f(x) = \sqrt{x - 2}$.

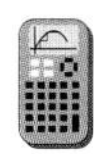

For Exercises 97–102, use your graphing calculator.

97. Check the graph drawn in Exercise 93.

98. Check the graph drawn in Exercise 95.

99. Determine whether the domain you gave in Exercise 81 is correct.

100. Determine whether the domain you gave in Exercise 82 is correct.

101. Graph $y = \sqrt[3]{x + 4}$.

102. Graph $f(x) = \sqrt[3]{2x - 3}$.

Group Activity

In this activity you will determine the conditions under which certain properties of radicals are true. We will discuss these properties later in this chapter. Discuss and answer these exercises as a group.

103. The property $\sqrt[n]{a} \cdot \sqrt[n]{b} = \sqrt[n]{ab}$, called the *multiplication property for radicals*, is true for certain real numbers a and b. By substituting values for a and b, determine under what conditions this property is true.

104. The property $\dfrac{\sqrt[n]{a}}{\sqrt[n]{b}} = \sqrt[n]{\dfrac{a}{b}}$, called the *division property for radicals*, is true for certain real numbers a and b. By substituting values for a and b, determine under what conditions this property is true.

105. Does $\sqrt{a} + \sqrt{b} = \sqrt{a + b}$? By substituting values for a and b, determine whether this statement is true.

Cumulative Review Exercises

[5.4–5.7] *Factor.*

106. $3y^2 - 18y + 27 - 3z^2$

107. $x^3 + \dfrac{1}{27}$

108. $(x + 2)^2 - (x + 2) - 12$

109. $2x^4 - 3x^3 - 6x^2 + 9x$

7.2 RATIONAL EXPONENTS

SSM VIDEO 7.2 CD Rom

1) **Change a radical expression to an exponential expression.**
2) **Simplify radical expressions.**
3) **Apply the rules of exponents to rational and negative exponents.**
4) **Factor expressions with rational exponents.**

1) Change a Radical Expression to an Exponential Expression

In this section we discuss changing radical expressions to exponential expressions and vice versa. When you see a rational exponent, you should realize that the expression can be written as a radical expression by using the following procedure.

$$\sqrt[n]{a} = a^{1/n}$$

When a is nonnegative, n can be any index.

When a is negative, n must be odd.

For the remainder of this chapter, unless you are instructed otherwise, assume that all variables in radicands represent nonnegative real numbers and that the radicand is nonnegative. With this assumption, we will not need to state that the variable is nonnegative whenever we have a radical with an even index. This will allow us to write many answers without absolute value signs.

EXAMPLE 1 Write each expression with rational exponents.

a) $\sqrt{6}$ **b)** $\sqrt[5]{7xy}$ **c)** $\sqrt[6]{\dfrac{3x^3}{5y}}$

Solution a) $\sqrt{6} = 6^{1/2}$ *Recall that the index of a square root is 2.*

b) $\sqrt[5]{7xy} = (7xy)^{1/5}$

c) $\sqrt[6]{\frac{3x^3}{5y}} = \left(\frac{3x^3}{5y}\right)^{1/6}$

Exponential expressions can be converted to radical expressions by reversing the procedure.

EXAMPLE 2 Write each expression without rational exponents.

a) $4^{1/2}$ b) $(-8)^{1/3}$ c) $y^{1/4}$ d) $(6x^2y)^{1/5}$

Solution a) $4^{1/2} = \sqrt{4} = 2$ b) $(-8)^{1/3} = \sqrt[3]{-8} = -2$

NOW TRY EXERCISE 21 c) $y^{1/4} = \sqrt[4]{y}$ d) $(6x^2y)^{1/5} = \sqrt[5]{6x^2y}$

2) Simplify Radical Expressions

We can expand the preceding rule so that radicals of the form $\sqrt[n]{a^m}$ can be written as exponential expressions. Consider $a^{2/3}$. We can write $a^{2/3}$ as $(a^{1/3})^2$ or $(a^2)^{1/3}$. This suggests $a^{2/3} = (\sqrt[3]{a})^2 = \sqrt[3]{a^2}$.

> For any nonnegative number a, and integers m and n,
>
> $$\sqrt[n]{a^m} = \left(\sqrt[n]{a}\right)^m = a^{m/n}$$
>
> (Power → m; Index → n)

This rule can be used to change an expression from radical form to exponential form and vice versa. When changing a radical expression to exponential form, the *power* is placed in the *numerator*, and the *index or root* is placed in the *denominator* of the rational exponent. Thus, for example, $\sqrt[3]{x^4}$ can be written $x^{4/3}$. Also $(\sqrt[5]{y})^2$ can be written $y^{2/5}$. Additional examples follow.

Examples

$\sqrt{y^3} = y^{3/2}$ $\sqrt[3]{z^2} = z^{2/3}$ $\sqrt[5]{2^8} = 2^{8/5}$

$(\sqrt{z})^3 = z^{3/2}$ $(\sqrt[4]{x})^3 = x^{3/4}$ $(\sqrt[4]{6})^3 = 6^{3/4}$

By this rule, for nonnegative values of the variable we can write

$$\sqrt[3]{x^4} = (\sqrt[3]{x})^4 \qquad (\sqrt[5]{y})^2 = \sqrt[5]{y^2}$$

EXAMPLE 3 Write each expression with rational exponents, then simplify.

a) $\sqrt[3]{x^9}$ b) $(\sqrt[4]{x})^{12}$

Solution a) $\sqrt[3]{x^9} = x^{9/3} = x^3$ b) $(\sqrt[4]{x})^{12} = x^{12/4} = x^3$

Exponential expressions with rational exponents can be converted to radical expressions by reversing the procedure. The *numerator* of the rational exponent is the *power*, and the *denominator* of the rational exponent is the *index or root* of the radical expression. Here are some examples.

Examples

$x^{1/2} = \sqrt{x}$	$5^{1/3} = \sqrt[3]{5}$
$6^{2/3} = \sqrt[3]{6^2}$ or $(\sqrt[3]{6})^2$	$y^{3/10} = \sqrt[10]{y^3}$ or $(\sqrt[10]{y})^3$
$x^{9/5} = \sqrt[5]{x^9}$ or $(\sqrt[5]{x})^9$	$y^{10/3} = \sqrt[3]{y^{10}}$ or $(\sqrt[3]{y})^{10}$

Notice that you may choose, for example, to write $6^{2/3}$ as either $\sqrt[3]{6^2}$ or $(\sqrt[3]{6})^2$.

EXAMPLE 4 Write each expression without rational exponents.

a) $x^{2/3}$ **b)** $(6ab)^{5/4}$

Solution **a)** $x^{2/3} = \sqrt[3]{x^2}$ or $(\sqrt[3]{x})^2$ **b)** $(6ab)^{5/4} = \sqrt[4]{(6ab)^5}$ or $(\sqrt[4]{6ab})^5$

EXAMPLE 5 Simplify. **a)** $25^{3/2}$ **b)** $\sqrt[6]{(16)^3}$ **c)** $\sqrt[4]{y^{12}}$ **d)** $(\sqrt[10]{z})^5$

Solution **a)** Sometimes an expression with a rational exponent can be simplified more easily by writing the expression as a radical, as illustrated.

$$\begin{aligned} 25^{3/2} &= (\sqrt{25})^3 && \textit{Write as a radical.} \\ &= (5)^3 \\ &= 125 \end{aligned}$$

b) Sometimes a radical expression can be simplified more easily by writing the expression with rational exponents, as illustrated in parts **b)** through **d)**.

$$\begin{aligned} \sqrt[6]{(16)^3} &= 16^{3/6} && \textit{Write with a rational exponent.} \\ &= 16^{1/2} && \textit{Reduce fraction.} \\ &= \sqrt{16} && \textit{Write as a radical.} \\ &= 4 && \textit{Simplify.} \end{aligned}$$

c) $\sqrt[4]{y^{12}} = y^{12/4} = y^3$

NOW TRY EXERCISE 43

d) $(\sqrt[10]{z})^5 = z^{5/10} = z^{1/2} = \sqrt{z}$

Now let's consider $\sqrt[5]{x^5}$. When written in exponential form, this is $x^{5/5} = x^1 = x$. This leads to the following rule.

> For any nonnegative real number a,
> $$\sqrt[n]{a^n} = (\sqrt[n]{a})^n = a^{n/n} = a$$

In the preceding box we specified that a was nonnegative. If n is an even index and a is a negative real number, $\sqrt[n]{a^n} = |a|$ and not a. For example, $\sqrt[6]{(-5)^6} = |-5| = 5$. *Since we are assuming, except where noted otherwise, that variables in radicands represent nonnegative real numbers*, we may write $\sqrt[6]{x^6} = x$ and not $|x|$. This assumption also lets us write $\sqrt{x^2} = x$ and $(\sqrt[4]{z})^4 = z$.

Examples

$\sqrt{5^2} = 5$	$\sqrt[4]{y^4} = y$
$\sqrt[8]{x^8} = x$	$(\sqrt[5]{z})^5 = z$

3) Apply the Rules of Exponents to Rational and Negative Exponents

In Section 1.5 we introduced and discussed the rules of exponents. In that section we only used exponents that were whole numbers. The rules still apply when the exponents are rational numbers. Let's review those rules now.

Rules of Exponents

For all real numbers a and b and all rational numbers m and n,

Product rule	$a^m \cdot a^n = a^{m+n}$
Quotient rule	$\dfrac{a^m}{a^n} = a^{m-n}, \quad a \neq 0$
Negative exponent rule	$a^{-m} = \dfrac{1}{a^m}, \quad a \neq 0$
Zero exponent rule	$a^0 = 1, \quad a \neq 0$
Raising a power to a power	$(a^m)^n = a^{m \cdot n}$
Raising a product to a power	$(ab)^m = a^m b^m$
Raising a quotient to a power	$\left(\dfrac{a}{b}\right)^m = \dfrac{a^m}{b^m}, \quad b \neq 0$

Using these rules, we will now work some problems in which the exponents are rational numbers.

EXAMPLE 6 Evaluate. **a)** $8^{-2/3}$ **b)** $(-27)^{-2/3}$

Solution **a)** Begin by using the negative exponent rule.

$$8^{-2/3} = \frac{1}{8^{2/3}} \qquad \textit{Negative exponent rule}$$

$$= \frac{1}{(\sqrt[3]{8})^2} \qquad \textit{Write the denominator as a radical.}$$

$$= \frac{1}{2^2} \qquad \textit{Simplify the denominator.}$$

$$= \frac{1}{4}$$

b) $$(-27)^{-2/3} = \frac{1}{(-27)^{2/3}} = \frac{1}{(\sqrt[3]{-27})^2} = \frac{1}{(-3)^2} = \frac{1}{9}$$

Note that Example 6 **a)** could have been evaluated as follows:

$$8^{-2/3} = \frac{1}{8^{2/3}} = \frac{1}{\sqrt[3]{8^2}} = \frac{1}{\sqrt[3]{64}} = \frac{1}{4}$$

NOW TRY EXERCISE 59

However, it is generally easier to evaluate the root before applying the power.

In Chapter 1 we indicated that

$$\left(\frac{a}{b}\right)^{-n} = \left(\frac{b}{a}\right)^n$$

We use this fact in the following example.

EXAMPLE 7 Evaluate $\left(\frac{4}{25}\right)^{-1/2}$.

Solution $\left(\frac{4}{25}\right)^{-1/2} = \left(\frac{25}{4}\right)^{1/2} = \sqrt{\frac{25}{4}} = \frac{5}{2}$

NOW TRY EXERCISE 63

HELPFUL HINT

How do the expressions $-25^{1/2}$ and $(-25)^{1/2}$ differ?

Recall that $-x^2$ means $-(x^2)$. The same principle applies here.

$$-25^{1/2} = -(25)^{1/2} = -\sqrt{25} = -5$$
$$(-25)^{1/2} = \sqrt{-25} \text{ This is not a real number.}$$

EXAMPLE 8 Simplify each expression and write the answer without negative exponents.

a) $x^{1/2} \cdot x^{-2/3}$ **b)** $(3x^2y^{-4})^{-1/2}$ **c)** $3.2x^{1/3}(2.4x^{1/2} + x^{-1/4})$ **d)** $\left(\frac{4x^{-3}y^{2/3}}{y^{-1/3}}\right)^{1/4}$

Solution **a)**
$$\begin{aligned} x^{1/2} \cdot x^{-2/3} &= x^{(1/2)-(2/3)} && \textit{Product rule} \\ &= x^{-1/6} && \textit{Find the LCD and subtract the exponents.} \\ &= \frac{1}{x^{1/6}} && \textit{Negative exponent rule} \end{aligned}$$

b)
$$\begin{aligned} (3x^2y^{-4})^{-1/2} &= 3^{-1/2}x^{2(-1/2)}y^{-4(-1/2)} && \textit{Raise the product to the power.} \\ &= 3^{-1/2}x^{-1}y^2 && \textit{Multiply the exponents.} \\ &= \frac{y^2}{3^{1/2}x} \text{ or } \frac{y^2}{x\sqrt{3}} && \textit{Negative exponent rule} \end{aligned}$$

c) Begin by using the distributive property.

$$\begin{aligned} 3.2x^{1/3}(2.4x^{1/2} + x^{-1/4}) &= (3.2x^{1/3})(2.4x^{1/2}) + (3.2x^{1/3})(x^{-1/4}) && \textit{Distributive property} \\ &= (3.2)(2.4)(x^{(1/3)+(1/2)}) + 3.2x^{(1/3)-(1/4)} && \textit{Product rule} \\ &= 7.68x^{5/6} + 3.2x^{1/12} \end{aligned}$$

d)
$$\begin{aligned} \left(\frac{4x^{-3}y^{2/3}}{y^{-1/3}}\right)^{1/4} &= (4x^{-3}y^{(2/3)-(-1/3)})^{1/4} && \textit{Quotient rule} \\ &= (4x^{-3}y)^{1/4} && \textit{Subtract the exponents.} \\ &= 4^{1/4}x^{-3(1/4)}y^{1/4} && \textit{Raise the product to a power.} \\ &= 4^{1/4}x^{-3/4}y^{1/4} && \textit{Multiply the exponents.} \\ &= \frac{4^{1/4}y^{1/4}}{x^{3/4}} && \textit{Negative exponent rule} \end{aligned}$$

NOW TRY EXERCISE 83

EXAMPLE 9 Simplify. **a)** $\sqrt[15]{(6x)^5}$ **b)** $\left(\sqrt[4]{a^2b^3c}\right)^{12}$ **c)** $\sqrt{\sqrt[3]{x}}$

Solution

a) $\sqrt[15]{(6x)^5} = (6x)^{5/15}$ *Write with a rational exponent.*

$= (6x)^{1/3}$ *Simplify the exponent.*

$= \sqrt[3]{6x}$ *Write as a radical.*

b) $\left(\sqrt[4]{a^2b^3c}\right)^{12} = \left(a^2b^3c\right)^{12/4}$ *Write with a rational exponent.*

$= \left(a^2b^3c\right)^3$

$= a^6b^9c^3$ *Raise the product to a power.*

c) $\sqrt{\sqrt[3]{x}} = \sqrt{x^{1/3}}$ *Write $\sqrt[3]{x}$ as $x^{1/3}$.*

$= \left(x^{1/3}\right)^{1/2}$ *Write with a rational exponent.*

$= x^{1/6}$ *Raise the power to a power.*

$= \sqrt[6]{x}$ *Write as a radical.*

NOW TRY EXERCISE 41

Using Your Calculator

Finding Roots or Expressions with Rational Exponents on a Scientific or Graphing Calculator

There are often many ways to evaluate an expression like $\left(\sqrt[5]{845}\right)^3$ or $845^{3/5}$ on a calculator. The procedure to use varies from calculator to calculator. One general method is to write the expression with a rational exponent and use the $\boxed{y^x}$ or $\boxed{\wedge}$ key with parentheses keys, as shown below.*

SCIENTIFIC CALCULATOR

To evaluate $845^{3/5}$, press

Answer displayed

845 $\boxed{y^x}$ $\boxed{(}$ 3 $\boxed{\div}$ 5 $\boxed{)}$ $\boxed{=}$ 57.03139903

To evaluate $845^{-3/5}$, which means $\frac{1}{845^{3/5}}$, press

Answer displayed

845 $\boxed{y^x}$ $\boxed{(}$ 3 $\boxed{+/-}$ $\boxed{\div}$ 5 $\boxed{)}$ $\boxed{=}$ 0.017534201

GRAPHING CALCULATOR

To evaluate $845^{3/5}$, press the following keys.

Answer displayed

845 $\boxed{\wedge}$ $\boxed{(}$ 3 $\boxed{\div}$ 5 $\boxed{)}$ $\boxed{\text{ENTER}}$ 57.03139903

To evaluate $845^{-3/5}$, press the following keys.

Answer displayed

845 $\boxed{\wedge}$ $\boxed{(}$ $\boxed{(-)}$ 3 $\boxed{\div}$ 5 $\boxed{)}$ $\boxed{\text{ENTER}}$ .0175342008

*Keystrokes to use vary from calculator to calculator. Read your calculator manual to learn how to evaluate exponential expressions.

4) Factor Expressions with Rational Exponents

In higher-level math courses, you may have the need to factor out a variable with a rational exponent. To factor a rational expression, factor out the term with the smallest (or most negative) exponent.

EXAMPLE 10 Factor $x^{2/5} + x^{-3/5}$.

Solution The smallest of the two exponents is $-3/5$. Therefore, we will factor $x^{-3/5}$ from both terms. To find the new exponent on the variable that originally had the greater exponent, we subtract the exponent that was factored out from the original exponent.

$$x^{\frac{2}{5}} + x^{-\frac{3}{5}} = x^{-\frac{3}{5}}\left(x^{\frac{2}{5}-\left(-\frac{3}{5}\right)} + 1\right)$$

$$= x^{-\frac{3}{5}}(x^1 + 1)$$

$$= x^{-\frac{3}{5}}(x + 1)$$

We can check our factoring by multiplying.

$$\begin{aligned} x^{-3/5}(x+1) &= x^{-3/5} \cdot x + x^{-3/5} \cdot 1 \\ &= x^{(-3/5)+1} + x^{-3/5} \\ &= x^{2/5} + x^{-3/5} \end{aligned}$$

NOW TRY EXERCISE 111 Since we obtained the original expression, the factoring is correct.

Exercise Set 7.2

Concept/Writing Exercises

1. a) Under what conditions is $\sqrt[n]{a}$ a real number?
   b) When $\sqrt[n]{a}$ is a real number, how can it be expressed with rational exponents?
2. a) Under what conditions is $\sqrt[n]{a^m}$ a real number?
   b) Under what conditions is $(\sqrt[n]{a})^m$ a real number?
   c) When $\sqrt[n]{a^m}$ is a real number, how can it be expressed with rational exponents?
3. a) Under what conditions is $\sqrt[n]{a^n}$ a real number?
   b) When n is even and $a \geq 0$, what is $\sqrt[n]{a^n}$ equal to?
   c) When n is odd, what is $\sqrt[n]{a^n}$ equal to?
   d) When n is even and a may be any real number, what is $\sqrt[n]{a^n}$ equal to?
4. a) Explain the difference between $-16^{1/2}$ and $(-16)^{1/2}$.
   b) Evaluate each expression in part **a)** if possible.

Practice the Skills

In this exercise set, assume that all variables represent positive real numbers. Write each expression in exponential form.

5. $\sqrt{x^5}$
6. $\sqrt{y^5}$
7. $\sqrt{8^5}$
8. $\sqrt[3]{z^2}$
9. $\sqrt[5]{x^4}$
10. $\sqrt[3]{z^5}$
11. $(\sqrt{x})^3$
12. $(\sqrt[3]{y})^2$
13. $\sqrt[4]{a^3b^5}$
14. $\sqrt[4]{4c^7d^5}$
15. $\sqrt[8]{5r^4s^9t^{12}}$
16. $\sqrt[17]{3a^5 + 4b}$

Write each expression in radical form.

17. $x^{1/2}$
18. $y^{2/3}$
19. $z^{3/2}$
20. $5^{1/2}$
21. $(24y^2)^{1/2}$
22. $(35c^2)^{5/2}$
23. $(19x^2y^4)^{-1/2}$
24. $(24xy^2)^{1/2}$
25. $(2m^2n^3)^{2/5}$
26. $(5r + s^2)^{1/4}$
27. $(a^2 - 4b^2)^{-2/3}$
28. $(3r^2 + 2m)^{-1/3}$

Simplify each radical expression by changing the expression to exponential form. Write the answer in radical form when appropriate.

29. $\sqrt{y^6}$
30. $\sqrt{x^{12}}$
31. $\sqrt{z^8}$
32. $\sqrt[3]{x^6}$

33. $\sqrt[3]{x^9}$ **34.** $\sqrt[6]{y^2}$ **35.** $\sqrt[10]{z^5}$ **36.** $\sqrt{2^4}$

37. $(\sqrt{5.1})^2$ **38.** $\sqrt[4]{(6.83)^4}$ **39.** $\sqrt[6]{y^6}$ **40.** $(\sqrt[5]{x})^5$

41. $(\sqrt[8]{xyz})^4$ **42.** $\sqrt[3]{4^6a^3}$ **43.** $(\sqrt[3]{xy^2})^9$ **44.** $(\sqrt[4]{a^4bc^3})^{40}$

45. $\sqrt{\sqrt{x}}$ **46.** $\sqrt{\sqrt{x^2y}}$ **47.** $\sqrt[3]{\sqrt{x^5}}$ **48.** $\sqrt[5]{\sqrt[3]{ab}}$

Evaluate if possible. If the expression is not a real number, so state.

49. $4^{1/2}$ **50.** $8^{2/3}$ **51.** $(-4)^{1/2}$ **52.** $\left(\frac{4}{9}\right)^{1/2}$

53. $\left(\frac{9}{25}\right)^{1/2}$ **54.** $\left(\frac{1}{8}\right)^{1/3}$ **55.** $-16^{1/2}$ **56.** $(-16)^{1/2}$

57. $-64^{1/3}$ **58.** $4^{-1/2}$ **59.** $64^{-1/3}$ **60.** $16^{-3/2}$

61. $4^{-3/2}$ **62.** $81^{-3/4}$ **63.** $-\left(\frac{4}{49}\right)^{-1/2}$ **64.** $\left(\frac{81}{16}\right)^{-3/4}$

65. $25^{1/2} + 169^{1/2}$ **66.** $25^{-1/2} + 36^{-1/2}$ **67.** $343^{-1/3} + 9^{-1/2}$ **68.** $16^{-1/2} - 625^{-3/4}$

Simplify. Write the answer in exponential form without negative exponents.

69. $x^5 \cdot x^{1/2}$ **70.** $x^{1/3} \cdot x^{3/8}$ **71.** $\frac{x^{1/2}}{x^{1/3}}$ **72.** $x^{-3/5}$

73. $(x^{1/2})^{-2}$ **74.** $(z^{-1/4})^{-1/2}$ **75.** $(6^{-1/3})^0$ **76.** $\frac{x^4}{x^{-1/2}}$

77. $\frac{5y^{-1/3}}{60y^{-2}}$ **78.** $x^{-1/2} \cdot x^{-3/5}$ **79.** $4x^{5/3} \cdot 2x^{-7/2}$ **80.** $(x^{-2/5})^{1/3}$

81. $\left(\frac{8}{64x}\right)^{1/3}$ **82.** $\left(\frac{81}{3y^4}\right)^{1/3}$ **83.** $\left(\frac{22x^{3/7}}{2x^{1/2}}\right)^2$ **84.** $\left(\frac{x^{-1/3}}{x^{-2}}\right)^{1/2}$

85. $\left(\frac{y^4}{4y^{-2/5}}\right)^{-3}$ **86.** $\left(\frac{81z^{1/2}y^3}{9z^{1/2}}\right)^{1/2}$ **87.** $\left(\frac{x^{3/4}y^{-2}}{x^{1/2}y^2}\right)^4$ **88.** $\left(\frac{250a^{-3/4}b^5}{2a^{-2}b^2}\right)^{2/3}$

Multiply.

89. $3z^{-1/2}(2z^4 - z^{1/2})$ **90.** $-2x^{-4/9}(2x^{1/9} - x^2)$ **91.** $5x^{-1}(x^{-4} + 2x^{-1/2})$

92. $-4a^{3/2}(a^{3/2} - a^{-3/2})$ **93.** $-4x^{5/3}(-2x^{1/2} + x^{1/3})$ **94.** $\frac{1}{2}x^{-2}(6x^{4/3} - 8x^{-1/2})$

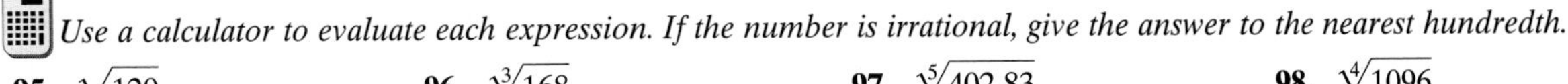

Use a calculator to evaluate each expression. If the number is irrational, give the answer to the nearest hundredth.

95. $\sqrt{120}$ **96.** $\sqrt[3]{168}$ **97.** $\sqrt[5]{402.83}$ **98.** $\sqrt[4]{1096}$

99. $45^{2/3}$ **100.** $697.2^{3/2}$ **101.** $1000^{-1/2}$ **102.** $8060^{-3/2}$

Problem Solving

103. Under what conditions will $\sqrt[n]{a^n} = (\sqrt[n]{a})^n = a$?

104. By selecting values for a and b, show that $(a^2 + b^2)^{1/2}$ *is not equal to* $a + b$.

105. By selecting values for a and b, show that $(a^{1/2} + b^{1/2})^2$ *is not equal to* $a + b$.

106. Determine whether $\sqrt[3]{\sqrt{x}} = \sqrt{\sqrt[3]{x}}$, $x \geq 0$.

Factor. Write the answer without negative exponents.

107. $x^{3/2} + x^{1/2}$ **108.** $x^{1/4} - x^{5/4}$ **109.** $y^{1/3} - y^{4/3}$

110. $x^{-1/2} + x^{1/2}$ **111.** $y^{-3/5} + y^{2/5}$ **112.** $a^{1/5} + a^{-4/5}$

In Exercises 113 through 118, use a calculator where appropriate.

113. The function, $E(t) = 2^{10} \cdot 2^t$ approximates the number of bacteria in a certain culture after t hours.

a) The initial number of bacteria is determined when $t = 0$. What is the initial number of bacteria?

b) How many bacteria are there after $\frac{1}{2}$ hour?

114. Carbon dating is used by scientists to find the age of fossils, bones, and other items. The formula used in carbon dating is $P = P_0 2^{-t/5600}$, where P_0 represents the original amount of carbon 14 (C_{14}) present and P represents the amount of C_{14} present after t years. If 10 milligrams (mg) of C_{14} is present in an animal bone recently excavated, how many milligrams will be present in 5000 years?

115. Each year more and more people contribute to their companies' 401(k) retirement plans. The total assets $A(t)$, in U.S. 401(k) plans, in billions of dollars, can be approximated by the function $A(t) = 2.69t^{3/2}$, where t is years since 1983 and $1 \leq t \leq 13$. (Therefore, this function holds from 1984 through 1997.) Estimate the total assets in U.S. 401(k) plans in **a)** 1990 and **b)** 1997.

116. Retail Internet sales are increasing annually. The total amount, $I(t)$, in billions of dollars, of Internet sales can be approximated by the function $I(t) = 0.25t^{5/3}$, where t is years since 1995 and $1 \leq t \leq 6$. Find the total Internet sales in **a)** 1996 and **b)** 1999.

117. Evaluate $(3^{\sqrt{2}})^{\sqrt{2}}$. Explain how you determined your answer.

118. **a)** On your calculator, evaluate 3^{π}.
b) Explain why your value from part **a)** does or does not make sense.

119. Find the domain of $f(x) = (x - 2)^{1/2}(x + 3)^{-1/2}$.

120. Find the domain of $f(x) = (x + 4)^{1/2}(x - 6)^{-1/2}$

Determine the index to be placed in the shaded area to make the statement true. Explain how you determined your answer.

121. $\sqrt[4]{\sqrt{\sqrt{x}}} = x^{1/24}$

122. $\sqrt[4]{\sqrt[5]{\sqrt{\sqrt[3]{z}}}} = z^{1/120}$

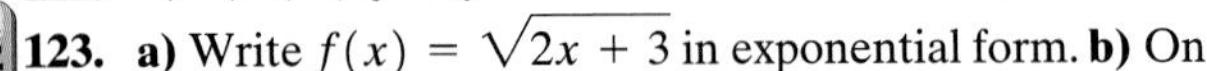

123. **a)** Write $f(x) = \sqrt{2x + 3}$ in exponential form. **b)** On your grapher, check that the answer you gave in part **a)** is correct by graphing both $f(x)$ as given and the function you gave in exponential form.

Cumulative Review Exercises

[3.2] **124.** Determine which of the following graphs are functions and which are relations.

a)

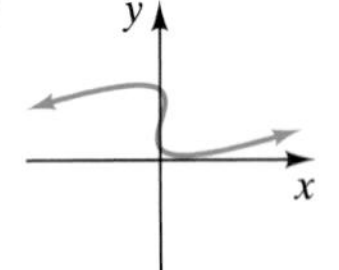

b) **c)**

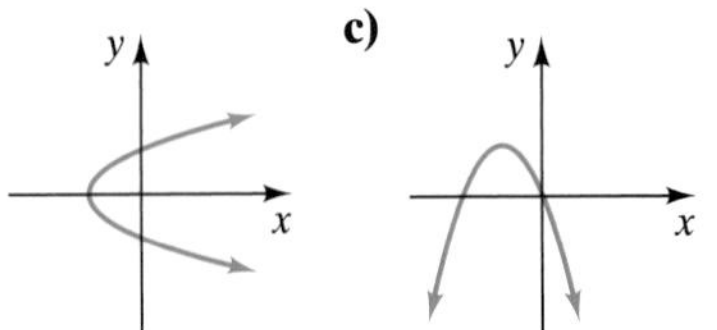

[6.3] **125.** Simplify $\dfrac{a^{-2} + ab^{-1}}{ab^{-2} - a^{-2}b^{-1}}$.

[6.4] **126.** Solve the equation $\dfrac{3x - 2}{x + 4} = \dfrac{2x + 1}{3x - 2}$.

[6.5] **127.** Amy Mayfield can fly her plane 500 miles against the wind in the same time it takes her to fly 560 miles with the wind. If the wind blows at 25 miles per hour, find the speed of the plane in still air.

7.3 MULTIPLYING AND SIMPLIFYING RADICALS

SSM VIDEO 7.3 CD Rom

1. Apply the product rule for radicals.
2. Simplify radicals whose radicands are natural numbers.
3. Simplify radicals.
4. Simplify the product of two radicals.

1 Apply the Product Rule for Radicals

In this section we will simplify radicals using the **product rule for radicals.** To introduce this rule, observe that $\sqrt{4} \cdot \sqrt{9} = 2 \cdot 3 = 6$. Also, $\sqrt{4 \cdot 9} = \sqrt{36} = 6$. We see that $\sqrt{4} \cdot \sqrt{9} = \sqrt{4 \cdot 9}$. This is one example of the product rule for radicals.

Product Rule for Radicals

For nonnegative real numbers a and b,

$$\sqrt[n]{a} \cdot \sqrt[n]{b} = \sqrt[n]{ab}$$

Examples of the Product Rule

$$\sqrt{20} = \begin{cases} \sqrt{1} \cdot \sqrt{20} \\ \sqrt{2} \cdot \sqrt{10} \\ \sqrt{4} \cdot \sqrt{5} \end{cases}$$

$\sqrt{20}$ can be factored into any of these forms

$$\sqrt[3]{20} = \begin{cases} \sqrt[3]{1} \cdot \sqrt[3]{20} \\ \sqrt[3]{2} \cdot \sqrt[3]{10} \\ \sqrt[3]{4} \cdot \sqrt[3]{5} \end{cases}$$

$\sqrt[3]{20}$ can be factored into any of these forms

$$\sqrt{x^7} = \begin{cases} \sqrt{x} \cdot \sqrt{x^6} \\ \sqrt{x^2} \cdot \sqrt{x^5} \\ \sqrt{x^3} \cdot \sqrt{x^4} \end{cases}$$

$\sqrt{x^7}$ can be factored into any of these forms

$$\sqrt[3]{x^7} = \begin{cases} \sqrt[3]{x} \cdot \sqrt[3]{x^6} \\ \sqrt[3]{x^2} \cdot \sqrt[3]{x^5} \\ \sqrt[3]{x^3} \cdot \sqrt[3]{x^4} \end{cases}$$

$\sqrt[3]{x^7}$ can be factored into any of these forms

2 Simplify Radicals Whose Radicands are Natural Numbers

To help clarify our explanations, we will introduce **perfect powers**. A number is a **perfect square** if it is the square of a natural number. A number is a **perfect cube** if it is a cube of a natural number. A number is a **perfect fourth power** if it is the fourth power of a natural number, and so on.

Examples of Perfect Squares and Perfect Cubes

Squares of natural numbers: $1^2, \; 2^2, \; 3^2, \; 4^2, \; 5^2, \; 6^2, \; 7^2, \; 8^2, \; 9^2, \ldots$

Perfect squares: 1, 4, 9, 16, 25, 36, 49, 64, 81,...

Cubes of natural numbers: $1^3, \; 2^3, \; 3^3, \; 4^3, \; 5^3, \; 6^3, \; 7^3, \; 8^3, \; 9^3, \ldots$

Perfect cubes: 1, 8, 27, 64, 125, 216, 343, 512, 729,...

Note that the square root of any perfect square is a whole number. For example,

$$\sqrt{36} = \sqrt{6^2} = 6^{2/2} = 6$$

Similarly, the cube root of any perfect cube is a whole number. For example,

$$\sqrt[3]{125} = \sqrt[3]{5^3} = 5^{3/3} = 5$$

The fourth root of any perfect fourth-power is a whole number, etc.

Now we will discuss how to simplify radicals whose radicands are natural numbers.

To Simplify Radicals Whose Radicands Are Natural Numbers

1. Write the radicand as the product of two numbers, one of which is the largest perfect power number for the given index.
2. Use the product rule to write the expression as a product of roots.
3. Find the roots of any perfect power numbers.

If we are simplifying a *square* root, we will write the radicand as the product of the largest *perfect square* and another number. If we are simplifying a *cube* root, we will write the radicand as the product of the largest *perfect cube* and another number, and so on.

EXAMPLE 1 Simplify. **a)** $\sqrt{32}$ **b)** $\sqrt{60}$ **c)** $\sqrt[3]{54}$

Solution **a)** Since we are evaluating a square root, we look for the largest perfect square that divides 32. The largest perfect square that divides, or is a factor of, 32 is 16.

$$\sqrt{32} = \sqrt{16 \cdot 2} = \sqrt{16}\sqrt{2} = 4\sqrt{2}$$

b) The largest perfect square that is a factor of 60 is 4.

$$\sqrt{60} = \sqrt{4 \cdot 15} = \sqrt{4}\sqrt{15} = 2\sqrt{15}$$

c) The largest perfect cube that is a factor of 54 is 27.

$$\sqrt[3]{54} = \sqrt[3]{27 \cdot 2} = \sqrt[3]{27}\sqrt[3]{2} = 3\sqrt[3]{2}$$

NOW TRY EXERCISE 11

HELPFUL HINT

In Example 1a), if you first thought that 4 was the largest perfect square that divided 32, you could proceed as follows:

$$\sqrt{32} = \sqrt{4 \cdot 8} = \sqrt{4}\sqrt{8} = 2\sqrt{8}$$
$$= 2\sqrt{4 \cdot 2} = 2\sqrt{4}\sqrt{2} = 2 \cdot 2\sqrt{2} = 4\sqrt{2}$$

Note that the final result is the same, but you must perform more steps. The chart on page 439 can help you determine the largest perfect square or perfect cube that is a factor of a radicand.

In Example 1b), $\sqrt{15}$ can be factored as $\sqrt{5}\sqrt{3}$; however, since neither 5 nor 3 is a perfect square, $\sqrt{15}$ cannot be simplified.

3 Simplify Radicals

Now we will discuss perfect powers of variables. The radicand x^n is a **perfect square** when n is a multiple of 2 (an even natural number). The radicand x^n is a **perfect cube** when n is a multiple of 3. In general, the radicand x^n is a **perfect power** when n is a *multiple of the index* of the radical (or when n is divisible by the index).

Examples

Perfect squares: $x^2, x^4, x^6, x^8, x^{10}, \ldots$

Perfect cubes: $x^3, x^6, x^9, x^{12}, x^{15}, \ldots$

Perfect fourth powers of x: $x^4, x^8, x^{12}, x^{16}, x^{20}, \ldots$

Perfect powers of x for index n: $x^n, x^{2n}, x^{3n}, x^{4n}, x^{5n}, \ldots$

HELPFUL HINT

A quick way to determine if a radicand x^n is a perfect power for an index is to determine if the exponent n is divisible by the index of the radical. For example, consider $\sqrt[5]{x^{20}}$. Since the exponent, 20, is divisible by the index, 5, x^{20} is a perfect fifth power. Now consider $\sqrt[6]{x^{20}}$. Since the exponent, 20, is not divisible by the index, 6, x^{20} is not a perfect sixth power. However, x^{18} and x^{24} are both perfect sixth powers since 6 divides both 18 and 24.

When the radicand is a perfect power for the index, the radical can be simplified by writing it in exponential form, as in Example 2.

EXAMPLE 2 Simplify. **a)** $\sqrt{x^4}$ **b)** $\sqrt[3]{x^{12}}$ **c)** $\sqrt[6]{y^{24}}$

Solution **a)** $\sqrt{x^4} = x^{4/2} = x^2$ **b)** $\sqrt[3]{x^{12}} = x^{12/3} = x^4$ **c)** $\sqrt[6]{y^{24}} = y^{24/6} = y^4$

Now we give a general procedure for simplifying radicals.

To Simplify Radicals

1. If the radicand contains a coefficient other than 1, write it as a product of two numbers, one of which is the largest perfect power for the index.
2. Write each variable factor as a product of two factors, one of which is the largest perfect power of the variable for the index.
3. Use the product rule to write the radical expression as a product of radicals. Place all the perfect powers (numbers and variables) under the same radical.
4. Simplify the radical containing the perfect powers.

EXAMPLE 3 Simplify. **a)** $\sqrt{x^9}$ **b)** $\sqrt[5]{x^{23}}$

Solution Because the radicands have coefficients of 1, we start with step 2 of the procedure.

a) The largest perfect square less than or equal to x^9 is x^8.

$$\sqrt{x^9} = \sqrt{x^8 \cdot x} = \sqrt{x^8} \cdot \sqrt{x} = x^{8/2}\sqrt{x} = x^4\sqrt{x}$$

b) The largest perfect fifth power less than or equal to x^{23} is x^{20}.

$$\sqrt[5]{x^{23}} = \sqrt[5]{x^{20} \cdot x^3} = \sqrt[5]{x^{20}}\sqrt[5]{x^3} = x^{20/5}\sqrt[5]{x^3} = x^4\sqrt[5]{x^3}$$

EXAMPLE 4 Simplify. **a)** $\sqrt{x^{12}y^{17}}$ **b)** $\sqrt[4]{x^6y^{23}}$

Solution **a)** x^{12} is a perfect square. The largest perfect square that is a factor of y^{17} is y^{16}. Write y^{17} as $y^{16} \cdot y$.

$$\begin{aligned}\sqrt{x^{12}y^{17}} = \sqrt{x^{12} \cdot y^{16} \cdot y} &= \sqrt{x^{12}y^{16}}\sqrt{y} \\ &= \sqrt{x^{12}}\sqrt{y^{16}}\sqrt{y} \\ &= x^{12/2}y^{16/2}\sqrt{y} \\ &= x^6y^8\sqrt{y}\end{aligned}$$

b) We begin by finding the largest perfect fourth power factors of x^6 and y^{23}. For an index of 4, the largest perfect power that is a factor of x^6 is x^4. The largest perfect power that is a factor of y^{23} is y^{20}.

$$\begin{aligned}\sqrt[4]{x^6y^{23}} &= \sqrt[4]{x^4 \cdot x^2 \cdot y^{20} \cdot y^3} \\ &= \sqrt[4]{x^4y^{20} \cdot x^2y^3} \\ &= \sqrt[4]{x^4y^{20}}\sqrt[4]{x^2y^3} \\ &= xy^5\sqrt[4]{x^2y^3}\end{aligned}$$

Often the steps where we change the radical expression to exponential form are done mentally, and those steps are not illustrated. For instance, in

Example 6 **b)** we changed $\sqrt[4]{x^4 y^{20}}$ to xy^5 mentally and did not show the intermediate steps.

EXAMPLE 5 Simplify. **a)** $\sqrt{80x^5 y^{12} z^3}$ **b)** $\sqrt[3]{54x^{17} y^{25}}$

Solution **a)** The largest perfect square that is a factor of 80 is 16. The largest perfect square that is a factor of x^5 is x^4. The expression y^{12} is a perfect square. The largest perfect square that is a factor of z^3 is z^2. Place all the perfect squares under the same radical, and then simplify.

$$\begin{aligned}\sqrt{80x^5 y^{12} z^3} &= \sqrt{16 \cdot 5 \cdot x^4 \cdot x \cdot y^{12} \cdot z^2 \cdot z}\\ &= \sqrt{16x^4 y^{12} z^2 \cdot 5xz}\\ &= \sqrt{16x^4 y^{12} z^2} \cdot \sqrt{5xz}\\ &= 4x^2 y^6 z\sqrt{5xz}\end{aligned}$$

b) The largest perfect cube that is a factor of 54 is 27. The largest perfect cube that is a factor of x^{17} is x^{15}. The largest perfect cube that is a factor of y^{25} is y^{24}.

$$\begin{aligned}\sqrt[3]{54x^{17} y^{25}} &= \sqrt[3]{27 \cdot 2 \cdot x^{15} \cdot x^2 \cdot y^{24} \cdot y}\\ &= \sqrt[3]{27x^{15} y^{24} \cdot 2x^2 y}\\ &= \sqrt[3]{27x^{15} y^{24}} \cdot \sqrt[3]{2x^2 y}\\ &= 3x^5 y^8 \sqrt[3]{2x^2 y}\end{aligned}$$

NOW TRY EXERCISE 27

HELPFUL HINT

In Example 4b) we showed that

$$\sqrt[4]{x^6 y^{23}} = xy^5 \sqrt[4]{x^2 y^3}$$

This radical can also be simplified by dividing the exponents on the variables in the radicand, 6 and 23, by the index, 4, and observing the quotients and remainders.

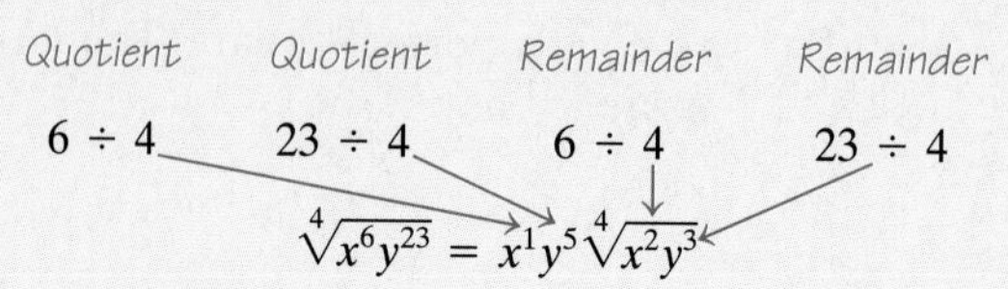

Can you explain why this procedure works? You may wish to use this procedure to work or check certain problems.

4 Simplify the Product of Two Radicals

To multiply radicals, we use the product rule given earlier. After multiplying we can often simplify the new radical (see Examples 6 and 7).

EXAMPLE 6 Multiply and simplify. **a)** $\sqrt[3]{2x}\,\sqrt[3]{4x^2}$ **b)** $\sqrt[4]{8x^{11} y}\,\sqrt[4]{8x^6 y^{22}}$

Solution **a)** $\sqrt[3]{2x}\,\sqrt[3]{4x^2} = \sqrt[3]{2x \cdot 4x^2}$

$= \sqrt[3]{8x^3}$ *$8x^3$ is a perfect cube.*

$= 2x$

b) $\sqrt[4]{8x^{11}y}\sqrt[4]{8x^6y^{22}} = \sqrt[4]{8x^{11}y \cdot 8x^6y^{22}}$

$= \sqrt[4]{64x^{17}y^{23}}$

$= \sqrt[4]{16x^{16}y^{20}}\sqrt[4]{4xy^3}$ *The largest perfect fourth root factors are 16, x^{16}, and y^{20}.*

$= 2x^4y^5\sqrt[4]{4xy^3}$

NOW TRY EXERCISE 43

When a radical is simplified, the radicand does not have any variable with an exponent greater than or equal to the index.

EXAMPLE 7 Multiply and simplify $\sqrt{2x}(\sqrt{8x} - \sqrt{32})$.

Solution Begin by using the distributive property.

$$\sqrt{2x}(\sqrt{8x} - \sqrt{32}) = (\sqrt{2x})(\sqrt{8x}) + (\sqrt{2x})(-\sqrt{32})$$
$$= \sqrt{16x^2} - \sqrt{64x}$$
$$= 4x - \sqrt{64}\sqrt{x}$$
$$= 4x - 8\sqrt{x}$$

Note in Example 7 that the same result could be obtained by first simplifying $\sqrt{8x}$ and $\sqrt{32}$ and then multiplying. You may wish to try this now.

EXAMPLE 8 If $f(x) = \sqrt[3]{x^2y}$ and $g(x) = \sqrt[3]{x^4y^2} + \sqrt[3]{xy}$, find $(f \cdot g)(x)$.

Solution From Section 3.6 we know that $(f \cdot g)(x) = f(x) \cdot g(x)$.

$$(f \cdot g)(x) = f(x) \cdot g(x)$$
$= \sqrt[3]{x^2y} \cdot (\sqrt[3]{x^4y^2} + \sqrt[3]{xy})$ *Substitute the given functions.*

$= \sqrt[3]{x^2y}\sqrt[3]{x^4y^2} + \sqrt[3]{x^2y}\sqrt[3]{xy}$ *Distributive property*

$= \sqrt[3]{x^6y^3} + \sqrt[3]{x^3y^2}$ *Product rule for radicals*

$= x^2y + x\sqrt[3]{y^2}$

NOW TRY EXERCISE 61

EXAMPLE 9 Simplify $f(x)$ if **a)** $f(x) = \sqrt{x+2}\sqrt{x+2}$, **b)** $f(x) = \sqrt{3x^2 - 30x + 75}$. Assume that the variable may be any real number.

Solution **a)** $f(x) = \sqrt{x+2}\sqrt{x+2}$

$= \sqrt{(x+2)(x+2)}$ *Product rule for Radicals*

$= \sqrt{(x+2)^2}$

$= |x+2|$

Since we are told the variable can be any real number we write the answer with absolute value signs. If we had been told that $x + 2$ was nonnegative, then we could have written our answer as $x + 2$.

b) $f(x) = \sqrt{3x^2 - 30x + 75}$

$= \sqrt{3(x^2 - 10x + 25)}$ Factor out 3.

$= \sqrt{3(x-5)^2}$ Write as the square of a binomial.

$= \sqrt{3}\sqrt{(x-5)^2}$ Product rule for radicals

$= \sqrt{3}|x-5|$

Since the variable could be any real number we write our answer with absolute value signs. If we had been told that $x - 5$ was nonnegative, then we could have written our answer as $\sqrt{3}(x-5)$.

NOW TRY EXERCISE 89

We will do additional multiplication of radicals in Sections 7.4 and 7.5.

Exercise Set 7.3

Concept/Writing Exercises

1. **a)** How do you obtain the numbers that are perfect squares?
   b) List the first six perfect squares.
2. **a)** How do you obtain the numbers that are perfect cubes?
   b) List the first six perfect cube numbers.
3. **a)** How do you obtain numbers that are perfect fifth powers?
   b) List the first five perfect fifth-power numbers.
4. In your own words, state the product rule for radicals.
5. We stated that for nonnegative real numbers a and b, $\sqrt[n]{a} \cdot \sqrt[n]{b} = \sqrt[n]{ab}$. Why is it necessary to specify that both a and b are nonnegative real numbers?
6. **a)** In your own words, explain how to simplify radicals.
   b) Simplify $\sqrt{32x^5y^4}$ using the procedure given in part **a)**.

Practice the Skills

In Exercises 7–86, assume that all variables represent nonnegative real numbers, except where indicated otherwise.

Simplify.

7. $\sqrt{75}$
8. $\sqrt{40}$
9. $\sqrt{32}$
10. $\sqrt{72}$
11. $\sqrt[3]{16}$
12. $\sqrt[3]{24}$
13. $\sqrt[3]{54}$
14. $\sqrt[4]{80}$
15. $-\sqrt{x^3}$
16. $\sqrt{y^5}$
17. $7\sqrt{x^{11}}$
18. $\sqrt{a^{30}}$
19. $\sqrt{b^{27}}$
20. $\sqrt[3]{y^7}$
21. $\sqrt[4]{b^{23}}$
22. $\sqrt{24x^3}$
23. $3\sqrt[3]{24c^{11}}$
24. $-3\sqrt[4]{16x^{10}}$
25. $\sqrt{x^3y^7}$
26. $2\sqrt{50xy^4}$
27. $\sqrt[3]{81a^6b^8}$
28. $\sqrt[3]{16x^3y^6}$
29. $4\sqrt[3]{54x^{12}y^{13}}$
30. $\sqrt[4]{x^9y^{12}z^{15}}$
31. $-\sqrt[5]{64x^{12}y^7}$
32. $\sqrt[3]{32c^4w^9z}$
33. $\sqrt[4]{32x^8y^9z^{19}}$
34. $\sqrt[3]{81x^7y^{21}z^{50}}$

Simplify.

35. $\sqrt{50}\sqrt{2}$
36. $\sqrt[3]{2}\sqrt[3]{4}$
37. $\sqrt[3]{2}\sqrt[3]{28}$
38. $\sqrt[3]{3}\sqrt[3]{54}$
39. $\sqrt{15xy^4}\sqrt{6xy^3}$
40. $(\sqrt{6xy^2})^2$
41. $\sqrt{9m^3n^7}\sqrt{3mn^4}$
42. $\sqrt[3]{5ab^2}\sqrt[3]{25a^4b^{12}}$
43. $\sqrt[3]{9x^7y^{12}}\sqrt[3]{6x^4y}$
44. $(\sqrt[3]{2x^3y^4})^2$
45. $(\sqrt[3]{5x^2y^6})^2$
46. $\sqrt[4]{3x^9y^{12}}\sqrt[4]{54x^4y^7}$
47. $\sqrt[5]{x^{24}y^{30}z^9}\sqrt[5]{x^{13}y^8z^7}$
48. $\sqrt[4]{8x^4yz^3}\sqrt[4]{2x^2y^3z^7}$

Simplify.

49. $\sqrt{2}(\sqrt{6} + \sqrt{2})$
50. $\sqrt{5}(\sqrt{5} + \sqrt{3})$
51. $\sqrt{3}(\sqrt{12} - \sqrt{6})$
52. $2(2\sqrt{8} - 3\sqrt{2})$
53. $\sqrt{2}(\sqrt{18} + \sqrt{8})$
54. $\sqrt[3]{x}(\sqrt[3]{x^2} + \sqrt[3]{x^5})$
55. $\sqrt{3y}(\sqrt{27y^2} - \sqrt{y})$
56. $2\sqrt[3]{x^4y^5}(\sqrt[3]{8x^{12}y^4} + \sqrt[3]{16xy^9})$
57. $\sqrt[4]{8x^3y^5}(\sqrt[4]{4x^5y^7} - \sqrt[4]{3x^7y^6})$
58. $\sqrt[5]{16x^7y^6}(\sqrt[5]{2x^6y^9} - \sqrt[5]{10x^3y^7})$

In Exercises 59–66, $f(x)$ and $g(x)$ are given. Find $(f \cdot g)(x)$.

59. $f(x) = \sqrt{2x},\ g(x) = \sqrt{8x} - \sqrt{32}$

60. $f(x) = \sqrt{5x},\ g(x) = \sqrt{5x} - \sqrt{10x}$

61. $f(x) = \sqrt[3]{xy^2},\ g(x) = \sqrt[3]{x^5y^2} + \sqrt[3]{x^2y^2}$

62. $f(x) = \sqrt[3]{2x^2y^2},\ g(x) = \sqrt[3]{4xy^5} + \sqrt[3]{32x}$

63. $f(x) = \sqrt[4]{3x^2y},\ g(x) = \sqrt[4]{9x^4y} - \sqrt[4]{x^7}$

64. $f(x) = \sqrt[4]{2x^3y^2},\ g(x) = \sqrt[4]{8x^5y^7} - \sqrt[4]{3x^5y^6}$

65. $f(x) = \sqrt[5]{8x^4y^6},\ g(x) = \sqrt[5]{4x^6y^9} - \sqrt[5]{10xy^7}$

66. $f(x) = \sqrt[6]{x^4y^5},\ g(x) = \sqrt[6]{x^9y^{12}} + \sqrt[6]{x^4y^3}$

Simplify. These exercises are a combination of the types of exercises presented earlier in this exercise set.

67. $\sqrt{24}$

68. $\sqrt{200}$

69. $\sqrt[3]{32}$

70. $\sqrt[3]{x^5}$

71. $\sqrt[3]{y^{13}}$

72. $\sqrt{36x^5}$

73. $\sqrt[3]{80x^{11}}$

74. $\sqrt[3]{x^9y^{11}z}$

75. $\sqrt[6]{128ab^{17}c^9}$

76. $\sqrt{75}\sqrt{6}$

77. $\sqrt[4]{8}\sqrt[4]{10}$

78. $\sqrt[5]{14x^4y^2}\sqrt[5]{3x^4y^3}$

79. $\sqrt{20xy^4}\sqrt{6x^5y^7}$

80. $\sqrt{6}(4 - \sqrt{2})$

81. $\sqrt{x}(\sqrt{x} + 3)$

82. $(\sqrt[3]{4x^5y^2})^2$

83. $\sqrt[3]{y}(2\sqrt[3]{y} - \sqrt[3]{y^8})$

84. $\sqrt[3]{2x^9y^6z}\sqrt[3]{12xy^4z^3}$

85. $\sqrt[3]{3ab^2}(\sqrt[3]{4a^4b^3} - \sqrt[3]{8a^5b^4})$

86. $\sqrt[4]{4st^2}(\sqrt[4]{2s^5t^6} + \sqrt[4]{5s^9t^2})$

Simplify the following. Assume that x can be any real number.

87. $f(x) = \sqrt{2x+5}\sqrt{2x+5}$

88. $g(a) = \sqrt{3a+4}\sqrt{3a+4}$

89. $h(r) = \sqrt{4r^2 - 32r + 64}$

90. $f(b) = \sqrt{20b^2 + 60b + 45}$

Problem Solving

91. Assume that x can be any real number. Simplify $\sqrt[n]{(x-4)^{2n}}$ if

a) n is even.

b) n is odd.

92. Consider $F = \sqrt{ab^2}$. How will the value of F change if

a) both a and b are doubled?

b) a is doubled and b is halved?

c) both a and b are halved?

Multiply each expression using the FOIL method. We will discuss problems like these in Section 7.4.

93. $(\sqrt{3} + \sqrt{2})(\sqrt{3} - \sqrt{2})$

94. $(2\sqrt{x} - 3\sqrt{y})(2\sqrt{x} + 3\sqrt{y})$

95. $(5\sqrt{a} - 4\sqrt{b})(2\sqrt{a} + 3\sqrt{b})$

96. Prove $\sqrt{a \cdot b} = \sqrt{a}\sqrt{b}$ by converting $\sqrt{a \cdot b}$ to exponential form.

97. Law enforcement officials sometimes use the formula $s = \sqrt{30FB}$ to determine a car's speed, s, in miles per hour, from a car's skid marks. The F in the formula represents the "road factor," which is determined by the road's surface, and the B represents the braking distance, in feet. Officer Jenkins is investigating an accident. Find the car's speed if the skid marks are 80 feet long and **a)** the road was dry asphalt, whose road factor is 0.85, **b)** the road was wet gravel, whose road factor is 0.52.

98. Many recent articles have discussed the increasing cost of long-term care in a nursing home. The average cost, $c(t)$, in dollars, for a year in a nursing home can be estimated by the function $c(t) = 345.0\sqrt{t^3}$, where t is years since 1964 and $1 \le t \le 34$. Estimate the cost of a year in a nursing home in **a)** 1970 and **b)** 1998.

99. The rate at which water flows through a particular fire hose, R, in gallons per minute, can be approximated by the formula $R = 28d^2\sqrt{P}$, where d is the diameter of the nozzle, in inches, and P is the nozzle pressure, in pounds per square inch. The Leadville, Colorado, fire department is fighting a fire. If the nozzle has a diameter of 2.5 inches and the nozzle pressure is 80 pounds per square inch, find the flow rate.

100. The number of applicants to Princeton University has increased from 1960 through 1990. The number of applicants $N(t)$, in thousands, can be estimated by the function $N(t) = \sqrt[3]{14.1t^2}$, where t is years since 1959 and $1 \le t \le 31$. Estimate the number of applications in **a)** 1960 and **b)** 1980.

101. In statistics, the standard deviation of the population is a measure of the spread of a set of data about the mean of the data. The greater the spread, the greater the standard deviation. There are a number of formulas used to find the standard deviation, σ, read "sigma". One formula is $\sigma = \sqrt{npq}$, where n represents the sample size, p represents the percent chance (or probability) that something specific happens, and q represents the percent chance (or probability) that the specific thing does not happen. In a sample of 600 people who purchase airline tickets, the percent that showed up for their flight, p, was 0.93, and the percent that did not show up for their flight, q, was 0.07. Use this information to find σ.

Cumulative Review Exercises

[1.2] **102.** What is a rational number?

[1.3] **103.** What is a real number?

104. What is an irrational number?

105. What is the definition of $|a|$?

[2.2] **106.** Solve the formula $E = \frac{1}{2}mv^2$ for m.

[2.5] **107.** Solve the inequality $-4 < 2x - 3 \le 5$ and indicate the solution **a)** on the number line; **b)** in internal notation; **c)** in set builder notation.

7.4 DIVIDING AND SIMPLIFYING RADICALS

SSM VIDEO 7.4 CD Rom

1. Apply the quotient rule for radicals.
2. Rationalize denominators.
3. Rationalize a denominator using the conjugate.
4. Know when a radical is simplified.

1 Apply the Quotient Rule for Radicals

In mathematics we sometimes need to divide one radical expression by another. To divide radicals, or to simplify radicals containing fractions, we use the **quotient rule for radicals**.

Quotient Rule for Radicals

For nonnegative real numbers a and b,

$$\frac{\sqrt[n]{a}}{\sqrt[n]{b}} = \sqrt[n]{\frac{a}{b}}, \quad b \neq 0$$

Examples 1 through 3 illustrate how to use the quotient rule to simplify radical expressions.

EXAMPLE 1 Simplify. **a)** $\dfrac{\sqrt{75}}{\sqrt{3}}$ **b)** $\dfrac{\sqrt[3]{24x}}{\sqrt[3]{3x}}$ **c)** $\dfrac{\sqrt[3]{x^4y^7}}{\sqrt[3]{xy^{-2}}}$

Solution In each part we use the quotient rule to write the quotient of radicals as a single radical. Then we simplify.

a) $\dfrac{\sqrt{75}}{\sqrt{3}} = \sqrt{\dfrac{75}{3}} = \sqrt{25} = 5$

b) $\dfrac{\sqrt[3]{24x}}{\sqrt[3]{3x}} = \sqrt[3]{\dfrac{24x}{3x}} = \sqrt[3]{8} = 2$

c) $\dfrac{\sqrt[3]{x^4y^7}}{\sqrt[3]{xy^{-2}}} = \sqrt[3]{\dfrac{x^4y^7}{xy^{-2}}}$ *Quotient rule for radicals*

$= \sqrt[3]{x^3y^9}$ *Simplify the radicand.*

$= xy^3$

EXAMPLE 2 Simplify. **a)** $\sqrt{\dfrac{9}{4}}$ **b)** $\sqrt[3]{\dfrac{8x^4y}{27xy^{10}}}$ **c)** $\sqrt[4]{\dfrac{15xy^5}{3x^9y}}$

Solution In each part we first simplify the radicand, if possible. Then we use the quotient rule to write the given radical as a quotient of radicals.

a) $\sqrt{\dfrac{9}{4}} = \dfrac{\sqrt{9}}{\sqrt{4}} = \dfrac{3}{2}$

b) $\sqrt[3]{\dfrac{8x^4y}{27xy^{10}}} = \sqrt[3]{\dfrac{8x^3}{27y^9}} = \dfrac{\sqrt[3]{8x^3}}{\sqrt[3]{27y^9}} = \dfrac{2x}{3y^3}$

c) $\sqrt[4]{\dfrac{15xy^5}{3x^9y}} = \sqrt[4]{\dfrac{5y^4}{x^8}} = \dfrac{\sqrt[4]{5y^4}}{\sqrt[4]{x^8}} = \dfrac{\sqrt[4]{y^4}\sqrt[4]{5}}{x^2} = \dfrac{y\sqrt[4]{5}}{x^2}$

NOW TRY EXERCISE 35

When dividing radicals that contain different indices we use rational exponents, as illustrated in Example 3.

EXAMPLE 3 Simplify. **a)** $\dfrac{\sqrt[5]{(m+n)^7}}{\sqrt[3]{(m+n)^4}}$ **b)** $\dfrac{\sqrt[3]{a^5b^4}}{\sqrt{a^2b}}$

Solution Begin by writing the numerator and denominator with rational exponents.

a) $\dfrac{\sqrt[5]{(m+n)^7}}{\sqrt[3]{(m+n)^4}} = \dfrac{(m+n)^{7/5}}{(m+n)^{4/3}}$ *Write with rational exponents.*

$= (m+n)^{(7/5)-(4/3)}$ *Quotient rule for exponents*

$= (m+n)^{1/15}$

$= \sqrt[15]{m+n}$ *Write as a radical.*

b) $\dfrac{\sqrt[3]{a^5b^4}}{\sqrt{a^2b}} = \dfrac{(a^5b^4)^{1/3}}{(a^2b)^{1/2}}$ *Write with rational exponents.*

$= \dfrac{a^{5/3}b^{4/3}}{ab^{1/2}}$ *Raise the product to a power.*

$= a^{(5/3)-1}b^{(4/3)-(1/2)}$ *Quotient rule for exponents*

$= a^{2/3}b^{5/6}$

$= a^{4/6}b^{5/6}$ *Write the fractions with denominator 6.*

$= (a^4b^5)^{1/6}$ *Rewrite using the laws of exponents.*

NOW TRY EXERCISE 107 $= \sqrt[6]{a^4b^5}$ *Write as a radical.*

2 Rationalize Denominators

When the denominator of a fraction contains a radical, we generally simplify the expression by **rationalizing the denominator**. To rationalize a denominator is to remove all radicals from the denominator. Denominators are rationalized because, without a calculator, it is often easier to evaluate a fraction with a whole-number denominator than one with a radical.

To Rationalize a Denominator

Multiply both the numerator and the denominator of the fraction by the denominator, or by a radical that will result in the radicand in the denominator becoming a perfect power.

When both the numerator and denominator are multiplied by the same radical expression, you are in effect multiplying the fraction by 1, which does not change its value.

EXAMPLE 4 Simplify. **a)** $\dfrac{1}{\sqrt{5}}$ **b)** $\dfrac{x}{3\sqrt{2}}$ **c)** $\dfrac{3}{\sqrt{2x}}$ **d)** $\dfrac{\sqrt[3]{16a^4}}{\sqrt[3]{b}}$

Solution To simplify each expression, we must rationalize the denominators.

a) $\dfrac{1}{\sqrt{5}} = \dfrac{1}{\sqrt{5}} \cdot \dfrac{\sqrt{5}}{\sqrt{5}} = \dfrac{\sqrt{5}}{\sqrt{25}} = \dfrac{\sqrt{5}}{5}$

b) $\dfrac{x}{3\sqrt{2}} = \dfrac{x}{3\sqrt{2}} \cdot \dfrac{\sqrt{2}}{\sqrt{2}} = \dfrac{x\sqrt{2}}{3 \cdot 2} = \dfrac{x\sqrt{2}}{6}$

c) There are two factors in the radicand, 2 and x. We must make each factor a perfect square. Since 2^2 or 4 is a perfect square and x^2 is a perfect square, we multiply both numerator and denominator by $\sqrt{2x}$.

$$\frac{3}{\sqrt{2x}} = \frac{3}{\sqrt{2x}} \cdot \frac{\sqrt{2x}}{\sqrt{2x}}$$

$$= \frac{3\sqrt{2x}}{\sqrt{4x^2}}$$

$$= \frac{3\sqrt{2x}}{2x}$$

d) There are no common factors in the numerator and denominator. Before we rationalize the denominator let's simplify the numerator.

$$\frac{\sqrt[3]{16a^4}}{\sqrt[3]{b}} = \frac{\sqrt[3]{8a^3}\,\sqrt[3]{2a}}{\sqrt[3]{b}} \qquad \text{Product rule for radicals}$$

$$= \frac{2a\sqrt[3]{2a}}{\sqrt[3]{b}} \qquad \text{Simplify the numerator.}$$

Now we rationalize the denominator. Since the denominator is a cube root, we need to make the radicand a perfect cube. Since the denominator contains b and we want b^3, we need two more factors of b. We therefore multiply both numerator and denominator by $\sqrt[3]{b^2}$.

$$= \frac{2a\sqrt[3]{2a}}{\sqrt[3]{b}} \cdot \frac{\sqrt[3]{b^2}}{\sqrt[3]{b^2}}$$

$$= \frac{2a\sqrt[3]{2ab^2}}{\sqrt[3]{b^3}}$$

$$= \frac{2a\sqrt[3]{2ab^2}}{b}$$

EXAMPLE 5 Simplify. **a)** $\sqrt{\dfrac{3}{5}}$ **b)** $\sqrt[3]{\dfrac{x}{2y^2}}$ **c)** $\sqrt[4]{\dfrac{32x^9y^6}{3z^2}}$

Solution In each part we will use the quotient rule to write the radical as a quotient of two radicals.

a) $\sqrt{\dfrac{3}{5}} = \dfrac{\sqrt{3}}{\sqrt{5}} \cdot \dfrac{\sqrt{5}}{\sqrt{5}} = \dfrac{\sqrt{15}}{\sqrt{25}} = \dfrac{\sqrt{15}}{5}$

b) $\sqrt[3]{\dfrac{x}{2y^2}} = \dfrac{\sqrt[3]{x}}{\sqrt[3]{2y^2}}$

The denominator is $\sqrt[3]{2y^2}$ and we want to change it to $\sqrt[3]{2^3y^3}$. We now multiply both the numerator and denominator by the cube root of an expression that will make the radicand in the denominator $\sqrt[3]{2^3y^3}$. Since $2 \cdot 2^2 = 2^3$ and $y^2 \cdot y = y^3$, we multiply both numerator and denominator by $\sqrt[3]{2^2y}$.

$$\frac{\sqrt[3]{x}}{\sqrt[3]{2y^2}} = \frac{\sqrt[3]{x}}{\sqrt[3]{2y^2}} \cdot \frac{\sqrt[3]{2^2y}}{\sqrt[3]{2^2y}}$$

$$= \frac{\sqrt[3]{x}\,\sqrt[3]{4y}}{\sqrt[3]{2^3y^3}}$$

$$= \frac{\sqrt[3]{4xy}}{2y}$$

c) After using the quotient rule, we simplify the numerator.

$$\sqrt[4]{\frac{32x^9y^6}{3z^2}} = \frac{\sqrt[4]{32x^9y^6}}{\sqrt[4]{3z^2}}$$ Quotient rule for radicals

$$= \frac{\sqrt[4]{16x^8y^4}\sqrt[4]{2xy^2}}{\sqrt[4]{3z^2}}$$ Product rule for radicals

$$= \frac{2x^2y\sqrt[4]{2xy^2}}{\sqrt[4]{3z^2}}$$ Simplify the numerator.

Now we rationalize the denominator. To make the radicand in the denominator a perfect fourth power, we need to get each factor to a power of 4. Since the denominator contains one factor of 3, we need 3 more factors of 3, or 3^3. Since there are two factors of z, we need 2 more factors of z, or z^2. Thus we will multiply both numerator and denominator by $\sqrt[4]{3^3z^2}$.

$$= \frac{2x^2y\sqrt[4]{2xy^2}}{\sqrt[4]{3z^2}} \cdot \frac{\sqrt[4]{3^3z^2}}{\sqrt[4]{3^3z^2}}$$

$$= \frac{2x^2y\sqrt[4]{2xy^2}\sqrt[4]{27z^2}}{\sqrt[4]{3z^2}\sqrt[4]{3^3z^2}}$$

$$= \frac{2x^2y\sqrt[4]{54xy^2z^2}}{\sqrt[4]{3^4z^4}}$$ Product rule for radicals

$$= \frac{2x^2y\sqrt[4]{54xy^2z^2}}{3z}$$

NOW TRY EXERCISE 49

Note: There are no perfect fourth power factors of 54, so we are finished.

3) Rationalize a Denominator Using the Conjugate

We will discuss adding and subtracting radicals in detail in Section 7.5. In this section we give only a brief introduction to this topic. Only radicals that have the same radicands and index may be added or subtracted. We add such radicals, called *like radicals*, by adding or subtracting their coefficients and multiplying this sum or difference by the like radical.

Examples

$3\sqrt{3} - 3\sqrt{3} = 0$ $\qquad$ $3\sqrt{x} + \sqrt{x} = 3\sqrt{x} + 1\sqrt{x} = 4\sqrt{x}$

$2\sqrt{5} - 9\sqrt{5} = -7\sqrt{5}$ $\qquad$ $3\sqrt{y} - 3\sqrt{y} = 0$

The terms $\sqrt{2} + \sqrt{3}$ cannot be added in radical form since the radicands are different.

When the denominator of a rational expression is a binomial that contains a radical, we rationalize the denominator. We do this by multiplying both the numerator and the denominator of the fraction by the **conjugate** of the denominator. The conjugate of a binomial is a binomial having the same two terms with the sign of the second term changed.

Expression	**Conjugate**
$3 + \sqrt{2}$	$3 - \sqrt{2}$
$2\sqrt{3} - \sqrt{5}$	$2\sqrt{3} + \sqrt{5}$
$\sqrt{x} + \sqrt{y}$	$\sqrt{x} - \sqrt{y}$
$a + \sqrt{b}$	$a - \sqrt{b}$

When a binomial is multiplied by its conjugate, the outer and inner products will sum to zero.

EXAMPLE 6 Multiply $(2 + \sqrt{3})(2 - \sqrt{3})$.

Solution Multiply using the FOIL method.

$$\overset{F}{2(2)} + \overset{O}{2(-\sqrt{3})} + \overset{I}{2(\sqrt{3})} + \overset{L}{\sqrt{3}(-\sqrt{3})} = 4 - 2\sqrt{3} + 2\sqrt{3} - \sqrt{9}$$
$$= 4 - \sqrt{9}$$
$$= 4 - 3 = 1$$

In Example 6, we would get the same result using the formula for the product of the sum and difference of the same two terms. The product results in the difference of two squares, $(a + b)(a - b) = a^2 - b^2$,

$$(2 + \sqrt{3})(2 - \sqrt{3}) = 2^2 - (\sqrt{3})^2$$
$$= 4 - 3 = 1$$

EXAMPLE 7 Multiply $(\sqrt{3} - \sqrt{5})(\sqrt{3} + \sqrt{5})$.

Solution We will multiply using the formula for the product of the sum and difference of the same two terms.

$$(\sqrt{3} - \sqrt{5})(\sqrt{3} + \sqrt{5}) = (\sqrt{3})^2 - (\sqrt{5})^2$$
$$= 3 - 5 = -2$$

EXAMPLE 8 Simplify. **a)** $\dfrac{5}{2 + \sqrt{3}}$ **b)** $\dfrac{6}{\sqrt{5} - \sqrt{2}}$ **c)** $\dfrac{x - \sqrt{y}}{x + \sqrt{y}}$

Solution In each part, we rationalize the denominator by multiplying the numerator and the denominator by the conjugate of the denominator.

a)
$$\frac{5}{2 + \sqrt{3}} = \frac{5}{2 + \sqrt{3}} \cdot \frac{2 - \sqrt{3}}{2 - \sqrt{3}}$$
$$= \frac{5(2 - \sqrt{3})}{(2 + \sqrt{3})(2 - \sqrt{3})}$$
$$= \frac{5(2 - \sqrt{3})}{4 - 3}$$
$$= 5(2 - \sqrt{3}) \quad \text{or} \quad 10 - 5\sqrt{3}$$

b)
$$\frac{6}{\sqrt{5}-\sqrt{2}} = \frac{6}{\sqrt{5}-\sqrt{2}} \cdot \frac{\sqrt{5}+\sqrt{2}}{\sqrt{5}+\sqrt{2}}$$
$$= \frac{6(\sqrt{5}+\sqrt{2})}{5-2}$$
$$= \frac{\overset{2}{\cancel{6}}(\sqrt{5}+\sqrt{2})}{\underset{1}{\cancel{3}}}$$
$$= 2(\sqrt{5}+\sqrt{2}) \quad \text{or} \quad 2\sqrt{5}+2\sqrt{2}$$

c)
$$\frac{x-\sqrt{y}}{x+\sqrt{y}} = \frac{x-\sqrt{y}}{x+\sqrt{y}} \cdot \frac{x-\sqrt{y}}{x-\sqrt{y}}$$
$$= \frac{x^2 - x\sqrt{y} - x\sqrt{y} + \sqrt{y^2}}{x^2 - y}$$
$$= \frac{x^2 - 2x\sqrt{y} + y}{x^2 - y}$$

NOW TRY EXERCISE 69 Remember that you cannot divide out x^2 or y because they are terms, not factors.

Avoiding Common Errors

The following simplifications are correct because the numbers and variables divided out are not within square roots.

CORRECT	CORRECT
$\frac{\overset{2}{\cancel{6}}\sqrt{2}}{\underset{1}{\cancel{3}}} = 2\sqrt{2}$	$\frac{\cancel{x}\sqrt{2}}{\cancel{x}} = \sqrt{2}$

An expression within a square root cannot be divided by an expression not within the square root.

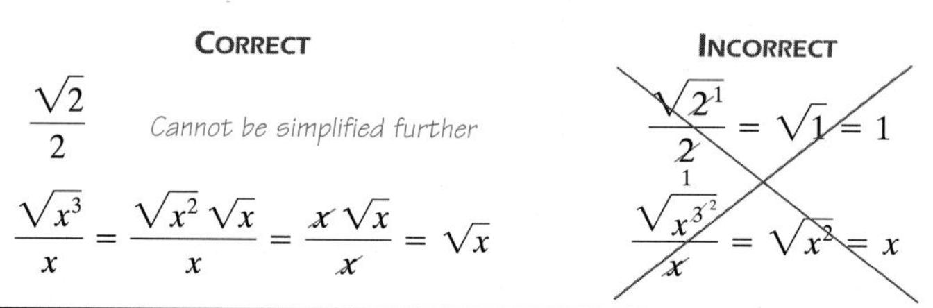

CORRECT		INCORRECT
$\frac{\sqrt{2}}{2}$	Cannot be simplified further	$\frac{\sqrt{\cancel{2}^{1}}}{\underset{1}{\cancel{2}}} = \sqrt{1} = 1$
$\frac{\sqrt{x^3}}{x} = \frac{\sqrt{x^2}\sqrt{x}}{x} = \frac{\cancel{x}\sqrt{x}}{\cancel{x}} = \sqrt{x}$		$\frac{\sqrt{x^{\cancel{3}^{2}}}}{\cancel{x}} = \sqrt{x^2} = x$

4) Know When a Radical is Simplified

After you have simplified a radical expression, you should check it to make sure that it is simplified as far as possible.

A Radical Expression is Simplified When the Following Are All True

1. No perfect powers are factors of the radicand.
2. No radicand contains a fraction.
3. No denominator contains a radical.

Exercise Set 7.4

Concept/Writing Exercises

1. Give the quotient rule for radicals.
2. a) What is the conjugate of a binomial?
   b) What is the conjugate of $x - \sqrt{3}$?
3. What does it mean to rationalize a denominator?
4. a) Explain how to rationalize a denominator that contains a radical expression of one term.
   b) Rationalize $\frac{4}{\sqrt{3y}}$ using the procedure you specified in part **a)**.
5. a) Explain how to rationalize a denominator that contains a binomial in which one or both terms is a radical expression.
   b) Rationalize $\frac{\sqrt{2}+\sqrt{5}}{\sqrt{2}-\sqrt{5}}$ using the procedure you specified in part **a)**.
6. What are the three conditions that must be met for a radical expression to be simplified?

Practice the Skills

In this exercise set, assume that all variables represent positive real numbers.

Simplify.

7. $\sqrt{\frac{36}{4}}$
8. $\sqrt{\frac{4}{25}}$
9. $\frac{\sqrt{3}}{\sqrt{27}}$
10. $\sqrt{\frac{81}{49}}$
11. $\sqrt[3]{\frac{2}{16}}$
12. $\frac{\sqrt[3]{108}}{\sqrt[3]{3}}$
13. $\frac{-\sqrt{24}}{\sqrt{3}}$
14. $\sqrt[3]{\frac{c^3}{27}}$
15. $\sqrt{\frac{r^4}{25}}$
16. $\frac{\sqrt[3]{2x^6}}{\sqrt[3]{16x^3}}$
17. $\sqrt{\frac{27x^6}{3x^2}}$
18. $\sqrt{\frac{72x^2y^5}{8x^2y^7}}$
19. $\sqrt[3]{\frac{7xy}{8x^{13}}}$
20. $\sqrt[3]{\frac{25x^2y^5}{5x^8y^{-2}}}$
21. $\sqrt[4]{\frac{20x^4}{81x^{-8}}}$
22. $\sqrt[4]{\frac{3a^6b^5}{16a^2b^{13}}}$

Simplify.

23. $\frac{1}{\sqrt{5}}$
24. $\frac{3}{\sqrt{3}}$
25. $\frac{\sqrt{m}}{\sqrt{2}}$
26. $\frac{5\sqrt{3}}{\sqrt{5}}$
27. $\frac{\sqrt{x}}{\sqrt{y}}$
28. $\frac{2\sqrt{3}}{\sqrt{w}}$
29. $\sqrt{\frac{5m}{8}}$
30. $\frac{2\sqrt{3}}{\sqrt{y^3}}$
31. $\frac{2n}{\sqrt{18n}}$
32. $\sqrt{\frac{120x}{4y^3}}$
33. $\sqrt{\frac{8x^6y}{2xz}}$
34. $\sqrt{\frac{5pq^4}{2r}}$
35. $\sqrt{\frac{20y^4z^3}{3xy^{-2}}}$
36. $\sqrt{\frac{5xy^6}{6z}}$
37. $\sqrt{\frac{18x^4y^3}{2z^3}}$
38. $\sqrt{\frac{45y^{12}z^{10}}{2x}}$

Simplify.

39. $\frac{1}{\sqrt[3]{2}}$
40. $\frac{2}{\sqrt[3]{4}}$
41. $\frac{1}{\sqrt[3]{5}}$
42. $\frac{5}{\sqrt[3]{x}}$
43. $\sqrt[3]{\frac{5x}{y}}$
44. $\sqrt[3]{\frac{1}{4x}}$
45. $-\sqrt[3]{\frac{5c}{9y^2}}$
46. $\frac{3}{\sqrt[4]{a}}$
47. $\frac{5m}{\sqrt[4]{2}}$
48. $\sqrt[4]{\frac{5}{3x^3}}$
49. $\sqrt[4]{\frac{2x^3}{4y^2}}$
50. $\sqrt[3]{\frac{3x^2}{2y^2}}$
51. $\sqrt[3]{\frac{15x^6y^7}{2z^2}}$
52. $\sqrt[3]{\frac{8xy^2}{2z^2}}$
53. $\sqrt[6]{\frac{r^4s^9}{2r^5}}$
54. $\sqrt[6]{\frac{2x^4y^8z}{3x^7}}$

Simplify.

55. $(5+\sqrt{5})(5-\sqrt{5})$

56. $(4+\sqrt{2})(4-\sqrt{2})$

57. $(7-\sqrt{5})(7+\sqrt{5})$

58. $(\sqrt{x}+5)(\sqrt{x}-5)$

59. $(\sqrt{6}+x)(\sqrt{6}-x)$

60. $(\sqrt{x}+y)(\sqrt{x}-y)$

61. $(\sqrt{x}+\sqrt{y})(\sqrt{x}-\sqrt{y})$

62. $(5-\sqrt{y})(5+\sqrt{y})$

63. $(2\sqrt{3}-\sqrt{2})(2\sqrt{3}+\sqrt{2})$

64. $(3\sqrt{a}-7\sqrt{b})(3\sqrt{a}+7\sqrt{b})$

Simplify.

65. $\dfrac{5}{\sqrt{2}+1}$

66. $\dfrac{1}{2+\sqrt{3}}$

67. $\dfrac{2}{5-\sqrt{6}}$

68. $\dfrac{4}{\sqrt{2}-7}$

69. $\dfrac{2}{\sqrt{2}+\sqrt{3}}$

70. $\dfrac{\sqrt{5}}{2\sqrt{5}-\sqrt{6}}$

71. $\dfrac{1}{\sqrt{17}-\sqrt{8}}$

72. $\dfrac{2}{6+\sqrt{x}}$

73. $\dfrac{3\sqrt{5}}{\sqrt{a}-3}$

74. $\dfrac{4\sqrt{x}}{\sqrt{x}-y}$

75. $\dfrac{\sqrt{8x}}{x+\sqrt{y}}$

76. $\dfrac{\sqrt{2}-2\sqrt{3}}{\sqrt{2}+4\sqrt{3}}$

77. $\dfrac{\sqrt{c}-\sqrt{2d}}{\sqrt{c}-\sqrt{d}}$

78. $\dfrac{\sqrt{a^3}+\sqrt{a^7}}{\sqrt{a}}$

79. $\dfrac{2\sqrt{xy}-\sqrt{xy}}{\sqrt{x}+\sqrt{y}}$

80. $\dfrac{2}{\sqrt{x+2}-3}$

Simplify. These exercises are a combination of the type of exercises presented earlier in this exercise set.

81. $\sqrt{\dfrac{x}{9}}$

82. $\sqrt[4]{\dfrac{x^4}{16}}$

83. $\sqrt{\dfrac{2}{5}}\dfrac{\sqrt{10}}{5}$

84. $\sqrt{\dfrac{a}{b}}$

85. $(\sqrt{5}+\sqrt{6})(\sqrt{5}-\sqrt{6})$

86. $\sqrt[3]{\dfrac{1}{3}}$

87. $\sqrt{\dfrac{24x^3y^6}{5z}}$

88. $\dfrac{6}{4-\sqrt{y}}$

89. $\sqrt{\dfrac{12xy^4}{2x^3y^4}}$

90. $\dfrac{4x}{\sqrt[3]{5y}}$

91. $(\sqrt{x}+3)(\sqrt{x}-3)$

92. $\dfrac{\sqrt{x}}{\sqrt{x}+5\sqrt{y}}$

93. $-\dfrac{7\sqrt{x}}{\sqrt{98}}$

94. $\sqrt{\dfrac{2xy^4}{18xy^2}}$

95. $\sqrt[4]{\dfrac{3y^2}{2x}}$

96. $\sqrt{\dfrac{25x^2y^5}{3z}}$

97. $\sqrt[3]{\dfrac{32y^{12}z^{10}}{2x}}$

98. $\dfrac{\sqrt{3}+\sqrt{4}}{\sqrt{2}+\sqrt{3}}$

99. $\dfrac{\sqrt{ar}}{\sqrt{a}-2\sqrt{r}}$

100. $\sqrt[4]{\dfrac{2}{9x}}$

101. $\dfrac{\sqrt[3]{6x}}{\sqrt[3]{5xy}}$

102. $(3\sqrt{y}-2\sqrt{x})(5\sqrt{y}+\sqrt{x})$

103. $\sqrt[4]{\dfrac{2x^7y^{12}z^4}{3x^9}}$

104. $\dfrac{3}{\sqrt{y+3}-\sqrt{y}}$

Simplify.

105. $\dfrac{\sqrt{(a+b)^4}}{\sqrt[3]{a+b}}$

106. $\dfrac{\sqrt[3]{c+2}}{\sqrt[4]{(c+2)^3}}$

107. $\dfrac{\sqrt[5]{(a+2b)^4}}{\sqrt[3]{(a+2b)^2}}$

108. $\dfrac{\sqrt[6]{(r+3)^5}}{\sqrt[3]{(r+3)^5}}$

109. $\dfrac{\sqrt[3]{r^2s^4}}{\sqrt{rs}}$

110. $\dfrac{\sqrt{a^2b^4}}{\sqrt[3]{ab^2}}$

111. $\dfrac{\sqrt[5]{x^4y^6}}{\sqrt[3]{(xy)^2}}$

112. $\dfrac{\sqrt[6]{4m^8n^4}}{\sqrt[4]{m^4n^2}}$

Problem Solving

113. Under certain conditions the formula

$$d = \sqrt{\frac{72}{I}}$$

is used to show the relationship between the illumination on an object, I, in lumens per meter, and the distance, d, in meters, the object is from the light source. If the illumination on a person standing near a light source is 5.3 lumens per meter, how far is the person from the light source?

114. When sufficient pressure is applied to a particular particle board, the particle board will break (or rupture). The thicker the particle board the greater will be the pressure that will need to be applied before the board breaks. The formula

$$T = \sqrt{\frac{0.05\,LB}{M}}$$

relates the thickness of a specific particle board, T, in inches, the board's length, L, in inches, the board's load, B, in pounds, that will cause the board to rupture, and M, the modulus of rupture, in pounds per square inch. The modulus of rupture is a constant determined by sample tests on the specific type of particle board.

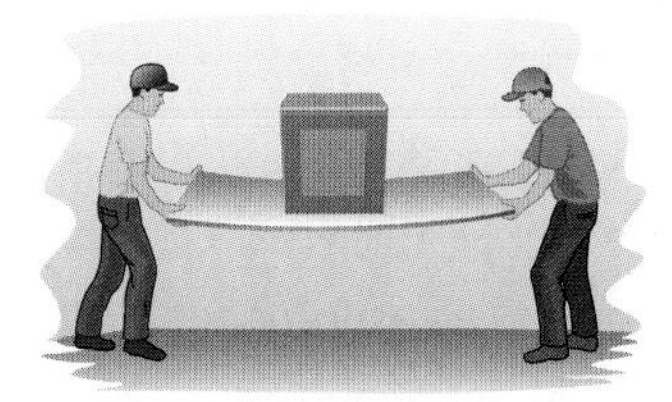

Find the thickness of a 36-inch-long particle board if the modulus of rupture is 2560 pounds per square inch and the board ruptures when 800 pounds are applied.

115. A new restaurant wants to have a spherical fish tank in its lobby. The radius, r, in inches, of a spherical tank can be found by the formula

$$r = \sqrt[3]{\frac{3V}{4\pi}}$$

where V is the volume of the tank in cubic inches. Find the radius of a spherical tank whose volume is 7238.23 cubic inches.

116. If we consider the set of consecutive natural numbers $1, 2, 3, 4, \ldots, n$ to be the population, the standard deviation, σ, which is a measure of the spread of the data from the mean, can be calculated by the formula

$$\sigma = \sqrt{\frac{n^2 - 1}{12}}$$

where n represents the number of natural numbers in the population. Find the standard deviation for the first 100 consecutive natural numbers.

117. The number of farms in the United States is declining annually (however, the size of the remaining farms is increasing). A function that can be used to estimate the number of farms, $N(t)$, in millions, is

$$N(t) = \frac{6.21}{\sqrt[4]{t}}$$

where t is years since 1949 and $1 \le t \le 48$. Estimate the number of farms in the United States in **a)** 1950 and **b)** 1997.

118. The U.S. infant mortality rate has been declining steadily. The infant mortality rate, $N(t)$, defined as deaths per 1000 live births, can be estimated by the function

$$N(t) = \frac{28.46}{\sqrt[3]{t^2}}$$

where t is years since 1959 and $1 \le t \le 37$. Estimate the infant mortality rate in **a)** 1960 and **b)** 1996.

119. We stated that for nonnegative real numbers a and b, $b \neq 0$, $\frac{\sqrt[n]{a}}{\sqrt[n]{b}} = \sqrt[n]{\frac{a}{b}}$. Why is it necessary to specify that both a and b are nonnegative real numbers?

120. **a)** Will $\frac{\sqrt[n]{x}}{\sqrt[n]{x}}$ always equal 1?

b) If your answer to part **a)** was no, under what conditions does $\frac{\sqrt[n]{x}}{\sqrt[n]{x}}$ equal 1?

121. Which is greater, $\frac{2}{\sqrt{2}}$ or $\frac{3}{\sqrt{3}}$? Explain.

122. Which is greater, $\frac{\sqrt{3}}{2}$ or $\frac{2}{\sqrt{3}}$? Explain.

123. Consider the functions $f(x) = x^{a/2}$ and $g(x) = x^{b/3}$.

a) List three values for a that will result in $x^{a/2}$ being a perfect square.

b) List three values for b that will result in $x^{b/3}$ being a perfect cube.

c) If $x \geq 0$, find $(f \cdot g)(x)$.

d) If $x \geq 0$, find $(f/g)(x)$.

Rationalize each denominator.

124. $\dfrac{1}{\sqrt{a+b}}$

125. $\dfrac{3}{\sqrt{2a-3b}}$

In higher math courses it may be necessary to rationalize the numerators of radical expressions. Rationalize the numerators of the following expressions. (Your answers will contain radicals in the denominators.)

126. $\dfrac{\sqrt{6}}{3}$

127. $\dfrac{5-\sqrt{5}}{6}$

128. $\dfrac{4\sqrt{x}-\sqrt{3}}{x}$

129. $\dfrac{\sqrt{x+h}-\sqrt{x}}{h}$

130. Prove $\sqrt[n]{\dfrac{a}{b}} = \dfrac{\sqrt[n]{a}}{\sqrt[n]{b}}$ by converting $\sqrt[n]{\dfrac{a}{b}}$ to exponential form.

Cumulative Review Exercises

[2.2] **131.** Solve the equation $A = \frac{1}{2}h(b_1 + b_2)$ for b_2.

[2.4] **132.** Two cars leave from West Point at the same time traveling in opposite directions. One travels 10 miles per hour faster than the other. If the two cars are 270 miles apart after 3 hours, find the speed of each car.

[5.2] **133.** Multiply $(4x^2 - 3x - 2)(2x - 3)$.

[6.4] **134.** Solve $(2x - 3)(x - 2) = 4x - 6$.

7.5 ADDING AND SUBTRACTING RADICALS

1) Add and subtract radicals.

SSM VIDEO 7.5 CD Rom

1) Add and Subtract Radicals

Like radicals are radicals having the same radicand and index. **Unlike radicals** are radicals differing in either the radicand or the index.

Examples of Like Radicals	Examples of Unlike Radicals	
$\sqrt{5}, 3\sqrt{5}$	$\sqrt{5}, \sqrt[3]{5}$	*indices differ*
$5\sqrt{7}, -2\sqrt{7}$	$\sqrt{5}, \sqrt{7}$	*radicands differ*
$\sqrt{x}, 5\sqrt{x}$	$\sqrt{x}, \sqrt{2x}$	*radicands differ*
$\sqrt[3]{2x}, -4\sqrt[3]{2x}$	$\sqrt{x}, \sqrt[3]{x}$	*indices differ*
$\sqrt[4]{x^2y^5}, -\sqrt[4]{x^2y^5}$	$\sqrt[3]{xy}, \sqrt[3]{x^2y}$	*radicands differ*

Like radicals are added and subtracted in exactly the same way that like terms are added or subtracted. To add or subtract like radicals, add or subtract their numerical coefficients and multiply this sum or difference by the like radical.

Examples of Adding and Subtracting Like Radicals

$$3\sqrt{5} + 2\sqrt{5} = (3 + 2)\sqrt{5} = 5\sqrt{5}$$

$$5\sqrt{x} - 7\sqrt{x} = (5 - 7)\sqrt{x} = -2\sqrt{x}$$

$$\sqrt[3]{4x^6} + 5\sqrt[3]{4x^6} = (1 + 5)\sqrt[3]{4x^6} = 6\sqrt[3]{4x^6}$$

$$4\sqrt{5x} - y\sqrt{5x} = (4 - y)\sqrt{5x}$$

EXAMPLE 1 Simplify. **a)** $6 + 4\sqrt{2} - \sqrt{2} + 3$ **b)** $2\sqrt[3]{x} + 5x + 4\sqrt[3]{x} - 3$

Solution **a)** $6 + 4\sqrt{2} - \sqrt{2} + 3 = 3\sqrt{2} + 9$ (or $9 + 3\sqrt{2}$)

b) $2\sqrt[3]{x} + 5x + 4\sqrt[3]{x} - 3 = 6\sqrt[3]{x} + 5x - 3$

It is sometimes possible to convert unlike radicals into like radicals by simplifying one or more of the radicals.

EXAMPLE 2 Simplify $\sqrt{3} + \sqrt{27}$.

Solution Since $\sqrt{3}$ and $\sqrt{27}$ are <u>un</u>like radicals, they cannot be added in their present form. We can simplify $\sqrt{27}$ to obtain like radicals.

$$\begin{aligned}\sqrt{3} + \sqrt{27} &= \sqrt{3} + \sqrt{9}\sqrt{3} \\ &= \sqrt{3} + 3\sqrt{3} = 4\sqrt{3}\end{aligned}$$

To Add or Subtract Radicals

1. Simplify each radical expression.
2. Combine like radicals (if there are any).

EXAMPLE 3 Simplify.

a) $4\sqrt{24} + \sqrt{54}$ **b)** $2\sqrt{45} - \sqrt{80} + \sqrt{20}$ **c)** $\sqrt[3]{27} + \sqrt[3]{81} - 4\sqrt[3]{3}$

Solution

a) $$\begin{aligned}4\sqrt{24} + \sqrt{54} &= 4 \cdot \sqrt{4} \cdot \sqrt{6} + \sqrt{9} \cdot \sqrt{6} \\ &= 4 \cdot 2\sqrt{6} + 3\sqrt{6} \\ &= 8\sqrt{6} + 3\sqrt{6} = 11\sqrt{6}\end{aligned}$$

b) $$\begin{aligned}2\sqrt{45} - \sqrt{80} + \sqrt{20} &= 2 \cdot \sqrt{9} \cdot \sqrt{5} - \sqrt{16} \cdot \sqrt{5} + \sqrt{4} \cdot \sqrt{5} \\ &= 2 \cdot 3\sqrt{5} - 4\sqrt{5} + 2\sqrt{5} \\ &= 6\sqrt{5} - 4\sqrt{5} + 2\sqrt{5} = 4\sqrt{5}\end{aligned}$$

c) $$\begin{aligned}\sqrt[3]{27} + \sqrt[3]{81} - 4\sqrt[3]{3} &= 3 + \sqrt[3]{27}\sqrt[3]{3} - 4\sqrt[3]{3} \\ &= 3 + 3\sqrt[3]{3} - 4\sqrt[3]{3} = 3 - \sqrt[3]{3}\end{aligned}$$

NOW TRY EXERCISE 21

EXAMPLE 4 Simplify. **a)** $\sqrt{x^2} - \sqrt{x^2 y} + x\sqrt{y}$ **b)** $\sqrt[3]{x^{10} y^2} - \sqrt[3]{x^4 y^8}$

Solution

a) $$\begin{aligned}\sqrt{x^2} - \sqrt{x^2 y} + x\sqrt{y} &= x - \sqrt{x^2}\sqrt{y} + x\sqrt{y} \\ &= x - x\sqrt{y} + x\sqrt{y} \\ &= x\end{aligned}$$

b) $$\begin{aligned}\sqrt[3]{x^{10} y^2} - \sqrt[3]{x^4 y^8} &= \sqrt[3]{x^9} \cdot \sqrt[3]{xy^2} - \sqrt[3]{x^3 y^6} \cdot \sqrt[3]{xy^2} \\ &= x^3\sqrt[3]{xy^2} - xy^2\sqrt[3]{xy^2}\end{aligned}$$

Now factor out the common factor $\sqrt[3]{xy^2}$.

NOW TRY EXERCISE 35

$$= (x^3 - xy^2)\sqrt[3]{xy^2}$$

EXAMPLE 5 Simplify $4\sqrt{2} - \frac{1}{\sqrt{8}} + \sqrt{32}$.

Solution Begin by rationalizing the denominator and by simplifying $\sqrt{32}$.

$$4\sqrt{2} - \frac{1}{\sqrt{8}} + \sqrt{32} = 4\sqrt{2} - \frac{1}{\sqrt{8}} \cdot \frac{\sqrt{2}}{\sqrt{2}} + \sqrt{16}\sqrt{2}$$

$$= 4\sqrt{2} - \frac{\sqrt{2}}{\sqrt{16}} + 4\sqrt{2}$$

$$= 4\sqrt{2} - \frac{1}{4}\sqrt{2} + 4\sqrt{2}$$

$$= \left(4 - \frac{1}{4} + 4\right)\sqrt{2}$$

NOW TRY EXERCISE 45

$$= \frac{31\sqrt{2}}{4}$$

Now that we have discussed adding and subtracting radical expressions in more depth, let us multiply a few more radicals.

EXAMPLE 6 Simplify. **a)** $(3\sqrt{6} - \sqrt{3})^2$ **b)** $(\sqrt[3]{x} - \sqrt[3]{2y^2})(\sqrt[3]{x^2} - \sqrt[3]{8y})$

Solution **a)** $(3\sqrt{6} - \sqrt{3})^2 = (3\sqrt{6} - \sqrt{3})(3\sqrt{6} - \sqrt{3})$

Now multiply the factors using the FOIL method.

$$(3\sqrt{6})(3\sqrt{6}) + (3\sqrt{6})(-\sqrt{3}) + (-\sqrt{3})(3\sqrt{6}) + (-\sqrt{3})(-\sqrt{3})$$

$$= 9(6) - 3\sqrt{18} - 3\sqrt{18} + 3$$

$$= 54 - 3\sqrt{18} - 3\sqrt{18} + 3$$

$$= 57 - 6\sqrt{18}$$

$$= 57 - 6\sqrt{9}\sqrt{2}$$

$$= 57 - 18\sqrt{2}$$

b) Multiply the factors using the FOIL method.

$$(\sqrt[3]{x})(\sqrt[3]{x^2}) + (\sqrt[3]{x})(-\sqrt[3]{8y}) + (-\sqrt[3]{2y^2})(\sqrt[3]{x^2}) + (-\sqrt[3]{2y^2})(-\sqrt[3]{8y})$$

$$= \sqrt[3]{x^3} - \sqrt[3]{8xy} - \sqrt[3]{2x^2y^2} + \sqrt[3]{16y^3}$$

NOW TRY EXERCISE 63

$$= x - 2\sqrt[3]{xy} - \sqrt[3]{2x^2y^2} + 2y\sqrt[3]{2}$$

HELPFUL HINT

The product rule and quotient rule for radicals presented in Sections 7.3 and 7.4 are

$$\sqrt[n]{a} \cdot \sqrt[n]{b} = \sqrt[n]{ab} \qquad \frac{\sqrt[n]{a}}{\sqrt[n]{b}} = \sqrt[n]{\frac{a}{b}}$$

Students often incorrectly assume similar properties exist for addition and subtraction. They do not. To illustrate this, let n be a square root (index 2), $a = 9$, and $b = 16$.

$$\sqrt[n]{a} + \sqrt[n]{b} \neq \sqrt[n]{a + b}$$
$$\sqrt{9} + \sqrt{16} \neq \sqrt{9 + 16}$$
$$3 + 4 \neq \sqrt{25}$$
$$7 \neq 5$$

Exercise Set 7.5

Concept/Writing Exercises

1. What are like radicals?
2. a) Explain how to add like radicals.

   b) Using the procedure in part a), add $\frac{3}{5}\sqrt{5} + \frac{5}{4}\sqrt{5}$.
3. Use a calculator to estimate $\sqrt{3} + 3\sqrt{2}$.
4. Use a calculator to estimate $2\sqrt{3} + \sqrt{5}$.
5. Does $\sqrt{a} + \sqrt{b} = \sqrt{a + b}$? Explain your answer and give an example supporting your answer.
6. Since $64 + 36 = 100$, does $\sqrt{64} + \sqrt{36} = \sqrt{100}$? Explain your answer.

Practice the Skills

In this exercise set, assume that all variables represent positive real numbers.

Simplify.

7. $6\sqrt{3} - 2\sqrt{3}$
8. $6\sqrt[3]{7} - 8\sqrt[3]{7}$
9. $2\sqrt{3} - 2\sqrt{3} - 4\sqrt{3} + 5$
10. $12\sqrt[3]{15} + 5\sqrt[3]{15} - 8\sqrt[3]{15}$
11. $3\sqrt[4]{y} - 6\sqrt[4]{y}$
12. $3\sqrt[5]{y} - \sqrt[5]{y} + 3$
13. $3\sqrt{5} - \sqrt[3]{x} + 4\sqrt{5} + 3\sqrt[3]{x}$
14. $\sqrt{a} + \sqrt{b} + a + 3\sqrt{b}$
15. $6 + 4\sqrt[3]{a} - 7\sqrt[3]{a}$
16. $5\sqrt{x} + 4 + 3\sqrt{x} + 2x - \sqrt{x}$

Simplify.

17. $\sqrt{8} - \sqrt{12}$
18. $\sqrt{75} + \sqrt{108}$
19. $-6\sqrt{75} + 4\sqrt{125}$
20. $3\sqrt{250} + 5\sqrt{160}$
21. $-4\sqrt{90} + 3\sqrt{40} + 2\sqrt{10}$
22. $3\sqrt{40x^2y} + 2x\sqrt{490y}$
23. $\sqrt{500xy^2} + y\sqrt{320x}$
24. $4\sqrt{32} - \sqrt{18} + 2\sqrt{128}$
25. $5\sqrt{8} + 2\sqrt{50} - 3\sqrt{72}$
26. $2\sqrt{5x} - 3\sqrt{20x} - 4\sqrt{45x}$
27. $3\sqrt{27c^2} - 2\sqrt{108c^2} - \sqrt{48c^2}$
28. $3\sqrt{50a^2} - 3\sqrt{72a^2} - 8a\sqrt{18}$
29. $\sqrt[3]{54} - \sqrt[3]{16}$
30. $4\sqrt[3]{5} - 5\sqrt[3]{40}$
31. $\sqrt[3]{108} + 2\sqrt[3]{32}$
32. $2\sqrt[3]{16} + \sqrt[3]{54}$
33. $\sqrt[3]{27} - 5\sqrt[3]{8}$
34. $3\sqrt{45x^3} + \sqrt{5x}$
35. $2\sqrt[3]{a^4b^2} + 3a\sqrt[3]{ab^2}$
36. $3y\sqrt[4]{48x^5} - x\sqrt[4]{3x^5y^4}$
37. $\sqrt{4r^7s^5} + 3r^2\sqrt{r^3s^5} - 2rs\sqrt{r^5s^3}$
38. $x\sqrt[3]{27x^5y^2} - x^2\sqrt[3]{x^2y^2} + 2\sqrt[3]{x^8y^2}$
39. $\sqrt[3]{128x^9y^{10}} - 2x^2y\sqrt[3]{16x^3y^7}$
40. $5\sqrt[3]{320x^5y^8} + 2x\sqrt[3]{135x^2y^8}$

Simplify.

41. $\frac{1}{\sqrt{2}} + \frac{\sqrt{2}}{2}$
42. $\frac{1}{\sqrt{3}} + \frac{\sqrt{3}}{3}$
43. $\sqrt{3} - \frac{1}{\sqrt{3}}$

44. $\sqrt{\frac{1}{6}} + \sqrt{24}$

45. $3\sqrt{2} - \frac{2}{\sqrt{8}} + \sqrt{50}$

46. $\sqrt{\frac{1}{2}} + 3\sqrt{2} + \sqrt{18}$

47. $4\sqrt{x} + \frac{1}{\sqrt{x}} + \sqrt{\frac{1}{x}}$

48. $\frac{1}{3} + \frac{1}{\sqrt{3}} + \sqrt{75}$

49. $\frac{1}{2}\sqrt{18} - \frac{3}{\sqrt{2}} - 3\sqrt{50}$

50. $\frac{\sqrt{3}}{3} + 2\sqrt{\frac{1}{3}} + \sqrt{12}$

51. $-2\sqrt{\frac{x}{y}} + 3\sqrt{\frac{y}{x}}$

52. $\sqrt{\frac{3}{8}} + \sqrt{\frac{3}{2}}$

Simplify.

53. $(\sqrt{3} + 4)(\sqrt{3} + 5)$

54. $(\sqrt{3} + 1)(\sqrt{3} - 6)$

55. $(1 + \sqrt{5})(6 + \sqrt{5})$

56. $(3 - \sqrt{2})(4 - \sqrt{8})$

57. $(4 - \sqrt{2})(5 + \sqrt{2})$

58. $(5\sqrt{6} + 3)(4\sqrt{6} - 2)$

59. $(4\sqrt{3} + \sqrt{2})(\sqrt{3} - \sqrt{2})$

60. $(\sqrt{3} + 4)^2$

61. $(2\sqrt{5} - 3)^2$

62. $(\sqrt{y} + \sqrt{6z})(\sqrt{2z} - \sqrt{8y})$

63. $(2\sqrt{3x} - \sqrt{y})(3\sqrt{3x} + \sqrt{y})$

64. $(\sqrt[3]{9} + \sqrt[3]{2})(\sqrt[3]{3} + \sqrt[3]{4})$

65. $(\sqrt[3]{4} - \sqrt[3]{6})(\sqrt[3]{2} - \sqrt[3]{36})$

66. $(\sqrt[3]{4x} - \sqrt[3]{2y})(\sqrt[3]{4x} + \sqrt[3]{10})$

Simplify. These exercises are a combination of the types of exercises presented earlier in this exercise set.

67. $\sqrt{5} + 2\sqrt{5}$

68. $-2\sqrt{x} - 3\sqrt{x}$

69. $\sqrt{125} + \sqrt{20}$

70. $3\sqrt{7} + 2\sqrt{63} - 2\sqrt{28}$

71. $\frac{\sqrt{6}}{2} + \frac{1}{\sqrt{6}}$

72. $-\sqrt[4]{x} + 6\sqrt[4]{x} - 2\sqrt[4]{x}$

73. $2\sqrt[3]{81} + 4\sqrt[3]{24}$

74. $4\sqrt{3} - \frac{3}{\sqrt{3}} + 2\sqrt{18}$

75. $(3\sqrt{2} - 4)(\sqrt{2} + 5)$

76. $(\sqrt{5} + \sqrt{2})(\sqrt{2} + \sqrt{20})$

77. $4\sqrt{3x^3} - \sqrt{12x}$

78. $2b\sqrt[4]{a^4 b} + ab\sqrt[4]{16b}$

79. $\frac{3}{\sqrt{y}} - \sqrt{\frac{9}{y}} + \sqrt{y}$

80. $2\sqrt[3]{24a^3 y^4} + 4a\sqrt[3]{81y^4}$

81. $(\sqrt[3]{x^2} - \sqrt[3]{y})(\sqrt[3]{x} - 2\sqrt[3]{y^2})$

82. $2x\sqrt[3]{xy} + 5y\sqrt[3]{x^4 y^4}$

83. $(\sqrt[3]{a} + 5)(\sqrt[3]{a^2} - 3)$

84. $\frac{2}{\sqrt{50}} - 3\sqrt{50} - \frac{1}{\sqrt{8}}$

85. $2\sqrt{\frac{8}{3}} - 4\sqrt{\frac{100}{6}}$

86. $-5\sqrt{\frac{x^2}{y^3}} + \frac{2x}{y}\sqrt{\frac{1}{y}}$

Problem Solving

Find the perimeter and area of the following figures. Write the area and perimeter in radical form with the radicals simplified.

87.

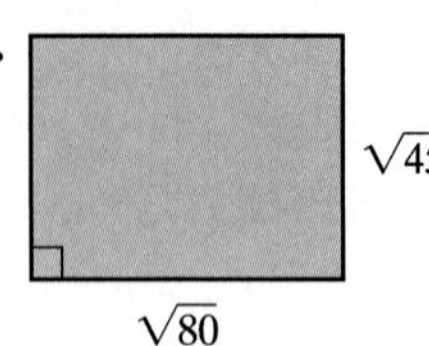

88.

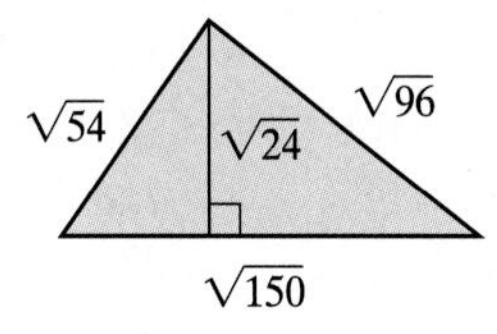

89.

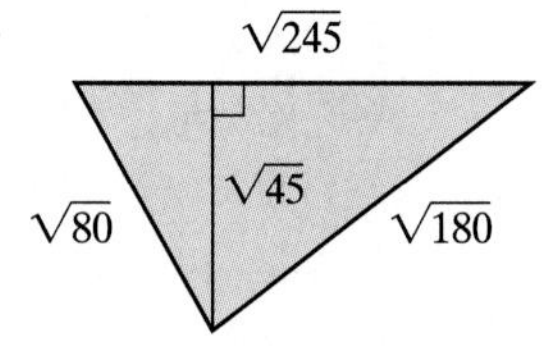

90.

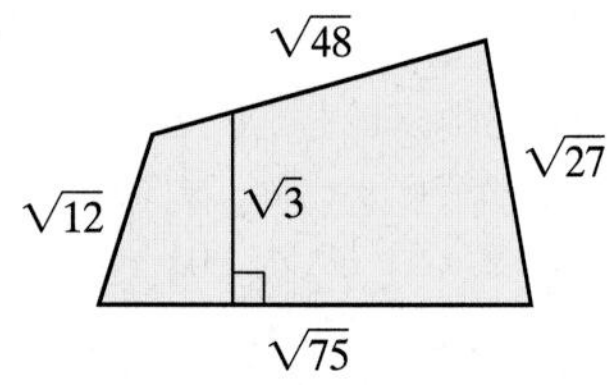

91. The function $f(t) = 3\sqrt{t} + 19$ can be used to approximate the median height, $f(t)$, in inches, for U.S. girls of age t, in months, where $1 \le t \le 60$. Estimate the median height of girls at age **a)** 36 months and **b)** 40 months.

92. On a chilly day when the wind is blowing, the temperature feels colder than it actually is. The temperature that it feels like is called the *wind chill temperature*. There are a number of formulas used to determine the wind chill temperature. One formula is

$$T_w = 33 - \frac{(10.45 + 10\sqrt{v} - v)(33 - T)}{22}$$

where T_w is the wind chill temperature in degrees Celsius, v is the wind speed in meters per second, and T is the actual temperature in degrees Celsius. Determine the wind chill temperature if the actual temperature is 4°C and the wind speed is 15 meters per second.

93. Which is greater, $\frac{1}{\sqrt{3} + 2}$ or $2 + \sqrt{3}$? (Do not use a calculator.) Explain how you determined your answer.

94. Which is greater, $\frac{1}{\sqrt{3}} + \sqrt{75}$ or $\frac{2}{\sqrt{12}} + \sqrt{48} + 2\sqrt{3}$? Do not use a calculator.) Explain how you determined your answer.

95. If $\sqrt{a + b} = \sqrt{c + d}$, must $a + b = c + d$? Explain and give an example supporting your answer.

96. If $\sqrt[3]{a + b} = \sqrt[3]{c + d}$ must $a + b = c + d$? Explain and give an example supporting your answer.

97. Will the sum of two radicals always be a radical? Give an example to support your answer.

98. Will the difference of two radicals always be a radical? Give an example to support your answer.

Factor each of the following by factoring out a radical expression.

99. $\sqrt{10} + \sqrt{5}$

100. $\sqrt{6} - \sqrt{18}$

*In Exercises 101 and 102, **a)** write each term as the square of an expression, **b)** then factor using the formula for the difference of two squares.*

101. $x^2 - 3$

102. $2x^2 - 7$

103. The graph of $f(x) = \sqrt{x}$ is shown.

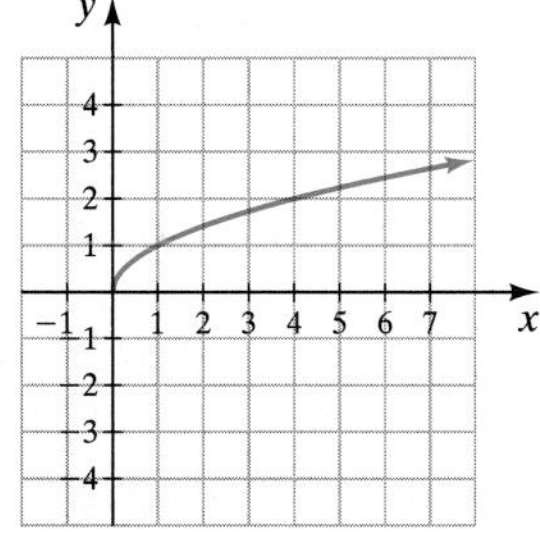

a) If $g(x) = 2$, sketch the graph of $(f + g)(x)$.
b) What effect does adding 2 have to the graph of $f(x)$?

104. The graph of $f(x) = -\sqrt{x}$ is shown.

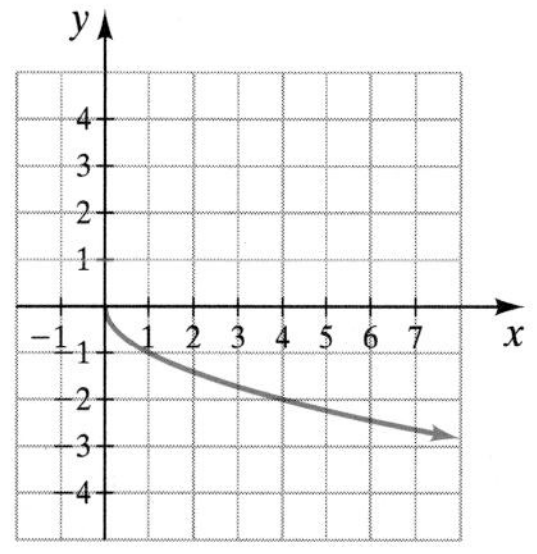

a) If $g(x) = 3$, sketch the graph of $(f + g)(x)$.
b) What effect does adding 3 have to the graph of $f(x)$?

105. You are given that $f(x) = \sqrt{x}$ and $g(x) = \sqrt{x} - 2$.
a) Sketch the graph of $(f - g)(x)$. Explain how you determined your answer.
b) What is the domain of $(f - g)(x)$?

106. You are given that $f(x) = \sqrt{x}$ and $g(x) = -\sqrt{x} - 3$.
a) Sketch the graph of $(f + g)(x)$. Explain how you determined your answer.
b) What is the domain of $(f + g)(x)$?

107. Graph the function $f(x) = \sqrt{x^2}$.

108. Graph the function $f(x) = \sqrt{x^2} - 4$.

Group Activity

The following two exercises will reinforce many of the concepts presented in this chapter. Work each problem as a group. Make sure each member of the group understands each step in obtaining the solution. The figures in each exercise are similar. For each exercise, use a proportion to find the length of side x. Write the answer in radical form with a rationalized denominator.

109. $5 + \sqrt{3}$, $\sqrt{12}$

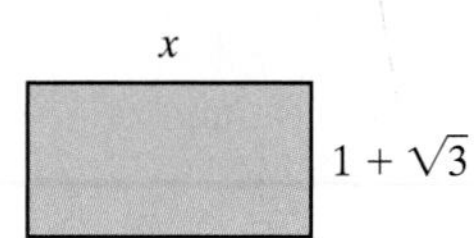

110.

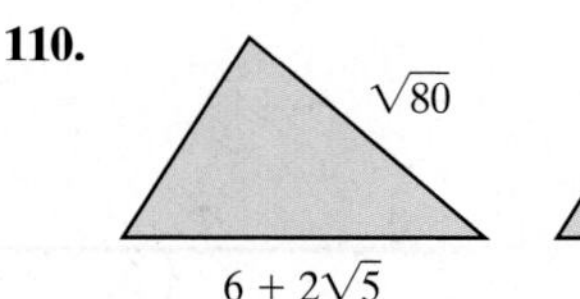

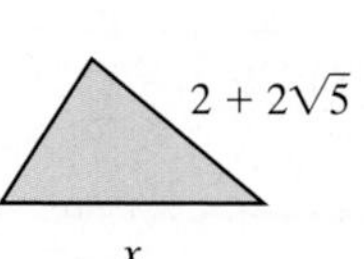

Cumulative Review Exercises

111. Simplify $\dfrac{(2x^{-2}y^3)^2}{(xy^2)^{-2}}$

[6.5] **112.** Solve the equation $20x^2 + 3x - 9 = 0$.

[8.2] **113.** Simplify $\left(\dfrac{x^{3/4}y^{2/3}}{x^{1/2}y}\right)^2$.

[8.3] **114.** Simplify $\sqrt[3]{3x^2y}\left(\sqrt[3]{9x^4y^3} - \sqrt[3]{x^{10}y^7}\right)$.

7.6 SOLVING RADICAL EQUATIONS

SSM VIDEO 7.6 CD Rom

1. Solve equations containing one radical.
2. Solve equations containing two radicals.
3. Solve equations containing two radical terms and a nonradical term.
4. Solve applications using radical equations.
5. Solve for a variable in a radicand.

1 Solve Equations Containing One Radical

A **radical equation** is an equation that contains a variable in a radicand.

Examples of Radical Equations

$$\sqrt{x} = 4, \qquad \sqrt[3]{y + 4} = 9, \qquad \sqrt{x - 2} = 4 + \sqrt{x + 8}$$

To Solve Radical Equations

1. Rewrite the equation so that one radical containing a variable is by itself (isolated) on one side of the equation.
2. Raise each side of the equation to a power equal to the index of the radical.
3. Combine like terms.
4. If the equation still contains a term with a variable in a radicand, repeat steps 1 through 3.
5. Solve the resulting equation for the variable.
6. Check all solutions in the original equation for extraneous solutions.

Recall from Section 6.4 that an extraneous solution is a number obtained when solving an equation that is not a solution to the original equation.

The following examples illustrate the procedure for solving radical equations.

EXAMPLE 1 Solve the equation $\sqrt{x} = 6$.

Solution The square root containing the variable is already by itself on one side of the equation. Square both sides of the equation.

$$\sqrt{x} = 6$$
$$(\sqrt{x})^2 = (6)^2$$
$$x = 36$$

CHECK:

$$\sqrt{x} = 6$$
$$\sqrt{36} \stackrel{?}{=} 6$$
$$6 = 6 \quad \textit{True}$$

EXAMPLE 2 Solve. **a)** $\sqrt{x + 4} - 6 = 0$ **b)** $\sqrt[3]{x} + 4 = 6$

Solution The first step in each part will be to isolate the term containing the radical.

a)

$$\sqrt{x + 4} - 6 = 0$$
$$\sqrt{x + 4} = 6 \quad \textit{Isolate the radical containing the variable.}$$
$$(\sqrt{x + 4})^2 = 6^2 \quad \textit{Square both sides.}$$
$$x + 4 = 36 \quad \textit{Solve for the variable.}$$
$$x = 32$$

A check will show that 32 is the solution.

b)

$$\sqrt[3]{x} + 4 = 6$$
$$\sqrt[3]{x} = 2 \quad \textit{Isolate the radical containing the variable.}$$
$$(\sqrt[3]{x})^3 = 2^3 \quad \textit{Cube both sides.}$$
$$x = 8$$

NOW TRY EXERCISE 17 A check will show that 8 is the solution.

HELPFUL HINT

Don't forget to check your solutions in the original equation. When you raise both sides of an equation to a power you may introduce extraneous solutions.

Consider the equation $x = 2$. Note what happens when you square both sides of the equation.

$$x = 2$$
$$x^2 = 2^2$$
$$x^2 = 4$$

Note that the equation $x^2 = 4$ has two solutions, $+2$ and -2. Since the original equation $x = 2$ has only one solution, 2, we introduced the extraneous solution, -2.

EXAMPLE 3 Solve $\sqrt{2x - 3} = x - 3$.

Solution Since the radical is already isolated, we square both sides of the equation.

$$(\sqrt{2x-3})^2 = (x-3)^2$$
$$2x - 3 = (x-3)(x-3)$$
$$2x - 3 = x^2 - 6x + 9$$
$$0 = x^2 - 8x + 12$$

Now we factor and use the zero-factor property.

$$x^2 - 8x + 12 = 0$$
$$(x-6)(x-2) = 0$$
$$x - 6 = 0 \quad \text{or} \quad x - 2 = 0$$
$$x = 6 \qquad\qquad x = 2$$

CHECK:

$x = 6$		$x = 2$	
$\sqrt{2x-3} = x - 3$		$\sqrt{2x-3} = x - 3$	
$\sqrt{2(6)-3} \stackrel{?}{=} 6 - 3$		$\sqrt{2(2)-3} \stackrel{?}{=} 2 - 3$	
$\sqrt{9} \stackrel{?}{=} 3$		$\sqrt{1} \stackrel{?}{=} -1$	
$3 = 3$	*True*	$1 = -1$	*False*

Thus, 6 is a solution, but 2 is not a solution to the equation. The 2 is an extraneous solution because 2 satisfies the equation $(\sqrt{2x-3})^2 = (x-3)^2$, but not the original equation, $\sqrt{2x-3} = x - 3$.

Using Your Graphing Calculator

In Example 3 we found the solution to $\sqrt{2x-3} = x - 3$ to be 6. If we let $y_1 = \sqrt{2x-3}$ and $y_2 = x - 3$ and graph y_1 and y_2 on a graphing calculator, we get Figure 7.3. Notice the graphs appear to intersect at $x = 6$, which is what we expect.

The table of values in Figure 7.4 shows that the y-coordinate at the point of intersection is 3. In the table, ERROR appears under y_1 for the values of x of 0 and 1. For any values less than $\frac{3}{2}$, the value of $2x - 3$ is negative and therefore $\sqrt{2x-3}$ is not a real number. The domain of function y_1 is $\{x \mid x \geq \frac{3}{2}\}$, which may be found by solving the inequality $2x - 3 \geq 0$.

You can use your grapher to either solve or check radical equations.

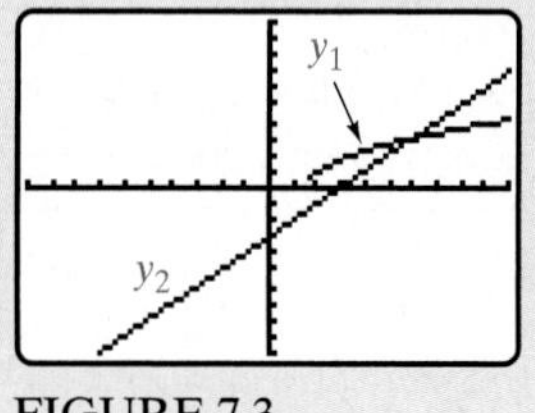

FIGURE 7.3

X	Y1	Y2
0	ERROR	-3
1	ERROR	-2
2	1	-1
3	1.7321	0
4	2.2361	1
5	2.6458	2
6	3	3

X=6

FIGURE 7.4

Exercises

Use your graphing calculator to determine whether the indicated value is the solution to the radical equation. If it is not the solution, use your grapher to determine the solution.

1. $\sqrt{2x+9} = 5(x-7), 8$

2. $\sqrt{3x+4} = \sqrt{x+12}, 6$

EXAMPLE 4 Solve $2x - 5\sqrt{x} - 3 = 0$.

Solution First, write the equation with the square root containing the variable by itself on one side of the equation.

$$\begin{aligned} 2x - 5\sqrt{x} - 3 &= 0 \\ -5\sqrt{x} &= -2x + 3 \\ \text{or}\quad 5\sqrt{x} &= 2x - 3 \end{aligned}$$

Now square both sides of the equation.

$$\begin{aligned} (5\sqrt{x})^2 &= (2x - 3)^2 \\ 25x &= (2x - 3)(2x - 3) \\ 25x &= 4x^2 - 12x + 9 \\ 0 &= 4x^2 - 37x + 9 \\ 0 &= (4x - 1)(x - 9) \end{aligned}$$

$$4x - 1 = 0 \quad \text{or} \quad x - 9 = 0$$

$$4x = 1 \qquad\qquad x = 9$$

$$x = \frac{1}{4}$$

CHECK: $x = \frac{1}{4}$

$$\begin{aligned} 2x - 5\sqrt{x} - 3 &= 0 \\ 2\left(\frac{1}{4}\right) - 5\sqrt{\frac{1}{4}} - 3 &\stackrel{?}{=} 0 \\ \frac{1}{2} - 5\left(\frac{1}{2}\right) - 3 &\stackrel{?}{=} 0 \\ -5 &= 0 \quad \textit{False} \end{aligned}$$

$x = 9$

$$\begin{aligned} 2x - 5\sqrt{x} - 3 &= 0 \\ 2(9) - 5\sqrt{9} - 3 &\stackrel{?}{=} 0 \\ 18 - 5(3) - 3 &\stackrel{?}{=} 0 \\ 18 - 15 - 3 &\stackrel{?}{=} 0 \\ 0 &= 0 \quad \textit{True} \end{aligned}$$

NOW TRY EXERCISE 33

The solution is 9. The value $\frac{1}{4}$ is an extraneous solution.

2) Solve Equations Containing Two Radicals

Now we will look at some equations that contain two radicals.

EXAMPLE 5 Solve $\sqrt{4x^2 + 16} = 2\sqrt{x^2 + 3x - 2}$.

Solution Since the two radicals appear on different sides of the equation, we square both sides of the equation.

$$(\sqrt{4x^2 + 16})^2 = (2\sqrt{x^2 + 3x - 2})^2 \quad \text{Square both sides.}$$
$$4x^2 + 16 = 4(x^2 + 3x - 2)$$
$$4x^2 + 16 = 4x^2 + 12x - 8$$
$$16 = 12x - 8$$
$$24 = 12x$$
$$2 = x$$

A check will show that 2 is the solution.

EXAMPLE 6 For $f(x) = 3(x - 2)^{1/3}$ and $g(x) = (17x - 14)^{1/3}$, find all values of x for which $f(x) = g(x)$.

Solution You should realize that alternate ways of writing $f(x)$ and $g(x)$ are $f(x) = 3\sqrt[3]{x - 2}$ and $g(x) = \sqrt[3]{17x - 4}$. We could therefore work this example using radicals, but we will work instead with rational exponents. We set the two functions equal to each other and solve for x.

$$f(x) = g(x)$$
$$3(x - 2)^{1/3} = (17x - 14)^{1/3}$$
$$[3(x - 2)^{1/3}]^3 = [(17x - 14)^{1/3}]^3 \quad \text{Cube both sides.}$$
$$3^3(x - 2) = 17x - 14$$
$$27(x - 2) = 17x - 14$$
$$27x - 54 = 17x - 14$$
$$10x - 54 = -14$$
$$10x = 40$$
$$x = 4$$

A check will show that the solution is 4.

In Example 6, if you solve the equation $3\sqrt[3]{x - 2} = \sqrt[3]{17x - 14}$ you will obtain the solution 4. For additional practice, do this now.

NOW TRY EXERCISE 61

3) Solve Equations Containing Two Radical Terms and a Nonradical Term

When a radical equation contains two radical terms and a third nonradical term, you will sometimes need to raise both sides of the equation to a given power twice to obtain the solution. First, isolate one radical term. Then raise both sides of the equation to the given power. This will eliminate one of the radicals. Next, isolate the remaining radical on one side of the equation. Then raise both sides of the equation to the given power a second time. This procedure is illustrated in Example 7.

EXAMPLE 7 Solve $\sqrt{5x - 1} - \sqrt{3x - 2} = 1$.

Solution We must isolate one radical term on one side of the equation. We will begin by adding $\sqrt{3x - 2}$ to both sides of the equation to isolate $\sqrt{5x - 1}$. Then we will square both sides of the equation and combine like terms.

$$\begin{aligned}
\sqrt{5x-1} &= 1+\sqrt{3x-2} && \text{Isolate } \sqrt{5x-1}. \\
(\sqrt{5x-1})^2 &= (1+\sqrt{3x-2})^2 && \text{Square both sides.} \\
5x-1 &= (1+\sqrt{3x-2})(1+\sqrt{3x-2}) && \text{Write as a product.} \\
5x-1 &= 1+\sqrt{3x-2}+\sqrt{3x-2}+(\sqrt{3x-2})^2 && \text{Multiply.} \\
5x-1 &= 1+2\sqrt{3x-2}+3x-2 && \text{Combine terms; simplify.} \\
5x-1 &= 3x-1+2\sqrt{3x-2} && \text{Combine terms.} \\
2x &= 2\sqrt{3x-2} && \text{Isolate the radical term.} \\
x &= \sqrt{3x-2} && \text{Divide both sides by 2.}
\end{aligned}$$

We have isolated the remaining radical term. We now square both sides of the equation again and solve for x.

$$\begin{aligned}
x &= \sqrt{3x-2} \\
x^2 &= (\sqrt{3x-2})^2 && \text{Square both sides.} \\
x^2 &= 3x-2 \\
x^2-3x+2 &= 0 \\
(x-2)(x-1) &= 0
\end{aligned}$$

$$x-2=0 \quad \text{or} \quad x-1=0$$
$$x=2 \qquad\qquad x=1$$

NOW TRY EXERCISE 51

A check will show that both 2 and 1 are solutions of the equation.

EXAMPLE 8 For $f(x)=\sqrt{5x-1}-\sqrt{3x-2}$, find all values of x for which $f(x)=1$.

Solution Substitute 1 for $f(x)$. This gives

$$1=\sqrt{5x-1}-\sqrt{3x-2}$$

Since this is the same equation we solved in Example 7, the answers are $x=2$ and $x=1$. Verify for yourself that $f(2)=1$ and $f(1)=1$.

4) Solve Applications Using Radical Equations

Now we will look at a few of the many applications of radicals.

EXAMPLE 9 A regulation baseball diamond is a square with 90 feet between bases. How far is second base from home plate?

Solution **Understand** Figure 7.5 illustrates the problem. We need to find the distance from second base to home plate, which is the hypotenuse of a right triangle.

FIGURE 7.5

Translate To solve the problem we use the Pythagorean Theorem, $\text{leg}^2 + \text{leg}^2 = \text{hyp}^2$, or $a^2 + b^2 = c^2$.

$$90^2 + 90^2 = c^2 \quad \text{Substitute known values.}$$

Carry Out:

$$8100 + 8100 = c^2$$
$$16{,}200 = c^2$$
$$\sqrt{16{,}200} = c \quad \text{*See footnote.}$$
$$127.28 \approx c$$

Answer The distance from second base to home plate is about 127.28 feet.

EXAMPLE 10 The length of time it takes for a pendulum to make one complete swing back and forth is called the *period* of the pendulum. See Figure 7.6. The period of a pendulum, T, in seconds, can be found by the formula $T = 2\pi\sqrt{\frac{L}{32}}$, where L is the length of the pendulum in feet. Find the period of a pendulum if its length is 4 feet.

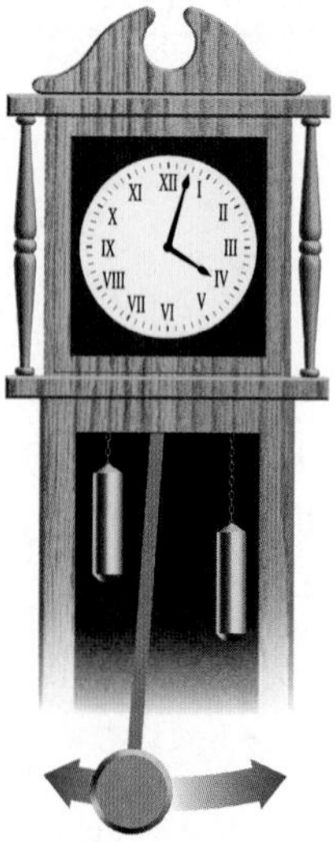

FIGURE 7.6

NOW TRY EXERCISE 77

Solution Substitute 4 for L and 3.14 for π in the formula. If you have a calculator that has π key, use it to enter π.

$$T = 2\pi\sqrt{\frac{L}{32}}$$
$$= 2(3.14)\sqrt{\frac{4}{32}}$$
$$= 2(3.14)\sqrt{0.125} \approx 2.22$$

Thus, the period is about 2.22 seconds. If you have a grandfather clock with a 4-foot pendulum, it will take about 2.22 seconds for it to swing back and forth.

5 Solve for a Variable in a Radicand

You may be given a formula and be asked to solve for a variable in a radicand. To do so, follow the same general procedure used to solve a radical equation. Begin by isolating the radical expression. Then raise both sides of the equation to the same power as the index of the radical. This procedure is illustrated in Example 11 **b)**.

EXAMPLE 11 A formula in statistics for finding the maximum error of estimation is $E = Z\frac{\sigma}{\sqrt{n}}$.

a) Find E if $Z = 1.28$, $\sigma = 5$, and $n = 36$.

b) Solve this equation for n.

---

* $c^2 = 16{,}200$ has two solutions: $c = \sqrt{16{,}200}$ and $c = -\sqrt{16{,}200}$. Since we are solving for a length, which must be a positive quantity, we use the positive root.

Solution **a)** $E = Z\dfrac{\sigma}{\sqrt{n}} = 1.28\left(\dfrac{5}{\sqrt{36}}\right) = 1.28\left(\dfrac{5}{6}\right) \approx 1.07$

b) First multiply both sides of the equation by $\sqrt{n}$ to eliminate fractions. Then isolate $\sqrt{n}$. Finally, solve for n by squaring both sides of the equation.

$$E = Z\frac{\sigma}{\sqrt{n}}$$

$$\sqrt{n}\,(E) = \left(Z\frac{\sigma}{\sqrt{n}}\right)\sqrt{n} \qquad \textit{Eliminate fractions.}$$

$$\sqrt{n}\,(E) = Z\sigma$$

$$\sqrt{n} = \frac{Z\sigma}{E} \qquad \textit{Isolate the radical term.}$$

$$(\sqrt{n})^2 = \left(\frac{Z\sigma}{E}\right)^2 \qquad \textit{Square both sides.}$$

$$n = \left(\frac{Z\sigma}{E}\right)^2 \quad \text{or} \quad n = \frac{Z^2\sigma^2}{E^2}$$

Exercise Set 7.6

Concept/Writing Exercises

1. **a)** Explain in your own words how to solve a radical equation.
   b) Solve $\sqrt{2x + 26} - 2 = 4$ using the procedure you gave in part **a)**.
2. Consider the equation $\sqrt{x + 3} = -\sqrt{2x - 1}$. Explain why this equation can have no real solution.
3. Consider the equation $-\sqrt{x^2} = \sqrt{(-x)^2}$. By studying the equation, can you determine its solution? Explain.
4. Consider the equation $\sqrt[3]{x^2} = -\sqrt[3]{x^2}$. By studying the equation, can you determine its solution? Explain.
5. Explain without solving the equation how you can tell that $\sqrt{x - 3} + 3 = 0$ has no solution.
6. Why is it necessary to check solutions to radical equations?

Practice the Skills

Solve and check your solution(s). If the equation has no real solution, so state.

7. $\sqrt{x} = 4$
8. $\sqrt{x} = 7$
9. $\sqrt[3]{x} = 2$
10. $\sqrt[3]{x} = 4$
11. $\sqrt[4]{x} = 3$
12. $\sqrt{a - 3} + 5 = 6$
13. $-\sqrt{2x + 4} = -6$
14. $\sqrt{x} + 3 = 5$
15. $\sqrt[3]{2x + 11} = 3$
16. $\sqrt[3]{6x - 3} = 3$
17. $\sqrt[3]{3x} + 4 = 7$
18. $2\sqrt{4x - 3} = 10$
19. $\sqrt{a - 8} = 2\sqrt{3a - 2}$
20. $\sqrt{8c - 4} = \sqrt{7c + 2}$
21. $\sqrt{5b + 10} = -\sqrt{3b + 8}$
22. $\sqrt[4]{x + 8} = \sqrt[4]{2x}$
23. $\sqrt[3]{6x + 1} = \sqrt[3]{2x + 5}$
24. $\sqrt[4]{3x + 1} = 2$
25. $\sqrt{x^2 + 9x + 3} = -x$
26. $\sqrt{x^2 + 3x + 9} = x$
27. $\sqrt{m^2 + 6m - 4} = m$
28. $\sqrt{5a + 1} - 11 = 0$
29. $\sqrt{x^2 - 2} = x + 4$
30. $\sqrt{x^2 + 3} = x + 1$
31. $-\sqrt{x} = 2x - 1$
32. $\sqrt{3x + 4} = x - 2$
33. $\sqrt{5x + 6} = 2x - 6$
34. $\sqrt[3]{3x - 1} + 4 = 0$
35. $\sqrt[3]{x - 12} = \sqrt[3]{5x + 16}$
36. $\sqrt{6x - 1} = 3x$
37. $\sqrt{8b - 15} + b = 10$
38. $\sqrt[3]{4x - 3} - 3 = 0$
39. $(2a + 9)^{1/2} - a + 3 = 0$
40. $(2x^2 + 4x + 6)^{1/2} = \sqrt{2x^2 + 6}$
41. $(r + 2)^{1/3} = (3r + 8)^{1/3}$
42. $\sqrt[4]{x + 5} = -3$
43. $(5x + 18)^{1/4} = (9x + 2)^{1/4}$
44. $(3x + 6)^{1/3} + 3 = 0$
45. $(x^2 + 4x + 4)^{1/2} - x - 3 = 0$
46. $(5a + 2)^{1/4} = (2a + 16)^{1/4}$

Solve. You will have to square both sides of the equation twice to eliminate all radicals.

47. $\sqrt{2a-3} = \sqrt{2a} - 1$
48. $\sqrt{x} + 2 = \sqrt{x+16}$
49. $\sqrt{x+1} = 2 - \sqrt{x}$
50. $\sqrt{x+3} = \sqrt{x} - 3$
51. $\sqrt{x+7} = 5 - \sqrt{x-8}$
52. $\sqrt{y+2} = 1 + \sqrt{y-3}$
53. $\sqrt{b-3} = 4 - \sqrt{b+5}$
54. $\sqrt{4x-3} = 2 + \sqrt{2x-5}$
55. $\sqrt{r+10} + 3 + \sqrt{r-5} = 0$
56. $\sqrt{y+1} = \sqrt{y+2} - 1$
57. $\sqrt{2x+4} - \sqrt{x+3} - 1 = 0$
58. $2 + \sqrt{x+8} = \sqrt{3x+12}$

For each pair of functions, find all real values of x where $f(x) = g(x)$.

59. $f(x) = \sqrt{x+4}, g(x) = \sqrt{2x-2}$
60. $f(x) = \sqrt{x^2-6x+10}, g(x) = \sqrt{x-2}$
61. $f(x) = \sqrt[3]{5x-12}, g(x) = \sqrt[3]{6x-16}$
62. $f(x) = (14x-12)^{1/2}, g(x) = 2(3x+1)^{1/2}$
63. $f(x) = 2(8x+24)^{1/3}, g(x) = 4(2x-2)^{1/3}$
64. $f(x) = 2\sqrt{x+7}, g(x) = 10 - \sqrt{x+14}$

Solve each formula for the indicated variable.

65. $p = \sqrt{2v}$, for v
66. $l = \sqrt{4r}$, for r
67. $v = \sqrt{2gh}$, for g
68. $v = \sqrt{\dfrac{2E}{m}}$, for E
69. $v = \sqrt{\dfrac{FR}{m}}$, for F
70. $\omega = \sqrt{\dfrac{a_0}{x_0}}$, for x_0
71. $x = \sqrt{\dfrac{m}{k}}V_0$, for m
72. $T = 2\pi\sqrt{\dfrac{L}{32}}$, for L

Problem Solving

73. A telephone pole is at a right, or 90°, angle with the ground as shown in the figure. Find the length of the wire that connects to the pole 40 feet above the ground and is anchored to the ground 20 feet from the base of the pole.

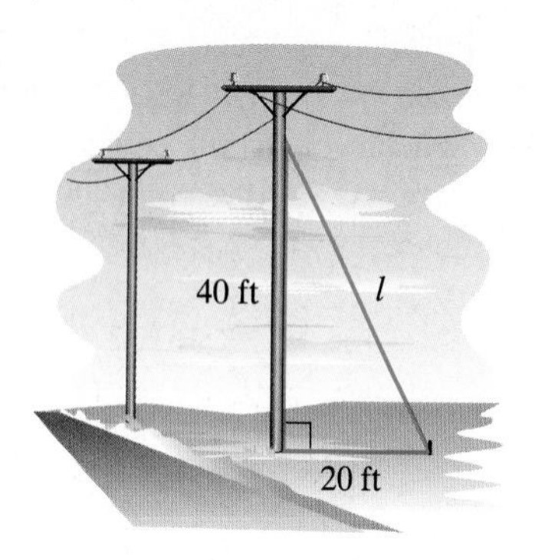

74. Ms. Song Tran places an extension ladder against her house. The base of the ladder is 2 meters from the house and the ladder rests against the house 6 meters above the ground. How far is her ladder extended?

75. When you are given the area of a square, the length of a side can be found by the formula $s = \sqrt{A}$. Find the side of Tom Kim's square garden if it has an area of 144 square feet.

76. When you are given the area of a circle, its radius can be found by the formula $r = \sqrt{A/\pi}$.

a) Find the radius of a basketball hoop if the area enclosed by the hoop is 254.47 square inches.

b) If the diameter of a basketball is 9 inches, what is the minimum distance possible between the hoop and the ball when the center of the ball is in the center of the hoop?

77. The formula for the period of a pendulum is

$$T = 2\pi\sqrt{\frac{l}{g}}$$

where T is the period in seconds, l is its length in feet, and g is the acceleration of gravity. On Earth, gravity is 32 ft/sec^2. The formula when used on Earth becomes

$$T = 2\pi\sqrt{\frac{l}{32}}$$

a) Find the period of a pendulum whose length is 6 feet.

b) If the length of a pendulum is doubled, what effect will this have on the period? Explain.

c) The gravity on the moon is 1/6 of that on Earth. If a pendulum has a period of 2 seconds on Earth, what will be the period of the same pendulum on the moon?

78. A formula for the length of a diagonal from the upper corner of a box to the opposite lower corner is $d = \sqrt{L^2 + W^2 + H^2}$, where L, W, and H are the length, width, and height, respectively.

a) Find the length of the diagonal of a suitcase of length 37 inches, width 15 inches, and height 9 inches.

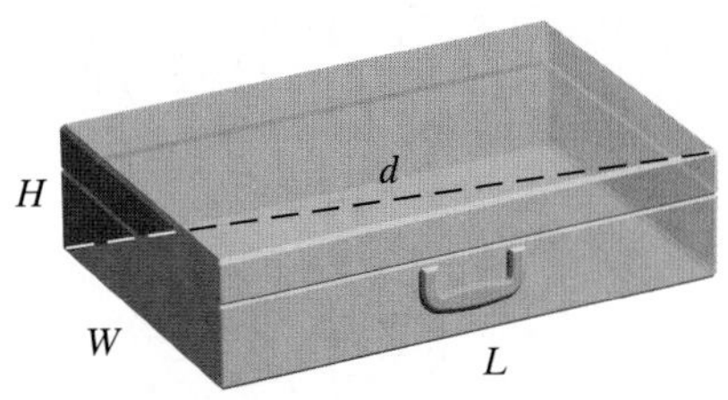

b) If the length, width, and height are all doubled, how will the diagonal change?

c) Solve the formula for W.

79. The formula

$$r = \sqrt[4]{\frac{8\mu l}{\pi R}}$$

is used in determining movement of blood through arteries. In the formula R represents the resistance to blood flow, μ is the viscosity of blood, l is the length of the artery, and r is the radius of the artery. Solve this equation for R.

80. The function

$$t = \frac{\sqrt{19.6s}}{9.8}$$

can be used to tell the time, t, in seconds, that an object has been falling if it has fallen s meters. Suppose Tamika Bidwell is bungee jumping and jumps from a 100-meter bridge (without pushing off upward or downward). If she travels a distance of 80 meters before the bungee cord begins to stretch, how long is she in free fall?

81. For any planet in our solar system, its "year" is the time it takes for the planet to revolve once around the sun. The number of Earth days in a given planet's year, N, is approximated by the formula $N = 0.2(\sqrt{R})^3$, where R is the mean distance of the planet to the sun in millions of kilometers. Find the number of Earth days in the year of the planet Earth, whose mean distance to the sun is 149.4 million kilometers.

82. Find the number of Earth days in the year of the planet Mercury, whose mean distance to the sun is 58 million kilometers.

83. When two forces, F_1 and F_2, pull at right angles to each other as illustrated below, the resultant, or the effective force, R, can be found by the formula $R = \sqrt{F_1^{\,2} + F_2^{\,2}}$. Two cars are trying to pull a third out of the mud, as illustrated. If car A is exerting a force of 600 pounds and car B is exerting a force of 800 pounds, find the resulting force on the car stuck in the mud.

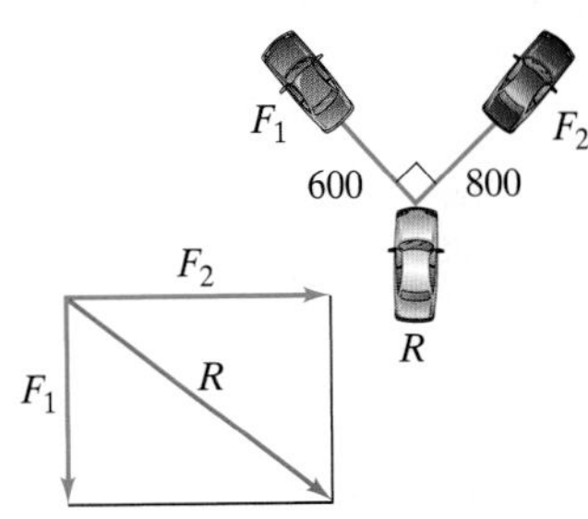

84. The escape velocity, or the velocity needed for a spacecraft to escape a planet's gravitational field, is found by the formula $v_e = \sqrt{2gR}$, where g is the force of gravity of the planet and R is the radius of the planet. Find the escape velocity for Earth, in meters per second, where $g = 9.75$ meters per second squared and $R = 6{,}370{,}000$ meters.

85. A formula used in the study of shallow-water wave motion is $c = \sqrt{gH}$, in which c is wave velocity, H is water depth, and g is the acceleration due to gravity. Find the wave velocity if the water's depth is 10 feet. (Use $g = 32$ ft/sec^2.)

86. When sound travels through air (or any gas), the velocity of the sound wave is dependent on the air (or gas) temperature. The velocity, v, in meters per second, at air temperature t, in degrees Celsius, can be found by the formula

$$v = 331.3\sqrt{1 + \frac{t}{273}}$$

Find the speed of sound in air whose temperature is 20°C (equivalent to 68°F).

A formula we have already mentioned and will be discussing in more detail shortly is the quadratic formula

$$x = \frac{-b \pm \sqrt{b^2 - 4ac}}{2a}$$

87. Find x when $a = 1, b = 0, c = -4$.

88. Find x when $a = 1, b = 1, c = -12$.

89. Find x when $a = 2, b = 5, c = -12$.

90. Find x when $a = -1, b = 4, c = 5$.

Given $f(x)$, find all values of x for which $f(x)$ is the indicated value.

91. $f(x) = \sqrt{x - 5}, f(x) = 4$

92. $f(x) = \sqrt[3]{2x + 3}, f(x) = 5$

93. $f(x) = \sqrt{3x^2 - 11} + 4, f(x) = 12$

94. $f(x) = 4 + \sqrt[3]{x^2 + 152}, f(x) = 10$

95. **a)** Consider the equation $\sqrt{4x - 12} = x - 3$. Setting each side of the equation equal to y yields the following system of equations.

$$y = \sqrt{4x - 12}$$

$$y = x - 3$$

The graphs of the equations in the system are illustrated in the figure.

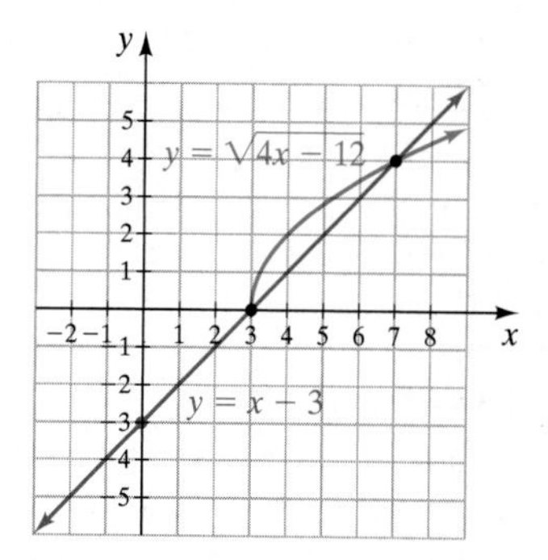

From the graphs, determine the values that appear to be solutions to the equation $\sqrt{4x - 12} = x - 3$. Explain how you determined your answer.

b) Substitute the values found in part **a)** into the original equation and determine whether they are the solutions to the equation.

c) Solve the equation $\sqrt{4x - 12} = x - 3$ algebraically and see if your solution agrees with the values obtained in part **a)**.

96. If the graph of a radical function, $f(x)$, does not intersect the x axis, then the equation $f(x) = 0$ has no real solutions. Explain why.

97. Suppose we are given a rational function $g(x)$. If $g(4) = 0$, then the graph of $g(x)$ must intersect the x axis at 4. Explain why.

98. The graph of the equation $y = \sqrt{x - 3} + 2$ is illustrated in the figure.

a) What is the domain of the function?

b) How many real solutions does the equation $\sqrt{x - 3} + 2 = 0$ have? List all the real solutions. Explain how you determined your answer.

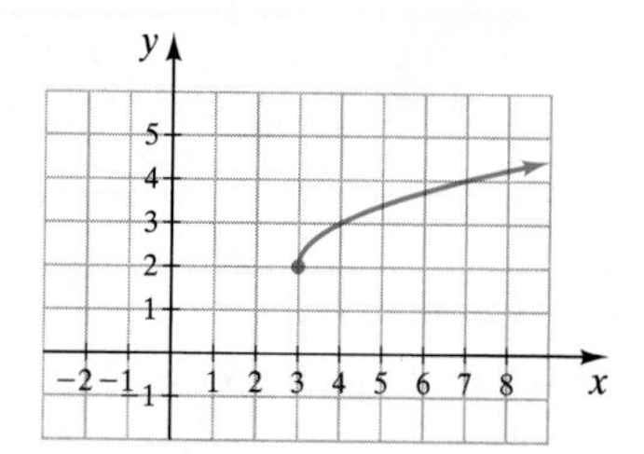

99. In statistics, a "confidence interval" is a range of values that is likely to contain the true value of the population. For a "95% confidence interval," the lower limit of the range, L_1, and the upper limit of the range, L_2, can be found by the formulas

$$L_1 = p - 1.96\sqrt{\frac{p(1 - p)}{n}}$$

$$L_2 = p + 1.96\sqrt{\frac{p(1 - p)}{n}}$$

where p represents the percent obtained from a sample and n is the size of the sample. Sylvia Dudley, a statistician, takes a sample of 36 families and finds that 60% of those surveyed used an answering machine in their home. She can therefore be 95% certain that the true percent of families that use an answering machine in their home is between L_1 and L_2. Find the values of L_1 and L_2. Use $p = 0.60$ and $n = 36$ in the formulas.

100. The *quadratic mean* (or *root mean square, RMS*) is often used in physical applications. In power distribution systems, for example, voltages and currents are usually referred to in terms of their RMS values. The quadratic mean of a set of scores is obtained by squaring each score and adding the results (signified by Σx^2), then dividing the value obtained by the number of scores, n, and then taking the square root of this value. We may express the formula as

$$\text{quadratic mean} = \sqrt{\frac{\Sigma x^2}{n}}$$

Find the quadratic mean of the numbers 2, 4, and 10.

Solve.

101. $\sqrt{x^2 + 9} = (x^2 + 9)^{1/2}$

102. $\sqrt{x^2 - 4} = (x^2 - 4)^{1/2}$

Use your graphing calculator to solve the equations. Round your solutions to the nearest tenth.

103. $\sqrt{x+8} = \sqrt{3x+5}$

104. $\sqrt{10x-16} - 15 = 0$

105. $\sqrt[3]{5x^2-6} - 4 = 0$

106. $\sqrt[3]{5x^2-10} = \sqrt[3]{4x+95}$

Challenge Problems

Solve.

107. $\sqrt{\sqrt{x+25} - \sqrt{x}} = 5$

108. $\sqrt{\sqrt{x+9} + \sqrt{x}} = 3$

109. $(3p-1)^{2/3} = (5p^2-p)^{1/3}$

Solve each equation for n.

110. $z = \dfrac{\overline{x} - \mu}{\dfrac{\sigma}{\sqrt{n}}}$

111. $z = \dfrac{p' - p}{\sqrt{\dfrac{pq}{n}}}$

Group Activity

Discuss and answer Exercise 112 as a group.

112. The area of a triangle is $A = \frac{1}{2}bh$. If the height is not known but we know the lengths of the three sides, we can use Heron's formula to find the area, A. Heron's formula is

$$A = \sqrt{S(S-a)(S-b)(S-c)}$$

where a, b, and c are the lengths of the three sides and

$$S = \frac{a+b+c}{2}$$

3 4 5

a) Have each group member use Heron's formula to find the area of a triangle whose sides are 3 inches, 4 inches, and 5 inches.

b) Compare your answers for part **a)**. If any member of the group did not get the correct answer, make sure they understand their error.

c) Have each member of the group do the following.

1) Draw a triangle on the following grid. Place each vertex of the triangle at the intersection of two grid lines.

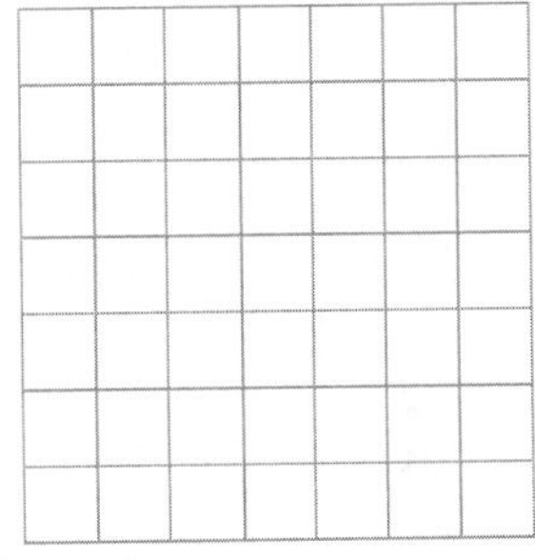

2) Measure with a ruler the length of each side of your triangle.

3) Use Heron's formula to find the area of your triangle.

4) Compare and discuss your work from part **c)**.

Cumulative Review Exercises

[2.2] **113.** Solve the formula $P_1P_2 - P_1P_3 = P_2P_3$ for P_2.

[6.1] **114.** Simplify $\dfrac{(2x+3)(3x-4)-(2x+3)(5x-1)}{(2x+3)}$.

Perform each indicated operation.

[6.1] **115.** $\dfrac{4a^2-9b^2}{4a^2+12ab+9b^2} \cdot \dfrac{6a^2b}{8a^2b^2-12ab^3}$

116. $(t^2-t-12) \div \dfrac{t^2-9}{t^2-3t}$

[6.2] **117.** $\dfrac{2}{x+3} - \dfrac{1}{x-3} + \dfrac{2x}{x^2-9}$

[6.4] **118.** Solve $2 + \dfrac{3x}{x-1} = \dfrac{8}{x-1}$.

7.7 COMPLEX NUMBERS

SSM VIDEO 7.7 CD Rom

1. Recognize a complex number.
2. Add and subtract complex numbers.
3. Multiply complex numbers.
4. Divide complex numbers.
5. Find powers of i.

1 Recognize a Complex Number

In Section 7.1 we stated that the square roots of negative numbers, such as $\sqrt{-4}$, are not real numbers. Numbers like $\sqrt{-4}$ are called **imaginary numbers**. Such numbers are called imaginary because when they were introduced many mathematicians refused to believe that they existed. Although they do not belong to the set of real numbers, the imaginary numbers do exist and are very useful in mathematics and science.

Every imaginary number has $\sqrt{-1}$, as a factor. The $\sqrt{-1}$, called the **imaginary unit**, is often denoted by the letter i.

$$i = \sqrt{-1}$$

We can therefore write

$$\sqrt{-4} = \sqrt{4}\sqrt{-1} = 2\sqrt{-1} = 2i$$
$$\sqrt{-9} = \sqrt{9}\sqrt{-1} = 3\sqrt{-1} = 3i$$
$$\sqrt{-7} = \sqrt{7}\sqrt{-1} = \sqrt{7}i \quad \text{or} \quad i\sqrt{7}$$

In this book we will generally write $i\sqrt{7}$ rather than $\sqrt{7}i$ to avoid confusion with $\sqrt{7i}$.

To help in writing square roots of negative numbers using i, we give the following rule.

For any positive real number n,

$$\sqrt{-n} = i\sqrt{n}$$

Examples

$$\sqrt{-4} = i\sqrt{4} = 2i \qquad \sqrt{-3} = i\sqrt{3}$$
$$\sqrt{-25} = i\sqrt{25} = 5i \qquad \sqrt{-10} = i\sqrt{10}$$

The real number system is a part of a larger number system, called the *complex number system*. Now we will discuss **complex numbers**.

Definition

Every number of the form

$$a + bi$$

where a and b are real numbers, is a **complex number**.

Every real number and every imaginary number are also complex numbers. A complex number has two parts: a real part, a, and an imaginary part, b.

$$a + bi$$

If $b = 0$, the complex number is a real number. If $a = 0$, the complex number is a *pure imaginary number.*

Examples of Complex Numbers

$3 + 4i$	$a = 3,\ b = 4$	
$5 - i\sqrt{3}$	$a = 5,\ b = -\sqrt{3}$	
5	$a = 5,\ b = 0$	(real number, $b = 0$)
$2i$	$a = 0,\ b = 2$	(imaginary number, $a = 0$)
$-i\sqrt{7}$	$a = 0,\ b = -\sqrt{7}$	(imaginary number, $a = 0$)

We stated that all real numbers and imaginary numbers are also complex numbers. The relationship between the various sets of numbers is illustrated in Figure 7.7.

Complex Numbers

Real Numbers		Nonreal Numbers
Rational numbers $\frac{1}{2}, -\frac{3}{5}, \frac{9}{4}$; Integers $-4, -9$, ; Whole numbers $0, 4, 12$	Irrational numbers $\sqrt{2}, \sqrt{3}$, $-\sqrt{7}, \pi$	$2 + 3i$, $6 - 4i$, $\sqrt{2} + i\sqrt{3}$, $i\sqrt{5}$, $6i$

FIGURE 7.7

EXAMPLE 1 Write each of the following complex numbers in the form $a + bi$.

a) $3 + \sqrt{-16}$ **b)** $5 - \sqrt{-12}$ **c)** 4 **d)** $\sqrt{-18}$ **e)** $6 + \sqrt{5}$

Solution

a) $3 + \sqrt{-16} = 3 + \sqrt{16}\sqrt{-1}$.
$= 3 + 4i$

b) $5 - \sqrt{-12} = 5 - \sqrt{12}\sqrt{-1}$
$= 5 - \sqrt{4}\sqrt{3}\sqrt{-1}$
$= 5 - 2\sqrt{3}i$ or $5 - 2i\sqrt{3}$

c) $4 = 4 + 0i$

d) $\sqrt{-18} = 0 + \sqrt{-18}$
$= 0 + \sqrt{9}\sqrt{2}\sqrt{-1}$
$= 0 + 3\sqrt{2}i$ or $0 + 3i\sqrt{2}$

NOW TRY EXERCISE 23

e) Both 6 and $\sqrt{5}$ are real numbers. The answer is $(6 + \sqrt{5}) + 0i$.

Complex numbers can be added, subtracted, multiplied, and divided. To perform these operations, we use the definitions that $i = \sqrt{-1}$ and

$$i^2 = -1$$

2) Add and Subtract Complex Numbers

We now explain how to add or subtract complex numbers.

To Add or Subtract Complex Numbers

1. Change all imaginary numbers to bi form.
2. Add (or subtract) the real parts of the complex numbers.
3. Add (or subtract) the imaginary parts of the complex numbers.
4. Write the answer in the form $a + bi$.

EXAMPLE 2 Add $(4 + 13i) + (-6 - 8i)$.

Solution
$$\begin{aligned}(4 + 13i) + (-6 - 8i) &= 4 + 13i - 6 - 8i \\ &= 4 - 6 + 13i - 8i && \text{Rearrange terms.} \\ &= -2 + 5i && \text{Combine like terms.}\end{aligned}$$

EXAMPLE 3 Subtract $(6 - \sqrt{-8}) - (4 + \sqrt{-18})$.

Solution
$$\begin{aligned}(6 - \sqrt{-8}) - (4 + \sqrt{-18}) &= (6 - \sqrt{8}\sqrt{-1}) - (4 + \sqrt{18}\sqrt{-1}) \\ &= (6 - \sqrt{4}\sqrt{2}\sqrt{-1}) - (4 + \sqrt{9}\sqrt{2}\sqrt{-1}) \\ &= (6 - 2i\sqrt{2}) - (4 + 3i\sqrt{2}) \\ &= 6 - 2i\sqrt{2} - 4 - 3i\sqrt{2} \\ &= 6 - 4 - 2i\sqrt{2} - 3i\sqrt{2} \\ &= 2 - 5i\sqrt{2}\end{aligned}$$

NOW TRY EXERCISE 35

3) Multiply Complex Numbers

Now let's discuss how to multiply complex numbers.

To Multiply Complex Numbers

1. Change all imaginary numbers to bi form.
2. Multiply the complex numbers as you would multiply polynomials.
3. Substitute -1 for each i^2.
4. Combine the real parts and the imaginary parts. Write the answer in $a + bi$ form.

EXAMPLE 4 Multiply. **a)** $3i(5 - 2i)$. **b)** $\sqrt{-4}\,(\sqrt{-2} + 7)$ **c)** $(3 - \sqrt{-8})(\sqrt{-2} + 5)$

Solution **a)** $3i(5 - 2i) = 3i(5) + (3i)(-2i)$ *Distributive Property*

$= 15i - 6i^2$

$= 15i - 6(-1)$ *Replace i^2 with -1*

$= 15i + 6$ or $6 + 15i$

b) $\sqrt{-4}\,(\sqrt{-2} + 7) = 2i(i\sqrt{2} + 7)$ *Change imaginary numbers to bi form*

$= (2i)(i\sqrt{2}) + (2i)(7)$ *Distributive Property*

$= 2i^2\sqrt{2} + 14i$

$= 2(-1)\sqrt{2} + 14i$ *Replace i^2 with -1*

$= -2\sqrt{2} + 14i$

c) $(3 - \sqrt{-8})(\sqrt{-2} + 5) = (3 - \sqrt{8}\sqrt{-1})(\sqrt{2}\sqrt{-1} + 5)$

$= (3 - 2i\sqrt{2})(i\sqrt{2} + 5)$

Now use the FOIL method to multiply.

$= (3)(i\sqrt{2}) + (3)(5) + (-2i\sqrt{2})(i\sqrt{2}) + (-2i\sqrt{2})(5)$

$= 3i\sqrt{2} + 15 - 2i^2(2) - 10i\sqrt{2}$

$= 3i\sqrt{2} + 15 - 2(-1)(2) - 10i\sqrt{2}$

$= 3i\sqrt{2} + 15 + 4 - 10i\sqrt{2}$

$= 19 - 7i\sqrt{2}$

NOW TRY EXERCISE 45

Avoiding Common Errors

What is $\sqrt{-4} \cdot \sqrt{-2}$?

CORRECT

$\sqrt{-4} \cdot \sqrt{-2} = 2i \cdot i\sqrt{2}$

$= 2i^2\sqrt{2}$

$= 2(-1)\sqrt{2}$

$= -2\sqrt{2}$

INCORRECT

~~$\sqrt{-4} \cdot \sqrt{-2} = \sqrt{8}$~~

~~$= \sqrt{4} \cdot \sqrt{2}$~~

~~$= 2\sqrt{2}$~~

Recall that $\sqrt{a} \cdot \sqrt{b} = \sqrt{ab}$ only for *nonnegative* real numbers a and b.

4) Divide Complex Numbers

The **conjugate of a complex number** $a + bi$ is $a - bi$. For example,

Complex Number	Conjugate
$3 + 4i$	$3 - 4i$
$1 - i\sqrt{3}$	$1 + i\sqrt{3}$
$2i$ (or $0 + 2i$)	$-2i$ (or $0 - 2i$)

When a complex number is multiplied by its conjugate using the FOIL method, the inner and outer products will sum to zero. For example,

$$\begin{aligned}(5 + 2i)(5 - 2i) &= 25 - 10i + 10i - 4i^2 \\ &= 25 - 4i^2 \\ &= 25 - 4(-1) \\ &= 25 + 4 = 29\end{aligned}$$

Now we explain how to divide complex numbers.

To Divide Complex Numbers

1. Change all imaginary numbers to bi form.
2. Rationalize the denominator by multiplying both the numerator and denominator by the conjugate of the denominator.
3. Write the answer in $a + bi$ form.

EXAMPLE 5 Divide $\dfrac{4 + i}{i}$.

Solution Begin by multiplying both numerator and denominator by $-i$, the conjugate of i.

$$\begin{aligned}\frac{4 + i}{i} \cdot \frac{-i}{-i} &= \frac{(4 + i)(-i)}{-i^2} \\ &= \frac{-4i - i^2}{-i^2} \\ &= \frac{-4i - (-1)}{-(-1)} \\ &= \frac{-4i + 1}{1} \\ &= 1 - 4i\end{aligned}$$

EXAMPLE 6 Divide $\dfrac{6 - 5i}{2 - i}$.

Solution Multiply both numerator and denominator by $2 + i$, the conjugate of $2 - i$.

$$\begin{aligned}\frac{6 - 5i}{2 - i} \cdot \frac{2 + i}{2 + i} &= \frac{12 + 6i - 10i - 5i^2}{4 - i^2} \\ &= \frac{12 - 4i - 5(-1)}{4 - (-1)} \\ &= \frac{17 - 4i}{5} \quad \text{or} \quad \frac{17}{5} - \frac{4}{5}i\end{aligned}$$

NOW TRY EXERCISE 65

EXAMPLE 7 A concept needed for the study of electronics is *impedance*. Impedance affects the current in a circuit. The impedance, Z, in a circuit is found by the formula $Z = \frac{V}{I}$, where V is voltage and I is current. Find Z when $V = 1.6 - 0.3i$ and $I = -0.2i$, where $i = \sqrt{-1}$.

Solution $Z = \dfrac{V}{I} = \dfrac{1.6 - 0.3i}{-0.2i}$. Now multiply both numerator and denominator by $0.2i$.

$$Z = \frac{1.6 - 0.3i}{-0.2i} \cdot \frac{0.2i}{0.2i} = \frac{0.32i - 0.06i^2}{-0.04i^2}$$

$$= \frac{0.32i + 0.06}{0.04}$$

$$= \frac{0.32i}{0.04} + \frac{0.06}{0.04}$$

$$= 8i + 1.5 \quad \text{or} \quad 1.5 + 8i$$

Most algebra books use i as the imaginary unit. However, most electronics books use j as the imaginary unit because i is often used to represent current.

5) Find Powers of i

Using $i = \sqrt{-1}$ and $i^2 = -1$, we can find other **powers of *i***. For example,

$$i^3 = i^2 \cdot i = -1 \cdot i = -i \qquad i^6 = i^4 \cdot i^2 = 1(-1) = -1$$

$$i^4 = i^2 \cdot i^2 = (-1)(-1) = 1 \qquad i^7 = i^4 \cdot i^3 = 1(-i) = -i$$

$$i^5 = i^4 \cdot i^1 = 1 \cdot i = i \qquad i^8 = i^4 \cdot i^4 = (1)(1) = 1$$

Note that successive powers of i rotate through the four values $i, -1, -i, 1$ (see Fig. 7.8).

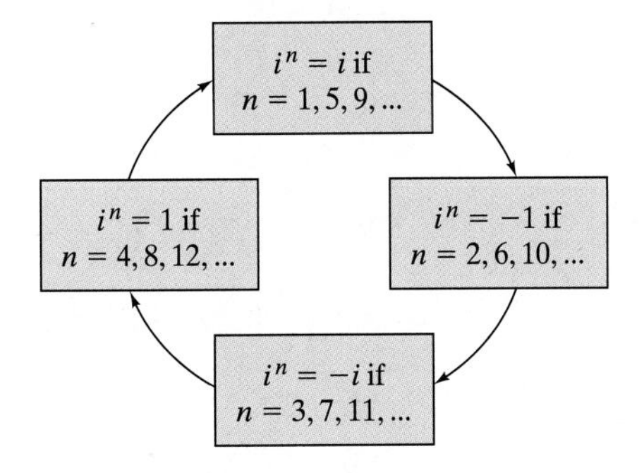

FIGURE 7.8

EXAMPLE 8 Evaluate. **a)** i^{35} **b)** i^{81}

Solution Write each expression as a product of factors such that the exponent of one factor is the largest multiple of 4 less than or equal to the given exponent. Then write this factor as i^4 raised to some power. Since i^4 has a value of 1, the expression i^4 raised to a power will also have a value of 1.

a) $i^{35} = i^{32} \cdot i^3 = \left(i^4\right)^8 \cdot i^3 = 1 \cdot i^3 = 1(-i) = -i$

NOW TRY EXERCISE 101

b) $i^{81} = i^{80} \cdot i^1 = \left(i^4\right)^{20} \cdot i = 1 \cdot i = i$

HELPFUL HINT

A quick way of evaluating i^n is to divide the exponent by 4 and observe the remainder.

If the remainder is 0, the value is 1.
If the remainder is 1, the value is i.
If the remainder is 2, the value is -1.
If the remainder is 3, the value is $-i$.

For Example 8 **a)**

$$\begin{array}{r} 8 \\ 4\overline{)35} \\ \underline{32} \\ 3 \end{array} \longleftarrow \text{answer is } -i$$

For 8 **b)**

$$\begin{array}{r} 20 \\ 4\overline{)81} \\ \underline{80} \\ 1 \end{array} \longleftarrow \text{answer is } i$$

Exercise Set 7.7

1. What does i equal?
2. Write $\sqrt{-n}$ using i.
3. Are all of the following complex numbers? If any are not complex numbers, explain why. **a)** 4 **b)** $-\frac{1}{6}$ **c)** $3 - \sqrt{-2}$ **d)** $5 - 3i$ **e)** $4.2i$ **f)** $3 + \sqrt{7}$
4. What does i^2 equal?
5. Is every real and every imaginary number a conplex number?
6. Is every complex number a real number?
7. What is the conjugate of $a + bi$?
8. Is $i \cdot i$ a real number? Explain.
9. List, if possible, a number that is *not*
   a) a rational number.
   b) an irrational number.
   c) a real number.
   d) an imaginary number.
   e) a complex number.
10. Write a paragraph or two explaining the relationship between the real numbers, imaginary numbers, and complex numbers. Include in your discussion how the various sets of numbers relate to each other.

Practice the Skills

Write each expression as a complex number in the form $a + bi$.

11. 3
12. $\sqrt{9}$
13. $3 + \sqrt{-4}$
14. $-\sqrt{5}$
15. $6 + \sqrt{3}$
16. $\sqrt{-8}$
17. $\sqrt{-25}$
18. $2 + \sqrt{-5}$
19. $4 + \sqrt{-12}$
20. $\sqrt{-4} + \sqrt{-16}$
21. $\sqrt{-25} - 2i$
22. $3 + \sqrt{-72}$
23. $9 - \sqrt{-9}$
24. $\sqrt{75} - \sqrt{-128}$
25. $2i - \sqrt{-80}$
26. $\sqrt{288} - \sqrt{-96}$

Add or subtract.

27. $(12 - 6i) + (3 + 2i)$
28. $(6 - 3i) - 2(2 - 4i)$
29. $\left(12 + \frac{5}{9}i\right) - \left(4 - \frac{3}{4}i\right)$
30. $\left(\frac{5}{8} + \sqrt{-4}\right) + \left(\frac{2}{3} + 7i\right)$
31. $(13 - \sqrt{-4}) - (-5 + \sqrt{-9})$
32. $(7 + \sqrt{5}) + (2\sqrt{5} + \sqrt{-5})$
33. $(\sqrt{3} + \sqrt{2}) + (3\sqrt{2} - \sqrt{-8})$
34. $(3 - \sqrt{-72}) + (4 - \sqrt{-32})$
35. $(19 + \sqrt{-147}) + (\sqrt{-75})$
36. $(13 + \sqrt{-108}) - (\sqrt{49} - \sqrt{-48})$
37. $(\sqrt{12} + \sqrt{-49}) - (\sqrt{49} - \sqrt{-12})$
38. $(\sqrt{20} - \sqrt{-12}) + (2\sqrt{5} + \sqrt{-75})$

Multiply.

39. $2(-3 - 2i)$
40. $-3(\sqrt{5} + 2i)$
41. $i(6 + i)$
42. $2i(2 - 5i)$
43. $-3.5i(6.4 - 1.8i)$
44. $\sqrt{-5}\,(2 + 3i)$
45. $\sqrt{-4}\,(\sqrt{3} + 2i)$
46. $\sqrt{-8}\,(\sqrt{2} - \sqrt{-2})$
47. $\sqrt{-6}\,(\sqrt{3} + \sqrt{-6})$
48. $-\sqrt{-2}\,(3 - \sqrt{-8})$
49. $(3 + 2i)(1 + i)$
50. $(3 - 4i)(6 + 5i)$
51. $(4 - 6i)(3 - i)$
52. $(3i + 4)(2i - 3)$
53. $\left(\frac{1}{4} + \sqrt{-3}\right)(2 + \sqrt{3})$
54. $(2 - 3i)(4 + \sqrt{-4})$
55. $(5 - \sqrt{-8})\left(\frac{1}{4} + \sqrt{-2}\right)$
56. $\left(\frac{3}{5} - \frac{1}{4}i\right)\left(\frac{2}{3} + \frac{2}{5}i\right)$

Divide.

57. $\dfrac{-5}{-3i}$
58. $\dfrac{2}{5i}$
59. $\dfrac{2 + 3i}{2i}$
60. $\dfrac{1 + i}{-3i}$
61. $\dfrac{2 + 5i}{5i}$
62. $\dfrac{6}{2 + i}$
63. $\dfrac{7}{7 - 2i}$
64. $\dfrac{4 + 2i}{1 + 3i}$

65. $\frac{6-3i}{4+2i}$ **66.** $\frac{4-3i}{4+3i}$ **67.** $\frac{4}{6-\sqrt{-4}}$ **68.** $\frac{5}{3+\sqrt{-5}}$

69. $\frac{\sqrt{6}}{\sqrt{3}-\sqrt{-9}}$ **70.** $\frac{\sqrt{2}}{5+\sqrt{-12}}$ **71.** $\frac{\sqrt{10}+\sqrt{-3}}{5-\sqrt{-20}}$ **72.** $\frac{12-\sqrt{-12}}{\sqrt{3}+\sqrt{-5}}$

73. $\frac{\sqrt{-60}}{\sqrt{-2}}$ **74.** $\frac{\sqrt{-150}}{\sqrt{6}}$ **75.** $\frac{\sqrt{-80}\sqrt{-5}}{\sqrt{-2}}$ **76.** $\frac{\sqrt{-32}}{\sqrt{-18}\sqrt{2}}$

Perform each indicated operation. These exercises are a combination of the types of exercises presented earlier in this exercise set.

77. $(4-2i)+(3-5i)$ **78.** $\left(\frac{1}{2}+i\right)-\left(\frac{3}{5}-\frac{2}{3}i\right)$ **79.** $(8-\sqrt{-6})-(2-\sqrt{-24})$

80. $(\sqrt{8}-\sqrt{2})-(\sqrt{-12}-\sqrt{-48})$ **81.** $5.2(4-3.2i)$ **82.** $\sqrt{-6}(\sqrt{3}-\sqrt{-10})$

83. $(5+2i)(3-5i)$ **84.** $(\sqrt{3}+2i)(\sqrt{6}-\sqrt{-8})$ **85.** $\frac{5-4i}{2i}$

86. $\frac{1}{2-3i}$ **87.** $\frac{4}{\sqrt{3}-\sqrt{-4}}$ **88.** $\frac{5-2i}{3+2i}$

89. $\left(5-\frac{5}{9}i\right)-\left(2-\frac{3}{5}i\right)$ **90.** $\frac{4}{7}\left(4-\frac{3}{5}i\right)$ **91.** $\left(\frac{2}{3}-\frac{1}{5}i\right)\left(\frac{3}{5}-\frac{3}{4}i\right)$

92. $\sqrt{\frac{4}{9}}\left(\sqrt{\frac{25}{36}}-\sqrt{-\frac{4}{25}}\right)$ **93.** $\frac{\sqrt{-96}}{\sqrt{-24}}$ **94.** $\frac{-1-2i}{2+\sqrt{-5}}$

95. $(5.23-6.41i)-(8.56-4.5i)-7.1i$ **96.** $(\sqrt{-6}+3)(\sqrt{-15}+5)$

For each imaginary number, indicate whether its value is i, −1, −i, or 1.

97. i^{10} **98.** i^{43} **99.** i^{200} **100.** i^{211}

101. i^{93} **102.** i^{103} **103.** i^{907} **104.** i^{1113}

Problem Solving

Answer true or false. Support your answer with an example.

105. The product of two pure imaginary numbers is always a real number.

106. The sum of two pure imaginary numbers is always an imaginary number.

107. The product of two complex numbers is always a real number.

108. The sum of two complex numbers is always a complex number.

109. What values of n will result in i^n being a real number? Explain.

110. What values of n will result in i^{2n} being a real number? Explain.

111. If $f(x)=x^2$, find $f(i)$.

112. If $f(x)=x^2$, find $f(3i)$.

113. If $f(x)=x^4-2x$, find $f(2i)$.

114. If $f(x)=x^3-3x^2$, find $f(5i)$.

115. If $f(x)=x^2-x$, find $f(3+i)$.

116. If $f(x)=\frac{x^2}{x-2}$, find $f(4-i)$.

Evaluate each expression for the given value of x.

117. $x^2-2x+5, x=1-2i$ **118.** $x^2-2x+5, x=1+2i$

119. $x^2+2x+8, x=-1+i\sqrt{5}$ **120.** $x^2+2x+8, x=-1-i\sqrt{5}$

Determine whether the given value of x is a solution to the equation.

121. $x^2-4x+5=0, x=2-i$ **122.** $x^2-4x+5=0, x=2+i$

123. $x^2-6x+12=0, x=-3+i\sqrt{3}$ **124.** $x^2-6x+12=0, x=3-i\sqrt{3}$

125. Find the impedance, Z, using the formula $Z = V/I$ when $V = 1.8 + 0.5i$ and $I = 0.6i$. See Example 7.

126. Refer to Exercise 125. Find the impedance when $V = 2.4 - 0.6i$ and $I = -0.4i$.

127. Under certain conditions, the total impedance, Z_T, of a circuit is given by the formula

$$Z_T = \frac{Z_1 Z_2}{Z_1 + Z_2}$$

Find Z_T when $Z_1 = 2 - i$ and $Z_2 = 4 + i$.

128. Refer to Exercise 127. Find Z_T when $Z_1 = 3 - i$ and $Z_2 = 5 + i$.

129. Determine whether i^{-1} is equal to $i, -1, -i$, or 1. Show your work.

130. Determine whether i^{-2} is equal to $i, -1, -i$, or 1. Show your work.

In Chapter 8 we will use the quadratic formula $x = \dfrac{-b \pm \sqrt{b^2 - 4ac}}{2a}$ *to solve equations of the form* $ax^2 + bx + c = 0$. ***(a)*** *Use the quadratic formula to solve the following quadratic equations.* ***(b)*** *Check each of the two solutions by substituting the values found for x (one at a time) back in the original equation. In these exercises, the ± (read "plus or minus") results in two distinct complex answers.*

131. $x^2 - 4x + 6 = 0$

132. $x^2 - 2x + 6 = 0$

Given the complex numbers $a = 3 + 2i\sqrt{3}$, $b = 1 + i\sqrt{3}$, *evaluate each expression.*

133. $a + b$

134. $a - b$

135. ab

136. $\dfrac{a}{b}$

Cumulative Review Exercises

[4.3] **137.** Josefina Saavedra, a grocer, has two coffees, one selling for \$5.50 per pound and the other for \$6.30 per pound. How many pounds of each type of coffee should she mix to make 40 pounds of coffee to sell for \$6.00 per pound?

Factor.

[6.2] **138.** $4x^4 + 12x^2 + 9$

139. $15r^2s^2 + rs - 6$

140. $8r^3 - 27s^6$

Summary

Key Terms and Phrases

7.1
Cube root
Even roots
Index
Odd roots
Principal square root
Radical expression
Radical sign
Radicand
Square root

7.2
Rational exponent

7.3
Perfect cubes
Perfect powers
Perfect squares
Product rule for radicals
Quotient rule for radicals
Rationalizing the denominator
Simplified radical

7.4
Conjugate

7.5
Like radicals
Unlike radicals

7.6
Radical equation

7.7
Complex number
Conjugate of a complex number
Imaginary number
Imaginary unit
Powers of i

IMPORTANT FACTS

If n is Even and $a \geq 0$:

$$\sqrt[n]{a} = b \text{ if } b \geq 0 \text{ and } b^n = a$$

If n is Odd:

$$\sqrt[n]{a} = b \text{ if } b^n = a$$

Rules of Radicals

$\sqrt{a^2} = |a|$

$\sqrt{a^2} = a, \quad a \geq 0$

$\sqrt[n]{a^n} = a, \quad a \geq 0$

$\sqrt[n]{a} = a^{1/n}, \quad a \geq 0$

$\sqrt[n]{a^m} = (\sqrt[n]{a})^m = a^{m/n}, \quad a \geq 0$

$\sqrt[n]{a}\sqrt[n]{b} = \sqrt[n]{ab}, \quad a \geq 0, b \geq 0$

$\dfrac{\sqrt[n]{a}}{\sqrt[n]{b}} = \sqrt[n]{\dfrac{a}{b}}, \quad a \geq 0, b > 0$

A Radical is Simplified When the Following are all True:

1. No perfect powers are factors of any radicand.
2. No radicand contains a fraction.
3. No denominator contains a radical.

Powers of i

$$i = \sqrt{-1}, \quad i^2 = -1, \quad i^3 = -i, \quad i^4 = 1$$

Review Exercises

[7.1] *Evaluate.*

1. $\sqrt{9}$ **2.** $\sqrt[3]{-27}$ **3.** $\sqrt[4]{256}$ **4.** $\sqrt[3]{-64}$

Use absolute value to evaluate.

5. $\sqrt{(-7)^2}$ **6.** $\sqrt{(-93.4)^2}$

Write as an absolute value.

7. $\sqrt{x^2}$ **8.** $\sqrt{(x-2)^2}$ **9.** $\sqrt{(x-y)^2}$ **10.** $\sqrt{(x^2-4x+12)^2}$

For the remainder of these review exercises, assume that all variables represent positive real numbers.

[7.2] *Write in exponential form.*

11. $\sqrt{x^5}$ **12.** $\sqrt[3]{x^5}$ **13.** $(\sqrt[4]{y})^{15}$ **14.** $\sqrt[7]{5^2}$

Write in radical form.

15. $a^{1/2}$ **16.** $y^{3/5}$ **17.** $(2m^2n)^{9/5}$ **18.** $(2a+3b)^{-3/4}$

Simplify each radical expression by changing the expression to exponential form. Write the answer in radical form.

19. $\sqrt[3]{3^6}$ **20.** $\sqrt{x^{10}}$ **21.** $(\sqrt[3]{4})^6$ **22.** $\sqrt[20]{x^4}$

Evaluate if possible. If the expression is not a real number, so state.

23. $-25^{1/2}$ **24.** $(-25)^{1/2}$ **25.** $\left(\frac{8}{27}\right)^{-1/3}$ **26.** $36^{-1/2} - 8^{-2/3}$

Simplify. Write the answer without negative exponents.

27. $x^{3/5} \cdot x^{-1/3}$ **28.** $\left(\frac{64}{y^6}\right)^{1/3}$ **29.** $\left(\frac{y^{-3/5}}{y^{1/5}}\right)^{2/3}$ **30.** $\left(\frac{30x^4y^{-2}}{5y^{1/2}}\right)^2$

Multiply.

31. $z^{1/3}(2z^{5/3} - 4z)$ **32.** $\frac{3}{4}r^{-2/3}\left(4r^{-3/2} + \frac{4}{3}r^{2/3}\right)$

Factor each expression. Write the answer without negative exponents.

33. $x^{1/5} + x^{6/5}$ **34.** $x^{-1/2} + x^{2/3}$

For each function, find the indicated value of the function. Use your calculator to evaluate irrational numbers. Round irrational numbers to the nearest thousandth.

35. If $f(x) = \sqrt{6x + 13}$, find $f(6)$. **36.** If $g(x) = \sqrt[3]{9r - 17}$, find $g(4)$.

Graph the following functions.

37. $f(x) = \sqrt{x}$ **38.** $f(x) = \sqrt{x} - 4$

[7.2–7.5] *Simplify.*

39. $\left(\frac{3r^2p^{1/3}}{r^{1/2}p^{4/3}}\right)^3$ **40.** $\left(\frac{4y^{2/5}z^{1/3}}{x^{-1}y^{3/5}}\right)^{-1}$ **41.** $\sqrt{80}$

42. $\sqrt[3]{54}$ **43.** $\sqrt{50x^3y^7}$ **44.** $\sqrt[3]{125x^7y^{10}}$

45. $(\sqrt[6]{x^{12}y^7z^{17}})^{42}$ **46.** $\sqrt{20}\sqrt{5}$ **47.** $\sqrt{5x}\sqrt{8x^5}$

48. $\sqrt[3]{2x^4y^5}\sqrt[3]{16x^4y^4}$ **49.** $(\sqrt[3]{5x^2y^3})^2$ **50.** $\sqrt[4]{8x^4y^7}\sqrt[4]{2x^5y^9}$

51. $\sqrt{3x}(\sqrt{12x} - \sqrt{20})$ **52.** $(\sqrt[5]{a^7b^{12}c^9})^{35}$ **53.** $\sqrt[3]{2x^2y}(\sqrt[3]{4x^4y^7} + \sqrt[3]{9x})$

54. $\sqrt{\sqrt{x^2y}}$ **55.** $\sqrt{\sqrt[3]{x^4y}}$ **56.** $\sqrt{\frac{36}{25}}$

57. $\sqrt[3]{\frac{x^3}{8}}$ **58.** $\frac{\sqrt[3]{2x^9}}{\sqrt[3]{16x^6}}$ **59.** $\sqrt{\frac{32x^2y^5}{2x^4y}}$

60. $\sqrt[3]{\frac{108x^3y^6}{2y^3}}$ **61.** $\frac{x}{\sqrt{7}}$ **62.** $\sqrt{\frac{2}{5}}$

63. $\sqrt{\frac{12x}{5y}}$ **64.** $\frac{2}{\sqrt[3]{x}}$ **65.** $\sqrt[3]{\frac{3x}{5y}}$

66. $\sqrt{\frac{3x^2}{y}}$ **67.** $\sqrt{\frac{18x^4y^5}{3z}}$ **68.** $\sqrt{\frac{125x^2y^5}{3z}}$

69. $\sqrt[4]{\frac{2x^2y^6}{8x^3}}$ **70.** $\sqrt[3]{\frac{4x^5y^3}{x^6}}$ **71.** $\sqrt[3]{\frac{y^6}{2x^2}}$

72. $(3 - \sqrt{2})(3 + \sqrt{2})$ **73.** $(\sqrt{x} + y)(\sqrt{x} - y)$ **74.** $(x - \sqrt{y})(x + \sqrt{y})$

75. $(\sqrt{3} + 5)^2$ **76.** $(\sqrt{x} - \sqrt{3y})(\sqrt{x} + \sqrt{5y})$ **77.** $(\sqrt[3]{2x} - \sqrt[3]{3y})(\sqrt[3]{3x} - \sqrt[3]{2y})$

78. $\dfrac{5}{2 + \sqrt{5}}$

79. $\dfrac{x}{3 + \sqrt{x}}$

80. $\dfrac{\sqrt{x}}{\sqrt{x} + \sqrt{y}}$

81. $\dfrac{\sqrt{x} - 2\sqrt{y}}{\sqrt{x} - \sqrt{y}}$

82. $\dfrac{4}{\sqrt{y + 2} - 3}$

83. $\sqrt[3]{x} + 3\sqrt[3]{x} - 2\sqrt[3]{x}$

84. $\sqrt{3} + \sqrt{27} - \sqrt{192}$

85. $\sqrt[3]{16} - 5\sqrt[3]{54} + 2\sqrt[3]{64}$

86. $4\sqrt{2} - \dfrac{3}{\sqrt{32}} + \sqrt{50}$

87. $3\sqrt{x^5y^6} - \sqrt{16x^7y^8}$

88. $2\sqrt[3]{x^7y^8} - \sqrt[3]{x^4y^2} + 3\sqrt[3]{x^{10}y^2}$

In Exercises 89 and 90, $f(x)$ and $g(x)$ are given. Find $(f \cdot g)(x)$.

89. $f(x) = \sqrt{3x}, g(x) = \sqrt{6x} - \sqrt{10}$

90. $f(x) = \sqrt[3]{2x^2}, g(x) = \sqrt[3]{4x^4} + \sqrt[3]{8x^5}$

Simplify. Assume the variable can be any real number.

91. $f(x) = \sqrt{2x + 4}\sqrt{2x + 4}$

92. $g(a) = \sqrt{20a^2 + 50a + 125}$

Simplify.

93. $\dfrac{\sqrt[3]{(x + 5)^5}}{\sqrt{(x + 5)^3}}$

94. $\dfrac{\sqrt[3]{a^3b^2}}{\sqrt[4]{a^4b}}$

95. The graph of $f(x) = \sqrt{x} + 2$ is given.
 a) For $g(x) = -3$, sketch the graph of $(f + g)(x)$.
 b) What is the domain of $(f + g)(x)$?

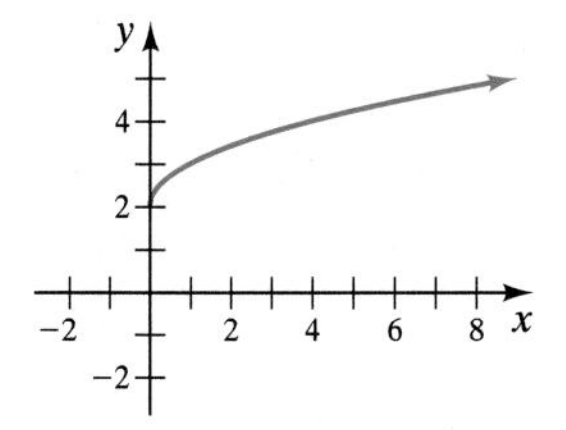

96. The graph of $f(x) = -\sqrt{x}$ is given.
 a) For $g(x) = \sqrt{x} + 2$, sketch the graph of $(f + g)(x)$.
 b) What is the domain of $(f + g)(x)$?

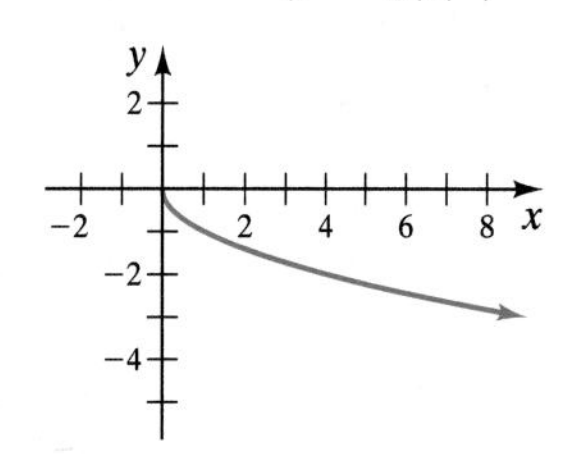

[7.6] *Solve each equation and check your solutions.*

97. $\sqrt[3]{x} = 4$

98. $\sqrt{3x + 4} = \sqrt{5x + 12}$

99. $2 + \sqrt[3]{x} = 4$

100. $\sqrt{x^2 + 2x - 4} = x$

101. $\sqrt[3]{x - 9} = \sqrt[3]{5x + 3}$

102. $(x^2 + 5)^{1/2} = x + 1$

103. $\sqrt{x} + 3 = \sqrt{3x + 9}$

104. $\sqrt{6x - 5} - \sqrt{2x + 6} - 1 = 0$

For each pair of functions, find all values of x for which $f(x) = g(x)$.

105. $f(x) = \sqrt{3x + 4}, g(x) = 2\sqrt{2x - 4}$

106. $f(x) = (4x + 3)^{1/3}, g(x) = (6x - 9)^{1/3}$

Solve the following for the indicated variable.

107. $V = \sqrt{\dfrac{2L}{w}}$, for L.

108. $r = \sqrt{\dfrac{A}{\pi}}$, for A.

Solve.

109. How long a wire does a phone company need to reach the top of a 5-meter telephone pole from a point on the ground 2 meters from the base of the pole?

110. Use the formula $v = \sqrt{2gh}$ to find the velocity of an object after it has fallen 20 feet ($g = 32$ ft/s^2).

111. Use the formula

$$T = 2\pi\sqrt{\frac{L}{32}}$$

to find the period of a pendulum, T, if its length, L, is 64 feet.

112. There are two types of energy: kinetic and potential. Potential energy is due to position and kinetic energy is due to motion. For example, if you hold a billiard ball above the ground it has potential energy. If you let go of the ball the potential energy is changed to kinetic energy as the ball drops. The formula

$$V = \sqrt{\frac{2K}{m}}$$

can be used to determine the velocity, V, in meters per second, when a mass, m, in kilograms, has a kinetic energy, K, joules. A 0.145-kg baseball is thrown. If the kinetic energy of the moving ball is 45 joules, at what speed is the ball moving?

113. Albert Einstein found that if an object at rest, with mass m_0, is made to travel close to the speed of light, its mass increases to m, where

$$m = \frac{m_0}{\sqrt{1 - \frac{v^2}{c^2}}}$$

In the formula, v is the velocity of the moving object and c is the speed of light.* In an accelerator used for cancer therapy, particles travel at speeds of $0.98c$, that is, at 98% the speed of light. At a speed of $0.98c$, determine a particle's mass, m, in terms of its rest mass, m_0. Use $v = 0.98c$ in the above formula.

[7.7] *Write each expression as a complex number in the form a + bi.*

114. 5 **115.** −6 **116.** $2 - \sqrt{-256}$ **117.** $3 + \sqrt{-16}$

Perform each indicated operation.

118. $(3 + 2i) + (4 - i)$
119. $(4 - 6i) - (3 - 4i)$
120. $(\sqrt{3} + \sqrt{-5}) + (2\sqrt{3} - \sqrt{-7})$
121. $\sqrt{-6}(\sqrt{6} + \sqrt{-6})$
122. $(4 + 3i)(2 - 3i)$
123. $(6 + \sqrt{-3})(4 - \sqrt{-15})$
124. $\frac{2}{3i}$
125. $\frac{2 + \sqrt{3}}{2i}$
126. $\frac{5}{3 + 2i}$
127. $\frac{\sqrt{3}}{5 - \sqrt{-6}}$

Evaluate each expression for the given value of x.

128. $x^2 - 2x + 9, x = 1 + 2i\sqrt{2}$
129. $x^2 - 2x + 12, x = 1 - 2i$

For each imaginary number, indicate whether its value is i, −1, −i, or 1.

130. i^{53} **131.** i^{19} **132.** i^{404} **133.** i^{5326}

Practice Test

1. Use absolute value to evaluate $\sqrt{(-26)^2}$.
2. Write $\sqrt{(3x - 4)^2}$ as an absolute value.
3. Simplify $\left(\frac{y^{2/3} \cdot y^{-1}}{y^{1/4}}\right)^2$
4. Factor $x^{3/5} + x^{-2/5}$.
5. Graph $f(x) = \sqrt{x}$.

Simplify. Assume that all variables represent positive real numbers.

6. $\sqrt{50x^5y^8}$
7. $\sqrt[3]{4x^5y^2}\sqrt[3]{10x^6y^8}$
8. $\sqrt{\frac{2x^4y^5}{8z}}$

*The speed of light is 3.00×10^8 meters per second. However, you do not need this information to solve this problem.

9. $\sqrt[3]{\frac{1}{x}}$

10. $\frac{\sqrt{2}}{2+\sqrt{8}}$

11. $\sqrt{27} + 2\sqrt{3} - 5\sqrt{75}$

12. $\sqrt[3]{8x^3y^5} + 2\sqrt[3]{x^6y^8}$

13. $(\sqrt{5} - 3)(2 - \sqrt{8})$

14. $\sqrt[3]{\sqrt{x^4y^2}}$

15. $\frac{\sqrt[4]{(7x+2)^5}}{\sqrt[3]{(7x+2)^2}}$

Solve.

16. $\sqrt{4x - 3} = 7$

17. $\sqrt{x^2 - x - 12} = x + 3$

18. $\sqrt{x - 15} = \sqrt{x} - 3$

19. For $f(x) = (9x + 37)^{1/3}$ and $g(x) = 2(2x + 2)^{1/3}$, find all values of x such that $f(x) = g(x)$.

20. Solve the formula $w = \frac{\sqrt{2gh}}{4}$ for h.

21. A ladder is placed against a house. If the base of the ladder is 5 feet from the house and the ladder rests on the house 12 feet above the ground, find the length of the ladder.

22. A formula used in the study of springs is
$$T = 2\pi\sqrt{\frac{m}{k}}$$
where T is the period of a spring (the time for the spring to stretch and return to its rest point), m is the mass on the spring, in kilograms, and k is the spring's constant, in newtons/meter. A mass of 1400 kilograms rests on a spring. Find the period of the spring if the spring's constant is 65,000 newtons/meter.

23. Multiply $(6 - \sqrt{-4})(3 + \sqrt{-2})$.

24. Divide $\frac{\sqrt{5}}{2 - \sqrt{-8}}$.

25. Evaluate $x^2 + 6x + 12$ for $x = -3 + i$.

Cumulative Review Test

1. Solve $\frac{1}{5}(x - 3) = \frac{3}{4}(x + 3) - x$.
2. Define **a)** a relation; **b)** a function.
3. **a)** State the domain of $f(x) = \sqrt{x} + 2$.
   b) Graph $f(x)$.
4. Solve the following system of equations.
$$x - 4y = 6$$
$$3x - y = 2$$
5. Simplify $\left(\frac{3x^2y^{-2}}{x^4y^{-5}}\right)^{-1}$.
6. For $f(x) = 3x^2 - 4x - 6$ and $g(x) = 2x - 5$, find $(f \cdot g)(x)$.
7. Divide $\frac{3x^2 + 10x + 10}{x + 2}$.
8. Graph $f(x) = x^3 - 2$.
9. Factor $2x^2 - 12x + 18 - 2y^2$.
10. Find the domain of $f(x) = \frac{x + 2}{3x - 5}$.
11. Simplify $\frac{(x + 2)(x - 4) + (x - 1)(x - 4)}{3(x - 4)}$.
12. Multiply $\frac{4x^2 + 8x + 3}{2x^2 - x - 1} \cdot \frac{x^2 - 1}{4x^2 + 12x + 9}$.
13. Subtract $\frac{x + 1}{x^2 + 2x - 3} - \frac{x}{2x^2 + 11x + 15}$.
14. Solve $4 - \frac{5}{y} = \frac{4y}{y + 1}$.
15. Simplify $\left(\frac{x^2y^{1/2}}{x^{1/4}}\right)^2$.
16. Simplify $\sqrt[3]{4x^{10}y^{20}} \cdot \sqrt[3]{4x^3y^9}$.
17. Solve $\sqrt{2x^2 + 7} + 3 = 8$.
18. Divide $\frac{2}{3 + \sqrt{-6}}$.
19. Jerrel Hocking by himself can paint the living room in his house in 2 hours. Jerrell's son Mike can paint the same room by himself in 3 hours. How long will it take them to paint the room if they work together?
20. A wire reaches from the top of a 30-foot telephone pole to a point on the ground 20 feet from the base of the pole. What is the length of the wire?

QUADRATIC FUNCTIONS

CHAPTER 8

Use the Angel Web site at www.prenhall.com/angel to be linked to an internet resource that will help you further explore the following application.

Life insurance companies base their rates on remaining life expectancy. The more years you are expected to live lower your premiums. On page 509, we use the quadratic formula to determine the age of a person with a remaining life expectancy of 14.3 years.

Preview and Perspective

We introduced quadratic functions in Section 5.1. Now we expand on the concepts. In Section 8.1 we introduce completing the square and in Section 8.2 we discuss the quadratic formula. After studying these sections, you will know three techniques for solving quadratic equations: factoring (when possible), completing the square, and the quadratic formula.

In Section 8.3 we study applications of quadratic equations. In Section 8.4 we introduce equations that can be expressed and solved as if they were quadratic equations. Such equations are called equations that are quadratic in form.

We discuss techniques for graphing quadratic functions in Section 8.5. Some of what we discuss, such as the translation of parabolas and completing the square, will be used again in Section 10.1 when we discuss parabolas further.

In Section 8.6 we solve quadratic and other nonlinear inequalities in one variable. We will draw on what we learned in Section 2.5, where we solved linear inequalities in one variable.

8.1 SOLVING QUADRATIC EQUATIONS BY COMPLETING THE SQUARE

SSM VIDEO 8.1 CD Rom

1) **Use the square root property to solve equations.**
2) **Write perfect square trinomials.**
3) **Solve quadratic equations by completing the square.**

In this section we introduce two concepts, the square root property and completing the square. The square root property will be used in several sections of this book.

In Section 5.8 we solved quadratic, or second degree, equations by factoring. Quadratic equations that cannot be solved by factoring can be solved by completing the square, or by the quadratic formula, which is presented in Section 8.2.

1) Use the Square Root Property to Solve Equations

In Section 7.1 we stated that every positive number has two square roots. Thus far, we have been using only the positive square root. In this section we use both the positive and negative square roots of a number.

Positive Square Root of 25	Negative Square Root of 25
$\sqrt{25} = 5$	$-\sqrt{25} = -5$

A convenient way to indicate the two square roots of a number is to use the plus or minus symbol, $\pm$. For example, the square roots of 25 can be indicated ± 5, read "plus or minus 5." The equation $x^2 = 25$ has two solutions, the two square roots of 25, which are ± 5. If you check each root, you will see that each value satisfies the equation. The **square root property** can be used to find the solutions to equations of the form $x^2 = a$.

Square Root Property

If $x^2 = a$, where a is a real number, then $x = \pm\sqrt{a}$.

EXAMPLE 1 Solve the following equations. **a)** $x^2 - 9 = 0$ **b)** $x^2 + 5 = 65$

Solution **a)** Add 9 to both sides of the equation to isolate the variable.

$$x^2 - 9 = 0$$
$$x^2 = 9 \qquad \textit{Isolate the variable.}$$
$$x = \pm\sqrt{9} \qquad \textit{Square root property}$$
$$= \pm 3$$

Check the solutions in the original equation.

$$x = 3 \qquad\qquad x = -3$$
$$x^2 - 9 = 0 \qquad\qquad x^2 - 9 = 0$$
$$3^2 - 9 \stackrel{?}{=} 0 \qquad\qquad (-3)^2 - 9 \stackrel{?}{=} 0$$
$$0 = 0 \quad \textit{True} \qquad\qquad 0 = 0 \quad \textit{True}$$

In both cases the check is true, which means that both 3 and −3 are solutions to the equation.

b)
$$x^2 + 5 = 65$$
$$x^2 = 60 \qquad \textit{Isolate the variable.}$$
$$x = \pm\sqrt{60} \qquad \textit{Square root property}$$
$$= \pm\sqrt{4}\sqrt{15} \qquad \textit{Simplify.}$$
$$= \pm 2\sqrt{15}$$

The solutions are $2\sqrt{15}$ and $-2\sqrt{15}$.

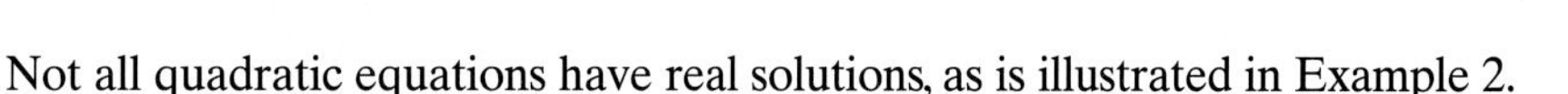

Not all quadratic equations have real solutions, as is illustrated in Example 2.

EXAMPLE 2 Solve the equation $x^2 + 7 = 0$.

Solution
$$x^2 + 7 = 0$$
$$x^2 = -7 \qquad \textit{Isolate the variable.}$$
$$x = \pm\sqrt{-7} \qquad \textit{Square root property}$$
$$= \pm i\sqrt{7}$$

The solutions are $i\sqrt{7}$ and $-i\sqrt{7}$.

EXAMPLE 3 Solve the equation $(a - 4)^2 = 32$.

Solution Begin by taking the square root of both sides of the equation.

$$(a - 4)^2 = 32$$
$$a - 4 = \pm\sqrt{32}$$
$$a = 4 \pm \sqrt{32} \qquad \textit{Add 4 to both sides.}$$
$$= 4 \pm \sqrt{16}\sqrt{2} \qquad \textit{Simplify.}$$
$$= 4 \pm 4\sqrt{2}$$

NOW TRY EXERCISE 19 The solutions are $4 + 4\sqrt{2}$ and $4 - 4\sqrt{2}$.

2) Write Perfect Square Trinomials

Now that we know the square root property we can focus our attention on completing the square. To understand this procedure, you need to know how to form perfect square trinomials. Perfect square trinomials were introduced in Section 5.6. Recall that a *perfect square trinomial* is a trinomial that can be expressed as the square of a binomial. Some examples follow.

Perfect Square Trinomials	Factors	Square of a Binomial
$x^2 + 8x + 16$	$= (x + 4)(x + 4)$	$= (x + 4)^2$
$x^2 - 8x + 16$	$= (x - 4)(x - 4)$	$= (x - 4)^2$
$x^2 + 10x + 25$	$= (x + 5)(x + 5)$	$= (x + 5)^2$
$x^2 - 10x + 25$	$= (x - 5)(x - 5)$	$= (x - 5)^2$

In a perfect square trinomial with a leading coefficient of 1, there is a relationship between the coefficient of the first-degree term and the constant term. In such trinomials the constant term is the square of one-half the coefficient of the first degree term.

Let's examine some perfect square trinomials for which the leading coefficient is 1.

$$x^2 + 8x + 16 = (x + 4)^2$$

$$\left[\tfrac{1}{2}(8)\right]^2 = (4)^2$$

$$x^2 - 10x + 25 = (x - 5)^2$$

$$\left[\tfrac{1}{2}(-10)\right]^2 = (-5)^2$$

When a perfect square trinomial with a leading coefficient of 1 is written as the square of a binomial, the constant in the binomial is one-half the coefficient of the first-degree term in the trinomial. For example,

$$x^2 + 8x + 16 = (x + 4)^2$$

$$\tfrac{1}{2}(8)$$

$$x^2 - 10x + 25 = (x - 5)^2$$

$$\tfrac{1}{2}(-10)$$

3) Solve Quadratic Equations by Completing the Square

Now we introduce completing the square. To solve a quadratic equation by **completing the square**, we add (or subtract) a constant to (or from) both sides of the equation so that the remaining trinomial is a perfect square trinomial. Then we use the square root property to solve the resulting equation. We will now summarize the procedure.

To Solve a Quadratic Equation by Completing the Square

1. Use the multiplication (or division) property of equality if necessary to make the leading coefficient 1.
2. Rewrite the equation with the constant by itself on the right side of the equation.
3. Take one-half the numerical coefficient of the first-degree term, square it, and add this quantity to both sides of the equation.
4. Replace the trinomial with the square of a binomial.
5. Use the square root property to take the square root of both sides of the equation.
6. Solve for the variable.
7. Check your solutions in the *original* equation.

EXAMPLE 4 Solve the equation $x^2 + 6x + 5 = 0$ by completing the square.

Solution Since the leading coefficient is 1, step 1 is not necessary.

Step 2: Move the constant, 5, to the right side of the equation by subtracting 5 from both sides of the equation.

$$x^2 + 6x + 5 = 0$$

$$x^2 + 6x = -5$$

Step 3: Determine the square of one-half the numerical coefficient of the first-degree term.

$$\frac{1}{2}(6) = 3, \qquad 3^2 = 9$$

Add this value to both sides of the equation.

$$x^2 + 6x + 9 = -5 + 9$$

$$x^2 + 6x + 9 = 4$$

Step 4: By following this procedure, we produce a perfect square trinomial on the left side of the equation. The expression $x^2 + 6x + 9$ is a perfect square trinomial that can be expressed as $(x + 3)^2$.

$\frac{1}{2}$ the numerical coefficient of the first-degree term is $\frac{1}{2}(6) = +3$

$$(x + 3)^2 = 4$$

Step 5: Take the square root of both sides of the equation.

$$x + 3 = \pm\sqrt{4}$$

$$x + 3 = \pm 2$$

Step 6: Finally, solve for x by subtracting 3 from both sides of the equation.

$$x + 3 - 3 = -3 \pm 2$$

$$x = -3 \pm 2$$

$$x = -3 + 2 \quad \text{or} \quad x = -3 - 2$$
$$x = -1 \qquad\qquad x = -5$$

Step 7: Check both solutions in the original equation.

$x = -1$		$x = -5$	
$x^2 + 6x + 5 = 0$		$x^2 + 6x + 5 = 0$	
$(-1)^2 + 6(-1) + 5 \stackrel{?}{=} 0$		$(-5)^2 + 6(-5) + 5 \stackrel{?}{=} 0$	
$1 - 6 + 5 \stackrel{?}{=} 0$		$25 - 30 + 5 \stackrel{?}{=} 0$	
$0 = 0$	*True*	$0 = 0$	*True*

NOW TRY EXERCISE 29 Since each number checks, both -1 and -5 are solutions to the original equation.

HELPFUL HINT

When solving the equation $x^2 + bx + c = 0$ by completing the square, we obtain $x^2 + bx + \left(\frac{b}{2}\right)^2$ on the left side of the equation and a constant on the right side of the equation. We then replace $x^2 + bx + \left(\frac{b}{2}\right)^2$ with $\left(x + \frac{b}{2}\right)^2$. In the figure that follows we show why

$$x^2 + bx + \left(\frac{b}{2}\right)^2 = \left(x + \frac{b}{2}\right)^2$$

The figure is a square with sides of length $x + \frac{b}{2}$. The area is therefore $\left(x + \frac{b}{2}\right)^2$. The area of the square can also be determined by adding the areas of the four sections as follows:

$$x^2 + \frac{b}{2}x + \frac{b}{2}x + \left(\frac{b}{2}\right)^2 = x^2 + bx + \left(\frac{b}{2}\right)^2$$

Comparing the areas, we see that $x^2 + bx + \left(\frac{b}{2}\right)^2 = \left(x + \frac{b}{2}\right)^2$.

x $\frac{b}{2}$

x x^2 $\frac{b}{2}x$

$\frac{b}{2}$ $\frac{b}{2}x$ $\left(\frac{b}{2}\right)^2$

The area of this piece represents the term we add to each side of the equation when we complete the square.

EXAMPLE 5 Solve the equation $-x^2 = -3x - 18$ by completing the square.

Solution The numerical coefficient of the squared term must be 1, not -1. Therefore, begin by multiplying both sides of the equation by -1 to make the coefficient of the squared term equal to 1.

$$-x^2 = -3x - 18$$
$$-1(-x^2) = -1(-3x - 18)$$
$$x^2 = 3x + 18$$

Now move all terms except the constant to the left side of the equation.

$$x^2 - 3x = 18$$

Take half the numerical coefficient of the x-term, square it, and add this product to both sides of the equation. Then rewrite the left side of the equation as the square of a binomial.

$$\frac{1}{2}(-3) = -\frac{3}{2} \qquad \left(-\frac{3}{2}\right)^2 = \frac{9}{4}$$

$$x^2 - 3x + \frac{9}{4} = 18 + \frac{9}{4} \qquad \textit{Complete the square.}$$

$$\left(x - \frac{3}{2}\right)^2 = 18 + \frac{9}{4} \qquad \textit{Rewrite the trinomial as the square of a binomial.}$$

$$\left(x - \frac{3}{2}\right)^2 = \frac{72}{4} + \frac{9}{4}$$

$$\left(x - \frac{3}{2}\right)^2 = \frac{81}{4}$$

$$x - \frac{3}{2} = \pm\sqrt{\frac{81}{4}} \qquad \textit{Square root property}$$

$$x - \frac{3}{2} = \pm\frac{9}{2}$$

$$x = \frac{3}{2} \pm \frac{9}{2} \qquad \textit{Add } \tfrac{3}{2} \textit{ to both sides.}$$

$$x = \frac{3}{2} + \frac{9}{2} \quad \text{or} \quad x = \frac{3}{2} - \frac{9}{2}$$

$$x = \frac{12}{2} = 6 \qquad\qquad x = -\frac{6}{2} = -3$$

NOW TRY EXERCISE 33

In the following examples we will not show some of the intermediate steps.

EXAMPLE 6 Solve the equation $x^2 - 6x + 17 = 0$.

Solution

$$x^2 - 6x + 17 = 0$$

$$x^2 - 6x = -17 \qquad \textit{Move the constant to the right side.}$$

$$x^2 - 6x + 9 = -17 + 9 \qquad \textit{Complete the square.}$$

$$(x - 3)^2 = -8 \qquad \textit{Write the trinomial as the square of a binomial.}$$

$$x - 3 = \pm\sqrt{-8} \qquad \textit{Square root property}$$

$$x - 3 = \pm 2i\sqrt{2} \qquad \textit{Simplify.}$$

$$x = 3 \pm 2i\sqrt{2} \qquad \textit{Solve for } x.$$

NOW TRY EXERCISE 37

The solutions are $3 + 2i\sqrt{2}$ and $3 - 2i\sqrt{2}$. Note that the solutions to the equation $x^2 - 6x + 17 = 0$ are not real. The solutions are complex numbers.

EXAMPLE 7 Solve the equation $-3m^2 + 6m + 24 = 0$ by completing the square.

Solution

$$-3m^2 + 6m + 24 = 0$$

$$-\frac{1}{3}(-3m^2 + 6m + 24) = -\frac{1}{3}(0) \quad \text{Multiply by } -\tfrac{1}{3} \text{ to obtain the leading coefficient of 1.}$$

$$m^2 - 2m - 8 = 0$$

Now proceed as before.

$$m^2 - 2m = 8 \quad \text{Move the constant to the right side.}$$

$$m^2 - 2m + 1 = 8 + 1 \quad \text{Complete the square}$$

$$(m - 1)^2 = 9 \quad \text{Write the trinomial as the square of a binomial.}$$

$$m - 1 = \pm 3 \quad \text{Square root property}$$

$$m = 1 \pm 3 \quad \text{Solve for } m.$$

$$m = 1 + 3 \quad \text{or} \quad m = 1 - 3$$

$$m = 4 \qquad\qquad m = -2$$

If you were asked to solve the equation $-\frac{1}{4}x^2 + 2x - 8 = 0$ by completing the square, what would you do first? If you answered "multiply both sides of the equation by -4 to make the leading coefficient 1," you answered correctly. To solve the equation $\frac{2}{3}x^2 + 3x - 5 = 0$, you would multiply both sides of the equation by $\frac{3}{2}$ to obtain a leading coefficient of 1.

Generally, quadratic equations that cannot be easily solved by factoring will be solved by the *quadratic formula*, which will be presented in the next section. We introduced completing the square because we use it to derive the quadratic formula in Section 8.2. We will use completing the square later in this chapter and in Chapter 10.

EXAMPLE 8 The compound interest formula $A = p\left(1 + \frac{r}{n}\right)^{nt}$ can be used to find the amount, A, when an initial principal, p, is invested at an annual interest rate, r, compounded n times a year for t years.

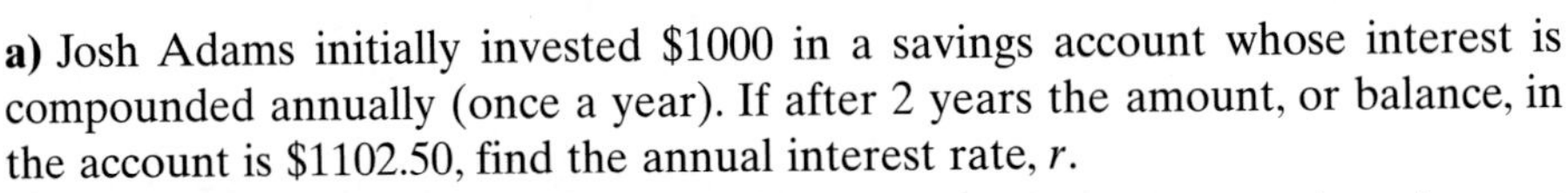

a) Josh Adams initially invested \$1000 in a savings account whose interest is compounded annually (once a year). If after 2 years the amount, or balance, in the account is \$1102.50, find the annual interest rate, r.

b) Trisha McDowell initially invested \$1000 in a savings account whose interest is compounded quarterly. If after 3 years the amount in the account is \$1195.62, find the annual interest rate, r.

Solution a) Understand We are given the following information:

$$p = \$1000, \quad A = \$1102.50, \quad n = 1, \quad t = 2$$

We are asked to find the annual rate, r. To do so, we substitute the appropriate values in the formula and solve for r.

Translate

$$A = p\left(1 + \frac{r}{n}\right)^{nt}$$

$$1102.50 = 1000\left(1 + \frac{r}{1}\right)^{1(2)}$$

Carry Out

$$1102.50 = 1000(1 + r)^2$$

$$1.10250 = (1 + r)^2 \qquad \textit{Divide both sides by 1000.}$$

$$\sqrt{1.10250} = 1 + r \qquad \textit{Square root property}$$

$$1.05 = 1 + r \qquad \textit{Subtract 1 from both sides.}$$

$$0.05 = r$$

Answer The interest rate is 0.05 or 5%.

b) Understand We are given

$$p = 1000, \quad A = \$1195.62, \quad n = 4, \quad t = 3.$$

To find r, we substitute the appropriate values in the formula and solve for r.

Translate

$$A = p\left(1 + \frac{r}{n}\right)^{nt}$$

$$1195.62 = 1000\left(1 + \frac{r}{4}\right)^{4(3)}$$

Carry out

$$1.19562 = \left(1 + \frac{r}{4}\right)^{12} \qquad \textit{Divide both sides by 1000.}$$

$$\sqrt[12]{1.19562} = 1 + \frac{r}{4} \qquad \textit{Take the 12th root of both sides (or raise both sides to the } \tfrac{1}{12} \textit{ power).}$$

$$1.015 = 1 + \frac{r}{4} \qquad \textit{Evaluate } \sqrt[12]{1.19562} \textit{ on a calculator.}$$

$$0.015 = \frac{r}{4} \qquad \textit{Subtract 1 from both sides.}$$

$$0.06 = r \qquad \textit{Multiply both sides by 4.}$$

NOW TRY EXERCISE 71

Answer The interest rate is 0.06 or 6%.

Exercise Set 8.1

Concept/Writing Exercises

1. Write the two square roots of 36.
2. Write the two square roots of 17.
3. Write the square root property.
4. What is the first step in completing the square?
5. Explain how to determine whether a trinomial is a perfect square trinomial.
6. Write a paragraph explaining in your own words how to construct a perfect square trinomial.

Practice the Skills

Use the square root property to solve each equation.

7. $x^2 = 49$
8. $x^2 = 18$
9. $y^2 = 48$
10. $x^2 - 11 = 19$
11. $z^2 + 12 = 40$
12. $y^2 + 15 = 80$
13. $(p - 4)^2 = 16$
14. $(y - 3)^2 = 45$

15. $(z + \frac{1}{3})^2 = \frac{4}{9}$
16. $(x - 0.2)^2 = 0.64$
17. $(x + 1.8)^2 = 0.81$
18. $(x + \frac{1}{2})^2 = \frac{4}{9}$
19. $(2a - 5)^2 = 12$
20. $(4y + 1)^2 = 8$
21. $(2y + \frac{1}{2})^2 = \frac{4}{25}$
22. $(3x - \frac{1}{4})^2 = \frac{9}{25}$

Solve each equation by completing the square.

23. $x^2 + 2x - 15 = 0$
24. $x^2 - 6x + 8 = 0$
25. $x^2 - 4x - 5 = 0$
26. $x^2 + 8x + 12 = 0$
27. $a^2 + 3a + 2 = 0$
28. $x^2 + 4x - 32 = 0$
29. $r^2 - 8r + 15 = 0$
30. $x^2 - 9x + 14 = 0$
31. $x^2 + 2x + 12 = 0$
32. $x^2 + 9x + 18 = 0$
33. $-z^2 + 9z - 20 = 0$
34. $2r^2 = -7r + 4$
35. $b^2 = 3b + 28$
36. $-4x = -x^2 + 10$
37. $x^2 + 3x + 6 = 0$
38. $x^2 - x - 3 = 0$
39. $-s^2 + 5s = -8$
40. $\frac{1}{4}x^2 + \frac{3}{4}x - \frac{3}{2} = 0$
41. $-\frac{1}{4}a^2 - \frac{1}{2}a = 0$
42. $2x^2 - 6x = 0$
43. $12a^2 - 4a = 0$
44. $6x^2 = 9x$
45. $-\frac{1}{2}p^2 - p + \frac{3}{2} = 0$
46. $2x^2 + 2x - 24 = 0$
47. $2x^2 + 18x + 4 = 0$
48. $2x^2 = 8x + 90$
49. $3x^2 + 33x + 72 = 0$
50. $3w^2 + 2w - 1 = 0$
51. $\frac{2}{3}x^2 + \frac{4}{3}x + 1 = 0$
52. $3c^2 - 8c + 4 = 0$
53. $-3x^2 + 6x = 6$
54. $2x^2 - x = -5$
55. $\frac{5}{2}x^2 + \frac{3}{2}x - \frac{5}{4} = 0$
56. $2x^2 - \frac{1}{3}x = -2$

Problem Solving

57. Solve $x^2 + 2ax + a^2 = k$ by completing the square.

58. Solve $(x \pm a)^2 - k = 0$ by completing the square.

59. The product of two consecutive positive odd integers is 63. Find the two odd integers.

60. The larger of two integers is 2 more than twice the smaller. Find the two numbers if their product is 12.

61. Donna Simm has marked off an area in her yard where she will plant tomato plants. Find the dimensions of the rectangular area if the length is 2 feet more than twice the width and the area is 60 square feet.

62. Ralph Evans is planning to blacktop his driveway. Find the dimensions of the rectangular driveway if its area is 381.25 square feet and its length is 18 feet greater than its width.

63. Daniella MacOvey is designing a house. One of the bedrooms is a square whose diagonal is 6 feet longer than the length of a side. Find the dimensions of the bedroom.

64. The Lakeside Hotel is planning to build a shallow wading pool for children. If the pool is to be square and the diagonal of the square is 10 feet longer than a side, find the dimensions of the pool.

65. When a triangle is inscribed in a semicircle where a diameter of the circle is a side of the triangle, the triangle formed is always a right triangle. If an isosceles triangle (two equal sides) is inscribed in a semicircle of radius 10 inches, find the length of the other two sides of the triangle.

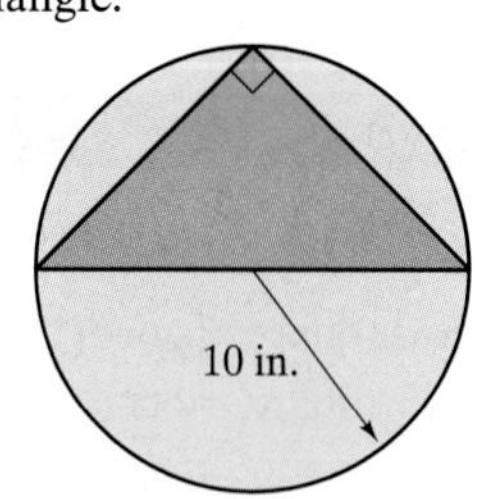

66. Refer to Exercise 65. Suppose a triangle is inscribed in a semicircle whose diameter is 12 meters. If one side of the inscribed triangle is 2 meters, find the third side.

67. The area of a circle is 24π square feet. Use the formula $A = \pi r^2$ to find the radius of the circle.

68. The area of a circle is 16.4π square meters. Find the radius of the circle.

Use the formula $A = p\left(1 + \frac{r}{n}\right)^{nt}$ *to answer Exercises 69–72.*

69. Kathy Hayes initially invested \$500 in a savings account whose interest is compounded annually. If after 2 years the amount in the account is \$551.25, find the annual interest rate.

70. Dileauth Newbar initially invested \$800 in a savings account whose interest is compounded annually. If after 2 years the amount in the account is \$898.88, find the annual interest rate.

71. Pedro Lewis initially invested \$1200 in a savings acount whose interest is compounded semiannually. If after 3 years the amount in the account is \$1432.86, find the annual interest rate.

72. Suki Kimm initially invested \$1500 in a savings account whose interest is compounded semiannually. If after 4 years the amount in the account is \$2052.85, find the annual interest rate.

73. In Chapter 10, we will start with an equation like $x^2 + 6x + y^2 - 10y = -18$ and change its form to $(x + 3)^2 + (y - 5)^2 = 16$. We do this by completing the square twice, once for the x-terms and once for the y-terms. Show how this is done.

74. Refer to Exercise 73. Change $4x^2 + 9y^2 - 48x + 72y = -144$ to the form $a(x - h)^2 + b(y - k)^2 = c$ by completing the square twice, once for the x-terms and once for the y-terms.

75. The surface area S and volume V of a right circular cylinder of radius r and height h are given by the formulas

$$S = 2\pi r^2 + 2\pi rh, \qquad V = \pi r^2 h$$

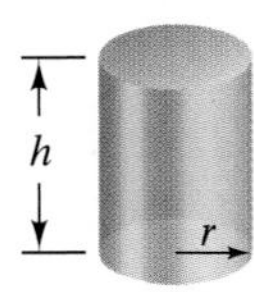

a) Find the surface area of the cylinder if its height is 10 inches and its volume is 160 cubic inches.

b) Find the radius if the height is 10 inches and the volume is 160 cubic inches.

c) Find the radius if the height is 10 inches and the surface area is 160 square inches.

Group Activity

Discuss and answer Exercise 76 as a group.

76. On the following grid, the points (x_1, y_1), (x_2, y_2), and (x_1, y_2) are plotted.

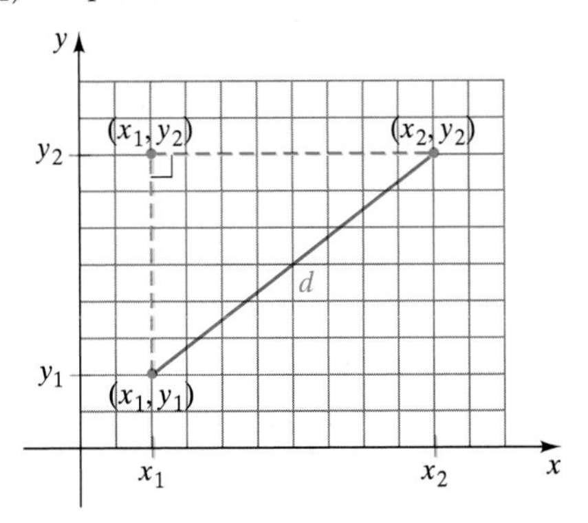

a) Explain why (x_1, y_2) is placed where it is and not somewhere else on the graph.

b) Express the length of the orange dashed line in terms of y_2 and y_1. Explain how you determined your answer.

c) Express the length of the green dashed line in terms of x_2 and x_1.

d) Using the Pythagorean theorem and the right triangle ABC, derive a formula for the distance, d, between points (x_1, y_1) and (x_2, y_2).* Explain how you determined the formula.

e) Use the formula you determined in part **(d)** to find the distance of the line segment between the points (1, 4) and (3, 7).

Cumulative Review Exercises

[2.2] **77.** Solve $2xy - 3yz = -xy + z$ for z.

[7.1] **78.** Express $\sqrt{(x^2 - 4x)^2}$ as an absolute value.

[7.2] **79.** Evaluate $25^{-1/2}$.

80. Simplify $\dfrac{x^{3/4}y^{1/2}}{x^{1/4}y^2}$.

*The distance formula will be discussed in Section 10.1.

8.2 SOLVING QUADRATIC EQUATIONS BY THE QUADRATIC FORMULA

SSM VIDEO 8.2 CD Rom

1. **Derive the quadratic formula.**
2. **Use the quadratic formula to solve equations.**
3. **Write a quadratic equation, given its solutions.**
4. **Use the discriminant to determine the number of real solutions to a quadratic equation.**
5. **Study applications that use quadratic equations.**

1) Derive the Quadratic Formula

The quadratic formula can be used to solve any quadratic equation. *It is the most useful and most versatile method of solving quadratic equations.* It is generally used in place of completing the square because of its efficiency.

The standard form of a quadratic equation is $ax^2 + bx + c = 0$, where a is the coefficient of the squared term, b is the coefficient of the first-degree term, and c is the constant.

Quadratic Equation in Standard Form	Values of Coefficients		
$x^2 - 3x + 4 = 0$	$a = 1,$	$b = -3,$	$c = 4$
$1.3x^2 - 4.2 = 0$	$a = 1.3,$	$b = 0,$	$c = -4.2$
$-\frac{5}{6}x^2 + \frac{3}{5}x = 0$	$a = -\frac{5}{6},$	$b = \frac{3}{5},$	$c = 0$

We can derive the quadratic formula by starting with a quadratic equation in standard form and completing the square, as discussed in the preceding section.

$$ax^2 + bx + c = 0$$

$$\frac{ax^2}{a} + \frac{b}{a}x + \frac{c}{a} = 0 \qquad \textit{Divide both sides by } a.$$

$$x^2 + \frac{b}{a}x = -\frac{c}{a} \qquad \textit{Subtract } c/a \textit{ from both sides.}$$

$$x^2 + \frac{b}{a}x + \frac{b^2}{4a^2} = -\frac{c}{a} + \frac{b^2}{4a^2}$$

Take 1/2 of b/a; that is, $b/2a$, and square it to get $b^2/4a^2$. Then add this expression to both sides.

$$\left(x + \frac{b}{2a}\right)^2 = \frac{b^2}{4a^2} - \frac{c}{a}$$

Rewrite the left side of the equation as the square of a binomial.

$$\left(x + \frac{b}{2a}\right)^2 = \frac{b^2 - 4ac}{4a^2}$$

Write the right side with a common denominator.

$$x + \frac{b}{2a} = \pm\sqrt{\frac{b^2 - 4ac}{4a^2}} \qquad \textit{Square root property}$$

$$x + \frac{b}{2a} = \pm\frac{\sqrt{b^2 - 4ac}}{2a} \qquad \textit{Quotient rule for radicals}$$

$$x = \frac{-b}{2a} \pm \frac{\sqrt{b^2 - 4ac}}{2a}$$ Subtract $b/2a$ from both sides.

$$x = \frac{-b \pm \sqrt{b^2 - 4ac}}{2a}$$ Write with a common denominator to get the quadratic formula

2) Use the Quadratic Formula to Solve Equations

Now that we have derived the quadratic formula, we can use it to solve quadratic equations.

To Solve a Quadratic Equation by the Quadratic Formula

1. Write the quadratic equation in standard form $ax^2 + bx + c = 0$, and determine the numerical values for a, b, and c.
2. Substitute the values for a, b, and c in the quadratic formula and then evaluate the formula to obtain the solution.

The Quadratic Formula

$$x = \frac{-b \pm \sqrt{b^2 - 4ac}}{2a}$$

EXAMPLE 1 Solve the equation $x^2 + 2x - 8 = 0$ by the quadratic formula.

Solution $a = 1, \quad b = 2, \quad c = -8.$

$$x = \frac{-b \pm \sqrt{b^2 - 4ac}}{2a}$$

$$x = \frac{-(2) \pm \sqrt{(2)^2 - 4(1)(-8)}}{2(1)}$$

$$= \frac{-2 \pm \sqrt{4 + 32}}{2}$$

$$= \frac{-2 \pm \sqrt{36}}{2}$$

$$= \frac{-2 \pm 6}{2}$$

$$x = \frac{-2 + 6}{2} \quad \text{or} \quad x = \frac{-2 - 6}{2}$$

$$x = \frac{4}{2} = 2 \qquad x = \frac{-8}{2} = -4$$

A check will show that both 2 and −4 are solutions to the equation.

The solution to Example 1 could also be obtained by factoring, as illustrated below.

$$x^2 + 2x - 8 = 0$$

$$(x + 4)(x - 2) = 0$$

$$x + 4 = 0 \quad \text{or} \quad x - 2 = 0$$

$$x = -4 \qquad x = 2$$

When you are given a quadratic equation to solve and the method to solve it has not been specified, you may try solving by factoring first (as we discussed in Section 5.8). If the equation cannot be easily factored, use the quadratic formula.

When solving a quadratic equation using the quadratic formula, the calculations may be easier if the leading coefficient, a, is positive. Thus, if solving the quadratic equation $-x^2 + 3x = 2$, you may wish to rewrite the equation as $x^2 - 3x + 2 = 0$.

Avoiding Common Errors

The entire numerator of the quadratic formula must be divided by $2a$.

CORRECT

$$x = \frac{-b \pm \sqrt{b^2 - 4ac}}{2a}$$

INCORRECT

$$x = -b \pm \frac{\sqrt{b^2 - 4ac}}{2a}$$

$$x = \frac{-b}{2a} \pm \sqrt{b^2 - 4ac}$$

EXAMPLE 2 Given $f(x) = 2x^2 + 4x$, find all real values of x for which $f(x) = 5$.

Solution We wish to determine all real values of x for which

$$2x^2 + 4x = 5$$

We can solve this equation with the quadratic formula. First we write the equation in standard form.

$$2x^2 + 4x - 5 = 0$$

Now we use the quadratic formula with $a = 2$, $b = 4$, and $c = -5$.

$$x = \frac{-b \pm \sqrt{b^2 - 4ac}}{2a}$$

$$= \frac{-4 \pm \sqrt{(4)^2 - 4(2)(-5)}}{2(2)} = \frac{-4 \pm \sqrt{56}}{4} = \frac{-4 \pm 2\sqrt{14}}{4}$$

Next we factor out 2 from both terms in the numerator, then we divide out the common factor.

$$x = \frac{\overset{1}{\cancel{2}}(-2 \pm \sqrt{14})}{\underset{2}{\cancel{4}}} = \frac{-2 \pm \sqrt{14}}{2}^*$$

Thus, the solutions are $\frac{-2 + \sqrt{14}}{2}$ and $\frac{-2 - \sqrt{14}}{2}$.

Note that the expression in Example 2, $2x^2 + 4x - 50$, is not factorable. Therefore, Example 2 could not be solved by factoring.

NOW TRY EXERCISE 53

* Solutions will be given in this form in the Answer Section.

Avoiding Common Errors

Some students use the quadratic formula correctly until the last step, where they make an error. Below are illustrated both the correct and incorrect procedures for simplifying an answer.

When *both* terms in the numerator *and* the denominator have a common factor, that common factor may be divided out, as follows:

Correct

$$\frac{2 + 4\sqrt{3}}{2} = \frac{\overset{1}{\cancel{2}}(1 + 2\sqrt{3})}{\underset{1}{\cancel{2}}} = 1 + 2\sqrt{3}$$

$$\frac{6 + 3\sqrt{3}}{6} = \frac{\overset{1}{\cancel{3}}(2 + \sqrt{3})}{\underset{2}{\cancel{6}}} = \frac{2 + \sqrt{3}}{2}$$

Below are some common errors. Study them carefully so you will not make them. Can you explain why each of the following procedures is incorrect?

Incorrect

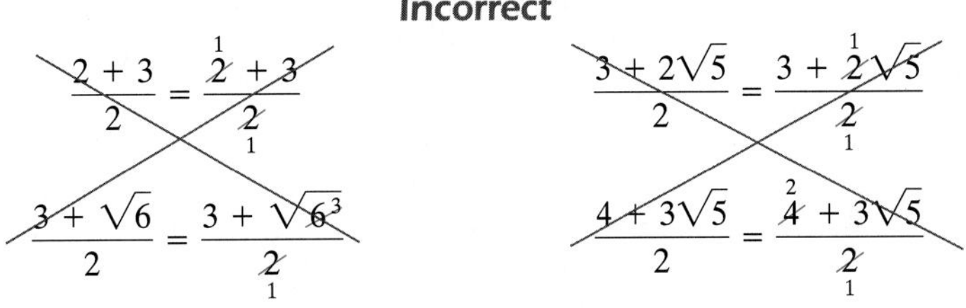

Note that $(2 + 3)/2$ simplifies to $5/2$. However, $(3 + 2\sqrt{5})/2$, $(3 + \sqrt{6})/2$, and $(4 + 3\sqrt{5})/2$ cannot be simplified any further.

EXAMPLE 3 Solve the equation $p^2 + \frac{2}{5}p + \frac{1}{3} = 0$ by the quadratic formula.

Solution Do not let the change in variable worry you. The quadratic formula is used exactly the same way as when x is the variable.

We could solve this equation using the quadratic formula with $a = 1$, $b = \frac{2}{5}$, and $c = \frac{1}{3}$. However, when a quadratic equation contains fractions, it is generally easier to begin by multiplying both sides of the equation by the least common denominator. In this example, the least common denominator is 15.

$$15\left(p^2 + \frac{2}{5}p + \frac{1}{3}\right) = 15(0)$$

$$15p^2 + 6p + 5 = 0$$

Now we can use the quadratic formula with $a = 15$, $b = 6$, and $c = 5$.

$$\begin{aligned} p &= \frac{-b \pm \sqrt{b^2 - 4ac}}{2a} \\ &= \frac{-6 \pm \sqrt{6^2 - 4(15)(5)}}{2(15)} \\ &= \frac{-6 \pm \sqrt{-264}}{30} \\ &= \frac{-6 \pm \sqrt{-4}\,\sqrt{66}}{30} \end{aligned}$$

$$= \frac{-6 \pm 2i\sqrt{66}}{30}$$

$$= \frac{\overset{1}{\cancel{2}}(-3 \pm i\sqrt{66})}{\underset{15}{\cancel{30}}}$$

$$= \frac{-3 \pm i\sqrt{66}}{15}$$

NOW TRY EXERCISE 43

The solutions are $\frac{-3 + i\sqrt{66}}{15}$ and $\frac{-3 - i\sqrt{66}}{15}$. Note that neither solution is real.

If all the numerical coefficients in a quadratic equation have a common factor, you should factor it out before using the quadratic formula. For example, consider the equation $3x^2 + 12x + 3 = 0$. Here $a = 3, b = 12$, and $c = 3$. If we use the quadratic formula, we would eventually obtain $x = -2 \pm \sqrt{3}$ as solutions. By factoring the equation before using the formula, we get

$$3x^2 + 12x + 3 = 0$$

$$3(x^2 + 4x + 1) = 0$$

If we consider $x^2 + 4x + 1 = 0$, then $a = 1, b = 4$, and $c = 1$. If we use these new values of a, b, and c in the quadratic formula, we will obtain the identical solution, $x = -2 \pm \sqrt{3}$. However, the calculations with these smaller values of a, b, and c are simplified. Solve both equations now using the quadratic formula to convince yourself.

3) Write a Quadratic Equation, Given Its Solutions

If we are given the solutions of an equation, we can find the equation by working backwards. This procedure is illustrated in Example 4.

EXAMPLE 4 Write an equation that has solutions.

a) -3 and 2 **b)** $3 + 2i$ and $3 - 2i$.

Solution **a)** If the solutions are -3 and 2, we write

$$x = -3 \quad \text{or} \quad x = 2$$

$$x + 3 = 0 \qquad x - 2 = 0 \qquad \textit{Set equations equal to 0.}$$

$$(x + 3)(x - 2) = 0 \qquad \textit{Zero factor property}$$

$$x^2 - 2x + 3x - 6 = 0 \qquad \textit{Multiply factors.}$$

$$x^2 + x - 6 = 0 \qquad \textit{Combine terms.}$$

Thus, the equation is $x^2 + x - 6 = 0$. Many other equations have solutions -3 and 2. In fact, any equation of the form $a(x^2 + x - 6) = 0$ has those solutions. Can you explain why?

b)

$$x = 3 + 2i \quad \text{or} \quad x = 3 - 2i$$

$$x - (3 + 2i) = 0 \qquad x - (3 - 2i) = 0 \qquad \textit{Set equations equal to 0.}$$

$$[x - (3 + 2i)][x - (3 - 2i)] = 0 \qquad \textit{Zero factor property}$$

$$x \cdot x - x(3 - 2i) - x(3 + 2i) + (3 + 2i)(3 - 2i) = 0 \quad \text{Multiply.}$$

$$x^2 - 3x + 2xi - 3x - 2xi + (9 - 4i^2) = 0 \quad \text{Distributive property; multiply}$$

$$x^2 - 6x + 9 - 4i^2 = 0 \quad \text{Combine terms.}$$

$$x^2 - 6x + 9 - 4(-1) = 0 \quad \text{Substitute } i^2 = -1.$$

$$x^2 - 6x + 13 = 0 \quad \text{Simplify.}$$

NOW TRY EXERCISE 59

The equation $x^2 - 6x + 13 = 0$ has complex roots, $3 + 2i$ and $3 - 2i$.

In Example 4a), the equation $x^2 + x - 6 = 0$ has solutions -3 and 2. Consider the graph of $f(x) = x^2 + x - 6$. The x-intercepts of the graph of $f(x)$ occur when $f(x) = 0$, or when $x^2 + x - 6 = 0$. Therefore the x-coordinates of the x-intercepts of the graph of $f(x) = x^2 + x - 6$ are $(-3, 0)$ and $(2, 0)$, as shown in Figure 8.1. In Example **4b)**, there are no real solutions. Thus the graph $f(x) = x^2 - 6x + 13$ has no x-intercepts. The graph of $f(x) = x^2 - 6x + 13$ is shown in Figure 8.2.

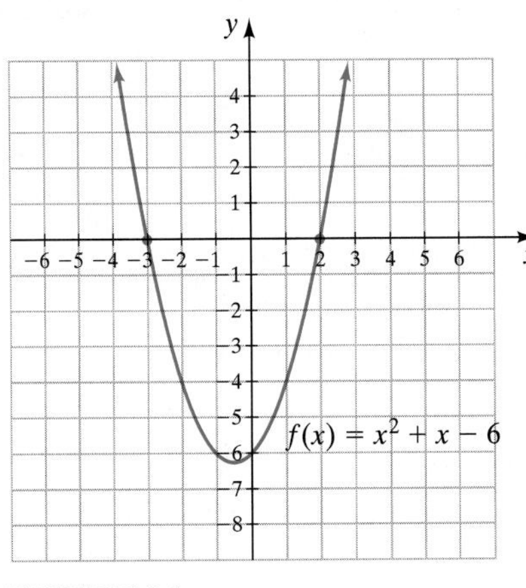

FIGURE 8.1

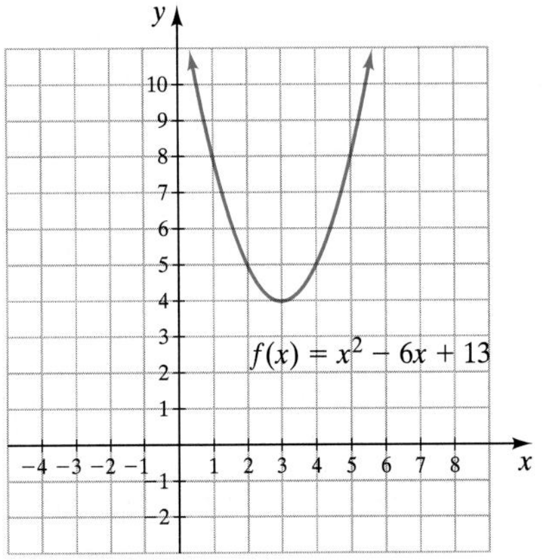

FIGURE 8.2

4) Use the Discriminant to Determine the Number of Real Solutions to a Quadratic Equation

The expression under the radical sign in the quadratic formula is called the **discriminant**.

$$\underbrace{b^2 - 4ac}_{\text{discriminant}}$$

The discriminant gives the number and nature of solutions of a quadratic equation.

Solutions of a Quadratic Equation

For a quadratic equation of the form $ax^2 + bx + c = 0$, $a \neq 0$:

If $b^2 - 4ac > 0$, the quadratic equation has two distinct real number solutions.

If $b^2 - 4ac = 0$, the quadratic equation has a single real number solution.

If $b^2 - 4ac < 0$, the quadratic equation has no real number solution.

EXAMPLE 5 **a)** Find the discriminant of the equation $x^2 - 8x + 16 = 0$.

b) How many real number solutions does the given equation have?

c) Use the quadratic formula to find the solution(s).

Solution **a)** $a = 1, \quad b = -8, \quad c = 16.$

$$b^2 - 4ac = (-8)^2 - 4(1)(16)$$
$$= 64 - 64 = 0$$

b) Since the discriminant equals zero, there is a single real number solution.

c)
$$x = \frac{-b \pm \sqrt{b^2 - 4ac}}{2a}$$
$$= \frac{-(-8) \pm \sqrt{0}}{2(1)} = \frac{8 \pm 0}{2} = \frac{8}{2} = 4$$

The only solution is 4.

EXAMPLE 6 Without actually finding the solutions, determine whether the following equations have two distinct real number solutions, a single real number solution, or no real number solution.

a) $2x^2 - 4x + 6 = 0$ **b)** $x^2 - 5x - 8 = 0$ **c)** $4x^2 - 12x = -9$

Solution We use the discriminant of the quadratic formula to answer these questions.

a) $b^2 - 4ac = (-4)^2 - 4(2)(6) = 16 - 48 = -32$

Since the discriminant is negative, this equation has no real number solution.

b) $b^2 - 4ac = (-5)^2 - 4(1)(-8) = 25 + 32 = 57$

Since the discriminant is positive, this equation has two distinct real number solutions.

c) First, rewrite $4x^2 - 12x = -9$ as $4x^2 - 12x + 9 = 0$.

$$b^2 - 4ac = (-12)^2 - 4(4)(9) = 144 - 144 = 0$$

NOW TRY EXERCISE 13

Since the discriminant is zero, this equation has a single real number solution.

The discriminant can be used to find the number of real solutions to an equation of the form $ax^2 + bx + c = 0$. Since the x-intercepts of a quadratic function, $f(x) = ax^2 + bx + c$, occur where $f(x) = 0$, the discriminant can also be used to find the number of x-intercepts of a quadratic function. Figure 8.3 shows the relationship between the discriminant and the number of x-intercepts for a function of the form $f(x) = ax^2 + bx + c$.

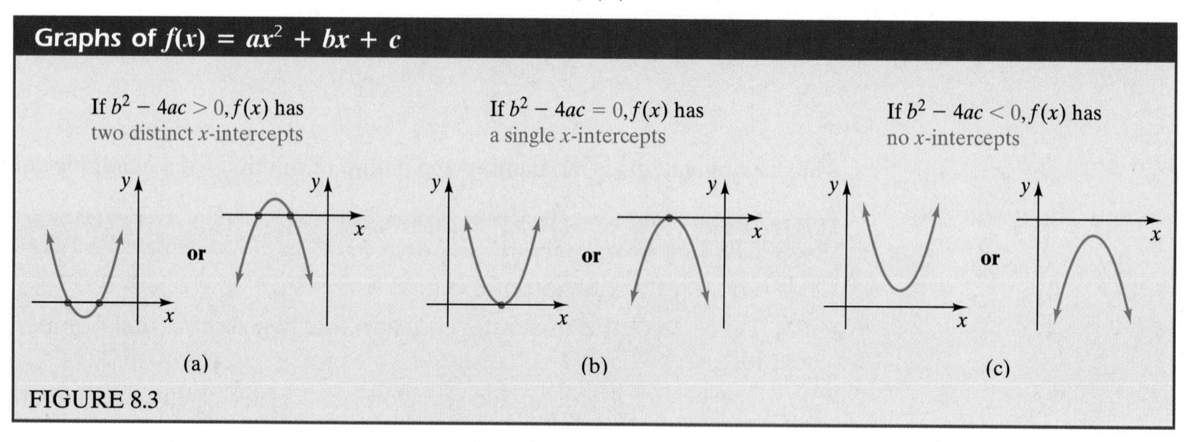

FIGURE 8.3

We will discuss graphing of quadratic functions in detail in Section 8.5.

5) Study Applications That Use Quadratic Equations

We will now look at some applications of quadratic equations.

EXAMPLE 7 Laserox, a startup company, projects that its annual profits, $p(n)$, in thousands of dollars, over the first 8 years of operation can be approximated by the function $p(n) = 1.2n^2 + 4n - 8$, where n is the number of years completed.

a) Estimate the profit (or loss) of the company after the first year.

b) Estimate the profit (or loss) of the company after 8 years.

c) Estimate the time needed for the company to break even.

Solution **a)** To estimate the profit after 1 year, we evaluate the function at 1.

$$p(n) = 1.2n^2 + 4n - 8$$
$$p(1) = 1.2(1)^2 + 4(1) - 8 = -2.8$$

Thus, at the end of the first year the company projects a loss of $2800.

b) $p(8) = 1.2(8)^2 + 4(8) - 8 = 100.8$
Thus, at the end of the eighth year the projected profit is $100,800.

c) Understand The company will break even when the profit is 0. Thus, to find the break-even point we solve the equation

$$1.2n^2 + 4n - 8 = 0$$

We can use the quadratic formula to solve this equation.

Translate $a = 1.2, \quad b = 4, \quad c = -8.$

$$n = \frac{-b \pm \sqrt{b^2 - 4ac}}{2a}$$
$$= \frac{-4 \pm \sqrt{4^2 - 4(1.2)(-8)}}{2(1.2)}$$

Carry Out

$$= \frac{-4 \pm \sqrt{16 + 38.4}}{2.4}$$
$$= \frac{-4 \pm \sqrt{54.4}}{2.4}$$
$$\approx \frac{-4 \pm 7.376}{2.4}$$

$$n = \frac{-4 + 7.376}{2.4} \approx 1.4 \qquad \text{or} \qquad n = \frac{-4 - 7.376}{2.4} \approx -4.74$$

Answer Since time cannot be negative, the break-even time is about 1.4 years.

An important formula in physics is $h = \frac{1}{2}gt^2 + v_0t + h_0$. When an object is projected upward from an initial height, h_0, with initial velocity of v_0, this formula can be used to find the height h of the object above the ground at any time, t. The g in the formula is the acceleration of gravity. Since the acceleration of Earth's gravity is -32 ft/sec^2, we use -32 for g in the formula when discussing Earth. This formula can also be used in describing projectiles on the moon and other planets, but the value of g in the formula will need to change for each planetary body. We will use this formula in Example 8.

EXAMPLE 8 Jennifer Posmoga is standing on top of a building and throws a ball upward from a height of 60 feet with an initial velocity of 30 feet per second. Use the formula $h = \frac{1}{2}gt^2 + v_0t + h_0$ to answer the following questions.

a) How long after the ball is thrown, to the nearest tenth of a second, will the ball be 25 feet from the ground?

b) How long after the ball is thrown will the ball strike the ground?

Solution **a) Understand** We will illustrate this problem with a diagram (see Fig. 8.4). Here $g = -32$, $v_0 = 30$, and $h_0 = 60$. We are asked to find the time, t, it takes for the ball to reach a height, h, of 25 feet above the ground. We substitute these values into the formula and then solve for t.

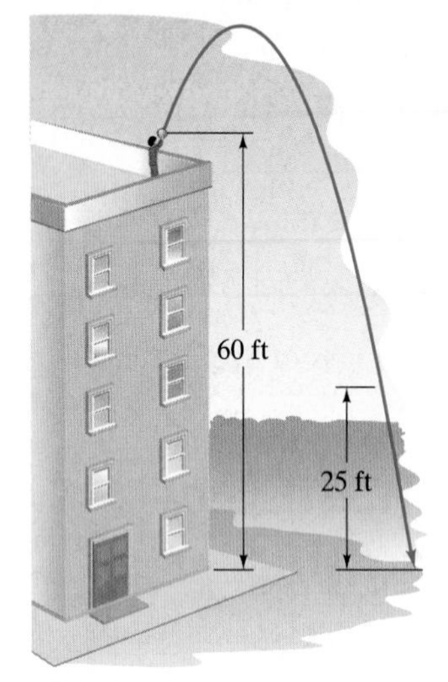

FIGURE 8.4

Translate

$$h = \frac{1}{2}gt^2 + v_0t + h_0$$

$$25 = \frac{1}{2}(-32)t^2 + 30t + 60$$

Carry Out Now we write the quadratic equation in standard form and solve for t by using the quadratic formula.

$$0 = -16t^2 + 30t + 35$$

$$\text{or} \quad -16t^2 + 30t + 35 = 0$$

$$a = -16, \quad b = 30, \quad c = 35$$

$$t = \frac{-b \pm \sqrt{b^2 - 4ac}}{2a}$$

$$= \frac{-30 \pm \sqrt{(30)^2 - 4(-16)(35)}}{2(-16)}$$

$$= \frac{-30 \pm \sqrt{3140}}{-32}$$

$$t = \frac{-30 + \sqrt{3140}}{-32} \quad \text{or} \quad t = \frac{-30 - \sqrt{3140}}{-32}$$

$$\approx -0.8 \qquad\qquad \approx 2.7$$

Answer Since time cannot be negative, the only acceptable solution is 2.7 seconds. Thus, about 2.7 seconds after the ball is thrown, it will be 25 feet above the ground.

b) Understand We wish to find the time at which the ball strikes the ground. When the ball strikes the ground, its distance above the ground is 0. We substitute $h = 0$ into the formula and solve for t.

Translate

$$h = \frac{1}{2}gt^2 + v_0t + h_0$$

$$0 = \frac{1}{2}(-32)t^2 + 30t + 60$$

Carry Out

$$0 = -16t^2 + 30t + 60$$

$$a = -16, \quad b = 30, \quad c = 60$$

$$t = \frac{-b \pm \sqrt{b^2 - 4ac}}{2a}$$

$$= \frac{-30 \pm \sqrt{(30)^2 - 4(-16)(60)}}{2(-16)}$$

$$= \frac{-30 \pm \sqrt{4740}}{-32}$$

$$t = \frac{-30 + \sqrt{4740}}{-32} \quad \text{or} \quad t = \frac{-30 - \sqrt{4740}}{-32}$$

$$\approx -1.2 \qquad\qquad \approx 3.1$$

Answer Since time cannot be negative, the only reasonable solution is 3.1 seconds. Thus, the ball will strike the ground approximately 3.1 seconds after it is thrown.

EXAMPLE 9 The function $N(t) = 0.0054t^2 - 1.46t + 95.11$ can be used to estimate the average number of years of life expectancy remaining for a person of age t years where $30 \le t \le 100$.

a) Estimate the remaining life expectancy of a person of age 50.

b) If a person has a remaining life expectancy of 14.3 years, estimate the age of the person.

Solution **a) Understand** We would expect that the older a person gets the shorter the remaining life expectancy. To determine the remaining life expectancy for a 50-year-old, we substitute 50 for t in the function and evaluate.

Translate

$$N(t) = 0.0054t^2 - 1.46t + 95.11$$

$$N(50) = 0.0054(50)^2 - 1.46(50) + 95.11$$

Carry Out

$$= 0.0054(2500) - 73.00 + 95.11$$

$$= 13.5 - 73.00 + 95.11$$

$$= 35.61$$

Answer and Check The answer appears reasonable. Thus, on the average, a 50-year-old can expect to live another 35.61 years to an age of 85.61 years.

b) Understand Here we are given the remaining life expectancy, $N(t)$, and asked to find the age of the person, t. To solve this problem, we substitute 14.3 for $N(t)$ and solve for t. To solve for t, we will use the quadratic formula.

Translate

$$N(t) = 0.0054t^2 - 1.46t + 95.11$$

$$14.3 = 0.0054t^2 - 1.46t + 95.11$$

Carry Out

$$0 = 0.0054t^2 - 1.46t + 80.81$$

$$a = 0.0054, \quad b = -1.46, \quad c = 80.81$$

$$t = \frac{-b \pm \sqrt{b^2 - 4ac}}{2a}$$

$$= \frac{-(-1.46) \pm \sqrt{(-1.46)^2 - 4(0.0054)(80.81)}}{2(0.0054)}$$

$$= \frac{1.46 \pm \sqrt{2.1316 - 1.745496}}{0.0108}$$

$$= \frac{1.46 \pm \sqrt{0.386104}}{0.0108}$$

$$\approx \frac{1.46 \pm 0.6214}{0.0108}$$

$$t = \frac{1.46 + 0.6214}{0.0108} \quad \text{or} \quad t = \frac{1.46 - 0.6214}{0.0108}$$

$$= 192.72 \qquad\qquad = 77.65$$

Answer Since 192.72 is not a reasonable age, we can exclude that as a possibility. Thus the average person who has a life expectancy of 14.3 years is about 77.65 years old.

NOW TRY EXERCISE 85

Exercise Set 8.2

Concept/Writing Exercises

1. Give the quadratic formula. (You should memorize this formula.)
2. Consider the two equations $-6x^2 + \frac{1}{2}x - 5 = 0$ and $6x^2 - \frac{1}{2}x + 5 = 0$? Must the solutions to these two equations be the same? Explain your answer.
3. Consider $12x^2 - 15x - 6 = 0$ and $3(4x^2 - 5x - 2) = 0$.
   a) Will the solution to the two equations be the same? Explain.
   b) Solve $12x^2 - 15x - 6 = 0$.
   c) Solve $3(4x^2 - 5x - 2) = 0$.
4. **a)** Explain how to find the discriminant.
   b) What is the discriminant for the equation $3x^2 - 6x + 20 = 0$?
5. Write a paragraph or two explaining the relationship between the value of the discriminant and the number of real solutions to a quadratic equation. In your paragraph, explain *why* the value of the discriminant determines the number of real solutions.
6. Write a paragraph or two explaining the relationship between the value of the discriminant and the number of x-intercepts of $f(x) = ax^2 + bx + c$. In your paragraph, explain when the function will have no, one, and two x-intercepts.

Practice the Skills

Use the discriminant to determine whether each equation has two distinct real solutions, a single real solution, or no real solution.

7. $x^2 - 5x + 6 = 0$
8. $4x^2 + 2x + 3 = 0$
9. $2a^2 - 4a + 7 = 0$
10. $-2x^2 + x - 8 = 0$
11. $5p^2 + 3p - 7 = 0$
12. $2x^2 = 16x - 32$
13. $-5x^2 + 5x - 6 = 0$
14. $4.1x^2 - 3.1x - 5.2 = 0$
15. $x^2 + 10.2x + 26.01 = 0$
16. $\frac{1}{2}x^2 + \frac{1}{3}x + 12 = 0$
17. $b^2 = -3b - \frac{9}{4}$
18. $\frac{x^2}{3} = \frac{2x}{5}$

Solve each equation by the quadratic formula.

19. $x^2 + 9x + 20 = 0$
20. $x^2 + 6x + 8 = 0$
21. $x^2 + 2x - 3 = 0$
22. $x^2 - 3x - 10 = 0$
23. $a^2 - 6a = -5$
24. $x^2 = 13x - 36$

25. $r^2 - 81 = 0$

26. $x^2 - 25 = 0$

27. $x^2 - 4x = 0$

28. $2x^2 = 4x + 1$

29. $3w^2 - 4w + 5 = 0$

30. $4s^2 - 8s + 6 = 0$

31. $x^2 + 6x = -2$

32. $x^2 - 2x - 1 = 0$

33. $-6x^2 + 21x = -27$

34. $-x^2 + 2x + 15 = 0$

35. $(2a + 3)(3a - 1) = 2$

36. $(2w - 6)(3w + 4) = -20$

37. $-2a^2 = a + 3$

38. $9x^2 + 6x + 1 = 0$

39. $2x^2 + 6x = 0$

40. $3x^2 - 5x = 0$

41. $m = \dfrac{-m + 6}{m - 4}$

42. $3p = \dfrac{5p + 6}{2p + 3}$

43. $\dfrac{1}{2}x^2 + 2x + \dfrac{2}{3} = 0$

44. $x^2 - \dfrac{x}{5} - \dfrac{1}{3} = 0$

45. $-x^2 + \dfrac{11}{3}x + \dfrac{10}{3} = 0$

46. $x^2 - \dfrac{7}{6}x + \dfrac{2}{3} = 0$

47. $0.1x^2 + 0.6x - 1.2 = 0$

48. $-2.3x^2 + 5.6x + 0.4 = 0$

49. $-1.62x^2 - 0.94x + 4.85 = 0$

50. $1.74x^2 - 2.04x + 6.2 = 0$

For each function, determine all real values of the variable for which the function has the value indicated.

51. $f(x) = x^2 - 3x + 3,\ f(x) = 3$

52. $g(r) = r^2 + 17r + 84,\ g(r) = 12$

53. $h(t) = 2t^2 - 7t + 1,\ h(t) = -3$

54. $f(x) = 2x^2 - 7x + 5,\ f(x) = 0$

55. $g(a) = 2a^2 - 3a + 16,\ g(a) = 14$

56. $h(x) = 6x^2 + x + 10,\ h(x) = -2$

Write a function that has the given solution.

57. $4, 6$

58. $-2, 5$

59. $3, -4$

60. $-1, -6$

61. $\dfrac{1}{2}, 3$

62. $-2, \dfrac{2}{3}$

63. $-\dfrac{3}{5}, \dfrac{2}{3}$

64. $\dfrac{3}{5}, \dfrac{1}{4}$

65. $\sqrt{3}, -\sqrt{3}$

66. $\sqrt{7}, -\sqrt{7}$

67. $2i, -2i$

68. $5i, -5i$

69. $3 + \sqrt{2}, 3 - \sqrt{2}$

70. $5 - \sqrt{3},\ 5 + \sqrt{3}$

71. $2 + 3i, 2 - 3i$

72. $5 - 4i, 5 + 4i$

Problem Solving

73. Give your own example of a quadratic equation that can be solved by the quadratic formula but not by factoring over the set of integers.

74. Are there any quadratic equations that **a)** can be solved by the quadratic formula that cannot be solved by completing the square? **b)** can be solved by completing the square that cannot be solved by factoring over the set of integers?

75. When solving a quadratic equation by the quadratic formula, if the discriminant is a perfect square, must the equation be factorable over the set of integers? Explain.

76. When solving a quadratic equation by the quadratic formula, if the discriminant is a natural number, must the equation be factorable over the set of integers? Explain.

In Exercises 77–93, use a calculator as needed to give the solution in decimal form. Round irrational numbers to the nearest hundredth.

77. Twice the square of a positive number increased by 3 times the number is 14. Find the number.

78. Three times the square of a positive number decreased by twice the number is 21. Find the number.

79. The length of a rectangular garden is 2 feet less than 3 times its width. Find the length and width if the area of the garden is 21 square feet.

80. Lora Wallman wishes to fence in a rectangular region along a river bank by constructing fencing as illustrated in the diagram. If she has only 400 feet of fencing and wishes to enclose an area of 15,000 square feet, find the dimensions of the rectangular region.

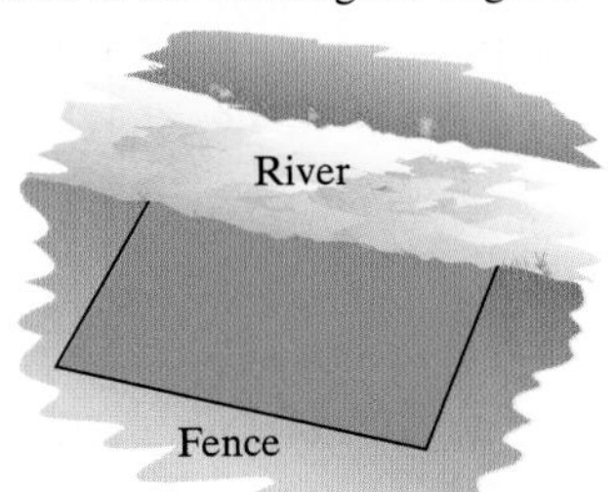

81. John Williams, a professional photographer, has a 6-inch by 8-inch photo. He wishes to reduce the photo by the same amount on each side so that the resulting photo will have half the area of the original photo. By how much will he have to reduce the length of each side?

82. The temperature, T, in degrees Fahrenheit, in a car's radiator during the first 4 minutes of driving is a function of time, t. The temperature can be found by the formula $T = 6.2t^2 + 12t + 32, 0 \leq t \leq 4$.

a) What is the car's radiator temperature at the instant the car is turned on?

b) What is the car's radiator temperature after the car has been driven for 1 minute?

c) How long after the car has begun operating will the car's radiator temperature reach 120°F?

83. A video store's weekly profit, P, in thousands of dollars, is a function of the rental price of the tapes, t. The profit equation is $P = 0.2t^2 + 1.5t - 1.2, 0 \leq t \leq 5$.

a) What is the store's weekly profit or loss if they charge \$1 per tape?

b) What is the weekly profit if they charge \$5 per tape?

c) At what tape rental price will their weekly profit be 1.6 thousand dollars?

84. The cost, C, of a ranch house in Norfolk, Virginia, is a function of the number of square feet of the house, s. The cost of a house can be approximated by the formula

$$C = -0.01s^2 + 80s + 20{,}000, \quad 1200 \leq s \leq 4000$$

a) Estimate the cost of a 1500-square-foot house.

b) How large a house can Bill Dodge purchase if he has \$150,000 to spend on a house?

85. The attendance at NASCAR Winston Cup Races has been growing steadily since 1987. The attendance, $A(t)$, in millions, can be estimated by the function $A(t) = 0.03t^2 + 0.02t + 2.69$, where t is years since 1987 and $1 \leq t \leq 20$.

a) Estimate the NASCAR attendance in 1990.

b) In what year was NASCAR attendance 5.3 million?

86. At a college, records show that the average person's grade point average, G, is a function of the number of hours he or she studies and does homework per week, h. The grade point average can be estimated by the equation $G = 0.01h^2 + 0.2h + 1.2, \ 0 \leq h \leq 8$.

a) What is the GPA of the average student who studies for 0 hours a week?

b) What is the GPA of an average student who studies 4 hours per week?

c) To obtain a 3.2 GPA, how many hours per week would the average student need to study?

87. The following data appeared in the April 20, 1998 issue of *Newsweek*.

Veterinary Expenditures

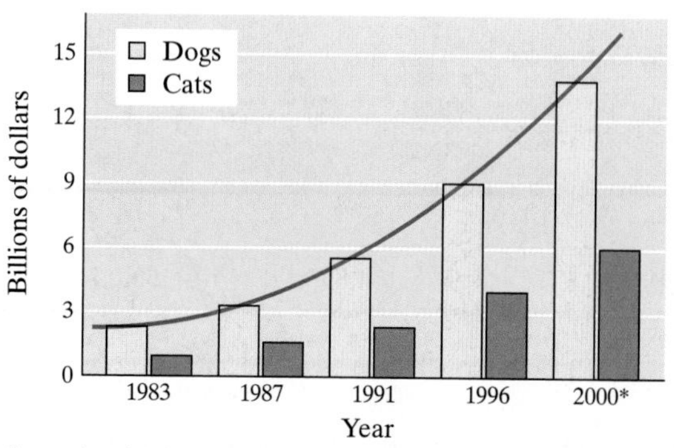

Source: American Veterinary Medical Association *Newsweek Projection

The annual amount spent in the United States on veterinary bills for dogs, $m(t)$, in billions of dollars, can be estimated by the function $m(t) = 0.05t^2 - 0.32t + 3.15$, where t represents the number of years since 1982 and $1 \leq t \leq 18$.

a) Use the function to estimate the amount spent on veterinary bills for dogs in 1985.

b) In what year was \$12 billion spent on veterinary bills for dogs?

88. According to the U.S. Department of Education, U.S. elementary and secondary school enrollment is expected to increase from 1990 to 2006 and level off about 2006. The function $N(t) = -0.043t^2 + 1.22t + 46.0$ can be used to estimate total elementary and secondary school enrollment, in millions, where t is years since 1989 and $1 \leq t \leq 17$.

a) Estimate total enrollment in 1995.

b) In what year will the total enrollment reach 54 million students?

89. An article in the November 24, 1997 issue of *Newsweek* stated that Americans buy 40% of the world's indigestion remedies. (Is it because we eat spicier foods or because we are a nervous nation?) The function $s(t) = 0.02t^2 - 0.02t + 1.20$ can be used to estimate U.S. sales of indigestion remedies, in billions of dollars, where t represents years since 1992 and $1 \leq t \leq 9$.

a) Estimate the amount spent in the United States on indigestion remedies in 1995.

b) In what year did U.S. residents spend \$2.0 billion on indigestion remedies?

90. The following graph summarizes the data provided in 1998 by the National Center on Addiction and Substance Abuse on the percent of students at various ages who say that their school is *not* drug-free.

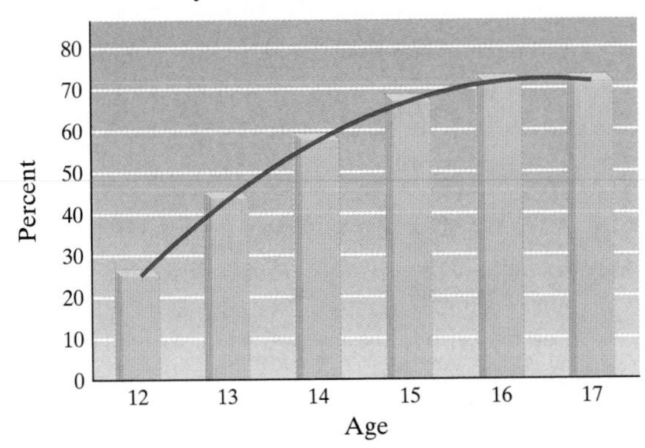

The function $f(a) = -2.32a^2 + 76.58a - 559.87$ can be used to estimate the percent of students who say their school is not drug-free. In the function, a represents the students' age, where $12 \le a \le 17$. Use the function to answer the following questions.

a) Estimate the percent of 13-year-olds say their school is not drug-free?

b) At what age do 70% of the students say their school is not drug-free?

91. The roller coaster at Busch Gardens in Williamsburg, Virginia, has a number of steep drops. One of its drops has a vertical distance of 62 feet. The speed, s, of the last car, in feet per second, t seconds after it has begun its drop can be calculated by the formula $s = 6.74t + 2.3, 0 \le t \le 4$. The height of the last car from the bottom of the drop t seconds after it has begun this drop can be found by the formula $h = -3.3t^2 - 2.3t + 62, 0 \le t \le 4$.

a) Find the time it takes for the last car to travel from the top of the drop to the bottom of the drop.

b) Find the speed of the last car when it reaches the bottom of the drop.

In Exercises 92 and 93, use the equation $h = \frac{1}{2}gt^2 + v_0t + h_0$ (refer to Example 8).

92. A horseshoe is thrown upward from an initial height of 80 feet with an initial velocity of 60 feet per second. How long after the horseshoe is projected upward

a) will it be 20 feet from the ground?

b) will it strike the ground?

93. Gravity on the moon is about one-sixth of that on Earth. Suppose Neil Armstrong is standing on a hill on the moon 60 feet high. If he jumps upward with a velocity of 40 feet per second, how long will it take for him to land on the ground below the hill?

Solve by the quadratic formula.

94. $x^2 - \sqrt{5}x - 10 = 0$

95. $x^2 + 5\sqrt{6}x + 36 = 0$

Challenge Problems

96. A metal cube expands when heated. If each edge increases 0.20 millimeter after being heated and the total volume increases by 6 cubic millimeters, find the original length of a side of the cube.

97. The equation $x^n = 1$ has n solutions (including the complex solutions). Find the six solutions to $x^6 = 1$. (*Hint:* Rewrite the equation as $x^6 - 1 = 0$, then factor using the formula for the difference of two squares.)

Group Activity

Discuss and answer Exercises 98 and 99 as a group.

98. The "golden ratio," which occurs often in mathematics, was introduced by the ancient Greeks. The golden ratio is used in art, architecture, advertising, and many other areas. To obtain the golden ratio, divide a line segment such that the ratio of the larger part to the whole is equal to the ratio of the smaller part to the larger part. Consider the following diagram.

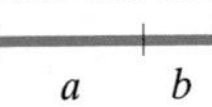

Using the diagram, the golden ratio will be $\dfrac{a}{a+b} = \dfrac{b}{a}$.

a) If we let the smaller part be 1, determine the value of a, the larger part.

b) Then substitute the values for a and b and use cross multiplication to show that $\frac{a}{a+b} = \frac{b}{a}$.

c) Write the answer in part a) as a decimal number rounded to the nearest thousandth.

d) The artist George Seurat (1859–1891) believed that certain "natural proportions" exist in nature and that those proportions are the ones most likely to please the human eye. You can see this within his painting *La Parade*, which is composed of rectangles within rectangles. Where in the painting are golden ratios used?

e) A golden rectangle is one whose ratio of length to width is $\frac{1+\sqrt{5}}{2}$.

$$\frac{\text{length}}{\text{width}} = \frac{a+b}{a} = \frac{a}{b} = \frac{1+\sqrt{5}}{2}$$

Where in the painting are golden rectangles used?

99. Travis Hawley is on the fourth floor of an eight-story building and Courtney Prenzlow is on the roof. Travis is 60 feet above the ground while Courtney is 120 feet above the ground.

a) Determine the time it takes for Travis's rock to strike the ground.

b) Determine the time it takes for Courtney's rock to strike the ground.

c) If Travis throws a rock upward with an initial velocity of 100 feet per second at the same time that Courtney throws a rock upward at 60 feet per second, whose rock will strike the ground first? Explain

d) Will the rocks ever be at the same distance above the ground? If so, at what times?

Cumulative Review Exercises

[7.3] **100.** Simplify $\sqrt[5]{64x^9y^{12}z^{20}}$.

101. Simplify $\sqrt[3]{4x^2y^8}(\sqrt[3]{2x^5y^4} + \sqrt[3]{6xy})$.

[7.4] **102.** Simplify $\frac{x+\sqrt{y}}{x-\sqrt{y}}$.

[7.6] **103.** Solve the equation $\sqrt{2x+4} - 1 = \sqrt{x+3}$.

8.3 QUADRATIC EQUATIONS: APPLICATIONS AND PROBLEM SOLVING

SSM VIDEO 8.3 CD Rom

1) Solve additional applications.
2) Solve for a variable in a formula.

1) Solve Additional Applications

We have already seen many applications of quadratic equations. In this section we will look at additional applications. In Example 1, we use the quadratic formula to solve a motion problem. In Example 2, we use the quadratic formula to solve a mixture problem. Then we will discuss solving for a variable in a formula.

Motion Problems

We first discussed motion problems in Section 2.4. The motion problem we give here is solved using the quadratic formula.

EXAMPLE 1 It is a beautiful day, so Richard Semmler decides to go for a slow and relaxing ride in his motorboat on the Potomac River. His trip starts near the Clara Barton Natural Historic Site in Bethesda, Maryland. He travels 12 miles until he enters the Arlington, Virginia/Washington, DC area, where he turns around and heads back to his starting point. The total time of his trip is 5 hours. Along the way, he passes a digital sign that indicates the river current is 2 miles per hour. If during the entire trip he did not touch the throttle (used to change the speed), find the speed the boat would have been traveling in still water.

Solution **Understand** We are asked to find the rate of the boat in still water. Let r = the rate of the boat in still water. We know that the total time of the trip is 5 hours. Thus, the time downriver plus the time upriver must sum to 5 hours. Since distance = rate · time, we can find the time by dividing the distance by the rate.

Direction	Distance	Rate	Time
Downriver (with current)	12	$r + 2$	$\frac{12}{r+2}$
Upriver (against current)	12	$r - 2$	$\frac{12}{r-2}$

Translate time downriver + time upriver = total time

$$\frac{12}{r+2} + \frac{12}{r-2} = 5$$

Carry Out

$$(r+2)(r-2)\left(\frac{12}{r+2} + \frac{12}{r-2}\right) = (r+2)(r-2)(5) \quad \text{Multiply by the LCD.}$$

$$\cancel{(r+2)}(r-2)\left(\frac{12}{\cancel{r+2}}\right) + (r+2)\cancel{(r-2)}\left(\frac{12}{\cancel{r-2}}\right) = (r+2)(r-2)(5) \quad \text{Distributive Property}$$

$$12(r-2) + 12(r+2) = 5(r^2 - 4)$$

$$12r - 24 + 12r + 24 = 5r^2 - 20 \quad \text{Distributive Property}$$

$$24r = 5r^2 - 20 \quad \text{Simplify}$$

$$\text{or} \quad 5r^2 - 24r - 20 = 0$$

Using the quadratic formula with $a = 5$, $b = -24$, and $c = -20$, we obtain

$$r = \frac{24 \pm \sqrt{976}}{10}$$

$$r \approx 5.5 \quad \text{or} \quad r \approx -0.7$$

Anwer Since the rate cannot be negative, the rate or speed of the boat in still water is about 5.5 miles per hour.

NOW TRY EXERCISE 17

Notice that in real-life situations most answers are not integral values.

Work Problems

Let's do an example involving a work problem. Work problems were discussed in Section 6.5. You may wish to review that section before studying the next example.

EXAMPLE 2 After Hurricane Bob, the Duvals needed to remove water from their flooded basement. They had one sump pump (used to remove water) and borrowed a second from their local fire department. The fire chief informed the Duvals he expects that with both pumps working together, their basement should empty in about 6 hours. He also informed them that the fire department's sump pump has a higher horsepower, and it would empty their basement by itself in 2 hours less time than the Duvals' pump could if it were working alone. How long would it take each pump to empty the basement if each were working alone?

Solution **Understand** Recall from Section 6.5 that the rate of work multiplied by the time worked gives the part of the task completed.

Let t = number of hours for slower pump to complete the job by itself

$t - 2$ = number of hours for the faster pump to complete the job by itself

Pump	Rate of Work	Time Worked	Part of Task Completed
Slower pump	$\frac{1}{t}$	6	$\frac{6}{t}$
Faster pump	$\frac{1}{t-2}$	6	$\frac{6}{t-2}$

Translate

$$\begin{pmatrix}\text{part of task}\\ \text{by slower pump}\end{pmatrix} + \begin{pmatrix}\text{part of task}\\ \text{by faster pump}\end{pmatrix} = 1$$

$$\frac{6}{t} + \frac{6}{t-2} = 1$$

Carry Out

$$t(t-2)\left(\frac{6}{t} + \frac{6}{t-2}\right) = t(t-2)(1)$$ *Multiply both sides by the LCD $t(t-2)$.*

$$\cancel{t}(t-2)\left(\frac{6}{\cancel{t}}\right) + t(\cancel{t-2})\left(\frac{6}{\cancel{t-2}}\right) = t^2 - 2t$$ *Distributive Property*

$$6(t-2) + 6t = t^2 - 2t$$

$$6t - 12 + 6t = t^2 - 2t$$

$$t^2 - 14t + 12 = 0$$

Using the quadratic formula, we obtain

$$t = \frac{14 \pm \sqrt{148}}{2}$$

$$t \approx 13.1 \quad \text{or} \quad t \approx 0.9$$

Answer Both 13.1 and 0.9 satisfy the equation $\frac{6}{t} + \frac{6}{t-2} = 1$ (with some round-off involved). However, if we accept 0.9 as a solution, then the faster pump could complete the task in a negative time ($t - 2 = 0.9 - 2 = -1.1$ hours), which is not possible. Therefore, 0.9 hour is not an acceptable solution. The only solution is 13.1 hours. The slower pump takes approximately 13.1 hours by itself, and the faster pump takes approximately $13.1 - 2$ or 11.1 hours by itself to empty the basement.

NOW TRY EXERCISE 23

2) Solve for a Variable in a Formula

When the square of a variable appears in a *formula*, you may need to use the square root property to solve for the variable. However, *when you use the square root property in most formulas, you will use only the principal or positive root*, because you are generally solving for a quantity that cannot be negative.

EXAMPLE 3 **a)** The formula for the area of a circle is $A = \pi r^2$. Solve this equation for r.

b) *Newton's law of universal gravity* states that every particle in the universe attracts every other particle with a force proportional to the product of their masses and inversely proportional to the square of the distance between them. We may represent Newton's law as

$$F = G\frac{m_1 m_2}{r^2}$$

Solve the equation for r.

Solution **a)**

$$A = \pi r^2$$

$$\frac{A}{\pi} = r^2 \qquad \text{Isolate } r^2 \text{ by dividing both sides by } \pi.$$

$$\sqrt{\frac{A}{\pi}} = r \qquad \text{Square root property}$$

b)

$$F = G\frac{m_1 m_2}{r^2}$$

$$r^2 F = G m_1 m_2 \qquad \text{Multiply both sides of formula by } r^2.$$

$$r^2 = \frac{G m_1 m_2}{F} \qquad \text{Isolate } r^2 \text{ by dividing both sides by } F.$$

$$r = \sqrt{\frac{G m_1 m_2}{F}} \qquad \text{Square root property}$$

NOW TRY EXERCISE 11

EXAMPLE 4 The diagonal of a box can be calculated by the formula

$$d = \sqrt{L^2 + W^2 + H^2}$$

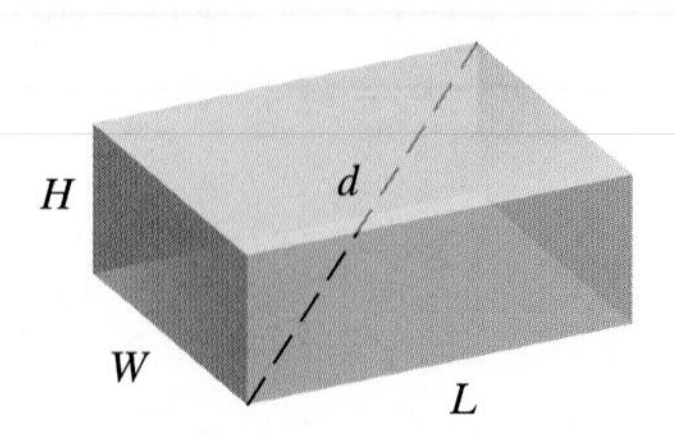

FIGURE 8.5

where L is the length, W is the width, and H is the height of the box. See Figure 8.5.

a) Find the diagonal of a suitcase of length 30 inches, width 15 inches, and height 10 inches.

b) Solve the equation for the width, W.

Solution a) Understand To find the diagonal we need to substitute the appropriate values into the formula and solve for the diagonal, d.

Translate

$$d = \sqrt{L^2 + W^2 + H^2}$$

$$d = \sqrt{(30)^2 + (15)^2 + (10)^2}$$

Carry Out

$$= \sqrt{900 + 225 + 100}$$

$$= \sqrt{1225}$$

$$= 35$$

Answer Thus, the diagonal of the suitcase is 35 inches.

b) Our first step in solving for W is to square both sides of the formula.

$$d = \sqrt{L^2 + W^2 + H^2}$$

$$d^2 = \left(\sqrt{L^2 + W^2 + H^2}\right)^2 \qquad \textit{Square both sides.}$$

$$d^2 = L^2 + W^2 + H^2 \qquad \textit{Use } \left(\sqrt{a}\right)^2 = a, a \geq 0.$$

$$d^2 - L^2 - H^2 = W^2 \qquad \textit{Isolate } W^2.$$

$$\sqrt{d^2 - L^2 - H^2} = W \qquad \textit{Square root property}$$

EXAMPLE 5 The surface area of a right circular cone is

$$s = \pi r\sqrt{r^2 + h^2}$$

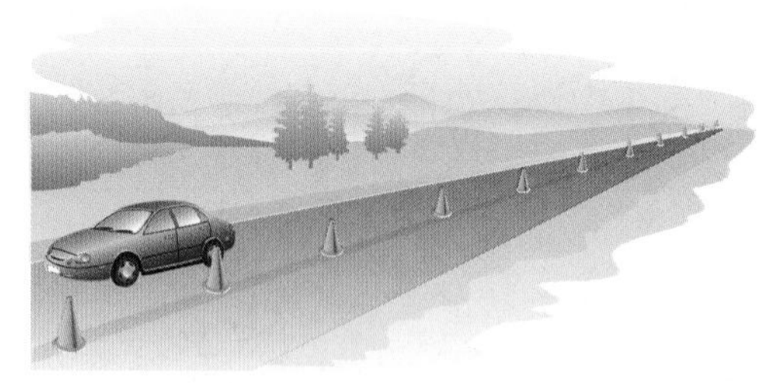

a) An orange safety cone used on roads is 18 inches high with a radius of 12 inches. Find the surface area of the cone.

b) Solve the formula for h.

Solution **a) Understand and Translate** To find the surface area, we substitute the appropriate values into the formula.

$$s = \pi r\sqrt{r^2 + h^2}$$

$$= \pi(12)\sqrt{(12)^2 + (18)^2}$$

Carry Out

$$= 12\pi\sqrt{144 + 324}$$
$$= 12\pi\sqrt{468}$$
$$\approx 815.56$$

Answer The surface area is about 815.56 square inches.

b) To solve for h we need to isolate h on one side of the equation. There are various ways to solve the equation for h.

$$s = \pi r\sqrt{r^2 + h^2}$$

$$\frac{s}{\pi r} = \sqrt{r^2 + h^2} \quad \textit{Divide both sides by } \pi r.$$

$$\left(\frac{s}{\pi r}\right)^2 = \left(\sqrt{r^2 + h^2}\right)^2 \quad \textit{Square both sides.}$$

$$\frac{s^2}{\pi^2 r^2} = r^2 + h^2 \quad \textit{Use } \left(\sqrt{a}\right)^2 = a,\ a \geq 0.$$

$$\frac{s^2}{\pi^2 r^2} - r^2 = h^2 \quad \textit{Subtract } r^2 \textit{ from both sides.}$$

$$\sqrt{\frac{s^2}{\pi^2 r^2} - r^2} = h \quad \textit{Square root property}$$

Other acceptable answers are $h = \sqrt{\dfrac{s^2 - \pi^2 r^4}{\pi^2 r^2}}$ and $h = \dfrac{\sqrt{s^2 - \pi^2 r^4}}{\pi r}$. Can you explain why?

NOW TRY EXERCISE 15

Exercise Set 8.3

Concept/Writing Exercises

1. In general, when solving for a variable in a formula, whether you use the square root property or the quadratic formula, you use only the positive square root. Explain why.

2. Suppose $P = ☺^2 + ⊡^2$ is a real formula. Solving for ☺ gives $☺ = \sqrt{P - ⊡^2}$. If ☺ is to be a real number, what relationship must exist between P and ⊡?

Practice the Skills

Solve for the indicated variable.

3. $A = s^2$, for s (area of a square)
4. $E = mc^2$, for c (Einstein's famous energy formula)
5. $E = i^2 r$, for i (current in electronics)
6. $A = 4\pi r^2$, for r (surface area of a sphere)
7. $V = \pi r^2 h$, for r (volume of a right circular cylinder)
8. $V = \frac{1}{3}\pi r^2 h$, for r (volume of a right circular cone)
9. $a^2 + b^2 = c^2$, for b (Pythagorean Theorem)
10. $P = \frac{1}{2}kx^2$, for x (potential energy of stretched spring)
11. $E = \dfrac{1}{2}mv^2$, for v (kinetic energy)
12. $f_x^2 + f_y^2 = f^2$, for f_x (forces acting on an object)
13. $a = \dfrac{v_2^2 - v_1^2}{2d}$, for v_1 (acceleration of a vehicle)
14. $x^2 + y^2 = r^2$, for x (formula for circle)
15. $v' = \sqrt{c^2 - v^2}$, for c (relativity, v' is read "v prime")
16. $L = L_0\sqrt{1 - \dfrac{v^2}{c^2}}$, for v (art, a painting's contraction)

Problem Solving

17. Julie Hildebrand, a long distance runner, who lives in Frankenmuth, Michigan, starts jogging at her house. She jogs 6 miles and then turns around and jogs back to her house. The first part of her jog is mostly uphill, so her speed averages 2 miles per hour less than her returning speed. If the total time she spent jogging was $1\frac{3}{4}$ hours, find her speed going and her speed returning.

18. Scott Barr, a truck driver, was transporting a heavy load of new cars on a car carrier from Detroit, Michigan, to Indianapolis, Indiana. On his return trip to Detroit, since his truck was lighter, he averaged 10 miles per hour faster than on his trip to Indianapolis. If the total distance traveled each way is 300 miles and the total time he spent driving is 11 hours, find his average speed going and returning.

19. Paul and Rema Jones, who live in Cedar Rapids, Iowa, decided that they want a well on their property. They hired the Ruth Cardiff Drilling Company to drill the well. The drilling company had to drill 64 feet to hit water. The drilling company informed the Joneses that they had just ordered new drilling equipment that drills at an average of 1 foot per hour faster, and that with their new equipment, they would have hit water in 3.2 hours less time. Find the rate at which their present equipment drills.

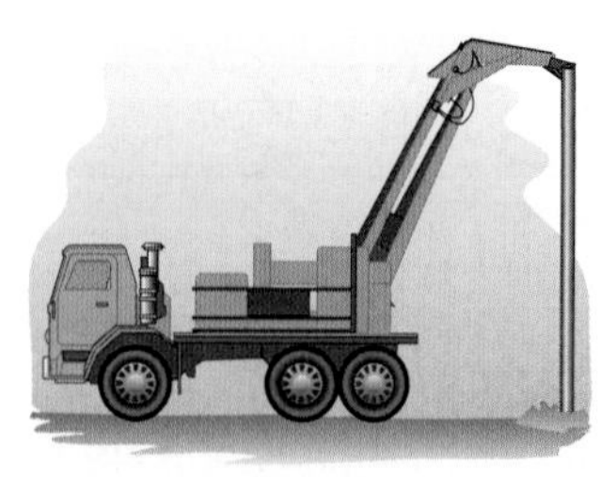

20. Susan Fackert traveled from near Garden of the Gods in Colorado Springs, Colorado, to just outside Grand Junction, Colorado. She took Interstate 25 to Denver, and then Interstate 70 to Grand Junction. The total distance she traveled was 300 miles. After she got to Grand Junction she figured out that had she averaged 10 miles per hour faster, she would have arrived 1 hour earlier. Find the average speed that Susan drove.

21. The Lewis Carroll family and their two friends Charles Dodgson and George Boole want to travel from Amarillo, Texas, to Las Vegas, Nevada. Since only one car is available and all the people would not fit comfortably in one car for the long trip, Charles and George decide to travel by Amtrak. The Carroll family and Charles and George squeeze into the car for the short trip to the Amtrak station. Once at the station, Charles and George board the train. The train and car leave Amarillo at the same time. During the trip Lewis and George speak by cellular phone, and Lewis informs George that they have just stopped for the evening after traveling for 500 miles. One and two-thirds hours later George calls Lewis and informs him that the train had just reached Las Vegas, a distance of 800 miles from Amarillo. Assuming the train averaged 20 miles per hour faster than the car, find the average speed of the car and the train.

22. Barry Harmon flew his single-engine Cessna airplane 80 miles with the wind from Jackson Hole, Wyoming, to above Blackfoot, Idaho. He then turned around and flew back to Jackson Hole against the wind. If the wind was a constant 30 miles per hour, and the total time going and returning was 1.3 hours, find the speed of the plane in still air.

23. Two mechanics, Bonita Rich and Pamela Pearson, take 6 hours to rebuild an engine when they work together. If each worked alone, Bonita, the more experienced mechanic, could complete the job 1 hour faster than Pamela. How long would it take each of them to rebuild the engine working alone?

24. After a small oil spill, two cleanup ships were sent to siphon off the oil floating in Baffin Bay. The newer ship can clean up the entire spill by itself in 3 hours less time than the older ship takes by itself. Working together the two ships can clean up the oil spill in 8 hours. How long will it take the newer ship by itself to clean up the spill?

25. The O'Connors own a small janitorial service. John requires $\frac{1}{2}$ hour more time to clean the Moose Club by himself than Chris does working by herself. If together they can clean the club in 6 hours, find the time required by each to clean the club.

26. A small electric heater requires 6 minutes longer to raise the temperature in an unheated garage to a comfortable level than does a larger electric heater. Together the two heaters can raise the garage temperature to a comfortable level in 42 minutes. How long would it take each heater by itself to raise the temperature in the garage to a comfortable level?

27. Lisa Coyle drove from Lubbock, Texas, to Plainview, Texas, a distance of 60 miles. She then stopped for 2.5 hours to see a friend in Plainview before continuing her journey from Plainview to Amarillo, Texas, a distance of 100 miles. If she drove 10 miles per hour faster from Lubbock to Plainview and the total time of the trip was 5.5 hours, find her average speed from Lubbock to Plainview.

28. Write your own motion problem and solve it.

29. Write your own work problem and solve it.

Challenge Problems

30. When an object is tossed downward with an initial velocity of v_0, the distance, s, it falls in t seconds can be found by the formula $s = \frac{1}{2}gt^2 - v_0 t$. This formula can be rewritten $\frac{1}{2}gt^2 - v_0 t - s = 0$.

Use the quadratic formula to solve this equation for t. Your answer will contain the letters g, t, and v.

Cumulative Review Exercises

[2.2] **31.** Solve $S = 2\pi rh + 2\pi r^2$ for h.

[3.2] **32.** Is the set of ordered pairs $\{(2,3),(4,5),(6,0),(2,1)\}$ a function? Explain.

33. Explain how the vertical line test may be used to determine whether a graph is a function.

34. If $f(x) = \frac{3}{2}x^3 - \frac{1}{2}x^2 + 3$, find $f(\frac{1}{3})$.

35. What is the domain of a function?

36. What is the range of a function?

37. Is $f(x) = \frac{1}{x}$ a polynomial function? Explain.

8.4 WRITING EQUATIONS IN QUADRATIC FORM

SSM VIDEO 8.4 CD Rom

1) **Solve equations that are quadratic in form.**

2) **Solve equations with rational exponents.**

1) Solve Equations That Are Quadratic in Form

Sometimes we need to solve an equation that is not a quadratic equation but can be rewritten in the form of a quadratic equation. We can then solve the equation in quadratic form by factoring, completing the square, or the quadratic formula.

Definition

An equation that can be written in the form $au^2 + bu + c = 0$ for $a \neq 0$ where u is an algebraic expression, is called **quadratic in form.**

When you are given an equation quadratic in form, make a substitution to get the equation in the form $au^2 + bu + c = 0$. For example,

Equation Quadratic in Form	Substitution	Equation with Substitution
$y^4 - y^2 - 6 = 0$	$u = y^2$	$u^2 - u - 6 = 0$
$2(x + 5)^2 - 5(x + 5) - 12 = 0$	$u = x + 5$	$2u^2 - 5u - 12 = 0$

To solve equations quadratic in form, we use the following procedure. We will illustrate this procedure in Example 1.

To Solve Equations Quadratic in Form

1. Make a substitution that will result in an equation of the form $au^2 + bu + c = 0$, $a \neq 0$, where u is a function of the original variable.
2. Solve the equation $au^2 + bu + c = 0$ for u.
3. Replace u with the function of the original variable from step 1 and solve the resulting equation for the original variable.
4. Check for extraneous solutions by substituting the apparent solutions in the original equation.

EXAMPLE 1 Solve the equation $p^4 + 2p^2 = 8$.

Solution

$$p^4 + 2p^2 - 8 = 0 \qquad \text{Set equation equal to 0.}$$

$$(p^2)^2 + 2p^2 - 8 = 0 \qquad \text{Write } p^4 \text{ as } (p^2)^2 \text{ to obtain equation in desired form.}$$

Now let $u = p^2$. This gives an equation that is quadratic in form.

$$u^2 + 2u - 8 = 0 \qquad \text{Substitute } u \text{ for } p^2.$$

$$(u + 4)(u - 2) = 0 \qquad \text{Solve the equation for } u.$$

$$u + 4 = 0 \quad \text{or} \quad u - 2 = 0$$

$$u = -4 \qquad u = 2$$

$$p^2 = -4 \qquad p^2 = 2 \qquad \text{Replace } u \text{ with } p^2.$$

$$p = \pm\sqrt{-4} \qquad p = \pm\sqrt{2} \qquad \text{Solve for } p.$$

$$p = \pm 2i$$

Check the four possible solutions in the *original* equation.

$p = 2i$

$$p^4 + 2p^2 = 8$$
$$(2i)^4 + 2(2i)^2 \stackrel{?}{=} 8$$
$$2^4 i^4 + 2(2^2)(i^2) \stackrel{?}{=} 8$$
$$16(1) + 8(-1) \stackrel{?}{=} 8$$
$$16 - 8 = 8$$

True

$p = -2i$

$$p^4 + 2p^2 = 8$$
$$(-2i)^4 + 2(-2i)^2 \stackrel{?}{=} 8$$
$$(-2)^4 i^4 + 2(-2)^2 i^2 \stackrel{?}{=} 8$$
$$16(1) + 8(-1) \stackrel{?}{=} 8$$
$$16 - 8 = 8$$

True

$p = \sqrt{2}$

$$p^4 + 2p^2 = 8$$
$$(\sqrt{2})^4 + 2(\sqrt{2})^2 \stackrel{?}{=} 8$$
$$4 + 2(2) \stackrel{?}{=} 8$$
$$8 = 8$$

True

$p = -\sqrt{2}$

$$p^4 + 2p^2 = 8$$
$$(-\sqrt{2})^4 + 2(-\sqrt{2})^2 \stackrel{?}{=} 8$$
$$4 + 2(2) \stackrel{?}{=} 8$$
$$8 = 8$$

True

Thus, the solutions are $2i$, $-2i$, $\sqrt{2}$, and $-\sqrt{2}$.

The solutions to equations like $p^4 + 2p^2 = 8$ will always check unless a mistake has been made. In equations like this, extraneous solutions will not be introduced. However, extraneous solutions *may* be introduced when working with rational exponents, as will be shown shortly.

NOW TRY EXERCISE 7

HELPFUL HINT

Students sometimes solve the equation for u but then forget to complete the problem by solving for the original variable. Remember that if the original equation is in x you must obtain values for x. If the original equation is in p (as in Example 1) you must obtain values for p, and so on.

EXAMPLE 2 Solve the equation $4(2w + 1)^2 - 16(2w + 1) + 15 = 0$.

Solution If we let $u = 2w + 1$, the equation becomes

$$4(2w + 1)^2 - 16(2w + 1) + 15 = 0$$

$$4u^2 - 16u + 15 = 0 \qquad \text{Substitute } u \text{ for } 2w + 1.$$

Now we can factor and solve.

$$(2u - 3)(2u - 5) = 0$$

$$2u - 3 = 0 \quad \text{or} \quad 2u - 5 = 0$$

$$2u = 3 \qquad\qquad 2u = 5$$

$$u = \frac{3}{2} \qquad\qquad u = \frac{5}{2}$$

We are not finished. Since the variable in the original equation is w, we must solve for w, not u. Therefore, we substitute back $2w + 1$ for u and solve for w.

$$u = \frac{3}{2} \qquad\qquad u = \frac{5}{2}$$

$$2w + 1 = \frac{3}{2} \qquad\qquad 2w + 1 = \frac{5}{2} \qquad \text{Substitute } 2w + 1 \text{ for } u.$$

$$2w = \frac{1}{2} \qquad\qquad 2w = \frac{3}{2}$$

$$w = \frac{1}{4} \qquad\qquad w = \frac{3}{4}$$

NOW TRY EXERCISE 17 A check will show that both $\frac{1}{4}$ and $\frac{3}{4}$ are solutions to the original equation.

EXAMPLE 3 Find the x-intercepts of the function $f(x) = 2x^{-2} + x^{-1} - 1$.

Solution The x-intercepts occur where $f(x) = 0$. Therefore, to find the x-intercepts we must solve the equation

$$2x^{-2} + x^{-1} - 1 = 0$$

This equation can be expressed as

$$2(x^{-1})^2 + x^{-1} - 1 = 0$$

When we let $u = x^{-1}$, the equation becomes

$$2u^2 + u - 1 = 0$$

$$(2u - 1)(u + 1) = 0$$

$$2u - 1 = 0 \quad \text{or} \quad u + 1 = 0$$
$$u = \frac{1}{2} \qquad\qquad u = -1$$

Now we substitute x^{-1} for u.

$$x^{-1} = \frac{1}{2} \quad \text{or} \quad x^{-1} = -1$$
$$\frac{1}{x} = \frac{1}{2} \qquad\qquad \frac{1}{x} = -1$$
$$x = 2 \qquad\qquad x = -1$$

A check will show that both 2 and -1 are solutions to the original equation. Thus, the x-intercepts are $(2, 0)$ and $(-1, 0)$.

The equation in Example 3 could also be expressed as

$$\frac{2}{x^2} + \frac{1}{x} - 1 = 0$$

A second method to solve this equation is to multiply both sides of the equation by the least common denominator, x^2, then simplify.

$$x^2\left(\frac{2}{x^2} + \frac{1}{x} - 1\right) = x^2 \cdot 0$$
$$2 + x - x^2 = 0$$
$$x^2 - x - 2 = 0$$
$$(x - 2)(x + 1) = 0$$
$$x - 2 = 0 \quad \text{or} \quad x + 1 = 0$$
$$x = 2 \qquad\qquad x = -1$$

Many of the equations solved in this section may be solved in more than one way.

NOW TRY EXERCISE 39

2) Solve Equations with Rational Exponents

When solving equations that are quadratic in form with rational exponents, we raise both sides of the equation to some power to eliminate the rational exponents. Recall that we did this in Section 7.6 when we solved radical equations. Whenever you raise both sides of an equation to a power, you may introduce extraneous solutions. **Therefore, whenever you raise both sides of an equation to a power, you must check all apparent solutions in the original equation to make sure that none are extraneous.** We will now work two examples showing how to solve equations that contain rational exponents. We use the same procedure as used earlier.

EXAMPLE 4 Solve the equation $x^{2/5} + x^{1/5} - 6 = 0$.

Solution This equation can be rewritten as

$$\left(x^{1/5}\right)^2 + x^{1/5} - 6 = 0$$

Let $u = x^{1/5}$. Then the equation becomes

$$u^2 + u - 6 = 0$$
$$(u + 3)(u - 2) = 0$$
$$u + 3 = 0 \quad \text{or} \quad u - 2 = 0$$
$$u = -3 \qquad\qquad u = -2$$

Now substitute $x^{1/5}$ for u and raise both sides of the equation to the fifth power to remove the rational exponents.

$$x^{1/5} = -3 \quad \text{or} \quad x^{1/5} = 2$$
$$\left(x^{1/5}\right)^5 = (-3)^5 \qquad \left(x^{1/5}\right)^5 = 2^5$$
$$x = -243 \qquad\qquad x = 32$$

The two *possible* solutions are −243 and 32. Remember that whenever you raise both sides of an equation to a power, as you did here, you need to check for extraneous solutions.

CHECK:

$x = -243$

$$x^{2/5} + x^{1/5} - 6 = 0$$
$$(-243)^{2/5} + (-243)^{1/5} - 6 \stackrel{?}{=} 0$$
$$\left(\sqrt[5]{-243}\right)^2 + \sqrt[5]{-243} - 6 \stackrel{?}{=} 0$$
$$(-3)^2 - 3 - 6 \stackrel{?}{=} 0$$
$$9 - 9 = 0$$

True

$x = 32$

$$x^{2/5} + x^{1/5} - 6 = 0$$
$$(32)^{2/5} + (32)^{1/5} - 6 \stackrel{?}{=} 0$$
$$\left(\sqrt[5]{32}\right)^2 + \sqrt[5]{32} - 6 \stackrel{?}{=} 0$$
$$2^2 + 2 - 6 \stackrel{?}{=} 0$$
$$4 + 2 - 6 = 0$$

True

NOW TRY EXERCISE 41

Since both values check, the solutions are −243 and 32.

EXAMPLE 5 Solve the equation $2p - \sqrt{p} - 10 = 0$.

Solution We can express this equation as

$$2p - p^{1/2} - 10 = 0$$
$$2\left(p^{1/2}\right)^2 - p^{1/2} - 10 = 0$$

If we let $u = p^{1/2}$, this equation is quadratic in form.

$$2u^2 - u - 10 = 0$$
$$(2u - 5)(u + 2) = 0$$
$$2u - 5 = 0 \quad \text{or} \quad u + 2 = 0$$
$$2u = 5 \qquad\qquad u = -2$$
$$u = \frac{5}{2}$$

However, since our original equation is in the variable p we must solve for p. We substitute $p^{1/2}$ for u.

$$p^{1/2} = \frac{5}{2} \qquad\qquad p^{1/2} = -2$$

Now we square both sides of the equation.

$$(p^{1/2})^2 = \left(\frac{5}{2}\right)^2 \qquad (p^{1/2})^2 = (-2)^2$$

$$p = \frac{25}{4} \qquad p = 4$$

We must now check both apparent solutions in the original equation.

CHECK:

$$p = \frac{25}{4} \qquad p = 4$$

$$2p - \sqrt{p} - 10 = 0 \qquad 2p - \sqrt{p} - 10 = 0$$

$$2\left(\frac{25}{4}\right) - \sqrt{\frac{25}{4}} - 10 \stackrel{?}{=} 0 \qquad 2(4) - \sqrt{4} - 10 \stackrel{?}{=} 0$$

$$\frac{25}{2} - \frac{5}{2} - 10 \stackrel{?}{=} 0 \qquad 8 - 2 - 10 \stackrel{?}{=} 0$$

$$0 = 0 \quad \textit{True} \qquad -4 = 0 \quad \textit{False}$$

NOW TRY EXERCISE 13

Since 4 does not check, it is an extraneous solution. The only solution is $\frac{25}{4}$.

Example 5 could also be solved by writing the equation as $\sqrt{p} = 2p - 10$ and squaring both sides of the equation. Try this now. If you have forgotten how to do this, review Section 7.6.

Exercise Set 8.4

Concept/Writing Exercises

1. Explain how you can determine whether a given equation can be expressed as an equation that is quadratic in form.

2. When solving an equation that is quadratic in form, when is it essential to check your answer for extraneous solutions? Explain why.

Practice the Skills

Solve each equation.

3. $x^4 - 10x^2 + 9 = 0$
4. $x^4 - 13x^2 + 36 = 0$
5. $x^4 - 7x^2 + 12 = 0$
6. $4x^4 - 17x^2 + 4 = 0$
7. $r^4 - 8r^2 = -15$
8. $9x^4 = 57x^2 - 18$
9. $x^4 + 2x^2 = 8$
10. $-r^4 - 4r^2 + 5 = 0$
11. $x + 4\sqrt{x} - 12 = 0$
12. $r = -\sqrt{r} + 2$
13. $x + \sqrt{x} = 6$
14. $2a - 7\sqrt{a} = 30$
15. $9x + 3\sqrt{x} = 2$
16. $2s + 6\sqrt{s} = 8$
17. $(x + 3)^2 + 2(x + 3) = 24$
18. $2(x + 1)^2 - 5(x + 1) - 3 = 0$
19. $6(x - 2)^2 = -19(x - 2) - 10$
20. $2(x + 2)^2 + 5(x + 2) - 3 = 0$
21. $(x^2 - 1)^2 - (x^2 - 1) - 6 = 0$
22. $(x^2 - 5)^2 + 3(x^2 - 5) - 10 = 0$
23. $18(x^2 - 5)^2 + 27(x^2 - 5) + 10 = 0$
24. $28(x^2 - 8)^2 - 23(x^2 - 8) - 15 = 0$
25. $r^{-2} + 6r^{-1} + 9 = 0$
26. $x^{-2} - x^{-1} = 20$
27. $2b^{-2} = 7b^{-1} - 3$

28. $6m^{-2} - m^{-1} - 12 = 0$

29. $x^{-2} + 6x^{-1} - 16 = 0$

30. $p^{-2} = -9p^{-1} + 10$

31. $x^{2/3} - 5x^{1/3} + 6 = 0$

32. $x^{2/3} + 11x^{1/3} + 28 = 0$

33. $c^{2/5} - 9c^{1/5} = -18$

34. $y^{2/5} + y^{1/5} + 16 = 0$

35. $-2a - 5a^{1/2} + 3 = 0$

36. $r^{2/3} - 7r^{1/3} + 6 = 0$

Find all x-intercepts of each function.

37. $f(x) = x - 5\sqrt{x} + 4$

38. $f(x) = 3x - 17\sqrt{x} + 10$

39. $g(x) = 4x^{-2} - 19x^{-1} - 5$

40. $g(x) = 4x^{-2} + 12x^{-1} + 9$

41. $f(x) = x^{2/3} + x^{1/3} - 6$

42. $f(x) = x^{1/2} + 4x^{1/4} - 5$

43. $g(x) = (x^2 - 3x)^2 + 2(x^2 - 3x) - 24$

44. $g(x) = (x^2 - 6x)^2 - 5(x^2 - 6x) - 24$

Problem Solving

45. Give a general procedure for solving an equation of the form $ax^4 + bx^2 + c = 0$.

46. Give a general procedure for solving an equation of the form $ax^{2n} + bx^n + c = 0$.

47. Write an equation of the form $ax^4 + bx^2 + c = 0$ that has solutions ± 2 and ± 4. Explain how you obtained your answer.

48. Write an equation of the form $ax^4 + bx^2 + c = 0$ that has solutions ± 3 and $\pm 2i$. Explain how you obtained your answer.

49. Is it possible for an equation of the form $ax^4 + bx^2 + c = 0$ to have exactly one real solution? Explain.

50. Is it possible for an equation of the form $ax^4 + bx^2 + c = 0$ to have exactly one imaginary solution? Explain.

51. Solve the equation $\frac{3}{x^2} - \frac{3}{x} = 60$ by

a) multiplying both sides of the equation by the LCD.

b) writing the equation with negative exponents.

52. Solve the equation $1 = \frac{2}{x} - \frac{2}{x^2}$ by

a) multiplying both sides of the equations by the LCD.

b) writing the equation with negative exponents.

Find all real solutions to each equation.

53. $15(r + 2) + 22 = -\frac{8}{r + 2}$

54. $2(p + 3) - 4 = \frac{3}{p + 3}$

55. $4 - (x - 2)^{-1} = 3(x - 2)^{-2}$

56. $3(x - 4)^{-2} = 16(x - 4)^{-1} + 12$

57. $x^6 - 9x^3 + 8 = 0$

58. $x^6 - 3x^3 - 40 = 0$

59. $(x^2 + 2x - 2)^2 - 7(x^2 + 2x - 2) + 6 = 0$

60. $(x^2 + 3x - 2)^2 - 10(x^2 + 3x - 2) + 16 = 0$

61. $2n^4 - 6n^2 - 3 = 0$

62. $3x^4 + 8x^2 - 1 = 0$

Cumulative Review Exercises

[2.2] **63.** Solve $P_1 = \frac{T_1 P_2}{T_2}$ for T_2.

[4.2] **64.** Solve the following system of equations.

$$a + 2b + 2c = 1$$

$$2a - b + c = 3$$

$$4a + b + 2c = 0$$

[5.2] **65. a)** Write expressions to represent each of the four areas shown in the figure.

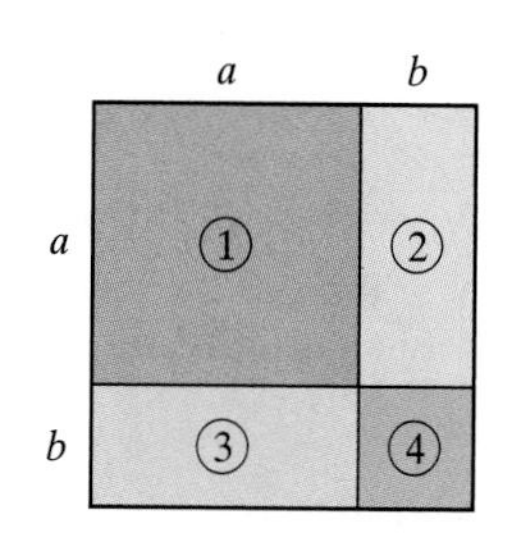

b) Express the total area shown as the square of a binomial.

[6.1] **66.** Simplify $\frac{6x^2 + 7x - 20}{4x^2 - 25}$.

8.5 GRAPHING QUADRATIC FUNCTIONS

1) Determine when a parabola opens upward or downward.
2) Find the axis of symmetry, vertex, and x-intercepts of a parabola.
3) Graph quadratic functions using the axis of symmetry, vertex, and intercepts.
4) Solve maximum and minimum problems.
5) Understand translations of parabolas.
6) Write functions in the form $f(x) = a(x - h)^2 + k$.

We graphed quadratic equations by plotting points in Section 3.2, and we had a brief discussion of the x-intercepts of the graphs of quadratic functions in Section 5.8. In this section we study the graphs of quadratic functions, called **parabolas**, in more depth. In objective 3 we will explain how to graph quadratic functions using the axis of symmetry, vertex, and intercepts. In objective 5 we will study patterns in the graphs of parabolas and use these patterns to determine translations, or shifts, that can be used to graph parabolas.

1) Determine When a Parabola Opens Upward or Downward

Parabolas have a shape that resembles, but are not the same as, the letter U. Parabolas may open upward or downward. For a quadratic function of the form $f(x) = ax^2 + bx + c$, the *sign* of the leading coefficient, a, determines whether a parabola opens upward or downward. *When $a > 0$, the parabola opens upward*, see Figure 8.6a. *When $a < 0$, the parabola opens downward*, see Figure 8.6b.

For a parabola that opens upward, the **vertex** is the lowest point on the curve. The minimum value of the function is the y-coordinate of the vertex. The minimum value is obtained when the x-coordinate of the vertex is substituted into the function. For a parabola that opens downward, the vertex is the highest point on the curve. The maximum value of the function is the y-coordinate of the vertex. The maximum value is obtained when the x-coordinate of the vertex is substituted into the function.

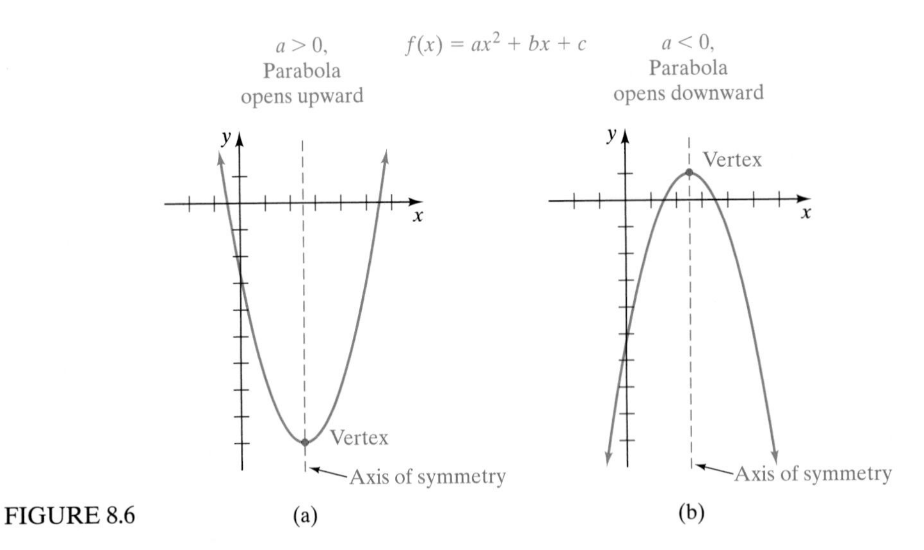

FIGURE 8.6

2) Find the Axis of Symmetry, Vertex, and x-Intercepts of a Parabola

Graphs of quadratic functions of the form $f(x) = ax^2 + bx + c$ will have **symmetry** about an imaginary line through the vertex. This means that if we fold the paper along this imaginary line, called the **axis of symmetry**, the right and left sides of the graph will coincide (see Fig. 8.6). We will now give the equation for finding the axis of symmetry.

> **To Find the Axis of Symmetry**
>
> For a function of the form $f(x) = ax^2 + bx + c$, the equation of the **axis of symmetry** of the parabola is
>
> $$x = -\frac{b}{2a}$$

Now we will derive the formula for the axis of symmetry, and find the coordinates of the vertex of a parabola, by beginning with a quadratic function of the form $f(x) = ax^2 + bx + c$ and completing the square on the first two terms.

$$f(x) = ax^2 + bx + c$$

$$= a\left(x^2 + \frac{b}{a}x\right) + c \qquad \textit{Factor out } a.$$

Half the coefficient of x is $\frac{b}{2a}$. Its square is $\frac{b^2}{4a^2}$. Add this term inside the parentheses and $-a\left(\frac{b^2}{4a^2}\right)$ outside the parentheses so that the sum added is 0.

$$f(x) = a\left[x^2 + \frac{b}{a}x + \left(\frac{b^2}{4a^2}\right)\right] - a\left(\frac{b^2}{4a^2}\right) + c$$

$$= a\left(x + \frac{b}{2a}\right)^2 - \frac{b^2}{4a} + c \qquad \textit{Replace the trinomial with the square of a binomial.}$$

$$= a\left(x + \frac{b}{2a}\right)^2 - \frac{b^2}{4a} + \frac{4ac}{4a} \qquad \textit{Write fractions with a common denominator.}$$

$$= a\left(x + \frac{b}{2a}\right)^2 + \frac{4ac - b^2}{4a} \qquad \textit{Combine the last two terms; write with the variable a first.}$$

$$= a\left[x - \left(-\frac{b}{2a}\right)\right]^2 + \frac{4ac - b^2}{4a}$$

The expression $\left[x - \left(-\frac{b}{2a}\right)\right]^2$ will always be greater than or equal to 0. (Why?) If $a > 0$, the parabola will open upward and have a minimum value. The minimum value will occur when $x = -\frac{b}{2a}$. If $a < 0$, the parabola will open downward and have a maximum value. The maximum value will occur when $x = -\frac{b}{2a}$. To determine the lowest, or highest, point on a parabola, substitute $-\frac{b}{2a}$ for x in the function to find y. The resulting ordered pair will be the vertex of the parabola. Since the axis of symmetry is the vertical line through the vertex, its equation is found using the x-coordinate of the ordered pair. Thus the equation of the axis of symmetry is $x = -\dfrac{b}{2a}$. Note that when $x = -\dfrac{b}{2a}$, the value of $f(x)$ is $\dfrac{4ac - b^2}{4a}$. Do you know why?

To Find the Vertex of a Parabola

The parabola represented by the function $f(x) = ax^2 + bx + c$ will have axis of symmetry $x = -\dfrac{b}{2a}$ and vertex

$$\left(-\frac{b}{2a}, \frac{4ac - b^2}{4a}\right)$$

Since we often find the y-coordinate of the vertex by substituting the x-coordinate of the vertex into $f(x)$, the vertex may also be designated as

$$\left(-\frac{b}{2a}, f\left(-\frac{b}{2a}\right)\right)$$

The parabola given by the function $f(x) = ax^2 + bx + c$ will open upward when a is greater than 0 and open downward when a is less than 0.

Recall that to find the x-intercept of the graph of $f(x) = ax^2 + bx + c$, we set $f(x) = 0$ and solve the equation

$$ax^2 + bx + c = 0.$$

This equation may be solved by factoring, the quadratic formula, or completing the square.

As we mentioned in Section 8.2 the discriminant, $b^2 - 4ac$, may be used to determine the *number of x-intercepts*. The following table summarizes information about the discriminant.

Discriminant, $b^2 - 4ac$	Number of x-Intercepts	Possible graphs of $f(x) = ax^2 + bx + c$
>0	Two	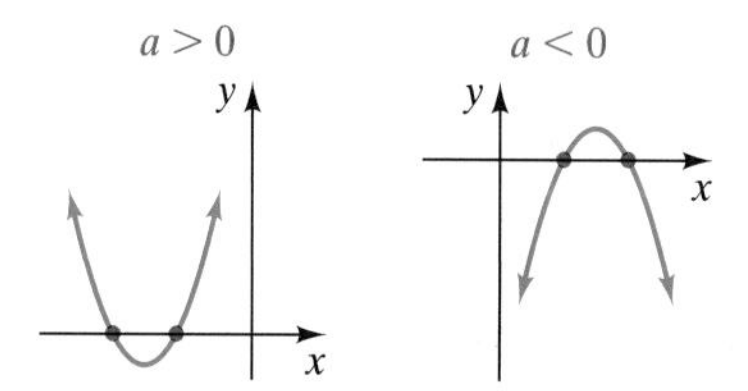 $a > 0$, $a < 0$ (x, y)
$=0$	One	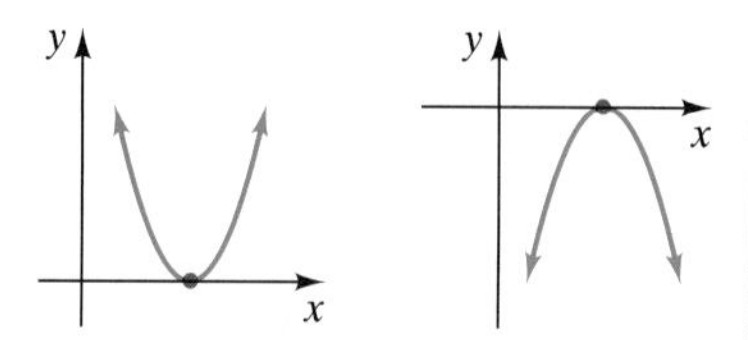 (x, y)
<0	None	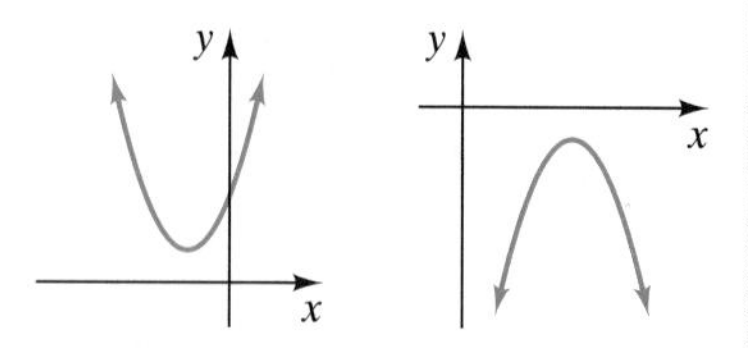 (x, y)

3) Graph Quadratic Functions Using the Axis of Symmetry, Vertex, and Intercepts

Now we will draw graphs of quadratic functions.

EXAMPLE 1 Consider the equation $y = -x^2 + 8x - 12$.

a) Determine whether the parabola opens upward or downward.

b) Find the y-intercept.

c) Find the vertex.

d) Find the x-intercepts, if any.

e) Draw the graph.

Solution **a)** Since a is -1, which is less than 0, the parabola opens downward.

b) To find the y-intercept, set $x = 0$ and solve for y.

$$y = -(0)^2 + 8(0) - 12 = -12$$

The y-intercept is $(0, -12)$.

c)

$$x = -\frac{b}{2a} = -\frac{8}{2(-1)} = 4$$

$$y = \frac{4ac - b^2}{4a} = \frac{4(-1)(-12) - 8^2}{4(-1)} = \frac{48 - 64}{-4} = 4$$

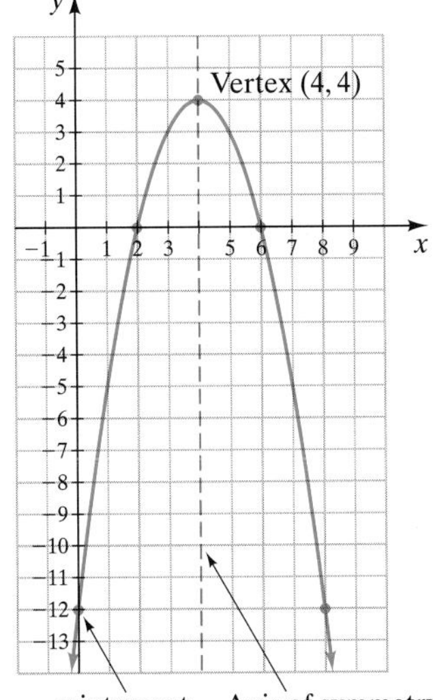

FIGURE 8.7

The vertex is at $(4, 4)$. The y-coordinate of the vertex could also be found by substituting 4 for x in the function and finding the corresponding value of y, which is 4.

d) To find the x-intercepts, set $y = 0$.

$$0 = -x^2 + 8x - 12$$

or

$$x^2 - 8x + 12 = 0$$

$$(x - 6)(x - 2) = 0$$

$$x - 6 = 0 \quad \text{or} \quad x - 2 = 0$$

$$x = 6 \quad \text{or} \quad x = 2$$

Thus, the x-intercepts are $(2, 0)$ and $(6, 0)$. These values could also be found by the quadratic formula (or by completing the square).

NOW TRY EXERCISE 13

e) Use all this information to draw the graph (Fig. 8.7).

If you obtain irrational values when finding x-intercepts by the quadratic formula, use your calculator to estimate these values, then plot these decimal values. For example, if you obtain $x = (2 \pm \sqrt{10})/2$, you would evaluate

$(2 + \sqrt{10})/2$ and $(2 - \sqrt{10})/2$ on your calculator and obtain 2.58 and −0.58, respectively, to the nearest hundredth. The x-intercepts would therefore be (2.58, 0) and (−0.58, 0).

EXAMPLE 2 Consider the function $f(x) = 2x^2 + 3x + 4$.

a) Determine whether the parabola opens upward or downward.

b) Find the y-intercept.

c) Find the vertex.

d) Find the x-intercepts, if any.

e) Draw the graph.

Solution **a)** Since a is 2, which is greater than 0, the parabola opens upward.

b) Since $f(x)$ is the same as y, to find the y-intercept, set $x = 0$ and solve for $f(x)$, or y.

$$f(x) = 2(0)^2 + 3(0) + 4 = 4.$$

The y-intercept is (0, 4).

c) $x = -\frac{b}{2a} = -\frac{3}{2(2)} = -\frac{3}{4}$

$$y = \frac{4ac - b^2}{4a} = \frac{4(2)(4) - 3^2}{4(2)} = \frac{32 - 9}{8} = \frac{23}{8}$$

The vertex is $\left(-\frac{3}{4}, \frac{23}{8}\right)$. The y-coordinate of the vertex can also be found by evaluating $f\left(-\frac{3}{4}\right)$.

d) To find the x-intercepts, set $f(x) = 0$.

$$0 = 2x^2 + 3x + 4$$

This trinomial cannot be factored. To determine whether this equation has any real solutions, evaluate the discriminant.

$$b^2 - 4ac = 3^2 - 4(2)(4) = 9 - 32 = -23$$

Since the discriminant is less than 0, this equation has no real solutions. You should have expected this answer because the y-coordinate of the vertex is a positive number and therefore above the x-axis. Since the parabola opens upward, it cannot intersect the x-axis.

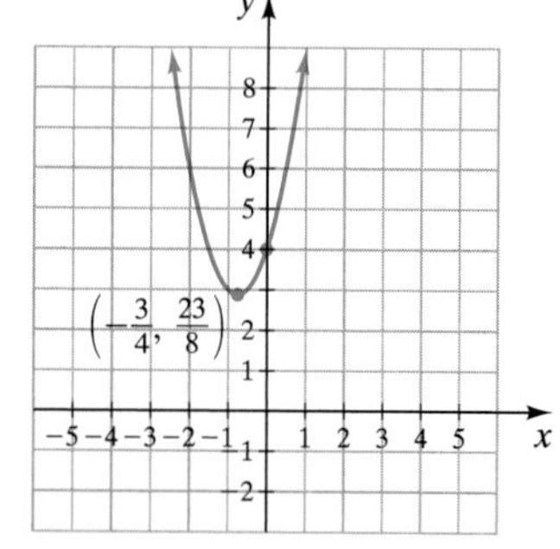

FIGURE 8.8

NOW TRY EXERCISE 31

e) The graph is given in Figure 8.8.

4 Solve Maximum and Minimum Problems

A parabola that opens upward has a **minimum value** at its vertex, as illustrated in Figure 8.9a. A parabola that opens downward has a **maximum value** at its vertex, as shown in Figure 8.9b. If you are given a function of the form $f(x) = ax^2 + bx + c$, the maximum or minimum value will occur at $-\frac{b}{2a}$, and the value will be $\dfrac{4ac - b^2}{4a}$. There are many real-life problems that require finding maximum and minimum values.

$y = ax^2 + bx + c$

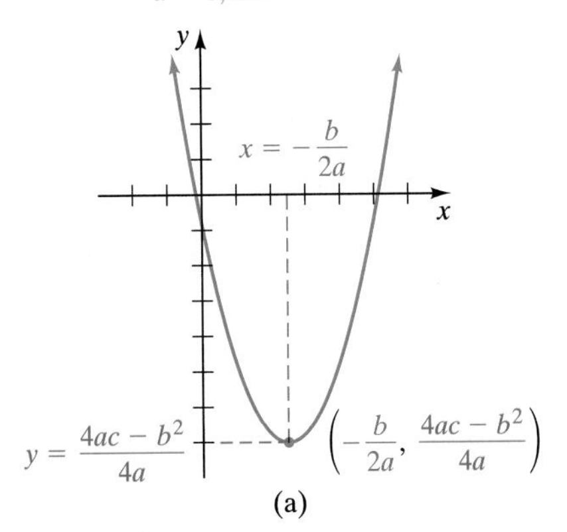

(a)

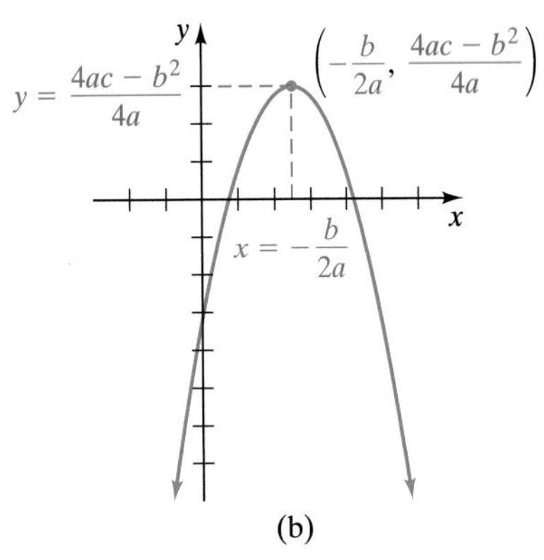

(b)

FIGURE 8.9

EXAMPLE 3 Each year in Gettysburg, Pennsylvania, a group of performers reenact the battle of Gettysburg that occurred in the Civil War. In the battle, many cannons were fired. If a cannon was fired from a height of 9.8 meters above the ground, at a certain angle, the height of the cannonball above the ground, h, in meters, at time, t, in seconds, is found by the function (see Figure 8.10).

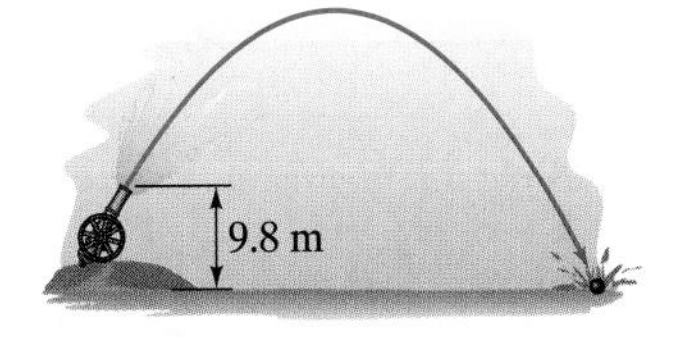

FIGURE 8.10

$$h(t) = -4.9t^2 + 24.5t + 9.8$$

a) Find the maximum height attained by the cannonball.

b) Find the time it takes for the cannonball to reach its maximum height.

c) Find the time it takes for the cannonball to strike the ground.

Solution **a) Understand** The cannonball will follow the path of a parabola that opens downward ($a < 0$). The cannonball will rise to a maximum height, then begin its fall back to the ground due to gravity. To find the maximum height, we use the formula $y = \dfrac{4ac - b^2}{4a}$.

Translate

$$y = \frac{4ac - b^2}{4a}$$

$$= \frac{4(-4.9)(9.8) - (24.5)^2}{4(-4.9)}$$

Carry Out

$$= \frac{-192.08 - 600.25}{-19.6}$$

$$= \frac{-792.33}{-19.6}$$

$$= 40.425 \text{ meters}$$

Answer Thus the maximum height obtained by the cannonball is about 40.4 meters.

b) The cannonball reaches its maximum height at

$$t = -\frac{b}{2a} = -\frac{24.5}{2(-4.9)} = 2.5 \text{ seconds}$$

c) Understand and Translate When the cannonball strikes the ground, its distance from the ground, $h(t)$, is 0. Thus to determine when the cannonball strikes the ground, we solve the equation

$$-4.9t^2 + 24.5t + 9.8 = 0$$

We use the quadratic formula with $a = -4.9$, $b = 24.5$, and $c = 9.8$.

Carry Out

$$t = \frac{-b \pm \sqrt{b^2 - 4ac}}{2a}$$

$$= \frac{-24.5 \pm \sqrt{(24.5)^2 - 4(-4.9)(9.8)}}{2(-4.9)}$$

$$= \frac{-24.5 \pm \sqrt{792.33}}{-9.8}$$

$$t \approx \frac{-24.5 - 28.15}{-9.8} \quad \text{or} \quad t \approx \frac{-24.5 + 28.15}{-9.8}$$

$$t \approx 5.37 \text{ sec} \qquad\qquad t \approx -0.37 \text{ sec}$$

Answer The only acceptable time is 5.37 seconds. Thus the cannonball strikes the ground in about 5.37 seconds. Notice in part **b)** that the time it takes the cannonball to reach its maximum height, 2.5 seconds, is not half of the total time the cannonball was in flight, 5.37 seconds. This is because the cannonball was fired from a height of 9.8 meters and not ground level.

EXAMPLE 4 John W. Brown is building a corral in the shape of a rectangle for newborn calves (see Fig. 8.11). If he has 100 meters of fencing, find the dimensions of the corral that will give the greatest area.

Solution **Understand** We are given the perimeter of the corral, 100 meters. The formula for the perimeter of a rectangle is $P = 2l + 2w$. For this problem, $100 = 2l + 2w$. We are asked to maximize the area, A, where

$$A = l \cdot w$$

FIGURE 8.11

We need to express the area in terms of one variable, not two. To express the area in terms of l, we solve $100 = 2l + 2w$ for w, then make a substitution.

Translate

$$100 = 2l + 2w$$

$$100 - 2l = 2w$$

$$50 - l = w$$

Carry Out Now we substitute $w = 50 - l$ in $A = l \cdot w$. This gives

$$A = lw$$
$$A = l(50 - l)$$
$$A = -l^2 + 50l$$

In this quadratic $a = -1$, $b = 50$, and $c = 0$. The maximum area will occur at

$$l = -\frac{b}{2a} = -\frac{50}{2(-1)} = 25$$

Answer The length that will give the largest area is 25 meters. The width, $w = 50 - l$, will also be 25 meters. Thus a square with dimensions 25 by 25 meters will give the largest area.

The largest area can be found by substituting $l = 25$ in the formula $A = l(50 - l)$ or by using $A = \frac{4ac - b^2}{4a}$. In either case, we obtain an area of 625 square meters.

In Example 4 when we obtained the equation $A = -l^2 + 50l$, we could have completed the square as follows:

$$\begin{aligned} A &= -(l^2 - 50l) \\ &= -(l^2 - 50l + 625) + 625 \\ &= -(l - 25)^2 + 625 \end{aligned}$$

From this equation we can determine that the maximum area, 625 square meters, occurs when the length is 25 meters.

NOW TRY EXERCISE 73

5 Understand Translations of Parabolas

Now we will look at another method used to graph parabolas. With this method, you start with a graph of the form $f(x) = ax^2$ and **translate**, or shift, the position of the graph to obtain the graph of the function you are seeking. As a reference, Figure 8.12 shows the graphs of $f(x) = x^2$, $g(x) = 2x^2$, and $h(x) = \frac{1}{2}x^2$. Figure 8.13 shows the graphs of $f(x) = -x^2$, $g(x) = -2x^2$, and $h(x) = -\frac{1}{2}x^2$.

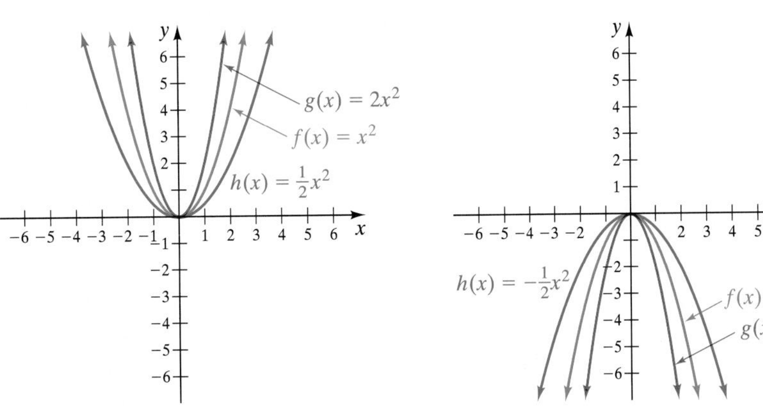

FIGURE 8.12

FIGURE 8.13

You can verify that each of the graphs is correct by plotting points. Notice that in Figures 8.12 and 8.13 the *value of a* in $f(x) = ax^2$ determines the width of the parabola. As $|a|$ gets larger, the parabola gets narrower and as $|a|$ gets smaller, the parabola gets wider.

Now let's consider the three functions $f(x) = x^2$, $g(x) = (x - 2)^2$, and $h(x) = (x + 2)^2$. These functions are graphed in Figure 8.14. (You can verify that these are graphs of the three functions by plotting points.)

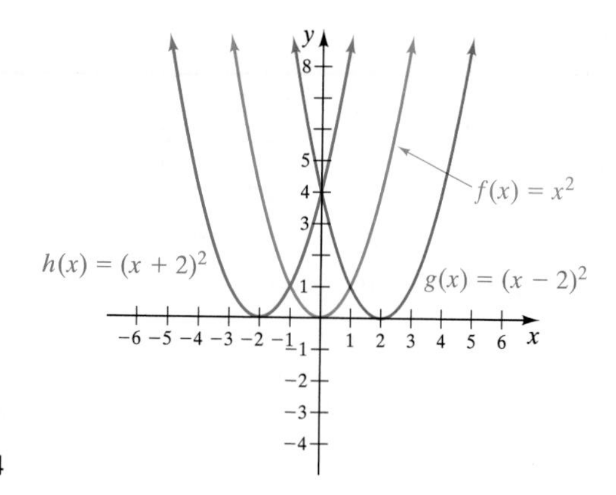

FIGURE 8.14

Notice that the graphs of $g(x)$ and $h(x)$ are identical to the graph of $f(x)$ except that $g(x)$ has been translated, or shifted, 2 units to the right and $h(x)$ has been translated 2 units to the left. In general, *the graph of* $g(x) = a(x - h)^2$ *is the graph of* $f(x) = ax^2$ *shifted h units to the right if h is a positive real number, and* $|h|$ *units to the left if h is a negative real number.*

Now consider the graphs of $f(x) = x^2$, $g(x) = x^2 + 3$ and $h(x) = x^2 - 3$ that are illustrated in Figure 8.15. You can verify that these are graphs of the three functions by plotting points.

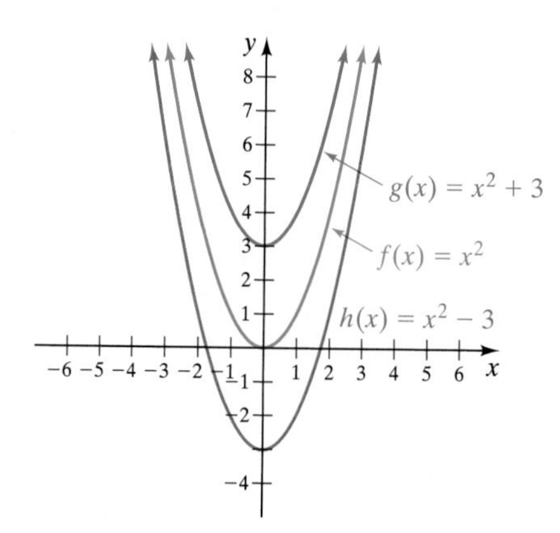

FIGURE 8.15

Notice that the graphs of $g(x)$ and $h(x)$ are identical to the graph of $f(x)$ except that $g(x)$ is translated up 3 units, and $h(x)$ is translated down 3 units. In general, *the graph of* $g(x) = ax^2 + k$ *is the graph of* $f(x) = ax^2$ *shifted k units up if k is a positive real number, and* $|k|$ *units down if k is a negative real number.*

Now consider the graphs of $f(x) = x^2$ and $g(x) = (x - 2)^2 + 3$, shown in Figure 8.16.

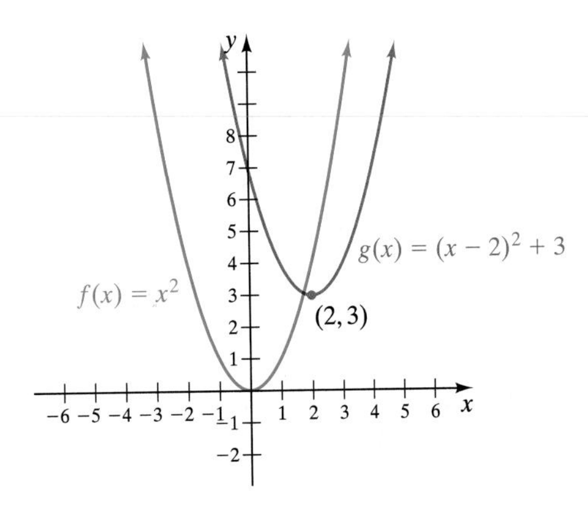

FIGURE 8.16

Notice the graph of $g(x)$ has the same general shape as that of $f(x)$. The graph of $g(x)$ is the graph of $f(x)$ translated 2 units to the right and 3 units up. This graph and the discussion preceding it lead to the following important fact.

For any function $f(x) = ax^2$, the graph of $g(x) = a(x - h)^2 + k$ will have the same shape as the graph of $f(x)$. The graph of $g(x)$ will be the graph of $f(x)$ shifted as follows:

- If h is a positive real number, the graph will be shifted h units to the right.
- If h is a negative real number, the graph will be shifted $|h|$ units to the left.
- If k is a positive real number, the graph will be shifted k units up.
- If k is a negative real number, the graph will be shifted $|k|$ units down.

Examine the graph of $g(x) = (x - 2)^2 + 3$ in Figure 8.16. Notice that its axis of symmetry is $x = 2$ and its vertex is $(2, 3)$.

The graph of any function of the form

$$f(x) = a(x - h)^2 + k$$

will be a parabola with axis of symmetry $x = h$ and vertex at (h, k).

Example	Axis of Symmetry	Vertex	Parabola Opens
$f(x) = 2(x - 4)^2 + 5$	$x = 4$	$(4, 5)$	upward, $a > 0$
$f(x) = -\frac{1}{2}(x - 6)^2 - 4$	$x = 6$	$(6, -4)$	downward, $a < 0$

Now consider $f(x) = 2(x + 5)^2 + 3$. We can rewrite this as $f(x) = 2[x - (-5)]^2 + 3$. Therefore, h has a value of -5 and k has a value of 3. The graph of this function has axis of symmetry $x = -5$ and vertex at $(-5, 3)$.

Example	Axis of Symmetry	Vertex	Parabola Opens
$f(x) = 3(x + 4)^2 - 2$	$x = -4$	$(-4, -2)$	upward, $a > 0$
$f(x) = -\frac{1}{2}(x + \frac{1}{3})^2 + \frac{1}{4}$	$x = -\frac{1}{3}$	$(-\frac{1}{3}, \frac{1}{4})$	downward, $a < 0$

Now we are ready to graph parabolas using translations.

EXAMPLE 5 The graph of $f(x) = -2x^2$ is illustrated in Figure 8.17. Using this graph as a guide, graph $g(x) = -2(x + 3)^2 - 4$.

Solution The function $g(x)$ may be written $g(x) = -2[x - (-3)]^2 - 4$. Therefore in the function, h has a value of -3 and k has a value of -4. The graph of $g(x)$ will therefore be the graph of $f(x)$ translated 3 units to the left (because $h = -3$) and 4 units down (because $k = -4$). The graphs of $f(x)$ and $g(x)$ are illustrated in Figure 8.18.

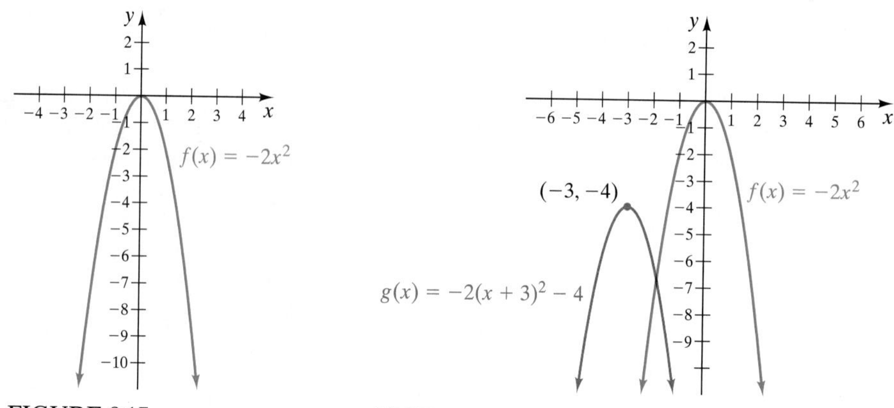

NOW TRY EXERCISE 37

FIGURE 8.17

FIGURE 8.18

In objective 2, we started with a function of the form $f(x) = ax^2 + bx + c$ and completed the square to obtain

$$f(x) = a\left[x - \left(-\frac{b}{2a}\right)\right]^2 + \frac{4ac - b^2}{4a}$$

We stated that the vertex of the parabola of this function was $\left(-\frac{b}{2a}, \frac{4ac - b^2}{4a}\right)$. Suppose we substitute h for $-\frac{b}{2a}$ and k for $\frac{4ac - b^2}{4a}$ in the function. We then get

$$f(x) = a(x - h)^2 + k$$

which we know is a parabola with vertex at (h, k). Therefore, both functions $f(x) = ax^2 + bx + c$ and $f(x) = a(x - h)^2 + k$ yield the same vertex and axis of symmetry for any given function.

6 Write Functions in the Form $f(x) = a(x - h)^2 + k$

If we wish to graph parabolas using translations, we need to change the form of a function from $f(x) = ax^2 + bx + c$ to $f(x) = a(x - h)^2 + k$. To do this, we *complete the square* as was discussed in Section 8.1. By completing the square we obtain a perfect square trinomial, which we can represent as the square of a binomial. Examples 6 and 7 explain the procedure. We will use this procedure again in Chapter 10, when we discuss conic sections.

EXAMPLE 6 Given $f(x) = x^2 - 4x + 7$.

a) Write $f(x)$ in the form $f(x) = a(x - h)^2 + k$.

b) Graph $f(x)$.

Solution **a)** We use the x^2 and $-4x$ terms to obtain a perfect square trinomial.

$$f(x) = (x^2 - 4x) + 7$$

Now we take half the coefficient of the x-term and square it.

$$\left[\frac{1}{2}(-4)\right]^2 = 4$$

We then add this value, 4, within the parentheses. Since we are adding 4 within parentheses, we add -4 outside parentheses. Adding 4 and -4 to an expression is the same as adding 0, which does not change the value of the expression.

$$f(x) = (x^2 - 4x + 4) - 4 + 7$$

By doing this we have created a perfect square trinomial within the parentheses, plus a constant outside the parentheses. We express the perfect square trinomial as the square of a binomial.

$$f(x) = (x - 2)^2 + 3$$

The function is now in the form we are seeking.

b) Since $a = 1$, which is greater than 0, the parabola opens upward. The axis of symmetry of the parabola is $x = 2$ and the vertex is at $(2, 3)$. The y-intercept can be easily obtained by substituting $x = 0$ and finding $f(x)$. When $x = 0$, $f(x) = (-2)^2 + 3 = 7$. Thus the y-intercept is at 7. By plotting the vertex, y-intercept, and a few other points, we obtain the graph in Figure 8.19. The figure also shows the graph of $y = x^2$ for comparison.

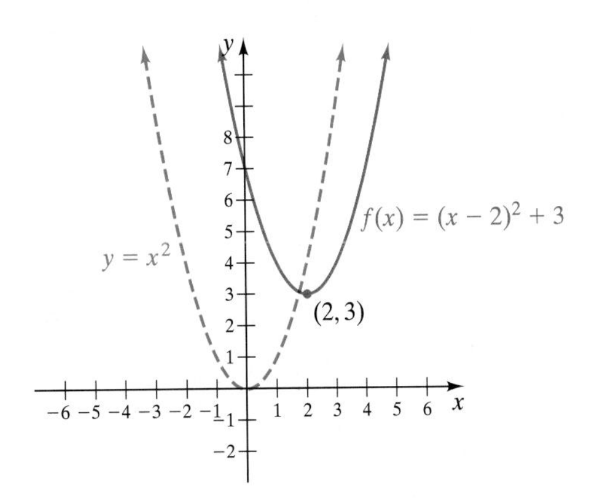

FIGURE 8.19

EXAMPLE 7 Given $f(x) = -2x^2 - 10x - 13$.

a) Write $f(x)$ in the form $f(x) = a(x - h)^2 + k$.

b) Graph $f(x)$.

Solution **a)** When the leading coefficient is not 1, we factor out the leading coefficient from the terms containing the variable.

$$f(x) = -2(x^2 + 5x) - 13$$

Now we complete the square.

Half of coefficient of x-term squared

$$\left[\frac{1}{2}(5)\right]^2 = \frac{25}{4}$$

If we add $\frac{25}{4}$ within the parentheses, we are actually adding $-2(\frac{25}{4})$ or $-\frac{25}{2}$, since each term in parentheses is multiplied by -2. Therefore, to compensate, we must add $\frac{25}{2}$ outside the parentheses.

$$f(x) = -2\left(x^2 + 5x + \frac{25}{4}\right) + \frac{25}{2} - 13$$

$$= -2\left(x + \frac{5}{2}\right)^2 - \frac{1}{2}$$

b) Since $a = -2$, the parabola opens downward. The axis of symmetry is $x = -\frac{5}{2}$ and the vertex is $(-\frac{5}{2}, -\frac{1}{2})$. The y-intercept is at $f(0) = -13$. We plot a few points and draw the graph in Figure 8.20. In the figure we also show the graph of $y = -2x^2$ for comparison.

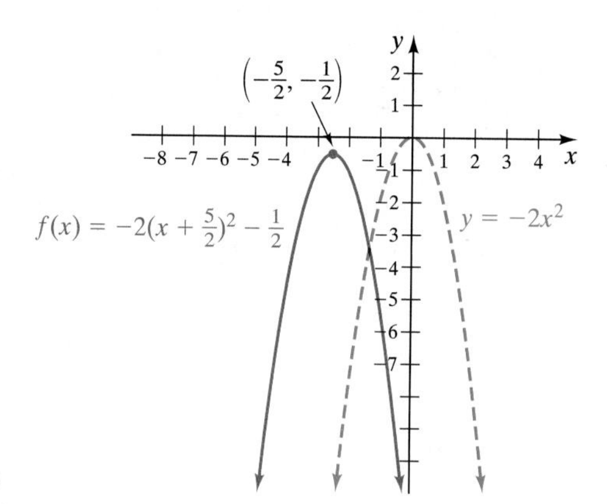

FIGURE 8.20

NOW TRY EXERCISE 51

Notice that $f(x) = -2(x + \frac{5}{2})^2 - \frac{1}{2}$ has no x-intercepts. Therefore there are no real values of x for which $f(x) = 0$.

A second way to change the equation from $f(x) = ax^2 + bx + c$ to $f(x) = a(x - h)^2 + k$ form is to let $h = -\dfrac{b}{2a}$ and $k = \dfrac{4ac - b^2}{4a}$. Find the values for h and k and then substitute the values obtained into $f(x) = a(x - h)^2 + k$. For example, for the function $f(x) = -2x^2 - 10x - 13$, in Example 7, $a = -2$, $b = -10$, and $c = -13$. Then

$$h = -\frac{b}{2a} = -\frac{-10}{2(-2)} = -\frac{5}{2}$$

$$k = \frac{4ac - b^2}{4a} = \frac{4(-2)(-13) - (-10)^2}{4(-2)} = -\frac{1}{2}$$

Therefore,

$$\begin{aligned} f(x) &= a(x-h)^2 + k \\ &= -2\left[x - \left(-\frac{5}{2}\right)\right]^2 - \frac{1}{2} \\ &= -2\left(x + \frac{5}{2}\right)^2 - \frac{1}{2} \end{aligned}$$

This answer checks with that obtained in Example 7.

Exercise Set 8.5

Concept/Writing Exercises

1. What is the graph of a quadratic equation called?
2. What is the vertex of a parabola?
3. What is the axis of symmetry of a parabola?
4. What is the equation of the axis of symmetry of the graph of $f(x) = ax^2 + bx + c$?
5. What is the vertex of the graph of $f(x) = ax^2 + bx + c$?
6. How many x-intercepts does a quadratic function have if the discriminant is **a)** <0, **b)** $=0$, **c)** >0?
7. For $f(x) = ax^2 + bx + c$, will $f(x)$ have a maximum or a minimum if **a)** $a > 0$, **b)** $a < 0$? Explain.
8. Explain how you may find the x-intercepts of the graph of a quadratic function.
9. Explain how you may find the y-intercept of the graph of a quadratic function.
10. Consider the graph of $f(x) = ax^2$. Explain how the shape of $f(x)$ changes as $|a|$ increases and as $|a|$ decreases.
11. Consider the graph of $f(x) = ax^2$. What is the general shape of $f(x)$ if **a)** $a > 0$, **b)** $a < 0$.
12. Will the graphs of $f(x) = ax^2$ and $g(x) = -ax^2$ have the same vertex for any nonzero real number a? Explain.

Practice the Skills

a) Determine whether the parabola opens upward or downward. b) Find the y-intercept. c) Find the vertex. d) Find the x-intercepts (if any). Use a calculator to find approximate values for the x-intercepts if they are irrational. e) Draw the graph.

13. $f(x) = x^2 + 8x + 15$
14. $f(x) = x^2 + 2x - 3$
15. $f(x) = x^2 - 6x + 4$
16. $f(x) = -x^2 + 4x - 5$
17. $f(x) = x^2 + 6x + 9$
18. $f(x) = -2x^2 + 4x - 8$
19. $y = 2x^2 - x - 6$
20. $y = -3x^2 + 6x - 9$
21. $y = 3x^2 + 4x + 3$
22. $f(x) = -3x^2 - 2x - 6$
23. $f(x) = -2x^2 - 6x + 4$
24. $y = x^2 + 4$
25. $y = x^2 + 4x$
26. $y = -x^2 + 6x$
27. $f(x) = 3x^2 + 10x$
28. $y = 3x^2 + 4x - 6$
29. $f(x) = -x^2 + 3x - 5$
30. $f(x) = -2x^2 - 6x + 5$
31. $f(x) = -4x^2 + 6x - 9$
32. $f(x) = -2x^2 + 5x + 4$

Using the graphs in Figures 8.12 through 8.15 as a guide, graph each function and label the vertex.

33. $f(x) = (x-3)^2$
34. $f(x) = (x+4)^2$
35. $f(x) = x^2 + 3$
36. $f(x) = x^2 - 4$
37. $f(x) = (x-2)^2 + 3$
38. $f(x) = (x-4)^2 - 2$
39. $g(x) = -(x+3)^2 - 2$
40. $g(x) = (x-1)^2 + 4$
41. $f(x) = (x+4)^2 + 4$
42. $h(x) = (x+4)^2 - 1$
43. $h(x) = -2(x+1)^2 - 3$
44. $y = -2(x-3)^2 + 1$
45. $y = -2(x-2)^2 + 2$
46. $f(x) = -(x-5)^2 + 2$

*In Exercises 47–56, **a)** Express each function in the form $f(x) = a(x - h)^2 + k$. **b)** Draw the graph of each function and label the vertex.*

47. $f(x) = x^2 - 6x + 8$ **48.** $f(x) = x^2 + 6x + 5$ **49.** $g(x) = x^2 - x - 12$ **50.** $g(x) = x^2 - 2x - 15$

51. $f(x) = -x^2 - 4x - 6$ **52.** $g(x) = x^2 - 4x - 5$ **53.** $h(x) = -x^2 + 4x - 12$ **54.** $f(x) = 2x^2 + 5x - 3$

55. $g(x) = -2x^2 - 9x - 9$ **56.** $f(x) = 2x^2 + 7x - 4$

Problem Solving

*Match the functions in Exercises 57–60 with the appropriate graphs labeled **a)** through **d)**.*

a)

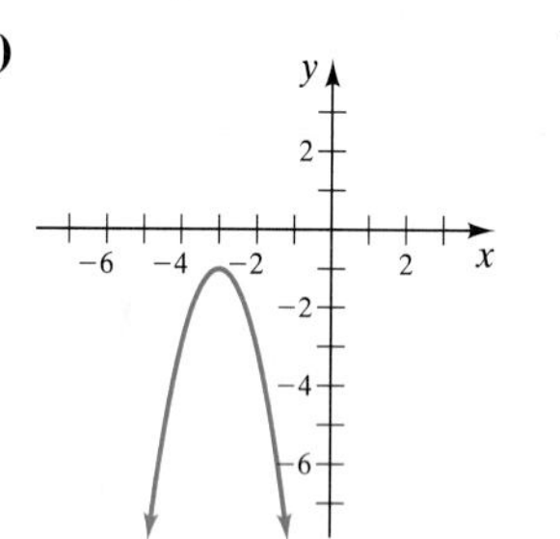

b)

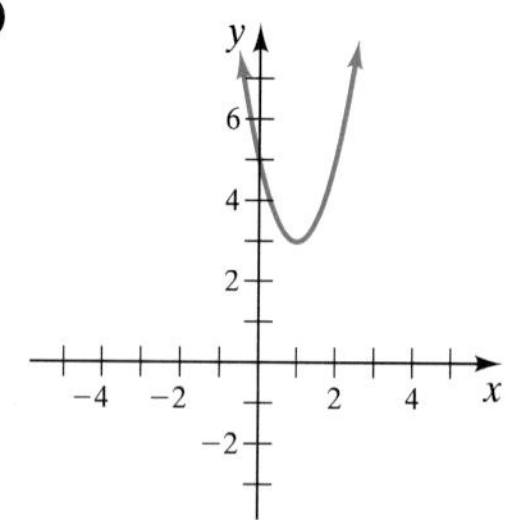

c) **d)**

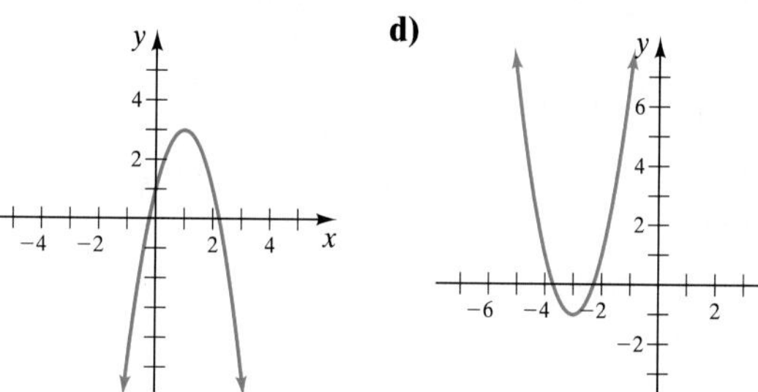

57. $f(x) = 2(x + 3)^2 - 1$

58. $f(x) = -2(x + 3)^2 - 1$

59. $f(x) = 2(x - 1)^2 + 3$

60. $f(x) = -2(x - 1)^2 + 3$

61. What is the distance between the vertices of the graphs of $f(x) = (x - 2)^2 + \frac{5}{2}$ and $g(x) = (x - 2)^2 - \frac{3}{2}$?

62. What is the distance between the vertices of the graphs of $f(x) = 2(x - 4)^2 - 3$ and $g(x) = -3(x - 4)^2 + 2$?

63. What is the distance between the vertices of the graphs of $f(x) = 2(x + 4)^2 - 3$ and $g(x) = -(x + 1)^2 - 3$?

64. What is the distance between the vertices of the graphs of $f(x) = -\frac{1}{3}(x - 3)^2 - 2$ and $g(x) = 2(x + 5)^2 - 2$?

65. Write the function that has the shape of $f(x) = 2x^2$ and a vertex at $(3, -2)$.

66. Write the function that has the shape of $f(x) = -\frac{1}{2}x^2$ and a vertex at $\left(\frac{2}{3}, -5\right)$.

67. Write the function that has the shape of $f(x) = -4x^2$ and a vertex at $\left(-\frac{3}{5}, -\sqrt{2}\right)$.

68. Write the function that has the shape of $f(x) = \frac{3}{5}x^2$ and a vertex at $(-\sqrt{3}, \sqrt{5})$.

69. Consider $f(x) = x^2 - 8x + 12$ and $g(x) = -x^2 + 8x - 12$.

a) Without graphing, can you explain how the graphs of the two functions compare?

b) Will the graphs have the same x-intercepts? Explain.

c) Will the graphs have the same vertex? Explain.

d) Graph both functions on the same axes.

70. By observing the leading coefficient in a quadratic function, and by determining the coordinates of the vertex of its graph, explain how you can determine the number of x-intercepts the parabola has.

71. The Riverside High School Theater Club is trying to set the price of tickets for a play. If the price is too low, they will not make enough money to cover expenses, and if the price is too high, not enough people will pay the price of a ticket. They estimate that their total income per concert, I, in hundreds of dollars, can be approximated by the formula

$$I = -x^2 + 24x - 44, 0 \le x \le 24$$

where x is the cost of a ticket.

a) Draw a graph of income versus the cost of a ticket.

b) Determine the minimum cost of a ticket for the theater club to break even.

c) Determine the maximum cost of a ticket that the theater club can charge and break even.

d) How much should they charge to receive the maximum income?

e) Find the maximum income.

72. An object is projected upward with an initial velocity of 192 feet per second. The object's distance above the ground, d, after t seconds may be found by the formula $d = -16t^2 + 192t$.

a) Find the object's distance above the ground after 3 seconds.

b) Draw a graph of distance versus time.

c) What is the maximum height the object will reach?

d) At what time will it reach its maximum height?

e) At what time will the object strike the ground?

73. The Holley-Berry Company earns a weekly profit according to the function $f(x) = -0.4x^2 + 80x - 200$, where x is the number of bird feeders built and sold.

a) Find the number of bird feeders that the company must sell in a week to obtain the maximum profit.

b) Find the maximum profit.

74. The Jane Nakote Company earns a weekly profit according to the function $f(x) = -1.2x^2 + 180x - 280$, where x is the number of rocking chairs built and sold.

a) Find the number of rocking chairs that the company must sell in a week to obtain the maximum profit.

b) Find the maximum profit.

75. Ramon Loomis throws a ball into the air with an initial velocity of 32 feet per second. The height of the ball at any time t is given by the formula $h = 32t - 16t^2$. At what time does the ball reach its maximum height? What is the maximum height?

76. When a baseball is thrown upward with a velocity of 64 feet per second from the top of a 160-foot-tall building, its distance from the ground, h, at any time, t, can be found by the formula $h = -16t^2 + 64t + 160$. Find the maximum height obtained by the baseball.

77. Susan Jackson is designing plans for her house. What is the maximum possible area of a room if its perimeter is to be 60 feet?

78. What are the dimensions of a garden that will have the greatest area if its perimeter is to be 70 feet?

79. What is the minimum product of two numbers that differ by 8? What are the numbers?

80. What is the minimum product of two numbers that differ by 10? What are the numbers?

81. What is the maximum product of two numbers that add to 40? What are the numbers?

82. What is the maximum product of two numbers that add to 9? What are the numbers?

83. The following graph from the February 4, 1998 issue of *USA Today* indicates that the percent of workers who feel insecure on the job is near an all-time high.

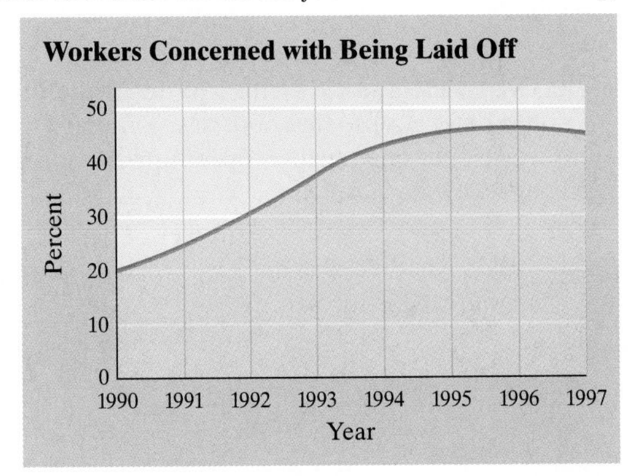

The function $p(x) = -0.007x^2 + 0.099x + 0.090$ can be used to estimate the percent of workers who feel insecure on the job, $p(x)$, where x is years since 1989 and $1 \le x \le 8$.

Use the function $p(x)$ to answer the following questions.

a) In what year was the percent of workers who felt insecure a maximum?

b) What was the maximum percent of workers that felt insecure on the job?

84. The following graph of the Bureau of Labor Statistics data indicates that the average annual hourly wage for private-sector workers, adjusted for inflation, is beginning to rise again.

Earnings of private-sector workers 1982 dollars

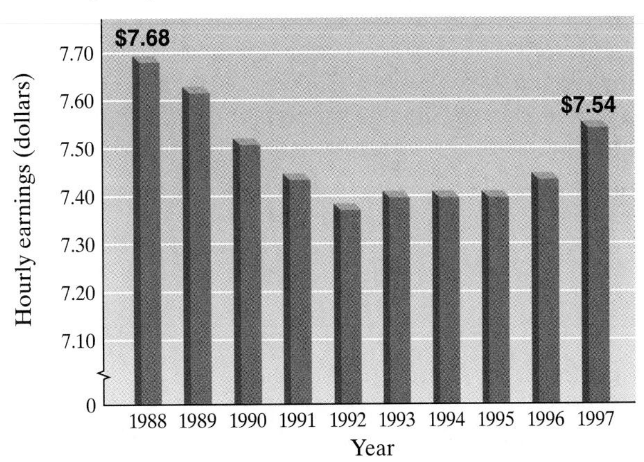

Source: Bureau of Labor Statistics

The average hourly wage, $w(t)$, in dollars, can be estimated by the function $w(t) = 0.011t^2 - 0.122t + 7.700$, where t is years since 1987 and $1 \le t \le 10$.

Use the function $w(t)$ to answer the following questions.

a) In what year was the hourly wage, adjusted for inflation, a minimum?

b) What was the wage?

85. Prior to the return of Hong Kong to China in 1997 there was a large emigration from Hong Kong, as shown in the following bar graph from *Newsweek*.

Emigration from Hong Kong

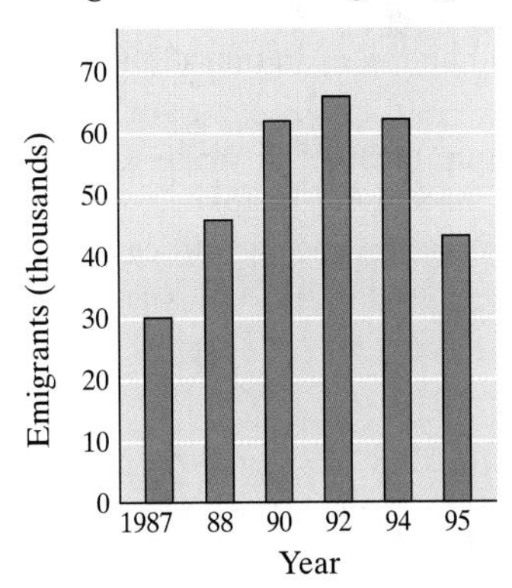

The number of people, in thousands, emigrating from Hong Kong can be estimated by the function $N(x) = -1.82x^2 + 20.08x + 11.86$, where x is the years since 1986 and $1 \le x \le 6$.

Use the function $N(x)$ to answer the following questions.

a) In what year was there the greatest emigration?

b) How many people emigrated?

86. In 1998, the Cigar Association of America published data that showed cigar consumption in the United States has had a resurgence. The number of cigars consumed, $N(t)$, in billions, can be estimated by the function $N(t) = 0.028t^2 - 0.366t + 4.623$, where t is years since 1984 and $1 \leq t \leq 12$.

a) In what year was the number of cigars consumed a minimum?

b) What was the minimum number of cigars consumed?

The profit of a company, in dollars, is the difference between the company's revenue and cost. Exercises 87 and 88 give cost, C, and revenue, R, equations for a particular company. The x represents the number of items produced and sold to distributors. Determine **a)** *the maximum profit of the company and* **b)** *the number of items that must be produced and sold to obtain the maximum profit.*

87. $C(x) = 2000 + 40x$
$R(x) = 800x - x^2$

88. $C(x) = 5000 + 12x$
$R(x) = 2000x - x^2$

Challenge Problems

89. In Example 3 on page 533 we used the function $h(t) = -4.9t^2 + 24.5t + 9.8$ to find that the maximum height, h, attained by a cannonball was 40.425 meters at a time, t, of 2.5 seconds after the cannonball was fired. Review Example 3 now.

a) Write $h(t)$ in the form $h(t) = a(t - h)^2 + k$ by completing the square. (Use the procedure we used in Example 7 on page 539.

b) Using the function you obtained in part **a)**, determine the maximum height attained by the cannonball, and the time after it was fired that the cannonball attained its maximum value.

c) Is the answer you obtained in part **b)** the same as the answer obtained in Example 3? If not, explain why not.

Group Activity

Discuss and answer Exercise 90 as a group.

90. a) Group member 1: Write two quadratic functions $f(x)$ and $g(x)$ so that the functions will not intersect.

b) Group member 2: Write two quadratic functions $f(x)$ and $g(x)$ so that neither function will have x-intercepts, and the vertices of the functions are on opposite sides of the x-axis.

c) Group member 3: Write two quadratic functions $f(x)$ and $g(x)$ so that both functions have the same vertex but one function opens upward and the other opens downward.

d) As a group, review each answer in parts **a)–c)** and decide whether each answer is correct. Correct any answer that is incorrect.

Cumulative Review Exercises

[2.2] **91.** Find the area shaded blue in the figure.

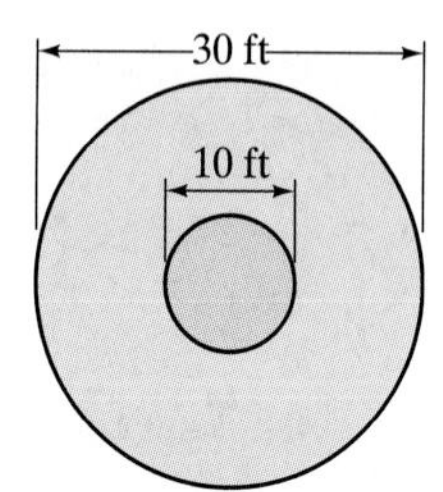

[4.3] **92.** Keon Duncan receives a weekly salary plus a commission of his sales. For one week, on sales of $6000 his total take-home pay was $1300. The next week on sales of $4000, his total take-home pay was $1000. Find his weekly salary and comission rate.

[5.4] **93.** Factor $x^3 + 2x - 5x^2 - 10$

[6.3] **94.** Simplify $\dfrac{3a^{-1} - b^{-1}}{(a - b)^{-1}}$.

8.6 QUADRATIC AND OTHER INEQUALITIES IN ONE VARIABLE

1) **Solve quadratic inequalities.**
2) **Solve other polynomial inequalities.**
3) **Solve rational inequalities.**

In Section 2.5 we discussed linear inequalities in one variable. Now we discuss quadratic inequalities in one variable.

When the equal sign in a quadratic equation of the form $ax^2 + bx + c = 0$ is replaced by an inequality sign, we get a **quadratic inequality**.

Examples of Quadratic Inequalities

$$x^2 + x - 12 > 0, \qquad 2x^2 - 9x - 5 \le 0$$

The **solution to a quadratic inequality** is the set of all values that make the inequality a true statement. For example, if we substitute 5 for x in $x^2 + x - 12 > 0$, we obtain

$$x^2 + x - 12 > 0$$
$$5^2 + 5 - 12 \overset{?}{>} 0$$
$$18 > 0 \quad \textit{True}$$

The inequality is true when x is 5, so 5 satisfies the inequality. However, 5 is not the only solution, for there are other values that satisfy (or are solutions to) the inequality. Does 4 satisfy the inequality? Does 2 satisfy the inequality?

1) Solve Quadratic Inequalities

A number of methods can be used to find the solutions to quadratic inequalities. We will begin by introducing a **sign graph**. Consider the function $f(x) = x^2 + x - 12$. Its graph is shown in Figure 8.21a. Figure 8.21b shows, in red, that when $x < -4$ or $x > 3$, $f(x) > 0$ or $x^2 + x - 12 > 0$. It also shows, in green, that when $-4 < x < 3$, $f(x) < 0$ or $x^2 + x - 12 < 0$.

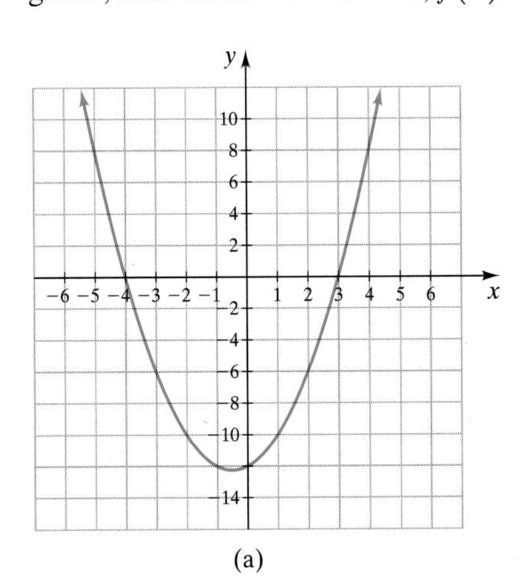

(a)

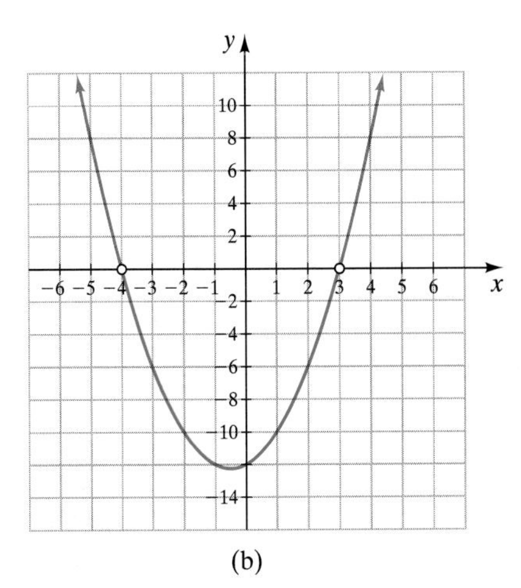

(b)

FIGURE 8.21

One way of finding the solutions to inequalities is to draw the graph and determine from the graph which values of the variable satisfy the inequality, as we just did. In many cases, it may be inconvenient or take too much time to draw the graph of a function, so we provide an alternate method to solve quadratic and other inequalities.

In Example 1 we will show how $x^2 + x - 12 > 0$ is solved using a number line. Then we will outline the procedure we used.

EXAMPLE 1 Solve $x^2 + x - 12 > 0$.

Solution Set the inequality equal to 0 and solve the equation.

$$x^2 + x - 12 = 0$$

$$(x + 4)(x - 3) = 0$$

$$x + 4 = 0 \quad \text{or} \quad x - 3 = 0$$

$$x = -4 \qquad\qquad x = 3$$

The numbers obtained are called **boundary values**. The boundary values are used to break a number line up into intervals. If the original inequality is $<$ or $>$, the boundary values are not part of the intervals. If the original inequality is $\le$ or $\ge$, the boundary values are part of the intervals.

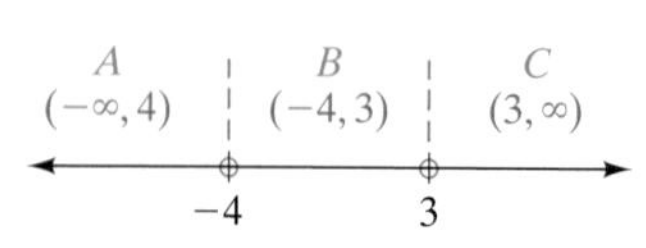

FIGURE 8.22

In Figure 8.22 we have labeled the intervals A, B, and C. Next, we select one test value in *each* interval. Then substitute each of those numbers, one at a time, in either $x^2 + x - 12 > 0$ or $(x + 4)(x - 3) > 0$ and determine whether they result in a true statement. If the test value results in a true statement, all values in that interval will also satisfy the inequality. If the test value results in a false statement, no numbers in that interval will satisfy the inequality.

In this example we will use the test values of −5 in interval A, 0 in interval B, and 4 in interval C (see Fig. 8.23).

FIGURE 8.23

Interval A, $x = -5$	Interval B, $x = 0$	Interval C, $x = 4$
Is $x^2 + x - 12 > 0$?	Is $x^2 + x - 12 > 0$?	Is $x^2 + x - 12 > 0$?
$(-5)^2 - 5 - 12 \overset{?}{>} 0$	$0^2 + 0 - 12 \overset{?}{>} 0$	$4^2 + 4 - 12 \overset{?}{>} 0$
$8 > 0$	$-12 > 0$	$8 > 0$
True	*False*	*True*

Since the test values in both intervals A and C satisfy the inequality, the solution is all real numbers in intervals A or C. The inequality symbol is $>$. The values −4 and 3 are not included in the solution since they make the inequality equal to 0.

The solution is $(-\infty, -4) \cup (3, \infty)$. See Figure 8.24. This solution is consistent with the graph in Figure 8.21b.

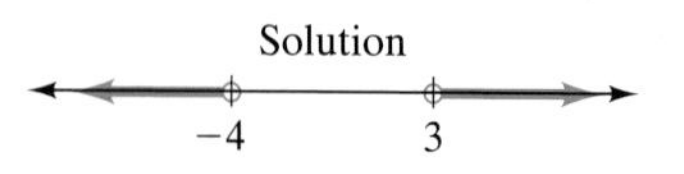

FIGURE 8.24

NOW TRY EXERCISE 9

To Solve Quadratic and Other Inequalities

1. Write the inequality as an equation and solve the equation.
2. If solving a rational inequality, determine the values that make any denominator 0.
3. Construct a number line. Mark each solution from step 1 and numbers obtained in step 2 on the number line. Mark the lowest value on the left, with values increasing from left to right.
4. Select a test value in each interval and determine whether it satisfies the inequality. Also test each boundary value.
5. Write the solution in the form requested by your instructor.

EXAMPLE 2 Solve the inequality $x^2 - 10x \geq -25$. Indicate the solution **a)** on a number line, **b)** in interval notation, and **c)** in set builder notation.

Solution Write the inequality as an equation, then solve the equation.

$$x^2 - 10x = -25$$
$$x^2 - 10x + 25 = 0$$
$$(x - 5)(x - 5) = 0$$
$$x - 5 = 0 \quad \text{or} \quad x - 5 = 0$$
$$x = 5 \qquad\qquad x = 5$$

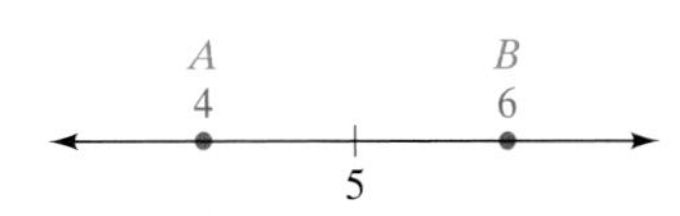

FIGURE 8.25

Since both factors are the same there is only one boundary value (see Fig. 8.25). Both test values, 4 and 6, result in true statements.

Interval A, $x = 4$		**Interval** B, $x = 6$	
$x^2 - 10x \geq -25$		$x^2 - 10x \geq -25$	
$16 - 40 \overset{?}{\geq} -25$		$36 - 60 \overset{?}{\geq} -25$	
$-24 \geq -25$	*True*	$-24 \geq -25$	*True*

The solution set includes both intervals, and is the set of real numbers, ℝ. The answers to parts a), b), and c) follow.

a)  (number line shaded over all real numbers, 0 marked) **b)** $(-\infty, \infty)$ **c)** $\{x | -\infty < x < \infty\}$

In Example 2, if we consider the inequality $x^2 - 10x + 25 \geq 0$ as $(x - 5)^2 \geq 0$ we can see that the solution must be the set of real numbers, since $(x - 5)^2$ must be greater than or equal to 0 for any real number x. The solution to $x^2 - 10x < -25$ is the empty set $\varnothing$. Can you explain why?

EXAMPLE 3 Solve the inequality $x^2 - 2x - 4 \leq 0$. Express the solution in interval notation.

Solution First we need to solve the equation $x^2 - 2x - 4 = 0$. Since this equation is not factorable, we use the quadratic formula to solve.

$$x = \frac{-b \pm \sqrt{b^2 - 4ac}}{2a}$$
$$= \frac{2 \pm \sqrt{4 - 4(1)(-4)}}{2(1)} = \frac{2 \pm \sqrt{20}}{2} = \frac{2 \pm 2\sqrt{5}}{2} = 1 \pm \sqrt{5}$$

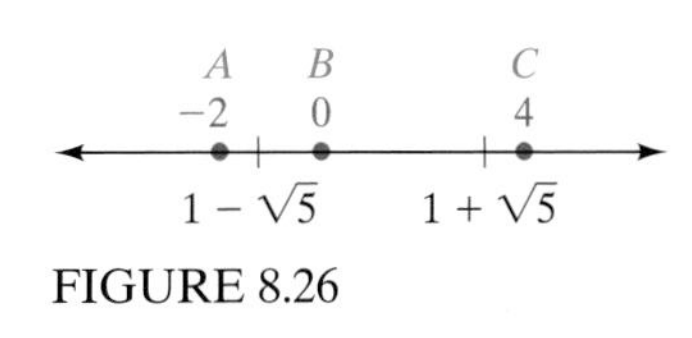

FIGURE 8.26

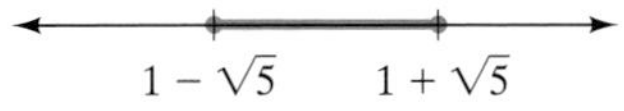

1 − √5 1 + √5

FIGURE 8.27

The value of $1 - \sqrt{5}$ is about −1.24 and the value of $1 + \sqrt{5}$ is about 3.24. We will select test values of −2, 0, and 4 (see Fig. 8.26)

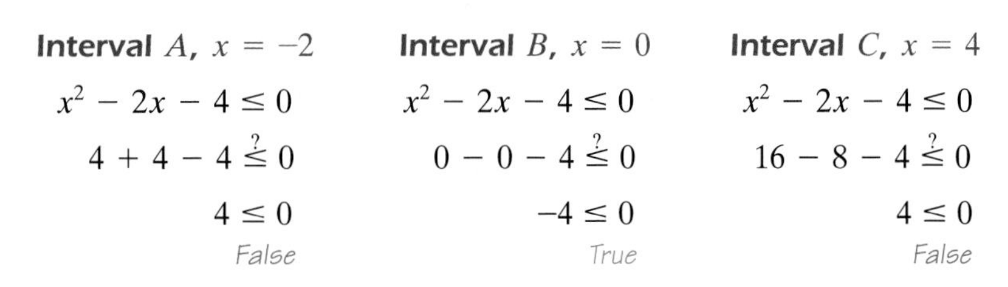

Interval A, $x = -2$	**Interval** B, $x = 0$	**Interval** C, $x = 4$
$x^2 - 2x - 4 \leq 0$	$x^2 - 2x - 4 \leq 0$	$x^2 - 2x - 4 \leq 0$
$4 + 4 - 4 \overset{?}{\leq} 0$	$0 - 0 - 4 \overset{?}{\leq} 0$	$16 - 8 - 4 \overset{?}{\leq} 0$
$4 \leq 0$	$-4 \leq 0$	$4 \leq 0$
False	*True*	*False*

The boundary values are part of the solution because the inequality symbol is $\leq$ and the boundary values make the inequality equal to 0. Thus, the solution in interval notation is $[1 - \sqrt{5}, 1 + \sqrt{5}]$. The solution is illustrated on the number line in Figure 8.27.

NOW TRY EXERCISE 13

HELPFUL HINT

If $ax^2 + bx + c = 0$ with $a > 0$ has two distinct real solutions, then:

Inequality of Form	Solution Is	Solution on Number Line
$ax^2 + bx + c \geq 0$	End intervals	
$ax^2 + bx + c \leq 0$	Center interval	

Example 1 is an inequality of the form $ax^2 + bx + c > 0$, and Example 3 is an inequality of the form $ax^2 + bx + c \leq 0$. Example 2 does not have two distinct real solutions, so this Helpful Hint does not apply.

2) Solve Other Polynomial Inequalities

A procedure similar to the one used earlier can be used to solve other **polynomial inequalities**, as illustrated in the following examples.

EXAMPLE 4 Solve the inequality $(3x - 2)(x + 3)(x + 5) < 0$. Illustrate the solution on a number line and write the solution in set builder notation.

Solution We use the zero factor property to solve the equation $(3x - 2)(x + 3)(x + 5) = 0$.

$$3x - 2 = 0 \quad \text{or} \quad x + 3 = 0 \quad \text{or} \quad x + 5 = 0$$

$$x = \frac{2}{3} \qquad\qquad x = -3 \qquad\qquad x = -5$$

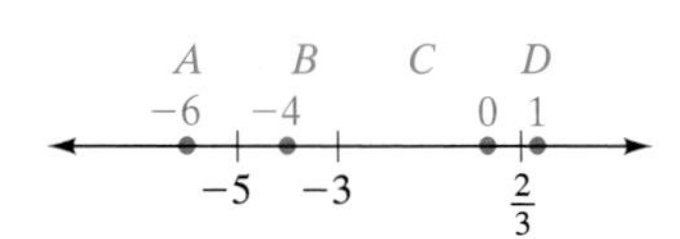

FIGURE 8.28

The solutions -5, -3, and $\frac{2}{3}$ break the number line into four intervals (see Fig. 8.28). The test values we will use are $-6, -4, 0$, and 1. We show the results in the following table.

Interval	Test Value	$(3x - 2)(x + 3)(x + 5)$	< 0
A	-6	-60	True
B	-4	14	False
C	0	-30	True
D	1	24	False

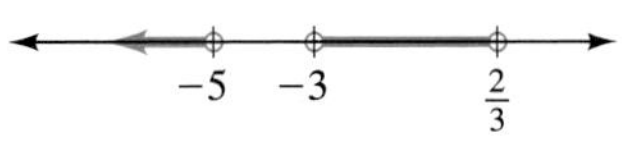

FIGURE 8.29

Since the original inequality symbol is <, the boundary values are not part of the solution. The solution, intervals A and C, is illustrated on the number line in Fig. 8.29. The solution in set builder notation is $\{x \mid x < -5 \text{ or } -3 < x < \frac{2}{3}\}$.

EXAMPLE 5 Given $f(x) = 3x^3 - 3x^2 - 6x$, find all values of x for which $f(x) \geq 0$.

Solution We need to solve the inequality

$$3x^3 - 3x^2 - 6x \geq 0$$

We start by solving the equation $3x^3 - 3x^2 - 6x = 0$.

$$3x(x^2 - x - 2) = 0$$

$$3x(x - 2)(x + 1) = 0$$

$$3x = 0 \quad \text{or} \quad x - 2 = 0 \quad \text{or} \quad x + 1 = 0$$

$$x = 0 \qquad\qquad x = 2 \qquad\qquad x = -1$$

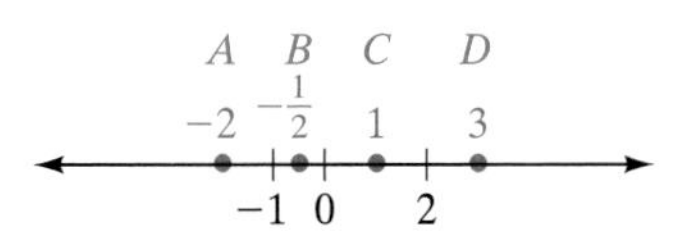

FIGURE 8.30

The solutions −1, 0, and 2 break the number line into four intervals (see Fig. 8.30). The test values that we will use are $-2, -\frac{1}{2}, 1$, and 3.

Interval	Test Value	$3x^3 - 3x^2 - 6x$	≥ 0
A	−2	−24	False
B	$-\frac{1}{2}$	$\frac{15}{8}$	True
C	1	−6	False
D	3	36	True

Since the original inequality is ≥, the boundary values are part of the solution. The solution, intervals B and D, is illustrated on the number line in Figure 8.31. The solution in interval notation is $[-1, 0] \cup [2, \infty)$. Figure 8.32 shows the graph of $f(x) = 3x^3 - 3x^2 - 6x$. Notice $f(x) \geq 0$ for $-1 \leq x \leq 0$ and for $x \geq 2$, which agrees with our solution.

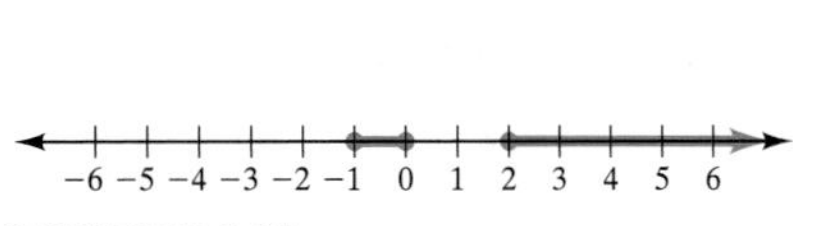

FIGURE 8.31

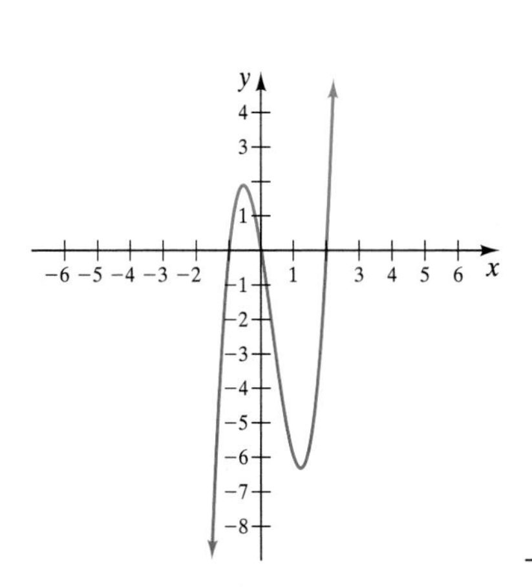

FIGURE 8.32

NOW TRY EXERCISE 31

3) Solve Rational Inequalities

In Examples 6 and 7 we solve **rational inequalities**, which are inequalities that contain a rational expression.

EXAMPLE 6 Solve the inequality $\frac{x-1}{x+3} \geq 2$ and graph the solution on a number line.

Solution Change the $\geq$ to $=$ and solve the resulting equation.

$$\frac{x-1}{x+3} = 2$$

$$\cancel{x+3} \cdot \frac{x-1}{\cancel{x+3}} = 2(x+3) \qquad \textit{Multiply both sides by } x+3.$$

$$x - 1 = 2x + 6$$

$$-1 = x + 6$$

$$-7 = x$$

When solving rational inequalities, we also need to determine the value or values that make the denominator 0. We set the denominator equal to 0 and solve.

$$x + 3 = 0$$

$$x = -3$$

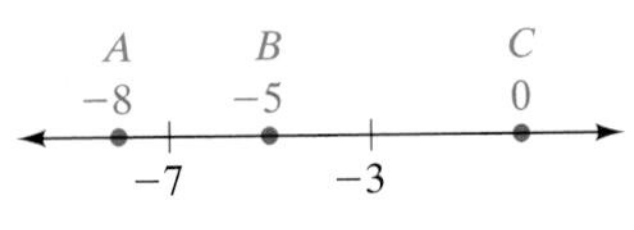

FIGURE 8.33

We use the solution to the equation, −7, and the values that make the denominator zero, −3, to determine the intervals, shown in Figure 8.33. We will use −8, −5, and 0 as our test values.

Interval A, $x = -8$

$$\frac{x-1}{x+3} \geq 2$$

$$\frac{-8-1}{-8+3} \overset{?}{\geq} 2$$

$$\frac{-9}{-5} \geq 2 \quad \textit{False}$$

Interval B, $x = -5$

$$\frac{x-1}{x+3} \geq 2$$

$$\frac{-5-1}{-5+3} \overset{?}{\geq} 2$$

$$3 \geq 2 \quad \textit{True}$$

Interval C, $x = 0$

$$\frac{x-1}{x+3} \geq 2$$

$$\frac{0-1}{0+3} \overset{?}{\geq} 2$$

$$-\frac{1}{3} \geq 2 \quad \textit{False}$$

Only interval B satisfies the inequality. Whenever we have a rational inequality we must be very careful to determine which boundary values are included in the solution. Remember we can never include in our solution any value that makes the denominator 0. Now check the boundary values −7 and −3. Since −7 results in the inequality being true, −7 is a solution. Since division by 0 is not permitted, −3 is not a solution. Thus the solution is $[-7, -3)$. The solution is illustrated on the number line in Figure 8.34.

FIGURE 8.34

In Example 6 we solved $\frac{x-1}{x+3} \geq 2$. Suppose we graphed $f(x) = \frac{x-1}{x+3}$. For what values of x would $f(x) \geq 2$? If you answered $-7 \leq x < -3$ you answered correctly. Figure 8.35 shows the graph of $f(x) = \frac{x-1}{x+3}$ and the graph of $y = 2$. Notice $f(x) \geq 2$ when $-7 \leq x < -3$.

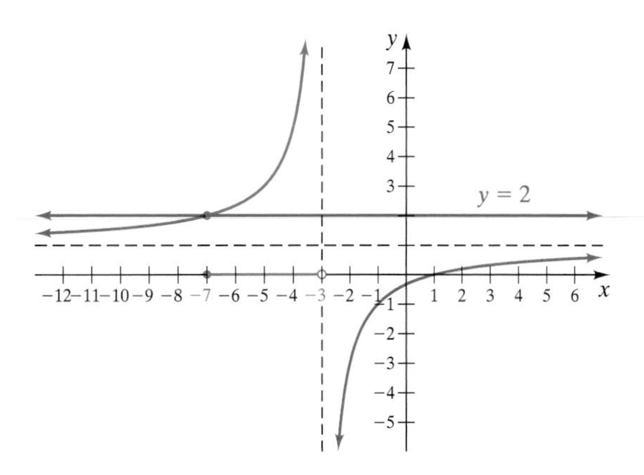

FIGURE 8.35

EXAMPLE 7 Solve the inequality $\dfrac{(x-3)(x+4)}{x+1} \geq 0$. Illustrate the solution on a number line and give the solution in interval notation.

Solution The solutions to the equation $\dfrac{(x-3)(x+4)}{x+1} = 0$ are 3 and -4 since these are the values that make the numerator equal to 0. The equation is not defined at -1. We therefore use the values 3, -4, and -1 to determine the intervals on the number line (see Fig. 8.36). Checking test values at -5, -2, 0, and 4, we find that the values in intervals B and D, $-4 < x < -1$ and $x > 3$, satisfy the inequality. Check the test values yourself to verify this. The values 3 and -4 make the inequality equal to 0 and are part of the solution. The inequality is not defined at -1, so -1 is not part of the solution. The solution is $[-4, -1) \cup [3, \infty)$. The solution is illustrated on the number line in Figure 8.37.

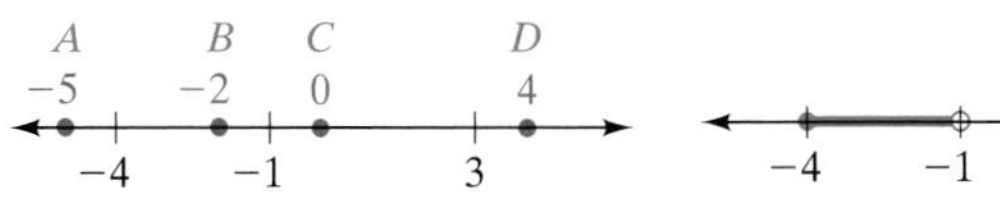

FIGURE 8.36

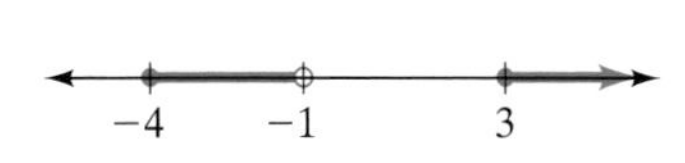

FIGURE 8.37

NOW TRY EXERCISE 51

Exercise Set 8.6

Concept/Writing Exercises

1. The graph of $f(x) = x^2 - 7x + 10$ is given. Find the solution to **a)** $f(x) > 0$ and **b)** $f(x) < 0$.

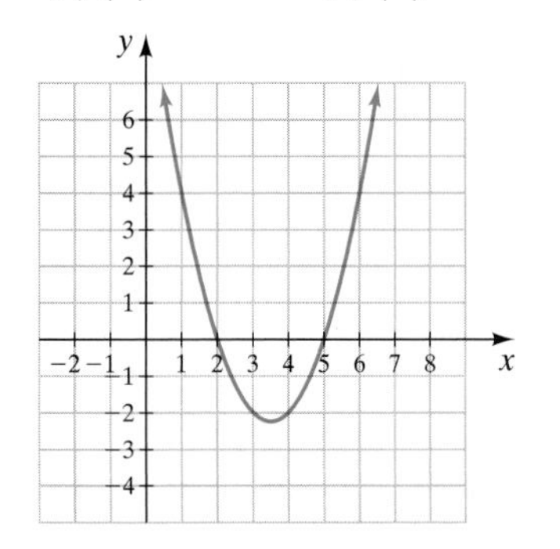

2. The graph of $f(x) = -x^2 - 4x + 5$ is given. Find the solution to **a)** $f(x) \geq 0$ and **b)** $f(x) \leq 0$.

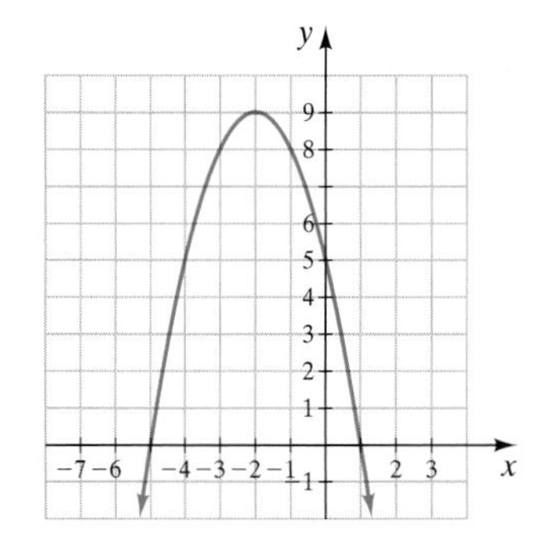

Practice the Skills

Solve each inequality and graph the solution on a number line.

3. $x^2 - 3x - 10 \geq 0$
4. $x^2 + 8x + 7 < 0$
5. $x^2 + 4x > 0$
6. $x^2 - 5x \geq 0$
7. $x^2 - 16 < 0$
8. $y^2 - 25 \leq 0$
9. $2x^2 + 5x - 3 \geq 0$
10. $3n^2 - 7n \leq 6$
11. $5x^2 + 19x \leq 4$
12. $3x^2 + 5x - 3 \leq 0$
13. $2x^2 - 12x + 9 \leq 0$
14. $5x^2 \leq -20x - 4$

Solve each inequality and give the solution in interval notation.

15. $(x - 1)(x + 1)(x + 4) > 0$
16. $(x - 3)(x + 2)(x + 5) \leq 0$
17. $x(x - 3)(2x + 6) \geq 0$
18. $(x - 3)(x + 4)(x - 2) \leq 0$
19. $(2c + 5)(3c - 6)(c + 6) > 0$
20. $(2p - 1)(p + 5)(3p + 6) \geq 0$
21. $(x + 2)(x + 2)(3x - 8) \geq 0$
22. $(x + 3)^2(4x - 5) \leq 0$
23. $x^3 - 4x^2 + 4x < 0$
24. $x^3 + 3x^2 - 18x > 0$

For each function provided, find all values of x for which f(x) satisfies the indicated conditions. Graph the solution on a number line.

25. $f(x) = x^2 + 4x, f(x) \geq 0$
26. $f(x) = x^2 - 9x, f(x) \leq 0$
27. $f(x) = x^2 - 14x + 48, f(x) < 0$
28. $f(x) = 4x^2 - 4x - 15, f(x) > 0$
29. $f(x) = 2x^2 + 9x - 4, f(x) \leq 2$
30. $f(x) = x^2 + 5x - 4, f(x) \leq 3$
31. $f(x) = 2x^3 + 9x^2 - 35x, f(x) \geq 0$
32. $f(x) = x^3 - 5x, f(x) \leq 0$

Solve each inequality and give the solution in set builder notation.

33. $\frac{x + 3}{x - 1} > 0$
34. $\frac{x - 5}{x + 2} < 0$
35. $\frac{y - 4}{y - 1} \leq 0$
36. $\frac{x + 6}{x + 2} \geq 0$
37. $\frac{2x - 4}{x - 1} < 0$
38. $\frac{3x + 6}{x + 4} \geq 0$
39. $\frac{3a + 6}{2a - 1} \geq 0$
40. $\frac{3x + 4}{2x - 1} < 0$
41. $\frac{x + 4}{x - 4} \leq 0$
42. $\frac{k + 3}{k} \geq 0$
43. $\frac{4x - 2}{2x - 4} > 0$
44. $\frac{3x + 5}{x - 2} \leq 0$

Solve each inequality and give the solution in interval notation.

45. $\frac{(x + 2)(x - 4)}{x + 6} < 0$
46. $\frac{(x - 3)(x - 6)}{x + 4} \geq 0$
47. $\frac{(w - 6)(w - 1)}{w - 3} \geq 0$
48. $\frac{x + 6}{(x - 2)(x + 4)} > 0$
49. $\frac{x - 6}{(x + 4)(x - 1)} \leq 0$
50. $\frac{x}{(x + 3)(x - 3)} \leq 0$
51. $\frac{(x - 3)(2x + 5)}{x - 6} > 0$
52. $\frac{r(r - 3)}{2r + 6} < 0$

Solve each inequality and graph the solution on a number line.

53. $\frac{2}{x - 3} \geq -1$
54. $\frac{3}{m - 1} \leq -1$
55. $\frac{4}{x - 2} \geq 2$
56. $\frac{2}{2a - 1} > 2$
57. $\frac{2p - 5}{p - 4} \leq 1$
58. $\frac{2x}{x + 1} > 1$
59. $\frac{w}{3w - 2} > -2$
60. $\frac{x - 1}{2x + 6} \leq -3$
61. $\frac{x + 8}{x + 2} > 1$
62. $\frac{4x + 2}{2x - 3} \geq 2$

63. The graph of $y = \dfrac{x^2 - 4x + 4}{x - 4}$ is illustrated. Determine the solution to the following inequalities.

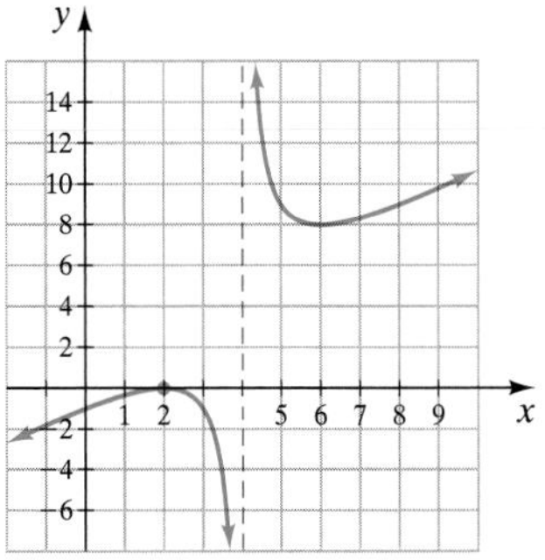

a) $\dfrac{x^2 - 4x + 4}{x - 4} > 0$

b) $\dfrac{x^2 - 4x + 4}{x - 4} < 0$

Explain how you determined your answer.

64. The graph of $y = \dfrac{x^2 + x - 6}{x - 4}$ is illustrated. Determine the solution to the following inequalities.

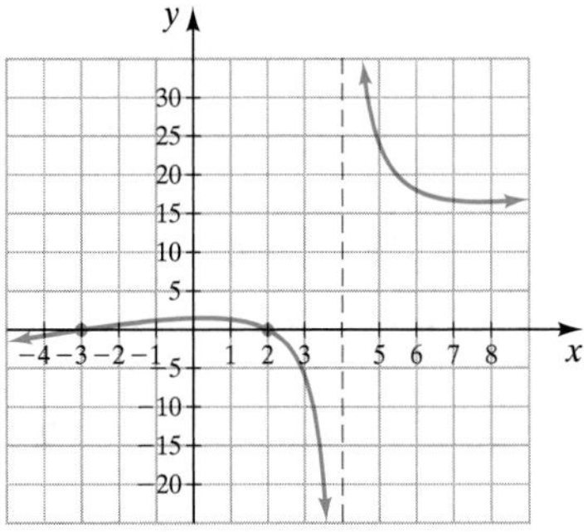

a) $\dfrac{x^2 + x - 6}{x - 4} \geq 0$

b) $\dfrac{x^2 + x - 6}{x - 4} < 0$

Explain how you determined your answer.

65. Write a quadratic inequality whose solution is,

.

66. Write a quadratic inequality whose solution is,

.

67. Write a rational inequality whose solution is

.

68. Write a rational inequality whose solution is

.

69. What is the solution to the inequality $(x + 3)^2(x - 1)^2 \geq 0$? Explain your answer.

70. What is the solution to the inequality $x^2(x - 3)^2(x + 4)^2 < 0$? Explain your answer.

71. What is the solution to the inequality $\dfrac{x^2}{(x + 1)^2} \geq 0$? Explain your answer.

72. What is the solution to the inequality $\dfrac{x^2}{(x - 3)^2} > 0$? Explain your answer.

73. If $f(x) = ax^2 + bx + c$ where $a > 0$ and the discriminant is negative, what is the solution to $f(x) < 0$? Explain.

74. If $f(x) = ax^2 + bx + c$ where $a < 0$ and the discriminant is negative, what is the solution to $f(x) > 2$? Explain.

Challenge Problems

Solve each inequality and graph the solution on the number line.

75. $(x + 1)(x - 3)(x + 5)(x + 9) \geq 0$

76. $\dfrac{(x - 4)(x + 2)}{x(x + 6)} \geq 0$

Write a quadratic inequality with the following solutions. Many answers are possible. Explain how you determined your answers.

77. $(-\infty, 0) \cup (3, \infty)$

78. $\{4\}$

79. $\varnothing$

80. $\mathbb{R}$

Group Activity

Discuss and answer Exercises 81 and 82 as a group.

81. Consider the number line on the right, where a, b, and c are distinct real numbers.

a) In which intervals will the real numbers satisfy the inequality $(x - a)(x - b)(x - c) > 0$? Explain.

b) In which intervals will the real numbers satisfy the inequality $(x - a)(x - b)(x - c) < 0$? Explain.

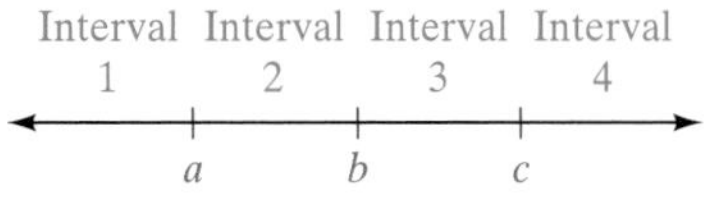

82. Consider the number line below where a, b, c, and d are distinct real numbers.

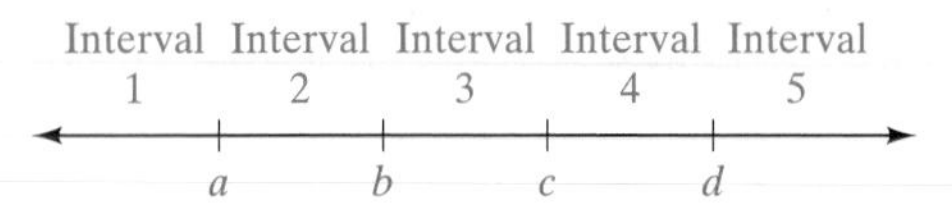

a) In which intervals do the real numbers satisfy the inequality $(x - a)(x - b)(x - c)(x - d) > 0$? Explain.

b) In which interval do the real numbers satisfy the inequality $(x - a)(x - b)(x - c)(x - d) < 0$? Explain.

Cumulative Review Exercises

In Exercises 83 and 84, find the domain of each function.

[6.1] **83.** $f(x) = \dfrac{3}{x^2 - 4}$

[8.1] **84.** $f(x) = \sqrt{x - 4}$.

[6.3] **85.** Simplify $\dfrac{ab^{-2} - a^{-1}b}{a^{-2} + ab^{-1}}$.

[7.7] **86.** Multiply $(\sqrt{-8} + \sqrt{2})(\sqrt{-2} - \sqrt{8})$.

SUMMARY

Key Words and Phrases

8.1
Completing the square
Perfect square trinomial

8.2
Discriminant

8.3
Solving for a variable in a formula

8.4
Equation quadratic in form

8.5
Axis of symmetry
Maximum value problem
Minimum value problem
Parabola
Symmetry
Translate a parabola
Vertex

8.6
Boundary values
Polynomial inequality
Quadratic inequality
Rational inequality
Sign graph
Solution to a quadratic inequality

IMPORTANT FACTS

Square Root Property: If $x^2 = a$, then $x = \pm\sqrt{a}$.

Quadratic formula: $x = \dfrac{-b \pm \sqrt{b^2 - 4ac}}{2a}$

Discriminant: $b^2 - 4ac$

For an equation of the form $ax^2 + bx + c = 0$.
If $b^2 - 4ac > 0$, the quadratic equation has two distinct real solutions.
If $b^2 - 4ac = 0$, the quadratic equation has one real solution.
If $b^2 - 4ac < 0$, the quadratic equation has no real solution.

For $f(x) = ax^2 + bx + c$, the vertex of the parabola is $\left(-\dfrac{b}{2a}, \dfrac{4ac - b^2}{4a}\right)$ or $\left(-\dfrac{b}{2a}, f\left(-\dfrac{b}{2a}\right)\right)$.

For $f(x) = a(x - h)^2 + k$, the vertex of the parabola is (h, k).

If $f(x) = ax^2 + bx + c$, $a > 0$, the function will have a minimum value of $\dfrac{4ac - b^2}{4a}$ at $x = -\dfrac{b}{2a}$.

If $f(x) = ax^2 + bx + c$, $a < 0$, the function will have a maximum value of $\dfrac{4ac - b^2}{4a}$ at $x = -\dfrac{b}{2a}$.

Review Exercises

[8.1] *Use the square root property to solve each equation for x.*

1. $(x - 4)^2 = 20$ **2.** $(3x - 4)^2 = 60$ **3.** $\left(x - \frac{2}{3}\right)^2 = \frac{1}{9}$ **4.** $\left(2x - \frac{1}{2}\right)^2 = 4$

Solve each equation by completing the square.

5. $x^2 - 8x + 15 = 0$ **6.** $x^2 - 3x - 54 = 0$ **7.** $x^2 = -5x + 6$

8. $x^2 + 2x - 5 = 0$ **9.** $-a^2 - 6a + 10 = 0$ **10.** $2r^2 - 8r = -64$

[8.2] *Determine whether each equation has two distinct real solutions, a single real solution, or no real solution.*

11. $3x^2 - 4x - 20 = 0$ **12.** $-3x^2 + 4x = 9$ **13.** $2x^2 + 6x + 7 = 0$

14. $n^2 - 12n = -36$ **15.** $-3x^2 - 4x + 8 = 0$ **16.** $x^2 - 9x + 6 = 0$

Soive each equation by the quadratic formula.

17. $5x^2 - 7x = 6$ **18.** $x^2 - 18 = 7x$ **19.** $x^2 - x + 30 = 0$

20. $6d^2 + d - 15 = 0$ **21.** $2x^2 + 4x - 3 = 0$ **22.** $-2x^2 + 3x + 6 = 0$

23. $x^2 - 6x + 7 = 0$ **24.** $3x^2 - 6x - 8 = 0$ **25.** $2x^2 - 5x = 0$

26. $1.2r^2 + 5.7r = 2.3$ **27.** $x^2 = \frac{5}{6}x + \frac{25}{6}$ **28.** $x^2 + \frac{5}{4}x = \frac{3}{8}$

For the given function, determine all real values of the variable for which the function has the value indicated.

29. $f(x) = x^2 - 4x - 45, f(x) = 15$ **30.** $g(x) = 6x^2 + 5x, g(x) = 6$

31. $h(r) = 5r^2 - 7r - 6, h(r) = -4$ **32.** $f(x) = -2x^2 + 6x + 5, f(x) = -4$

Write a function that has the given solutions.

33. $3, -2$ **34.** $\frac{2}{3}, -3$ **35.** $-\sqrt{5}, \sqrt{5}$ **36.** $3 - 2i, 3 + 2i$

[8.1–8.3]

37. Paula Ryba is designing a rectangular storage shed. If the area is to be 63 square feet and the length is to be 2 feet greater than the width, find the dimensions of the shed.

38. Find the length of side x in the following figure.

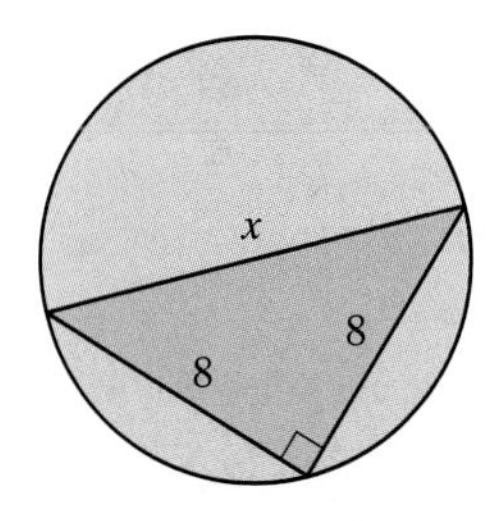

39. Katherine Chu invested \$800 in a savings account whose interest is compounded annually. If after 2 years the amount in the account is \$882, find the annual interest rate.

40. The larger of two positive numbers is 4 greater than the smaller. Find the two numbers if their product is 45.

41. The length of a rectangle is 1 inch less than twice its width. Find the sides of the rectangle if its area is 66 square inches.

42. The value, V, in dollars per acre of a wheat crop d days after planting is given by the formula $V = 12d - 0.05d^2$, $20 < d < 80$. Find the value of an acre of wheat 50 days after it has been planted.

43. The distance, d, in feet, that an object is from the ground t seconds after being dropped from an airplane is given by the formula $d = -16t^2 + 1800$.

a) Find the distance the object is from the ground 3 seconds after it has been dropped.

b) When will the object hit the ground?

44. If an object is thrown upward from the top of a 100-foot-tall building, its height above the ground, $h(t)$, at any time, t, can be found by the function $h(t) = -16t^2 + 16t + 100$.

a) Find the height of the object at 2 seconds.

b) When will the object hit the ground?

45. A tractor has an oil leak. The amount of oil, $L(t)$, in milliliters per hour that leaks out is a function of the tractor's operating temperature, t, in degrees Celsius. The function is

$$L(t) = 0.0004t^2 + 0.16t + 20, 100°\text{C} \leq t \leq 160°\text{C}$$

a) How many milliliters of oil will leak out in 1 hour if the operating temperature of the tractor is 100°C?

b) If oil is leaking out at 53 milliliters per hour, what is the operating temperature of the tractor?

46. Two molding machines can complete an order in 12 hours. The larger machine could complete the order by itself in 1 hour less time than the smaller machine could by itself. How long will it take each machine to complete the order working by itself?

47. Bryce Dodson drove 20 miles at a constant speed, then increased his speed by 10 miles per hour for the next 30 miles. If the time required to travel the 50 miles was 0.9 hours, find the speed he drove during the first 20 miles.

48. Rachel Dorushka canoed downstream going with the current for 3 miles, then turned around and canoed upstream against the current to her starting point. If the total time she spent canoeing was 4 hours and the current was 0.4 miles per hour, what is the speed she canoes in still water?

Solve each equation for the variable indicated.

49. $V_x^2 + V_y^2 = V^2$, for V_y (study of vectors)

50. $a = \dfrac{v_2^2 - v_1^2}{2d}$, for v_2 (acceleration of a vehicle)

[8.4] *Solve each equation.*

51. $p^4 - 5p^2 = 24$

52. $6m^{-2} + 11m^{-1} - 10 = 0$

53. $4x + 23\sqrt{x} - 6 = 0$

54. $2m^{2/3} - 7m^{1/3} = -6$

55. $10(p + 2) + 7 = \dfrac{12}{p + 2}$

56. $6(x - 2)^{-2} = -13(x - 2)^{-1} + 8$

Find all x-intercepts of the given function.

57. $f(x) = 30x + 13\sqrt{x} - 10$

58. $g(x) = 12(x^2 - 4x)^2 + 16(x^2 - 4x) - 3$

[8.5] ***a)** Determine whether the parabola opens upward or downward. **b)** Find the y-intercept. **c)** Find the vertex. **d)** Find the x-intercepts if they exist. **e)** Draw the graph.*

59. $y = x^2 + 6x$

60. $y = x^2 + 2x - 8$

61. $y = -x^2 - 9$

62. $y = -2x^2 - x + 15$

63. The Hornell High School Theater estimates that their total income, I, in hundreds of dollars, for their production of the play *The King and I* can be approximated by the formula $I = -x^2 + 22x - 30, 2 \le x \le 20$, where x is the cost of a ticket.

a) How much should they charge to maximize their profit?

b) What is the maximum profit?

64. Jasonna Waters tosses a ball upward from the top of a 60-foot building. The height, $s(t)$, of the ball at any time t can be determined by the function $s(t) = -16t^2 + 80t + 60$.

a) At what time will the ball attain its maximum height?

b) What is the maximum height?

Graph each function.

65. $f(x) = (x - 3)^2$

66. $f(x) = -(x + 2)^2 - 3$

67. $g(x) = -2(x + 4)^2 - 1$

68. $h(x) = \frac{1}{2}(x - 1)^2 + 3$

[8.6] *Graph the solution to each inequality on a number line.*

69. $x^2 + 6x + 5 \ge 0$

70. $x^2 + 2x - 15 \le 0$

71. $x^2 \le 11x - 20$

72. $3x^2 + 8x > 16$

73. $4x^2 - 9 \le 0$

74. $5x^2 - 25 > 0$

Solve each inequality and give the solution in set builder notation.

75. $\dfrac{x + 2}{x - 3} > 0$

76. $\dfrac{x - 5}{x + 2} \le 0$

77. $\dfrac{2x - 4}{x + 1} \ge 0$

78. $\dfrac{3x+5}{x-6} < 0$

79. $(x+3)(x+1)(x-2) > 0$

80. $x(x-3)(x-5) \leq 0$

Solve each inequality and give the solution in interval notation.

81. $(3x+4)(x-1)(x-3) \geq 0$

82. $2x(x+2)(x+5) < 0$

83. $\dfrac{x(x-4)}{x+2} > 0$

84. $\dfrac{(x-2)(x-5)}{x+3} < 0$

85. $\dfrac{x-3}{(x+2)(x-5)} \geq 0$

86. $\dfrac{x(x-5)}{x+3} \leq 0$

Solve each inequality and graph the solution on a number line.

87. $\dfrac{3}{x+4} \geq -1$

88. $\dfrac{2x}{x-2} \leq 1$

89. $\dfrac{2x+3}{3x-5} < 4$

Practice Test

Solve by completing the square.

1. $x^2 = -x + 12$

2. $4x^2 + 8x = -12$

Solve by the quadratic formula.

3. $x^2 - 5x - 6 = 0$

4. $x^2 + 5 = -8x$

Solve by the method of your choice.

5. $3x^2 - 5x = 0$

6. $-2x^2 = 9x - 5$

7. Write a function that has x-intercepts $3, -\frac{5}{2}$.

8. Solve the formula $K = \frac{1}{2}mv^2$ for v.

9. The cost, c, of a house in Duquoin, Illinois, is a function of the number of square feet, s, of the house. The cost of the house can be approximated by

$$c = -0.01s^2 + 78s + 22{,}000, \quad 1300 \leq s \leq 3900$$

a) Estimate the cost of a 2000-square-foot house.

b) How large a house can Clarissa Skocy purchase if she wishes to spend \$160,000 on a house?

10. Jacob Zwick drove his 4-wheel drive Jeep from Anchorage, Alaska, to the Chena River State Recreation area, a distance of 520 miles. Had he averaged 15 miles per hour faster, the trip would have taken 2.4 hours less. Find the average speed that Jacob drove.

Solve.

11. $10m^4 + 21m^2 = 10$

12. $3r^{2/3} + 11r^{1/3} - 42 = 0$

13. Find all x-intercepts of $f(x) = 16x - 40\sqrt{x} + 25$.

Graph each function.

14. $f(x) = (x-3)^2 + 2$

15. $h(x) = -\frac{1}{2}(x-2)^2 - 2$

16. Determine whether $5x^2 = 4x + 2$ has two distinct real solutions, a single solution, or no real solution. Explain your answer.

17. Consider the quadratic equation $y = x^2 - 2x - 8$.

a) Determine whether the parabola opens upward or downward.

b) Find the y-intercept.

c) Find the vertex.

d) Find the x-intercepts (if they exist)

e) Draw the graph.

18. Write a quadratic function whose x-intercepts are $(-6, 0), \left(\frac{1}{2}, 0\right)$.

Solve each inequality and graph the solution on a number line.

19. $x^2 - x \geq 42$

20. $\dfrac{(x+3)(x-4)}{x+1} \geq 0$

*Solve the following inequality. Write the answer in **a)** interval notation and **b)** set builder notation.*

21. $\dfrac{x + 3}{x + 2} \leq -1$

22. The length of a rectangular Persian carpet is 4 feet greater than twice its width. Find the length and width of the carpet if its area is 48 square feet.

23. Jose Ramirez throws a ball upward from the top of a building. The distance, d, of the ball from the ground at any time t is $d = -16t^2 + 64t + 80$. How long will it take for the ball to strike the ground?

24. The Leigh Ann Sims Company earns a weekly profit according to the function $f(x) = -1.4x^2 + 56x - 60$, where x is the number of wood carvings made and sold each week.

a) Find the number of carvings the company must sell in a week to maximize its profit.

b) What is its maximum weekly profit?

25. The cost, C, and revenue, R, equations for a company are given. The x represents the number of items produced and sold. Determine

a) the number of items that must be sold to maximize profit.

b) the maximum profit of the company.

$$C(x) = 8000 + 20x$$

$$R(x) = 500x - x^2$$

Cumulative Review Test

1. Evaluate $(|-3| - 5) - (4 \cdot |-6|)$.

2. The circle graph shows the U.S. households by size, in percents, in 1996. The total number of U.S. households in 1996 was 9.96×10^7

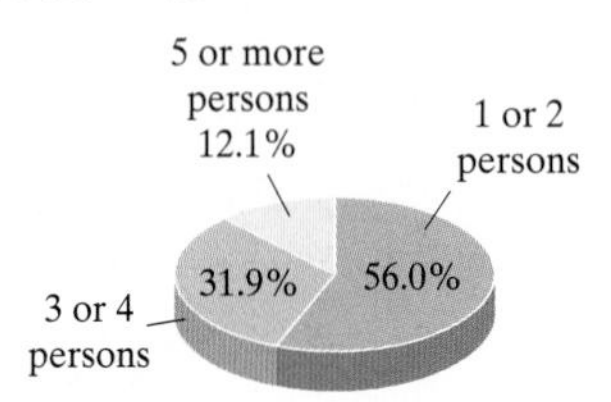

a) Find the number of households with 1 or 2 persons.

b) Find the number of households with 5 or more persons.

c) How many more households with 1 or 2 persons were there than households with 5 or more persons?

3. Solve the equation $-2(x + 5) = 3\{[5 - (x - 5)] - 6x\}$.

4. Solve the following inequality.

$$\left|\frac{x - 5}{3}\right| < 8$$

5. Solve the following equation.

$$\left|\tfrac{1}{3}a + 3\right| = \left|\tfrac{2}{3}a - 1\right|$$

6. Determine whether the set of ordered pairs is a function. Explain your answer.

$$\{(-5, 2), (-4, 6), (3, 5), (2, 6), (5, 0)\}$$

7. a) Determine whether the following graph is a function. Explain your answer.

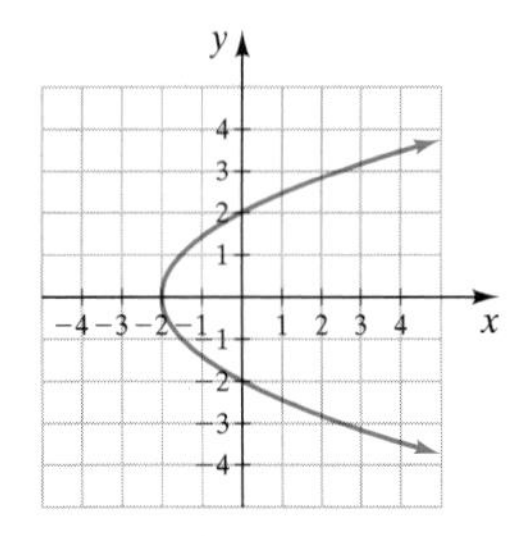

b) Determine the domain and range of the function or relation.

8. Graph each equation.

a) $x = -4$

b) $y = 2$

9. a) Write $\frac{1}{2}y = 2(x - 3) + 4$ in standard form.

b) Graph $\frac{1}{2}y = 2(x - 3) + 4$.

10. Determine the equation of a line perpendicular to the graph of $4y = -3x + 7$ that passes through $(2, -1)$. Write the equation in point–slope form.

11. Solve the following system of equations using matrices.

$$x - 2y = 8$$
$$2x + y = 6$$

12. Evaluate the following determinant.

$$\begin{vmatrix} 4 & 0 & -2 \\ 3 & 5 & 1 \\ 1 & -1 & 7 \end{vmatrix}$$

13. Factor $3p^4 - 12p^2q + 9p^2q - 36q^2$.

14. If $f(x) = x^3 - 13x - 12$ and $g(x) = x - 4$, find

a) $(f/g)(x)$

b) $(f/g)(6)$.

15. If $f(x) = 2x^2 + 4x - 6$ and $g(x) = 3x - 4$, find $(f \cdot g)(x)$.

16. One square has a side 4 inches longer than the side of a second square. If the area of the larger square is 121 square inches, find the length of a side of each square.

17. Solve the following equation.

$$\frac{1}{a - 2} = \frac{4a - 1}{a^2 + 5a - 14} + \frac{2}{a + 7}$$

18. The wattage rating of an appliance, w, varies jointly as the square of the current, I, and the resistance, R. If the wattage is 12 when the current is 2 amperes and the resistance is 100 ohms, find the wattage when the current is 0.8 amperes and the resistance is 600 ohms.

19. Simplify $\dfrac{3 - 4i}{2 + 3i}$.

20. Solve $4x^2 = -3x - 12$.

EXPONENTIAL AND LOGARITHMIC FUNCTIONS

CHAPTER

9

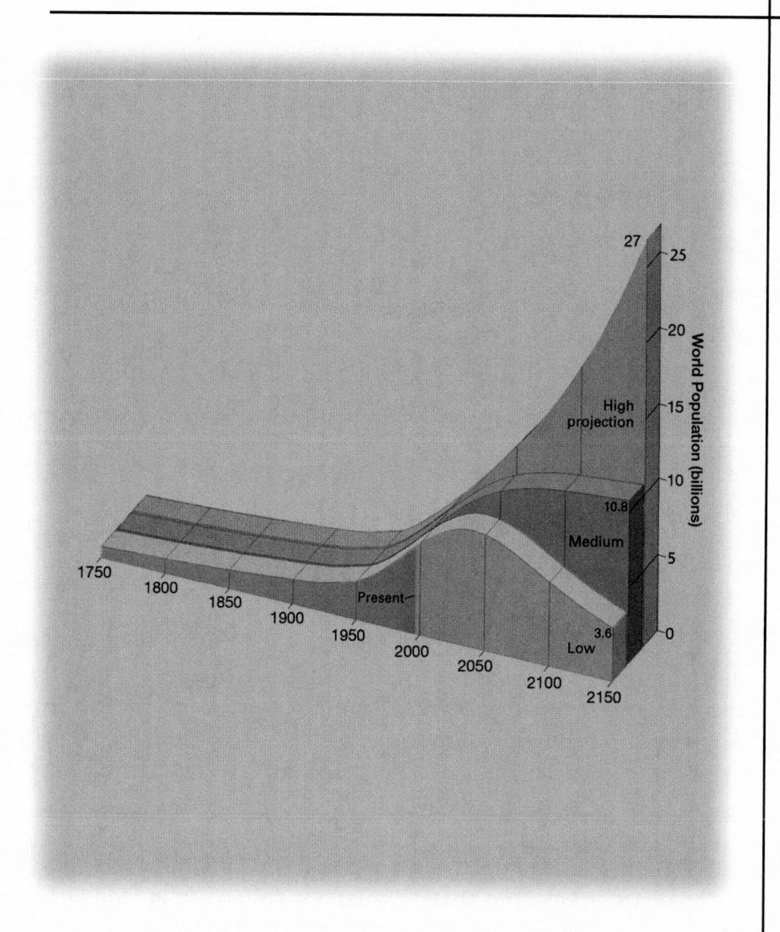

Use the Angel Web site at www.prenhall.com/angel to be linked to an internet resource that will help you further explore the following application.

The Population Division of the Department of Economic and Social Affairs uses mathematical models to make estimates and projections of the world population. Future projections are made by making an assumption about the world fertility level, that is, the average number of births per woman. The world fertility levels have been declining over the last 25 years. In the 1950s, it was 5 births per woman, it is now only 2.7 births per woman.

Currently the world population is growing at 1.33% per year. Because the population is growing by a percentage rather than a fixed amount, it is modeled by an exponential function rather than a linear one. On page 614, you will investigate the doubling time for world population.

Preview and Perspective

Exponential and logarithmic functions have a wide variety of uses, some of which you will see as you read through this chapter. You often read in newspaper and magazine articles that health care spending, the federal deficit, and the world population, to list just a few, are growing exponentially. By the time you finish this chapter you should have a clear understanding of just what this means.

In Section 9.1 we introduce composite and inverse functions. The remainder of the chapter is spent discussing logarithmic and exponential functions. Common logarithms, that is, logarithms to the base 10, are discussed in Section 9.5.

In Section 9.6 we introduce very special functions, the natural exponential function and the natural logarithmic function. In both of these functions the base is e, an irrational number whose value is approximately 2.7183. Many natural phenomena, such as carbon dating, radioactive decay, and the growth of savings invested in an account compounding interest continuously, can be described by natural exponential functions.

9.1 COMPOSITE AND INVERSE FUNCTIONS

SSM

VIDEO 9.1

CD Rom

1. **Find composite functions.**
2. **Understand one-to-one functions.**
3. **Find inverse functions.**
4. **Find the composition of a function and its inverse.**

The focus of this chapter is logarithms. However, before we can study logarithms we need to discuss composite functions, one-to-one functions, and inverse functions. Let's start with composite functions.

1 Find Composite Functions

Often we come across situations where one quantity is a function of one variable. That variable, in turn, is a function of some other variable. For example, the cost of advertising on a television show may be a function of the Nielson rating of the show. The Nielson rating, in turn, is a function of the number of people who watch the show. So, in the final outcome, the cost of advertising may be affected by the number of people who watch the show. Functions like this are called *composite functions*.

Let's consider another example. Suppose that 1 U.S. dollar can be converted to 1.30 Canadian dollars and that 1 Canadian dollar can be converted to 6.2 Mexican pesos. Using this information, we can convert 20 U.S. dollars to Mexican pesos. We have the following functions.

$$g(x) = 1.30x \text{ (U.S. dollars to Canadian dollars)}$$

$$f(x) = 6.2x \text{ (Canadian dollars to Mexican pesos)}$$

If we let $x = 20$, for \$20 U.S., then it can be converted into \$26 Canadian using function g:

$$g(x) = 1.30x$$

$$g(20) = 1.30(20) = \$26 \text{ Canadian}$$

The \$26 Canadian can, in turn, be converted into 161.20 Mexican pesos using function f:

$$f(x) = 6.2x$$
$$f(26) = 6.2(26) = 161.20 \text{ Mexican pesos}$$

Is there a way of finding this conversion without performing this string of calculations? The answer is yes. One U.S. dollar can be converted to Mexican pesos by substituting the $1.3x$ found in function $g(x)$ for the x in $f(x)$. This gives a new function, h, which converts U.S. dollars directly into Mexican pesos.

$$g(x) = 1.30x, \qquad f(x) = 6.2x$$
$$h(x) = f[g(x)]$$
$$= 6.2(1.30x) \quad \text{Substitute } g(x) \text{ for } x \text{ in } f(x)$$
$$= 8.06x$$

Thus, for each U.S. dollar, x, we get 8.06 Mexican pesos. If we substitute \$20 for x, we get 161.20 pesos, which is what we expected.

$$h(x) = 8.06x$$
$$h(20) = 8.06(20) = 161.20$$

Function h, called a **composition of f with g**, is denoted $(f \circ g)$ and is read "f composed with g" or "f circle g." Figure 9.1 shows how the composite function h relates to functions f and g.

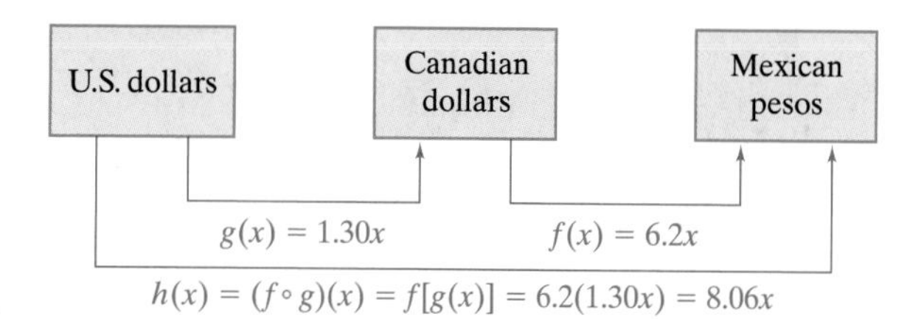

FIGURE 9.1

We now define **composite function**.

Definition

Composite Function

The **composite function** $f \circ g$ is defined as

$$(f \circ g)(x) = f[g(x)]$$

When we are given $f(x)$ and $g(x)$, to find $(f \circ g)(x)$ we substitute $g(x)$ for x in $f(x)$ to get $f[g(x)]$.

EXAMPLE 1 Given $f(x) = x^2 - 2x + 3$ and $g(x) = x - 5$, find

a) $f(a)$ **b)** $(f \circ g)(x)$ **c)** $(f \circ g)(3)$

Solution **a)** To find $f(a)$, we substitute a for each x in $f(x)$.

$$f(x) = x^2 - 2x + 3$$
$$f(a) = a^2 - 2a + 3$$

b) $(f \circ g)(x) = f[g(x)]$. To find $(f \circ g)(x)$, we substitute $g(x)$, which is $x - 5$, for each x in $f(x)$.

$$f(x) = x^2 - 2x + 3, \qquad g(x) = x - 5$$
$$\begin{aligned} f[g(x)] &= (x-5)^2 - 2(x-5) + 3 \\ &= (x-5)(x-5) - 2x + 10 + 3 \\ &= x^2 - 10x + 25 - 2x + 13 \\ &= x^2 - 12x + 38 \end{aligned}$$

Therefore, the composite function of f with g is $x^2 - 12x + 38$.

$$(f \circ g)(x) = f[g(x)] = x^2 - 12x + 38$$

c) To find $(f \circ g)(3)$, we substitute 3 for x in $(f \circ g)(x)$.

$$(f \circ g)(x) = x^2 - 12x + 38$$
$$(f \circ g)(3) = 3^2 - 12(3) + 38 = 11$$

How do you think we would determine $(g \circ f)(x)$ or $g[f(x)]$? If you answered "substitute $f(x)$ for each x in $g(x)$," you answered correctly. Using $f(x)$ and $g(x)$ as given in Example 1, we find $(g \circ f)(x)$ as follows.

$$g(x) = x - 5, \qquad f(x) = x^2 - 2x + 3$$
$$\begin{aligned} g[f(x)] &= (x^2 - 2x + 3) - 5 \\ &= x^2 - 2x + 3 - 5 \\ &= x^2 - 2x - 2 \end{aligned}$$

Therefore, the composite function of g with f is $x^2 - 2x - 2$.

$$(g \circ f)(x) = g[f(x)] = x^2 - 2x - 2$$

By comparing the illustrations above, we see that $f[g(x)] \neq g[f(x)]$.

EXAMPLE 2 Given $f(x) = x^2 + 4$ and $g(x) = \sqrt{x-2}$, find

a) $(f \circ g)(x)$ **b)** $(g \circ f)(x)$

Solution **a)** To find $(f \circ g)(x)$, we substitute $g(x)$, which is $\sqrt{x-2}$, for each x in $f(x)$. You should realize that $\sqrt{x-2}$ is a real number only when $x \geq 2$.

$$f(x) = x^2 + 4$$
$$(f \circ g)(x) = f[g(x)] = \left(\sqrt{x-2}\right)^2 + 4 = x - 2 + 4 = x + 2, \quad x \geq 2.$$

Since values of $x < 2$ are not in the domain of $g(x)$, values of $x < 2$ are not in the domain of $(f \circ g)(x)$.

b) To find $(g \circ f)(x)$, we substitute $f(x)$, which is $x^2 + 4$, for each x in $g(x)$.

$$g(x) = \sqrt{x - 2}$$
$$(g \circ f)(x) = g[f(x)] = \sqrt{(x^2 + 4) - 2} = \sqrt{x^2 + 2}$$

EXAMPLE 3 Given $f(x) = x - 1$ and $g(x) = x + 7$, find

a) $(f \circ g)(x)$ **b)** $(f \circ g)(2)$ **c)** $(g \circ f)(x)$ **d)** $(g \circ f)(2)$

Solution **a)**

$$f(x) = x - 1$$
$$(f \circ g)(x) = f[g(x)] = (x + 7) - 1 = x + 6$$

b) We find $(f \circ g)(2)$ by substituting 2 for each x in $(f \circ g)(x)$.

$$(f \circ g)(x) = x + 6$$
$$(f \circ g)(2) = 2 + 6 = 8$$

c)
$$g(x) = x + 7$$
$$(g \circ f)(x) = g[f(x)] = (x - 1) + 7 = x + 6$$

d) Since $(g \circ f)(x) = x + 6$, $(g \circ f)(2) = 2 + 6 = 8$.

In general, $(f \circ g)(x) \neq (g \circ f)(x)$. In Example 3, $(f \circ g)(x) = (g \circ f)(x)$, but this is only due to the specific functions used.

NOW TRY EXERCISE 11

HELPFUL HINT

Do not confuse finding the product of two functions with finding a composite function.

Product of functions f and g:	$(fg)(x)$	or	$(f \cdot g)(x)$
Composite function of f with g:	$(f \circ g)(x)$		

When multiplying functions f and g, we can use a dot between the f and g. When finding the composite function of f with g, we use a small *open* circle.

2) Understand One-to-One Functions

Consider the following two sets of ordered pairs.

$$A = \{(1, 2), (3, 5), (4, 6), (-2, 1)\}$$
$$B = \{(1, 2), (3, 5), (4, 6), (-2, 5)\}$$

Both sets of ordered pairs, A and B, are functions since each value of x has a unique value of y. In set A, each value of y also has a unique value of x, as shown in Figure 9.2. In set B, each value of y does not have a unique value of x. In the ordered pairs $(3, 5)$ and $(-2, 5)$, the y-value 5 corresponds with two values of x, as shown in Figure 9.3.

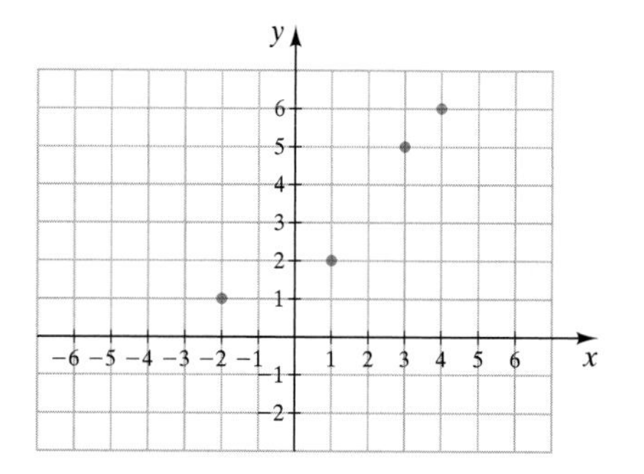

FIGURE 9.2

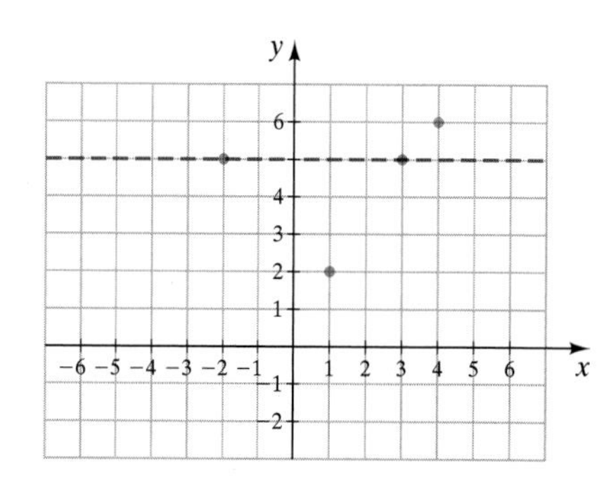

FIGURE 9.3

The set of ordered pairs in A is an example of a *one-to-one function*. The set of ordered pairs in B is not a one-to-one function. In a **one-to-one function**, each value in the range has a unique value in the domain. Thus, if y is a one-to-one function of x, in addition to each x-value having a unique y-value (the definition of a function), each y-value must also have a unique x-value.

Definition

One-to-One Function

A function is a **one-to-one function** if each value in the range corresponds with exactly one value in the domain.

For a function to be a one-to-one function, it must pass not only a **vertical line test** (the test to ensure that it is a function) but also a **horizontal line test** (to test the one-to-one criteria).

Consider the function $f(x) = x^2$ (see Fig. 9.4). Note that it is a function since it passes the vertical line test. For each value of x, there is a unique value of y. Does each value of y also have a unique value of x? The answer is no, as illustrated in Figure 9.5. Note that for the indicated value of y there are two values of x, namely x_1 and x_2. If we limit the domain of $f(x) = x^2$ to values of x-greater than or equal to 0, then each x-value has a unique y-value and each y-value also has a unique x-value (see Fig. 9.6). The function $f(x) = x^2, x \geq 0$, Figure 9.6, is an example of a one-to-one function.

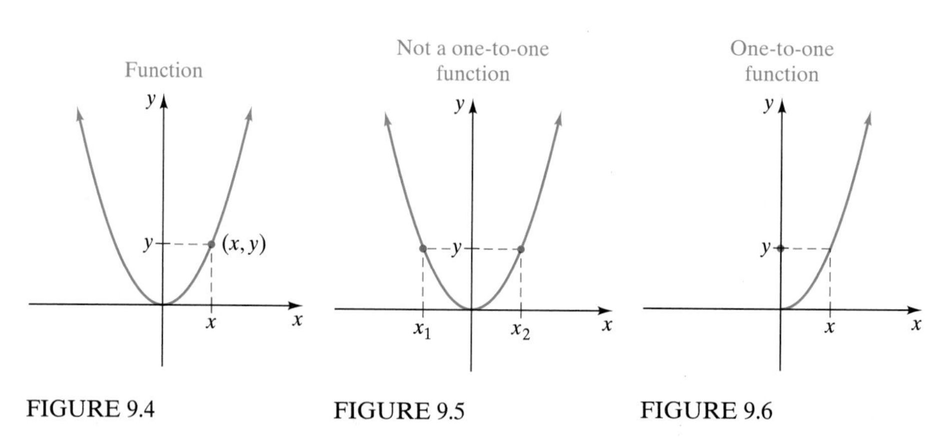

FIGURE 9.4 FIGURE 9.5 FIGURE 9.6

In Figure 9.7 all the graphs are functions since they all pass the vertical line test. However, only the graphs in parts (a), (d), and (e) are one-to-one functions since they also pass the horizontal line test.

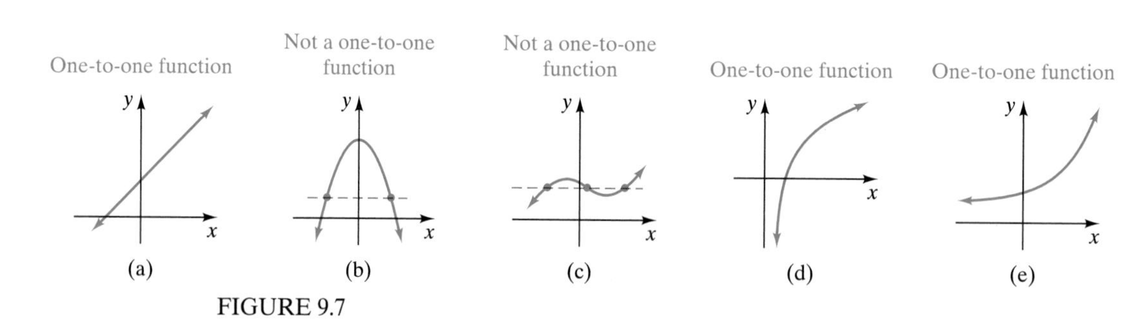

FIGURE 9.7

EXAMPLE 4 Determine which of the following functions are one-to-one functions.

a)

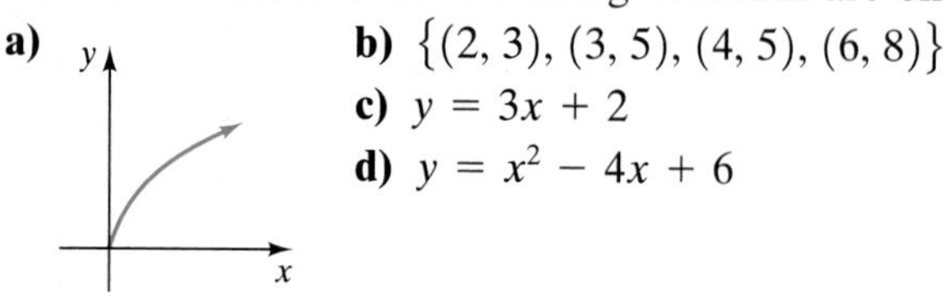

b) $\{(2, 3), (3, 5), (4, 5), (6, 8)\}$

c) $y = 3x + 2$

d) $y = x^2 - 4x + 6$

Solution **a)** Yes, the graph is a one-to-one function because it passes the horizontal line test.

b) No, the set of ordered pairs is not a one-to-one function. Note that the y-value 5 is associated with two x-values, 3 and 4.

c) Yes, the graph of this function is a straight line, and straight lines (except for horizontal lines) pass the horizontal line test.

d) No, the graph of this function is a parabola, and parabolas do not pass the horizontal line test.

NOW TRY EXERCISE 29

3) Find Inverse Functions

Now that we have discussed one-to-one functions, we can introduce inverse functions. You must be aware that **only one-to-one functions have inverse functions.** If a function is one-to-one, its **inverse function** may be obtained by interchanging the first and second coordinates in each ordered pair of the function. Thus, for each ordered pair (x, y) in the function, the ordered pair (y, x) will be in the inverse function. For example,

Function: $\{(1, 4), (2, 0), (3, 7), (-2, 1), (-1, -5)\}$

Inverse function: $\{(4, 1), (0, 2), (7, 3), (1, -2), (-5, -1)\}$

Note that the domain of the function becomes the range of the inverse function, and the range of the function is the domain of the inverse function.

If we graph the points in the function and the points in the inverse function (Fig. 9.8), we see that the points are symmetric with respect to the line $y = x$.

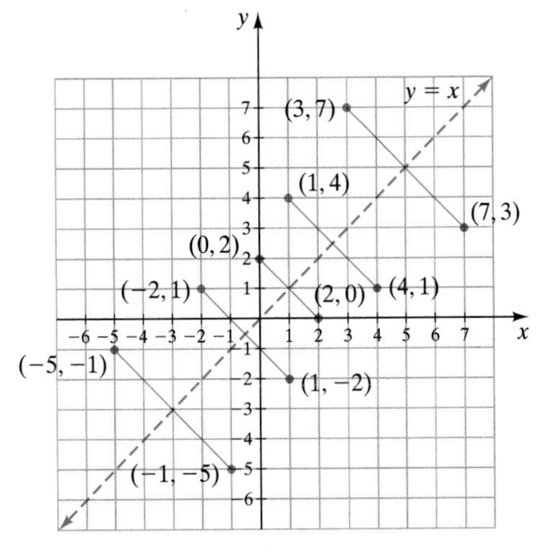

FIGURE 9.8

For a function $f(x)$, the notation $f^{-1}(x)$ represents its inverse function. Note that the -1 in the notation is *not* an exponent.

Definition

Inverse Function

If $f(x)$ is a one-to-one function with ordered pairs of the form (x, y), its **inverse function**, $f^{-1}(x)$, is a one-to-one function with ordered pairs of the form (y, x).

When a function $f(x)$ and its inverse function $f^{-1}(x)$ are graphed on the same axes, $f(x)$ *and* $f^{-1}(x)$ are *symmetric about the line* $y = x$ as seen in Figure 9.8.

When a one-to-one function is given as an equation, its inverse function can be found by the following procedure.

To Find the Inverse Function of a One-to-One Function

1. Replace $f(x)$ with y.
2. Interchange the two variables x and y. (This gives the inverse function.)
3. Solve the equation for y.
4. Replace y with $f^{-1}(x)$ (the inverse function notation).

The following example will illustrate the procedure.

EXAMPLE 5 **a)** Find the inverse function of $f(x) = 4x + 2$.

b) On the same axes, graph both $f(x)$ and $f^{-1}(x)$.

Solution **a)** We will follow the four-step procedure.

$$f(x) = 4x + 2$$

Step 1 $y = 4x + 2$ *Replace $f(x)$ with y.*

Step 2 $x = 4y + 2$ *Interchange x and y.*

Step 3 $x - 2 = 4y$ *Solve for y.*

$$\frac{x-2}{4} = y$$

$$y = \frac{x-2}{4}$$

Step 4 $f^{-1}(x) = \dfrac{x-2}{4}$ *Replace y with $f^{-1}(x)$.*

b) Below we show tables of values for $f(x)$ and $f^{-1}(x)$. The graphs of $f(x)$ and $f^{-1}(x)$ are shown in Figure 9.9.

x	$y = f(x)$
0	2
1	6

x	$y = f^{-1}(x)$
2	0
6	1

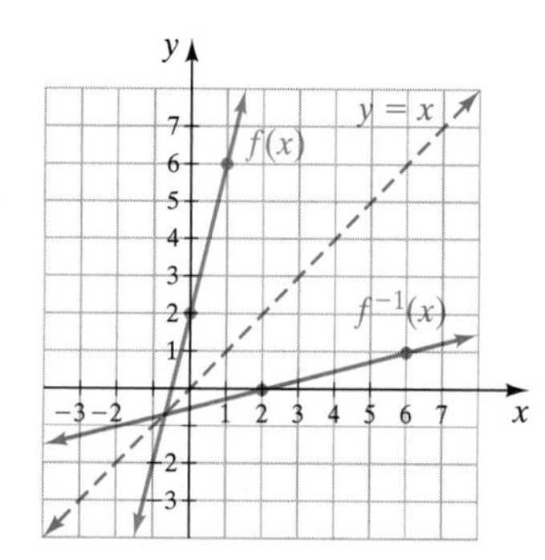

FIGURE 9.9

Note the symmetry of $f(x)$ and $f^{-1}(x)$ about the line $y = x$. Also note that both the domain and range of both $f(x)$ and $f^{-1}(x)$ are the set of real numbers, $\mathbb{R}$.

NEW TRY EXERCISE 53

In Chapter 7 when we solved equations containing cube roots, we cubed each side of the equation. To solve cubic equations, we raise each side of the equation to the one-third power, which is equivalent to taking the cube root of each side of the equation. Recall from Chapter 7 that $\sqrt[3]{a^3} = a$ for any real number a.

EXAMPLE 6 **a)** Find the inverse function of $f(x) = x^3 + 2$.

b) On the same axes, graph both $f(x)$ and $f^{-1}(x)$.

Solution **a)** This function is one-to-one, therefore we will follow the four-step procedure to find its inverse.

$$f(x) = x^3 + 2$$

Step 1 $\quad y = x^3 + 2 \quad$ *Replace $f(x)$ with y.*

Step 2 $\quad x = y^3 + 2 \quad$ *Interchange x and y.*

Step 3 $\quad x - 2 = y^3 \quad$ *Solve for y.*

$$\sqrt[3]{x-2} = \sqrt[3]{y^3}$$

$$\sqrt[3]{x-2} = y$$

$$y = \sqrt[3]{x-2}$$

Step 4 $\quad f^{-1}(x) = \sqrt[3]{x-2} \quad$ *Replace y with $f^{-1}(x)$.*

b) Below we show tables of values for $f(x)$ and $f^{-1}(x)$.

x	$y = f(x)$
−2	−6
−1	1
0	2
1	3
2	10

x	$y = f^{-1}(x)$
−6	−2
1	−1
2	0
3	1
10	2

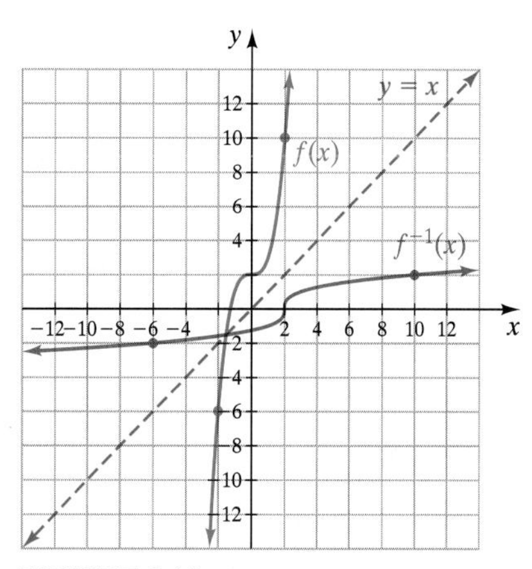

FIGURE 9.10

Notice for each point (a, b) on the graph of $f(x)$, the point (b, a) appears on the graph of $f^{-1}(x)$. The graphs of $f(x)$ and $f^{-1}(x)$ are shown in Figure 9.10.

NOW TRY EXERCISE 47

Using Your Graphing Calculator

In Example 6, we were given $f(x) = x^3 + 2$ and found that $f^{-1}(x) = \sqrt[3]{x - 2}$. The graphs of these functions are symmetric about the line $y = x$, although it may not appear that way on a graphing calculator. Figure 9.11 shows the two graphs using a standard calculator window.

Standard window

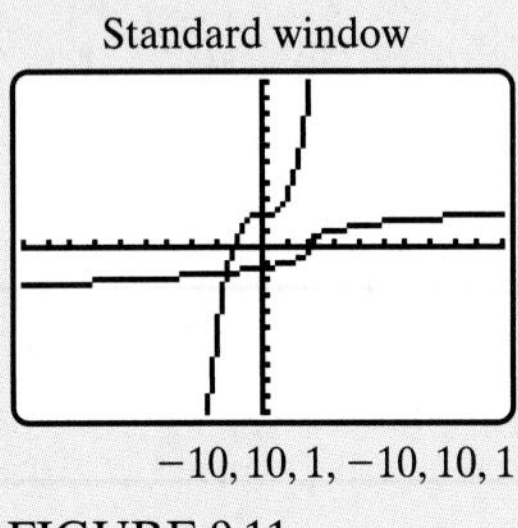

$-10, 10, 1, -10, 10, 1$

FIGURE 9.11

ZSquared window

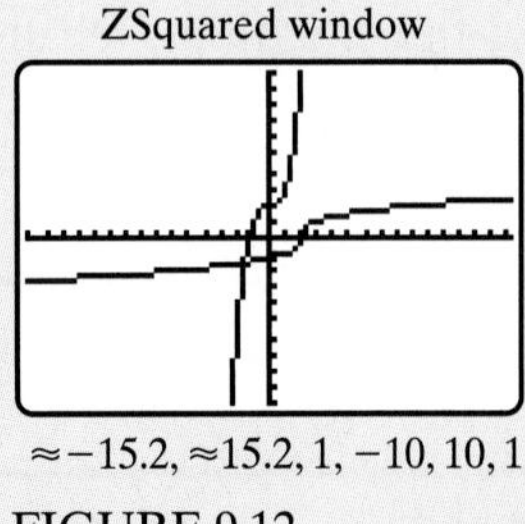

$\approx -15.2, \approx 15.2, 1, -10, 10, 1$

FIGURE 9.12

ZDecimal window

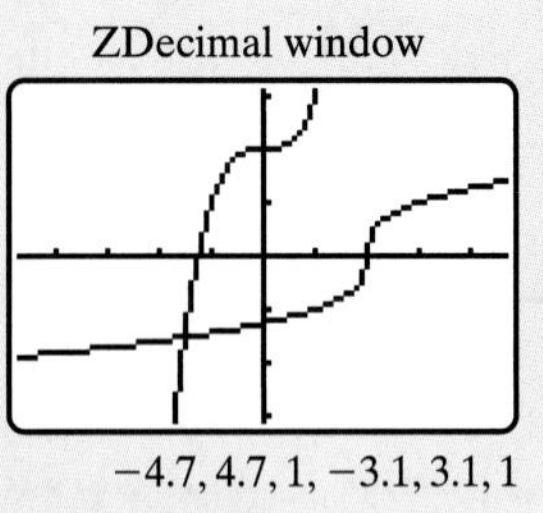

$-4.7, 4.7, 1, -3.1, 3.1, 1$

FIGURE 9.13

Since the horizontal axis is longer than the vertical axis, and both axes have ten tick marks, the graphs appear distorted. Many calculators have a feature to "square the axes." When this feature is used, the window is still rectangular, but the distance between the tick marks is equalized. To equalize the spacing between the tick marks on the vertical and horizontal axes on the TI-82 and TI-83, press ZOOM then select option 5, ZSquare. Figure 9.12 shows the graphs after this option is selected. A third illustration of the graphs can be obtained using ZOOM, option 4, ZDecimal. This option resets the x-axis to go from -4.7 to 4.7 and the y-axis to go from -3.1 to 3.1, as shown in Figure 9.13.

4 Find the Composition of a Function and Its Inverse

If two functions $f(x)$ and $f^{-1}(x)$ are inverses of each other, $(f \circ f^{-1})(x) = x$ and $(f^{-1} \circ f)(x) = x$.

EXAMPLE 7 In Example 5 we determined that for $f(x) = 4x + 2$, $f^{-1}(x) = \dfrac{x - 2}{4}$. Show that

a) $(f \circ f^{-1})(x) = x$, **b)** $(f^{-1} \circ f)(x) = x$

Solution **a)** To determine $(f \circ f^{-1})(x)$, substitute $f^{-1}(x)$ for each x in $f(x)$.

$$f(x) = 4x + 2$$

$$(f \circ f^{-1})(x) = 4\left(\frac{x - 2}{4}\right) + 2$$

$$= x - 2 + 2 = x$$

b) To determine $(f^{-1} \circ f)(x)$, substitute $f(x)$ for each x in $f^{-1}(x)$.

$$f^{-1}(x) = \frac{x - 2}{4}$$

$$(f^{-1} \circ f)(x) = \frac{4x + 2 - 2}{4}$$

$$= \frac{4x}{4} = x$$

Thus, $(f \circ f^{-1})(x) = (f^{-1} \circ f)(x) = x$.

EXAMPLE 8 In Example 6 we determined that $f(x) = x^3 + 2$ and $f^{-1}(x) = \sqrt[3]{x-2}$ are inverse functions. Show that

a) $(f \circ f^{-1})(x) = x$ **b)** $(f^{-1} \circ f)(x) = x$

Solution **a)** To determine $(f \circ f^{-1})(x)$, substitute $f^{-1}(x)$ for each x in $f(x)$.

$$f(x) = x^3 + 2$$
$$(f \circ f^{-1})(x) = \left(\sqrt[3]{x-2}\right)^3 + 2$$
$$= x - 2 + 2 = x$$

b) To determine $(f^{-1} \circ f)(x)$, substitute $f(x)$ for each x in $f^{-1}(x)$.

$$f^{-1}(x) = \sqrt[3]{x - 2}$$
$$(f^{-1} \circ f)(x) = \sqrt[3]{(x^3 + 2) - 2}$$
$$= \sqrt[3]{x^3} = x$$

Thus, $(f \circ f^{-1})(x) = (f^{-1} \circ f)(x) = x$.

Because a function and its inverse "undo" each other, the composite of a function with its inverse results in the given value from the domain. For example, for any function $f(x)$ and its inverse $f^{-1}(x)$, $(f^{-1} \circ f)(3) = 3$, and $(f \circ f^{-1})\left(-\frac{1}{2}\right) = -\frac{1}{2}$.

NEW TRY EXERCISE 61

Exercise Set 9.1

Concept/Writing Exercises

1. Explain how to find $(f \circ g)(x)$ when you are given $f(x)$ and $g(x)$.
2. Explain how to find $(g \circ f)(x)$ when you are given $f(x)$ and $g(x)$.
3. **a)** What are one-to-one functions?
   b) Explain how you may determine whether a graph is a one-to-one function.
4. Do all functions have inverse functions? If not, which functions do?
5. Consider the set of ordered pairs $\{(3, 5), (4, 2), (-1, 3), (0, -2)\}$.
   a) Is this set of ordered pairs a function? Explain.
   b) Does this function have an inverse? Explain.
   c) If this function has an inverse, give the inverse function. Explain how you determined your answer.
6. Suppose $f(x)$ and $g(x)$ are inverse functions.
   a) What is $(f \circ g)(x)$ equal to?
   b) What is $(g \circ f)(x)$ equal to?
7. What is the relationship between the domain and range of a function and the domain and range of its inverse function?
8. What is the value of $(f \circ f^{-1})(6)$? Explain.

Practice the Skills

For each pair of functions, find **a)** $(f \circ g)(x)$, **b)** $(f \circ g)(4)$, **c)** $(g \circ f)(x)$, *and* **d)** $(g \circ f)(4)$.

9. $f(x) = x^2 + 5,\ g(x) = x - 4$
10. $f(x) = x + 3,\ g(x) = x^2 + x - 4$
11. $f(x) = x + 2,\ g(x) = x^2 + 4x - 2$
12. $f(x) = x + 1,\ g(x) = \frac{1}{x}$
13. $f(x) = \frac{3}{x},\ g(x) = x^2 + 1$
14. $f(x) = x^2 + 3,\ g(x) = x^2 - 6$
15. $f(x) = x - 4,\ g(x) = \sqrt{x + 5};\ x \geq -5$
16. $f(x) = \sqrt{x + 4},\ x \geq -4,\ g(x) = x + 5$

Determine whether each function is a one-to-one function.

17. **18.** **19.** **20.**

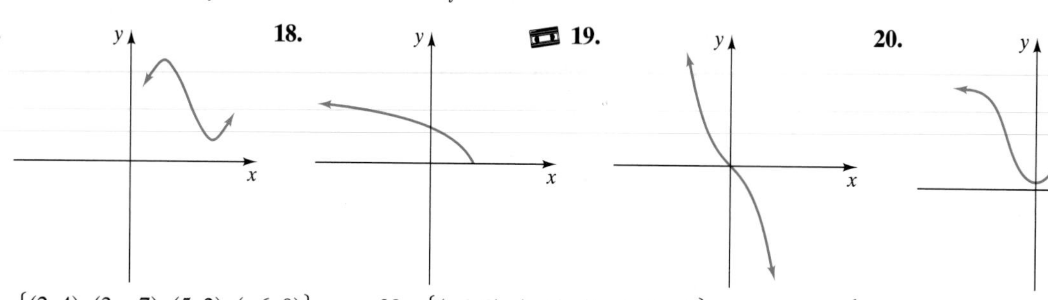

21. $\{(2, 4), (3, -7), (5, 3), (-6, 0)\}$

22. $\{(-4, 2), (2, 3), (4, 1), (0, 4)\}$

23. $\{(-4, 2), (5, 3), (0, 2), (3, 7)\}$

24. $\{(0, 5), (1, 4), (-3, 5), (4, 2)\}$

25. $y = 2x + 4$

26. $y = 3x - 6$

27. $y = x^2 - 4$

28. $y = x^2 - 2x + 4$

29. $y = x^2 - 4, x \geq 0$

30. $y = x^2 - 4, x \leq 0$

31. $y = |x|$

32. $y = \sqrt{x}$

For the given function, find the domain and range of both f(x) and $f^{-1}(x)$.

33. $\{(4, 0), (9, 3), (2, 7), (-1, 6), (-2, 4)\}$

34. $\left\{(-2, -3), (-4, 0), (5, 3), (6, 2), \left(2, \frac{1}{2}\right)\right\}$

35.

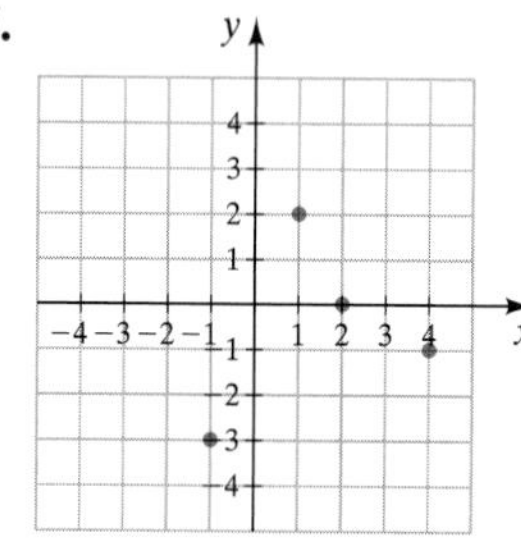

36.

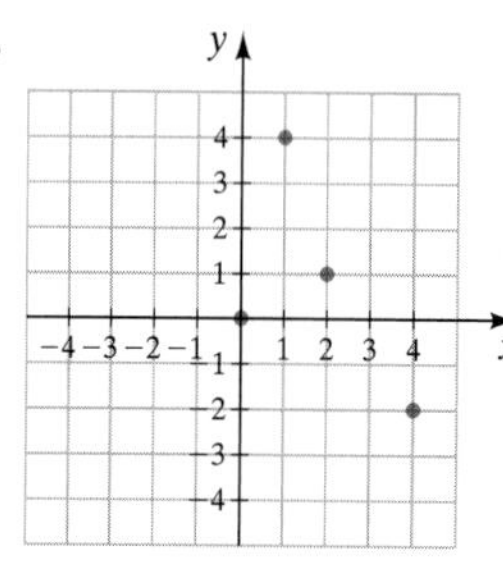

37. **38.**

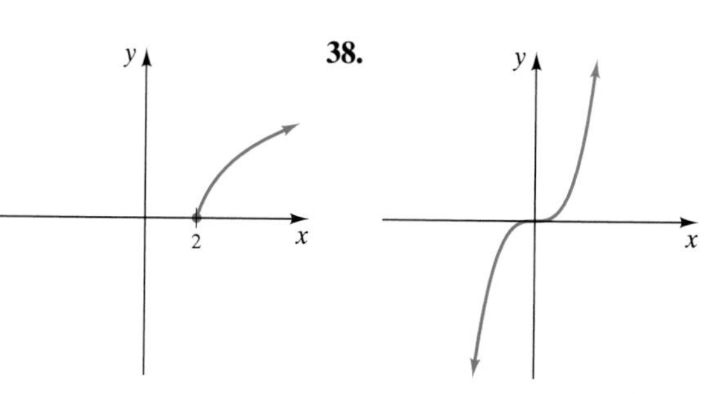

*For each function **a)** determine whether it is one-to-one; **b)** if it is one-to-one, find its inverse function.*

39. $f(x) = x - 5$

40. $g(x) = 2x - 6$

41. $h(x) = 4x$

42. $f(x) = x^2$

43. $g(x) = \frac{1}{x}$

44. $h(x) = \frac{2}{x}$

45. $f(x) = x^2 + 4$

46. $g(x) = x^3 + 6$

47. $g(x) = x^3 - 5$

48. $f(x) = \sqrt{x}, x \geq 0$

49. $g(x) = \sqrt{x + 2}, x \geq -2$

50. $f(x) = x^2 - 1, x \geq 0$

51. $h(x) = x^2 - 4, x \geq 0$

52. $h(x) = |x|$

*For each one-to-one function, **a)** find $f^{-1}(x)$ and **b)** graph f(x) and $f^{-1}(x)$ on the same axes.*

53. $f(x) = 2x + 8$

54. $f(x) = -3x + 6$

55. $f(x) = \sqrt{x}, x \geq 0$

56. $f(x) = \sqrt{x + 1}, x \geq -1$

57. $f(x) = \sqrt[3]{x}$

58. $f(x) = \sqrt[3]{x + 1}$

59. $f(x) = \frac{1}{x}, x > 0$

60. $f(x) = \frac{1}{x}$

For each pair of inverse functions, show that $(f \circ f^{-1})(x) = x$ and $(f^{-1} \circ f)(x) = x$.

61. $f(x) = x - 4, f^{-1}(x) = x + 4$

62. $f(x) = 3x + 2, f^{-1}(x) = \frac{x - 2}{3}$

63. $f(x) = \sqrt[3]{x - 2}, f^{-1}(x) = x^3 + 2$

64. $f(x) = \sqrt[3]{x + 8}, f^{-1}(x) = x^3 - 8$

65. $f(x) = \dfrac{2}{x}, f^{-1}(x) = \dfrac{2}{x}$

66. $f(x) = \sqrt{x + 1}, f^{-1}(x) = x^2 - 1, x \geq 0$

Problem Solving

67. Is $(f \circ g)(x) = (g \circ f)(x)$ for all values of x? Explain and give an example to support your answer.

68. Consider the functions $f(x) = \sqrt{x + 5}, x \geq -5$, and $g(x) = x^2 - 5, x \geq 0$.

a) Show that $(f \circ g)(x) = (g \circ f)(x)$ for $x \geq 0$.

b) Explain why we need to stipulate that $x \geq 0$ for part **a)** to be true.

69. Consider the functions $f(x) = x^3 + 2$ and $g(x) = \sqrt[3]{x - 2}$.

a) Show that $(f \circ g)(x) = (g \circ f)(x)$.

b) What are the domains of $f(x)$, $g(x)$, $(f \circ g)(x)$, and $(g \circ f)(x)$? Explain.

70. For the function $f(x) = x^3$, $f(2) = 2^3 = 8$. Explain why $f^{-1}(8) = 2$.

71. For the function $f(x) = x^4, x > 0, f(2) = 16$. Explain why $f^{-1}(16) = 2$.

72. The function $f(x) = 12x$ converts feet, x, into inches. Find the inverse function that converts inches into feet. In the inverse function, what do x and $f^{-1}(x)$ represent?

73. The function $f(x) = 3x$ converts yards, x, into feet. Find the inverse function that converts feet into yards. In the inverse function, what do x and $f^{-1}(x)$ represent?

74. The function $f(x) = \frac{22}{15}x$ converts miles per hour, x, into feet per second. Find the inverse function that converts feet per second to miles per hour. Explain how the inverse function is used.

75. The function $f(x) = \frac{5}{9}(x - 32)$ converts degrees Fahrenheit, x, to degrees Celsius. Find the inverse function that changes degrees Celsius into degrees Fahrenheit. Explain how the inverse function is used.

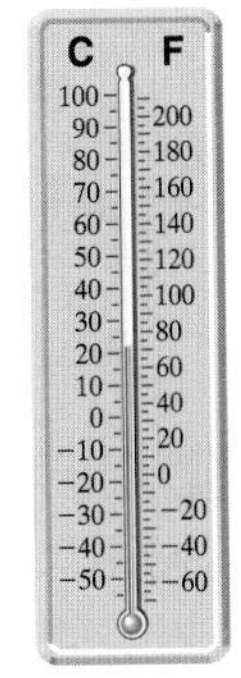

76. **a)** Does the function $f(x) = |x|$ have an inverse? Explain why.

b) If the domain is limited to $x \geq 0$, does the function have an inverse? Explain.

c) Find the inverse function of $f(x) = |x|, x \geq 0$.

Use your graphing calculator to determine whether the following functions are inverses.

77. $f(x) = 3x - 4, g(x) = \dfrac{x}{3} + \dfrac{4}{3}$

78. $f(x) = \sqrt{4 - x^2}, g(x) = \sqrt{4 - 2x}$

79. $f(x) = x^3 - 12, g(x) = \sqrt[3]{x + 12}$

80. $f(x) = x^5 + 5, g(x) = \sqrt[5]{x - 5}$

Challenge Problems

81. When a pebble is thrown into a pond, the circle formed by the pebble hitting the water expands with time. The surface area of the expanding circle may be found by the formula $A = \pi r^2$. The radius, r, of the circle, in feet, is a function of time, t, in seconds. Suppose that the function is $r(t) = 2t$.

a) Find the radius of the circle at 3 seconds.

b) Find the surface area of the circle at 3 seconds.

c) Express the surface area as a function of time by finding $A \circ r$.

d) Using the function found in part **c)**, find the surface area of the circle at 3 seconds.

e) Do your answers in parts **b)** and **d)** agree? If not, explain why.

82. The surface area, S, of a spherical balloon of radius r, in inches, is found by $S(r) = 4\pi r^2$. If the balloon is being blown up at a constant rate by a machine, then the radius of the balloon is a function of time. Suppose that this function is $r(t) = 1.2t$, where t is in seconds.

a) Find the radius of the balloon at 2 seconds.

b) Find the surface area at 2 seconds.

c) Express the surface area as a function of time by finding $S \circ r$.

d) Using the function found in part **b)**, find the surface area after 2 seconds.

e) Do your answers in parts **b)** and **d)** agree? If not, explain why.

Group Activity

Discuss and answer Exercise 83 as a group.

83. Consider the function $f(x) = 2^x$. This is an example of an *exponential function*, which we will discuss in the next section.

a) Graph this function by substituting values for x and finding the corresponding values of $f(x)$.

b) Do you think this function has an inverse? Explain your answer.

c) Using the graph in part **a)**, draw the inverse function, $f^{-1}(x)$ on the same axes.

d) Explain how you obtained the graph of $f^{-1}(x)$.

Cumulative Review Exercises

[4.2] **84.** Solve the following system of equations.

$$2x + 3y - 4z = 18$$
$$x - y - z = 3$$
$$x - 2y - 2z = 2$$

[5.3] **85.** Divide using synthetic division.

$$(x^3 + 6x^2 + 6x - 8) \div (x + 2)$$

[7.4] **86.** Simplify $\sqrt{\dfrac{24x^3y^2}{3xy^3}}$.

[8.1] **87.** Solve the equation $x^2 + 2x - 6 = 0$ by completing the square.

9.2 EXPONENTIAL FUNCTIONS

1. Graph exponential functions
2. Solve applications of exponential functions

SSM VIDEO 9.2 CD Rom

1 Graph Exponential Functions

We often read about things that are growing exponentially. For example, you may read that the world population is growing exponentially or that the use of e-mail is growing exponentially. What does this indicate? The graph in Figure 9.14 shows present and projected future population growth worldwide. The graph in Figure 9.15 shows the present and projected number of Internet kiosks. Both graphs have the same general shape, as indicated by the brown curve. Both graphs are exponential functions and rise rapidly.

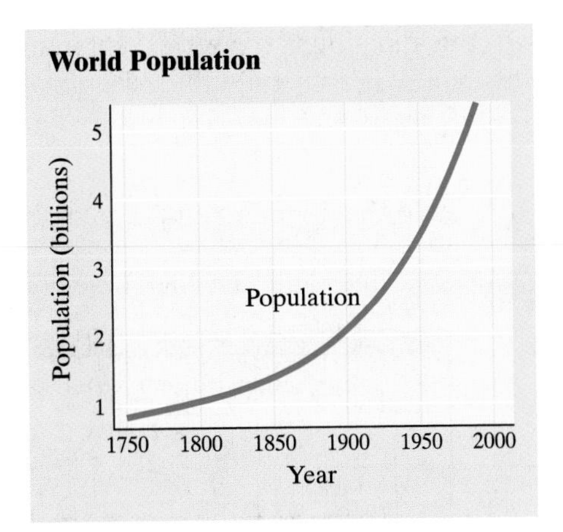

FIGURE 9.14

Number Of Internet Kiosks

Number of kiosks

0

1,000

2,000

3,000

4,000

5,000

1997 1998 1999 2000 2001

Year

Source: The Yankee Group. Adapted from *Newsweek*, May 25, 1998, p. 14.

*1998-2001 are projections

FIGURE 9.15

In the quadratic function $f(x) = x^2$, the variable is the base and the exponent is a constant. In the function $f(x) = 2^x$, the constant is the base and the variable is the exponent. The function $f(x) = 2^x$ is an example of an *exponential function*, which we now define.

Definition

Exponential Function

For any real number $a > 0$ and $a \neq 1$,

$$f(x) = a^x$$

is an **exponential function.**

An exponential function is a function of the form $f(x) = a^x$, where a is a positive real number not equal to 1.

Examples of Exponential Functions

$$f(x) = 2^x, \qquad f(x) = 5^x, \qquad f(x) = \left(\frac{1}{2}\right)^x$$

Since $y = f(x)$, functions of the form $y = a^x$ are also exponential functions. Exponential functions can be graphed by selecting values for x, finding the corresponding values of y [or $f(x)$], and plotting the points.

EXAMPLE 1 Graph the exponential function $y = 2^x$. State the domain and range of the function.

Solution First, construct a table of values.

x	−4	−3	−2	−1	0	1	2	3	4
y	$\frac{1}{16}$	$\frac{1}{8}$	$\frac{1}{4}$	$\frac{1}{2}$	1	2	4	8	16

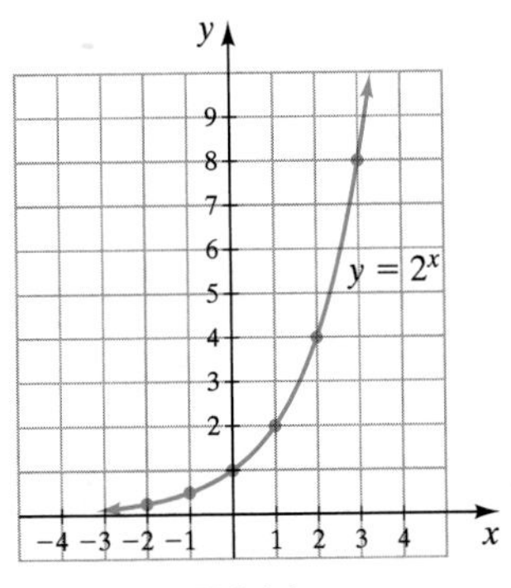

FIGURE 9.16

Now plot these points and connect them with a smooth curve (Fig. 9.16).

Domain: $\mathbb{R}$

Range: $\{y \mid y > 0\}$

The domain of this function is the set of real numbers, ℝ. The range is the set of values greater than 0. If you study the equation $y = 2^x$, you should realize that y must always be positive because 2 is positive.

EXAMPLE 2 Graph $y = \left(\frac{1}{2}\right)^x$. State the domain and range of the function.

Solution Construct a table of values and plot the curve (Fig. 9.17).

x	−4	−3	−2	−1	0	1	2	3	4
y	16	8	4	2	1	$\frac{1}{2}$	$\frac{1}{4}$	$\frac{1}{8}$	$\frac{1}{16}$

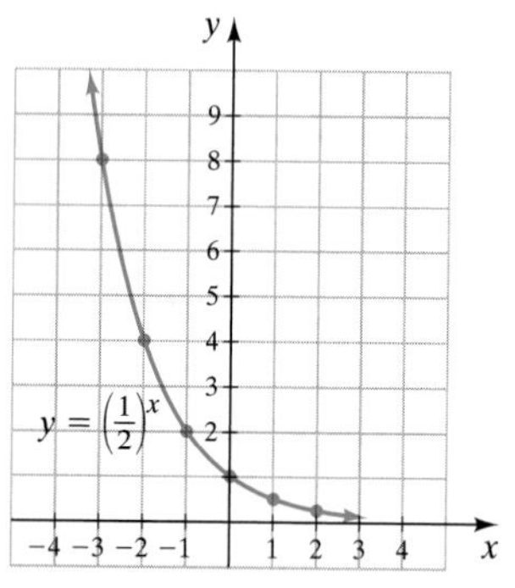

FIGURE 9.17

The domain is the set of real numbers ℝ. The range is $\{y \mid y > 0\}$.

NOW TRY EXERCISE 13

Note that the graphs in Figures 9.16 and 9.17 are both one-to-one functions. *The graphs of exponential functions of the form $y = a^x$ are similar to Figure 9.16 when $a > 1$ and similar to Figure 9.17 when $0 < a < 1$.* Note that $y = 1^x$ is not a one-to-one function, so we exclude it from our discussion of exponential functions.

What will the graph of $y = 2^{-x}$ look like? Remember that 2^{-x} means $\frac{1}{2^x}$ or $\left(\frac{1}{2}\right)^x$. Thus, the graph of $y = 2^{-x}$ will be identical to the graph in Figure 9.17. Now consider the equation $y = \left(\frac{1}{2}\right)^{-x}$. This equation may be rewritten as $y = 2^x$ since $\left(\frac{1}{2}\right)^{-x} = 2^x$. Thus, the graph of $y = \left(\frac{1}{2}\right)^{-x}$ will be identical to the graph in Figure 9.16.

Using Your Graphing Calculator

In Figure 9.18 we show the graph of the function $y = 2^x$ on the standard window of a graphing calculator. In this chapter we will sometimes use equations like $y = 2000(1.08)^x$. If you were to graph this function on a standard calculator window, you would not see any of the graph. Can you explain why? By observing the function, can you determine the y-intercept of the graph? To determine the y-intercept, substitute 0 for x. When you do so, you find the y-intercept is at $2000(1.08)^0 = 2000(1) = 2000$. In Figure 9.19 we show the graph of $y = 2000(1.08)^x$.

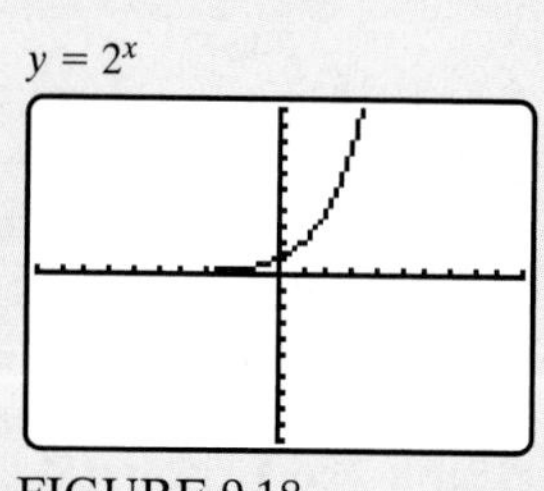

FIGURE 9.18

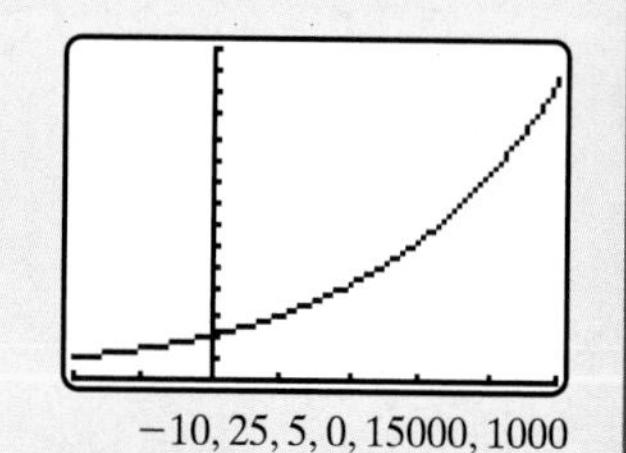

FIGURE 9.19

2) Solve Applications of Exponential Functions

Exponential functions are often used to describe the growth and decay of certain quantities. Example 3 illustrates an exponential equation used in genetics.

EXAMPLE 3 The number of gametes (or reproductive cells), g, in a certain species of plant is determined by the function $g = 2^n$, where n is the number of cells that an individual of the species commonly has. Determine the number of gametes if the individual has 12 cells.

Solution By evaluating 2^{12} on a calculator, we can determine that an individual of this species, with 12 cells, has 4096 gametes.

EXAMPLE 4 We have seen the *compound interest formula* $A = p\left(1 + \frac{r}{n}\right)^{nt}$ in earlier chapters. When interest is compounded periodically (yearly, monthly, quarterly), this formula can be used to find the amount, A.

In the formula, r is the interest rate, p is the principal, n is the number of compounding periods per year, and t is the number of years. Suppose that \$10,000 is invested at 8% interest compounded quarterly for 6 years. Find the amount in the account after 6 years.

Solution **Understand** We are given that the principal, p, is \$10,000. We are also given that the interest rate, r, is 8%. Because the interest is compounded quarterly, the number of compounding periods, n, is 4. The money is invested for 6 years. Therefore, t is 6.

Translate Now we substitute these values into the formula

$$A = p\left(1 + \frac{r}{n}\right)^{nt}$$

$$= 10{,}000\left(1 + \frac{0.08}{4}\right)^{4(6)}$$

Carry Out

$$= 10{,}000(1 + 0.02)^{24}$$

$$= 10{,}000(1.02)^{24}$$

$$\approx 10{,}000(1.608437) \quad \text{From a calculator}$$

$$\approx 16{,}084.37$$

Answer The original \$10,000 has grown to \$16,084.37 after 6 years.

NEW TRY EXERCISE 31

EXAMPLE 5 Carbon 14 dating is used by scientists to find the age of fossils and other artifacts. The formula used in carbon dating is

$$A = A_0 \cdot 2^{-t/5600}$$

where A_0 represents the amount of carbon 14 present when the fossil was formed and A represents the amount of carbon 14 present after t years. If 500 grams of carbon 14 were present when an organism died, how many grams will be found in the fossil 2000 years later?

Solution **Understand** When the fossil died, 500 grams of carbon 14 were present. Therefore $A_0 = 500$. To find out how many grams of carbon 14 will be present 2000 years later, we substitute 2000 for t in the formula.

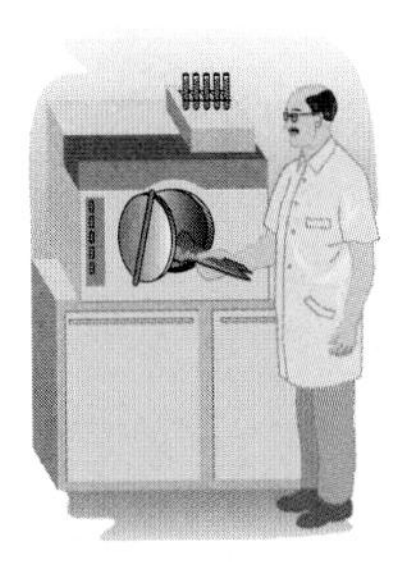

Translate

$$A = A_0 \cdot 2^{-t/5600}$$

$$= 500(2)^{-2000/5600}$$

Carry out

$$\approx 500(0.7807092) \quad \text{From a calculator}$$

$$\approx 390.35 \text{ grams}$$

Answer After 2000 years, about 390.35 of the original 500 grams of carbon 14 are still present.

NOW TRY EXERCISE 33

EXAMPLE 6 The graph in Figure 9.20 shows the number of students in the United States taking an advanced placement test from 1955 through 1995.

An exponential function that closely approximates this curve is $f(x) = 10.72(1.10)^x$. In the function, $f(x)$ represents the number who took the test, in thousands, and x is years since 1954. Using the given function, estimate the number of AP exams given in **a)** 1975 and **b)** 1995.

Students Taking AP Exams

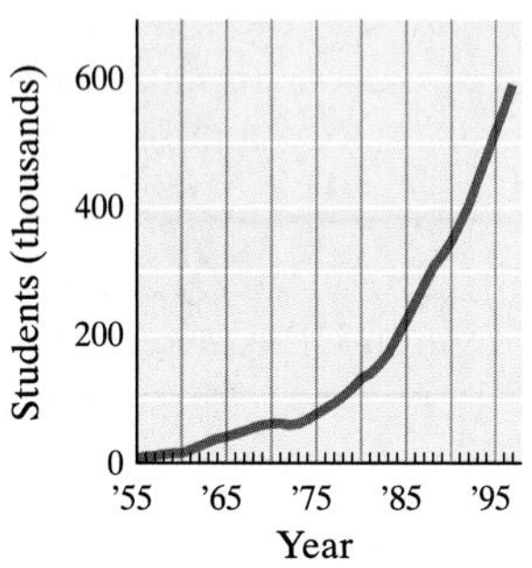

Source: Advanced Placement Program

FIGURE 9.20

Solution a) Understand In the function, x is years since 1954. Therefore, 1955 would be represented by $x = 1$, 1965 by $x = 11$, and so on. Since 1975 is 21 years since 1954, to find the number of students who took the AP test in 1975 we need to evaluate the function for $x = 21$.

Translate and $\quad f(x) = 10.72(1.10)^x$

Carry Out $\quad f(21) = 10.72(1.10)^{21} \approx 79.3306794$ *From a calculator*

Answer Therefore, about 79,331 students took an AP test in 1975.

b) Since 1995 is 41 years since 1954, to find the number of students who took the AP test in 1995 we need to evaluate the function for $x = 41$.

$$f(x) = 10.72(1.10)^x$$

$$f(41) = 10.72(1.10)^{41} \approx 533.6971417$$ *From a calculator*

NOW TRY EXERCISE 41 Therefore, about 533,697 students took a test in 1995.

Exercise Set 9.2

Concept/Writing Exercises

1. What are exponential functions?
2. Consider the exponential function $y = 2^x$.
   a) As x increases, what happens to y?
   b) Can y ever be 0? Explain.
   c) Can y ever be negative? Explain.
3. Consider the exponential function $y = \left(\frac{1}{2}\right)^x$.
   a) As x increases, what happens to y?
   b) Can y ever be 0? Explain.
   c) Can y ever be negative? Explain.
4. Consider the exponential function $y = 2^{-x}$. Write an equivalent exponential function that does not contain a negative sign in the exponent. Explain how you obtained your answer.
5. Consider the equations $y = 2^x$ and $y = 3^x$.
   a) Will both graphs have the same or different y-intercepts? Explain and determine their y-intercepts.
   b) How will the graphs of the two functions compare?
6. Consider the equation $y = \left(\frac{1}{3}\right)^x$.
   a) What is the y-intercept of the graph?
   b) How will the graphs of $y = \left(\frac{1}{3}\right)^x$ and $y = 3^{-x}$ compare? Explain.

Practice the Skills

Graph each exponential function.

7. $y = 2^x$
8. $y = 3^x$
9. $y = \left(\frac{1}{2}\right)^x$
10. $y = \left(\frac{1}{3}\right)^x$
11. $y = 4^x$
12. $y = 5^x$
13. $y = 3^{-x}$
14. $y = \left(\frac{1}{3}\right)^{-x}$
15. $y = 2^{x-1}$
16. $y = 3^{x-1}$
17. $y = 2^x - 1$
18. $y = 3^x - 1$
19. $y = 2^{2x} - 4$
20. $y = 3^{x-3}$

Problem Solving

21. We stated earlier that, for exponential functions $f(x) = a^x$, the value of a cannot equal 1.
    a) What does the graph of $f(x) = a^x$ look like when $a = 1$?
    b) Is $f(x) = a^x$ a function when $a = 1$?
    c) Does $f(x) = a^x$ have an inverse function when $a = 1$? Explain your answer.
22. How will the graphs of $y = a^x$ and $y = a^x + k, k > 0$, compare? Explain.
23. How will the graphs of $y = a^x$ and $y = a^x - k, k > 0$, compare?
24. For $a > 1$, how will the graphs of $y = a^x$ and $y = a^{x+1}$ compare?
25. For $a > 1$, how will the graphs of $y = a^x$ and $y = a^{x+2}$ compare?
26. a) Is $y = x^\pi$ an exponential function? Explain.
    b) Is $y = \pi^x$ an exponential function? Explain.
27. The following graph shows world population growth through history. The world population has been growing exponentially since the beginning of the industrial revolution. The following table indicates the population in 1650 (about 0.5 billion people) and each later year for which the world population doubled.

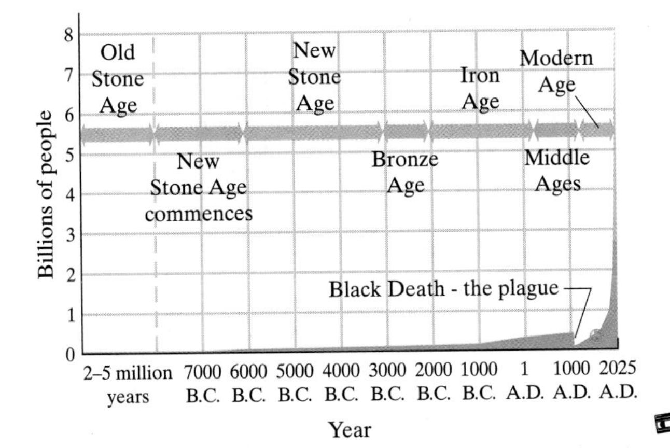

Year	Population (billions)
1650	0.5
1850	1
1930	2
1976	4
2016	8

a) Draw a graph of the data provided in the table. Place the year on the horizontal axis.
b) How long did it take for the population to double (called the *doubling time*) from 0.5 billion to 1 billion people?
c) What is the doubling time for the population to increase from 4 billion to the projected 8 billion people?

28. The following graph indicates linear growth of $100 invested at 7% simple interest and exponential growth at 7% interest compounded annually. In the formulas, A represents the amount in dollars and t represents the time in years.

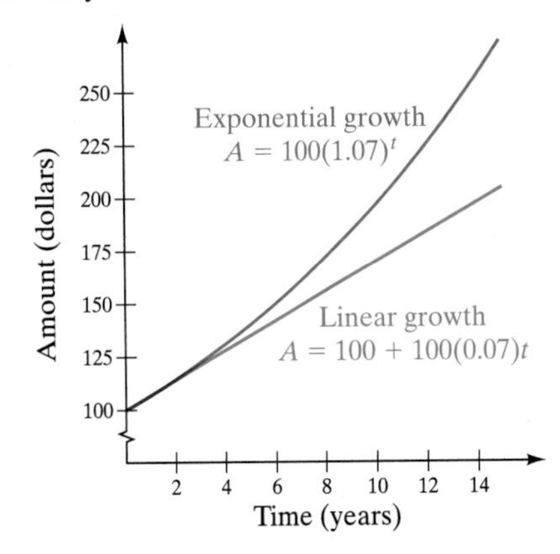

a) Use the graph to estimate the doubling time for $100 invested at 7% simple interest.
b) Estimate the doubling time for $100 invested at 7% interest compounded annually?
c) Estimate the difference in amounts after 10 years for $100 invested by each method.
d) Most banks compound interest daily instead of annually. What effect does this have on the total amount? Explain.

29. Use the formula $g = 2^n$ to determine the number of gametes a plant has if it has 8 cells (see Example 3).
30. Determine the number of gametes a plant has if it has 10 cells.
31. If Don Gecewicz invests $5000 at 6% interest compounded quarterly, find the amount after 4 years (see Example 4).
32. If Bridget Hunt invests $8000 at 8% interest compounded quarterly, find the amount after 5 years.
33. If 12 grams of carbon 14 are originally present in a certain animal bone, how much will remain at the end of 1000 years? Use $A = A_0 \cdot 2^{-t/5600}$ (see Example 5).
34. If 60 grams of carbon 14 are originally present in the fossil Tim Jonas found at an archeological site, how much will remain after 10,000 years?

35. The amount of a radioactive substance present, in grams, at time t in years is given by the formula $y = 80(2)^{-0.4t}$. Find the number of grams present in **a)** 10 years; **b)** 100 years.

36. The expected future population of Ackworth, which presently has 2000 residents, can be approximated by the formula $y = 2000\ (1.2)^{0.1x}$, where x is the number of years in the future. Find the expected population of the town in **a)** 10 years; **b)** 100 years.

37. The number of a certain type of bacteria present in a culture is determined by the equation $y = 5000(3)^x$, where x is the number of days the culture has been growing. Find the number of bacteria in **a)** 5 days; **b)** 7 days.

38. The average U.S. resident used about 116,000 gallons of water in 1998. Suppose that each year after 1998 the average resident is able to reduce the amount of water used by 5%. The amount of water used by the average resident t years after 1998 could then be found by the formula $A = 116{,}000(0.95)^t$.

a) Explain why this formula may be used to find the amount of water used.

b) What would be the average amount of water used in the year 2003?

39. Presently about $\frac{2}{3}$ of all aluminum cans are recycled each year, while about $\frac{1}{3}$ are disposed of in landfills. The recycled aluminum is used to make new cans. Americans used about 190,000,000 aluminum cans in 1998. The number of new cans made each year from recycled 1998 aluminum cans n years later can be estimated by the formula $A = 190{,}000{,}000\left(\frac{2}{3}\right)^n$.

a) Explain why the formula may be used to estimate the number of the cans made from recycled aluminum cans n years after 1998.

b) How many cans will be made from 1998 recycled aluminum cans in 2001?

40. Since 1979 the number of mutual fund accounts has been growing exponentially. The function $f(x) = 12.48(1.16)^x$ can be used to estimate the number of mutual fund accounts, in millions, where x is years since 1979 and $1 \le x \le 16$. Use the given function to estimate the number of mutual fund accounts in 1992.

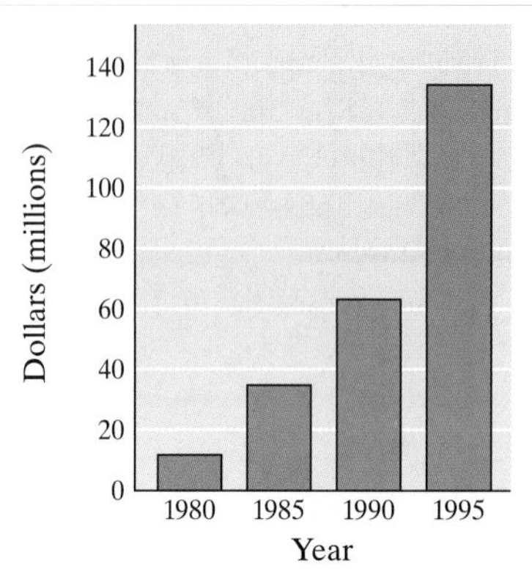

41. Atmospheric pressure varies with altitude. The greater the altitude the lower the pressure, as shown in the following graph.

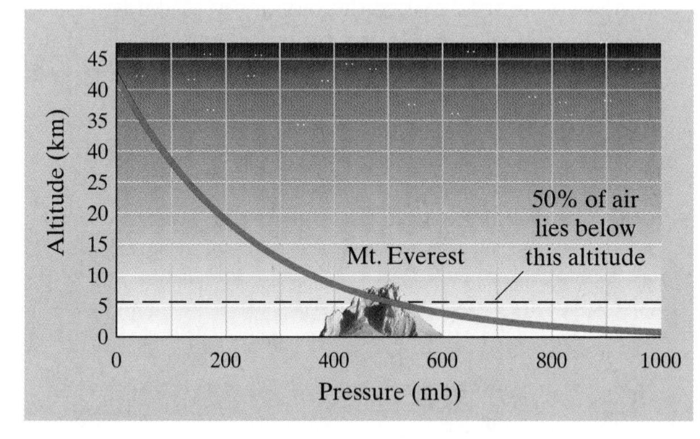

The equation $A = 41.97(0.996)^x$ can be used to estimate the altitude, A, in kilometers, for a given pressure, x, in millibars (mb). If the atmospheric pressure on top of Mt. Everest is about 389 mb, estimate the altitude of the top of Mt. Everest.

42. The number of e-mails internationally is expected to increase exponentially from 1997 to 2002. The number of e-mails per day, in millions, can be estimated by the function $f(x) = 2.56(2.49)^x$, where x is years since 1996 and $1 \le x \le 6$. Estimate the number of e-mails in 2002.

43. The cost of long-term care is increasing exponentially. The function $f(x) = 3318.10(1.08)^x$ can be used to estimate the cost of one year in a nursing home from 1964 through 2001. In the function, x represents years since 1964 and $1 \le x \le 37$. Estimate the cost of one year in a long-term care facility in 1998.

44. Do you think that the world's population can continue to grow exponentially forever? Explain.

Write both sides of each equation with the same base, then solve the equation. We will discuss problems like this in Section 9.6.

45. $2^{3x+2} = 16$

46. $16^x = 64$

47. In Exercise 28 we graphed the amount for various years when \$100 is invested at 7% simple interest and at 7% interest compounded annually.

a) Use the compound interest formula given in Example 4 to determine the amount if \$100 is compounded daily at 7% for 10 years (assume 365 days per year).

b) Estimate the difference in the amount in 10 years for the \$100 invested at 7% simple interest versus the 7% interest compounded daily.

48. Graph $y = 2^x$ and $y = 3^x$ on the same window.

49. **a)** Graph $y = 3^{x-5}$. **b)** Use your grapher to solve the equation $4 = 3^{x-5}$. Round your answer to the nearest hundredth.

50. **a)** Graph $y = \left(\frac{1}{2}\right)^{2x+3}$.

b) Use your grapher to solve the equation $-3 = \left(\frac{1}{2}\right)^{2x+3}$. Round your answer to the nearest hundredth.

Group Activity

51. Suppose Bob Jenkins gives Carol Dantuma \$1 on day 1, \$2 on day 2, \$4 on day 3, \$8 on day 4, and continues this doubling process for 30 days.

a) Group member 1: Determine how much Bob will give Carol on day 12.

b) Group member 2: Determine how much Bob will give Carol on day 15.

c) Group member 3: Determine how much Bob will give Carol on day 20.

As a group, answer parts **d), e),** and **f).**

d) Express the amount, using exponential form, that Bob gives Carol on day n.

e) How much, in dollars, will Bob give Carol on day 30? Write the amount in exponential form. Then use a calculator to evaluate.

f) Express the total amount Bob gives Carol over the 30 days as a sum of exponential terms. (Do not find the actual value.)

52. Functions that are exponential or are approximately exponential are commonly seen.

a) Have each member of the group individually determine a function not given in this section that may approximate an exponential function. You may use newspapers, books, or other sources.

b) As a group, discuss one another's functions. Determine whether each function presented is an exponential function.

c) As a group, write a paper that discusses each of the exponential functions and state why you believe each function is exponential.

Cumulative Review Exercises

[5.1] **53.** Consider the polynomial

$$2.3x^4y - 6.2x^6y^2 + 9.2x^5y^2.$$

a) Write the polynomial in descending order of the variable x.

b) What is the degree of the polynomial?

c) What is the leading coefficient?

[5.2] **54.** If $f(x) = x + 3$, $g(x) = x^2 - 2x + 4$, find $(f \cdot g)(x)$.

[7.1] **55.** Write $\sqrt{a^2 - 8a + 16}$ as an absolute value.

[7.4] **56.** Simplify $\sqrt[4]{\dfrac{32x^5y^9}{2y^3z}}$.

9.3 LOGARITHMIC FUNCTIONS

SSM VIDEO 9.3 CD Rom

1) Convert from exponential form to logarithmic form.
2) Graph logarithmic functions.
3) Compare the graphs of exponential and logarithmic functions.
4) Solve applications of logarithmic functions.

1) Convert from Exponential Form to Logarithmic Form

Now we are ready to introduce **logarithms.** Consider the exponential function $y = 2^x$. Recall from Section 9.1 that to find the inverse function, we interchange x and y and solve the equation for y. Interchanging x and y gives the equation

$x = 2^y$, which is the inverse of $y = 2^x$. But at this time we have no way of solving the equation $x = 2^y$ for y. To solve this equation for y, we introduce a new definition.

Definition

Logarithm

For all positive numbers a, where $a \neq 1$,

$$y = \log_a x \quad \text{means} \quad x = a^y$$

In the equation $y = \log_a x$, the word *log* is an abbreviation for the word *logarithm*; $y = \log_a x$ is read "y is the logarithm of x to the base a." The letter y represents the logarithm, the letter a represents the base, and the letter x represents the number.

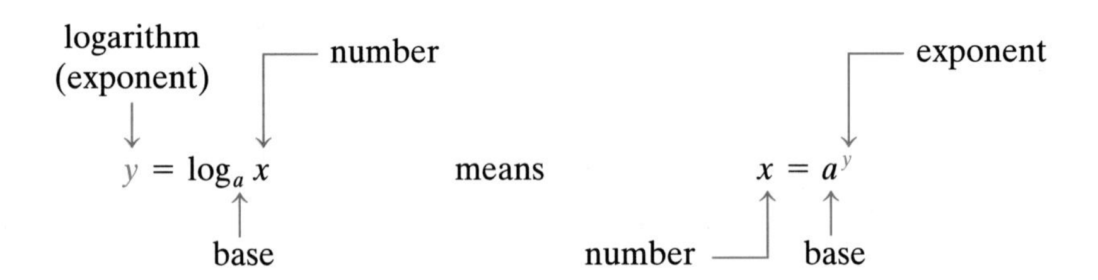

In words, the logarithm of the number x to the base a is the *exponent* to which the base a must be raised to equal the number x. In short, a logarithm is an exponent. For example,

$$2 = \log_{10} 100 \quad \text{means} \quad 100 = 10^2$$

In $\log_{10} 100 = 2$, the logarithm is 2, the base is 10, and the number is 100. The logarithm, 2, is the *exponent* to which the base, 10, must be raised to equal the number, 100. Note $10^2 = 100$.

Following are some examples of how an exponential expression can be converted to a logarithmic expression.

Exponential Form	**Logarithmic Form**
$10^0 = 1$	$\log_{10} 1 = 0$
$4^2 = 16$	$\log_4 16 = 2$
$\left(\frac{1}{2}\right)^5 = \frac{1}{32}$	$\log_{1/2} \frac{1}{32} = 5$
$5^{-2} = \frac{1}{25}$	$\log_5 \frac{1}{25} = -2$

Now let's do a few examples involving conversion from exponential form to logarithmic form, and vice versa.

EXAMPLE 1 Write each equation in logarithmic form.

a) $3^4 = 81$ **b)** $\left(\frac{1}{5}\right)^3 = \frac{1}{125}$ **c)** $2^{-4} = \frac{1}{16}$

NOW TRY EXERCISE 25

Solution **a)** $\log_3 81 = 4$ **b)** $\log_{1/5} \frac{1}{125} = 3$ **c)** $\log_2 \frac{1}{16} = -4$

EXAMPLE 2 Write each equation in exponential form.

a) $\log_6 36 = 2$ **b)** $\log_3 9 = 2$ **c)** $\log_{1/3} \frac{1}{81} = 4$

Solution **a)** $6^2 = 36$ **b)** $3^2 = 9$ **c)** $\left(\frac{1}{3}\right)^4 = \frac{1}{81}$

EXAMPLE 3 Write each equation in exponential form; then find the unknown value.

a) $y = \log_5 25$ **b)** $2 = \log_a 16$ **c)** $3 = \log_{1/2} x$

Solution **a)** $5^y = 25$. Since $5^2 = 25$, $y = 2$.

b) $a^2 = 16$. Since $4^2 = 16$, $a = 4$.

NOW TRY EXERCISE 55

c) $\left(\frac{1}{2}\right)^3 = x$. Since $\left(\frac{1}{2}\right)^3 = \frac{1}{8}$, $x = \frac{1}{8}$.

2) Graph Logarithmic Functions

Now that we know how to convert from exponential form to logarithmic form and vice versa, we can graph logarithmic functions. Equations of the form $y = \log_a x$, $a > 0$, $a \neq 1$, and $x > 0$, are called **logarithmic functions** since their graphs pass the vertical line test. To graph a logarithmic function, change it to exponential form and then plot points. This procedure is illustrated in Examples 4 and 5.

EXAMPLE 4 Graph $y = \log_2 x$. State the domain and range of the function.

Solution $y = \log_2 x$ means $x = 2^y$. Using $x = 2^y$, construct a table of values. The table will be easier to develop by selecting values for y and finding the corresponding values for x.

x	$\frac{1}{16}$	$\frac{1}{8}$	$\frac{1}{4}$	$\frac{1}{2}$	1	2	4	8	16
y	−4	−3	−2	−1	0	1	2	3	4

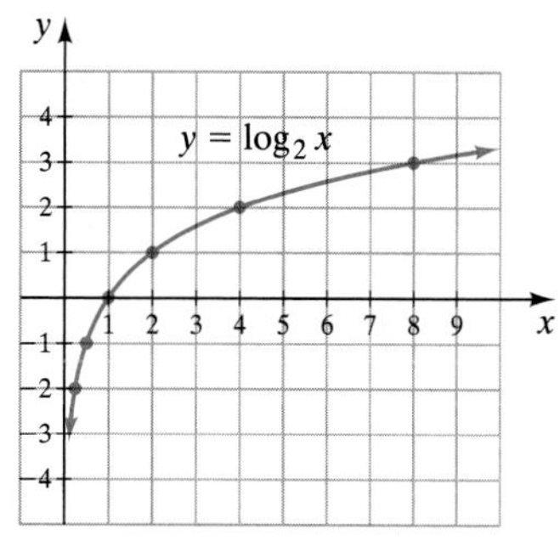

FIGURE 9.21

Now draw the graph (Fig. 9.21). The domain, the set of x-values, is $\{x \mid x > 0\}$. The range, the set of y-values, is all real numbers, $\mathbb{R}$.

EXAMPLE 5 Graph $y = \log_{1/2} x$. State the domain and range of the function.

Solution $y = \log_{1/2} x$ means $x = \left(\frac{1}{2}\right)^y$. Construct a table of values by selecting values for y and finding the corresponding values of x.

x	16	8	4	2	1	$\frac{1}{2}$	$\frac{1}{4}$	$\frac{1}{8}$	$\frac{1}{16}$
y	−4	−3	−2	−1	0	1	2	3	4

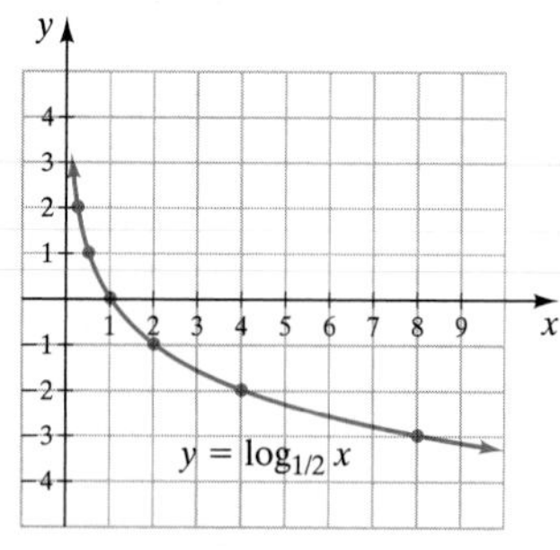

FIGURE 9.22

The graph is illustrated in Fig. 9.22. The domain is $\{x | x > 0\}$. The range is the set of real numbers, $\mathbb{R}$.

If we study the domains in Examples 4 and 5, we see that the domains of both $y = \log_2 x$ and $y = \log_{1/2} x$ are $\{x | x > 0\}$. In fact, **for any logarithmic function $y = \log_a x$, the domain is $\{x | x > 0\}$.** Also note that the graphs in Examples 4 and 5 are both graphs of one-to-one functions.

3 Compare the Graphs of Exponential and Logarithmic Functions

Recall that to find inverse functions we switch x and y and solve the resulting equation for y. Consider $y = a^x$. If we switch x and y, we get $x = a^y$. By our definition of logarithm this function may be rewritten as $y = \log_a x$, which is an equation solved for y. Therefore, $y = a^x$ and $y = \log_a x$ are inverse functions. In Figure 9.23, we show general graphs of $y = a^x$ and $y = \log_a x$, $a > 1$, on the same axes. Notice they are symmetric about the line $y = x$.

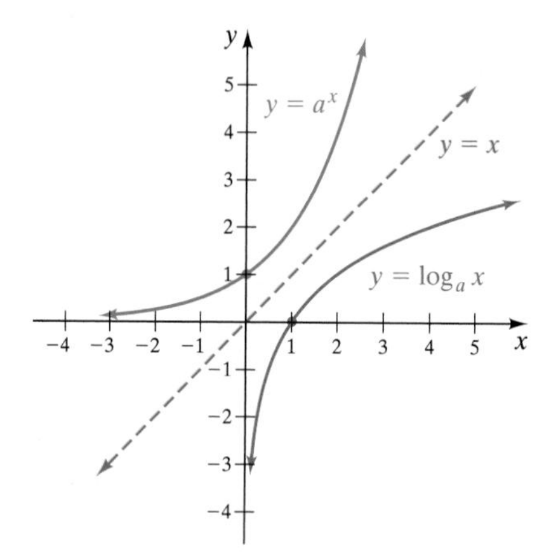

FIGURE 9.23

The graphs of $y = 2^x$ and $y = \log_2 x$ are illustrated in Figure 9.24. The graphs of $y = \left(\frac{1}{2}\right)^x$ and $y = \log_{1/2} x$ are illustrated in Figure 9.25. In each figure, the graphs are inverses of each other and are symmetric with respect to the line $y = x$.

NOW TRY EXERCISE 13

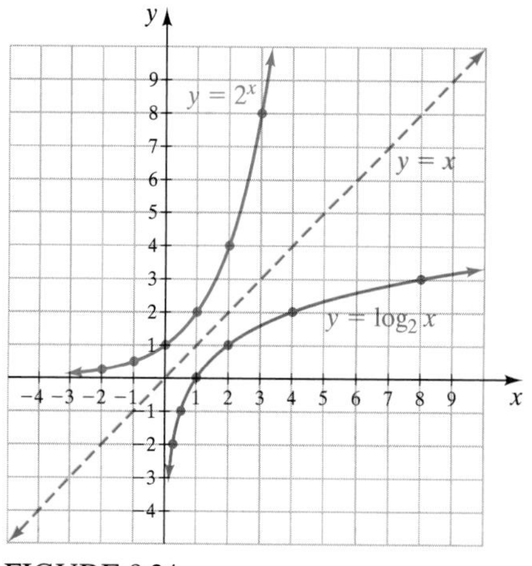

FIGURE 9.24

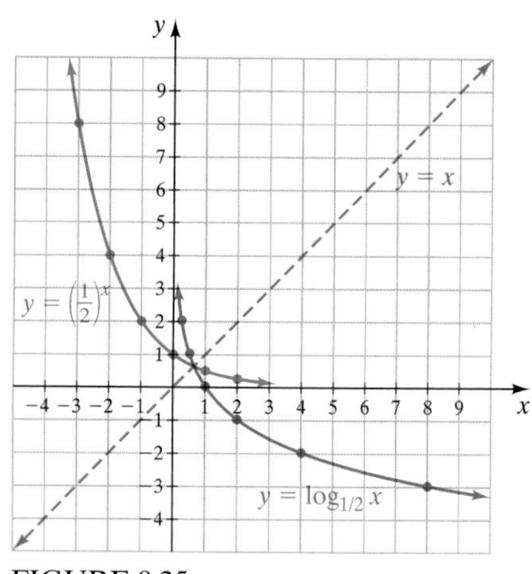

FIGURE 9.25

4 Solve Applications of Logarithmic Functions

We will see many applications of logarithms later, but let's look at one application now.

EXAMPLE 6 Logarithms are used to measure the magnitude of earthquakes. The Richter scale for measuring earthquakes was developed by Charles R. Richter. The magnitude, R, of an earthquake on the Richter scale is given by the formula

$$R = \log_{10} I$$

where I represents the number of times greater (or more intense) the earthquake is than the smallest measurable activity that can be measured on a seismograph.

a) If an earthquake measures 4 on the Richter scale, how many times more intense is it than the smallest measurable activity?

b) How many times more intense is an earthquake that measures 5 than an earthquake that measures 4?

Solution a) Understand The Richter number, R, is 4. To find how many times more intense the earthquake is than the smallest measurable activity, I, we substitute $R = 4$ in the formula and solve for I.

Translate

$$R = \log_{10} I$$

$$4 = \log_{10} I$$

Carry Out

$$10^4 = I \qquad \textit{Change to exponential form.}$$

$$10{,}000 = I$$

Answer Therefore, an earthquake that measures 4 is 10,000 times more intense than the smallest measurable activity.

b)

$$5 = \log_{10} I$$

$$10^5 = I \qquad \textit{Change to exponential form.}$$

$$100{,}000 = I$$

Since $(10{,}000)(10) = 100{,}000$, an earthquake that measures 5 is 10 times more intense than an earthquake that measures 4.

NOW TRY EXERCISE 91

Exercise Set 9.3

Concept/Writing Exercises

1. Consider the logarithmic function $y = \log_a x$.
   a) What are the restrictions on a?
   b) What is the domain of the function?
   c) What is the range of the function?
2. Write $y = \log_a x$ in exponential form.
3. If some points on the graph of the exponential function $f(x) = a^x$ are $\left(-3, \frac{1}{27}\right), \left(-2, \frac{1}{9}\right), \left(-1, \frac{1}{3}\right), (0, 1), (1, 3), (2, 9)$, and $(3, 27)$, list some points on the graph of the logarithmic function $g(x) = \log_a x$. Explain how you determined your answer.
4. For the logarithmic function $y = \log_a(x - 3)$, what must be true about x? Explain.
5. Discuss the relation between the graphs $y = a^x$ and $y = \log_a x$ for $a > 0$ and $a \neq 1$.
6. What is the x-intercept of the graph of an equation of the form $y = \log_a x$?

Practice the Skills

Graph the logarithmic function.

7. $y = \log_2 x$
8. $y = \log_3 x$
9. $y = \log_{1/2} x$
10. $y = \log_{1/3} x$
11. $y = \log_5 x$
12. $y = \log_{1/4} x$

Graph each pair of functions on the same axes.

13. $y = 2^x, y = \log_{1/2} x$
14. $y = \left(\frac{1}{2}\right)^x, y = \log_2 x$
15. $y = 2^x, y = \log_2 x$
16. $y = \left(\frac{1}{2}\right)^x, y = \log_{1/2} x$

Write each equation in logarithmic form.

17. $2^3 = 8$
18. $3^5 = 243$
19. $9^{1/2} = 3$
20. $8^{1/3} = 2$
21. $\left(\frac{1}{2}\right)^5 = \frac{1}{32}$
22. $\left(\frac{1}{4}\right)^2 = \frac{1}{16}$
23. $2^{-3} = \frac{1}{8}$
24. $5^{-2} = \frac{1}{25}$
25. $4^{-3} = \frac{1}{64}$
26. $64^{1/3} = 4$
27. $16^{-1/2} = \frac{1}{4}$
28. $36^{1/2} = 6$
29. $8^{-1/3} = \frac{1}{2}$
30. $81^{-1/4} = \frac{1}{3}$
31. $10^{0.6990} = 5$
32. $10^{1.0792} = 12$
33. $e^2 = 7.3891$
34. $e^{-1} = 0.3679$
35. $c^b = w$
36. $r^n = A$

Write each equation in exponential form.

37. $\log_2 8 = 3$
38. $\log_3 9 = 2$
39. $\log_{1/3} \frac{1}{9} = 2$
40. $\log_{1/2} \frac{1}{16} = 4$
41. $\log_5 \frac{1}{125} = -3$
42. $\log_9 3 = \frac{1}{2}$
43. $\log_{125} 5 = \frac{1}{3}$
44. $\log_8 \frac{1}{64} = -2$
45. $\log_{27} \frac{1}{3} = -\frac{1}{3}$
46. $\log_{10} 100 = 2$
47. $\log_{10} 1000 = 3$
48. $\log_6 216 = 3$
49. $\log_{10} 8 = 0.9031$
50. $\log_{10} 0.62 = -0.2076$
51. $\log_e 6.52 = 1.8749$
52. $\log_e 30 = 3.4012$
53. $\log_r c = -a$
54. $\log_w s = -p$

Write each equation in exponential form; then find the unknown value.

55. $\log_4 16 = y$
56. $\log_a 81 = 4$
57. $\log_2 x = 5$
58. $\log_2 \frac{1}{8} = y$
59. $\log_a \frac{1}{27} = -3$
60. $\log_{1/2} x = 2$
61. $\log_a \frac{1}{64} = 3$
62. $\log_4 \frac{1}{16} = y$

Evaluate the following.

63. $\log_{10} 100$
64. $\log_{10} 10$
65. $\log_{10} 1$
66. $\log_{10} 1000$
67. $\log_{10} 10{,}000$
68. $\log_{10} 100{,}000$
69. $\log_4 64$
70. $\log_3 \frac{1}{27}$
71. $\log_8 \frac{1}{64}$
72. $\log_7 1$
73. $\log_9 9$
74. $\log_5 5$
75. $\log_5 1$
76. $\log_4 1024$

Problem Solving

77. Between which two integers must $\log_{10} 425$ lie? Explain.

78. Between which two integers must $\log_{10} 0.672$ lie? Explain.

79. Between which two integers must $\log_3 62$ lie? Explain.

80. Between which two integers must $\log_5 0.3256$ lie? Explain.

81. For $x > 1$, which will grow faster as x increases, 2^x or $\log_{10} x$? Explain.

82. For $x > 1$, which will grow faster as x increases, x or $\log_{10} x$? Explain.

Change to exponential form, then solve for x. We will discuss rules for solving problems like this in Section 9.4.

83. $x = \log_{10} 10^5$ **84.** $x = \log_7 7^9$ **85.** $x = \log_b b^3$ **86.** $x = \log_e e^5$

Change to logarithmic form, then solve for x. We will discuss rules for solving problems like this in Section 9.4.

87. $x = 10^{\log_{10} 8}$ **88.** $x = 6^{\log_6 4}$ **89.** $x = b^{\log_b 9}$ **90.** $x = c^{\log_c 2}$

91. If the magnitude of an earthquake is 7 on the Richter scale, how many times more intense is the earthquake than the smallest measurable activity? Use $R = \log_{10} I$ (see Example 6).

92. How many times more intense is an earthquake that measures 3 on the Richter scale than an earthquake that measures 1?

93. Graph $y = \log_2(x - 1)$.

94. Graph $y = \log_3(x - 2)$.

Cumulative Review Exercises

[5.4–5.7] *Factor.*

95. $24x^2 - 6xy + 16xy - 4y^2$

96. $2(a - 3)^2 + 7(a - 3) - 15$

97. $4x^4 - 36x^2$

98. $8x^3 + \frac{1}{27}$

9.4 PROPERTIES OF LOGARITHMS

SSM VIDEO 9.4 CD Rom

1. **Use the product rule for logarithms.**
2. **Use the quotient rule for logarithms.**
3. **Use the power rule for logarithms.**
4. **Use additional properties of logarithms.**

1 Use the Product Rule for Logarithms

When finding the logarithm of an expression, the expression is called the **argument** of the logarithm. For example, in $\log_{10} 3$ the 3 is the argument, and in $\log_{10}(2x + 4)$ the $(2x + 4)$ is the argument. When the argument contains a variable, we assume that the argument represents a positive value. *Remember, only logarithms of positive numbers exist.*

To be able to do calculations using logarithms, you must understand their properties. The first property we discuss is the product rule for logarithms.

Product Rule for Logarithms

For positive real numbers x, y, and a, $a \neq 1$,

$$\log_a xy = \log_a x + \log_a y \qquad \text{Property 1}$$

To prove this property, we let $\log_a x = m$ and $\log_a y = n$. Remember, logarithms are exponents. Now we write each logarithm in exponential form.

$$\log_a x = m \quad \text{means} \quad a^m = x$$

$$\log_a y = n \quad \text{means} \quad a^n = y$$

By substitution and using the rules of exponents, we see that

$$xy = a^m \cdot a^n = a^{m+n}$$

We can now convert $xy = a^{m+n}$ to logarithmic form.

$$xy = a^{m+n} \quad \text{means} \quad \log_a xy = m + n$$

Finally, substituting $\log_a x$ for m and $\log_a y$ for n, we obtain

$$\log_a xy = \log_a x + \log_a y$$

which is property 1.

Examples of Property 1

$$\log_3(5 \cdot 7) = \log_3 5 + \log_3 7$$

$$\log_4 3x = \log_4 3 + \log_4 x$$

$$\log_8 x^2 = \log_8(x \cdot x) = \log_8 x + \log_8 x \text{ or } 2 \log_8 x$$

2) Use the Quotient Rule for Logarithms

Now we give the quotient rule for logarithms, which we refer to as property 2.

Quotient Rule for Logarithms

For positive real numbers x, y, and a, $a \neq 1$,

$$\log_a \frac{x}{y} = \log_a x - \log_a y \qquad \text{Property 2}$$

Examples of Property 2

$$\log_3 \frac{12}{4} = \log_3 12 - \log_3 4$$

$$\log_6 \frac{x}{3} = \log_6 x - \log_6 3$$

$$\log_5 \frac{x}{x + 2} = \log_5 x - \log_5 (x + 2)$$

3) Use the Power Rule for Logarithms

The next property we discuss is the power rule for logarithms.

Power Rule for Logarithms

If x and a are positive real numbers, $a \neq 1$, and n is any real number, then

$$\log_a x^n = n \log_a x \qquad \text{Property 3}$$

Examples of Property 3

$$\log_2 4^3 = 3 \log_2 4$$

$$\log_{10} x^2 = 2 \log_{10} x$$

$$\log_5 \sqrt{12} = \log_5 (12)^{1/2} = \frac{1}{2} \log_5 12$$

$$\log_8 \sqrt[5]{x + 3} = \log_8 (x + 3)^{1/5} = \frac{1}{5} \log_8 (x + 3)$$

Properties 2 and 3 can be proved in a manner similar to that given for property 1 (see Exercises 69 and 70).

EXAMPLE 1 Use properties 1 through 3 to expand.

a) $\log_8 \frac{27}{43}$ **b)** $\log_4(64 \cdot 180)$ **c)** $\log_{10}(32)^{1/5}$

Solution

a) $\log_8 \frac{27}{43} = \log_8 27 - \log_8 43$ *Quotient rule*

b) $\log_4(64 \cdot 180) = \log_4 64 + \log_4 180$ *Product rule*

c) $\log_{10}(32)^{1/5} = \frac{1}{5}\log_{10} 32$ *Power rule*

NOW TRY EXERCISE 9

Often we will have to use two or more of these properties in the same problem.

EXAMPLE 2 Expand.

a) $\log_{10} 4(x+2)^3$ **b)** $\log_5 \frac{(4-x)^2}{3}$

c) $\log_5\left(\frac{4-x}{3}\right)^2$ **d)** $\log_5 \frac{[x(x+4)]^3}{2}$

Solution

a) $\log_{10} 4(x+2)^3 = \log_{10} 4 + \log_{10}(x+2)^3$ *Product rule*

$= \log_{10} 4 + 3\log_{10}(x+2)$ *Power rule*

b) $\log_5 \frac{(4-x)^2}{3} = \log_5(4-x)^2 - \log_5 3$ *Quotient rule*

$= 2\log_5(4-x) - \log_5 3$ *Power rule*

c) $\log_5\left(\frac{4-x}{3}\right)^2 = 2\log_5\left(\frac{4-x}{3}\right)$ *Power rule*

$= 2[\log_5(4-x) - \log_5 3]$ *Quotient rule*

$= 2\log_5(4-x) - 2\log_5 3$ *Distributive property*

d) $\log_5 \frac{[x(x+4)]^3}{2} = \log_5[x(x+4)]^3 - \log_5 2$ *Quotient rule*

$= 3\log_5 x(x+4) - \log_5 2$ *Power rule*

$= 3[\log_5 x + \log_5(x+4)] - \log_5 2$ *Product rule*

$= 3\log_5 x + 3\log_5(x+4) - \log_5 2$ *Distributive property*

NOW TRY EXERCISE 17

Note that the product rule can be expanded to evaluate the product of 3 or more quantities. For example, $\log_5 xyz = \log_5 x + \log_5 y + \log_5 z$.

HELPFUL HINT

In Example 2**b)**, when we expanded $\log_5 \frac{(4-x)^2}{3}$, we first used the quotient rule. In Example 2**c)**, when we expanded $\log_5\left(\frac{4-x}{3}\right)^2$, we first used the power rule. Do you see the difference in the two problems? In $\log_5 \frac{(4-x)^2}{3}$, just the numerator of the argument is squared; therefore, we use the quotient rule first. In $\log_5\left(\frac{4-x}{3}\right)^2$, the entire argument is squared, so we use the power rule first.

EXAMPLE 3 Write each of the following as the logarithm of a single expression.

a) $3\log_8(x+2) - \log_8 x$

b) $\log_7(x+1) + 2\log_7(x+4) - 3\log_7(x-5)$

Solution **a)**
$$3\log_8(x+2) - \log_8 x = \log_8(x+2)^3 - \log_8 x \quad \text{Power rule}$$
$$= \log_8 \frac{(x+2)^3}{x} \quad \text{Quotient rule}$$

b)
$$\log_7(x+1) + 2\log_7(x+4) - 3\log_7(x-5)$$
$$= \log_7(x+1) + \log_7(x+4)^2 - \log_7(x-5)^3 \quad \text{Power rule}$$
$$= \log_7(x+1)(x+4)^2 - \log_7(x-5)^3 \quad \text{Product rule}$$
$$= \log_7 \frac{(x+1)(x+4)^2}{(x-5)^3} \quad \text{Quotient rule}$$

NOW TRY EXERCISE 29

HELPFUL HINT

The Correct Rules Are

$$\log_a xy = \log_a x + \log_a y$$
$$\log_a \frac{x}{y} = \log_a x - \log_a y$$

Note that:

$$\log_a(x+y) \neq \log_a x + \log_a y \qquad \log_a(xy) \neq (\log_a x)(\log_a y)$$
$$\log_a(x-y) \neq \log_a x - \log_a y \qquad \log_a(x/y) \neq \frac{\log_a x}{\log_a y}$$

4) Use Additional Properties of Logarithms

The last properties we discuss in this section will be used to solve equations in Section 9.6.

Additional Properties of Logarithms

If $a > 0$, and $a \neq 1$, then

$$\log_a a^x = x \qquad \text{Property 4}$$

and
$$a^{\log_a x} = x \quad (x > 0) \qquad \text{Property 5}$$

Examples of Property 4

$\log_6 6^5 = 5$

$\log_6 6^x = x$

Examples of Property 5

$3^{\log_3 7} = 7$

$5^{\log_5 x} = x \ (x > 0)$

EXAMPLE 4 Evaluate. **a)** $\log_5 25$ **b)** $\sqrt{16}^{\log_4 9}$

Solution **a)** $\log_5 25$ may be written as $\log_5 5^2$. By property 4,
$$\log_5 25 = \log_5 5^2 = 2$$

b) $\sqrt{16}^{\log_4 9}$ may be written $4^{\log_4 9}$. By property 5,
$$\sqrt{16}^{\log_4 9} = 4^{\log_4 9} = 9$$

NOW TRY EXERCISE 45

Exercise Set 9.4

Concept/Writing Exercises

1. In your own words, explain the product rule for logarithms.
2. In your own words, explain the quotient rule for logarithms.
3. In your own words, explain the power rule for logarithms.
4. Explain why we need to stipulate that x and y are positive real numbers when discussing the product and quotient rules.

Practice the Skills

Use properties 1–3 to expand.

5. $\log_4(6 \cdot 9)$
6. $\log_5(3 \cdot 7)$
7. $\log_8 7(x + 3)$
8. $\log_{10} x(x + 8)$
9. $\log_6 \dfrac{27}{5}$
10. $\log_9 \dfrac{\sqrt{x}}{12}$
11. $\log_{10} \dfrac{\sqrt{x}}{x - 9}$
12. $\log_5 3^{12}$
13. $\log_8 x^4$
14. $\log_5(r + 4)^3$
15. $\log_{10} 3(8)^2$
16. $\log_8 x^2(x - 2)$
17. $\log_4 \sqrt{\dfrac{a^5}{a + 4}}$
18. $\log_{10}(x - 3)^2 x^3$
19. $\log_{10} \dfrac{d^4}{(a + 2)^3}$
20. $\log_7 x^2(x - 2)$
21. $\log_8 \dfrac{y(y + 2)}{y^3}$
22. $\log_{10}\left(\dfrac{x}{6}\right)^2$
23. $\log_{10} \dfrac{2m}{3n}$
24. $\log_5 \dfrac{\sqrt{a}\,\sqrt[3]{b}}{\sqrt[4]{c}}$

Write as a logarithm of a single expression.

25. $2\log_{10} x - \log_{10}(x - 5)$
26. $3\log_8 x + 2\log_8(x + 1)$
27. $2(\log_5 a - \log_5 3)$
28. $\dfrac{1}{2}[\log_6(x - 1) - \log_6 3]$
29. $\log_{10} n + \log_{10}(n - 3) - \log_{10}(n + 1)$
30. $2\log_5 x + \log_5(x - 4) + \log_5(x - 2)$
31. $\dfrac{1}{2}[\log_5(x - 4) - \log_5 x]$
32. $5\log_7(a + 3) + 2\log_7(a - 1) - \dfrac{1}{2}\log_7 a$
33. $2\log_9 5 + \dfrac{1}{3}\log_9(r - 6) - \dfrac{1}{2}\log_9 r$
34. $5\log_6(x + 3) - [2\log_6(x - 4) + 3\log_6 x]$
35. $4\log_6 3 - [2\log_6(x + 3) + 4\log_6 x]$
36. $2\log_7(m - 4) + 3\log_7(m + 3) - [5\log_7 2 + 3\log_7(m - 2)]$

Find the value by writing each argument using the numbers 2 and/or 5 and using the values $\log_{10} 2 = 0.3010$ and $\log_{10} 5 = 0.6990$.

37. $\log_{10} 10$
38. $\log_{10} 0.4$
39. $\log_{10} 2.5$
40. $\log_{10} 4$
41. $\log_{10} 25$
42. $\log_{10} 8$

Evaluate (see Example 4).

43. $5^{\log_5 10}$
44. $\log_5 5$
45. $(2^3)^{\log_8 5}$
46. $\log_8 64$
47. $\log_3 27$
48. $2\log_9 \sqrt{9}$
49. $5(\sqrt[3]{27})^{\log_3 5}$
50. $\dfrac{1}{2}\log_6 \sqrt[3]{6}$

Problem Solving

51. For $x > 0$ and $y > 0$, is $\log_a \frac{x}{y} = \log_a xy^{-1} = \log_a x + \log_a y^{-1} = \log_a x + \log_a \frac{1}{y}$?

52. Read Exercise 51. By the quotient rule, $\log_a \frac{x}{y} = \log_a x - \log_a y$. Can we therefore conclude that $\log_a x - \log_a y = \log_a x + \log_a \frac{1}{y}$?

53. Use the product rule to show that

$$\log_a \frac{x}{y} = \log_a x + \log_a \frac{1}{y}$$

54. a) Explain why

$$\log_a \frac{3}{xy} \neq \log_a 3 - \log_a x + \log_a y$$

b) Expand $\log_a \frac{3}{xy}$ correctly.

55. Express $\log_a(x^2 - 4) - \log_a(x + 2)$ as a single logarithm and simplify.

56. Express $\log_a(x - 8) - \log_a(x^2 - 4x - 32)$ as a single logarithm and simplify.

57. Is $\log_a(x^2 + 8x + 16) = 2\log_a(x + 4)$? Explain.

58. Is $\log_a(4x^2 - 20x + 25) = 2\log_a(2x - 5)$? Explain.

If $\log_{10} x = 0.4320$, find the following.

59. $\log_{10} x^2$

60. $\log_{10} \sqrt{x}$

61. $\log_{10} \sqrt[4]{x}$

62. $\log_{10} x^{10}$

If $\log_{10} x = 0.5000$ and $\log_{10} y = 0.2000$, find the following.

63. $\log_{10} xy$

64. $\log_{10}\left(\frac{x}{y}\right)$

65. Using the information given in the instructions for Exercises 63 and 64, is it possible to find $\log_{10}(x + y)$? Explain.

66. Are the graphs of $y = \log_b x^2$ and $y = 2\log_b x$ the same? Explain your answer by discussing the domains of each equation.

Use properties 1–3 to expand.

67. $\log_2 \frac{\sqrt[4]{xy}\,\sqrt[3]{a}}{\sqrt[5]{a - b}}$

68. $\log_3 \left[\frac{(a^2 + b^2)(c^2)}{(a - b)(b + c)(c + d)}\right]^2$

69. Prove the quotient rule for logarithms.

70. Prove the power rule for logarithms.

Group Activity

Discuss and answer Exercise 71 as a group.

71. Consider $\log_a \frac{\sqrt{x^4 y}}{\sqrt{xy^3}}$, where $x > 0$ and $y > 0$.

a) Group member 1: Expand the expression using the quotient rule.

b) Group member 2: Expand the expression using the product rule.

c) Group member 3: First simplify $\frac{\sqrt{x^4 y}}{\sqrt{xy^3}}$, then expand the resulting logarithm.

d) Check each other's work and make sure all answers are correct. Can this expression be simplified by all three methods?

Cumulative Review Exercises

Perform each indicated operation.

[6.1] **72.** $\frac{2x + 5}{x^2 - 7x + 12} \div \frac{x - 4}{2x^2 - x - 15}$

[6.2] **73.** $\frac{2x + 5}{x^2 - 7x + 12} - \frac{x - 4}{2x^2 - x - 15}$

[6.5] **74.** Mike Eisen can paint a house by himself in 4 days and Jill McGhee can paint the same house by herself in 5 days. How long would it take them to paint the house together?

[7.3] **75.** Multiply and then simplify

$$\sqrt[3]{4x^4 y^7} \cdot \sqrt[3]{12x^7 y^{10}}.$$

9.5 COMMON LOGARITHMS

SSM VIDEO 9.5 CD Rom

1) **Find common logarithms of powers of 10.**
2) **Find common logarithms.**
3) **Find antilogarithms.**

1) Find Common Logarithms of Powers of 10

The properties discussed in Section 9.4 can be used with any valid base (a real number greater than 0 and not equal to 1). However, since we are used to working in base 10, we will often use the base 10 when computing with logarithms. **Base 10 logarithms** are called **common logarithms**. When we are working with common logarithms, it is not necessary to list the base. Thus, $\log x$ means $\log_{10} x$.

The properties of logarithms written as common logarithms follow. For positive real numbers x and y, and any real number n,

1. $\log xy = \log x + \log y$

2. $\log \dfrac{x}{y} = \log x - \log y$

3. $\log x^n = n \log x$.

The logarithms of most numbers are irrational numbers. Even the values given by calculators are usually only approximations of the actual values. Even though we are working with approximations when evaluating most logarithms, we generally write the logarithm with an equal sign. Thus, rather than writing $\log 6 \approx 0.77815$, we will write $\log 6 = 0.77815$.

In Chapter 1 we learned that 1 can be expressed as 10^0 and 10 can be expressed as 10^1. Since, for example, 5 is between 1 and 10, it must also be between 10^0 and 10^1.

$$1 < 5 < 10$$

$$10^0 < 5 < 10^1$$

The number 5 can be expressed as the base 10 raised to an exponent between 0 and 1. The number 5 is approximately equal to $10^{0.69897}$. The common logarithm of 5 is about 0.69897.

$$\log 5 = 0.69897$$

Definition

Common Logarithm

The **common logarithm** of a positive real number is the *exponent* to which the base 10 is raised to obtain the number.

$$\text{If } \log N = L, \quad \text{then} \quad 10^L = N.$$

For example, if $\log 5 = 0.69897$, then $10^{0.69897} = 5$.

Now consider the number 50.

$$10 < 50 < 100$$

$$10^1 < 50 < 10^2$$

The number 50 can be expressed as the base 10 raised to an exponent between 1 and 2. The number $50 = 10^{1.69897}$; thus $\log 50 = 1.69897$.

2) Find Common Logarithms

To find common logarithms of numbers, we can use a calculator that has a logarithm key, LOG.

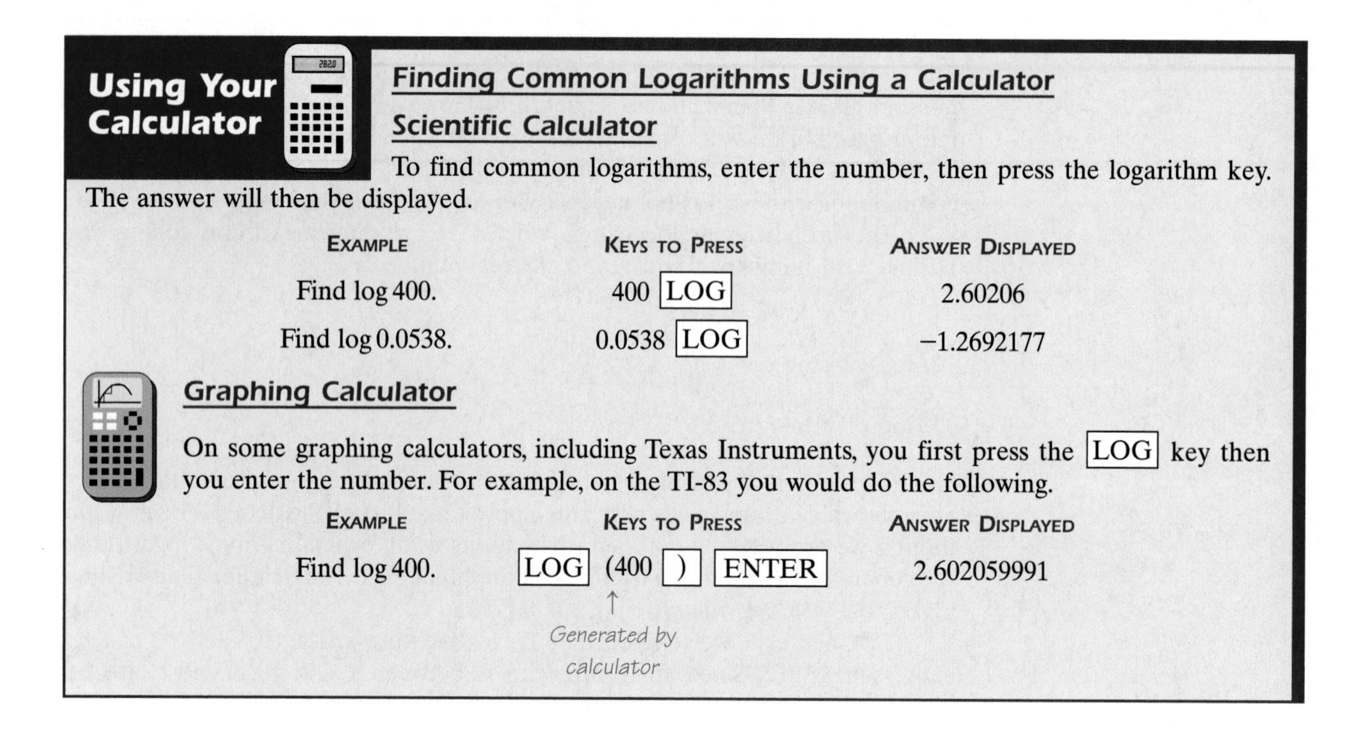

Using Your Calculator
Finding Common Logarithms Using a Calculator
Scientific Calculator
To find common logarithms, enter the number, then press the logarithm key. The answer will then be displayed.
EXAMPLE | KEYS TO PRESS | ANSWER DISPLAYED
Find log 400. | 400 LOG | 2.60206
Find log 0.0538. | 0.0538 LOG | −1.2692177
Graphing Calculator
On some graphing calculators, including Texas Instruments, you first press the LOG key then you enter the number. For example, on the TI-83 you would do the following.
EXAMPLE | KEYS TO PRESS | ANSWER DISPLAYED
Find log 400. | LOG (400) ENTER | 2.602059991
Generated by calculator

EXAMPLE 1 Find the exponent to which the base 10 must be raised to obtain the number 43,600.

Solution *We are asked to find the exponent, which is a logarithm.* We need to determine log 43,600. Using a calculator, we find

$$\log 43{,}600 = 4.6394865$$

NOW TRY EXERCISE 7

Thus, the exponent is 4.6394865. Note that $10^{4.6394865} = 43{,}600$.

3) Find Antilogarithms

The question that should now be asked is, "if we know the common logarithm of a number, how do we find the number?" For example, if $\log N = 3.406$, what is N? To find N, the number, we need to determine the value of $10^{3.406}$. Since

$$10^{3.406} = 2546.830253$$

$N = 2546.830253$. The number 2546.830253 is the *antilogarithm* of 3.406.

When we find the value of the number from the logarithm, we say we are finding the **antilogarithm** or **inverse logarithm**. If the logarithm of N is L, then N is the antilogarithm or inverse logarithm of L.

Definition

Antilogarithm

$$\text{If } \log N = L, \quad \text{then} \quad N = \text{antilog } L.$$

When we are given the common logarithm, which is the exponent on the base 10, the *antilog is the number* obtained when the base 10 is raised to that exponent.

Examples

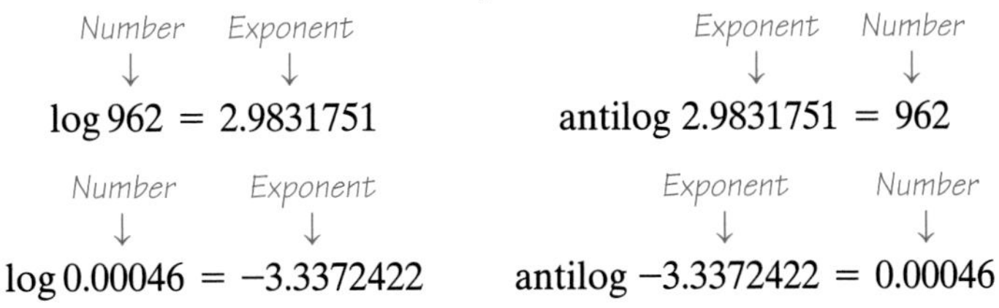

When finding an antilog, we start with the logarithm, or the exponent, and end with the number equal to 10 raised to that logarithm or exponent. If antilog $-3.3372422 = 0.00046$ then $10^{-3.3372422} = 0.00046$.

Using Your Calculator

Finding Antilogarithms

Scientific Calculator

To find antilogarithms on a scientific calculator, enter the logarithm, and press the [2nd], [INV], or [SHIFT] key, depending upon which of these keys your calculator has. Then press the [LOG] key. After the [LOG] key is pressed, the antilog will be displayed.

Example	Keys to Press	Answer Displayed
Find antilog 2.9831751.	2.9831751 [INV] [LOG]	962.00006*
Find antilog −3.3372422.	3.3372422 [+/-] [INV] [LOG]	0.00046**

When you are finding the antilog of a negative value, enter the value and then press the [+/-] key before pressing the inverse and logarithm keys.

* Some calculators give slightly different answers, depending on their electronics.

** Some calculators may display answers in scientific notation form.

Graphing Calculator

On most graphing calculators you press the [2nd] then [LOG] key before you enter the logarithm. On the TI-83 and on certain other calculators, 10^x is printed directly above the [LOG] key. The antilog is actually the value of 10^x, where x is the logarithm. When you press [2nd] [LOG], the TI-83 displays 10^(. You then enter the logarithm followed by the [)] key. After you press [ENTER], the antilog is displayed.

Example	Keys to Press	Answer Displayed
Find antilog 2.9831751.	[2nd] [LOG]* (2.9831751[)] [ENTER]	962.0000619
Find antilog −3.3372422.	[2nd] [LOG] ([(−)] 3.3372422[)] [ENTER]	4.599999664E⁻4**

* Left parenthesis is generated by the TI-83.

** Recall from scientific notation this number is 0.0004599999664.

Since we generally do not need the accuracy given by most calculators, in the exercise set that follows we will round logarithms to four decimal places and antilogarithms to three **significant digits**. In a number written in decimal form, any zeros preceding the first nonzero digit are not significant digits. The first nonzero digit in a number, moving from left to right, is the first significant digit.

	Examples
0.0063402	First significant digit is shaded.
3.0424080	First three significant digits are shaded.
0.0000138483	First three significant digits are shaded.
206,435.05	First four significant digits are shaded.

EXAMPLE 2 Find the value obtained when the base 10 is raised to the -1.052 power.

Solution We are asked to find the value of $10^{-1.052}$. Since we are given the exponent, or logarithm, we can find the value by taking the antilog of -1.052.

$$\text{antilog } -1.052 = 0.0887156$$

NOW TRY EXERCISE 55

Thus, $10^{-1.052} = 0.0887$ rounded to three significant digits.

EXAMPLE 3 Find N if $\log N = 3.742$.

Solution We are given the logarithm and asked to find the antilog, or the number N.

$$\text{antilog } 3.742 = 5520.7744$$

NOW TRY EXERCISE 33

Thus, $N = 5520.7744$.

EXAMPLE 4 Find the following antilogs and round to three significant digits.

a) antilog 6.827 **b)** antilog -2.35

Solution **a)** Using a calculator, we find antilog $6.827 = 6{,}714{,}288.5$. Rounding to three significant digits, we get antilog $6.827 = 6{,}710{,}000$.

b) Using a calculator, we find antilog $-2.35 = 0.0044668$. Rounding to three significant digits, we get antilog $-2.35 = 0.00447$.

Exercise Set 9.5

Concept/Writing Exercises

1. What are common logarithms?
2. Write $\log N = L$ in exponential form.
3. What are antilogarithms?
4. If $\log 652 = 2.8142$, what is antilog 2.8142?

Practice the Skills

Find the common logarithm of each number. Round the answer to four decimal places.

5. 45
6. 8
7. 19,200
8. 1000
9. 0.0000857
10. 27,700
11. 100
12. 0.000835
13. 3.75
14. 0.375
15. 0.000472
16. 0.00872

Find the antilog of each logarithm. Round the answer to three significant digits.

17. 0.6325 **18.** 2.6464 **19.** 4.6283 **20.** 5.8149

21. −1.0585 **22.** −2.3382 **23.** 0.0000 **24.** 5.5922

25. 2.5011 **26.** −4.4306 **27.** −0.1543 **28.** −1.2549

Find each number N. Round N to three significant digits.

29. $\log N = 2.0000$ **30.** $\log N = 1.6730$ **31.** $\log N = -2.103$ **32.** $\log N = 1.9330$

33. $\log N = 4.5202$ **34.** $\log N = 2.7404$ **35.** $\log N = -1.06$ **36.** $\log N = -1.1469$

37. $\log N = -0.3686$ **38.** $\log N = 1.5159$ **39.** $\log N = -0.3936$ **40.** $\log N = -1.3206$

To what exponent must the base 10 be raised to obtain each value? Round your answer to four decimal places.

41. 3560 **42.** 817,000 **43.** 0.0727 **44.** 0.00612

45. 102 **46.** 8.92 **47.** 0.00128 **48.** 73,700,000

Find the value obtained when 10 is raised to the following exponents. Round to three significant digits.

49. 2.4360 **50.** 3.7118 **51.** −0.158 **52.** −2.2351

53. −1.6091 **54.** 4.8537 **55.** 1.3503 **56.** −2.1918

By changing the logarithm to exponential form, evaluate the common logarithm without the use of a calculator.

57. $\log 1$ **58.** $\log 100$ **59.** $\log 0.1$ **60.** $\log 1000$

61. $\log 0.01$ **62.** $\log 10$ **63.** $\log 0.001$ **64.** $\log 10{,}000$

In Section 9.4 we stated that for $a > 0$, and $a \neq 1$, $\log_a a^x = x$ and $a^{\log_a x} = x$ ($x > 0$). Rewriting these properties using common logarithms ($a = 10$), we obtain $\log 10^x = x$ and $10^{\log x} = x$ ($x > 0$), respectively. Use these properties to evaluate the following.

65. $\log 10^5$ **66.** $\log 10^{6.7}$ **67.** $10^{\log 7}$ **68.** $10^{\log 8.3}$

69. $6 \log 10^{5.2}$ **70.** $8 \log 10^{4.6}$ **71.** $5(10^{\log 9.4})$ **72.** $2.3(10^{\log 5.2})$

Problem Solving

73. On your calculator, you find $\log 462$ and obtain the value 1.6646. Can this value be correct? Explain.

74. On your calculator, you find $\log 6250$ and obtain the value 2.7589. Can this value be correct? Explain.

75. On your calculator, you find $\log 0.163$ and obtain the value −2.7878. Can this value be correct? Explain.

76. On your calculator, you find $\log -1.23$ and obtain the value 0.08991. Can this value be correct? Explain.

77. Is $\log \frac{y}{3x} = \log y - \log 3 + \log x$? Explain.

78. Is $\log \frac{5x^2}{2} = 2(\log 5 + \log x) - \log 2$? Explain.

If $\log 25 = 1.3979$ and $\log 5 = 0.6990$, find the answer if possible. If it is not possible to find the answer, explain why. Do not look the logarithms up on your calculator except to check answers.

79. $\log 125$ **80.** $\log \frac{1}{5}$ **81.** $\log 30$

82. $\log 625$ **83.** $\log \frac{1}{25}$ **84.** $\log \sqrt{5}$

Solve each problem. Round your answers to the nearest hundredth.

85. The magnitude of an earthquake on the Richter scale is given by the formula $R = \log I$, where I is the number of times more intense the quake is than the minimum level for comparison.

a) Find the Richter scale number for an earthquake that is 12,000 times more intense than the minimum level for comparison.

b) If the Richter scale number of an earthquake is 4.29, how many times more intense is the earthquake than the minimum level for comparison?

86. In astronomy, a formula used to find the diameter, in kilometers, of minor planets (also called asteroids) is $\log d = 3.7 - 0.2g$, where g is a quantity called the absolute magnitude of the minor planet. Find the diameter of a minor planet if its absolute magnitude is **a)** 11 and **b)** 20. **c)** Find the absolute magnitude of the minor planet whose diameter is 5.8 kilometers.

87. The movie *Titanic* was released in 1997. It was the number one movie in box office receipts for many weeks. During each of the first eight weeks its box office receipts, $f(x)$, in millions of dollars, could be approximated by the function $f(x) = 30 - 5\log x$, where $x = 1$ represents the first week of its release, $x = 2$ the second week, and so on, and $1 \le x \le 8$. Use the function to estimate the *Titanic*'s weekly box office receipts during **a)** the first week of its release and **b)** the eighth week of its release?

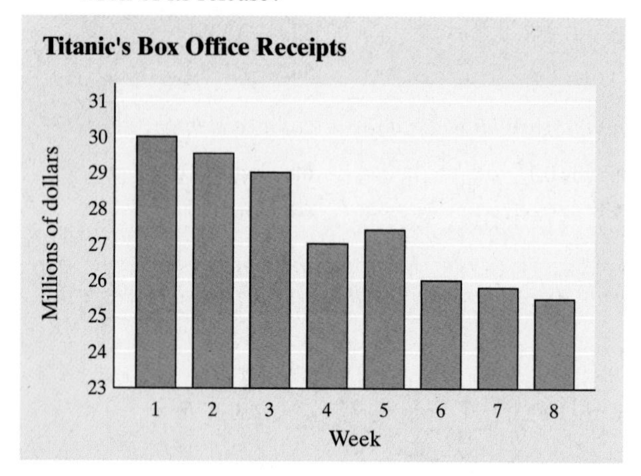

88. The average score on a standardized test is a function of the number of hours studied for the test. The average score, $f(x)$, in points, can be approximated by $f(x) = \log 0.3x + 1.8$, where x is the number of hours studied for the test. The maximum possible score on the test is 4.0. Find the score received by the average person who studied for **a)** 15 hours and **b)** 55 hours.

89. A formula sometimes used to estimate the seismic energy released by an earthquake is $\log E = 11.8 + 1.5m_s$, where E is the seismic energy and m_s is the surface wave magnitude.

a) Find the energy released in an earthquake whose surface wave magnitude is 6.

b) If the energy released during an earthquake is 1.2×10^{15}, what is the magnitude of the surface wave?

90. The sound pressure level, s_p, is given by the formula $s_p = 20\log\frac{p_r}{0.0002}$, where p_r is the sound pressure in dynes/cm^2.

a) Find the sound pressure level if the sound pressure is 0.0036 dynes/cm^2.

b) If the sound pressure level is 10.0, find the sound pressure.

91. The Richter scale, used to measure the strength of earthquakes, relates the magnitude, M, of the earthquake to the release of energy, E, in ergs, by the formula

$$M = \frac{\log E - 11.8}{1.5}$$

An earthquake releases 1.259×10^{21} ergs of energy. What is the magnitude of such an earthquake on the Richter scale?

92. The pH is a measure of the acidity or alkalinity of a solution. The pH of water, for example, is 7. In general, acids have pH numbers less than 7 and alkaline solutions have pH numbers greater than 7. The pH of a solution is defined as $\text{pH} = -\log[H_3O^+]$, where H_3O^+ represents the hydronium ion concentration of the solution. Find the pH of a solution whose hydronium ion concentration is 2.8×10^{-3}.

Group Activity

93. In section 9.7 we introduce the *change of base formula*, $\log_a x = \frac{\log_b x}{\log_b a}$, where a and b are bases and x is a positive number.

a) Group member 1: Use the change of base formula to evaluate $\log_3 45$. (*Hint*: Let $b = 10$.)

b) Group member 2: Repeat part **a)** for $\log_5 30$.

c) Group member 3: Repeat part **a)** for $\log_6 40$.

d) As a group, use the fact that $\log_a x = \frac{\log_b x}{\log_b a}$, where $b = 10$, to graph the equation $y = \log_2 x$ for $x > 0$. Use a grapher if available.

Cumulative Review Exercises

[8.2] **94.** Solve the following quadratic equation using the quadratic formula.

$$-3x^2 - 4x - 8 = 0$$

[8.3] **95.** In 4 hours the Simpsons traveled 15 miles downriver in their motorboat, and then turned around and returned home. If the river current is 5 miles per hour, find the speed of their boat in still water.

[8.4] **96.** Graph the solution to $\frac{2x - 3}{5x + 10} < 0$ on a number line.

[8.5] **97.** Draw the graph of $y = (x - 2)^2 + 1$.

9.6 EXPONENTIAL AND LOGARITHMIC EQUATIONS

SSM VIDEO 9.6 CD Rom

1) **Solve exponential and logarithmic equations.**
2) **Solve applications.**

1) Solve Exponential and Logarithmic Equations

In Sections 9.2 and 9.3 we introduced **exponential** and **logarithmic equations**. In this section we give more examples of their use and discuss further procedures for solving such equations.

To solve exponential and logarithmic equations, we often use the following properties 6a through 6d.

Properties for Solving Exponential and Logarithmic Equations

a. If $x = y$, then $a^x = a^y$.
b. If $a^x = a^y$, then $x = y$.
c. If $x = y$, then $\log x = \log y \quad (x > 0, y > 0)$.
d. If $\log x = \log y$, then $x = y \quad (x > 0, y > 0)$. Properties 6a–6d

We will be referring to these properties when explaining the solutions to the examples in this section.

EXAMPLE 1 Solve the equation $8^x = \frac{1}{2}$.

Solution To solve this equation, we will write both sides of the equation with the same base, 2, then use property 6b.

$$8^x = \frac{1}{2}$$

$$(2^3)^x = \frac{1}{2} \qquad \text{Write 8 as } 2^3.$$

$$2^{3x} = 2^{-1} \qquad \text{Write } 1/2 \text{ as } 2^{-1}.$$

Using property 6b, we can write

$$3x = -1$$

$$x = -\frac{1}{3}$$

NOW TRY EXERCISE 7

When both sides of the exponential equation cannot be written as a power of the same base, we often begin by taking the logarithm of both sides of the equation, as in Example 2. In the following examples, we will round logarithms to the nearest ten-thousandth.

EXAMPLE 2 Solve the equation $5^n = 20$.

Solution Take the logarithm of both sides of the equation and solve for n.

$$\log 5^n = \log 20$$
$$n \log 5 = \log 20 \quad \text{Power rule}$$
$$n = \frac{\log 20}{\log 5} \quad \text{Divide both sides by } \log 5.$$
$$\approx \frac{1.3010}{0.6990} \approx 1.8612$$

NOW TRY EXERCISE 15

Some logarithmic equations can be solved by expressing the equation in exponential form. **It is necessary to check logarithmic equations for extraneous solutions.** When checking a solution, if you obtain the logarithm of a nonpositive number, the solution is extraneous.

EXAMPLE 3 Solve the equation $\log_2(x + 1)^3 = 4$.

Solution Write the equation in exponential form.

$$(x + 1)^3 = 2^4 \quad \text{Write in exponential form.}$$
$$(x + 1)^3 = 16$$
$$x + 1 = \sqrt[3]{16} \quad \text{Take the cube root of both sides.}$$
$$x = -1 + \sqrt[3]{16} \quad \text{Solve for } x.$$

CHECK:

$$\log_2(x + 1)^3 = 4$$
$$\log_2[(-1 + \sqrt[3]{16}) + 1]^3 \overset{?}{=} 4$$
$$\log_2(\sqrt[3]{16})^3 \overset{?}{=} 4$$
$$\log_2 16 \overset{?}{=} 4 \quad (\sqrt[3]{16})^3 = 16$$
$$2^4 \overset{?}{=} 16 \quad \text{Write in exponential form.}$$
$$16 = 16 \quad \text{True}$$

NOW TRY EXERCISE 31

Other logarithmic equations can be solved using the properties of logarithms given in earlier sections.

EXAMPLE 4 Solve the equation $\log(3x + 2) + \log 9 = \log(x + 5)$.

Solution

$$\log(3x + 2) + \log 9 = \log(x + 5)$$
$$\log(3x + 2)(9) = \log(x + 5) \quad \text{Product rule}$$
$$(3x + 2)(9) = (x + 5) \quad \text{Property 6d}$$
$$27x + 18 = x + 5$$
$$26x + 18 = 5$$
$$26x = -13$$
$$x = -\frac{1}{2}$$

Check for yourself that the solution is $-\frac{1}{2}$.

EXAMPLE 5 Solve the equation $\log x + \log(x + 1) = \log 12$.

Solution

$$\log x + \log(x + 1) = \log 12$$

$$\log x(x + 1) = \log 12 \qquad \textit{Product rule}$$

$$x(x + 1) = 12 \qquad \textit{Property 6d}$$

$$x^2 + x = 12$$

$$x^2 + x - 12 = 0$$

$$(x + 4)(x - 3) = 0$$

$$x + 4 = 0 \quad \text{or} \quad x - 3 = 0$$

$$x = -4 \qquad\qquad x = 3$$

CHECK: $x = -4$

$$\log x + \log(x + 1) = \log 1a$$

$$\log(-4) + \log(-3) \stackrel{?}{=} \log 12$$

Stop. ↑ ↑
Logarithms of negative numbers are not real numbers.

$x = 3$

$$\log x + \log(x + 1) = \log 12g$$

$$\log 3 + \log 4 \stackrel{?}{=} \log 12 +$$

$$\log(3)(4) \stackrel{?}{=} \log 12$$

$$\log 12 = \log 12 \qquad \textit{True}$$

Thus, −4 is an extraneous solution. The only solution is 3.

Using Your Graphing Calculator

We have indicated how equations in one variable may be solved graphically. Logarithmic and exponential equations may also be solved graphically by graphing each side of the equation and finding the x-coordinate of the point of intersection of the two graphs. In Example 5 we found that the solution to the equation $\log x + \log(x + 1) = \log 12$ was $x = 3$. Figure 9.26 shows the graphical solution to this equation. The horizontal line is the graph of $y = \log 12$ since $\log 12$ is a constant. Notice that the x-coordinate of the point of intersection of the two graphs, 3, is the solution to the equation.

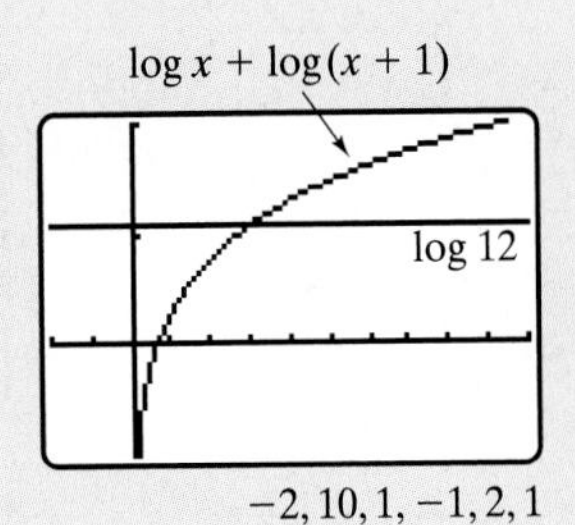

FIGURE 9.26

EXAMPLE 6 Solve the equation $\log(3x - 5) - \log 5x = 1.23$.

Solution

$$\log(3x - 5) - \log 5x = 1.23$$

$$\log \frac{3x - 5}{5x} = 1.23 \qquad \textit{Quotient rule}$$

$$\frac{3x - 5}{5x} = \text{antilog } 1.23$$

$$\frac{3x - 5}{5x} = 17.0 \qquad \textit{Rounded to 3 significant digits}$$

$$3x - 5 = 5x(17.0)$$

$$3x - 5 = 85x$$

$$-5 = 82x$$

$$x = -\frac{5}{82} \approx -0.061$$

CHECK:

$$\log(3x - 5) - \log 5x = 1.23$$

$$\log[3(-0.061) - 5] - \log[(5)(-0.061)] \stackrel{?}{=} 1.23$$

$$\log(-5.183) - \log(-0.305) \stackrel{?}{=} 1.23 \quad \text{Stop.}$$

Since we have the logarithms of negative numbers, -0.061 is an extraneous solution. Thus, this equation has no solution; that is, its solution is the empty set, $\varnothing$.

NOW TRY EXERCISE 51

2) Solve Applications

In Section 9.2 we introduced and worked problems using the compound interest formula, $A = p\left(1 + \frac{r}{n}\right)^{nt}$. The value of A can also be found by using logarithms, as in Example 7.

EXAMPLE 7 The amount of money, A, accumulated in a savings account for a given principal, p, interest rate, r, number of compounding periods, n, and number of years, t, can be found by the formula

$$A = p\left(1 + \frac{r}{n}\right)^{nt}$$

Suppose \$1000 is invested in a savings account at 8% interest compounded annually for 5 years. The amount accumulated at the end of 5 years is

$$\begin{aligned} A &= 1000\left(1 + \frac{0.08}{1}\right)^{5(1)} \\ &= 1000(1 + 0.08)^5 \\ &= 1000(1.08)^5 \end{aligned}$$

Use logarithms to find the amount accumulated.

Solution Begin by taking the logarithms of both sides of the equation,

$$\begin{aligned} \log A &= \log\left[1000(1.08)^5\right] \\ &= \log 1000 + \log(1.08)^5 && \text{Product rule} \\ &= \log 1000 + 5\log 1.08 && \text{Power rule} \\ &= 3.00 + 5(0.0334) \\ &= 3.00 + 0.167 \\ &= 3.167 \\ A &= \text{antilog } 3.167 \\ A &\approx 1469 \end{aligned}$$

In 5 years \$1000 would grow to about \$1469. This amount includes the \$1000 principal and about \$469 interest.

NOW TRY EXERCISE 55

EXAMPLE 8 If there are initially 1000 bacteria in a culture, and the number of bacteria doubles each hour, the number of bacteria after t hours can be found by the formula

$$N = 1000(2)^t$$

How long will it take for the culture to grow to 30,000 bacteria?

Solution

$$N = 1000(2)^t$$

$$30{,}000 = 1000(2)^t \qquad \textit{Substitute 30,000 for N.}$$

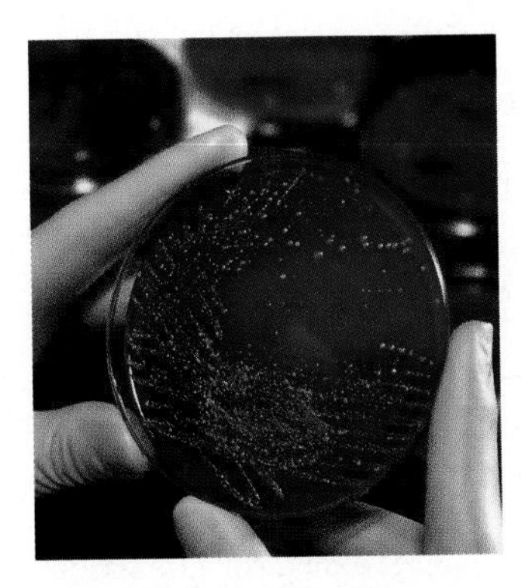

Now take the logarithm of both sides of the equation.

$$\log 30{,}000 = \log\left[1000(2)^t\right]$$

$$\log 30{,}000 = \log 1000 + \log 2^t \qquad \textit{Product rule}$$

$$\log 30{,}000 = \log 1000 + t(\log 2) \qquad \textit{Power rule}$$

$$4.4771 = 3.000 + t(0.3010)$$

$$4.4771 = 3.000 + 0.3010t$$

$$1.4771 = 0.3010t$$

$$\frac{1.4771}{0.3010} = t$$

$$4.91 \approx t$$

In approximately 4.91 hours, there will be 30,000 bacteria in the culture.

Exercise Set 9.6

Concept/Writing Exercises

1. Use properties 6a–6d to write two possible answers to complete the following statement: If $m = n$, then ...
2. If $\log c = \log d$, then what is the relationship between c and d?
3. If $c^r = c^s$, then what is the relationship between r and s?
4. In properties 6c and 6d, we specify that both x and y must be positive. Explain why.
5. After solving a logarithmic equation, what must you do?
6. How can you tell quickly that $\log(x + 4) = \log(-2)$ has no real solution?

Practice the Skills

Solve each exponential equation without using a calculator.

7. $5^x = 125$
8. $3^x = 243$
9. $16^x = \frac{1}{4}$
10. $5^{-x} = \frac{1}{25}$
11. $2^{3x-2} = 16$
12. $64^x = 4^{4x+1}$
13. $27^x = 3^{2x+3}$
14. $\left(\frac{1}{2}\right)^x = 8$

Use a calculator to solve each equation. Round your answers to the nearest hundredth.

15. $7^x = 50$
16. $1.05^x = 15$
17. $4^{x-1} = 20$
18. $2.3^{x-1} = 5.6$
19. $1.63^{x+1} = 25$
20. $4^x = 9^{x-2}$
21. $3^{x+4} = 6^x$
22. $5^x = 2^{x+5}$

Solve each logarithmic equation. Use a calculator where appropriate. If the answer is irrational round the answer to the nearest hundredth.

23. $\log_9 x = \frac{1}{2}$

24. $\log_2 x = -3$

25. $\log_5 x = -2$

26. $\log x = 1$

27. $\log_2 (5 - 3x) = 3$

28. $\log_4 (2x + 2) = 3$

29. $\log_5 (x + 2)^3 = 3$

30. $\log_3 (a - 2)^2 = 2$

31. $\log_2 (a + 4)^2 = 4$

32. $\log_2 x + \log_2 3 = 1$

33. $\log (2x - 3)^3 = 3$

34. $\log_3 2x + \log_3 x = 3$

35. $\log (r + 2) = \log (3r - 1)$

36. $\log 2a = \log (1 - a)$

37. $\log (2x + 1) + \log 4 = \log (7x + 8)$

38. $\log (x + 3) + \log x = \log 4$

39. $\log n + \log (3n - 5) = \log 2$

40. $\log (x + 4) - \log x = \log (x + 1)$

41. $\log 5 + \log y = 0.72$

42. $\log (x + 4) - \log x = 1.22$

43. $2 \log x - \log 4 = 2$

44. $\log 6000 - \log (x + 2) = 3.15$

45. $\log x + \log (x - 3) = 1$

46. $2 \log_2 x = 2$

47. $\log x = \frac{1}{3} \log 27$

48. $\log_7 x = \frac{3}{2} \log_7 64$

49. $\log_8 x = 3 \log_8 2 - \log_8 4$

50. $\log_4 x + \log_4 (6x - 7) = \log_4 5$

51. $\log_5 (x + 3) + \log_5 (x - 2) = \log_5 6$

52. $\log_7 (x + 6) - \log_7 (x - 3) = \log_7 4$

53. $\log_2 (x + 3) - \log_2 (x - 6) = \log_2 4$

54. $\log (x - 7) - \log (x + 3) = \log 6$

Problem Solving

Solve each problem. Round your answers to the nearest hundredth.

55. Find the amount accumulated if Marlina Zuhl puts \$1200 in a savings account offering 6% interest compounded annually for 5 years (see Example 7).

56. If the initial number of bacteria in the culture in Example 8 is 4500, when will the number of bacteria in the culture reach 50,000? Use $N = 4500(2)^t$.

57. If after 4 hours the culture in Example 8 contains 2224 bacteria, how many bacteria were present initially?

58. The amount, A, of 200 grams of a certain radioactive material remaining after t years can be found by the equation $A = 200(0.800)^t$. When will 40 grams remain?

59. The infant mortality rate (deaths per 1000 live births) in the United States has been decreasing since before 1959. Although it has fallen significantly, it is still higher than in many other nations (in 1996 Japan had the lowest rate at 4.0). The U.S. infant mortality rate can be approximated by the function $f(x) = 26 - 12 \log x$, where x is years since 1959 and $1 \le x \le 37$. Use the function to estimate the U.S. infant mortality rate in **a)** 1960 and **b)** 1996.

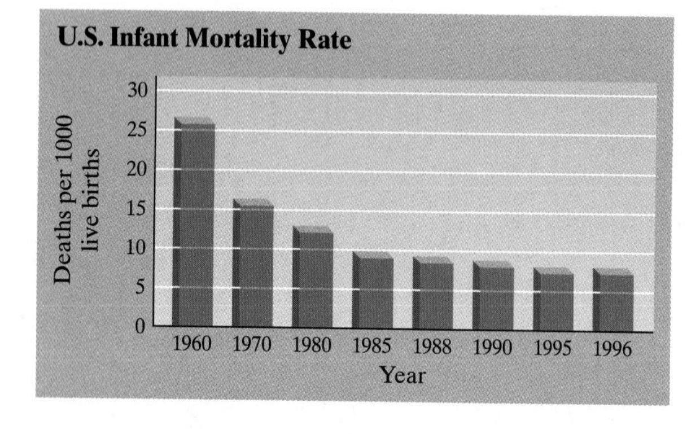

60. Since 1985, the percent of the U.S. population covered by traditional medical care has fallen each year while the percent covered by managed medical care has increased each year. The function

$$f(x) = -8 + 35 \log 4.4x$$

can be used to approximate the percent of the U.S. population covered by managed care, where x is years since 1984 and $1 \le x \le 11$. Use the function to estimate the percent of the U.S. population covered by managed care in **a)** 1985 and **b)** 1995.

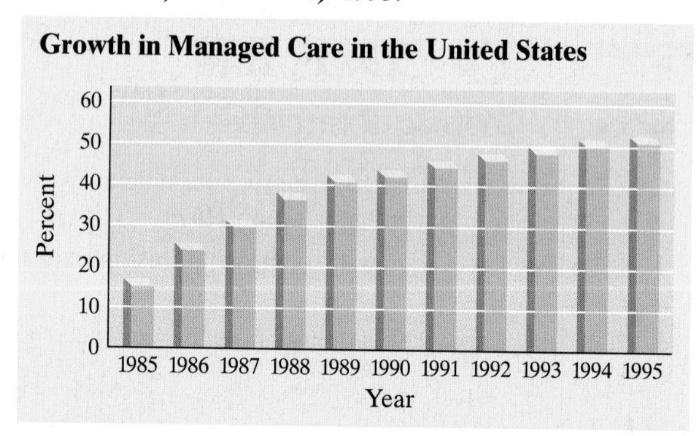

61. A machine purchased for business can be depreciated to reduce income tax. The value of the machine at the end of its useful life is called its *scrap value*. When the machine depreciates by a constant percentage annually, its scrap value, S, is $S = c(1 - r)^n$, where c is the original cost, r is the annual rate of depreciation as a decimal, and n is the useful life in years. Find the scrap value of a machine that costs \$50,000, has a useful life of 12 years, and has an annual depreciation rate of 15%.

62. If the machine in Exercise 61 costs \$100,000, has a useful life of 15 years, and has an annual depreciation rate of 8%, find its scrap value.

63. The power gain, P, of an amplifier is defined as

$$P = 10\log\left(\frac{P_{out}}{P_{in}}\right)$$

where P_{out} is the output power in watts and P_{in} is the input power in watts. If an amplifier has an output power of 12.6 watts and an input power of 0.146 watts, find the power gain.

64. Measured on the Richter scale, the magnitude, R, of an earthquake of intensity I is defined by $R = \log I$, where I is the number of times more intense the earthquake is than the minimum level for comparison.

a) How many times more intense was the 1906 San Francisco earthquake, which measured 8.25 on the Richter scale, than the minimum level for comparison?

b) How many times more intense is an earthquake that measures 6.4 on the Richter scale than one that measures 4.7?

65. The decibel scale is used to measure the magnitude of sound. The magnitude d, in decibels, of a sound is defined to be $d = 10\log I$, where I is the number of times greater (or more intense) the sound is than the minimum intensity of audible sound.

a) An airplane engine (nearby) measures 120 decibels. How many times greater than the minimum level of audible sound is the airplane engine?

b) The intensity of the noise in a busy city street is 70 decibels. How many times greater is the intensity of the sound of the airplane engine than the sound of the city street?

66. In the following procedure, we begin with a true statement and end with a false statement. Can you find the error?

$2 < 3$	*True*
$2\log(0.1) < 3\log(0.1)$	*Multiply both sides by* $\log(0.1)$.
$\log(0.1)^2 < \log(0.1)^3$	*Property 3*
$(0.1)^2 < (0.1)^3$	*Property 6d*
$0.01 < 0.001$	*False*

67. Solve $8^x = 16^{x-2}$.

68. Solve $27^x = 81^{x-3}$.

69. Use equations that are quadratic in form to solve the equation. $2^{2x} - 6(2^x) + 8 = 0$.

70. Use equations that are quadratic in form to solve the equation $2^{2x} - 18(2^x) + 32 = 0$.

Change the exponential or logarithmic equation to the form $ax + by = c$, then solve the system of equations.

71. $2^x = 8^y$
$x + y = 4$

72. $3^{2x} = 9^{y+1}$
$x - 2y = -3$

73. $\log(x + y) = 2$
$x - y = 8$

74. $\log(x + y) = 3$
$2x - y = 5$

Use your calculator to estimate the solutions to the nearest tenth. If a real solution does not exist, so state.

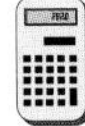

75. $\log(x + 3) + \log x = \log 16$

76. $\log(3x + 5) = 2.3x - 6.4$

77. $5.6\log(5x - 12) = 2.3\log(x - 5.4)$

78. $5.6\log(x + 12.2) - 1.6\log(x - 4) = 20.3\log(2x - 6)$

Cumulative Review Exercises

[2.5] **79.** Solve the inequality $\frac{x-4}{2} - \frac{2x-5}{5} > 3$ and indicate the solution in **a)** set builder notation and **b)** interval notation.

[3.2] **80.** Indicate which of the graphs are functions. If the graph is a function, is it a one-to-one function?

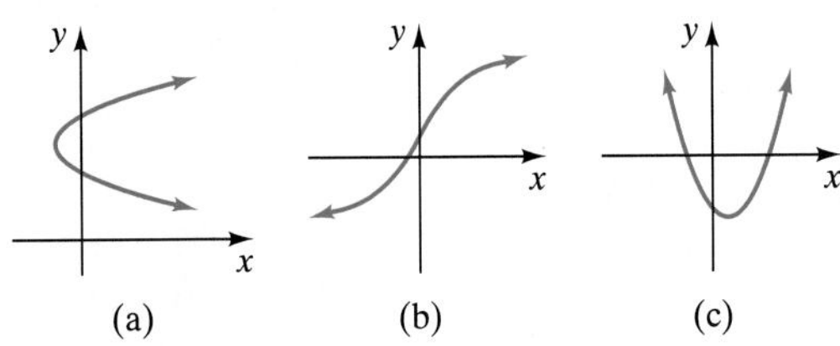

[5.2] **81.** Multiply $(x^2 - 4x + 3)(2x - 3)$.

[5.3] **82.** Divide $2x^2 + 11x + 15$ by $x + 4$.

9.7 NATURAL EXPONENTIAL AND NATURAL LOGARITHMIC FUNCTIONS

SSM VIDEO 9.7 CD Rom

1) Identify the natural exponential function.
2) Identify the natural logarithmic function.
3) Find values on a calculator.
4) Find natural logarithms using the change of base formula.
5) Solve natural logarithmic and natural exponential equations.
6) Solve applications.

The natural exponential function and *its inverse*, the natural logarithmic function, are exponential functions and logarithmic functions of the type presented in the previous sections. They share all the properties of exponential functions and logarithmic functions discussed earlier. The importance of these special functions lies in the many varied applications in real life of a unique irrational number designated by the letter e.

1) Identify the Natural Exponential Function

In Section 9.2 we indicated that exponential functions were of the form $f(x) = a^x$, $a > 0$ and $a \neq 1$. Now we introduce a very special exponential function. It is called the **natural exponential function**, and it uses the number ***e***. Like the irrational number π, the number e is an irrational number whose value can only be approximated by a decimal number. The number e plays a very important role in higher-level mathematics courses. The value of e is approximately 2.7183. Now we define the natural exponential function.

Definition

The Natural Exponential Function

The Natural Exponential function is

$$f(x) = e^x$$

where $e \approx 2.7183$.

2) Identify the Natural Logarithmic Function

We discussed common logarithms in Section 9.4. Now we will discuss natural logarithms.

Definition

Natural Logarithms

Natural logarithms are logarithms to the base e. Natural logarithms are indicated by the letters ln.

$$\log_e x = \ln x$$

The notation $\ln x$ is read the "natural logarithm of x."

You must remember that the base of the natural logarithm is e. Thus, when you change a natural logarithm to exponential form, the base of the exponential expression will be e.

For $x > 0$ if $y = \ln x$, then $e^y = x$.

EXAMPLE 1 Find the value by changing the natural logarithm to exponential form.

a) $\ln 1$ **b)** $\ln e$

Solution **a)** Let $y = \ln 1$; then $e^y = 1$. Since any nonzero value to the 0th power equals 1, y must equal 0. Thus, $\ln 1 = 0$.

b) Let $y = \ln e$; then $e^y = e$. For e^y to equal e, y must equal 1. Thus, $\ln e = 1$.

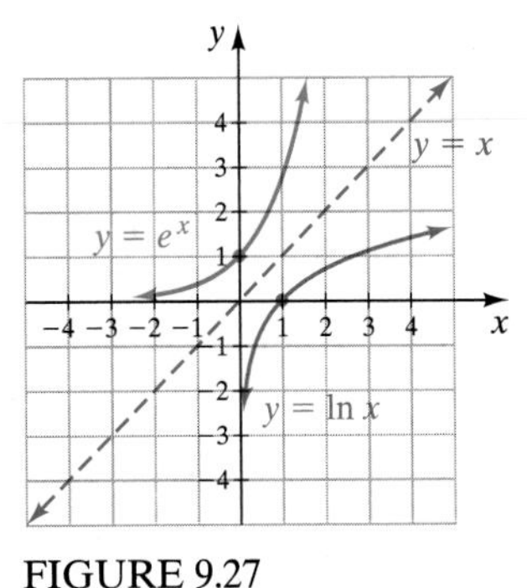

FIGURE 9.27

The functions $y = a^x$ and $y = \log_a x$ are inverse functions. Similarly, the functions $y = e^x$ and $y = \ln x$ are inverse functions. (Remember, $y = \ln x$ means $y = \log_e x$.) The graphs of $y = e^x$ and $y = \ln x$ are illustrated in Figure 9.27. Notice that the graphs are symmetric about the line $y = x$, which is what we expect of inverse functions.

Note that the graph of $y = e^x$ is similar to graphs of the form $y = a^x$, $a > 1$, and that the graph of $y = \ln x$ is similar to graphs of the form $y = \log_a x$, $a > 1$.

3) Find Values on a Calculator

Now we will learn how to find natural logarithms on a calculator.

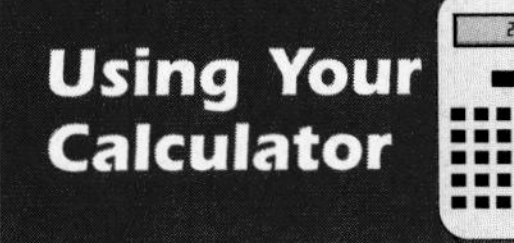

Find Natural Logarithms on a Calculator

Natural logarithms can be found using a calculator that has a [LN] key. Natural logarithms are found in the same manner that we found common logarithms on a calculator, except we use the natural log key, [LN], instead of the common log key, [LOG].

Scientific Calculator

EXAMPLE	KEYS TO PRESS	ANSWER DISPLAYED
Find ln 242.	242 [LN]	5.4889377
Find ln 0.85.	.85 [LN]	−0.1625189

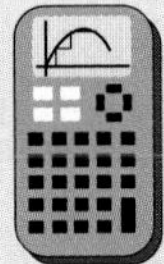

Graphing Calculator*

On the TI-83, after the [LN] key is pressed, the calculator displays ln(on the screen.

EXAMPLE	KEYS TO PRESS	ANSWER DISPLAYED
Find ln 242.	[LN] (242 [ )] [ENTER]	5.488937726
Find ln 0.85.	[LN] (.85 [ )] [ENTER]	−.1625189295

*The keys are for a TI-83. Read your manual for instructions on how to find natural logarithms on your calculator.

When finding the natural logarithm of a number, we are finding an exponent. The natural logarithm of a number is the exponent to which the base e must be raised to obtain that number. For example,

$$\text{if } \ln 242 = 5.4889377, \text{ then } e^{5.4889377} = 242.$$

$$\text{if } \ln 0.85 = -0.1625189, \text{ then } e^{-0.1625189} = 0.85.$$

Since $y = \ln x$ and $y = e^x$ are inverse functions, we can use the inverse key [INV] in combination with the natural log key [LN] to obtain values for e^x.

Using Your Calculator

Finding Values of e^x on a Calculator

Scientific Calculator

To find values of e^x, first enter the exponent on e. Then press either, [SHIFT], [2nd], or [INV], depending on your calculator. Then press the natural log key, [LN]. After the [LN] key is pressed, the value of e^x will be displayed.

EXAMPLE	KEYS TO PRESS	ANSWER DISPLAYED
Find $e^{5.24}$.	5.24 [INV] [LN]	188.6701
Find $e^{-1.639}$.	1.639 [+/-] [INV] [LN]	0.1941741

Graphing Calculator*

On the TI-83, after the [2nd] [LN] is pressed, the calculator displays e^(on the screen.

EXAMPLE	KEYS TO PRESS	ANSWER DISPLAYED
Find $e^{5.24}$.	[2nd] [LN] (5.24 [)] [ENTER]	188.6701024
Find $e^{-1.639}$.	[2nd] [LN] ([(−)] 1.639 [)] [ENTER]	.1941741194

*Keys are for a TI-83. Read your manual for instructions on how to evaluate natural exponential expressions on your calculator.

Since e is about 2.7183, if we evaluated $(2.7183)^{5.24}$ we would obtain a value close to 188.6701. Also, if we found $\ln 188.6701$ on a calculator, we would obtain a value close to 5.24. What do you think we would get if we evaluated $\ln 0.1941741$ on a calculator? If you answered a value close to −1.639, you answered correctly.

EXAMPLE 2 Find N if **a)** $\ln N = 4.92$ and **b)** $\ln N = -0.0253$.

Solution **a)** If we write $\ln N = 4.92$ in exponential form, we get $e^{4.92} = N$. Thus, we simply need to evaluate $e^{4.92}$ to determine N.

$$e^{4.92} = 137.00261 \quad \text{From a calculator}$$

Thus, $N = 137.00261$.

b) If we write $\ln N = -0.0253$ in exponential form, we get $e^{-0.0253} = N$.

$$e^{-0.0253} = 0.9750174 \quad \text{From a calculator}$$

NOW TRY EXERCISE 17 Thus, $N = 0.9750174$.

4) Find Natural Logarithms Using the Change of Base Formula

If you are given a logarithm in a base other than 10 or e you will not be able to evaluate it on your calculator directly. When this occurs, you can use the **change of base formula**.

Change of Base Formula

For any logarithm bases a and b, and positive number x,

$$\log_a x = \frac{\log_b x}{\log_b a}$$

In the change of base formula, 10 is often used in place of base b because we can find common logarithms on a calculator. Replacing base b with 10, we get

$$\log_a x = \frac{\log_{10} x}{\log_{10} a} \quad \text{or} \quad \log_a x = \frac{\log x}{\log a}$$

EXAMPLE 3 Use the change of base formula to find $\log_3 24$.

Solution If we substitute 3 for a and 24 for x in $\log_a x = \dfrac{\log x}{\log a}$, we obtain

$$\log_3 24 = \frac{\log 24}{\log 3} \approx \frac{1.3802}{0.4771} \approx 2.8929$$

NOW TRY EXERCISE 23 Note that $3^{2.8929} \approx 24$.

We can use the same procedure as in Example 3 to find natural logarithms using the change of base formula. For example, to evaluate $\ln 20$ (or $\log_e 20$), we can substitute e for a and 20 for x in the formula $\log_a x = \dfrac{\log x}{\log a}$.

$$\log_e 20 = \frac{\log 20}{\log e} \approx \frac{1.3010}{0.4343} \approx 2.9956$$

Thus, $\ln 20 \approx 2.9956$. If you find $\ln 20$ on a calculator, you will obtain a very close value.

Since $\log e \approx 0.4343$, to evaluate natural logarithms using common logarithms, we use the formula

$$\ln x = \frac{\log x}{\log e} \approx \frac{\log x}{0.4343}$$

EXAMPLE 4 Use the change of base formula to find $\ln 95$.

Solution

$$\ln 95 = \frac{\log 95}{\log e} \approx \frac{1.9778}{0.4343} \approx 4.5540$$

5 Solve Natural Logarithmic and Natural Exponential Equations

The properties of logarithms discussed in Section 9.4 still hold true for natural logarithms. Following is a summary of these properties in the notation of natural logarithms.

Properties for Natural Logarithms

$\ln xy = \ln x + \ln y$	$(x > 0 \text{ and } y > 0)$	*Product rule*
$\ln \frac{x}{y} = \ln x - \ln y$	$(x > 0 \text{ and } y > 0)$	*Quotient rule*
$\ln x^n = n \ln x$	$(x > 0)$	*Power rule*

Consider the expression $\ln e^x$, which means $\log_e e^x$. From property 4 on page 588, $\log_e e^x = x$. Thus $\ln e^x = x$. Similarly, $e^{\ln x} = e^{\log_e x} = x$ by property 5. Although $\ln e^x = x$ and $e^{\ln x} = x$ are just special cases of properties 4 and 5, respectively, we will call these properties 7 and 8 so that we can make reference to them.

Additional Properties for Natural Logarithms and Natural Exponential Expressions

$\ln e^x = x$	Property 7
$e^{\ln x} = x, \quad x > 0$	Property 8

Using property 7, $\ln e^x = x$, we can state, for example, that $\ln e^{kt} = kt$, and $\ln e^{-2.06t} = -2.06t$. Using property 8, $e^{\ln x} = x$, we can state, for example, that $e^{\ln(t+2)} = t + 2$ and $e^{\ln kt} = kt$.

EXAMPLE 5 Solve the equation $\ln y - \ln(x + 6) = t$ for y.

Solution

$$\ln y - \ln(x + 6) = t$$

$$\ln \frac{y}{x + 6} = t \qquad \textit{Quotient rule}$$

To eliminate the natural logarithm on the left side of the equation, we rewrite both sides of the equation in exponential form, with the base e. By property 6a given on page 597 (if $x = y$, then $a^x = a^y$), we write

$$e^{\ln[y/(x+6)]} = e^t \qquad \textit{Property 6a}$$

By Property 8, $e^{\ln x} = x$, we obtain

$$\frac{y}{x + 6} = e^t \qquad \textit{Property 8}$$

Now we solve the equation for y.

$$y = (x + 6)e^t$$

In Example 5 we could have changed $\ln \frac{y}{x + 6} = t$ to $\frac{y}{x + 6} = e^t$ using a different procedure, as shown below.

$$\ln \frac{y}{x + 6} = t \quad \text{means} \quad \log_e \frac{y}{x + 6} = t$$

Changing the logarithm on the right to exponential form gives

$$e^t = \frac{y}{x + 6}$$

NOW TRY EXERCISE 31

The procedure used in Example 5 reinforces some properties of logarithms.

EXAMPLE 6 Solve the equation $225 = 450e^{-0.4t}$ for t.

Solution Begin by dividing both sides of the equation by 450 to isolate $e^{-0.4t}$.

$$\frac{225}{450} = \frac{\cancel{450}\, e^{-0.4t}}{\cancel{450}}$$

$$0.5 = e^{-0.4t}$$

Now take the natural logarithm of both sides of the equation to eliminate the exponential expression on the right side of the equation.

$$\ln 0.5 = \ln e^{-0.4t}$$

$$\ln 0.5 = -0.4t \qquad \textit{Property 7}$$

$$-0.6931472 = -0.4t$$

$$\frac{-0.6931472}{-0.4} = t$$

$$1.732868 = t$$

EXAMPLE 7 Solve the equation $P = P_0 e^{kt}$ for t.

Solution We can follow the same procedure as used in Example 6.

$$P = P_0 e^{kt}$$

$$\frac{P}{P_0} = \frac{P_0 e^{kt}}{P_0} \qquad \textit{Divide both sides by } P_0$$

$$\frac{P}{P_0} = e^{kt}$$

$$\ln \frac{P}{P_0} = \ln e^{kt} \qquad \textit{Take natural log of both sides.}$$

$$\ln P - \ln P_0 = \ln e^{kt} \qquad \textit{Quotient Property}$$

$$\ln P - \ln P_0 = kt \qquad \textit{Property 7}$$

$$\frac{\ln P - \ln P_0}{k} = t \qquad \textit{Solve for } t.$$

NOW TRY EXERCISE 51

6 Solve Applications

Now let us look at some applications that involve the natural exponential function and natural logarithms.

When a quantity increases or decreases at an *exponential rate*, a formula often used to find the value of P after time t is

$$P = P_0 e^{kt}$$

where P_0 is the initial or starting value and k is the constant rate of growth or decay. We will refer to this formula as the **exponential growth (or decay) formula**. In the formula, other letters may be used in place of P. When $k > 0$, P increases as t increases. When $k < 0$, P decreases and gets closer to 0 as t increases.

EXAMPLE 8 Banks often credit compound interest on a continuous basis. When interest is compounded continuously, the balance, P, in the account at any time, t, can be calculated by the exponential growth formula $P = P_0 e^{kt}$, where P_0 is the principal initially invested and k is the interest rate.

a) Suppose the interest rate is 6% compounded continuously and \$1000 is initially invested. Determine the balance in the account after 1 year.

b) How long will it take the account to double in value?

Solution **a) Understand and Translate** We are told that the principal initially invested, P_0, is \$1000. We are also given that the time, t, is 1 year and that the interest rate, k, is 6% or 0.06. We substitute these values in the given formula and solve for P.

$$P = P_0e^{kt}$$

$$P = 1000e^{(0.06)(1)}$$

Carry out

$$= 1000e^{0.06}$$

$$= 1000(1.0618365) \quad \text{From a calculator}$$

$$= 1061.8365$$

Answer After 1 year the balance in the account is \$1061.84.

b) Understand and Translate For the value of the account to double, the balance in the account would have to reach \$2000. Therefore, we substitute 2000 for P and solve for t.

$$P = P_0e^{kt}$$

$$2000 = 1000e^{0.06t}$$

$$2 = e^{0.06t}$$

Carry Out

$$\ln 2 = \ln e^{0.06t} \quad \text{Take natural log of both sides.}$$

$$\ln 2 = 0.06t \quad \text{Property 7}$$

$$\frac{\ln 2}{0.06} = t$$

$$\frac{0.6931472}{0.06} = t$$

$$11.552453 = t$$

Answer Thus, with an interest rate of 6% compounded continuously, the account would double in about 11.6 years.

EXAMPLE 9 Assume that the value of the island of Manhattan has grown at an exponential rate of 8% per year since 1626, when Peter Minuit of the Dutch West India Company purchased the island for \$24. What is the value of the island of Manhattan in 2000?

Solution **Understand** Since the value is increasing exponentially, we can use the exponential growth formula. We will use V_0 to represent initial value, and V to represent the value in 2000. By subtracting 1626 from 2000 we find that the time, t, is 374 years.

Translate

$$V = V_0e^{kt}$$

$$= 24e^{(0.08)(374)}$$

Carry Out

$$= 24e^{29.92}$$
$$\approx 24(9.86485937 \times 10^{12})$$
$$\approx 2.367566249 \times 10^{14}$$
$$\approx 236{,}756{,}624{,}900{,}000$$

Answer If the value grew exponentially at 8%, the value of Manhattan after 374 years is about $236,756,624,900,000. This example illustrates the effect of compounding interest over a long period of time.

EXAMPLE 10 Strontium 90 is a radioactive isotope that decays exponentially at 2.8% per year. Suppose there are initially 1000 grams of strontium 90 in a substance.

a) Find the number of grams of strontium 90 left after 50 years.

b) Find the half-life of strontium 90.

Solution **a) Understand** Since the strontium 90 is decaying over time, the value of k in the formula $P = P_0 e^{kt}$ is negative. Since the rate of decay is 2.8% per year, we use $k = -0.028$. Therefore the formula we use is $P = P_0 e^{-0.028t}$.

Translate

$$P = P_0 e^{-0.028t}$$
$$= 1000e^{-0.028(50)}$$

Carry Out

$$= 1000e^{-1.4}$$
$$= 1000(0.246597)$$
$$= 246.597$$

Answer Thus after 50 years, 246.597 grams of strontium 90 remain.

b) To find the half-life, we need to determine when 500 grams of strontium 90 are left.

$$P = P_0 e^{-0.028t}$$
$$500 = 1000e^{-0.028t}$$
$$0.5 = e^{-0.028t}$$
$$\ln 0.5 = \ln e^{-0.028t} \quad \text{Take natural log of both sides.}$$
$$-0.6931472 = -0.028t \quad \text{Property 7}$$
$$\frac{-0.6931472}{-0.028} = t$$
$$24.755257 \approx t$$

NOW TRY EXERCISE 71 Thus, the half-life of strontium 90 is about 24.8 years.

EXAMPLE 11 The formula for estimating the number, N, of a particular toy sold is $N = 400 + 250 \ln a$, where a is the amount of money spent on advertising the toy.

a) If \$2000 is spent on advertising, what are the expected sales of the toy?

b) How much money should be spent on advertising to sell 1500 toys?

Solution **a)**

$$\begin{aligned} N &= 400 + 250 \ln a \\ &= 400 + 250 \ln 2000 && \text{Substitute 2000 for } a. \\ &= 400 + 250(7.6009025) \\ &= 2300.2256 \end{aligned}$$

Thus, approximately 2300 toys are expected to be sold.

b) Understand and Translate We are asked to find the amount of money to be spent on advertising, a, to sell 1500 toys. We substitute the appropriate values in the given equation and solve for a.

$$N = 400 + 250 \ln a$$

$$1500 = 400 + 250 \ln a \qquad \text{Substitute 1500 for } N.$$

Carry Out

$$1100 = 250 \ln a$$

$$\frac{1100}{250} = \ln a$$

$$4.4 = \ln a$$

$$e^{4.4} = e^{\ln a} \qquad \text{Write both sides with base } e, \text{ Property 6a}$$

$$81.45 = a \qquad \text{Property 8}$$

NOW TRY EXERCISE 67

Answer Thus, about \$81 should be spent on advertising to sell 1500 toys.

Using Your Graphing Calculator

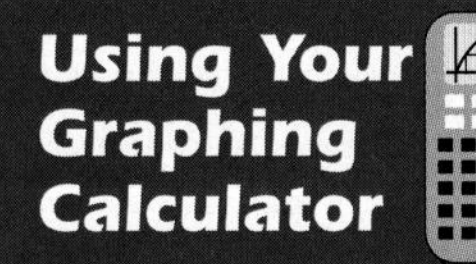

Equations containing natural logarithms and natural exponential functions can be solved on your graphing calculator. For example, to solve the equation $\ln x + \ln(x + 3) = \ln 8$ we set

$$y_1 = \ln x + \ln(x + 3)$$

$$y_2 = \ln 8$$

and find the intersection of the graphs, as shown in Figure 9.28.

In Figure 9.28 we used the CALC, INTERSECT option to find the intersection of the graphs. The solution is the x-coordinate of the intersection. The solution to the equation is $x = 1.7012$, to the nearest thousandth.

To solve the equation $4e^{0.3x} - 5 = x + 3$, we set

$$y_1 = 4e^{0.3x} - 5$$

$$y_2 = x + 3$$

and find the intersection of the equations, as shown in Figure 9.29. This equation has two solutions since there are two intersections.

The solutions to the equation are approximately $x = -7.5896$ and $x = 3.5284$. In Exercises 78 through 82, we use the grapher to check or solve equations.

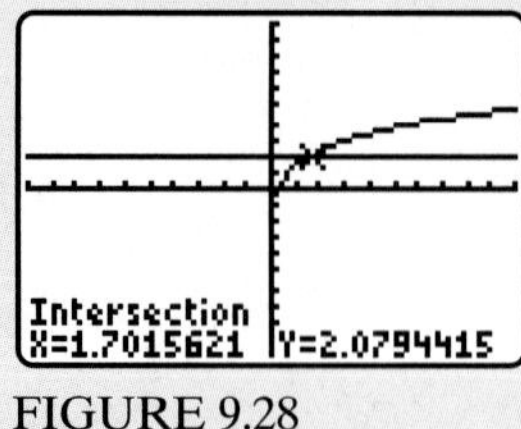

FIGURE 9.28

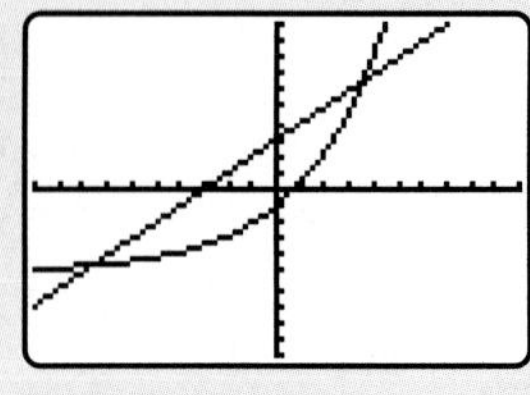

FIGURE 9.29

Exercise Set 9.7

Concept/Writing Exercises

1. **a)** What is the base in the natural exponential function?
   b) What is the approximate value of e?
2. What is another way of writing $\log_e x$?
3. What is the domain of $\ln x$?
4. Under what conditions will $\ln x < 0$?
5. Give the change of base formula.
6. Is $n \log_e x = \ln x^n$? Explain.
7. To what is $\ln e^x$ equal?
8. To what is $e^{\ln x}$ equal?
9. Under what circumstances will P in the formula $P = P_0 e^{kt}$ increase when t increases?
10. Under what conditions will P in the formula $P = P_0 e^{kt}$ decrease when t increases?

Practice the Skills

Find the following values. Round values to four decimal places.

11. $\ln 35$
12. $\ln 0.432$
13. $\ln 302$
14. $\ln 0.0038$

Find the value of N. Round values to three significant digits.

15. $\ln N = 1.6$
16. $\ln N = 4.96$
17. $\ln N = -2.72$
18. $\ln N = 0.632$

Use the change of base formula to find the value of the following logarithms. Do not round logarithms in the change of base formula. Write the answer rounded to the nearest ten-thousandths.

19. $\ln 60$
20. $\ln 562$
21. $\ln 0.046$
22. $\ln 198$
23. $\log_3 25$
24. $\log_7 96$
25. $\log_2 32$
26. $\log_5 0.463$
27. $\ln 2700$
28. $\log_6 4000$
29. $\log_3 0.0049$
30. $\ln 8462$

Solve the following logarithmic equations.

31. $\ln x + \ln(x - 1) = \ln 12$
32. $\ln(x + 3) + \ln(x + 2) = \ln 6$
33. $\ln x = 5 \ln 2 - \ln 8$
34. $\ln x + \ln(x - 1) = \ln 2$
35. $\ln(x^2 - 4) - \ln(x + 2) = \ln 1$
36. $\ln x = \frac{3}{2} \ln 4$

Each of the following equations is in the form $P = P_0 e^{kt}$. Solve each equation for the remaining variable. Remember, e is a constant.

37. $P = 700e^{1.7(1.2)}$
38. $1000 = V_0 e^{0.6(2)}$
39. $50 = P_0 e^{-0.05(3)}$
40. $120 = 60e^{2t}$
41. $90 = 30e^{1.4t}$
42. $20 = 40e^{-0.5t}$
43. $80 = 40e^{k(3)}$
44. $10 = 50e^{k(4)}$
45. $20 = 40e^{k(2.4)}$
46. $100 = A_0 e^{-0.02(3)}$
47. $A = 6000e^{-0.08(3)}$
48. $75 = 100e^{-0.04t}$

Solve for the indicated variable.

49. $V = V_0 e^{kt}$, for V_0
50. $P = P_0 e^{kt}$, for P_0
51. $P = 150e^{4t}$, for t
52. $200 = P_0 e^{kt}$, for t
53. $A = A_0 e^{kt}$, for k
54. $140 = R_0 e^{kt}$, for k
55. $\ln y - \ln x = 2.3$, for y
56. $\ln y + 5 \ln x = \ln 2$, for y
57. $\ln y - \ln(x + 3) = 6$, for y
58. $\ln(x + 2) - \ln(y - 1) = \ln 5$, for y

Problem Solving

Use a calculator to solve.

59. If $e^x = 12.183$, find the value of x. Explain how you obtained your answer.
60. To what exponent must the base e be raised to obtain the value 184.93? Explain how you obtained your answer.
61. If $5000 is invested at 8% compounded continuously, **a)** determine the balance in the account after 2 years, and **b)** how long would it take the value of the account to double? (See Example 8)

62. Refer to Example 10. Determine the amount of strontium 90 remaining after 20 years if there were originally 70 grams.

63. The supply of caviar has been decreasing at an alarming rate. In 1986 there were about 33 thousand tons of caviar produced and sold. By 1996 less than 7 thousand tons were produced and sold. The amount of caviar available, $f(x)$, in thousands of tons, can be approximated by the function $f(x) = 30 - 15 \ln 0.4x$, where x is years since 1984 and $3 \le x \le 18$. Use the function to approximate the amount of caviar available in **a)** 1989 and **b)** 2000.

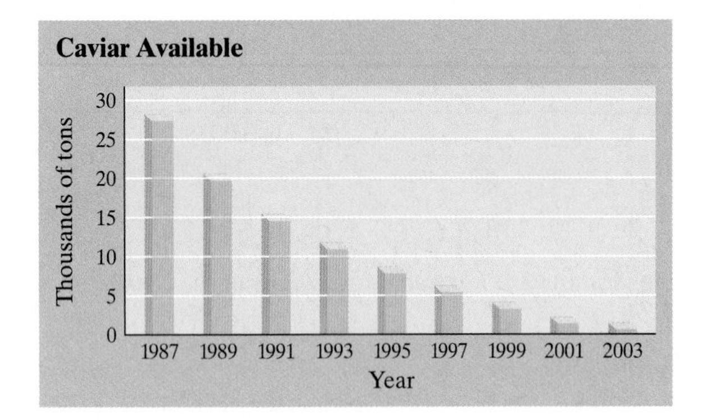

64. The following graph shows the increase in the number of social security beneficiaries from 1940 projected to 2070. The graph also shows the number of workers per beneficiary from 1940 through 2070. The number of workers per beneficiary from 1940 through 1960 can be approximated by the function $f(x) = 44.2 - 12.8 \ln x$, where x is years since 1939 and $1 \le x \le 26$. Use the given function to estimate the number of workers per beneficiary in **a)** 1940 and **b)** 1960.

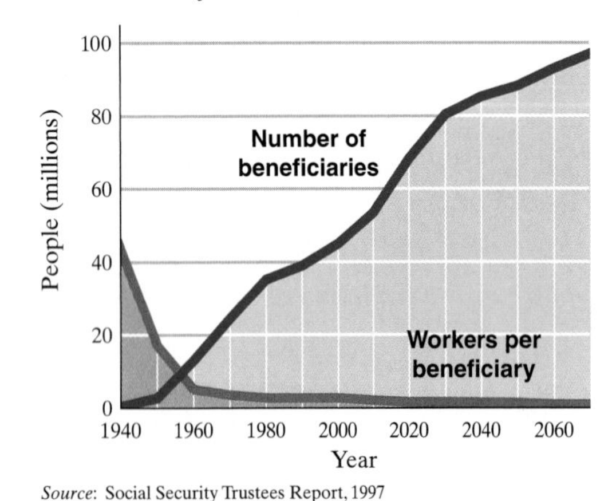

Source: Social Security Trustees Report, 1997

65. For a certain soft drink, the percent of a target market, $f(t)$, that buys the soft drink is a function of the number of days, t that the soft drink is advertised. The function that describes this relationship is $f(t) = 1 - e^{-0.04t}$.

a) What percent of the target market buys the soft drink after 50 days of advertising?

b) How many days of advertising are needed if 75% of the target market is to buy the soft drink?

66. For a certain type of tie, the number of ties sold, $N(a)$, is a function of the dollar amount spent on advertising, a (in thousands of dollars). The function that describes this relationship is $N(a) = 800 + 300 \ln a$.

a) How many ties were sold after \$1500 (or \$1.5 thousand) was spent on advertising?

b) How much money must be spent on advertising to sell 1000 ties?

67. It was found in a psychological study that the average walking speed, $f(P)$, of a person living in a city is a function of the population of the city. For a city of population P, the average walking speed in feet per second is given by $f(P) = 0.37 \ln P + 0.05$. The population of Nashville, Tennessee, is 972,000.

a) What is the average walking speed of a person living in Nashville?

b) What is the average walking speed of a person living in New York City, population 8,567,000?

c) If the average walking speed of the people in a certain city is 5.0 feet per second, what is the population of the city?

68. The percent of doctors who accept and prescribe a new drug is given by the formula $P(t) = 1 - e^{-0.22t}$, where t is the time in months since the drug was placed on the market. What percentage of doctors accept a new drug 2 months after it is placed on the market?

69. The world population in 1999 was estimated at about 5.98 billion people. It is estimated that the world's population continues to grow exponentially at the rate of 1.33% per year. The world's expected population, in billions, in t years is given by the formula $P(t) = 5.98e^{0.0133t}$ where t is years since 1999.

a) Find the expected world's population in the year 2006.

b) In how many years will the world's population double?

70. Plutonium, which is commonly used in nuclear reactors, decays at a rate of 0.003% per year. The formula $A = A_0e^{kt}$ can be used to find the amount of plutonium remaining from an initial amount, A_0, after t years. In the formula the k is replaced with -0.00003.

a) If 1000 grams of plutonium is present in 2000, how many grams of plutonium will remain in the year 2100, 100 years later?

b) Find the half-life of plutonium.

71. Carbon dating is used to estimate the age of ancient plants and objects. The radioactive element, carbon 14, is most often used for this purpose. Carbon 14 decays at a rate of 0.01205% per year. The amount of carbon 14 remaining in an object after t years can be found by the function $f(t) = v_0e^{-0.0001205t}$, where v_0 is the initial amount present.

a) If an ancient animal bone originally had 20 grams of carbon 14, and when found it had 9 grams of carbon 14, how old is the bone?

b) How old is an item that has 50% of its original carbon 14 remaining?

72. Determine the age of a fossil if it contains 80% of its amount of carbon 14 (see Exercise 71).

73. At what rate, compounded continuously, must a sum of money be invested if it is to double in 6 years?

74. How much money must be deposited today to become $20,000 in 18 years if invested at 6% compounded continuously?

75. The power supply of a satellite is a radioisotope. The power P, in watts, remaining in the power supply is a function of the time the satellite is in space.

a) If there are 50 grams of the isotope originally, the power remaining after t days is $P = 50e^{-0.002t}$. Find the power remaining after 100 days.

b) When will the power remaining drop to 10 watts?

76. During the nuclear accident at Chernobyl in the Ukraine in 1986, two of the radioactive materials that escaped into the atmosphere were cesium 137, with a decay rate of 2.3% and strontium 90, with a decay rate of 2.8%.

a) Which material will decompose more quickly? Explain.

b) What percentage of the cesium will remain in 2036, 50 years after the accident?

77. In the study of radiometric dating (using radioactive isotopes to determine the age of items), the formula

$$t = \frac{t_h}{0.693}\ln\left(\frac{N_0}{N}\right)$$

is often used. In the formula, t is the age of the item, t_h is the half-life of the radioactive isotope used, N_0 is the original number of radioactive atoms present, and N is the number remaining at time t. Suppose a rock originally contained 5×10^{12} atoms of uranium 238. Uranium 238 has a half-life of 4.5×10^9 years. If at present there are 4×10^{12} atoms present, how old is the rock?

In Exercises 78–82, use your graphing calculator. In Exercises 80–82, round your answers to the nearest thousandth.

78. Check your answer to Exercise 31.

79. Check your answer to Exercise 33.

80. Solve the equation $e^{x-4} = 12\ln(x + 2)$

81. Solve the equation $\ln(4 - x) = 2\ln x + \ln 2.4$

82. Solve the equation $3x - 6 = 2e^{0.2x} - 12$.

Challenge Exercises

In Exercises 83–86, when you solve for the given variable, write the answer without using the natural logarithm.

83. The intensity of light as it passes through a certain medium is found by the formula $x = k(\ln I_0 - \ln I)$. Solve this equation for I_0.

84. The distance traveled by a train initially moving at velocity v_0 after the engine is shut off can be calculated by the formula $x = \frac{1}{k}\ln(kv_0t + 1)$. Solve this equation for v_0.

85. A formula used in studying the action of a protein molecule is $\ln M = \ln Q - \ln(1 - Q)$. Solve this equation for Q.

86. An equation relating the current and time in an electric circuit is $\ln i - \ln I = \frac{-t}{RC}$. Solve this equation for i.

Cumulative Review Exercises

[1.5] **87.** Simplify $(x^2y^{-2})^{-1}(4xy^3)^2$

[6.3] **88.** Simplify $\dfrac{\frac{3}{x^2} - \frac{2}{x}}{\frac{x}{4}}$.

[6.4] **89.** Solve the formula $\frac{1}{f} = \frac{1}{p} + \frac{1}{q}$ for q.

[7.3] **90.** Simplify $\sqrt[3]{128x^7y^9z^{13}}$.

SUMMARY

Key Words and Phrases

9.1
Composite fuctions
Composition of functions
Horizontal line test
Inverse function
One-to-one function
Vertical line test

9.2
Exponential function

9.3
Logarithm
Logarithmic function

9.4
Power rule for logarithms
Product rule for logarithms
Quotient rule for logarithms

9.5
Antilogarithms
Base 10 logarithm
Common logarithm
Inverse logarithm
Significant digits

9.6
Exponential equation
Logarithmic equation

9.7
Change of base formula
e
Exponential growth (or decay) formula
Natural exponential function
Natural logarithm
Natural logarithmic function

IMPORTANT FACTS

Composite function of f with g: $(f \circ g)(x) = f[g(x)]$
Composite function of g with f: $(g \circ f)(x) = g[f(x)]$
If $f(x)$ and $g(x)$ are inverse functions, then $(f \circ g)(x) = (g \circ f)(x) = x$.
$y = a^x$ and $y = \log_a x$ are inverse functions.
$y = e^x$ and $y = \ln x$ are inverse functions.
The domain of a logarithmic function of the form $y = \log_a x$ is $x > 0$.

Properties of Logarithms

1. $\log_a xy = \log_a x + \log_a y$
2. $\log_a \dfrac{x}{y} = \log_a x - \log_a y$
3. $\log_a x^n = n \log_a x$
4. $\log_a a^x = x$
5. $a^{\log_a x} = x\ (x > 0)$

Change of Base Formula

$$\log_a x = \frac{\log_b x}{\log_b a}$$

To solve exponential and logarithmic equations, we also use these properties:

6. a) If $x = y$, then $a^x = a^y$.
   b) If $a^x = a^y$, then $x = y$.
   c) If $x = y$, then $\log x = \log y$ $(x > 0, y > 0)$.
   d) If $\log x = \log y$, then $x = y$ $(x > 0, y > 0)$.

Natural Logarithms

7. $\ln e^x = x$
8. $e^{\ln x} = x, x > 0$

Review Exercises

[9.1] *Given $f(x) = x^2 - 3x + 4$ and $g(x) = 2x - 5$, find the following.*

1. $(f \circ g)(x)$
2. $(f \circ g)(2)$
3. $(g \circ f)(x)$
4. $(g \circ f)(-3)$

Given $f(x) = 3x + 2$ and $g(x) = \sqrt{x-4}$, $x \geq 4$, find the following.

5. $(f \circ g)(x)$
6. $(g \circ f)(x)$

Determine whether each function is a one-to-one function.

7. 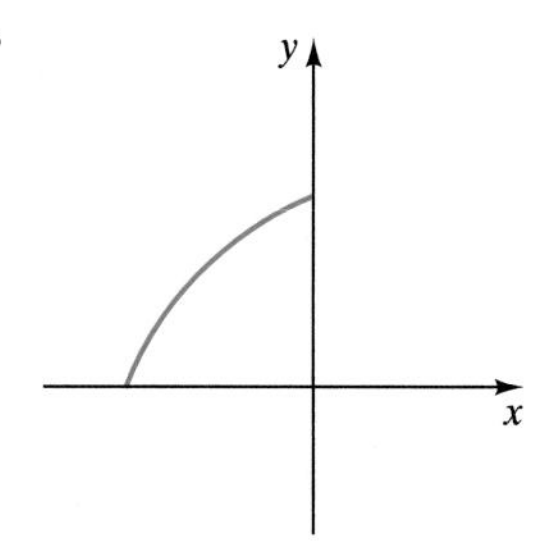

8. 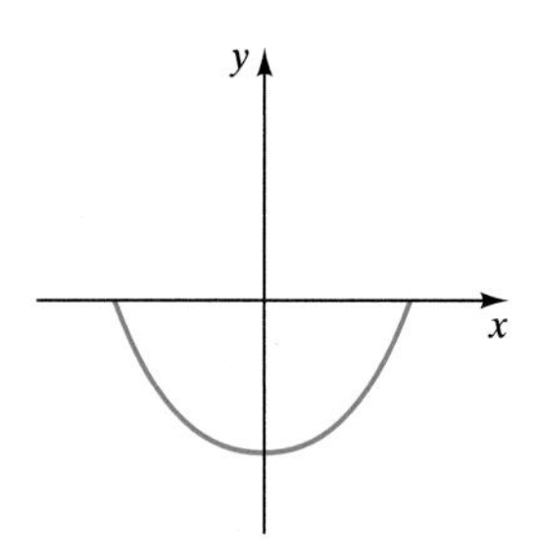

9. $\{(2, 3), (4, 0), (-5, 7), (3, 8)\}$

10. $\left\{(0, -2), (5, 6), (3, -2), \left(\frac{1}{2}, 4\right)\right\}$

11. $y = \sqrt{x+1}$, $x \geq -1$

12. $y = x^2 - 9$

For each function, find the domain and range of both $f(x)$ and $f^{-1}(x)$.

13. $\{(5, 3), (6, 2), (-4, -3), (0, 7)\}$

14. 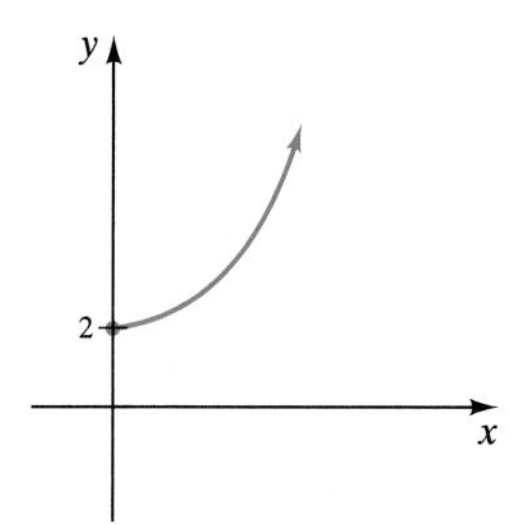

Find $f^{-1}(x)$ and graph $f(x)$ and $f^{-1}(x)$ on the same axes.

15. $y = f(x) = 4x - 2$

16. $y = f(x) = \sqrt[3]{x-1}$

[9.2] *Graph the following functions.*

17. $y = 2^x$

18. $y = \left(\frac{1}{2}\right)^x$

19. The following graph shows the worldwide growth in the use of computer CD-ROM drives.

Hard-driving Hardware (worldwide)

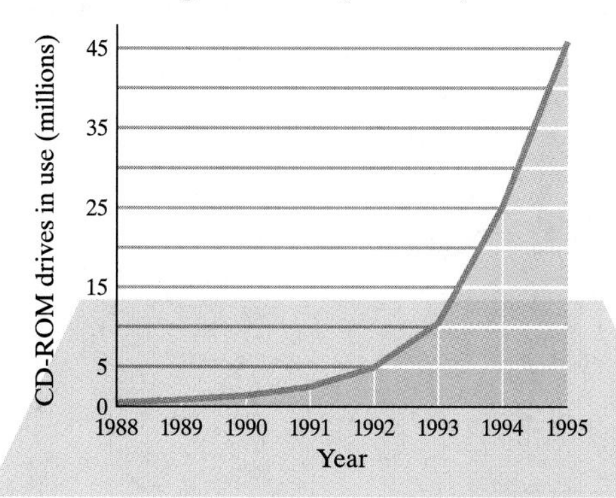

a) Was the growth more nearly linear or exponential from 1988 through 1990? Explain.

b) In the period 1988–1995, is the graph more nearly linear or exponential? Explain.

c) How many measurements were used in constructing this graph?

d) Estimate the time for the number of drives to double from the number used in 1992.

e) Estimate the time for the number of drives to double from the number used in 1993.

f) Estimate when 35 million drives were in use worldwide.

[9.3] *Write each equation in logarithmic form.*

20. $4^2 = 16$ **21.** $8^{1/3} = 2$ **22.** $6^{-2} = \frac{1}{36}$

Write each equation in exponential form.

23. $\log_5 25 = 2$ **24.** $\log_{1/3} \frac{1}{9} = 2$ **25.** $\log_3 \frac{1}{9} = -2$

Write each equation in exponential form and find the missing value.

26. $3 = \log_4 x$ **27.** $3 = \log_a 8$ **28.** $-3 = \log_{1/4} x$

Graph the following functions.

29. $y = \log_2 x$ **30.** $y = \log_{1/2} x$

[9.4] *Use the properties of logarithms to expand each expression.*

31. $\log_8 \sqrt{12}$ **32.** $\log (x - 8)^5$

33. $\log \dfrac{2(x - 3)}{x}$ **34.** $\log \dfrac{x^4}{39(2x + 8)}$

Write the following as the logarithm of a single expression.

35. $2 \log x - 3 \log (x + 1)$ **36.** $3(\log 2 + \log x) - \log y$

37. $\frac{1}{2}[\ln x - \ln (x + 2)] - \ln 2$ **38.** $3 \ln x + \frac{1}{2} \ln (x + 1) - 3 \ln (x + 4)$

Evaluate.

39. $8^{\log_8 9}$ **40.** $\log_4 4^5$ **41.** $3 \log_7 49$ **42.** $4^{\log_8 \sqrt{8}}$

[9.5] *Use a calculator to find the common logarithm of each number. Round your answer to four decimal places.*

43. 8200 **44.** 0.000716

Use a calculator to find the antilog of each number. Give the antilog to three significant digits.

45. 2.9186 **46.** −1.3747

Use a calculator to find N. Round your answer to three significant digits.

47. $\log N = 2.3304$ **48.** $\log N = -1.2262$

Evaluate.

49. $\log 10^4$ **50.** $10^{\log 3}$ **51.** $7.5 \log 10^{4.2}$ **52.** $3(10^{\log 1.7})$

[9.6] *Solve without using a calculator.*

53. $9 = 3^x$ **54.** $49^x = \frac{1}{7}$ **55.** $2^{2x+3} = 32$ **56.** $27^x = 3^{2x+5}$

Solve using a calculator. Round your answers to the nearest thousandth.

57. $4^x = 37$ **58.** $3.2^x = 187$ **59.** $10.9^{x+1} = 492$ **60.** $3^{x+2} = 8^x$

Solve the logarithmic equation.

61. $\log_5 (x + 2) = 3$ **62.** $\log x - \log(3x - 5) = \log 2$

63. $\log_3 x + \log_3 (2x + 1) = 1$ **64.** $\ln (x + 1) - \ln (x - 2) = \ln 4$

[9.7] *Solve each exponential equation for the remaining variable.*

65. $40 = 20e^{0.6t}$ **66.** $100 = A_0 e^{-0.42(3)}$

Solve for the indicated variable.

67. $A = A_0 e^{kt}$, for t **68.** $150 = 600e^{kt}$, for k

69. $\ln y - \ln x = 2$, for y

70. $\ln(y + 3) - \ln(x + 1) = \ln 5$, for y

Use the change of base formula to evaluate.

71. $\ln 450$

72. $\log_3 50$

[9.2–9.7]

73. Find the amount of money accumulated if Justine Elwood puts \$12,000 in a savings account yielding 10% interest per year for 8 years. Use $A = p(1 + r)^n$.

74. Plutonium is a radioactive element that decays over time. If there were originally 1000 mg of plutonium, the amount remaining, R, after t years is $R = 1000(0.5)^{0.000041t}$. Calculate the amount of plutonium present after 20,000 years.

75. The bacteria *Escherichia coli* are commonly found in the bladders of humans. Suppose that 2000 bacteria are present at time 0. Then the number of bacteria present t minutes later may be found by the formula $N(t) = 2000(2)^{0.05t}$.

a) When will 50,000 bacteria be present?

b) Suppose that a human bladder infection is classified as a condition with 120,000 bacteria. When would a person develop a bladder infection if he started with 2000 bacteria?

76. A class of history students is given a final exam at the end of the course. As part of a research project, the students are also given equivalent forms of the exam each month for n months. The average grade of the class after n months may be found by the formula $A(n) = 72 - 18\log(n + 1), \quad n \geq 0$.

a) What was the class average when the students took the original exam ($n = 0$)?

b) What was the class average for the exam given 2 months later?

c) After how many months was the class average 59.4?

77. The atmospheric pressure, P, in pounds per square inch at an elevation of x feet above sea level can be found by the formula $P = 14.7e^{-0.00004x}$. Find the atmospheric pressure at an elevation of 12,000 feet.

78. If \$10,000 is placed in a savings account paying 7% interest compounded continuously, find the time needed for the account to double in value.

Practice Test

1. **a)** Determine whether the following function is a one-to-one function.

$$\{(4, 2), (-3, 8), (-1, 3), (5, 7)\}$$

b) List the set of ordered pairs in the inverse function.

2. Given $f(x) = x^2 - 4$ and $g(x) = x + 3$, find **a)** $(f \circ g)(x)$ and **b)** $(f \circ g)(3)$.

3. Given $f(x) = x^2 + 5$ and $g(x) = \sqrt{x - 4}$, $x \geq 4$, find **a)** $(g \circ f)(x)$ and **b)** $(g \circ f)(6)$.

*In Exercises 4 and 5, **a)** find $f^{-1}(x)$ and **b)** graph $f(x)$ and $f^{-1}(x)$ on the same axes.*

4. $y = f(x) = -3x - 5$

5. $y = f(x) = \sqrt{x - 1}, x \geq 1$

6. What is the domain of $y = \log_a x$?

7. Evaluate $\log_8 \frac{1}{64}$.

8. Graph $y = 2^x$.

9. Graph $y = \log_2 x$.

10. Write $4^{-3} = \frac{1}{64}$ in logarithmic form.

11. Write $\log_3 243 = 5$ in exponential form.

Write in exponential form and find the missing value.

12. $4 = \log_2 x$

13. $y = \log_{27} 3$

14. Expand $\log_3 \dfrac{x(x - 4)}{x^2}$.

15. Write as the logarithm of a single expression.

$$3\log_8(x - 4) + 2\log_8(x + 1) - \frac{1}{2}\log_8 x$$

16. Evaluate $2\log_9 \sqrt{9}$.

17. **a)** Find log 4620.
b) Find log 0.000638.

18. Solve $3^x = 123$ for x.

19. Solve $\log 4x = \log(x + 3) + \log 2$ for x.

20. Solve $\log(x + 5) - \log(x - 2) = \log 6$ for x.

21. Find N if $\ln N = 3.52$.

22. Evaluate $\log_6 40$ using the change of base formula.

23. Solve $200 = 500e^{-0.03t}$ for t.

24. What amount of money accumulates if Say-Chun Ling puts \$3500 in a savings account yielding 6% interest compounded quarterly for 10 years?

25. The amount of carbon 14 remaining after t years is found by the formula $v = v_0 e^{-0.0001205t}$, where v_0 is the original amount of carbon 14. If a fossil originally contained 60 grams of carbon 14, and now contains 40 grams of carbon 14, how old is the fossil?

Cumulative Review Test

1. Evaluate $\dfrac{6 - |-18| \div 3^2 - 6}{4 - |-8| \div 2^2}$.

2. Simplify $\left(\dfrac{3x^4y^{-3}}{6xy^4z^2}\right)^{-3}$.

3. Jason Sykora jogs at 4 mph while Kendra Rathbun jogs at 5 mph. Kendra starts jogging $\frac{1}{2}$ hour after Jason and jogs along the same path.

a) How long after Kendra starts will the two meet?

b) How far from the starting point will they be when they meet?

4. Solve $\frac{1}{2}(2x + 12) \geq 3x - 6 + x$.

5. Find the domain and range of the function shown

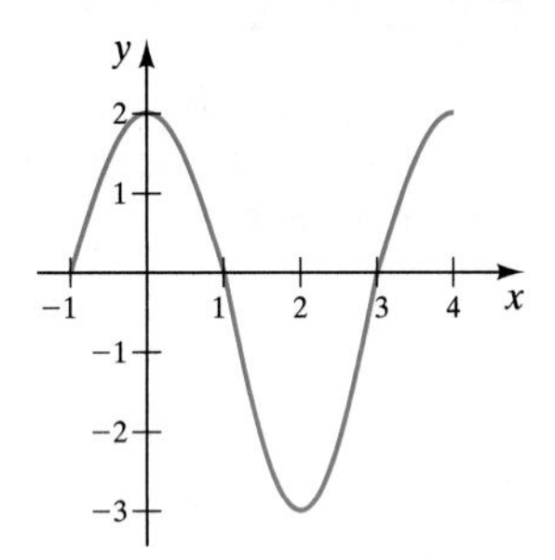

6. Use the points indicated in the graph to answer this problem.

a) Determine the slope of the line shown.

b) Find the equation of the line. Write the equation in standard form.

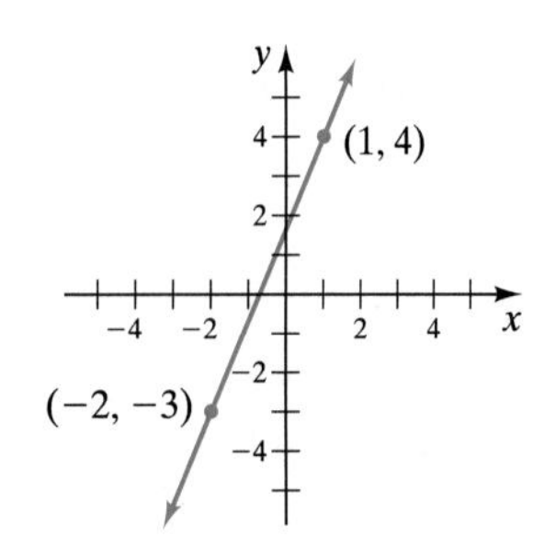

7. Graph $y < \frac{2}{3}x + 3$.

8. Solve the following system.

$$0.4x + 0.6y = 3.2$$
$$1.4x - 0.3y = 1.6$$

9. Solve the following system using matrices.

$$x + y = 6$$
$$-2x + y = 3$$

10. Evaluate the following determinant.

$$\begin{vmatrix} 3 & 0 & -1 \\ 2 & 5 & 3 \\ -1 & 4 & 6 \end{vmatrix}$$

11. Factor $12x^2 - 5xy - 3y^2$.

12. Given $f(x) = 3x^2 - 7x - 11$, find all values of x for which $f(x) = 9$.

13. Divide $2x^2 - x - 26$ by $x - 4$.

14. Solve the following equation.

$$\frac{x + 1}{x + 2} + \frac{x - 2}{x - 3} = \frac{x^2 - 4}{x^2 - x - 6}$$

15. Simplify $\dfrac{x^{-3} + x^{-2}}{x^{-1} - x^{-2}}$.

16. Simplify $\sqrt[3]{\dfrac{27x^4y^5}{5xy^7}}$.

17. Solve $\sqrt[3]{5x + 1} + 4 = 0$.

18. Graph $y = (x - 4)^2 + 1$.

19. Given $f(x) = x^3 - 6x^2 + 5x$, find all values of x for which $f(x) \geq 0$. Write the answer in interval notation.

20. Strontium 90 is a radioactive substance that decays exponentially at 2.8% per year. Suppose there are originally 600 grams of strontium 90. **a)** Find the number of grams left after 60 years. **b)** What is the half-life of strontium 90?

CONIC SECTIONS

CHAPTER 10

Use the Angel Web site at www.prenhall.com/angel to be linked to an internet resource that will help you further explore the following application.

The shape of an ellipse gives it an unusual feature. Anything bounced off the wall of an elliptical shape from one focal point will ricochet to the other focal point. This feature has been used in architecture and medicine. One example is the National Statuary Hall in the Capitol Building which has an elliptically shaped domed ceiling. If you whisper at one focal point, your whisper can be heard at the other focal point. The trick only works at the focal points which can be determined if you know the length and the width of the ellipse. Similarly, a ball hit from one focus of an elliptical billiard table would rebound to the other focal point. On page 638, you will determine the location of the foci of an elliptical billiard table.

Preview and Perspective

The focus of this chapter is on graphing conic sections. These include the circle, ellipse, parabola, and hyperbola. We have already discussed parabolas. In this chapter we will learn more about them.

We solved linear systems of equations graphically in Section 4.1. In Section 10.4 we solve nonlinear systems of equations. The systems we solve will contain the equation of at least one conic section. We will provide both graphical and algebraic solutions to the systems.

Conic sections are important in higher mathematics courses and in certain science and engineering courses. If you take additional mathematics courses, you will probably cover this material again, but in more depth.

10.1 THE PARABOLA AND THE CIRCLE

1. **Identify and describe the conic sections.**
2. **Review parabolas.**
3. **Graph parabolas of the form $x = a(y - k)^2 + h$.**
4. **Learn the distance and midpoint formulas.**
5. **Graph circles with centers at the origin.**
6. **Graph circles with centers at (h, k).**

1 Identify and Describe Conic Sections

In previous chapters we discussed parabolas. A **parabola** is one type of conic section. Parabolas will be discussed further in this section. Other conic sections are circles, ellipses, and hyperbolas. Each of these shapes is called a conic section because each can be made by slicing a cone and observing the shape of the slice. The methods used to slice the cone to obtain each conic section are illustrated in Figure 10.1.

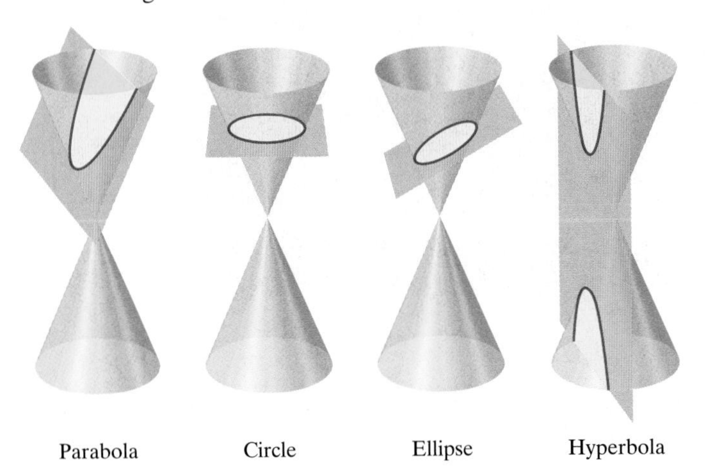

FIGURE 10.1

2 Review Parabolas

We discussed parabolas in Sections 8.5. Example 1 will refresh your memory on how to graph parabolas in the forms $y = ax^2 + bx + c$ and $y = a(x - h)^2 + k$.

EXAMPLE 1 Consider $y = 2x^2 + 4x - 6$.

a) Write the equation in $y = a(x - h)^2 + k$ form.

b) Determine whether the parabola opens upward or downward.

c) Determine the vertex of the parabola.

d) Determine the y-intercept of the parabola.

e) Determine the x-intercepts of the parabola.

f) Graph the parabola.

Solution **a)** First, factor 2 from the two terms containing the variable to make the coefficient of the squared term 1. (Do not factor 2 from the constant, −6.) Then complete the square.

$$\begin{aligned} y &= 2x^2 + 4x - 6 \\ &= 2(x^2 + 2x) - 6 \\ &= 2(x^2 + 2x + 1) - 2 - 6 && \textit{Complete the square.} \\ &= 2(x + 1)^2 - 8 \end{aligned}$$

b) The parabola opens upward since $a = 2$, which is greater than 0.

c) The vertex of the graph of an equation in the form $y = a(x - h)^2 + k$ is (h, k). Therefore the vertex of the graph of $y = 2(x + 1)^2 - 8$ is $(-1, -8)$. The vertex of a parabola can also be found using

$$\left(-\frac{b}{2a}, \frac{4ac - b^2}{4a}\right) \quad \text{or} \quad \left(-\frac{b}{2a}, f\left(-\frac{b}{2a}\right)\right)$$

Show that both of these procedures give $(-1, -8)$ as the vertex of the parabola now.

d) To determine the y-intercept, let $x = 0$ and solve for y.

$$\begin{aligned} y &= 2(x + 1)^2 - 8 \\ &= 2(0 + 1)^2 - 8 \\ &= -6 \end{aligned}$$

The y-intercept is $(0, -6)$.

e) To determine the x-intercepts, let $y = 0$ and solve for x.

$$\begin{aligned} y &= 2(x + 1)^2 - 8 \\ 0 &= 2(x + 1)^2 - 8 && \textit{Substitute 0 for y.} \\ 8 &= 2(x + 1)^2 && \textit{Add 8 to both sides.} \\ 4 &= (x + 1)^2 && \textit{Divide both sides by 2.} \\ \pm 2 &= x + 1 && \textit{Square root property} \\ -1 \pm 2 &= x && \textit{Subtract 1 from both sides.} \end{aligned}$$

$$x = -1 - 2 \quad \text{or} \quad x = -1 + 2$$
$$x = -3 \qquad\qquad x = 1$$

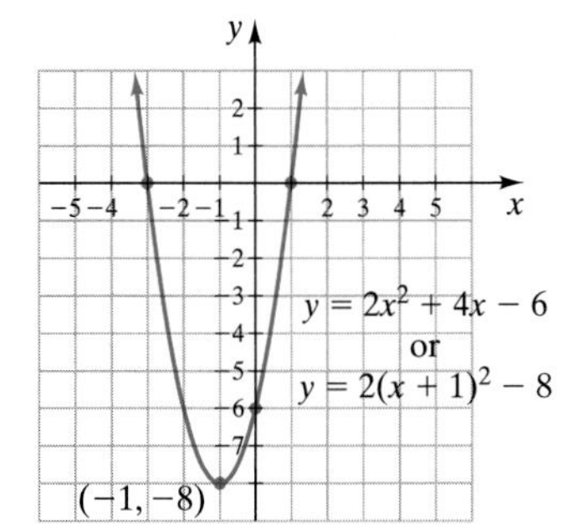

FIGURE 10.2

The x-intercepts are $(-3, 0)$ and $(1, 0)$. The x-intercepts could also be found by substituting 0 for y in $y = 2x^2 + 4x - 6$ and solving for x using factoring or the quadratic formula. Do this now and show that you get the same x-intercepts.

f) We use the vertex and the x- and y-intercepts to draw the graph, which is shown in Figure 10.2.

NOW TRY EXERCISE 11

3 Graph Parabolas of the Form $x = a(y - k)^2 + h$

Parabolas can also open to the right or left. The graph of an equation of the form $x = a(y - k)^2 + h$ will be a parabola whose vertex is at the point (h, k). If a is a positive number, the parabola will open to the right, and if a is a negative number, the parabola will open to the left. The four different forms of a parabola are shown in Figure 10.3.

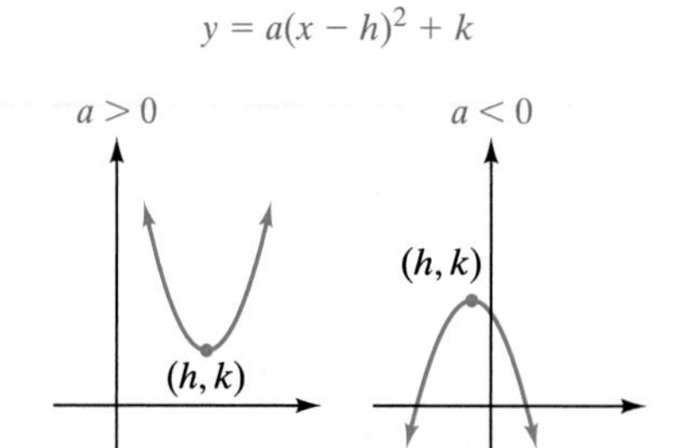

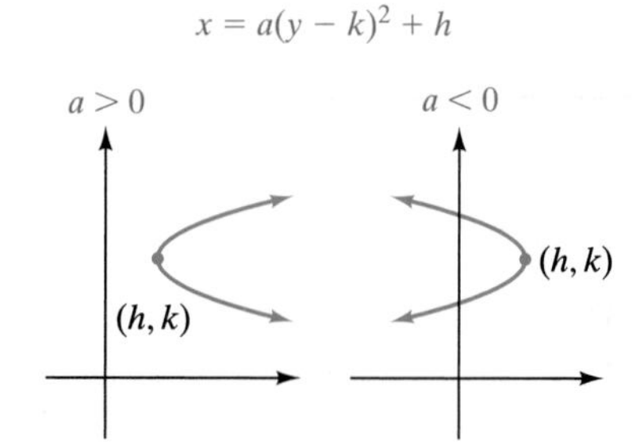

FIGURE 10.3

Parabola with Vertex at (h, k)

1. $y = a(x - h)^2 + k$, $a > 0$ (opens upward)
2. $y = a(x - h)^2 + k$, $a < 0$ (opens downward)
3. $x = a(y - k)^2 + h$, $a > 0$ (opens to the right)
4. $x = a(y - k)^2 + h$, $a < 0$ (opens to the left)

Note that equations of the form $y = a(x - h)^2 + k$ are functions since their graphs pass the vertical line test. However, equations of the form $x = a(y - k)^2 + h$ are not functions since their graphs do not pass the vertical line test.

EXAMPLE 2 Sketch the graph of $x = -2(y + 4)^2 - 1$.

Solution The graph opens to the left since the equation is of the form $x = a(y - k)^2 + h$ and $a = -2$, which is less than 0. The equation can be expressed as $x = -2[y - (-4)]^2 - 1$. Thus, $h = -1$ and $k = -4$. The vertex of the graph is $(-1, -4)$ See Fig. 10.4. If we set $y = 0$, we see that the x-intercept is at $-2(0 + 4)^2 - 1 = -2(16) - 1$ or -33. By substitutive values for y you can find corresponding values of x. When $y = -2$, $x = -9$, and when $y = -6$, $x = -9$. These points are marked on the graph.

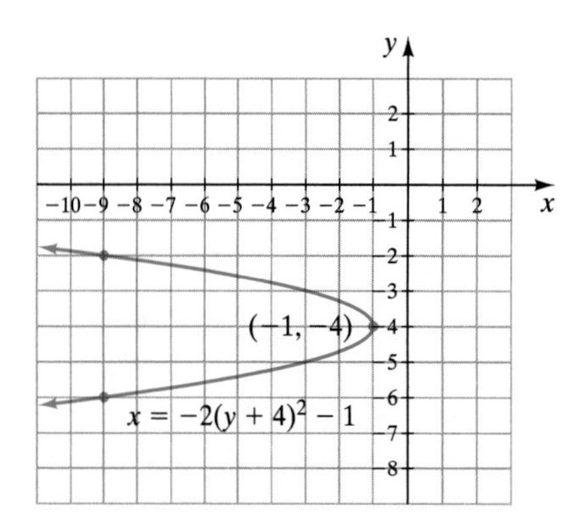

FIGURE 10.4

EXAMPLE 3 **a)** Write the equation $x = -2y^2 + 4y + 5$ in the form $x = a(y - k)^2 + h$.
b) Graph $x = -2y^2 + 4y + 5$.

Solution **a)** First factor -2 from the first two terms. Then complete the square on the expression within parentheses.

$$x = -2y^2 + 4y + 5$$
$$= -2(y^2 - 2y) + 5$$
$$= -2(y^2 - 2y + 1) + 2 + 5$$
$$= -2(y - 1)^2 + 7$$

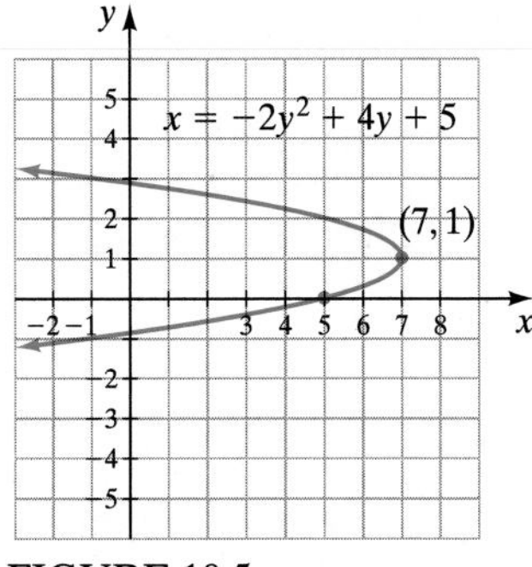

FIGURE 10.5

b) Since $a < 0$, this parabola opens to the left. Note that when $y = 0$, $x = 5$. Therefore, the x-intercept is $(5, 0)$. The vertex of the parabola is $(7, 1)$. Using the quadratic formula, we can determine that the y-intercepts are about $(0, 2.87)$ and $(0, -0.87)$. The graph is shown in Figure 10.5.

NOW TRY EXERCISE 27

4) Learn the Distance and Midpoint Formulas

Now we will derive a formula to find the **distance** between two points on a line. We will use this formula shortly to develop the formula for a circle. Consider Figure 10.6.

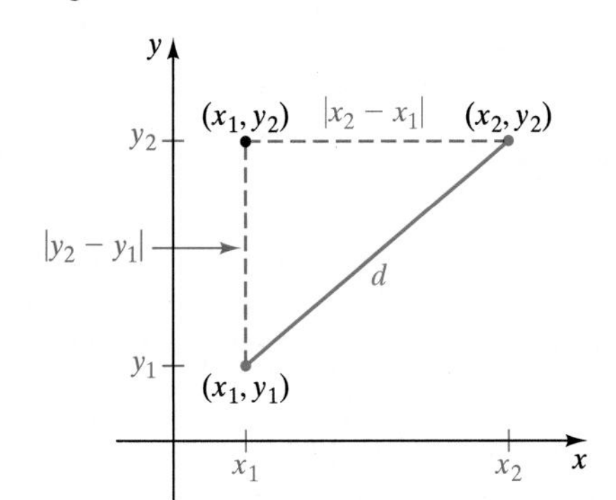

FIGURE 10.6

The horizontal distance between the two points (x_1, y_2) and (x_2, y_2), indicated by the red dashed line, is $|x_2 - x_1|$. We use the absolute value because we want the distance to be positive. If x_1 was larger than x_2, then $x_2 - x_1$ would be negative. The vertical distance between the points (x_1, y_1) and (x_1, y_2), indicated by the green dashed line, is $|y_2 - y_1|$. Using the Pythagorean theorem where d is the distance between the two points, we get

$$d^2 = |x_2 - x_1|^2 + |y_2 - y_1|^2$$

Since any nonzero number squared is positive, we do not need absolute value signs. We can therefore write

$$d^2 = (x_2 - x_1)^2 + (y_2 - y_1)^2$$

Using the square root property, with the principal square root, we get the distance between the points (x_1, y_1) and (x_2, y_2), which is $d = \sqrt{(x_2 - x_1)^2 + (y_2 - y_1)^2}$.

Distance Formula

The distance, d, between any two points (x_1, y_1) and (x_2, y_2) can be found by the distance formula

$$d = \sqrt{(x_2 - x_1)^2 + (y_2 - y_1)^2}$$

The distance between any two points will always be a positive number. Can you explain why? When finding the distance, it makes no difference which point we designate as point 1, (x_1, y_1), or point 2, (x_2, y_2). Note that the square of any real number will always be greater than or equal to zero. For example, $(5 - 2)^2 = (2 - 5)^2 = 9$.

EXAMPLE 4 Determine the distance between the points (2, 3) and (−4, 1).

Solution As an aid we plot the points (Fig. 10.7). Label (2, 3) point 1 and (−4, 1) point 2. Thus, (x_2, y_2) represents (−4, 1) and (x_1, y_1) represents (2, 3). Now use the distance formula to find the distance, d.

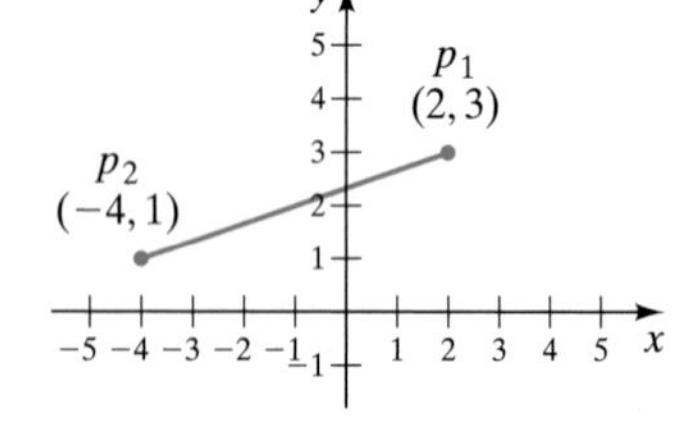

FIGURE 10.7

$$\begin{aligned} d &= \sqrt{(x_2 - x_1)^2 + (y_2 - y_1)^2} \\ &= \sqrt{(-4 - 2)^2 + (1 - 3)^2} \\ &= \sqrt{(-6)^2 + (-2)^2} \\ &= \sqrt{36 + 4} \\ &= \sqrt{40} \quad \text{or} \quad \approx 6.32 \end{aligned}$$

NOW TRY EXERCISE 37 Thus, the distance between the points (2, 3) and (−4, 1) is about 6.32 units.

HELPFUL HINT

Students will sometimes begin finding the distance correctly using the distance formula but will forget to take the square root of the sum $(x_2 - x_1)^2 + (y_2 - y_1)^2$ to obtain the correct answer. When taking the square root, remember that $\sqrt{a^2 + b^2} \neq a + b$.

It is often necessary to find the **midpoint** of a line segment between two given endpoints. To do this, we use the midpoint formula.

Midpoint Formula

Given any two points (x_1, y_1) and (x_2, y_2), the point halfway between the given points can be found by the midpoint formula:

$$\text{midpoint} = \left(\frac{x_1 + x_2}{2}, \frac{y_1 + y_2}{2}\right)$$

To find the midpoint, we take the average (the mean) of the x-coordinates and of the y-coordinates.

EXAMPLE 5 A line segment through the center of a circle intersects the circle at the points $(-3, 6)$ and $(4, 1)$. Find the center of the circle.

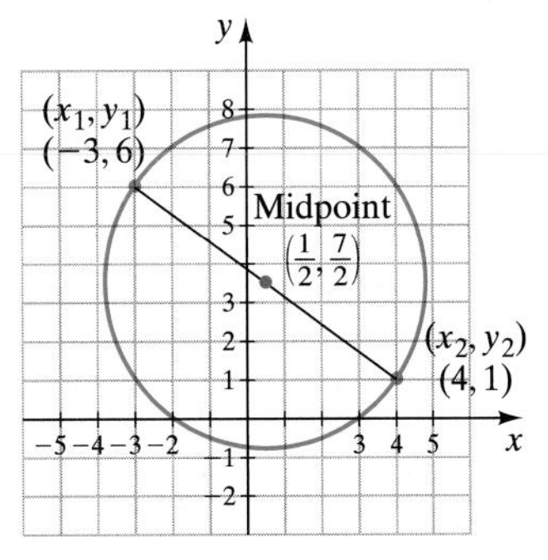

FIGURE 10.8

Solution To find the center of the circle, we find the midpoint of the line segment between $(-3, 6)$ and $(4, 1)$. It makes no difference which points we label (x_1, y_1) and (x_2, y_2). We will let $(-3, 6)$ be (x_1, y_1) and $(4, 1)$ be (x_2, y_2), see Figure 10.8.

$$\text{midpoint} = \left(\frac{x_1 + x_2}{2}, \frac{y_1 + y_2}{2}\right)$$

$$= \left(\frac{-3 + 4}{2}, \frac{6 + 1}{2}\right) = \left(\frac{1}{2}, \frac{7}{2}\right)$$

The point $\left(\frac{1}{2}, \frac{7}{2}\right)$ is halfway between the points $(-3, 6)$ and $(4, 1)$. It is also the center of the circle.

5) Graph Circles with Centers at the Origin

A **circle** may be defined as the set of points in a plane that are the same distance from a fixed point called its **center**.

The formula for the *standard form* of a circle whose center is at the origin may be derived using the distance formula. Let (x, y) be a point on a circle of radius r with center at $(0, 0)$, see Figure 10.9. Using the distance formula, we have

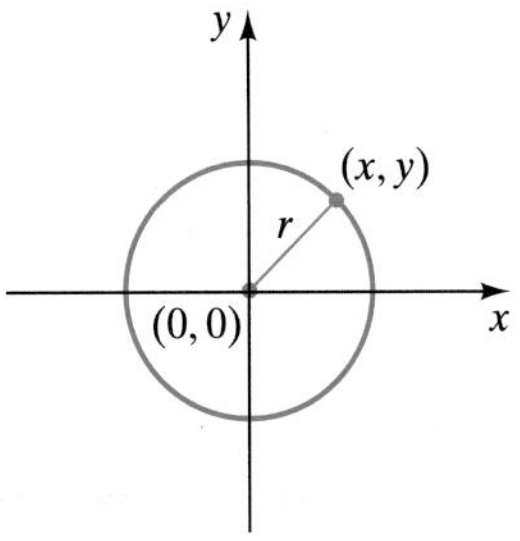

FIGURE 10.9

$$d = \sqrt{(x_2 - x_1)^2 + (y_2 - y_1)^2} \quad \textit{Distance formula}$$

$$\text{or} \quad r = \sqrt{(x - 0)^2 + (y - 0)^2} \quad \textit{Subsitute } r \textit{ for } d, (x, y) \textit{ for } (x_2, y_2), \textit{ and } (0, 0) \textit{ for } (x_1, y_1).$$

$$r = \sqrt{x^2 + y^2} \quad \textit{Simplify the radicand.}$$

$$r^2 = x^2 + y^2 \quad \textit{Square both sides.}$$

Circle with Its Center at the Origin and Radius r

$$x^2 + y^2 = r^2$$

For example, $x^2 + y^2 = 16$ is a circle with its center at the origin and radius 4, and $x^2 + y^2 = 7$ is a circle with its center at the origin and radius $\sqrt{7}$. Note that $4^2 = 16$ and $(\sqrt{7})^2 = 7$.

EXAMPLE 6 Graph the following equations.

a) $x^2 + y^2 = 64$ **b)** $y = \sqrt{64 - x^2}$ **c)** $y = -\sqrt{64 - x^2}$

Solution **a)** If we rewrite the equation as

$$x^2 + y^2 = 8^2$$

we see that the radius is 8. The graph is illustrated in Figure 10.10.

b) If we solve the equation $x^2 + y^2 = 64$ for y, we obtain

$$y^2 = 64 - x^2$$

$$y = \pm\sqrt{64 - x^2}$$

The graph of $y = \sqrt{64 - x^2}$, where y is the principal square root, lies above and on the x-axis. For any value of x in the domain of the function, the value of y must be greater than or equal to 0. Why? The graph is the semicircle shown in Figure 10.11.

c) The graph of $y = -\sqrt{64 - x^2}$ is also a semicircle. However, this graph lies below and on the x-axis. For any value of x in the domain of the function, the value of y must be less than or equal to 0. Why? The graph is shown in Figure 10.12.

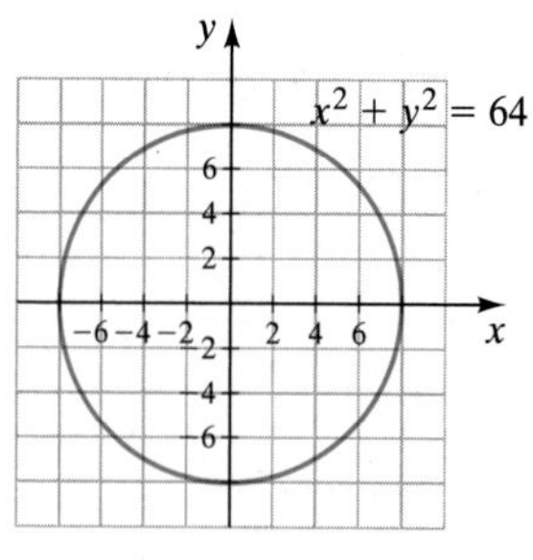

FIGURE 10.10

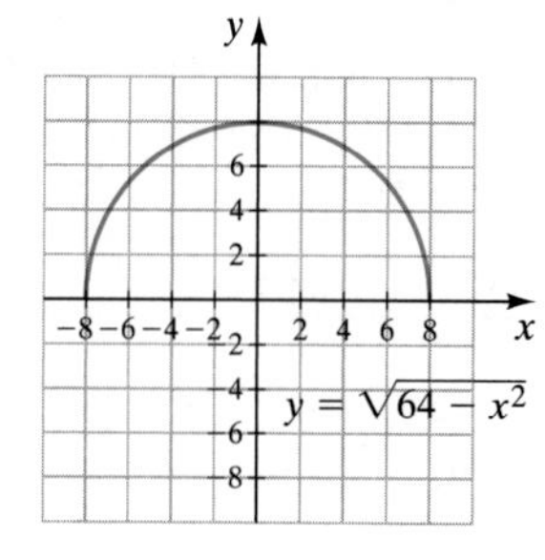

FIGURE 10.11

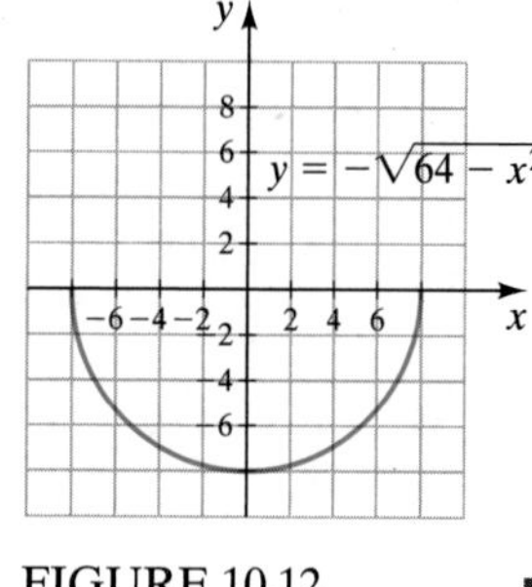

FIGURE 10.12

NOW TRY EXERCISE 81

Consider the equations $y = \sqrt{64 - x^2}$ and $y = -\sqrt{64 - x^2}$ in Example 1**b)** and 1**c)**. If you square both sides of the equations and rearrange the terms, you will obtain $x^2 + y^2 = 64$. Try this now and see.

Using Your Graphing Calculator

When using your calculator, you insert the function you wish to graph to the right of $y =$. Circles are not functions since they do not pass the vertical line test. To graph the equation $x^2 + y^2 = 64$, which is a circle of radius 8, we solve the equation for y to obtain $y = \pm\sqrt{64 - x^2}$. We then graph the two functions $y_1 = \sqrt{64 - x^2}$ and $y_2 = -\sqrt{64 - x^2}$ on the same axes to obtain the circle. These graphs are illustrated in Figure 10.13. Because of the distortion (described in the Using Your Graphing Calculator box in Section 9.1), the graph does not appear to be a circle. When you use the SQUARE *feature* of the calculator, the figure appears as a circle. (see Fig. 10.14)

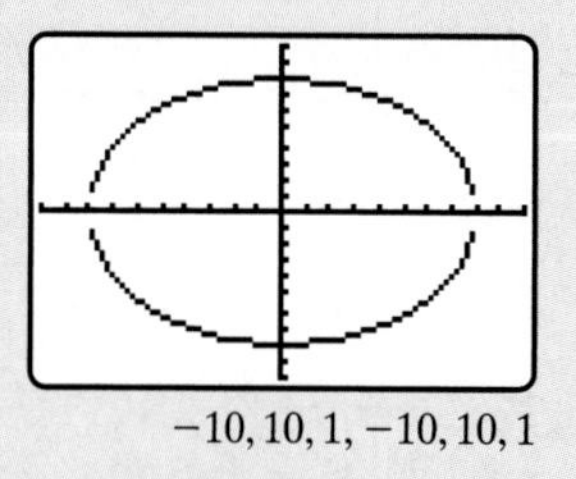

$-10, 10, 1, -10, 10, 1$

FIGURE 10.13

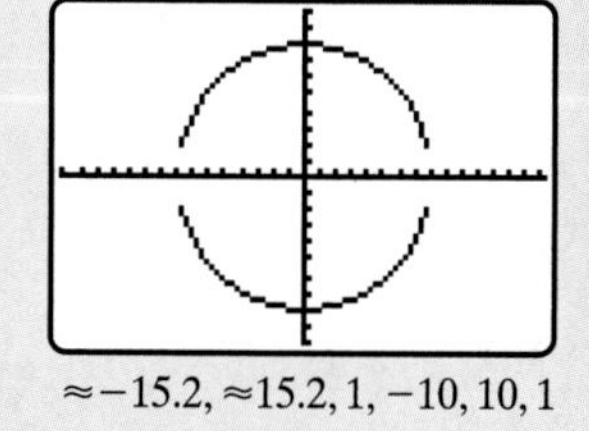

$\approx -15.2, \approx 15.2, 1, -10, 10, 1$

FIGURE 10.14

6 Graph Circles with Centers at (h, k)

The standard form of a circle with center at (h, k) and radius r can be derived using the distance formula. Let (h, k) be the center of the circle and let (x, y) be any point on the circle (see Fig. 10.15). If the radius r represents the distance between points (x, y) and (h, k), then by the distance formula

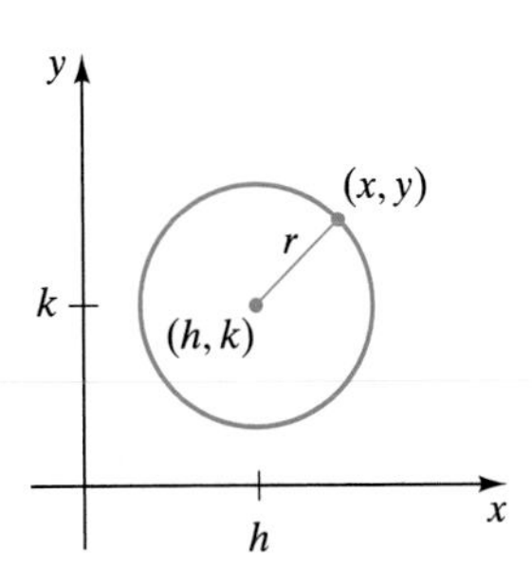

FIGURE 10.15

$$r = \sqrt{(x - h)^2 + (y - k)^2}$$

We now square both sides of the equation to obtain the standard form of a circle with center at (h, k) and radius r.

$$r^2 = (x - h)^2 + (y - k)^2$$

Circle with Its Center at (h, k) and Radius r

$$(x - h)^2 + (y - k)^2 = r^2$$

EXAMPLE 7 Determine the equation of the circle shown in Figure 10.16.

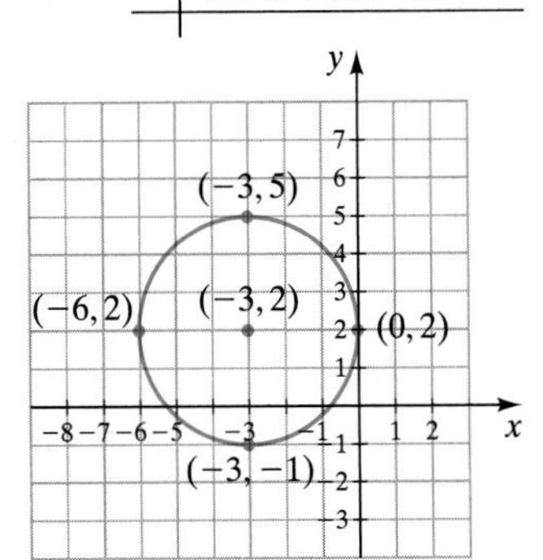

FIGURE 10.16

Solution The center is $(-3, 2)$ and the radius is 3.

$$(x - h)^2 + (y - k)^2 = r^2$$
$$[x - (-3)]^2 + (y - 2)^2 = 3^2$$
$$(x + 3)^2 + (y - 2)^2 = 9$$

EXAMPLE 8 **a)** Show that the graph of the equation $x^2 + y^2 + 6x - 2y + 6 = 0$ is a circle.
b) Determine the center and radius of the circle and then draw the circle.

Solution **a)** We will write this equation in standard form by completing the square. First we rewrite the equation, placing all the terms containing like variables together.

$$x^2 + 6x + y^2 - 2y + 6 = 0$$

Then we move the constant to the right side of the equation.

$$x^2 + 6x + y^2 - 2y = -6$$

Now we complete the square twice, once for each variable. We will first work with the variable x.

$$x^2 + 6x + 9 + y^2 - 2y = -6 + 9$$

Now we work with the variable y.

$$x^2 + 6x + 9 + y^2 - 2y + 1 = -6 + 9 + 1$$

or

$$\underbrace{x^2 + 6x + 9}_{(x+3)^2} + \underbrace{y^2 - 2y + 1}_{(y-1)^2} = 4$$
$$(x + 3)^2 + (y - 1)^2 = 4$$
$$(x + 3)^2 + (y - 1)^2 = 2^2$$

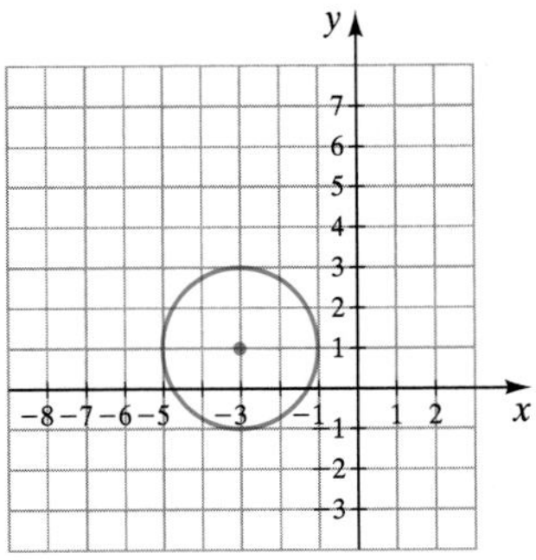

FIGURE 10.17

b) The center of the circle is at $(-3, 1)$ and the radius is 2. The circle is sketched in Figure 10.17.

NOW TRY EXERCISE 93

Exercise Set 10.1

Concept/Writing Exercises

1. Name the four conic sections. Draw a picture showing how each is formed.
2. Explain how to determine the direction a parabola will open by examining the equation.
3. Will all parabolas of the form $y = a(x - h)^2 + k$, $a > 0$ be functions? Explain. What will be the domain and range of $y = a(x - h)^2 + k, a > 0$?
4. Will all parabolas of the form $x = a(y - k)^2 + h$, $a > 0$ be functions? Explain. What will be the domain and range of $x = a(y - k)^2 + h, a > 0$?
5. How will the graphs of $y = 2(x - 3)^2 + 4$ and $y = -2(x - 3)^2 + 4$ compare?
6. Give the distance formula.
7. When the distance between two different points is found by using the distance formula, why must the distance always be a positive number?
8. Give the midpoint formula.
9. What is the definition of a circle?
10. What is the equation of a circle with center at (h, k)?

Practice the Skills

Graph each equation.

11. $y = (x - 2)^2 + 3$
12. $x = (y + 6)^2 + 1$
13. $x = (y - 4)^2 - 3$
14. $x = -(y - 5)^2 + 4$
15. $y = 2(x + 6)^2 - 4$
16. $y = -3(x - 5)^2 + 3$
17. $x = -5(y + 3)^2 - 6$
18. $x = -(y - 7)^2 + 8$
19. $y = -2\left(x + \frac{1}{2}\right)^2 + 6$
20. $y = -\left(x - \frac{5}{2}\right)^2 + \frac{1}{2}$

a) *In Exercises 21–32, write each equation in the form* $y = a(x - h)^2 + k$ *or* $x = a(y - k)^2 + h$. ***b)*** *Graph the equation.*

21. $y = x^2 + 2x$
22. $y = x^2 - 4x$
23. $x = y^2 + 6y$
24. $y = x^2 - 6x + 8$
25. $y = x^2 + 2x - 15$
26. $x = y^2 + 8y + 7$
27. $x = -y^2 + 6y - 9$
28. $y = -x^2 + 4x - 4$
29. $y = x^2 + 7x + 10$
30. $x = -y^2 + 3y - 4$
31. $y = 2x^2 - 4x - 4$
32. $x = 3y^2 - 12y - 36$

Determine the distance between each pair of points. Use a calculator where appropriate and round your answer to the nearest hundredth.

33. $(2, -2)$ and $(2, -5)$
34. $(-5, 5)$ and $(-5, 1)$
35. $(-4, 3)$ and $(5, 3)$
36. $(-1, -1)$ and $(3, 2)$
37. $(-3, -5)$ and $(6, -2)$
38. $(5, 3)$ and $(-5, -3)$
39. $(0, 6)$ and $(5, -1)$
40. $(4.2, -3.6)$ and $(-2.6, 2.3)$
41. $(-1.6, 3.5)$ and $(-4.3, -1.7)$
42. $(3, -1)$ and $\left(\frac{1}{2}, 4\right)$
43. $\left(\frac{3}{4}, 2\right)$ and $\left(-\frac{1}{2}, 6\right)$
44. $(4, 0)$ and $\left(-\frac{3}{5}, -4\right)$
45. $(\sqrt{5}, -\sqrt{2})$ and $(0, 0)$
46. $(\sqrt{7}, -\sqrt{3})$ and $(0, 0)$
47. $(\sqrt{3}, \sqrt{7})$ and $(\sqrt{2}, \sqrt{3})$
48. $(\sqrt{6}, \sqrt{5})$ and $(\sqrt{2}, -\sqrt{3})$

Determine the midpoint of the line segment between each pair of points.

49. $(1, 4)$ and $(2, 6)$
50. $(-5, 3)$ and $(5, -3)$
51. $(0, 8)$ and $(4, -6)$
52. $(-2, -8)$ and $(-6, -2)$
53. $(4, 7)$ and $(1, -3)$
54. $(15.3, -6.2)$ and $(8.2, -12.4)$
55. $(-9.62, 12.58)$ and $(3.52, 6.57)$
56. $\left(3, \frac{1}{2}\right)$ and $(2, -4)$
57. $\left(\frac{5}{2}, 3\right)$ and $\left(2, \frac{9}{2}\right)$
58. $\left(-\frac{5}{2}, -\frac{11}{2}\right)$ and $\left(-\frac{7}{2}, \frac{3}{2}\right)$
59. $(\sqrt{5}, 1)$ and $(\sqrt{3}, 4)$
60. $(-\sqrt{2}, 3)$ and $(3, -\sqrt{5})$

Write the equation of each circle with the given center and radius.

61. Center $(0, 0)$, radius 3
62. Center $(0, 0)$, radius 5
63. Center $(3, 0)$, radius 1
64. Center $(0, -2)$, radius 7
65. Center $(-6, 5)$, radius 5
66. Center $(-4, -1)$, radius 4
67. Center $(4, 7)$, radius $\sqrt{8}$
68. Center $(0, -2)$, radius $\sqrt{12}$

Write the equation of each circle. Assume the radius is a whole number.

69.

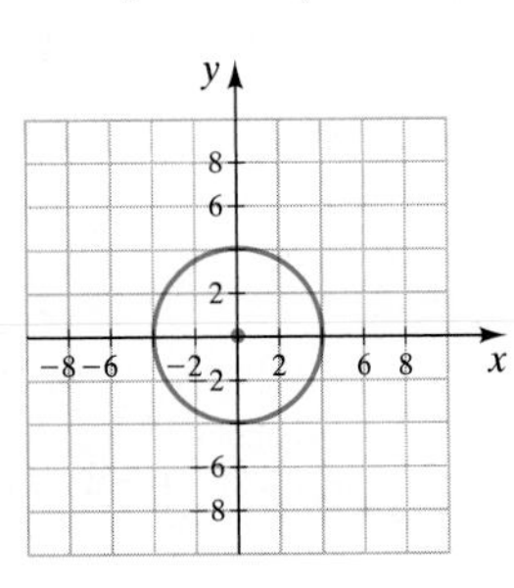

70.

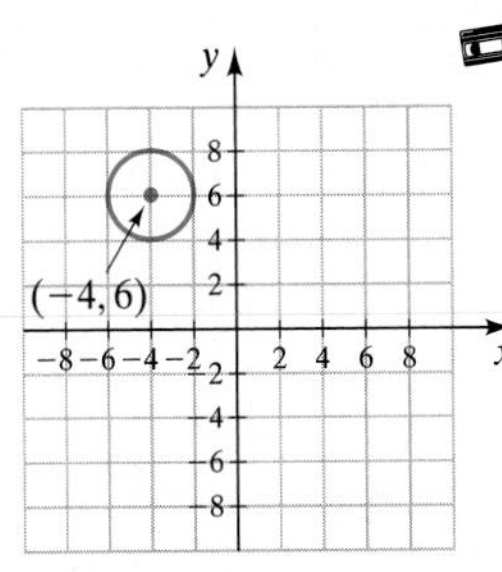

71.

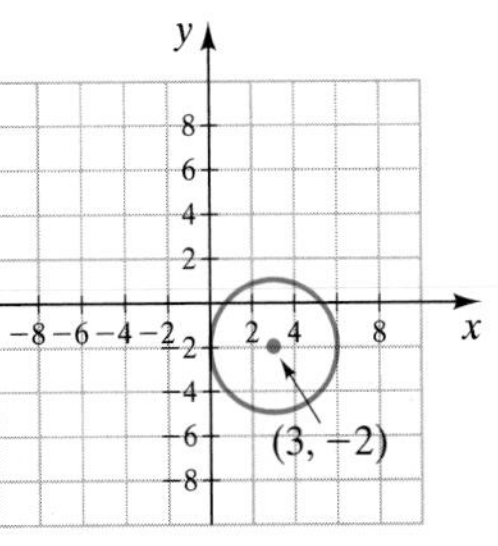

72.

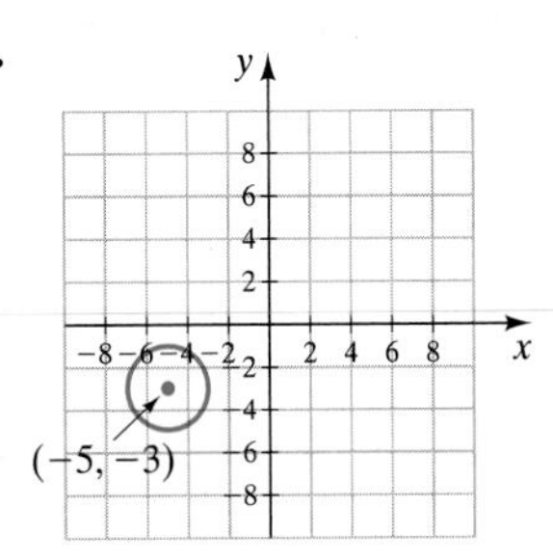

Graph each equation

73. $x^2 + y^2 = 16$

74. $x^2 + y^2 = 3$

75. $x^2 + y^2 = 10$

76. $x^2 + (y - 3)^2 = 4$

77. $(x + 4)^2 + y^2 = 25$

78. $(x - 2)^2 + (y + 3)^2 = 16$

79. $(x + 8)^2 + (y + 2)^2 = 9$

80. $(x + 1)^2 + (y - 4)^2 = 36$

81. $y = \sqrt{16 - x^2}$

82. $y = \sqrt{25 - x^2}$

83. $y = -\sqrt{4 - x^2}$

84. $y = -\sqrt{49 - x^2}$

a) *Use the procedure for completing the square to write each equation in standard form.* ***b)*** *Draw the graph.*

85. $x^2 + y^2 + 10y - 75 = 0$

86. $x^2 + y^2 - 4y = 0$

87. $x^2 + 8x - 9 + y^2 = 0$

88. $x^2 + y^2 + 6x - 4y + 9 = 0$

89. $x^2 + y^2 + 2x - 4y - 4 = 0$

90. $x^2 + y^2 + 4x - 6y - 3 = 0$

91. $x^2 + y^2 + 6x - 2y + 6 = 0$

92. $x^2 + y^2 + 8x - 4y + 4 = 0$

93. $x^2 + y^2 - 8x + 2y + 13 = 0$

94. $x^2 - x + y^2 + 3y - \frac{3}{2} = 0$

Problem Solving

Find the x- and y-intercept, if they exist, of the graph of each equation.

95. $x = y^2 - 6y - 7$

96. $x = -y^2 + 8y - 12$

97. $x = 2(y - 5)^2 + 6$

98. $x = -(y + 2)^2 - 8$

99. If you know the midpoint of a line segment, is it possible to determine the length of the line segment? Explain.

100. If you know one endpoint of a line segment and the length of the line segment, can you determine the other endpoint? Explain.

101. Find the length of the line segment whose midpoint is (4, −6) with one endpoint at (7, −2).

102. Find the length of the line segment whose midpoint is (−2, 4) with one endpoint at (3, 6).

103. Find the equation of a circle with center at (−5, 2) that is tangent to the x-axis (that is, the circle touches the x-axis at only one point).

104. Find the equation of a circle with center at (−3, −4) that is tangent to the y-axis.

In Exercises 105 and 106, find ***a)*** *the radius of the circle whose diameter is along the line shown,* ***b)*** *the center of the circle, and* ***c)*** *the equation of the circle.*

105.

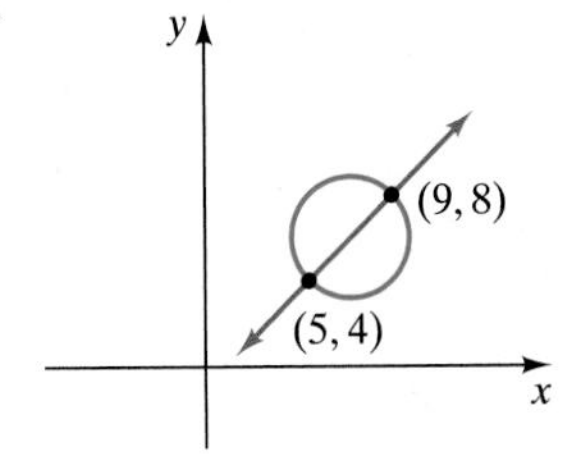

106.

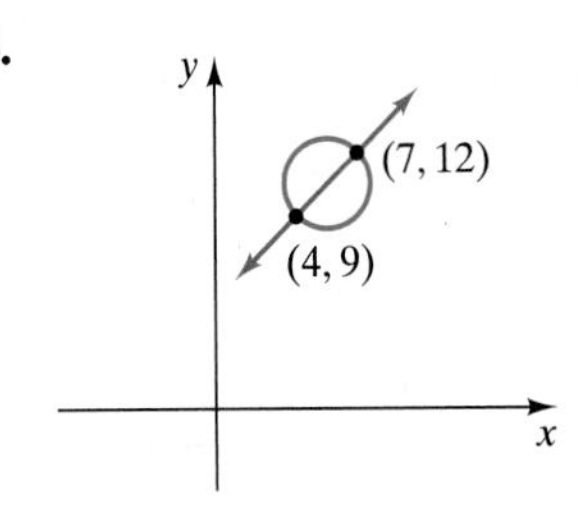

107. What is the maximum number and the minimum number of points of intersection possible for the graphs of $y = a(x - h_1)^2 + k_1$ and $x = a(y - k_2)^2 + h_2$? Explain.

108. Consider the figure below.

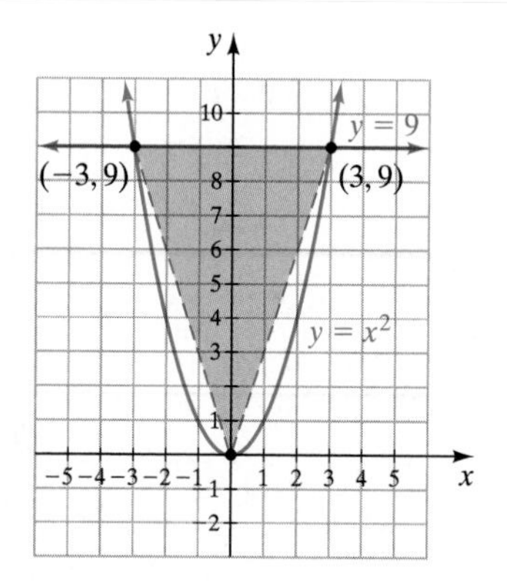

a) Find the area of the triangle outlined in green.

b) When a triangle is inscribed within a parabola, as in the figure, the area within the parabola from the base of the triangle is $\frac{4}{3}$ the area of the triangle. Find the area within the parabola from $x = -3$ to $x = 3$.

109. The Ferris wheel at Navy Pier in Chicago is 150 feet tall. The radius of the wheel itself is 68.2 feet.

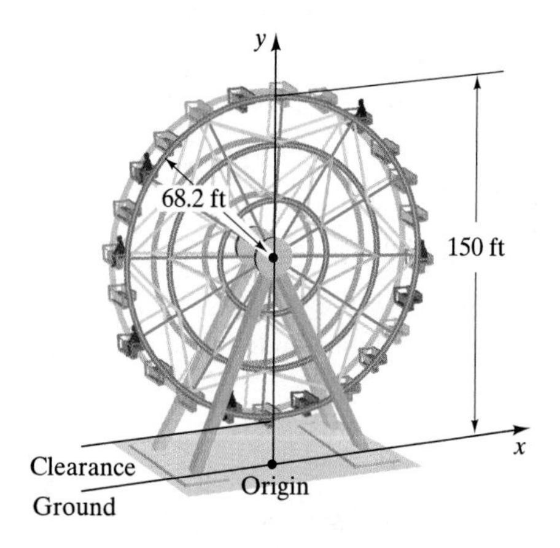

a) What is the clearance below the wheel?

b) How high is the center of the wheel from the ground?

c) Find the equation of the wheel. Assume the origin is on the ground directly below the center of the wheel.

110. The photo shows the bridge on Route 255 that crosses the Mississippi River near Jefferson Barracks, Missouri. The arch of the bridge is a parabola with equation $y = -\dfrac{x^2}{1135}$. The road directly under the arch has a length of 909 feet. The maximum vertical distance that the arch is from the road below is 182 feet.

a) What is the distance from either end of the road under the arch to the midpoint of the road?

b) In the figure below we set the vertex of the arch at the origin. Use the given formula to find values of y if x is 0 feet, ±100 feet, ±200 feet, ±300 feet, ±400 feet, and ±454.5 feet. Explain what each ordered pair represents, and explain how the given equation is used in the design of the arch.

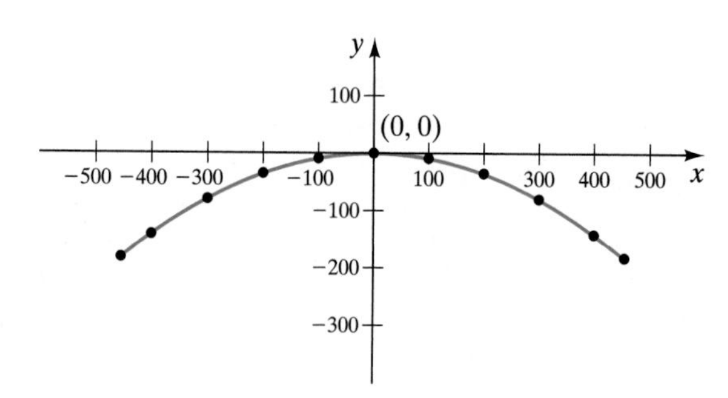

111. Consider the figure below. Write an equation for **a)** the blue circle, **b)** the red circle, and **c)** the green circle. **d)** Find the shaded area.

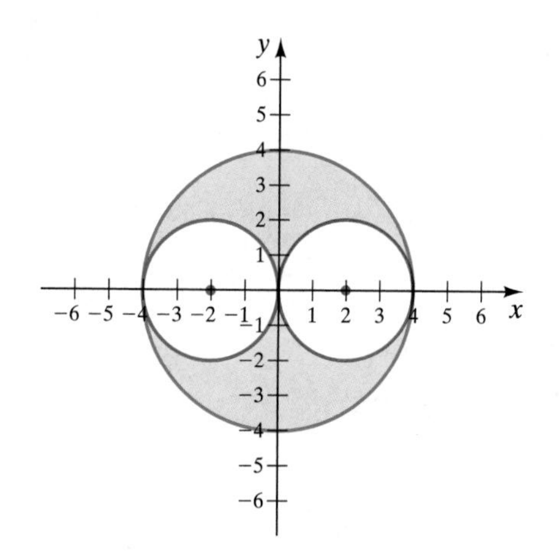

112. Find the shaded area of the square in the figure. The equation of the circle is $x^2 + y^2 = 9$.

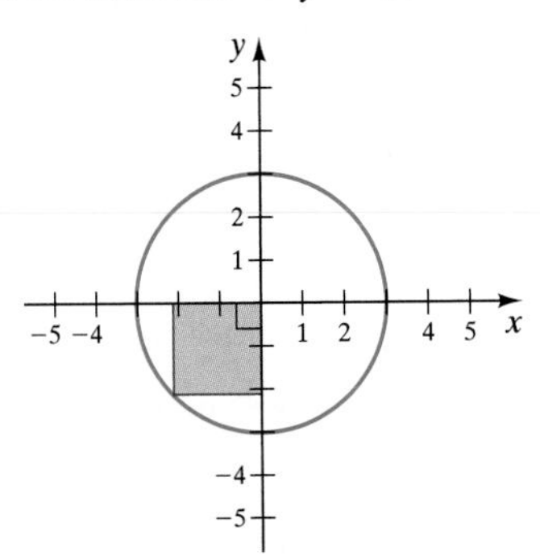

113. Consider the equations $x^2 + y^2 = 16$ and $(x - 2)^2 + (y - 2)^2 = 16$. By considering the center and radius of each circle, determine the number of points of intersection of the two circles.

114. Find the area between the two concentric circles whose equations are $(x - 2)^2 + (y + 4)^2 = 16$ and $(x - 2)^2 + (y + 4)^2 = 64$. *Concentric circles* are circles that have the same center.

115. A highway department is planning to construct a semicircular tunnel through a mountain. The tunnel is to be large enough so that a truck 8 feet wide and 10 feet tall will pass through the center of the tunnel with 1 foot to spare when driving down the center of the tunnel (as shown in the figure below.) Determine the minimum radius of the tunnel.

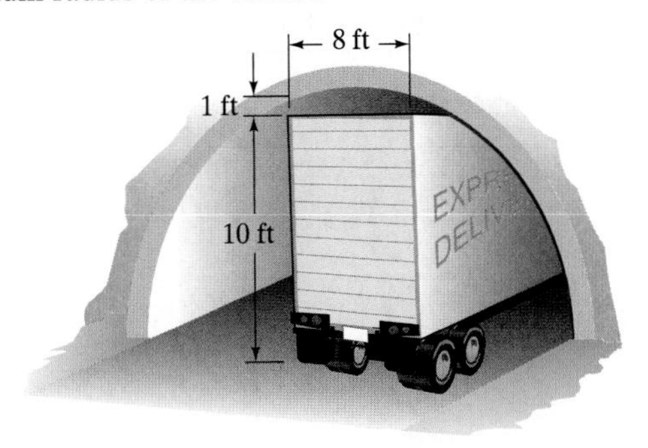

Group Activity

Discuss and answer Exercise 116 as a group.

116. The equation of a parabola can be found if three points on the parabola are known. To do so, start with $y = ax^2 + bx + c$. Then substitute the x- and y-coordinates of the first point into the equation. This will result in an equation in a, b, and c. Repeat the procedure for the other two points. This process yields a system of 3 equations in 3 variables. Next solve the system for a, b, and c. To find the equation of the parabola, substitute the values found for a, b, and c into the equation $y = ax^2 + bx + c$.

Three points on a parabola are $(0, 12)$, $(3, -3)$, and $(-2, 32)$.

a) Individually, find a system of equations in three variables that can be used to find the equation of the parabola. Then compare your answers. If each member of the group does not have the same system, determine why.

b) Individually, solve the system and determine the values of a, b, and c. Then compare your answers.

c) Individually, write the equation of the parabola passing through $(0, 12)$, $(3, -3)$, and $(-2, 32)$. Then compare your answers.

d) Individually, write the equation in

$$y = a(x - h)^2 + k$$

form. Then compare your answers.

e) Individually, graph the equation in part **d)**. Then compare your answers.

Cumulative Review Exercises

[3.4] **117.** Write the equation, in slope–intercept form, of the graph that passes through the points $(-6, 4)$ and $(-2, 2)$.

[4.5] **118.** Evaluate the following determinant.

$$\begin{vmatrix} 4 & 0 & 3 \\ 5 & 2 & -1 \\ 3 & 6 & 4 \end{vmatrix}$$

[5.1] **119.** If $P(x) = -2x^2 - x + 2$, find $P\left(\frac{1}{3}\right)$.

[6.6] **120.** T varies jointly as m_1 and m_2 and inversely as the square of R. If $T = \frac{3}{2}$ when $m_1 = 6$, $m_2 = 4$, and $R = 4$, find T when $m_1 = 6$, $m_2 = 10$, and $R = 2$.

10.2 THE ELLIPSE

SSM

VIDEO 10.2

CD Rom

1) Graph ellipses.
2) Graph ellipses with centers at (h, k).

1) Graph Ellipses

An **ellipse** may be defined as a set of points in a plane, the sum of whose distances from two fixed points is a constant. The two fixed points are called the **foci** (each is a focus) of the ellipse (see Fig. 10.18).

FIGURE 10.18

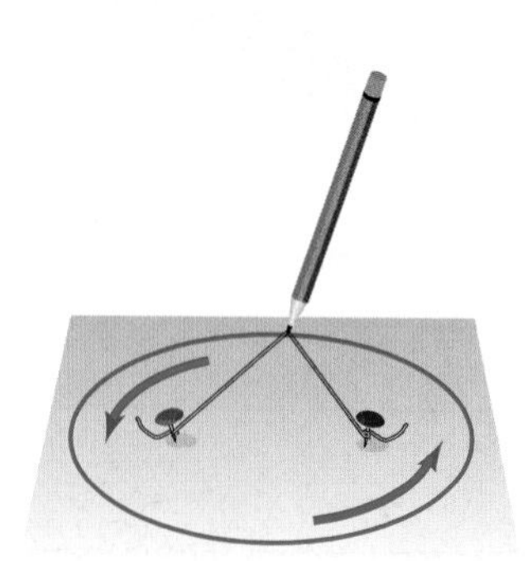

FIGURE 10.19

We can construct an ellipse using a length of string and two thumbtacks. Place the two thumbtacks fairly close together (Fig. 10.19). Then tie the ends of the string to the thumbtacks. With a pencil or pen pull the string taut, and, while keeping the string taut, draw the ellipse by moving the pencil around the thumbtacks.

The standard form of an ellipse with its center at the origin (Fig. 10.20) follows.

Ellipse with Its Center at the Origin

$$\frac{x^2}{a^2} + \frac{y^2}{b^2} = 1$$

where $(a, 0)$ and $(-a, 0)$ are the x-intercepts and $(0, b)$ and $(0, -b)$ are the y-intercepts.

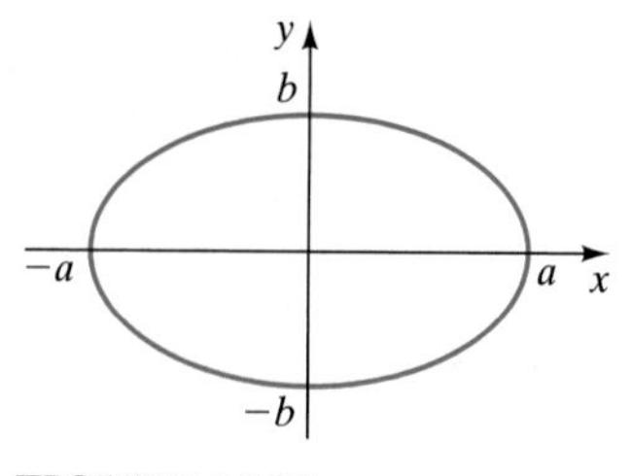

FIGURE 10.20

In Figure 10.20, the line segment from $-a$ to a on the x-axis is the *longer* or **major axis** and the line segment from $-b$ to b is the *shorter* or **minor axis** of the ellipse.

Graph $\frac{x^2}{9} + \frac{y^2}{4} = 1$.

Solution We can rewrite the equation as

$$\frac{x^2}{3^2} + \frac{y^2}{2^2} = 1$$

Thus, $a = 3$ and the x-intercepts are $(3, 0)$ and $(-3, 0)$. Since $b = 2$, the y-intercepts are $(0, 2)$ and $(0, -2)$. The ellipse is illustrated in Figure 10.21.

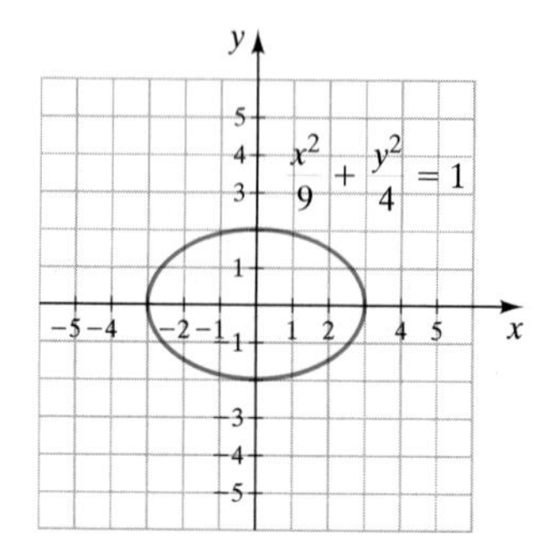

FIGURE 10.21

An equation may be camouflaged so that it may not be obvious that its graph is an ellipse. This is illustrated in Example 2.

EXAMPLE 2 Graph $20x^2 + 9y^2 = 180$.

Solution To make the right side of the equation equal to 1, we divide both sides of the equation by 180. We then obtain an equation that we can recognize as an ellipse.

$$\frac{20x^2 + 9y^2}{180} = \frac{180}{180}$$

$$\frac{20x^2}{180} + \frac{9y^2}{180} = 1$$

$$\frac{x^2}{9} + \frac{y^2}{20} = 1$$

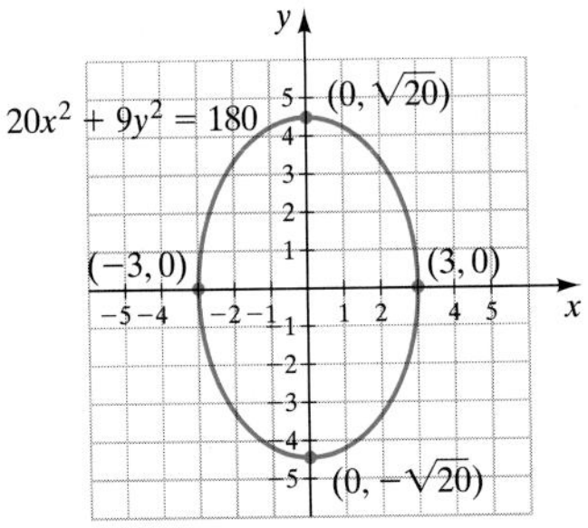

The equation can now be recognized as an ellipse in standard form.

$$\frac{x^2}{a^2} + \frac{y^2}{b^2} = 1$$

Since $a^2 = 9$, $a = 3$. We know that $b^2 = 20$; thus $b = \sqrt{20}$ (or approximately 4.47).

$$\frac{x^2}{3^2} + \frac{y^2}{(\sqrt{20})^2} = 1$$

FIGURE 10.22

The x-intercepts are $(3, 0)$ and $(-3, 0)$. The y-intercepts are $(0, -\sqrt{20})$ and $(0, \sqrt{20})$. The graph is illustrated in Figure 10.22. Note that the major axis lies along the y-axis instead of along the x-axis.

NOW TRY EXERCISE 15

EXAMPLE 3 Write the equation of the ellipse illustrated in Figure. 10.23.

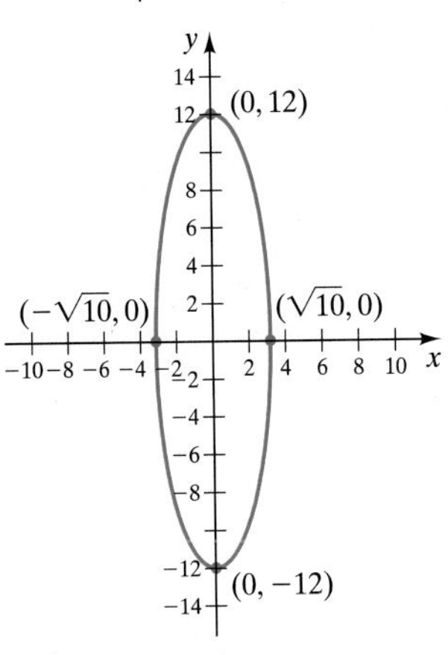

FIGURE 10.23

Solution The x-intercepts are $(-\sqrt{10}, 0)$ and $(\sqrt{10}, 0)$; thus $a = \sqrt{10}$ and $a^2 = 10$. The y-intercepts are $(0, -12)$ and $(0, 12)$; thus, $b = 12$ and $b^2 = 144$.

$$\frac{x^2}{a^2} + \frac{y^2}{b^2} = 1$$

$$\frac{x^2}{10} + \frac{y^2}{144} = 1$$

2) Graph Ellipses with Centers at (h, k)

Horizontal and vertical translations, similar to those used in Chapter 8, may be used to obtain the equation of an ellipse with center at (h, k).

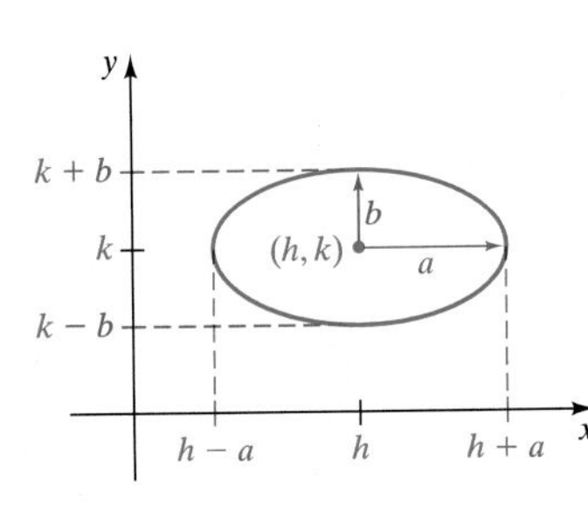

FIGURE 10.24

Ellipse with Center at (h, k)

$$\frac{(x-h)^2}{a^2} + \frac{(y-k)^2}{b^2} = 1$$

In the formula, the h shifts the graph left or right from the origin and k shifts the graph up or down from the origin, as shown in Figure 10.24.

EXAMPLE 4 Graph $\frac{(x-2)^2}{25} + \frac{(y+3)^2}{16} = 1$.

Solution This is the graph of $\frac{x^2}{25} + \frac{y^2}{16} = 1$ or $\frac{x^2}{5^2} + \frac{y^2}{4^2} = 1$ translated so that its center is at $(2, -3)$. Note that $a = 5$ and $b = 4$. The graph is shown in Figure 10.25.

NOW TRY EXERCISE 23

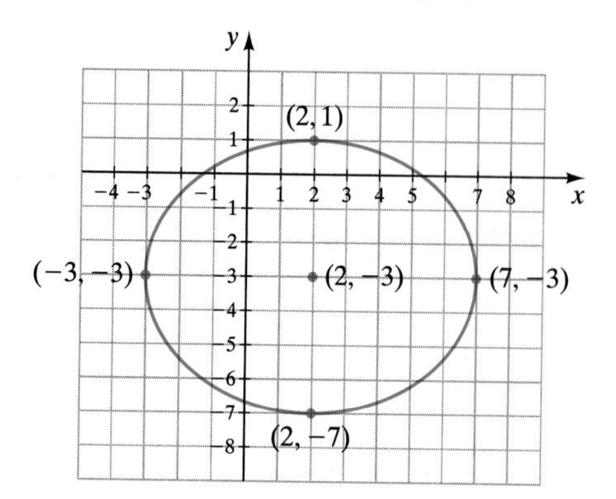

FIGURE 10.25

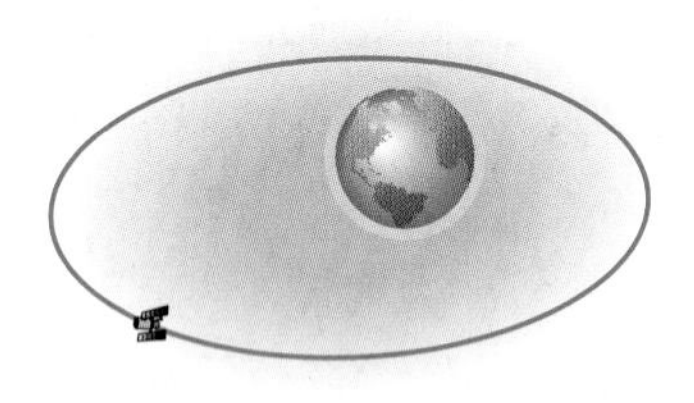

FIGURE 10.26

An understanding of ellipses is useful in many areas. Astronomers know that planets revolve in elliptical orbits around the sun. Communications satellites move in elliptical orbits around Earth (see Fig. 10.26).

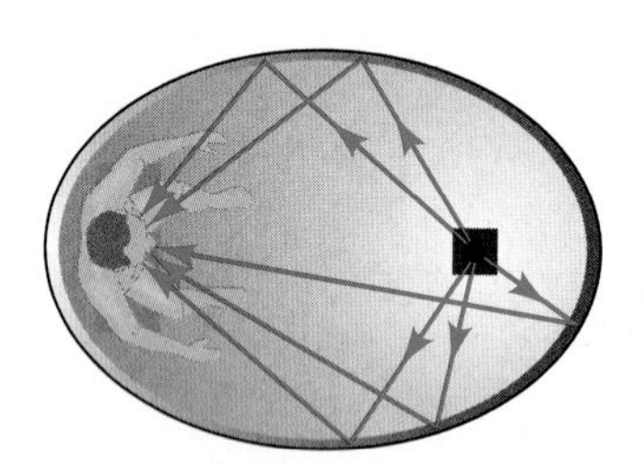

FIGURE 10.27

Ellipses are used in medicine to smash kidney stones. When a signal emerges from one focus of an ellipse, the signal is reflected to the other focus. In kidney stone machines, the person is situated so that the stone to be smashed is at one focus of an elliptical shaped chamber called a lithotripter (see Fig. 10.27 and Exercises 37 and 39).

In certain buildings with ellipsoidal ceilings, a person standing at one focus can whisper something and a person standing at the other focus can clearly hear what the person whispered. There are many other uses for ellipses, including lamps that are made to concentrate light at a specific point.

Using Your Graphing Calculator

Ellipses are not functions. To graph ellipses on a graphing calculator, we solve the equation for y. This will give the two equations that we use to graph the ellipse.

In Example 1, we graphed $\frac{x^2}{9} + \frac{y^2}{4} = 1$. Solving this equation for y, we get

$$\frac{x^2}{9} + \frac{y^2}{4} = 1$$

$$36 \cdot \frac{x^2}{9} + 36 \cdot \frac{y^2}{4} = 1 \cdot 36 \quad \text{Multiply by the LCD.}$$

$$4x^2 + 9y^2 = 36$$

$$9y^2 = 36 - 4x^2$$

$$y^2 = \frac{36 - 4x^2}{9}$$

$$y^2 = \frac{4(9 - x^2)}{9} \quad \text{Factor 4 from the numerator.}$$

$$y = \pm\frac{2}{3}\sqrt{9 - x^2} \quad \text{Square root property.}$$

To graph the ellipse, we let $y_1 = \frac{2}{3}\sqrt{9 - x^2}$ and $y_2 = -\frac{2}{3}\sqrt{9 - x^2}$ and graph both equations. The graphs of y_1 and y_2 are illustrated in Figure10.28.

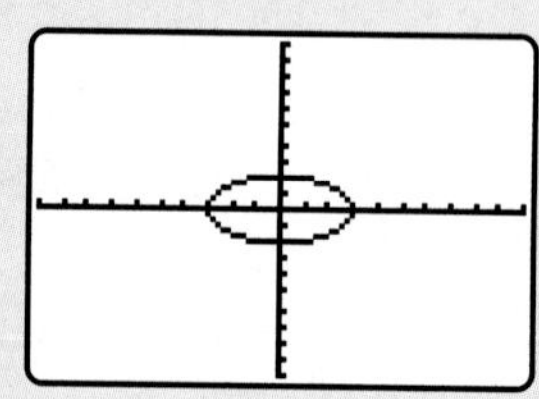

FIGURE 10.28

Exercise Set 10.2

Concept/Writing Exercises

1. What is the definition of an ellipse?
2. What is the equation of an ellipse with its center at the origin?
3. What is the equation of an ellipse with its center at (h, k)?
4. In your own words, discuss the graphs of $\frac{x^2}{a^2} + \frac{y^2}{b^2} = 1$ when $a > b$, $a < b$, and $a = b$.
5. Explain why the circle is a special case of the ellipse.
6. In the formula $\frac{x^2}{a^2} + \frac{y^2}{b^2} = 1$, what do the a and b represent?

Practice the Skills

Graph each equation.

7. $\frac{x^2}{4} + \frac{y^2}{1} = 1$
8. $\frac{x^2}{9} + \frac{y^2}{4} = 1$
9. $\frac{x^2}{4} + \frac{y^2}{9} = 1$
10. $\frac{x^2}{25} + \frac{y^2}{9} = 1$
11. $\frac{x^2}{16} + \frac{y^2}{25} = 1$
12. $\frac{x^2}{9} + \frac{y^2}{121} = 1$
13. $9x^2 + 12y^2 = 108$
14. $9x^2 + 4y^2 = 36$
15. $100x^2 + 25y^2 = 400$
16. $x^2 + 36y^2 = 36$
17. $9x^2 + 25y^2 = 225$
18. $x^2 + 2y^2 = 8$
19. $\frac{x^2}{16} + \frac{(y-2)^2}{9} = 1$
20. $\frac{(x-4)^2}{9} + \frac{(y+3)^2}{25} = 1$
21. $\frac{(x+1)^2}{9} + \frac{(y-2)^2}{4} = 1$
22. $\frac{(x-3)^2}{16} + \frac{(y-4)^2}{25} = 1$
23. $4(x-2)^2 + 9(y+2)^2 = 36$
24. $(x+3)^2 + 9(y+1)^2 = 81$
25. $12(x+4)^2 + 3(y-1)^2 = 48$
26. $16(x-2)^2 + 4(y+3)^2 = 16$

Problem Solving

27. How many points are on the graph of $9x^2 + 36y^2 = 0$? Explain.
28. Consider the graph of the equation $\frac{x^2}{a^2} + \frac{y^2}{b^2} = 1$. Explain what will happen to the shape of the graph as the value of b gets closer to the value of a. What is the shape of the graph when $a = b$?

Find the equation of the ellipse that contains the following points.

29. $(2, 0), (-2, 0), (0, 3), (0, -3)$
30. $(7, 0), (-7, 0), (0, 4), (0, -4)$
31. How many points of intersection will the graphs of the equations $x^2 + y^2 = 16$ and $\frac{x^2}{4} + \frac{y^2}{9} = 1$ have? Explain.
32. How many points of intersection will the graphs of the equations $y = 2(x-2)^2 - 3$ and $\frac{(x-2)^3}{4} + \frac{(y+3)^2}{9} = 1$ have? Explain.

Write the following equations in standard form. Determine the center of each ellipse.

33. $x^2 + 4y^2 - 4x - 8y - 92 = 0$
34. $x^2 + 4y^2 + 6x + 16y - 11 = 0$
35. An art gallery has an elliptical hall. The maximum distance from one focus to the wall is 90.2 feet and the minimum distance is 20.7 feet. Find the distance between the foci.
36. A space shuttle transported a communications satellite to space. The satellite travels in an elliptical orbit around Earth. The maximum distance of the satellite from Earth is 23,200 miles and the minimum distance is 22,800 miles. Earth is at one focus of the ellipse. Find the distance from Earth to the other focus.
37. Suppose the lithotripter machine described on page 636 is 6 feet long and 4 feet wide. Describe the location of the foci.

38. An elliptical billiard table is 8 feet long by 5 feet wide. Determine the location of the foci. On such a table, if a ball is put at each focus and one ball is hit with enough force, it would hit the ball at the other focus no matter where it banks on the table.

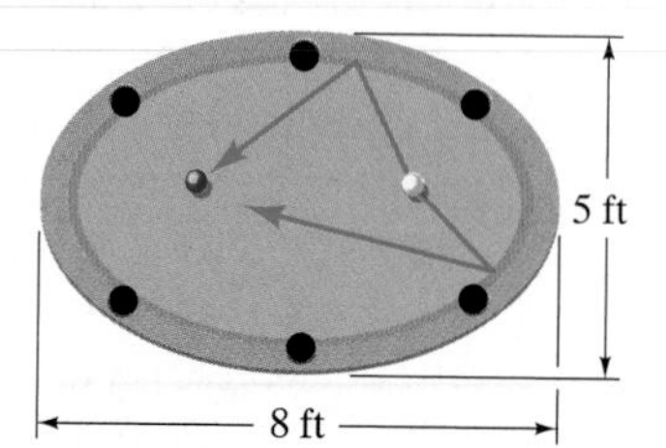

39. On page 636 we gave a brief introduction to the lithotripter, which uses ultrasound waves to shatter kidney stones. Do research and write a detailed report describing the procedure used to shatter kidney stones. Make sure that you explain how the waves are directed on the stone.

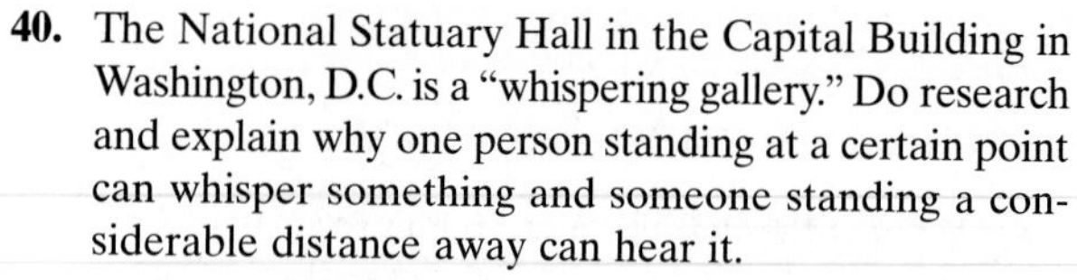

40. The National Statuary Hall in the Capital Building in Washington, D.C. is a "whispering gallery." Do research and explain why one person standing at a certain point can whisper something and someone standing a considerable distance away can hear it.

41. Check your answer to Exercise 7 on your grapher.

42. Check your answer to Exercise 15 on your grapher.

Challenge Problems

Determine the equation of the ellipse that contains the following points.

43. $(-4, 3), (2, 3), (-1, 5), (-1, 1)$

44. $(2, 2), (6, 2), (4, 5), (4, -1)$

Group Activity

Work Exercise 45 individually. Then compare your answers.

45. The photo shows an elliptical tunnel (with the bottom part of the ellipse not shown) near Rockefeller Center in New York City. The maximum width of the tunnel is 18 feet and the maximum height *from the ground to the top* is 10.5 feet.

a) If the *completed ellipse* would have a maximum height of 15 feet, how high from the ground is the center of the elliptical tunnel?

b) Consider the following graph, which could be used to represent the tunnel.

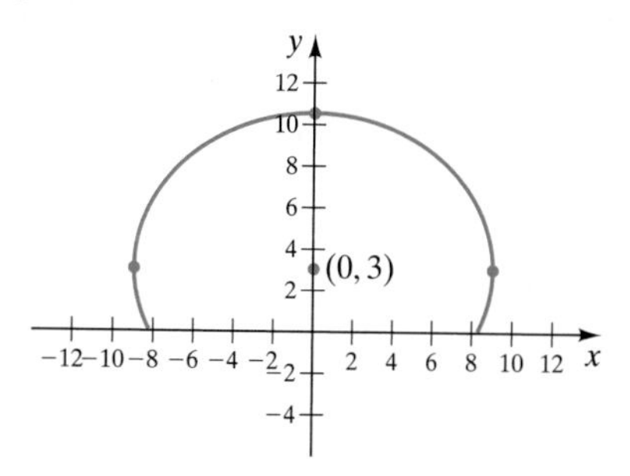

If the ellipse were continued, what would be the other y-intercept of the graph?

c) Write the equation of the ellipse, if completed, in part **b)**.

Cumulative Review Exercises

Solve the following equations or inequalities. Indicate the solution on a number line.

[2.5] **46.** $-3 \le 4 - \frac{1}{2}x < 6$

[2.6] **47.** $|2x - 4| = 8$

48. $|2x - 4| \le 8$

49. $|2x - 4| > 8$

10.3 THE HYPERBOLA

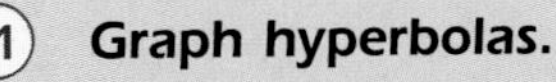

1) **Graph hyperbolas.**
2) **Review conic sections.**

SSM VIDEO 10.3 CD Rom

1) Graph Hyperbolas

A **hyperbola** is the set of points in a plane the difference of whose distances from two fixed points (called foci) is a constant. A hyperbola may look like a pair of parabolas (see Figure 10.29), however, the shapes are actually quite different. A hyperbola has two **vertices**. The point halfway between the two vertices is the **center** of the hyperbola. The line through the vertices is called the **transverse axis**. In Figure 10.29a, the transverse axis lies along the x-axis, and in Figure 10.29b, the transverse axis lies along the y-axis.

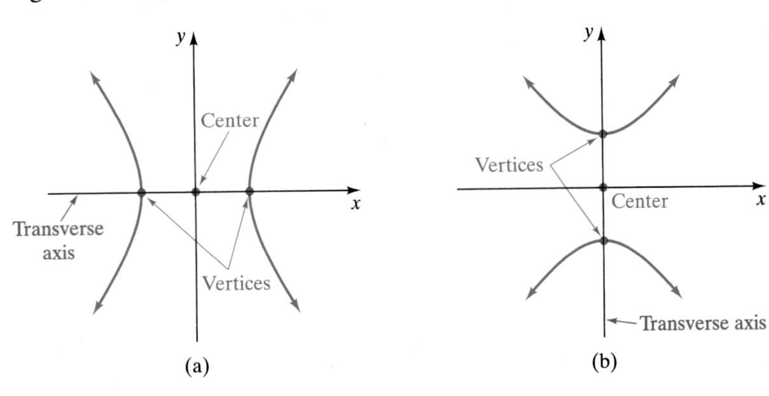

FIGURE 10.29

The dashed lines in Figure 10.30 are called **asymptotes**. The asymptotes are not a part of the hyperbola but are used as an aid in graphing hyperbolas. (We will discuss asymptotes shortly.) Also given in Figure 10.30 is the standard form of the equation for each hyperbola. In Figure 10.30a, both vertices are a units from the origin. In Figure 10.30b, both vertices are b units from the origin. Note that in the standard form of the equation the denominator of the x^2 term is always a^2 and the denominator of the y^2 term is always b^2.

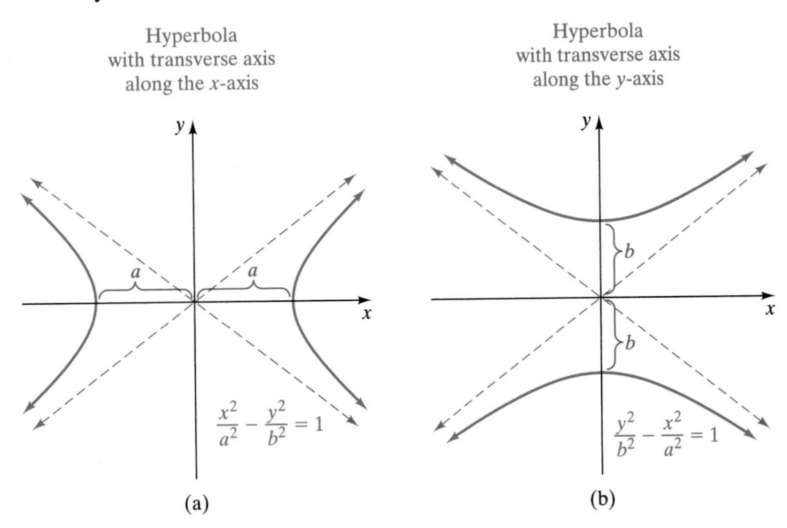

FIGURE 10.30

A hyperbola centered at the origin whose transverse axis is parallel to one of the coordinate axes has either x-intercepts (Figure 10.30a) or y-intercepts (Figure 10.30b), but not both. When a hyperbola is centered at the origin, the intercepts are the vertices of the hyperbola. When written in standard form, the intercepts will be on the axis indicated by the variable with the positive coefficient. The intercepts will be the positive and the negative square root of the denominator of the positive term.

Examples	Intercepts on	Intercepts
$\dfrac{x^2}{25} - \dfrac{y^2}{9} = 1$	x-axis	$(-5, 0)$ and $(5, 0)$
$\dfrac{y^2}{9} - \dfrac{x^2}{25} = 1$	y-axis	$(0, -3)$ and $(0, 3)$

Asymptotes can help you graph hyperbolas. The asymptotes are two straight lines that go through the center of the hyperbola (see Fig. 10.30). As the values of x and y get larger, the graph of the hyperbola approaches the asymptotes. The equations of the asymptotes of a hyperbola whose center is the origin are

$$y = \frac{b}{a}x \qquad \text{and} \qquad y = -\frac{b}{a}x$$

The asymptotes can be drawn quickly by plotting the four points (a, b), $(-a, b)$, $(a, -b)$, and $(-a, -b)$, then connecting these points with dashed lines to form a rectangle. Next draw dashed lines through the opposite corners of the rectangle to obtain the asymptotes.

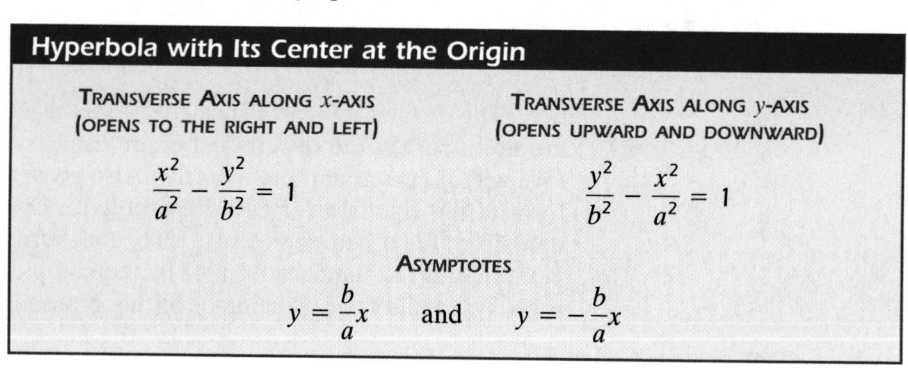

Hyperbola with Its Center at the Origin

TRANSVERSE AXIS ALONG x-AXIS (OPENS TO THE RIGHT AND LEFT)	TRANSVERSE AXIS ALONG y-AXIS (OPENS UPWARD AND DOWNWARD)
$\dfrac{x^2}{a^2} - \dfrac{y^2}{b^2} = 1$	$\dfrac{y^2}{b^2} - \dfrac{x^2}{a^2} = 1$

ASYMPTOTES

$$y = \frac{b}{a}x \qquad \text{and} \qquad y = -\frac{b}{a}x$$

EXAMPLE 1 **a)** Determine the equations of the asymptotes of the hyperbola with equation

$$\frac{x^2}{9} - \frac{y^2}{16} = 1$$

b) Draw the hyperbola using the asymptotes.

Solution **a)** The value of a^2 is 9; the positive square root of 9 is 3. The value of b^2 is 16; the positive square root of 16 is 4. The asymptotes are

$$y = \frac{b}{a}x \qquad \text{and} \qquad y = -\frac{b}{a}x$$

or

$$y = \frac{4}{3}x \qquad \text{and} \qquad y = -\frac{4}{3}x$$

b) To graph the hyperbola, we first graph the asymptotes. To graph the asymptotes, we can plot the points (3, 4), (−3, 4), (3, −4), and (−3, −4) and draw the rectangle as illustrated in Figure 10.31. The asymptotes are the dashed lines through the opposite corners of the rectangle

Since the x-term in the original equation is positive, the graph intersects the x-axis. Since the denominator of the positive term is 9, the vertices are at (3, 0) and (−3, 0). Now draw the hyperbola by letting the hyperbola approach its asymptotes (Fig. 10.32). Note that the asymptotes are drawn using dashed lines since they are not part of the hyperbola. They are used merely to help draw the graph.

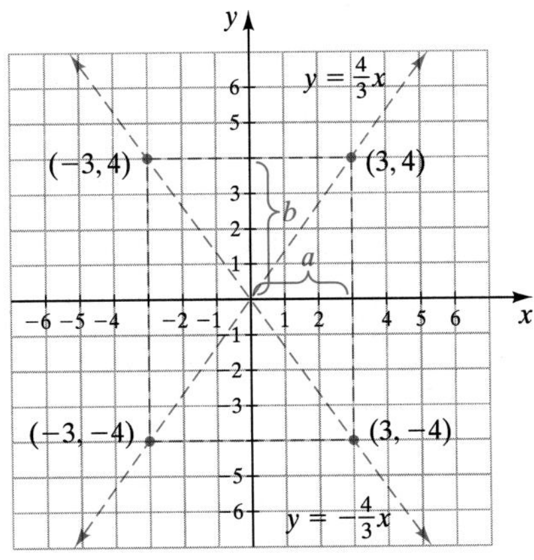

NOW TRY EXERCISE 9

FIGURE 10.31

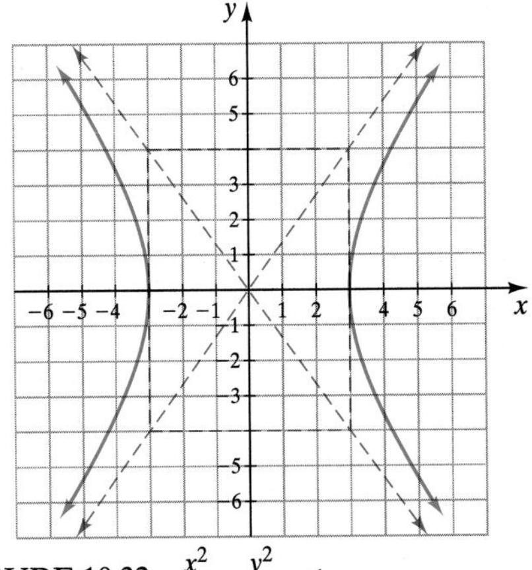

FIGURE 10.32 $\frac{x^2}{9} - \frac{y^2}{16} = 1$

EXAMPLE 2 **a)** Show that the equation $-25x^2 + 4y^2 = 100$ is a hyperbola by expressing the equation in standard form.

b) Determine the equations of the asymptotes of the graph.

c) Draw the graph.

Solution **a)** We divide both sides of the equation by 100 to obtain 1 on the right side of the equation.

$$\frac{-25x^2 + 4y^2}{100} = \frac{100}{100}$$

$$\frac{-25x^2}{100} + \frac{4y^2}{100} = 1$$

$$\frac{-x^2}{4} + \frac{y^2}{25} = 1$$

Rewriting the equation in standard form (positive term first), we get

$$\frac{y^2}{25} - \frac{x^2}{4} = 1$$

b) The equations of the asymptotes are

$$y = \frac{5}{2}x \quad \text{and} \quad y = -\frac{5}{2}x$$

c) The graph intersects the y-axis at $(0, 5)$ and $(0, -5)$. Figure 10.33a illustrates the asymptotes, and Figure 10.33b illustrates the hyperbola.

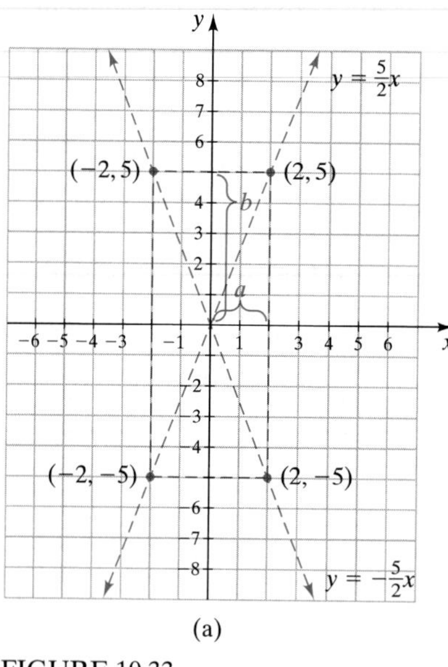

(a)

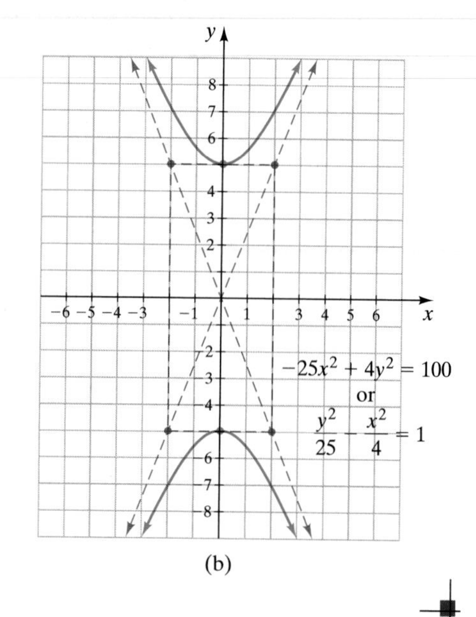

(b)

NOW TRY EXERCISE 17

FIGURE 10.33

Using Your Graphing Calculator

We can graph hyperbolas just as we did circles and ellipses. To graph hyperbolas on the graphing calculator, solve the equation for y and graph each part. Consider Example 1,

$$\frac{x^2}{9} - \frac{y^2}{16} = 1$$

Show that if you solve this equation for y you get $y = \pm\frac{4}{3}\sqrt{x^2 - 9}$. Let $y_1 = \frac{4}{3}\sqrt{x^2 - 9}$ and $y_2 = -\frac{4}{3}\sqrt{x^2 - 9}$. Figure 10.34a, 10.34b, 10.34c, and 10.34d give the graphs of y_1 and y_2 for different window settings. The window settings used are indicated above each graph.

Standard setting

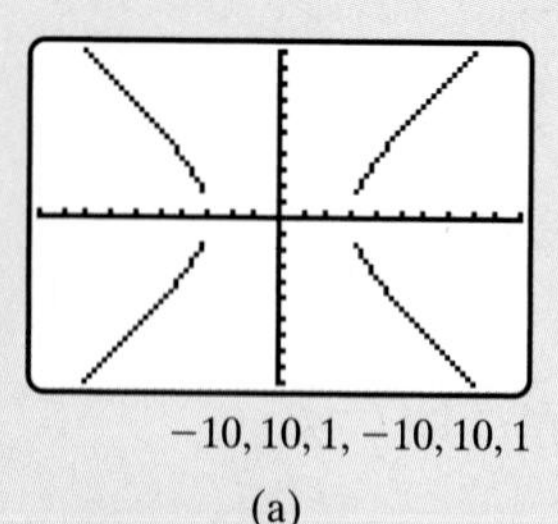

$-10, 10, 1, -10, 10, 1$

(a)

ZOOM: option 5 ZSquare setting

$\approx -15.2, \approx 15.2, 1, -10, 10, 1$

(b)

ZOOM: option 4, ZDecimal setting

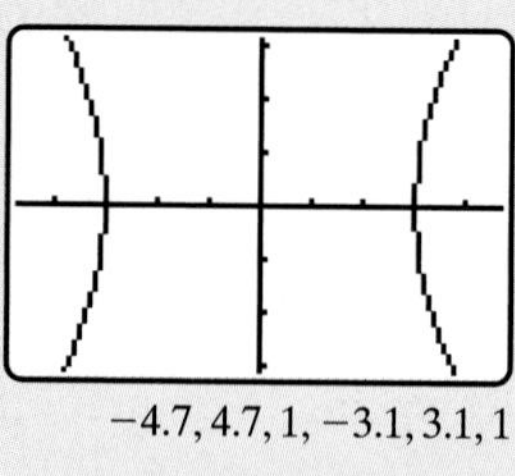

$-4.7, 4.7, 1, -3.1, 3.1, 1$

(c)

Set window as shown below figure (called "friendly window settings")

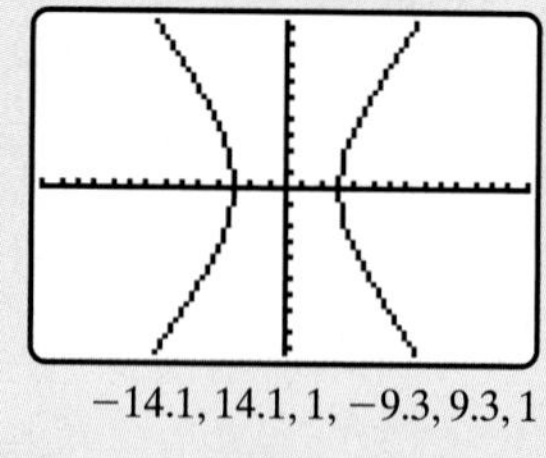

$-14.1, 14.1, 1, -9.3, 9.3, 1$

(d)

FIGURE 10.34

In part (d), the "friendly window setting," the ratio of the length of the x-axis (28.2 units) to the length of the y-axis (18.6 units) is about 1.516. This is the same ratio as the length to the width of the display window of the calculator on the TI-82 and TI-83.

2) Review Conic Sections

The following chart summarizes conic sections.

Parabola	Circle	Ellipse	Hyperbola
$y = a(x - h)^2 + k$ or $y = ax^2 + bx + c$; $a > 0$; $a < 0$	$x^2 + y^2 = r^2$	$\frac{x^2}{a^2} + \frac{y^2}{b^2} = 1$	$\frac{x^2}{a^2} - \frac{y^2}{b^2} = 1$
$x = a(y - k)^2 + h$ or $x = ay^2 + by + c$; $a > 0$; $a < 0$	$(x - h)^2 + (y - k)^2 = r^2$	$\frac{(x - h)^2}{a^2} + \frac{(y - k)^2}{b^2} = 1$	$\frac{y^2}{b^2} - \frac{x^2}{a^2} = 1$
			Asymptotes: $y = \frac{b}{a}x$ and $y = -\frac{b}{a}x$

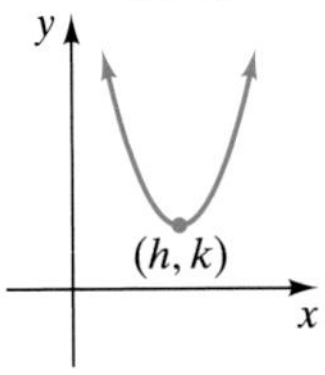

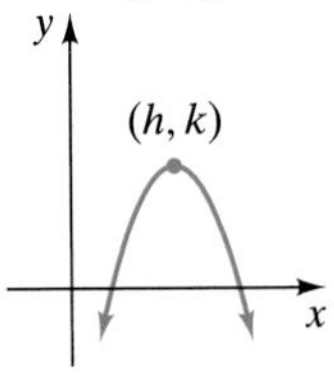

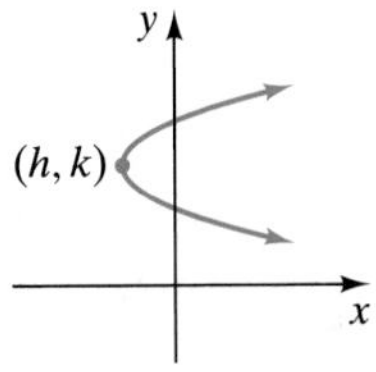

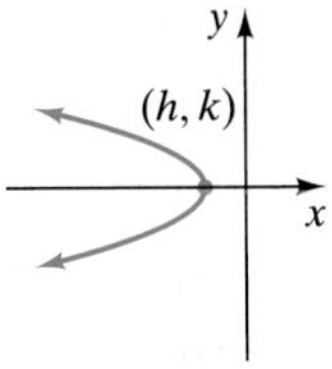

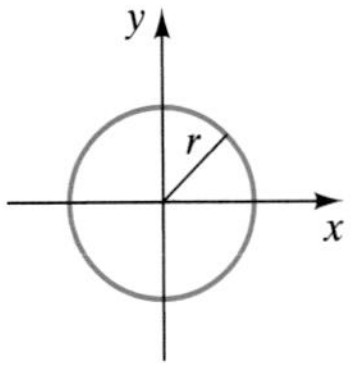

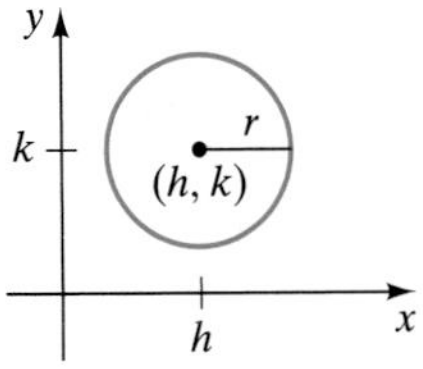

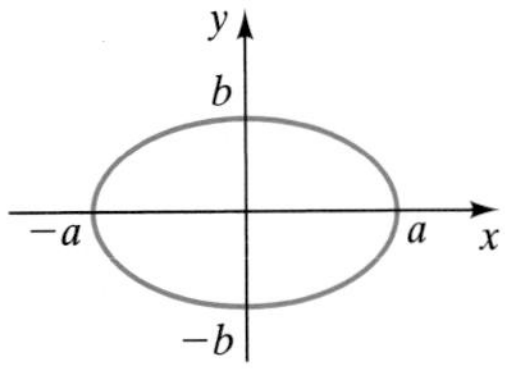

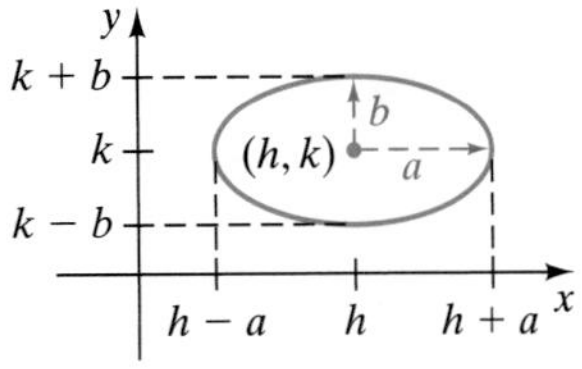

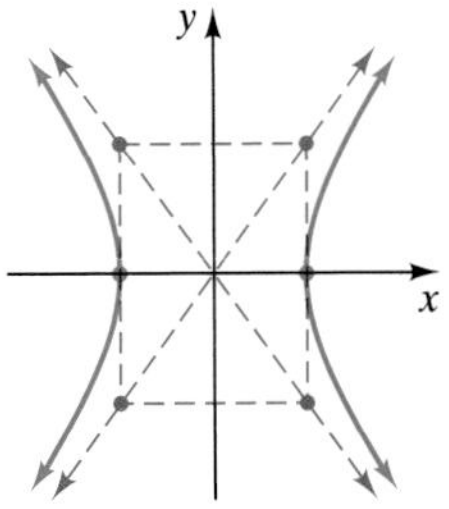

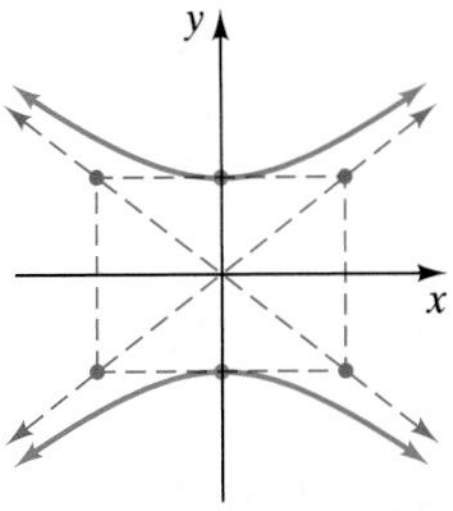

NOW TRY EXERCISE 35

Exercise Set 10.3

Concept/Writing Exercises

1. What is the definition of a hyperbola?

2. What are asymptotes? How do you find the equations of the asymptotes of a hyperbola?

3. Discuss the graph of $\frac{x^2}{a^2} - \frac{y^2}{b^2} = 1$ for nonzero real numbers a and b. Include the transverse axis, vertices, and asymptotes.

4. Discuss the graph of $\frac{y^2}{b^2} - \frac{x^2}{a^2} = 1$ for nonzero real numbers a and b. Include transverse axis, vertices, and asymptotes.

Practice the Skills

a) Determine the equations of the asymptotes for each equation. b) Graph the equation.

5. $\frac{x^2}{4} - \frac{y^2}{1} = 1$
6. $\frac{x^2}{9} - \frac{y^2}{4} = 1$
7. $\frac{y^2}{9} - \frac{x^2}{16} = 1$
8. $\frac{y^2}{25} - \frac{x^2}{4} = 1$
9. $\frac{y^2}{25} - \frac{x^2}{36} = 1$
10. $\frac{x^2}{9} - \frac{y^2}{25} = 1$
11. $\frac{x^2}{4} - \frac{y^2}{4} = 1$
12. $\frac{y^2}{49} - \frac{x^2}{100} = 1$
13. $\frac{y^2}{16} - \frac{x^2}{81} = 1$
14. $\frac{x^2}{25} - \frac{y^2}{16} = 1$
15. $\frac{y^2}{25} - \frac{x^2}{16} = 1$
16. $\frac{y^2}{4} - \frac{x^2}{36} = 1$

a) Write each equation in standard form and determine the equations of the asymptotes. b) Draw the graph.

17. $16x^2 - 4y^2 = 64$
18. $25x^2 - 16y^2 = 400$
19. $9y^2 - x^2 = 9$
20. $4y^2 - 25x^2 = 100$
21. $4y^2 - 36x^2 = 144$
22. $x^2 - 25y^2 = 25$
23. $25x^2 - 9y^2 = 225$
24. $64y^2 - 25x^2 = 1600$

Indicate whether the graph of each equation is a parabola, a circle, an ellipse, or a hyperbola. In equations that contain both x^2 and y^2 terms, divide both sides of the equation by the constant term to put the equation in a more recognizable form.

25. $4x = 6x^2 + y + 3$
26. $6x^2 + 6y^2 = 24$
27. $4x^2 - 4y^2 = 29$
28. $9x^2 - 16y^2 = 36$
29. $x = y^2 + 6y - 7$
30. $x^2 - 4y^2 = 36$
31. $-2x^2 + 4y^2 = 16$
32. $3x^2 + 3y^2 = 12$
33. $5x^2 + 10y^2 = 12$
34. $7x^2 + 15y^2 = 144$
35. $x + y = 2y^2 + 9$
36. $6x^2 - 9y^2 = 36$
37. $6x^2 + 6y^2 = 36$
38. $-y^2 + 4x^2 = 16$
39. $-3x^2 - 3y^2 = -27$
40. $-6x^2 + 2y^2 = -6$
41. $-6y^2 + x^2 = -9$
42. $4x^2 - 9y^2 = 36$

Problem Solving

43. Find the equation of the hyperbola whose intercepts are (0, 2) and (0, −2) and whose asymptotes are $y = \frac{1}{2}x$ and $y = -\frac{1}{2}x$.

44. Find the equation of the hyperbola whose intercepts are (−3, 0) and (3, 0) and whose asymptotes are $y = 2x$ and $y = -2x$.

45. Find an equation of a hyperbola whose transverse axis is along the x-axis and whose equations of the asymptotes are $y = \frac{5}{3}x$ and $y = -\frac{5}{3}x$. Is this the only possible answer? Explain.

46. Find an equation of a hyperbola whose transverse axis is along the y-axis and whose equations of the asymptotes are $y = \frac{2}{3}x$ and $y = -\frac{2}{3}x$. Is this the only possible answer? Explain.

47. Are any hyperbolas of the form $\frac{x^2}{a^2} - \frac{y^2}{b^2} = 1$ functions? Explain.

48. Are any hyperbolas of the form $\frac{y^2}{b^2} - \frac{x^2}{a^2} = 1$ functions? Explain.

49. Considering the graph of $\frac{x^2}{9} - \frac{y^2}{4} = 1$, determine the domain and range of the relation.

50. Considering the graph of $\frac{y^2}{16} - \frac{x^2}{9} = 1$, determine the domain and range of the relation.

51. If the equation $\frac{x^2}{a^2} - \frac{y^2}{b^2} = 1$, where $a > b$, is graphed, and then the values of a and b are interchanged, and the new equation is graphed, how will the two graphs compare? Explain your answer.

52. If the equation $\frac{x^2}{a^2} - \frac{y^2}{b^2} = 1$, where $a > b$, is graphed, and then the signs of each term on the left side of the equation are changed, and the new equation is graphed, how will the two graphs compare? Explain your answer.

53. Check your answer to Exercise 7 on your grapher.

54. Check your answer to Exercise 9 on your grapher.

Cumulative Review Exercises

[2.1] **55.** Solve the equation $\frac{x}{2} + \frac{2}{3}(x - 6) = x + 4$.

[3.2] **56.** What are the range and domain of a function?

[4.6] **57.** Determine the solution to the following system of inequalities graphically.

$$6x - 2y < 12$$
$$y \geq -2x + 3$$

[8.2] **58.** Solve the quadratic equation $-2x^2 + 6x - 5 = 0$.

10.4 NONLINEAR SYSTEMS OF EQUATIONS AND THEIR APPLICATIONS

SSM VIDEO 10.4 CD Rom

1) **Solve nonlinear systems using substitution.**
2) **Solve nonlinear systems using addition.**
3) **Solve applications.**

1) Solve Nonlinear Systems Using Substitution

In Chapter 4 we discussed systems of linear equations. Here we discuss nonlinear systems of equations. A **nonlinear system of equations** is a system of equations in which at least one equation is not linear (that is, one whose graph is not a straight line).

The solution to a system of equations is the point or points that satisfy all equations in the system. Consider the system of equations

$$x^2 + y^2 = 25$$
$$3x + 4y = 0$$

Both equations are graphed on the same axes in Figure 10.35. Note that the graphs appear to intersect at the points $(-4, 3)$ and $(4, -3)$. The check shows that these points satisfy both equations in the system and are therefore solutions to the system.

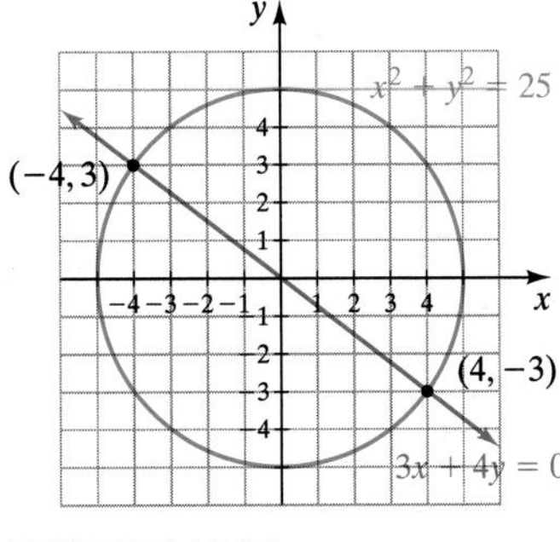

FIGURE 10.35

CHECK: $(-4, 3)$

$$x^2 + y^2 = 25 \qquad 3x + 4y = 0$$
$$(-4)^2 + 3^2 \stackrel{?}{=} 25 \qquad 3(-4) + 4(3) \stackrel{?}{=} 0$$
$$16 + 9 \stackrel{?}{=} 25 \qquad -12 + 12 \stackrel{?}{=} 0$$
$$25 = 25 \quad \text{True} \qquad 0 = 0 \quad \text{True}$$

CHECK: $(4, -3)$

$$4^2 + (-3)^2 = 25 \qquad 3(4) + 4(-3) = 0$$
$$16 + 9 \stackrel{?}{=} 25 \qquad 12 - 12 \stackrel{?}{=} 0$$
$$25 = 25 \quad \text{True} \qquad 0 = 0 \quad \text{True}$$

The graphical procedure for solving a system of equations may be inaccurate since we have to estimate the point or points of intersection. An exact answer may be obtained algebraically.

To solve a system of equations algebraically, we often solve one or more of the equations for one of the variables and then use substitution. This procedure is illustrated in Examples 1 and 2.

EXAMPLE 1 Solve the previous system of equations algebraically using substitution.

$$x^2 + y^2 = 25$$
$$3x + 4y = 0$$

Solution We first solve the linear equation $3x + 4y = 0$ for either x or y. We will solve for y.

$$3x + 4y = 0$$
$$4y = -3x$$
$$y = -\frac{3x}{4}$$

Now we substitute $-\frac{3x}{4}$ for y in the equation $x^2 + y^2 = 25$ and solve for the remaining variable, x.

$$x^2 + y^2 = 25$$
$$x^2 + \left(-\frac{3x}{4}\right)^2 = 25$$
$$x^2 + \frac{9x^2}{16} = 25$$
$$16\left(x^2 + \frac{9x^2}{16}\right) = 16(25)$$
$$16x^2 + 9x^2 = 400$$
$$25x^2 = 400$$
$$x^2 = \frac{400}{25} = 16$$
$$x = \pm\sqrt{16} = \pm 4$$

Next, we find the corresponding value of y for each value of x by substituting each value of x (one at a time) in the equation solved for y.

$$x = 4 \qquad\qquad x = -4$$
$$y = -\frac{3x}{4} \qquad\qquad y = -\frac{3x}{4}$$
$$= -\frac{3(4)}{4} \qquad\qquad = -\frac{3(-4)}{4}$$
$$= -3 \qquad\qquad = 3$$

The solutions are $(4, -3)$ and $(-4, 3)$. This checks with the solution obtained graphically in Figure 10.35.

NOW TRY EXERCISE 3

Note that our objective in using substitution is to obtain a single equation containing only one variable.

EXAMPLE 2 Solve the following system of equations using substitution.

$$y = x^2 - 3$$
$$x^2 + y^2 = 9$$

Solution Since both equations contain x^2, we will solve one of the equations for x^2. We will choose to solve $y = x^2 - 3$ for x^2.

$$y = x^2 - 3$$
$$y + 3 = x^2$$

Now substitute $y + 3$ for x^2 in the equation $x^2 + y^2 = 9$.

$$x^2 + y^2 = 9$$
$$(y + 3) + y^2 = 9$$
$$y^2 + y + 3 = 9$$
$$y^2 + y - 6 = 0$$
$$(y + 3)(y - 2) = 0$$
$$y + 3 = 0 \quad \text{or} \quad y - 2 = 0$$
$$y = -3 \qquad\qquad y = 2$$

Now find the corresponding values of x by substituting the values found for y.

$$y = -3 \qquad\qquad y = 2$$
$$y = x^2 - 3 \qquad\qquad y = x^2 - 3$$
$$-3 = x^2 - 3 \qquad\qquad 2 = x^2 - 3$$
$$0 = x^2 \qquad\qquad 5 = x^2$$
$$0 = x \qquad\qquad \pm\sqrt{5} = x$$

This system has three solutions: $(0, -3)$, $(\sqrt{5}, 2)$, and $(-\sqrt{5}, 2)$.

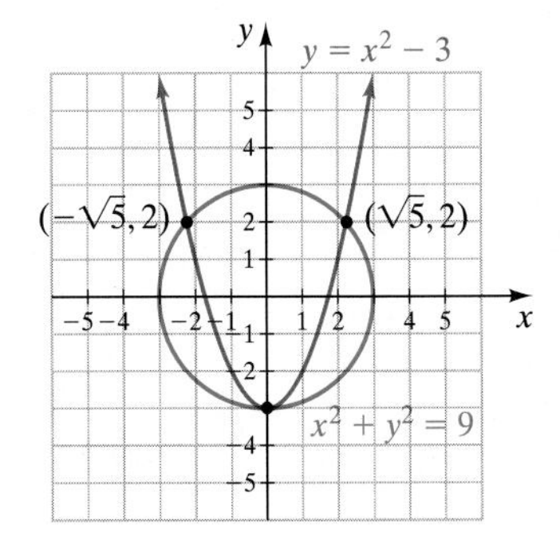

FIGURE 10.36

NOW TRY EXERCISE 11

Note that in Example 2 the graph of the equation $y = x^2 - 3$ is a parabola and the graph of the equation $x^2 + y^2 = 9$ is a circle. The graphs of both equations are illustrated in Figure 10.36.

HELPFUL HINT

Students will sometimes solve for one variable and assume that they have the solution. Remember that the solution, if one exists, to a system in two variables consists of one or more ordered pairs.

2 Solve Nonlinear Systems Using Addition

We can often solve systems of equations more easily using the addition method that was discussed in Section 4.1. As with the substitution method, our objective is to obtain a single equation containing only one variable.

EXAMPLE 3 Solve the system of equations using the addition method.

$$x^2 + y^2 = 9$$
$$2x^2 - y^2 = -6$$

Solution If we add the two equations, we will obtain one equation containing only one variable.

$$\begin{aligned} x^2 + y^2 &= -9 \\ 2x^2 - y^2 &= -6 \\ \hline 3x^2 - y^2 &= -3 \\ x^2 &= -1 \\ x &= \pm 1 \end{aligned}$$

Now solve for the variable y by substituting $x = \pm 1$ in either of the original equations.

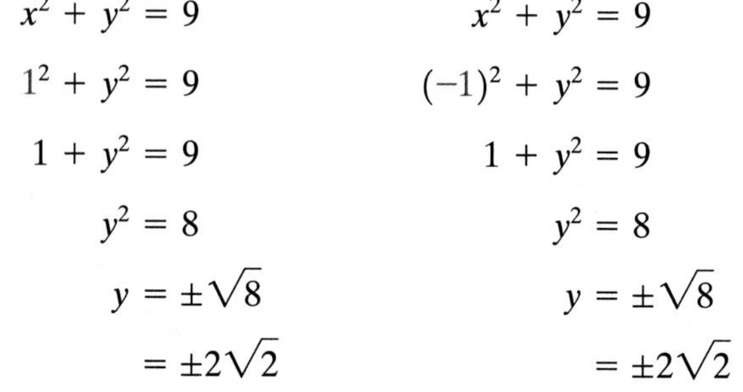

$$\begin{aligned} x &= 1 \\ x^2 + y^2 &= 9 \\ 1^2 + y^2 &= 9 \\ 1 + y^2 &= 9 \\ y^2 &= 8 \\ y &= \pm\sqrt{8} \\ &= \pm 2\sqrt{2} \end{aligned} \qquad \begin{aligned} x &= -1 \\ x^2 + y^2 &= 9 \\ (-1)^2 + y^2 &= 9 \\ 1 + y^2 &= 9 \\ y^2 &= 8 \\ y &= \pm\sqrt{8} \\ &= \pm 2\sqrt{2} \end{aligned}$$

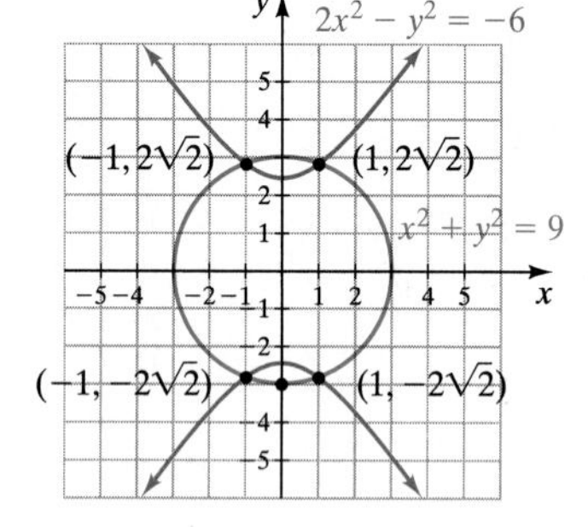

FIGURE 10.37

There are four solutions to this system of equations:

$$(1, 2\sqrt{2}), (1, -2\sqrt{2}), (-1, 2\sqrt{2}), \text{ and } (-1, -2\sqrt{2})$$

The graphs of the equations in the system solved in Example 3 are given in Figure 10.37. Notice the four points of intersection of the two graphs.

It is possible that a system of equations has no real solution (therefore, the graphs do not intersect). Example 4 illustrates such a case.

EXAMPLE 4 Solve the system of equations using the addition method.

$$x^2 + 4y^2 = 16 \qquad (eq.\ 1)$$
$$x^2 + y^2 = 1 \qquad (eq.\ 2)$$

Solution Multiply *(eq. 2)* by −1 and add the resulting equation to *(eq. 1)*.

$$\begin{aligned} x^2 + 4y^2 &= 16 \\ -x^2 - y^2 &= -1 \qquad (eq.\ 2) \text{ multiplied by } -1. \\ \hline 3y^2 &= 15 \\ y^2 &= 5 \\ y &= \pm\sqrt{5} \end{aligned}$$

Now solve for x.

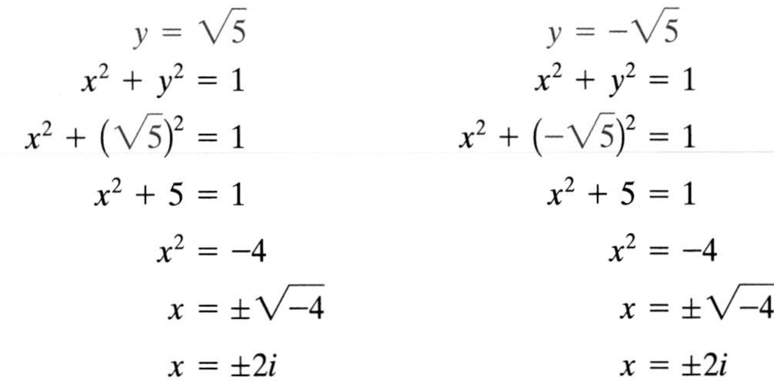

$y = \sqrt{5}$	$y = -\sqrt{5}$
$x^2 + y^2 = 1$	$x^2 + y^2 = 1$
$x^2 + (\sqrt{5})^2 = 1$	$x^2 + (-\sqrt{5})^2 = 1$
$x^2 + 5 = 1$	$x^2 + 5 = 1$
$x^2 = -4$	$x^2 = -4$
$x = \pm\sqrt{-4}$	$x = \pm\sqrt{-4}$
$x = \pm 2i$	$x = \pm 2i$

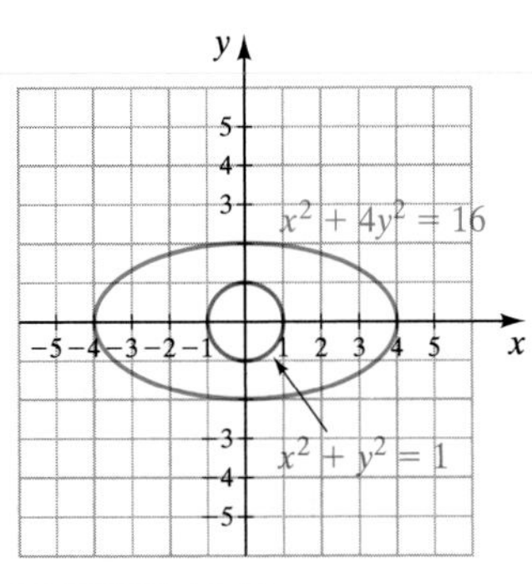

FIGURE 10.38

Since x is an imaginary number for both values of y, this system of equations has no real solution.

The graphs of the equations in Example 4 are shown in Figure 10.38. Notice that the two graphs do not intersect; therefore, there is no real solution. This agrees with the answer we obtained in Example 4.

NOW TRY EXERCISE 17

3) Solve Applications

Now we will study some applications of nonlinear systems.

EXAMPLE 5 The area, A, of a rectangular playground is 8000 square feet and its perimeter, P, is 360 feet. Find the length and width of the rectangle.

Solution **Understand and Translate** We begin by drawing a sketch (see Fig. 10.39).

$$\text{Let } x = \text{length}$$
$$y = \text{width}$$

Since $A = xy$ and $P = 2x + 2y$, the system of equations is

$$xy = 8000$$
$$2x + 2y = 360$$

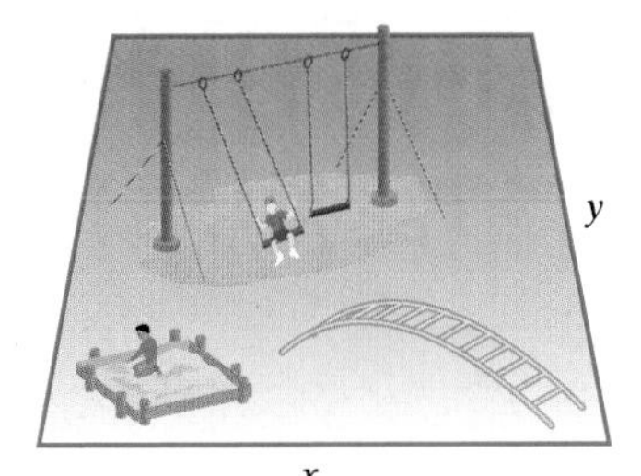

FIGURE 10.39

Carry Out We will solve this system using substitution. Since $2x + 2y = 360$ is a linear equation, we solve this equation for y (we could also solve for x).

$$2x + 2y = 360$$
$$2y = 360 - 2x$$
$$y = \frac{360 - 2x}{2} = \frac{360}{2} - \frac{2x}{2} = 180 - x$$

Now we substitute $180 - x$ for y in $xy = 8000$.

$$xy = 8000$$
$$x(180 - x) = 8000$$
$$180x - x^2 = 8000$$
$$x^2 - 180x + 8000 = 0$$
$$(x - 100)(x - 80) = 0$$

$$x - 100 = 0 \qquad \text{or} \qquad x - 80 = 0$$
$$x = 100 \qquad\qquad x = 80$$

Answer If x is 100, then $y = 180 - x = 180 - 100 = 80$. And if $x = 80$, then $y = 180 - 80 = 100$. Thus, the dimensions of the playground are 80 feet by 100 feet.

In Example 5 the graph of $xy = 8000$ is a hyperbola, and the graph of $2x + 2y = 360$ is a straight line. If you graph these two equations, the graphs will intersect at (80, 100) and (100, 80).

NOW TRY EXERCISE 27

EXAMPLE 6

Hike 'n' Bike Company produces and sells bicycles. Its weekly cost equation is $C = 50x + 400, 0 \le x \le 160$, and its weekly revenue equation is $R = 100x - 0.3x^2, 0 \le x \le 160$, where x is the number of bicycles produced and sold each week. Find the number of bicycles that must be produced and sold for Hike 'n' Bike to break even.

Solution

Understand and Translate A company breaks even when its cost equals its revenue. When its cost is greater than its revenue, the company has a loss. When its revenue exceeds its cost, the company makes a profit.

The system of equations is

$$C = 50x + 400$$
$$R = 100x - 0.3x^2$$

For Hike 'n' Bike to break even, its cost must equal its revenue. Thus, we write

$$C = R$$
$$50x + 400 = 100x - 0.3x^2$$

Carry Out Writing this quadratic equation in standard form, we obtain

$$0.3x^2 - 50x + 400 = 0, \quad 0 \le x \le 160$$

We will solve this equation using the quadratic formula.

$$a = 0.3, \qquad b = -50, \qquad c = 400$$

$$x = \frac{-b \pm \sqrt{b^2 - 4ac}}{2a}$$
$$= \frac{-(-50) \pm \sqrt{(-50)^2 - 4(0.3)(400)}}{2(0.3)}$$
$$= \frac{50 \pm \sqrt{2020}}{0.6}$$

$$x = \frac{50 + \sqrt{2020}}{0.6} \approx 158.2 \qquad \text{or} \qquad x = \frac{50 - \sqrt{2020}}{0.6} \approx 8.4$$

Answer The cost will equal the revenue and the company will break even when approximately 8 bicycles are sold. The cost will also equal the revenue when approximately 158 bicycles are sold. The company will make a profit when between 9 and 158 bicycles are sold. When fewer than 9 or more than 158 bicycles are sold, the company will have a loss (see Fig. 10.40).

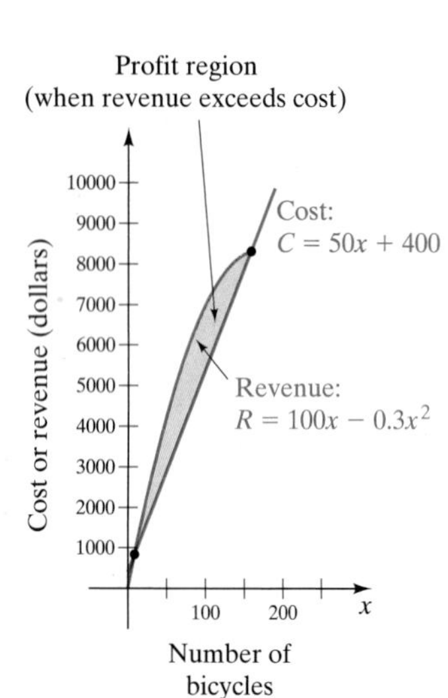

FIGURE 10.40

Using Your Graphing Calculator

To solve nonlinear systems of equations graphically, graph the equations and find the intersections of the graphs. Consider the system in Example 1, $x^2 + y^2 = 25$ and $3x + 4y = 0$. To graph $x^2 + y^2 = 25$, we use $y_1 = \sqrt{25 - x^2}$ and $y_2 = -\sqrt{25 - x^2}$. Therefore, to solve this system we find the intersection of

$$y_1 = \sqrt{25 - x^2}$$

$$y_2 = -\sqrt{25 - x^2}$$

$$y_3 = -\frac{3}{4}x$$

The system is graphed, using the ZOOM: 5 (ZSquared) in Figure 10.41a.* In Figure 10.41b we graph the same two equations using the "friendly numbers" shown below the figure. Using the calculator with either the TRACE and ZOOM features, the TABLE feature, or the INTERSECT feature, you will find that the solutions are $(4, -3)$ and $(-4, 3)$.

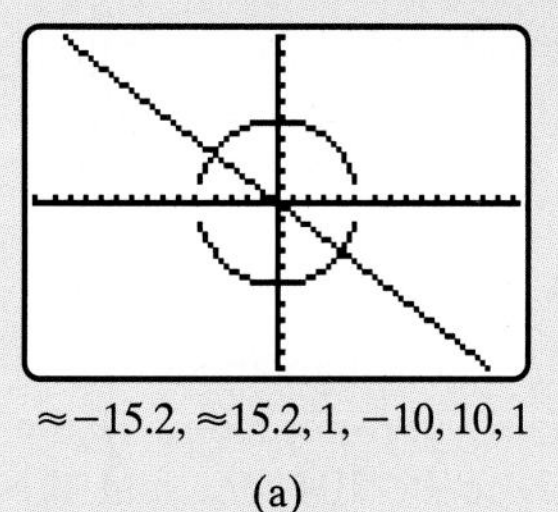

$\approx -15.2, \approx 15.2, 1, -10, 10, 1$

(a)

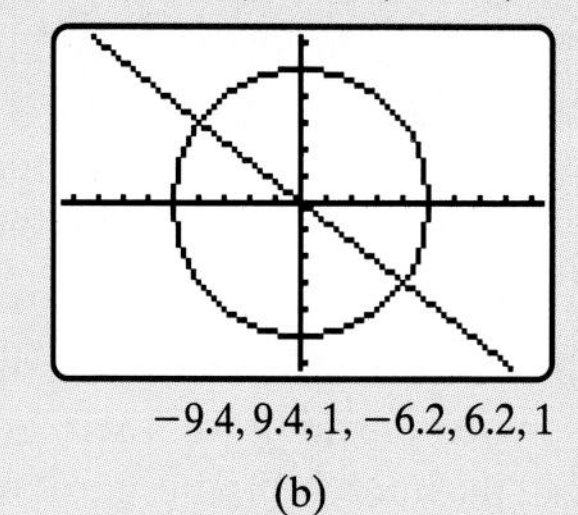

$-9.4, 9.4, 1, -6.2, 6.2, 1$

(b)

FIGURE 10.41

* Start with the standard window, then select ZOOM: 5 to get this graph.

Exercise Set 10.4

Concept/Writing Exercises

1. What is a nonlinear system of equations?

2. Explain how nonlinear systems of equations may be solved graphically.

Practice the Skills

Find all real solutions to each system of equations by substitution.

3. $x^2 + y^2 = 9$, $x + 2y = 3$

4. $x^2 + y^2 = 4$, $x - 2y - 4 = 0$

5. $y = x^2 - 5$, $3x + 2y = 10$

6. $x + y = 4$, $x^2 - y^2 = 4$

7. $2x^2 - y^2 = -8$, $x - y = 6$

8. $y^2 = -x + 4$, $x^2 + y^2 = 6$

9. $x^2 - 4y^2 = 16$, $x^2 + y^2 = 1$

10. $2x^2 + y^2 = 16$, $x^2 - y^2 = -4$

11. $y = x^2 - 3$, $x^2 + y^2 = 9$

12. $x^2 + y^2 = 25$, $x - 3y = -5$

Find all real solutions to each system of equations using the addition method.

13. $x^2 - y^2 = 4$, $x^2 + y^2 = 4$

14. $x^2 + y^2 = 25$, $x^2 - 2y^2 = 7$

15. $x^2 + y^2 = 13$, $2x^2 + 3y^2 = 30$

16. $3x^2 - y^2 = 4$, $x^2 + 4y^2 = 10$

17. $4x^2 + 9y^2 = 36$, $2x^2 - 9y^2 = 18$

18. $5x^2 - 2y^2 = -13$, $3x^2 + 4y^2 = 39$

19. $2x^2 + 3y^2 = 21$, $x^2 + 2y^2 = 12$

20. $2x^2 + y^2 = 11$, $x^2 + 3y^2 = 28$

21. $-x^2 - 2y^2 = 6$, $5x^2 + 15y^2 = 20$

22. $x^2 - 2y^2 = 7$, $x^2 + y^2 = 34$

23. $x^2 + y^2 = 9$, $16x^2 - 4y^2 = 64$

24. $3x^2 + 4y^2 = 35$, $2x^2 + 5y^2 = 42$

Problem Solving

25. Make up your own nonlinear system of equations whose solution is the empty set. Explain how you know the system has no solution.

26. If a system of equations consists of an ellipse and a hyperbola, what is the maximum number of points of intersection? Make a sketch to illustrate this.

27. Elizabeth Shepherd wants to build a dance floor at her restaurant. If the perimeter of the dance floor is to be 68 feet and the area of the dance floor is to be 240 square feet, find the dimensions of the dance floor.

28. A garden club is designing a rectangular garden in the entrance of their housing development. If the perimeter of the garden is to be 104 feet and the area is to be 480 square feet, find the dimensions of the garden.

29. A country has recently decided to change its currency. One of its new bills is designed to have an area of 112 square centimeters with a diagonal of $\sqrt{260}$ centimeters. Find the length and width of the new bill.

30. A rectangular ice rink has an area of 3000 square feet. If the diagonal across the rink is 85 feet, find the dimensions of the rink.

31. Judy Stamm, a carpenter, has a rectangular piece of wood. When she measures the diagonal it measures 17 inches. When she cuts the wood along the diagonal, the perimeter of each triangle formed is 40 inches. Find the dimensions of the original piece of wood.

32. A sail on a sailboat is shaped like a right triangle with a perimeter of 36 meters and a hypotenuse 15 meters. Find the length of the legs of the triangle.

33. Maureen Smith fences in a rectangular area along a river bank as illustrated. If 20 feet of fencing encloses an area of 48 square feet, find the dimensions of the enclosed area.

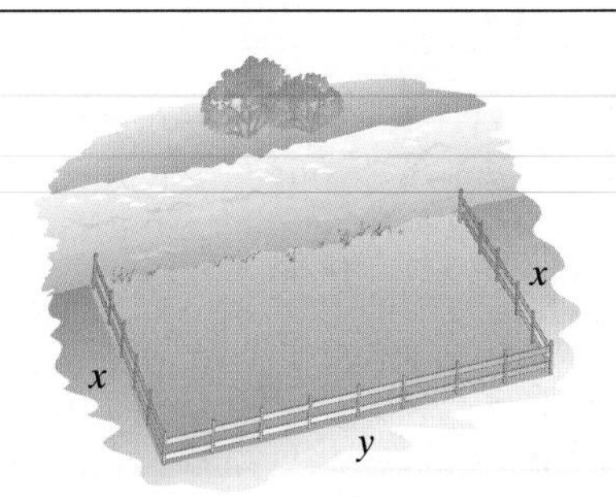

34. A rectangular area is to be fenced along a river as illustrated in Exercise 33. If 40 feet of fencing encloses an area of 200 square feet, find the dimensions of the enclosed area.

35. Paul Martin throws a football upward from the ground. Its height above the ground at any time t is given by the formula $d = -16t^2 + 64t$. At the same time that the football is thrown, Shannon Ryan throws a baseball upward from the top of an 80-foot-tall building. Its height above the ground at any time t is given by the formula $d = -16t^2 + 16t + 80$. Find the time at which the two balls will be the same height above the ground.

36. Michael Sutton throws a tennis ball downward from a helicopter flying at a height of 1000 feet. The height of the ball above the ground at any time t is found by the formula $d = -16t^2 - 10t + 1000$. At the instant the ball is thrown from the helicopter, Tiashana Thompson throws a snowball upward from the top of a 800-foot-tall building. The height above the ground of the snowball at any time t is found by the formula $d = -16t^2 + 80t + 800$. At what time will the ball and snowball pass each other? (Neglect air resistance.)

37. Simple interest is calculated using the simple interest formula, interest = principal · rate · time or $i = prt$. If Seana Hayden invests a certain principal at a specific interest rate for 1 year, the interest she obtains is \$7.50. If she increases the principal by \$25 and the interest rate is decreased by 1%, the interest remains the same. Find the principal and the interest rate.

38. If Evan Girard invests a certain principal at a specific interest rate for 1 year, the interest she obtains is \$72. If she increases the principal by \$120 and the interest rate is decreased by 2%, the interest remains the same. Find the principal and the interest rate. Use $i = prt$.

For the given cost and revenue equations, find the break-even point(s).

39. $C = 10x + 300, R = 30x - 0.1x^2$

40. $C = 12.6x + 150, R = 42.8x - 0.3x^2$

41. $C = 80x + 900, R = 120x - 0.2x^2$

42. $C = 0.6x^2 + 9, R = 12x - 0.2x^2$

Solve the following systems using your graphing calculator. Round your answers to the nearest hundredth.

43. $3x - 5y = 12$
$x^2 + y^2 = 10$

44. $y = 2x^2 - x + 2$
$4x^2 + y^2 = 36$

Challenge Problems

45. The intersection of three roads forms a right triangle, as shown in the figure.

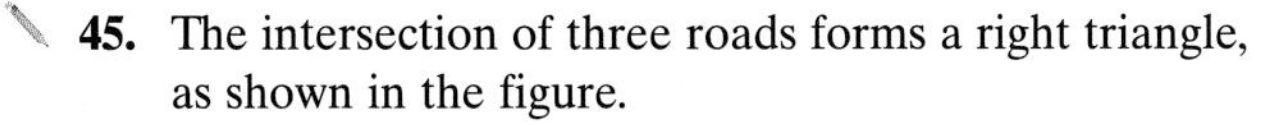

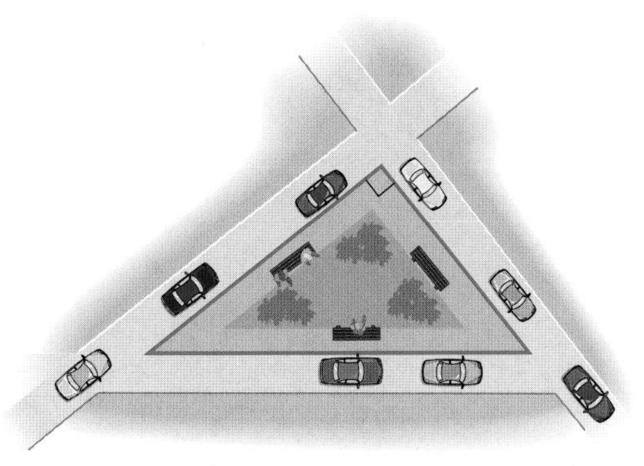

If the hypotenuse is 26 yards and the area is 120 square yards, find the length of the two legs of the triangle.

46. In the figure shown, R represents the radius of the larger circle and r represents the radius of the smaller circle. If $R = 2r$ and if the shaded area is 122.5π, find r and R.

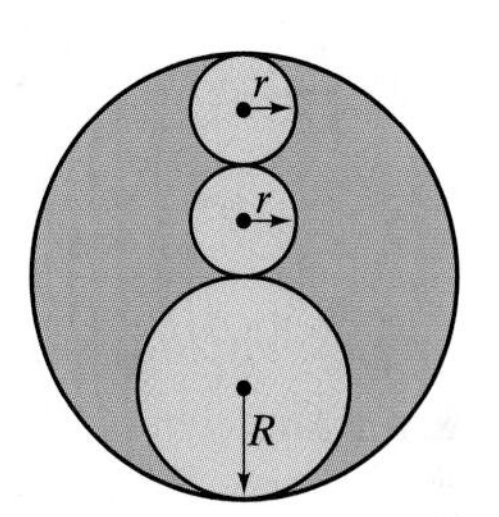

Cumulative Review Exercises

[1.4] **47.** List the order of operations we follow when evaluating an expression.

48. Solve $\frac{3}{5}(2x - y) = \frac{3}{4}(2x - 3y) + 6$ for y.

[2.2] **49.** Use the compound interest formula $A = p\left(1 + \frac{r}{n}\right)^{nt}$ to find the amount, A, when the principal, p, is \$5000, the rate r is 8%, the number of compounding periods, n, is 2, and the number of years, t, is 2.

[2.5] **50.** Solve the inequality $\frac{3 - 4y}{3} \geq \frac{2y - 6}{4} - \frac{7}{6}$.

[7.1] **51.** Graph $f(x) = \sqrt{x + 2}$.

[7.6] **52.** Given $f(x) = \sqrt{x^2 - 3x - 9}$ and $g(x) = \sqrt{2x - 3}$, find all values of x for which $f(x) = g(x)$ and $f(x)$ and $g(x)$ are real numbers.

SUMMARY

Key Words and Phrases

10.1

Center of a circle
Circle
Distance
Midpoint
Parabola

10.2

Ellipse
Foci
Major axis
Minor axis

10.3

Asymptotes
Center of a hyperbola
Hyperbola
Transverse axis
Vertices of a hyperbola

10.4

Nonlinear system of equations

IMPORTANT FACTS

Distance Formula: $d = \sqrt{(x_2 - x_1)^2 + (y_2 - y_1)^2}$

Midpoint Formula: $\left(\frac{x_1 + x_2}{2}, \frac{y_1 + y_2}{2}\right)$

Circle	Ellipse
$(x - h)^2 + (y - k)^2 = r^2$ center at (h, k) radius r	$\frac{(x - h)^2}{a^2} + \frac{(y - k)^2}{b^2} = 1$ center at (h, k)

Parabola	Hyperbola
$y = a(x - h)^2 + k$ vertex at (h, k) opens upward when $a > 0$ opens downward when $a < 0$ $x = a(y - k)^2 + h$ vertex at (h, k) opens right when $a > 0$ opens left when $a < 0$	x-axis transverse axis $\frac{x^2}{a^2} - \frac{y^2}{b^2} = 1$ y-axis transverse axis $\frac{y^2}{b^2} - \frac{x^2}{a^2} = 1$ asymptotes: $y = \frac{b}{a}x$ or $y = -\frac{b}{a}x$

Review Exercises

[10.1] *Find the length and the midpoint of the line segment between each pair of points.*

1. $(0, 0), (3, -4)$ **2.** $(6, 2), (2, -1)$ **3.** $(-2, -3), (3, 9)$ **4.** $(-4, 3), (-2, 5)$

Graph each equation.

5. $y = (x - 3)^2 + 4$ **6.** $y = (x + 4)^2 - 5$ **7.** $x = (y - 1)^2 + 4$ **8.** $x = -2(y + 4)^2 - 3$

a) *Write each equation in the form $y = a(x - h)^2 + k$ or $x = a(y - k)^2 + h$.* ***b)*** *Graph the equation.*

9. $y = x^2 - 6x$ **10.** $x = -y^2 - 2y + 8$ **11.** $x = y^2 + 5y + 4$ **12.** $y = 2x^2 - 8x - 24$

a) *In Exercises 13–18, write the equation of each circle in standard form.* ***b)*** *Draw the graph.*

13. Center $(0, 0)$, radius 5 **14.** Center $(-3, 4)$, radius 3 **15.** $x^2 + y^2 - 4y = 0$
16. $x^2 + y^2 - 2x + 6y + 1 = 0$ **17.** $x^2 - 8x + y^2 - 10y + 40 = 0$ **18.** $x^2 + y^2 - 4x + 10y + 17 = 0$

Graph each equation.

19. $y = \sqrt{16 - x^2}$ **20.** $y = -\sqrt{25 - x^2}$

Determine the equation of each circle.

21.

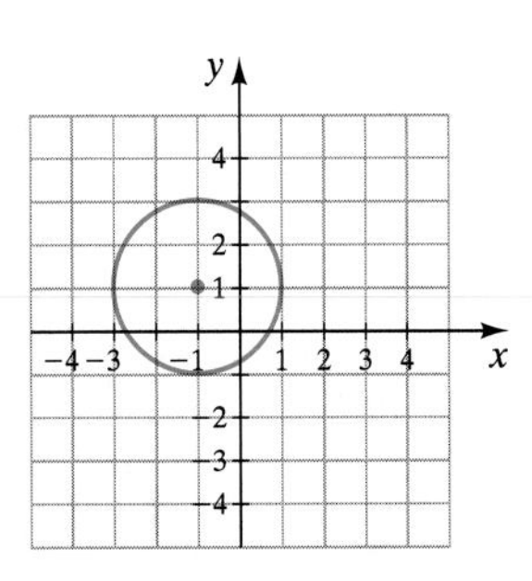

22.

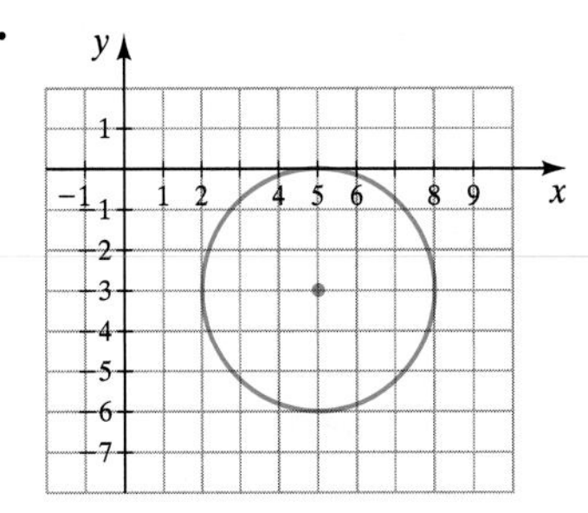

[10.2] *Graph each equation.*

23. $\frac{x^2}{9} + \frac{y^2}{4} = 1$

24. $\frac{x^2}{9} + \frac{y^2}{64} = 1$

25. $4x^2 + 9y^2 = 36$

26. $9x^2 + 16y^2 = 144$

27. $\frac{(x-3)^2}{16} + \frac{(y+2)^2}{4} = 1$

28. $\frac{(x+3)^2}{9} + \frac{y^2}{25} = 1$

29. $25(x-2)^2 + 9(y-1)^2 = 225$

30. $4(x+3)^2 + 25(y-2)^2 = 100$

[10.3] ***a)*** *Determine the equations of the asymptotes for each equation.* ***b)*** *Draw the graph.*

31. $\frac{x^2}{4} - \frac{y^2}{9} = 1$

32. $\frac{y^2}{16} - \frac{x^2}{4} = 1$

33. $\frac{y^2}{9} - \frac{x^2}{25} = 1$

34. $\frac{x^2}{4} - \frac{y^2}{36} = 1$

a) *Write each equation in standard form.* ***b)*** *Determine the equations of the asymptotes.* ***c)*** *Draw the graph.*

35. $9y^2 - 4x^2 = 36$

36. $x^2 - 16y^2 = 16$

37. $25x^2 - 16y^2 = 400$

38. $49y^2 - 9x^2 = 441$

[10.1–10.3] *Identify the graph of each equation as a circle, ellipse, parabola, or hyperbola.*

39. $\frac{x^2}{4} - \frac{y^2}{25} = 1$

40. $4x^2 + 9y^2 = 144$

41. $4x^2 + 4y^2 = 16$

42. $x^2 - 25y^2 = 25$

43. $\frac{x^2}{16} + \frac{y^2}{9} = 1$

44. $y = (x-3)^2 + 4$

45. $4x^2 + 9y^2 = 36$

46. $x = -y^2 - 6y - 7$

[10.4] *Find all real solutions to each system of equations by substitution.*

47. $x^2 + y^2 = 9$
$y = 3x + 9$

48. $x^2 - y^2 = 4$
$2x + 2y = 8$

49. $x^2 + y^2 = 4$
$x^2 - y^2 = 4$

50. $x^2 + 4y^2 = 4$
$x^2 - 6y^2 = 12$

Find all real solutions to each system of equations using the addition method.

51. $x^2 + y^2 = 16$
$x^2 - y^2 = 16$

52. $x^2 + y^2 = 25$
$x^2 - 2y^2 = -2$

53. $-4x^2 + y^2 = -12$
$8x^2 + 2y^2 = -8$

54. $-2x^2 - 3y^2 = -6$
$5x^2 + 4y^2 = 15$

55. Richard Spencer's dining room table has an area of 32 square feet and a perimeter of 24 feet. Find the length and width of the table.

56. The Dip and Dap Company has a cost equation of $C = 20.3x + 120$ and a revenue equation of $R = 50.2x - 0.2x^2$, where x is the number of bottles of glue sold. Find the number of bottles of glue the company must sell to break even.

57. Tanya Richardson just purchased a rectangular Persian carpet. If the carpet has an area of 300 square feet and the diagonal of the carpet is 25 feet, find the length and width of the carpet.

58. If Willis Bilderback invests a certain principal at a specific interest rate for 1 year, the interest is \$250. If he increases the principal by \$1250 and the interest rate is decreased by 1%, the interest remains the same. Find the principal and interest rate. Use $i = prt$.

Practice Test

1. Why are parabolas, circles, ellipses, and hyperbolas called conic sections?
2. Determine the length of the line segment whose endpoints are $(-4, 5)$ and $(3, 4)$.
3. Determine the midpoint of the line segment whose endpoints are $(4, 2)$ and $(-6, 5)$.
4. Determine the vertex of the graph of $y = -2(x - 3)^2 + 4$, then graph the equation.
5. Graph $x = y^2 - 2y + 4$.
6. Write the equation $x = -3y^2 + 12y - 8$ in the form $x = a(y - k)^2 + h$, then draw the graph.
7. Write the equation of a circle with center at $(-3, -1)$ and radius 4, then draw the graph.
8. What is the range and domain of the graph of $(x + 1)^2 + (y - 1)^2 = 9$?
9. Write the equation of the circle shown.

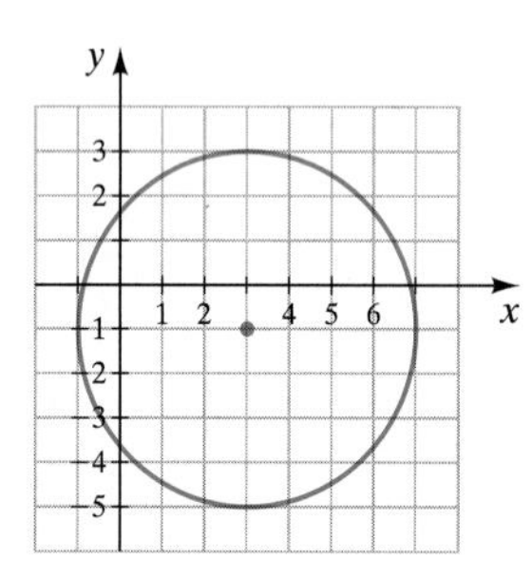

10. Graph $y = -\sqrt{16 - x^2}$.
11. Write the equation $x^2 + y^2 - 2x - 6y + 1 = 0$ in standard form, then draw the graph.
12. Graph $9x^2 + 16y^2 = 144$.
13. Is the graph below the graph of $\frac{(x + 2)^2}{4} + \frac{(y + 1)^2}{16} = 1$? Explain your answer.

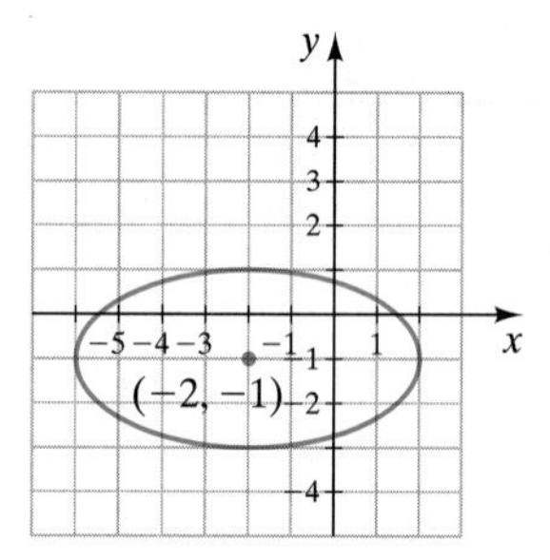

14. Graph $4(x - 3)^2 + 16(y + 1)^2 = 64$.
15. Determine the center of the ellipse given by $x^2 + y^2 + 4x + 8y - 12 = 0$.
16. Explain how to determine whether the transverse axis of a hyperbola lies on the x- or y-axis.
17. What are the equations of the asymptotes of the graph of $\frac{x^2}{16} - \frac{y^2}{36} = 1$?
18. Graph $\frac{y^2}{25} - \frac{x^2}{1} = 1$.
19. Graph $\frac{x^2}{4} - \frac{y^2}{9} = 1$.

In Exercises 20 and 21, determine whether the graph of the equation is a parabola, circle, ellipse, or hyperbola. Explain how you determined your answer.

20. $4x^2 - 16y^2 = 48$
21. $16x^2 + 4y^2 = 64$

Solve each system of equations.

22. $x^2 + y^2 = 6$
    $2x^2 - y^2 = 3$
23. $x + y = 5$
    $x^2 + y^2 = 4$
24. Mary Choi owns a rectangular plot of land that has an area of 6000 square feet. Find the dimensions of the plot if its perimeter is 320 feet.
25. David Gillespie owns a truck. The bed of the truck has an area of 60 square feet, and the diagonal across the bed measures 13 feet. Find the dimensions of the bed of the truck.

Cumulative Review Test

1. Solve the following system of equations algebraically.

$$2x - y = 6$$
$$3x + 2y = 4$$

2. If $f(x) = x^2 + 2x + 5$, find $f(a + 3)$.

3. Multiply $\dfrac{6x^2 + 5x - 4}{2x^2 - 3x + 1} \cdot \dfrac{4x^2 - 1}{8x^3 + 1}$.

4. Subtract $\dfrac{x}{x + 3} - \dfrac{x + 1}{2x^2 - 2x - 24}$.

5. Solve $\dfrac{y + 1}{y + 3} + \dfrac{y - 3}{y - 2} = \dfrac{2y^2 - 15}{y^2 + y - 6}$.

6. Simplify $\sqrt{\dfrac{12x^5y^3}{8z}}$.

7. Simplify $\dfrac{6}{\sqrt{3} - \sqrt{5}}$.

8. Solve $3\sqrt[3]{2x + 2} = \sqrt[3]{80x - 24}$.

9. Evaluate $(5 - 4i)(5 + 4i)$.

10. Solve $(x - 3)^2 = 28$.

11. Solve $3x^2 - 4x - 8 = 0$ by the quadratic formula.

12. Solve $3p^{2/3} + 14p^{1/3} - 24 = 0$.

13. Solve the inequality $\dfrac{3x - 2}{x + 4} \geq 0$ and graph the solution on a number line.

14. If $f(x) = x^2 + 6x$ and $g(x) = 2x - 3$, find
a) $(f \circ g)(x)$ **b)** $(g \circ f)(x)$

15. Graph $9x^2 + 4y^2 = 36$.

16. Graph $\dfrac{y^2}{25} - \dfrac{x^2}{16} = 1$.

17. Solve the equation $\log(3x - 4) + \log 4 = \log(x + 6)$.

18. Solve the equation $250 = 500e^{-0.3t}$ for t.

19. The price of a suit is reduced by 20% and then reduced an additional \$25. If the sale price of the suit is \$155, find the original cost of the suit.

20. The Nut Shop sells cashews for \$7 per pound and peanuts for \$5 per pound. If a customer wants a 4-pound mixture of these two nuts and wants the mixture to cost \$25 before tax, how many pounds of each nut should be mixed?

SEQUENCES, SERIES, AND THE BINOMIAL THEOREM

CHAPTER 11

Use the Angel Web site at www.prenhall.com/angel to be linked to an internet resource that will help you further explore the following application.

Have you ever noticed that the rebound height of a ball depends on the height from which it was dropped? A ball's rebound height is a percentage of the height from which it is dropped. If a ball rebounds 4 feet when dropped from 6 feet, it has rebounded $66\frac{2}{3}\%$ of its original height. Theoretically, every rebound will have a rebound and the ball will never stop bouncing. Would it be possible to calculate the total vertical distance traveled by a ball that never stops bouncing? In this chapter you will learn how to make this calculation, that is, you will learn how to sum an infinite geometric sequence. On page 684 you estimate the total vertical distance a dropped ball travels.

Preview and Perspective

Sequences and series are discussed in this chapter. A sequence is a list of numbers in a specific order and a series is the sum of the numbers in a sequence. In this book we discuss two types of sequences and series: arithmetic and geometric. Series can be used to solve many real-life problems as illustrated in this chapter. If you take higher-level mathematics courses, you may use sequences to approximate some irrational numbers. Other types of sequences not discussed in this book, including the Fibonacci sequence, may be discussed in other mathematics courses.

In Section 11.1 we use the summation symbol, Σ. If you take a course in statistics you will use this symbol often.

We introduce the binomial theorem for expanding an expression of the form $(a + b)^n$ in Section 11.4.

11.1 SEQUENCES AND SERIES

1) **Find the terms of a sequence.**
2) **Write a series.**
3) **Find partial sums of a series.**
4) **Use summation notation, Σ.**

1) Find the Terms of a Sequence

Many times we see patterns in numbers. For example, suppose you are given a job offer with a starting salary of $25,000. You are given two options for your annual salary increases. One option is an annual salary increase of $2000 per year. The salary you would receive under this option is shown below.

Year	1	2	3	4	...
	↓	↓	↓	↓	
Salary	$25,000	$27,000	$29,000	$31,000	...

Each year the salary is $2000 greater than the previous year. The three dots on the right of the lists of numbers indicate that the list continues in the same manner.

The second option is a 5% salary increase each year. The salary you would receive under this option is shown below.

Year	1	2	3	4	...
	↓	↓	↓	↓	
Salary	$25,000	$26,250	$27,562.50	$28,940.63	...

With this option, the salary in a given year after year 1 is 5% greater than the previous year's salary.

The two lists of numbers that illustrate the salaries are examples of sequences. A **sequence** of numbers is a list of numbers arranged in a specific order. Consider the list of numbers given below, which is a sequence.

$$5, 10, 15, 20, 25, 30, \ldots$$

The first term is 5. We indicate this by writing $a_1 = 5$. Since the second term is 10, $a_2 = 10$, and so on. The three dots indicate that the sequence continues indefinitely and is an **infinite sequence.**

Definition

An **infinite sequence** is a function whose domain is the set of natural numbers.

Consider the infinite sequence 5, 10, 15, 20, 25, 30, 35, ...

$$\begin{array}{lllllllllll} \text{Domain:} & \{1, & 2, & 3, & 4, & 5, & 6, & 7, & \ldots, & n, & \ldots\} \\ & \downarrow & \downarrow & \downarrow & \downarrow & \downarrow & \downarrow & \downarrow & & \downarrow & \\ \text{Range:} & \{5, & 10, & 15, & 20, & 25, & 30, & 35, & \ldots, & 5n, & \ldots\} \end{array}$$

Note that the terms of the sequence 5, 10, 15, 20, ... are found by multiplying each natural number by 5. For any natural number n, the corresponding term in the sequence is $5 \cdot n$ or $5n$. The **general term of the sequence**, a_n, which defines the sequence, is $a_n = 5n$.

$$a_n = f(n) = 5n$$

To find the twelfth term of the sequence, substitute 12 for n in the general term of the sequence, $a_{12} = 5 \cdot 12 = 60$. Thus, the twelfth term of the sequence is 60. Note that the terms in the sequence are the function values, or the numbers in the range of the function. When writing the sequence, we do not use set braces. The general form of a sequence is

$$a_1, a_2, a_3, a_4, \ldots, a_n, \ldots$$

For the infinite sequence 2, 4, 8, 16, 32, ..., 2^n, ... we can write

$$a_n = f(n) = 2^n$$

Notice that when $n = 1$, $a_1 = 2^1 = 2$; when $n = 2$, $a_2 = 2^2 = 4$; when $n = 3$, $a_3 = 2^3 = 8$; when $n = 4$, $a_4 = 2^4 = 16$; and so on. What is the seventh term of this sequence? The answer is $a_7 = 2^7 = 128$.

A sequence may also be **finite**.

Definition

A **finite sequence** is a function whose domain includes only the first n natural numbers.

A finite sequence has only a finite number of terms.

Examples of Finite Sequences

5, 10, 15, 20	domain is $\{1, 2, 3, 4\}$
2, 4, 8, 16, 32	domain is $\{1, 2, 3, 4, 5\}$

EXAMPLE 1 Write the finite sequence defined by $a_n = 2n + 1$, for $n = 1, 2, 3, 4$.

Solution

$$a_n = 2n + 1$$
$$a_1 = 2(1) + 1 = 3$$
$$a_2 = 2(2) + 1 = 5$$
$$a_3 = 2(3) + 1 = 7$$
$$a_4 = 2(4) + 1 = 9$$

Thus, the sequence is 3, 5, 7, 9.

Since each term of the sequence in Example 1 is larger than the preceding term, it is called an **increasing sequence**.

EXAMPLE 2 Given $a_n = (2n + 3)/n^2$.

a) Find the first term in the sequence.

b) Find the third term in the sequence.

c) Find the fifth term in the sequence.

Solution

a) When $n = 1$, $a_1 = \dfrac{2(1) + 3}{1^2} = \dfrac{5}{1} = 5.$

b) When $n = 3$, $a_3 = \dfrac{2(3) + 3}{3^2} = \dfrac{9}{9} = 1.$

c) When $n = 5$, $a_5 = \dfrac{2(5) + 3}{5^2} = \dfrac{13}{25}.$

Note in Example 2 that since there is no restriction on n, a_n is the general term of an infinite sequence.

In Example 2, since each term of the sequence generated by $a_n = (2n + 3)/n^2$ will be smaller than the preceding term, the sequence is called a **decreasing sequence.**

EXAMPLE 3 Find the first four terms of the sequence whose general term is $a_n = (-1)^n(n)$.

Solution

$$a_n = (-1)^n(n)$$
$$a_1 = (-1)^1(1) = -1$$
$$a_2 = (-1)^2(2) = 2$$
$$a_3 = (-1)^3(3) = -3$$
$$a_4 = (-1)^4(4) = 4$$

If we write the sequence in Example 3, we get $-1, 2, -3, 4, \ldots, (-1)^n(n)$. Notice that each term alternates in sign. We call this an **alternating sequence.**

NOW TRY EXERCISE 21

2) Write a Series

A **series** is the sum of the terms of a sequence. A series may be finite or infinite, depending on whether the sequence it is based on is finite or infinite.

Examples

FINITE SEQUENCE

$$a_1, a_2, a_3, a_4, a_5$$

FINITE SERIES

$$a_1 + a_2 + a_3 + a_4 + a_5$$

INFINITE SEQUENCE

$$a_1, a_2, a_3, a_4, a_5, \ldots, a_n, \ldots$$

INFINITE SERIES

$$a_1 + a_2 + a_3 + a_4 + a_5 + \cdots + a_n + \cdots$$

EXAMPLE 4 Write the first eight terms of the sequence; then write the series that represents the sum of that sequence if

a) $a_n = \left(\frac{1}{2}\right)^n$ **b)** $a_n = \left(\frac{1}{2}\right)^{n-1}$

Solution **a)** We begin with $n = 1$; thus, the first eight terms of the sequence whose general term is $a_n = \left(\frac{1}{2}\right)^n$ are

$$\left(\frac{1}{2}\right)^1, \left(\frac{1}{2}\right)^2, \left(\frac{1}{2}\right)^3, \left(\frac{1}{2}\right)^4, \left(\frac{1}{2}\right)^5, \left(\frac{1}{2}\right)^6, \left(\frac{1}{2}\right)^7, \left(\frac{1}{2}\right)^8$$

or

$$\frac{1}{2}, \frac{1}{4}, \frac{1}{8}, \frac{1}{16}, \frac{1}{32}, \frac{1}{64}, \frac{1}{128}, \frac{1}{256}$$

The series that represents the sum of the sequence is

$$\frac{1}{2} + \frac{1}{4} + \frac{1}{8} + \frac{1}{16} + \frac{1}{32} + \frac{1}{64} + \frac{1}{128} + \frac{1}{256} = \frac{255}{256}$$

b) We again begin with $n = 1$; thus, the first eight terms of the sequence whose general term is $a_n = \left(\frac{1}{2}\right)^{n-1}$ are

$$\left(\frac{1}{2}\right)^{1-1}, \left(\frac{1}{2}\right)^{2-1}, \left(\frac{1}{2}\right)^{3-1}, \left(\frac{1}{2}\right)^{4-1}, \left(\frac{1}{2}\right)^{5-1}, \left(\frac{1}{2}\right)^{6-1}, \left(\frac{1}{2}\right)^{7-1}, \left(\frac{1}{2}\right)^{8-1}$$

or

$$\left(\frac{1}{2}\right)^0, \left(\frac{1}{2}\right)^1, \left(\frac{1}{2}\right)^2, \left(\frac{1}{2}\right)^3, \left(\frac{1}{2}\right)^4, \left(\frac{1}{2}\right)^5, \left(\frac{1}{2}\right)^6, \left(\frac{1}{2}\right)^7$$

or

$$1, \frac{1}{2}, \frac{1}{4}, \frac{1}{8}, \frac{1}{16}, \frac{1}{32}, \frac{1}{64}, \frac{1}{128}$$

The series that represents the sum of this sequence is

$$1 + \frac{1}{2} + \frac{1}{4} + \frac{1}{8} + \frac{1}{16} + \frac{1}{32} + \frac{1}{64} + \frac{1}{128} = \frac{255}{128} \quad \text{or} \quad 1\frac{127}{128}$$

3) Find Partial Sums of a Series

A **partial sum of a series** is the sum of a finite number of consecutive terms of the series, beginning with the first term.

$s_1 = a_1$ *First partial sum*

$s_2 = a_1 + a_2$ *Second partial sum*

$s_3 = a_1 + a_2 + a_3$ *Third partial sum*

$\vdots$

$s_n = a_1 + a_2 + a_3 + \cdots + a_n$ *nth partial sum*

EXAMPLE 5 Given the infinite sequence defined by $a_n = (1 + n^2)/n$, find

a) the first partial sum **b)** the third partial sum

Solution **a)** $s_1 = a_1 = \frac{1 + 1^2}{1} = \frac{1 + 1}{1} = 2$

b) $s_3 = a_1 + a_2 + a_3$

$$= \frac{1 + 1^2}{1} + \frac{1 + 2^2}{2} + \frac{1 + 3^2}{3}$$

$$= 2 + \frac{5}{2} + \frac{10}{3} = \frac{47}{6} \quad \text{or} \quad 7\frac{5}{6}$$

NOW TRY EXERCISE 33

4) Use Summation Notation, Σ

When the general term of a sequence is known, the Greek letter **sigma**, Σ, can be used to write a series. For example, the sum of the first five terms of the sequence 5, 7, 9, 11, 13, ..., $2n + 3$, ..., can be represented using the **summation notation**.

$$\sum_{n=1}^{5} (2n + 3)$$

This notation is read "the sum as n goes from 1 to 5 of $2n + 3$."

To evaluate the series represented by $\sum_{n=1}^{5} (2n + 3)$, we first substitute 1 for n in $2n + 3$ and list the value obtained. Then we substitute 2 for n in $2n + 3$ and list the value. We follow this procedure for the values 1 through 5. We then sum these values to obtain the series value.

$$\sum_{n=1}^{5} (2n + 3) = (2 \cdot 1 + 3) + (2 \cdot 2 + 3) + (2 \cdot 3 + 3) + (2 \cdot 4 + 3) + (2 \cdot 5 + 3)$$

$$= 5 + 7 + 9 + 11 + 13$$

$$= 45$$

The letter n used in the summation notation is called the **index of summation** or simply the *index*. Any letter can be used for the index. The 1 below the summation symbol is called the **lower limit**, and the 5 above the summation symbol is called the **upper limit** of the summation.

EXAMPLE 6 Write out the series $\sum_{k=1}^{6} (k^2 + 1)$ and evaluate it.

Solution

$$\sum_{k=1}^{6} (k^2 + 1) = (1^2 + 1) + (2^2 + 1) + (3^2 + 1) + (4^2 + 1) + (5^2 + 1) + (6^2 + 1)$$

$$= 2 + 5 + 10 + 17 + 26 + 37$$

$$= 97$$

NOW TRY EXERCISE 53

EXAMPLE 7 Consider the general term of a sequence $a_n = 2n^2 - 4$. Represent the third partial sum in summation notation.

Solution The third partial sum will be the sum of the first three terms, $a_1 + a_2 + a_3$. We can represent the third partial sum as $\sum_{n=1}^{3}(2n^2 - 4)$.

In the summation $\sum_{i=1}^{5} x_i$, the index is i. Note that

$$\sum_{i=1}^{5} x_i = x_1 + x_2 + x_3 + x_4 + x_5$$

We use this notation in Example 8.

EXAMPLE 8 For the following set of values $x_1 = 3, x_2 = 4, x_3 = 5, x_4 = 6$, and $x_5 = 7$, does $\sum_{i=1}^{5} x_i^2 = \left(\sum_{i=1}^{5} x_i\right)^2$?

Solution

$$\begin{aligned}\sum_{i=1}^{5} x_i^2 &= x_1^2 + x_2^2 + x_3^2 + x_4^2 + x_5^2 \\ &= 3^2 + 4^2 + 5^2 + 6^2 + 7^2 \\ &= 9 + 16 + 25 + 36 + 49 = 135\end{aligned}$$

$$\begin{aligned}\left(\sum_{i=1}^{5} x_i\right)^2 &= (x_1 + x_2 + x_3 + x_4 + x_5)^2 \\ &= (3 + 4 + 5 + 6 + 7)^2 = (25)^2 = 625\end{aligned}$$

NOW TRY EXERCISE 65 Since $135 \neq 625$, $\sum_{i=1}^{5} x_i^2 \neq \left(\sum_{i=1}^{5} x_i\right)^2$.

When a summation symbol is written without any upper and lower limits, it means that all the given data are to be summed.

EXAMPLE 9 A formula used to find the arithmetic mean, $\bar{x}$ (read x bar), of a set of data is $\bar{x} = \frac{\Sigma x}{n}$, where n is the number of pieces of data.

Joan Sally's five test grades are 70, 90, 83, 74, and 92. Find the arithmetic mean of her grades.

Solution

$$\bar{x} = \frac{\Sigma x}{n} = \frac{70 + 90 + 83 + 74 + 92}{5} = \frac{409}{5} = 81.8$$

Exercise Set 11.1

Concept/Writing Exercises

1. What is a sequence?
2. What is an infinite sequence?
3. What is a finite sequence?
4. What is an increasing sequence?
5. What is a decreasing sequence?
6. What is an alternating sequence?
7. What is a series?
8. What is the nth partial sum of a series?
9. Write the following notation in words: $\sum_{n=1}^{5}(n + 2)$.
10. Consider the summation $\sum_{k=1}^{5}(k + 2)$. **a)** What is the 1 called? **b)** What is the 5 called? **c)** What is the k called?

Practice the Skills

Write the first five terms of the sequence whose nth term is shown.

11. $a_n = 3n$
12. $a_n = 2n + 3$
13. $a_n = \dfrac{n+4}{n}$
14. $a_n = \dfrac{1}{n}$
15. $a_n = \dfrac{3}{n^2}$
16. $a_n = n^2 - n$
17. $a_n = \dfrac{n+2}{n+1}$
18. $a_n = \dfrac{n+2}{n+3}$
19. $a_n = (-1)^n$
20. $a_n = (-1)^{2n}$
21. $a_n = (-2)^{n+1}$
22. $a_n = n(n+2)$

Find the indicated term of the sequence whose nth term is shown.

23. $a_n = 2n + 7$, twelfth term
24. $a_n = 2^n$, seventh term
25. $a_n = 2n - 4$, fifth term
26. $a_n = (-1)^n$, eighth term
27. $a_n = (-2)^n$, fourth term
28. $a_n = n(n+5)$, eighth term
29. $a_n = \dfrac{n^2}{2n+1}$, ninth term
30. $a_n = \dfrac{n(n+1)}{n^2}$, tenth term

Find the first and third partial sums, s_1 and s_3, for each sequence.

31. $a_n = 2n + 3$
32. $a_n = \dfrac{3n}{n+2}$
33. $a_n = 2^n + 1$
34. $a_n = \dfrac{n-1}{n+1}$
35. $a_n = (-1)^{2n}$
36. $a_n = \dfrac{2n^2}{n+4}$
37. $a_n = \dfrac{n^2}{2}$
38. $a_n = \dfrac{n+3}{2n}$

Write the next three terms of each sequence.

39. 2, 4, 8, 16, 32, ...
40. $\frac{1}{2}, \frac{1}{3}, \frac{1}{4}, \frac{1}{5}, \ldots$
41. 5, 7, 9, 11, 13, ...
42. 5, 10, 15, 20, 25, ...
43. $1, \frac{1}{2}, \frac{1}{3}, \frac{1}{4}, \frac{1}{5}, \ldots$
44. $\frac{2}{3}, \frac{3}{4}, \frac{4}{5}, \frac{5}{6}, \frac{6}{7}, \ldots$
45. −1, 1, −1, 1, −1, ...
46. −10, −20, −30, −40, ...
47. $1, \frac{1}{3}, \frac{1}{9}, \frac{1}{27}, \ldots$
48. $\frac{1}{3}, \frac{2}{3}, \frac{3}{3}, \frac{4}{3}, \ldots$
49. $1, -\frac{1}{2}, \frac{1}{4}, -\frac{1}{8}, \ldots$
50. $\frac{2}{3}, \frac{1}{3}, \frac{1}{6}, \frac{1}{12}, \ldots$
51. 7, −1, −9, −17, ...
52. 37, 32, 27, 22, ...

Write out each series, then evaluate it.

53. $\sum_{n=1}^{5} (3n - 1)$
54. $\sum_{k=1}^{4} (k^2 - 1)$
55. $\sum_{k=1}^{6} (2k^2 - 3)$
56. $\sum_{i=1}^{4} \frac{i^2}{2}$
57. $\sum_{n=2}^{4} \frac{n^2 + n}{n+1}$
58. $\sum_{n=2}^{5} \frac{n^3}{n+1}$

For the given general term a_n, write an expression using Σ to represent the indicated partial sum.

59. $a_n = n + 3$, fifth partial sum
60. $a_n = n^2 + 1$, fourth partial sum
61. $a_n = \dfrac{n^2}{4}$, third partial sum
62. $a_n = \dfrac{n^2+1}{n+1}$, third partial sum

For the set of values $x_1 = 2$, $x_2 = 3$, $x_3 = 5$, $x_4 = -1$, and $x_5 = 4$, find each of the following.

63. $\sum_{i=1}^{5} x_i$
64. $\sum_{i=1}^{5} x_i^2$
65. $\left(\sum_{i=1}^{5} x_i\right)^2$
66. $\sum_{i=1}^{4} (x_i^2 + 3)$

Find the arithmetic mean $\overline{x}$, of the following sets of data.

67. 15, 20, 25, 30, 35
68. 16, 20, 96, 18, 25
69. 72, 83, 4, 60, 18, 20
70. 5, 12, 9, 12, 17, 36, 70

Problem Solving

71. Create your own sequence that is an increasing sequence and list the first five terms.

72. Create your own sequence that is a decreasing sequence and list the first five terms.

73. Create your own sequence that is an alternating sequence and list the first five terms.

74. Write **a)** $\sum_{i=1}^{n} x_i$ as a sum of terms and **b)** $\sum_{j=1}^{n} x_j$ as a sum of terms. **c)** For a given set of values of x, from x_1 to x_n, will $\sum_{i=1}^{n} x_i = \sum_{j=1}^{n} x_j$? Explain.

75. Solve $\bar{x} = \dfrac{\Sigma x}{n}$ for Σx.

76. Solve $\bar{x} = \dfrac{\Sigma x}{n}$ for n.

77. Is $\sum_{i=1}^{n} 2x_i = 2\sum_{i=1}^{n} x_i$? Illustrate your answer with an example.

78. Is $\sum_{i=1}^{n} \dfrac{x_i}{2} = \dfrac{1}{2}\sum_{i=1}^{n} x_i$? Illustrate your answer with an example.

79. Let $x_1 = 3, x_2 = 5, x_3 = 2$, and $y_1 = 4, y_2 = 1, y_3 = 6$. Find the following. Note that $\Sigma x = x_1 + x_2 + x_3$, $\Sigma y = y_1 + y_2 + y_3$, and $\Sigma xy = x_1 y_1 + x_2 y_2 + x_3 y_3$.
a) Σx **b)** Σy **c)** $\Sigma x \cdot \Sigma y$ **d)** Σxy **e)** Is $\Sigma x \cdot \Sigma y = \Sigma xy$?

Group Activity

Exercises 80 through 85 will give you more practice using Σ. It may also be helpful if you plan on taking a statistics course.

When no upper and lower limits are placed on a summation symbol, it indicates that the sum of all values is to be found. Consider the following values:

$$x_1 = 1, \quad x_2 = 3, \quad x_3 = 5, \quad x_4 = 7, \quad x_5 = 9$$
$$f_1 = 3, \quad f_2 = 4, \quad f_3 = 5, \quad f_4 = 0, \quad f_5 = 2$$

Then

$$\Sigma x = x_1 + x_2 + x_3 + x_4 + x_5 = 1 + 3 + 5 + 7 + 9 = 25$$
$$\Sigma f = f_1 + f_2 + f_3 + f_4 + f_5 = 3 + 4 + 5 + 0 + 2 = 14$$
$$\Sigma xf = x_1 f_1 + x_2 f_2 + x_3 f_3 + x_4 f_4 + x_5 f_5$$
$$= 1(3) + 3(4) + 5(5) + 7(0) + 9(2) = 58$$

A 10-point quiz is given to a statistics class of 20 students. The results of the quiz are indicated in the table. The grade on the quiz is represented by x and the frequency of the grade (or the number of students who obtained that grade) is represented by f. For example, from the table we see that two students received a grade of 10.

x	6	7	8	9	10
f	1	5	7	5	2

As a group, using the values in the table, find

80. Σf

81. Σxf

82. $\Sigma x^2 f$

83. $(\Sigma xf)^2$

84. The mean, $\bar{x}$, of a set of data may be found by the formula $\bar{x} = \dfrac{\Sigma xf}{n}$, where $n = \Sigma f$. As a group, use this formula to find the mean of this set of data.

85. Before working this exercise, work Exercises 80 through 83. The **standard deviation** of a set of data is a measure of the spread of the data from the mean of the set of data. A formula used to calculate standard deviation in statistics is

$$s = \sqrt{\frac{n(\Sigma x^2 f) - (\Sigma xf)^2}{n(n-1)}}$$

where $n = \Sigma f$.

a) Individually, find the standard deviation of the set of 20 quiz grades given in the table for Exercises 80 through 83. Then compare your answers.

b) Some calculators will determine the standard deviation of a set of data. Read your calculator manual and see if it has the ability to find standard deviations. If so, check your answer to part **a)** on your calculator.

Cumulative Review Exercises

[5.8] **86.** Solve the equation $2x^2 + 15 = 13x$ by factoring.

[8.2] **87.** How many real solutions does the equation $6x^2 - 3x - 4 = 2$ have? Explain how you obtained your answer.

[10.2] **88.** Graph the equation $\dfrac{x^2}{4} + \dfrac{y^2}{1} = 1$.

[10.4] **89.** Solve the following system of equations.

$$x^2 + y^2 = 5$$
$$x = 2y$$

11.2 ARITHMETIC SEQUENCES AND SERIES

SSM VIDEO 11.2 CD Rom

1) **Find the common difference in an arithmetic sequence.**
2) **Find the *n*th term of an arithmetic sequence.**
3) **Find the *n*th partial sum of an arithmetic series.**

1) Find the Common Difference in an Arithmetic Sequence

In the previous section we started our discussion by assuming you got a job with a starting salary of \$25,000. One option for salary increases was an increase of \$2000 each year. This would result in the sequence

$$\$25{,}000, \$27{,}000, \$29{,}000, \$31{,}000, \ldots$$

This is an example of an arithmetic sequence.

Definition

An **arithmetic sequence** is a sequence in which each term after the first differs from the preceding term by a constant amount.

The constant amount by which each pair of successive terms differs is called the **common difference**, d. The common difference can be found by subtracting any term from the term that directly follows it.

Arithmetic Sequence	Common Difference
$1, 4, 7, 10, 13, 16, \ldots$	$d = 4 - 1 = 3$
$-7, -2, 3, 8, 13, \ldots$	$d = -2 - (-7) = -2 + 7 = 5$
$\frac{7}{2}, \frac{2}{2}, -\frac{3}{2}, -\frac{8}{2}, -\frac{13}{2}, -\frac{18}{2}, \ldots$	$d = \frac{2}{2} - \frac{7}{2} = -\frac{5}{2}$

EXAMPLE 1 Write the first five terms of the arithmetic sequence with

a) first term 6 and common difference 3

b) first term 3 and common difference −2

Solution **a)** Start with 6 and keep adding 3. The sequence is 6, 9, 12, 15, 18.

b) 3, 1, −1, −3, −5

2) Find the *n*th Term of an Arithmetic Sequence

In general, an arithmetic sequence with first term a_1 and common difference d has the following terms:

$$a_1 = a_1, \quad a_2 = a_1 + d, \quad a_3 = a_1 + 2d, \quad a_4 = a_1 + 3d, \quad \text{and so on.}$$

If we continue this process, we can see that the nth term, a_n, can be found by the following formula:

nth Term of an Arithmetic Sequence

$$a_n = a_1 + (n - 1)d$$

EXAMPLE 2 **a)** Write an expression for the general (or nth) term, a_n, of the arithmetic sequence whose first term is -3 and whose common difference is 4.

b) Find the twelfth term of the sequence.

Solution **a)** The nth term of the sequence is $a_n = a_1 + (n-1)d$. Substituting $a_1 = -3$ and $d = 4$, we obtain

$$\begin{aligned} a_n &= a_1 + (n-1)d \\ &= -3 + (n-1)4 \\ &= -3 + 4(n-1) \\ &= -3 + 4n - 4 \\ &= 4n - 7 \end{aligned}$$

Thus, $a_n = 4n - 7$.

b) $a_n = 4n - 7$

NOW TRY EXERCISE 5

$$a_{12} = 4(12) - 7 = 48 - 7 = 41.$$

The twelfth term in the sequence is 41.

EXAMPLE 3 Find the number of terms in the arithmetic sequence 5, 9, 13, 17, …, 41.

Solution The first term, a_1, is 5; the nth term is 41, and the common dfference, d, is 4. Substitute the appropriate values in the formula for the nth term and solve for n.

$$\begin{aligned} a_n &= a_1 + (n-1)d \\ 41 &= 5 + (n-1)4 \\ 41 &= 5 + 4n - 4 \\ 41 &= 4n + 1 \\ 40 &= 4n \\ 10 &= n \end{aligned}$$

NOW TRY EXERCISE 17

The sequence has 10 terms.

3 Find the nth Partial Sum of an Arithmetic Series

An **arithmetic series** is the sum of the terms of an arithmetic sequence. A finite arithmetic series can be written

$$s_n = a_1 + (a_1 + d) + (a_1 + 2d) + (a_1 + 3d) + \cdots + (a_n - 2d) + (a_n - d) + a_n$$

If we consider the last term as a_n, the term before the last term will be $a_n - d$, the second before the last term will be $a_n - 2d$, and so on.

A formula for the nth partial sum, s_n, can be obtained by adding the reverse of s_n to itself.

$$\begin{aligned} s_n &= a_1 + (a_1 + d) + (a_1 + 2d) + \cdots + (a_n - 2d) + (a_n - d) + a_n \\ s_n &= a_n + (a_n - d) + (a_n - 2d) + \cdots + (a_1 + 2d) + (a_1 + d) + a_1 \\ \hline 2s_n &= (a_1 + a_n) + (a_1 + a_n) + (a_1 + a_n) + \cdots + (a_1 + a_n) + (a_1 + a_n) + (a_1 + a_n) \end{aligned}$$

Since the right side of the equation contains n terms of $(a_1 + a_n)$, we can write

$$2s_n = n(a_1 + a_n)$$

Now divide both sides of the equation by 2 to obtain the following formula.

nth Partial Sum of an Arithmetic Series

$$s_n = \frac{n(a_1 + a_n)}{2}$$

EXAMPLE 4 Find the sum of the first 25 natural numbers.

Solution The arithmetic sequence is 1, 2, 3, 4, 5, 6, ..., 25. The first term, a_1, is 1; the last term, a_n, is 25. There are 25 terms; thus, $n = 25$. Using the formula for the nth partial sum, we have

$$s_n = \frac{n(a_1 + a_n)}{2} = \frac{25(1 + 25)}{2} = \frac{25(26)}{2} = 25(13) = 325$$

The sum of the first 25 natural numbers is 325. Thus, $s_{25} = 325$.

EXAMPLE 5 The first term of an arithmetic sequence is 4, and the last term is 31. If $s_n = 175$, find the number of terms in the sequence and the common difference.

Solution We substitute the appropriate values, $a_1 = 4$, $a_n = 31$, and $s_n = 175$, in the formula for the nth partial sum and solve for n.

$$s_n = \frac{n(a_1 + a_n)}{2}$$

$$175 = \frac{n(4 + 31)}{2}$$

$$175 = \frac{35n}{2}$$

$$350 = 35n$$

$$10 = n$$

There are 10 terms in the sequence. We can now find the common difference.

$$a_n = a_1 + (n - 1)d$$

$$31 = 4 + (10 - 1)d$$

$$31 = 4 + 9d$$

$$27 = 9d$$

$$3 = d$$

NOW TRY EXERCISE 37

The common difference is 3. The sequence is 4, 7, 10, 13, 16, 19, 22, 25, 28, 31.

Examples 6 and 7 illustrate some applications of arithmetic sequences and series.

EXAMPLE 6 Lori Sullivan is given a starting salary of \$25,000 and is promised a \$1200 raise after each of the next 8 years. Find her salary during her eighth year of work.

Solution **Understand** Her salaries during the first few years would be

$$\$25{,}000, \$26{,}200, \$27{,}400, \$28{,}600, \ldots$$

Since we are adding a constant amount each year, this is an arithmetic sequence. The general term of an arithmetic sequence is $a_n = a_1 + (n - 1)d$.

Translate In this example, $a_1 = 25{,}000$ and $d = 1200$. Thus, for $n = 8$, Lori's salary would be

$$a_8 = 25{,}000 + (8 - 1)1200$$

Carry Out

$$= 25{,}000 + 7(1200)$$
$$= 25{,}000 + 8400$$
$$= 33{,}400$$

Answer During her eighth year of work Lori's salary would be \$33,400. If we listed all the salaries for the 8-year period, they would be \$25,000, \$26,200, \$27,400, \$28,600, \$29,800, \$31,000, \$32,200, \$33,400.

EXAMPLE 7 Each swing of a pendulum (left to right or right to left) is 3 inches shorter than the preceding swing. The first swing is 8 feet.

a) Find the length of the twelfth swing.

b) Determine the distance traveled by the pendulum during the first 12 swings.

Solution **a) Understand** Since each swing is decreasing by a constant amount, this problem can be represented as an arithmetic series. Since the first swing is given in feet and the decrease in swing in inches, we will change 3 inches to 0.25 feet ($3 \div 12 = 0.25$). The twelfth swing can be considered a_{12}. The difference, d, is negative since the distance is decreasing with each swing.

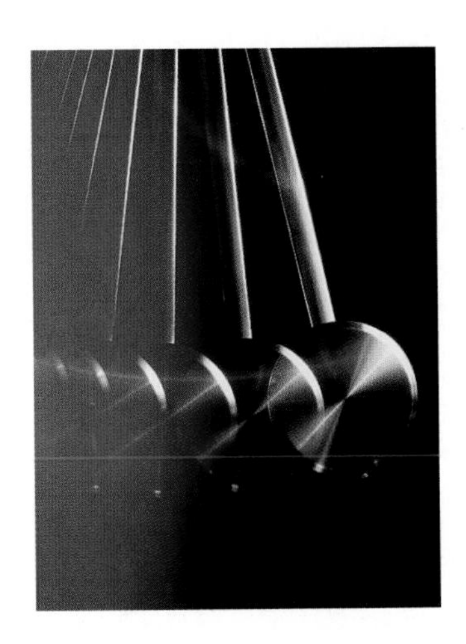

Translate

$$a_n = a_1 + (n - 1)d$$
$$a_{12} = 8 + (12 - 1)(-0.25)$$

Carry Out

$$= 8 + 11(-0.25)$$
$$= 8 - 2.75$$
$$= 5.25 \text{ feet}$$

Answer The twelfth swing is 5.25 feet.

b) Understand and Translate The distance traveled during the first 12 swings can be found using the formula for the nth partial sum The first swing, a_1, is 8 feet and the twelfth swing, a_{12}, is 5.25 feet.

$$s_n = \frac{n(a_1 + a_n)}{2}$$

$$s_{12} = \frac{12(a_1 + a_{12})}{2}$$

Carry Out

$$= \frac{12(8 + 5.25)}{2} = \frac{12(13.25)}{2} = 6(13.25) = 79.5 \text{ feet}$$

NOW TRY EXERCISE 59 **Answer** The pendulum travels 79.5 feet during its first 12 swings.

Exercise Set 11.2

Concept/Writing Exercises

1. What is an arithmetic sequence?
2. What is an arithmetic series?
3. What do we call the constant amount by which each pair of successive terms in an arithmetic sequence differs?
4. How can the common difference in an arithmetic sequence be found?

Practice the Skills

Write the first five terms of the arithmetic sequence with the given first term and common difference. Write the expression for the general (or nth) term, a_n, of the arithmetic sequence.

5. $a_1 = 4, d = 3$
6. $a_1 = 7, d = 2$
7. $a_1 = -5, d = 2$
8. $a_1 = -8, d = -3$
9. $a_1 = \frac{1}{2}, d = \frac{3}{2}$
10. $a_1 = -\frac{5}{3}, d = -\frac{1}{3}$
11. $a_1 = 100, d = -5$
12. $a_1 = \frac{5}{4}, d = -\frac{3}{4}$

Find the indicated quantity of the arithmetic sequence.

13. $a_1 = 5, d = 3$; find a_4
14. $a_1 = 8, d = -2$; find a_6
15. $a_1 = -6, d = -1$; find a_{18}
16. $a_1 = -15, d = 3$; find a_{20}
17. $a_1 = -2, d = \frac{5}{3}$; find a_{10}
18. $a_1 = 5, a_8 = -21$; find d
19. $a_1 = 3, a_9 = 19$; find d
20. $a_1 = \frac{1}{2}, a_7 = \frac{19}{2}$; find d
21. $a_1 = 4, a_n = 28, d = 3$; find n
22. $a_1 = -2, a_n = -20, d = -3$; find n
23. $a_1 = -\frac{7}{3}, a_n = -\frac{17}{3}, d = -\frac{2}{3}$; find n
24. $a_1 = 100, a_n = 60, d = -8$; find n

Find the sum, s_n, and common difference, d, of each sequence.

25. $a_1 = 1, a_{10} = 19, n = 10$
26. $a_1 = -5, a_7 = 13, n = 7$
27. $a_1 = \frac{3}{5}, a_8 = 2, n = 8$
28. $a_1 = 12, a_8 = -23, n = 8$
29. $a_1 = \frac{12}{5}, a_5 = \frac{28}{5}, n = 5$
30. $a_1 = -3, a_6 = 15.5, n = 6$
31. $a_1 = 7, a_{11} = 67, n = 11$
32. $a_1 = 14.25, a_{31} = 18.75, n = 31$

Write the first four terms of each sequence; then find a_{10} and s_{10}.

33. $a_1 = 5, d = 3$
34. $a_1 = -4, d = -2$
35. $a_1 = -8, d = -5$
36. $a_1 = \frac{7}{2}, d = \frac{5}{2}$
37. $a_1 = 100, d = -7$
38. $a_1 = -15, d = 4$
39. $a_1 = \frac{9}{5}, d = \frac{3}{5}$
40. $a_1 = 35, d = 3$

Find the number of terms in each sequence and find s_n.

41. $1, 4, 7, 10, \ldots, 43$
42. $-8, -6, -4, -2, \ldots, 42$

43. $-9, -5, -1, 3, \ldots, 27$

44. $\frac{1}{2}, \frac{2}{2}, \frac{3}{2}, \frac{4}{2}, \frac{5}{2}, \ldots, \frac{17}{2}$

45. $-\frac{5}{6}, -\frac{7}{6}, -\frac{9}{6}, -\frac{11}{6}, \ldots, -\frac{21}{6}$

46. $7, 14, 21, 28, \ldots, 63$

47. $-12, -16, -20, \ldots, -52$

48. $9, 12, 15, 18, \ldots, 93$

Problem Solving

49. Find the sum of the first 50 natural numbers.

50. Find the sum of the first 50 even numbers.

51. Find the sum of the first 50 odd numbers.

52. Find the sum of the first 30 multiples of 5.

53. Find the sum of the first 20 multiples of 3.

54. Find the sum of the numbers between 50 and 200, inclusive.

55. Determine how many numbers between 7 and 1610 are divisible by 6.

56. Determine how many numbers between 12 and 1470 are divisible by 8.

57. Karl Friedrich Gauss (1777–1855), a famous mathematician, as a child found the sum of the first 100 natural numbers quickly in his head $(1 + 2 + 3 + \cdots + 100)$. Explain how he might have done this and find the sum of the first 100 natural numbers as you think Gauss might have. (*Hint:* $1 + 100 = 101$, $2 + 99 = 101$, etc.)

58. Find a formula for the sum of the first n consecutive odd numbers starting with 1.

$$1 + 3 + 5 + \cdots + (2n - 1)$$

59. A long vine is attached to the branch of a tree. Jane Hyatt swings from the vine, and each swing (left to right or right to left) is $\frac{1}{2}$ foot less than the previous swing. If her first swing is 22 feet, find **a)** the length of the seventh swing and **b)** the distance traveled during her seven swings.

60. Each swing of a pendulum is 2 inches shorter than the previous swing (left to right or right to left). The first swing is 6 feet. Find **a)** the length of the eighth swing and **b)** the distance traveled by the pendulum during the eight swings.

61. Beth Casey drops a ball from a second-story window. Each time the ball bounces, the height attained is 6 inches less than on the previous bounce. If the first bounce reaches a height of 6 feet, find the height attained on the eleventh bounce.

62. Suppose when a ping-pong ball falls from the table it bounces to a height of 3 feet. If each successive bounce is 3 inches less than the previous bounce, find the height attained on the twelfth bounce.

63. Yamil Bermudez stacks logs so that there are 20 logs in the bottom layer, and each layer contains one log less than the layer below it. How many logs are in the pile?

64. Suppose Yamil, in Exercise 63, stopped stacking the logs after completing the layer containing 8 logs. How many logs are in the pile?

65. If Craig Campanella saves \$1 on day 1, \$2 on day 2, \$3 on day 3, and so on, how much money, in total, will he have saved on day 31?

66. If Dan Currier saves 50 cents on day 1, \$1.00 on day 2, \$1.50 on day 3, and so on, how much, in total, will he have saved by the end of one year (365 days)?

67. Susan Cottendon recently retired and met with her financial planner. She arranged to receive \$32,000 the first year. Because of inflation, each year she will get \$400 more than she received the previous year.

a) What income will she receive in her tenth year of retirement?

b) How much money will she have received in total during her first 10 years of retirement?

68. Christie Catalano is given a starting salary of $20,000 and is told she will receive a $1000 raise at the end of each year.

a) Find her salary during year 12.

b) How much will she receive in total during her first 12 years?

69. The sum of the interior angles of a triangle, a quadrilateral, a pentagon, and a hexagon are 180°, 360°, 540°, and 720°, respectively. Use the pattern here to find the formula for the sum of the interior angles of a polygon with n sides.

70. Another formula that may be used to find the nth partial sum of an arithmetic series is

$$s_n = \frac{n}{2}[2a_1 + (n-1)d]$$

Derive this formula using the two formulas presented in this section.

Group Activity

In calculus, a topic of importance is limits. Consider $a_n = \frac{1}{n}$. The first five terms of this sequence are $\frac{1}{1}, \frac{1}{2}, \frac{1}{3}, \frac{1}{4}, \frac{1}{5}$. Since the value of $\frac{1}{n}$ gets closer and closer to 0 as n gets larger and larger, we say that the limit of $\frac{1}{n}$ as n approaches infinity is 0. We write this as $\lim_{n\to+\infty} \frac{1}{n} = 0$ or $\lim_{n\to+\infty} a_n = 0$. Notice that $\frac{1}{n}$ can never equal 0, but its value approaches 0 as n gets larger and larger.

a) *Group member 1: Find $\lim_{n\to+\infty} a_n$ for Exercises 71 and 72.*

b) *Group member 2: Find $\lim_{n\to+\infty} a_n$ for Exercises 73 and 74.*

c) *Group member 3: Find $\lim_{n\to+\infty} a_n$ for Exercises 75 and 76.*

d) *Exchange work and check each other's answers.*

71. $a_n = \dfrac{1}{n-2}$

72. $a_n = \dfrac{n}{n+1}$

73. $a_n = \dfrac{1}{n^2+2}$

74. $a_n = \dfrac{2n+1}{n}$

75. $a_n = \dfrac{4n-3}{3n+1}$

76. $a_n = \dfrac{n^2}{n+1}$

Cumulative Review Exercises

[4.1] **77.** Consider the following system of equations.

$$2x + 3y = -4$$
$$-x - y = -1$$

Without solving the system, determine how many solutions the system has. Explain.

[4.1] **78.** Solve the system of equations in Exercise 77.

[3.7] **79.** Graph $|x - 2| < 4$ in the Cartesian coordinate system.

[9.4] **80.** Graph $\dfrac{(x+2)^2}{4} + \dfrac{(y-3)^2}{9} = 1$.

11.3 GEOMETRIC SEQUENCES AND SERIES

SSM

VIDEO 11.3

CD Rom

1. Find the common ratio in a geometric sequence.
2. Find the nth term of a geometric sequence.
3. Find the nth partial sum of a geometric series.
4. Identify infinite geometric series.
5. Find the sum of an infinite geometric series.
6. Study applications of geometric series.

1) Find the Common Ratio in a Geometric Sequence

In Section 11.1 we assumed you got a job with a starting salary of $25,000. We also mentioned that an option for salary increases was a 5% salary increase each year. This would result in the following sequence.

$25,000, $26,250, $27,562.50, $28,940.63, . . .

This is an example of a geometric sequence.

Definition

A **geometric sequence** is a sequence in which each term after the first is a multiple of the preceding term.

The common multiple is called the **common ratio.**

The common ratio, r, in any geometric sequence can be found by dividing any term, except the first, by the preceding term. The common ratio of the previous geometric sequence is 1.05 (or 105%).

Consider the geometric sequence

$$1, 3, 9, 27, 81, \ldots, 3^{n-1}, \ldots$$

The common ratio is 3 since $3 \div 1 = 3$ (or $9 \div 3 = 3$, and so on).

Geometric Sequence	Common Ratio
$4, 8, 16, 32, 64, \ldots, 4(2^{n-1}), \ldots$	2
$3, 12, 48, 192, 768, \ldots, 3(4^{n-1}), \ldots$	4
$7, \frac{7}{2}, \frac{7}{4}, \frac{7}{8}, \frac{7}{16}, \ldots, 7\left(\frac{1}{2}\right)^{n-1}, \ldots$	$\frac{1}{2}$

EXAMPLE 1 Determine the first five terms of the geometric sequence if $a_1 = 4$ and $r = \frac{1}{2}$.

Solution $a_1 = 4, \quad a_2 = 4 \cdot \frac{1}{2} = 2, \quad a_3 = 2 \cdot \frac{1}{2} = 1, \quad a_4 = 1 \cdot \frac{1}{2} = \frac{1}{2}, \quad a_5 = \frac{1}{2} \cdot \frac{1}{2} = \frac{1}{4}$

Thus, the first five terms of the geometric sequence are

$$4, 2, 1, \frac{1}{2}, \frac{1}{4}$$

2) Find the nth Term of a Geometric Sequence

In general, a geometric sequence with first term a_1 and common ratio r has the following terms:

$$a_1, \quad a_1 r, \quad a_1 r^2, \quad a_1 r^3, \quad a_1 r^4, \ldots, a_1 r^{n-1}, \ldots$$

a_1	$a_1 r$	$a_1 r^2$	$a_1 r^3$	$a_1 r^4$	$a_1 r^{n-1}$
↑	↑	↑	↑	↑	↑
1st term	2nd term	3rd term	4th term	5th term	nth term

Thus, we can see that the nth term of a geometric sequence is given by the following formula:

nth Term of a Geometric Sequence

$$a_n = a_1 r^{n-1}$$

EXAMPLE 2 **a)** Write an expression for the general (or nth) term, a_n, of the geometric sequence with $a_1 = 3$ and $r = -2$.

b) Find the twelfth term of this sequence.

Solution **a)** The nth term of the sequence is $a_n = a_1 r^{n-1}$. Substituting $a_1 = 3$ and $r = -2$, we obtain

$$a_n = a_1 r^{n-1} = 3(-2)^{n-1}$$

Thus, $a_n = 3(-2)^{n-1}$.

b)

$$a_n = 3(-2)^{n-1}$$

$$a_{12} = 3(-2)^{12-1} = 3(-2)^{11} = 3(-2048) = -6144$$

The twelfth term of the sequence is −6144. The first twelve terms of the sequence are 3, −6, 12, −24, 48, −96, 192, −384, 768, −1536, 3072, −6144.

NOW TRY EXERCISE 19

EXAMPLE 3 Find r and a_1 for the geometric sequence with $a_2 = 24$ and $a_5 = 648$.

Solution The sequence can be represented with blanks for the missing terms.

$$__,\ \underset{\underset{a_2}{\uparrow}}{24},\ __,\ __,\ \underset{\underset{a_5}{\uparrow}}{648}$$

If we assume that a_2 is the first term of a sequence with the same common ratio, we obtain

$$\underset{\underset{\substack{\text{1st}\\\text{term}}}{\uparrow}}{24},\ __,\ __,\ \underset{\underset{\substack{\text{4th}\\\text{term}}}{\uparrow}}{648}$$

Now we use the formula. We let the first term, a_1, be 24 and the number of terms, n, be 4.

$$a_n = a_1 r^{n-1}$$

$$648 = 24r^{4-1}$$

$$648 = 24r^3$$

$$\frac{648}{24} = r^3$$

$$27 = r^3$$

$$3 = r$$

Thus, the common ratio is 3.

The first term of the original sequence must be $24 \div 3$ or 8. Thus, $a_1 = 8$. The first term, a_1, could also be found using the formula with $a_n = 648$, $r = 3$, and $n = 5$. Find a_1 by the formula now.

NOW TRY EXERCISE 61

3) Find the nth Partial Sum of a Geometric Series

A **geometric series** is the sum of the terms of a geometric sequence. The sum of the first n terms, s_n, of a geometric sequence can be expressed as

$$s_n = a_1 + a_1 r + a_1 r^2 + a_1 r^3 + \cdots + a_1 r^{n-2} + a_1 r^{n-1} \qquad (eq.\ 1)$$

If we multiply both sides of the equation by r, we obtain

$$rs_n = a_1 r + a_1 r^2 + a_1 r^3 + \cdots + a_1 r^{n-1} + a_1 r^n \qquad (eq.\ 2)$$

Now we subtract the corresponding sides of $(eq.\ 2)$ from $(eq.\ 1)$. The red colored terms drop out, leaving

$$s_n - rs_n = a_1 - a_1r^n$$

Now we solve the equation for s_n.

$$s_n(1 - r) = a_1(1 - r^n) \qquad \text{Factor.}$$

$$s_n = \frac{a_1(1 - r^n)}{1 - r} \qquad \text{Divide both sides by } 1 - r.$$

Thus, we have the following formula for the nth partial sum of a geometric series.

nth Partial Sum of a Geometric Series

$$s_n = \frac{a_1(1 - r^n)}{1 - r}, \qquad r \neq 1$$

EXAMPLE 4 Find the seventh partial sum of a geometric series whose first term is 8 and whose common ratio is $\frac{1}{2}$.

Solution

$$s_n = \frac{a_1(1 - r^n)}{1 - r}$$

$$s_7 = \frac{8\left[1 - \left(\frac{1}{2}\right)^7\right]}{1 - \frac{1}{2}} = \frac{8\left(1 - \frac{1}{128}\right)}{\frac{1}{2}} = \frac{8\left(\frac{127}{128}\right)}{\frac{1}{2}} = \frac{127}{16} \cdot \frac{2}{1} = \frac{127}{8}$$

Thus, $s_7 = \frac{127}{8}$.

EXAMPLE 5 Given $s_n = 93$, $a_1 = 3$, and $r = 2$, find n.

Solution

$$s_n = \frac{a_1(1 - r^n)}{1 - r}$$

$$93 = \frac{3(1 - 2^n)}{1 - 2}$$

$$93 = \frac{3(1 - 2^n)}{-1}$$

$$-93 = 3(1 - 2^n)$$

$$-31 = 1 - 2^n$$

$$-32 = -2^n$$

$$32 = 2^n$$

$$2^5 = 2^n$$

Therefore, $n = 5$.

4 Identify Infinite Geometric Series

All the geometric sequences that we have examined thus far have been finite since they have had a last term. The following sequence is an example of an infinite geometric sequence.

$$1, \frac{1}{2}, \frac{1}{4}, \frac{1}{8}, \frac{1}{16}, \ldots, \left(\frac{1}{2}\right)^{n-1}, \ldots$$

Note that the three dots at the end of the sequence indicate that the sequence continues indefinitely. The sum of the terms in an infinite geometric sequence forms an **infinite geometric series**. For example,

$$1 + \frac{1}{2} + \frac{1}{4} + \frac{1}{8} + \frac{1}{16} + \cdots + \left(\frac{1}{2}\right)^{n-1} + \cdots$$

is an infinite geometric series. Let's find some partial sums.

Partial Sum	Series	Sum
Second	$1 + \frac{1}{2}$	1.5
Third	$1 + \frac{1}{2} + \frac{1}{4}$	1.75
Fourth	$1 + \frac{1}{2} + \frac{1}{4} + \frac{1}{8}$	1.875
Fifth	$1 + \frac{1}{2} + \frac{1}{4} + \frac{1}{8} + \frac{1}{16}$	1.9375
Sixth	$1 + \frac{1}{2} + \frac{1}{4} + \frac{1}{8} + \frac{1}{16} + \frac{1}{32}$	1.96875

With each successive partial sum, the amount being added is less than with the previous partial sum. Also, the sum seems to be getting closer and closer to 2. In Example 6, we will show that the sum of this infinite geometric series is indeed 2.

5) Find the Sum of an Infinite Geometric Series

Consider the formula for the sum of the first n terms of an infinite geometric series:

$$s_n = \frac{a_1(1 - r^n)}{1 - r}, \qquad r \neq 1$$

What happens to r^n if $|r| < 1$ and n gets larger and larger? Suppose that $r = \frac{1}{2}$; then

$$\left(\frac{1}{2}\right)^1 = 0.5, \quad \left(\frac{1}{2}\right)^2 = 0.25, \quad \left(\frac{1}{2}\right)^3 = 0.125, \quad \left(\frac{1}{2}\right)^{20} \approx 0.000001$$

We can see that when $|r| < 1$ the value of r^n gets exceedingly close to 0 as n gets larger and larger. Thus, when considering the sum of an infinite geometric series, symbolized s_∞, the expression r^n approaches 0 when $|r| < 1$. Therefore, replacing r^n with 0 in the formula $s_n = \frac{a_1(1 - r^n)}{1 - r}$ leads to the following formula.

Sum of an Infinite Geometric Series

$$s_\infty = \frac{a_1}{1 - r} \quad \text{where} \quad |r| < 1$$

EXAMPLE 6 Find the sum of the terms of the infinite sequence $1, \frac{1}{2}, \frac{1}{4}, \frac{1}{8}, \ldots$.

Solution $a_1 = 1$ and $r = \frac{1}{2}$. Note that $\left|\frac{1}{2}\right| < 1$.

$$s_\infty = \frac{a_1}{1-r} = \frac{1}{1-\frac{1}{2}} = \frac{1}{\frac{1}{2}} = 2$$

Thus, $1 + \frac{1}{2} + \frac{1}{4} + \frac{1}{8} + \frac{1}{16} + \cdots + \left(\frac{1}{2}\right)^{n-1} + \cdots = 2.$

EXAMPLE 7 Find the sum of the infinite geometric series

$$5 - 2 + \frac{4}{5} - \frac{8}{25} + \frac{16}{125} - \frac{32}{625} + \cdots$$

Solution The terms of the corresponding sequence are $5, -2, \frac{4}{5}, -\frac{8}{25}, \ldots$. Note that $a_1 = 5$.

$$r = -2 \div 5 = -\frac{2}{5}$$

Since $\left|-\frac{2}{5}\right| < 1$,

$$s_\infty = \frac{a_1}{1-r}$$

$$= \frac{5}{1-\left(-\frac{2}{5}\right)} = \frac{5}{1+\frac{2}{5}} = \frac{5}{\frac{7}{5}} = \frac{25}{7}$$

NOW TRY EXERCISE 47

EXAMPLE 8 Write 0.343434 ... as a ratio of integers.

Solution We can write this decimal as

$$0.34 + 0.0034 + 0.000034 + \cdots + (0.34)(0.01)^{n-1} + \cdots$$

This is an infinite geometric series with $r = 0.01$. Since $|r| < 1$,

$$s_\infty = \frac{a_1}{1-r} = \frac{0.34}{1-0.01} = \frac{0.34}{0.99} = \frac{34}{99}$$

NOW TRY EXERCISE 57

If you divide 34 by 99 on a calculator, you will see .34343434 displayed.

What is the sum of a geometric series when $|r| > 1$? Consider the geometric sequence in which $a_1 = 1$ and $r = 2$.

$$1, 2, 4, 8, 16, 32, \ldots, 2^{n-1}, \ldots$$

The sum of its terms is

$$1 + 2 + 4 + 8 + 16 + 32 + \cdots + 2^{n-1} + \cdots$$

What is the sum of this series? As n gets larger and larger, the sum gets larger and larger. We therefore say that the sum "does not exist." For $|r| > 1$, the sum of an infinite geometric series does not exist.

6 Study Applications of Geometric Series

Now let's look at some applications of geometric series.

EXAMPLE 9 Todd Morency invests \$1000 at 8% interest compounded annually in a savings account. Determine the amount in his account and the amount of interest earned at the end of 6 years.

Solution **Understand** Suppose we let P represent any principal invested. At the beginning of the second year the amount grows to $P + 0.08P$ or $1.08P$. This amount will be the principal invested for year 2. At the beginning of the third year the second year's principal will grow by 8% to $(1.08P)(1.08)$, or $(1.08)^2P$. The amount in Todd's account at the beginning of successive years is

Year 1	Year 2	Year 3	Year 4
P	$1.08P$	$(1.08)^2P$	$(1.08)^3P$

and so on. This is a geometric series with $r = 1.08$. The amount in his account at the end of 6 years will be the same as the amount in his account at the beginning of year 7. We will therefore use the formula

$$a_n = a_1 r^{n-1}, \quad \text{with } r = 1.08 \text{ and } n = 7$$

Translate We have a geometric series with $a_1 = 1000$, $r = 1.08$, and $n = 7$. Substituting these values into the formula, we obtain the following.

$$a_n = a_1 r^{n-1}$$

$$a_n = 1000(1.08)^{n-1}$$

Carry Out

$$a_7 = 1000(1.08)^{7-1}$$

$$= 1000(1.08)^6$$

$$\approx 1000(1.58687)$$

$$\approx 1586.87$$

Answer After 6 years the amount in the account is \$1586.87. The amount of interest earned is \$1586.87 − \$1000 = \$586.87.

EXAMPLE 10 Suppose someone offered you \$1000 a day for each day of a 30-day month. Or, you could elect to take a penny on day 1, 2¢ on day 2, 4¢ on day 3, 8¢ on day 4, and so on. The amount would continue to double each day for 30 days.

a) Without doing any calculations, take a guess at which of the two offerings would provide the greater total return for 30 days.

b) Calculate the total amount you would receive by selecting \$1000 a day for 30 days.

c) Calculate the amount you would receive on day 30 by selecting 1¢ on day 1 and doubling the amount each day for 30 days.

d) Calculate the total amount you would receive for 30 days by selecting 1¢ on day 1 and doubling the amount each day for 30 days.

Solution **a)** Each of you will have your own answer to part **a)**.

b) If you received \$1000 a day for 30 days, you would receive 30(\$1000) = \$30,000.

c) Understand Since the amount is doubled each day, this represents a geometric sequence with $r = 2$. The chart that follows shows the amount you would receive in each of the first 7 days. We also show the amounts written with base 2, the common ratio.

Day	1	2	3	4	5	6	7
Amount (cents)	1	2	4	8	16	32	64
Amount (cents)	2^0	2^1	2^2	2^3	2^4	2^5	2^6

Notice that for any given day, the exponent on 2 is 1 less than the given day. For example, on day 7, the amount is 2^6. In general, the amount on day n is 2^{n-1}.

Translate To find the amount received on day 30, we evaluate $a_n = a_1 r^{n-1}$ for $n = 30$.

$$a_n = a_1 r^{n-1}$$
$$a_{30} = 1(2)^{30-1}$$

Carry Out

$$a_{30} = 1(2)^{29}$$
$$= 1(536{,}870{,}912)$$
$$= 536{,}870{,}912$$

Answer On day 30, the amount that you would receive is 536,870,912 cents or $5,368,709.12

d) Understand and Translate To find the total amount received over the 30 days, we find the 30th partial sum.

$$s_n = \frac{a_1(1 - r^n)}{1 - r}$$
$$s_{30} = \frac{1(1 - 2^{30})}{1 - 2}$$

Carry Out

$$= \frac{1(1 - 1{,}073{,}741{,}824)}{-1}$$
$$= 1{,}073{,}741{,}823$$

Answer Therefore, over 30 days the total amount you would receive by this method would be 1,073,741,823 cents or $10,737,418.23. The amount received by this method greatly surpasses the $30,000 received by selecting $1000 a day for 30 days.

EXAMPLE 11

On each swing (left to right or right to left), a certain pendulum travels 90% as far as on its previous swing. For example, if the swing to the right is 10 feet, the swing back to the left is $0.9 \times 10 = 9$ feet (see Fig. 11.1). If the first swing is 10 feet, determine the total distance traveled by the pendulum by the time it comes to rest.

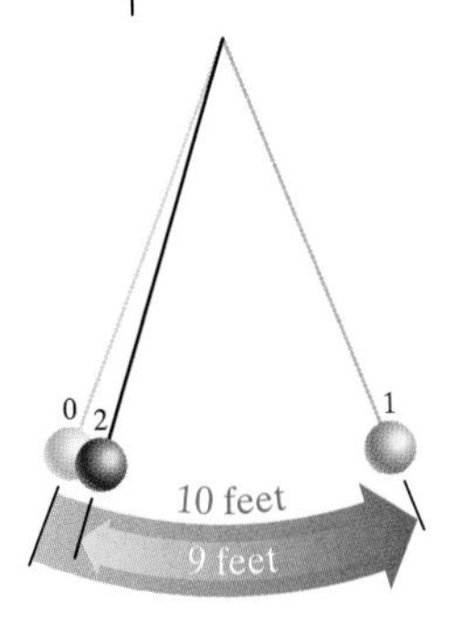

FIGURE 11.1

Solution **Understand** This problem may be considered an infinite geometric series with $a_1 = 10$ and $r = 0.9$. We can therefore use the formula $s_\infty = \dfrac{a_1}{1 - r}$ to find the total distance traveled by the pendulum.

Translate and Carry Out

$$s_\infty = \frac{a_1}{1 - r} = \frac{10}{1 - 0.9} = \frac{10}{0.1} = 100 \text{ feet}$$

Answer By the time the pendulum comes to rest it has traveled 100 feet.

NOW TRY EXERCISE 75

Exercise Set 11.3

Concept/Writing Exercises

1. What is a geometric sequence?
2. What is a geometric series?
3. Explain how to find the common ratio in a geometric sequence.
4. What is an infinite geometric series?
5. In a geometric series, if $|r| < 1$, what does r^n approach as n gets larger and larger?
6. What is the sum of an infinite geometric series when $|r| > 1$?

Practice the Skills

Determine the first five terms of each geometric sequence.

7. $a_1 = 4, r = 3$
8. $a_1 = 6, r = \frac{1}{2}$
9. $a_1 = 90, r = \frac{1}{3}$
10. $a_1 = -12, r = -1$
11. $a_1 = -5, r = -2$
12. $a_1 = 1, r = -\frac{1}{2}$
13. $a_1 = 3, r = \frac{3}{2}$
14. $a_1 = 60, r = \frac{1}{3}$

Find the indicated term of each geometric sequence.

15. $a_1 = 4, r = 2$; find a_6
16. $a_1 = -12, r = \frac{1}{2}$; find a_{10}
17. $a_1 = 15, r = 3$; find a_7
18. $a_1 = -20, r = -2$; find a_{10}
19. $a_1 = 2, r = \frac{1}{2}$; find a_8
20. $a_1 = 5, r = \frac{2}{3}$; find a_9
21. $a_1 = -3, r = -2$; find a_{12}
22. $a_1 = 80, r = \frac{1}{3}$; find a_{12}

Find the indicated sum.

23. $a_1 = 4, r = 5$; find s_6
24. $a_1 = 9, r = \frac{1}{2}$; find s_6
25. $a_1 = 80, r = 2$; find s_7
26. $a_1 = 1, r = -2$; find s_{12}
27. $a_1 = -30, r = -\frac{1}{2}$; find s_9
28. $a_1 = \frac{3}{5}, r = 3$; find s_7
29. $a_1 = -9, r = \frac{2}{5}$; find s_5
30. $a_1 = 35, r = \frac{1}{5}$; find s_{12}

For each geometric sequence, find the common ratio, r, and then write an expression for the general (or nth) term, a_n.

31. $5, \frac{5}{2}, \frac{5}{4}, \frac{5}{8}, \ldots$
32. $3, 9, 27, 81, \ldots$
33. $2, -6, 18, -54, \ldots$
34. $\frac{3}{4}, \frac{1}{2}, \frac{1}{3}, \frac{2}{9}$
35. $-1, -3, -9, -18, \ldots$
36. $\frac{4}{3}, \frac{8}{3}, \frac{16}{3}, \frac{32}{3}, \ldots$

Find the sum of the terms in each geometric sequence.

37. $6, 3, \frac{3}{2}, \frac{3}{4}, \frac{3}{8}, \ldots$
38. $\frac{1}{3}, \frac{1}{9}, \frac{1}{27}, \frac{1}{81}, \ldots$
39. $5, 2, \frac{4}{5}, \frac{8}{25}, \ldots$
40. $-\frac{4}{3}, -\frac{4}{9}, -\frac{4}{27}, -\frac{4}{81}, \ldots$
41. $\frac{1}{3}, \frac{4}{15}, \frac{16}{75}, \frac{64}{375}, \ldots$
42. $6, -2, \frac{2}{3}, -\frac{2}{9}, \frac{2}{27}, \ldots$
43. $9, -1, \frac{1}{9}, -\frac{1}{81}, \ldots$
44. $\frac{5}{3}, 1, \frac{3}{5}, \frac{9}{25}, \ldots$

Find the sum of each infinite geometric series.

45. $1 + \frac{1}{2} + \frac{1}{4} + \frac{1}{8} + \cdots$
46. $4 + 2 + 1 + \frac{1}{2} + \cdots$
47. $8 + \frac{16}{3} + \frac{32}{9} + \frac{64}{27} + \cdots$

48. $10 - 5 + \frac{5}{2} - \frac{5}{4} + \cdots$

49. $-60 + 20 - \frac{20}{3} + \frac{20}{9} - \cdots$

50. $4 + \frac{8}{3} + \frac{16}{9} + \frac{32}{27} + \cdots$

51. $-12 - \frac{12}{5} - \frac{12}{25} - \frac{12}{125} - \cdots$

52. $5 - 1 + \frac{1}{5} - \frac{1}{25} + \cdots$

Write each repeating decimal as a ratio of integers.

53. 0.2727...

54. 0.454545...

55. 0.7777...

56. 0.375375...

57. 0.515151...

58. 0.742742...

Problem Solving

59. In a geometric series, $a_3 = 28$ and $a_5 = 112$; find r and a_1.

60. In a geometric series, $a_2 = 27$ and $a_5 = 1$; find r and a_1.

61. In a geometric series, $a_2 = 15$ and $a_5 = 405$; find r and a_1.

62. In a geometric series, $a_2 = 12$ and $a_5 = -324$; find r and a_1.

63. A loaf of bread currently costs \$1.40. Determine the cost of a loaf of bread in 8 years if inflation were to grow at a constant rate of 3% per year. *Hint:* In year 2, the cost of a loaf of bread is \$1.40(1.03).

64. A specific bicycle currently costs \$400. Determine the cost of the bicycle in 12 years if inflation were to grow at a constant rate of 5% per year.

65. A substance loses half its mass each day. If there are initially 300 grams of the substance, find

a) the number of days after which only 37.5 grams of the substance remain

b) the amount of the substance remaining after 8 days

66. The number of a certain type of bacteria doubles every hour. If there are initially 1000 bacteria, how long will it take for the number of bacteria to reach 64,000?

67. In January, 1999, the population of the United States was about 271 million. If the population grows at a rate of 2.4% per year, find

a) The population in 12 years

b) the number of years for the population to double

68. A piece of farm equipment that costs \$75,000 decreases in value by 15% per year. Find the value of the equipment after 4 years.

69. The amount of light filtering through a lake diminishes by one-half for each meter of depth.

a) Write a sequence indicating the amount of light remaining at depths of 1, 2, 3, 4, and 5 meters.

b) What is the general term for this sequence?

c) What is the remaining light at a depth of 7 meters?

70. You invest \$10,000 in a savings account paying 6% interest annually. Find the amount in your account at the end of 8 years.

71. A tracer dye is injected into Mark Shermak for medical reasons. After each hour, two-thirds of the previous hour's dye remains. How much dye remains in Mark's system after 10 hours?

72. On each swing (left to right or right to left), a pendulum travels 80% as far as on its previous swing. If the first swing is 8 feet, determine the total distance traveled by the pendulum by the time it comes to rest.

73. Shawna Kelly goes bungee jumping off a bridge above water. On the initial jump, the bungee cord stretches to 220 feet. Assume the first bounce reaches a height of 60% of the original jump and that each additional bounce reaches a height of 60% of the previous bounce.

a) What will be the height of the third bounce?

b) Theoretically, Shawna would never stop bouncing, but realistically, she will. Use the infinite geometric series to estimate the total distance Shawna travels in a *downward* direction.

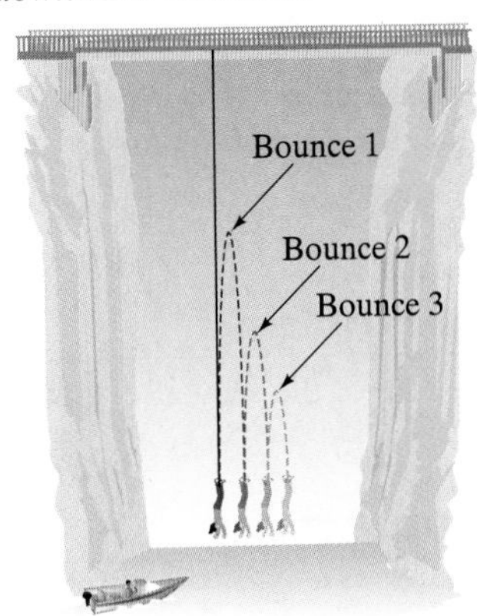

74. Repeat Exercise 73 **b)**, but this time find the total distance traveled in an *upward* direction.

75. A ping-pong ball falls off a table 30 inches high. Assume that the first bounce reaches a height of 70% of the distance the ball fell and each additional bounce reaches a height of 70% of the previous bounce.

a) How high will the ball bounce on the fourth bounce?

b) Theoretically, the ball would never stop bouncing, but realistically, it will. Estimate the total distance the ball travels in the *downward* direction.

76. Repeat Exercise 75 **b)**, but this time find the total distance traveled in the *upward* direction.

77. Suppose that you form stacks of blue chips such that there is one blue chip in the first stack and in each successive stack you double the number of chips. Thus you have stacks of 1, 2, 4, 8, and so on blue chips. You also form stacks of red chips, starting with one red chip and then tripling the number in each successive stack. Thus the stacks will contain 1, 3, 9, 27 and so on red chips. How many more would the sixth stack of red chips have than the sixth stack of blue chips?

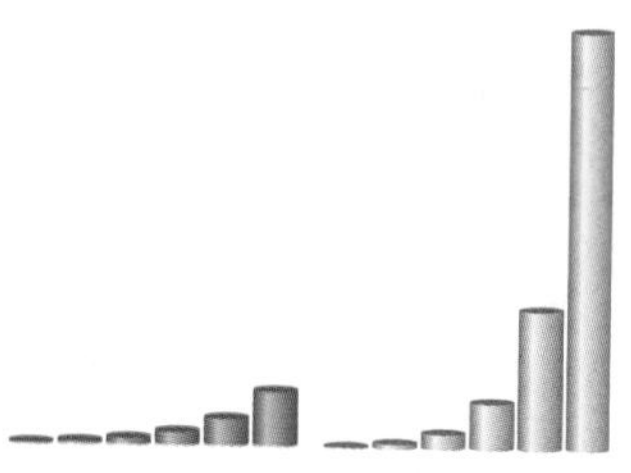

78. If you start with $1 and double your money each day, how many days will it take to surpass $1,000,000?

79. One method of depreciating an item on an income tax return is the declining balance method. With this method, a given percentage of the cost of the item is depreciated each year. Suppose that an item has a 5-year life and is depreciated using the declining balance method. Then at the end of its first year it loses $\frac{1}{5}$ of its value and $\frac{4}{5}$ of its value remains. At the end of the second year it loses $\frac{1}{5}$ of the remaining $\frac{4}{5}$ of its value, and so on. A car has a 5-year life expectancy and costs $9800.

a) Write a sequence showing the value of the car remaining for each of the first 3 years.

b) What is the general term of this sequence?

c) Find the value of the car at the end of 5 years.

80. In Exercise Set 9.6, Exercise 61, a formula for scrap value was given. The scrap value, S, is found by $S = c(1 - r)^n$ where c is the original cost, r is the annual depreciation rate and n is the number of years the object is depreciated.

a) If you have not already done so, do Exercise 79 above to find the value of the car remaining at the end of 5 years.

b) Use the formula given to find the scrap value of the car at the end of 5 years and compare this answer with the answer found in part **a)**.

81. A ball is dropped from a height of 10 feet. The ball bounces to a height of 9 feet. On each successive bounce the ball rises to 90% of its previous height. Find the *total vertical distance* traveled by the ball when it comes to rest.

82. A particle follows the path indicated by the wave shown. Find the *total vertical distance* traveled by the particle.

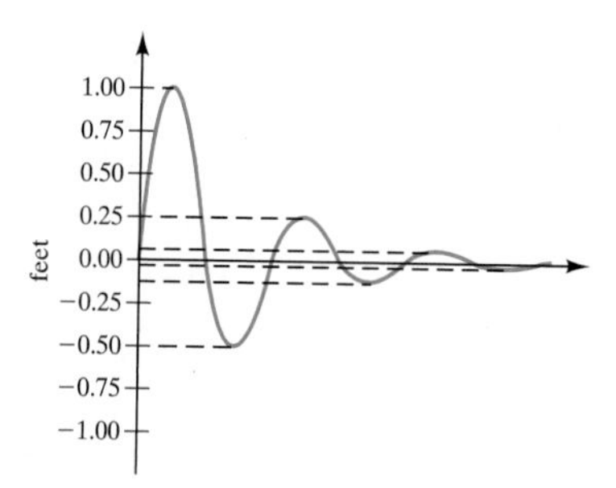

83. The formula for the nth term of a geometric sequence is $a_n = a_1 r^{n-1}$. If $a_1 = 1$, $a_n = r^{n-1}$.

a) How do you think the graphs of $y_1 = 2^{n-1}$ and $y_2 = 3^{n-1}$ will compare?

b) Graph both y_1 and y_2 and determine whether your answer to part **a)** was correct.

84. Use your grapher to decide the value of n to the nearest hundredth, where $100 = 3 \cdot 2^{n-1}$.

Challenge Problem

85. Find the sum of the sequence, 1, 2, 4, 8, ..., 1,048,576 and the number of terms in the sequence.

Group Activity

Discuss and answer Exercise 86 as a group.

86. The two bar graphs show annual revenues for two companies. The TECO Corporation's revenue grew at an arithmetic rate, and the Tom Malek Corporation revenue grew at a geometric rate.

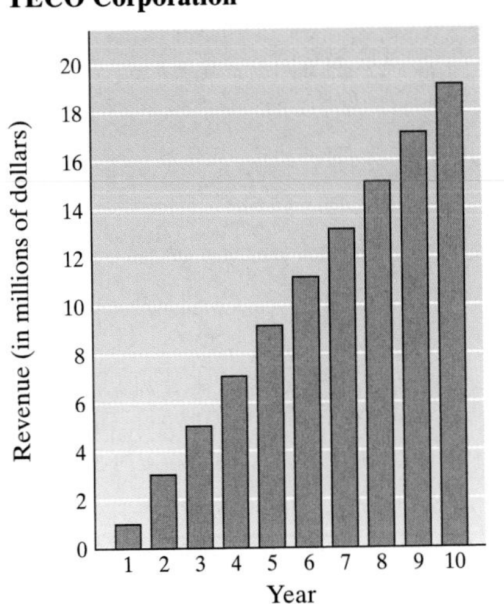

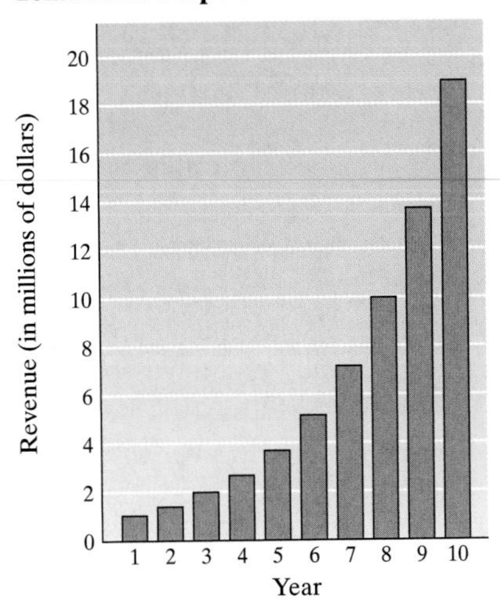

a) Using the first and tenth year's revenues, estimate the common difference in the arithmetic rate.
b) Using the first and tenth year's revenues, estimate the common ratio in the geometric rate.
c) Over the 10 years shown, estimate the total revenue received by the TECO Corporation.
d) Over the 10 years shown, estimate the total revenue received by the Tom Malek Corporation.
e) Assuming past trends continue, estimate the revenue from the TECO Corporation in year 11.
f) Estimate the revenue from the Tom Malek Corporation in year 11.
g) In which of the two companies would you rather own stock? Explain.

Cumulative Review Exercises

[5.2] **87.** Multiply $(4x^2 - 3x - 6)(2x - 3)$.

[5.3] **88.** Divide $(16x^2 + 10x - 18) \div (2x + 5)$.

[6.5] **89.** It takes Mrs. Donovan twice as long to load a truck as it takes Mr. Donovan. If together they can load the truck in 8 hours, how long would it take Mr. Donovan to load the truck by himself?

[7.2] **90.** Evaluate $\left(\frac{9}{100}\right)^{-1/2}$.

[7.3] **91.** Simplify $\sqrt[3]{9x^2y}\left(\sqrt[3]{3x^4y^6} - \sqrt[3]{8xy^4}\right)$.

[7.5] **92.** Simplify $x\sqrt{y} - 2\sqrt{x^2y} + \sqrt{4x^2y}$.

[7.6] **93.** Solve $\sqrt{a^2 + 9a + 3} = -a$.

11.4 THE BINOMIAL THEOREM

SSM

VIDEO 11.4

CD Rom

1) **Evaluate factorials.**
2) **Use Pascal's triangle.**
3) **Use the binomial theorem.**

1) Evaluate Factorials

To understand the binomial theorem, you must have an understanding of what **factorials** are. The notation $n!$ is read "n factorial." Its definition follows.

Definition

***n* Factorial**

$$n! = n(n-1)(n-2)(n-3)\cdots(1)$$

for any positive integer n.

Examples

$6! = 6 \cdot 5 \cdot 4 \cdot 3 \cdot 2 \cdot 1 = 720$

$8! = 8 \cdot 7 \cdot 6 \cdot 5 \cdot 4 \cdot 3 \cdot 2 \cdot 1 = 40{,}320$

Note that by definition **0! is 1**.

Using Your Calculator

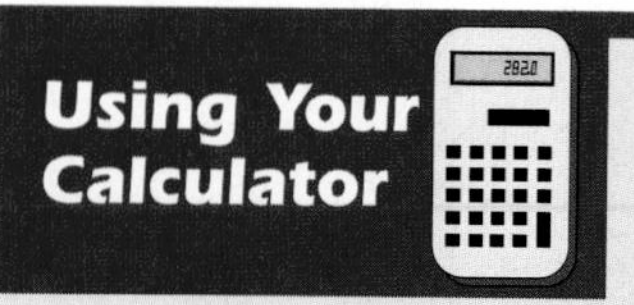

Factorials can be found on calculators that contain an $\boxed{n!}$ or $\boxed{x!}$ key. Often, the factorial key is a second function key. In the following, the answers appear after $\boxed{n!}$.

Evaluate 6!   6 $\boxed{\text{2nd}}$ $\boxed{n!}$ 720

Evaluate 9!   9 $\boxed{\text{2nd}}$ $\boxed{n!}$ 362880

Graphing calculators do not have a factorial key. On most graphing calculators, factorials are found under the $\boxed{\text{MATH}}$, Probability function menu.

2) Use Pascal's Triangle

Using polynomial multiplication, we can obtain the following expansions of the binomial $a + b$:

$$(a + b)^0 = 1$$

$$(a + b)^1 = a + b$$

$$(a + b)^2 = a^2 + 2ab + b^2$$

$$(a + b)^3 = a^3 + 3a^2b + 3ab^2 + b^3$$

$$(a + b)^4 = a^4 + 4a^3b + 6a^2b^2 + 4ab^3 + b^4$$

$$(a + b)^5 = a^5 + 5a^4b + 10a^3b^2 + 10a^2b^3 + 5ab^4 + b^5$$

$$(a + b)^6 = a^6 + 6a^5b + 15a^4b^2 + 20a^3b^3 + 15a^2b^4 + 6ab^5 + b^6$$

Note that when expanding a binomial of the form $(a + b)^n$:

1. There are $n + 1$ terms in the expansion.
2. The first term is a^n and the last term is b^n.
3. Reading from left to right, the exponents on a decrease by 1 from term to term, while the exponents on b increase by 1 from term to term.
4. The sum of the exponents on the variables in each term is n.
5. The coefficients of the terms equidistant from the ends are the same.

If we examine just the variables in $(a + b)^5$, we have a^5, a^4b, a^3b^2, a^2b^3, ab^4, and b^5.

The numerical coefficients of each term in the expansion of $(a + b)^n$ can be found by using **Pascal's triangle**, named after Blaise Pascal, a seventeenth century French mathematician. For example, if $n = 5$, we can determine the numerical coefficients of $(a + b)^5$ as follows.

Exponent on Binomial	Pascal's Triangle
$n = 0$	1
$n = 1$	1 1
$n = 2$	1 2 1
$n = 3$	1 3 3 1
$n = 4$	1 4 6 4 1
$n = 5$	1 5 10 10 5 1
$n = 6$	1 6 15 20 15 6 1

Examine row 5 ($n = 4$) and row 6 ($n = 5$).

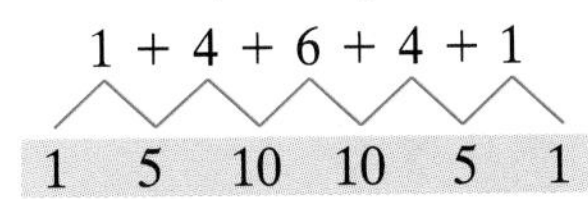

Notice that the first and last numbers in each row are 1, and the inner numbers are obtained by adding the two numbers in the row above (to the right and left). The numerical coefficients of $(a + b)^5$ are 1, 5, 10, 10, 5, and 1. Thus we can write the expansion of $(a + b)^5$ by using the information in **1–5** on page 686 for the variables and their exponents, and by using Pascal's triangle for the coefficients.

$$(a + b)^5 = a^5 + 5a^4b + 10a^3b^2 + 10a^2b^3 + 5ab^4 + b^5$$

This method of expanding a binomial is not practical when n is large.

3) Use the Binomial Theorem

We will shortly introduce a more practical method, called the binomial theorem, to expand expressions of the form $(a + b)^n$. However, before we introduce this formula, we need to explain how to find binomial coefficients of the form $\binom{n}{r}$.

Binomial Coefficients

For n and r nonnegative integers, $n > r$,

$$\binom{n}{r} = \frac{n!}{r! \cdot (n - r)!}$$

The binomial coefficient $\binom{n}{r}$ is read "the number of *combinations* of n items taken r at a time." Combinations are used in the many areas of mathematics, including the study of probability.

EXAMPLE 1 Evaluate $\binom{5}{2}$.

Solution Using the definition, if we substitute 5 for n and 2 for r, we obtain

$$\binom{5}{2} = \frac{5!}{2! \cdot (5 - 2)!} = \frac{5!}{2! \cdot 3!} = \frac{5 \cdot 4 \cdot \cancel{3 \cdot 2 \cdot 1}}{(2 \cdot 1) \cdot \cancel{(3 \cdot 2 \cdot 1)}} = 10$$

Thus, $\binom{5}{2}$ equals 10.

EXAMPLE 2 Evaluate. **a)** $\binom{7}{4}$ **b)** $\binom{4}{4}$ **c)** $\binom{5}{0}$

Solution

a) $\binom{7}{4} = \frac{7!}{4! \cdot (7-4)!} = \frac{7!}{4! \cdot 3!} = \frac{7 \cdot 6 \cdot 5 \cdot \cancel{4 \cdot 3 \cdot 2 \cdot 1}}{(\cancel{4 \cdot 3 \cdot 2 \cdot 1})(3 \cdot 2 \cdot 1)} = 35$

b) $\binom{4}{4} = \frac{4!}{4! \cdot (4-4)!} = \frac{\cancel{4!}}{\cancel{4!} \cdot 0!} = \frac{1}{1} = 1$ *Remember that* $0! = 1$.

c) $\binom{5}{0} = \frac{5!}{0! \cdot (5-0)!} = \frac{\cancel{5!}}{0! \cdot \cancel{5!}} = \frac{1}{1} = 1$

By studying Examples 2(b) and (c), you can reason that, for any positive integer n,

NOW TRY EXERCISE 11

$$\binom{n}{n} = 1 \quad \text{and} \quad \binom{n}{0} = 1$$

Using Your Graphing Calculator

All graphing calculators can evaluate binomial coefficients. On most graphers, the notation ${}_nC_r$ is used instead of $\binom{n}{r}$. Thus $\binom{7}{4}$ would be represented as ${}_7C_4$ on a grapher. Read your graphing calculator manual to learn how to evaluate combinations.

Now we introduce the binomial theorem.

Binomial Theorem

For any positive integer n,

$$(a+b)^n = \binom{n}{0}a^n b^0 + \binom{n}{1}a^{n-1}b^1 + \binom{n}{2}a^{n-2}b^2 + \binom{n}{3}a^{n-3}b^3 + \cdots + \binom{n}{n}a^0 b^n$$

Notice in the binomial theorem that the sum of the exponents on the variables in each term is n, and the bottom number in the combination is always the same as the exponent on the second variable in the term. For example, if we consider the term $\binom{n}{3}a^{n-3}b^3$, the sum of the exponents on the variables is $(n-3)+3=n$. Also, the exponent on the variable b is 3, and the bottom number in the combination is also 3.

Now we will expand $(a+b)^5$ using the binomial theorem and see if we get the same expression as we did when we used polynomial multiplication and Pascal's triangle to obtain the expansion.

$$\begin{aligned}(a+b)^5 &= \tbinom{5}{0}a^5b^0 + \tbinom{5}{1}a^{5-1}b^1 + \tbinom{5}{2}a^{5-2}b^2 + \tbinom{5}{3}a^{5-3}b^3 + \tbinom{5}{4}a^{5-4}b^4 + \tbinom{5}{5}a^{5-5}b^5 \\ &= \tbinom{5}{0}a^5b^0 + \tbinom{5}{1}a^4b^1 + \tbinom{5}{2}a^3b^2 + \tbinom{5}{3}a^2b^3 + \tbinom{5}{4}a^1b^4 + \tbinom{5}{5}a^0b^5 \\ &= \frac{5!}{0! \cdot 5!}a^5 + \frac{5!}{1! \cdot 4!}a^4b + \frac{5!}{2! \cdot 3!}a^3b^2 + \frac{5!}{3! \cdot 2!}a^2b^3 + \frac{5!}{4! \cdot 1!}ab^4 + \frac{5!}{0! \cdot 5!}b^5 \\ &= a^5 + 5a^4b + 10a^3b^2 + 10a^2b^3 + 5ab^4 + b^5\end{aligned}$$

This is the same expression as we obtained earlier.

In the binomial theorem, the first and last terms of the expansion contain a factor raised to the zero power. Since any nonzero number raised to the 0th power equals 1, we could have omitted those factors. These factors were included so that you could see the pattern better.

EXAMPLE 3 Use the binomial theorem to expand $(2x + 3)^6$.

Solution If we use $2x$ for a and 3 for b, we obtain

$$(2x+3)^6 = \binom{6}{0}(2x)^6(3)^0 + \binom{6}{1}(2x)^5(3)^1 + \binom{6}{2}(2x)^4(3)^2 + \binom{6}{3}(2x)^3(3)^3 + \binom{6}{4}(2x)^2(3)^4 + \binom{6}{5}(2x)^1(3)^5 + \binom{6}{6}(2x)^0(3)^6$$

$$= 1(2x)^6 + 6(2x)^5(3) + 15(2x)^4(9) + 20(2x)^3(27) + 15(2x)^2(81) + 6(2x)(243) + 1(3)^6$$

$$= 64x^6 + 576x^5 + 2160x^4 + 4320x^3 + 4860x^2 + 2916x + 729$$

NOW TRY EXERCISE 13

EXAMPLE 4 Use the binomial theorem to expand $(5x - 2y)^4$.

Solution Write $(5x - 2y)^4$ as $[5x + (-2y)]^4$. Use $5x$ in place of a and $-2y$ in place of b in the binomial theorem.

$$[5x + (-2y)]^4 = \binom{4}{0}(5x)^4(-2y)^0 + \binom{4}{1}(5x)^3(-2y)^1 + \binom{4}{2}(5x)^2(-2y)^2 + \binom{4}{3}(5x)^1(-2y)^3 + \binom{4}{4}(5x)^0(-2y)^4$$

$$= 1(5x)^4 + 4(5x)^3(-2y) + 6(5x)^2(-2y)^2 + 4(5x)(-2y)^3 + 1(-2y)^4$$

$$= 625x^4 - 1000x^3y + 600x^2y^2 - 160xy^3 + 16y^4$$

NOW TRY EXERCISE 17

Exercise Set 11.4

Concept/Writing Exercises

1. Explain how to construct Pascal's triangle. Construct the first five rows of Pascal's triangle.
2. Explain in your own words how to find $n!$ for any whole number n.

Practice the Skills

Evaluate each combination.

3. $\binom{5}{2}$
4. $\binom{6}{3}$
5. $\binom{5}{5}$
6. $\binom{9}{3}$
7. $\binom{7}{0}$
8. $\binom{10}{8}$
9. $\binom{8}{4}$
10. $\binom{12}{8}$
11. $\binom{8}{2}$
12. $\binom{7}{5}$

Use the binomial theorem to expand each expression.

13. $(x + 4)^3$
14. $(2x + 3)^3$
15. $(a - b)^4$
16. $(2r + s^2)^4$
17. $(3a - b)^5$
18. $(x + 2y)^5$
19. $\left(2x + \frac{1}{2}\right)^4$
20. $\left(\frac{2}{3}x + \frac{3}{2}\right)^4$
21. $\left(\frac{x}{2} - 3\right)^4$
22. $(3x^2 + y)^5$

Write the first four terms of each expansion.

23. $(x + 10)^{10}$
24. $(2x + 3)^8$
25. $(3x - y)^7$
26. $(3p + 2q)^{11}$
27. $(x^2 - 3y)^8$
28. $\left(2x + \frac{y}{5}\right)^9$

Problem Solving

29. Is $n!$ equal to $n \cdot (n-1)!$? Explain and give an example to support your answer.

30. Is $(n+1)!$ equal to $(n+1) \cdot n!$? Explain and give an example to support your answer.

31. Is $(n-3)!$ equal to $(n-3)(n-4)(n-5)!$ for $n \geq 5$? Explain and give an example to support your answer.

32. Under what conditions will $\binom{n}{m}$, where n and m are nonnegative integers, have a value of 1?

33. What are the first, second, next to last, and last terms of the expansion $(x+3)^8$?

34. What are the first, second, next to last, and last terms of the expansion $(2x+5)^6$?

35. Write the binomial theorem using summation notation.

36. Prove that $\binom{n}{r} = \binom{n}{n-r}$ for any whole numbers n and r, and $r \leq n$.

Group Activity

When the probability of a specific thing happening on any specific trial is p, and the probability that the specific thing fails to happen on any specific trial is q, then the probability of getting exactly r successes in n tries, P(r), is found using the formula $P(r) = \binom{n}{r} p^r q^{n-r}$. *For example, the probability of rolling a three on a die (one of a pair of dice) is* $\frac{1}{6}$, *and the probability of not rolling a three is* $\frac{5}{6}$. *The probability of rolling exactly 2 threes out of 5 rolls is found as follows.*

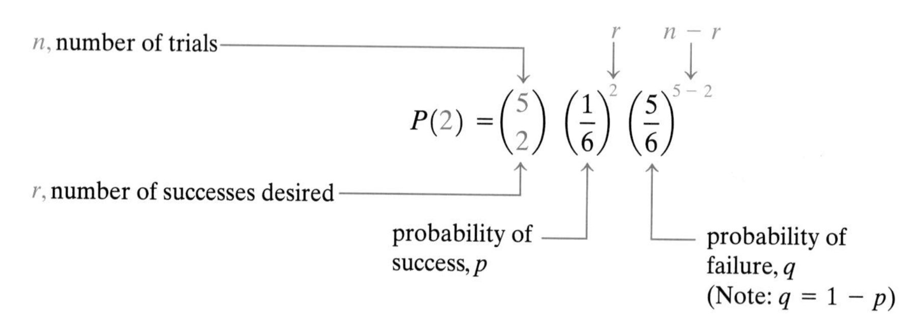

Probabilities are numbers between 0 and 1.00. A probability of 0 indicates that the event cannot happen and a probability of 1.00 indicates that the event must happen.

Discuss and answer Exercises 37 and 38 as a group.

37. Suppose that 40% of all people who are exposed to a particular virus catch the virus ($p = 0.4$). Determine the probability that exactly 3 of the next 6 people who are exposed to the virus catch it.

38. Matthew McCulley's probability of making a free throw from the foul line when playing basketball is 0.80. Find the probability that he makes exactly 4 of his next 7 free throws.

Cumulative Review Exercises

[2.2] *Solve* $s_n - s_n r = a_1 - a_1 r^n$ *for the indicated variable.*

39. s_n

40. a_1

Factor.

[5.5] **41.** $16x^2 - 8x - 3.$

[5.7] **42.** $3a^2b^2 - 24ab^2 + 48b^2.$

SUMMARY

Key Words and Phrases

11.1

Alternating sequence
Decreasing sequence
Finite sequence
General term of a sequence
Increasing sequence
Index of summation
Infinite sequence
Lower limit of a summation
Partial sum of a series
Sequence
Series
Sigma
Summation notation
Upper limit of a summation

11.2

Arithmetic sequence
Arithmetic series
Common difference

11.3

Common ratio
Geometric sequence
Geometric series
Infinite geometric series

11.4

Binomial coefficient
Binomial theorem
Factorial
Pascal's triangle

IMPORTANT FACTS

$$\sum_{i=1}^{n} x_i = x_1 + x_2 + x_3 + \cdots + x_n$$

nth Partial Sum of an Arithmetic Series

$$s_n = \frac{n(a_1 + a_n)}{2}$$

Pascal's Triangle

$$\begin{array}{ccccccccc} & & & & 1 & & & & \\ & & & 1 & & 1 & & & \\ & & 1 & & 2 & & 1 & & \\ & 1 & & 3 & & 3 & & 1 & \\ 1 & & 4 & & 6 & & 4 & & 1 \end{array}$$

nth Term of a Geometric Sequence

$$a_n = a_1 r^{n-1}$$

nth Partial Sum of a Geometric Series

$$s_n = \frac{a_1(1 - r^n)}{1 - r}, r \neq 1$$

n Factorial

$$n! = n(n - 1)(n - 2) \cdots (2)(1)$$

Sum of an Infinite Geometric Series

$$s_\infty = \frac{a_1}{1 - r}, |r| < 1$$

nth Term of an Arithmetic Sequence

$$a_n = a_1 + (n - 1)d$$

Binomial Coefficients

$$\binom{n}{r} = \frac{n!}{r! \cdot (n - r)!}$$

Binomial Theorem

$$(a + b)^n = \binom{n}{0}a^n b^0 + \binom{n}{1}a^{n-1} b^1 + \binom{n}{2}a^{n-2} b^2 + \binom{n}{3}a^{n-3} b^3 + \cdots + \binom{n}{n}a^0 b^n$$

Review Exercises

[11.1] *Write the first five terms of each sequence.*

1. $a_n = n + 2$

2. $a_n = \frac{1}{n}$

3. $a_n = n(n + 1)$

4. $a_n = \frac{n^2}{n + 4}$

Find the indicated term of each sequence.

5. $a_n = 3n + 4$, seventh term

6. $a_n = (-1)^n + 3$, seventh term

7. $a_n = \frac{n + 7}{n^2}$, ninth term

8. $a_n = (n)(n - 3)$, eleventh term

For each sequence, find the first and third partial sums, s_1 and s_3.

9. $a_n = 3n + 2$

10. $a_n = 2n^2$

11. $a_n = \frac{n + 3}{n + 2}$

12. $a_n = (-1)^n(n + 2)$

Write the next three terms of each sequence. Then write an expression for the general term, a_n.

13. 1, 2, 4, 8, ...

14. −8, 4, −2, 1, ...

15. $\frac{2}{3}, \frac{4}{3}, \frac{8}{3}, \frac{16}{3}, \ldots$

16. 9, 6, 3, 0, ...

Write out each series. Then find the sum of the series.

17. $\sum_{n=1}^{3} (n^2 + 2)$

18. $\sum_{k=1}^{4} k(k + 2)$

19. $\sum_{k=1}^{5} \frac{k^2}{3}$

20. $\sum_{n=1}^{4} \frac{n}{n + 1}$

For the set of values $x_1 = 3, x_2 = 9, x_3 = 5, x_4 = 10$, evaluate the indicated sum.

21. $\sum_{i=1}^{4} x_i$

22. $\sum_{i=1}^{4} x_i^2$

23. $\sum_{i=2}^{3} (x_i^2 + 1)$

24. $\left(\sum_{i=1}^{4} x_i\right)^2$

[11.2] *Write the first five terms of the arithmetic sequence with indicated first term and common difference.*

25. $a_1 = 5, d = 2$

26. $a_1 = \frac{1}{2}, d = -2$

27. $a_1 = -12, d = -\frac{1}{2}$

28. $a_1 = -100, d = \frac{1}{5}$

For each arithmetic sequence, find the indicated value.

29. $a_1 = 2, d = 3$; find a_9

30. $a_1 = 50, a_5 = 34$; find d

31. $a_1 = -3, a_7 = 0$; find d

32. $a_1 = 12, a_n = -13, d = -5$; find n

Find s_n and d for each arithmetic sequence.

33. $a_1 = 7, a_8 = 21, n = 8$

34. $a_1 = -12, a_7 = -48, n = 7$

35. $a_1 = \frac{3}{5}, a_7 = 3, n = 7$

36. $a_1 = -\frac{10}{3}, a_9 = -6, n = 9$

Write the first four terms of each arithmetic sequence. Then find a_{10} and s_{10}.

37. $a_1 = 2, d = 4$

38. $a_1 = -8, d = -3$

39. $a_1 = \frac{5}{6}, d = \frac{2}{3}$

40. $a_1 = -80, d = 4$

Find the number of terms in each arithmetic sequence. Then find s_n.

41. $3, 8, 13, \ldots, 53$

42. $-16, -11, -6, -1, \ldots, 24$

43. $\frac{6}{10}, \frac{9}{10}, \frac{12}{10}, \frac{15}{10}, \ldots, \frac{36}{10}$

44. $-5, 0, 5, 10, \ldots, 85$

[11.3] *Determine the first five terms of each geometric sequence.*

45. $a_1 = 5, r = 2$

46. $a_1 = -12, r = \frac{1}{2}$

47. $a_1 = 20, r = -\frac{2}{3}$

48. $a_1 = -100, r = \frac{1}{5}$

Find the indicated term of each geometric sequence.

49. $a_1 = 12, r = \frac{1}{3}$; find a_7

50. $a_1 = 25, r = 2$; find a_9

51. $a_1 = -8, r = -2$; find a_9

52. $a_1 = \frac{5}{12}, r = \frac{2}{3}$; find a_8

Find each sum.

53. $a_1 = 12, r = 2$; find s_8

54. $a_1 = \frac{3}{5}, r = \frac{5}{3}$; find s_7

55. $a_1 = -84, r = -\frac{1}{4}$; find s_5

56. $a_1 = 9, r = \frac{3}{2}$; find s_9

For each geometric sequence, find the common ratio, r, and then write an expression for the general term, a_n.

57. $6, 12, 24, \ldots$

58. $8, \frac{8}{3}, \frac{8}{9}, \ldots$

59. $-4, -20, -100, \ldots$

60. $\frac{9}{5}, \frac{18}{15}, \frac{36}{45}, \ldots$

Find the sum of the terms in each infinite geometric sequence.

61. $7, \frac{7}{2}, \frac{7}{4}, \frac{7}{8}, \ldots$

62. $-8, \frac{8}{3}, -\frac{8}{9}, \frac{8}{27}, \ldots$

63. $-5, -\frac{10}{3}, -\frac{20}{9}, -\frac{40}{27}, \ldots$

64. $\frac{7}{2}, 1, \frac{2}{7}, \frac{4}{49}, \ldots$

Find the sum of each infinite series.

65. $2 + 1 + \frac{1}{2} + \frac{1}{4} + \cdots$

66. $7 + \frac{7}{3} + \frac{7}{9} + \frac{7}{27} + \cdots$

67. $-12 - \frac{24}{3} - \frac{48}{9} - \frac{96}{27} - \cdots$

68. $5 - 1 + \frac{1}{5} - \frac{1}{25} + \cdots$

Write the repeating decimal as a ratio of integers.

69. $0.5252\ldots$

70. $0.531531\ldots$

[11.4] *Use the binomial theorem to expand the expression.*

71. $(3x + y)^4$

72. $(2x - 3y^2)^3$

Write the first four terms of the expansion.

73. $(x - 2y)^9$

74. $(2a^2 + 3b)^8$

[11.1] **75.** Find the sum of the integers between 100 and 200, inclusive.

[11.3] **76.** Wayne Siegert is offered a job with a starting salary of $30,000 with the agreement that his salary will increase by $1000 at the end of each of the next 7 years.

a) Write a sequence showing his salary for the first 5 years.

b) What is the general term of this sequence?

c) If this process were continued, what would his salary be after 9 years?

77. You begin with $100, double that to get $200, double that again to get $400, and so on. How much will you have after you perform this process ten times?

78. If the inflation rate was a constant 8% per year (each year the cost of living is 8% greater than the previous year), how much would a product that costs $200 now cost after 12 years?

79. On each swing (left to right or right to left), a pendulum travels 92% as far as on its previous swing. If the first swing is 8 feet, find the distance traveled by the pendulum by the time it comes to rest.

Practice Test

1. **a)** What is an infinite sequence?

b) What is a finite sequence?

2. What is a series?

3. **a)** What is an arithmetic sequence?

b) What is a geometric sequence?

4. Are the following sequences arithmetic, geometric, or neither? Explain your answer.

a) $1, 1, 2, 3, 5, 8, 13, \ldots$

b) $4, 1, -2, -5, -8, \ldots$

c) $4, -2, 1, -\frac{1}{2}, \frac{1}{4}, -\frac{1}{8}, \ldots$

5. Write the first five terms of the sequence with $a_n = \dfrac{n+2}{n^2}$.

6. Find the first and third partial sums of $a_n = \dfrac{2n+3}{n}$.

7. Write out the following series and find the sum of the series.

$$\sum_{n=1}^{5} (2n^2 + 3)$$

8. For $x_1 = 4$, $x_2 = 2$, $x_3 = 8$, and $x_4 = 12$, find $\sum_{i=1}^{4} x_i^2$.

9. Write the general term for the following arithmetic sequence.

$$\frac{1}{3}, \frac{2}{3}, \frac{3}{3}, \frac{4}{3}, \ldots$$

10. Write the general term for the following geometric sequence.

$$5, 10, 20, 40, \ldots$$

Write the first four terms of each sequence.

11. $a_1 = 12$, $d = -3$

12. $a_1 = \dfrac{5}{8}$, $r = \dfrac{2}{3}$

13. Find a_8 when $a_1 = 100$ and $d = -12$.

14. Find s_8 for the arithmetic series with $a_1 = 3$ and $a_8 = -11$.

15. Find the number of terms in the arithmetic sequence $-4, -16, -28, \ldots, -148$.

16. Find a_7 when $a_1 = 8$ and $r = \dfrac{2}{3}$.

17. Find s_7 when $a_1 = \dfrac{3}{5}$ and $r = -5$.

18. Find the common ratio and write an expression for the general term of the sequence $12, 6, 3, \frac{3}{2}, \ldots$.

19. Find the sum of the following infinite geometric series.

$$3 + 2 + \frac{4}{3} + \frac{8}{9} + \cdots$$

20. Write $0.6262\ldots$ as a ratio of integers.

21. Evaluate $\binom{7}{2}$.

22. Use the binomial theorem to expand $(x + 2y)^4$.

23. Paul Misselwitz's five test grades are 74, 93, 83, 87, and 68. Use $\bar{x} = \frac{\Sigma x}{n}$ to find the arithmetic mean of Paul's grades.

24. To save for retirement, Jamie Monroe plans to save \$1000 the first year, \$2000 the second year, \$3000 the third year, and to increase the amount saved by \$1000 in each successive year. How much will she have saved by the end of her twentieth year of savings?

25. The number of bacteria in a culture is tripling every hour. If there are initially 500 bacteria in the culture, how many bacteria will be in the culture by the end of the eighth hour?

Cumulative Review Test

1. Solve the equation $\frac{1}{2}x + \frac{1}{3}(x - 2) = \frac{3}{4}(x - 5)$.

2. Solve the inequality $|2x - 3| - 4 > 10$.

3. Graph $3x - 5y > 10$.

4. Solve the following system of equations.

$$5x - 2y = 8$$
$$x - y = 4$$

5. Solve the formula $A = \frac{pt}{p + t}$ for p.

6. Add $\frac{x}{x^2 + x - 12} + \frac{x + 2}{3x^2 + 16x + 16}$.

7. Simplify $\frac{\sqrt[3]{24x^6 y^3}}{\sqrt[3]{2x^2 y^5}}$.

8. Simplify $\sqrt{28} - 3\sqrt{7} + \sqrt{63}$.

9. Solve $\sqrt{5x + 1} - \sqrt{2x - 2} = 2$.

10. Simplify $\frac{5 - 2i}{3 + 4i}$.

11. Solve $-4x^2 - 2x + 8 = 0$.

12. Solve the inequality $\frac{(2x - 3)(x - 4)}{x + 1} < 0$.

13. Graph $4x^2 + 4y^2 = 36$.

14. **a)** Write the equation $y = x^2 + 2x - 3$ in the form $y = a(x - h)^2 + k$. **b)** Then graph the equation.

15. Graph $y = 2^x$ and $y = \log_2 x$ on the same axes.

16. Solve the following equation.

$$\log(4x - 1) + \log 3 = \log(8x + 13)$$

17. In an arithmetic series, if $a_1 = 8$ and $a_6 = 28$, find d.

18. In a geometric series, if $a_1 = 5$ and $r = 3$, find s_3.

19. A train and car leave from South Point heading for the same destination. The car averages 40 miles per hour and the train averages 60 miles per hour. If the train arrives 2 hours ahead of the car, find the distance between the two towns.

20. The Donovans' garden is in the shape of a rectangle. If the area of their garden is 300 square feet and the length of the garden is 20 feet greater than the width, find the dimensions of their garden.

Appendix

Geometric Formulas

Areas and perimeters			
Figure	**Sketch**	**Area**	**Perimeter**
Square	s	$A = s^2$	$P = 4s$
Rectangle	w, l	$A = lw$	$P = 2l + 2w$
Parallelogram	h, w, l	$A = lh$	$P = 2l + 2w$
Trapezoid	b_1, s_1, h, s_2, b_2	$A = \frac{1}{2}h(b_1 + b_2)$	$P = s_1 + s_2 + b_1 + b_2$
Triangle	s_1, h, s_2, b	$A = \frac{1}{2}bh$	$P = s_1 + s_2 + b$

Area and Circumference of Circle			
Circle	r	$A = \pi r^2$	$C = 2\pi r$

Volumes and Surface Areas of Three-Dimensional Figures			
Figure	**Sketch**	**Volume**	**Surface Area**
Rectangular solid	h, w, l	$V = lwh$	$s = 2lh + 2wh + 2wl$
Right circular cylinder	h, r	$V = \pi r^2 h$	$s = 2\pi rh + 2\pi r^2$
Sphere	r	$V = \frac{4}{3}\pi r^3$	$s = 4\pi r^2$
Right circular cone	h, r	$V = \frac{1}{3}\pi r^2 h$	$s = \pi r\sqrt{r^2 + h^2}$
Square or rectangular pyramid	h, w, l	$V = \frac{1}{3}lwh$	

Answers

CHAPTER 1

Exercise Set 1.1 **1–11.** Answers will vary. **13.** Do all the homework and preview the new material to be covered in class. **15.** See the steps on page 5 of your text. **17.** The more you put into the course, the more you will get out of it. **19.** Answers will vary.

Exercise Set 1.2 **1.** A letter used to represent various numbers **3.** A collection of objects **5.** Infinite, since there is no largest element. **7.** $>$, is greater than; $\geq$, is greater than or equal to; $<$, is less than; $\leq$, is less than or equal to; $\neq$, is not equal to **9.** $\{5, 6, 7, 8\}$ **11.** An integer can be written with a denominator of 1. **13.** True **15.** False **17.** False **19.** True **21.** True **23.** $<$ **25.** $<$ **27.** $>$ **29.** $<$ **31.** $>$ **33.** $<$ **35.** $>$ **37.** $>$ **39.** $A = \{6\}$ **41.** $C = \{6, 8\}$ **43.** $E = \{0, 1, 2, 3, 4, 5, 6\}$ **45.** $H = \{0, 5, 10, 15, ...\}$ **47.** $J = \{-5\}$ **49. a)** 4 **b)** 4, 0 **c)** $-3, 4, 0$ **d)** $-3, 4, \frac{1}{2}, \frac{5}{9}, 0, -1.23, \frac{99}{100}$ **e)** $\sqrt{2}, \sqrt{8}$ **f)** $-3, 4, \frac{1}{2}, \frac{5}{9}, 0, \sqrt{2}, \sqrt{8}, -1.23, \frac{99}{100}$ **51.** $A \cup B = \{5, 6, 7, 8\}; A \cap B = \{6, 7\}$ **53.** $A \cup B = \{-1, -2, -4, -5, -6\}; A \cap B = \{-2, -4\}$ **55.** $A \cup B = \{0, 1, 2, 3\}; A \cap B = \varnothing$ **57.** $A \cup B = \{0, 1, 2, 3, 4, 5, 6, 7, 8\}; A \cap B = \varnothing$ **59.** $A \cup B = \{0.1, 0.2, 0.3, 0.4, ...\}; A \cap B = \{0.2, 0.3\}$ **61.** The set of natural numbers. **63.** The set of even natural numbers greater than or equal to 8 and less than or equal to 30. **65.** The set of odd integers. **67. a)** Set A is the set of all x such that x is a natural number less than 8. **b)** $A = \{1, 2, 3, 4, 5, 6, 7\}$ **69.** 

71. 

73. 

75. 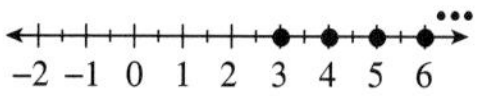

77. 

79. $\{x|x > -4\}$ **81.** $\{x|x \geq -3 \text{ and } x \in I\}$ or $\{x|x > -4 \text{ and } x \in I\}$ **83.** $\{x|-2 < x \leq 4.6\}$ **85.** $\left\{x\middle|x \leq \frac{37}{4}\right\}$ **87.** $\{x|-1 \leq x \leq 3 \text{ and } x \in I\}$ **89.** Yes **91.** Yes **93.** No **95.** Yes **97.** One example is $\{0.1, 0.2, 0.3, 0.4, 0.5\}$ **99.** One example is $A = \{3, 5, 7, 8, 9\}, B = \{5, 7\}$ **101. a)** {Levi's, Lands' End, L.L. Bean, Nike, Gold Toe, London Fog, Reebok, Hanes, Hanes Her Way, Arizona} **b)** {Levi's, L.L. Bean, Reebok} **c)** Union; asks for the brands in either category **d)** Intersection; asks for the brands that are common to both categories **103. a)** {Housing, Food & drinks, Transportation} **b)** {Apparel & upkeep, Entertainment} **105. a)** $\{1, 3, 4, 5, 6, 7\}$ **b)** $\{2, 3, 4, 6, 8\}$ **c)** $\{1, 2, 3, 4, 5, 6, 7, 8\}$ **d)** $\{3, 4, 6\}$ **107. a)** $\{x|x > 1\}$ includes fractions and decimal numbers which the other set does not contain. **b)** $\{2, 3, 4, 5, ...\}$ **c)** No, since it is not possible to list all real numbers greater than 1 in roster form. **109. a)** $0.\overline{1}; 0.\overline{2}; 0.\overline{3}$ **b)** $\frac{4}{9}; \frac{5}{9}; \frac{6}{9}$ or $\frac{2}{3}$ **c)** Based on a) and b), we deduce that $0.\overline{9} = \frac{9}{9} = 1$.

Exercise Set 1.3 **1.** Two numbers that sum to 0 **3.** $|a| = \begin{cases} a \text{ if } a \geq 0 \\ -a \text{ if } a < 0 \end{cases}$ **5.** Since a and $-a$ are the same distance from 0 on a number line, $|a| = |-a|$ for all real numbers, $\mathbb{R}$. **7.** Since $|a| = \begin{cases} a, & a \geq 0 \\ -a, & a \leq 0 \end{cases}$, then $|a| = -a$ only when $a \leq 0$. **9.** Since $|5| = 5$ and $|-5| = 5$, the desired values for a are 5 and -5. **15.** $-\frac{a}{b}$ or $\frac{-a}{b}$ **17. a)** $(ab)c = a(bc)$ **19.** Consider $2 + (3 \cdot 4)$ and $(2 + 3) \cdot (2 + 4)$. The left side is $2 + (3 \cdot 4) = 2 + 12 = 14$ and the right side is $(2 + 3) \cdot (2 + 4) = 5 \cdot 6 = 30$. **21.** 3 **23.** 6

25. $\frac{3}{4}$ **27.** 0 **29.** −7 **31.** $-\frac{5}{9}$ **33.** = **35.** < **37.** = **39.** > **41.** < **43.** < **45.** −1, 2, |3|, |−5|, 6

47. −32, −|4|, 4, |−7|, 15 **49.** −|2.9|, −2.4, −2.1, −2, |−2.8| **51.** $-2, \frac{1}{3}, \left|-\frac{1}{2}\right|, \left|\frac{3}{5}\right|, \left|-\frac{3}{4}\right|$ **53.** 1 **55.** −11

57. −3 **59.** −5.78 **61.** $-\frac{17}{45}$ **63.** −2.4 **65.** −17.2 **67.** 4 **69.** −2 **71.** $\frac{17}{20}$ **73.** −48 **75.** $\frac{5}{4}$ **77.** 12

79. 235.9192 **81.** 10 **83.** 1 **85.** 8 **87.** $\frac{5}{3}$ **89.** −3 **91.** 5 **93.** −16 **95.** −402.738 **97.** −21.6 **99.** −4

101. $\frac{81}{16}$ **103.** −1 **105.** $-\frac{17}{45}$ **107.** 77 **109.** Commutative property of addition
111. Multiplicative property of zero **113.** Associative property of addition **115.** Identity property of multiplication
117. Associative property of multiplication **119.** Distributive property **121.** Identity property of addition
123. Inverse property of addition **125.** Inverse property of multiplication **127.** Double negative property

129. $-4; \frac{1}{4}$ **131.** $\frac{2}{3}; -\frac{3}{2}$ **133.** 49°F **135.** −148.2 feet or 148.2 feet below the starting point

137. a) Owe $12,400 to the publisher **b)** Receive $27,500 from the publisher **139. a)** Refund **b)** $1875
143. a) $5399 **b)** $2483 **c)** $4578 **d)** $258 **145.** −50 **147.** −1 **149.** True **150.** {1, 2, 3, 4, ...}

151. a) 3, 4, −2, 0 **b)** $3, 4, -2, \frac{5}{6}, 0$ **c)** $\sqrt{3}$ **d)** $3, 4, -2, \frac{5}{6}, \sqrt{3}, 0$ **152. a)** {1, 4, 7, 9, 12, 15} **b)** {4, 7}

153. (number line: open circle at −4, closed circle at 6, segment between)
−4 6

Exercise Set 1.4

1. a) Base **b)** Exponent **3. a)** Index **b)** Radicand **5.** Positive number whose square equals the radicand **7.** A positive number raised to an odd power is a positive number. **11. b)** 16 **13.** 16

15. 25 **17.** −16 **19.** $-\frac{81}{625}$ **21.** $-\frac{16}{81}$ **23.** −6 **25.** −5 **27.** 0.1 **29.** 0.013 **31.** 0.053 **33.** −557.060

35. 1.710 **37.** 4.160 **39.** −0.723 **41. a)** 9 **b)** −9 **43. a)** 1 **b)** −1 **45. a)** 1 **b)** −1 **47. a)** $\frac{1}{9}$

b) $-\frac{1}{9}$ **49. a)** 27 **b)** −27 **51. a)** −27 **b)** 27 **53. a)** −8 **b)** 8 **55. a)** $\frac{8}{27}$ **b)** $-\frac{8}{27}$ **57.** 30

59. −19 **61.** −13.09 **63.** $-\frac{49}{72}$ **65.** 26 **67.** 14 **69.** $\frac{23}{4}$ **71.** $\frac{81}{40}$ **73.** 23 **75.** 294 **77.** 64 **79.** 16

81. $\frac{27}{5}$ **83.** Undefined **85.** −4 **87.** 0 **89.** $-\frac{10}{3}$ **91.** $\frac{242}{5}$ **93.** $\frac{1}{4}$ **95.** −7 **97.** −40 **99.** $\frac{46}{9}$

101. −80 **103.** 33 **105.** −3 **107.** $\frac{3}{2}$ **109.** $(3x + 6)^2$; 225 **111.** $6(3x + 6) - 9$; 81 **113.** $\left(\frac{x + 3}{2y}\right)^2 - 3$; 1

115. a) 0.425 million metric tons **b)** 3.95 million metric tons **117. a)** 7.62% **b)** 21.78%
119. a) 2.3 billion operations per second **b)** 2623.5 billion operations per second **121. a)** 2.30 million
b) 39.96 million **122. a)** $\{b, c, f\}$ **b)** $\{a, b, c, d, f, g, h\}$ **123.** all real numbers or ℝ **124.** $a \geq 0$ **125.** 4; −4
126. −|6|, −4, −|−2|, 0, |−5| **127.** Associative property of addition

Exercise Set 1.5

1. a) $a^m \cdot a^n = a^{m+n}$ **3. a)** $a^{-m} = \frac{1}{a^m}, a \neq 0$ **5. a)** $(a^m)^n = a^{mn}$ **7. a)** $\left(\frac{a}{b}\right)^m = \frac{a^m}{b^m}, b \neq 0$

9. $x = \frac{1}{5}$ **11. a)** The opposite of x is $-x$; the reciprocal of x is $\frac{1}{x}$ **b)** $x^{-1}; \frac{1}{x}$ **c)** $-x$ **13.** 27 **15.** 81 **17.** 16

19. 16 **21.** $\frac{1}{16}$ **23.** $-\frac{1}{16}$ **25.** $\frac{1}{125}$ **27.** $-\frac{1}{125}$ **29.** $\frac{5}{y^3}$ **31.** x^4 **33.** $2ab^3$ **35.** $\frac{5}{2m^2n^3}$ **37.** $\frac{5z^4}{x^2y^3}$

39. $\frac{1}{6xy}$ **41.** 1 **43.** −2 **45.** −1 **47.** 7 **49.** $\frac{1}{5}$ **51.** x^4 **53.** 9 **55.** $\frac{1}{49}$ **57.** $\frac{1}{x^3}$ **59.** $5w^5$ **61.** $\frac{12}{x^7}$

63. $3p$ **65.** $-10r^7$ **67.** $8x^7y^2$ **69.** $\frac{3x^2}{y^6}$ **71.** $-\frac{3x^3z^2}{y^5}$ **73.** 64 **75.** $\frac{1}{64}$ **77.** x^8 **79.** $-x^3$ **81.** $-\frac{8}{x^6}$

83. $\frac{9}{25}$ **85.** $\frac{49}{4}$ **87.** $\frac{9}{4x^2}$ **89.** $\frac{16x^4}{y^4}$ **91.** $\frac{1}{8a^9b^3}$ **93.** $-\frac{x^9}{64y^{15}}$ **95.** $\frac{36x^2}{y^4}$ **97.** $8r^6s^{15}$ **99.** $\frac{y^6}{64x^3}$ **101.** $64x^9y^3$

103. $\dfrac{z^3}{8x^3y^3}$ **105.** $\dfrac{x^{16}}{y^{10}}$ **107.** $\dfrac{x^4y^8}{4z^{12}}$ **109.** $-\dfrac{8b^{12}}{a^6c^3}$ **111.** $\dfrac{27}{8x^{21}y^9}$ **113.** x^{7a+4} **115.** w^{7b+1} **117.** x^{w+7} **119.** x^{5p+2}
121. x^{2m+2} **123.** $\dfrac{5m^{2b}}{n^{2a}}$ **125. a)** $x < 0$ or $x > 1$ **b)** $0 < x < 1$ **c)** $x = 0$ or $x = 1$ **d)** Not true for $0 \le x \le 1$
127. a) An even number of negative factors is positive **b)** An odd number of negative factors is negative
129. a) Yes **b)** Yes, because $x^{-2} = \dfrac{1}{x^2}$ and $(-x)^{-2} = \dfrac{1}{(-x)^2} = \dfrac{1}{x^2}$. **131.** −3 **133.** −1; 3 **135.** $x^{9/8}$
137. $\dfrac{1}{x^{9/2}y^{19/6}}$ **140. a)** $A \cup B = \{1, 2, 3, 4, 5, 6, 8\}$ **b)** $A \cap B = \varnothing$ **141.** 

142. −6 **143.** −5

Exercise Set 1.6

1. A number greater than or equal to 1 and less than 10 multiplied by a power of 10
3. No; $10^{-n} = \dfrac{1}{10^n}$ which is positive. **5.** 7.3×10^3 **7.** 4.7×10^{-2} **9.** 1.9×10^4 **11.** 1.86×10^{-6}
13. 5.78×10^6 **15.** 1.01×10^{-4} **17.** 6400 **19.** 0.0000213 **21.** 0.312 **23.** 9,000,000 **25.** 207,000
27. 1,000,000 **29.** 150,000,000 **31.** 0.021 **33.** 13,000,000 **35.** 10,660 **37.** 40 **39.** 0.0000005733
41. 4.5×10^{-7} **43.** 2.0×10^3 **45.** 2.13×10^{-7} **47.** 1.645×10^{12} **49.** 3.2×10^5 **51.** 3.0×10^0
53. 6.974×10^{12} **55.** 9.905×10^2 **57.** 5.337×10^2 **59.** 1.536×10^8 **61.** 7.047×10^{-27} **63.** 3.333×10^{60}
65. a) Add 1 to the exponent. **b)** Add 2 to the exponent. **c)** Add 6 to the exponent. **d)** 7.59×10^{13}
67. a) 1.00×10^4 or 10,000 **b)** 4.725×10^5 or 472,500 **c)** error in part b); because 472,500 is greater than 10,000
69. a) 1,602,739.7 miles **b)** ≈ 8347.6 mph **71.** ≈ 5.816×10^9 people **73. a)** 2.1×10^8 pounds
b) 3.99×10^9 pounds **75. a)** ≈ 3.00×10^{10} passengers **b)** ≈ 1.47×10^{10} passengers **c)** ≈ 2.04 times greater
77. a) 945 million **b)** ≈ 20.96% **c)** ≈ 327 people per square mile **d)** ≈ 73.2 people per square mile
79. a) ≈ 5.87×10^{12} miles **b)** 500 seconds or $8\frac{1}{3}$ minutes **c)** 6.72×10^{11} seconds or 21,309 years

Review Exercises

1. {3, 4, 5, 6} **2.** {0, 3, 6, 9, ...} **3.** Yes **4.** Yes **5.** Yes **6.** No **7.** 4, 6 **8.** 4, 6, 0
9. −2, 4, 6, 0 **10.** $-2, 4, 6, \frac{1}{2}, 0, \frac{15}{27}, -\frac{1}{5}, 1.47$ **11.** $\sqrt{7}, \sqrt{3}$ **12.** $-2, 4, 6, \frac{1}{2}, \sqrt{7}, \sqrt{3}, 0, \frac{15}{27}, -\frac{1}{5}, 1.47$ **13.** False
14. True **15.** True **16.** True **17.** $A \cup B = \{1, 2, 3, 4, 5\}$; $A \cap B = \{2, 3, 4, 5\}$
18. $A \cup B = \{2, 3, 4, 5, 6, 7, 8, 9\}$; $A \cap B = \varnothing$ **19.** $A \cup B = \{1, 2, 3, 4, ...\}$; $A \cap B = \{2, 4, 6, ...\}$
20. $A \cup B = \{3, 4, 5, 6, 9, 10, 11, 12\}$; $A \cap B = \{9, 10\}$ **21.** 

22. 

23. 

24. 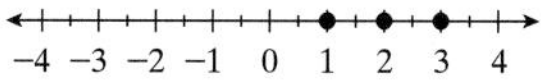

25. < **26.** < **27.** < **28.** = **29.** < **30.** >
31. > **32.** > **33.** −5, −2, 4, |7| **34.** $0, \frac{3}{5}, 2.3, |-3|$ **35.** −2, 3, |−5|, |−7| **36.** −4, −|3|, −2.1, −2
37. −4, −|−3|, 5, 6 **38.** −3, 0, |1.6|, |−2.3| **39.** Distributive property **40.** Commutative property of multiplication
41. Associative property of addition **42.** Identity property of addition **43.** Associative property of multiplication
44. Double negative property **45.** Multiplicative property of zero **46.** Inverse property of addition
47. Inverse property of multiplication **48.** Identity property of multiplication **49.** 3 **50.** 11 **51.** 8 **52.** −1
53. 0 **54.** 21 **55.** 19 **56.** −47 **57.** 15 **58.** 31 **59.** 6 **60.** 512 **61.** undefined **62.** $\dfrac{8}{3}$ **63.** 15
64. $\dfrac{26}{3}$ **65. (a)** \$816.37 million **(b)** \$4064.32 million **66. (a)** 944.53 ton-miles **(b)** 1413.7 ton-miles **67.** 64
68. x^8 **69.** x^4 **70.** y^9 **71.** x^7 **72.** $\dfrac{1}{x^3}$ **73.** $\dfrac{1}{27}$ **74.** 3 **75.** $9n^4$ **76.** $\dfrac{3}{2}$ **77.** $\dfrac{16}{9}$ **78.** $\dfrac{y^2}{x}$ **79.** $-21x^3y^9$
80. $\dfrac{8}{x^2y}$ **81.** $\dfrac{3y^7}{x^5}$ **82.** $\dfrac{3}{xy^9}$ **83.** $\dfrac{a^5}{b^4c^6}$ **84.** $\dfrac{4p^2}{qr^6}$ **85.** $125a^3b^3$ **86.** $\dfrac{x^{10}}{9y^2}$ **87.** x^6y^8 **88.** $-\dfrac{125y^3}{x^6z^9}$ **89.** $\dfrac{z^4}{36x^2y^6}$
90. $\dfrac{m^9}{27}$ **91.** $\dfrac{x^{12}}{16y^8}$ **92.** $-\dfrac{64y^3}{x^3z^{15}}$ **93.** $\dfrac{9x^{10}}{4y^{14}z^{12}}$ **94.** $-\dfrac{x^6z^2}{8y^2}$ **95.** 7.42×10^{-5} **96.** 2.6×10^5 **97.** 1.83×10^5
98. 1.0×10^{-6} **99.** 30,000 **100.** 0.02 **101.** 200,000,000 **102.** 2000 **103. a)** \$$2.32 \times 10^7$ or \$23,200,000
b) ≈ 1.24 times larger **104. a)** 10,400,000,000 **b)** 10.4 billion km **c)** 5.2×10^8 km or 520,000,000 km
d) 6.24×10^9 miles or 6,240,000,000 miles

Practice Test **1.** $A = \{6, 7, 8, 9, \ldots\}$ **2.** True **3.** True **4.** $-\frac{3}{5}, 2, -4, 0, \frac{19}{12}, 2.57, -1.92$

5. $-\frac{3}{5}, 2, -4, 0, \frac{19}{12}, 2.57, \sqrt{8}, \sqrt{2}, -1.92$ **6.** $A \cup B = \{5, 7, 8, 9, 10, 11, 14\}; A \cap B = \{8, 10\}$
7. $A \cup B = \{1, 3, 5, 7, \ldots\}; A \cap B = \{3, 5, 7, 9, 11\}$ **8.** 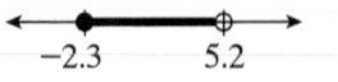

9. 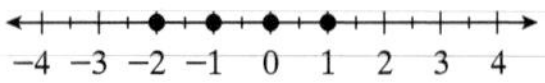

10. $-|4|, -2, |3|, 6$ **11.** Associative property of addition **12.** Commutative property of addition **13.** -5
14. 23 **15.** Undefined **16.** $-\frac{37}{22}$ **17.** 17 **18. a)** \$0.99 **b)** \$3.36 **19.** $\frac{1}{9}$ **20.** $\frac{9}{x^4y^2}$ **21.** $\frac{y^5z^7}{3x^3}$
22. $-\frac{y^{21}}{27x^{12}}$ **23.** 2.42×10^8 **24.** 260,000,000 **25. a)** $\approx 5.528 \times 10^7$
b) $\approx 4.672 \times 10^7$ **c)** $\approx 8.56 \times 10^6$

CHAPTER 2

Using Your Calculator, 2.1 **1.** No **2.** Yes

Exercise Set 2.1 **1.** The terms of an expression are the parts added. **3. a)** $\frac{1}{4}$ **b)** -1 **c)** $-\frac{3}{5}$
5. a) Like terms have the same variables and exponents. **b)** No; the exponent on x is different for each term.
7. No **9.** If $a = b$, then $a + c = b + c$ **11. a)** An identity is an equation that is always true. **b)** $\mathbb{R}$
13. b) -4 **15.** Symmetric property **17.** Transitive property **19.** Reflexive property
21. Addition property of equality **23.** Multiplication property of equality **25.** Multiplication property of equality
27. First **29.** Second **31.** Zero **33.** First **35.** Thirteenth **37.** Twelfth **39.** $15x - 5$ **41.** $5x^2 - x - 5$
43. $8.7c^2 + 3.6c$ **45.** Cannot be simplified **47.** $4xy + y^2 - 2$ **49.** $-3.56x - 42.76$ **51.** $\frac{8}{3}x + \frac{13}{2}$
53. $-17x - 4$ **55.** $6x - 3y$ **57.** $-9b + 93$ **59.** $4r^2 - 2rs + 3r + 4s$ **61.** 4 **63.** $\frac{3}{2}$ **65.** 2 **67.** 15
69. 5 **71.** 1 **73.** 1 **75.** -2 **77.** 1 **79.** -1 **81.** 5 **83.** 5 **85.** 2 **87.** $-\frac{1}{2}$ **89.** 6 **91.** 2 **93.** 0
95. -36 **97.** -4 **99.** 12 **101.** -4 **103.** $\frac{15}{16}$ **105.** 3.86 **107.** 1.18 **109.** $-15{,}000.00$ **111.** 1701.39
113. 47,800 **115.** $\varnothing$; contradiction **117.** $\{0\}$; conditional **119.** $\mathbb{R}$; identity **121.** $\mathbb{R}$; identity
123. $\varnothing$; contradiction **125. a)** 33 **b)** 1997 **127. a)** \$60 billion **b)** 1999
129. Answers may vary. One possible answer is $2x = 8, x + 3 = 7$, and $x - 2 = 2$. **131.** Answers may vary. One possible answer is $x + 5 = x + 5$. Make sure that the equation is always true.
133. Answers may vary. One possible answer is $4 + 3x + 1 = x + 9$. **135.** Substitute 6 for x and solve for n; 18
137. $\triangle = \frac{\odot + \square}{*}$ **139.** $\triangle = \frac{\otimes}{\odot + \square}$ **141. b)** $|a| = \begin{cases} a \text{ if } a > 0 \\ -a \text{ if } a < 0 \end{cases}$ **142. a)** -9 **b)** 9 **143.** $-\frac{27}{64}$ **144.** -4

Exercise Set 2.2 **1.** A mathematical model is an expression or equation that represents a real life situation.
3. 1. Understand; 2. Translate; 3. Carry out; 4. Check; 5. Answer **5. a)** $w = 1$ **b)** $w = \frac{P - 2l}{2}$ **c)** No
d) They are the same because the formula and the equation are equivalent when $P = 10$ and $l = 4$. **7.** 70 **9.** 42
11. 201.06 **13.** 250 **15.** 150 **17.** $\frac{7}{4}$ **19.** $\sqrt{145} \approx 12.04$ **21.** 4 **23.** 119.10 **25.** $y = -3x + 5$
27. $y = 2x + 5$ **29.** $y = \frac{5x + 4}{3}$ **31.** $y = \frac{-x + 12}{4}$ **33.** $y = x + 2$ **35.** $y = \frac{3x - 17}{6}$ **37.** $t = \frac{d}{r}$
39. $l = \frac{A}{w}$ **41.** $w = \frac{P - 2l}{2}$ **43.** $h = \frac{V}{lw}$ **45.** $b = \frac{2A}{h}$ **47.** $y = \frac{C - Ax}{B}$ **49.** $m = \frac{y - b}{x}$
51. $m = \frac{y - y_1}{x - x_1}$ **53.** $\mu = x - z\sigma$ **55.** $T_2 = \frac{T_1P_2}{P_1}$ **57.** $h = \frac{2A}{b_1 + b_2}$ **59.** $n = \frac{2S}{f + l}$ **61.** $F = \frac{9}{5}C + 32$
63. $m_1 = \frac{Fd^2}{km_2}$ **65. a)** $m = 1.15k$ **b)** The quotient of 6076 and 5280 is about 1.15. **c)** ≈ 23.58 mph **67.** \$90
69. 4 years **71.** 8 feet **73.** 1256.64 square feet **75. a)** ≈ 667.98 cubic inches **b)** ≈ 2.89 gallons

c) ≈ 2.89 ounces **77.** \$11,264.93 **79.** $4\frac{1}{4}\%$ tax-free **81. a)** 146.25 beats per minute **b)** 34.29 years

83. a) $d = \dfrac{\frac{g}{3}}{u}$ **b)** Every 71.67 days **85. a)** $\text{BMI} = \dfrac{w}{h^2}$ **b)** $\text{BMI} = \dfrac{w(705)}{h^2}$ **c)** Answers will vary. **87.** -22

88. $\dfrac{3}{2}$ **89.** 15 **90.** 78

Exercise Set 2.3

1. $A + 4A = 180; A = 36°; B = 144°$ **3.** $x + x + (20 + 2x) = 180$; 40°, 60°, 80° **5.** $p - 0.25p = 187.50$; \$250 **7.** $12.50x = 940$; 75.2 weeks **9.** $p + 0.073p = 22{,}600$; \$21,062.44 **11.** $0.20n + 35 = 80$; 225 miles **13.** $0.5t + 20 = 2t$; more than 13 times **15.** $t + 4.45t = 622{,}608$; aircraft carrier, 114,240 tons; *Jahre Viking*, 508,368 tons **17.** $g + 2g - 250 = 5000$; \$1750 to the global equities fund; \$3250 to the stock fund **19.** $x + (2x - 5) + (2x + 2) = 57$; 12 grasses; 19 weeds; 26 trees **21.** $460(0.10)x = 260$; 5.65 years **23.** $x + 0.07x + 0.15x = 15.75$; \$12.91 **25.** $m + 0.251m = 50.8$; 40.6 hours **27.** $d + (d + 428{,}000) + (d + 428{,}000 + 812{,}000) + (2d - 115{,}000) = 10{,}508{,}000$; Dell: 1,791,000; IBM: 2,219,000; Packard Bell-NEC: 3,031,000; Compaq: 3,467,000 **29. a)** $563.50x = 538.30x + 0.02(70{,}000) + 200$; ≈ 63.49 months **b)** First National **31. a)** $510x = 420.50x + 2500$; ≈ 28 months **b)** Yes **33.** $7x = 91$; 13 meters by 13 meters **35.** $6w + 4(w + 1) = 114$; width is 11 meters; length is 12 meters **37.** $p - 0.10p - 5 = 49$; \$60 **39.** $50x = 810 + 0.10(50x)$; 18 **41.** $0.07(375 - x) = 17.50$; \$125

43. $\frac{5}{8}x + \frac{3}{8}x + 0.15x = 184.60$; Newton family: \$100.33; Lee family: \$84.27

45. a) $\dfrac{70 + 83 + 97 + 84 + 74 + x + x}{7} = 80$; 76 points **b)** No, he would need a score of 111.

47. Answers will vary. **49.** 220 miles **51.** $\dfrac{13}{5}$ **52.** -2.7 **53.** $\dfrac{5}{16}$ **54.** -5 **55.** $\dfrac{y^{18}}{8x^{12}}$

Exercise Set 2.4

1. $5(1.2) + 4.5(1.2) = d$; 11.4 miles **3.** $16t - 14t = 4$; 2 hours **5.** $3(r + 18) = (3 + 1.2)r$; Freight train travels at 45 mph; passenger train travels at 63 mph **7. a)** $2(4r) - 2r = 18$; 12 mph **b)** 24 miles **c)** \$36 **9. a)** $3.4t = 1.2(12 - t)$; 3.13 hours **b)** 21.3 miles **11.** $400t + 600(t + 2) = 15{,}000$; 13.8 hours **13.** $0.04p + 0.05(11{,}000 - p) = 530$; \$2000 at 4%; \$9000 at 5% **15. a)** $108.75m + 27.25(4m) = 10{,}000$; Microsoft, 45 shares; Hilton, 180 shares **b)** \$201.25 **17.** $6.20x + 5.80(18) = 6.10(x + 18)$; 54 pounds **19.** $0.12x + 0.05(40) = 0.08(x + 40)$; 30 ounces **21.** $0.115c + 0(32) = 0.05(c + 32)$; about 24.6 ounces **23.** $0.76(16) + 12x = 0.82(28)$; 90% **25.** $28{,}200 - x = 32{,}450 - (6400 - x)$; Mr. is \$1075; Mrs. is \$5325 **27.** $4r + 4(r + 10) = 480$; Bob at 55 mph; Julie at 65 mph **29.** $0.06x = 0.10(8000 - x)$; \$5000 was invested at 6%; \$3000 was invested at 10% **31.** $500t + 550(2t) = 5200$; 9.75 hours **33. a)** $800t = 520(t + 2)$; ≈ 3.71 hours **b)** ≈ 2971.4 miles **35.** $6.00x + 6.50(18 - x) = 114$; 6 hours at \$6.00; 12 hours at \$6.50 **37.** $1(r) + 1(r - 16) = 100$; 58 mph **39.** $0.80x + 0(128 - x) = 0.06(128)$; 9.6 ounces of the 80% solution; 118.4 ounces of water **41.** $0.05(400) + 0.015x = 0.02(x + 400)$; 2400 quarts **43.** $70t = 50t + 200$; 10 minutes **45.** $D - 6(5) - 6(5) = 20$; 80 feet **47.** Answers will vary. **49.** ≈ 149 miles **51.** 6 quarts **52.** -5.7

53. $\dfrac{18}{5}$ **54.** $y = \dfrac{x - 42}{30}$ **55.** 140 miles

Exercise Set 2.5

1. $x \le 7$ includes 7 and $x < 7$ does not include 7. **3. a)** Use open circles when the endpoints are not included. **b)** Use closed circles when the endpoints are included. **c)** Answers will vary. One possible answer is $x > 4$. **d)** Answers will vary. One possible answer is $x \ge 4$. **5.** No real number is both greater than 4 and less than 2. **7. a)** -4 **b)** $(-\infty, -4)$ **c)** $\{x | x < -4\}$

9. a) 5.2 **b)** $[5.2, \infty)$ **c)** $\{x | x \ge 5.2\}$ **11. a)** 2, $\frac{12}{5}$ **b)** $\left[2, \frac{12}{5}\right)$ **c)** $\left\{x \middle| 2 \le x < \frac{12}{5}\right\}$

13. a) -7, -4 **b)** $(-7, -4]$ **c)** $\{x | -7 < x \le -4\}$ **15.** 6 **17.** 7

19. 3.6 **21.** 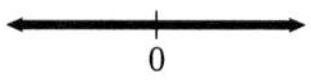 0 **23.**  -15 **25.** 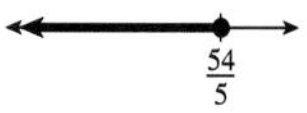 $\frac{54}{5}$ **27.** $(-\infty, 18)$

29. $\left(-\infty, \frac{3}{2}\right)$ **31.** $\varnothing$ **33.** $(-\infty, \infty)$ **35.** $(-2, 6)$ **37.** $\left(-\frac{3}{5}, \frac{8}{5}\right]$ **39.** $\left[4, \frac{11}{2}\right)$ **41.** $(2, 5.8]$

43. $\left\{x \middle| \frac{5}{2} < x \le \frac{13}{2}\right\}$ **45.** $\{x|0 < x \le 1\}$ **47.** $\{x|3 < x \le 33\}$ **49.** $\{x|0 < x < 4\}$ **51.** $\varnothing$

53. $\{x|-5 < x < 2\}$ **55.** $(-\infty, 4)$ **57.** $[0, 2]$ **59.** $(-\infty, 0) \cup (6, \infty)$ **61. a)** $l + w + d \le 61$ **b)** $10\frac{1}{2}$ inches

63. $70x \le 800$; 11 boxes **65.** $4.25 + 0.45x \le 9.50$; 14 minutes **67.** $10{,}025 + 1.09x < 6.42x$; 1881 books

69. $85 + 0.256x < 0.33x$; 1149 pieces **71.** $8(90) + 6.25x \le 2000$; 204 hours **73.** $\frac{90 + 87 + 96 + 79 + x}{5} \ge 90$; 98

75. $80 \le \frac{87 + 92 + 70 + 75 + x}{5} < 90$; $76 \le x \le 100$ **77.** $7.2 < \frac{7.48 + 7.15 + x}{3} < 7.8$; $6.97 < x < 8.77$

79. a) 1990, 1991, and 1992 **b)** 1991 to 1996 **81.** No, $-1 > -2$ but $(-1)^2 < (-2)^2$. **83.** Answers will vary.
85. $84 \le x \le 100$ **87. a)** Answers will vary. **b)** $(-3, \infty)$ **89. a)** $A \cup B = \{1, 2, 3, 4, 5, 6, 8, 9\}$
b) $A \cap B = \{1, 8\}$ **90. a)** 4 **b)** 0, 4 **c)** $-3, 4, \frac{5}{2}, 0, -\frac{29}{80}$ **d)** $-3, 4, \frac{5}{2}, \sqrt{7}, 0, -\frac{29}{80}$

91. Associative property of addition **92.** Commutative property of addition **93.** $V = \frac{R - L + Dr}{r}$

Exercise Set 2.6

1. Set $x = a$ or $x = -a$. **3.** $\varnothing$; There is no real number whose absolute value is negative.
5. Write $-a < x < a$. **7.** Write $x < -a$ or $x > a$. **9. a)** $m < x < n$; 
b) $x < m$ or $x > n$;  **11. a)** No solution **b)** One **c)** Two **13. a)** C **b)** A **c)** D
d) B **e)** E **15.** $\{-5, 5\}$ **17.** $\{-12, 12\}$ **19.** $\varnothing$ **21.** $\{-12, 2\}$ **23.** $\{-16, 4\}$ **25.** $\left\{\frac{3}{2}, \frac{11}{6}\right\}$ **27.** $\{-17, 23\}$

29. $\{3\}$ **31.** $\{y|-5 \le y \le 5\}$ **33.** $\{x|-2 \le x \le 16\}$ **35.** $\left\{z \middle| 0 \le z \le \frac{10}{3}\right\}$ **37.** $\{x|-9 \le x \le 6\}$

39. $\{x|-1.9 \le x \le 2.7\}$ **41.** $\varnothing$ **43.** $\left\{x \middle| -4 < x < \frac{52}{3}\right\}$ **45.** $\left\{\frac{1}{4}\right\}$ **47.** $\{x|x < -3 \text{ or } x > 3\}$

49. $\{x|x < -9 \text{ or } x > 1\}$ **51.** $\left\{y \middle| y \le -\frac{4}{3} \text{ or } y \ge 4\right\}$ **53.** $\left\{w \middle| w \le -\frac{35}{3} \text{ or } w \ge 15\right\}$ **55.** $\{x|x < 2 \text{ or } x > 6\}$

57. $\{x|x \le -18 \text{ or } x \ge 2\}$ **59.** $\mathbb{R}$ **61.** $\{x|x < 2 \text{ or } x > 2\}$ **63.** $\left\{\frac{4}{3}, 5\right\}$ **65.** $\{-3, 1\}$ **67.** $\left\{-\frac{12}{5}, 28\right\}$

69. $\left\{\frac{2}{5}\right\}$ **71.** $\{-7, 7\}$ **73.** $\{x|-2 < x < 8\}$ **75.** $\{x|x < -14 \text{ or } x > 4\}$ **77.** $\left\{-\frac{11}{4}, \frac{7}{4}\right\}$

79. $\left\{x \middle| x < -\frac{5}{2} \text{ or } x > -\frac{5}{2}\right\}$ **81.** $\left\{x \middle| -\frac{13}{3} \le x \le \frac{5}{3}\right\}$ **83.** $\varnothing$ **85.** $\{w|-16 < w < 8\}$ **87.** $\mathbb{R}$ **89.** $\left\{2, \frac{22}{3}\right\}$

91. $\left\{-\frac{3}{2}, \frac{9}{7}\right\}$ **93. a)** $[3.5, 4.5]$ **b)** Between 3.5 feet and 4.5 feet, inclusive **95.** $|x| = 5$ **97.** $|x| \ge 5$

99. $x = -\frac{b}{a}$ **101. a)** Set $ax + b = -c$ or $ax + b = c$ and solve each equation for x.

b) $\left\{x \middle| x = \frac{-c - b}{a} \text{ or } x = \frac{c - b}{a}\right\}$ **103. a)** Write $ax + b < -c$ or $ax + b > c$ and solve each inequality for x.

b) $\left\{x \middle| x < \frac{-c - b}{a} \text{ or } x > \frac{c - b}{a}\right\}$ **105.** $\mathbb{R}$ **107.** $\{x|x \ge 0\}$ **109.** $\{2\}$ **111.** $\{x|x \le 2\}$ **113.** $\{3\}$ **115.** $\varnothing$

117. $\frac{29}{72}$ **118.** 25 **119.** 1.33 miles **120.** $\{x|x < 4\}$

Review Exercises

1. Tenth **2.** First **3.** Seventh **4.** Cannot be simplified **5.** $7x^2 + 2xy - 4$ **6.** 8
7. $4x - 3y + 6$ **8.** $\frac{49}{6}$ **9.** 20 **10.** $-\frac{13}{3}$ **11.** 0 **12.** $-\frac{9}{2}$ **13.** No solution **14.** $\mathbb{R}$ **15.** 200

16. $\frac{1}{4}$ **17.** 176 **18.** -4 **19.** $l = \frac{A}{w}$ **20.** $h = \frac{A}{\pi r^2}$ **21.** $w = \frac{P - 2l}{2}$ **22.** $r = \frac{d}{t}$ **23.** $m = \frac{y - b}{x}$

24. $y = \frac{2x - 5}{3}$ **25.** $V_2 = \frac{P_1V_1}{P_2}$ **26.** $a = \frac{2S - b}{3}$ **27.** $l = \frac{K - 2d}{2}$ **28.** $x - 0.60x = 20$; \$50
29. $4750 + 350x = 5800$; 3 years **30.** $300 + 0.06x = 650$; \$5833.33

31. $3.30x = 27.50$; 9 or more **32.** $x - 0.40x - 20 = 120$; \$233.33 **33.** ≈ 30.6 rolls of film each hour **34.** \$6000 at 8%; \$4000 at 5% **35.** $2\frac{2}{3}$ hours **36. a)** 3000 mph **b)** 16,500 miles **37.** 15 pounds of \$6.00 coffee; 25 pounds of \$6.80 coffee **38.** \$25 **39. a)** 1 hour **b)** 14.4 miles **40.** 40°; 65°; 75° **41.** 300 gallons per hour; 450 gallons per hour **42.** 24; 25 **43.** 7.5 ounces **44.** \$4500 at 10%; \$7500 at 6% **45.** more than 5 **46.** 90 mph **47.** 7 **48.** −3 **49.**  $\frac{5}{2}$ **50.**  $\frac{21}{4}$ **51.**  $-\frac{9}{2}$ **52.**  −10 **53.**  $\frac{2}{5}$ **54.**  $\frac{20}{9}$ **55.** $468 + 80x \le 1525$; 13 boxes **56.** $4.50 + 0.95x \le 8.65$; 7 minutes **57.** $3 + 1.5x \ge 27$; 17 weeks **58.** $(5, 11)$ **59.** $[-3, 3)$ **60.** $\left(\frac{7}{2}, 6\right)$ **61.** $\left(\frac{8}{3}, 6\right)$ **62.** $\left[-\frac{1}{2}, \frac{23}{2}\right)$ **63.** $(2, 14)$ **64.** $\{x|81 \le x \le 100\}$ **65.** $\{x|-3 < x < 3\}$ **66.** $\mathbb{R}$ **67.** $\{x|x \le -4\}$ **68.** $\left\{x\middle|x \le -4 \text{ or } x > \frac{17}{5}\right\}$ **69.** $\{-4, 4\}$ **70.** $\{x|-3 < x < 3\}$ **71.** $\{x|x \le -4 \text{ or } x \ge 4\}$ **72.** $\{-5, 13\}$ **73.** $\{x|x \le -3 \text{ or } x \ge 7\}$ **74.** $\left\{-\frac{1}{2}, \frac{9}{2}\right\}$ **75.** $\{x|-2 < x < 5\}$ **76.** $\{-1, 4\}$ **77.** $\{x|-14 < x < 22\}$ **78.** $\left\{\frac{1}{4}, \frac{7}{2}\right\}$ **79.** $\mathbb{R}$ **80.** $\varnothing$ **81.** $(4, 7]$ **82.** $\left(-\frac{17}{2}, \frac{27}{2}\right]$ **83.** $[-2, 4]$ **84.** $(-\infty, \infty)$ **85.** $\left(\frac{2}{3}, 8\right]$

Practice Test **1.** Sixth **2.** $16p - 3q - 4pq$ **3.** $x - 7$ **4.** $\frac{27}{7}$ **5.** $-\frac{34}{5}$ **6.** 2 **7.** $\varnothing$ **8.** $\mathbb{R}$ **9.** $\frac{13}{3}$ **10.** $b = \frac{a - 2c}{3}$ **11.** $b_2 = \frac{2A - hb_1}{h}$ **12.** $x + 0.07x = 668.75$; \$625 **13.** $35 + 0.15x = 65$; 200 miles **14.** $4(1.25) + (5.25)(1.25) = x$; ≈ 11.56 miles **15.** $0.12x + 0.25(10) = 0.20(x + 10)$; 6.25 liters **16.** $0.08x + 0.07(12{,}000 - x) = 910$; \$7000 at 8%; \$5000 at 7% **17.**  −3 **18.** 33 **19.** $\left(\frac{9}{2}, 7\right]$ **20.** $(-12, 12)$ **21.** $\{-1, 9\}$ **22.** $\left\{-\frac{14}{3}, \frac{26}{5}\right\}$ **23.** $\left\{\frac{2}{3}\right\}$ **24.** $\{x|x < -1 \text{ or } x > 4\}$ **25.** $\left\{x\middle|\frac{1}{2} \le x \le \frac{5}{2}\right\}$

Cumulative Review Test **1. a)** $A \cup B = \{1, 2, 3, 4, 5, 6, 7, 9, 10, 12\}$ **b)** $A \cap B = \{4, 6, 9, 12\}$ **2. a)** Commutative property of addition **b)** Associative property of multiplication **c)** Distributive property **3.** < **4.** −80 **5.** −15 **6.** −29 **7.** 7 **8.** $\frac{1}{4x^8y^6}$ **9.** $\frac{9m^{10}}{n^{12}}$ **10.** r^{5m-8} **11.** 4.06×10^7 **12.** $\$1.95 \times 10^9$ or \$1,950,000,000 **13.** $\frac{1}{5}$ **14.** 1.15 **15.** $-\frac{56}{33}$ **16.** 10 **17.** A conditional equation is true only under specific conditions. An identity is true for all values of the variable. An inconsistent equation is never true. **18.** 3 **19.** $t = \frac{A - p}{pr}$ **20. a)** −2, $\frac{8}{5}$ **b)** $\left\{x\middle|-2 < x < \frac{8}{5}\right\}$ **c)** $\left(-2, \frac{8}{5}\right)$ **21.** $\{-5, 1\}$ **22.** $\{x|x \le -10 \text{ or } x \ge 14\}$ **23.** \$2250 **24.** 40 mph; 50 mph **25.** $1\frac{1}{3}$ liters of the 20% solution; $\frac{2}{3}$ liter of the 50% solution

CHAPTER 3

Exercise Set 3.1 **1. a)** A straight line **b)** Two; Two points uniquely determine a straight line. **3.** 0; $\frac{1}{0}$ is undefined. **5.** $A(3, 1), B(-6, 0), C(2, -4), D(-2, -4), E(0, 3), F(-8, 1), G\left(\frac{3}{2}, -1\right)$

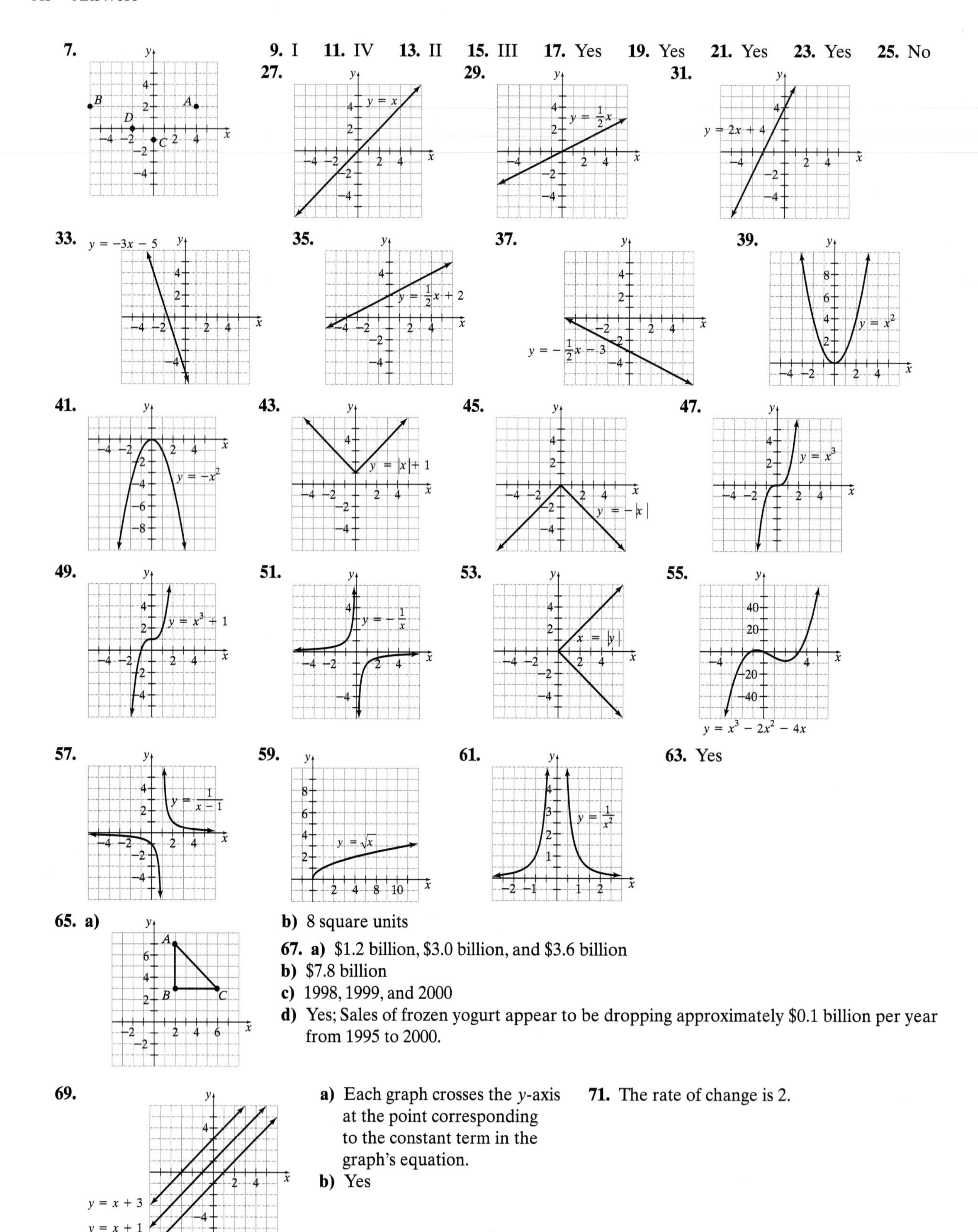

9. I **11.** IV **13.** II **15.** III **17.** Yes **19.** Yes **21.** Yes **23.** Yes **25.** No

63. Yes

65. b) 8 square units

67. a) \$1.2 billion, \$3.0 billion, and \$3.6 billion
b) \$7.8 billion
c) 1998, 1999, and 2000
d) Yes; Sales of frozen yogurt appear to be dropping approximately \$0.1 billion per year from 1995 to 2000.

69. a) Each graph crosses the y-axis at the point corresponding to the constant term in the graph's equation.
b) Yes

71. The rate of change is 2.

73. The rate of change is 3. **75.** Answers may vary. One possible answer is the points (2, −1) and (3, 2). **77.** c
79. a **81.** b **83.** a

85. 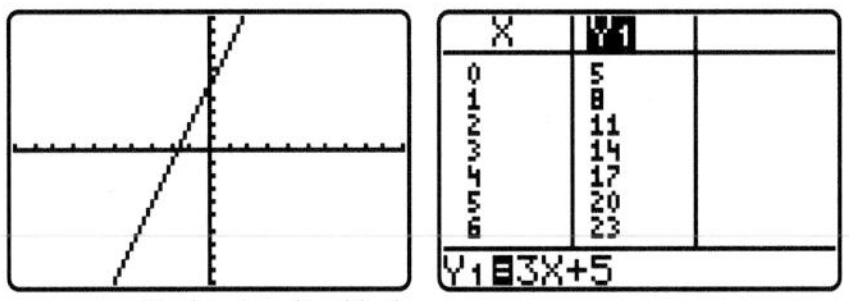

−10, 10, 1, −10, 10, 1

87. 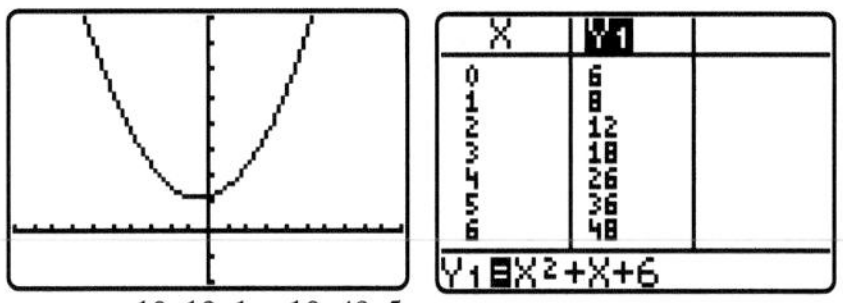

−10, 10, 1, −10, 40, 5

89. 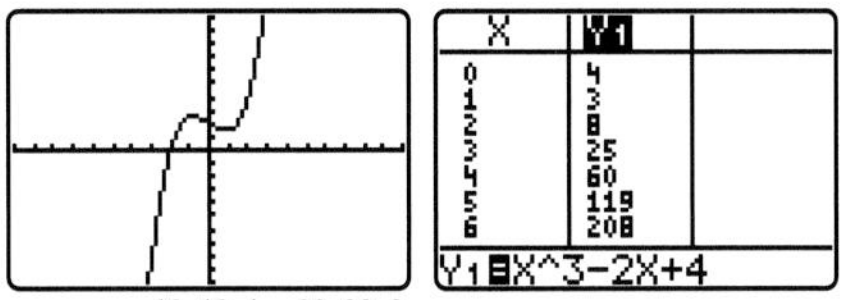

−10, 10, 1, −20, 20, 2

91. 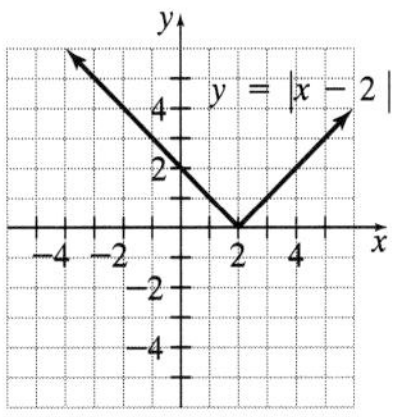

95. $\frac{3}{2}$
96. 140 miles
97. $\{x|-2 < x \le 4\}$
98. $\left\{x \middle| x < -\frac{7}{3} \text{ or } x > 1\right\}$

Exercise Set 3.2

1. Any set of ordered pairs **3.** No, a relation can have two ordered pairs with the same first element but a function cannot. **5.** If each vertical line drawn through any part of the graph intersects the graph in at most one point, the graph represents a function. **7.** The set of values for the dependent variable
9. Domain: $\mathbb{R}$; Range: $\mathbb{R}$ **11.** If y depends on x, then y is the dependent variable. **13.** f of x **15. a)** A function **b)** Domain: {3, 5, 10}; Range: {6, 10, 20} **17. a)** A function **b)** Domain: {Ron, Jayne, Cecilia}; Range: {18, 19}
19. a) Not a function **b)** Domain: {1990, 1996, 1999}; Range: {20, 32, 33} **21. a)** A function **b)** Domain: {1, 2, 3, 4, 5}; Range: {1, 2, 3, 4, 5} **23. a)** A function **b)** Domain: {1, 2, 3, 4, 5, 7}; Range: {−1, 0, 2, 4, 5} **25. a)** Not a function **b)** Domain: {1, 2, 3, 5}; Range: {−4, −1, 0, 1, 2}
27. a) Not a function **b)** Domain: {0, 1, 2}; Range: {−7, −1, 2, 3} **29. a)** A function **b)** Domain: $\mathbb{R}$; Range: $\mathbb{R}$ **c)** $x = 2$ **31. a)** Not a function **b)** Domain: {−2}; Range: $\mathbb{R}$ **c)** $x = -2$ **33. a)** Not a function **b)** Domain: $\{x|-4 \le x \le 4\}$; Range: $\{y|-2 \le y \le 2\}$ **c)** $x = 0$ **35. a)** Not a function **b)** Domain: $\mathbb{R}$; Range: $\mathbb{R}$ **c)** $x = 2$ **37. a)** A function **b)** Domain: {1, 2, 3}; Range: {1} **c)** No values of x **39. a)** A function **b)** Domain: $\{x|-20 \le x \le 10\}$; Range: $\{y|-2 \le y \le 2\}$ **c)** $x = -17.5$ or $x = -7.5$ or $x = 2.5$ **41. a)** 21 **b)** 0
43. a) −6 **b)** −4 **45. a)** 12 **b)** −13 **47. a)** 2 **b)** 9.6 **49. a)** 2 **b)** 5 **51. a)** 0 **b)** −5
53. a) 240 miles **b)** 720 miles **55. a)** $C(r) = 2\pi r$ **b)** ≈ 56.5 feet **57. a)** $F(C) = \frac{9}{5}C + 32$ **b)** 68°F
59. a) 18.23°C **b)** 27.68°C **61. a)** 78.32° **b)** 73.04° **63. a)** 91 oranges **b)** 204 oranges
65. Answers will vary. One possible interpretation: The man walks on level ground, about 30 feet above sea level, for 5 minutes. For the next 5 minutes he walks uphill to 45 feet above sea level. For 5 minutes he walks on level ground, then walks quickly downhill for 3 minutes to an elevation of 20 feet above sea level. For 7 minutes he walks on level ground. Then he walks quickly uphill.
67. Answers will vary. One possible interpretation: A woman drives in stop-and-go traffic for 5 minutes. Then she drives on the highway for 15 minutes, gets off onto a country road for a few minutes, and returns to stop-and-go traffic.
69. a) Yes **b)** The number of active U.S. military personnel in all branches from 1975 to 1997 **c)** 1,700,000 **d)** 34% **71. a)** Yes **b)** No **c)** Yes **d)** 1994 **e)** 1991
73. a) 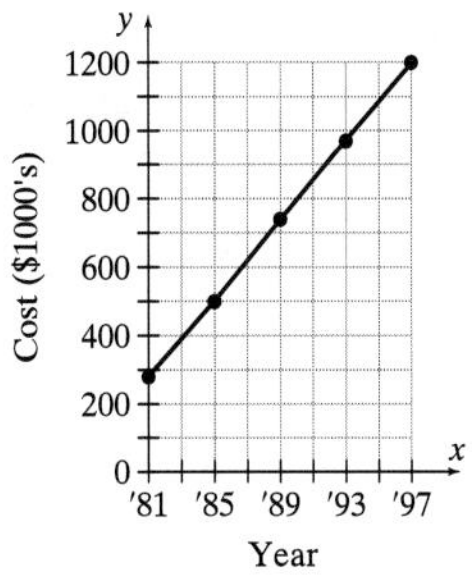

b) Yes **c)** $1,100,000 **75. a)** 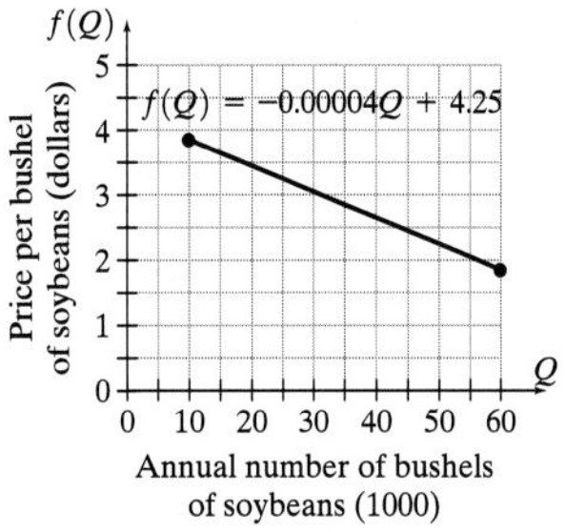

b) $2.65 per bushel

78. $x = \frac{1}{2}$ **79.** $p^2 = \frac{E - a_1p_1 - a_3p_3}{a_2}$ **80. a)** 

b) $(3, \infty)$ **c)** $\{x|x > 3\}$
81. $x = -2$ or $x = 10$

Exercise Set 3.3 **1.** $ax + by = c$ **3.** To find the x-intercept, set $y = 0$ and solve for x. To find the y-intercept, set $x = 0$ and solve for y. **5.** Horizontal line **7.** Vertical line **9.** Graph both sides of the equation. The solution is the x-coordinate of the intersection. **11.** $3x - y = 2$ **13.** $2x - 3y = -4$ **15.** $3x - 4y = -14$

17.

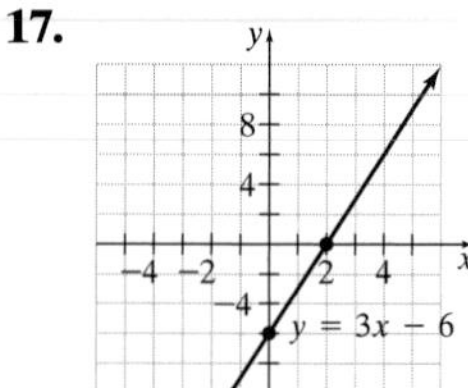

19.

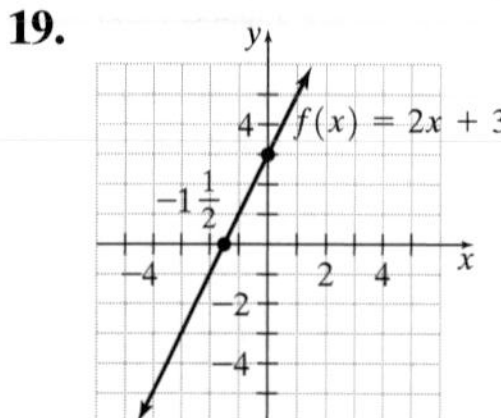

21.

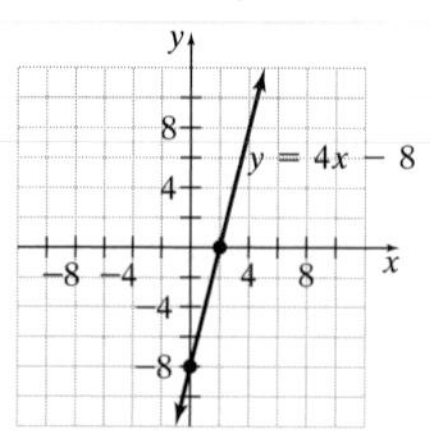

23.

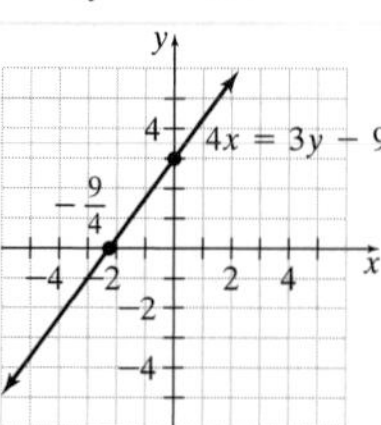

25.

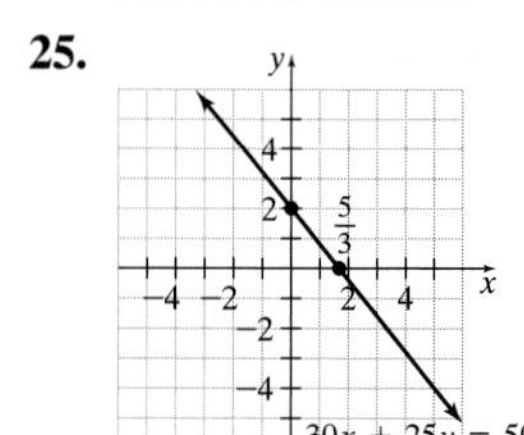

27.

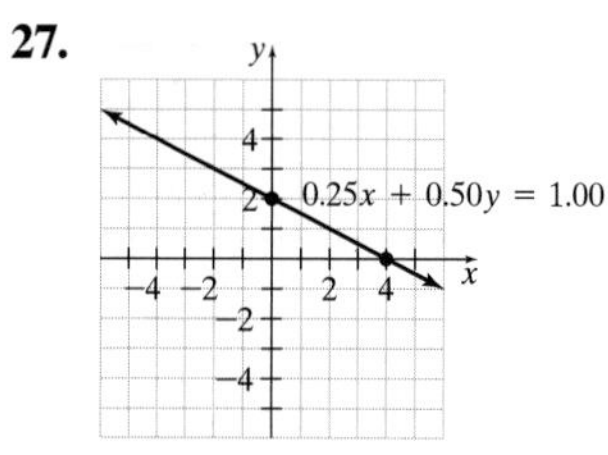

29.

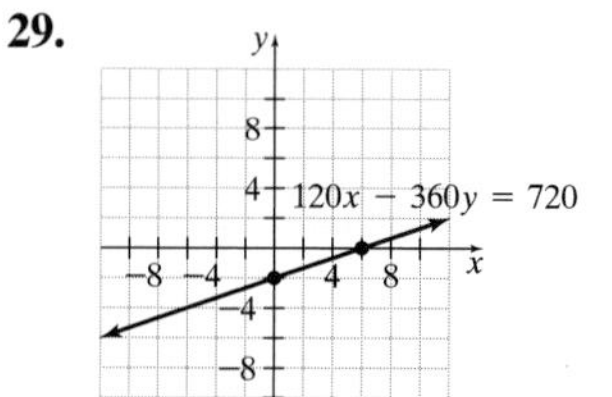

31.

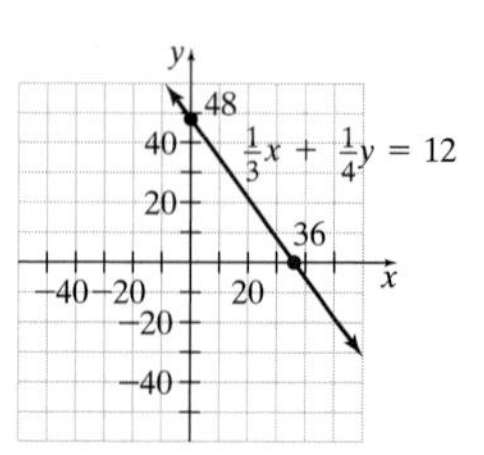

33.

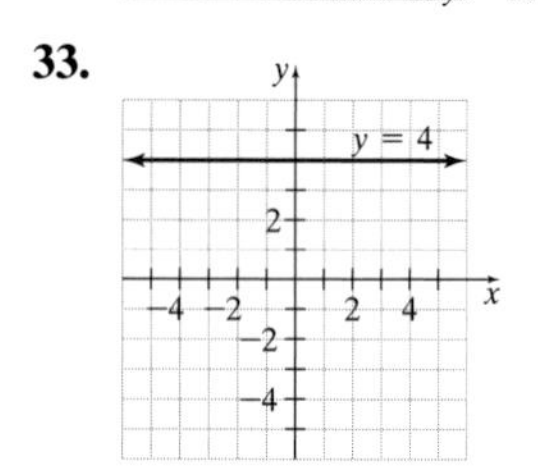

35.

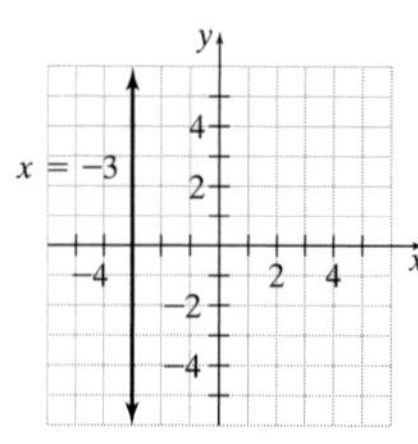

37.

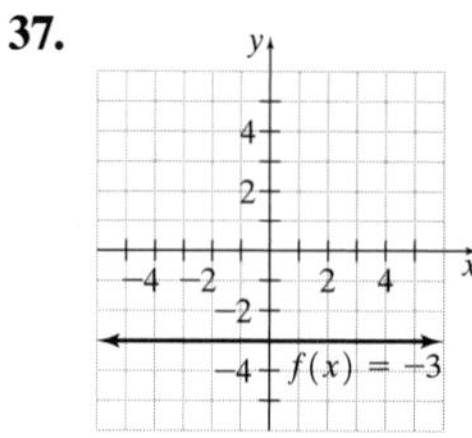

39.

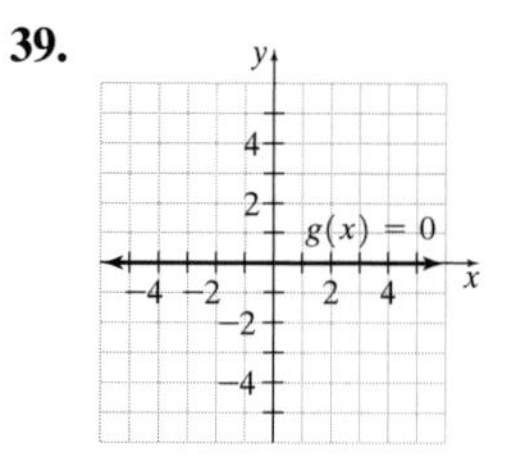

41.

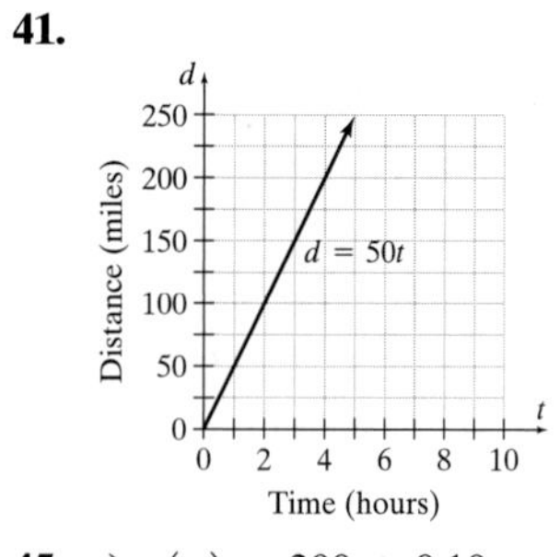

43. a)

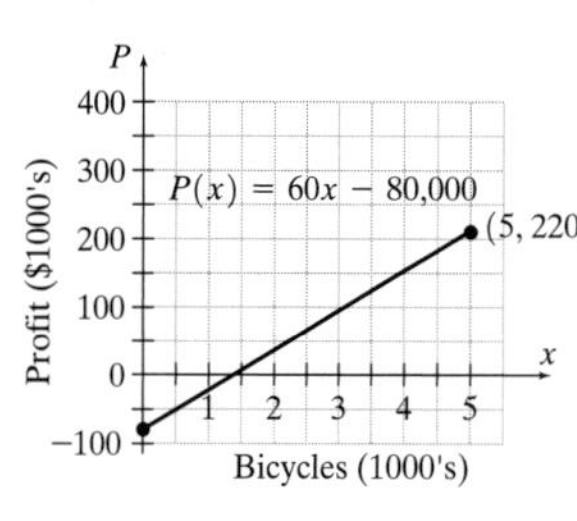

b) About 1300 bicycles
c) About 3800 bicycles

45. a) $s(x) = 300 + 0.10x$
b)

c) \$700
d) \$3000

47. a) There is only one y-value for each x-value.
b) Length; weight
c) Yes
d) 11.5 kilograms
e) 65 centimeters
f) 12.0–15.5 kilograms
g) Increases

49. When the graph goes through the origin
51. Answers will vary. One possible answer is $f(x) = 4$.
53. Both intercepts will be at 0.

55. a)

b) 2 (or −2) units **c)** 4 (or −4) units **d)** $\frac{1}{2}$; slope

57. $x = -2$ **59.** $x = -1.5$ **61.** (1.2, 0); (0, −3.6) **63.** (−2, 0); (0, −2.5)
65. a) Answers will vary. **b)** $x = a + b$ or $x = a - b$
66. a) Answers will vary. **b)** $a - b < x < a + b$
67. a) Answers will vary. **b)** $x < a - b$ or $x > a + b$
68. $\{-2, 2\}$

Exercise Set 3.4 **1.** Select two points on the line; find $\frac{\Delta y}{\Delta x}$. **3.** The line falls going from left to right. **5.** We cannot divide by zero. **7.** Solve for y. **9. a)** Moved up 3 units **b)** $(0, 1)$ **11.** The change in y for a unit change in x **13.** 2 **15.** $-\frac{1}{2}$ **17.** -1 **19.** Undefined **21.** 0 **23.** $-\frac{2}{3}$ **25.** $a = 3$ **27.** $b = 2$ **29.** $x = 6$ **31.** $x = 0$ **33.** $m = -3; y = -3x$ **35.** Slope is undefined; $x = -2$ **37.** $m = -\frac{1}{3}; y = -\frac{1}{3}x + 2$ **39.** $m = -\frac{3}{2}; y = -\frac{3}{2}x + 15$ **41.** $y = -x + 2; m = -1; (0, 2)$ **43.** $y = \frac{2}{3}x - 2; m = \frac{2}{3}; (0, -2)$

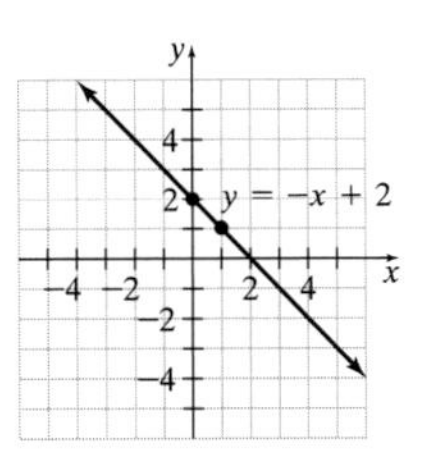

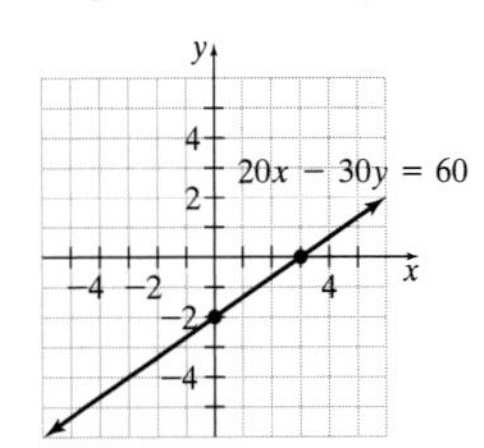

45. $y = \frac{5}{2}x + 2; m = \frac{5}{2}; (0, 2)$

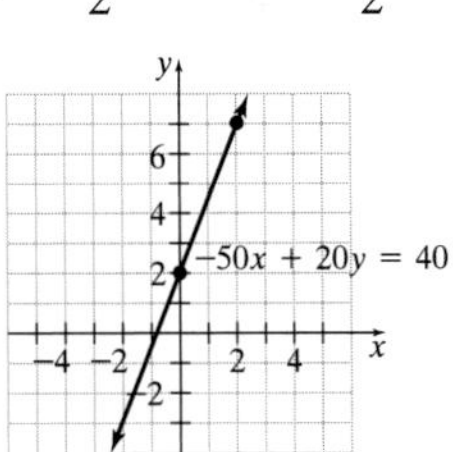

47.

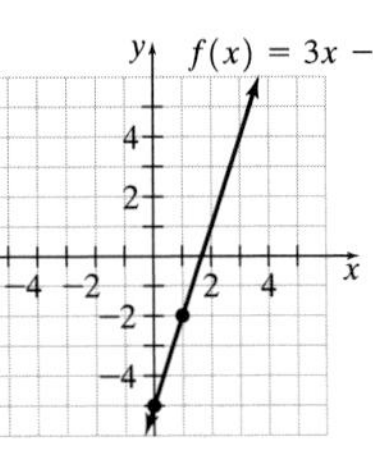

49.

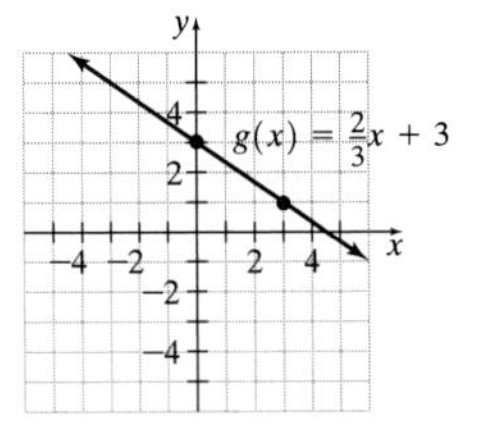

51. a) 2 **b)** 4 **c)** 1 **d)** 3 **53.** If the slopes are the same and the y-intercepts are different, the lines are parallel. **55.** $(0, -5)$ **57. a)** $y = 3x + 1$ **b)** $y = 3x - 5$ **59. a)** 2 **b)** $(0, 4)$ **c)** $y = 2x + 4$ **61.** $y = \frac{3}{2}x - 7$ **63.** 0.2 **65. a–b)**

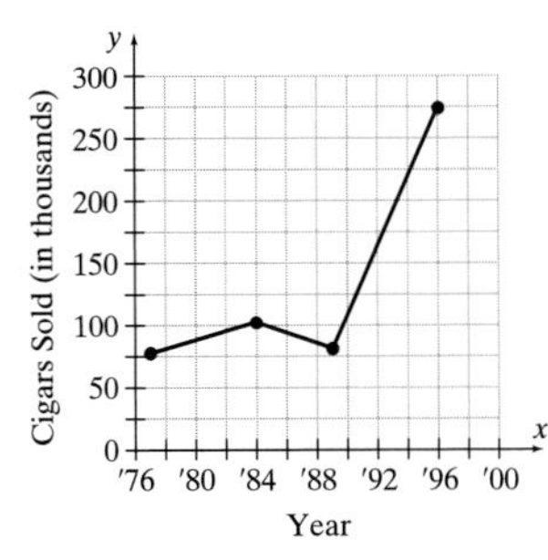

c) $\approx 3.43; -4.2; \approx 27.57$ **d)** 1989 to 1996 **67. a)** $h(x) = -x + 200$ **b)** 186 beats per minute **69. a)** $s(t) = -0.95t + 100$ **b)** 76.2% **71.** The y-intercept is wrong. **73.** The slope is wrong. **75.** Height: 14.2 inches; width: 6.4 inches **78.** 19 **79.** $x = -\frac{92}{5}$ **80.** $x = 2.4$ **81.** First: 75 mph; second: 60 mph **82. a)** $x < -2$ or $x > 1$ **b)** $-2 < x < 1$

Exercise Set 3.5 **1.** $y - y_1 = m(x - x_1)$ **3.** Two lines are perpendicular if their slopes are negative reciprocals or if one line is vertical and the other is horizontal. **5.** $y = 3x - 2$ **7.** $y = x - 3$ **9.** $y = \frac{1}{2}x - \frac{9}{2}$ **11.** $y = -\frac{3}{2}x$ **13.** Parallel **15.** Perpendicular **17.** Neither **19.** Perpendicular **21.** Parallel **23.** Parallel **25.** Neither **27.** Perpendicular **29.** Perpendicular **31.** Neither **33.** $y = 2x + 1$ **35.** $y = -\frac{3}{2}x - \frac{11}{30}$ **37.** $f(x) = -2x - \frac{16}{3}$ **39.** $f(x) = -3x + 11$ **41.** $y = -\frac{2}{3}x + 6$ **43. a)** $n(t) = 1.42t + 63.1$ **b)** About 112.8 thousand **45. a)** $C(s) = 45.7s + 95.8$ **b)** About 324.3 calories **47. a)** $m(t) = 0.14t + 22.8$ **b)** About 28.3 years **49. a)** $n(t) = -0.032t + 2.8$ **b)** About 0.41 million

51. a–b)

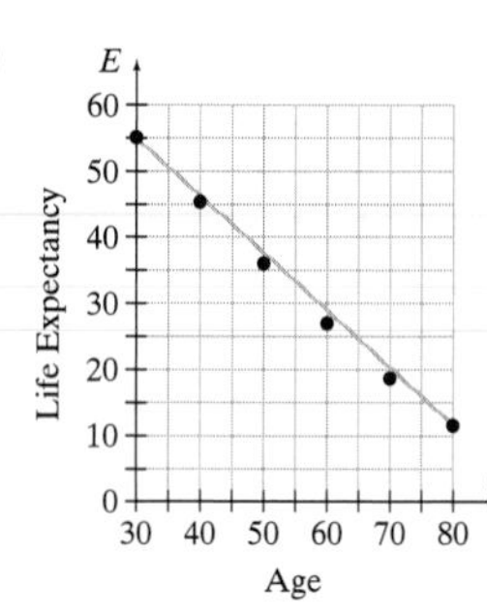

c) $E(a) = -0.87a + 81.2$ **d)** About 24.65 years
53. a) $V(t) = 3.57t + 40$ **b)** ≈ 157.81 million
55. a) $N(t) = -0.5t + 69$ **b)** $H(t) = 1.83t + 43$
57. a) $n(t) = -0.58t + 23.2$; about 12,760 **b)** Yes; Answers may vary.
c) No; Answers may vary. **59.** $\left(-\infty, \frac{2}{5}\right)$
60. Reverse the direction of the inequality. **61. a)** Any set of ordered pairs
b) A correspondence where each member of the domain corresponds to a unique member in the range **c)** Answers may vary.
62. Domain: $\{3, 4, 5, 6\}$; Range: $\{-2, -1, 2, 3\}$

Exercise Set 3.6

1. Yes, this is how addition of functions is defined. **3.** $g(x) \neq 0$ since division by zero is undefined. **5.** No, subtraction is not commutative. One example is $5 - 3 = 2 \neq 3 - 5 = -2$ **7. a)** 2 **b)** -8 **c)** -15 **d)** $-\frac{3}{5}$ **9. a)** $x^2 - x + 4$ **b)** $a^2 - a + 4$ **c)** 6 **11. a)** $x^3 + x^2 - 3x + 5$ **b)** $a^3 + a^2 - 3a + 5$ **c)** 11 **13. a)** $4x^3 - x + 4$ **b)** $4a^3 - a + 4$ **c)** 34 **15.** -1 **17.** -13 **19.** -60 **21.** $-\frac{12}{17}$ **23.** 13 **25.** $-\frac{3}{4}$ **27.** $2x^2 - 6$ **29.** 26 **31.** 66 **33.** -189 **35.** $-\frac{3}{7}$ **37.** $-\frac{1}{45}$ **39.** $-2x^2 + 2x - 6$ **41.** 3 **43.** -4 **45.** -3 **47.** Undefined **49.** 4 **51.** 3 **53.** -3 **55.** -6
57. a) 4.9 million **b)** 4.7 million **c)** 2.9 million **d)** 1.6 million **e)** 2.0 million **f)** Answers will vary.
59. a) 32 million **b)** 100 million **c)** The number of PC's shipped outside of the U.S. **d)** 68 million
61. $f(a)$ and $g(a)$ must either be opposites or both be equal to 0. **63.** $f(a) = g(a)$
65. $f(a)$ and $g(a)$ must have opposite signs.

67.

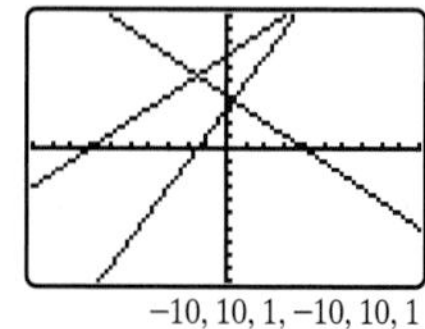

$-10, 10, 1, -10, 10, 1$

69.

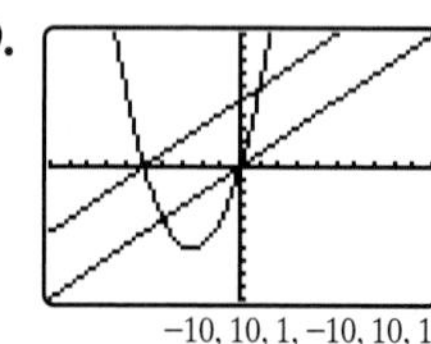

$-10, 10, 1, -10, 10, 1$

72. $h = \frac{2A}{b}$ **73.** \$450 **74.**

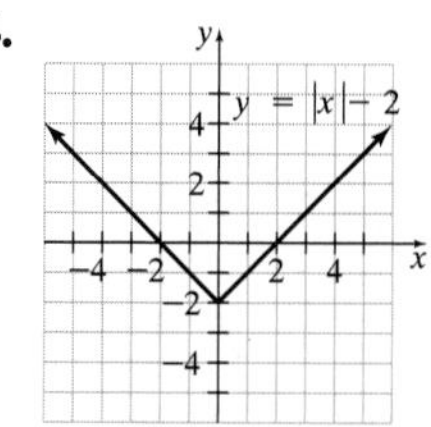

75.

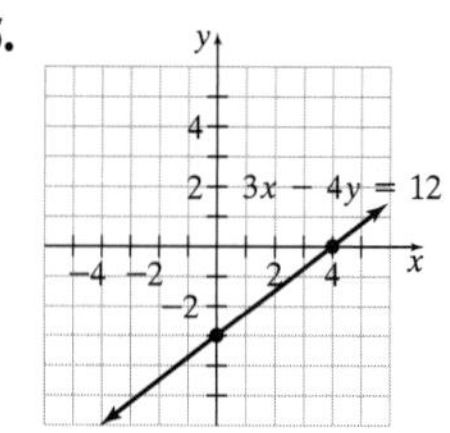

Exercise Set 3.7

1. The inequalities $\geq$ and $\leq$ include the corresponding equation; the points on the line satisfy the equation. **3.** $(0, 0)$ cannot be used as a check point if the line passes through the origin.

5.

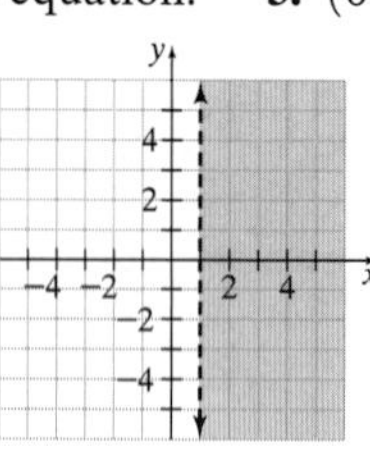

7.

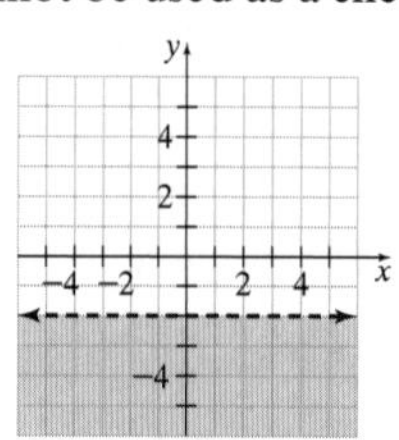

9.

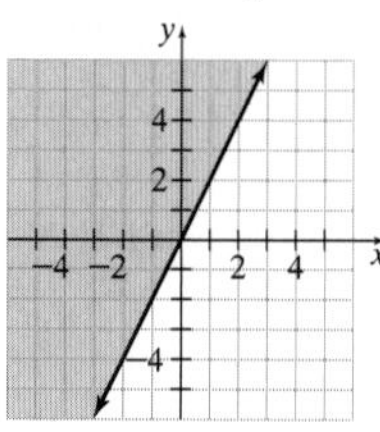

11.

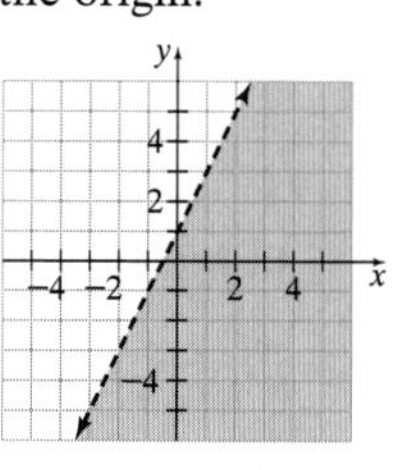

13.

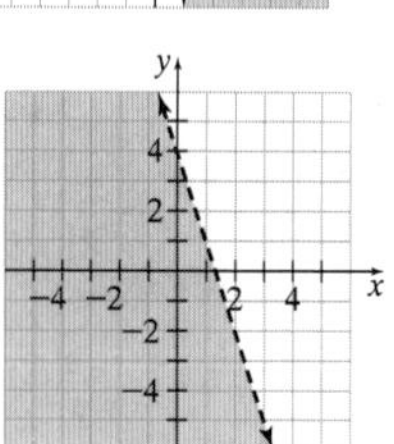

15.

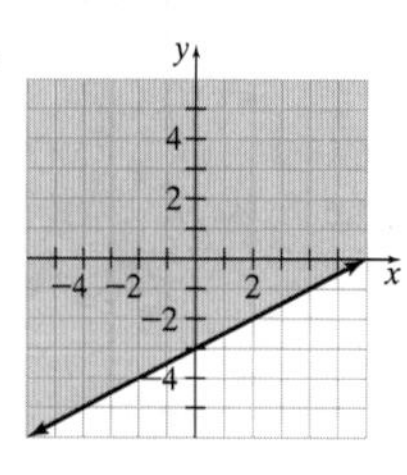

17.

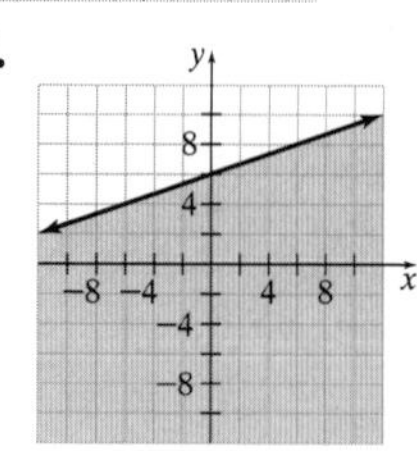

19.

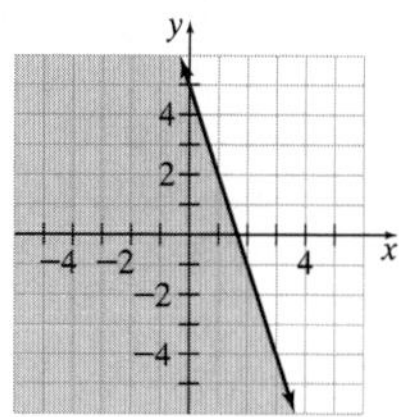

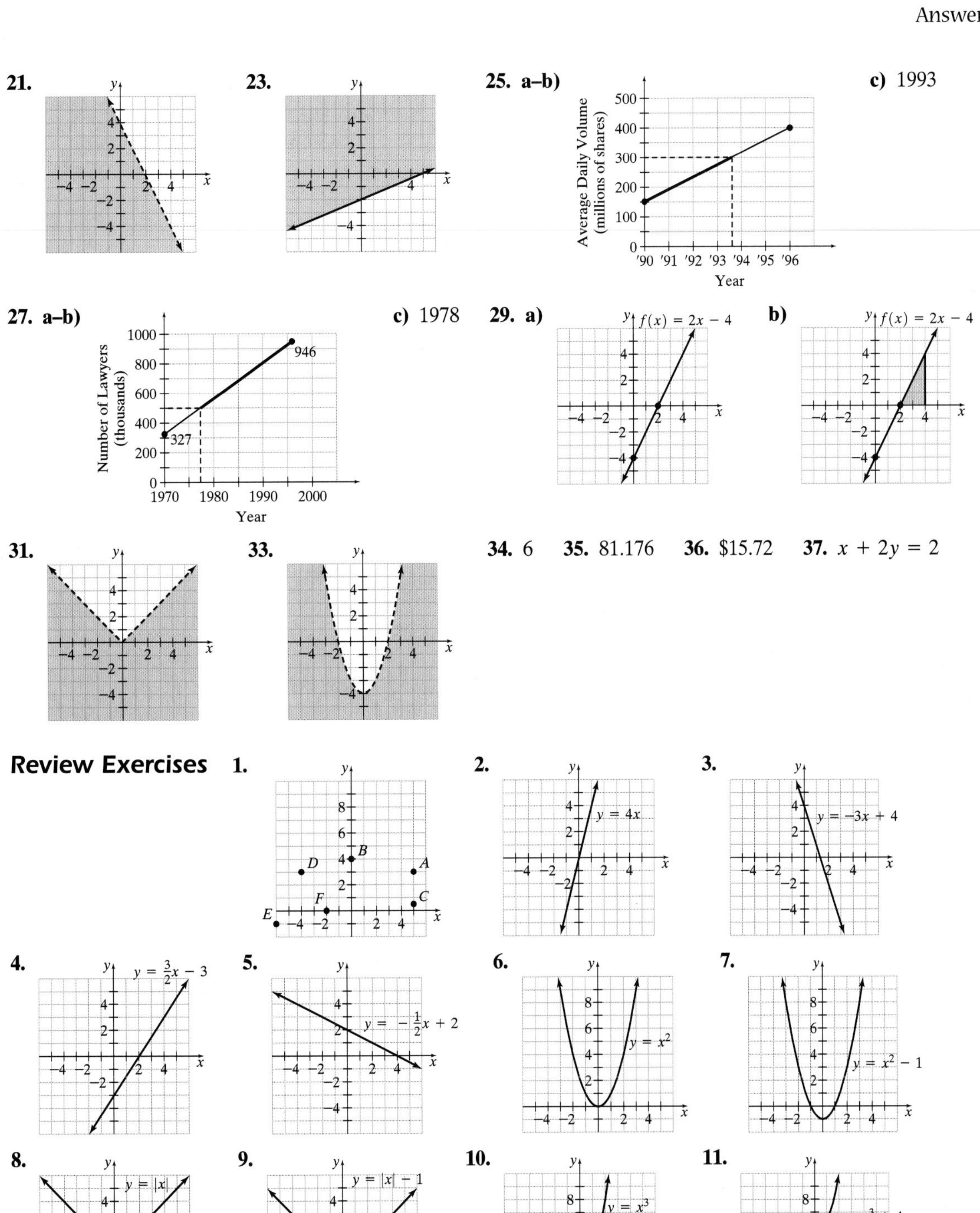

34. 6 **35.** 81.176 **36.** $15.72 **37.** $x + 2y = 2$

12. A function is a correspondence where each member of the domain corresponds to exactly one member of the range. **13.** No, every relation is not a function. $\{(4, 2), (4, -2)\}$ is a relation but not a function. Yes, every function is a relation because it is a set of ordered pairs. **14.** Yes, each member of the domain corresponds to exactly one member of the range. **15.** No, the domain element 4 corresponds to more than one member of the range (2 and 0).
16. a) No, the relation is not a function **b)** Domain: $\{x \mid -1 \le x \le 1\}$; Range: $\{y \mid -1 \le y \le 1\}$
17. a) No, the relation is not a function. **b)** Domain: $\{x \mid -2 \le x \le 2\}$; Range: $\{y \mid -1 \le y \le 1\}$

18. a) Yes, the relation is a function. **b)** Domain: $\mathbb{R}$; Range: $\{y \mid y \leq 0\}$
19. a) Yes, the relation is a function. **b)** Domain: $\mathbb{R}$; Range: $\mathbb{R}$ **20. a)** 10 **b)** $a^2 + 2a - 5$ **21. a)** -10 **b)** 46
22. Answers will vary. **23. a)** 720 **b)** 1500 **24. a)** 84 feet **b)** 36 feet

25.

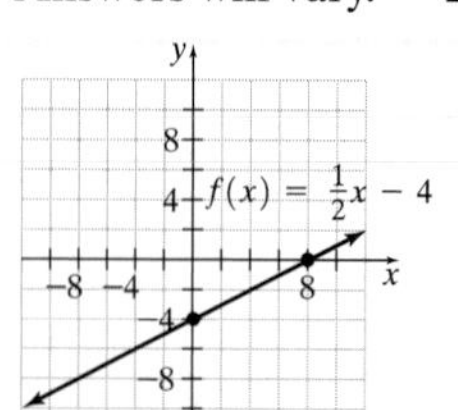

26. **27.** **28.**

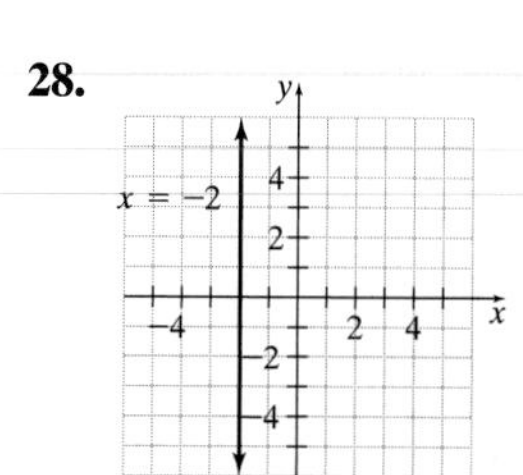

29. a)

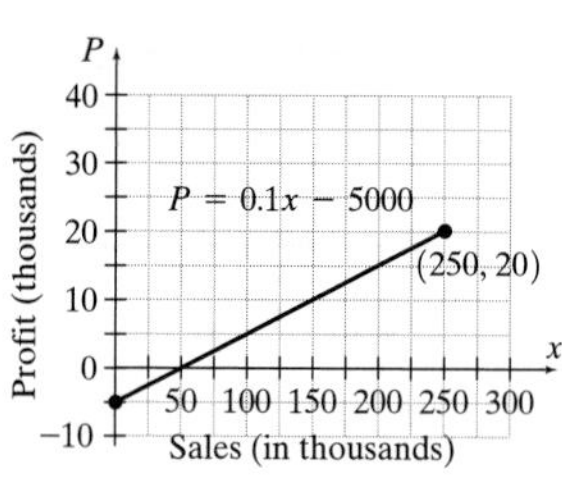

b) Approximately 50,000 bagels **c)** Approximately 250,000 bagels

30.

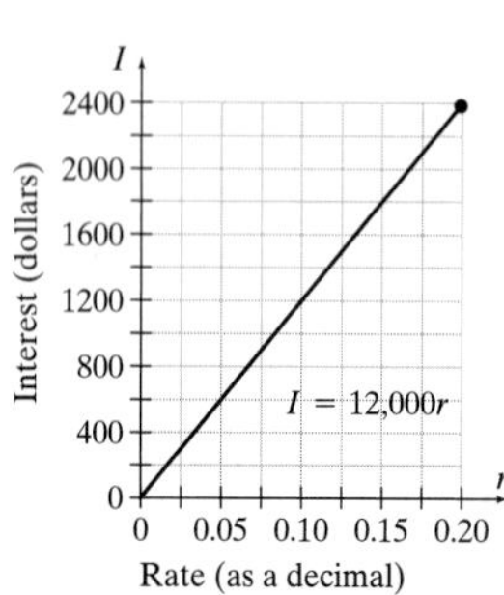

31. $m = -1, (0, 5)$ **32.** $m = -4, \left(0, \frac{1}{2}\right)$ **33.** $m = -\frac{3}{5}, \left(0, \frac{12}{5}\right)$ **34.** $m = -\frac{9}{7}, \left(0, \frac{15}{7}\right)$ **35.** m is undefined; no y-intercept **36.** $m = 0, (0, 6)$ **37.** -7 **38.** $-\frac{1}{3}$ **39.** $m = 0; y = 3$ **40.** m is undefined; $x = 2$ **41.** $m = -\frac{1}{2}; y = -\frac{1}{2}x + 2$ **42.** $-\frac{3}{5}$ or -0.6 **43.** $(0, 0)$

44. a)

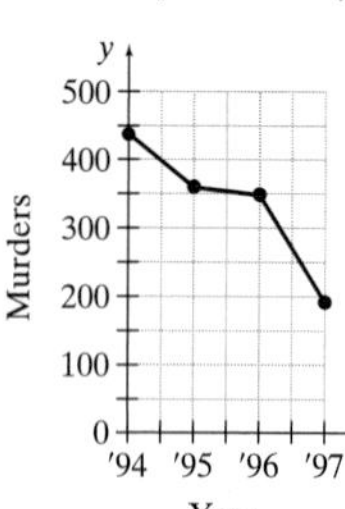

b) 1994 to 1995: -77; 1995 to 1996: -11; 1996 to 1997: -157
45. $n(t) = 0.7t + 35.6$ **46.** Parallel **47.** Perpendicular **48.** Neither **49.** $y = -\frac{2}{3}x + 4$ **50.** $y = x - 1$ **51.** $y = 3x + 20$ **52.** $y = \frac{2}{5}x - \frac{18}{5}$ **53.** $y = -\frac{5}{3}x - 4$ **54.** $y = -\frac{1}{2}x + 4$ **55.** Neither **56.** Parallel **57.** Perpendicular **58.** Neither **59. a)** $b(t) = -39t + 206$ **b)** 167 **60. a)** $p(t) = 61.6t + 522$ **b)** 953,200 **c)** Yes **61.** $x^2 - x - 1$ **62.** 5

63. $-x^2 + 5x - 9$ **64.** -15 **65.** -56 **66.** 70 **67.** $-\frac{2}{3}$ **68.** -2 **69. a)** 6000 **b)** 13,000 **c)** 19,000
70. a) 720 **b)** 900 **c)** 1620

71.

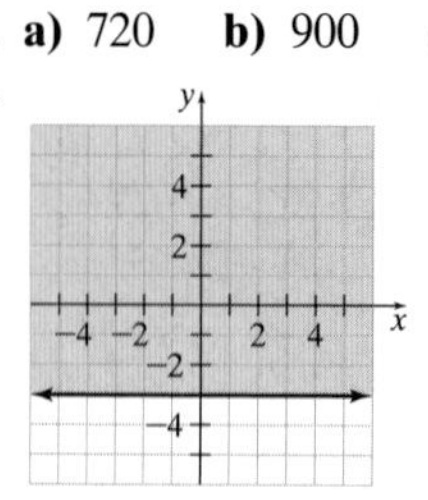

72.

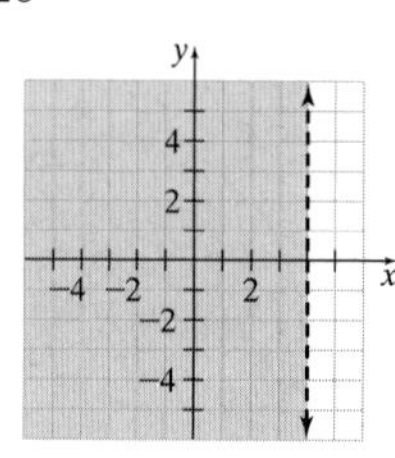

73.

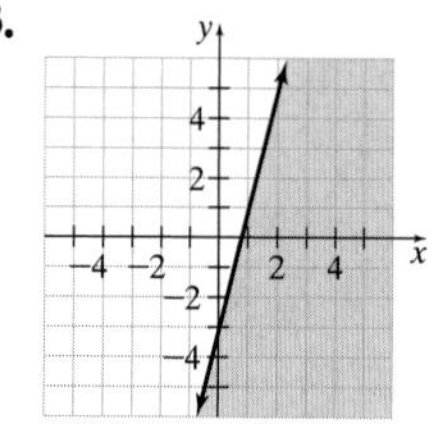

74.

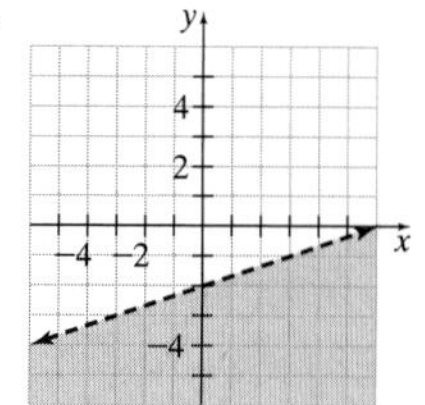

Practice Test **1.**

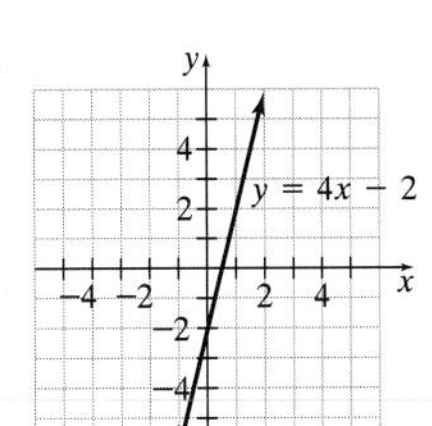

2.

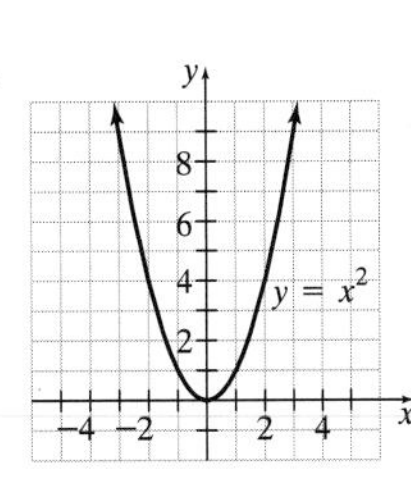

3.

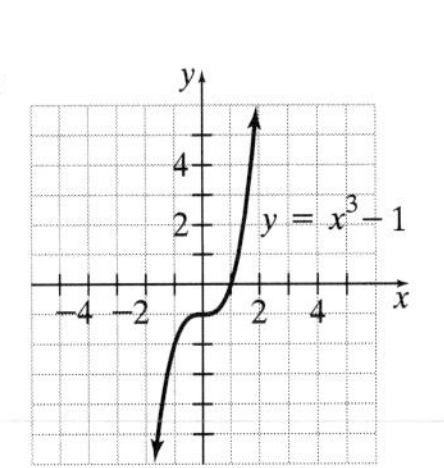

4.

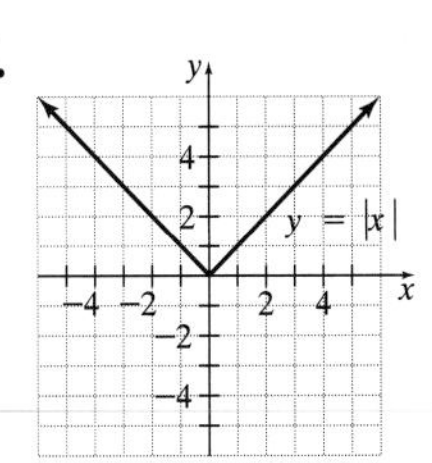

5. A function is a correspondence where each member in the domain corresponds with exactly one member in the range. **6.** Yes, because each member in the domain corresponds to exactly one member in the range. **7.** Yes; Domain: $\mathbb{R}$; Range: $\{y|y \le 4\}$ **8.** No; Domain: $\{x|-3 \le x \le 3\}$; Range: $\{y|-2 \le y \le 2\}$ **9.** 26

10. $100y + 200x = 400$

11.

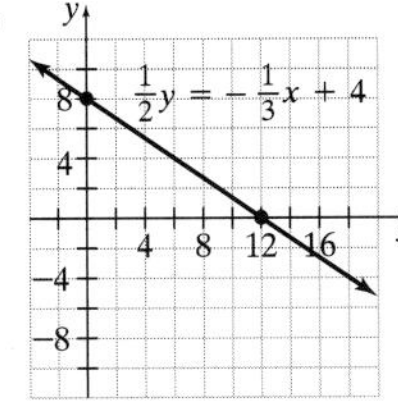

12.

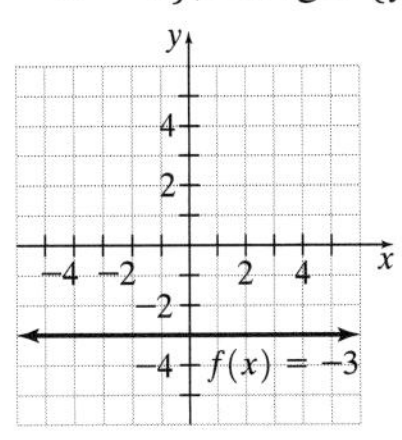

13.

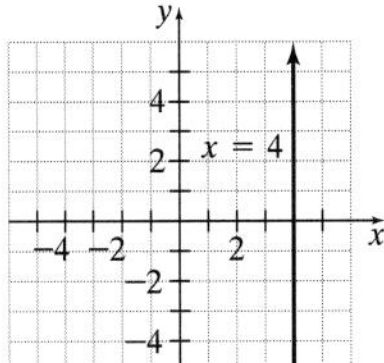

14. a)

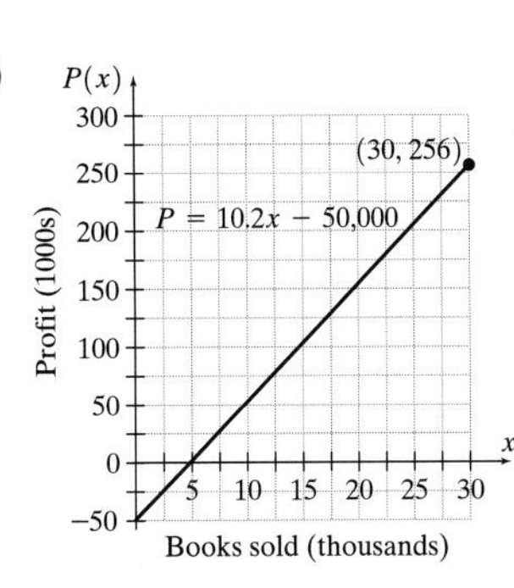

b) 4900 books **c)** 14,700 books **15.** $m = \frac{4}{3}, (0, -3)$ **16.** $-\frac{3}{10}$ **17.** $p(t) = 2.386t + 274.634$ **18.** Perpendicular **19.** $y = -\frac{2}{3}x - 2$ **20. a)** $n(t) = 4545.6t + 6851$ **b)** 29,579 **21.** 13 **22.** $-\frac{1}{2}$ **23.** $2a^2 - a$ **24. a)** 44 million tons **b)** 18 million tons **c)** 26 million tons

25.

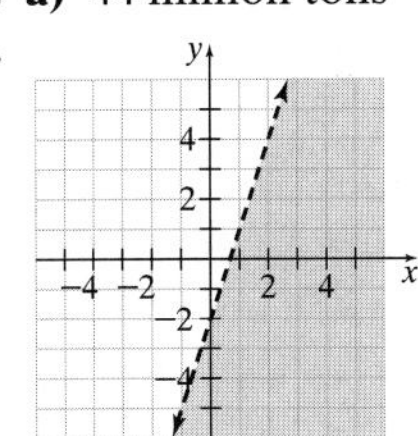

Cumulative Review Test **1. a)** $A \cap B = \{2, 4, 6\}$ **b)** $A \cup B = \{1, 2, 3, 4, 5, 6, 8\}$ **2. a)** None **b)** $-6, -4, -\sqrt{2}, 0, \frac{1}{3}, \sqrt{3}, 4.67, \frac{37}{2}$ **3.** 92 **4.** $16x^4y^6$ **5.** $\frac{x^9}{8y^{15}}$ **6. a)** 4.9938×10^8 or 499,380,000 **b)** 3.5916×10^8 or 359,160,000 **c)** 1.4022×10^8 or 140,220,000 **7.** $\frac{13}{2}$ **8.** $-\frac{138}{5}$ **9.** $7x - 7$ **10.** $b_1 = \frac{2A}{h} - b_2$ **11.** 24 liters **12.** $x > -\frac{10}{3}$ **13.** $1 < x < 5$ **14.** $\left\{x \middle| x < -\frac{1}{2} \text{ or } x > \frac{7}{2}\right\}$ **15.** $\left\{\frac{4}{3}, \frac{12}{5}\right\}$

16.

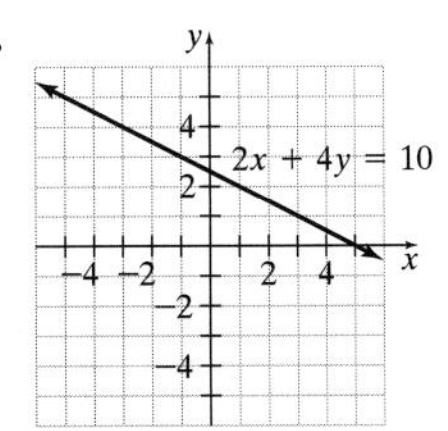

17. Neither **18. a)** Not a function **b)** Domain: $\{x|x \le 2\}$; Range: $\mathbb{R}$ **19.** $x^2 + 7x - 8$ **20.** 260

CHAPTER 4

Using Your Graphing Calculator, 4.1 **1.** (2.76, 0.82) **2.** (13.29, 9.57) **3.** (−4.67, −4.66) **4.** (−2.25, 10.52)

Exercise Set 4.1 **1.** The solution to a system of linear equations is the point(s) that satisfy all equations in the system. **3.** A consistent system of equations has a solution. **5.** An inconsistent system of equations is a system of equations that has no solution. **7.** Compare the slopes and y-intercepts of the equations. If the slopes are different, the sytem is consistent. If the slopes and y-intercepts are the same, the system is dependent. If the slopes are the same and the y-intercepts are different, the system is inconsistent. **9.** You will get a false statement, like $6 = 0$. **11.** None **13.** b) **15.** b) **17.** Consistent; one solution **19.** Dependent; infinite number of solutions **21.** Inconsistent; no solution **23.** Inconsistent; no solution **25.** $(-1, 4)$ **27.** Inconsistent **29.** Dependent **31.** $(1, 1)$ **33.** $(1, 0)$ **35.** Inconsistent **37.** $(3, 1)$ **39.** $(5, 5)$ **41.** No solution **43.** $\left(\frac{1}{2}, -\frac{39}{2}\right)$ **45.** Infinite number of solutions **47.** $(1, -3)$ **49.** $\left(-\frac{19}{5}, -3\right)$ **51.** $(8, 6)$ **53.** $(2, -2)$ **55.** $\left(2, \frac{9}{2}\right)$ **57.** No solution **59.** $(4, -2)$ **61.** $(4, -1)$ **63.** Infinite number of solutions **65.** $\left(\frac{14}{5}, -\frac{12}{5}\right)$ **67.** $\left(\frac{37}{7}, \frac{19}{7}\right)$ **69.** $(3, 2)$ **71.** $(4, 0)$ **73.** $(4, 3)$ **75.** $(10, 4)$ **77.** Answers may vary. The system should involve a variable that has a coefficient of 1. **79.** 2.61 years after 1995 or in 1997 **81.** Multiply the first equation by -2 and notice that the x- and y-terms have the same coefficients but the constant terms are different. **83. a)** Infinite number **b)** $m = -\frac{1}{2}$; $y = -\frac{1}{2}x + 5$; $(0, 5)$ **c)** Yes **85.** One example is: $x + y = 1, 2x + 2y = 2$ **87. a)** One example is: $x + y = 7, x - y = -3$. **b)** Choose coefficients for x and y, then use the given coordinates to find the constants. **89.** $A = 2$ and $B = 5$ **91.** $m = 4, b = -2$ **93.** The system is dependent or one graph is not in the viewing window. **95.** $(8, -1)$ **97.** $(-1, 2)$ **99.** $\left(\frac{1}{a}, 5\right)$ **102.** Rational numbers can be expressed as quotients of two integers. Irrational numbers cannot. **103. a)** Yes, the set of real numbers includes the set of rational numbers. **b)** Yes, the set of real numbers includes the set of irrational numbers. **104.** $\mathbb{R}$ **105.** 520 **106.** No, the points $(-3, 4)$ and $(-3, 2)$ have the same first coordinate but different second coordinates.

Exercise Set 4.2 **1.** The graph will be a plane. **3.** $(3, 2, 5)$ **5.** $\left(-7, -\frac{35}{4}, -3\right)$ **7.** $(0, 3, 6)$ **9.** $(-2, 2, 3)$ **11.** $(-1, 3, 2)$ **13.** $(3, 1, -2)$ **15.** $(2, -1, 3)$ **17.** $\left(\frac{2}{3}, -\frac{1}{3}, 1\right)$ **19.** $(4, -1, -2)$ **21.** $\left(-\frac{11}{17}, \frac{7}{34}, -\frac{49}{17}\right)$ **23.** $(0, 0, 0)$ **25.** $(3, -1, 2)$ **27.** $(1, 1, 2)$ **29.** Inconsistent **31.** Dependent **33.** Inconsistent **35.** No point is common to all three planes. Therefore, the system is inconsistent. **37.** A line is common to all three planes. Therefore, there are an infinite number of points common to all three planes and the system is dependent. **39. a)** Yes, if two or more of the planes are parallel, there will be no solution. **b)** Yes, three planes may intersect at a single point. **c)** No, the possibilities are no solution, one solution, or infinitely many solutions. **41.** $A = 3, B = -2, C = 1$; $3x - 2y + z = -2$ **43.** One example is $x + y + z = 10, x + 2y + z = 11, x + y + 2z = 16$; Choose coefficients for x, y, and z, then use the given coordinates to find the constants. **45. a)** $a = 1, b = 2, c = -4$ **b)** $y = x^2 + 2x - 4$ **47.** $(-1, 2, 1, 5)$ **49. a)** $\frac{5}{12}$ hour or 25 minutes **b)** 1.25 miles **50.** $\left\{x \middle| x < -\frac{3}{2} \text{ or } x > \frac{27}{2}\right\}$ **51.** $\left\{x \middle| -\frac{8}{3} < x < \frac{16}{3}\right\}$ **52.** $\varnothing$

Exercise Set 4.3 **1. a)** About \$46.7 million **b)** \$37 million **c)** \$120 million **3.** Hamburger has 21 grams; fries have 67 grams **5.** Expos is \$80.42; Braves is \$134.16 **7.** 52°, 128° **9.** Salary: \$500; commission rate: 4% **11.** Fixed charge: \$300; charge for each flier: \$0.25 **13.** 3.75 gallons of 20% solution; 6.25 gallons of 4% solution **15.** 10 gallons of concentrate; 190 gallons of water **17.** 60 regular orders; 74 jumbo orders **19.** 70 gallons of the 5% butterfat milk; 30 gallons of skim milk **21.** 5.96 ounces of heavy cream; 10.04 ounces of half-and-half **23.** 4.8 ounces of apple juice; 3.2 ounces of raspberry juice **25. a)** $\frac{2400}{17} \approx 141.2$ cm **b)** The rat will pull away. **27.** 182 small cones; 78 large cones **29.** Slower machine works 5 minutes; faster machines works 8 minutes **31.** About 0.39 liters of pure antifreeze must be added to about 15.61 liters of the mixture. **33.** 8 Model A chairs; 12 Model B chairs **35.** Tom traveled at 60 mph; Melissa traveled at 75 mph. **37.** Approximately 7 years after 1992 or in 1999 **39. a)** Plan 1: $c = 0.15m$; Plan 2: $c = 0.10m + 4.95$

b)

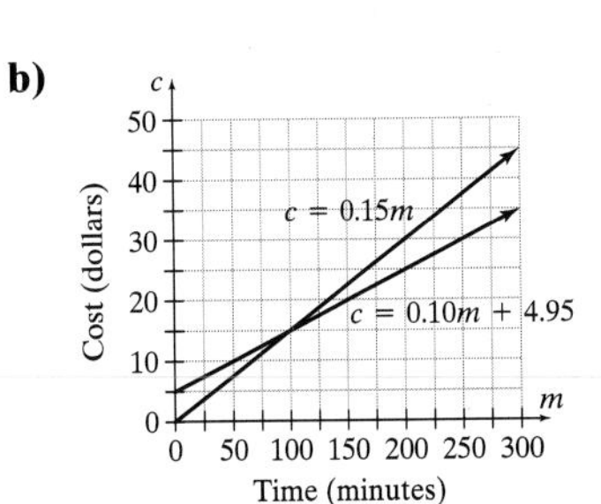

c) Approximately 100 minutes **d)** 99 minutes; the graph cannot be read with enough accuracy to distinguish between 99 and 100 minutes. **41. a)** $x + y + z = 141; x = y + 4; z = 8y - 3$ **b)** 14 commissioned officers; 18 chief petty officers; 109 other enlisted men **43. a)** $x + y + z = 41; z = 3x - 14; y = 2x - 5$ **b)** Iraq: 10 million; Angola: 15 million; Iran: 16 million **45. a)** $x + y + z = 180; x = \frac{2}{3}y; z = 3y - 30$ **b)** 30°, 45°, 105° **47. a)** $x + y + z = 8; 0.10x + 0.12y + 0.20z = (0.13)8; z = x - 2$ **b)** 4 liters of the 10% solution; 2 liters of the 12% solution; 2 liters of the 20% solution **49. a)** $5x + 4y + 7z = 154; 3x + 2y + 5z = 94; 2x + 2y + 4z = 76$ **b)** 10 children's chairs; 12 standard chairs; 8 executive chairs **51.** $I_A = \frac{27}{38}; I_B = -\frac{15}{38}; I_C = -\frac{6}{19}$ **54.** $-\frac{35}{8}$ **55.** 4 **56.** Use the vertical line test. **57.** $y = x - 10$

Exercise Set 4.4

1. Same number of rows and columns **3.** Change the −1 in the second row to 1 **5. a)** A row of numbers contains all 0's. **b)** All the numbers in a row on the left side of the augmented matrix are 0's, but the number on the right side is not 0. **7.** $\left[\begin{array}{cc|c} 1 & -\frac{1}{2} & -\frac{5}{4} \\ 3 & 5 & -1 \end{array}\right]$ **9.** $\left[\begin{array}{ccc|c} 4 & 0 & 3 & 8 \\ -\frac{5}{7} & 1 & -\frac{2}{7} & -2 \\ -1 & 3 & 5 & 12 \end{array}\right]$ **11.** $\left[\begin{array}{cc|c} 1 & 3 & 12 \\ 0 & 17 & 30 \end{array}\right]$ **13.** $\left[\begin{array}{ccc|c} 1 & 0 & 8 & \frac{1}{4} \\ 0 & 2 & -38 & -\frac{13}{4} \\ 6 & -3 & 1 & 0 \end{array}\right]$ **15.** (3, 0) **17.** (3, 1) **19.** $\left(0, \frac{1}{2}\right)$ **21.** Dependent system **23.** (−2, 1) **25.** Inconsistent system **27.** $\left(\frac{2}{3}, \frac{1}{4}\right)$ **29.** $\left(\frac{4}{5}, -\frac{7}{8}\right)$ **31.** (2, 0, 1) **33.** (3, 1, 2) **35.** $\left(-2, \frac{1}{3}, 0\right)$ **37.** Dependent system **39.** $\left(\frac{1}{2}, 2, 4\right)$ **41.** Inconsistent system **43.** $\left(5, \frac{1}{3}, -\frac{1}{2}\right)$ **45.** No, this is the same as switching the order of the equations. **47.** $\angle x = 30°, \angle y = 65°, \angle z = 85°$ **49.** 26% by Chiquita, 25% by Dole, 14% by Del Monte, 35% by other **51. a)** {1, 2, 3, 4, 5, 6, 9, 10} **b)** {4, 6} **52. a)** (number line from −2, open, to 4, closed) **b)** $\{x | -2 < x \le 4\}$ **c)** (−2, 4] **53.** A graph is an illustration of the set of points whose coordinates satisfy an equation. **54.** −76

Exercise Set 4.5

1. Answers will vary. **3.** If $D = 0$, and D_x, D_y, and D_z are all equal to 0, the system is dependent. **5.** 23 **7.** −8 **9.** −12 **11.** 44 **13.** (3, 1) **15.** (3, 2) **17.** $\left(\frac{60}{17}, -\frac{11}{17}\right)$ **19.** $\left(\frac{1}{2}, -4\right)$ **21.** (2, −3) **23.** (2, 5) **25.** (2, −1, 3) **27.** $\left(-\frac{1}{2}, \frac{1}{2}, -\frac{3}{2}\right)$ **29.** $\left(\frac{165}{14}, -\frac{153}{14}, -\frac{6}{7}\right)$ **31.** (−1, 0, 2) **33.** Dependent system; infinite number of solutions **35.** (1, −1, 2) **37.** Dependent system; infinite number of solutions **39.** Inconsistent system; no solution **41.** (−1, 5, −2) **43.** It has the opposite sign **45.** 0 **47.** Yes, the determinant will become the opposite of the original value. **49.** Yes, the determinant will become the opposite of the original value. **51.** 5 **53.** 2 **55. a)** $x = \frac{c_1b_2 - c_2b_1}{a_1b_2 - a_2b_1}$ **b)** $y = \frac{a_1c_2 - a_2c_1}{a_1b_2 - a_2b_1}$ **56.** $\left(-\infty, \frac{14}{11}\right)$

57.

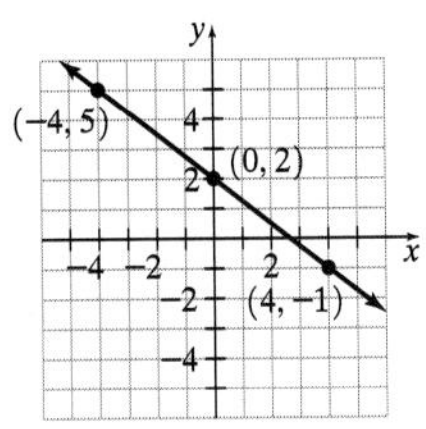

58.

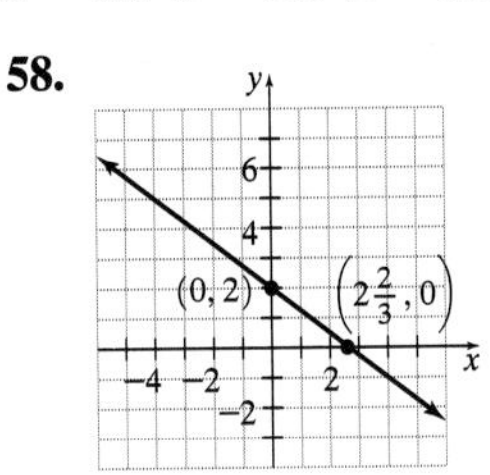

59.

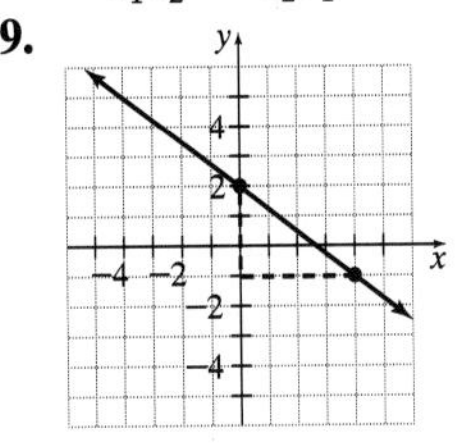

Exercise Set 4.6

1. Answers will vary.

3.

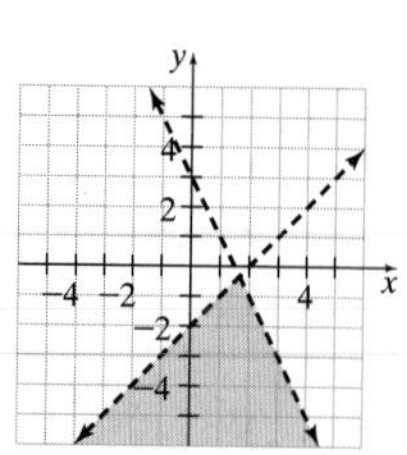

5.

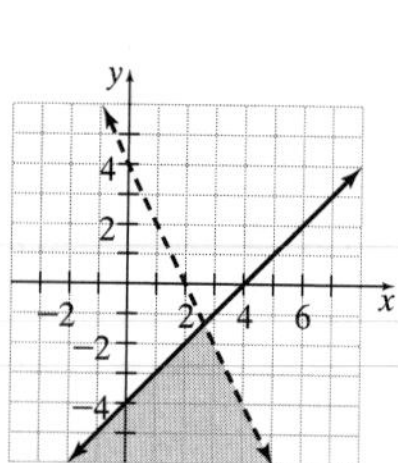

7.

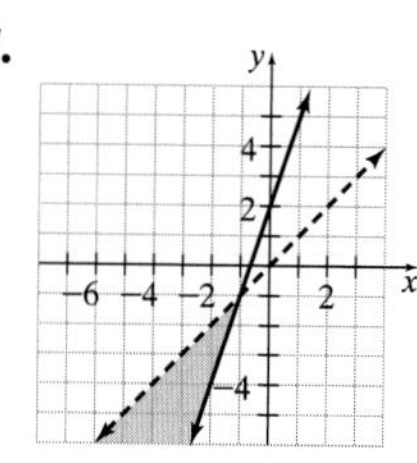

9.

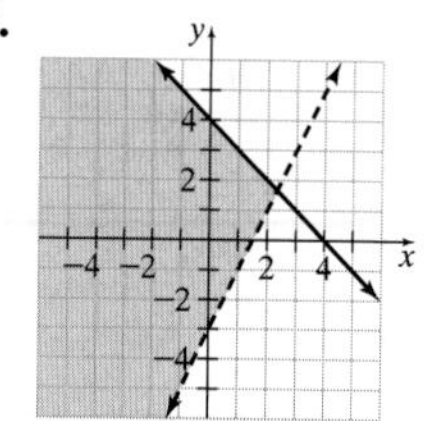

11.

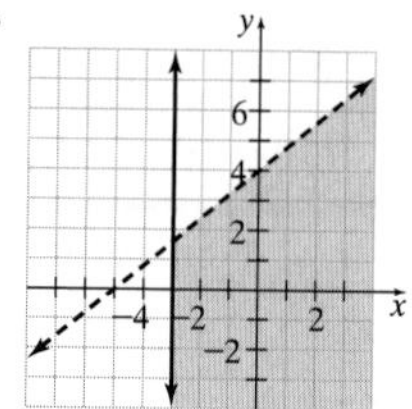

13.

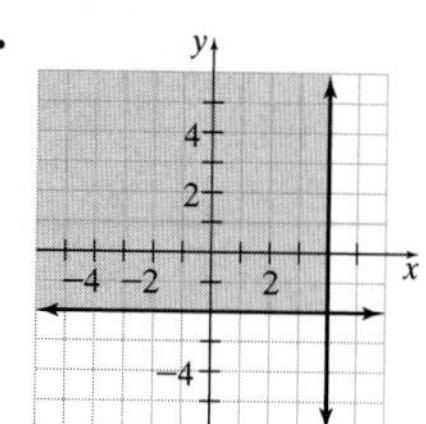

15.

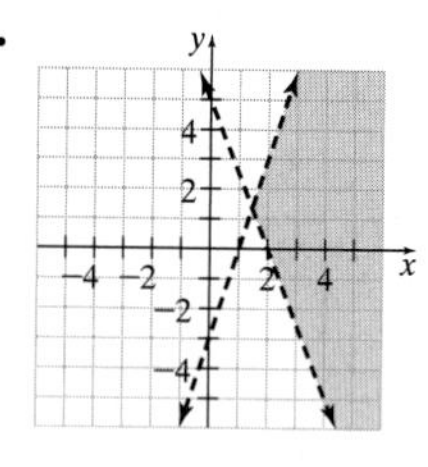

17. No solution

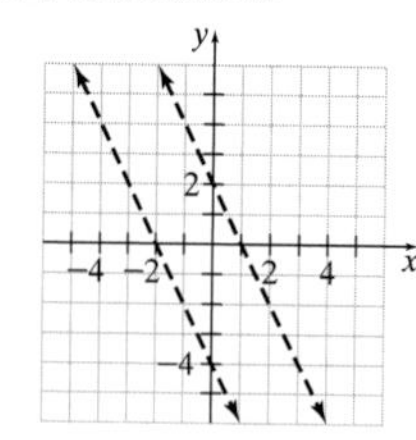

19.

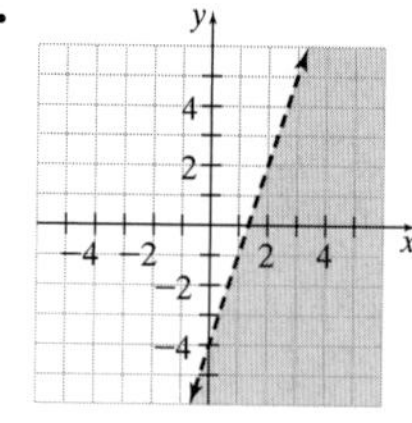

21.

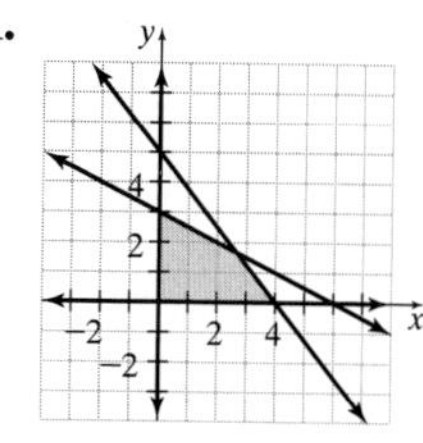

23.

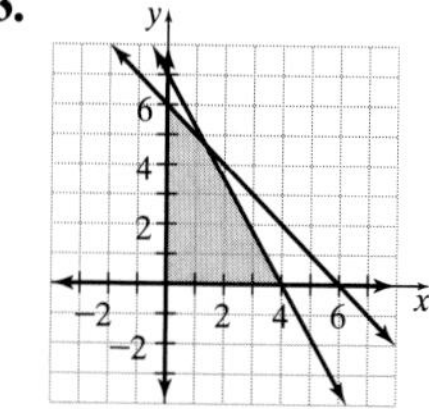

25.

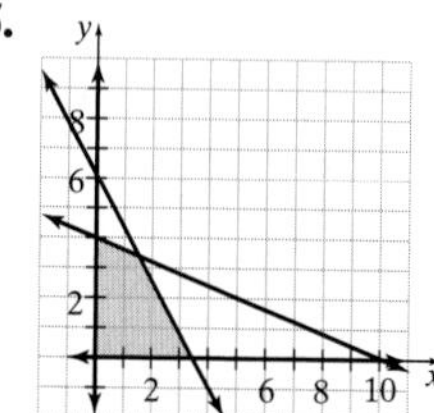

27.

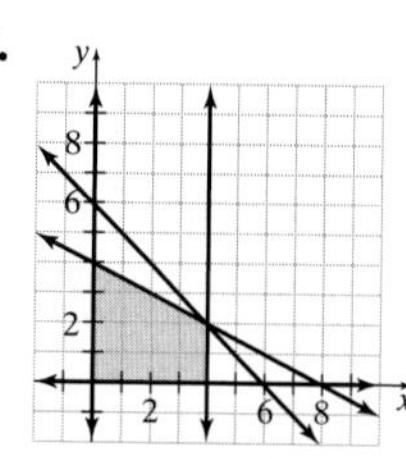

29.

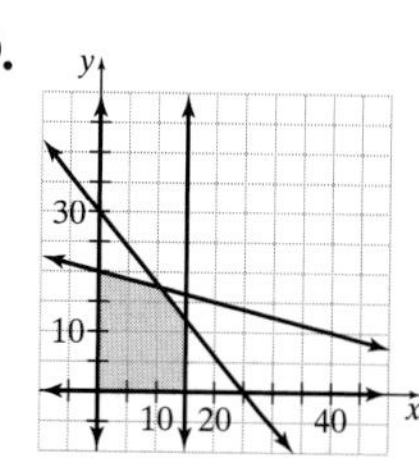

31.

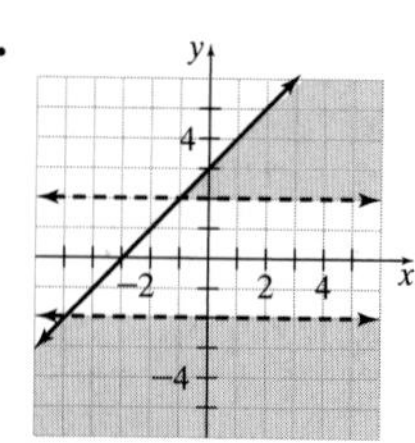

33.

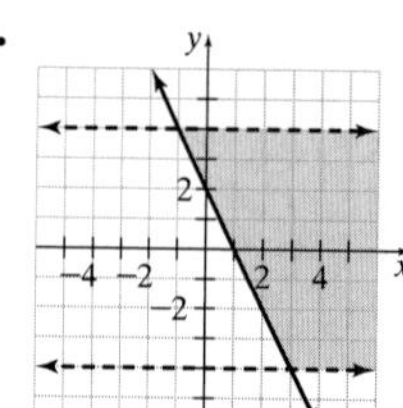

35.

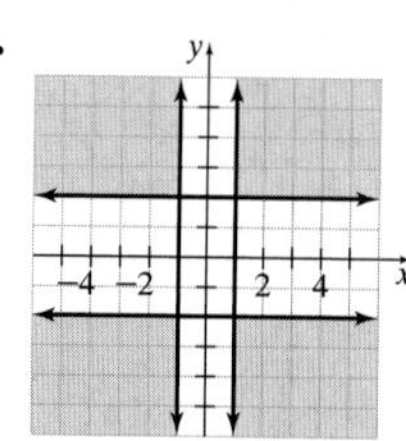

37.

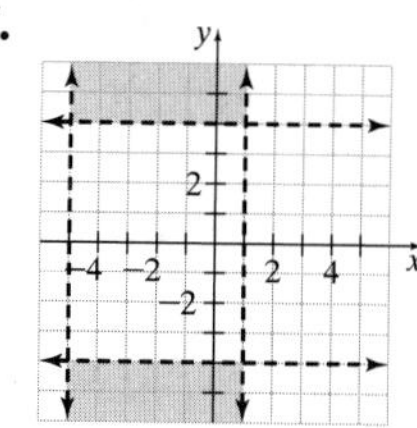

39.

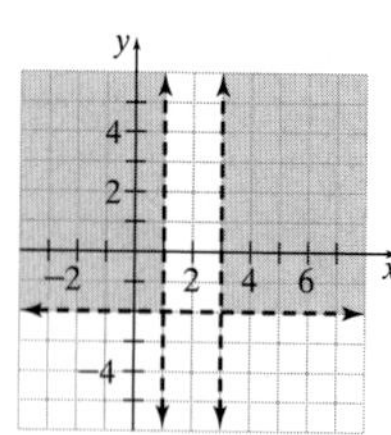

41. If the boundary lines are parallel, there may be no solution.

43. There are no solutions. Opposite sides of the same line are being shaded, but not the line itself.

45. There are an infinite number of solutions. Both inequalities include the line $5x - 2y = 3$.

47. There are an infinite number of solutions. The lines are parallel, but the same side of each line is being shaded.

49.

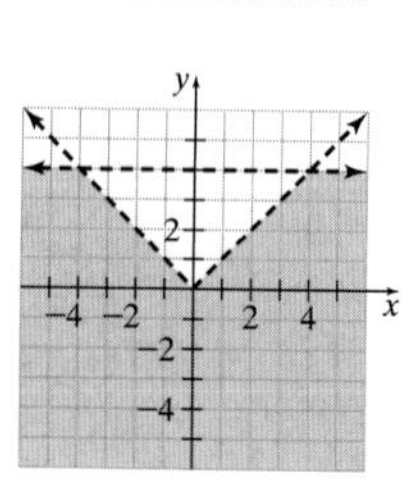

51. $f_2 = \dfrac{f_3 d_3 - f_1 d_1}{d_2}$

52. Domain: $\{-1, 0, 4, 5\}$; Range: $\{-5, -2, 2, 3\}$

53. Domain: $\mathbb{R}$; Range: $\mathbb{R}$

54. Domain: $\mathbb{R}$; Range: $\{y | y \geq -1\}$

Review Exercises **1.** Inconsistent; no solution **2.** Consistent; one solution **3.** Consistent; one solution **4.** Consistent; one solution **5.** $(-2, 1)$ **6.** $(-2, 3)$ **7.** $(0, 4)$ **8.** Dependent **9.** $(3, 7)$ **10.** $(2, 3)$ **11.** $(2, 5)$ **12.** $(5, 2)$ **13.** $(8, -2)$ **14.** $(1, -2)$ **15.** $(26, -16)$ **16.** $(5, -2)$ **17.** $\left(\frac{32}{13}, \frac{8}{13}\right)$ **18.** $\left(-1, \frac{13}{3}\right)$ **19.** $(1, 2)$ **20.** $\left(\frac{7}{5}, \frac{13}{5}\right)$ **21.** $(6, -2)$ **22.** $\left(-\frac{78}{7}, -\frac{48}{7}\right)$ **23.** Infinite number of solutions **24.** No solution **25.** $\left(2, 5, \frac{34}{5}\right)$ **26.** $\left(5, -\frac{15}{4}, -2\right)$ **27.** $(1, 2, -1)$ **28.** $(3, 1, 2)$ **29.** $\left(\frac{8}{3}, \frac{2}{3}, 3\right)$ **30.** $(0, 2, -3)$ **31.** Inconsistent system; no solution **32.** Dependent system; infinite number of solutions **33.** The brother is 24 years old and the niece is 6 years old. **34.** The speed of the plane is 565 mph and the speed of the wind is 35 mph. **35.** James should combine 2 liters of the 20% acid solution to 4 liters of the 50% acid solution. **36.** 410 adult tickets and 240 children's tickets were sold. **37.** His ages were 41 years and 77 years. **38.** \$20,000 was invested at 7%, \$15,000 was invested at 5%, and \$5000 was invested at 3%. **39.** $\left(-1, \frac{1}{3}\right)$ **40.** 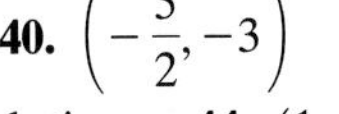 $\left(-\frac{5}{2}, -3\right)$ **41.** Dependent system; infinite number of solutions **42.** $(2, 1, -2)$ **43.** Inconsistent system; no solution **44.** $(1, -1, 3)$ **45.** $(4, -1)$ **46.** $(1, -1)$ **47.** $(-1, 2)$ **48.** $(4, 1, 3)$ **49.** $(-1, 5, -2)$ **50.** Inconsistent system; no solution

51.
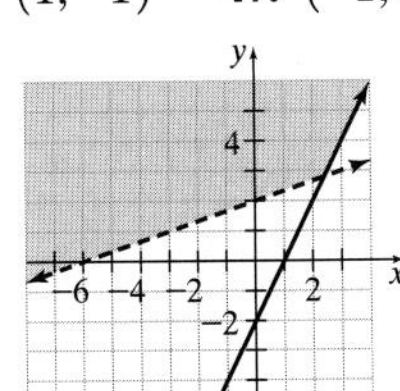

52.
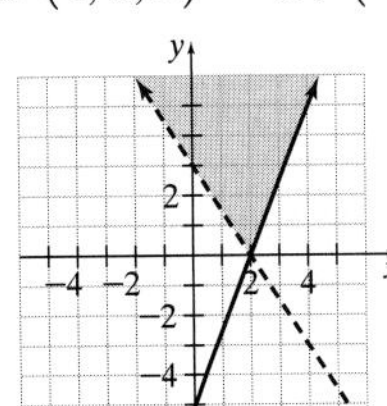

53.
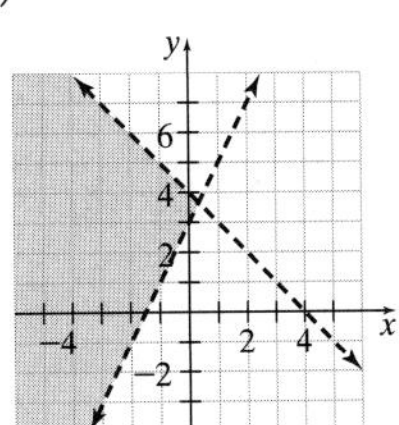

54. No solution
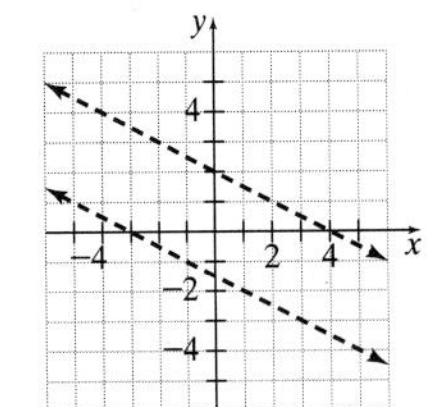

55.
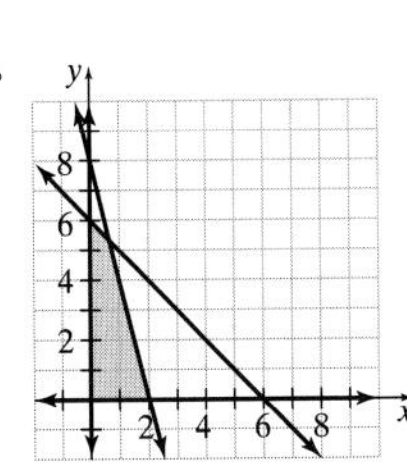

56.
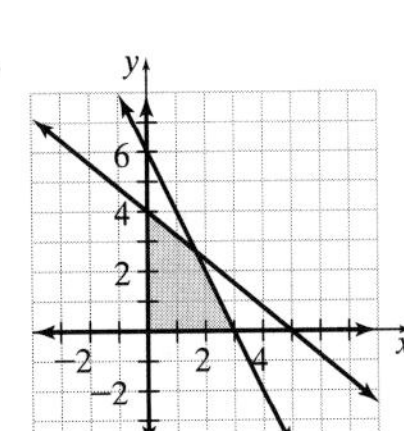

57.
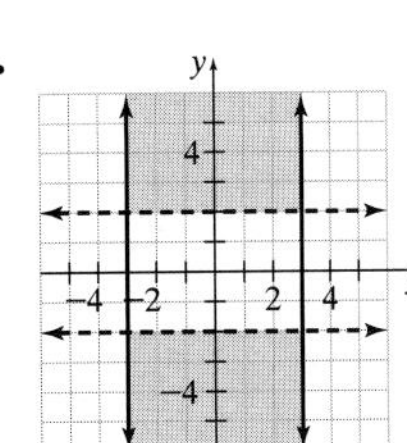

58.
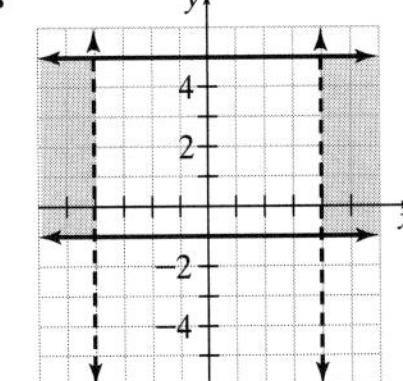

Practice Test **1.** Answers will vary. **2.** Consistent; one solution **3.** Dependent; infinite number **4.** Inconsistent; no solution **5.** $(2, 4)$ **6.** $(1, 5)$ **7.** $(2, 0)$ **8.** $(3, -1)$ **9.** $\left(\frac{5}{2}, 0\right)$ **10.** Dependent; infinite number **11.** $\left(\frac{44}{19}, \frac{48}{19}\right)$ **12.** $(1, -1, 2)$ **13.** $\left[\begin{array}{rrr|r} 4 & -5 & 3 & 2 \\ 2 & -1 & -2 & 4 \\ 3 & 2 & -1 & -3 \end{array}\right]$ **14.** $\left[\begin{array}{rrr|r} 6 & -2 & 4 & 4 \\ 0 & 5 & -3 & 12 \\ 2 & -1 & 4 & -3 \end{array}\right]$ **15.** $(3, 1)$ **16.** $(3, -1, 2)$ **17.** 32 **18.** 165 **19.** $(1, 9)$ **20.** $(-1, -1, 2)$ **21.** $6\frac{2}{3}$ pounds cashews; $13\frac{1}{3}$ pounds peanuts **22.** $6\frac{2}{3}$ liters 6% solution; $3\frac{1}{3}$ liters 15% solution **23.** 4, 9, and 12

24.
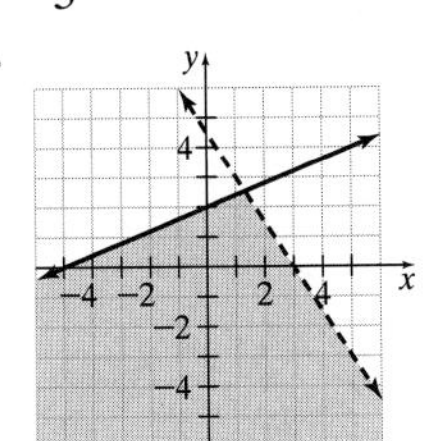

25.
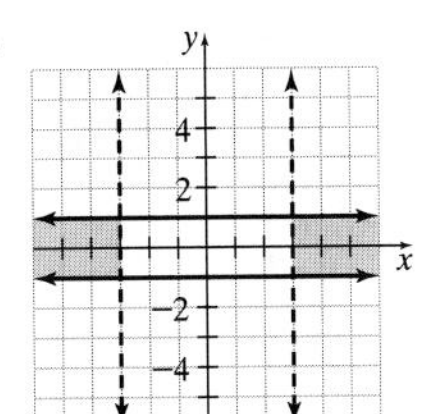

Cumulative Review Test **1.** 0 **2. a)** 9, 1 **b)** $\frac{1}{2}, -4, 9, 0, -4.63, 1$ **c)** $\frac{1}{2}, -4, 9, 0, \sqrt{3}, -4.63, 1$ **3.** $-|-8|, -1, \frac{5}{8}, \frac{3}{4}, |-4|, |-10|$ **4.** 7 **5.** -15 **6.** $-\frac{5}{4}, \frac{11}{4}$ **7.** $b = \frac{R - 3a}{3}$ **8.** $\left\{x \middle| \frac{2}{3} < x \leq \frac{34}{3}\right\}$

9. 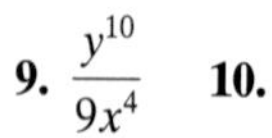 $\frac{y^{10}}{9x^4}$ **10.** 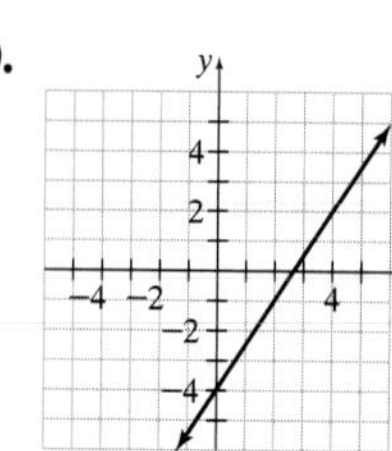 **11.** $y = \frac{2}{3}x + \frac{5}{3}$ **12.** 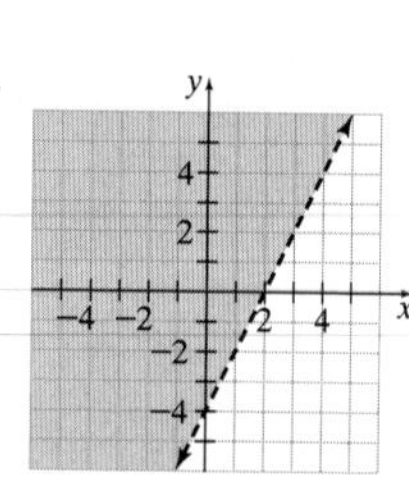 **13. a)** Function **b)** Function **c)** Not a function **14. a)** 14 **b)** $c^2 - 3c + 4$ **15.** (1, 3) **16.** (6, −5) **17.** (2, 1, 3) **18.** 10°, 80°, 90° **19.** 1 hour **20.** 600 of the \$20 tickets; 400 of the \$16 tickets

CHAPTER 5

Using Your Graphing Calculator, 5.1 **1.** 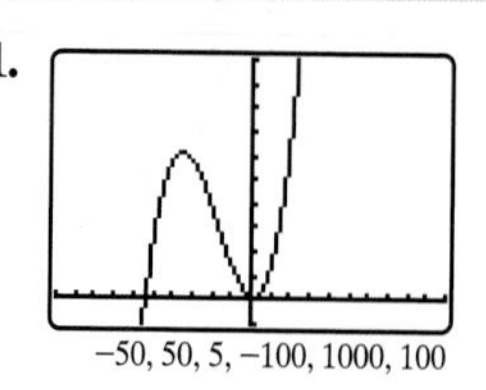
−50, 50, 5, −100, 1000, 100

2. 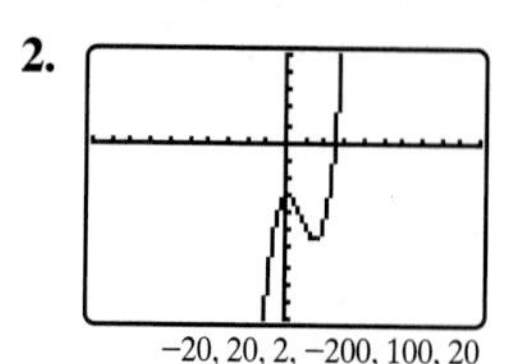
−20, 20, 2, −200, 100, 20

Exercise Set 5.1 **1.** The terms are the parts that are added. **3.** A polynomial is a finite sum of terms in which all variables have whole number exponents and no variable appears in a denominator. **5.** The leading coefficient is the coefficient of the leading term. **7. a)** It is the same as that of the highest-degree term. **b)** 7 **9. a)** A polynomial is linear if its degree is 0 or 1. **b)** Answers will vary. One example is $x + 4$. **11. a)** A polynomial is cubic if it has degree 3 and is in one variable. **b)** Answers will vary. One example is $x^3 + x - 4$. **13.** Answers will vary. One example is $x^5 + x + 1$. **15.** Monomial **17.** Monomial **19.** Trinomial **21.** Negative exponent; not a polynomial **23.** $-x^2 - 4x - 8$; second **25.** $10x^2 + 3xy + 6y^2$; second **27.** In descending order; fourth **29. a)** 6th **b)** 3 **31. a)** 6th **b)** 9 **33. a)** 17th **b)** $-\frac{1}{2}$ **35.** −4 **37.** −7 **39.** −2.0312 **41.** $x^2 + 11x - 9$ **43.** $x^2 - 8x - 2$ **45.** $2y^2 + 6y - 9$ **47.** $-\frac{2}{3}a^2 - \frac{29}{36}a + 3$ **49.** $-3.5x^2 - 3.1x - 27.5$ **51.** $-\frac{4}{3}x^3 - \frac{1}{4}x^2y + 4xy^2$ **53.** $5a - 10b + 13c$ **55.** $8a^2b - 10ab + 11b^2$ **57.** $2r^2 - 4rt - 3t^2$ **59.** $10x^2 - 5x - 6$ **61.** $-3w^2 + 6w$ **63.** $-x + 11$ **65.** $-3x^2 + 2x - 12$ **67.** $-4.4a^2 - 5.7a - 26.4$ **69.** $-\frac{11}{2}x^2y + 6xy^2 + \frac{2}{45}$ **71.** $5x^{2r} - 8x^r + 3$ **73.** $-x^{2s} - 2x^s + 7$ **75.** $7b^{4n} - 5b^{3n} - 2b^{2n} - 1$ **77.** No **79.** No **81.** $A \approx 113.10$ in^2 **83.** 674 feet **85. a)** $P(x) = 2x^2 + 360x - 4050$ **b)** \$51,950 **87.** c) **89.** c) **91. a)** 883,060 **b)** Yes **93. a)** 182,970 **b)** 636,970 **95. a)** 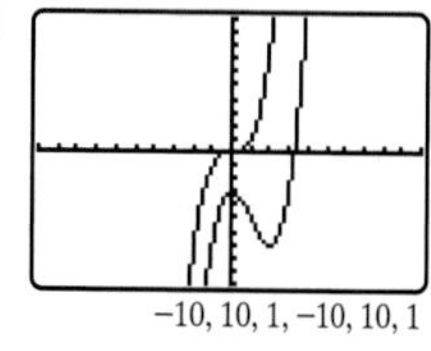
−10, 10, 1, −10, 10, 1

b) Increase **c)** Answers will vary. **d)** Decrease **e)** Answers will vary. **97.** b) **101.** (−8, 2) **102.** (−8, 2) **103.** (−8, 2) **104.** 1, 5, and 6

Exercise Set 5.2 **1.** Answers will vary. **3. a)** Answers will vary. **b)** $x^3 - 2x^2 - 21x + 12$ **5. a)** Answers will vary. **b)** Answers will vary. One possible answer is $(x + 4)(x - 4)$. **c)** Answers will vary. **d)** Answers will vary. One possible answer is $x^2 - 16$. **7.** Yes **9.** $24x^2y^5$ **11.** $\frac{1}{9}x^7y^8z^2$ **13.** $6x^6y^3 - 9x^3y^4 - 12x^2y$ **15.** $2xyz + \frac{8}{3}y^2z - 6y^3z$ **17.** $2.85a^{11}b^5 - 1.38a^9b^7 + 0.36a^6b^9$ **19.** $12x^2 - 38x + 30$ **21.** $-2x^3 + 8x^2 - 3x + 12$ **23.** $\frac{2}{15}x^2 + \frac{1}{3}xz - \frac{1}{5}z^2$ **25.** $12.88a^2 + 1.82ab - 5.88b^2$ **27.** $x^3 - 7x^2 + 14x - 8$ **29.** $4x^3 + x^2 - 20x + 4$ **31.** $2a^3 - 7a^2b + 5ab^2 - 6b^3$ **33.** $27x^3 - 27x^2 + 9x - 1$ **35.** $6a^4 - 28a^3 + 35a^2 - 3a - 6$ **37.** $10r^4 - 2r^3s - r^2s^2 + rs^3 - 2s^4$ **39.** $x^2 + 4x + 4$ **41.** $4x^2 - 12xy + 9y^2$ **43.** $25m^4 - 4n^2$ **45.** $y^2 + 8y - 4xy + 16 - 16x + 4x^2$

47. $16 - 8x + 24y + x^2 - 6xy + 9y^2$ **49.** $a^2 - b^2 - 4b - 4$ **51.** $6y^2 - y - 12$ **53.** $4x^2 - \frac{9}{16}$

55. $\frac{2}{5}x^3y^7 - \frac{1}{6}x^6y^5 + \frac{4}{3}x^3y^7z^5$ **57.** $2x^3 + 10x^2 + 9x - 9$ **59.** $6x^3 - x^2y - 16xy^2 + 6y^3$
61. $x^3 + 9x^2 + 27x + 27$ **63.** $w^2 - 9x^2 - 24x - 16$ **65.** $a^4 - 2a^2b^2 + b^4$ **67.** $2x^3 - 4x^2 - 64x + 192$
69. a) $x^2 - x - 20$ **b)** -8 **71. a)** $-3x^3 + 12x^2 - 2x + 8$ **b)** 0 **73. a)** $10x^3 + 36x^2 - 2x - 12$ **b)** 1196
75. a) $x^2 + 8x + 15$ **b)** $x^2 + 8x + 15$ **77.** $36 - x^2$ **79.** $x^2 + y^2$ **81. a)** $11x + 12$
b) 117 square inches; 50 square inches **83.** $(x - 6)(x + 6)$ **85.** $(x + 6)(x + 6)$ **87.** $a(x - n)(x - n)(x - n)$
89. a) Answers may vary. **b)** $a^2 + 2ab + b^2$ **c)** $a^2 + 2ab + b^2$ **d)** Same **91. a)** $A = P(1 + r)^t$
b) $1123.60 **93.** $a^2 + 2ab + b^2 + 3a + 3b + 4$ **95.** $15x^{3t-1} + 12x^{4t}$ **97.** $12x^{3m} - 18x^m - 10x^{2m} + 15$
99. $y^{a^2-b^2}$ **101.** $x^4 - 12x^3y + 54x^2y^2 - 108xy^3 + 81y^4$ **103. a)** Answers will vary. **b)** It is correct.
105. $y^2 - 2y - 2xy + 2x + x^2 + 1$ **107.** Yes; no **108. a)** $ax + by = c$ **b)** $y = mx + b$
c) $y - y_1 = m(x - x_1)$ **109.** $(2, -1, 6)$ **110.** 150 first class; 400 bulk postage

Exercise Set 5.3

1. a) Answers will vary. **b)** $\frac{5}{3}x^3 - 2x^2 - \frac{4}{3}x - 4 + \frac{7}{3x}$ **3.** Yes; answers will vary.

5. Place them in descending order of the variable. **7. a)** Answers will vary. **b)** $x + 8 + \frac{36}{x - 5}$ **9.** $3x + 4$

11. $2x + 1$ **13.** $3x^2 - x - 2$ **15.** $x^3 - \frac{3}{2}x^2 + 3x - 2$ **17.** $2x^2 - 4xy + \frac{3}{2}y^2$ **19.** $\frac{3x}{y} - 6x^2 + \frac{9y}{2x}$

21. $\frac{z}{2} + z^2 - \frac{3}{2}x^2y^4z^7$ **23.** $x + 3$ **25.** $2x + 3$ **27.** $2a + 5 + \frac{1}{a - 2}$ **29.** $3x + 2$ **31.** $2r + 3$

33. $3x^2 - 3x + 1 + \frac{2}{3x + 2}$ **35.** $-x^2 - 7x - 5 - \frac{8}{x - 1}$ **37.** $2y^2 + 3y - 1$ **39.** $2x^2 + x - 2 - \frac{2}{2x - 1}$

41. $3x^3 + 6x + 4$ **43.** $x + 4$ **45.** $3x^3 - 8x^2 - 5$ **47.** $x + 3$ **49.** $x + 8 + \frac{12}{x - 3}$ **51.** $3x + 5 + \frac{10}{x - 4}$

53. $4x^2 + x + 3 + \frac{3}{x - 1}$ **55.** $3x^2 - 2x + 2 + \frac{6}{x + 3}$ **57.** $y^3 + y^2 + y + 1$ **59.** $y^4 - \frac{10}{y + 1}$

61. $3x^2 + 3x - 3$ **63.** $2x^3 + 2x - 2$ **65.** 10 **67.** 0; factor **69.** $-\frac{19}{4}$ or -4.75 **71.** $3x + 2$ **73.** 3 times

75. No **77.** If the remainder is 0, $x - a$ is a factor. **79.** $x^2 - 2x - 8$ **81.** $x^2 - x - 8$ **83.** $2x^2 + 3xy - y^2$

85. $x + \frac{5}{2} + \frac{11}{2(2x - 3)}$ **87.** $w = r + 1$ **89.** $x^3 - 6x^2 + 13x - 10$ **91.** $2x + 1 - \frac{3}{2x} - \frac{1}{2x^2}$ **93.** Not a factor

95. Factor **97. a)** $3x^2 - 2x + 5 - \frac{13}{3x + 5}$ **b)** Because we are expressing the remainder in terms of $3x + 5$ rather than $x + \frac{5}{3}$, the denominator of the remainder is altered rather than the numerator. **98.** $\frac{x^4}{16y^6}$

99. 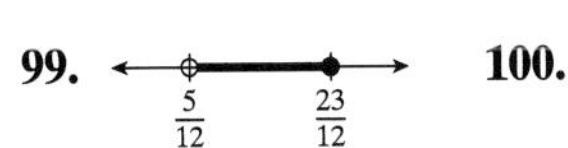

100. $\frac{2}{21}$ **101.** 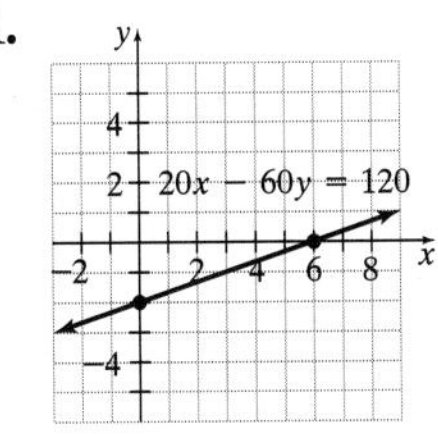

Exercise Set 5.4

1. Determine if all the terms contain a greatest common factor and, if so, factor it out.
3. a) Answers will vary. **b)** $2x^2y$ **c)** $2x^2y(3y^4 - x + 6x^7y^2)$ **5. a)** Answers will vary. **b)** $(2x + y^2)(3x^2 - y^3)$
7. xy^4 **9.** $8(n + 1)$ **11.** $3(2x^2 + x - 3)$ **13.** $2(4p^2 - 3p + 2)$ **15.** $x^3(7x^2 - 9x + 3)$
17. $-3y(8y^{14} - 3y^2 + 1)$ **19.** $3xy(x + 2xy + 1)$ **21.** $4a^2c(10a^3b^4 - 2a^2b^2c + 1)$
23. $3pq^2r(3p^3q^3 - pr + 2q^3r^2)$ **25.** $-2(26x^2y^2 + 8xy^3 - 13z)$ **27.** $-2(3x - 1)$ **29.** $-(w^2 - 7w + 5)$
31. $-3(r^2 + 2r - 3)$ **33.** $-2rs^3(3r^3 - 2rs - s^2)$ **35.** $-a^2b(a^2bc - 5ac^2 - 1)$ **37.** $(x + 1)(a - 2)$
39. $(4a - 3)(x - 1)$ **41.** $-(x - 2)(2x - 9)$ **43.** $-2(a + 2)(a + 2)$ **45.** $(x - 5)(x + 3)$
47. $2(2n - 5)(2n - 1)$ **49.** $(a + b)(x + y)$ **51.** $(x^2 + 4)(x - 3)$ **53.** $(2m - 5n)(5m - 6n)$

55. $6(x^2 - 2)(x + 3)$ **57.** $a^2(a^2 + 1)(a - 1)$ **59. a)** 240 feet **b)** $h(t) = -16t(t - 8)$ **c)** 240 feet
61. a) $\approx$ 2856.64 ft^2 **b)** $A = r(\pi r + 2l)$ **c)** $A \approx$ 2856.64 ft^2 **63. a)** $(1 - 0.06)(x + 0.06x) = (0.94)(1.06x)$
b) $0.9964x$; slightly lower than the price of the 1999 model (99.64%)
65. a) $(x + 0.15x) - 0.20(x + 0.15x) = 0.80(x + 0.15x)$ **b)** $0.92x$; 92% of the regular price
67. $(3x - 2)^4(15ax - 10a + 4)$ **69.** $2(x - 3)(2x^4 - 12x^3 + 15x^2 + 9x + 2)$ **71.** $(x^2 + 2x - 3)(a + b)$
73. $x^{4m}(x^{2m} - 2)$ **75.** $x^{2m}(3x^{2m} - 2x^m + 1)$ **77.** $(b^r - d^r)(a^r + c^r)$ **79. a)** Yes **b)** 0 **c)** Answers will vary.
81. a) They should be the same graph.
b)

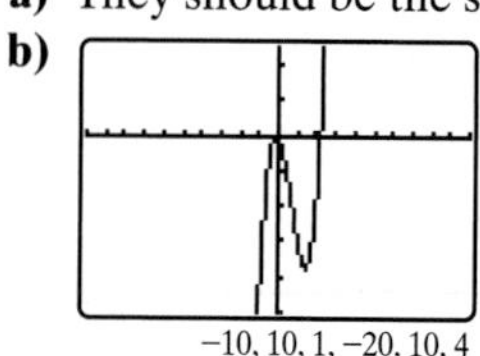

−10, 10, 1, −20, 10, 4

c) Answers will vary. **d)** Factoring is not correct.
83. $-\frac{5}{24}$ **84.** $\left\{x \middle| -\frac{4}{3} \le x < \frac{14}{3}\right\}$
85. \$4000 was invested at 5%; \$6000 was invested at 6%
86. \$4000 was invested at 5%; \$6000 was invested at 6%
87. $3x - 4y = 8$

Using Your Graphing Calculator, 5.5

1. Yes **2.** No

Exercise Set 5.5

1. Factor out the greatest common factor if it is present. **3. a)** Answers will vary.
b) $(2x - 3)(3x + 4)$ **5.** Both are +. **7.** One is +, one is −. **9.** $(x + 6)(x + 1)$ **11.** $(y - 1)(y - 11)$
13. $(x - 8)^2$ **15.** $(x + 2)(x - 15)$ **17.** $(x - y)(x - 3y)$ **19.** $(z - 2y)(z - 5y)$ **21.** $4(x + 4)(x - 1)$
23. $x(x - 6)(x + 3)$ **25.** $(p - 1)(5p - 3)$ **27.** $3(x - 2)(x + 1)$ **29.** $(3x + 7)(2x - 9)$
31. $(10x - 7)(3x - 5)$ **33.** $(16x - 3)(2x - 1)$ **35.** $2(3w + 4z)(3w - z)$ **37.** $(4x - 3y)(2x + 9y)$
39. $10(5b - 2)(2b - 1)$ **41.** $ab^5(a - 4)(a + 3)$ **43.** $3b^2c(b - 3c)^2$ **45.** $4m^6n^3(m + 2n)(2m - 3n)$
47. $(6x - 5)(5x + 4)$ **49.** $8x^2y^4(x + 4)(x - 1)$ **51.** $(x^2 + 3)(x^2 - 2)$ **53.** $(x^2 + 2)(x^2 + 3)$
55. $(2a^2 + 5)(3a^2 - 5)$ **57.** $(2x + 5)(2x + 3)$ **59.** $(3a + 1)(2a + 5)$ **61.** $(ab - 3)(ab - 5)$
63. $(3xy - 5)(xy + 1)$ **65.** $(2a - 5)(a - 1)(5 - a)$ **67.** $(x + 2)(x - 3)(2x + 3)$ **69.** $(y^2 - 10)(y^2 + 3)$
71. $(x + 2)(x + 1)(x + 3)$ **73.** $a^3b^2(5a - 3b)(a - b)$ **75.** $2x^2 - 5xy - 12y^2$ **77.** Divide; $x + 6$
79. a) Answers will vary. **b)** $(6x - 5)(5x + 8)$; $(7x - 1)(7x - 13)$ **81.** $\pm 3, \pm 9$ **83.** 6 or −6 **85. a)** 4
b) $(x - 3)(x - 5)$ **87. a)** −8 **b)** Not factorable **89.** Answers will vary. One example is $x^2 + 2x + 1$.
91. $(2a^n + 3)(2a^n - 5)$ **93.** $(x + y)^2(x - 4y)(x - 3y)$ **95.** $(x^n - 2)(x^n + 5)$
97. a) Answers will vary. **b)** Correct **99.** 9.0×10^{10} **100.** 0, the change in y is 0. **101.** Undefined
102. $y = -\frac{3}{2}x + \frac{11}{2}$ **103.** $-x^2y - 8xy^2 + 6$

Exercise Set 5.6

1. a) Answers will vary. **b)** $(x + 4)(x - 4)$ **3.** Answers will vary.
5. $a^3 + b^3 = (a + b)(a^2 - ab + b^2)$ **7.** $(x + 9)(x - 9)$ **9.** $(x + 4)(x - 4)$ **11.** $(1 + 3a)(1 - 3a)$
13. $(5 + 4y^2)(5 - 4y^2)$ **15.** $\left(\frac{1}{4} + x\right)\left(\frac{1}{4} - x\right)$ **17.** $(ab + 7c)(ab - 7c)$ **19.** $(0.2x + 0.3)(0.2x - 0.3)$
21. $x(12 - x)$ **23.** $(a + 3b + 2)(a - 3b - 2)$ **25.** $(x + 5)^2$ **27.** $(2 + x)^2$ **29.** $(2x - 5y)^2$
31. $(0.9x - 0.2)^2$ **33.** $(w^2 + 8)^2$ **35.** $(x + y + 1)^2$ **37.** $(a + b)^2(a - b)^2$ **39.** $(x + 3 + y)(x + 3 - y)$
41. $(x + 7)(-x + 3)$ **43.** $(3a - 2b + 3)(3a - 2b - 3)$ **45.** $4xy$ **47.** $(x - 3)(x^2 + 3x + 9)$
49. $(x + y)(x^2 - xy + y^2)$ **51.** $(4 - a)(16 + 4a + a^2)$ **53.** $(3y - 2x)(9y^2 + 6xy + 4x^2)$
55. $3(2x - 3y)(4x^2 + 6xy + 9y^2)$ **57.** $5(x - 5y)(x^2 + 5xy + 25y^2)$ **59.** $(x + 2)(x^2 + x + 1)$
61. $(x - y - 3)(x^2 - 2xy + y^2 + 3x - 3y + 9)$ **63.** $-9(b^2 + 3b + 3)$ **65.** $(11y^2 + 7x)(11y^2 - 7x)$
67. $(4y + 9x)(4y - 9x)$ **69.** $(5x^2 + 9y^3)(5x^2 - 9y^3)$ **71.** $(x - 4)(x^2 + 4x + 16)$ **73.** $(3xy + 4)^2$
75. $(a^2 + b^2)^2$ **77.** $(x - 1 + y)(x - 1 - y)$ **79.** $(x + y + 1)(x^2 + 2xy + y^2 - x - y + 1)$
81. $3m(-m + 2n)$ **83. a)** $a^2 - b^2$ **b)** $(a + b)(a - b)$ **85. a)** $6a^3 - 6ab^2$ **b)** $6a(a + b)(a - b)$
87. a) $\frac{4}{3}\pi R^3 - \frac{4}{3}\pi r^3$ **b)** $\frac{4}{3}\pi(R - r)(R^2 + Rr + r^2)$ **89.** 12; −12 **91.** $c = 4$
93. a) Find an expression whose square is $25x^2 - 30x + 9$ **b)** $s(x) = 5x - 3$ **c)** 7
95. $(x^2 + 4x + 8)(x^2 - 4x + 8)$ **97.** $h(2a + h)$ **99. a)** 16 **b)** $x^2 + 8x + 16$ **c)** $(x + 4)^2$
101. $(8x^{2a} + 3y^{3a})(8x^{2a} - 3y^{3a})$ **103.** $(a^n - 8)^2$ **105.** $(x^n - 2)(x^{2n} + 2x^n + 4)$ **107.** Correct
109. a) $(x^3 - 1)(x^3 + 1)$ **b)** $(x^2 - 1)(x^4 + x^2 + 1)$ **111. a)** 3, 6 **b)** $-2, \frac{5}{9}, -1.67, 0, 3, 6$ **c)** $\sqrt{3}, -\sqrt{6}$
d) $-2, \frac{5}{9}, -1.67, 0, \sqrt{3}, -\sqrt{6}, 3, 6$ **112.** $\{4, 5, 6, 7\}$ **113.** $\{4, 5, 6, 7, \ldots\}$ **114.** 2 **115.** Width is 3 feet; length is 8 feet.

Exercise Set 5.7 **1.** Answers will vary. **3.** $4(x + 4)(x - 3)$ **5.** $(5s - 3)(2s + 5)$ **7.** $-(4r - 3)(2r - 5)$ **9.** $0.4(x + 0.3)(x - 0.3)$ **11.** $5x(x^2 + 3)(x^2 - 3)$ **13.** $3x^4(x - 4)(x - 1)$ **15.** $5x^2y^2(x - 3)(x + 4)$ **17.** $x^2(x + y)(x - y)$ **19.** $x^4y^2(x - 1)(x^2 + x + 1)$ **21.** $x(x^2 + 4)(x + 2)(x - 2)$ **23.** $4(x^2 + 2y)(x^4 - 2x^2y + 4y^2)$ **25.** $2(a + b + 3)(a + b - 3)$ **27.** $6(x + 3y)^2$ **29.** $x(x + 4)$ **31.** $(2a + b)(a - 2b)$ **33.** $(y + 5)^2$ **35.** $5a^2(3a - 1)^2$ **37.** $\left(x + \frac{1}{3}\right)\left(x^2 - \frac{1}{3}x + \frac{1}{9}\right)$ **39.** $(x + 3)(x - 3)(3x + 2)$ **41.** $ab(a + 4b)(a - 4b)$ **43.** $(9 + x + y)(9 - x - y)$ **45.** $2(4x - 3)(3x - 2)$ **47.** $(8x + 3)(2x - 5)$ **49.** $(x^2 + 9)(x + 3)(x - 3)$ **51.** $(5c - 6y)(b - 2x)$ **53.** $(x^2 + 1)(3x^2 - 4)$ **55.** $(y + x - 4)(y - x + 4)$ **57.** $(2y + 5)(y + 8)$ **59.** $(a + 6b + 4c)(a + 6b - 4c)$ **61.** $3x^2y(x + 3)(2x - 1)$ **63.** $(x + y)^2(x - y)^2$ **65.** e **67.** a **69.** f **71.** c **73. a)** $a(a + b) - b(a + b) = a^2 - b^2$ **b)** $(a - b)(a + b)$ **75. a)** $a^2 + 2ab + b^2$ **b)** $(a + b)^2$ **77. a)** $2a(a - b) + 2b(a - b)$ **b)** $2(a + b)(a - b)$ **79. a)** Answers will vary. **b)** Answers will vary. **81. a)** $x^{-5}(x^2 - 2x - 3)$ **b)** $x^{-5}(x - 3)(x + 1)$ **83. a)** $x^{-3/2}(5x^2 + 2x - 3)$ **b)** $x^{-3/2}(5x - 3)(x + 1)$ **84.** -8 **85.** No solution **86.** $\{x|x > 6\}$ **87.** $\mathbb{R}$

Using Your Graphing Calculator, 5.8 **1.** $y = x^2 - 6x + 5$ **2.** $y = x^2 - x - 6$ **3.** $y = x^2 + 4x$

Exercise Set 5.8 **1.** The degree of a polynomial function is the same as the degree of the leading term. **3.** $ax^2 + bx + c = 0$ **5. a)** The zero factor property only holds when one side of the equation is 0. **b)** -2 and -5 **7. a)** Answers will vary. **b)** $\frac{4}{3}, -\frac{5}{4}$ **9. a)** Legs **b)** Hypotenuse **11.** -8 and -2 **13.** $0, -5$ **15.** $0, -9$ **17.** $-\frac{5}{2}, 3, -2$ **19.** 3 **21.** $0, 12$ **23.** $0, -2$ **25.** $-4, 3$ **27.** $10, 2$ **29.** $-6, 3$ **31.** $6, -4$ **33.** $0, -21, 2$ **35.** $-6, -5$ **37.** $2, -\frac{5}{6}$ **39.** $11, 1$ **41.** $0, -8$ **43.** $0, -4, 3$ **45.** $0, \frac{8}{3}$ **47.** $0, -\frac{4}{5}, \frac{4}{5}$ **49.** $-8, 0$ **51.** $5, -3$ **53.** $\frac{4}{3}, \frac{5}{2}$ **55.** $-\frac{1}{3}, 2$ **57.** $0, -5, -3$ **59.** $-\frac{1}{3}, -2$ **61.** $\frac{3}{5}, \frac{5}{2}$ **63.** $-\frac{5}{4}, 6$ **65.** $(-6, 0), (4, 0)$ **67.** $(-7, 0)$ **69.** $(0, 0), \left(\frac{4}{3}, 0\right), \left(\frac{5}{2}, 0\right)$ **71.** d) **73.** b) **75.** $y = x^2 - x - 12$ **77.** $y = 2x^2 - x - 15$ **79.** Width is 2 feet; length is 5 feet **81.** Base is 10 feet; height is 16 feet **83.** 2.5 cm **85.** 3 feet **87.** 2 seconds **89.** 13 feet **91.** 60 bicycles **93.** 13 inches by 13 inches **95. a)** $V = a^3 - ab^2$ **b)** $V = a(a + b)(a - b)$ **c)** $b = 3$ inches **97. a)** $f(x) = x^2 + 7x + 10$ **b)** $x^2 + 7x + 10 = 0$ **c)** An infinite number; any function of the form $f(x) = a(x^2 + 7x + 10), a \neq 0$ **d)** An infinite number; any equation of the form $a(x^2 + 7x + 10) = 0, a \neq 0$ **99. a)** Answers will vary. Examples are:

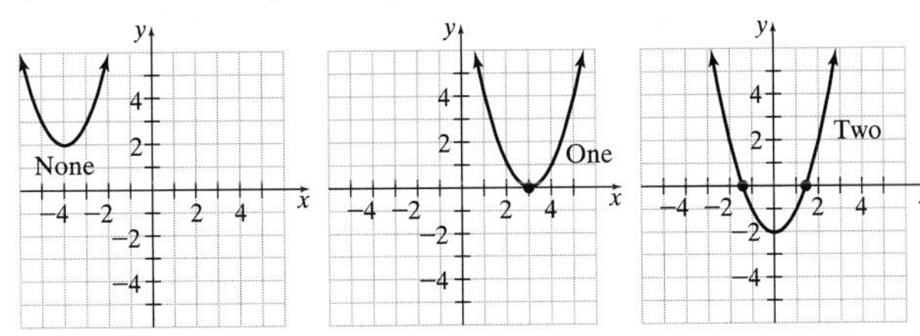

b) No x-intercepts, one x-intercept, or two x-intercepts **101.** 73.721949 mph **103.** $\pm 2, \pm 3$ **108. a)** Carmen, 4.8 mph; Bob, 6 mph **b)** 24 miles **109.**

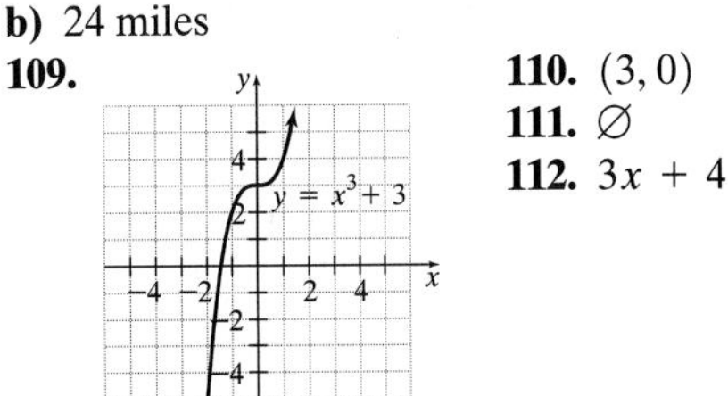

110. $(3, 0)$ **111.** $\varnothing$ **112.** $3x + 4$

Review Exercises **1. a)** Trinomial **b)** $4x^2 + 5x - 3$ **c)** Second degree **2. a)** Trinomial **b)** $x^2 + xy - y^2$ **c)** Second degree **3. a)** Polynomial **b)** $2x^4 - 9x^2y + 6xy^3 - 3$ **c)** Fourth degree **4.** Not a polynomial **5.** $2x^2 + 10x + 3$ **6.** $4x^3 + 8x^2 + 12x$ **7.** $x^2 + 10x + 25$ **8.** $7y^2 + 2$ **9.** $25x^2 - 9$ **10.** $6x^2y + 9xy^2 + 2x + 3y$ **11.** $2x - 3$ **12.** $2x^3 - 8x^2 - 9$ **13.** $-2x^4y^2 - 2x^3y^7 + 12xy^3$ **14.** $9x^2 - 12xy + 4y^2$ **15.** $3x^2y + 3xy - 9y^2$ **16.** $25x^2y^2 - 36$ **17.** $3x - 2y + 1$ **18.** $4x^2 - 25y^4$ **19.** $x + 4 - \frac{5}{x - 3}$ **20.** $\frac{x^2}{2y} + x + \frac{3y}{2}$ **21.** $x^2 + 6xy + 9y^2 + 4x + 12y + 4$ **22.** $x^2 + 6xy + 9y^2 - 4$

23. $-9xy - 9x + 5y^2$ **24.** $6x^3 - x^2 - 24x + 18$ **25.** $2x^3 + x^2 - 3x - 4$ **26.** $4x^4 + 12x^3 + 6x^2 + 13x - 15$ **27.** $2x^2 + 3x - 4 + \dfrac{2}{2x + 3}$ **28.** $x^3y + 6x^2y + 7xy^2 + x^2y^2 + y^3$ **29.** 57 **30.** −149 **31. a)** \$780.46 billion **b)** Yes **32. a)** \$773.13 billion **b)** Yes **33. a)** $x^2 - x - 6$ **b)** 0 **34. a)** $2x^3 - 4x^2 - 6x + 12$ **b)** 12 **35. a)** $x^3 - x^2 - 5x + 6$ **b)** 9 **36. a)** $x^4 - 4$ **b)** 77 **37.** $3x^2 + 7x + 21 + \dfrac{73}{x - 3}$

38. $2y^4 - 2y^3 - 8y^2 + 8y - 7 + \dfrac{6}{y + 1}$ **39.** $x^4 + 2x^3 + 4x^2 + 8x + 16 + \dfrac{12}{x - 2}$ **40.** $2x^2 + 2x + 6$

41. 3 **42.** −236 **43.** $-\dfrac{53}{9}$ or $-5.\overline{8}$ **44.** 0; factor **45.** $4(3x^2 + x + 2)$ **46.** $6x^4(10 + x^5 - 3xy^2)$

47. $(2x - 1)(10x - 3)$ **48.** $3xy^2z^2(4y^2z + 2xy - 5x^2z)$ **49.** $(x + 4y)(5x - y)$ **50.** $(4x + 5y)(3x - 2y)$ **51.** $(3x - y)(-4x + 9y)$ **52.** $3(a^2 + 3b)(a^2 - 4b)$ **53.** $(x + 5)(x + 3)$ **54.** $-(x - 15)(x + 3)$ **55.** $(x - 18y)(x + 3y)$ **56.** $x(4x - 5)(2x + 5)$ **57.** $x(4x - 5)(x - 1)$ **58.** $x(12x + 1)(x + 5)$ **59.** $(x^2 + 4)(x^2 - 5)$ **60.** $(x + 9)(x + 11)$ **61.** $(x + 6)(x - 6)$ **62.** $(x^2 + 9)(x - 3)(x + 3)$ **63.** $(x - 1)(x - 5)$ **64.** $(2x - 3)^2$ **65.** $(3y + 4)^2$ **66.** $(w^2 - 8)^2$ **67.** $(a + 3b + 2c)(a + 3b - 2c)$ **68.** $(x - 2)(x^2 + 2x + 4)$ **69.** $(2x + 3)(4x^2 - 6x + 9)$ **70.** $(x - 1)(x^2 + 4x + 7)$ **71. a)** $a^2 - 4b^2$ **b)** $(a + 2b)(a - 2b)$ **72. a)** $4a^3 - 4ac^2$ **b)** $4a(a + c)(a - c)$ **73.** $y^2(x + 3)(x - 5)$ **74.** $3x(x - 4)(x - 2)$ **75.** $3xy^4(x - 2)(x + 6)$ **76.** $3y(y^2 + 3)(y^2 - 3)$ **77.** $2y(x + 2)(x^2 - 2x + 4)$ **78.** $5x^2y(x + 2)^2$ **79.** $3x(2x + 1)(x - 4)$ **80.** $(x + 5 + y)(x + 5 - y)$ **81.** $3(x + 2y)(x^2 - 2xy + 4y^2)$ **82.** $(x + 4)^2(x - 1)$ **83.** $(4x + 1)(4x + 5)$ **84.** $(2x^2 - 1)(2x^2 + 3)$ **85.** $(x - 2)(x + 1)(x - 1)$ **86.** $(3x + 4y)(3a - b)$ **87.** $(2pq - 3)(3pq + 2)$ **88.** $(3x^2 - 2)^2$ **89.** $(2y + x + 2)(2y - x - 2)$ **90.** $2(3a + 5)(4a + 3)$

91. $3x^2y^4(x + 3)(2x - 3)$ **92.** $\left(x - \dfrac{2}{3}y^2\right)\left(x^2 + \dfrac{2}{3}xy^2 + \dfrac{4}{9}y^4\right)$ **93. a)** $a^2 - 4b^2$ **b)** $(a + 2b)(a - 2b)$

94. a) $2ab + 2b^2$ **b)** $2b(a + b)$ **95. a)** $2a(a + 3b) + b(a + 3b)$ **b)** $(2a + b)(a + 3b)$

96. a) $a^2 + 2ab + b^2$ **b)** $(a + b)^2$ **97.** $-\dfrac{2}{3}, 5$ **98.** $0, \dfrac{3}{2}$ **99.** $-\dfrac{4}{3}, 0$ **100.** −4, 6 **101.** −5, −3 **102.** −4, 2

103. −2, −5 **104.** 0, 2, 4 **105.** $0, -\dfrac{1}{4}, \dfrac{4}{3}$ **106.** $-\dfrac{3}{2}, \dfrac{1}{4}$ **107.** −2, 2 **108.** −3, 2 **109.** (6, 0), (−5, 0)

110. $\left(\dfrac{6}{5}, 0\right), \left(\dfrac{5}{4}, 0\right)$ **111.** $\left(\dfrac{7}{2}, 0\right), \left(-\dfrac{4}{7}, 0\right)$ **112.** Width is 7 feet; length is 9 feet

113. Height is 4 feet; base is 11 feet **114.** 5 inches for the smaller square; 9 inches for the larger square
115. 9 seconds **116.** 9

Practice Test

1. a) Trinomial **b)** $-6x^4 - 4x^2 + 2x$ **c)** Degree is 4 **d)** −6 **2.** $4x^5 - 2y^2 + \dfrac{5}{x}$

3. $-6x^2 + xy + y^2$ **4.** $4x^3 + 8x^2y - 9xy^2 - 6y^3$ **5.** $x - 5 + \dfrac{25}{2x + 3}$ **6.** $2x^2y + 3x + 7y^2$

7. $2x^2 + 5x + 20 + \dfrac{53}{x - 3}$ **8.** $-6x^7y^6 + 18x^4y^7 - 9x^3y^4$ **9.** $3x^3 + 3x^2 + 15x + 15 + \dfrac{79}{x - 5}$ **10.** −89

11. $3xy^2(3x^2 + 4xy^3 - 9y^2)$ **12.** $3x(x - 3)(x + 1)$ **13.** $(2x + 3y)(x + 2y)$ **14.** $5(x - 2)(x + 1)$ **15.** $y^6(3x - 2)(9x^2 + 6x + 4)$ **16.** $(x + 2)(x + 6)$ **17.** $(2x^2 + 9)(x^2 - 2)$ **18. a)** $3x^2 - 19x + 20$ **b)** −6 **19.** $-\dfrac{3}{4}, 6$ **20.** 0, −5, 1 **21.** $c = \dfrac{b_n - b_1}{a - 1}$ **22.** $\left(\dfrac{1}{4}, 0\right), \left(-\dfrac{3}{2}, 0\right)$ **23.** $y = x^2 - 8x + 12$

24. Height is 4 meters; base is 14 meters **25.** 7 seconds

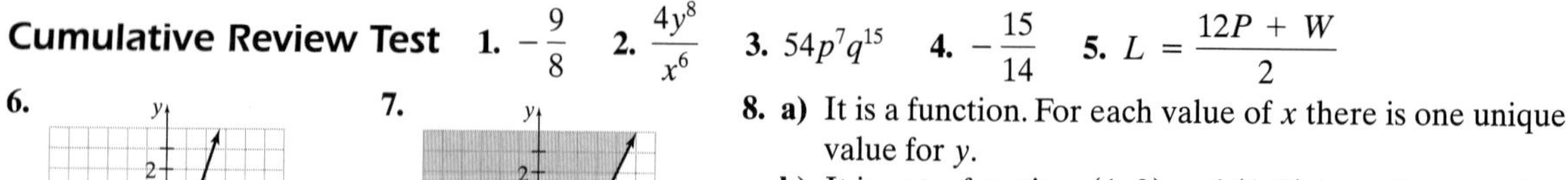

Cumulative Review Test

1. $-\dfrac{9}{8}$ **2.** $\dfrac{4y^8}{x^6}$ **3.** $54p^7q^{15}$ **4.** $-\dfrac{15}{14}$ **5.** $L = \dfrac{12P + W}{2}$

6.

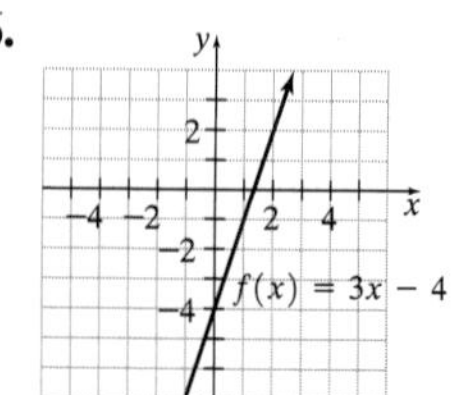

7.

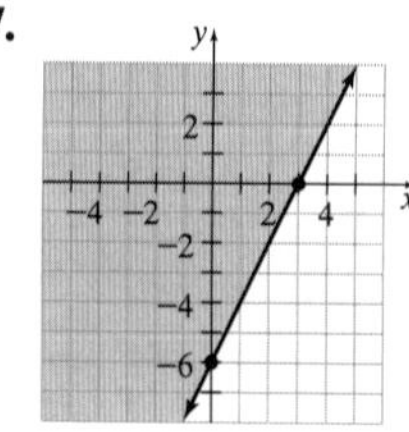

8. a) It is a function. For each value of x there is one unique value for y.
b) It is not a function; (1, 2) and (1, 0) have the same first coordinate.

9. $\left(\dfrac{20}{11}, -\dfrac{14}{11}\right)$

10.

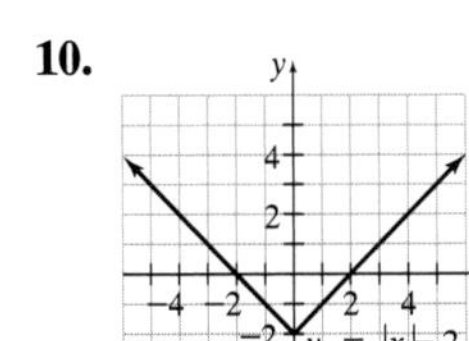

11. $7x^2 - 9x$ **12.** $2x^3 - 11x^2 + 3x + 30$ **13.** $3xy^4 - \frac{8y^3}{3} - \frac{4}{x}$ **14.** -5 **15.** $x(x^2 + 2)(x - 3)$ **16.** $3y(x - 2)(4x - 1)$ **17.** $(y + 2)(y - 2)(y^2 + 6)$ **18.** $(2x - 3y^2)(4x^2 + 6xy^2 + 9y^4)$ **19.** 620 pages **20.** $34 \le x < 84$

CHAPTER 6

Exercise Set 6.1

1. a) A rational expression is an expression of the form $\frac{p}{q}$, p and q polynomials, $q \neq 0$ **b)** Answers will vary. **3. a)** A rational function is a function of the form $f(x) = \frac{p}{q}$, p and q polynomials, $q \neq 0$ **b)** Answers will vary. **5. a)** The domain of a rational function is the set of values that can replace the variable **b)** $\{x|x \text{ is a real number}, x \neq -3, x \neq 3\}$ **7. a)** Answers will vary. **b)** -1 **9. a)** Answers will vary. **b)** $\frac{1}{r + 4}$ **11.** 4 **13.** 5 and $\frac{5}{2}$ **15.** No excluded values **17.** $\{p|p \text{ is a real number and } p \neq 4\}$ **19.** $\{x|x \text{ is a real number and } x \neq -3, x \neq 2\}$ **21.** $\left\{a \middle| a \text{ is a real number and } a \neq \frac{1}{2}, a \neq -2\right\}$ **23.** $1 - y$ **25.** $\frac{x - 2y}{5}$ **27.** -1 **29.** $-(p + 4)$ **31.** $\frac{a - 5}{a + 3}$ **33.** $4x^2 + 10xy + 25y^2$ **35.** $\frac{2x - 5}{2}$ **37.** $\frac{x - w}{x + w}$ **39.** $\frac{a^2 + ab + b^2}{a + b}$ **41.** $\frac{xy^2}{9}$ **43.** $12x^3y^2$ **45.** 1 **47.** 1 **49.** $\frac{1}{x - 8}$ **51.** $\frac{1}{x - 3}$ **53.** $\frac{(a + 1)^2}{9(a + b)^2}$ **55.** $\frac{x - 4}{4x + 1}$ **57.** $\frac{(x + 2)(x - 2)}{(x^2 + 2x + 4)(x^2 + 4)}$ **59.** $3x^2$ **61.** $\frac{x^3}{x + 2}$ **63.** $\frac{1}{r + 3}$ **65.** $\frac{x^2(2x - 1)}{(x - 1)(x - 3)}$ **67.** $\frac{r + 5s}{2r + 5s}$ **69.** $\frac{2p - q^2}{p^2 + q^2}$ **71.** One possible answer is $\frac{1}{(x - 2)(x + 3)}$. **73.** The numerator is never 0. **75. a)** 4 **b)** 2 and -2 **77.** One possible answer is $f(x) = \frac{x - 2}{(x - 3)(x + 1)}$. **79.** $x + 5$ **81.** $y^2 - 4y - 5$ **83.** $x^2 + x - 2$; factors must be $(x - 1)(x + 2)$ **85.** $2x^2 + x - 6$; factors must be $(x + 2)(2x - 3)$ **87.** $2(a - b)$ **89.** $\frac{(x + 2)(3x + 1)}{(2x - 3)(x + 1)}$ **91.** $\frac{x - 1}{x + 3}$ **93.** $\frac{1}{x^2(x - p)^n}$ **95.** x^y **97. a)** $\{x|x \text{ is a real number}, x \neq 2\}$ **b)**

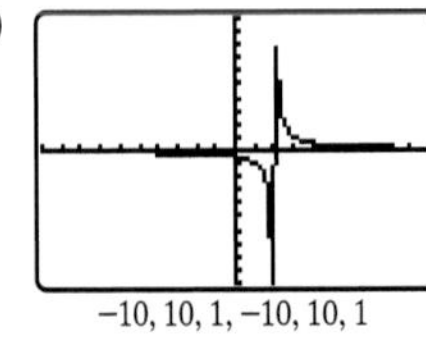

−10, 10, 1, −10, 10, 1

c) Decreasing **d)** Increasing **99. a)** $\{x|x \text{ is a real number}, x \neq 2\}$ **b)**

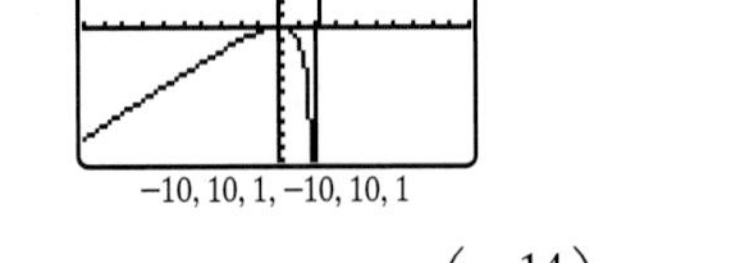

−10, 10, 1, −10, 10, 1

c) Decreasing **d)** Increasing **103.** $\frac{y}{3x}$ **104.** $h = \frac{3V}{4\pi r^2}$ **105.** $(0, 4)$ **106.** Slope is $-\frac{1}{3}$ and the y-intercept is $\left(0, \frac{14}{3}\right)$ **107.** $\left(-\frac{2}{7}, \frac{15}{7}\right)$ **108.** $3x^2y - 7xy - 4y^2 - 2x$

Exercise Set 6.2

1. a) Answers will vary. **b)** Answers will vary. **c)** $(8x + 11)(8x - 11)(x - 2)$ **3. a)** The entire numerator was not subtracted. **b)** $\frac{x^2 - 4x - x^2 - x + 2}{(x + 3)(x - 2)}$ **5.** $\frac{7x - 11}{x - 2}$ **7.** $\frac{x - 3}{x - 1}$ **9.** $\frac{5}{r - 2}$ **11.** $\frac{x + 5}{x + 3}$ **13.** $(x + 3)(x + 2)$ **15.** $48x^3y$ **17.** $z - 4$ **19.** $(a - 8)(a + 3)(a + 8)$

21. $(x - 1)(x + 4)(4x + 9)$ **23.** $\frac{8}{3x}$ **25.** $\frac{10y + 9}{12y^2}$ **27.** $\frac{25y^2 - 8x^2}{40x^4y^3}$ **29.** $\frac{4 + 6y}{3y}$ **31.** $\frac{3a}{a - b}$ **33.** $\frac{-4x^2 + 25x - 36}{(x + 3)(x - 3)}$ **35.** $\frac{2m^2 + 5m - 2}{(m - 5)(m + 2)}$ **37.** $\frac{1}{x - 5}$ **39.** $\frac{-2p - 7}{(p - 4)(p + 4)(2p - 3)}$ **41.** $\frac{5x^2 + 14x - 49}{(x + 5)(x - 2)}$ **43.** $\frac{3a - 1}{4a + 1}$ **45.** $\frac{2x^2 - 4xy + 4y^2}{(x - 2y)^2(x + 2y)}$ **47.** $\frac{16}{r - 4}$ **49.** 0 **51.** 1 **53.** $\frac{x - 6}{x - 4}$ **55.** $\frac{x^2 - 18x - 30}{(5x + 6)(x - 2)}$ **57.** $\frac{8m^2 + 5mn}{(2m + 3n)(3m + 2n)(2m + n)}$ **59.** 0 **61.** $\frac{1}{2x + 3y}$ **63.** No, they should be added first, then factored. **65.** Yes, factor -1 from the numerator and the denominator. **67. a)** $\{x|x \text{ is a real number}, x \neq 3\}$ **b)** $\{x|x \text{ is a real number}, x \neq -4\}$ **c)** $\frac{2x^2 + 3x + 8}{(x - 3)(x + 4)}$ **d)** $\{x|x \text{ is a real number}, x \neq 3, x \neq -4\}$ **69.** Domain: $\{x|x \text{ is a real number}, x \neq 2\}$; range: $\{y|y \text{ is a real number}, y \neq 1\}$ **71.** $\frac{x^2 + 6x + 6}{(x + 2)(x - 2)(x + 3)}$ **73.** $\frac{3x}{x^4 + x^3 - 10x^2 - 4x + 24}$ **75.** $\frac{a}{b} + \frac{c}{d} = \frac{a}{b} \cdot \frac{d}{d} + \frac{c}{d} \cdot \frac{b}{b} = \frac{ad}{bd} + \frac{cb}{db} = \frac{ad + cb}{bd}$ or $\frac{ad + bc}{bd}$ **77. a)** 4 **b)** $\frac{a^2 - b^2}{a^2}$ **79.** $7x^2 - 6x + 6$ **81.** $\frac{3x + 10}{x - 2}$ **83.** $\frac{1}{(a - 5)(a + 3)}$ **85.** $-x^2 + 4x + 5$ **87. a)** $\frac{ax + bn - bx}{n}$ **b)** 79.2 **89.** $\frac{a - b + 1}{(a - b)^2}$ **91.** No **93. a)** $\frac{x + 1}{x}$ **b)** $\frac{x^2 + x + 1}{x^2}$ **c)** $\frac{x^5 + x^4 + x^3 + x^2 + x + 1}{x^5}$ **d)** $\frac{x^n + x^{n-1} + x^{n-2} + \cdots + 1}{x^n}$ **94.** \$180 **95. a)** 6 minutes **b)** 960 bottles **96.** $3x^3y^2 - \frac{x^2}{y^2} + \frac{5y}{3}$ **97.** $2x - 1$

Exercise Set 6.3

1. One that has a fractional expression in the numerator or the denominator or both **3.** $\frac{y - x}{xy}$ **5.** $\frac{75a}{b^5}$ **7.** $\frac{12x^3}{y^6}$ **9.** $\frac{x(y - 1)}{1 + x}$ **11.** $\frac{5}{3a^2}$ **13.** $\frac{x - y}{y}$ **15.** $\frac{m + 2}{2m^2 + 1}$ **17.** $x^2(x - y)$ **19.** $\frac{-a + 1}{(a + 1)(2a + 1)}$ **21.** $\frac{x - 1}{x + 1}$ **23.** $\frac{a^2 + 1}{2a}$ **25.** $\frac{x}{5(x - 3)}$ **27.** $\frac{x^2 + 5x - 6}{x(x - 2)}$ **29.** $\frac{3x(x - 3)}{(x + 4)(3x - 5)}$ **31.** $\frac{3 + a^2b}{a^2}$ **33.** $\frac{ab}{b + a}$ **35.** $\frac{b(1 + a)}{a(1 - b)}$ **37.** $\frac{y - x}{x + y}$ **39.** $\frac{(a + b)^2}{ab}$ **41.** $\frac{6y - x}{3xy}$ **43.** $\frac{2y - 3xy + 5y^2}{3y^2 - 2x}$ **45. a)** $\frac{2}{7}$ **b)** $\frac{4}{13}$ **47.** $R_T = \frac{R_1R_2R_3}{R_2R_3 + R_1R_3 + R_1R_2}$ **49.** a **51. a)** $\frac{1}{x + h}$ **b)** $\frac{-h}{x(x + h)}$ **c)** $-\frac{1}{x(x + h)}$ **53. a)** $\frac{2}{(x + h)^2}$ **b)** $\frac{-4xh - 2h^2}{x^2(x + h)^2}$ **c)** $\frac{-4x - 2h}{x^2(x + h)^2}$ **55.** $\frac{4a^2 + 1}{4a(2a^2 + 1)}$ **57.** $\frac{5}{12}$ **58.** $\frac{13}{48}$ **59.** $\left\{x \middle| x \le -\frac{5}{2} \text{ or } x \ge \frac{13}{2}\right\}$ **60.** **61.** $x^2 - 5x - 23 - \frac{37}{x - 2}$

Exercise Set 6.4

1. A number obtained when solving an equation that is not a true solution to the original equation **3. a)** Multiply both sides of the equation by 12 to remove fractions. **b)** -24 **c)** Write each term with the LCD of 12 so that the fractions can be added. **d)** $\frac{-x + 24}{12}$ **5.** Similar figures are figures whose corresponding angles are the same and whose corresponding sides are in proportion. **7.** -2 **9.** 1 **11.** 3 **13.** All real numbers **15.** $\frac{1}{4}$ **17.** $\frac{14}{3}$ **19.** No solution **21.** $\frac{6}{5}$ **23.** ≈ -1.63 **25.** 8 **27.** $-\frac{4}{3}; 1$ **29.** 3.76 **31.** $-3, -2$ **33.** -5 **35.** $-\frac{5}{2}$ **37.** 5 **39.** No solution **41.** No solution **43.** $\frac{17}{4}$ **45.** 12 and 2 **47.** 12 and 4 **49.** $-2, 1$

51. $-5, \frac{1}{3}$ **53.** 3 **55.** $P_2 = \frac{V_1P_1}{V_2}$ **57.** $V_2 = \frac{V_1P_1}{P_2}$ **59.** $r = 1 - \frac{a}{S}$ or $r = \frac{S-a}{S}$ **61.** $s = \frac{x - \bar{x}}{z}$ **63.** $w = \frac{fl - df}{d}$ **65.** $q = \frac{pf}{p - f}$ **67.** $a = \frac{v_2 - v_1}{t_2 - t_1}$ **69.** $d = \frac{a_n - a_1}{n - 1}$ **71. a)** $\frac{2x + 7}{(x-2)(x+2)}$ **b)** $-\frac{7}{2}$ **73. a)** $\frac{4}{b + 5}$ **b)** No solution **75.** $c \neq 0$; division by 0 is undefined. **77.** $f(x)$ is graph b) and $g(x)$ is graph a); $f(x)$ is not defined for $x = 3$ **79. a)** \$6250 **b)** $R = \frac{AC}{0.80I}$ **81. a)** 20 ft/min^2 **b)** $t_1 = t_2 + \frac{v_1 - v_2}{a}$ **83.** 35 units **85.** $\approx$ 176.47 ohms **87.** $\approx$ 0.101m **89. a)** $\approx$ 9.71% **b)** Tax Free Money Market since 9.71% > 7.68% **91.** One possible answer is $\frac{1}{x-4} + \frac{1}{x+2} = 0$ **93.** One possible answer is $\frac{1}{x} + \frac{1}{x} = \frac{2}{x}$ **95.** $-1 < x \leq 3$ **96.** $\frac{3}{2}$ **97.** $-9x^3 + 24x^2 - 8$ **98.** $8(x - 2y^2)(x^2 + 2xy^2 + 4y^4)$

Exercise Set 6.5

1. Equal to $\frac{1}{2}$, since each paints $\frac{1}{2}$ the wall in $\frac{1}{2}$ the time. **3.** $\frac{x}{3} + \frac{x}{6} = 1$; 2 hours **5.** $\frac{x}{4} + \frac{x}{6} = 1$; 2.4 hours **7.** $\frac{x}{3} - \frac{x}{4} = 1$; 12 hours **9.** $\frac{8}{12} + \frac{8}{x} = 1$; 24 hours **11.** $\frac{50}{60} + \frac{50}{x} = 1$; 300 minutes **13.** $\frac{x}{8} + \frac{x}{12} - \frac{x}{7} = 1$; $\frac{168}{11}$ or $\approx$ 15.27 minutes **15.** $\frac{x}{6} + \frac{x}{5} + \frac{x}{4} = 1$; $\frac{60}{37}$ or $\approx$ 1.62 hours **17.** $\frac{6}{15} + \frac{x}{20} = 1$; 12 hours **19.** $\frac{4x}{3 + x} = \frac{5}{2}$; 5 **21.** $\frac{1}{x} + \frac{1}{2x} = \frac{3}{4}$; 2 and 4 **23.** $\frac{1}{x} + \frac{1}{x+2} = \frac{5}{12}$; 4 and 6 **25.** $\frac{2}{x} + 3 = \frac{31}{10}$; 20 **27.** $3x + \frac{2}{x} = 5$; $\frac{2}{3}$ and 1 **29.** $\frac{400}{x} = \frac{600}{x + 40}$; car: 80 km/hour; train: 120 km/hour **31.** $\frac{120}{r + 2} = \frac{52}{r}$; $\approx$ 1.53 feet/second **33.** $\frac{5.7}{x + 2.9} = \frac{3.4}{x}$; $\approx$ 4.29 mph **35.** $\frac{40}{x} + 2\left(\frac{40}{x}\right) = 4$; 30 km/hour **37.** $\frac{x}{14.7} + \frac{80 - x}{5.8} = 10.6$; $\approx$ 30.6 yards **39.** $\frac{16.4}{x} = \frac{24.2}{x + 5.2}$; local $\approx$ 10.9 mph, express $\approx$ 16.1 mph **41.** $\frac{x}{70} + 2 = \frac{x}{50}$; 350 miles **43.** $\frac{x}{3.6} + 0.2 = \frac{x}{2.4}$; 1.44 miles **45.** $\frac{600}{r} + 2 = \frac{600}{r - 10}$; 60 mph **47.** $\frac{x}{20{,}000} + 0.6 = \frac{x}{18{,}000}$; 108,000 miles **49.** Answers will vary. **51.** Answers will vary. **53.** $-\frac{5}{2}x^2 + \frac{7}{5}xy - 6y^2$ **54.** $12x^3 - 34x^2 + 21x + 4$ **55.** $4x + 5 + \frac{22}{3x - 2}$ **56.** $(4x + 3)(2x + 5)$

Exercise Set 6.6

1. a) As one increases, the other increases. **b)** Answers will vary. **c)** Answers will vary. **3.** One quantity varies as a product of two or more quantities **5.** Direct **7.** Inverse **9.** Inverse **11.** Direct **13.** Inverse **15.** Direct **17.** Inverse **19. a)** $x = ky$ **b)** 72 **21. a)** $y = kR$ **b)** 306 **23. a)** $R = \frac{k}{W}$ **b)** $\frac{1}{20} = 0.05$ **25. a)** $A = \frac{kB}{C}$ **b)** 9 **27. a)** $x = ky$ **b)** 18 **29. a)** $y = kR^2$ **b)** 20 **31. a)** $C = \frac{k}{J}$ **b)** $\approx$ 0.41 **33. a)** $F = \frac{kM_1M_2}{d}$ **b)** 40 **35.** Doubled **37.** Halved **39.** 3096 mg **41.** $\approx$ 139.35 minutes **43.** 240 miles **45.** 6400 cc **47.** 0.32 seconds **49.** $\approx$ 8.18 kg **51.** \$438.90 **53.** 4600 tapes **55.** 25 ohms **57.** \$124.92 **59. a)** $F = \frac{km_1m_2}{d^2}$ **b)** The new force is 24 times the original force. **61.** $d = -5$

62.

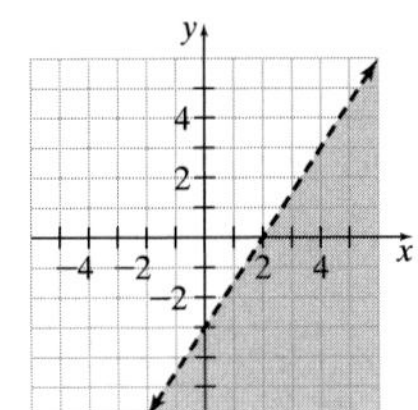

63. The base salary is \$400 and the commission rate is 3%. **64.**

Review Exercises **1.** 4 **2.** -1 **3.** No excluded values **4.** $\{x|x$ is a real number and $x \neq -3\}$ **5.** $\{x|x$ is a real number and $x \neq 0\}$ **6.** $\{x|x$ is a real number and $x \neq 5$ and $x \neq -2\}$ **7.** x **8.** $x - 3$ **9.** -1 **10.** $\dfrac{x-1}{x-2}$ **11.** $\dfrac{x-3}{x+1}$ **12.** $\dfrac{a^2 + 2ab + 4b^2}{a + 2b}$ **13.** $x(x+1)$ **14.** $(x+y)(x-y)$ **15.** $(x+7)(x-5)(x+2)$ **16.** $(x+2)^2(x-2)(x+1)$ **17.** $6xz^2$ **18.** $-\dfrac{1}{2}$ **19.** $8x^3z^5$ **20.** $\dfrac{3}{x}$ **21.** $\dfrac{(x+y)y^2}{2x^3}$ **22.** $3x + 2$ **23.** 1 **24.** $\dfrac{1}{3x+5}$ **25.** $\dfrac{5x^2 - 12y}{3x^2y}$ **26.** $\dfrac{7x+12}{x+2}$ **27.** $\dfrac{5x+12}{x+3}$ **28.** $\dfrac{x+y}{x}$ **29.** $\dfrac{1}{3(a+3)}$ **30.** $\dfrac{a^2+c^2}{ac}$ **31.** $\dfrac{x+1}{2x-1}$ **32.** 1 **33.** $2x(x-5y)$ **34.** $\dfrac{-3x^2+2x-4}{3x(x-2)}$ **35.** $\dfrac{x^2-2x-5}{(x+5)(x-5)}$ **36.** $\dfrac{3(x-1)}{(x+3)(x-3)}$ **37.** $\dfrac{1}{a+2}$ **38.** $\dfrac{16(x-2y)}{3(x+2y)}$ **39.** $\dfrac{4}{(x+2)(x-3)(x-2)}$ **40.** $\dfrac{2}{x(x+2)}$ **41.** $\dfrac{2(x-4)}{(x-3)(x-5)}$ **42.** $\dfrac{22x+5}{(x-5)(x-10)(x+5)}$ **43.** $\dfrac{x^3+y^2}{x^3-y^2}$ **44.** $-\dfrac{1}{x-3}$ **45.** $\dfrac{(x+3)(x-1)}{4(x+1)}$ **46.** $\dfrac{x+3}{x+5}$ **47.** $\dfrac{x-4}{x-6}$ **48.** $\dfrac{x^2+6x-24}{(x-1)(x+9)}$ **49. a)** $\{x|x$ is a real number, $x \neq -3\}$ **b)** $\{x|x$ is a real number, $x \neq -5\}$ **c)** $\dfrac{2x^2+4x-20}{(x+3)(x+5)}$ **d)** $\{x|x$ is a real number, $x \neq -3, x \neq -5\}$ **50. a)** $\{x|x$ is a real number, $x \neq -3, x \neq 3\}$ **b)** $\{x|x$ is a real number, $x \neq 3\}$ **c)** $\dfrac{x^2+6x+6}{(x+3)(x-3)}$ **d)** $\{x|x$ is a real number, $x \neq -3, x \neq 3\}$ **51.** $\dfrac{5yz}{6}$ **52.** $\dfrac{xy+1}{y^3}$ **53.** $\dfrac{2(2x+1)}{x(6x-1)}$ **54.** $\dfrac{2a+1}{2}$ **55.** $\dfrac{x+1}{-x+1}$ **56.** $\dfrac{3x^2-29x+68}{4x^2-6x-54}$ **57.** 9 **58.** 2.8 **59.** 6 **60.** $-7.\overline{5}$ **61.** 52 **62.** 2.4 **63.** 18 **64.** $\dfrac{1}{2}$ **65.** -6 **66.** -18 **67.** $E = \dfrac{I(R+nr)}{n}$ **68.** $a = \dfrac{S(1-r)}{1-r^n}$ **69.** 120 ohms **70.** 900 ohms and 1800 ohms **71.** 3 cm **72.** 15 cm **73.** $\dfrac{x}{3} + \dfrac{x}{4} = 1$; $\dfrac{12}{7}$ or $\approx$ 1.71 hour **74.** $\dfrac{40}{75} + \dfrac{40}{x} = 1$; $\dfrac{600}{7}$ or $\approx$ 85.71 hours **75.** $\dfrac{5 \cdot x}{x+8} = 1; 2$ **76.** $1 - \dfrac{1}{2x} = \dfrac{1}{3x}; \dfrac{5}{6}$ **77.** $\dfrac{20}{15+x} = \dfrac{10}{15-x}$; 5 mph **78.** $\dfrac{450}{3x} + 6 = \dfrac{450}{x}$; car is 50 mph; plane is 150 mph **79.** 20 **80.** 20 **81.** $\approx$ 426.7 **82.** 5 inches **83.** \$119.88 **84.** 400 feet **85.** 200.96 square units **86.** 2.38 minutes

Practice Test **1.** -4 and 7 **2.** $\left\{x \middle| x \text{ is a real number}, x \neq \dfrac{5}{4}, x \neq -\dfrac{5}{2}\right\}$ **3.** $4x^5y + 8 + 9xy^2$ **4.** $\dfrac{2x+5y}{2x+y}$ **5.** $\dfrac{1}{x^4y^2}$ **6.** $a + 3$ **7.** $\dfrac{x-3y}{3}$ **8.** $x^2 + y^2$ **9.** $\dfrac{10x+3}{2x^2}$ **10.** $-\dfrac{1}{(x+4)(x-4)}$ **11.** $\dfrac{x(x+10)}{(2x-1)^2(x+3)}$ **12.** $\dfrac{m(6m+n)}{(6m+5n)(2m-n)(2m+3n)}$ **13.** $\dfrac{1}{r-2}$ **14.** $\dfrac{3x^2+2x-9}{(x+5)(2x+3)}$ **15.** $\dfrac{(x-9)(x+1)}{(x+5)(2x+3)}$ **16.** $\left\{x \middle| x \text{ is a real number}, x \neq -5, x \neq -\dfrac{3}{2}\right\}$ **17.** $\dfrac{y+x}{y-x}$ **18.** $\dfrac{x^2(y+1)}{y+x}$ **19.** 60 **20.** 12 **21.** $s = \dfrac{R}{I} - \dfrac{w}{2}$ **22.** 0.75 watt **23.** 1.125 **24.** $\dfrac{40}{13}$ or $\approx$ 3.08 hour **25.** $6\dfrac{2}{3}$ miles

Cumulative Review Test **1.** 

2. $-27\dfrac{3}{4}$ **3.** $\dfrac{1}{4a^{10}b^2}$ **4. a)** $\approx$ \$7.486 $\times 10^{11}$ **b)** $\approx$ \$1.925 $\times 10^{11}$ **c)** $\approx$ \$5.561 $\times 10^{11}$ **5.** 20 **6.** 8 **7. a)** \$25,000 **b)** Plan 2 **8.** 3400 bottles per hour

9.

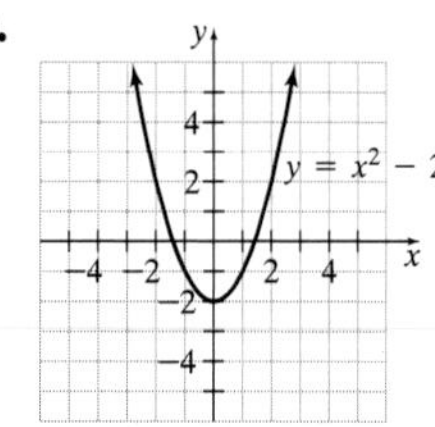

10.

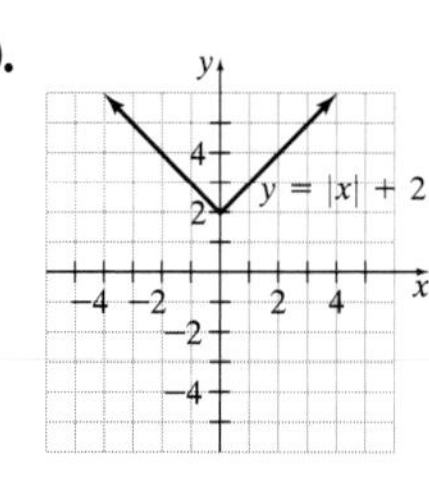

11. a) and c) **12.** $y = \frac{3}{5}x - 2$ **13.** $y = -\frac{3}{4}x + \frac{1}{4}$ **14. a)** 158,015 **b)** 36,609 **c)** Answers will vary. **15.** $(2, -1, 3)$ **16.** 3.2 liters of 10% solution; 0.8 liter of 20% solution **17. a)** $x^3 - 16x^2 + 85x - 150$ **b)** $x - 6$ **c)** $\{x|x \text{ is a real number}, x \neq 5\}$ **18.** $2pq(2p + 3)(6p - 5)$ **19.** $-\frac{5}{3}$ **20.** No solution

CHAPTER 7

Exercise Set 7.1

1. a) Two, positive and negative **b)** 6, −6 **c)** Principal square root **d)** 6 **3.** There is no real number which, when squared, results in −49. **5.** No; if the radicand is negative, the answer is not a real number. **7.** 7 **9.** −4 **11.** 5 **13.** Not a real number **15.** 6 **17.** −7 **19.** 2 **21.** Not a real number **23.** −4.68 **25.** $\frac{5}{3}$ **27.** Not a real number **29.** 1.75 **31.** 4 **33.** 1 **35.** 43 **37.** 235.23 **39.** 0.03 **41.** 32.4 **43.** $|a - 9|$ **45.** $|x - 3|$ **47.** $|3x + 5|$ **49.** $|6 - 3x|$ **51.** $|y^2 - 4y + 3|$ **53.** $|8a - b|$ **55.** $|a^4|$ **57.** $|a^{12}|$ **59.** $|a - 4|$ **61.** $|x - 4|$ **63.** 2 **65.** 9.381 **67.** 5.192 **69.** −3.583 **71.** $x < \frac{1}{2}$ **73.** $x \geq 1$ **75.** $x \geq 3$ **77. a)** All real numbers **b)** $a \geq 0$ **79.** If n is even, we are finding the even root of a positive number. If n is odd, the expression is also real. **81.** $\{x|x \text{ is a real number } x > -3\}$ **83.** d) **85.** a) **87.** One answer is $f(x) = \sqrt{x - 4}$ **89. a)** No **b)** Yes **c)** Yes; answers will vary. **91. a)** $\sqrt{1288} \approx 35.89$ feet/second **b)** $\sqrt{2576} \approx 50.75$ feet/second

93.

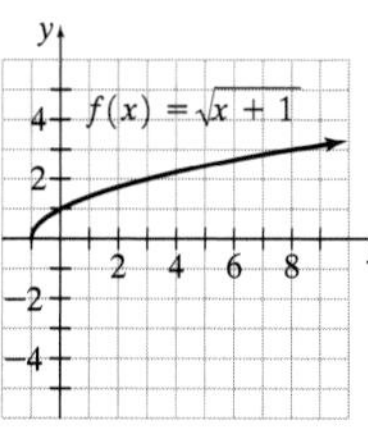

95.

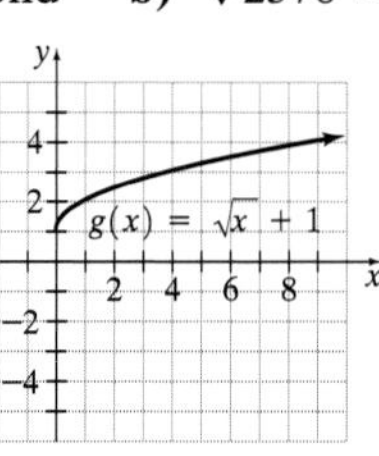

101.

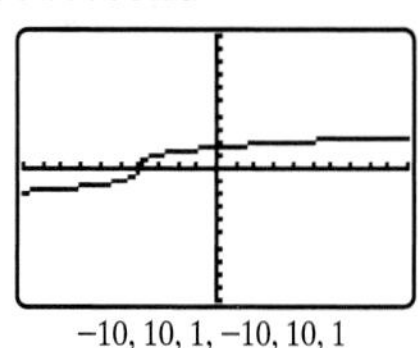

106. $3(y - 3 + z)(y - 3 - z)$ **107.** $\left(x + \frac{1}{3}\right)\left(x^2 - \frac{1}{3}x + \frac{1}{9}\right)$ **108.** $(x - 2)(x + 5)$ **109.** $x(x^2 - 3)(2x - 3)$

Exercise Set 7.2

1. a) When n is even and $a \geq 0$, or n is odd **b)** $a^{1/n}$ **3. a)** Always real **b)** a **c)** a **d)** $|a|$ **5.** $x^{5/2}$ **7.** $8^{5/2}$ **9.** $x^{4/5}$ **11.** $x^{3/2}$ **13.** $(a^3b^5)^{1/4}$ **15.** $(5r^4s^9t^{12})^{1/8}$ **17.** $\sqrt{x}$ **19.** $\sqrt{z^3}$ **21.** $\sqrt{24y^2}$ **23.** $\frac{1}{\sqrt{19x^2y^4}}$ **25.** $(\sqrt[5]{2m^2n^3})^2$ **27.** $\frac{1}{(\sqrt[3]{a^2 - 4b^2})^2}$ **29.** y^3 **31.** z^4 **33.** x^3 **35.** $\sqrt{z}$ **37.** 5.1 **39.** y **41.** $\sqrt{xyz}$ **43.** x^3y^6 **45.** $\sqrt[4]{x}$ **47.** $\sqrt[6]{x^5}$ **49.** 2 **51.** Not a real number **53.** $\frac{3}{5}$ **55.** −4 **57.** −4 **59.** $\frac{1}{4}$ **61.** $\frac{1}{8}$ **63.** $-\frac{7}{2}$ **65.** 18 **67.** $\frac{10}{21}$ **69.** $x^{11/12}$ **71.** $x^{1/6}$ **73.** $\frac{1}{x}$ **75.** 1 **77.** $\frac{y^{5/3}}{12}$ **79.** $\frac{8}{x^{11/6}}$ **81.** $\frac{1}{2x^{1/3}}$ **83.** $\frac{121}{x^{1/7}}$ **85.** $\frac{64}{y^{66/5}}$ **87.** $\frac{x}{y^{16}}$ **89.** $6z^{7/2} - 3$ **91.** $\frac{5}{x^5} + \frac{10}{x^{3/2}}$ **93.** $8x^{13/6} - 4x^2$ **95.** 10.95 **97.** 3.32 **99.** 12.65 **101.** 0.03 **103.** When n is odd, or n is even with $a \geq 0$ **105.** Answers will vary. One example is $(4^{1/2} + 9^{1/2})^2 \neq 4 + 9$ or $(2 + 3)^2 \neq 4 + 9$ or $5^2 \neq 4 + 9$ or $25 \neq 13$. **107.** $x^{1/2}(x + 1)$ **109.** $y^{1/3}(1 - y)$ **111.** $\frac{1 + y}{y^{3/5}}$ **113. a)** 2^{10} **b)** $2^{10}\sqrt{2} \approx 1448$ bacteria **115. a)** $2.69\sqrt{7^3} \approx \$49.82$ billion **b)** $2.69\sqrt{14^3} \approx \$140.91$ billion **117.** 9 **119.** $\{x|x \text{ is a real number}, x \geq 2\}$ **121.** 3 **123. a)** $f(x) = (2x + 3)^{1/2}$ **b)** a) is correct because the graphs are the same.

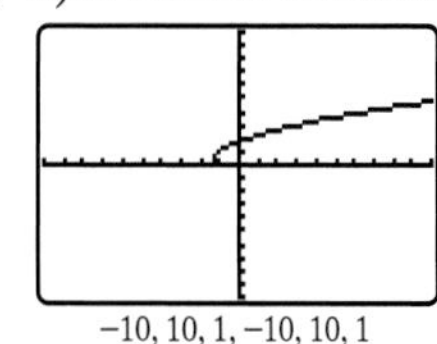

124. c) is a function; a), b), and c) are relations. **125.** $\frac{b^2 + a^3b}{a^3 - b}$ **126.** 0, 3 **127.** $\approx$ 441.67 mph

Exercise Set 7.3 **1. a)** Square the natural numbers **b)** 1, 4, 9, 16, 25, 36 **3. a)** Raise natural numbers to the fifth power. **b)** 1, 32, 243, 1024, 3125 **5.** If n is even and a or b is negative, the numbers are not real numbers and the rule does not apply. **7.** $5\sqrt{3}$ **9.** $4\sqrt{2}$ **11.** $2\sqrt[3]{2}$ **13.** $3\sqrt[3]{2}$ **15.** $-x\sqrt{x}$ **17.** $7x^5\sqrt{x}$ **19.** $b^{13}\sqrt{b}$
21. $b^5\sqrt[4]{b^3}$ **23.** $6c^3\sqrt[3]{3c^2}$ **25.** $xy^3\sqrt{xy}$ **27.** $3a^2b^2\sqrt[3]{3b^2}$ **29.** $12x^4y^4\sqrt[3]{2y}$ **31.** $-2x^2y\sqrt[5]{2x^2y^2}$ **33.** $2x^2y^2z^4\sqrt[4]{2yz^3}$
35. 10 **37.** $2\sqrt[3]{7}$ **39.** $3xy^3\sqrt{10y}$ **41.** $3m^2n^5\sqrt{3n}$ **43.** $3x^3y^4\sqrt[3]{2x^2y}$ **45.** $xy^4\sqrt[3]{25x}$ **47.** $x^7y^7z^3\sqrt[5]{x^2y^3z}$
49. $2\sqrt{3} + 2$ **51.** $6 - 3\sqrt{2}$ **53.** 10 **55.** $9y\sqrt{y} - y\sqrt{3}$ **57.** $2x^2y^3\sqrt[4]{2} - x^2y^2\sqrt[4]{24x^2y^3}$ **59.** $4x - 8\sqrt{x}$
61. $x^2y\sqrt[3]{y} + xy\sqrt[3]{y}$ **63.** $x\sqrt[4]{27x^2y^2} - x^2\sqrt[4]{3xy}$ **65.** $2x^2y^3 - xy^2\sqrt[5]{80y^3}$ **67.** $2\sqrt{6}$ **69.** $2\sqrt[3]{4}$ **71.** $y^4\sqrt[3]{y}$
73. $2x^3\sqrt[3]{10x^2}$ **75.** $2b^2c\sqrt[6]{2ab^5c^3}$ **77.** $2\sqrt[4]{5}$ **79.** $2x^3y^5\sqrt{30y}$ **81.** $x + 3\sqrt{x}$ **83.** $2\sqrt[3]{y^2} - y^3$
85. $ab\sqrt[3]{12a^2b^2} - 2a^2b^2\sqrt[3]{3}$ **87.** $|2x + 5|$ **89.** $2|r - 4|$ **91. a)** $(x - 4)^2$ **b)** $(x - 4)^2$ **93.** 1
95. $10a + 7\sqrt{ab} - 12b$ **97. a)** ≈ 45.17 mph **b)** ≈ 35.33 mph **99.** ≈ 1565.25 gallons per minute **101.** ≈ 6.25
102. A rational number is a number that can be represented as the quotient of two integers, denominator not zero.
103. A real number is a number that can be represented on a real number line
104. An irrational number is a number that cannot be expressed as a quotient of two integers.
105. $|a| = \begin{cases} a \text{ if } a \ge 0 \\ -a \text{ if } a < 0 \end{cases}$ **106.** $m = \dfrac{2E}{v^2}$ **107. a)** 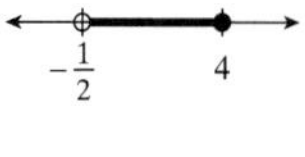

b) $\left(-\frac{1}{2}, 4\right]$ **c)** $\left\{x \,\middle|\, -\frac{1}{2} < x \le 4\right\}$

Exercise Set 7.4 **1.** $\dfrac{\sqrt[n]{a}}{\sqrt[n]{b}} = \sqrt[n]{\dfrac{a}{b}}, a \ge 0, b > 0$ **3.** To remove radicals from a denominator
5. a) Answers will vary. **b)** $-\dfrac{7 + 2\sqrt{10}}{3}$ **7.** 3 **9.** $\dfrac{1}{3}$ **11.** $\dfrac{1}{2}$ **13.** $-2\sqrt{2}$ **15.** $\dfrac{r^2}{5}$ **17.** $3x^2$ **19.** $\dfrac{\sqrt[3]{7y}}{2x^4}$
21. $\dfrac{x^3\sqrt[4]{20}}{3}$ **23.** $\dfrac{\sqrt{5}}{5}$ **25.** $\dfrac{\sqrt{2m}}{2}$ **27.** $\dfrac{\sqrt{xy}}{y}$ **29.** $\dfrac{\sqrt{10m}}{4}$ **31.** $\dfrac{\sqrt{2n}}{3}$ **33.** $\dfrac{2x^2\sqrt{xyz}}{z}$ **35.** $\dfrac{2y^3z\sqrt{15xz}}{3x}$
37. $\dfrac{3x^2y\sqrt{yz}}{z^2}$ **39.** $\dfrac{\sqrt[3]{4}}{2}$ **41.** $\dfrac{\sqrt[3]{25}}{5}$ **43.** $\dfrac{\sqrt[3]{5xy^2}}{y}$ **45.** $-\dfrac{\sqrt[3]{15cy}}{3y}$ **47.** $\dfrac{5m\sqrt[4]{8}}{2}$ **49.** $\dfrac{\sqrt[4]{8x^3y^2}}{2y}$ **51.** $\dfrac{x^2y^2\sqrt[3]{60yz}}{2z}$
53. $\dfrac{s\sqrt[6]{32r^5s^3}}{2r}$ **55.** 20 **57.** 44 **59.** $6 - x^2$ **61.** $x - y$ **63.** 10 **65.** $5\sqrt{2} - 5$ **67.** $\dfrac{10 + 2\sqrt{6}}{19}$
69. $2\sqrt{3} - 2\sqrt{2}$ **71.** $\dfrac{\sqrt{17} + 2\sqrt{2}}{9}$ **73.** $\dfrac{3\sqrt{5a} + 9\sqrt{5}}{a - 9}$ **75.** $\dfrac{2x\sqrt{2x} - 2\sqrt{2xy}}{x^2 - y}$ **77.** $\dfrac{c + \sqrt{cd} - \sqrt{2cd} - d\sqrt{2}}{c - d}$
79. $\dfrac{x\sqrt{y} - y\sqrt{x}}{x - y}$ **81.** $\dfrac{\sqrt{x}}{3}$ **83.** $\dfrac{\sqrt{10}}{5}$ **85.** -1 **87.** $\dfrac{2xy^3\sqrt{30xz}}{5z}$ **89.** $\dfrac{\sqrt{6}}{x}$ **91.** $x - 9$ **93.** $-\dfrac{\sqrt{2x}}{2}$
95. $\dfrac{\sqrt[4]{24x^3y^2}}{2x}$ **97.** $\dfrac{2y^4z^3\sqrt[3]{2x^2z}}{x}$ **99.** $\dfrac{a\sqrt{r} + 2r\sqrt{a}}{a - 4r}$ **101.** $\dfrac{\sqrt[3]{150y^2}}{5y}$ **103.** $\dfrac{y^3z\sqrt[4]{54x^2}}{3x}$ **105.** $\sqrt[3]{(a + b)^5}$
107. $\sqrt[15]{(a + 2b)^2}$ **109.** $\sqrt[6]{rs^5}$ **111.** $\sqrt[15]{x^2y^8}$ **113.** ≈ 3.69 meters **115.** 12 inches **117. a)** 6.21 million **b)** 2.36 million **119.** If n is even and a or b is negative, the numbers are not real and this rule does not apply.
121. $\dfrac{3}{\sqrt{3}}$ **123. a)** 4, 8, 12 **b)** 9, 18, 27 **c)** $x^{(3a+2b)/6}$ **d)** $x^{(3a-2b)/6}$ **125.** $\dfrac{3\sqrt{2a} - 3b}{2a - 3b}$ **127.** $\dfrac{10}{15 + 3\sqrt{5}}$
129. $\dfrac{1}{\sqrt{x + h} + \sqrt{x}}$ **131.** $b_2 = \dfrac{2A}{h} - b_1$ **132.** 40 mph, 50 mph **133.** $8x^3 - 18x^2 + 5x + 6$ **134.** $4, \dfrac{3}{2}$

Exercise Set 7.5 **1.** Radicals with the same radicands and index **3.** ≈ 5.97 **5.** No; answers will vary.
7. $4\sqrt{3}$ **9.** $5 - 4\sqrt{3}$ **11.** $-3\sqrt[4]{y}$ **13.** $7\sqrt{5} + 2\sqrt[3]{x}$ **15.** $6 - 3\sqrt[3]{a}$ **17.** $2(\sqrt{2} - \sqrt{3})$
19. $-30\sqrt{3} + 20\sqrt{5}$ **21.** $-4\sqrt{10}$ **23.** $18y\sqrt{5x}$ **25.** $2\sqrt{2}$ **27.** $-7c\sqrt{3}$ **29.** $\sqrt[3]{2}$ **31.** $7\sqrt[3]{4}$ **33.** -7
35. $5a\sqrt[3]{ab^2}$ **37.** $3r^3s^2\sqrt{rs}$ **39.** 0 **41.** $\sqrt{2}$ **43.** $\dfrac{2\sqrt{3}}{3}$ **45.** $\dfrac{15\sqrt{2}}{2}$ **47.** $2\sqrt{x}\left(2 + \dfrac{1}{x}\right)$ **49.** $-15\sqrt{2}$
51. $\left(-\dfrac{2}{y} + \dfrac{3}{x}\right)\sqrt{xy}$ **53.** $23 + 9\sqrt{3}$ **55.** $11 + 7\sqrt{5}$ **57.** $18 - \sqrt{2}$ **59.** $10 - 3\sqrt{6}$ **61.** $29 - 12\sqrt{5}$
63. $18x - \sqrt{3xy} - y$ **65.** $8 - 2\sqrt[3]{18} - \sqrt[3]{12}$ **67.** $3\sqrt{5}$ **69.** $7\sqrt{5}$ **71.** $\dfrac{2\sqrt{6}}{3}$ **73.** $14\sqrt[3]{3}$ **75.** $-14 + 11\sqrt{2}$
77. $(4x - 2)\sqrt{3x}$ **79.** $\sqrt{y}$ **81.** $x - 2\sqrt[3]{x^2y^2} - \sqrt[3]{xy} + 2y$ **83.** $a - 3\sqrt[3]{a} + 5\sqrt[3]{a^2} - 15$ **85.** $-\dfrac{16\sqrt{6}}{3}$

87. $P = 14\sqrt{5}; A = 60$ **89.** $P = 17\sqrt{5}; A = 52.5$ **91. a)** 37 inches **b)** $\approx$ 37.97 inches **93.** $2 + \sqrt{3}$ **95.** Yes **97.** No **99.** $\sqrt{5}(\sqrt{2} + 1)$ **101. a)** $(x)^2 - (\sqrt{3})^2$ **b)** $(x + \sqrt{3})(x - \sqrt{3})$

103. a)

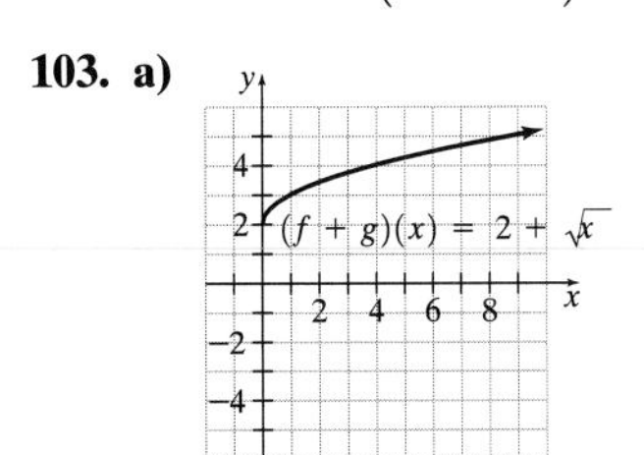

b) Raises the graph 2 units

105. a)

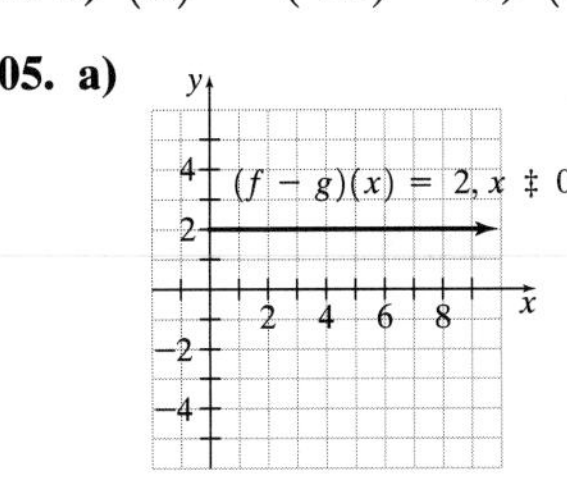

b) $\{x|x$ is a real number, $x \geq 0\}$

107.

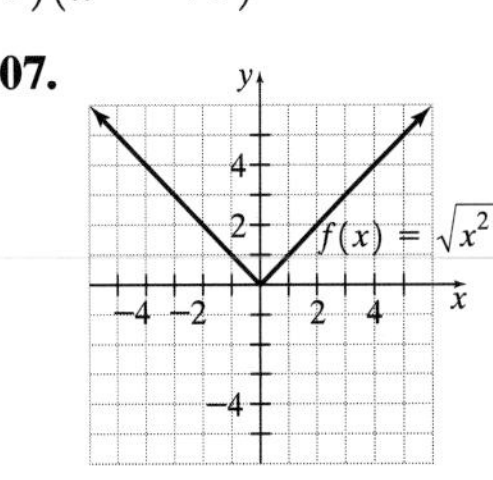

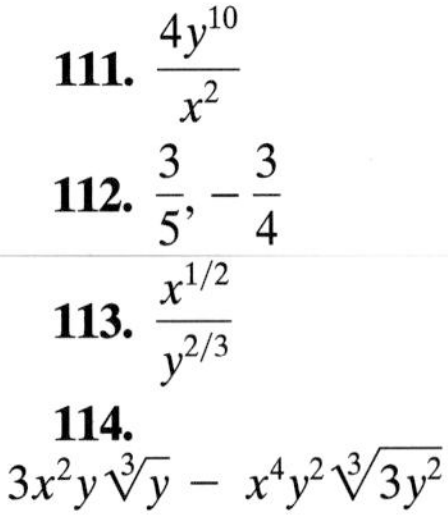

111. $\dfrac{4y^{10}}{x^2}$

112. $\dfrac{3}{5}, -\dfrac{3}{4}$

113. $\dfrac{x^{1/2}}{y^{2/3}}$

114. $3x^2y\sqrt[3]{y} - x^4y^2\sqrt[3]{3y^2}$

Using Your Graphing Calculator, 7.6

1.

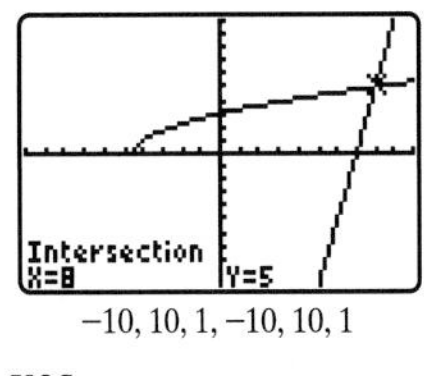

−10, 10, 1, −10, 10, 1

yes

2.

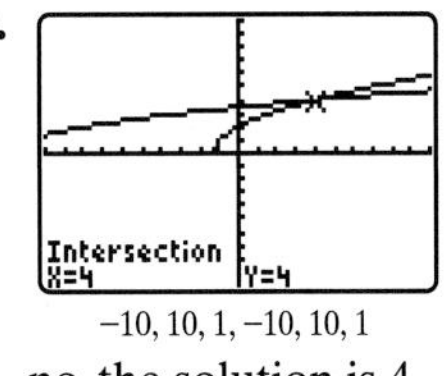

−10, 10, 1, −10, 10, 1

no, the solution is 4

Exercise Set 7.6

1. a) Answers will vary. **b)** 5 **3.** 0 **5.** Answers will vary. **7.** 16 **9.** 8 **11.** 81 **13.** 16 **15.** 8 **17.** 9 **19.** No solution **21.** No solution **23.** 1 **25.** $-\dfrac{1}{3}$ **27.** $\dfrac{2}{3}$ **29.** $-\dfrac{9}{4}$ **31.** $\dfrac{1}{4}$ **33.** 6 **35.** −7 **37.** 5 **39.** 8 **41.** −3 **43.** 4 **45.** $-\dfrac{5}{2}$ **47.** 2 **49.** $\dfrac{9}{16}$ **51.** 9 **53.** 4 **55.** No solution **57.** 6 **59.** 6 **61.** 4 **63.** 5 **65.** $v = \dfrac{p^2}{2}$ **67.** $g = \dfrac{v^2}{2h}$ **69.** $F = \dfrac{mv^2}{R}$ **71.** $m = \dfrac{x^2k}{V_0^2}$ **73.** $\approx$ 44.7 feet **75.** 12 feet **77. a)** $\approx$ 2.72 seconds **b)** $\sqrt{2T}$ **c)** $\approx$ 4.90 seconds **79.** $R = \dfrac{8\mu l}{\pi r^4}$ **81.** $\approx$ 365.2 days **83.** 1000 pounds **85.** $\approx$ 17.89 feet/second **87.** 2, −2 **89.** $\dfrac{3}{2}, -4$ **91.** 21 **93.** ±5 **95. a)** 3, 7 **b)** Yes **c)** Agrees **97.** At $x = 4$; $g(x) = 0$ or $y = 0$. Therefore, the graph must have an x-intercept at 4. **99.** $L_1 \approx 0.44$; $L_2 \approx 0.76$ **101.** All real numbers **103.** 1.5 **105.** −3.7, 3.7 **107.** No solution **109.** $\dfrac{1}{4}, 1$ **111.** $n = \dfrac{z^2pq}{(p' - p)^2}$ **113.** $P_2 = \dfrac{P_1P_3}{P_1 - P_3}$ **114.** $-2x - 3$ **115.** $\dfrac{3a}{2b(2a + 3b)}$ **116.** $t(t - 4)$ **117.** $\dfrac{3}{x + 3}$ **118.** 2

Exercise Set 7.7

1. $\sqrt{-1}$ **3.** All are complex numbers. **5.** Yes **7.** $a - bi$ **9.** Answers will vary. One example is: **a)** $\sqrt{2}$ **b)** 1 **c)** $\sqrt{-3}$ or $2i$ **d)** 6 **e)** Every number we have studied is a complex number. **11.** $3 + 0i$ **13.** $3 + 2i$ **15.** $(6 + \sqrt{3}) + 0i$ **17.** $0 + 5i$ **19.** $4 + 2i\sqrt{3}$ **21.** $0 + 3i$ **23.** $9 - 3i$ **25.** $0 + (2 - 4\sqrt{5})i$ **27.** $15 - 4i$ **29.** $8 + \dfrac{47}{36}i$ **31.** $18 - 5i$ **33.** $(4\sqrt{2} + \sqrt{3}) - 2i\sqrt{2}$ **35.** $19 + 12i\sqrt{3}$ **37.** $(2\sqrt{3} - 7) + (7 + 2\sqrt{3})i$ **39.** $-6 - 4i$ **41.** $-1 + 6i$ **43.** $-6.3 - 22.4i$ **45.** $-4 + 2i\sqrt{3}$ **47.** $-6 + 3i\sqrt{2}$ **49.** $1 + 5i$ **51.** $6 - 22i$ **53.** $\left(\dfrac{1}{2} + \dfrac{\sqrt{3}}{4}\right) + (2\sqrt{3} + 3)i$ **55.** $\dfrac{21}{4} + \dfrac{9}{2}i\sqrt{2}$ **57.** $-\dfrac{5i}{3}$ **59.** $\dfrac{3 - 2i}{2}$ **61.** $\dfrac{5 - 2i}{5}$ **63.** $\dfrac{49 + 14i}{53}$ **65.** $\dfrac{9 - 12i}{10}$ **67.** $\dfrac{3 + i}{5}$ **69.** $\dfrac{\sqrt{2} + i\sqrt{6}}{4}$ **71.** $\dfrac{(5\sqrt{10} - 2\sqrt{15}) + (10\sqrt{2} + 5\sqrt{3})i}{45}$ **73.** $\sqrt{30}$ **75.** $10i\sqrt{2}$ **77.** $7 - 7i$ **79.** $6 + i\sqrt{6}$ **81.** $20.8 - 16.64i$ **83.** $25 - 19i$ **85.** $\dfrac{-4 - 5i}{2}$ **87.** $\dfrac{4\sqrt{3} + 8i}{7}$ **89.** $3 + \dfrac{2}{45}i$ **91.** $\dfrac{1}{4} - \dfrac{31}{50}i$ **93.** 2 **95.** $-3.33 - 9.01i$ **97.** −1 **99.** 1 **101.** i **103.** $-i$ **105.** True **107.** False **109.** Even values **111.** −1 **113.** $16 - 4i$ **115.** $5 + 5i$ **117.** 0 **119.** 2 **121.** Yes **123.** No **125.** $\approx 0.83 - 3i$ **127.** $\approx 1.5 - 0.33i$ **129.** $-i$ **131.** $2 \pm i\sqrt{2}$ **133.** $4 + 3i\sqrt{3}$ **135.** $-3 + 5i\sqrt{3}$

137. 15 pounds of the \$5.50 per pound coffee; 25 pounds of the \$6.30 per pound coffee **138.** $(2x^2 + 3)^2$
139. $(3rs + 2)(5rs - 3)$ **140.** $(2r - 3s^2)(4r^2 + 6rs^2 + 9s^4)$

Review Exercises

1. 3 **2.** -3 **3.** 4 **4.** -4 **5.** 7 **6.** 93.4 **7.** $|x|$ **8.** $|x - 2|$ **9.** $|x - y|$
10. $|x^2 - 4x + 12|$ **11.** $x^{5/2}$ **12.** $x^{5/3}$ **13.** $y^{15/4}$ **14.** $5^{2/7}$ **15.** $\sqrt{a}$ **16.** $\sqrt[5]{y^3}$ **17.** $(\sqrt[5]{2m^2n})^9$
18. $\dfrac{1}{(\sqrt[4]{2a + 3b})^3}$ **19.** 9 **20.** x^5 **21.** 16 **22.** $\sqrt[5]{x}$ **23.** -5 **24.** Not a real number **25.** $\dfrac{3}{2}$ **26.** $-\dfrac{1}{12}$
27. $x^{4/15}$ **28.** $\dfrac{4}{y^2}$ **29.** $\dfrac{1}{y^{8/15}}$ **30.** $\dfrac{36x^8}{y^5}$ **31.** $2z^2 - 4z^{4/3}$ **32.** $\dfrac{3}{r^{13/6}} + 1$ **33.** $x^{1/5}(1 + x)$ **34.** $\dfrac{1 + x^{7/6}}{x^{1/2}}$
35. 7 **36.** 2.668

37.

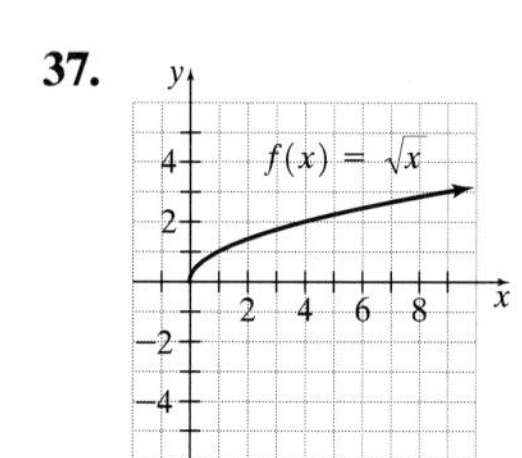

38.

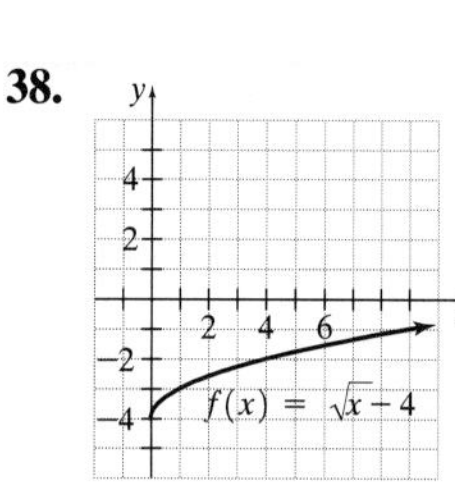

39. $\dfrac{27r^{9/2}}{p^3}$ **40.** $\dfrac{y^{1/5}}{4xz^{1/3}}$ **41.** $4\sqrt{5}$ **42.** $3\sqrt[3]{2}$ **43.** $5xy^3\sqrt{2xy}$
44. $5x^2y^3\sqrt[3]{xy}$ **45.** $x^{84}y^{49}z^{119}$ **46.** 10 **47.** $2x^3\sqrt{10}$
48. $2x^2y^3\sqrt[3]{4x^2}$ **49.** $xy^2\sqrt[3]{25x}$ **50.** $2x^2y^4\sqrt[4]{x}$
51. $6x - 2\sqrt{15x}$ **52.** $a^{49}b^{84}c^{63}$ **53.** $2x^2y^2\sqrt[3]{y^2} + x\sqrt[3]{18y}$
54. $\sqrt[4]{x^2y}$ **55.** $\sqrt[6]{x^4y}$ **56.** $\dfrac{6}{5}$ **57.** $\dfrac{x}{2}$ **58.** $\dfrac{x}{2}$ **59.** $\dfrac{4y^2}{x}$

60. $3xy\sqrt[3]{2}$ **61.** $\dfrac{x\sqrt{7}}{7}$ **62.** $\dfrac{\sqrt{10}}{5}$ **63.** $\dfrac{2\sqrt{15xy}}{5y}$ **64.** $\dfrac{2\sqrt[3]{x^2}}{x}$ **65.** $\dfrac{\sqrt[3]{75xy^2}}{5y}$ **66.** $\dfrac{x\sqrt{3y}}{y}$ **67.** $\dfrac{x^2y^2\sqrt{6yz}}{z}$
68. $\dfrac{5xy^2\sqrt{15yz}}{3z}$ **69.** $\dfrac{y\sqrt[4]{4x^3y^2}}{2x}$ **70.** $\dfrac{y\sqrt[3]{4x^2}}{x}$ **71.** $\dfrac{y^2\sqrt[3]{4x}}{2x}$ **72.** 7 **73.** $x - y^2$ **74.** $x^2 - y$ **75.** $28 + 10\sqrt{3}$
76. $x + \sqrt{5xy} - \sqrt{3xy} - y\sqrt{15}$ **77.** $\sqrt[3]{6x^2} - \sqrt[3]{4xy} - \sqrt[3]{9xy} + \sqrt[3]{6y^2}$ **78.** $-10 + 5\sqrt{5}$ **79.** $\dfrac{3x - x\sqrt{x}}{9 - x}$
80. $\dfrac{x - \sqrt{xy}}{x - y}$ **81.** $\dfrac{x - \sqrt{xy} - 2y}{x - y}$ **82.** $\dfrac{4\sqrt{y + 2} + 12}{y - 7}$ **83.** $2\sqrt[3]{x}$ **84.** $-4\sqrt{3}$ **85.** $8 - 13\sqrt[3]{2}$ **86.** $\dfrac{69\sqrt{2}}{8}$
87. $(3x^2y^3 - 4x^3y^4)\sqrt{x}$ **88.** $(2x^2y^2 - x + 3x^3)\sqrt[3]{xy^2}$ **89.** $3x\sqrt{2} - \sqrt{30x}$ **90.** $2x^2 + 2x^2\sqrt[3]{2x}$ **91.** $|2x + 4|$
92. $\sqrt{5}|2a + 5|$ **93.** $\sqrt[6]{x + 5}$ **94.** $\sqrt[12]{b^5}$

95. a)

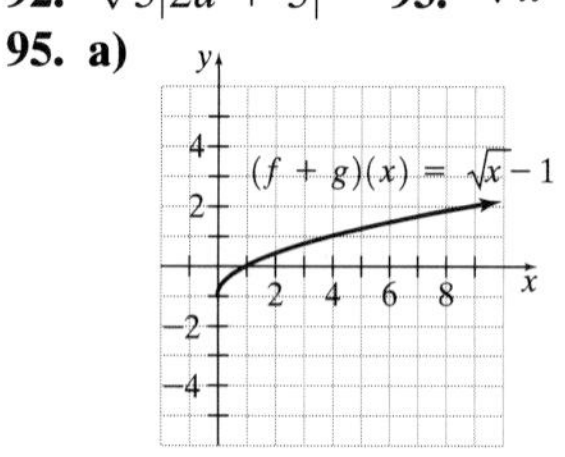

b) $\{x|x \text{ is a real number}, x \geq 0\}$

96. a)

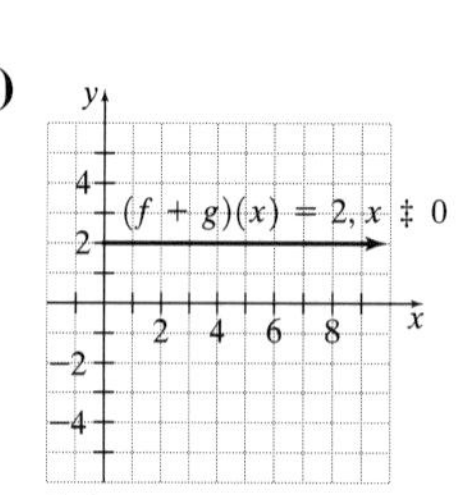

b) $\{x|x \text{ is a real number}, x \geq 0\}$

97. 64 **98.** No real solution **99.** 8
100. 2 **101.** -3 **102.** 2
103. 0 and 9 **104.** 5 **105.** 4
106. 6 **107.** $L = \dfrac{V^2w}{2}$
108. $A = \pi r^2$
109. $\sqrt{29} \approx 5.39$ meters

110. $\sqrt{1280} \approx 35.78$ feet/second **111.** $2\pi\sqrt{2} \approx 8.89$ seconds **112.** ≈ 25 meters per second
113. $\approx 5m_o$; It is about 5 times its original mass. **114.** $5 + 0i$ **115.** $-6 + 0i$ **116.** $2 - 16i$ **117.** $3 + 4i$
118. $7 + i$ **119.** $1 - 2i$ **120.** $3\sqrt{3} + (\sqrt{5} - \sqrt{7})i$ **121.** $-6 + 6i$ **122.** $17 - 6i$
123. $(24 + 3\sqrt{5}) + (4\sqrt{3} - 6\sqrt{15})i$ **124.** $-\dfrac{2i}{3}$ **125.** $\dfrac{(-2 - \sqrt{3})i}{2}$ **126.** $\dfrac{15 - 10i}{13}$ **127.** $\dfrac{5\sqrt{3} + 3i\sqrt{2}}{31}$
128. 0 **129.** 7 **130.** i **131.** $-i$ **132.** 1 **133.** -1

Practice Test

1. 26 **2.** $|3x - 4|$ **3.** $\dfrac{1}{y^{7/6}}$ **4.** $x^{-2/5}(x + 1)$

5.

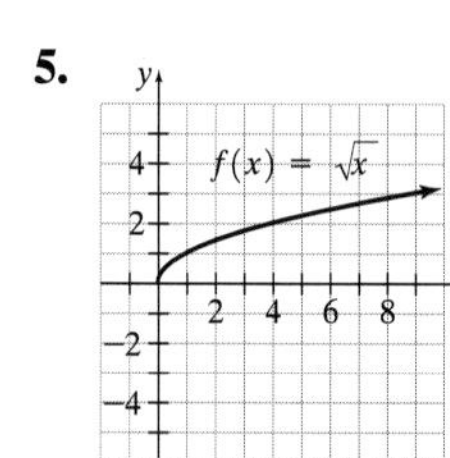

6. $5x^2y^4\sqrt{2x}$ **7.** $2x^3y^3\sqrt[3]{5x^2y}$ **8.** $\dfrac{x^2y^2\sqrt{yz}}{2z}$ **9.** $\dfrac{\sqrt[3]{x^2}}{x}$ **10.** $\dfrac{2 - \sqrt{2}}{2}$
11. $-20\sqrt{3}$ **12.** $(2xy + 2x^2y^2)\sqrt[3]{y^2}$ **13.** $2\sqrt{5} - 2\sqrt{10} - 6 + 6\sqrt{2}$ **14.** $\sqrt[6]{x^4y^2}$
15. $\sqrt[12]{(7x + 2)^7}$ **16.** 13 **17.** -3 **18.** 16 **19.** 3 **20.** $h = \dfrac{8w^2}{g}$ **21.** 13 feet
22. ≈ 0.92 seconds **23.** $(18 + 2\sqrt{2}) + (6\sqrt{2} - 6)i$ **24.** $\dfrac{\sqrt{5} + i\sqrt{10}}{6}$ **25.** 2

Cumulative Review Test **1.** $\frac{57}{9}$ **2. a)** Any set of ordered pairs
b) A correspondence such that each element of the domain corresponds with exactly one member of the range.
3. a) $\{x|x \geq 0\}$
b)

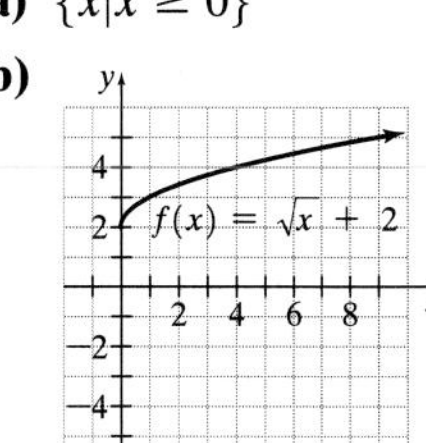

4. $\left(\frac{2}{11}, -\frac{16}{11}\right)$ **5.** $\frac{x^2}{3y^3}$ **6.** $6x^3 - 23x^2 + 8x + 30$ **7.** $3x + 4 + \frac{2}{x + 2}$
8.
9. $2(x - 3 + y)(x - 3 - y)$
10. $\left\{x \middle| x \text{ is a real number, } x \neq \frac{5}{3}\right\}$ **11.** $\frac{2x + 1}{3}$
12. $\frac{x + 1}{2x + 3}$ **13.** $\frac{x^2 + 8x + 5}{(x + 3)(x - 1)(2x + 5)}$ **14.** -5
15. $x^{7/2}y$ **16.** $2x^4y^9\sqrt[3]{2xy^2}$ **17.** -3 and 3
18. $\frac{6 - 2i\sqrt{6}}{15}$ **19.** $1\frac{1}{5}$ hours **20.** $\sqrt{1300} \approx 36.1$ feet

CHAPTER 8

Exercise Set 8.1 **1.** ± 6 **3.** If $x^2 = a$, where a is a real number, then $x = \pm\sqrt{a}$. **5.** $\left(\frac{b}{2}\right)^2$ must equal c.
7. ± 7 **9.** $\pm 4\sqrt{3}$ **11.** $\pm 2\sqrt{7}$ **13.** $8, 0$ **15.** $\frac{1}{3}, -1$ **17.** $-0.9, -2.7$ **19.** $\frac{5 \pm 2\sqrt{3}}{2}$ **21.** $-\frac{1}{20}, -\frac{9}{20}$ **23.** $3, -5$
25. $5, -1$ **27.** $-2, -1$ **29.** $5, 3$ **31.** $-1 \pm i\sqrt{11}$ **33.** $5, 4$ **35.** $-4, 7$ **37.** $\frac{-3 \pm i\sqrt{15}}{2}$ **39.** $\frac{5 \pm \sqrt{57}}{2}$
41. $0, -2$ **43.** $0, \frac{1}{3}$ **45.** $1, -3$ **47.** $\frac{-9 \pm \sqrt{73}}{2}$ **49.** $-3, -8$ **51.** $\frac{-2 \pm i\sqrt{2}}{2}$ **53.** $1 \pm i$ **55.** $\frac{-3 \pm \sqrt{59}}{10}$
57. $-a \pm \sqrt{k}$ **59.** 7 and 9 **61.** 5 feet by 12 feet **63.** ≈ 14.49 feet by 14.49 feet **65.** $10\sqrt{2} \approx 14.14$ inches
67. $2\sqrt{6} \approx 4.90$ feet **69.** 5% **71.** 6% **73.** $x^2 + 6x + 9 + y^2 - 10y + 25 = -18 + 9 + 25;$
$(x + 3)^2 + (y - 5)^2 = 16$ **75. a)** $S = 32 + 80\sqrt{\pi} \approx 173.80$ square inches **b)** $r = \frac{4\sqrt{\pi}}{\pi} \approx 2.26$ inches
c) $r = -5 + \sqrt{\frac{80 + 25\pi}{\pi}} \approx 2.1$ inches **77.** $\frac{3xy}{3y + 1}$ **78.** $|x^2 - 4x|$ **79.** $\frac{1}{5}$ **80.** $\frac{x^{1/2}}{y^{3/2}}$

Exercise Set 8.2 **1.** $x = \frac{-b \pm \sqrt{b^2 - 4ac}}{2a}$ **3. a)** Yes; $12x^2 - 15x - 6 = 3(4x^2 - 5x - 2)$ **b)** $\frac{5 \pm \sqrt{57}}{8}$
c) $\frac{5 \pm \sqrt{57}}{8}$ **5.** If $b^2 - 4ac > 0$, then the quadratic equation will have two distinct real solutions. Since there is a positive number under the radical sign in the quadratic formula, the value of the radical will be real and there will be two real solutions. If $b^2 - 4ac = 0$, then the equation has the single real solution $-\frac{b}{2a}$. If $b^2 - 4ac < 0$, the expression under the radical sign in the quadratic formula is negative. Thus the equation has no real solution.
7. Two real solutions **9.** No real solution **11.** Two real solutions **13.** No real solution **15.** One real solution
17. One real solution **19.** $-4, -5$ **21.** $1, -3$ **23.** $1, 5$ **25.** $9, -9$ **27.** $4, 0$ **29.** $\frac{2 \pm i\sqrt{11}}{3}$ **31.** $-3 \pm \sqrt{7}$
33. $\frac{9}{2}, -1$ **35.** $\frac{1}{2}, -\frac{5}{3}$ **37.** $\frac{-1 \pm i\sqrt{23}}{4}$ **39.** $0, -3$ **41.** $\frac{3 \pm \sqrt{33}}{2}$ **43.** $\frac{-6 \pm 2\sqrt{6}}{3}$ **45.** $\frac{11 \pm \sqrt{241}}{6}$
47. $\frac{-0.6 \pm \sqrt{0.84}}{0.2}$ or $-3 \pm \sqrt{21}$ **49.** $\frac{-0.94 \pm \sqrt{32.3116}}{3.24}$ or $\frac{-47 \pm \sqrt{80{,}779}}{162}$ **51.** $3, 0$ **53.** $\frac{7 \pm \sqrt{17}}{4}$
55. No real numbers **57.** $f(x) = x^2 - 10x + 24$ **59.** $f(x) = x^2 + x - 12$ **61.** $f(x) = 2x^2 - 7x + 3$
63. $f(x) = 15x^2 - x - 6$ **65.** $f(x) = x^2 - 3$ **67.** $f(x) = x^2 + 4$ **69.** $f(x) = x^2 - 6x + 7$
71. $f(x) = x^2 - 4x + 13$ **73.** Answers will vary. **75.** Yes, if the discriminant is a perfect square, the simplified expression will not contain a radical and the quadratic equation can be solved by factoring. **77.** 2
79. Width: 3 feet; length: 7 feet **81.** 2 inches **83. a)** \$0.5 thousand **b)** \$11.3 thousand **c)** \$1.55
85. a) 3.02 million **b)** 9 years after 1987, or in 1996 **87. a)** \$2.64 billion **b)** 16.9 years after 1982, or in 1998

89. a) \$1.32 billion **b)** 6.84 years after 1992, or in 1998 **91. a)** 4 seconds **b)** 29.26 feet per second **93.** $\approx$ 16.37 seconds **95.** $-3\sqrt{6}, -2\sqrt{6}$ **97.** $\pm 1, \frac{-1 \pm i\sqrt{3}}{2}, \frac{1 \pm i\sqrt{3}}{2}$ **100.** $2xy^2z^4\sqrt[5]{2x^4y^2}$ **101.** $2x^2y^4\sqrt[3]{x} + 2xy^3\sqrt[3]{3}$ **102.** $\frac{x^2 + 2x\sqrt{y} + y}{x^2 - y}$ **103.** 6

Exercise Set 8.3

1. Answers will vary. **3.** $s = \sqrt{A}$ **5.** $i = \sqrt{\frac{E}{r}}$ **7.** $r = \sqrt{\frac{V}{\pi h}}$ **9.** $b = \sqrt{c^2 - a^2}$ **11.** $v = \sqrt{\frac{2E}{m}}$ **13.** $v_1 = \sqrt{v_2^2 - 2ad}$ **15.** $c = \sqrt{(v')^2 + v^2}$ **17.** Going rate is 6 mph; return rate is 8 mph **19.** 4 feet/hour **21.** Car: 60 mph; train: 80 mph (car: 100 mph; train: 120 mph is not reasonable) **23.** Bonita: 11.52 hours; Pamela: 12.52 hours **25.** Chris: $\approx$ 11.76 hours; John: $\approx$ 12.26 hours **27.** 60 mph **29.** Answers will vary. **31.** $h = \frac{S - 2\pi r^2}{2\pi r}$ **32.** No. In a function each x-value must have a unique y-value. This is not the case with the points (2, 3) and (2, 1). **33.** If a vertical line can be drawn to intersect the graph at more than one point, the graph is not a function. **34.** 3 **35.** The domain is the set of values that can be used for the independent variable. **36.** The range is the set of values that are obtained for the dependent variable. **37.** No. $f(x) = \frac{1}{x}$ is not a polynomial since a polynomial function must have whole number exponents on the variable and, in this case, $f(x) = \frac{1}{x} = x^{-1}$ and -1 is not a whole number.

Exercise Set 8.4

1. The equation can be written in the form $au^2 + bu + c = 0$ **3.** $\pm 3, \pm 1$ **5.** $\pm 2, \pm\sqrt{3}$ **7.** $\pm\sqrt{3}, \pm\sqrt{5}$ **9.** $\pm\sqrt{2}, \pm 2i$ **11.** 4 **13.** 4 **15.** $\frac{1}{9}$ **17.** $1, -9$ **19.** $\frac{4}{3}, -\frac{1}{2}$ **21.** $\pm i, \pm 2$ **23.** $\pm\frac{\sqrt{39}}{3}, \pm\frac{5\sqrt{6}}{6}$ **25.** $-\frac{1}{3}$ **27.** $\frac{1}{3}, 2$ **29.** $-\frac{1}{8}, \frac{1}{2}$ **31.** 27, 8 **33.** 243, 7776 **35.** $\frac{1}{4}$ **37.** (1, 0), (16, 0) **39.** $(-4, 0), \left(\frac{1}{5}, 0\right)$ **41.** $(-27, 0), (8, 0)$ **43.** $(4, 0), (-1, 0)$ **45.** Let $u = x^2$ **47.** $0 = x^4 - 20x^2 + 64 = (x + 2)(x - 2)(x + 4)(x - 4)$ **49.** No, it can have no real solution, two real solutions, or four real solutions. **51. a)** $\frac{1}{5}, -\frac{1}{4}$ **b)** $\frac{1}{5}, -\frac{1}{4}$ **53.** $-\frac{14}{5}, -\frac{8}{3}$ **55.** $\frac{5}{4}, 3$ **57.** 2, 1 **59.** $-4, 2, -3, 1$ **61.** $\pm\sqrt{\frac{3 \pm \sqrt{15}}{2}}$ **63.** $T_2 = \frac{T_1P_2}{P_1}$ **64.** $(-1, -2, 3)$ **65. a)** 1: a^2; 2: ab; 3: ab; 4: b^2 **b)** $(a + b)^2$ **66.** $\frac{3x - 4}{2x - 5}$

Exercise Set 8.5

1. The graph of a quadratic equation is called a parabola. **3.** The axis of symmetry of a parabola is the line where, if the graph is folded, the two sides overlap. **5.** $\left(-\frac{b}{2a}, \frac{4ac - b^2}{4a}\right)$ **7. a)** When $a > 0$, $f(x)$ will have a minimum since the graph opens upward. **b)** When $a < 0$, $f(x)$ will have a maximum since the graph opens downward. **9.** Set $x = 0$ and solve for y **11. a)** **b)** **13. a)** Upward **b)** (0, 15) **c)** $(-4, -1)$ **d)** $(-5, 0), (-3, 0)$ **e)** $f(x) = x^2 + 8x + 15$ **15. a)** Upward **b)** (0, 4) **c)** $(3, -5)$ **d)** $(3 + \sqrt{5}, 0), (3 - \sqrt{5}, 0)$ **e)**

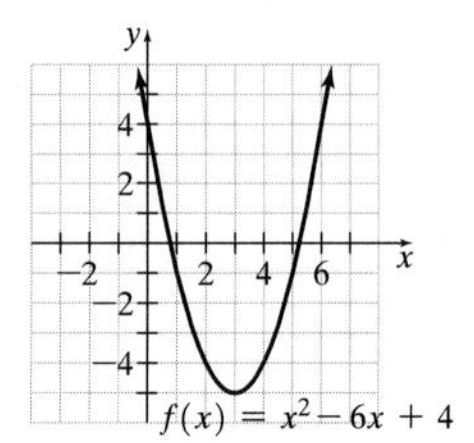

17. a) Upward **b)** $(0, 9)$ **c)** $(-3, 0)$ **d)** $(-3, 0)$

e)

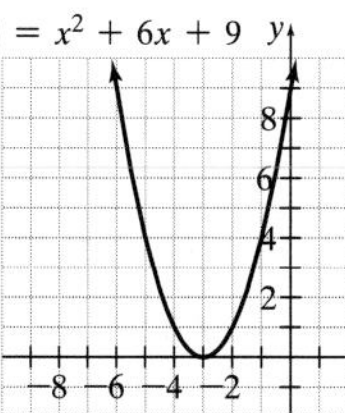

19. a) Upward **b)** $(0, -6)$ **c)** $\left(\frac{1}{4}, -\frac{49}{8}\right)$ **d)** $\left(-\frac{3}{2}, 0\right), (2, 0)$

e)

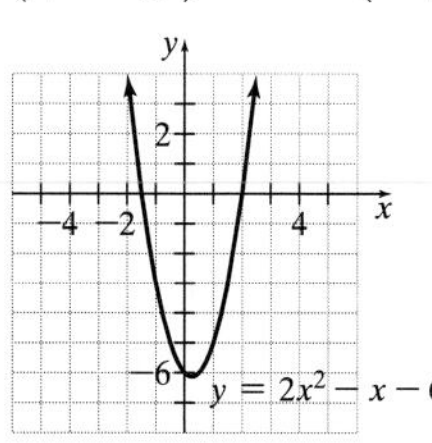

21. a) Upward **b)** $(0, 3)$ **c)** $\left(-\frac{2}{3}, \frac{5}{3}\right)$ **d)** No x-intercepts

e)

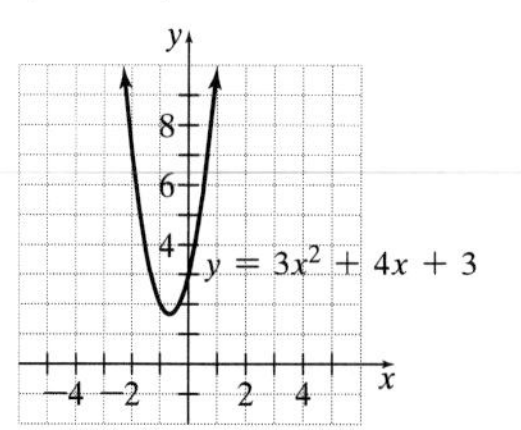

23. a) Downward **b)** $(0, 4)$ **c)** $\left(-\frac{3}{2}, \frac{17}{2}\right)$

d) $\left(\frac{-3 + \sqrt{17}}{2}, 0\right), \left(\frac{-3 - \sqrt{17}}{2}, 0\right)$

e) $f(x) = -2x^2 - 6x + 4$

25. a) Upward **b)** $(0, 0)$ **c)** $(-2, -4)$ **d)** $(0, 0), (-4, 0)$

e)

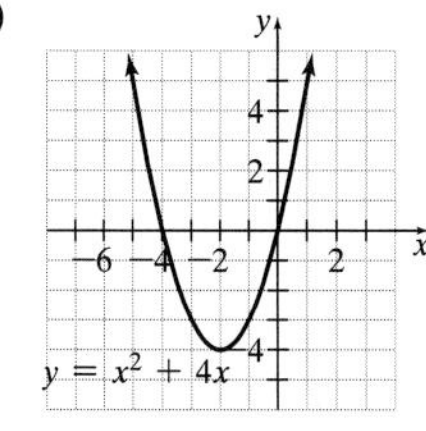

27. a) Upward **b)** $(0, 0)$ **c)** $\left(-\frac{5}{3}, -\frac{25}{3}\right)$ **d)** $(0, 0), \left(-\frac{10}{3}, 0\right)$

e)

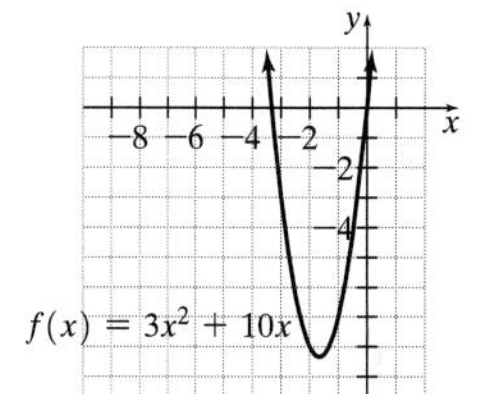

29. a) Downward **b)** $(0, -5)$ **c)** $\left(\frac{3}{2}, -\frac{11}{4}\right)$ **d)** No x-intercepts

e)

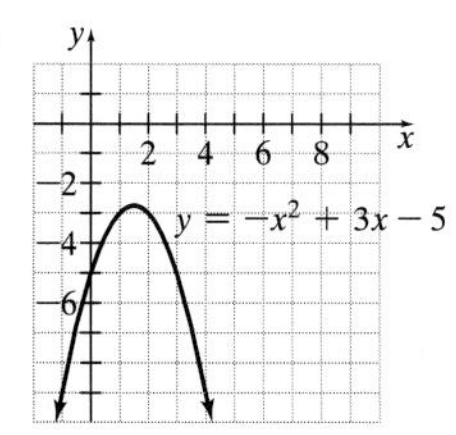

31. a) Downward **b)** $(0, -9)$ **c)** $\left(\frac{3}{4}, -\frac{27}{4}\right)$ **d)** No x-intercepts

e)

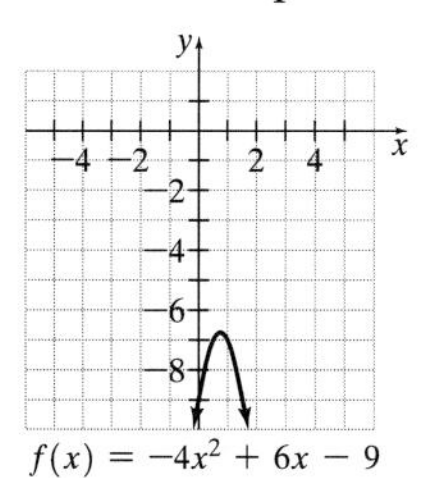

33.

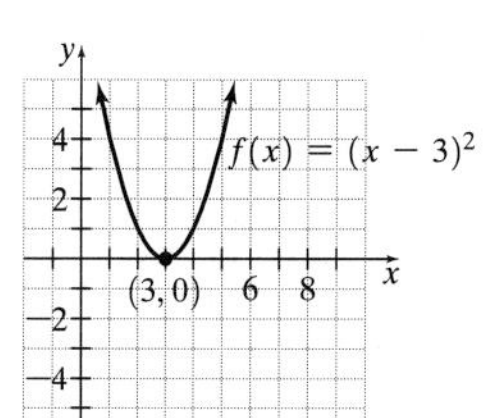

35.

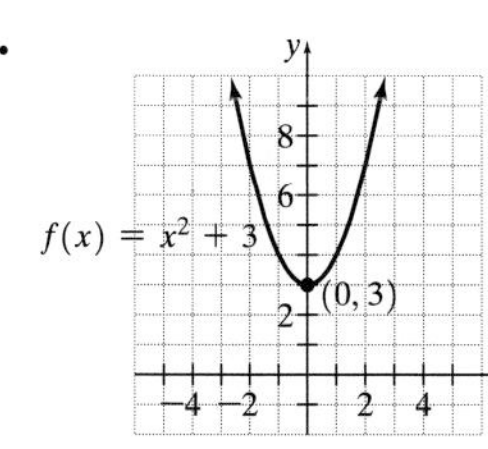

37.

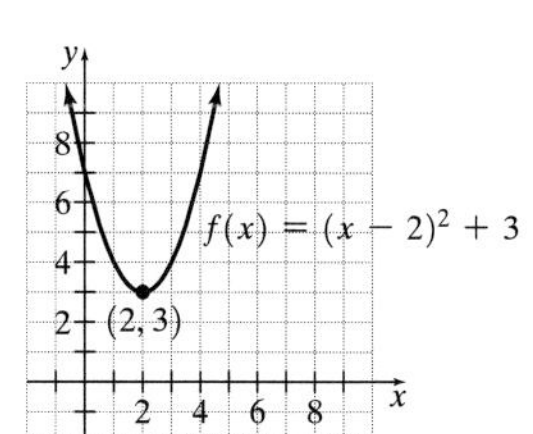

39.

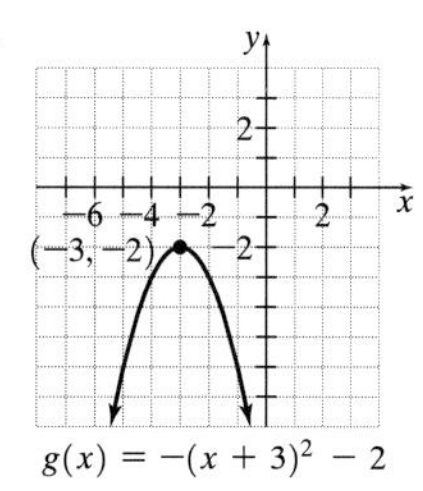

41.

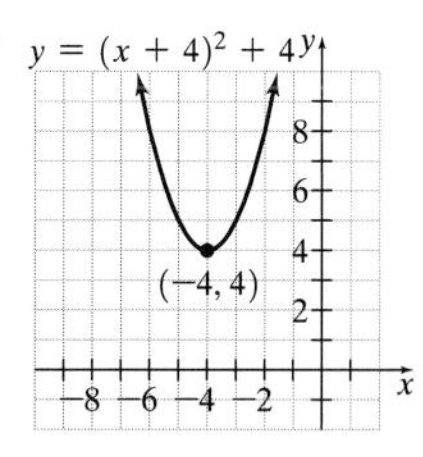

43.

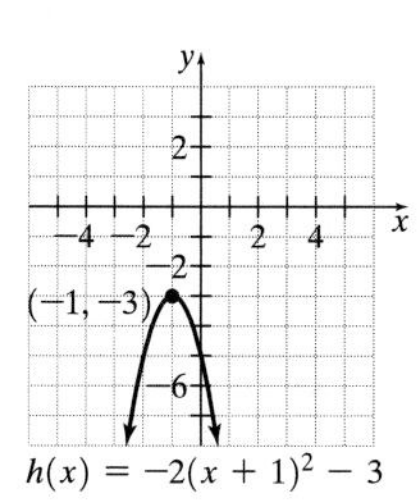

45.

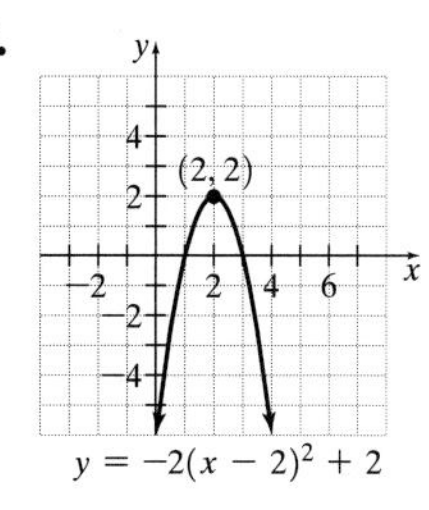

47. a) $f(x) = (x - 3)^2 - 1$

b)

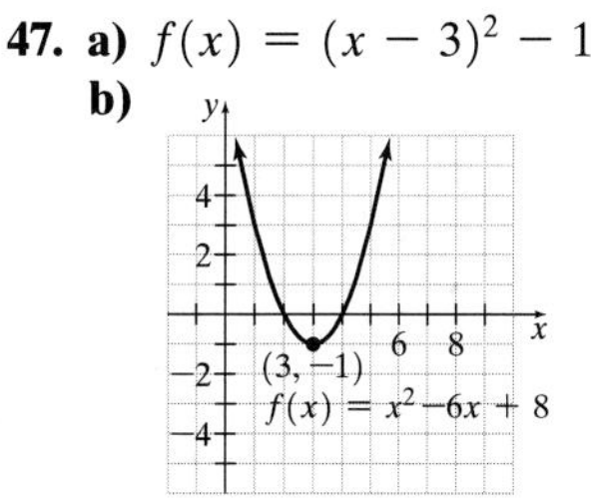

49. a) $g(x) = \left(x - \frac{1}{2}\right)^2 - \frac{49}{4}$

b)

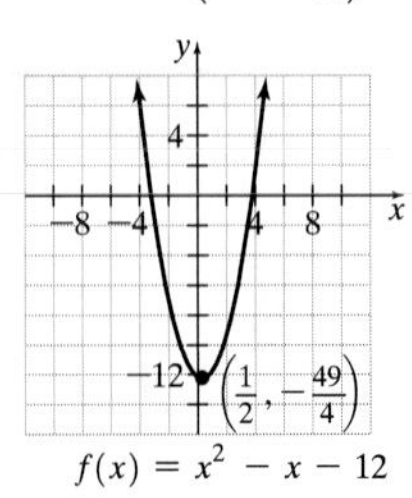

51. a) $f(x) = -(x + 2)^2 - 2$

b)

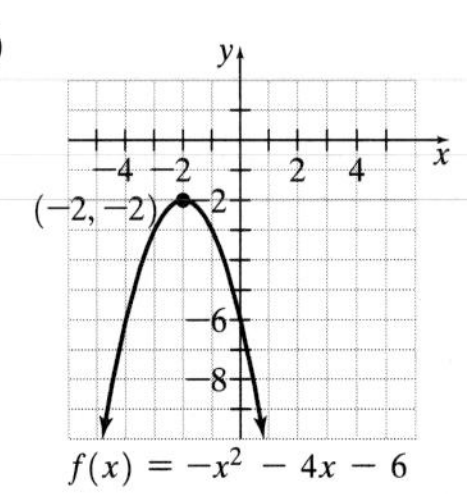

53. a) $h(x) = -(x - 2)^2 - 8$

b)

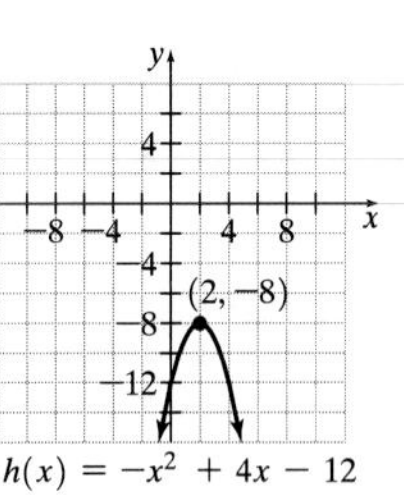

55. a) $g(x) = -2\left(x + \frac{9}{4}\right)^2 + \frac{9}{8}$

b)

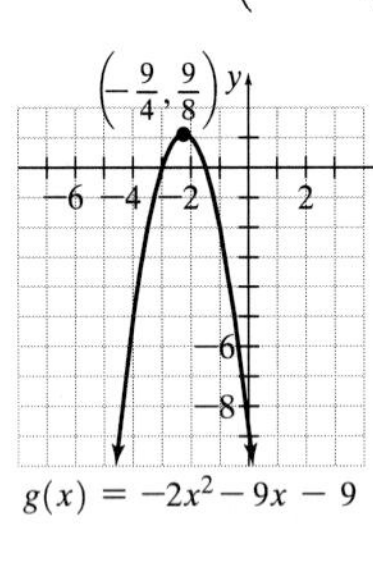

57. d) **59.** b) **61.** 4 units **63.** 3 units **65.** $f(x) = 2(x - 3)^2 - 2$

67. $f(x) = -4\left(x + \frac{3}{5}\right)^2 - \sqrt{2}$

69. a) The graphs will have the same x-intercepts but $f(x) = x^2 - 8x + 12$ will open upward and $g(x) = -x^2 + 8x - 12$ will open downward.
b) Yes **c)** No
d)

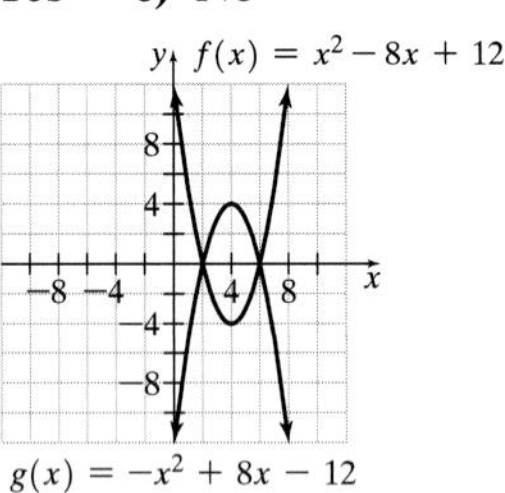

71. a)

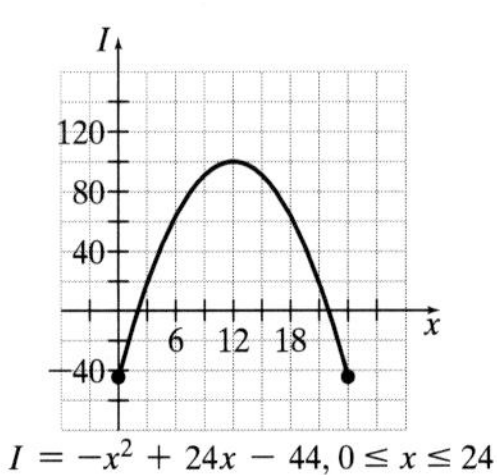

b) \$2
c) \$22
d) \$12
e) \$10,000

73. a) 100 **b)** \$3800 **75.** 1 second; 16 feet **77.** 225 ft^2 **79.** −16; −4 and 4 **81.** 400; 20 and 20
83. a) ≈ 7.07 years after 1989 or in 1996 **b)** ≈ 44% **85. a)** ≈ 5.52 years after 1986 or in 1991 **b)** ≈ 67,245
87. a) \$142,400 **b)** 380 **89. a)** $h(t) = -4.9(t - 2.5)^2 + 40.425$ **b)** 40.425 meters; 2.5 seconds **c)** Same
91. 200π ft^2 **92.** Salary: \$400; commission: 15% **93.** $(x - 5)(x^2 + 2)$ **94.** $\frac{(3a - b)(a - b)}{ab}$

Exercise Set 8.6

1. a) $x < 2$ or $x > 5$ **b)** $2 < x < 5$ **3.** −2, 5 **5.**

7. −4, 4 **9.** −3, $\frac{1}{2}$ **11.** −4, $\frac{1}{5}$ **13.**

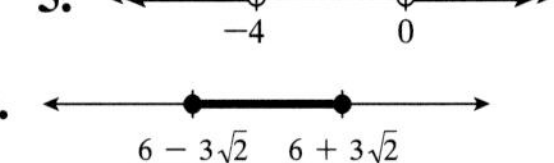

15. $(-4, -1) \cup (1, \infty)$ **17.** $[-3, 0] \cup [3, \infty)$ **19.** $\left(-6, -\frac{5}{2}\right) \cup (2, \infty)$ **21.** $\left[\frac{8}{3}, \infty\right)$ **23.** $(-\infty, 0)$

25. −4, 0 **27.** 6, 8 **29.** $\frac{-9 - \sqrt{129}}{4}$, $\frac{-9 + \sqrt{129}}{4}$ **31.** −7, 0, $\frac{5}{2}$

33. $\{x | x < -3 \text{ or } x > 1\}$ **35.** $\{y | 1 < y \le 4\}$ **37.** $\{x | 1 < x < 2\}$ **39.** $\left\{a \middle| a \le -2 \text{ or } a > \frac{1}{2}\right\}$

41. $\{x | -4 \le x < 4\}$ **43.** $\left\{x \middle| x < \frac{1}{2} \text{ or } x > 2\right\}$ **45.** $(-\infty, -6) \cup (-2, 4)$ **47.** $[1, 3) \cup [6, \infty)$

49. $(-\infty, -4) \cup (1, 6]$ **51.** $\left(-\frac{5}{2}, 3\right) \cup (6, \infty)$ **53.** 1, 3 **55.** 2, 4

57. 

59. 

61. 

63. a) $(4, \infty)$ **b)** $(-\infty, 2) \cup (2, 4)$ **65.** $x^2 + 2x - 8 > 0$ **67.** $\dfrac{x+3}{x-4} \geq 0$ **69.** All real numbers

71. All real numbers except -1 **73.** No solution **75.** 

77. One possible answer is: $x^2 - 3x > 0$. Use a parabola that opens upward and has x-intercepts of $(0, 0)$ and $(3, 0)$. The x-values for which the parabola lies above the x-axis are $(-\infty, 0) \cup (3, \infty)$.
79. One possible answer is: $x^2 < 0$.Use a parabola that opens upward and has its vertex on or above the x-axis. Then there are no x-values for which the parabola lies below the x-axis.

83. Domain: $\{x|x$ is a real number and $x \neq -2, x \neq 2\}$ **84.** $\{x|x$ is a real number and $x \geq 4\}$ **85.** $\dfrac{a^3 - ab^3}{b^2 + a^3b}$

86. $-8 - 6i$

Review Exercises

1. $4 \pm 2\sqrt{5}$ **2.** $\dfrac{4 \pm 2\sqrt{15}}{3}$ **3.** $1, \dfrac{1}{3}$ **4.** $\dfrac{5}{4}, -\dfrac{3}{4}$ **5.** 5, 3 **6.** 9, −6 **7.** 1, −6
8. $-1 \pm \sqrt{6}$ **9.** $-3 \pm \sqrt{19}$ **10.** $2 \pm 2i\sqrt{7}$ **11.** Two real solutions **12.** No real solutions **13.** No real solutions
14. One real solution **15.** Two real solutions **16.** Two real solutions **17.** $2, -\dfrac{3}{5}$ **18.** 9, −2 **19.** $\dfrac{1 \pm i\sqrt{119}}{2}$
20. $\dfrac{3}{2}, -\dfrac{5}{3}$ **21.** $\dfrac{-2 \pm \sqrt{10}}{2}$ **22.** $\dfrac{3 \pm \sqrt{57}}{4}$ **23.** $3 \pm \sqrt{2}$ **24.** $\dfrac{3 \pm \sqrt{33}}{3}$ **25.** $\dfrac{5}{2}, 0$ **26.** $\dfrac{-5.7 \pm \sqrt{43.53}}{2.4}$
27. $\dfrac{5}{2}, -\dfrac{5}{3}$ **28.** $\dfrac{1}{4}, -\dfrac{3}{2}$ **29.** 10, −6 **30.** $-\dfrac{3}{2}, \dfrac{2}{3}$ **31.** $\dfrac{7 \pm \sqrt{89}}{10}$ **32.** $\dfrac{3 \pm 3\sqrt{3}}{2}$ **33.** $f(x) = x^2 - x - 6$
34. $f(x) = 3x^2 + 7x - 6$ **35.** $f(x) = x^2 - 5$ **36.** $f(x) = x^2 - 6x + 13$ **37.** Width = 7 feet; length = 9 feet
38. $x = 8\sqrt{2} \approx 11.31$ **39.** 5% **40.** 5; 9 **41.** Width = 6 inches; length = 11 inches **42.** \$475
43. a) 1656 feet **b)** ≈ 10.61 seconds **44. a)** 68 feet **b)** ≈ 3.05 seconds **45. a)** 40 ml **b)** 150°C
46. Smaller machine: 24.51 hours; larger machine: 23.51 hours **47.** 50 mph **48.** 1.6 mph **49.** $V_y = \sqrt{V^2 - V_x^2}$
50. $v_2 = \sqrt{v_1^2 + 2ad}$ **51.** $\pm 2\sqrt{2}, \pm i\sqrt{3}$ **52.** $\dfrac{3}{2}, -\dfrac{2}{5}$ **53.** $\dfrac{1}{16}$ **54.** $\dfrac{27}{8}, 8$ **55.** $-\dfrac{6}{5}, -\dfrac{7}{2}$ **56.** $4, \dfrac{13}{8}$
57. $\left(\dfrac{4}{25}, 0\right)$ **58.** $\left(\dfrac{4 \pm \sqrt{10}}{2}, 0\right), \left(\dfrac{12 \pm 5\sqrt{6}}{6}, 0\right)$

59. a) Upward **b)** $(0, 0)$
c) $(-3, -9)$ **d)** $(0, 0), (-6, 0)$
e)

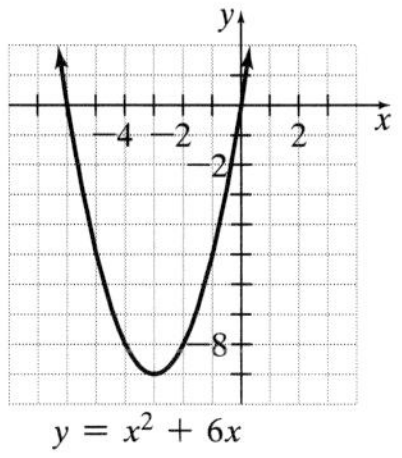

60. a) Upward **b)** $(0, -8)$
c) $(-1, -9)$ **d)** $(-4, 0), (2, 0)$
e)

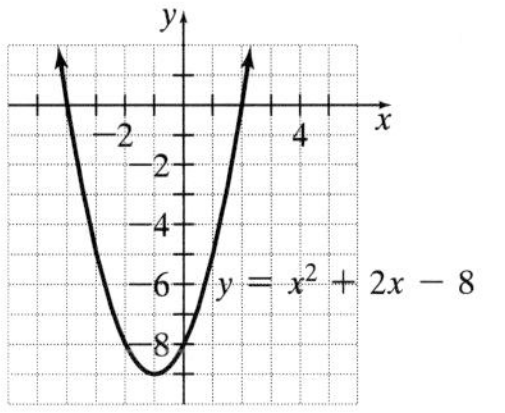

61. a) Downward **b)** $(0, -9)$
c) $(0, -9)$ **d)** No x-intercepts
e)

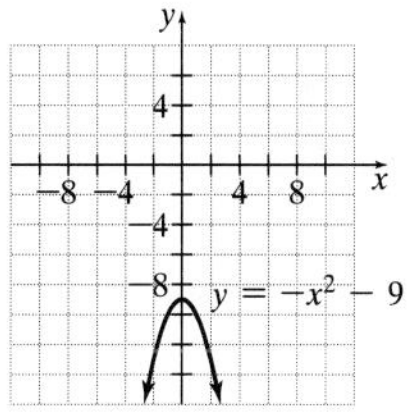

62. a) Downward **b)** $(0, 15)$
c) $\left(-\dfrac{1}{4}, \dfrac{121}{8}\right)$ **d)** $(-3, 0), \left(\dfrac{5}{2}, 0\right)$
e)

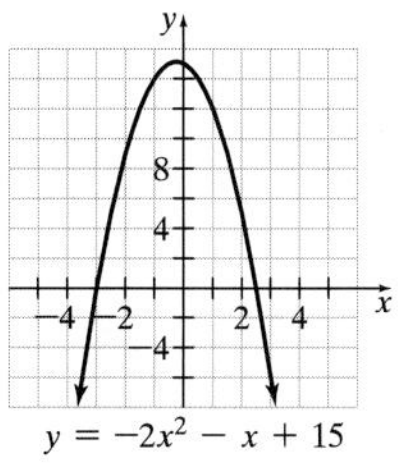

63. a) \$11 **b)** \$9100 **64. a)** 2.5 seconds **b)** 160 feet

65.

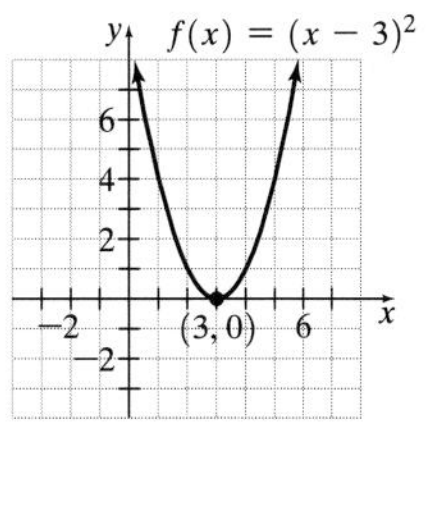

66.

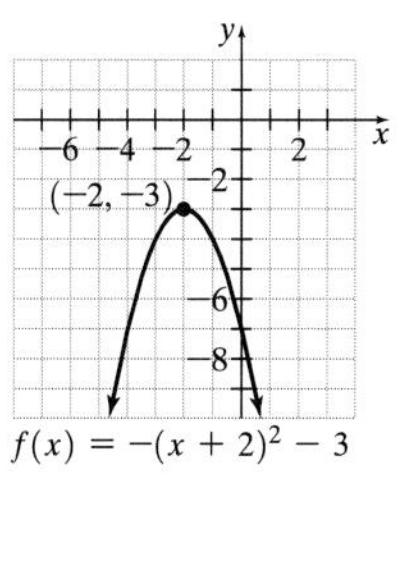

67.
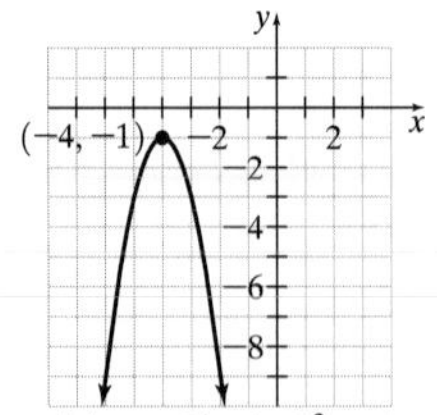

$g(x) = -2(x + 4)^2 - 1$

68.
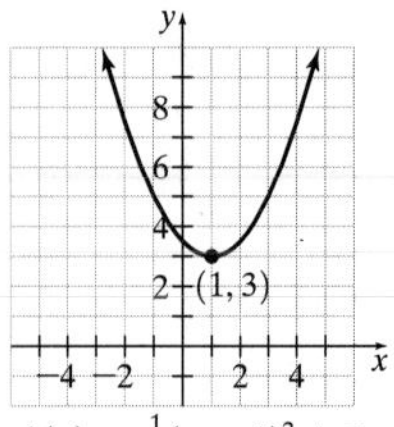

$h(x) = \frac{1}{2}(x - 1)^2 + 3$

69.
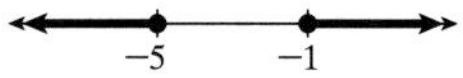

70.

71.
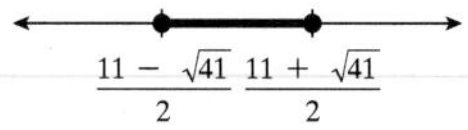

72.
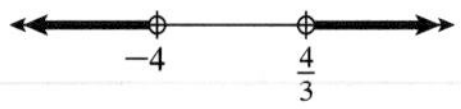

73.

74.

75. $\{x|x < -2 \text{ or } x > 3\}$ **76.** $\{x|-2 < x \le 5\}$

77. $\{x|x < -1 \text{ or } x \ge 2\}$ **78.** $\left\{x\middle|-\frac{5}{3} < x < 6\right\}$ **79.** $\{x|-3 < x < -1 \text{ or } x > 2\}$ **80.** $\{x|x \le 0 \text{ or } 3 \le x \le 5\}$

81. $\left[-\frac{4}{3}, 1\right] \cup [3, \infty)$ **82.** $(-\infty, -5) \cup (-2, 0)$ **83.** $(-2, 0) \cup (4, \infty)$ **84.** $(-\infty, -3) \cup (2, 5)$ **85.** $(-2, 3] \cup (5, \infty)$

86. $(-\infty, -3) \cup [0, 5]$ **87.**
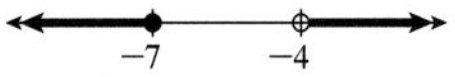

88.

89.
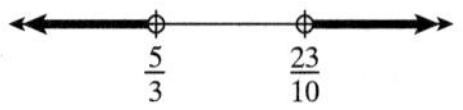

Practice Test

1. 3, −4 **2.** $-1 \pm i\sqrt{2}$ **3.** 6, −1 **4.** $-4 \pm \sqrt{11}$ **5.** $0, \frac{5}{3}$ **6.** $\frac{1}{2}, -5$

7. $f(x) = 2x^2 - x - 15$ **8.** $v = \sqrt{\frac{2K}{m}}$ **9. a)** \$138,000 **b)** 2712.57 square feet **10.** 50 mph

11. $\pm\frac{\sqrt{10}}{5}, \pm\frac{i\sqrt{10}}{2}$ **12.** $\frac{343}{27}, -216$ **13.** $\left(\frac{25}{16}, 0\right)$

14.
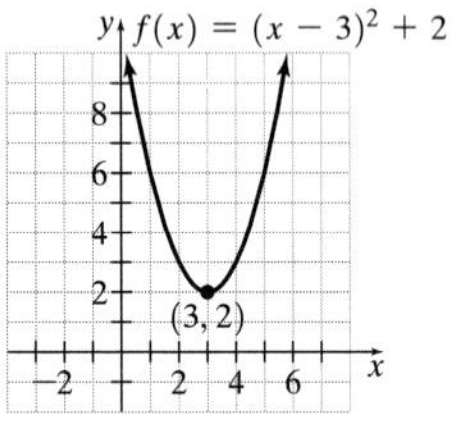

15.
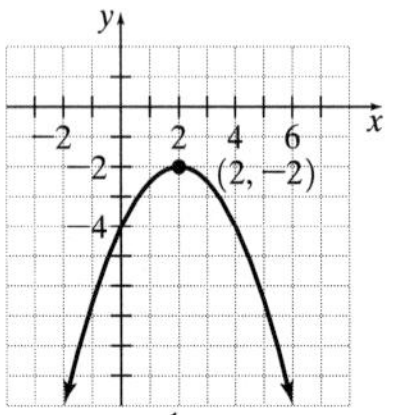

$h(x) = -\frac{1}{2}(x - 2)^2 - 2$

16. Two real solutions **17. a)** Upward **b)** (0, −8) **c)** (1, −9) **d)** (−2, 0), (4, 0)

e)
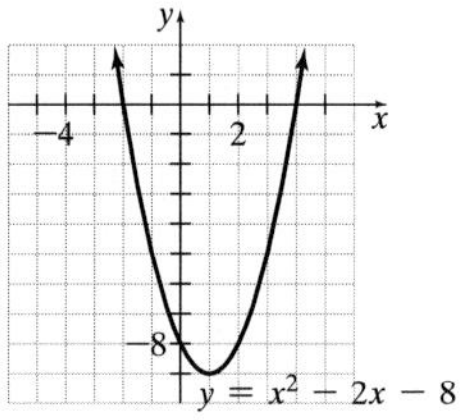

18. $f(x) = 2x^2 + 11x - 6$ **19.**

20.

21. a) $\left[-\frac{5}{2}, -2\right)$ **b)** $\left\{x\middle|-\frac{5}{2} \le x < -2\right\}$ **22.** Width = 4 feet; length = 12 feet **23.** 5 seconds

24. a) 20 **b)** \$500 **25. a)** 240 **b)** \$49,600

Cumulative Review Test

1. −26 **2. a)** 5.5776×10^7 **b)** 1.20516×10^7 **c)** 4.37244×10^7

3. $\frac{40}{19}$ **4.** $-19 < x < 29$ **5.** $\{12, -2\}$ **6.** Yes **7. (a)** No **(b)** Domain: $\{x|x \ge -2\}$; Range: $\mathbb{R}$

8. a)
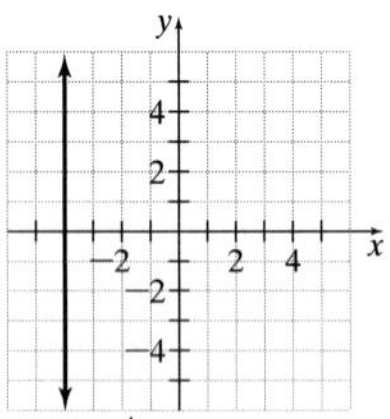

$x = -4$

b)
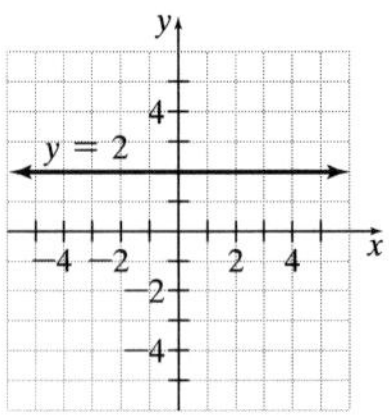

9. a) $4x - y = 4$ **b)**
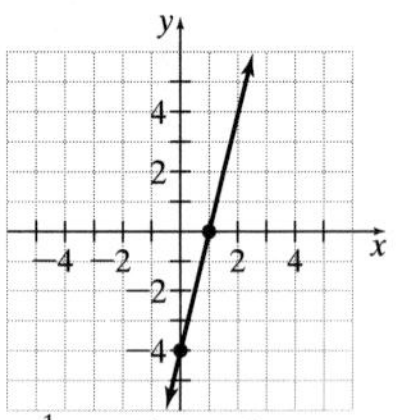

$\frac{1}{2}y = 2(x - 3) + 4$

10. $y + 1 = \frac{4}{3}(x - 2)$ **11.** (4, −2) **12.** 160 **13.** $3(p^2 + 3q)(p^2 - 4q)$ **14. a)** $x^2 + 4x + 3; x \ne 4$ **b)** 63

15. $6x^3 + 4x^2 - 34x + 24$ **16.** 7 inches; 11 inches **17.** $\frac{12}{5}$ **18.** 11.52 watts **19.** $\frac{-6 - 17i}{13}$ **20.** $\frac{-3 \pm i\sqrt{183}}{8}$

CHAPTER 9

Exercise Set 9.1 **1.** To find $(f \circ g)(x)$, substitute $g(x)$ for x in $f(x)$ **3. a)** Each y has a unique x **b)** Use the horizontal line test **5. a)** Yes; each first coordinate is paired with only one second coordinate. **b)** Yes; each second coordinate is paired with only one first coordinate. **c)** $\{(5, 3), (2, 4), (3, -1), (-2, 0)\}$; reverse each ordered pair. **7.** The domain of f is the range of f^{-1} and the range of f is the domain of f^{-1}. **9. a)** $x^2 - 8x + 21$ **b)** 5 **c)** $x^2 + 1$ **d)** 17 **11. a)** $x^2 + 4x$ **b)** 32 **c)** $x^2 + 8x + 10$ **d)** 58 **13. a)** $\dfrac{3}{x^2 + 1}$ **b)** $\dfrac{3}{17}$ **c)** $\dfrac{x^2 + 9}{x^2}$ **d)** $\dfrac{25}{16}$ **15. a)** $\sqrt{x + 5} - 4$ **b)** -1 **c)** $\sqrt{x + 1}$ **d)** $\sqrt{5}$ **17.** Not one-to-one **19.** One-to-one **21.** Yes **23.** No **25.** Yes **27.** No **29.** Yes **31.** No **33.** $f(x)$: Domain: $\{-2, -1, 2, 4, 9\}$; Range: $\{0, 3, 4, 6, 7\}$; $f^{-1}(x)$: Domain: $\{0, 3, 4, 6, 7\}$; Range: $\{-2, -1, 2, 4, 9\}$ **35.** $f(x)$: Domain: $\{-1, 1, 2, 4\}$; Range: $\{-3, -1, 0, 2\}$; $f^{-1}(x)$: Domain: $\{-3, -1, 0, 2\}$; Range: $\{-1, 1, 2, 4\}$ **37.** $f(x)$: Domain: $\{x|x \geq 2\}$; Range: $\{y|y \geq 0\}$; $f^{-1}(x)$: Domain: $\{x|x \geq 0\}$; Range: $\{y|y \geq 2\}$ **39. a)** Yes **b)** $f^{-1}(x) = x + 5$ **41. a)** Yes **b)** $h^{-1}(x) = \dfrac{x}{4}$ **43. a)** Yes **b)** $g^{-1}(x) = \dfrac{1}{x}$ **45. a)** No **47. a)** Yes **b)** $g^{-1}(x) = \sqrt[3]{x + 5}$ **49. a)** Yes **b)** $g^{-1}(x) = x^2 - 2, x \geq 0$ **51. a)** Yes **b)** $h^{-1}(x) = \sqrt{x + 4}, x \geq -4$

53. a) $f^{-1}(x) = \dfrac{x - 8}{2}$ **b)**

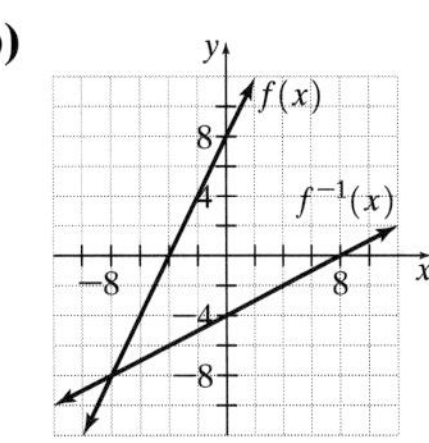

55. a) $f^{-1}(x) = x^2, x \geq 0$ **b)**

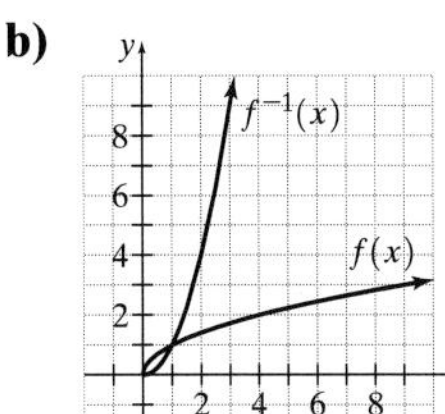

57. a) $f^{-1}(x) = x^3$ **b)**

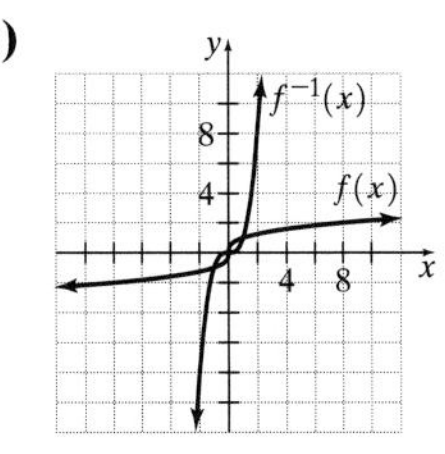

59. a) $f^{-1}(x) = \dfrac{1}{x}, x > 0$ **b)**

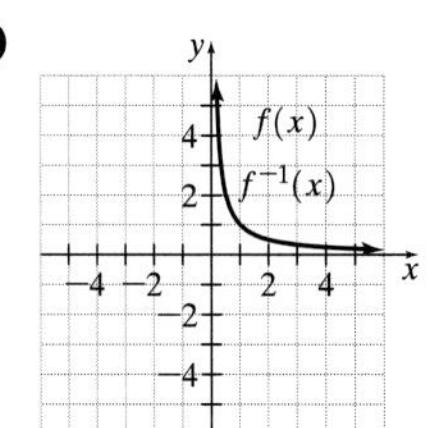

67. No, composition of functions is not commutative. Let $f(x) = x^2$ and $g(x) = x + 1$. Then $(f \circ g)(x) = x^2 + 2x + 1$, while $(g \circ f)(x) = x^2 + 1$. **69. a)** $(f \circ g)(x) = x$; $(g \circ f)(x) = x$ **b)** The Domain is $\mathbb{R}$ for all of them. **71.** The range of $f^{-1}(x)$ is the domain of $f(x)$. **73.** $f^{-1}(x) = \dfrac{x}{3}$; x is feet and $f^{-1}(x)$ is yards **75.** $f^{-1}(x) = \dfrac{9}{5}x + 32$; The inverse function converts Celsius to Fahrenheit. **77.** Yes **79.** Yes **81. a)** 6 feet **b)** $36\pi \approx 113.10$ square feet **c)** $A(t) = 4\pi t^2$ **d)** $36\pi \approx 113.10$ square feet **e)** Answers should agree. **84.** $(4, 2, -1)$ **85.** $x^2 + 4x - 2 - \dfrac{4}{x + 2}$ **86.** $\dfrac{2x\sqrt{2y}}{y}$ **87.** $x = -1 \pm \sqrt{7}$

Exercise Set 9.2 **1.** Functions of the form $f(x) = a^x, a > 0, a \neq 1$ **3. a)** As x increases, y decreases. **b)** No, $\left(\dfrac{1}{2}\right)^x$ can never be 0. **c)** No, $\left(\dfrac{1}{2}\right)^x$ is never negative. **5. a)** Same; (0, 1) **b)** $y = 3^x$ will be steeper than $y = 2^x$ for $x > 0$.

7.

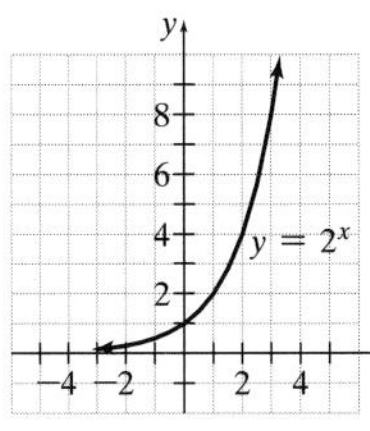

9.

11.

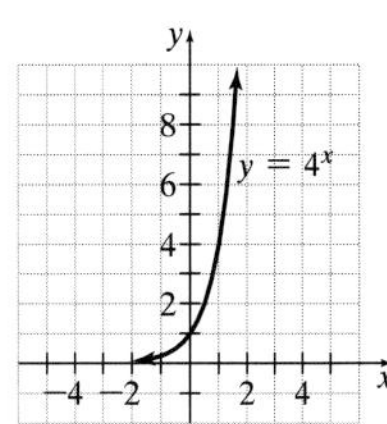

13.

15.

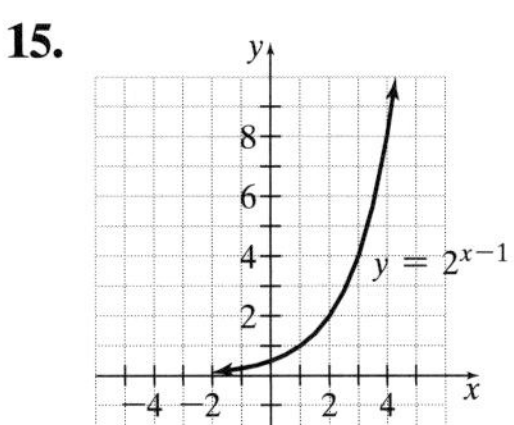

17.

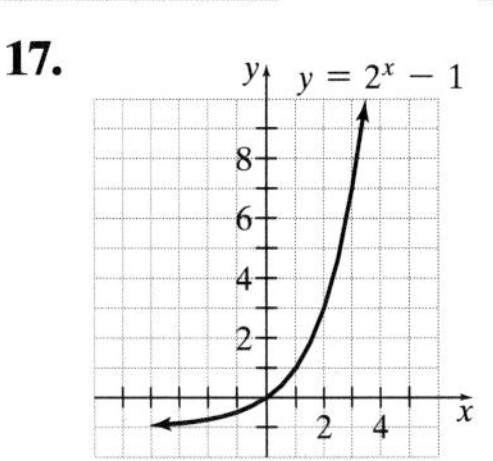

19.

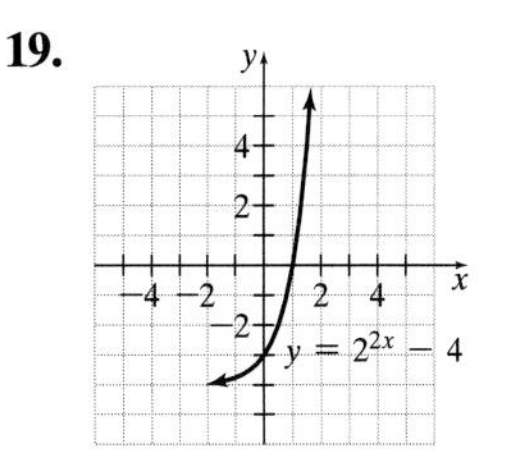

21. a) The horizontal line $y = 1$ **b)** Yes, it passes the vertical line test. **c)** No, it fails the horizontal line test. **23.** $y = a^x - k$ will be k units lower than $y = a^x$. **25.** The graph of $y = a^{x+2}$ is the graph of $y = a^x$ shifted 2 units to the left. **27. a)**

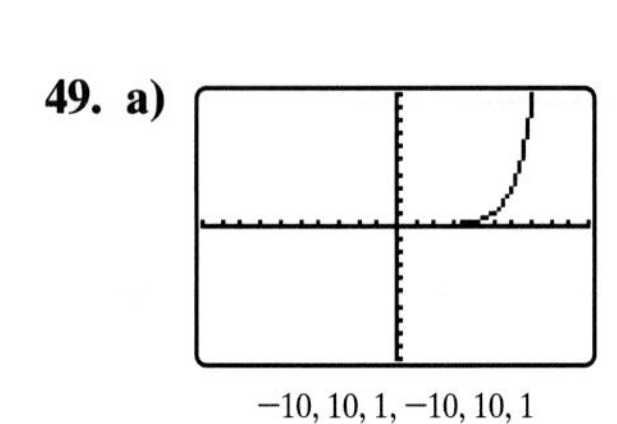

b) 200 years **c)** 40 years **29.** 256 gametes **31.** $6344.93 **33.** ≈ 10.6 grams **35. a)** 5 grams **b)** ≈ 7.28×10^{-11} grams **37. a)** 1,215,000 bacteria **b)** 10,935,000 bacteria **39. a)** Answers will vary. **b)** ≈ 56.3 million cans **41.** ≈ 8.83 km **43.** ≈ $45,425.23 **45.** $\frac{2}{3}$ **47. a)** $201.36 **b)** $31.36

49. a)

−10, 10, 1, −10, 10, 1

b) 6.26 **53. a)** $-6.2x^6y^2 + 9.2x^5y^2 + 2.3x^4y$ **b)** 8 **c)** −6.2 **54.** $x^3 + x^2 - 2x + 12$ **55.** $|a - 4|$ **56.** $\frac{2xy\sqrt[4]{xy^2z^3}}{z}$

Exercise Set 9.3

1. a) $a > 0$ and $a \neq 1$ **b)** $\{x|x \text{ is a real number and } x > 0\}$ **c)** $\mathbb{R}$

3. $\left(\frac{1}{27}, -3\right)\left(\frac{1}{9}, -2\right), \left(\frac{1}{3}, -1\right)(1, 0), (3, 1), (9, 2)$, and $(27, 3)$; the functions $f(x) = a^x$ and $g(x) = \log_a x$ are inverses.

5. The functions $y = a^x$ and $y = \log_a x$ for $a \neq 1$ are inverses of each other, thus the graphs are symmetric with respect to the line $y = x$. For each ordered pair (x, y) on the graph of $y = a^x$, the ordered pair (y, x) is on the graph of $y = \log_a x$.

7.

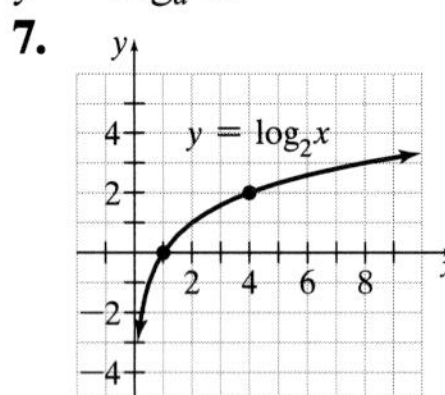

9.

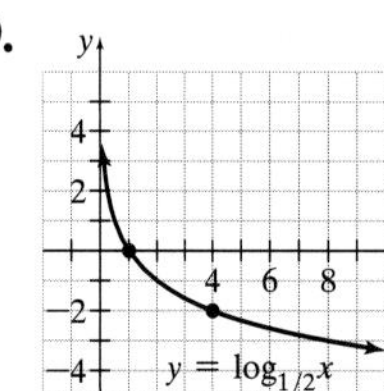

11.

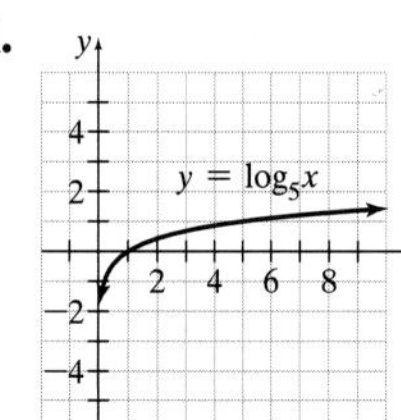

13.

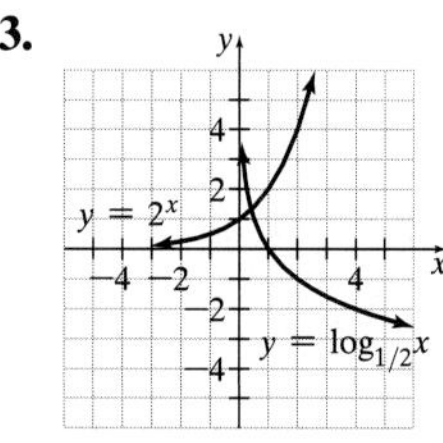

15.

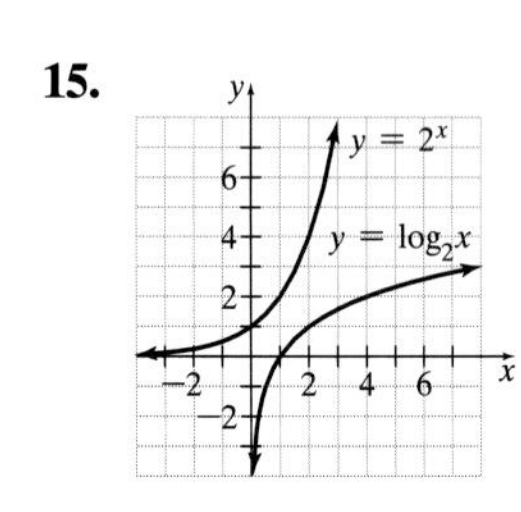

17. $\log_2 8 = 3$ **19.** $\log_9 3 = \frac{1}{2}$ **21.** $\log_{1/2} \frac{1}{32} = 5$ **23.** $\log_2 \frac{1}{8} = -3$ **25.** $\log_4 \frac{1}{64} = -3$ **27.** $\log_{16} \frac{1}{4} = -\frac{1}{2}$ **29.** $\log_8 \frac{1}{2} = -\frac{1}{3}$ **31.** $\log_{10} 5 = 0.6990$ **33.** $\log_e 7.3891 = 2$ **35.** $\log_c w = b$ **37.** $2^3 = 8$ **39.** $\left(\frac{1}{3}\right)^2 = \frac{1}{9}$ **41.** $5^{-3} = \frac{1}{125}$ **43.** $125^{1/3} = 5$ **45.** $27^{-1/3} = \frac{1}{3}$ **47.** $10^3 = 1000$ **49.** $10^{0.9031} = 8$ **51.** $e^{1.8749} = 6.52$ **53.** $r^{-a} = c$

55. 2 **57.** 32 **59.** 3 **61.** $\frac{1}{4}$ **63.** 2 **65.** 0 **67.** 4 **69.** 3 **71.** −2 **73.** 1 **75.** 0

77. 2 and 3, since 425 lies between $10^2 = 100$ and $10^3 = 1000$. **79.** 3 and 4, since 62 lies between $3^3 = 27$ and $3^4 = 81$. **81.** 2^x; Note that for $x = 10, 2^x = 1024$ while $\log_{10} x = 1$. **83.** 5 **85.** 3 **87.** 8 **89.** 9 **91.** 10,000,000

93.

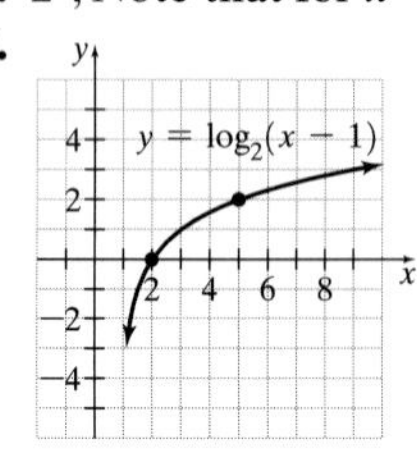

95. $2(3x + 2y)(4x - y)$ **96.** $(2a - 9)(a + 2)$ **97.** $4x^2(x + 3)(x - 3)$ **98.** $\left(2x + \frac{1}{3}\right)\left(4x^2 - \frac{2}{3}x + \frac{1}{9}\right)$

Exercise Set 9.4 **1.** Answers will vary. **3.** Answers will vary. **5.** $\log_4 6 + \log_4 9$ **7.** $\log_8 7 + \log_8(x + 3)$ **9.** $\log_6 27 - \log_6 5$ **11.** $\frac{1}{2}\log_{10} x - \log_{10}(x - 9)$ **13.** $4 \log_8 x$ **15.** $\log_{10} 3 + 2 \log_{10} 8$ **17.** $\frac{5}{2}\log_4 a - \frac{1}{2}\log_4(a + 4)$ **19.** $4 \log_{10} d - 3 \log_{10}(a + 2)$ **21.** $-2 \log_8 y + \log_8(y + 2)$ **23.** $\log_{10} 2 + \log_{10} m - \log_{10} 3 - \log_{10} n$ **25.** $\log_{10} \frac{x^2}{x - 5}$ **27.** $\log_5\left(\frac{a}{3}\right)^2$ **29.** $\log_{10} \frac{n(n - 3)}{n + 1}$ **31.** $\log_5 \sqrt{\frac{x - 4}{x}}$ **33.** $\log_9 \frac{25\sqrt[3]{r} - 6}{\sqrt{r}}$ **35.** $\log_6 \frac{3^4}{(x + 3)^2 x^4}$ **37.** 1 **39.** 0.3980 **41.** 1.3980 **43.** 10 **45.** 5 **47.** 3 **49.** 25 **51.** Yes **53.** $\log_a \frac{x}{y} = \log_a x \cdot \frac{1}{y} = \log_a x + \log_a \frac{1}{y}$ **55.** $\log_a(x - 2)$ **57.** Yes **59.** 0.8640 **61.** 0.1080 **63.** 0.7000 **65.** No **67.** $\frac{1}{4}\log_2 x + \frac{1}{4}\log_2 y + \frac{1}{3}\log_2 a - \frac{1}{5}\log_2(a - b)$ **72.** $\frac{(2x + 5)^2}{(x - 4)^2}$ **73.** $\frac{(3x + 1)(x + 9)}{(x - 4)(x - 3)(2x + 5)}$ **74.** $2\frac{2}{9}$ days **75.** $2x^3y^5\sqrt[3]{6x^2y^2}$

Exercise Set 9.5 **1.** Common logarithms are logarithms with base 10. **3.** Antilogarithms are numbers obtained by taking 10 to the power of the logarithm. **5.** 1.6532 **7.** 4.2833 **9.** −4.0670 **11.** 2.0000 **13.** 0.5740 **15.** −3.3261 **17.** 4.29 **19.** 42,500 **21.** 0.0874 **23.** 1.00 **25.** 317 **27.** 0.701 **29.** 100 **31.** 0.00789 **33.** 33,100 **35.** 0.0871 **37.** 0.428 **39.** 0.404 **41.** 3.5514 **43.** −1.1385 **45.** 2.0086 **47.** −2.8928 **49.** 273 **51.** 0.695 **53.** 0.0246 **55.** 22.4 **57.** 0 **59.** −1 **61.** −2 **63.** −3 **65.** 5 **67.** 7 **69.** 31.2 **71.** 47 **73.** No; $10^2 = 100$ and since $462 > 100$, log 462 must be greater than 2. **75.** No; $10^0 = 1$ and $10^{-1} = 0.1$ and since $1 > 0.163 > 0.1$, log 0.163 must be between 0 and −1. **77.** No; $\log \frac{y}{3x} = \log y - \log 3 - \log x$ **79.** 2.0970 **81.** Not possible **83.** −1.3979 **85. a)** 4.08 **b)** 19,500 **87. a)** \$30 million **b)** \$25.48 million **89. a)** 6.31×10^{20} **b)** 2.19 **91.** ≈ 6.2 **94.** $\frac{-2 \pm 2i\sqrt{5}}{3}$ **95.** 10 mph

96.

97.

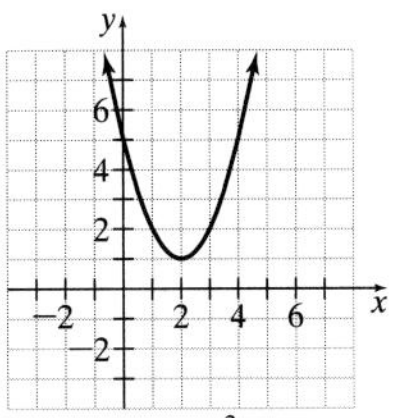

$y = (x - 2)^2 + 1$

Exercise Set 9.6 **1.** $a^m = a^n$, $\log m = \log n$ **3.** $r = s$ **5.** Check for extraneous roots. **7.** 3 **9.** $-\frac{1}{2}$ **11.** 2 **13.** 3 **15.** 2.01 **17.** 3.16 **19.** 5.59 **21.** 6.34 **23.** 3 **25.** $\frac{1}{25}$ **27.** −1 **29.** 3 **31.** 0, −8 **33.** $\frac{13}{2}$ **35.** $\frac{3}{2}$ **37.** 4 **39.** 2 **41.** 1.05 **43.** 20 **45.** 5 **47.** 3 **49.** 2 **51.** 3 **53.** 9 **55.** \$1605.87 **57.** 139 **59. a)** 26 **b)** 7.18 **61.** \$7112.09 **63.** 19.36 **65. a)** 1,000,000,000,000 **b)** 100,000 times greater **67.** 8 **69.** $x = 2$ and $x = 1$ **71.** (3, 1) **73.** (54, 46) **75.** 2.8 **77.** No solution **79. a)** $\{x|x > 40\}$ **b)** $(40, \infty)$ **80.** b) and c) are functions; only b) is one-to-one. **81.** $2x^3 - 11x^2 + 18x - 9$ **82.** $2x + 3 + \frac{3}{x + 4}$

Exercise Set 9.7 **1. a)** e **b)** 2.7183 **3.** $\{x|x$ is a real number and $x > 0\}$ **5.** $\log_a x = \frac{\log_b x}{\log_b a}$ **7.** x **9.** $k > 0$ **11.** 3.5553 **13.** 5.7104 **15.** 4.95 **17.** 0.0659 **19.** 4.0943 **21.** −3.0791 **23.** 2.9300 **25.** 5.0000 **27.** 7.9010 **29.** −4.8411 **31.** 4 **33.** 4 **35.** 3 **37.** $P \approx 5383.43$ **39.** $P_0 \approx 58.09$ **41.** $t \approx 0.7847$ **43.** $k \approx 0.2310$ **45.** $k \approx -0.2888$ **47.** $A \approx 4719.77$ **49.** $V_0 = \frac{V}{e^{kt}}$ **51.** $t = \frac{\ln P - \ln 150}{4}$

53. $k = \dfrac{\ln A - \ln A_0}{t}$ **55.** $y = xe^{2.3}$ **57.** $y = (x + 3)e^6$ **59.** $x = \ln 12.183 \approx 2.5000$ **61. a)** \$5867.55 **b)** 8.66 years **63. a)** 19.60 thousand tons **b)** 2.16 thousand tons **65. a)** 86.47% **b)** 34.66 days **67. a)** 5.15 feet/second **b)** 5.96 feet/second **c)** 646,000 **69. a)** $\approx$ 6.56 billion **b)** $\approx$ 52.12 years **71. a)** 6626.62 years **b)** 5752.26 years **73.** $\approx$ 11.55% **75. a)** 40.94 watts **b)** 804.72 days **77.** 1.45×10^9 years **81.** 1.099 **83.** $I_0 = Ie^{x/k}$ **85.** $Q = \dfrac{M}{1 + M}$ **87.** $16y^8$ **88.** $\dfrac{12 - 8x}{x^3}$ **89.** $q = \dfrac{fp}{p - f}$ **90.** $4x^2y^3z^4\sqrt[3]{2xz}$

Review Exercises

1. $4x^2 - 26x + 44$ **2.** 8 **3.** $2x^2 - 6x + 3$ **4.** 39 **5.** $3\sqrt{x - 4} + 2$ **6.** $\sqrt{3x - 2}, x \geq \dfrac{2}{3}$ **7.** One-to-one **8.** Not one-to-one **9.** One-to-one **10.** Not one-to-one **11.** One-to-one **12.** Not one-to-one **13.** $f(x)$: Domain: $\{-4, 0, 5, 6\}$; Range: $\{-3, 2, 3, 7\}$; $f^{-1}(x)$: Domain: $\{-3, 2, 3, 7\}$; Range: $\{-4, 0, 5, 6\}$ **14.** $f(x)$: Domain: $\{x|x \geq 0\}$; Range: $\{y|y \geq 2\}$; $f^{-1}(x)$: Domain: $\{x|x \geq 2\}$; Range: $\{y|y \geq 0\}$

15. $f^{-1}(x) = \dfrac{x + 2}{4}$;

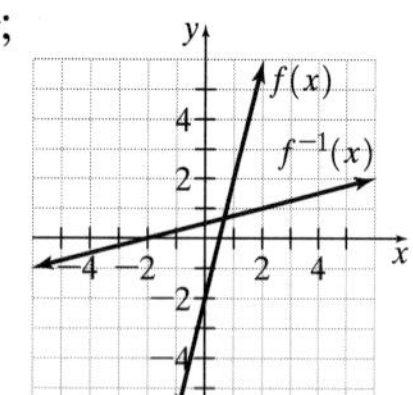

16. $f^{-1}(x) = x^3 + 1$;

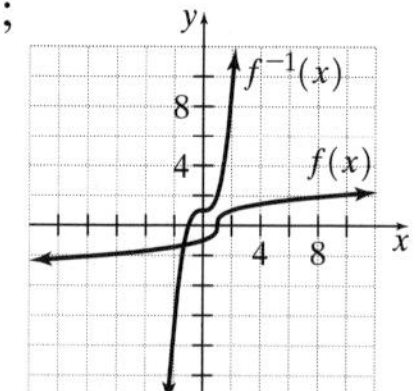

17.

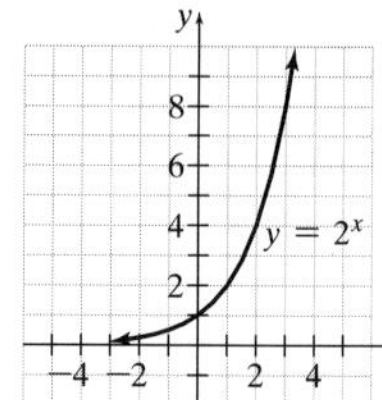

18.

19. a) Linear **b)** Exponential **c)** 8 **d)** 1 year **e)** Less than 1 year **f)** Mid-1994 **20.** $\log_4 16 = 2$ **21.** $\log_8 2 = \dfrac{1}{3}$ **22.** $\log_6 \dfrac{1}{36} = -2$ **23.** $5^2 = 25$ **24.** $\left(\dfrac{1}{3}\right)^2 = \dfrac{1}{9}$ **25.** $3^{-2} = \dfrac{1}{9}$ **26.** $x = 4^3$; 64 **27.** $a^3 = 8$; 2 **28.** $x = \left(\dfrac{1}{4}\right)^{-3}$; 64

29.

30.

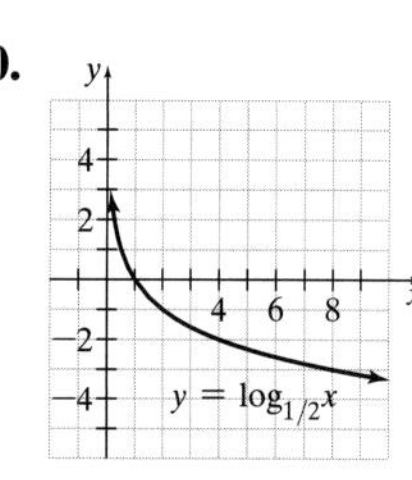

31. $\dfrac{1}{2}\log_8 12$ **32.** $5\log(x - 8)$ **33.** $\log 2 + \log(x - 3) - \log x$ **34.** $4\log x - \log 39 - \log(2x + 8)$ **35.** $\log \dfrac{x^2}{(x + 1)^3}$ **36.** $\log \dfrac{(2x)^3}{y}$ **37.** $\ln \dfrac{\sqrt{\frac{x}{x+2}}}{2}$ **38.** $\ln \dfrac{x^3\sqrt{x + 1}}{(x + 4)^3}$ **39.** 9 **40.** 5 **41.** 6 **42.** 2 **43.** 3.9138 **44.** −3.1451 **45.** 829 **46.** 0.0422 **47.** 214 **48.** 0.0594 **49.** 4 **50.** 3 **51.** 31.5 **52.** 5.1 **53.** 2 **54.** $-\dfrac{1}{2}$ **55.** 1 **56.** 5 **57.** 2.605 **58.** 4.497 **59.** 1.595 **60.** 2.240 **61.** 123 **62.** 2 **63.** 1 **64.** 3 **65.** $t \approx 1.1552$ **66.** $A_0 \approx 352.54$ **67.** $t = \dfrac{\ln A - \ln A_0}{k}$ **68.** $k = \dfrac{\ln 0.25}{t}$ **69.** $y = xe^2$ **70.** $y = 5x + 2$ **71.** 6.1092 **72.** 3.5609 **73.** \$25,723.07 **74.** 566.4 mg **75. a)** 92.88 minutes **b)** 118.14 minutes **76. a)** 72 **b)** $\approx$ 63.4 **c)** $\approx$ 4 months **77.** 9.10 lb/in^2 **78.** 9.90 years

Practice Test

1. a) Yes **b)** $\{(2, 4), (8, -3), (3, -1), (7, 5)\}$ **2. a)** $x^2 + 6x + 5$ **b)** 32 **3. a)** $\sqrt{x^2 + 1}$ **b)** $\sqrt{37}$ **4. a)** $f^{-1}(x) = -\dfrac{1}{3}(x + 5)$ **b)**

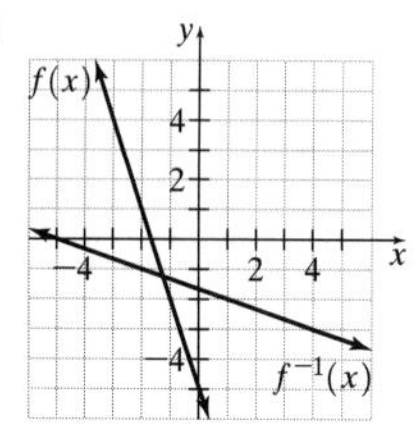

5. $f^{-1}(x) = x^2 + 1, x \geq 0$ **b)**

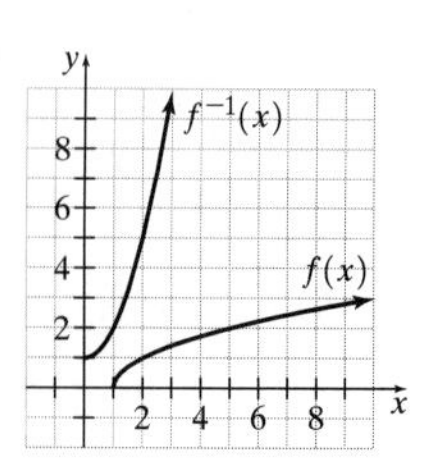

6. $\{x|x$ is a real number and $x > 0\}$ **7.** −2

8. 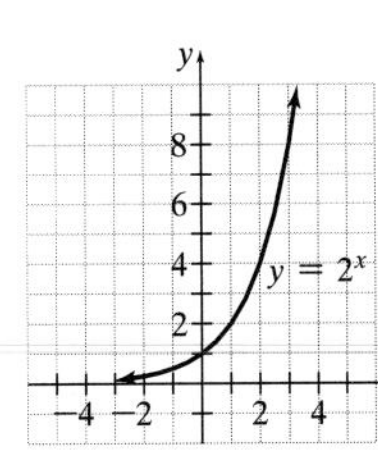

9. 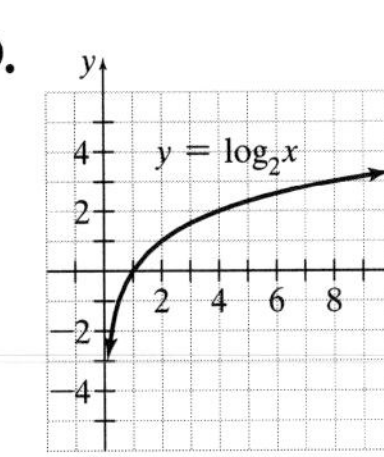

10. $\log_4 \frac{1}{64} = -3$ **11.** $3^5 = 243$ **12.** $2^4 = x; 16$ **13.** $3 = 27^y; \frac{1}{3}$ **14.** $\log_3 x + \log_3(x - 4) - 2\log_3 x$ **15.** $\log_8 \frac{(x-4)^3(x+1)^2}{\sqrt{x}}$ **16.** 1 **17. a)** 3.6646 **b)** −3.1952 **18.** 4.38 **19.** 3 **20.** $\frac{17}{5}$ **21.** 33.7844 **22.** 2.0588 **23.** 30.5430 **24.** $6349.06 **25.** 3364.86 years old

Cumulative Review Test

1. −1 **2.** $\frac{8y^{21}z^6}{x^9}$ **3. a)** 2 hours **b)** 10 miles **4.** $x \le 4$
5. Domain: $\{x|x \text{ is a real number}, -1 \le x \le 4\}$; Range: $\{y|y \text{ is a real number}, -3 \le y \le 2\}$
6. a) $\frac{7}{3}$ **b)** $7x - 3y = -5$ **7.** 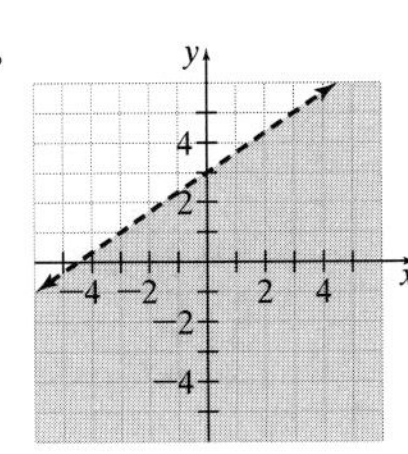

8. $(2, 4)$ **9.** $(1, 5)$ **10.** 41 **11.** $(4x - 3y)(3x + y)$ **12.** $-\frac{5}{3}, 4$ **13.** $2x + 7 + \frac{2}{x - 4}$ **14.** −1 **15.** $\frac{x + 1}{x^2 - x}$
16. $\frac{3x\sqrt[3]{25y}}{5y}$ **17.** −13 **18.** 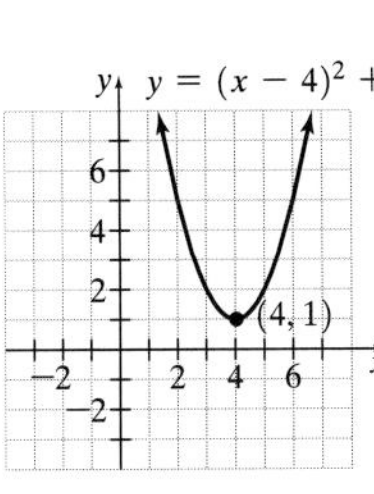

19. $[0, 1] \cup [5, \infty)$ **20. a)** 111.82 grams **b)** 24.8 years

CHAPTER 10

Exercise Set 10.1

1. Parabola, circle, ellipse, and hyperbola; for an illustration, see page 622.
3. Yes, because each value of x corresponds to only one value of y. The domain is ℝ, and the range is $\{y|y \ge k\}$.
5. The graphs have the same vertex, $(3, 4)$. The first graph opens upward, and the second graph opens downward.
7. The distance is always a positive number because both distances are squared and we use the principal square root.
9. A circle is the set of all points in a plane that are the same distance from a fixed point.

11. 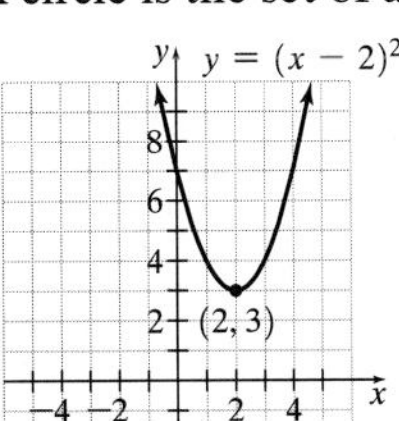

13. 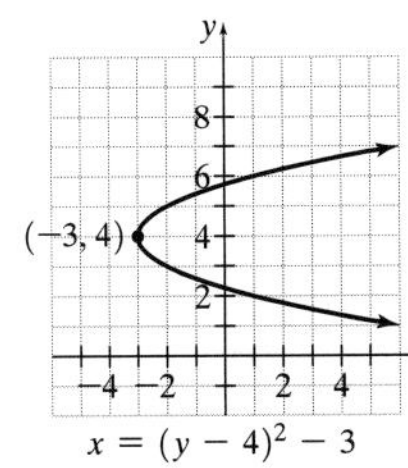

15. 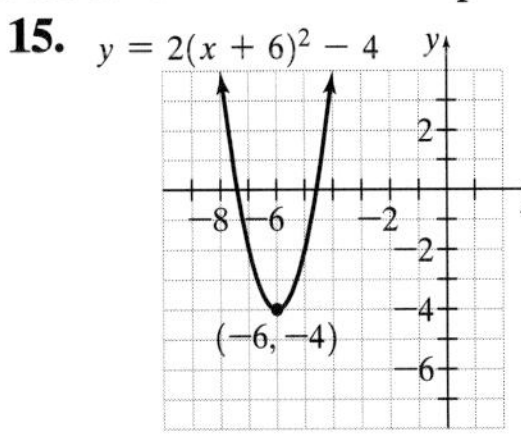

17. 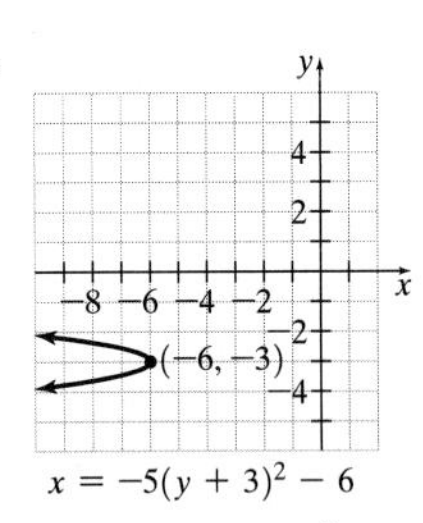

19. 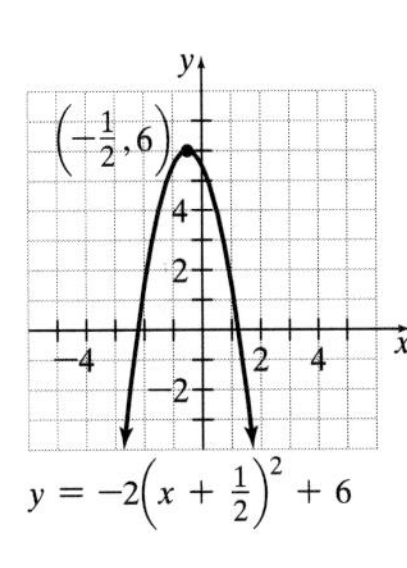

21. a) $y = (x + 1)^2 - 1$ **b)** 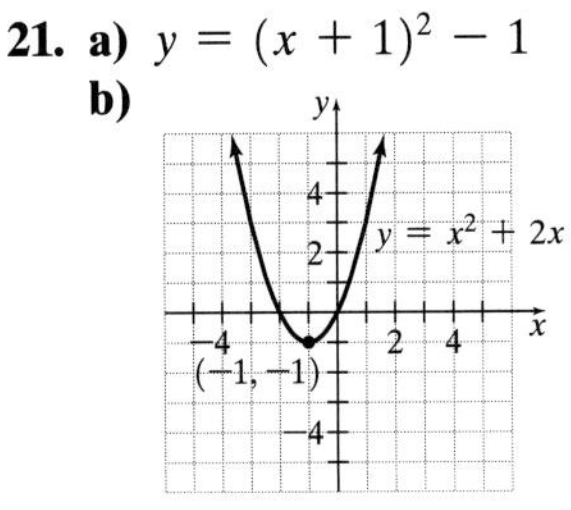

23. a) $x = (y + 3)^2 - 9$ **b)** 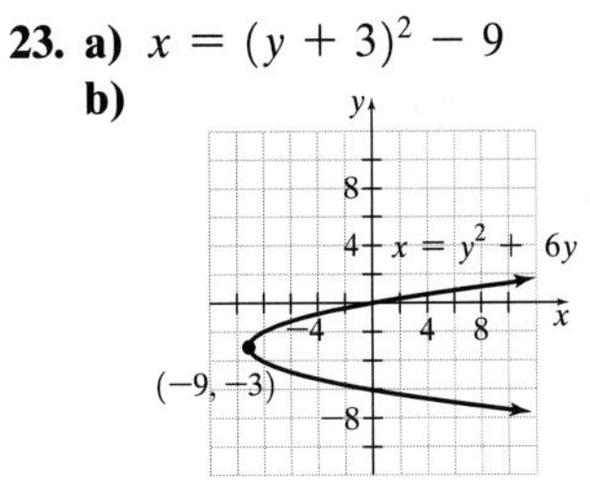

25. a) $y = (x + 1)^2 - 16$ **b)** 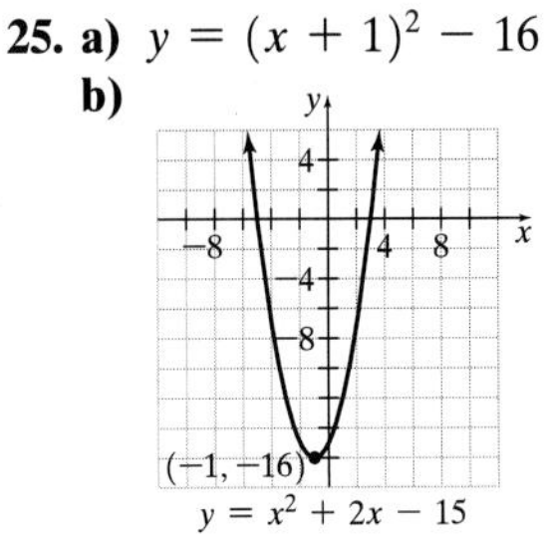

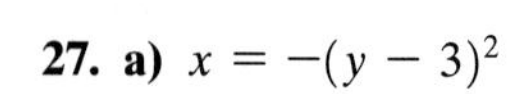

27. a) $x = -(y - 3)^2$
b)

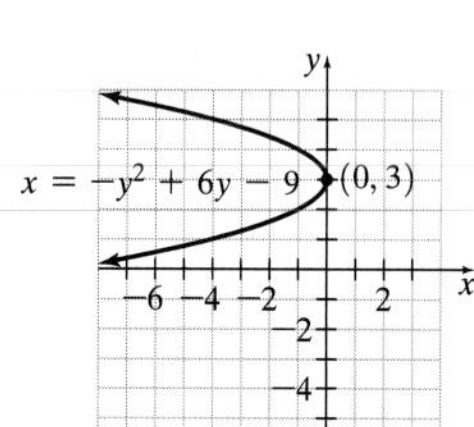

29. a) $y = \left(x + \frac{7}{2}\right)^2 - \frac{9}{4}$
b)

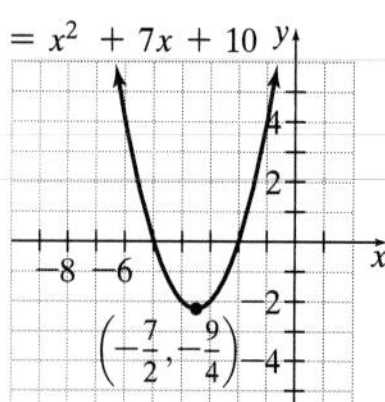

31. a) $y = 2(x - 1)^2 - 6$
b)

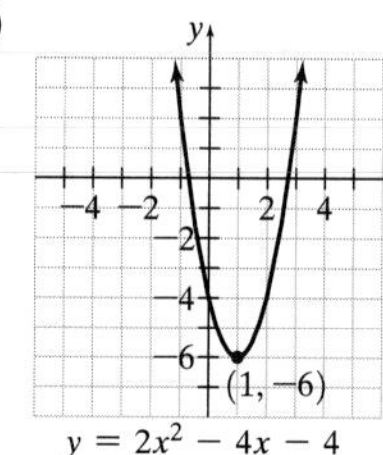

33. 3 **35.** 9
37. $\sqrt{90} \approx 9.49$
39. $\sqrt{74} \approx 8.60$
41. $\sqrt{34.33} \approx 5.86$
43. $\sqrt{\frac{281}{16}} \approx 4.19$
45. $\sqrt{7} \approx 2.65$

47. $\sqrt{15 - 2\sqrt{6} - 2\sqrt{21}} \approx 0.97$ **49.** $\left(\frac{3}{2}, 5\right)$ **51.** $(2, 1)$ **53.** $\left(\frac{5}{2}, 2\right)$ **55.** $(-3.05, 9.575)$ **57.** $\left(\frac{9}{4}, \frac{15}{4}\right)$

59. $\left(\frac{\sqrt{3} + \sqrt{5}}{2}, \frac{5}{2}\right)$ **61.** $x^2 + y^2 = 9$ **63.** $(x - 3)^2 + y^2 = 1$ **65.** $(x + 6)^2 + (y - 5)^2 = 25$

67. $(x - 4)^2 + (y - 7)^2 = 8$ **69.** $x^2 + y^2 = 16$ **71.** $(x - 3)^2 + (y + 2)^2 = 9$

73.

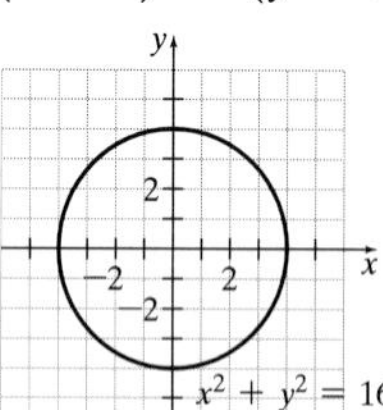

75.

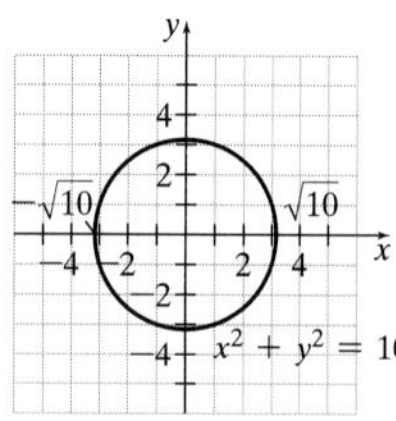

77.

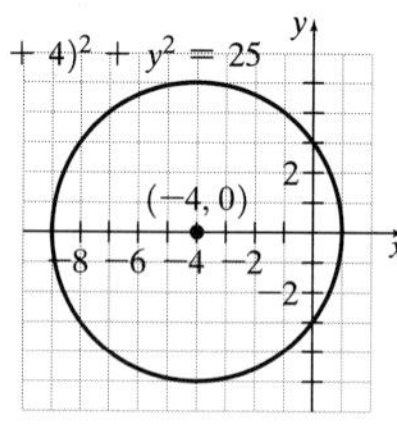

79.

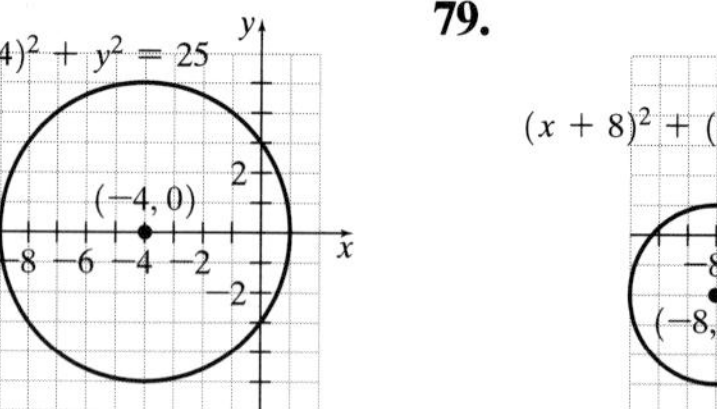

81.

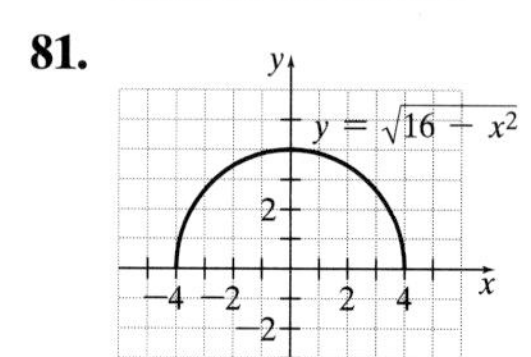

83.

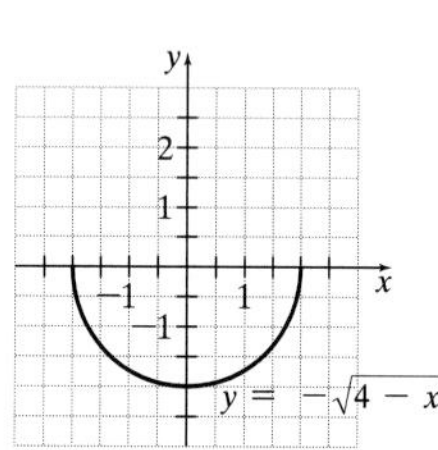

85. a) $x^2 + (y + 5)^2 = 10^2$
b)

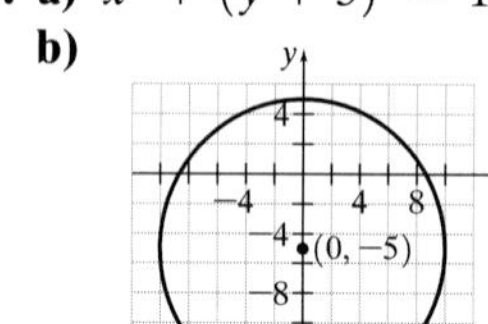

87. a) $(x + 4)^2 + y^2 = 5^2$
b)

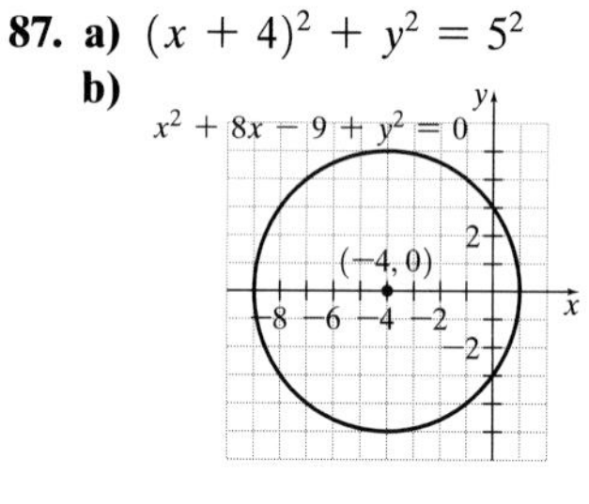

89. a) $(x + 1)^2 + (y - 2)^2 = 3^2$
b)

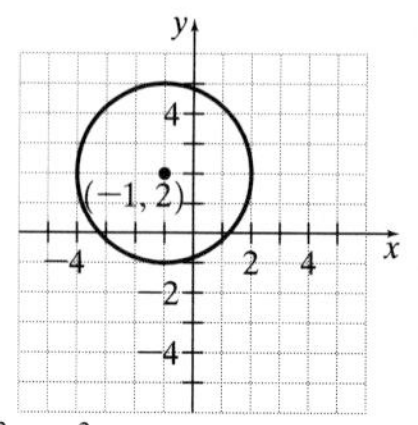

91. a) $(x + 3)^2 + (y - 1)^2 = 2^2$
b)

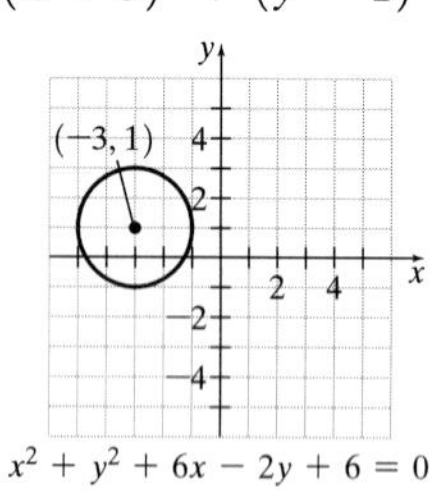

93. a) $(x - 4)^2 + (y + 1)^2 = 2^2$
b)

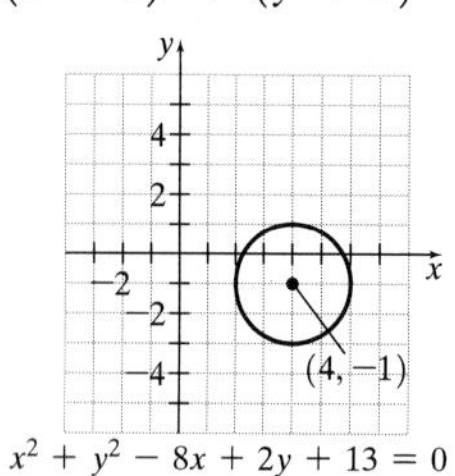

95. x-intercept: $(-7, 0)$; y-intercepts: $(0, -1)$, $(0, 7)$
97. x-intercept: $(56, 0)$; no y-intercept

99. No. For example, the origin is the midpoint of both the segment from $(1, 1)$ to $(-1, -1)$ and the segment from $(2, 2)$ to $(-2, -2)$, but these segments have different lengths. **101.** 10 **103.** $(x + 5)^2 + (y - 2)^2 = 2^2$ **105. a)** $2\sqrt{2}$ **b)** $(7, 6)$ **c)** $(x - 7)^2 + (y - 6)^2 = 8$ **107.** 4; 0 **109. a)** 13.6 feet **b)** 81.8 feet **c)** $x^2 + (y - 81.8)^2 = 68.2^2$ **111. a)** $x^2 + y^2 = 16$ **b)** $(x - 2)^2 + y^2 = 4$ **c)** $(x + 2)^2 + y^2 = 4$ **d)** 8π

113. 2 **115.** $1 + \sqrt{116} \approx 11.77$ feet **117.** $y = -\frac{1}{2}x + 1$ **118.** 128 **119.** $\frac{13}{9}$ **120.** 15

Exercise Set 10.2

1. An ellipse is a set of points in a plane, the sum of whose distances from two fixed points is constant. **3.** $\dfrac{(x-h)^2}{a^2}+\dfrac{(y-k)^2}{b^2}=1$ **5.** If $a=b$, the formula for a circle is obtained.

7.

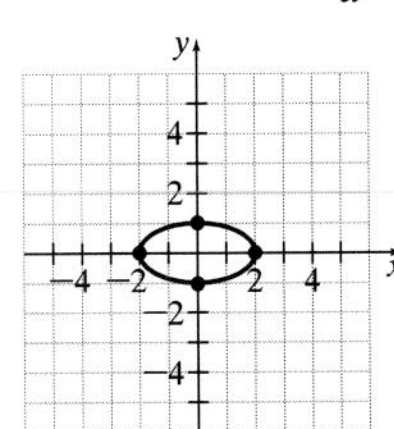

9.

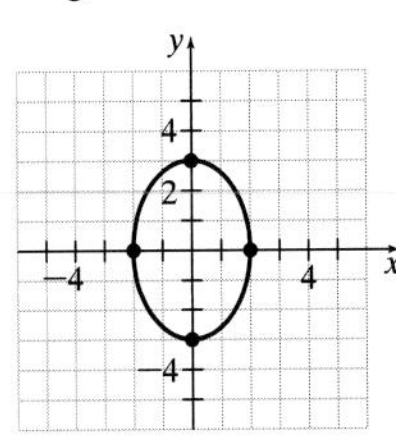

11.

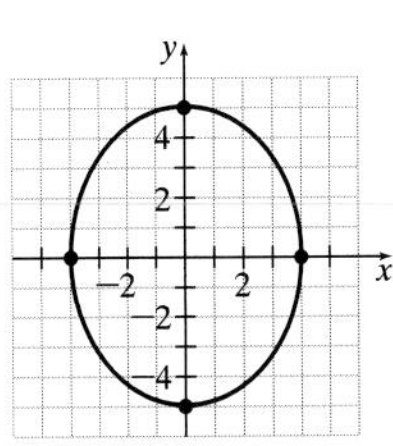

13.

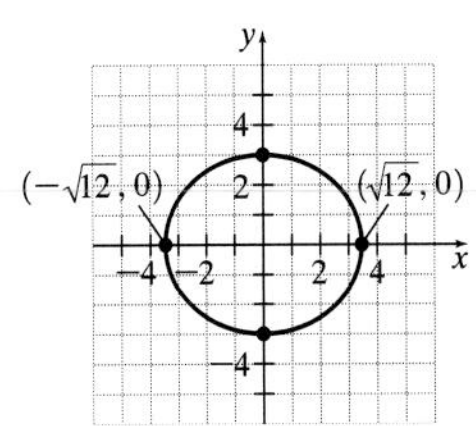

15.

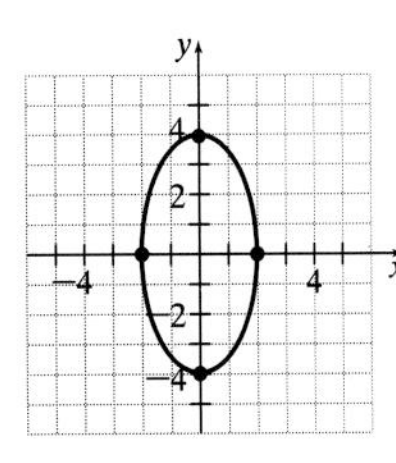

17.

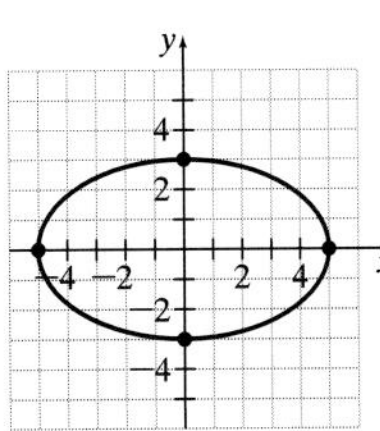

19.

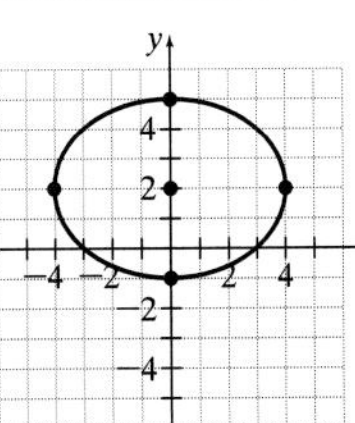

21.

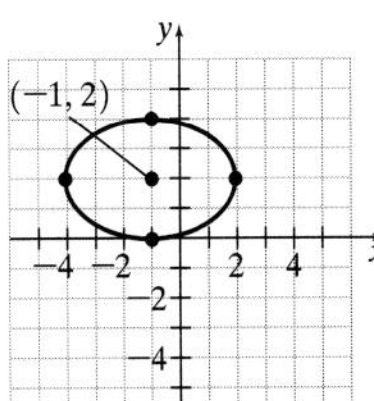

23.

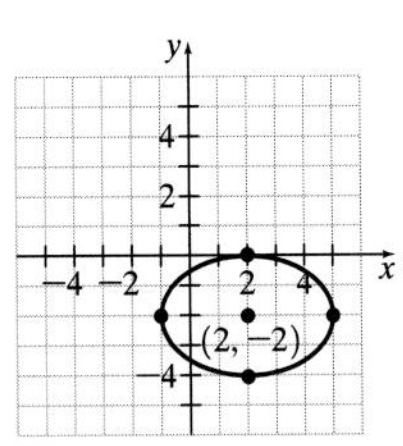

25.

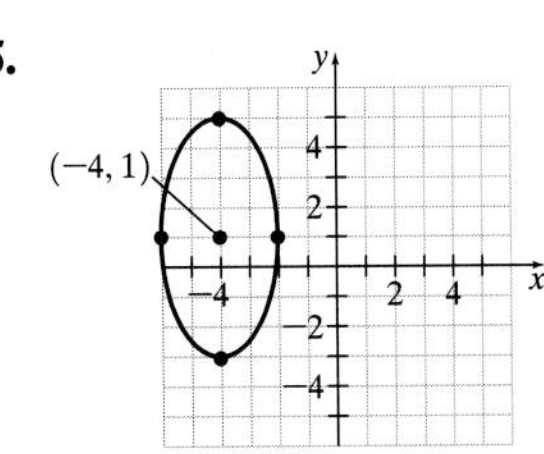

27. There is only one point, at $(0, 0)$. **29.** $\dfrac{x^2}{4}+\dfrac{y^2}{9}=1$

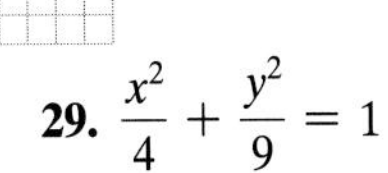

31. There are no points of intersection, because the ellipse with $a=2$ and $b=3$ is completely inside the circle of radius 4.

33. $\dfrac{(x-2)^2}{100}+\dfrac{(y-1)^2}{25}=1$; $(2, 1)$ **35.** 69.5 feet

37. $\sqrt{5}\approx 2.24$ feet, in both directions, from the center of the ellipse along the major axis **39.** Answers will vary.

43. $\dfrac{(x+1)^2}{9}+\dfrac{(y-3)^2}{4}=1$ **46.** −4 14 **47.** −2 6 **48.**

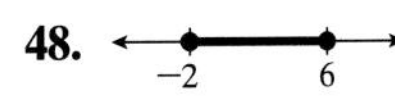

49.

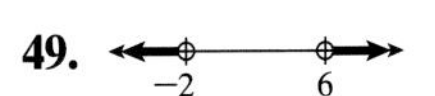

Exercise Set 10.3

1. A hyperbola is the set of points in a plane the differences of whose distances from two fixed points is a constant. **3.** The graph of $\dfrac{x^2}{a^2}-\dfrac{y^2}{b^2}=1$ is a hyperbola with vertices at $(a, 0)$ and $(-a, 0)$. Its transverse axis lies along the x-axis. The asymptotes are $y=\pm\dfrac{b}{a}x$.

5. a) $y=\pm\dfrac{1}{2}x$

b)

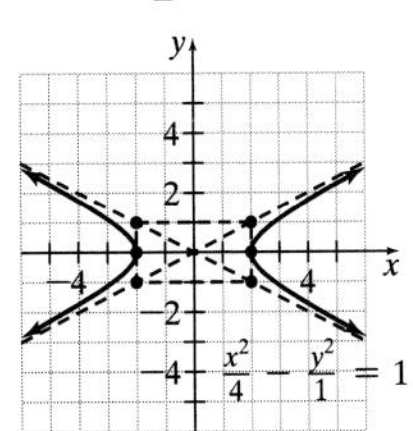

7. a) $y=\pm\dfrac{3}{4}x$

b)

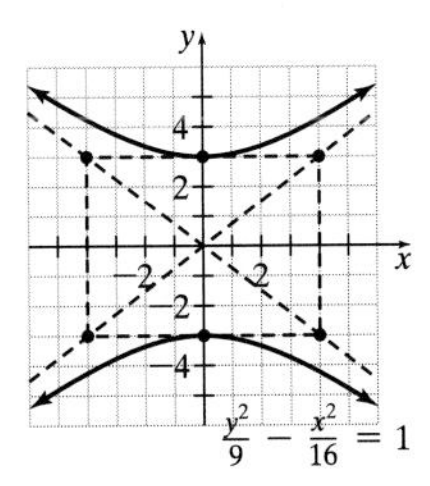

9. a) $y=\pm\dfrac{5}{6}x$

b)

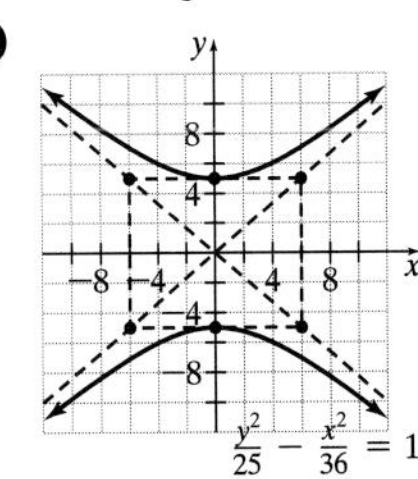

11. a) $y=\pm x$

b)

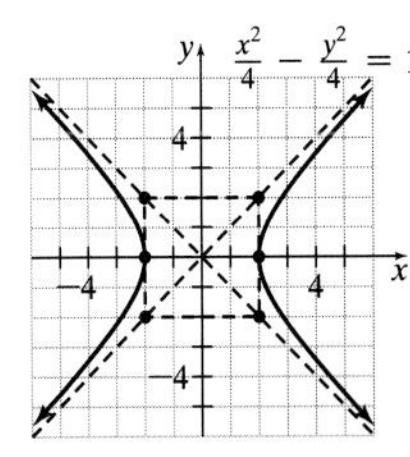

13. a) $y=\pm\dfrac{4}{9}x$

b)

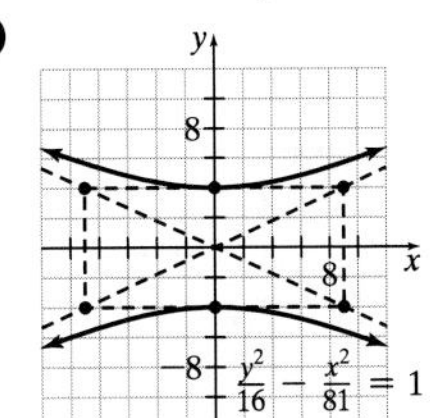

15. a) $y=\pm\dfrac{5}{4}x$

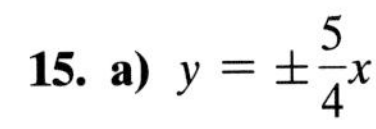

b)

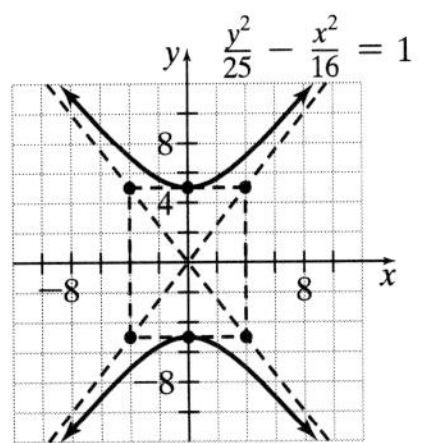

17. a) $\dfrac{x^2}{4}-\dfrac{y^2}{16}=1$; $y=\pm 2x$

b)

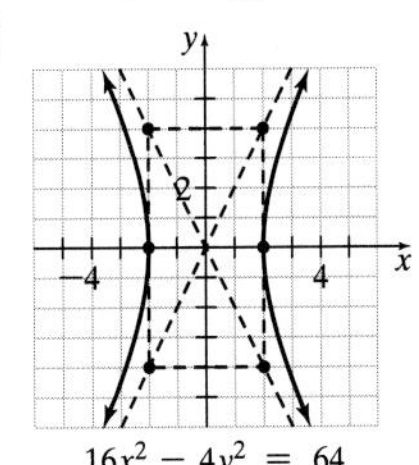

19. a) $\frac{y^2}{1} - \frac{x^2}{9} = 1; y = \pm\frac{1}{3}x$ **21. a)** $\frac{y^2}{36} - \frac{x^2}{4} = 1; y = \pm 3x$ **23. a)** $\frac{x^2}{9} - \frac{y^2}{25} = 1; y = \pm\frac{5}{3}x$

b)

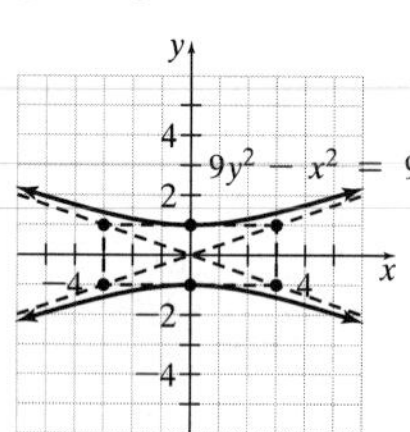

b)

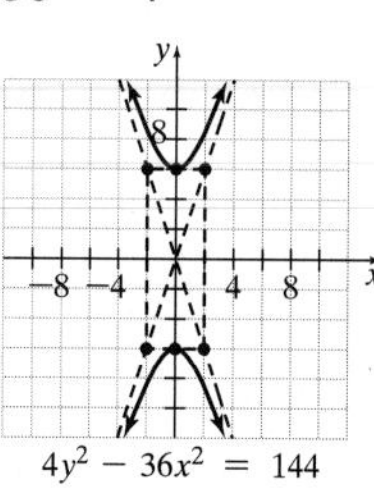

b)

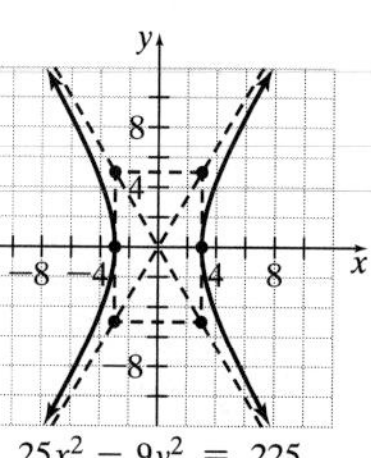

25. Parabola **27.** Hyperbola **29.** Parabola **31.** Hyperbola **33.** Ellipse **35.** Parabola **37.** Circle **39.** Circle **41.** Hyperbola **43.** $\frac{y^2}{4} - \frac{x^2}{16} = 1$ **45.** $\frac{x^2}{9} - \frac{y^2}{25} = 1$; No, $\frac{x^2}{18} - \frac{y^2}{50} = 1$ and others will also work.
47. No, for certain values of x, there are 2 possible values of y. **49.** Domain: $(-\infty, -3] \cup [3, \infty)$; Range: $\mathbb{R}$
51. Both graphs have a transverse axis along the x-axis. The vertices of the second graph will be closer to the origin, at $(\pm b, 0)$ instead of $(\pm a, 0)$. The second graph will open wider. **55.** 48 **56.** The domain of a function is the set of values for the independent variable. The range is the set of values obtained for the dependent variable.

57.

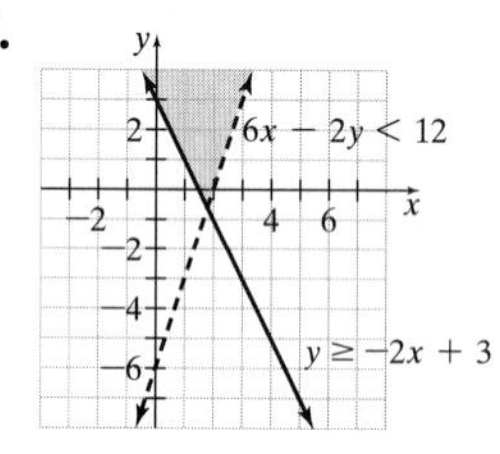

58. $\frac{3 \pm i}{2}$

Exercise Set 10.4

1. A nonlinear system of equations is a system in which at least one equation is nonlinear.
3. $(3, 0), \left(-\frac{9}{5}, \frac{12}{5}\right)$ **5.** $(-4, 11), \left(\frac{5}{2}, \frac{5}{4}\right)$ **7.** $(2, -4), (-14, -20)$ **9.** No real solution
11. $(0, -3), (\sqrt{5}, 2), (-\sqrt{5}, 2)$ **13.** $(2, 0), (-2, 0)$ **15.** $(3, 2), (3, -2), (-3, 2), (-3, -2)$ **17.** $(3, 0), (-3, 0)$
19. $(\sqrt{6}, \sqrt{3}), (\sqrt{6}, -\sqrt{3}), (-\sqrt{6}, \sqrt{3}), (-\sqrt{6}, -\sqrt{3})$ **21.** No real solution
23. $(\sqrt{5}, 2), (\sqrt{5}, -2), (-\sqrt{5}, 2), (-\sqrt{5}, -2)$ **25.** Answers will vary. **27.** 10 feet by 24 feet **29.** 8 cm by 14 cm
31. 8 inches by 15 inches **33.** 6 feet by 8 feet or 4 feet by 12 feet **35.** $\approx$ 1.67 seconds
37. Principal: \$125; Rate: 6% **39.** $\approx$ 16 and $\approx$ 184 **41.** $\approx$ 26 and $\approx$ 174 **43.** $(-1, -3), (3.12, -0.53)$
45. 10 yards, 24 yards **47.** Parentheses, exponents, multiplication or division, addition or subtraction
48. $y = \frac{2x + 40}{11}$ **49.** \$5849.29 **50.** $y \leq 2$ **51.**

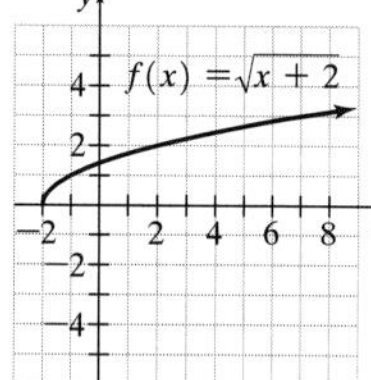

52. 6

Review Exercises

1. 5; $\left(\frac{3}{2}, -2\right)$ **2.** 5; $\left(4, \frac{1}{2}\right)$ **3.** 13; $\left(\frac{1}{2}, 3\right)$ **4.** $\sqrt{8} \approx 2.83$; $(-3, 4)$

5.

6.

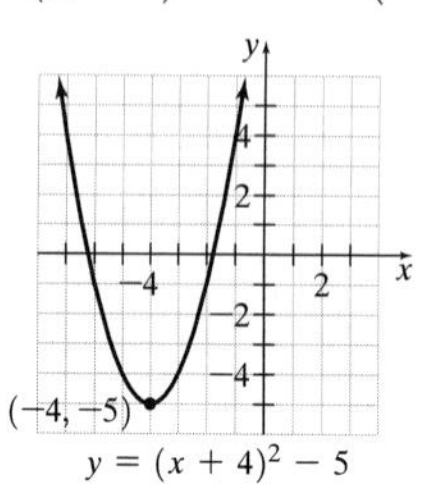

7.

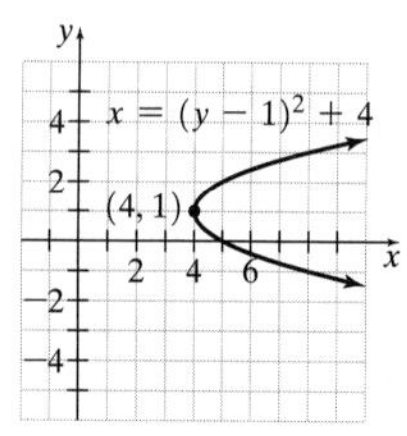

8.

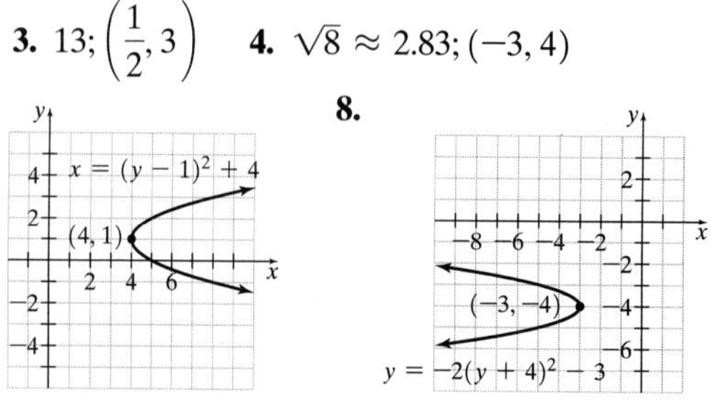

9. a) $y = (x - 3)^2 - 9$

b)

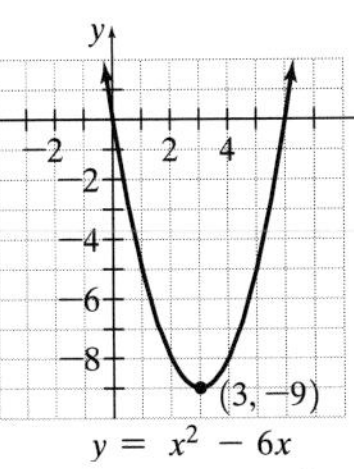

10. a) $x = -(y + 1)^2 + 9$

b)

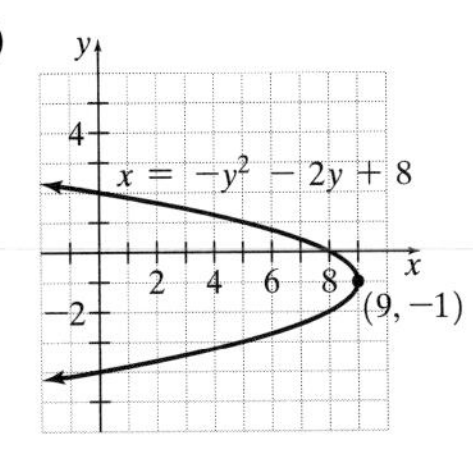

11. a) $x = \left(y + \frac{5}{2}\right)^2 - \frac{9}{4}$

b)

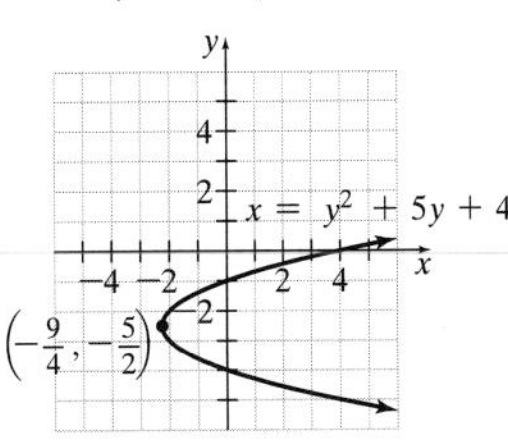

12. a) $y = 2(x - 2)^2 - 32$

b)

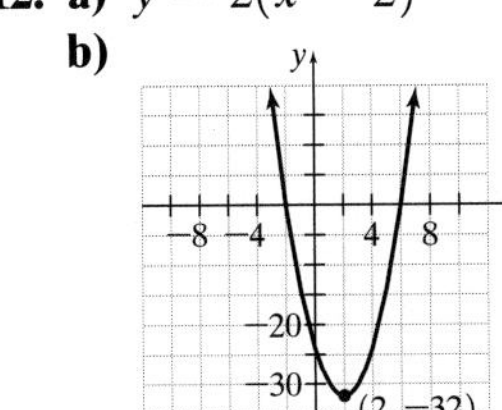

13. a) $x^2 + y^2 = 5^2$

b)

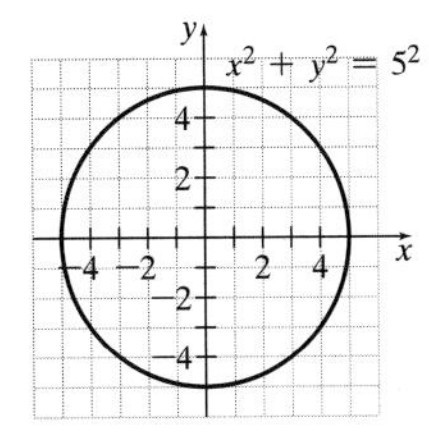

14. a) $(x + 3)^2 + (y - 4)^2 = 3^2$

b)

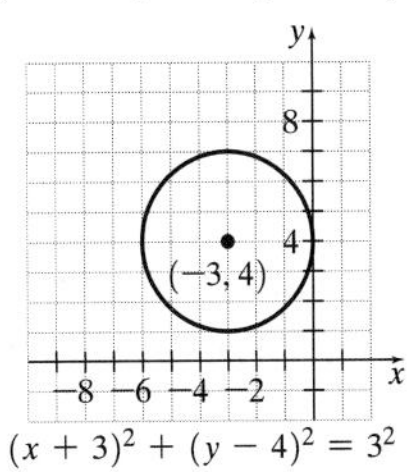

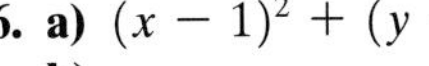

15. a) $x^2 + (y - 2)^2 = 2^2$

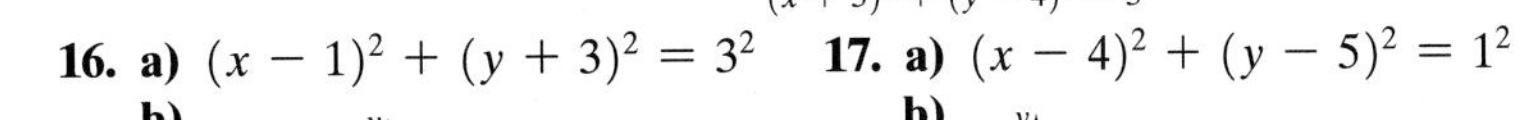

16. a) $(x - 1)^2 + (y + 3)^2 = 3^2$

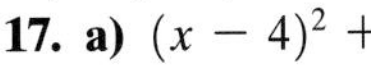

17. a) $(x - 4)^2 + (y - 5)^2 = 1^2$

15. b)

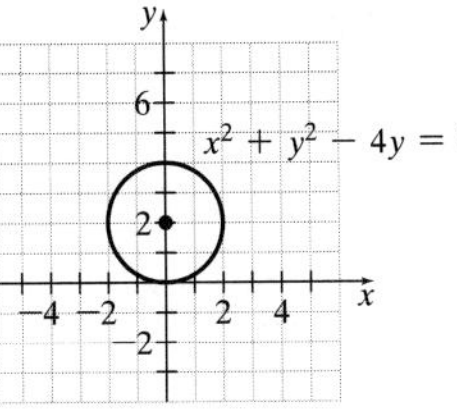

16. b)

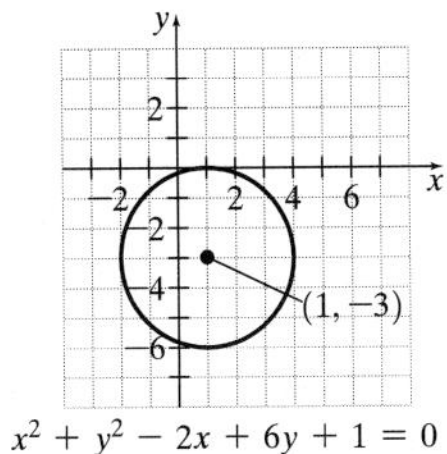

17. b)

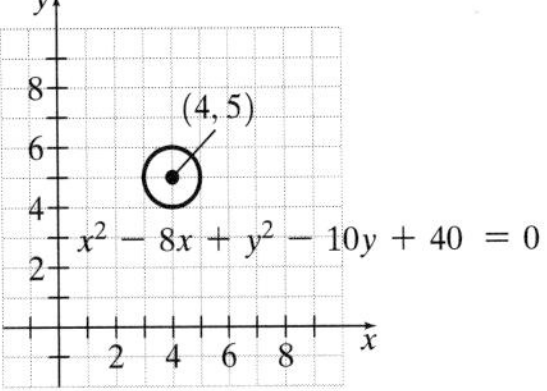

18. a) $(x - 2)^2 + (y + 5)^2 = (\sqrt{12})^2$

b)

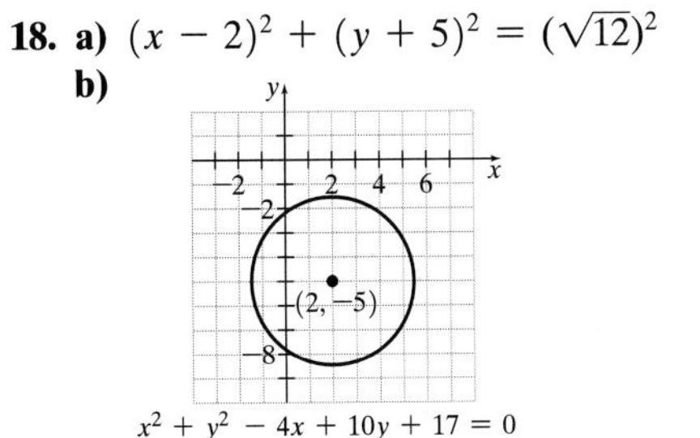

19.

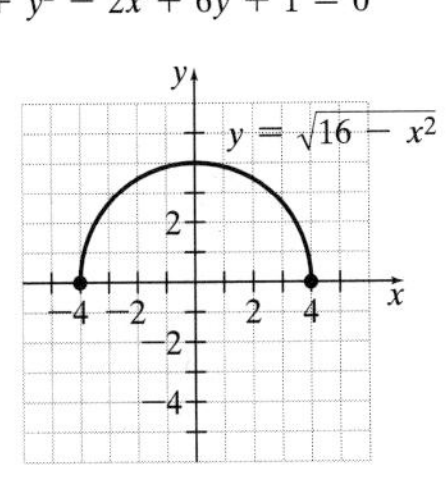

20.

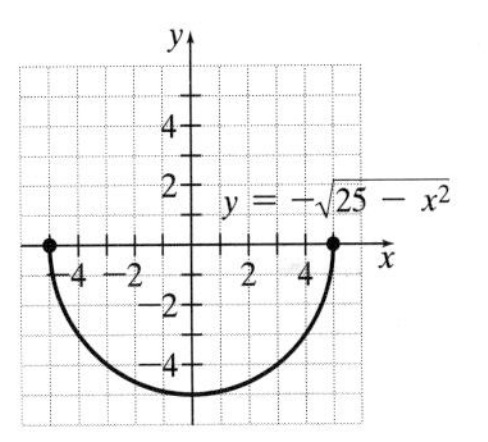

21. $(x + 1)^2 + (y - 1)^2 = 4$ **22.** $(x - 5)^2 + (y + 3)^2 = 9$

23.

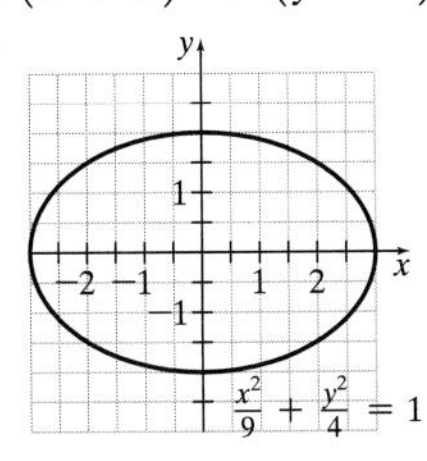

24.

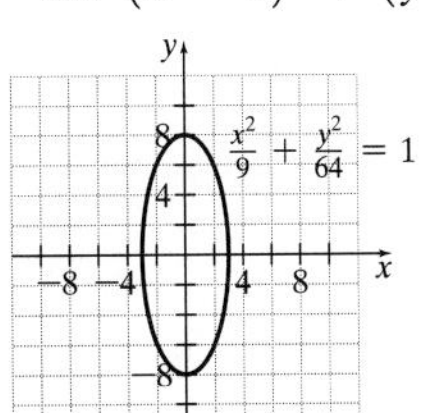

25.

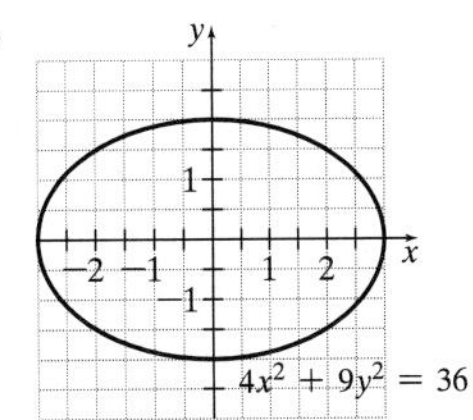

26.

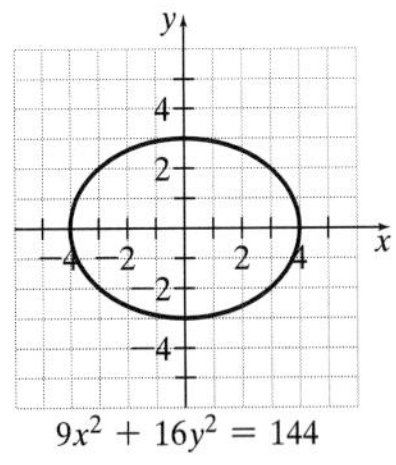

27.

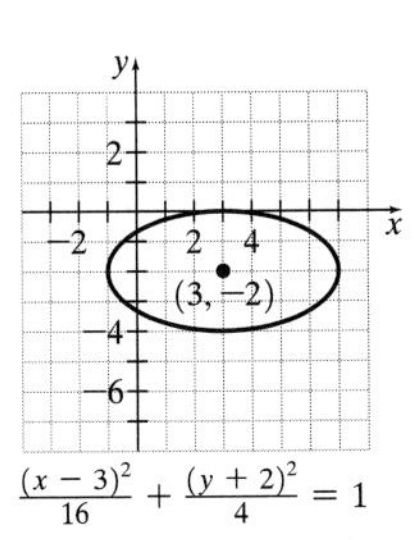

28.

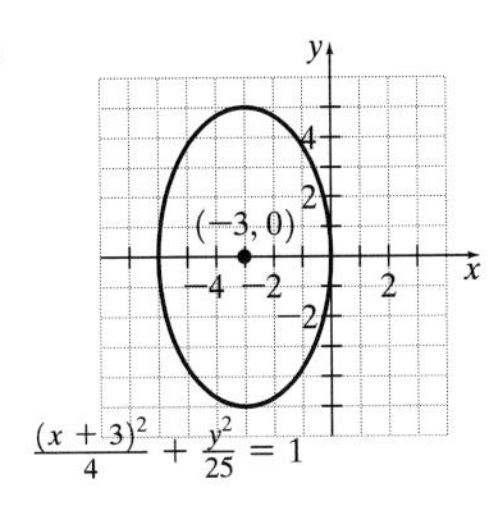

29.

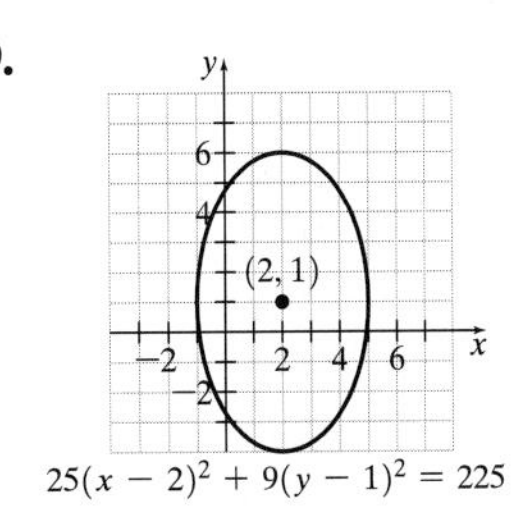

30.

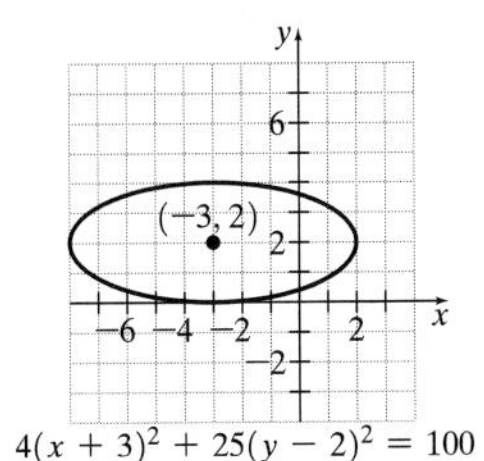

31. a) $y = \pm\frac{3}{2}x$

b)

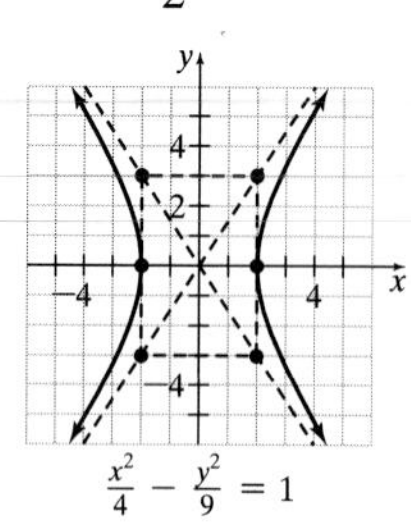

32. a) $y = \pm 2x$

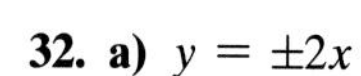

b)

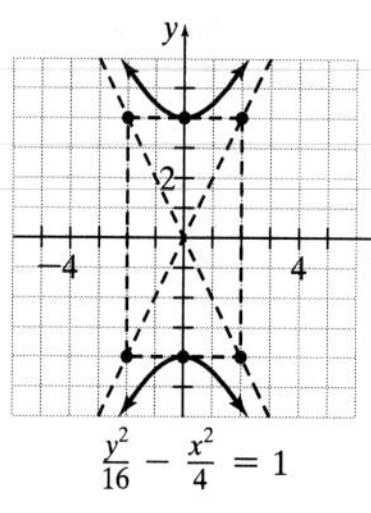

33. a) $y = \pm\frac{3}{5}x$

b)

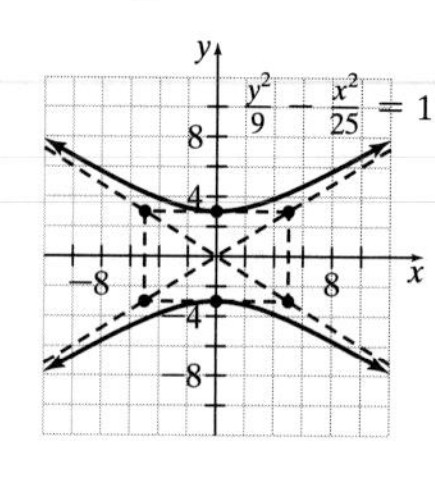

34. a) $y = \pm 3x$

b)

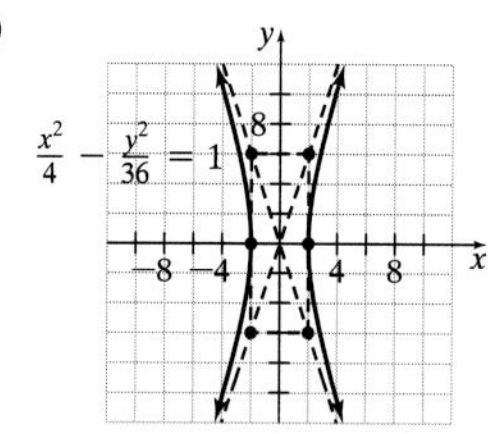

35. a) $\frac{y^2}{4} - \frac{x^2}{9} = 1$

b) $y = \pm\frac{2}{3}x$

c)

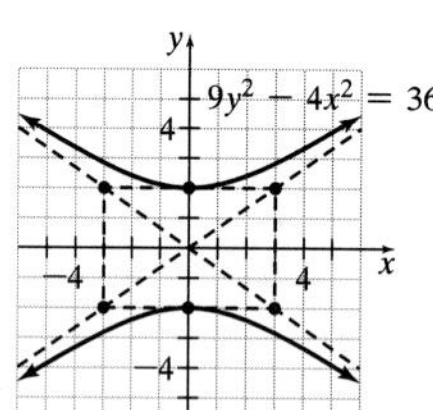

36. a) $\frac{x^2}{16} - \frac{y^2}{1} = 1$

b) $y = \pm\frac{1}{4}x$

c)

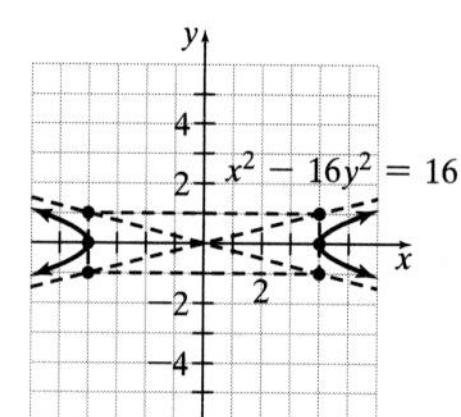

37. a) $\frac{x^2}{16} - \frac{y^2}{25} = 1$

b) $y = \pm\frac{5}{4}x$

c)

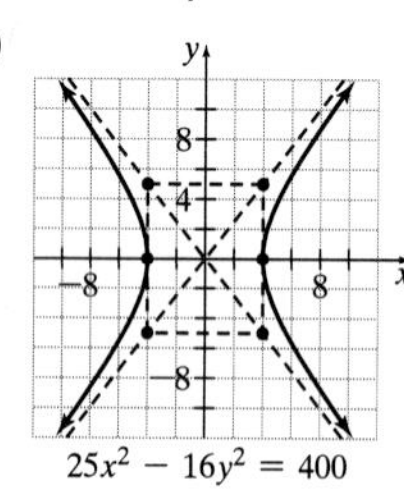

38. a) $\frac{y^2}{9} - \frac{x^2}{49} = 1$

b) $y = \pm\frac{3}{7}x$

c)

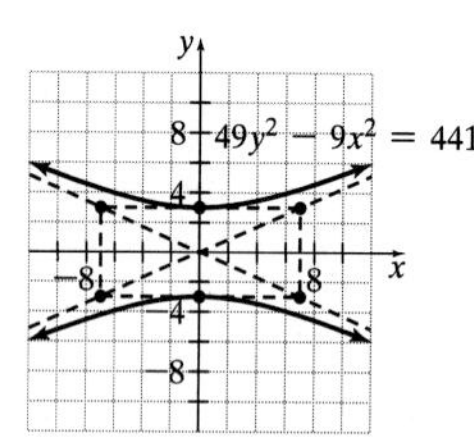

39. Hyperbola **40.** Ellipse **41.** Circle
42. Hyperbola **43.** Ellipse **44.** Parabola
45. Ellipse **46.** Parabola
47. $(-3, 0), \left(-\frac{12}{5}, \frac{9}{5}\right)$ **48.** $\left(\frac{5}{2}, \frac{3}{2}\right)$
49. $(2, 0), (-2, 0)$ **50.** No real solution
51. $(4, 0), (-4, 0)$
52. $(4, 3), (4, -3), (-4, 3), (-4, -3)$
53. No real solution
54. $(\sqrt{3}, 0), (-\sqrt{3}, 0)$
55. 4 feet by 8 feet **56.** ≈ 4 and ≈ 145
57. 15 feet by 20 feet
58. Principal: \$5000; rate: 5%

Practice Test

1. They are formed by cutting a cone or pair of cones. **2.** $\sqrt{50} \approx 7.07$ units

3. $\left(-1, \frac{7}{2}\right)$

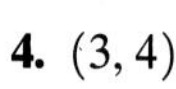

4. $(3, 4)$

$y = -2(x-3)^2 + 4$

5.

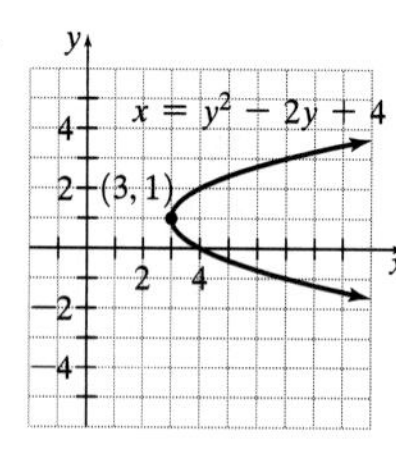

6. $x = -3(y-2)^2 + 4$

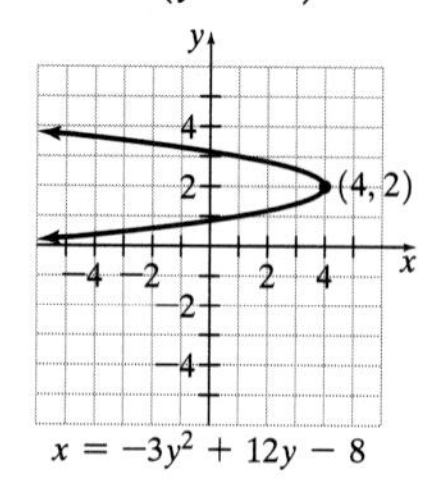

7. $(x + 3)^2 + (y + 1)^2 = 16$

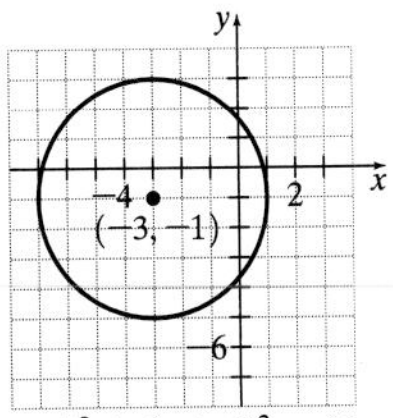

$(x + 3)^2 + (y + 1)^2 = 16$

8. Domain: $[-4, 2]$, Range: $[-2, 4]$

9. $(x - 3)^2 + (y + 1)^2 = 4^2$

10.

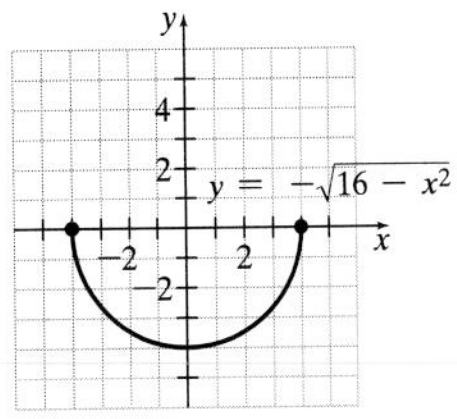

11. $(x - 1)^2 + (y - 3)^2 = 9$

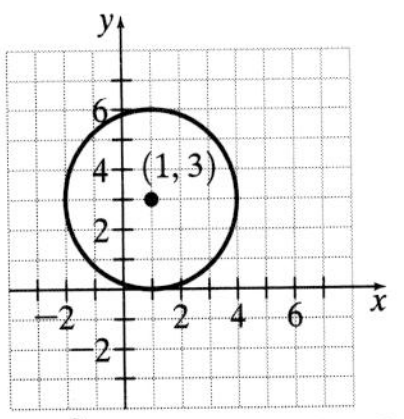

$x^2 + y^2 - 2x - 6y + 1 = 0$

12.

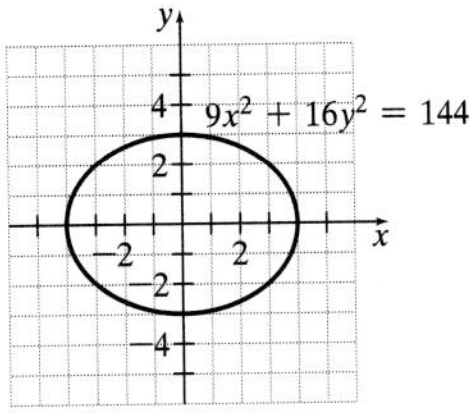

13. No, the values of a^2 and b^2 are switched.

14.

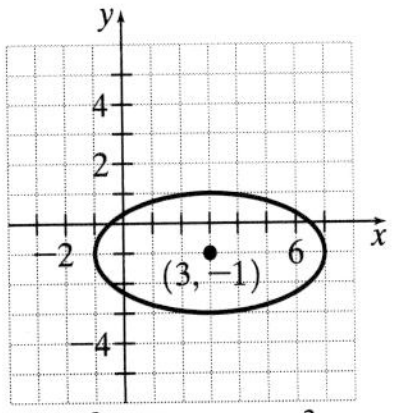

$4(x - 3)^2 + 16(y + 1)^2 = 64$

15. $(-2, -4)$

16. The transverse axis lies along the axis corresponding to the positive term of the equation in standard form.

17. $y = \pm\frac{3}{2}x$

18.

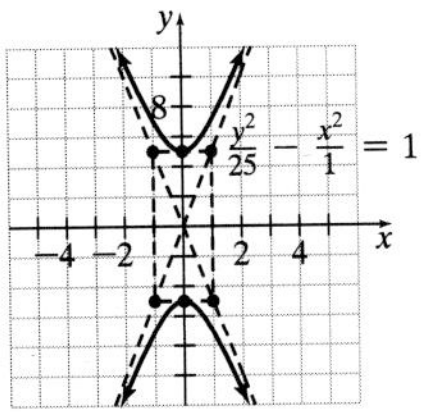

19.

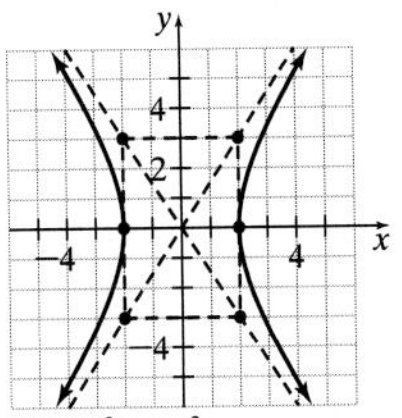

$\frac{x^2}{4} - \frac{y^2}{9} = 1$

20. Hyperbola

21. Ellipse

22. $(\sqrt{3}, \sqrt{3}), (\sqrt{3}, -\sqrt{3}), (-\sqrt{3}, \sqrt{3}), (-\sqrt{3}, -\sqrt{3})$

23. No real solution

24. 60 feet by 100 feet

25. 5 feet by 12 feet

Cumulative Review Test

1. $\left(\frac{16}{7}, -\frac{10}{7}\right)$ **2.** $a^2 + 8a + 20$ **3.** $\frac{(3x + 4)(2x - 1)}{(x - 1)(4x^2 - 2x + 1)}$

4. $\frac{2x^2 - 9x - 1}{2(x + 3)(x - 4)}$ **5.** 4 **6.** $\frac{x^2y\sqrt{6xyz}}{2z}$ **7.** $-3(\sqrt{3} + \sqrt{5})$ **8.** 3 **9.** 41 **10.** $3 \pm 2\sqrt{7}$ **11.** $\frac{2 \pm 2\sqrt{7}}{3}$

12. $\frac{64}{27}, -216$ **13.**

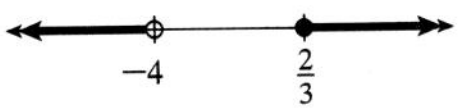

14. a) $4x^2 - 9$ **b)** $2x^2 + 12x - 3$

15.

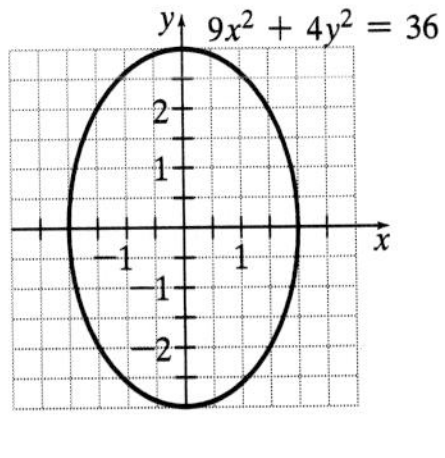

16.

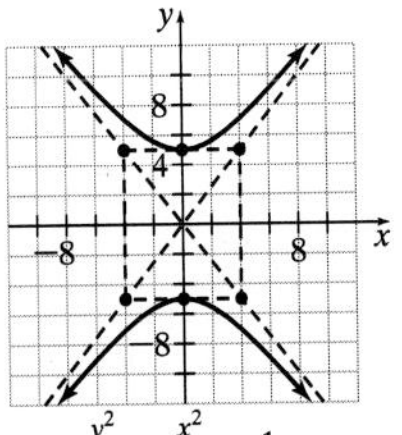

$\frac{y^2}{25} - \frac{x^2}{16} = 1$

17. 2

18. $t \approx 2.31$

19. \$225

20. $2\frac{1}{2}$ lb cashews, $1\frac{1}{2}$ lb peanuts

CHAPTER 11

Exercise Set 11.1 **1.** A sequence is a list of numbers arranged in a specific order **3.** A finite sequence is a function whose domain includes only the first n natural numbers **5.** In decreasing sequence, the terms decrease. **7.** A series is the sum of the terms of a sequence **9.** The sum as n goes from 1 to 5 of $n + 2$. **11.** 3, 6, 9, 12, 15 **13.** $5, 3, \frac{7}{3}, 2, \frac{9}{5}$ **15.** $3, \frac{3}{4}, \frac{1}{3}, \frac{3}{16}, \frac{3}{25}$ **17.** $\frac{3}{2}, \frac{4}{3}, \frac{5}{4}, \frac{6}{5}, \frac{7}{6}$ **19.** −1, 1, −1, 1, −1 **21.** 4, −8, 16, −32, 64 **23.** 31

25. 6 **27.** 16 **29.** $\frac{81}{19}$ **31.** $s_1 = 5; s_3 = 21$ **33.** $s_1 = 3; s_3 = 17$ **35.** $s_1 = 1; s_3 = 3$ **37.** $s_1 = \frac{1}{2}; s_3 = 7$
39. 64, 128, 256 **41.** 15, 17, 19 **43.** $\frac{1}{6}, \frac{1}{7}, \frac{1}{8}$ **45.** 1, −1, 1 **47.** $\frac{1}{81}, \frac{1}{243}, \frac{1}{729}$ **49.** $\frac{1}{16}, -\frac{1}{32}, \frac{1}{64}$ **51.** −25, −33, −41
53. $2 + 5 + 8 + 11 + 14 = 40$ **55.** $-1 + 5 + 15 + 29 + 47 + 69 = 164$ **57.** $2 + 3 + 4 = 9$
59. $\sum_{n=1}^{5} (n + 3)$ **61.** $\sum_{n=1}^{3} \frac{n^2}{4}$ **63.** 13 **65.** 169 **67.** 25 **69.** ≈ 42.83 **71.** Answers will vary.
73. Answers will vary. **75.** $\sum x = n\bar{x}$ **77.** Yes **79. a)** 10 **b)** 11 **c)** 110 **d)** 29 **e)** No **86.** $5, \frac{3}{2}$
87. Two, since $b^2 - 4ac$ is greater than zero. **88.**

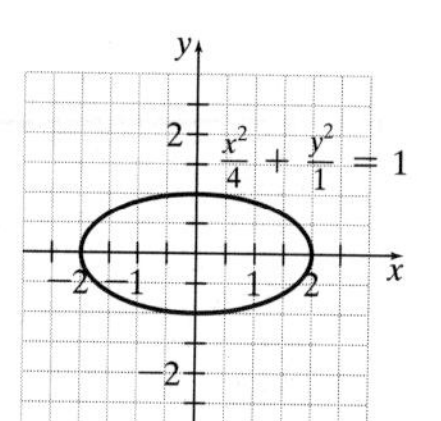

89. (2, 1), (−2, −1)

Exercise Set 11.2 **1.** In an arithmetic sequence, each term differs by a constant amount.
3. It is called the common difference. **5.** 4, 7, 10, 13, 16; $a_n = 3n + 1$ **7.** −5, −3, −1, 1, 3; $a_n = 2n - 7$
9. $\frac{1}{2}, 2, \frac{7}{2}, 5, \frac{13}{2}; a_n = \frac{3}{2}n - 1$ **11.** 100, 95, 90, 85, 80; $a_n = -5n + 105$ **13.** 14 **15.** −23 **17.** 13 **19.** 2
21. 9 **23.** 6 **25.** $s_{10} = 100; d = 2$ **27.** $s_8 = \frac{52}{5}; d = \frac{1}{5}$ **29.** $s_5 = 20; d = \frac{4}{5}$ **31.** $s_{11} = 407; d = 6$
33. 5, 8, 11, 14; $a_{10} = 32; s_{10} = 185$ **35.** −8, −13, −18, −23; $a_{10} = -53; s_{10} = -305$
37. 100, 93, 86, 79; $a_{10} = 37; s_{10} = 685$ **39.** $\frac{9}{5}, \frac{12}{5}, 3, \frac{18}{5}; a_{10} = \frac{36}{5}; s_{10} = 45$ **41.** $n = 15; s_{15} = 330$
43. $n = 10; s_{10} = 90$ **45.** $n = 9; s_9 = -\frac{39}{2}$ **47.** $n = 11; s_{11} = -352$ **49.** 1275 **51.** 2500 **53.** 630 **55.** 267
57. 50(101) = 5050 **59. a)** 19 feet **b)** 143.5 feet **61.** 1 foot **63.** 210 logs **65.** \$496 **67. a)** \$35,600
b) \$338,000 **69.** $a_n = 180(n - 2)$ **77.** One solution; slopes of the two lines are different **78.** (7, −6)
79. $|x - 2| < 4$ **80.** $\frac{(x + 2)^2}{4} + \frac{(y - 3)^2}{9} = 1$; (−2, 3)

Exercise Set 11.3 **1.** A geometric sequence is a sequence in which each term after the first is the same multiple of the preceding term **3.** To find the common ratio, take any term except the first and divide by the term that precedes it. **5.** 0 **7.** 4, 12, 36, 108, 324 **9.** $90, 30, 10, \frac{10}{3}, \frac{10}{9}$ **11.** −5, 10, −20, 40, −80 **13.** $3, \frac{9}{2}, \frac{27}{4}, \frac{81}{8}, \frac{243}{16}$
15. 128 **17.** 10,935 **19.** $\frac{1}{64}$ **21.** 6144 **23.** 15,624 **25.** 10,160 **27.** $-\frac{2565}{128}$ **29.** $-\frac{9279}{625}$
31. $r = \frac{1}{2}; a_n = 5\left(\frac{1}{2}\right)^{n-1}$ **33.** $r = -3; a_n = 2(-3)^{n-1}$ **35.** $r = 3; a_n = -1(3)^{n-1}$ **37.** 12 **39.** $\frac{25}{3}$ **41.** $\frac{5}{3}$
43. $\frac{81}{10}$ **45.** 2 **47.** 24 **49.** −45 **51.** −15 **53.** $\frac{3}{11}$ **55.** $\frac{7}{9}$ **57.** $\frac{17}{33}$ **59.** $r = 2$ or $r = -2; a_1 = 7$
61. $r = 3; a_1 = 5$ **63.** \$1.72 **65. a)** 3 days **b)** ≈ 1.172 grams **67. a)** 360.22 million people **b)** ≈ 29.23 years
69. a) $\frac{1}{2}, \frac{1}{4}, \frac{1}{8}, \frac{1}{16}, \frac{1}{32}$ **b)** $a_n = \frac{1}{2}\left(\frac{1}{2}\right)^{n-1} = \left(\frac{1}{2}\right)^n$ **c)** $\frac{1}{128}$ or ≈ 0.78% **71.** $\left(\frac{2}{3}\right)^{10}$ or ≈ 1.7%
73. a) 47.52 feet **b)** 550 feet **75. a)** 7.203 inches **b)** 100 inches **77.** 211

79. a) \$7840, \$6272, \$5017.60 **b)** $a_n = 7840\left(\frac{4}{5}\right)^{n-1}$ **c)** \$3211.26 **81.** 190 feet

83. a) y_2 goes up more steeply. **b)**

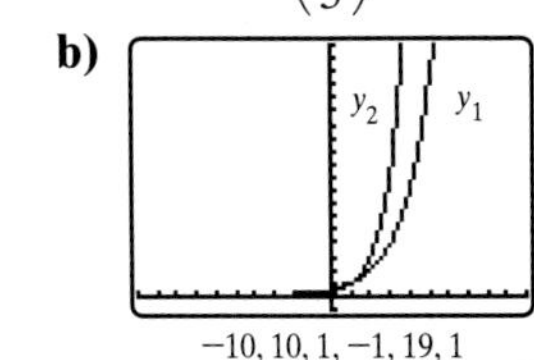

−10, 10, 1, −1, 19, 1

85. $n = 21; s_n = 2{,}097{,}151$ **87.** $8x^3 - 18x^2 - 3x + 18$ **88.** $8x - 15 + \frac{57}{2x+5}$ **89.** 12 hours **90.** $\frac{10}{3}$ **91.** $3x^2y^2\sqrt[3]{y} - 2xy\sqrt[3]{9y^2}$ **92.** $x\sqrt{y}$ **93.** $-\frac{1}{3}$

Exercise Set 11.4

1. The first and last numbers in each row are 1 and the inner numbers are obtained by adding the two numbers in the row above (to the right and left). For a diagram, see page 687. **3.** 10 **5.** 1 **7.** 1 **9.** 70 **11.** 28 **13.** $x^3 + 12x^2 + 48x + 64$ **15.** $a^4 - 4a^3b + 6a^2b^2 - 4ab^3 + b^4$

17. $243a^5 - 405a^4b + 270a^3b^2 - 90a^2b^3 + 15ab^4 - b^5$ **19.** $16x^4 + 16x^3 + 6x^2 + x + \frac{1}{16}$

21. $\frac{x^4}{16} - \frac{3x^3}{2} + \frac{27x^2}{2} - 54x + 81$ **23.** $x^{10} + 10x^9y + 45x^8y^2 + 120x^7y^3$

25. $2187x^7 - 5103x^6y + 5103x^5y^2 - 2835x^4y^3$ **27.** $x^{16} - 24x^{14}y + 252x^{12}y^2 - 1512x^{10}y^3$ **29.** Yes; $4! = 4 \cdot 3!$

31. Yes; $(7-3)! = (7-3)(7-4)(7-5)!$ **33.** x^8; $24x^7$; $17{,}496x$; 6561 **35.** $(a+b)^n = \sum_{i=0}^{n} \binom{n}{i} a^{n-i}b^i$

39. $s_n = \frac{a_1 - a_1r^n}{1-r} = \frac{a_1(1-r^n)}{1-r}$ **40.** $a_1 = \frac{s_n - s_nr}{1-r^n} = \frac{s_n(1-r)}{1-r^n}$ **41.** $(4x-3)(4x+1)$ **42.** $3b^2(a-4)^2$

Review Exercises

1. 3, 4, 5, 6, 7 **2.** $1, \frac{1}{2}, \frac{1}{3}, \frac{1}{4}, \frac{1}{5}$ **3.** 2, 6, 12, 20, 30 **4.** $\frac{1}{5}, \frac{2}{3}, \frac{9}{7}, 2, \frac{25}{9}$ **5.** 25 **6.** 2 **7.** $\frac{16}{81}$

8. 88 **9.** $s_1 = 5; s_3 = 24$ **10.** $s_1 = 2; s_3 = 28$ **11.** $s_1 = \frac{4}{3}; s_3 = \frac{227}{60}$ **12.** $s_1 = -3; s_3 = -4$

13. 16, 32, 64; $a_n = 2^{n-1}$ **14.** $-\frac{1}{2}, \frac{1}{4}, -\frac{1}{8}; a_n = -8\left(-\frac{1}{2}\right)^{n-1}$ **15.** $\frac{32}{3}, \frac{64}{3}, \frac{128}{3}; a_n = \frac{2}{3}(2)^{n-1}$

16. $-3, -6, -9; a_n = 9 - 3(n-1) = 12 - 3n$ **17.** $3 + 6 + 11 = 20$ **18.** $3 + 8 + 15 + 24 = 50$

19. $\frac{1}{3} + \frac{4}{3} + \frac{9}{3} + \frac{16}{3} + \frac{25}{3} = \frac{55}{3}$ **20.** $\frac{1}{2} + \frac{2}{3} + \frac{3}{4} + \frac{4}{5} = \frac{163}{60}$ **21.** 27 **22.** 215 **23.** 108 **24.** 729

25. 5, 7, 9, 11, 13 **26.** $\frac{1}{2}, -\frac{3}{2}, -\frac{7}{2}, -\frac{11}{2}, -\frac{15}{2}$ **27.** $-12, -\frac{25}{2}, -13, -\frac{27}{2}, -14$ **28.** $-100, -\frac{499}{5}, -\frac{498}{5}, -\frac{497}{5}, -\frac{496}{5}$

29. 26 **30.** −4 **31.** $\frac{1}{2}$ **32.** 6 **33.** $s_8 = 112; d = 2$ **34.** $s_7 = -210; d = -6$ **35.** $s_7 = \frac{63}{5}; d = \frac{2}{5}$

36. $s_9 = -42; d = -\frac{1}{3}$ **37.** 2, 6, 10, 14; $a_{10} = 38; s_{10} = 200$ **38.** $-8, -11, -14, -17; a_{10} = -35; s_{10} = -215$

39. $\frac{5}{6}, \frac{3}{2}, \frac{13}{6}, \frac{17}{6}; a_{10} = \frac{41}{6}; s_{10} = \frac{115}{3}$ **40.** $-80, -76, -72, -68; -44; a_{10} = -44; s_{10} = -620$ **41.** $n = 11; s_{11} = 308$

42. $n = 9; s_9 = 36$ **43.** $n = 11; s_{11} = \frac{231}{10}$ **44.** $n = 19; s_{19} = 760$ **45.** 5, 10, 20, 40, 80 **46.** $-12, -6, -3, -\frac{3}{2}, -\frac{3}{4}$

47. $20, -\frac{40}{3}, \frac{80}{9}, -\frac{160}{27}, \frac{320}{81}$ **48.** $-100, -20, -4, -\frac{4}{5}, -\frac{4}{25}$ **49.** $\frac{4}{243}$ **50.** 6400 **51.** −2048 **52.** $\frac{160}{6561}$

53. 3060 **54.** $\frac{37{,}969}{1215}$ **55.** $-\frac{4305}{64}$ **56.** $\frac{172{,}539}{256}$ **57.** $r = 2; a_n = 6(2)^{n-1}$ **58.** $r = \frac{1}{3}; a_n = 8\left(\frac{1}{3}\right)^{n-1}$

59. $r = 5; a_n = -4(5)^{n-1}$ **60.** $r = \frac{2}{3}; a_n = \frac{9}{5}\left(\frac{2}{3}\right)^{n-1}$ **61.** 14 **62.** −6 **63.** −15 **64.** $\frac{49}{10}$ **65.** 4 **66.** $\frac{21}{2}$

67. −36 **68.** $\frac{25}{6}$ **69.** $\frac{52}{99}$ **70.** $\frac{59}{111}$ **71.** $81x^4 + 108x^3y + 54x^2y^2 + 12xy^3 + y^4$

72. $8x^3 - 36x^2y^2 + 54xy^4 - 27y^6$ **73.** $x^9 - 18x^8y + 144x^7y^2 - 672x^6y^3$

74. $256a^{16} + 3072a^{14}b + 16{,}128a^{12}b^2 + 48{,}384a^{10}b^3$ **75.** 15,150 **76. a)** \$30,000; \$31,000; \$32,000; \$33,000; **b)** $a_n = 29{,}000 + 1000n$ **c)** \$39,000 **77.** \$102,400 **78.** \$503.63 **79.** 100 feet

Practice Test **1. a)** An infinite sequence is a function whose domain is the set of natural numbers. **b)** A finite sequence is a function whose domain includes only the first n natural numbers. **2.** A series is the sum of the terms of a sequence **3. a)** An arithmetic sequence is one whose terms differ by a constant amount. **b)** A geometric sequence is one whose terms differ by a common multiple. **4. a)** This sequence is neither arithmetic or geometric because the terms do not differ by a constant amount nor by a common multiple. **b)** This sequence is arithmetic because the terms differ by -3. **c)** This sequence is geometric because the terms differ by the multiple $-\frac{1}{2}$. **5.** $3, 1, \frac{5}{9}, \frac{3}{8}, \frac{7}{25}$ **6.** $s_1 = 5$; $s_3 = \frac{23}{2}$ **7.** $5 + 11 + 21 + 35 + 53 = 125$ **8.** 228 **9.** $a_n = \frac{1}{3} + \left(\frac{1}{3}\right)(n-1) = \frac{1}{3}n$ **10.** $a_n = 5(2)^{n-1}$ **11.** 12, 9, 6, 3 **12.** $\frac{5}{8}, \frac{5}{12}, \frac{5}{18}, \frac{5}{27}$ **13.** 16 **14.** -32 **15.** 13 **16.** $\frac{512}{729}$ **17.** $\frac{39{,}063}{5}$ **18.** $r = \frac{1}{2}$; $a_n = 12\left(\frac{1}{2}\right)^{n-1}$ **19.** 9 **20.** $\frac{62}{99}$ **21.** 21 **22.** $x^4 + 8x^3y + 24x^2y^2 + 32xy^3 + 16y^4$ **23.** 81 **24.** \$210,000 **25.** 3,280,500

Cumulative Review Test **1.** -37 **2.** $x < -\frac{11}{2}$ or $x > \frac{17}{2}$ **3.** **4.** $(0, -4)$ **5.** $p = \frac{At}{t - A}$ **6.** $\frac{4x^2 + 3x - 6}{(x + 4)(x - 3)(3x + 4)}$ **7.** $\frac{x\sqrt[3]{12xy}}{y}$ **8.** $2\sqrt{7}$ **9.** $3; \frac{11}{9}$ **10.** $\frac{7 - 26i}{25}$ **11.** $\frac{-1 \pm \sqrt{33}}{4}$ **12.** $x < -1$ or $\frac{3}{2} < x < 4$ **13.**

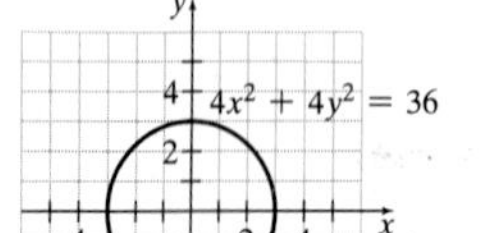

14. a) $y = (x + 1)^2 - 4$ **b)**

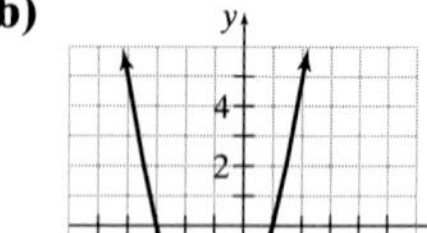

15.

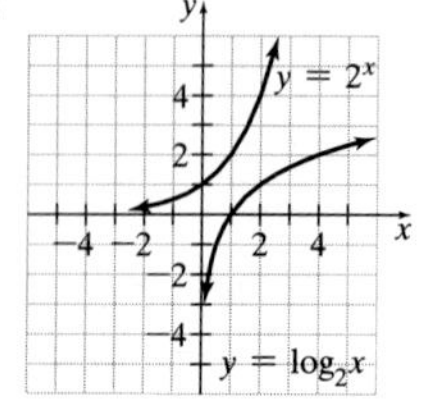

16. 4 **17.** 4 **18.** 65 **19.** 240 miles **20.** Width is 10 feet; length is 30 feet

INDEX

T

U

Photo credits:

1. Dimitra Lavrakas/Alaska Stock. **5.** Richard Hutchings/PhotoEdit. **27.** Chris Oxley/Corbis. **34.** Kevin Fleming/Corbis. **55.** Billy E. Barnes/PhotoEdit. **61.** D.Young-Wolff/PhotoEdit. **74.** AP/Wide World Photos. **76.** M.Antman/The Image Works. **89.** Michael T. Sedam/Corbis. **93.** M. Richards/PhotoEdit. **95.** Simon Battensby/Tony Stone Images. **102.** J.C.Leacock/Adventure Photo & Film. **103.** Vincent Dewitt/Stock Boston. **104.** Michael Newman/PhotoEdit. **105.** Allen R. Angel. **119.** Allen R. Angel. **137.** Gary Buss/FPG International LLC. **138.** Sheila Terry/Science Photo Library/Photo Researchers, Inc. **161.** Jim Steinberg/Photo Researchers, Inc. **165.** Susan Van Etten/PhotoEdit. **191.** Allen R. Angel. **221.** Steve Hansen/Stock Boston. **248.** Pacha/Corbis. **250.** Richard Hamilton Smith/Corbis. **240.** John P. Kelly/The Image Bank. **281.** Jeff Greenberg/PhotoEdit. **315.** Michael Newman/PhotoEdit. **349.** Neil Strassberg/Archive Photos. **348.** Allen R. Angel. **357.** NASA/Science Photo Library/Photo Researchers, Inc. **377.** Richard Pasley/Stock Boston. **396.** G.Biss/Masterfile Corporation. **402.** Dennis Marsico/Corbis. **403.** Allen R. Angel. **404.** Allen R. Angel. **404.** SuperStock, Inc. **405.** Kit Houghton/Corbis. **405.** Allen R. Angel. **406.** Greg Smith/Photographic Illustrators. **421.** James Shaffer/PhotoEdit. **438.** S.McBrady/PhotoEdit. **461.** Everett Johnson/ Tony Stone Images. **467.** Walter Schmid/Tony Stone Images. **471.** Allen R. Angel. **489.** David Young-Wolff/PhotoEdit. **513.** NASA/Masterfile Corporation. **514.** Francis G. Mayer/Corbis. Georges Pierre Seurat, "Invitation to the Side-Show" 1887-1888 Oil on canvas. 39 1/4 x 59 in. (99.7 x 149.9 cm). Metropolitan Museum of Art, New York, NY, USA.61.101.16. **557.** Chris Arend/Alaska Stock. **559.** National Geographic Society NGS Image Collection. **571.** Damir Frkovic/Masterfile Corporation. **578.** Dept. of Energy/Photo Researchers, Inc. **583.** Deborah Davis/Tony Stone Images. **601.** Charles Gupton/Tony Stone Images. **611.** Fred George/Tony Stone Images. **611.** Mae Scanlan Photography. **632.** Srenco Photography (c) 1996 Srenco Photography. **638.** Santi Visalli/Archive Photos Inc. **638.** Allen R. Angel. **659.** Richard Megna/Fundamental Photographs (c) Richard Megna/Fundamental Photographs, NYC. **671.** William James Warren/Corbis. **673.** Corbis. **690.** Kevin Fleming/Corbis

Chapter 7 Roots, Radicals, and Complex Numbers

If n is even and $a \geq 0$: $\sqrt[n]{a} = b$ if $b \geq 0$ and $b^n = a$

If n is odd: $\sqrt[n]{a} = b$ if $b^n = a$

Rules of radicals

$\sqrt{a^2} = |a|$

$\sqrt{a^2} = a,\ a \geq 0$

$\sqrt[n]{a^n} = a,\ a \geq 0$

$\sqrt[n]{a} = a^{1/n},\ a \geq 0$

$\sqrt[n]{a^m} = \left(\sqrt[n]{a}\right)^m = a^{m/n},\ a \geq 0$

$\sqrt[n]{a}\sqrt[n]{b} = \sqrt[n]{ab},\ a \geq 0, b \geq 0$

$\dfrac{\sqrt[n]{a}}{\sqrt[n]{b}} = \sqrt[n]{\dfrac{a}{b}},\ a \geq 0, b \geq 0$

A radical is simplified when the following are all true:

1. No perfect powers are factors of any radicand.
2. No radicand contains a fraction.
3. No denominator contains a radical.

Complex numbers: numbers of the form $a + bi$.

Powers of i: $i = \sqrt{-1}, i^2 = -1, i^3 = -i, i^4 = 1$

Chapter 8 Quadratic Functions

Square Root Property:

If $x^2 = a$, where a is a real number, then $x = \pm\sqrt{a}$.

A quadratic equation may be solved by factoring, completing the square, or the quadratic formula

Quadratic Formula: $x = \dfrac{-b \pm \sqrt{b^2 - 4ac}}{2a}$

Discriminant: $b^2 - 4ac$

If $b^2 - 4ac > 0$ then equation has two distinct real number solutions.

If $b^2 - 4ac = 0$ then equation has a single real number solution.

If $b^2 - 4ac < 0$ then equation has no real number solution.

Parabolas

For $f(x) = ax^2 + bx + c$, the vertex of the parabola is $\left(-\dfrac{b}{2a}, \dfrac{4ac - b^2}{4a}\right)$ or $\left(-\dfrac{b}{2a}, f\left(-\dfrac{b}{2a}\right)\right)$.

For $f(x) = a(x - h)^2 + k$, the vertex of the parabola is (h, k).

If $f(x) = ax^2 + bx + c, a > 0$, the function will have a minimum value of $\dfrac{4ac - b^2}{4a}$ at $x = -\dfrac{b}{2a}$.

If $f(x) = ax^2 + bx + c, a < 0$, the function will have a maximum value of $\dfrac{4ac - b^2}{4a}$ at $x = -\dfrac{b}{2a}$.

Chapter 9 Exponential and Logarithmic Functions

Composite function of function f with function g: $(f \circ g)(x) = f[g(x)]$

To find the **inverse function**, $f^{-1}(x)$, interchange all x's and y's and solve the resulting equation for y.

If $f(x)$ and $g(x)$ are inverse functions, then $(f \circ g)(x) = (g \circ f)(x) = x$.

Exponential function: $f(x) = a^x, a > 0, a \neq 1$

Logarithm: $y = \log_a x$ means $x = a^y, a > 0, a \neq 1$

Properties of Logarithms:

$\log_a xy = \log_a x + \log_a y$

$\log_a (x/y) = \log_a x - \log_a y$

$\log_a x^n = n \log_a x$

$\log_a a^x = x$

$a^{\log_a x} = x, x > 0$

Common Logarithms are logarithms to the base 10.

Natural Logarithms are logarithms to the base e, where $e \approx 2.7183$.

Antilogarithm: If $\log N = L$ then $N = \text{antilog } L$

Change of base formula: $\log_a x = \dfrac{\log_b x}{\log_b a}$

Natural exponential function: $f(x) = e^x$

To solve Exponential and Logarithmic equations we also use these properties:

If $x = y$, then $a^x = a^y$

If $a^x = a^y$, then $x = y$

If $x = y$, then $\log x = \log y$ $(x > 0, y > 0)$

If $\log x = \log y$, then $x = y$ $(x > 0, y > 0)$

$\ln e^x = x$

$e^{\ln x} = x, x > 0$

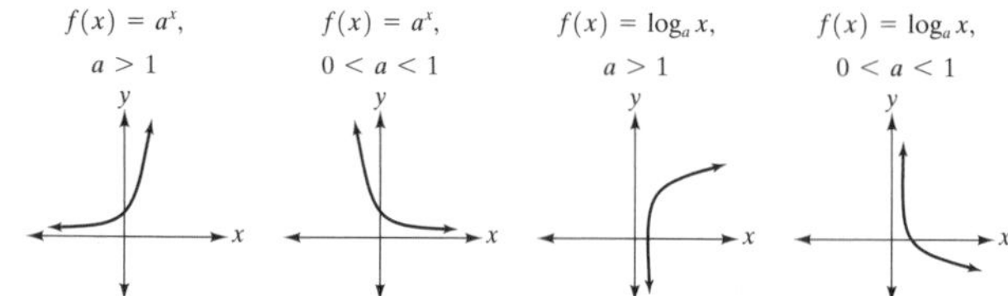